Lecture Notes in Computer Science 12696

More information about this subseries at http://www.springer.com/series/7410

Anne Canteaut · François-Xavier Standaert (Eds.)

Advances in Cryptology – EUROCRYPT 2021

40th Annual International Conference on the Theory and Applications of Cryptographic Techniques
Zagreb, Croatia, October 17–21, 2021
Proceedings, Part I

Editors
Anne Canteaut
Inria
Paris, France

François-Xavier Standaert
UCLouvain
Louvain-la-Neuve, Belgium

ISSN 0302-9743 ISSN 1611-3349 (electronic)
Lecture Notes in Computer Science
ISBN 978-3-030-77869-9 ISBN 978-3-030-77870-5 (eBook)
https://doi.org/10.1007/978-3-030-77870-5

LNCS Sublibrary: SL4 – Security and Cryptology

This Springer imprint is published by the registered company Springer Nature Switzerland AG
The registered company address is: Gewerbestrasse 11, 6330 Cham, Switzerland

Preface

Eurocrypt 2021, the 40th Annual International Conference on the Theory and Applications of Cryptographic Techniques, was held in Zagreb, Croatia, during October 17–21, 2021.[1] The conference was sponsored by the International Association for Cryptologic Research (IACR). Lejla Batina (Radboud University, The Netherlands) and Stjepan Picek (Delft University of Technology, The Netherlands) were responsible for the local organization.

We received a total of 400 submissions. Each submission was anonymized for the reviewing process and was assigned to at least three of the 59 Program Committee (PC) members. PC members were allowed to submit at most two papers. The reviewing process included a rebuttal round for all submissions. After extensive deliberations the PC accepted 78 papers. The revised versions of these papers are included in this three-volume proceedings.

The PC decided to give Best Paper Awards to the papers "*Non-Interactive Zero Knowledge from Sub-exponential DDH*" by Abhishek Jain and Zhengzhong Jin, "*On the (in)security of ROS*" by Fabrice Benhamouda, Tancrède Lepoint, Julian Loss, Michele Orrù, and Mariana Raykova and "*New Representations of the AES Key Schedule*" by Gaëtan Leurent and Clara Pernot. The authors of these three papers received an invitation to submit an extended version of their work to the *Journal of Cryptology*. The program also included invited talks by Craig Gentry (Algorand Foundation) and Sarah Meiklejohn (University College London).

We would like to thank all the authors who submitted papers. We know that the PC's decisions can be very disappointing, especially rejections of good papers which did not find a slot in the sparse number of accepted papers. We sincerely hope that these works will eventually get the attention they deserve.

We are indebted to the PC and the external reviewers for their voluntary work. Selecting papers from 400 submissions covering the many areas of cryptologic research is a huge workload. It has been an honor to work with everyone. We owe a big thank you to Kevin McCurley for his continuous support in solving all the minor issues we had with the HotCRP review system, to Gaëtan Leurent for sharing his MILP programs which made the papers assignments much easier, and to Simona Samardjiska who acted as Eurocrypt 2021 webmaster.

Finally, we thank all the other people (speakers, sessions chairs, rump session chairs…) for their contribution to the program of Eurocrypt 2021. We would also like to thank the many sponsors for their generous support, including the Cryptography Research Fund that supported student speakers.

April 2021

Anne Canteaut
François-Xavier Standaert

[1] This preface was written before the conference took place, under the assumption that it will take place as planned in spite of travel restrictions due to COVID-19.

Eurocrypt 2021

The 40th Annual International Conference on the Theory and Applications of Cryptographic Techniques

Sponsored by the *International Association for Cryptologic Research*
Zagreb, Croatia
October 17–21, 2021

General Co-chairs

Lejla Batina	Radboud University, The Netherlands
Stjepan Picek	Delft University of Technology, The Netherlands

Program Committee Chairs

Anne Canteaut	Inria, France
François-Xavier Standaert	UCLouvain, Belgium

Program Committee

Shweta Agrawal	IIT Madras, India
Joël Alwen	Wickr, USA
Foteini Baldimtsi	George Mason University, USA
Marshall Ball	Columbia University, USA
Begül Bilgin	Rambus - Cryptography Research, The Netherlands
Nir Bitansky	Tel Aviv University, Israel
Joppe W. Bos	NXP Semiconductors, Belgium
Christina Boura	University of Versailles, France
Wouter Castryck	KU Leuven, Belgium
Kai-Min Chung	Academia Sinica, Taiwan
Jean-Sébastien Coron	University of Luxembourg, Luxembourg
Véronique Cortier	LORIA, CNRS, France
Geoffroy Couteau	CNRS, IRIF, Université de Paris, France
Luca De Feo	IBM Research Europe, Switzerland
Léo Ducas (Area Chair: Public-Key Crypto)	CWI, Amsterdam, The Netherlands
Orr Dunkelman	University of Haifa, Israel
Stefan Dziembowski (Area Chair: Theory)	University of Warsaw, Poland
Thomas Eisenbarth	University of Lübeck, Germany
Dario Fiore	IMDEA Software Institute, Spain
Marc Fischlin	TU Darmstadt, Germany

Benjamin Fuller	University of Connecticut, USA
Adrià Gascón	Google, UK
Henri Gilbert	ANSSI, France
Shai Halevi	Algorand Foundation, USA
Annelie Heuser	Univ Rennes, CNRS, IRISA, France
Naofumi Homma	Tohoku University, Japan
Kristina Hostáková	ETH Zürich, Switzerland
Tetsu Iwata	Nagoya University, Japan
Marc Joye	Zama, France
Pascal Junod (Area Chair: Real-World Crypto)	Snap, Switzerland
Pierre Karpman	Université Grenoble-Alpes, France
Gregor Leander (Area Chair: Symmetric Crypto)	Ruhr-Universität Bochum, Germany
Benoît Libert	CNRS and ENS de Lyon, France
Julian Loss	University of Maryland, College Park, USA
Christian Majenz	CWI, Amsterdam, The Netherlands
Daniel Masny	Visa Research, USA
Bart Mennink	Radboud University, The Netherlands
Tarik Moataz	Aroki Systems, USA
Amir Moradi	Ruhr-Universität Bochum, Germany
Michael Naehrig	Microsoft Research, USA
María Naya-Plasencia	Inria, France
Claudio Orlandi	Aarhus University, Denmark
Elisabeth Oswald (Area Chair: Implementations)	University of Klagenfurt, Austria
Dan Page	University of Bristol, UK
Rafael Pass	Cornell Tech, USA
Thomas Peyrin	Nanyang Technological University, Singapore
Oxana Poburinnaya	University of Rochester and Ligero Inc., USA
Matthieu Rivain	CryptoExperts, France
Adeline Roux-Langlois	Univ Rennes, CNRS, IRISA, France
Louis Salvail	Université de Montréal, Canada
Yu Sasaki	NTT Laboratories, Japan
Tobias Schneider	NXP Semiconductors, Austria
Yannick Seurin	ANSSI, France
Emmanuel Thomé	LORIA, Inria Nancy, France
Vinod Vaikuntanathan	MIT, USA
Prashant Nalini Vasudevan	UC Berkeley, USA
Daniele Venturi	Sapienza University of Rome, Italy
Daniel Wichs	Northeastern University and NTT Research Inc., USA
Yu Yu	Shanghai Jiao Tong University, China

Additional Reviewers

Mark Abspoel
Hamza Abusalah
Alexandre Adomnicai
Archita Agarwal
Divesh Aggarwal
Shashank Agrawal
Gorjan Alagic
Martin R. Albrecht
Ghada Almashaqbeh
Bar Alon
Miguel Ambrona
Ghous Amjad
Prabhanjan Ananth
Toshinori Araki
Victor Arribas
Gilad Asharov
Roberto Avanzi
Melissa Azouaoui
Christian Badertscher
Saikrishna Badrinarayanan
Karim Baghery
Victor Balcer
Laasya Bangalore
Magali Bardet
James Bartusek
Balthazar Bauer
Carsten Baum
Christof Beierle
James Bell
Fabrice Benhamouda
Iddo Bentov
Olivier Bernard
Sebastian Berndt
Pauline Bert
Ward Beullens
Benjamin Beurdouche
Ritam Bhaumik
Erica Blum
Alexandra Boldyreva
Jonathan Bootle
Nicolas Bordes
Katharina Boudgoust
Florian Bourse
Xavier Boyen
Elette Boyle
Zvika Brakerski
Lennart Braun
Gianluca Brian
Marek Broll
Olivier Bronchain
Chris Brzuska
Benedikt Bünz
Chloe Cachet
Matteo Campanelli
Federico Canale
Ignacio Cascudo
Gaëtan Cassiers
Avik Chakraborti
Benjamin Chan
Eshan Chattopadhyay
Panagiotis Chatzigiannis
Shan Chen
Yanlin Chen
Yilei Chen
Yu Chen
Alessandro Chiesa
Ilaria Chillotti
Seung Geol Choi
Arka Rai Choudhuri
Michele Ciampi
Daniel Coggia
Benoît Cogliati
Ran Cohen
Andrea Coladangelo
Sandro Coretti-Drayton
Craig Costello
Daniele Cozzo
Ting Ting Cui
Debajyoti Das
Poulami Das
Bernardo David
Alex Davidson
Gareth Davies
Lauren De Meyer
Thomas Debris-Alazard
Leo de Castro
Thomas Decru
Jean Paul Degabriele
Akshay Degwekar
Amit Deo
Patrick Derbez
Itai Dinur
Christoph Dobraunig
Yevgeniy Dodis
Jack Doerner
Jelle Don
Benjamin Dowling
Eduoard Dufour Sans
Yfke Dulek
Frédéric Dupuis
Sylvain Duquesne
Avijit Dutta
Ehsan Ebrahimi
Kasra Edalat Nejdat
Naomi Ephraim
Thomas Espitau
Andre Esser
Grzegorz Fabiański
Xiong Fan
Antonio Faonio
Sebastian Faust
Serge Fehr
Patrick Felke
Rune Fiedler
Ben Fisch
Matthias Fitzi
Antonio Flórez-Gutiérrez
Cody Freitag
Georg Fuchsbauer
Ariel Gabizon
Nicolas Gama
Chaya Ganesh
Rachit Garg
Pierrick Gaudry
Romain Gay
Peter Gaži
Nicholas Genise
Craig Gentry

Marilyn George
Adela Georgescu
David Gerault
Essam Ghadafi
Satrajit Ghosh
Irene Giacomelli
Aarushi Goel
Junqing Gong
Alonso González
S. Dov Gordon
Louis Goubin
Marc Gourjon
Rishab Goyal
Lorenzo Grassi
Elijah Grubb
Cyprien de Saint Guilhem
Aurore Guillevic
Aldo Gunsing
Chun Guo
Qian Guo
Felix Günther
Iftach Haitner
Mohammad Hajiabadi
Mathias Hall-Andersen
Ariel Hamlin
Lucjan Hanzlik
Patrick Harasser
Dominik Hartmann
Eduard Hauck
Phil Hebborn
Javier Herranz
Amir Herzberg
Julia Hesse
Shoichi Hirose
Martin Hirt
Akinori Hosoyamada
Kathrin Hövelmanns
Andreas Hülsing
Ilia Iliashenko
Charlie Jacomme
Christian Janson
Stanislaw Jarecki
Ashwin Jha
Dingding Jia
Daniel Jost
Kimmo Järvinen
Guillaume Kaim
Chethan Kamath
Pritish Kamath
Fredrik Kamphuis
Ioanna Karantaidou
Shuichi Katsumata
Jonathan Katz
Tomasz Kazana
Marcel Keller
Mustafa Khairallah
Louiza Khati
Hamidreza Khoshakhlagh
Dakshita Khurana
Ryo Kikuchi
Eike Kiltz
Elena Kirshanova
Agnes Kiss
Karen Klein
Michael Klooß
Alexander Koch
Lisa Kohl
Vladimir Kolesnikov
Dimitris Kolonelos
Ilan Komargodski
Yashvanth Kondi
Venkata Koppula
Adrien Koutsos
Hugo Krawczyk
Stephan Krenn
Ashutosh Kumar
Ranjit Kumaresan
Po-Chun Kuo
Rolando L. La Placa
Thijs Laarhoven
Jianchang Lai
Virginie Lallemand
Baptiste Lambin
Eran Lambooij
Philippe Lamontagne
Rio Lavigne
Jooyoung Lee
Alexander Lemmens
Nikos Leonardos
Matthieu Lequesne
Antonin Leroux
Gaëtan Leurent
Jyun-Jie Liao
Damien Ligier
Huijia Lin
Benjamin Lipp
Maciej Liskiewicz
Qipeng Liu
Shengli Liu
Tianren Liu
Yanyi Liu
Chen-Da Liu-Zhang
Alex Lombardi
Patrick Longa
Vadim Lyubashevsky
Fermi Ma
Mimi Ma
Urmila Mahadev
Nikolaos Makriyannis
Giulio Malavolta
Damien Marion
Yoann Marquer
Giorgia Marson
Chloe Martindale
Ange Martinelli
Michael Meyer
Pierre Meyer
Andrew Miller
Brice Minaud
Ilya Mironov
Tal Moran
Saleet Mossel
Tamer Mour
Pratyay Mukherjee
Marta Mularczyk
Pierrick Méaux
Yusuke Naito
Joe Neeman
Patrick Neumann
Khoa Nguyen
Ngoc Khanh Nguyen
Phong Nguyen

Tuong-Huy Nguyen
Jesper Buus Nielsen
Ryo Nishimaki
Abderrahmane Nitaj
Anca Nitulescu
Lamine Noureddine
Adam O'Neill
Maciej Obremski
Cristina Onete
Michele Orru
Emmanuela Orsini
Carles Padro
Mahak Pancholi
Omer Paneth
Dimitris Papachristoudis
Sunoo Park
Anat Paskin-Cherniavsky
Alice Pellet-Mary
Olivier Pereira
Léo Perrin
Thomas Peters
Duy-Phuc Pham
Krzyszof Pietrzak
Jérôme Plût
Bertram Poettering
Yuriy Polyakov
Antigoni Polychroniadou
Alexander Poremba
Thomas Prest
Cassius Puodzius
Willy Quach
Anaïs Querol
Rahul Rachuri
Hugues Randriam
Adrian Ranea
Shahram Rasoolzadeh
Deevashwer Rathee
Mayank Rathee
Divya Ravi
Christian Rechberger
Michael Reichle
Jean-René Reinhard
Joost Renes
Nicolas Resch
João Ribeiro
Silas Richelson
Tania Richmond
Doreen Riepel
Peter Rindal
Miruna Rosca
Michael Rosenberg
Mélissa Rossi
Yann Rotella
Alex Russell
Théo Ryffel
Carla Ràfols
Paul Rösler
Rajeev Anand Sahu
Olga Sanina
Pratik Sarkar
Alessandra Scafuro
Christian Schaffner
Peter Scholl
Tobias Schmalz
Phillipp Schoppmann
André Schrottenloher
Jörg Schwenk
Adam Sealfon
Okan Seker
Jae Hong Seo
Karn Seth
Barak Shani
Abhi Shelat
Omri Shmueli
Victor Shoup
Hippolyte Signargout
Tjerand Silde
Mark Simkin
Luisa Siniscalchi
Daniel Slamanig
Benjamin Smith
Fang Song
Jana Sotáková
Pierre-Jean Spaenlehauer
Nicholas Spooner
Akshayaram Srinivasan
Damien Stehlé
Marc Stevens
Siwei Sun
Mehrdad Tahmasbi
Quan Quan Tan
Stefano Tessaro
Florian Thaeter
Aishwarya Thiruvengadam
Mehdi Tibouchi
Radu Titiu
Oleksandr Tkachenko
Yosuke Todo
Junichi Tomida
Ni Trieu
Eran Tromer
Daniel Tschudi
Giorgos Tsimos
Ida Tucker
Michael Tunstall
Akin Ünal
Dominique Unruh
Bogdan Ursu
Christine van Vredendaal
Wessel van Woerden
Marc Vauclair
Serge Vaudenay
Muthu Venkitasubramaniam
Damien Vergnaud
Gilles Villard
Fernando Virdia
Satyanarayana Vusirikala
Riad Wahby
Hendrik Waldner
Alexandre Wallet
Haoyang Wang
Hoeteck Wee
Weiqiang Wen
Benjamin Wesolowski
Jan Wichelmann
Luca Wilke
Mary Wootters
David Wu
Jiayu Xu
Sophia Yakoubov

Shota Yamada
Takashi Yamakawa
Sravya Yandamuri
Kang Yang
Lisa Yang
Kevin Yeo
Eylon Yogev
Greg Zaverucha
Mark Zhandry
Jiayu Zhang
Ruizhe Zhang
Yupeng Zhang
Vassilis Zikas
Paul Zimmermann
Dionysis Zindros

Contents – Part I

Best Papers

Public-Key Cryptography

Isogenies

Post-Quantum Cryptography

Lattices

Homomorphic Encryption

Symmetric Cryptanalysis

Contents – Part II

Masking and Secret-Sharing

Leakage, Faults and Tampering

Quantum Constructions and Proofs

Contents – Part III

Property-Preserving Hash Functions and ORAM

Blockchain

Privacy and Law Enforcement

Best Papers

Non-interactive Zero Knowledge from Sub-exponential DDH

Abhishek Jain(✉) and Zhengzhong Jin

Johns Hopkins University, Baltimore, MD, USA
{abhishek,zzjin}@cs.jhu.edu

Abstract. We provide the first constructions of non-interactive zero-knowledge and Zap arguments for NP based on the sub-exponential hardness of Decisional Diffie-Hellman against polynomial time adversaries (without use of groups with pairings).

Central to our results, and of independent interest, is a new notion of *interactive trapdoor hashing protocols.*

1 Introduction

Zero-knowledge (ZK) proofs [31] are a central object in the theory and practice of cryptography. A ZK proof allows a prover to convince a verifier about the validity of a statement without revealing any other information. ZK proofs have found wide applications in cryptography in all of their (interactive) avatars, but especially so in the non-interactive form where a proof consists of a single message from the prover to the verifier. This notion is referred to as *non-interactive zero knowledge* (NIZK) [22]. Applications of NIZKs abound and include advanced encryption schemes [23,46], signature schemes [4,7], blockchains [6], and more.

Since NIZKs for non-trivial languages are impossible in the plain model, the traditional (and de facto) model for NIZKs allows for a trusted setup that samples a common reference string (CRS) and provides it to the prover and the verifier algorithms. Starting from the work of [22], a major line of research has been dedicated towards understanding the assumptions that are sufficient for constructing NIZKs in the CRS model [5,9,14,16,18,19,21,28,30,34,35,50,53]. By now, NIZKs for NP are known from most of the standard assumptions known to imply public-key encryption – this includes factoring related assumptions [9,28], bilinear maps [18,34,35], and more recently, learning with errors (LWE) [14,50].

Notable exceptions to this list are standard assumptions related to the discrete-logarithm problem such as the Decisional Diffie-Hellman (DDH) assumption. In particular, the following question has remained open for three decades:

Do there exist NIZKs for NP based on DDH?

From a conceptual viewpoint, an answer to the above question would shed further light on the cryptographic complexity of NIZKs relative to public-key encryption. It would also improve our understanding of the power of groups with bilinear maps relative to non-pairing groups in cryptography. There are (at least)

A. Canteaut and F.-X. Standaert (Eds.): EUROCRYPT 2021, LNCS 12696, pp. 3–32, 2021.
https://doi.org/10.1007/978-3-030-77870-5_1

two prominent examples where bilinear maps have traditionally had an edge – advanced encryption schemes such as identity-based [10] and attribute-based encryption [33,54] (and more broadly, functional encryption [11,48,54]), and NIZKs. For the former, the gap has recently started to narrow in some important cases; see, e.g., [24]. We seek to understand whether such gap is inherent for NIZKs based on standard assumptions.[1]

A recent beautiful work of Brakerski et al. [13] demonstrates that this gap disappears if we additionally rely on the hardness of the learning parity with noise (LPN) problem. Namely, they construct NIZKs assuming that DDH and LPN are *both* hard. NIZKs based on the *sole* hardness of DDH, however, still remain elusive.

Zaps. Dwork and Naor [26] introduced the notion of *Zaps*, aka two-round public-coin proof systems in the plain model (i.e., without a trusted setup) that achieve a weaker form of privacy known as witness-indistinguishability (WI) [29]. Roughly speaking, WI guarantees that a proof for a statement with multiple witnesses does not reveal which of the witnesses was used in the computation of the proof.

Despite this seeming weakness, [26] proved that (assuming one-way functions) Zaps are equivalent to statistically-sound NIZKs in the common *random* string model. This allows for porting some of the known results for NIZKs to Zaps; specifically, those based on factoring assumptions and bilinear maps. Subsequently, alternative constructions of Zaps were proposed based on indistinguishability obfuscation [8]. Very recently, computationally-sound Zaps, aka Zap *arguments* were constructed based on quasi-polynomial LWE [2,32,42].

As in the case of NIZKs, constructing Zaps (or Zap arguments) for NP based on standard assumptions related to discrete-logarithm remains an open problem. Moreover, if we require *statistical* privacy, i.e., statistical Zap arguments [2,32], curiously, even bilinear maps have so far been insufficient.[2] In contrast, statistical NIZKs based on bilinear maps are known [34,35].

1.1 Our Results

In this work, we construct (statistical) NIZK and Zap arguments for NP based on the sub-exponential hardness of DDH against polynomial-time adversaries in standard groups.

Theorem 1 (Main Result – Informal). *Assuming sub-exponential hardness of DDH against polynomial-time attackers, there exist:*

- *(Statistical) NIZK arguments for NP in the common random string model.*
- *Statistical Zap arguments for NP.*

[1] If we allow for non-standard assumptions (albeit those not known to imply public-key encryption), then this gap is not inherent, as demonstrated by [16,21].

[2] A variant of statistical Zap arguments where the verifier is private-coin but the proofs are publicly verifiable is known from standard assumptions on bilinear maps [43].

Our NIZK achieves adaptive, multi-theorem statistical zero knowledge and non-adaptive soundness. By relaxing the zero-knowledge guarantee to be computational, we can achieve adaptive soundness. Our Zap argument achieves adaptive statistical witness indistinguishability and non-adaptive soundness.[3]

Our results rely on the assumption that *polynomial-time* adversaries cannot distinguish Diffie-Hellman tuples from random tuples with better than sub-exponentially small advantage. To the best of our knowledge, this assumption is unaffected by known attacks on the discrete logarithm problem.[4]

While our primary focus is on constructing NIZKs and Zap arguments from DDH, we note that our constructions enjoy certain properties that have previously not been achieved even using bilinear maps:

- Our NIZK constructions rely on a common *random* string setup unlike prior schemes based on bilinear maps that require a common *reference* string for achieving statistical ZK [34,35].
- Our statistical Zap argument is the first group-based construction (irrespective of whether one uses bilinear maps or not). Known constructions of Zaps from bilinear maps only achieve computational WI [34,35].

In particular, statistical NIZKs in the common random string model were previously only known from LWE (or circular-secure FHE) [14,50], and statistical Zap arguments were previously only known from (quasi-polynomial) LWE [2,32].

Interactive Trapdoor Hashing Protocols. Towards obtaining our results, we introduce the notion of *interactive trapdoor hashing protocols* (ITDH). An ITDH for a function family F is an interactive protocol between two parties – a sender and a receiver – where the sender holds an input x and the receiver holds a function $f \in F$. At the end of the protocol, the parties obtain an additive secret-sharing of $f(x)$. An ITDH must satisfy the following key properties:

- The sender must be *laconic* in that the length of each of its messages (consisting of a hash value) is independent of the input length.
- The receiver's messages must *hide* the function f.

ITDH generalizes and extends the recent notion of trapdoor hash functions (TDH) [25] to *multi-round interactive protocols.* Indeed, (ignoring some syntactic differences) a TDH can be viewed as an ITDH where both the receiver and the sender send a *single* message to each other.

[3] Following [43], by standard complexity leveraging, our statistical NIZK and Zap arguments can be upgraded (without changing our assumption) to achieve adaptive soundness for all instances of a priori (polynomially) bounded size. For the "unbounded-size" case, [49] proved the impossibility of statistical NIZKs where adaptive soundness is proven via a black-box reduction to falsifiable assumptions [44].

[4] There are well-known attacks for discrete logarithm over $\mathbb{Z}_q^*$ that require sub-exponential time and achieve constant success probability [1,20]. However, as observed in [16], a 2^t time algorithm with constant successful probability does not necessarily imply a polynomial time attack with 2^{-t} successful probability.

Our primary motivation for the study of ITDH is to explore the feasibility of a richer class of computations than what can be supported by known constructions of TDH. Presently, TDH constructions are known for a small class of computations such as linear functions and constant-degree polynomials (based on various assumptions such as DDH, Quadratic Residuosity, and LWE) [13,25]. We demonstrate that ITDH can support a much broader class of computations.

Assuming DDH, we construct a constant-round ITDH protocol for TC^0 circuits. While ITDH for TC^0 suffices for our main application, our approach can be generalized to obtain a polynomial-round ITDH for P/poly.

Theorem 2 (Informal). *Assuming DDH, there exists a constant-round ITDH for* TC^0.

We view ITDH as a natural generalization of TDH that might allow for a broader pool of applications. While our present focus is on the class of computations, it is conceivable that the use of interaction might enable additional properties in the future that are not possible (or harder to achieve) in the non-interactive setting.

Our Approach: Round Collapsing, *Twice*. We follow the correlation intractability framework for NIZKs implemented in a recent remarkable sequence of works [13,14,16,21,36,50]. The central idea of this framework is to instantiate the random oracle in the Fiat-Shamir paradigm [29] by so-called *correlation intractable hash functions* (CIH) [17]. In particular, given a CIH for all efficiently searchable relations, this approach can be used to collapse the rounds of so-called trapdoor sigma protocols [14] to obtain NIZKs in the CRS model.

The works of [14,50] used (leveled) fully homomorphic encryption to construct CIH for all efficiently searchable relations and therefore required LWE-related assumptions. Recently, Brakerski et al. [13] demonstrated a new approach for constructing CIH via (rate-1) TDH by crucially exploiting the laconic sender property of the latter. This raises hope for potential instantiations of CIH – ideally for all efficiently searchable relations – from other standard assumptions. So far, however, this approach has yielded CIH only for relations that can be approximated by constant-degree polynomials over $\mathbb{Z}_2$ due to limitations of known results for TDH. This severely restricts the class of compatible trapdoor sigma protocols that can be used for constructing NIZKs via the CIH framework. Indeed, Brakerski et al. rely crucially on LPN to construct such sigma protocols.

Somewhat counter-intuitively, we use *interaction* to address the challenge of constructing NIZKs solely from DDH. Specifically, we show that by using interaction – via the abstraction of ITDH – we can expand the class of functions that can be computed with a laconic sender. Furthermore, if an ITDH is sufficiently function-private (where the amount of security required depends on the round complexity), then we can *collapse its rounds* to construct CIH. Using this approach, we construct a CIH for TC^0 based on sub-exponential DDH.

Theorem 3 (Informal). *Assuming sub-exponential hardness of DDH against polynomial-time attackers, there exists a CIH for* TC^0.

Expanding the class of relations for CIH in turn expands the class of compatible trapdoor sigma protocols. In particular, we show that trapdoor sigma protocols for NP compatible with CIH from the above theorem can be built from DDH. This allows us to construct NIZK and Zap arguments in Theorem 1.

Overall, our approach for constructing NIZKs involves **two stages of round collapsing** – we first collapse rounds of ITDH to construct CIH, and then use CIH to collapse rounds of trapdoor sigma protocols to obtain NIZKs. Our construction of Zaps follows a similar blueprint, where the first step is the same as in the case of NIZKs and the second round-collapsing step is similar to the recent works of Badrinarayanan et al. [2] and Goyal et al. [32].

1.2 Guide to the Paper

We present the technical overview in Sect. 2 and the necessary preliminaries in Sect. 3. We define and construct ITDH in Sects. 4 and Sect. 5 respectively, and construct CIH for TC^0 in Sect. 6.

Due to page limits, we defer our constructions of NIZKs and Zap arguments to the full version.

2 Technical Overview

Our constructions rely on the correlation-intractability framework for instantiating the Fiat-Shamir paradigm. We start by recalling this framework.

Fiat-Shamir via Correlation Intractability. A family of hash functions defined by a tuple of algorithms $(\mathsf{Gen}, \mathsf{Hash})$ is said to be *correlation intractable* (CI) for a relation class $\mathcal{R}$ if for any $R \in \mathcal{R}$, given a hash key k sampled by Gen, an adversary cannot find an input x such that $(x, \mathsf{Hash}(\mathsf{k}, x)) \in R$. In the sequel, we focus on searchable relations where R is associated with a circuit C and $(x, y) \in R$ if and only if $y = C(x)$.

The CI framework instantiates the random oracle in the Fiat-Shamir paradigm for NIZKs via a family of CIH $(\mathsf{Gen}, \mathsf{Hash})$. Let Σ be a sigma protocol for a language $\mathcal{L}$ where the messages are denoted as α, β and γ. To obtain a NIZK in the CRS model, we collapse the rounds of Σ by computing β as the output of $\mathsf{Hash}(k, \alpha)$ for a key k sampled by Gen and fixed as part of CRS.

We now recall the argument for soundness of the resulting scheme. From the special soundness of Σ, for any $x \notin \mathcal{L}$ and any α, there exists a *bad challenge function* BadC such that the only possible accepting transcript (α, β, γ) must satisfy $\beta = \mathsf{BadC}(\alpha)$. In other words, any cheating prover must find an α such that $\beta = \mathsf{Hash}(\mathsf{k}, \alpha) = \mathsf{BadC}(\alpha)$. However, if $(\mathsf{Gen}, \mathsf{Hash})$ is CI for the relation searchable by BadC, then such an adversary must not exist.

Note that in general, BadC may not be efficiently computable. However, for *trapdoor sigma protocols*, BadC is efficiently computable given a "trapdoor" associated with the protocol. In this case, we only require CI for efficiently searchable relations.

Prior Work. A sequence of works [15,16,21,36,38] have constructed CIH for various classes of (not necessarily efficiently searchable) relations from well-defined, albeit strong assumptions that are not well understood. Recently, Canetti et al. [14] constructed CIH for all efficiently searchable relations from circular-secure fully homomorphic encryption. Subsequently, Peikert and Shiehian [50] obtained a similar result based on standard LWE.

Very recently, Brakerski et al. [13] leveraged the compactness properties of (rate-1) trapdoor hash functions to build CIH from standard assumptions. Specifically, assuming DDH (or other standard assumptions such as Quadratic Residuosity or LWE), they construct CIH for functions that can be approximated by a distribution on constant-degree polynomials. While this is a small class, [13] show that it nevertheless suffices for constructing NIZKs for NP. Specifically, they show that by relying on the LPN assumption, it is possible to construct trapdoor sigma protocols where the bad challenge function has probabilistic constant-degree representation. By collapsing the rounds of this protocol, they obtain NIZKs for NP.

Main Challenges. We now briefly discuss the main conceptual challenges in buildings NIZKs based only on DDH (in light of the work of [13]).

On the one hand, (non-pairing) group-based assumptions seem to have less structure than lattice assumptions; for example, we can only exploit linear homomorphisms. Hence it is not immediately clear how to construct rate-1 trapdoor hash functions from DDH beyond (probabilistic) linear functions or constant-degree polynomials (a constant-degree polynomial is also a linear function of its monomials).[5] On the other hand, it seems that we need CIH for more complicated functions in order to build NIZKs from (only) DDH via the CIH framework.

Indeed, the bad challenge function in trapdoor sigma protocols involves (at least) *extraction* from the commitment scheme used in the protocol, and it is unclear whether such extraction can be represented by probablistic constant-degree polynomials when the commitment scheme is constructed from standard group-based assumptions. For example, the decryption circuit for the ElGamal encryption scheme [27] (based on DDH) is in a higher complexity class, and is not known to have representation by probabilistic constant-degree polynomials. Indeed, there are known lower-bounds for functions that can be approximated by probabilistic polynomials. Specifically, [41,47,55,56] proved that approximating a n fan-in majority gate by probabilistic polynomials over binary field with a small constant error requires degree at least $\Omega(\sqrt{n})$.

Roadmap. We overcome the above dilemma by exploiting the power of interaction.

- In Sect. 2.1, we introduce the notion of interactive trapdoor hashing protocols (ITDH) – a generalization of TDH to multi-round interactive protocols.

[5] The breakthrough work of [12] shows that in the case of homomorphic secret-sharing, it is in fact possible to go beyond linear homomorphisms in traditional groups. The communication complexity of the sender in their scenario, however, grows with the input length and is *not* compact as in the case of TDH.

We show that despite increased interaction, ITDH can be used to build CIH. Namely, we devise a round-collapsing approach to construct CIH from ITDH.
- We next show that ITDH can capture a larger class of computations than what can be supported by known constructions of TDH. Namely, we construct a constant-round ITDH protocol for TC^0 where the sender is laconic (Sect. 2.2).
- Finally, we demonstrate that using DDH, it is possible to construct trapdoor sigma protocols where the bad challenge function can be computed in low depth. Using such sigma protocols, we build multi-theorem (statistical) NIZK and statistical Zap arguments for NP (Sects. 2.3 and 2.4, respectively).

2.1 Interactive Trapdoor Hashing Protocols

We start by providing an informal definition of ITDH and then describe our strategy for constructing CIH from ITDH.

Defining ITDH. An L-level ITDH is an interactive protocol between a "sender" and a "receiver", where the receiver's input is a circuit f and the sender's input is a string x. The two parties jointly compute $f(x)$ by multiple rounds of communication that are divided into L *levels.* Each level $\ell \in [L]$ consists of two consecutive protocol messages – a receiver's message, followed by the sender's response:

- First, the receiver uses f (and prior protocol information) to compute a key k_ℓ and trapdoor td_ℓ. It sends the key k_ℓ to the sender.
- Upon receiving this message, the sender computes a hash value h_ℓ together with an encoding e_ℓ. The sender sends h_ℓ to the receiver but keeps e_ℓ to herself. (The encoding e_ℓ can be viewed as sender's "private state" used for computing the next level message.)

Upon receiving the level L (i.e., final) message h_L from the sender, the receiver computes a decoding value d using the trapdoor. The function output $f(x)$ can be recovered by computing $\mathsf{e} \oplus \mathsf{d}$, where e is the *final* level encoding computed by the sender. We require the following properties from ITDH:

- **Compactness**: The sender's message in every level must be *compact.* Specifically, for every level $\ell \in [L]$, the size of the hash value h_ℓ is bounded by the security parameter, and is independent of the length of the sender's input x and the size of the circuit f.
- **Approximate Correctness**: For an overwhelming fraction of the random tapes for the receiver, for any input x, the Hamming distance between $\mathsf{e} \oplus \mathsf{d}$ and $f(x)$ must be small. Note that this is an *adaptive* definition in that the input x is chosen after the randomness for the receiver is fixed.
- **Leveled Function Privacy**: The receiver's messages computationally hide the circuit f. Specifically, we require that the receiver's message in every level can be simulated without knowledge of the circuit f. Moreover, we allow the privacy guarantee to be *different* for each level by use of different security parameters for different levels.

As we discuss in Sect. 4.1, barring some differences in syntax, trapdoor hash functions can be viewed as 1-level ITDH. We refer the reader to the technical sections for a formal definition of ITDH.

CIH from ITDH. We now describe our round-collapsing strategy for constructing CIH from ITDH. Given an L-level ITDH for a circuit family $\mathcal{C}$, we construct a family of CIH for relations searchable by $\mathcal{C}$ as follows:

- **Key Generation:** The key generation algorithm uses the function-privacy simulator for ITDH to compute a simulated receiver message for *every* level. It outputs a key k consisting of L simulated receiver messages (one for each level) as well as a random mask mask.
- **Hash Function:** Given a key k and an input x, the hash function uses the ITDH sender algorithm on input x to perform an ITDH protocol execution "in its head." Specifically, for every level $\ell \in [L]$, it reads the corresponding receiver message in the key k and uses it to computes the hash value and the encoding for that level. By proceeding in a level-by-level fashion, it obtains the final level encoding e. It outputs $\mathsf{e} \oplus \mathsf{mask}$.

We now sketch the proof for correlation intractability. For simplicity, we first consider the case when $L = 1$. We then extend the proof strategy to the multi-level case.

For $L = 1$, the proof of correlation intractability resembles the proof in [13]. We first switch the simulated receiver message in the CIH key to a "real" message honestly computed using a circuit $C \in \mathcal{C}$. Now, suppose that the adversary finds an x such that $\mathsf{Hash}(\mathsf{k}, x) = C(x)$. Then by approximate correctness of ITDH, $C(x) \approx \mathsf{e} \oplus \mathsf{d}$, where the "$\approx$" notation denotes closeness in Hamming distance. This implies that $\mathsf{e} \oplus \mathsf{d} \approx \mathsf{e} \oplus \mathsf{mask}$, and thus $\mathsf{d} \approx \mathsf{mask}$. However, once we fix the randomness used by the receiver, d *only depends on* h. Since h is compact, the value d is exponentially "sparse" in its range. Therefore, the probability that $\mathsf{d} \approx \mathsf{mask}$ is exponentially small, and thus such an input x exists with only negligible probability.

Let us now consider the multi-level case. Our starting idea is to switch the simulated receiver messages in the CIH key to "real" messages in a level-by-level manner. However, note that the honest receiver message at each level depends on the hash value sent by the sender in the previous level, and at the time of the key generation of the CIH, the sender's input has not been determined. Hence, it is not immediately clear how to compute the honest receiver message at each level without knowing the sender's input.

To get around this issue, at each level ℓ, we first simply *guess* the sender's hash value $\mathsf{h}_{\ell-1}$ in the previous level $(\ell - 1)$, and then switch the simulated receiver message in level ℓ to one computed honestly using the ITDH receiver algorithm on input $\mathsf{h}_{\ell-1}$. To ensure this guessing succeeds with high probability, we rely on the *compactness* of the hash values. Specifically, let λ_ℓ denote the security parameter for the ℓth level in ITDH (as mentioned earlier, we allow the security parameters for each level to be different). Then the guessing of the level $(\ell - 1)$ hash value succeeds with probability $2^{-\lambda_{\ell-1}}$. We set $\lambda_{\ell-1}$ to be sublinear in λ, where λ is the security parameter for CIH. Then, when we reach the final level, all our guesses are successful with probability $2^{-(\lambda_1+\lambda_2+\ldots+\lambda_L)}$, which is sub-exponential in λ. Since the probability of $\mathsf{d} \approx \mathsf{mask}$ can be exponentially small in λ, we can still get a contradiction.

However, the above argument assumes the function privacy is perfect, which is not the case. Indeed, at every level, we must also account for the adversary's distinguishing advantage when we switch a simulated message to a real message. In order to make the above argument go through, we need the distinguishing advantage to be a magnitude smaller than $2^{-\lambda_{\ell-1}}$ (for every ℓ). That is, we require ITDH to satisfy sub-exponential leveled functional privacy. Now, the distinguishing advantage can be bounded by $2^{-\lambda_\ell^c}$, where $0 < c < 1$ is a constant. Once we choose λ_ℓ large enough, then $2^{-\lambda_\ell^c}$ can be much smaller than $2^{-\lambda_{\ell-1}}$, and thus the above argument goes through as long as L is not too large.

In particular, there is room for trade-off between the number of levels in ITDH that we can collapse and the amount of leveled function privacy required. If we wish to rely on *polynomial time* and sub-exponential advantage assumptions, then the above transformation requires the number of levels to be *constant.* If we allow for *sub-exponential time* (and sub-exponential advantage) assumptions, then the above transformation can work for up to $O(\log\log\lambda)$ levels. We refer the reader to Sect. 6.2 for more details.

2.2 Constructing ITDH

We now provide an overview of our construction of constant-round ITDH for TC^0. Let *not-threshold* gate be a gate that computes a threshold gate and then outputs its negation. Since not-threshold gates are universal for threshold circuits, it suffices for our purposes to consider circuits that consist of only not-threshold gates.

At a high-level, we implement the following two-step blueprint for constructing ITDH:

- **Step 1 (Depth-1 Circuits):** First, we build an ITDH for a simple circuit family $\mathcal{T}$ where each circuit is simply a *single* layer of layer of not-threshold gates.
- **Step 2 (Sequential Composition):** Next, to compute circuits with larger depth, we *sequentially compose* multiple instances of ITDH from the first step, where the output of the ith ITDH is used as an input in the $(i+1)^{\text{th}}$ ITDH.

Overall, our construction uses only one cryptographic tool, namely, TDH for linear functions. As we will see later, we will use additional ideas to introduce non-linearity in the computation.

In the following, we elaborate on each of these steps. We first focus on step 2, namely, the sequential composition step, and discuss the main challenges therein. We will later describe how we implement step 1.

Controlling the Error. Recall that ITDH guarantees only approximate correctness, i.e., the xor of the final-level encoding e and decoding d is "close" (in terms of Hamming distance) to the true function output. Then, in a sequential composition of an ITDH protocol, *each* execution only guarantees approximate correctness. This means that the errors could *spread* across the executions, ultimately causing *every output bit of the final execution to be incorrect.* For example, suppose a coordinate of the output for an intermediate execution is flipped and later, the computation of every output bit depends on this flipped output bit. In this case, every output bit could be incorrect.

To overcome this issue, we observe that any circuit can be converted to a new circuit that satisfies a "parallel structure" demonstrated in Fig. 1.

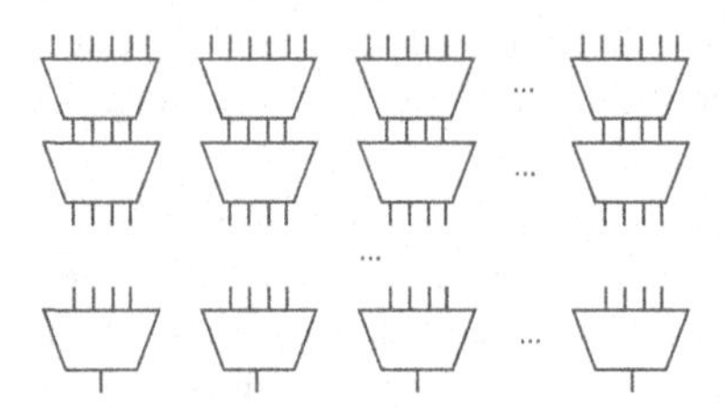

Fig. 1. Parallel structure. The top (resp., bottom) layer corresponds to input (resp., output) wires.

In such circuits, each output bit only depends on the input to *one* parallel repetition. Hence, the spreading of one Hamming error is controlled in one parallel execution. We leverage this observation to prove approximate correctness of the sequential composition.

Input Passing. Recall that the protocol output in any ITDH execution is "secret shared" between the sender and the receiver, where the sender holds the final level encoding e, and the receiver holds the decoding d. Then, a plausible way to implement Step 2 is for the receiver to simply send the decoding in the ith ITDH to the sender so that the latter can compute the output, and then use it as input in the $(i+1)^{\text{th}}$ ITDH. However, this leaks intermediate wire values (of the TC^0 circuit that we wish to compute) to the sender, thereby compromising function privacy. Note that the reverse strategy of requiring the sender to send the encoding to the receiver (to allow output computation) also does not work since it violates the compactness requirement on the sender's messages to the receiver.

To resolve this issue, we keep the secret-sharing structure of the output in every ITDH intact. Instead, we extend the functionality of the ITDH in Step 1 so that the output of the ith ITDH can be computed *within* the $(i+1)^{\text{th}}$ ITDH. Specifically, in Step 1, we construct an ITDH for a circuit family $\mathcal{T}^{\oplus}$ where every circuit consists of a single layer of Xor-*then*-Not-Threshold gates, namely, gates that first XOR the input with a pre-hardwired string and then compute the not-threshold operation on the resulting value. This allows for resolving the above problem as follows: the final-level encoding from the ith ITDH constitutes the sender's input in the $(i+1)^{\text{th}}$ ITDH. On the other hand, the decoding in the ith ITDH is used as the pre-hardwired string in the circuit computed by the $(i+1)^{\text{th}}$ ITDH.

Putting together these ideas in a careful manner, we are able to implement Step 2. We refer the reader to the technical section for more details on this step.

ITDH for $\mathcal{T}^{\oplus}$. We now discuss how we implement *revised* step 1, namely constructing an ITDH for $\mathcal{T}^{\oplus}$, where every circuit consists of a single layer of Xor-*then*-Not-Threshold gates. At a high-level, we proceed in the following three steps:

- We first "decompose" a circuit in $\mathcal{T}^{\oplus}$ as the composition of two linear functions.
- Next, we use a 1-level ITDH, which is implied by TDH, to compute each of these linear functions.
- Finally, we "compose" the two ITDH executions sequentially to obtain a 2-level ITDH for $\mathcal{T}^{\oplus}$.

An observant reader may wonder how we decompose the computation of threshold gates into linear functions. Indeed, composition of linear functions is still a linear function, while a threshold gate involves non-linear computation. As we will soon see, our decomposition strategy crucially relies on some "offline" processing by the parties on the intermediate encoding and decoding values between different TDH executions. This introduces the desired non-linearity in the computation.

For simplicity, let us focus on the simpler goal of computing a *single* Xor-*then*-Not-Threshold gate. Our ideas easily extend to the more general setting. To compute such a gate, we proceed in three simple steps.

- First, bitwise xor the input string x with another string y, where y is hard-wired in the circuit description.
- Next, sum the elements in the string $x \oplus y$.
- Finally, compare the summation with the threshold t (defined by the gate).

For the *first* step, let a and b be two bits at (say) the ith coordinate of x and y, respectively. Then $a \oplus b = 1$ if and only if $a = 0 \wedge b = 1$ or $a = 1 \wedge b = 0$. Hence, $a \oplus b = (1-a) \cdot b + a \cdot (1-b)$. Since b is part of the circuit description, the right hand side is a linear function of a over $\mathbb{Z}$. For the *second* step, we simply sum over the result of step 1 on all coordinates. Combining the first step and the second step, this summation is still a linear function of x over $\mathbb{Z}$, and thus we can use a TDH for linear functions to compute such a summation. We note, however, that the known construction of TDH in [13,25] is only for linear functions over $\mathbb{Z}_2$. We therefore extend the TDH construction in [13,25] to arbitrary polynomial modulus. In our case, since the summation cannot be more than n, it suffices to choose the modulo $(n+1)$.

We now proceed to express the comparison in the *third* step as a linear function. We start with a simpler case. Suppose that the summation obtained from the second step is $\mathsf{sum} \in \{0, 1, 2, \ldots, n\}$ and we want to compare it with a threshold t. Let $\mathbb{1}_{\mathsf{sum}}$ denote the indicator vector of x, i.e., $\mathbb{1}_{\mathsf{sum}} =$

$(0, 0, \ldots, 0, 1, 0, \ldots, 0)$, where the $(\mathsf{sum}+1)^{\text{th}}$ coordinate is 1, and all other coordinates are 0. Then, we have that

$$\mathsf{sum} < t \iff \langle \mathbb{1}_{\mathsf{sum}}, \mathbb{1}_{<t} \rangle = 1,$$

where $\mathbb{1}_{<t} = (1, 1, \ldots, 1, 0, \ldots, 0)$ is a vector with 1's on the first t coordinates, and 0's on the remaining coordinates. We can therefore express the comparison as the inner product between $\mathbb{1}_{\mathsf{sum}}$ and $\mathbb{1}_{<t}$, which is a linear function over $\mathbb{1}_{\mathsf{sum}}$. Hence, such a comparison can be computed by a TDH for linear functions over $\mathbb{Z}_2$.

The above discussion is oversimplified, however, since the sender and the receiver do not have the value sum. Instead, at the end of the "previous" TDH execution, the sender and the receiver only obtained encoding e and a decoding d, respectively, such that $(\mathsf{e}+\mathsf{d}) \bmod R = \mathsf{sum}$. Fortunately, we can still express the comparison $(\mathsf{e}+\mathsf{d}) \bmod R < t$ as

$$(\mathsf{e}+\mathsf{d}) \bmod R < t \iff \langle \mathbb{1}_{\mathsf{e}}, \mathbb{1}_{j,<t} \rangle = 1,$$

where $\mathbb{1}_{\mathsf{e}}$ is the indicator vector for e and $\mathbb{1}_{j,<t} = \sum_{j=0}^{t-1} \mathbb{1}_{(j-\mathsf{d}) \bmod R}$. This expression works because comparing $(\mathsf{e}+\mathsf{d}) \bmod R < t$ is equivalent to checking if there exists a $0 \leq j < t$ such that $(\mathsf{e}+\mathsf{d}) \mod R = j$, which is equivalent to checking whether $\mathsf{e} = (j - \mathsf{d}) \bmod R$. Note that the right hand side of this formula is a linear function of $\mathbb{1}_{\mathsf{e}}$, and can thus be computed using a TDH for linear functions over $\mathbb{Z}_2$.

In the above two executions of TDH, the sender processes e from the first TDH execution to obtain $\mathbb{1}_{\mathsf{e}}$, and uses it as the input to the second TDH. The receiver processes d from the first TDH execution to obtain $\mathbb{1}_{j,<t}$, and uses it as the function for the second TDH execution. Note that this intermediate processing is non-linear, since computing the indicator vector can be done by several equality checks, and equality check is not a linear function. Hence, it introduces the necessary non-linearity in the computation, but is done "outside" of the TDH executions.

2.3 Constructing NIZKs

Armed with our construction of CIH, we now sketch the main ideas underlying our construction of (statistical) multi-theorem NIZK for NP. We proceed in the following two steps:

1. First, using CIH for TC^0, we construct a non-interactive witness indistinguishable (NIWI) argument for NP in the common random string model. Our construction satisfies either statistical WI and non-adaptive soundness, or computational WI and adaptive soundness.
2. We then transform the above NIWI into an adaptive, *multi-theorem* NIZK for NP in the common random string model via a variant of the Feige-Lapidot-Shamir (FLS) "OR-trick" [28].[6] Our NIZK satisfies either statistical ZK and

[6] By using "programmable" CIH, one could directly obtain NIZKs in the first step. However, the resulting NIZK only achieves *single-theorem* ZK; hence an additional step is still required to obtain multi-theorem NIZKs.

non-adaptive soundness, or computational ZK and adaptive soundness. Crucially, our transformation does *not* require "CRS switching" in the security proof and hence works for both cases seamlessly while preserving the distribution of the CRS in the underlying NIWI.

Statistical NIZKs. In the remainder of this section, we focus on the construction of *statistical* NIZKs. We briefly discuss the steps necessary for obtaining the computational variant (with adaptive soundness) at the end of the section.

Towards implementing the first of the above two steps, we first build the following two ingredients:

- A lossy public key encryption scheme with an additional property that we refer to as *low-depth decryption*, from DDH. Roughly speaking, this property requires that there exists a TC^0 circuit Dec that takes as input any ciphertext ct and a secret key sk, and outputs the correct plaintext.
- A trapdoor sigma protocol for NP with bad challenge function in TC^0 from the above lossy public key encrytion scheme. We also require the trapdoor sigma protocol to satisfy an additional "knowledge extraction" property, which can be viewed as an analogue of special soundness for trapdoor sigma protocols. Looking ahead, we use this property to construct NIWIs with argument of knowledge property, which in turn is required for our FLS variant.

Lossy Public Key Encryption. The lossy public key encryption we use is essentially the same as in [3,40,51]. We start by briefly describing the scheme.

A public key $\mathsf{pk} = \begin{bmatrix} g^1 & g^b \\ g^a & g^c \end{bmatrix}$ is a matrix of elements in a group $\mathbb{G}$. When the matrix $\begin{bmatrix} 1 & b \\ a & c \end{bmatrix}$ is singular (i.e., $c = ab$), then the public key is in the "injective mode" and the secret key is $\mathsf{sk} = a$; when the matrix is non-singular (i.e., $c \neq ab$), then the public key is in the "lossy mode." The encryption algorithm is described as follows:

$$\mathsf{Enc}\left(\mathsf{pk}, m \in \{0,1\}; r = \begin{bmatrix} r_1 \\ r_2 \end{bmatrix}\right) = \begin{bmatrix} (g^1)^{r_1} \cdot (g^b)^{r_2} \\ (g^a)^{r_1} \cdot (g^c)^{r_2} \cdot g^m \end{bmatrix} = g^{\begin{bmatrix} 1 & b \\ a & c \end{bmatrix}\begin{bmatrix} r_1 \\ r_2 \end{bmatrix} + \begin{bmatrix} 0 \\ m \end{bmatrix}}.$$

Let us now argue the low-depth decryption property. Let $[c_1, c_2]^T$ denote the ciphertext obtained by encrypting a message m using an injective mode public key pk with secret key $\mathsf{sk} = a$. To decrypt the ciphertext, we can compute $c_1^{-a} \cdot c_2 = g^m$ and then comparing with $1_{\mathbb{G}}$ to recover m. However, it is not known whether c_1^{-a} can be computed in TC^0 (recall that a depends on the security parameter).

Towards achieving the low-depth decryption property, we use the following observation. Let $a_0, a_1, \ldots a_\lambda$ be the binary representation of a. Then, we have that

$$\left(c_1^{-2^0}\right)^{a_0} \cdot \left(c_1^{-2^1}\right)^{a_1} \cdot \left(c_1^{-2^2}\right)^{a_2} \cdot \ldots \cdot \left(c_1^{-2^\lambda}\right)^{a_\lambda} \cdot c_2 = g^m.$$

Note that given $[c_1, c_2]^T$, one can "precompute" $c_1^{-2^0}, c_1^{-2^1}, \ldots, c_1^{-2^\lambda}$ without using the secret key sk. In our application to NIZKs and Zaps, such precomputation can be performed by the prover and the verifier.

We leverage this observation to slightly modify the definition of low-depth decryption to allow for a *deterministic* polynomial-time "precomputation" algorithm PreComp. Specifically, we require that the output of $\mathsf{Dec}(\mathsf{PreComp}(1^\lambda, \mathsf{ct}), \mathsf{sk})$ is the correct plaintext m. We set $\mathsf{PreComp}(1^\lambda, c) = (c_1^{-2^0}, c_1^{-2^1}, \ldots, c_1^{-2^\lambda}, c_2)$, and allow the circuit Dec to receive $c_1^{-2^0}, c_1^{-2^1}, \ldots, c_1^{-2^\lambda}, c_2$ and $a_0, a_1, \ldots, a_\lambda$ as input. The decryption circuit Dec proceeds in the following steps:

- For each $i = 0, 1, \ldots, \lambda$, it chooses g_i to be either $1_{\mathbb{G}}$ or $c_1^{-2^i}$, such that $g_i = (c_1^{-2^i})^{a_i}$. This computation can be done in constant depth, and is hence in TC^0.
- Multiply the values $g_0, g_2, \ldots, g_\lambda$ and c_2. From [52], this iterative multiplication can be computed in TC^0 when we instantiate $\mathbb{G}$ as a subgroup of $\mathbb{Z}_q^*$.
- Compare the resulting value with $1_{\mathbb{G}}$. If they are equal, then output 0. Otherwise output 1.

Since each of the above steps can be computed in TC^0, we have that Dec is also in TC^0.

Trapdoor Sigma Protocol for NP. Recently, Brakerski et al. [13] constructed a "commit-and-open" style trapdoor sigma protocol where the only cryptographic primitive used is a commitment scheme. Crucially, the bad challenge function for their protocol involves the following two computations: extraction from the commitment, and a post-extraction verification using 3-CNF. By exploiting the specific form of their bad challenge function, we construct a trapdoor sigma protocol for NP with our desired properties by simply instantiating the commitment scheme in their protocol with the above lossy encryption scheme.

Let us analyze the bad challenge function of the resulting trapdoor sigma protocol. Since our lossy public key encryption satisfies the low-depth decryption property, the first step of the bad challenge computation can be done in TC^0. Next, note that the second step of the bad challenge computation is also in TC^0 since it involves evaluation of 3-CNF which can be computed in AC^0. Thus, the bad challenge function is in TC^0.

We observe that our protocol also satisfies a *knowledge extraction* property which requires that one can efficiently extract a witness from a *single* accepting transcript (α, β, γ) by using a trapdoor (namely, the secret key of the lossy public key encryption), if β does not equal to the output of the bad challenge function evaluated on α. We use this property to construct NIWIs with argument of knowledge property.

NIWI from Fiat-Shamir via CIH. We construct NIWI arguments in the CRS model by using CIH to collapse the rounds of our trapdoor sigma protocol repeated λ times in parallel. The CRS of the resulting construction contains a public-key of lossy public key encryption scheme from above and a CIH key.

When the public key is in lossy mode, the NIWI achieves statistical WI property and non-adaptive argument of knowledge property.

To prove the argument of knowledge property, we observe that for any accepting transcript $(\{\alpha_i\}_{i\in[\lambda]}, \{\beta_i\}_{i\in[\lambda]}, \{\gamma_i\}_{i\in[\lambda]})$, it follows from correlation intractability of the CIH that $\{\beta_i\}_{i\in[\lambda]}$ is *not* equal to the outputs of the bad challenge function evaluated on $\{\alpha_i\}_{i\in[\lambda]}$. Hence, there exists at least one index i^* such that β_{i^*} is not equal to the output of the bad challenge function on α_{i^*}. We can now extract a witness by relying on the knowledge extraction property of the i^*-th parallel execution of the trapdoor sigma protocol.

From NIWI to Multi-theorem NIZK. The FLS "OR-trick" [28] is a standard methodology to transform NIWIs (or single-theorem NIZKs) into multi-theorem NIZKs. Roughly speaking, the trick involves supplementing the CRS with an instance (say) y of a hard-on-average decision problem and requiring the prover to prove that either the "original" instance (say) x *or* y is true. This methodology involves switching the CRS either in the proof of soundness or zero-knowledge, which can potentially result in a degradation of security. E.g., in the former case, one may end up with non-adaptive (computational) soundness while in the latter case, one may end up with computational ZK even if the underlying scheme achieves statistical privacy. The instance y also needs to be chosen carefully depending on the desired security and whether one wants the resulting CRS to be a reference string or a random string.

We consider a variant of the "OR-trick" that does not require CRS switching and preserves the distribution of the CRS of the underlying scheme. We supplement the CRS with an instance of average-hard *search* problem, where the instance is subjected to the *uniform* distribution. For our purposes, the discrete logarithm problem suffices. The ZK simulator simply uses the secret exponent of the discrete-log instance in the CRS to simulate the proof. On the other hand, soundness can be argued by relying on the computational hardness of the discrete-log problem. One caveat of this transformation is that the proof of soundness requires the underlying NIWI to satisfy argument of knowledge property. We, note, however, that this property is usually easy to achieve (in the CRS model).

Using this approach, we obtain statistical multi-theorem NIZK arguments in the common *random* string model from sub-exponential DDH. Previously, group-based statistical NIZKs were known only in the common reference string model [34].

We remark that the above idea can be easily generalized to other settings. For example, starting from LWE-based single-theorem statistical NIZKs [50], one can embed the Shortest Integer Solution (SIS) problem in the CRS to build *multi-theorem* statistical NIZKs in the common random string model. This settles an open question stated in the work of [50].

Computational NIZKs with Adaptive Soundness. Using essentially the same approach as described above, we can also construct computational NIZKs for NP with adaptive soundness. The main difference is that instead of using lossy public-key encryption scheme in the construction of trapdoor sigma protocols, we use ElGamal encryption scheme [27]. Using the same ideas as for our lossy public-key encryption scheme, we observe that the ElGamal encryption scheme also satisfies *low-depth decryption* property. This allows us to follow the same sequence of steps as described above to obtain a computational NIZK for NP with adaptive soundness in the common *random* string model.[7]

2.4 Constructing Zaps

At a high-level, we follow a similar recipe as in the recent works of [2,32] who construct statistical Zap arguments from quasi-polynomial LWE.

The main idea in these works is to replace the (non-interactive) commitment scheme in a trapdoor sigma protocol with a *two-round* statistical-hiding commitment scheme in the plain model and then collapse the rounds of the resulting protocol using CIH, as in the case of NIZKs. Crucially, unlike the non-interactive commitment scheme that only allows for extraction in the CRS model, the two-round commitment scheme must support extraction in the plain model. The key idea for achieving such an extraction property (in conjunction with statistical-hiding property) is to allow for successful extraction with only negligible but still much larger than sub-exponential probability (for example, $2^{-\log^2 \lambda}$) [37]. By carefully using complexity leveraging, one can prove soundness of the resulting argument system.

Statistical-Hiding Commitment with Low-depth Extraction. We implement this approach by replacing the lossy public-key encryption scheme in our NIWI construction (from earlier) with a two-round statistical hiding commitment scheme. Since we need the bad challenge function of the sigma protocol to be in TC^0, we require the commitment scheme to satisfy an additional *low-depth extraction* property.

To construct such a scheme, we first observe that the construction of (public-coin) statistical-hiding extractable commitments in [2,32,37,39] only makes black-box use of a two-round oblivious transfer (OT) scheme. We instantiate this generic construction via the Naor-Pinkas OT scheme based on DDH [45]. By exploiting the specific structure of the generic construction as well as the fact that Naor-Pinkas OT decryption can be computed in TC^0, we are able to show that the extraction process can also be performed in TC^0. We refer the reader to the full version for more details.

3 Preliminaries

For any positive integer $N \in \mathbb{Z}, N > 0$, denote $[N] = \{1, 2, \dots, N\}$. For any integer $R > 0$, and $x \in \mathbb{Z}_R$, $0 \leq x < R$, the indicator vector $\mathbb{1}_x$ of x is a vector

[7] We note that one could obtain computational NIZKs with adaptive soundness by simply "switching the CRS" in our construction of statistical NIZKs. However, the resulting scheme in this case is in the common *reference* string model.

in $\{0,1\}^R$, where the $(x+1)^{\text{th}}$ position is 1, and all other coordinates are zero. A binary relation $\mathcal{R}$ is a subset of $\{0,1\}^* \times \{0,1\}^*$.

Statistical Distance. For any two discrete distributions P, Q, the statistical distance between P and Q is defined as $\mathsf{SD}(P,Q) = \sum_i \left| \Pr[P=i] - \Pr[Q=i] \right| / 2$ where i takes all the values in the support of P and Q.

Hamming Distance. Let n be an integer, and S be a set, and $x = (x_1, x_2, \ldots, x_n)$ and $(y_1, y_2, \ldots, y_n)$ be two tuples in S^n, the Hamming distance $\mathsf{Ham}(x,y)$ is defined as $\mathsf{Ham}(x,y) = |\{i \mid x_i \neq y_i\}|$.

Threshold Gate. Let $x_1, x_2, \ldots, x_n$ be n binary variables. A threshold gate is defined as the following function:

$$\mathsf{Th}_t(x_1, x_2, \ldots, x_n) = \begin{cases} 1 & \sum_{i \in [n]} x_i \geq t \\ 0 & \text{Otherwise} \end{cases}$$

Not-threshold Gate. A not-threshold gate $\overline{\mathsf{Th}}_t$ is the negation of a threshold gate.

Threshold Circuits and TC^0. A threshold circuit is a directed acyclic graph, where each node either computes a threshold gate of unbounded fan-in or a negation gate.

In this work, for any constant L, we use TC^0_L to denote the class of L-depth polynomial-size threshold circuits. When the depth L is not important or is clear from the context, we omit it and simply denote the circuit class TC^0_L as TC^0. The not-threshold gate is universal for TC^0, since we can convert any threshold circuit of constant depth to a constant depth circuit that only contains not-threshold gates. The conversion works as follows: for each negation gate, we convert it to a not-threshold gate with a single input and threshold $t = 1$. For each threshold gate, we convert it to a not-threshold gate with the same input and threshold and then compose it with a negation gate, where the negation gate can be implemented as a not-threshold gate.

We defer more preliminaries to the full version.

4 Interactive Trapdoor Hashing Protocols

In this section, we define interactive trapdoor hashing protocols (ITDH). At a high-level, ITDH is a generalization of trapdoor hash functions – which can be viewed as two-round two-party protocols with specific structural and communication efficiency properties – to multi-round protocols.

More specifically, an interactive trapdoor hashing protocol involves two parties – a sender and a receiver. The sender has an input x, while the receiver has a circuit f. The two parties jointly compute $f(x)$ over several rounds of interaction. We structure the protocols in multiple *levels*, where a level consists of the following two successive rounds:

- The receiver generates a key k and a trapdoor td using a key generation algorithm KGen, which takes as input the circuit f, the level number, and some additional internal state of the receiver. Then it sends k to the sender.
- Upon receiving a key k, the sender computes a hash value h and an encoding e using the algorithm $\mathsf{Hash\&Enc}$, which takes as input x, the key k, the level number, and the previous level encoding. Then it sends the hash h to the receiver, and keeps e as an internal state.

Finally, there is a decoding algorithm Dec that takes the internal state of the receiver after the last level as input, and outputs a decoding value d. Ideally, we want the output $f(x)$ to be $\mathsf{e} \oplus \mathsf{d}$.

In the following, we proceed to formally define this notion and its properties.

Per-level Security Parameter. In our formal definition of ITDH, we allow the security parameter to be *different* for every level. This formulation is guided by our main application, namely, constructing correlation-intractable hash functions (see Sect. 6). Nevertheless, we note that ITDH could also be meaningfully defined w.r.t. a single security parameter for the entire protocol.

4.1 Definition

Let $\mathcal{C} = \{\mathcal{C}_{n,u}\}_{n,u}$ be a family of circuits, where each circuit $f \in \mathcal{C}_{n,u}$ is a circuit of input length n and output length u. An L-level interactive trapdoor hashing protocol for the circuit family $\mathcal{C}$ is a tuple of algorithms $\mathsf{ITDH} = (\mathsf{KGen}, \mathsf{Hash\&Enc}, \mathsf{Dec})$ that are described below.

We use $\lambda_1, \ldots, \lambda_L$ to denote the security parameters for different levels. Throughout this work, these parameters are set so that they are polynomially related. That is, there exists a λ such that $\lambda_1, \ldots, \lambda_L$ are polynomials in λ.

- $\mathsf{KGen}(1^{\lambda_\ell}, \ell, f, \mathsf{h}_{\ell-1}, \mathsf{td}_{\ell-1})$: The key generation algorithm takes as input a security parameter λ_ℓ (that varies with the level number), a level number ℓ, a circuit $f \in \mathcal{C}_{n,u}$, a level $(\ell - 1)$ hash value $\mathsf{h}_{\ell-1}$ and trapdoor $\mathsf{td}_{\ell-1}$ (for $\ell = 1$, $\mathsf{h}_{\ell-1} = \mathsf{td}_{\ell-1} = \bot$). It outputs an ℓth level key k_ℓ and a trapdoor td_ℓ.
- $\mathsf{Hash\&Enc}(\mathsf{k}_\ell, x, \mathsf{e}_{\ell-1})$: The hash-and-encode algorithm takes as input a level ℓ hash key k_ℓ, an input x, and a level $(\ell - 1)$ encoding $\mathsf{e}_{\ell-1}$. It outputs an ℓth level hash value h_ℓ and an encoding $\mathsf{e}_\ell \in \{0,1\}^u$. When $\ell = 1$, we let $\mathsf{e}_{\ell-1} = \bot$.
- $\mathsf{Dec}(\mathsf{td}_L, \mathsf{h}_L)$: The decoding algorithm takes as input a level L trapdoor td_L and hash value h_L, and outputs a value $\mathsf{d} \in \{0,1\}^u$.

We require ITDH to satisfy the following properties:

- **Compactness:** For each level $\ell \in [L]$, the bit length of h_ℓ is at most λ_ℓ.
- **(Δ, ϵ)-Approximate Correctness:** For any $n, u \in \mathbb{N}$, any circuit $f \in \mathcal{C}_{n,u}$ and any sequence of security parameters $(\lambda_1, \ldots, \lambda_L)$, we have

$$\Pr_{r_1, r_2, \ldots, r_L}[\forall x \in \{0,1\}^n, \mathsf{Ham}(\mathsf{e} \oplus \mathsf{d}, f(x)) < \Delta(u)] > 1 - \epsilon(u, \lambda_1, \ldots, \lambda_L),$$

where e, d are obtained by the following procedure: Let $\mathsf{h}_0 = \mathsf{td}_0 = \mathsf{e}_0 = \bot$. For $\ell = 1, 2, \ldots, L$,

- Compute $(\mathsf{k}_\ell, \mathsf{td}_\ell) \leftarrow \mathsf{KGen}(1^{\lambda_\ell}, \ell, f, \mathsf{h}_{\ell-1}, \mathsf{td}_{\ell-1}; r_\ell)$ using random coins r_ℓ.
- Hash and encode the input x: $(\mathsf{h}_\ell, \mathsf{e}_\ell) \leftarrow \mathsf{Hash\&Enc}(\mathsf{k}_\ell, x, \mathsf{e}_{\ell-1})$.

Finally, let $\mathsf{e} = \mathsf{e}_L$ be the encoding at the final level, and $\mathsf{d} = \mathsf{Dec}(\mathsf{td}_L, \mathsf{h}_L)$.

- **Leveled Function Privacy:** There exist a simulator Sim and a negligible function $\nu(\cdot)$ such that for any level $\ell \in [L]$, any polynomials $n(\cdot)$ and $u(\cdot)$ in the security parameter, any circuit $f \in \mathcal{C}_{n,u}$, any trapdoor $\mathsf{td}' \in \{0,1\}^{|\mathsf{td}_{\ell-1}|}$, any hash value $\mathsf{h}' \in \{0,1\}^{|\mathsf{h}_{\ell-1}|}$, and any n.u. PPT distinguisher $\mathcal{D}$,

$$\Big| \Pr\left[(\mathsf{k}_\ell, \mathsf{td}_\ell) \leftarrow \mathsf{KGen}(1^{\lambda_\ell}, \ell, f, \mathsf{h}', \mathsf{td}') : \mathcal{D}(1^{\lambda_\ell}, \mathsf{k}_\ell) = 1\right] - \Pr\left[\widetilde{\mathsf{k}}_\ell \leftarrow \mathsf{Sim}(1^{\lambda_\ell}, 1^n, 1^u, \ell) : \mathcal{D}(1^{\lambda_\ell}, \widetilde{\mathsf{k}}_\ell) = 1\right] \Big| \leq \nu(\lambda_\ell).$$

We say that the ITDH satisfies sub-exponential leveled function privacy, if there exists a constant $0 < c < 1$ such that for any n.u. PPT distinguisher, $\nu(\lambda_\ell)$ is bounded by $2^{-\lambda_\ell^c}$ for any sufficiently large λ_ℓ.

Note that since the security parameters for different levels are polynomially related, $n(\cdot)$ and $u(\cdot)$ are polynomials in λ_ℓ iff they are polynomials in λ.

Relationship with Trapdoor Hash Functions. A 1-level ITDH is essentially the same as TDH, except that in TDH, there are two kinds of keys: a hash key and an encoding key. In particular, a hash value is computed using the hash key and can be reused with different encoding keys for different functions. In 1-level ITDH, however, the receiver's message only consists of one key that is used by the sender for computing both the hash value and the encoding. Therefore, the hash value is not reusable for different functions.

We choose the above formulation of ITDH for the sake of a simpler and cleaner definition. Moreover, if we consider multi-bit output functions, then the above difference disappears, since we can combine multiple functions into one multi-bit output function and encode it using one key.

5 Construction of ITDH

In this section, we construct an interactive trapdoor hashing protocol (ITDH) for TC^0 circuits. We refer the reader to Sect. 2 for a high-level overview of our approach. The remainder of this section is organized as follows:

- **Depth-1 Circuits:** In Sect. 5.1, we first construct a 2-level ITDH protocol for $\mathcal{T}^{\oplus}$ – roughly speaking, a family of depth-1 Xor-*then*-Not-Threshold circuits (see below for the precise definition of $\mathcal{T}^{\oplus}$).
- **Sequential Composition:** Next, in Sect. 5.2, we present a sequential composition theorem for ITDH where we show how to compose L instances of a 2-level ITDH for some circuit family to obtain a $2L$-level ITDH for a related circuit family.
- **Construction for TC^0:** Finally, in Sect. 5.3, we put these two constructions together to obtain an ITDH for TC^0.

5.1 ITDH for $\mathcal{T}^{\oplus}$

We start by introducing some notation and definitions.

XOR-*then*-Compute Circuits. Let $\mathcal{C} = \{\mathcal{C}_{n,u}\}_{n,u}$ be a circuit family, where for any n and u, $\mathcal{C}_{n,u}$ contains circuits with n-bit inputs and u-bit outputs. For any $\mathcal{C}$, we define an Xor-*then*-Compute circuit family $\mathcal{C}^{\oplus} = \{\mathcal{C}^{\oplus}_{n,u}\}_{n,u}$ consisting of circuits that *first* compute a bit-wise xor operation on the input with a fixed string and *then* compute a circuit in $\mathcal{C}$ on the resulting value.

Specifically, $\mathcal{C}^{\oplus}_{n,u}$ contains all the circuit $\mathrm{C}^{\oplus y} : \{0,1\}^n \to \{0,1\}^u$, where $y \in \{0,1\}^n$ and there exists a $C \in \mathcal{C}_{n,u}$ such that for every $x \in \{0,1\}^n$,

$$\mathrm{C}^{\oplus y}(x) = C(x \oplus y).$$

Circuit Families $\mathcal{T}$ and $\mathcal{T}^{\oplus}$. We define a circuit family $\mathcal{T} = \{\mathcal{T}_{n,u}\}_{n,u}$ consisting of depth-1 not-threshold circuits, i.e., a single layer of not-threshold gates (see Sect. 3). Specifically, $\mathcal{T}_{n,u}$ contains all circuits $\mathrm{T}_{\vec{t},\vec{I}} : \{0,1\}^n \to \{0,1\}^u$ where $\vec{t} = \{t_1, \dots, t_u\}$ is a set of positive integers, and $\vec{I} = \{I_1, \dots, I_u\}$ is a collection of sets $I_j \subseteq [n]$ s.t. for any $x \in \{0,1\}^n$,

$$\mathrm{T}_{\vec{t},\vec{I}}(x) = \left(\overline{\mathsf{Th}}_{t_1}(x[I_1]), \dots, \overline{\mathsf{Th}}_{t_u}(x[I_u])\right),$$

where for any index set $I_j = \{i_1, i_2, \dots, i_w\} \subseteq [n]$, we denote $x[I_j] = (x_{i_1}, x_{i_2}, \dots, x_{i_w})$ as the projection of string x to the set I_j.

The function family $\mathcal{T}^{\oplus} = \{\mathcal{T}^{\oplus}_{n,u}\}_{n,u}$ is defined as the Xor-*then*-Compute family corresponding to $\mathcal{T}$. We denote the circuits in $\mathcal{T}^{\oplus}_{n,u}$ as $\mathrm{T}^{\oplus y}_{\vec{t},\vec{I}}$, where $\vec{t}$, $\vec{I}$ and y are as defined above.

For a high-level overview of our construction, see Sect. 2.2. We now proceed to give a formal description of our construction.

Construction of ITDH for $\mathcal{T}^{\oplus}$. We construct a 2-level interactive trapdoor hashing protocol $\mathsf{ITDH} = (\mathsf{KGen}, \mathsf{Hash\&Enc}, \mathsf{Dec})$ for the circuit family $\mathcal{T}^{\oplus}$ as defined above. Our construction relies on the following ingredient: a trapdoor hash function $\mathsf{TDH} = (\mathsf{TDH.HKGen}, \mathsf{TDH.EKGen}, \mathsf{TDH.Hash}, \mathsf{TDH.Enc}, \mathsf{TDH.Dec})$ for the linear function family $\mathcal{F} = \{\mathcal{F}_{n,R}\}_{n,R}$ that achieves τ-enhanced correctness and function privacy.

For ease of exposition, we describe the algorithms of ITDH *separately* for each level. The first level algorithms of ITDH internally use TDH to evaluate a circuit (defined below) with input length $n_1 = n$ and modulus $R_1 = n + 1$. The second level algorithms of ITDH internally use TDH to evaluate another circuit (defined below) with input length $n_2 = R_1 \cdot u$ and modulus $R_2 = 2$. We use λ_1 and λ_2 to denote the security parameters input to the first and second level algorithms, respectively.

- **Level 1** $\mathsf{KGen}(1^{\lambda_1}, 1, \mathrm{T}^{\oplus y}_{\vec{t},\vec{I}}, \mathsf{h}_0 = \bot, \mathsf{td}_0 = \bot)$**:**
  - Sample a hash key of TDH w.r.t. security parameter λ_1, input length $n_1 = n$ and modulus $R_1 = n + 1$

$$\mathsf{hk}_1 \leftarrow \mathsf{TDH.HKGen}(1^{\lambda_1}, 1^{n_1=n}, 1^{R_1=n+1})$$

  - Parse $\vec{I} = \{I_1, \ldots, I_u\}$. For every $i \in [u]$, sample an encoding key:

$$(\mathsf{ek}_{1,i}, \mathsf{td}_{1,i}) \leftarrow \mathsf{TDH.EKGen}(\mathsf{hk}_1, \mathsf{XorSum}_{I_i,y})$$

    where for any set $I \subseteq [n]$, $\mathsf{XorSum}_{I,y}$ is the linear function described in Fig. 2.
  - Output $(\mathsf{k}_1, \mathsf{td}_1)$ where $\mathsf{k}_1 = (1, \mathsf{hk}_1, \{\mathsf{ek}_{1,i}\}_{i\in[u]})$ and $\mathsf{td}_1 = \{\mathsf{td}_{1,i}\}_{i\in[u]}$.
- **Level 1** $\mathsf{Hash\&Enc}(\mathsf{k}_1, x, \mathsf{e}_0 = \bot)$**:**
  - Parse $\mathsf{k}_1 = (1, \mathsf{hk}_1, \{\mathsf{ek}_{1,i}\}_{i\in[u]})$.
  - Compute "first level" hash over x: $\mathsf{h}_1 \leftarrow \mathsf{TDH.Hash}(\mathsf{hk}_1, x)$
  - For every $i \in [u]$, compute a "first level" encoding: $\mathsf{e}_{1,i} \leftarrow \mathsf{TDH.Enc}(\mathsf{ek}_{1,i}, x)$
  - Output $(\mathsf{h}_1, \mathsf{e}_1)$, where $\mathsf{e}_1 = \{\mathsf{e}_{1,i}\}_{i\in[u]}$.
- **Level 2** $\mathsf{KGen}(1^{\lambda_2}, 2, \mathrm{T}^{\oplus y}_{\vec{t},\vec{I}}, \mathsf{h}_1, \mathsf{td}_1)$**:**
  - Parse $\mathsf{td}_1 = \{\mathsf{td}_{1,i}\}_{i\in[u]}$. For every $i \in [u]$, decode h_1: $\mathsf{d}_{1,i} \leftarrow \mathsf{TDH.Dec}(\mathsf{td}_{1,i}, \mathsf{h}_1)$
  - Sample a new hash key of TDH w.r.t. security parameter λ_2, input length $n_2 = R_1 \cdot u$ and modulus $R_2 = 2$,

$$\mathsf{hk}_2 \leftarrow \mathsf{TDH.HKGen}(1^{\lambda_2}, 1^{n_2=R_1\cdot u}, 1^{R_2=2}).$$

  - Parse $\vec{t} = \{t_1, \ldots, t_u\}$. For each $i \in [u]$, sample a new encoding key

$$(\mathsf{ek}_{2,i}, \mathsf{td}_{2,i}) \leftarrow \mathsf{TDH.EKGen}(\mathsf{hk}_2, \mathsf{AddTh}_{i,t_i,\mathsf{d}_{1,i}}),$$

    where for any index $i \in [u]$, positive integer t and value $\mathsf{d} \in \mathbb{Z}_{R_1}$, $\mathsf{AddTh}_{i,t,\mathsf{d}}$ is the linear function defined in the Fig. 3.
  - Output $(\mathsf{k}_2, \mathsf{td}_2)$, where $\mathsf{k}_2 = (2, \mathsf{hk}_2, \{\mathsf{ek}_{2,i}\}_{i\in[u]})$ and $\mathsf{td}_2 = \{\mathsf{td}_{2,i}\}_{i\in[u]}$.
- **Level 2** $\mathsf{Hash\&Enc}(\mathsf{k}_2, x, \mathsf{e}_1)$**:**
  - Parse $\mathsf{k}_2 = (2, \mathsf{hk}_2, \{\mathsf{ek}_{2,i}\}_{i\in[u]})$, and $\mathsf{e}_1 = \{\mathsf{e}_{1,i}\}_{i\in[u]}$.
  - Compute "second level" hash over $\{\mathbb{1}_{\mathsf{e}_{1,i}}\}_{i\in[u]}$, where $\mathbb{1}_{\mathsf{e}}$ is the indicator vector for any e.

$$\mathsf{h}_2 \leftarrow \mathsf{TDH.Hash}(\mathsf{hk}_2, \{\mathbb{1}_{\mathsf{e}_{1,i}}\}_{i\in[u]})$$

  - For any $i \in [u]$, compute "second level" encoding: $\mathsf{e}_{2,i} \leftarrow \mathsf{TDH.Enc}(\mathsf{ek}_{2,i}, \{\mathbb{1}_{\mathsf{e}_{1,j}}\}_{j\in[u]})$.
  - Output $(\mathsf{h}_2, \mathsf{e}_2)$, where $\mathsf{e}_2 = \{\mathsf{e}_{2,i}\}_{i\in[u]})$.
- **Decoding** $\mathsf{Dec}(\mathsf{td}_2, \mathsf{h}_2)$**:**
  - Parse $\mathsf{td}_2 = \{\mathsf{td}_{2,i}\}_{i\in[u]}$. For every $i \in [u]$, decode h_2: $\mathsf{d}_{2,i} \leftarrow \mathsf{TDH.Dec}(\mathsf{td}_{2,i}, \mathsf{h}_2)$.
  - Output $\mathsf{d} = \{\mathsf{d}_{2,i}\}_{i\in[u]}$.

This completes the description of ITDH. We defer the proof of approximate correctness and leveled function privacy to the full version.

Linear Function $\mathsf{XorSum}_{I,y}(x_1, \ldots, x_n)$ over $\mathbb{Z}_{R_1}$

- Let $y = (y_1, y_2, \ldots, y_n)$.
- Compute and output $\sum_{i \in I} x_i \cdot (1 - y_i) + (1 - x_i) \cdot y_i$.

Fig. 2. Description of the linear function $\mathsf{XorSum}_{I,y}$. This function computes the sum over $\mathbb{Z}_{R_1}$ of I values obtained by bit-wise XOR of $y[I]$ and $x[I]$, where $x = (x_1, \ldots, x_n)$.

Linear Function $\mathsf{AddTh}_{i,t,\mathsf{d}}(\vec{\mathbb{e}})$ over $\mathbb{Z}_2$

- Let $\vec{\mathbb{e}} = (\mathbb{e}_1, \ldots, \mathbb{e}_u)$, where $\mathbb{e}_j \in \{0,1\}^{R_1}$ for every $j \in [u]$.
- Compute and output the inner product: $\langle \mathbb{e}_i, \mathbb{f} \rangle \bmod 2$, where $\mathbb{f} = \sum_{j=0}^{t-1} \mathbb{1}_{(j-\mathsf{d}) \bmod R_1}$ is the sum of indicator vectors for $(j - \mathsf{d}) \bmod R_1$, for $0 \leq j < t$.

Fig. 3. Description of the linear function $\mathsf{AddTh}_{i,t,\mathsf{d}}$. For any $\mathsf{e}_1, \mathsf{e}_2, \ldots, \mathsf{e}_u \in \mathbb{Z}_{R_1}$, this function computes whether $(\mathsf{e}_i + \mathsf{d}) \bmod R_1$ is less than the threshold t. The actual input $\vec{\mathbb{e}}$ to the function is such that $\mathbb{e}_i$ is the indicator vector for e_i.

5.2 ITDH Composition

In this section, we establish a sequential composition theorem for ITDH. Roughly speaking, we show how a 2-level ITDH for an "Xor-*then*-Compute" circuit family can be executed sequentially L times to obtain an ITDH for a related circuit family (the exact transformation is more nuanced; see below). The main benefit of sequential composition is that it can be used to increase the depth of circuits that can be computed by ITDH.

We start by introducing some notation and terminology for circuit composition that we shall use in the sequel.

Parallel Composition. Let w be a positive integer. Informally, an w-parallel composition of a circuit f' is a new circuit f that computes w copies of f' in parallel. More formally, for any circuit family $\mathcal{C}$, we define a corresponding parallel-composition circuit family as follows:

Definition 1 (Parallel Composition). *For any circuit family $\mathcal{C}$ and any polynomial $w = w(n)$, we say that $\mathcal{C}[\overrightarrow{w}] = \{\mathcal{C}[\overrightarrow{w}]_{n,u}\}_{n,u}$ is a family of w-parallel composition circuits if for every $f \in \mathcal{C}[\overrightarrow{w}]_{n,u}$, there exists a sequence of circuits $f'_1, f'_2, \ldots, f'_w \in \mathcal{C}_{n',u'}$ such that $n = n' \cdot w(n)$ and $u = u' \cdot w(n)$, and for any input $x = (x_1, x_2, \ldots, x_w) \in \{0,1\}^{n' \cdot w}$ (where every $x_i \in \{0,1\}^{n'}$), we have*

$$f(x_1, x_2, \ldots, x_w) = (f'_1(x_1), f'_2(x_2), \ldots, f'_w(x_w)).$$

Parallel-and-Sequential-Composition. For any circuit family $\mathcal{C}$, we now define another circuit family obtained via parallel *and* sequential composition of circuits in $\mathcal{C}$.

Informally speaking, for any polynomials $w(n)$ and $L(n)$ and an integer s, a w-parallel-and-L-sequential-composition of a circuit family $\mathcal{C}$ is a new circuit family $\mathcal{C}[\overrightarrow{w}_{\downarrow L}] = \{\mathcal{C}[\overrightarrow{w}_{\downarrow L}]_{n,s}\}_{n,s}$, where each circuit $f \in \mathcal{C}[\overrightarrow{w}_{\downarrow L}]_{n,s}$ is computed by a sequence of circuits $f_1, f_2, \ldots, f_L$. For any input x, to compute $f(x)$, we firstly evaluate f_1 on input x, then use the output $f_1(x)$ as the input to the circuit f_2, and so on, such that the output of f_L is the output of f. Furthermore, we require that for every $\ell \in [L]$, f_ℓ is an m-parallel composition of some sequence of circuits $f'_{\ell,1}, f'_{\ell,2}, \ldots, f'_{\ell,w} \in \mathcal{C}$. For the ease of presentation, we fix the output length of the circuit f_ℓ for every $\ell < L$ as s, and the output length of f as w.

Definition 2 (Parallel-and-Sequential-Composition). *Let $\mathcal{C} = \{\mathcal{C}_{n,u}\}_{n,u}$ be a circuit family, where each circuit in $\mathcal{C}_{n,u}$ has input length n and output length u. For any polynomials $w = w(n), L = L(n)$, and integer s, we say that $\mathcal{C}[\overrightarrow{w}_{\downarrow L}] = \{\mathcal{C}[\overrightarrow{w}_{\downarrow L}]_{n,s}\}_{n,s}$ is a family of* w-parallel-and-L-sequential-composition circuits *if every circuit $f \in \mathcal{C}[\overrightarrow{w}_{\downarrow L}]_{n,s}$ is of the form*

$$f = f_L \circ f_{L-1} \circ \ldots \circ f_1$$

where for every $\ell \in [L]$, $f_\ell : \{0,1\}^{n_\ell} \to \{0,1\}^{n_{\ell+1}}$ satisfies $n_1 = n, n_2 = n_3 = \ldots = n_{L-1} = s, n_L = w$. Furthermore, there exists a sequence of integers $\{n'_\ell\}_\ell$ and circuits $\{f'_{\ell,j}\}_{\ell\in[L],j\in[w]}$, where $f'_{\ell,j} \in \mathcal{C}_{n'_\ell, n'_{\ell+1}}$, and $n_\ell = n'_\ell \cdot w$,

$$f_\ell(x_1, \ldots, x_w) = \left(f'_{\ell,1}(x_1), f'_{\ell,2}(x_2), \ldots, f'_{\ell,w}(x_w)\right)$$

for every $x = (x_1, \ldots, x_w) \in \{0,1\}^{n'_\ell \cdot w}$, where $x_i \in \{0,1\}^{n'_\ell}$ for every $i \in [w]$.

Construction of ITDH for $\mathcal{C}[\overrightarrow{w}_{\downarrow L}]$. Let $\mathcal{C} = \{\mathcal{C}_{n,u}\}_{n,u}$ be any circuit family, and let $\mathcal{C}[\overrightarrow{w}]$ be the corresponding w-parallel composition circuit family. Let $\mathcal{C}[\overrightarrow{w}]^\oplus = \{\mathcal{C}[\overrightarrow{w}]^\oplus_{n,u}\}_{n,u}$ be the "Xor-*then*-Compute" circuit family defined w.r.t. $\mathcal{C}[\overrightarrow{w}]$. Let $\mathsf{ITDH} = (\mathsf{ITDH.KGen}, \mathsf{ITDH.Hash\&Enc}, \mathsf{ITDH.Dec})$ be a 2-level interactive trapdoor hashing protocol for $\mathcal{C}[\overrightarrow{w}]^\oplus = \{\mathcal{C}[\overrightarrow{w}]^\oplus_{n,u}\}_{n,u}$ with (Δ, ϵ)-approximate correctness and leveled function privacy.

Given ITDH, we construct a $2L$-level interactive trapdoor hashing protocol $\mathsf{ITDH}' = (\mathsf{KGen}, \mathsf{Hash\&Enc}, \mathsf{Dec})$ for the circuit family $\mathcal{C}[\overrightarrow{w}_{\downarrow L}]$ as defined above. For ease of exposition, we describe the algorithms of ITDH' for "odd" and "even" levels separately.

- **Level $\ell' = 2\ell - 1$, $\mathsf{KGen}(1^{\lambda_{\ell'}}, \ell', f, \mathsf{h}_{\ell'-1}, \mathsf{td}_{\ell'-1})$:**
  - If $\ell = 1$, set d_0 to be an all zero string of length n.
  - If $\ell \geq 2$, decode $\mathsf{h}_{\ell'-1}$: $\mathsf{d}_{\ell-1} \leftarrow \mathsf{ITDH.Dec}(\mathsf{td}_{\ell'-1}, \mathsf{h}_{\ell'-1})$
  - Let $f_1, \ldots, f_L$ be such that $f = f_L \circ f_{L-1} \circ \ldots \circ f_1$ (as defined above), where f_ℓ has input length n_ℓ and output length $n_{\ell+1}$.
  - Compute a key w.r.t. security parameter $\lambda_{\ell'}$ and the "Xor-*then*-Compute" circuit $f_\ell^{\oplus \mathsf{d}_{\ell-1}} \in \mathcal{C}[\overrightarrow{w}]^\oplus_{n_\ell, n_{\ell+1}}$

$$(\mathsf{k}_{\ell,1}, \mathsf{td}_{\ell,1}) \leftarrow \mathsf{ITDH.KGen}(1^{\lambda_{\ell'}}, 1^{n_\ell}, 1, f_\ell^{\oplus \mathsf{d}_{\ell-1}}, \bot, \bot).$$

- Output $(\mathsf{k}_{\ell'}, \mathsf{td}_{\ell'})$ where $\mathsf{k}_{\ell'} = (\ell', \mathsf{k}_{\ell,1})$ and $\mathsf{td}_{\ell'} = \mathsf{td}_{\ell,1}$.

- **Level** $\ell' = 2\ell - 1$, $\mathsf{Hash\&Enc}(\mathsf{k}_{\ell'}, x, \mathsf{e}_{\ell'-1})$:
  - If $\ell = 1$, let $x_\ell = x$, otherwise, let $x_\ell = \mathsf{e}_{\ell'-1}$. Execute

$$(\mathsf{h}_{\ell,1}, \mathsf{e}_{\ell,1}) \leftarrow \mathsf{ITDH.Hash\&Enc}(\mathsf{k}_{\ell,1}, x, \bot)$$

  - Output $(\mathsf{h}_\ell = \mathsf{h}_{\ell,1}, \mathsf{e}_\ell = (x_\ell, \mathsf{e}_{\ell,1}))$.
- **Level** $\ell' = 2\ell$, $\mathsf{KGen}(1^{\lambda_{\ell'}}, \ell', f, \mathsf{h}_{\ell'-1}, \mathsf{td}_{\ell'-1})$:
  - Parse $\mathsf{h}_{\ell'-1} = \mathsf{h}_{\ell,1}$, and $\mathsf{td}_{\ell'-1} = \mathsf{td}_{\ell,1}$.

$$(\mathsf{k}_{\ell,2}, \mathsf{td}_{\ell,2}) \leftarrow \mathsf{ITDH.KGen}(1^{\lambda_{\ell'}}, 1^{n_\ell}, 2, f_\ell^{\oplus \mathsf{d}_{\ell-1}}, \mathsf{h}_{\ell,1}, \mathsf{td}_{\ell,1})$$

  - Output $(\mathsf{k}_{\ell'}, \mathsf{td}_{\ell'})$, where $\mathsf{k}_{\ell'} = (\ell', \mathsf{k}_{\ell,2})$, and $\mathsf{td}_{\ell'} = \mathsf{td}_{\ell,2}$.
- **Level** $\ell' = 2\ell$, $\mathsf{Hash\&Enc}(\mathsf{k}_{\ell'}, x, \mathsf{e}_{\ell'-1})$:
  - Parse $\mathsf{e}_{\ell'-1} = (x_\ell, \mathsf{e}_{\ell,1}), \mathsf{k}_{\ell'} = \mathsf{k}_{\ell,2}$.
  - Output $(\mathsf{h}_{\ell'}, \mathsf{e}_{\ell'}) \leftarrow \mathsf{Hash\&Enc}(\mathsf{k}_{\ell,2}, x_\ell, \mathsf{e}_{\ell,1})$.
- **Decoding** $\mathsf{Dec}(\mathsf{td}_{2L}, \mathsf{h}_{2L})$:
  - Output $\mathsf{d} \leftarrow \mathsf{ITDH.Dec}(\mathsf{td}_{2L}, \mathsf{h}_{2L})$.

This completes the description of ITDH'. We defer the proof of approximate correctness and leveled function privacy to the full version.

5.3 ITDH for TC^0

We now describe how we can put the above constructions together to obtain an ITDH for TC^0. Recall that, we use the notation TC^0_L to denote the class of L-depth TC^0 circuits.

Let $\mathcal{T}[\overrightarrow{\downarrow L}^{w}]$ be the circuit family obtained by w-parallel-and-L-sequential composition of the circuit family $\mathcal{T}$, as per Definition 2. We first show that any circuit in TC^0_L can be converted to a circuit in $\mathcal{T}[\overrightarrow{\downarrow L}^{w}]$.

Lemma 1. TC^0_L *can be computed in* $\mathcal{T}[\overrightarrow{\downarrow L}^{w}]$. *Specifically, for any circuit* $f \in \mathsf{TC}^0_L$ *with* n *bit input and* w *output bits, we convert it in polynomial time to a circuit* $f' \in \mathcal{T}[\overrightarrow{\downarrow L}^{w}]$ *such that, for any* $x \in \{0,1\}^n$, $f(x) = f'(x, x, \ldots, x)$.

We defer the proof to the full version.

Next, we combine the construction of ITDH for the circuit family $\mathcal{T}^{\oplus}$ from Sect. 5.1 together with the sequential composition theorem in Sect. 5.2 to obtain an ITDH for the circuit family $\mathcal{T}[\overrightarrow{\downarrow L}^{w}]$, and therefore an ITDH for TC^0_L.

Theorem 4. *If for any inverse polynomial* τ *in the security parameter, there exists a trapdoor hash function* TDH *for linear function family* $\mathcal{F}$ *with* τ*-enhanced correctness and sub-exponential function privacy, then for any constants* $L = O(1)$, $\alpha = O(1)$, *and any polynomial* w *in the security parameter, there exists a* $2L$*-level interactive trapdoor hashing protocol for* TC^0_L *that achieves* (Δ, ϵ)*-approximate correctness and sub-exponential function privacy, where* $\Delta(w) = \alpha \cdot w$ *and for any* $\lambda_1 < \lambda_2 < \ldots < \lambda_{2L} < w/2L$, $\epsilon(w, \lambda_1, \ldots, \lambda_L) = 2^{-2w+O(1)}$.

We defer the proof to the full version.

ITDH for P/poly. Since any circuit in P/poly can be converted to a layered circuit as in Lemma 1, the above construction of ITDH for TC^0 can be naturally extended to obtain a polynomial-level ITDH for P/poly.

6 Correlation Intractable Hash Functions for TC^0

In this section, we build correlation intractable hash functions for the circuit family TC^0.

6.1 Definition

Correlation intractable hash (CIH) function is a tuple of algorithms $\mathsf{CIH} = (\mathsf{Gen}, \mathsf{Hash})$ described as follows:

- $\mathsf{Gen}(1^\lambda)$: It takes as input a security parameter λ and outputs a key k.
- $\mathsf{Hash}(\mathsf{k}, x)$: It takes as input a hash key k and a string x, and outputs a binary string y of length $w = w(\lambda)$.

We require CIH to satisfy the following property:

- **Correlation Intractability:** Recall that, a binary relation R is a subset of $\{0,1\}^* \times \{0,1\}^*$. We say that CIH is correlation intractable for a class of binary relations $\{\mathcal{R}_\lambda\}_\lambda$ if there exists a negligible function $\nu(\lambda)$ such that, for any $\lambda \in \mathbb{N}$, any n.u. PPT adversary $\mathcal{A}$, and any $R \in \mathcal{R}_\lambda$,

$$\Pr\left[\mathsf{k} \leftarrow \mathsf{Gen}(1^\lambda), x \leftarrow \mathcal{A}(1^\lambda, \mathsf{k}) : (x, \mathsf{Hash}(\mathsf{k}, x)) \in R\right] \le \nu(\lambda)$$

We say that the CIH is sub-exponential correlation intractable, if there exists a constants c such that for any n.u. PPT adversary, its successful probability is bounded by $2^{-\lambda^c}$ for any sufficiently large λ.

Definition 3 (CIH for TC^0). *Let $n(\lambda), w(\lambda)$ be polynomials. Let $L = O(1)$ be a constant. Recall that, we use TC^0_L to denote the class of L-depth threshold circuits. We say that CIH is a CIH for TC^0_L, if CIH is correlation intractable for the class of relations $\{\mathcal{R}_\lambda\}_\lambda$, where $\mathcal{R}_\lambda = \{R_{f,\lambda} \mid f \in \mathsf{TC}^0_L\}$, and*

$$R_{f,\lambda} = \{(x, y) \in \{0,1\}^{n(\lambda)} \times \{0,1\}^{w(\lambda)} \mid y = f(x)\}$$

6.2 Our Construction

For any $L = O(1)$, we show a generic transformation from an L-level ITDH for TC^0_L to a CIH for the same circuit family.

CIH for TC^0. Let $\mathsf{ITDH} = (\mathsf{ITDH.KGen}, \mathsf{ITDH.Hash\&Enc}, \mathsf{ITDH.Dec})$ be an L-level interactive trapdoor hashing protocol for the circuit class TC^0_L that satisfies the following properties:

- $(0.01w, 2^{-2w+O(1)})$-approximate correctness.
- Sub-exponential leveled function privacy. Let Sim be the leveled function privacy simulator. Let c be the constant in the sub-exponential security definition.

Correlation Intractable Hash CIH

- $\mathsf{Gen}(1^\lambda)$:
  - For each $\ell \in [L]$, set $\lambda_\ell = \lambda^{\frac{1}{2}(\frac{c}{2})^{L-\ell}}$.
  - Compute simulated receiver's messages for ITDH:

$$\forall \ell \in [L], \mathsf{k}_\ell \leftarrow \mathsf{ITDH.Sim}(1^{\lambda_\ell}, 1^n, 1^w, \ell)$$

  - Sample a mask $\mathsf{mask} \leftarrow \{0,1\}^w$ uniformly at random.
  - Output $\mathsf{k} = \left(\{\mathsf{k}_\ell\}_{\ell \in [L]}, \mathsf{mask}\right)$.
- $\mathsf{Hash}(\mathsf{k}, x)$:
  - Parse $\mathsf{k} = (\{\mathsf{k}_\ell\}_{\ell \in [L]}, \mathsf{mask})$.
  - Let $\mathsf{e}_0 = \perp$. Compute hash values and encodings for ITDH:

$$\forall \ell \in [L], (\mathsf{h}_\ell, \mathsf{e}_\ell) \leftarrow \mathsf{ITDH.Hash\&Enc}(\mathsf{k}_\ell, x, \mathsf{e}_{\ell-1}).$$

  - Output $\mathsf{e} \oplus \mathsf{mask}$, where $\mathsf{e} = \mathsf{e}_L$.

Fig. 4. Description of CIH.

We construct a correlation intractable hash function $\mathsf{CIH} = (\mathsf{CIH.Gen}, \mathsf{CIH.Hash})$ for TC^0_L in Fig. 4.

Theorem 5 (Correlation Intractability). *If $w = \Omega(\lambda)$, the construction in Fig. 4 is sub-exponential correlation intractable for the circuit class TC^0_L.*

Acknowledgements. The authors were supported in part by an NSF CNS grant 1814919, NSF CAREER award 1942789 and Johns Hopkins University Catalyst award. The first author was additionally supported in part by Office of Naval Research grant N00014-19-1-2294.

References

1. Adleman, L.: A subexponential algorithm for the discrete logarithm problem with applications to cryptography. In: 20th Annual Symposium on Foundations of Computer Sciences, pp. 55–60 (1979)
2. Badrinarayanan, S., Fernando, R., Jain, A., Khurana, D., Sahai, A.: Statistical ZAP arguments. In: Canteaut, A., Ishai, Y. (eds.) EUROCRYPT 2020, Part III. LNCS, vol. 12107, pp. 642–667. Springer, Cham (2020). https://doi.org/10.1007/978-3-030-45727-3_22
3. Bellare, M., Hofheinz, D., Yilek, S.: Possibility and impossibility results for encryption and commitment secure under selective opening. In: Joux, A. (ed.) EUROCRYPT 2009. LNCS, vol. 5479, pp. 1–35. Springer, Heidelberg (2009). https://doi.org/10.1007/978-3-642-01001-9_1

4. Bellare, M., Micciancio, D., Warinschi, B.: Foundations of group signatures: formal definitions, simplified requirements, and a construction based on general assumptions. In: Biham, E. (ed.) EUROCRYPT 2003. LNCS, vol. 2656, pp. 614–629. Springer, Heidelberg (2003). https://doi.org/10.1007/3-540-39200-9_38
5. Bellare, M., Yung, M.: Certifying cryptographic tools: the case of trapdoor permutations. In: Brickell, E.F. (ed.) CRYPTO 1992. LNCS, vol. 740, pp. 442–460. Springer, Heidelberg (1993). https://doi.org/10.1007/3-540-48071-4_31
6. Ben-Sasson, E., et al.: Zerocash: decentralized anonymous payments from bitcoin. In: 2014 IEEE Symposium on Security and Privacy, SP 2014, Berkeley, CA, USA, 18–21 May 2014, pp. 459–474. IEEE Computer Society (2014)
7. Bender, A., Katz, J., Morselli, R.: Ring signatures: stronger definitions, and constructions without random oracles. In: Halevi, S., Rabin, T. (eds.) TCC 2006. LNCS, vol. 3876, pp. 60–79. Springer, Heidelberg (2006). https://doi.org/10.1007/11681878_4
8. Bitansky, N., Paneth, O.: ZAPs and non-interactive witness indistinguishability from indistinguishability obfuscation. In: Dodis, Y., Nielsen, J.B. (eds.) TCC 2015, Part II. LNCS, vol. 9015, pp. 401–427. Springer, Heidelberg (2015). https://doi.org/10.1007/978-3-662-46497-7_16
9. Blum, M., Feldman, P., Micali, S.: Non-interactive zero-knowledge and its applications (extended abstract). In: 20th ACM STOC, Chicago, IL, USA, 2–4 May 1988, pp. 103–112. ACM Press (1988). https://doi.org/10.1145/62212.62222
10. Boneh, D., Franklin, M.: Identity-based encryption from the Weil pairing. In: Kilian, J. (ed.) CRYPTO 2001. LNCS, vol. 2139, pp. 213–229. Springer, Heidelberg (2001). https://doi.org/10.1007/3-540-44647-8_13
11. Boneh, D., Sahai, A., Waters, B.: Functional encryption: definitions and challenges. In: Ishai, Y. (ed.) TCC 2011. LNCS, vol. 6597, pp. 253–273. Springer, Heidelberg (2011). https://doi.org/10.1007/978-3-642-19571-6_16
12. Boyle, E., Gilboa, N., Ishai, Y.: Breaking the circuit size barrier for secure computation under DDH. In: Robshaw, M., Katz, J. (eds.) CRYPTO 2016, Part I. LNCS, vol. 9814, pp. 509–539. Springer, Heidelberg (2016). https://doi.org/10.1007/978-3-662-53018-4_19
13. Brakerski, Z., Koppula, V., Mour, T.: NIZK from LPN and trapdoor hash via correlation intractability for approximable relations. In: Micciancio, D., Ristenpart, T. (eds.) CRYPTO 2020, Part III. LNCS, vol. 12172, pp. 738–767. Springer, Cham (2020). https://doi.org/10.1007/978-3-030-56877-1_26
14. Canetti, R., et al.: Fiat-Shamir: from practice to theory. In: Charikar, M., Cohen, E. (eds.) 51st ACM STOC, Phoenix, AZ, USA, 23–26 June 2019, pp. 1082–1090. ACM Press (2019). https://doi.org/10.1145/3313276.3316380
15. Canetti, R., Chen, Y., Reyzin, L.: On the correlation intractability of obfuscated pseudorandom functions. In: Kushilevitz, E., Malkin, T. (eds.) TCC 2016, Part I. LNCS, vol. 9562, pp. 389–415. Springer, Heidelberg (2016). https://doi.org/10.1007/978-3-662-49096-9_17
16. Canetti, R., Chen, Y., Reyzin, L., Rothblum, R.D.: Fiat-Shamir and correlation intractability from strong KDM-secure encryption. In: Nielsen, J.B., Rijmen, V. (eds.) EUROCRYPT 2018, Part I. LNCS, vol. 10820, pp. 91–122. Springer, Cham (2018). https://doi.org/10.1007/978-3-319-78381-9_4
17. Canetti, R., Goldreich, O., Halevi, S.: The random oracle methodology, revisited. J. ACM **51**(4), 557–594 (2004). https://doi.org/10.1145/1008731.1008734
18. Canetti, R., Halevi, S., Katz, J.: A forward-secure public-key encryption scheme. In: Biham, E. (ed.) EUROCRYPT 2003. LNCS, vol. 2656, pp. 255–271. Springer, Heidelberg (2003). https://doi.org/10.1007/3-540-39200-9_16

19. Canetti, R., Lichtenberg, A.: Certifying trapdoor permutations, revisited. In: Beimel, A., Dziembowski, S. (eds.) TCC 2018, Part I. LNCS, vol. 11239, pp. 476–506. Springer, Cham (2018). https://doi.org/10.1007/978-3-030-03807-6_18
20. Coppersmith, D., Odlyzko, A.M., Schroeppel, R.: Discrete logarithms in gf(p). Algorithmica **1**(1), 1–15 (1986). https://doi.org/10.1007/BF01840433
21. Couteau, G., Katsumata, S., Ursu, B.: Non-interactive zero-knowledge in pairing-free groups from weaker assumptions. In: Canteaut, A., Ishai, Y. (eds.) EUROCRYPT 2020, Part III. LNCS, vol. 12107, pp. 442–471. Springer, Cham (2020). https://doi.org/10.1007/978-3-030-45727-3_15
22. De Santis, A., Micali, S., Persiano, G.: Non-interactive zero-knowledge proof systems. In: Pomerance, C. (ed.) CRYPTO 1987. LNCS, vol. 293, pp. 52–72. Springer, Heidelberg (1988). https://doi.org/10.1007/3-540-48184-2_5
23. Dolev, D., Dwork, C., Naor, M.: Non-malleable cryptography (extended abstract). In: 23rd ACM STOC, New Orleans, LA, USA, 6–8 May 1991, pp. 542–552. ACM Press (1991). https://doi.org/10.1145/103418.103474
24. Döttling, N., Garg, S.: Identity-based encryption from the Diffie-Hellman assumption. In: Katz, J., Shacham, H. (eds.) CRYPTO 2017, Part I. LNCS, vol. 10401, pp. 537–569. Springer, Cham (2017). https://doi.org/10.1007/978-3-319-63688-7_18
25. Döttling, N., Garg, S., Ishai, Y., Malavolta, G., Mour, T., Ostrovsky, R.: Trapdoor hash functions and their applications. In: Boldyreva, A., Micciancio, D. (eds.) CRYPTO 2019, Part III. LNCS, vol. 11694, pp. 3–32. Springer, Cham (2019). https://doi.org/10.1007/978-3-030-26954-8_1
26. Dwork, C., Naor, M.: Zaps and their applications. In: 41st FOCS, Redondo Beach, CA, USA, 12–14 November 2000, pp. 283–293. IEEE Computer Society Press (2000). https://doi.org/10.1109/SFCS.2000.892117
27. Elgamal, T.: A public key cryptosystem and a signature scheme based on discrete logarithms. IEEE Trans. Inf. Theory **31**(4), 469–472 (1985)
28. Feige, U., Lapidot, D., Shamir, A.: Multiple non-interactive zero knowledge proofs based on a single random string (extended abstract). In: 31st FOCS, St. Louis, MO, USA, 22–24 October 1990. pp. 308–317. IEEE Computer Society Press (1990). https://doi.org/10.1109/FSCS.1990.89549
29. Fiat, A., Shamir, A.: How to prove yourself: practical solutions to identification and signature problems. In: Odlyzko, A.M. (ed.) CRYPTO 1986. LNCS, vol. 263, pp. 186–194. Springer, Heidelberg (1987). https://doi.org/10.1007/3-540-47721-7_12
30. Goldreich, O., Rothblum, R.D.: Enhancements of trapdoor permutations. J. Cryptol. **26**(3), 484–512 (2013). https://doi.org/10.1007/s00145-012-9131-8
31. Goldwasser, S., Micali, S., Rackoff, C.: The knowledge complexity of interactive proof-systems (extended abstract). In: 17th ACM STOC, Providence, RI, USA, 6–8 May 1985, pp. 291–304. ACM Press (1985). https://doi.org/10.1145/22145.22178
32. Goyal, V., Jain, A., Jin, Z., Malavolta, G.: Statistical zaps and new oblivious transfer protocols. In: Canteaut, A., Ishai, Y. (eds.) EUROCRYPT 2020, Part III. LNCS, vol. 12107, pp. 668–699. Springer, Cham (2020). https://doi.org/10.1007/978-3-030-45727-3_23
33. Goyal, V., Pandey, O., Sahai, A., Waters, B.: Attribute-based encryption for fine-grained access control of encrypted data. In: Juels, A., Wright, R.N., De Capitani di Vimercati, S. (eds.) ACM CCS 2006, Alexandria, Virginia, USA, 30 October–3 November 2006, pp. 89–98. ACM Press (2006). https://doi.org/10.1145/1180405.1180418. Available as Cryptology ePrint Archive Report 2006/309

34. Groth, J., Ostrovsky, R., Sahai, A.: Non-interactive zaps and new techniques for NIZK. In: Dwork, C. (ed.) CRYPTO 2006. LNCS, vol. 4117, pp. 97–111. Springer, Heidelberg (2006). https://doi.org/10.1007/11818175_6
35. Groth, J., Ostrovsky, R., Sahai, A.: Perfect non-interactive zero knowledge for NP. In: Vaudenay, S. (ed.) EUROCRYPT 2006. LNCS, vol. 4004, pp. 339–358. Springer, Heidelberg (2006). https://doi.org/10.1007/11761679_21
36. Holmgren, J., Lombardi, A.: Cryptographic hashing from strong one-way functions (or: One-way product functions and their applications). In: Thorup, M. (ed.) 59th FOCS, Paris, France, 7–9 October 2018, pp. 850–858. IEEE Computer Society Press (2018). https://doi.org/10.1109/FOCS.2018.00085
37. Kalai, Y.T., Khurana, D., Sahai, A.: Statistical witness indistinguishability (and more) in two messages. In: Nielsen, J.B., Rijmen, V. (eds.) EUROCRYPT 2018, Part III. LNCS, vol. 10822, pp. 34–65. Springer, Cham (2018). https://doi.org/10.1007/978-3-319-78372-7_2
38. Kalai, Y.T., Rothblum, G.N., Rothblum, R.D.: From obfuscation to the security of Fiat-Shamir for proofs. In: Katz, J., Shacham, H. (eds.) CRYPTO 2017, Part II. LNCS, vol. 10402, pp. 224–251. Springer, Cham (2017). https://doi.org/10.1007/978-3-319-63715-0_8
39. Khurana, D., Sahai, A.: How to achieve non-malleability in one or two rounds. In: Umans, C. (ed.) 58th FOCS, Berkeley, CA, USA, 15–17 October 2017, pp. 564–575. IEEE Computer Society Press (2017). https://doi.org/10.1109/FOCS.2017.58
40. Kol, G., Naor, M.: Cryptography and game theory: designing protocols for exchanging information. In: Canetti, R. (ed.) TCC 2008. LNCS, vol. 4948, pp. 320–339. Springer, Heidelberg (2008). https://doi.org/10.1007/978-3-540-78524-8_18
41. Kopparty, S.: AC^0 lower bounds and pseudorandomness. Lecture notes for 'Topics in Complexity Theory and Pseudorandomness' (2013). https://sites.math.rutgers.edu/~sk1233/courses/topics-S13/lec4.pdf
42. Lombardi, A., Vaikuntanathan, V., Wichs, D.: 2-message publicly verifiable WI from (subexponential) LWE. Cryptology ePrint Archive, Report 2019/808 (2019). https://eprint.iacr.org/2019/808
43. Lombardi, A., Vaikuntanathan, V., Wichs, D.: Statistical ZAPR arguments from bilinear maps. In: Canteaut, A., Ishai, Y. (eds.) EUROCRYPT 2020, Part III. LNCS, vol. 12107, pp. 620–641. Springer, Cham (2020). https://doi.org/10.1007/978-3-030-45727-3_21
44. Naor, M.: On cryptographic assumptions and challenges. In: Boneh, D. (ed.) CRYPTO 2003. LNCS, vol. 2729, pp. 96–109. Springer, Heidelberg (2003). https://doi.org/10.1007/978-3-540-45146-4_6
45. Naor, M., Pinkas, B.: Efficient oblivious transfer protocols. In: Kosaraju, S.R. (ed.) 12th SODA, Washington, DC, USA, 7–9 January 2001, pp. 448–457. ACM-SIAM (2001)
46. Naor, M., Yung, M.: Public-key cryptosystems provably secure against chosen ciphertext attacks. In: 22nd ACM STOC, Baltimore, MD, USA, 14–16 May 1990, pp. 427–437. ACM Press (1990). https://doi.org/10.1145/100216.100273
47. Oliveira, I.C., Santhanam, R., Srinivasan, S.: Parity helps to compute majority. In: Shpilka, A. (ed.) 34th Computational Complexity Conference (CCC 2019). Leibniz International Proceedings in Informatics (LIPIcs), vol. 137, pp. 23:1–23:17. Schloss Dagstuhl-Leibniz-Zentrum fuer Informatik, Dagstuhl, Germany (2019). https://doi.org/10.4230/LIPIcs.CCC.2019.23. http://drops.dagstuhl.de/opus/volltexte/2019/10845
48. O'Neill, A.: Definitional issues in functional encryption. IACR Cryptol. ePrint Arch. 2010/556 (2010). http://eprint.iacr.org/2010/556

49. Pass, R.: Unprovable security of perfect NIZK and non-interactive non-malleable commitments. In: Sahai, A. (ed.) TCC 2013. LNCS, vol. 7785, pp. 334–354. Springer, Heidelberg (2013). https://doi.org/10.1007/978-3-642-36594-2_19
50. Peikert, C., Shiehian, S.: Noninteractive zero knowledge for NP from (plain) learning with errors. In: Boldyreva, A., Micciancio, D. (eds.) CRYPTO 2019, Part I. LNCS, vol. 11692, pp. 89–114. Springer, Cham (2019). https://doi.org/10.1007/978-3-030-26948-7_4
51. Peikert, C., Vaikuntanathan, V., Waters, B.: A framework for efficient and composable oblivious transfer. In: Wagner, D. (ed.) CRYPTO 2008. LNCS, vol. 5157, pp. 554–571. Springer, Heidelberg (2008). https://doi.org/10.1007/978-3-540-85174-5_31
52. Reif, J.H., Tate, S.R.: On threshold circuits and polynomial computation. SIAM J. Comput. **21**(5), 896–908 (1992)
53. Sahai, A., Waters, B.: How to use indistinguishability obfuscation: deniable encryption, and more. In: Shmoys, D.B. (ed.) 46th ACM STOC, New York, NY, USA, 31 May–3 June 2014, pp. 475–484. ACM Press (2014). https://doi.org/10.1145/2591796.2591825
54. Sahai, A., Waters, B.: Fuzzy identity-based encryption. In: Cramer, R. (ed.) EUROCRYPT 2005. LNCS, vol. 3494, pp. 457–473. Springer, Heidelberg (2005). https://doi.org/10.1007/11426639_27
55. Smolensky, R.: Algebraic methods in the theory of lower bounds for Boolean circuit complexity. In: Aho, A. (ed.) 19th ACM STOC, New York City, NY, USA, 25–27 May 1987, pp. 77–82. ACM Press (1987). https://doi.org/10.1145/28395.28404
56. Smolensky, R.: On representations by low-degree polynomials. In: 34th FOCS, Palo Alto, CA, USA, 3–5 November 1993, pp. 130–138. IEEE Computer Society Press (1993). https://doi.org/10.1109/SFCS.1993.366874

On the (in)security of ROS

Fabrice Benhamouda[1(✉)], Tancrède Lepoint[2], Julian Loss[3], Michele Orrù[4], and Mariana Raykova[2]

[1] Algorand Foundation, Singapore, Singapore
[2] Google, Mountain View, USA
{tancrede,marianar}@google.com
[3] University of Maryland, College Park, USA
[4] UC Berkeley, Berkeley, USA
michele.orru@berkeley.edu

Abstract. We present an algorithm solving the ROS (Random inhomogeneities in a Overdetermined Solvable system of linear equations) problem mod p in polynomial time for $\ell > \log p$ dimensions. Our algorithm can be combined with Wagner's attack, and leads to a sub-exponential solution for any dimension ℓ with best complexity known so far.

When concurrent executions are allowed, our algorithm leads to practical attacks against unforgeability of blind signature schemes such as Schnorr and Okamoto–Schnorr blind signatures, threshold signatures such as GJKR and the original version of FROST, multisignatures such as CoSI and the two-round version of MuSig, partially blind signatures such as Abe–Okamoto, and conditional blind signatures such as ZGP17. Schemes for e-cash and anonymous credentials (such as Anonymous Credentials Light) inspired from the above are also affected.

1 Introduction

One of the most fundamental concepts in cryptanalysis is the *birthday paradox*. Roughly, it states that among $O(\sqrt{p})$ random elements from the range $[0, p-1]$ (where p is a prime), there exist two elements a and b such that $a = b$, with high probability. In a seminal work, Wagner gave a generalization of the birthday paradox to ℓ dimensions which asks to find $x_i \in L_i, i \in [0, \ell-1]$ such that $x_0 + \cdots + x_{\ell-1} = 0 \pmod p$, where L_i are lists of random elements.

His work also showed a simple and elegant algorithm to solve the problem in subexponential time $O((\ell+1) \cdot 2^{\lceil \log p \rceil / (1 + \lfloor \log(\ell+1) \rfloor)})$ and explained how it could be applied to perform cryptanalysis on various schemes. Among the most important applications of Wagner's technique is a subexponential solution to the ROS (Random inhomogeneities in a Overdetermined Solvable system of linear equations) problem [Sch01,FPS20], which is defined as follows. Given a prime number p and access to a random oracle $\mathrm{H}_{\mathrm{ros}}$ with range in $\mathbb{Z}_p$, the ROS problem (in dimension ℓ) asks to find $(\ell+1)$ affine functions $\boldsymbol{\rho}_i$ for $i = 0, \ldots, \ell$, $(\ell+1)$ bit strings $\mathsf{aux}_i \in \{0,1\}^*$ (with $i \in [0, \ell]$), and a vector $\boldsymbol{c} = (c_0, \ldots, c_{\ell-1})$ such that:

$$\mathrm{H}_{\mathrm{ros}}(\boldsymbol{\rho}_i, \mathsf{aux}_i) = \boldsymbol{\rho}_i(\boldsymbol{c}) \qquad \text{for all } i \in [0, \ell].$$

A. Canteaut and F.-X. Standaert (Eds.): EUROCRYPT 2021, LNCS 12696, pp. 33–53, 2021.
https://doi.org/10.1007/978-3-030-77870-5_2

This problem was originally studied by Schnorr [Sch01] in the context of blind signature schemes. Using a solver for the ROS problem, Wagner showed that the unforgeability of the Schnorr and Okamoto-Schnorr blind signature schemes can be attacked in subexponential time whenever more than $O(\log p)$ signatures are issued concurrently. In this work, we revisit the ROS problem and its applications. We make the following contributions.

- We give the first polynomial time solution to the ROS problem for $\ell > \log p$ dimensions.
- We show how the above solution can be combined with Wagner's techniques to yield an improved subexponential algorithm for dimensions lower than $\log p$. The resulting construction offers a smooth trade-off between the work and the dimension needed to solve the ROS problem. It outperforms the runtime of Wagner's algorithm for a broad range of dimensions.
- Finally, we describe how to apply our new attack to an extensive list of schemes. These include: blind signatures [PS00, Sch01], threshold signatures [GJKR07, KG20a], multisignatures [STV+16, MPSW18a], partially blind signatures [AO00], conditionally blind signatures [ZGP17, GPZZ19], and anonymous credentials [BL13, Bra94] in a concurrent setting with $\ell > \log p$ parallel executions. While our attacks do not contradict the security arguments of those schemes (which are restricted only to sequential or bounded number of executions), it proves that these schemes are unpractical for some real-world applications (cf. Sect. 7).

1.1 Technical Overview

Let $\mathsf{Pgen}(1^\lambda)$ be a parameter generation algorithm that given as input the security parameter λ in unary form, outputs a prime p of length $\lambda = \lceil \log p \rceil$. In this work, we prove the following main theorem:

Theorem 1 (ROS attack). *If $\ell > \lambda$, then there exists a (probabilistic) adversary that runs in expected polynomial time and solves the ROS problem relative to* Pgen *with dimension ℓ with probability* 1.

Let $B(\boldsymbol{x}) := \sum_{i=0}^{\lambda-1} 2^i \boldsymbol{\rho}_i(x_i)$ for functions ρ_i where $i \in [0, \lambda-1]$. If we can set $\rho_i(x_i)$ to be the multivariate polynomials that evaluate to 0 at the point c_i^0 and to 1 at the point c_i^1 (for $i \in [0, \ell-1]$), then we can write any value $y \in [0, p-1]$ as $y = B(c_0^{b_0}, \ldots, c_{\ell-1}^{b_{\ell-1}})$, where the b_i values are such that $y = \sum_{i=0}^{\lambda-1} 2^i b_i$. Using this idea, we first define all the functions $\boldsymbol{\rho}_0, \ldots, \boldsymbol{\rho}_{\ell-1}$ along with the corresponding pairs of points c_i^0, c_i^1 that are obtained as $c_i^b := \mathrm{H}_{\mathrm{ros}}(\boldsymbol{\rho}_i, b)$. In a second step, we choose $\boldsymbol{\rho}_\ell(\boldsymbol{x}) := B(\boldsymbol{x})$, and query $y := \mathrm{H}_{\mathrm{ros}}(\boldsymbol{\rho}_\ell, \mathsf{aux}_\ell)$. Now, we can write $y = \sum_{i=0}^{\lambda-1} 2^i b_i$ which determines a point $c_i^{b_i}$ from every pair. We can output the chosen points in $\boldsymbol{c}$ along with the vector of affine functions $(\boldsymbol{\rho}_0, \ldots, \boldsymbol{\rho}_\ell)$ as a solution to the ROS problem. (Note that $\boldsymbol{\rho}_\ell = B(\boldsymbol{x})$ is also affine.) This attack runs in expected polynomial time (since with small probability, $\mathrm{H}_{\mathrm{ros}}$ produces collisions, in which case steps need to be repeated) and works whenever $\ell > \log p$. This requirement ensures that it is always possible to write any value with ℓ

terms in binary representation. To circumvent the restriction $\ell > \log p$, we prove a second theorem:

Theorem 2 (Generalized ROS attack). *Let $L \geq 0$ be an integer and $w \geq 0$ be a real number. If $\ell \geq \max\{2^w - 1, \lceil 2^w - 1 + \lambda - (w+1) \cdot L \rceil\}$, then there exists a (probabilistic) adversary that runs in expected time $O(2^{w+L})$ and solves the ROS problem relative to* Pgen *and dimension ℓ with probability* 1.

The idea of this attack is to combine the technique from the first attack with the basic subexponential attack of Wagner. Instead of writing y entirely in binary as above, which requires ℓ dimensions, we first find a sum s of 2^w values which include y, but satisfies $|s| \in [0, \frac{p}{2^{(w+1) \cdot L}} - 1] \pmod{p}$. Note that s can be represented with $\lambda - (w+1) \cdot L$ many bits in binary representation. This approach requires, in total, $\lceil 2^w + \lambda - (w+1) \cdot L - 1 \rceil$ dimensions and 2^{w+L} overall work. As illustrated in Fig. 4, this leads to improvements over Wagner's attack relatively quickly as the dimension ℓ of the ROS problem increases. We remark that, while in our first attack we give a concrete probability of failure, our second attack is based on the conjecture that Wagner's algorithm for $\mathbb{Z}_p$ succeeds with constant probability. While we are not aware of any formal analysis of Wagner's algorithm over $\mathbb{Z}_p$, we remark that it is considered a standard cryptanalytic tool [DEF+19]. Our attack can be seen as strictly improving over its (conjectured) performance when applied to solve the ROS problem.

1.2 Impact of the Attacks

Any cryptographic construction that bases its security guarantees on the hardness of the ROS problem is affected by our attacks.

Blind Signatures. An immediate consequence of our findings is the first polynomial-time attack against Schnorr blind signatures [Sch01] and Okamoto–Schnorr blind signatures [PS00] in the concurrent setting with $\ell > \log p$ parallel executions.[1] Structurally, our attack builds on the one shown by Schnorr [Sch01], who showed that a solver to the ROS problem can be turned into an attacker against one-more unforgeability of blind Schnorr and Okamoto-Schnorr signatures. As a concrete example, the attack in Sect. 5 breaks one-more unforgeability of blind Schnorr signatures over 256-bit elliptic curves in a few seconds (when implemented in Sage [S+20]), provided that the attacker can open 256 concurrent sessions.

Other Affected Constructions. Our attack can be adapted to an extensive list of schemes which include threshold signatures [GJKR07, KG20a], multisignatures [STV+16, MPSW18a], partially blind signatures [AO00], conditionally blind signatures [ZGP17, GPZZ19], blind anonymous group signatures [CFLW04], blind identity-based signcryption [YW05], and blind signature schemes from bilinear

[1] Okamoto–Schnorr signatures are proven secure only for ℓ parallel executions s.t. $Q^\ell/p \ll 1$, where Q is the number of queries to $\mathrm{H}_{\mathrm{ros}}$. Our attack does not contradict their analysis as our attack requires $\ell > \log_2 p > \log_Q p$.

Game $\mathrm{ROS}_{\mathsf{Pgen},\mathsf{A},\ell}(\lambda)$	Oracle $\mathrm{H}_{\mathrm{ros}}(\boldsymbol{\rho}, \mathsf{aux})$
$p \leftarrow \mathsf{Pgen}(1^\lambda)$	**if** $\mathrm{T}_{\mathrm{ros}}[\boldsymbol{\rho}, \mathsf{aux}] = \perp$ **then**
$\mathrm{T}_{\mathrm{ros}} := [\,]$	$\quad \mathrm{T}_{\mathrm{ros}}[\boldsymbol{\rho}, \mathsf{aux}] \leftarrow_\$ \mathbb{Z}_p$
$((\boldsymbol{\rho}_i, \mathsf{aux}_i)_{i \in [0,\ell]}, (c_j)_{j \in [0,\ell-1]}) \leftarrow \mathsf{A}^{\mathrm{H}_{\mathrm{ros}}}(p)$	**return** $\mathrm{T}_{\mathrm{ros}}[\boldsymbol{\rho}, \mathsf{aux}]$
return $(\ \forall i \neq j \in [0,\ell],\ \ (\boldsymbol{\rho}_i, \mathsf{aux}_i) \neq (\boldsymbol{\rho}_j, \mathsf{aux}_j)$	
$\quad \wedge\ \forall i \in [0,\ell],\ \ \sum_{j=0}^{\ell-1} c_j \rho_{i,j} + \rho_{i,\ell} = \mathrm{H}_{\mathrm{ros}}(\boldsymbol{\rho}_i, \mathsf{aux}_i))$	

Fig. 1. The $\mathrm{ROS}_{\mathsf{Pgen},\mathcal{A},\ell}(\lambda)$ game. Above, $\rho_{i,j}$ is the j-th coefficient of the polynomial $\boldsymbol{\rho}_i$, i.e., $\boldsymbol{\rho}_i(\boldsymbol{x}) = \sum_{j=0}^{\ell-1} \rho_{i,j} x_i + \rho_{i,\ell}$.

pairings [CHYC05]. We note that some of the previous works claim security only for non-concurrent executions or with a bounded number of executions; therefore, our attacks do not contradict their security claims but render these schemes unsuitable for a broad range of real-world use cases.

Scope of Our Attacks and Countermeasures. Our attacks do not extend to the modified-ROS [FPS20] and the generalized-ROS [HKLN20] problems. The concrete hardness of both problems remains an intriguing open question.

2 Preliminaries

In this work, we assume that logarithm is always base 2. Let again $\mathsf{Pgen}(1^\lambda)$ be a parameter generation algorithm that given as input the security parameter λ in unary outputs a prime p of length $\lambda = \lceil \log p \rceil$. The ROS problem for ℓ dimensions, displayed in Fig. 1, is *hard* if no adversary can solve the ROS problem in time polynomial in the security parameter λ. i.e.:

$$\mathsf{Adv}^{\mathrm{ros}}_{\mathsf{Pgen},\mathcal{A},\ell}(\lambda) := \Pr\left[\mathrm{ROS}_{\mathsf{Pgen},\mathcal{A},\ell}(\lambda) = 1\right] = \mathsf{negl}(\lambda)\,.$$

Alternative Formulations of ROS. Fuchsbauer et al. [FPS20, Fig. 7] present a variant of $\mathrm{ROS}_{\mathsf{Pgen},\mathcal{A},\ell}(\lambda)$ the gamewith linear instead of affine functions $\boldsymbol{\rho}_i$ (i.e., where $\rho_{i,\ell} = 0$). Hauck et al. [HKL19, Fig. 3] allow only for linear functions, and do not allow for auxiliary information aux within $\mathrm{H}_{\mathrm{ros}}$ (i.e., where $\mathsf{aux}_i = \perp$).[2] These formulations are all equivalent.

First, any adversary $\mathcal{A}$ for ROS with affine functions as per Fig. 1 can be reduced to an adversary $\mathcal{B}$ for ROS with linear functions as per [FPS20]: $\mathcal{B}$ runs $\mathcal{A}$ and for every query of the form $((\rho_{i,0}, \ldots, \rho_{i,\ell}), \mathsf{aux}_i)$ to the oracle $\mathrm{H}_{\mathrm{ros}}$ (made by $\mathcal{A}$), it returns $\mathrm{H}_{\mathrm{ros}}((\rho_{i,0}, \ldots, \rho_{i,\ell-1}), (\rho_{i,\ell} \| \mathsf{aux}_i)) - \rho_{i,\ell}$. Finally, $\mathcal{B}$ modifies accordingly the solution output by $\mathcal{A}$ by concatenating $\rho_{i,\ell}$ to the corresponding aux_i.

Second, any adversary $\mathcal{A}$ for ROS with linear functions can be reduced to an adversary $\mathcal{B}$ for ROS with linear functions and without auxiliary information as per [HKL19]. We assume without loss of generality that $\mathcal{A}$ never

[2] Our attacks only apply to the case where the scalar set $\mathcal{S}$ is a finite field.

makes twice the same query. Then $\mathcal{B}$ runs $\mathcal{A}$ and for every query of the form $((\rho_{i,0},\ldots,\rho_{i,\ell-1},0),\mathsf{aux}_i)$ to the oracle (made by $\mathcal{A}$), it picks a random scalar $r \in \mathbb{Z}_p^*$ and returns $\mathrm{H}_{\mathrm{ros}}((r\cdot\rho_{i,0},\ldots,r\cdot\rho_{i,\ell-1}),\bot)\cdot r^{-1} \bmod p$. When $\mathcal{A}$ outputs a solution $(\boldsymbol{\rho}_i,\mathsf{aux}_i)_{i\in[0,\ell]},(c_j)_{j\in[0,\ell-1]}$, $\mathcal{B}$ outputs $(r\cdot\boldsymbol{\rho}_i)_{i\in[0,\ell]},(c_j)_{j\in[0,\ell-1]}$. The simulation of the oracle $\mathrm{H}_{\mathrm{ros}}$ is perfect unless there is a collision in the scalar r, which happens with negligible probability in λ.

3 Attack

In this section, we prove Theorem 1. We abuse notation and $\boldsymbol{\rho}_i$ denotes both the vector $\boldsymbol{\rho}_i = (\rho_{i,0},\ldots,\rho_{i,\ell}) \in \mathbb{Z}_p^{\ell+1}$ and the corresponding affine function $\boldsymbol{\rho}_i(\boldsymbol{x}) = \sum_{j=0}^{\ell-1}\rho_{i,j}\cdot x_j + \rho_{i,\ell}$ (where $\boldsymbol{x} = (x_0,\ldots,x_{\ell-1})$).

Proof (of Theorem 1). We construct an adversary for $\mathrm{ROS}_{\mathsf{Pgen},\mathcal{A},\ell}(\lambda)$, where $\ell > \log p$. Recall that to simplify the description of the attack, we use a polynomial formulation of ROS, i.e., we represent vectors $\boldsymbol{\rho}_i = (\rho_{i,0},\ldots,\rho_{i,\ell})$ as linear multivariate polynomials in $\mathbb{Z}_p[x_0,\ldots,x_{\ell-1}]$:

$$\boldsymbol{\rho}_i(x_0,\ldots,x_{\ell-1}) = \rho_{i,0}x_0 + \cdots + \rho_{i,\ell-1}x_{\ell-1} + \rho_{i,\ell}\ . \tag{1}$$

The goal for the adversary $\mathcal{A}$ is to output $(\boldsymbol{\rho}_i,\mathsf{aux}_i)_{i\in[0,\ell]}$ and $\boldsymbol{c} = (c_0,\ldots,c_{\ell-1})$ such that:

$$\boldsymbol{\rho}_i(\boldsymbol{c}) = \mathrm{H}_{\mathrm{ros}}(\boldsymbol{\rho}_i,\mathsf{aux}_i) \qquad \text{for all } i\in[0,\ell].$$

Define:

$$\boldsymbol{\rho}_i := x_i \qquad \text{for } i = 0,\ldots,\ell-1,$$

and find two strings aux_i^0 and aux_i^1 such that $c_i^b := \mathrm{H}_{\mathrm{ros}}(\boldsymbol{\rho}_i,\mathsf{aux}_b)$ are different for $b=0$ and $b=1$.[3] Then, let:

$$x_i' := \frac{x_i - c_i^0}{c_i^1 - c_i^0}$$

for all $i = 0,\ldots,\ell-1$. We remark that, if $x_i = c_i^b$, then $x_i' = b$ (for $b = 0,1$). Define $\boldsymbol{\rho}_\ell := \sum_{i=0}^{\ell-1} 2^i x_i'$, and query $y := \mathrm{H}_{\mathrm{ros}}(\boldsymbol{\rho}_\ell,\bot)$.Finally, write y in binary as:

$$y = \sum_{i=0}^{\ell-1} 2^i b_i \pmod p.$$

(As $2^\ell > p$, it is possible to write y this way, and this implicitly defines the b_i's.) The adversary $\mathcal{A}$ outputs the solution $(\boldsymbol{\rho}_0,\mathsf{aux}_0^{b_0}),\ldots,(\boldsymbol{\rho}_{\ell-1},\mathsf{aux}_{\ell-1}^{b_{\ell-1}}),(\boldsymbol{\rho}_\ell,\bot)$ and

[3] This step is the reason why the algorithm is expected polynomial time instead of polynomial time. Note that, since $\mathsf{aux} \in \{0,1\}^*$, there will always be two values $\mathsf{aux}_i^0,\mathsf{aux}_i^1 \in \{0,1\}^*$ so that $c_i^0 \neq c_i^1$.

$\boldsymbol{c} := (c_0^{b_0}, \ldots, c_{\ell-1}^{b_{\ell-1}})$. We have indeed that, for $i \in [0, \ell-1]$, $\boldsymbol{\rho}_i(\boldsymbol{c}) = c_i^{b_i} = \mathrm{H}_{\mathrm{ros}}(\boldsymbol{\rho}_i, \mathsf{aux}_i^{b_i})$ and:

$$\boldsymbol{\rho}_\ell(\boldsymbol{c}) = \sum_{i=0}^{\ell-1} 2^i x_i'(\boldsymbol{c}) = \sum_{i=0}^{\ell-1} 2^i b_i = y = \mathrm{H}_{\mathrm{ros}}(\boldsymbol{\rho}_\ell, \perp) \ .$$

□

Remark 1. In [FPS20, Sec. 5], Fuchsbauer, Plouviez, and Seurin proposed a variant of ROS, called modified ROS. The attack above does not apply to modified ROS.

4 Generalized Attack

We present a combination of Wagner's subexponential k-list attack and the polynomial time attack from Sect. 3. This combined attack yields a subexponentially efficient algorithm against ROS which requires fewer dimensions than the attack in the previous section (i.e., less than $\lambda = \lceil \log p \rceil$). However, for some practical cases, the attack significantly outperforms Wagner's attack in terms of work, for the same number of dimensions. At a very high level, our attack works as follows. We set $k_1 = 2^w - 1$, $k_2 = \max(0, \lceil \lambda - (w+1) \cdot L \rceil)$, and the dimension $\ell = k_1 + k_2$, for some integer w and some real number $L > 0$.

First, we use a generalization of Wagner's algorithm to find a "small" sum $s = y_{k_2}^* + \cdots + y_\ell^*$ of k_1 values $y_i^* := -\mathrm{H}_{\mathrm{ros}}(\boldsymbol{\rho}_i, \mathsf{aux}_i)$, where the polynomials $\boldsymbol{\rho}_i(\boldsymbol{x})$ are chosen to make the second step of the attack work.[4] As we describe below, we can obtain that $|s| < 2^{k_2 - 1}$ using $O(2^{w+L})$ hash queries and space $O(w2^L)$. Then, we use the technique from the previous section in order to represent the sum s as a binary sum of at most k_2 terms. Finally, we subtract the $k_1 - 1$ terms $y_{k_2}^*, \ldots, y_{k_2+k_1-1}^* = y_{\ell-1}^*$ to extract the term y_ℓ^*. This solves the ROS problem. The attack runs in overall time $O(2^{w+L})$, space $O(w2^L)$, and requires $\ell = \max(2^w - 1, \lceil 2^w - 1 + \lambda - (w+1) \cdot L \rceil)$ dimensions.

We remark that the attack is a generalization of both Wagner's attack and our polynomial-time attack from Sect. 3. Wagner's attack corresponds to the case where $L = \lambda/(w+1)$ and $\ell = 2^w - 1$. Our polynomial-time attack corresponds to the case $w = 0$, $L = 0$, $\ell = \lambda$.

Examples. For a prime p of $\lambda = 256$ bits, a concrete example yields $w = 5, L = 15$, i.e., $\ell = 32 + 256 - 6 \cdot 15 - 1 = 197$ dimensions and time roughly 2^{20} and space roughly $5 \cdot 2^{15}$ (elements of $\mathbb{Z}_p$). On the other hand, Wagner's algorithm for 197 dimensions requires time roughly $2^{\lfloor \log 197 \rfloor} \cdot 2^{\frac{256}{\lfloor \log 197 \rfloor + 1}} = 2^7 \cdot 2^{32} = 2^{39}$ and space roughly $\lfloor \log 197 \rfloor \cdot 2^{\frac{256}{\lfloor \log 197 \rfloor + 1}} = 7 \cdot 2^{32}$.

For a 512 bit modulus, a concrete example yields $w = 6, L = 46$, i.e., $\ell = 64 + 512 - 7 \cdot 46 - 1 = 253$ dimensions and time roughly 2^{53} and space roughly

[4] In the actual attack, part of the second step is executed before to allow to choose these polynomials properly.

$6 \cdot 2^{46}$. Wagner's algorithm for 254 dimensions requires time roughly $2^{\lfloor \log 254 \rfloor} \cdot 2^{\frac{512}{\lfloor \log 255 \rfloor + 1}} = 2^7 \cdot 2^{64} = 2^{71}$ and space roughly $\lfloor \log 254 \rfloor \cdot 2^{\frac{512}{\lfloor \log 255 \rfloor + 1}} = 7 \cdot 2^{64}$.[5]

4.1 Generalized k-List Algorithm

In this section, we write elements $\mathbb{Z}_p$ as signed integers in $[-\frac{p-1}{2}, \frac{p-1}{2}]$. Let w and L be two positive integers. We define the following integer intervals:

$$I_i := \left[- \left\lfloor \frac{p-1}{2^{(w-i) \cdot L + 1}} \right\rfloor, \left\lfloor \frac{p-1}{2^{(w-i) \cdot L + 1}} \right\rfloor \right] .$$

Remark that $\mathbb{Z}_p = I_w$.

We now describe the k-list algorithm, which is the core of the Wagner's algorithm. We generalize it to match our needs and to output elements that sum to something in I_{-1} rather than to exactly 0. (This essentially corresponds to executing Wagner's attack as usual, but stopping earlier.) The algorithm is defined relative to random oracle $\mathrm{H}_{\mathrm{ros}}$. It takes as input $(w, L, \boldsymbol{\rho}_1, \ldots, \boldsymbol{\rho}_k)$ and outputs $(\mathsf{aux}_1^*, \ldots, \mathsf{aux}_k^*)$ with $k = 2^w$ such that:

$$s := y_1^* + \cdots + y_k^* \in I_{-1} \qquad \text{where } y_i^* := \mathrm{H}_{\mathrm{ros}}(\boldsymbol{\rho}_i, \mathsf{aux}_i^*) .$$

The high-level idea of the algorithm is to use $2^{w+1} - 1$ lists of about 2^L values organized as a tree, as depicted in Fig. 2, and to ensure that lists $\mathfrak{L}_i^w$ at level i contains elements from the set I_i.

- **Setup/Leaves:** $\mathsf{k\text{-}List}$ fills the lists $\mathfrak{L}_i^w$ in the leaves with 2^L points of the form $\mathrm{H}_{\mathrm{ros}}(\boldsymbol{\rho}_i, \mathsf{aux}) \in \mathbb{Z}_p = I_w$, for $\mathsf{aux} \in [1, 2^L]$.
- **Collisions/Join:** The algorithm now proceeds to find collisions in levels from w to 1. At level i, process the 2^{i-1} pairs of lists $(\mathfrak{L}_1^i, \mathfrak{L}_2^i), \ldots, (\mathfrak{L}_{2^i-1}, \mathfrak{L}_{2^i})$ into 2^{i-1} lists $\mathfrak{L}_1^{i-1}, \ldots, \mathfrak{L}_{2^w-1}^{i-1}$ as follows:

  $$\mathfrak{L}_j^{i-1} := \{a + b \quad : \quad a \in \mathfrak{L}_{2j-1}^i,\ b \in \mathfrak{L}_{2j}^i,\ a + b \in I_i\} .$$

  (Remember that $a, b \in \mathbb{Z}_p$ and $a + b$ is computed modulo p.) Moreover, we implicitly assume that the algorithm stores back pointers to a and b s.t. they can efficiently be recovered at a later point.
- **Output:** Let $\mathfrak{L}^0 = \mathfrak{L}_1^0$ denote the (only) list created at level 1. The algorithm finds an element $s \in \mathfrak{L}^0$ such that $s \in I_{-1}$. If no such element exists, it returns $\perp$. Otherwise, it recovers $k = 2^w$ strings $\mathsf{aux}_1^*, \ldots, \mathsf{aux}_k^*$ such that $y_i^* = \mathrm{H}_{\mathrm{ros}}(\boldsymbol{\rho}_i, \mathsf{aux}_i^*) \in \mathfrak{L}_i^w$ and $s = y_1^* + \cdots + y_k^*$. It returns $(\mathsf{aux}_1^*, \ldots, \mathsf{aux}_k^*)$.

We formally write the algorithm $\mathsf{k\text{-}List}$ in Fig. 3.

[5] Indeed, when considering the exact values of the constants in the asymptotics, the actual complexity of Wagner's attack is $2^{\lfloor \log(\ell+1) \rfloor} \cdot 2^{\frac{p}{\lfloor \ell+1 \rfloor + 1}}$.

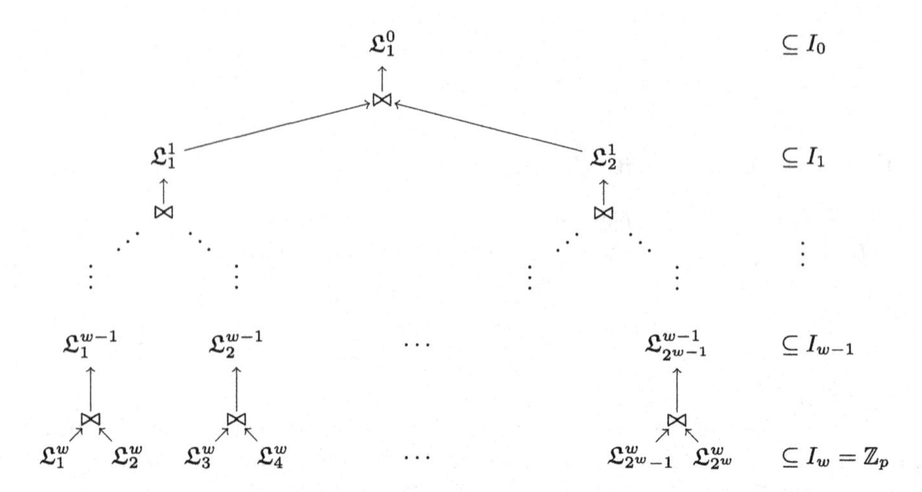

Fig. 2. Tree of lists for the k-list algorithm ($\bowtie$ represents the join operation in the algorithm; the sets in the right handside are the sets to which the elements of the lists of a given level belong).

Correctness. We do not prove correctness of k-List in this work, since our algorithm's correctness is implied by the correctness of Wagner's original algorithm. More precisely, our algorithm performs identical steps as Wagner's, but stops upon finding a sum of values with a suitably small absolute value, i.e., one that falls into I_0. On the other hand, Wagner's algorithm keeps continuing with more levels until it finds values who sum to 0. However, we remark that we are not aware of a formal analysis of Wagner's algorithm for values in $\mathbb{Z}_p$. The work of Minder and Sinclair [MS09] analyses the case of finding a weighted sum of *vectors* of $\mathbb{Z}_p$ values that sum to zero in each component, but uses a different technique from the one presented in Wagner's paper (and used here). Our attack can be seen as working under the assumption that Wagner's algorithm works correctly, i.e., has constant failure probability (see below). We can repeat the attack until it succeeds, which makes the resulting algorithm expected polynomial time. Formally analyzing the failure probability of Wagner's algorithm over $\mathbb{Z}_p$ remains an important open problem.

Complexity. Overall, the algorithm runs in time $O(2^{w+L})$ and is conjectured to succeed with constant probability. (As described [Wag02], this running time is made possible using an optimized join operation such as Hash Join or Merge Join). The algorithm uses space $O(2^{w+L})$, but by evaluating the collisions/joins in postfix order (in the tree), this can be reduced to $O(w2^L)$.

Algorithm $\text{k-List}^{\mathrm{H_{ros}}}(w, L, \boldsymbol{\rho}_1, \ldots, \boldsymbol{\rho}_{2^w})$

// Setup

$L_i^w := \{\mathrm{H_{ros}}(\boldsymbol{\rho}_i, \mathsf{aux})\}_{\mathsf{aux} \in [1, 2^L]}$ **for** $i \in [1, 2^w]$

// Collisions

for $i = w$ **downto** 1:

  for $j \in [1, 2^{i-1}]$:

    $\mathfrak{L}_j^{i-1} = \{a + b : a \in \mathfrak{L}_{2j-1}^i,\ b \in \mathfrak{L}_{2j}^i,\ a + b \in I_i\}$

// Output

look for an element $s = y_1^* + \cdots + y_k^* \in \mathfrak{L}^0 \cap I_{-1}$

if such an element does not exists **then return** $\perp$

return $(\mathsf{aux}_1^*, \ldots, \mathsf{aux}_k^*)$ such that $y_i^* = \mathrm{H_{ros}}(\boldsymbol{\rho}_i, \mathsf{aux}_i^*)$

Fig. 3. The k-list algorithm.

4.2 Combined Attack

We now prove Theorem 2.

Proof. Recall that $k_1 = 2^w - 1$ and $k_2 = \max(0, \lceil \lambda - (w+1) \cdot L \rceil)$. Set $\ell = k_1 + k_2$. For all $i \in [0, \ell - 1]$, define:

$$\boldsymbol{\rho}_i := x_i\,,$$

and find two strings aux_i^0 and aux_i^1 with different hash values $c_i^0 = \mathrm{H_{ros}}(\boldsymbol{\rho}_i, \mathsf{aux}_i^0)$ and $c_i^1 = \mathrm{H_{ros}}(\boldsymbol{\rho}_i, \mathsf{aux}_i^1)$. Then, let:

$$x_i' := \frac{x_i - c_i^0}{c_i^1 - c_i^0}$$

for all $i \in [0, k_2 - 1]$. We remark that, if $x_i = c_i^b$, then $x_i' = b$ (for $b = 0, 1$). Define:

$$\boldsymbol{\rho}_\ell := \sum_{i=0}^{k_2-1} 2^i x_i' - \left\lfloor \frac{p-1}{2^{(w+1)\cdot L+1}} \right\rfloor - \sum_{i=k_2}^{k_1+k_2-1} x_i\ .$$

Run $(\mathsf{aux}_{k_2}, \ldots, \mathsf{aux}_\ell) := \text{k-List}^{\mathrm{H_{ros}}}(w, L, \boldsymbol{\rho}_{k_2}, \ldots, \boldsymbol{\rho}_\ell)$ (where $k = k_1 + 1 = 2^w$) and define for $i \in [k_2, \ell]$:

$$y_i^* := \mathrm{H_{ros}}(\boldsymbol{\rho}_i, \mathsf{aux}_i^*)\ ,$$

and $c_i := y_i^*$ for $i \in [k_2, \ell - 1]$. Set:

$$s := \sum_{i=k_2}^{\ell} y_i^* \quad \in I_{-1} = \left[-\left\lfloor \frac{p-1}{2^{(w+1)\cdot L+1}} \right\rfloor, \left\lfloor \frac{p-1}{2^{(w+1)\cdot L+1}} \right\rfloor \right]\ . \tag{2}$$

Write $s + \lfloor (p-1)/2^{(w+1)\cdot L+1} \rfloor$ in binary as:

$$s + \left\lfloor \frac{p-1}{2^{(w+1)\cdot L+1}} \right\rfloor = \sum_{i=0}^{k_2-1} 2^i b_i \quad \in \left[0, \left\lfloor \frac{p-1}{2^{(w+1)\cdot L}} \right\rfloor \right]\,, \tag{3}$$

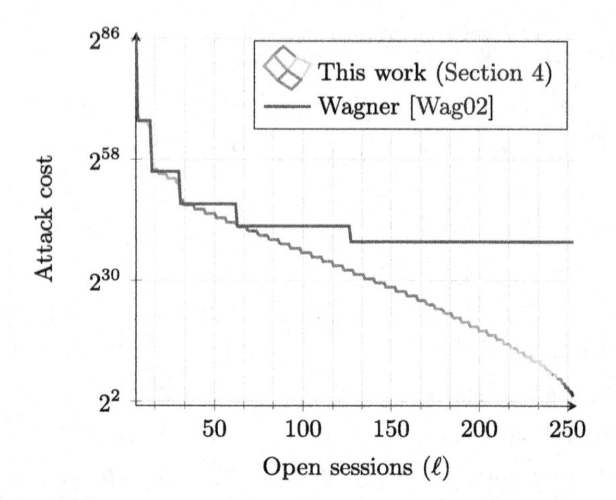

Fig. 4. Concrete cost of our combined attack compared to Wagner's [Wag02] for $\lambda = 256$ and $\ell < 256$. The color key indicates the different values of w used to estimate the cost. For $\ell \geq 256$, the attack of Sect. 3 applies.

which is possible since $p < 2^\lambda$, $k_2 = \lambda - (w+1) \cdot L$, hence $(p-1)/2^{(w+1)\cdot L} < 2^{k_2}$. Define:

$$\mathsf{aux}_i = \begin{cases} \mathsf{aux}_i^{b_i} & \text{for } i \in [0, k_2 - 1] \ , \\ \mathsf{aux}_i^* & \text{for } i \in [k_2, k_1 + k_2] \text{ from } \mathsf{k\text{-}List}. \end{cases}$$

$\mathcal{A}$ outputs: $(\boldsymbol{\rho}_0, \mathsf{aux}_0), \ldots, (\boldsymbol{\rho}_\ell, \mathsf{aux}_\ell)$ and:

$$\boldsymbol{c} := (c_0^{b_0}, \ldots, c_{k_2}^{b_{k_2}}, c_{k_2+1}, \ldots c_{k_2+k_1-1}) \ .$$

We have indeed that:

$$\boldsymbol{\rho}_i(\boldsymbol{c}) = c_i = \begin{cases} c_i^{b_i} = \mathrm{H}_{\mathrm{ros}}(\boldsymbol{\rho}_i, \mathsf{aux}_i^{b_i}) & \text{for } i \in [0, k_2 - 1] \ , \\ y_i^* = \mathrm{H}_{\mathrm{ros}}(\boldsymbol{\rho}_i, \mathsf{aux}_i^*) & \text{for } i \in [k_2, k_1 + k_2 - 1] \ . \end{cases}$$

and:

$$\begin{aligned} \boldsymbol{\rho}_\ell(\boldsymbol{c}) &= \sum_{i=0}^{k_2-1} 2^i x_i'(\boldsymbol{c}) - \left\lfloor \frac{p-1}{2^{(w+1)\cdot L+1}} \right\rfloor - \sum_{i=k_2}^{k_1+k_2-1} x_i(\boldsymbol{c}) \\ &= \sum_{i=0}^{k_2-1} 2^i b_i - \left\lfloor \frac{p-1}{2^{(w+1)\cdot L+1}} \right\rfloor - \sum_{i=k_2}^{k_1+k_2-1} y_i^* \\ &= s - \sum_{i=k_2}^{k_1+k_2-1} y_i^* = y_{k_2+k_1}^* = \mathrm{H}_{\mathrm{ros}}(\boldsymbol{\rho}_\ell, \mathsf{aux}_\ell^*) \ , \end{aligned}$$

where the third equality comes from Eq. 2 while the fourth equality comes from Eq. 3. The attack requires $k_1 + k_2 = \max\{2^w - 1, \lceil 2^w - 1 + \lambda - (w+1) \cdot L \rceil\}$ dimensions, runs in time $O(2^{w+L})$, and in space $O(w2^L)$. □

5 Affected Blind Signatures

For simplicity and clarity of exposition, we implement only the attack presented in Sect. 3. Our attack can be easily adapted for the one presented in Sect. 4.

Throughout the remaining of this manuscript, we will assume the existence of a group generator algorithm $\mathsf{GrGen}(1^\lambda)$ that, given as input the security parameter in unary form outputs the description $\Gamma = (\mathbb{G}, p, G)$ of a group $\mathbb{G}$ of prime order p generated by G. Similarly to Sect. 2, we assume that the prime p is of length λ. We use additive notation for the group law.

5.1 Schnorr Blind Signatures

A Schnorr blind signature [Sch01,FPS20] for a message $m \in \{0,1\}^*$ consists of a pair $(R, s) \in \mathbb{G} \times \mathbb{Z}_p$ such that $sG - cX = R$, where $c := \mathsf{H}(R, m)$ and $X \in \mathbb{G}$ is the verification key. A formal description of the protocol can be found in [FPS20, Fig. 6], using the same notation employed here.

We construct a probabilistic (expected) polynomial-time adversary $\mathcal{A}$ that is able to produce $\ell + 1$ signatures after opening $\ell \geq \lceil \log p \rceil = \lambda$ parallel sessions. $\mathcal{A}$ selects a message $m_\ell \in \{0,1\}^*$ for which a signature will be forged. It opens ℓ parallel sessions, querying $\textsc{Sign}_0()$ and receiving $\boldsymbol{R} = (R_0, \ldots, R_{\ell-1}) \in \mathbb{G}^\ell$. Let m_i^b be a random message and $c_i^b := \mathsf{H}(R_i, m_i^b)$ for $i \in [0, \ell-1]$ and $b \in \{0,1\}$. If $c_i^0 = c_i^1$, two different messages m_i^0 and m_i^1 are chosen until $c_i^0 \neq c_i^1$. Define $\boldsymbol{\rho}_\ell := \sum_i 2^i x_i'$ as per Sect. 3, that is:

$$\boldsymbol{\rho}_\ell(x_0, \ldots, x_{\ell-1}) := \sum_{i=0}^{\ell-1} 2^i \cdot \frac{x_i - c_i^0}{c_i^1 - c_i^0} = \sum_{i=0}^{\ell-1} \rho_{\ell,i} x_i + \rho_{\ell,\ell} \ . \tag{4}$$

Let $R_\ell := \boldsymbol{\rho}_\ell(\boldsymbol{R}) - \rho_{\ell,\ell} \cdot X$, where $\boldsymbol{\rho}_\ell(\boldsymbol{R})$ denotes the evaluation of the affine function $\boldsymbol{\rho}_\ell$ over $(R_0, \ldots R_{\ell-1})$. Define $c_\ell := \mathsf{H}(R_\ell, m_\ell) = \sum_{i=0}^{\ell-1} 2^i b_i$ and let $\boldsymbol{c} = (c_0^{b_0}, \ldots, c_{\ell-1}^{b_{\ell-1}})$. Complete the ℓ opened sessions querying $\textsc{Sign}_1(i, c_i^{b_i})$, for $i \in [0, \ell-1]$. The adversary thus obtains responses $\boldsymbol{s} := (s_0, \ldots, s_{\ell-1}) \in \mathbb{Z}_p^\ell$ satisfying:

$$s_i G - c_i^{b_i} X = R_i, \qquad \text{for } i \in [0, \ell-1].$$

Let $s_\ell := \boldsymbol{\rho}_\ell(\boldsymbol{s})$. Then $(m_\ell, (R_\ell, s_\ell))$ is a valid forgery. In fact, by perfect correctness of Schnorr blind signatures, we have:

$$\begin{aligned}
R_\ell = \boldsymbol{\rho}_\ell(\boldsymbol{R}) - \rho_{\ell,\ell} X &= \sum_{i=0}^{\ell-1} \rho_{\ell,i} \cdot R_i + \rho_{\ell,\ell} \cdot (G - X) \\
&= \sum_{i=0}^{\ell-1} \rho_{\ell,i} \cdot (s_i G - c_i^{b_i} X) + \rho_{\ell,\ell} \cdot (G - X) \\
&= \boldsymbol{\rho}_\ell(\boldsymbol{s}) \cdot G - \boldsymbol{\rho}_\ell(\boldsymbol{c}) \cdot X \\
&= s_\ell G - c_\ell X,
\end{aligned}$$

where $c_\ell = \mathsf{H}(R_\ell, m_\ell) = \boldsymbol{\rho}_\ell(\boldsymbol{c})$ by Eq. 4. Let $m_i := m_i^{b_i}$ for $i \in [0, \ell-1]$. The adversary outputs $(m_i, (R_i, s_i))$ for $i \in [0, \ell]$.

Remark 2. The attack does not apply to the clause blind Schnorr signature scheme [FPS20, Sec. 5], which relies on the modified ROS problem.

5.2 Okamoto–Schnorr Blind Signatures

An Okamoto–Schnorr blind signature [PS00] for a message m consists of a tuple $(R, s, t) \in \mathbb{G} \times \mathbb{Z}_p^2$ such that $sG + tH - cX = R$, where $c := \mathsf{H}(R, m)$, and (G, H) are two nothing-up-my-sleeve generators of $\mathbb{G}$. The attack of the previous section directly extends to Okamoto–Schnorr signatures: $\mathcal{A}$ operates exactly as before until Eq. 4. Then, the forgery is constructed as:

$$\big(R_\ell := \boldsymbol{\rho}_\ell(\boldsymbol{R}) + \rho_{\ell,\ell} H - \rho_{\ell,\ell} X, \quad s_\ell := \boldsymbol{\rho}_\ell(\boldsymbol{s}), \quad t_\ell := \boldsymbol{\rho}_\ell(\boldsymbol{t})\big).$$

We stress again that this does not contradict the security analysis of Stern and Pointcheval [PS00], whose security was reduced to $\mathrm{DLOG}_{\mathsf{GrGen},\mathcal{A}}(\lambda)$ for a $\mathsf{polylog}(\lambda)$ number of queries.

6 Other Constructions Affected

In this section, we overview how the attacks presented in Sects. 3 and 4 apply to a number of other cryptographic primitives. To simplify exposition, we focus on adapting the attack of Sect. 3. We note that, in some cases (e.g., multi-signatures), we break the security claims of the papers, while for other primitives (e.g., threshold signatures), our attack illustrates the tightness of the security theorems, which assume either non-concurrent setting, or up to a logarithmic number of concurrent executions.

6.1 Multi-signatures

A multi-signature scheme allows a group of signers $S_1, \ldots, S_n$, each having their own key pair $(\mathsf{pk}_j, \mathsf{sk}_j)$, to collaboratively sign a message m. The resulting signature can be verified given the message and the set of public keys of all signers.

CoSi. CoSi is a multi-signature scheme introduced by Syta et al. [STV+16], that features a two-round signing protocol. The signers are organized in a tree structure, where S_1 is the root of the tree. A signature for a message $m \in \{0,1\}^*$ consists of a pair $(c, s) \in \mathbb{Z}_p^2$ such that $c = \mathsf{H}(sG - c \cdot \mathsf{pk}, m)$, where $\mathsf{pk} = \sum_{j=1}^n \mathsf{pk}_j \in \mathbb{G}$ is the aggregated verification key. A formal description of the protocol can be found in [DEF+19, Sec. 2.5]; we use the same notation, except that we employ additive notation xG instead of multiplicative notation g^x.

Attack. We present an attack for a two-node tree where the attacker controls the root S_1. The attack can easily be extended to other settings, similarly to [DEF+19, Sec. 4.2]. Our attack allows the signer S_1 to forge one signature, for an arbitrary message $m_\ell \in \{0,1\}^*$, after performing $\ell \geq \lceil \log p \rceil = \lambda$ interactions with the honest signer S_2. Recall that $\mathsf{pk} = \mathsf{pk}_1 + \mathsf{pk}_2$ where $\mathsf{pk}_i = \mathsf{sk}_i G$. The signing protocol proceeds as follows. First, S_1 obtains a commitment $t_2 = r_2 G$ from S_2, and computes $\bar{t} = t_1 = r_1 G + t_2$ for a random r_1. Then, S_1 computes the challenge $c = \mathsf{H}(\bar{t}, m)$, and sends $(\bar{t}, c)$ to S_2. Next, S_2 returns $s_2 := r_2 + c \cdot \mathsf{sk}_2$. Finally, S_1 computes $s := s_2 + r_1 + c \cdot \mathsf{sk}_1$ and outputs the signature (c, s) for the message m.

The attack proceeds as follows. S_1 opens ℓ parallel sessions with ℓ arbitrary distinct messages $m_0, \ldots, m_{\ell-1} \in \{0,1\}^*$. For each session, S_1 gets the commitments $t_i = r_i G$ from S_2 at the end of the first round of signing. Now, it samples two random values $r_{i,0}, r_{i,1}$ for each $i \in [0, \ell-1]$, defines $\bar{t}_i^0 = r_{i,0} G + t_i$ and $\bar{t}_i^1 = r_{i,1} G + t_i$, and computes $c_i^b = \mathsf{H}(\bar{t}_i^b, m_i)$. (As usual, if $c_i^0 = c_i^1$, S_1 samples again $r_{i,0}$ and $r_{i,1}$ until $c_i^0 \neq c_i^1$.) S_1 then defines the polynomial $\boldsymbol{\rho} := \sum_{i=0}^{\ell-1} 2^i x_i / (c_i^1 - c_i^0)$, computes $t_\ell := \boldsymbol{\rho}(t_0, \ldots, t_{\ell-1})$ and $c_\ell := \mathsf{H}(t_\ell, m_\ell)$. S_1 computes $d_\ell = c_\ell - \boldsymbol{\rho}(c_0^0, \ldots, c_{\ell-1}^0)$ and writes this value in binary as $d_\ell = \sum_{i=0}^{\ell-1} 2^i b_i$. It then closes the ℓ sessions by using $\bar{t}_i = \bar{t}_i^{b_i}$ and $c_i = c_i^{b_i}$. At the last step of the signing sessions, S_1 obtains values $s_i = r_i + c_i \cdot \mathsf{sk}_2$ from S_2, and closes the sessions honestly using r_{i,b_i}. Finally, S_1 concludes its forgery by defining $s_\ell := \boldsymbol{\rho}(\boldsymbol{s}) + c_\ell \cdot \mathsf{sk}_1$: the pair (c_ℓ, s_ℓ) is a valid signature for m_ℓ. In fact:

$$
\begin{aligned}
s_\ell G - c_\ell \cdot \mathsf{pk} &= (\boldsymbol{\rho}(\boldsymbol{s}) + c_\ell \cdot \mathsf{sk}_1) G - c_\ell \cdot \mathsf{pk} \\
&= \sum_{i=0}^{\ell-1} \frac{2^i s_i}{c_i^1 - c_i^0} G - c_\ell \cdot \mathsf{pk}_2 \\
&= \sum_{i=0}^{\ell-1} \frac{2^i (r_i + c_i^{b_i} \cdot \mathsf{sk}_2)}{c_i^1 - c_i^0} G - c_\ell \cdot \mathsf{pk}_2 \\
&= \sum_{i=0}^{\ell-1} \frac{2^i r_i}{c_i^1 - c_i^0} G + \left(\sum_{i=0}^{\ell-1} \frac{2^i c_i^{b_i}}{c_i^1 - c_i^0} - c_\ell \right) \cdot \mathsf{pk}_2 \\
&= \sum_{i=0}^{\ell-1} \frac{2^i t_i}{c_i^1 - c_i^0} + \left(\sum_{i=0}^{\ell-1} 2^i b_i + \sum_{i=0}^{\ell-1} \frac{2^i c_i^0}{c_i^1 - c_i^0} - c_\ell \right) \cdot \mathsf{pk}_2 \\
&= \sum_{i=0}^{\ell-1} \frac{2^i t_i}{c_i^1 - c_i^0} + \underbrace{\left(\sum_{i=0}^{\ell-1} 2^i b_i + \boldsymbol{\rho}(c_0^0, \ldots, c_{\ell-1}^0) - c_\ell \right)}_{= d_\ell - d_\ell = 0} \cdot \mathsf{pk}_2 \\
&= \boldsymbol{\rho}(t_0, \ldots, t_{\ell-1}) = t_\ell \,,
\end{aligned}
$$

and $c_\ell = \mathsf{H}(t_\ell, m_\ell)$ by definition.

Two-Round MuSig. As in [DEF+19], the above technique (with some minor modifications) can be applied to the two-round MuSig as initially proposed by Maxwell et al. [MPSW18a], as the main difference between CoSi and two-round MuSig is in how the public key is aggregated in order to avoid rogue-key attacks. Our attack does not apply to the updated MuSig that uses a 3-round signing algorithm [MPSW18b].

6.2 Threshold Signatures

A (t, n)-threshold signature scheme assumes that the secret signing key is split among n parties $\mathsf{P}_1, \ldots, \mathsf{P}_n$ in a way that allows any subset of at least t out of the n parties to produce a valid signature. As long as the adversary corrupts less than the threshold number of parties, it is not possible to forge signatures or learn any information about the signing key.

GJKR07. Gennaro, Jarecki, Krawczyk, Rabin proposed a threshold signature scheme based on Pedersen's distributed key generation (DKG) protocol in [GJKR07, Section 5.2]. At a very high level, Pedersen's DKG protocol allows to generate a random group element $X = \chi G$ so that its discrete logarithm χ is shared both additively and according to Feldman secret sharing [Fel87] scheme, between a set of "qualified" parties. For the attack we present below, all parties $\mathsf{P}_1, \ldots, \mathsf{P}_n$ (included the ones that are controlled by the adversary) will remain qualified.[6] We denote by χ_j the additive share of party P_j. We have $\chi = \sum_{j=1}^{n} \chi_j$. Importantly for the attack, the adversary controlling for example P_1, can see all the group elements $\chi_2 G, \ldots, \chi_n G$ and then can choose its value χ_1. This is due to the way the Feldman secret sharing is performed.

In the threshold signature scheme of Gennaro et al. [GJKR07], the parties execute a distributed key generation procedure to produce a verification key $\mathsf{pk} := \mathsf{sk} \cdot G \in \mathbb{G}$, where the secret key sk is additively shared between the parties: each party P_j has an additive share sk_j, so that $\mathsf{sk} = \sum_{j=1}^{n} \mathsf{sk}_j$. A signature (R, s) for a message $m \in \{0,1\}^*$ is generated as follows. The participants run once again the distributed key generation protocol to produce a commitment $t = rG \in \mathbb{G}$, where r is additively shared between the parties: each party P_j has a share r_j, so that $r = \sum_{j=1}^{n} r_j$. Then, each party computes a share of the response:

$$s_j = r_j + c \cdot \mathsf{sk}_j, \qquad \text{where } c := \mathsf{H}(t, m). \tag{5}$$

Let $s := \sum_{j=1}^{n} s_j$. Then (c, s) is a valid signature on m. In fact:

$$sG = \sum_{j=1}^{n} r_j G + c \cdot \sum_{j=1}^{n} \mathsf{sk}_j \cdot G = t + c \cdot \mathsf{pk}, \tag{6}$$

where $c = \mathsf{H}(t, m)$.

[6] We do not use the fact that only a threshold $t+1$ of the parties are required to sign in our attack. We assume that all the parties come to sign, to simplify the description of the attack.

Concurrent Setting Insecurity. Gennaro et al. [GJKR07] proved the security of the scheme in a standalone *sequential* setting, where no two instances of the protocol can be run in parallel. We remark that if an adversary is allowed to start $\ell \geq \lceil \log p \rceil$ sessions in parallel, the attack against CoSi in Sect. 6.1 can be directly adapted to attack this threshold signature scheme for $n = 2$. The attack of both schemes use the fact that the adversary P_1 (or signer S_1 in CoSi) can see the commitment $t_2 = r_2 G$ of the honest party P_2 (or honest signed S_2) and only then choose r_1 that defines the commitment $t = r_1 G + t_2$. The generalization to any $n \geq 2$ is straightforward.

Scope of the Attack. Our attack is an attack against the proposed threshold signature scheme when instantiated with Pedersen's DKG, but not an attack against Perdersen's DKG itself (i.e., JF-DKG from [GJKR07, Fig. 1]). Furthermore, the attack does not work when Perdersen's DKG is replaced by the new DKG protocol from [GJKR07, Fig. 2].

Original Version of FROST. Komlo and Goldberg FROST [KG20a] proposed an extension of the above threshold signature scheme that was similarly affected by the above concurrent attack. On 19 July 2020, they updated the signing algorithm [KG20b] in a way that is no more susceptible to the above issue: each party now shares (D_j, E_j) and the commitment is computed as $R = \sum_j D_j + h_j E_j$, where $h_j := \mathsf{H}((D_j, E_j, j)_{j \in [t]})$. We direct the reader to [KG20b, Fig. 3] for a more detailed illustration of the problem and the fix.

6.3 Partially Blind Signatures

Partially blind signatures [AO00] are an extension of blind signature schemes that allow the signer to include some public metadata (e.g., expiration date, collateral conditions, server name, etc.) in the resulting signature. The original construction [AO00], as well as schemes inspired from it, such as Anonymous Credentials Light [BL13] and restrictive partially-blind signatures from bilinear pairings [CZMS06], might not provide the desired security properties.

Abe–Okamoto. Abe and Okamoto [AO00, Fig. 1] propose a partially blind signature scheme inspired from Schnorr blind signatures. Given a verification key $X := xG$ and some public information info that is hashed into the group $Z := \mathsf{H}(\mathsf{info})$, a partially blind signature for the message $m \in \{0,1\}^*$ is a tuple $(r, c, s, d) \in \mathbb{Z}_p$ where $c + d = \mathsf{H}(rG + cX,\ sG + dZ,\ Z,\ m)$.

Attack. The security of the above partially blind signature is proved up to a poly-logarithmic number of parallel open sessions in the security parameter [AO00]. We show that the security claim is tight by showing that there exists a poly-time attacker against one-more unforgeability in the setting where the adversary can have $\ell = O(\lambda)$ open sessions using the same metadata info. The attack follows essentially the same strategy of Sect. 5.1. First, the attacker opens ℓ parallel sessions and obtains the commitments $(A_i, B_i) \in \mathbb{G}^2$ for $i \in [0, \ell-1]$. It then constructs the polynomial $\boldsymbol{\rho}_\ell$ as per Eq. 4. The forged signature for an arbitrary message m^* is computed using the challenge:

$$e_\ell := \mathsf{H}(\boldsymbol{\rho}_\ell(\boldsymbol{A}) + \rho_{\ell,\ell}X,\ \boldsymbol{\rho}_\ell(\boldsymbol{B}) + \rho_{\ell,\ell}Z,\ Z,\ m^*) - \rho_{\ell,\ell}$$

and closing the ℓ sessions as in Sect. 5.1, i.e., by using the challenges $e_i^{b_i}$ where b_i is the i-th bit of the canonical representation of e_ℓ. Given the signatures $(r_i, c_i^{b_i}, s_i, d_i)$ for $i \in [0, \ell-1]$, the attacker can finally create its forgery $(\boldsymbol{\rho}(\boldsymbol{r}), \boldsymbol{\rho}(\boldsymbol{c}), \boldsymbol{\rho}(\boldsymbol{s}), \boldsymbol{\rho}(\boldsymbol{d}))$. The forgery is indeed correct because:

$$\begin{aligned}\boldsymbol{\rho}(\boldsymbol{c}) + \boldsymbol{\rho}(\boldsymbol{d}) &= \sum_i \rho_i(c_i^{b_i} + d_i) + \rho_{\ell,\ell} + \rho_{\ell,\ell} \\ &= \boldsymbol{\rho}(e_0^{b_0}, \ldots, e_{\ell-1}^{b_{\ell-1}}) + \rho_{\ell,\ell} \\ &= \mathsf{H}(\boldsymbol{\rho}_\ell(\boldsymbol{r})G + \boldsymbol{\rho}_\ell(\boldsymbol{c})X,\ \boldsymbol{\rho}_\ell(\boldsymbol{s})G + \boldsymbol{\rho}_\ell(\boldsymbol{d})Z,\ Z,\ m^*)\,.\end{aligned}$$

Anonymous Credentials Light. Inspired from Abe's blind signature [Abe01], Baldimitsi and Lysyanskaya [BL13] developed anonymous credentials light (ACL). The security proof of their scheme is under standard assumptions in the sequential settings. The public parameters are a so-called *real public key* $Y = xG$ and a *tag public key* $Z = wG$ (using the paper's notation). During the signing protocol, the signer produces two shares Z_1, Z_2 of Z such that $Z_1 + Z_2 = Z$, and proves either knowledge of Y (referred to as y-side), or of Z_1, Z_2 (so-called z-side). The discrete log of Z_1, Z_2 is never known by the signer, and the z-branch is inherited by Abe's blind signature and is necessary for the proof of security.

The essential difference between ACL and Abe's blind signature is the computation of Z_1: while in Abe's scheme it is computed invoking the random oracle over a random string (so that neither the user nor the signer know its discrete logarithm), in ACL it is computed starting from the user's commitment $C = \sum_{i=0}^{n} l_i H_i + rH$ (where $l_0, \ldots, l_n$) is the list of attributes) and the user could know a discrete-log relation across multiple sessions. This difference is fatal in the concurrent settings.

Attack. The attacker $\mathcal{A}$ opens ℓ parallel sessions, all with the same commitment C, and will provide a one-more forgery for an arbitrary message m^* on the same commitment C.

After opening the ℓ concurrent sessions, the attacker proves in zero-knowledge (as per protocol issuance) that the attributes required are valid, following the *reigistration phase* as prescribed in the protocol. Let $d_0, \ldots, d_{\ell-1}$ denote the randomization key used by the server to re-randomize the commitment C (displayed in [BL13, Fig. 1] as rnd) and sent to the user at the end of the registration phase. Upon receiving $A_i \in \mathbb{G}$ (the commitment of the y-side) and $A'_{1,i}, A'_{2,i}$ (the commitment of the z-side), for $i \in [0, \ell]$, the attacker computes the polynomial $\boldsymbol{\rho}_\ell$ defined in Sect. 3 (using the commitments and the message of the previous sessions), and computes the commitment forgeries:

$$A_\ell := \rho_\ell(A_0, \ldots, A_{\ell-1}) + \rho_{\ell,\ell} Y$$
$$A_{1,\ell} := \rho_\ell(A'_{1,0}, \ldots, A'_{1,\ell-1}) + \rho_{\ell,\ell} C$$
$$A_{2,\ell} := \rho_\ell(A'_{2,0}, \ldots, A'_{2,\ell-1}) + \rho_{\ell,\ell}(Z - C)$$

For simplicity, we assume that the re-randomization of Z is not performed by the attacker, i.e. $\tau = 1$, and that no blinding is performed: the attacker simply hashes the values, as they are received from the adversary. $\mathcal{A}$ sends the challenges according to the bits of $\mathsf{H}(Z, C, A_\ell, A_{1,\ell}, A_{2,\ell})$, similarly to Sect. 5, and receives the responses $(c_i, r_i, c'_i, r'_{1,i}, r'_{2,i}) \in \mathbb{Z}_p^5$, for $i \in [0, \ell]$. The adversary $\mathcal{A}$ computes the forged responses for the y-side:

$$c_\ell := \boldsymbol{\rho}(\boldsymbol{c}) = \sum_{i=0}^{\ell-1} \rho_{i,\ell} c_i + \rho_{\ell,\ell}$$
$$c'_\ell := \boldsymbol{\rho}(\boldsymbol{c}') = \sum_{i=0}^{\ell-1} \rho_{i,\ell} c'_i + \rho_{\ell,\ell}$$
$$r_\ell := \boldsymbol{\rho}(\boldsymbol{r}) = \sum_{i=0}^{\ell-1} \rho_{i,\ell} r_i + \rho_{\ell,\ell}$$
$$r'_{1,\ell} := \boldsymbol{\rho}(\boldsymbol{r}'_1 + \boldsymbol{c}' \circ \boldsymbol{d}) = \sum_{i=0}^{\ell-1} \rho_{i,\ell}(r'_{1,i} + c'_i d_i) + \rho_{\ell,\ell}$$
$$r'_{2,\ell} := \boldsymbol{\rho}(\boldsymbol{r}'_2 - \boldsymbol{c}' \circ \boldsymbol{d}) = \sum_{i=0}^{\ell-1} \rho_{i,\ell}(r'_{2,i} - c'_i d_i) + \rho_{\ell,\ell}$$

In fact, it holds that:

$$r_\ell G + c_\ell Y = \sum_{i=0}^{\ell} \rho_{i,\ell}\big(r_i G + c_i Y\big) + \rho_{\ell,\ell}(Y + G) = A_\ell$$
$$r'_{1,\ell} G + c'_\ell C = \sum_{i=0}^{\ell-1} \rho_{i,\ell}\big(r'_{1,i} G + c'_i(C + d_i G)\big) + \rho_{\ell,\ell}(C + G) = A_{1,\ell}$$
$$r'_{2,\ell} G + c'_\ell(Z - C) = \sum_{i=0}^{\ell-1} \rho_{i,\ell}\big(r'_{2,i} G + c'_i(Z - C - d_i G)\big) + \rho_{\ell,\ell}(Z - C) = A_{2,\ell}$$

And the verification of the re-randomization τ is trivially satisfied.

6.4 Conditional Blind Signatures

Conditional blind signatures (CBS), introduced by Grontas et al. [ZGP17], allow a user to request a blind signature on messages of their choice, and the server has a secret boolean input which determines if it will issue a valid signature or not. CBS only allow a *designated* verifier to check the validity of the signature; the user will not able to distinguish between valid and invalid signatures. Conditional blind signature have application in e-voting schemes [GPZZ19].

ZGP17. Zacharakis et al. [ZGP17] propose an instantiation of CBS as an extension of Okamoto–Schnorr blind signatures, where the (designated) verifier holds a secret verification key $k \in \mathbb{Z}_p$ and publishes $K = kG$ as public information. During the execution of Okamoto–Schnorr, one of the two responses (s, t) will be computed in $\mathbb{G}$ rather than $\mathbb{Z}_p$, using K as a generator. Only the designated verifier, who knows the discrete log of K can now check the verification equation.

The attack from Sect. 5.2 directly applies also to their scheme, and leads to a poly-time adversary that with λ queries to the signing oracle for the same bit $b = 1$ can produce one-more forgery with overwhelming probability. This attack does not invalidate the security claims of [ZGP17], which are argued only for a poly-logarithmic number of parallel open sessions.

6.5 Other Schemes

The following papers prove rely on the hardness of the ROS problem for their security proofs, and henceforth may not provide the expected security guarantees: blind anonymous group signatures [CFLW04]; blind identity-based signcryption [YW05]; blind signature schemes from bilinear pairings [CHYC05].

7 Conclusions

Our work provides a polynomial attack against $\text{ROS}_\ell(\lambda)$ when $\ell > \log p$, and a sub-exponential attack for $\ell < \log p$. This impacts the one-more unforgeability property of Schnorr and Okamoto–Schnorr blind signatures, plus a number of cryptographic schemes derived from them. Our attacks run in polynomial time only in the concurrent setting, and only for $\ell > \log p$ parallel signing sessions.

Concretely, the cost of the attack and the number of sessions required are rather small: for today's security parameters, the attack could be already mounted with $\ell = 9$ parallel open sessions. As already pointed out by [FPS20], even just $\ell = 16$ open sessions could lead to a forgery in time $O(2^{55})$. For $\ell = 128$, our attack of Sect. 4 leads to a forgery in time $O(2^{32})$. For $\ell = 256$, our attack of Sect. 3 produces a forgery in a matter of seconds on commodity hardware. Although 256 parallel signing sessions might seem at first unrealistic, modern large-scale web servers must handle more than 10 million concurrent sessions[7]. Given our attack, the main takeaway of our work is that blind Schnorr signatures are unsuitable for wide-scale deployments.

The easiest countermeasure to our attack could be to allow only for sequential signing sessions, as Schnorr blind signatures are unforgeable in the algebraic

[7] For further information, read the C10K problem ('99) and the C10M problem ('11).

group model for polynomially many sessions [KLRX]. Another countermeasure to our attack could be to employ (much) larger security parameters, require the signer to enforce strong ratio limits, *and* perform frequent key rotations, accepting the tradeoffs given by our attacks. Finally, Fuchsbauer et al. [FPS20] recently introduced a variant of blind Schnorr signatures (the *clause* version) which is unaffected by our attack. Unfortunately, it relies on the conjectured hardness of the so-called *modified ROS problem*, which is still relatively new and has not been subject to any significant cryptanalysis.

To conclude, other blind signature schemes are to this day considered secure and should be considered as alternatives: blind RSA [Cha82], blind BLS [Bol03], and Abe's blind signature scheme [Abe01, KLRX].

References

[Abe01] Abe, M.: A secure three-move blind signature scheme for polynomially many signatures. In: Pfitzmann, B. (ed.) EUROCRYPT 2001. LNCS, vol. 2045, pp. 136–151. Springer, Heidelberg (2001). https://doi.org/10.1007/3-540-44987-6_9

[AO00] Abe, M., Okamoto, T.: Provably secure partially blind signatures. In: Bellare, M. (ed.) CRYPTO 2000. LNCS, vol. 1880, pp. 271–286. Springer, Heidelberg (2000). https://doi.org/10.1007/3-540-44598-6_17

[BL13] Baldimtsi, F., Lysyanskaya, A.: Anonymous credentials light. In: Sadeghi, A.-R., Gligor, V.D., Yung, M. (eds.) ACM CCS 2013, pp. 1087–1098. ACM Press, November 2013

[Bol03] Boldyreva, A.: Threshold signatures, multisignatures and blind signatures based on the gap-diffie-hellman-group signature scheme. In: Desmedt, Y.G. (ed.) PKC 2003. LNCS, vol. 2567, pp. 31–46. Springer, Heidelberg (2003). https://doi.org/10.1007/3-540-36288-6_3

[Bra94] Brands, S.: Untraceable off-line cash in wallet with observers. In: Stinson, D.R. (ed.) CRYPTO 1993. LNCS, vol. 773, pp. 302–318. Springer, Heidelberg (1994). https://doi.org/10.1007/3-540-48329-2_26

[CFLW04] Chan, T.K., Fung, K., Liu, J.K., Wei, V.K.: Blind spontaneous anonymous group signatures for ad hoc groups. In: Castelluccia, C., Hartenstein, H., Paar, C., Westhoff, D. (eds.) ESAS 2004. LNCS, vol. 3313, pp. 82–94. Springer, Heidelberg (2005). https://doi.org/10.1007/978-3-540-30496-8_8

[Cha82] Chaum, D.: Blind signatures for untraceable payments. In: Chaum, D., Rivest, R.L., Sherman, A.T. (eds.) CRYPTO 1982, pp. 199–203. Plenum Press, New York (1982)

[CHYC05] Chow, S.S.M., Hui, L.C.K., Yiu, S.M., Chow, K.P.: Two improved partially blind signature schemes from bilinear pairings. In: Boyd, C., González Nieto, J.M. (eds.) ACISP 2005. LNCS, vol. 3574, pp. 316–328. Springer, Heidelberg (2005). https://doi.org/10.1007/11506157_27

[CZMS06] Chen, X., Zhang, F., Mu, Y., Susilo, W.: Efficient provably secure restrictive partially blind signatures from bilinear pairings. In: Di Crescenzo, G., Rubin, A. (eds.) FC 2006. LNCS, vol. 4107, pp. 251–265. Springer, Heidelberg (2006). https://doi.org/10.1007/11889663_21

[DEF+19] Drijvers, M., et al.: On the security of two-round multi-signatures. In: 2019 IEEE Symposium on Security and Privacy, pp. 1084–1101. IEEE Computer Society Press, May 2019

[Fel87] Feldman, P.: A practical scheme for non-interactive verifiable secret sharing. In: 28th FOCS, pp. 427–437. IEEE Computer Society Press, October 1987

[FPS20] Fuchsbauer, G., Plouviez, A., Seurin, Y.: Blind Schnorr signatures and signed ElGamal encryption in the algebraic group model. In: Canteaut, A., Ishai, Y. (eds.) EUROCRYPT 2020. LNCS, vol. 12106, pp. 63–95. Springer, Cham (2020). https://doi.org/10.1007/978-3-030-45724-2_3

[GJKR07] Gennaro, R., Jarecki, S., Krawczyk, H., Rabin, T.: Secure distributed key generation for discrete-log based cryptosystems. J. Cryptol. **20**(1), 51–83 (2007)

[GPZZ19] Grontas, P., Pagourtzis, A., Zacharakis, A., Zhang, B.: Towards everlasting privacy and efficient coercion resistance in remote electronic voting. In: Zohar, A., et al. (eds.) FC 2018. LNCS, vol. 10958, pp. 210–231. Springer, Heidelberg (2019). https://doi.org/10.1007/978-3-662-58820-8_15

[HKL19] Hauck, E., Kiltz, E., Loss, J.: A modular treatment of blind signatures from identification schemes. In: Ishai, Y., Rijmen, V. (eds.) EUROCRYPT 2019, Part III. LNCS, vol. 11478, pp. 345–375. Springer, Cham (2019). https://doi.org/10.1007/978-3-030-17659-4_12

[HKLN20] Hauck, E., Kiltz, E., Loss, J., Nguyen, N.K.: Lattice-based blind signatures, revisited. In: Micciancio, D., Ristenpart, T. (eds.) CRYPTO 2020, Part II. LNCS, vol. 12171, pp. 500–529. Springer, Cham (2020). https://doi.org/10.1007/978-3-030-56880-1_18

[KG20a] Komlo, C., Goldberg, I.: FROST: flexible round-optimized Schnorr threshold signatures (2020). https://crysp.uwaterloo.ca/software/frost/frost-extabs.pdf. Version from 7 January 2020. Accessed 04 Oct 2020

[KG20b] Komlo, C., Goldberg, I.: FROST: Flexible round-optimized Schnorr threshold signatures. Cryptology ePrint Archive, Report 2020/852 (2020). https://eprint.iacr.org/2020/852

[KLRX] Kastner, J., Loss, J., Xu, J.: On pairing-free blind signature schemes in the algebraic group model. Cryptology ePrint Archive, Report 2020/1071 (2020)

[MPSW18a] Maxwell, G., Poelstra, A., Seurin, Y., Wuille, P.: Simple Schnorr multi-signature with applications to Bitcoin. Cryptology ePrint Archive, Report 2018/068, Revision 20180118:124757 (2018). https://eprint.iacr.org/2018/068/20180118:124757

[MPSW18b] Maxwell, G., Poelstra, A., Seurin, Y., Wuille, P.: Simple Schnorr multi-signature with applications to Bitcoin. Cryptology ePrint Archive, Report 2018/068, Revision 20180520:191909 (2018). https://eprint.iacr.org/2018/068/20180520:191909

[MS09] Minder, L., Sinclair, A.: The extended k-tree algorithm. In: Mathieu, C. (ed.) 20th SODA, pp. 586–595. ACM-SIAM, January 2009

[PS00] Pointcheval, D., Stern, J.: Security arguments for digital signatures and blind signatures. J. Cryptol. **13**(3), 361–396 (2000)

[S+20] Stein, W.A., et al.: Sage Mathematics Software (Version 9.1). The Sage Development Team (2020). http://www.sagemath.org

[Sch01] Schnorr, C.P.: Security of blind discrete log signatures against interactive attacks. In: Qing, S., Okamoto, T., Zhou, J. (eds.) ICICS 2001. LNCS, vol. 2229, pp. 1–12. Springer, Heidelberg (2001). https://doi.org/10.1007/3-540-45600-7_1

[STV+16] Syta, E., et al.: Keeping authorities "honest or bust" with decentralized witness cosigning. In: 2016 IEEE Symposium on Security and Privacy, pp. 526–545. IEEE Computer Society Press, May 2016

[Wag02] Wagner, D.: A generalized birthday problem. In: Yung, M. (ed.) CRYPTO 2002. LNCS, vol. 2442, pp. 288–304. Springer, Heidelberg (2002). https://doi.org/10.1007/3-540-45708-9_19

[YW05] Yuen, T.H., Wei, V.K.: Fast and proven secure blind identity-based signcryption from pairings. In: Menezes, A. (ed.) CT-RSA 2005. LNCS, vol. 3376, pp. 305–322. Springer, Heidelberg (2005). https://doi.org/10.1007/978-3-540-30574-3_21

[ZGP17] Zacharakis, A., Grontas, P., Pagourtzis, A.: Conditional blind signatures. Cryptology ePrint Archive, Report 2017/682 (2017). http://eprint.iacr.org/2017/682

New Representations of the AES Key Schedule

Gaëtan Leurent(✉) and Clara Pernot

Inria, Paris, France
{gaetan.leurent,clara.pernot}@inria.fr

Abstract. In this paper we present a new representation of the AES key schedule, with some implications to the security of AES-based schemes. In particular, we show that the AES-128 key schedule can be split into four independent parallel computations operating on 32 bits chunks, up to linear transformation. Surprisingly, this property has not been described in the literature after more than 20 years of analysis of AES. We show two consequences of our new representation, improving previous cryptanalysis results of AES-based schemes.

First, we observe that iterating an odd number of key schedule rounds results in a function with short cycles. This explains an observation of Khairallah on mixFeed, a second-round candidate in the NIST lightweight competition. Our analysis actually shows that his forgery attack on mixFeed succeeds with probability 0.44 (with data complexity 220 GB), breaking the scheme in practice. The same observation also leads to a novel attack on ALE, another AES-based AEAD scheme.

Our new representation also gives efficient ways to combine information from the first subkeys and information from the last subkeys, in order to reconstruct the corresponding master keys. In particular we improve previous impossible differential attacks against AES-128.

Keywords: AES · Key schedule · mixFeed · ALE · Impossible differential attack

1 Introduction

The AES [1,17] is the most widely used block cipher today, designed by Daemen and Rijmen in 1999 and selected for standardization by NIST. Like all symmetric cryptography primitives, the security of the AES can only be evaluated with cryptanalysis, and there is a constant effort to study its resistance again old and new attacks, and to evaluate its security margin. There are three versions of AES, with different key sizes, and different number of rounds: AES-128 with 10 rounds, AES-192 with 12 rounds, and AES-256 with 14 rounds. After twenty years of cryptanalysis, many different attacks have been applied to AES, and we have a strong confidence in its security: the best attacks against AES-128 in the single-key setting reach only 7 rounds out of 10. The best attacks known so far are either impossible differential attacks (following a line of work starting with [2]) or meet-in-the-middle attacks (with a line of work starting from [18]), as listed in Table 2.

A. Canteaut and F.-X. Standaert (Eds.): EUROCRYPT 2021, LNCS 12696, pp. 54–84, 2021.
https://doi.org/10.1007/978-3-030-77870-5_3

Table 1. Comparison of attacks against ALE.

Attack		Enc.	Verif.	Time	Ref.
Existential Forgery	Known Plaintext	$2^{110.4}$	2^{102}	$2^{110.4}$	[34]
Existential Forgery	Known Plaintext	2^{103}	2^{103}	2^{104}	[30]
Existential Forgery	Known Plaintext	1	2^{120}	2^{120}	[30]
State Recovery, Almost Univ. Forgery	Known Plaintext	1	2^{121}	2^{121}	[30]
State Recovery, Almost Univ. Forgery	Chosen Plaintext	$2^{57.3}$	0	$2^{104.4}$	**New**

Table 2. Best single-key attacks against 7-round AES-128.

Attack	Data	Time	Mem.	Ref.	Note
Meet-in-the-middle	2^{97}	2^{99}	2^{98}	[19]	
	2^{105}	2^{105}	2^{90}	[19]	
	2^{105}	2^{105}	2^{81}	[9]	
	2^{113}	2^{113}	2^{74}	[9]	
Impossible differential	2^{113}	2^{113}	2^{74}	[13]	Using 4 out. diff. and state-test
	$2^{105.1}$	2^{113}	$2^{74.1}$	[13][a]	Using 4 out. diff
	$2^{106.1}$	$2^{112.1}$	$2^{73.1}$		Variant of [13] using 1 out. diff.
	$2^{104.9}$	$2^{110.9}$	$2^{71.9}$	**New**	Using 1 out. diff.

[a]The time complexity is incorrectly given as $2^{106.88}$ in [13].

1.1 Our Results

The key schedule is arguably the weakest part of the AES, and it is well known to cause issues in the related-key setting [5–7]. In this paper, we focus on the key schedule of AES, and we show a surprising alternative representation, where the key schedule is split into several independent chunks, and the actual subkeys are just linear combinations of the chunks.

Application to mixFeed and ALE. This representation is motivated by an observation made by Khairallah [29] on the AEAD scheme mixFeed: when the 11-round AES-128 key schedule is iterated there are apparently many short cycles of length roughly 2^{34}. Our representation explains this observation, and proves that the forgery attack of Khairallah against mixFeed actually succeeds with a very high probability. It only requires the encryption of one known message of length at least $2^{33.7}$ blocks, and generates a forgery with probability 0.44, making it a practical break of the scheme.

We also apply the same observation to ALE, another AES-based scheme that iterates the AES key schedule. We obtain a novel attack against ALE, with a much lower data complexity than previous attacks, but we need chosen plaintexts rather than known plaintexts (see Table 1).

Key recovery attack against AES-128. We also improve key recovery attacks against AES-128 based on impossible differential cryptanalysis. This type of attacks targets bytes of the first subkey and of the last subkey, and excludes some values that are shown impossible. Then, the attacker must iterate over the remaining candidates, and reconstruct the corresponding master keys. Using our new representation of the key schedule, we make the reconstruction of the master key more efficient. Therefore we can start from a smaller data set: we identify fewer impossible keys, but we process the larger number of key candidates without making this step the bottleneck.

While the improvement is quite modest (see Table 2), it is notable that we improve this attack in a non-negligible way, because cryptanalysis of AES has achieved a high level of technicality, and attacks are already thoroughly optimized. In particular, we obtain the best attack so far when the amount of memory is limited (*e.g.* below 2^{75}).

1.2 Organisation of the Paper

We start with a description of the AES-128 key schedule and describe our alternative representation in Sect. 2, before presenting applications to mixFeed (Sect. 3), ALE (Sect. 4) and impossible differential attacks against AES-128 (Sect. 5). We then describe an alternative representation of the AES-192 and AES-256 key schedules in Sect. 6, and some properties of the AES key schedules that might be useful in future works in Sect. 7.

2 A New Representation of the AES-128 Key Schedule

In AES-128, the key schedule is an iterative process to derive 11 subkeys from one master key. To start with, the 128 bits of the master key are divided into 4 words of 32 bits each: w_i for $0 \leq i \leq 3$. The following notations are used within the algorithm:

RotWord performs a cyclic permutation of one byte to the left.
SubWord applies the AES Sbox to each of the 4 bytes of a word.
RCon(i) is a round constant defined as $[x^{i-1}, 0, 0, 0]$ in the field $\mathbb{F}_{2^8}$ described in [1]. For simplicity, we denote x^{i-1} as c_i.

In order to construct w_i for $i \geq 4$, one applies the following steps:

- if $i \equiv 0 \bmod 4$, $w_i = \mathsf{SubWord}(\mathsf{RotWord}(w_{i-1})) \oplus \mathsf{RCon}(i/4) \oplus w_{i-4}$.
- else, $w_i = w_{i-1} \oplus w_{i-4}$.

The subkey at round r is the concatenation of the words w_{4r} to w_{4r+3}. We can also express the key schedule at the byte level, using k_i^r with $0 \leq i < 16$ to denote byte i of the round-r subkey (we use $k^r_{\langle i,j,\dots\rangle}$ as a shorthand for $k_i^r, k_j^r, \dots$). The subkey is typically represented as a 4×4 matrix with the AES byte ordering, with $w_i = k^{i/4}_{4(i \bmod 4)} \| k^{i/4}_{4(i \bmod 4)+1} \| k^{i/4}_{4(i \bmod 4)+2} \| k^{i/4}_{4(i \bmod 4)+3}$:

$$\begin{bmatrix} k_0^r & k_4^r & k_8^r & k_{12}^r \\ k_1^r & k_5^r & k_9^r & k_{13}^r \\ k_2^r & k_6^r & k_{10}^r & k_{14}^r \\ k_3^r & k_7^r & k_{11}^r & k_{15}^r \end{bmatrix} = \begin{bmatrix} \\ w_{4r} & w_{4r+1} & w_{4r+2} & w_{4r+3} \\ \\ \end{bmatrix}$$

The key schedule can be written as follows, with k the key schedule state, k'_i the state after one round of key schedule, and S the AES Sbox (see Fig. 1 and 3):

$$
\begin{aligned}
k'_0 &= k_0 \oplus S(k_{13}) \oplus c_i & k'_8 &= k_8 \oplus k_4 \oplus k_0 \oplus S(k_{13}) \oplus c_i \\
k'_1 &= k_1 \oplus S(k_{14}) & k'_9 &= k_9 \oplus k_5 \oplus k_1 \oplus S(k_{14}) \\
k'_2 &= k_2 \oplus S(k_{15}) & k'_{10} &= k_{10} \oplus k_6 \oplus k_2 \oplus S(k_{15}) \\
k'_3 &= k_3 \oplus S(k_{12}) & k'_{11} &= k_{11} \oplus k_7 \oplus k_3 \oplus S(k_{12}) \\
k'_4 &= k_4 \oplus k_0 \oplus S(k_{13}) \oplus c_i & k'_{12} &= k_{12} \oplus k_8 \oplus k_4 \oplus k_0 \oplus S(k_{13}) \oplus c_i \\
k'_5 &= k_5 \oplus k_1 \oplus S(k_{14}) & k'_{13} &= k_{13} \oplus k_9 \oplus k_5 \oplus k_1 \oplus S(k_{14}) \\
k'_6 &= k_6 \oplus k_2 \oplus S(k_{15}) & k'_{14} &= k_{14} \oplus k_{10} \oplus k_6 \oplus k_2 \oplus S(k_{15}) \\
k'_7 &= k_7 \oplus k_3 \oplus S(k_{12}) & k'_{15} &= k_{15} \oplus k_{11} \oplus k_7 \oplus k_3 \oplus S(k_{12})
\end{aligned}
$$

Invariant subspaces. Recently, several lightweight block ciphers have been analyzed using *invariant subspace* attacks. This type of attack was first proposed on PRINTcipher by Leander *et al.* [31]; the basic idea is to identify a linear subspace V and an offset u such that the round function F of a cipher satisfies $F(u+V) = F(u) + V$. At Eurocrypt 2015, Leander, Minaud and Rønjom [32] introduced an algorithm in order to detect such invariant subspaces. By applying this algorithm to four rounds of the AES-128 key schedule, we find invariant subspaces of dimension four over $\mathbb{F}_{2^8}$, and this implies a decomposition of the key schedule.

First, let's recall the generic algorithm for a permutation $F : \mathbb{F}_2^n \to \mathbb{F}_2^n$:

1. Guess an offset $u \in \mathbb{F}_2^n$ and a one-dimensional subspace V_0.
2. Compute $V_{i+1} = span\{(F(u+V_i) - F(u)) \cup V_i\}$.
3. If the dimension of V_{i+1} equals the dimension of V_i, we found an invariant subspace: $F(u+V) = F(u) + V$.
4. Else, we go on step 2.

In the case of the AES-128 key schedule, we use subspaces of $\mathbb{F}_{2^8}^{16}$ over the field $\mathbb{F}_{2^8}$ rather than over $\mathbb{F}_2$. If we apply this algorithm with the permutation F corresponding to 4 rounds of key schedule, with any key state u, and with V_0 the vector space generated by one of the first four bytes, we obtain 4 invariant affine subspaces whose linear parts are:

$$
\begin{aligned}
E_0 &= \{(a,b,c,d, \quad 0,b,0,d, \quad a,0,0,d, \quad 0,0,0,d) \quad \text{for } a,b,c,d \in \mathbb{F}_{2^8}\} \\
E_1 &= \{(a,b,c,d, \quad a,0,c,0, \quad 0,0,c,d, \quad 0,0,c,0) \quad \text{for } a,b,c,d \in \mathbb{F}_{2^8}\} \\
E_2 &= \{(a,b,c,d, \quad 0,b,0,d, \quad 0,b,c,0, \quad 0,b,0,0) \quad \text{for } a,b,c,d \in \mathbb{F}_{2^8}\} \\
E_3 &= \{(a,b,c,d, \quad a,0,c,0, \quad a,b,0,0, \quad a,0,0,0) \quad \text{for } a,b,c,d \in \mathbb{F}_{2^8}\}
\end{aligned}
$$

When we consider a single round R of the key schedule, the subspaces are not invariant, but are images of each other. We have the following relations, with u_0 an element in $(\mathbb{F}_{2^8})^{16}$ and $u_i = R^i(u_0)$, for $(1 \le i < 5)$:

$$
\begin{aligned}
R(E_0 + u_0) &= E_1 + u_1, & R(E_1 + u_1) &= E_2 + u_2, \\
R(E_2 + u_2) &= E_3 + u_3, & R(E_3 + u_3) &= E_0 + u_4
\end{aligned}
$$

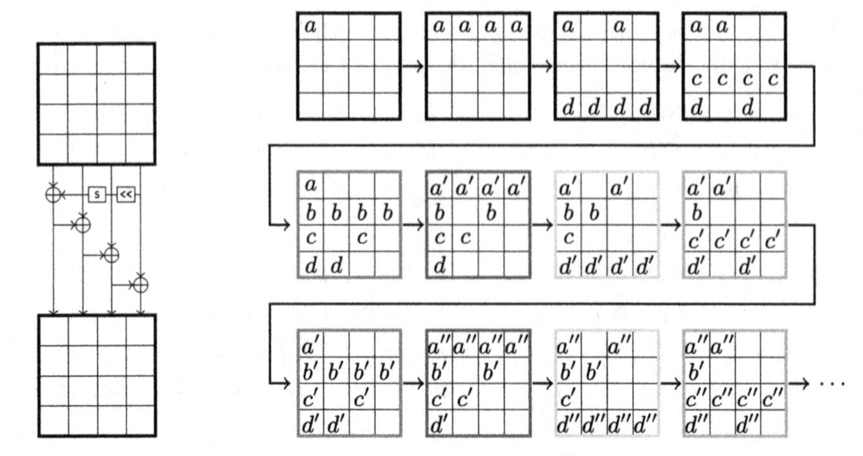

Fig. 1. AES key schedule. (figure adapted from [28])

Fig. 2. Evolution of a difference located on the first byte after several rounds of key schedule.

In other words, if the difference pattern between two states is in E_i, then after r rounds of key schedule, the difference pattern will be in $E_{(i+r)\%4}$.

This can be verified by tracking the differences in the key schedule, starting from a difference $(a, 0, 0, 0,\ 0, 0, 0, 0,\ 0, 0, 0, 0,\ 0, 0, 0, 0)$, as shown in Fig. 2. After four rounds we reach a difference $(a, b, c, d,\ 0, b, 0, d,\ 0, b, c, 0,\ 0, b, 0, 0)$, with differential transitions $a \to d$, $d \to c$, and $c \to b$ through the Sbox. Next, we obtain a difference $(a', b, c, d,\ a', 0, c, 0,\ a', b, 0, 0,\ a', 0, 0, 0)$, after an Sbox transition $b \to a \oplus a'$. Surprisingly, the dimension of the difference state does not increase, because there is a single active Sbox in each round, and it affects a difference that is already independent of the rest of the state. Therefore we have the four transitions given above, and the spaces are indeed invariant.

New representation from invariant subspaces. We actually have a much stronger property than just invariant spaces: the full space is the direct sum of those four vector spaces, with parallel invariant subspaces for any offset u:

$$(\mathbb{F}_{2^8})^{16} = E_0 \oplus E_1 \oplus E_2 \oplus E_3$$
$$\forall u, \forall i,\ F(u \oplus E_i) = F(u) \oplus E_i.$$

This implies that we can split the internal state according to those vector spaces. Indeed, there exists unique linear projections $\pi_i : (\mathbb{F}_{2^8})^{16} \to E_i$ for $0 \leq i < 4$ such that $\forall x \in E_i$, $\pi_i(x) = x$, and $\pi_i(E_j) = 0$ for $i \neq j$. In particular, we have $\forall x$, $x = \pi_0(x) \oplus \pi_1(x) \oplus \pi_2(x) \oplus \pi_3(x)$. This implies:

$$\begin{aligned} F(x) &= F\big(\pi_0(x) \oplus \pi_1(x) \oplus \pi_2(x) \oplus \pi_3(x)\big) \\ &\in F\big(\pi_0(x) \oplus \pi_1(x) \oplus \pi_2(x)\big) \oplus E_3 \\ &\in F\big(\pi_0(x) \oplus \pi_1(x)\big) \oplus E_3 \oplus E_2 \\ &\in F\big(\pi_0(x)\big) \oplus E_3 \oplus E_2 \oplus E_1 \end{aligned}$$

Therefore $\pi_0(F(x)) = \pi_0\big(F(\pi_0(x))\big)$. Similarly, $\pi_i(F(x)) = \pi_i\big(F(\pi_i(x))\big)$, and finally we can split the permutation in four independent 32-bit computations:

$$F(x) = \pi_0\big(F(\pi_0(x))\big) \oplus \pi_1\big(F(\pi_1(x))\big) \oplus \pi_2\big(F(\pi_2(x))\big) \oplus \pi_3\big(F(\pi_3(x))\big).$$

To obtain a representation that makes the 4 subspaces appear clearly, we perform a change of basis. Let $\{e_0, e_1, \dots, e_{15}\}$ be our new basis of $(\mathbb{F}_{2^8})^{16}$ defined as follows:

$$\text{Base of } E_0 \begin{cases} e_0 &= (0,0,0,1,0,0,0,1,0,0,0,1,0,0,0,1) \\ e_1 &= (0,0,1,0,0,0,0,0,0,0,0,0,0,0,0,0) \\ e_2 &= (0,1,0,0,0,1,0,0,0,0,0,0,0,0,0,0) \\ e_3 &= (1,0,0,0,0,0,0,0,1,0,0,0,0,0,0,0) \end{cases}$$

$$\text{Base of } E_1 \begin{cases} e_4 &= (0,0,1,0,0,0,1,0,0,0,1,0,0,0,1,0) \\ e_5 &= (0,1,0,0,0,0,0,0,0,0,0,0,0,0,0,0) \\ e_6 &= (1,0,0,0,1,0,0,0,0,0,0,0,0,0,0,0) \\ e_7 &= (0,0,0,1,0,0,0,0,0,0,0,1,0,0,0,0) \end{cases}$$

$$\text{Base of } E_2 \begin{cases} e_8 &= (0,1,0,0,0,1,0,0,0,1,0,0,0,1,0,0) \\ e_9 &= (1,0,0,0,0,0,0,0,0,0,0,0,0,0,0,0) \\ e_{10} &= (0,0,0,1,0,0,0,1,0,0,0,0,0,0,0,0) \\ e_{11} &= (0,0,1,0,0,0,0,0,0,0,1,0,0,0,0,0) \end{cases}$$

$$\text{Base of } E_3 \begin{cases} e_{12} &= (1,0,0,0,1,0,0,0,1,0,0,0,1,0,0,0) \\ e_{13} &= (0,0,0,1,0,0,0,0,0,0,0,0,0,0,0,0) \\ e_{14} &= (0,0,1,0,0,0,1,0,0,0,0,0,0,0,0,0) \\ e_{15} &= (0,1,0,0,0,0,0,0,0,1,0,0,0,0,0,0) \end{cases}$$

Let $s_0, s_1, \dots, s_{15}$ be the coordinates in the new basis. They can be obtained by multiplying the original coordinates $(k_0, \dots, k_{15})$ with the matrix $A = C_0^{-1}$, where the columns of the transition matrix C_0 are the coordinates of the vectors $e_0, e_1, \dots, e_{15}$ expressed in the old basis (canonical basis):

$$C_0 = \begin{pmatrix}
0&0&0&1&0&0&1&0&0&1&0&0&1&0&0&0 \\
0&0&1&0&0&1&0&0&1&0&0&0&0&0&0&1 \\
0&1&0&0&1&0&0&0&0&0&0&1&0&0&1&0 \\
1&0&0&0&0&0&0&1&0&0&1&0&0&1&0&0 \\
0&0&0&0&0&0&1&0&0&0&0&0&1&0&0&0 \\
0&0&1&0&0&0&0&0&1&0&0&0&0&0&0&0 \\
0&0&0&0&1&0&0&0&0&0&0&0&0&0&1&0 \\
1&0&0&0&0&0&0&0&0&0&1&0&0&0&0&0 \\
0&0&0&1&0&0&0&0&0&0&0&0&1&0&0&0 \\
0&0&0&0&0&0&0&0&1&0&0&0&0&0&0&1 \\
0&0&0&0&1&0&0&0&0&0&0&1&0&0&0&0 \\
1&0&0&0&0&0&0&1&0&0&0&0&0&0&0&0 \\
0&0&0&0&0&0&0&0&0&0&0&0&1&0&0&0 \\
0&0&0&0&0&0&0&0&1&0&0&0&0&0&0&0 \\
0&0&0&0&1&0&0&0&0&0&0&0&0&0&0&0 \\
1&0&0&0&0&0&0&0&0&0&0&0&0&0&0&0
\end{pmatrix}
\qquad
A = \begin{pmatrix}
0&0&0&0&0&0&0&0&0&0&0&0&0&0&0&1 \\
0&0&1&0&0&0&1&0&0&0&1&0&0&0&1&0 \\
0&0&0&0&0&1&0&0&0&0&0&0&0&1&0&0 \\
0&0&0&0&0&0&0&0&1&0&0&0&1&0&0&0 \\
0&0&0&0&0&0&0&0&0&0&0&0&0&0&1&0 \\
0&1&0&0&0&1&0&0&0&1&0&0&0&1&0&0 \\
0&0&0&0&1&0&0&0&0&0&0&0&1&0&0&0 \\
0&0&0&0&0&0&0&0&0&0&0&1&0&0&0&1 \\
0&0&0&0&0&0&0&0&0&0&0&0&0&1&0&0 \\
1&0&0&0&1&0&0&0&1&0&0&0&1&0&0&0 \\
0&0&0&0&0&0&0&1&0&0&0&0&0&0&0&1 \\
0&0&0&0&0&0&0&0&0&0&1&0&0&0&1&0 \\
0&0&0&0&0&0&0&0&0&0&0&0&1&0&0&0 \\
0&0&0&1&0&0&0&1&0&0&0&1&0&0&0&1 \\
0&0&0&0&0&0&1&0&0&0&0&0&0&0&1&0 \\
0&0&0&0&0&0&0&0&0&1&0&0&0&1&0&0
\end{pmatrix}$$

Therefore, we use:

$$\begin{array}{llll} s_0 = k_{15} & s_1 = k_{14} \oplus k_{10} \oplus k_6 \oplus k_2 & s_2 = k_{13} \oplus k_5 & s_3 = k_{12} \oplus k_8 \\ s_4 = k_{14} & s_5 = k_{13} \oplus k_9 \oplus k_5 \oplus k_1 & s_6 = k_{12} \oplus k_4 & s_7 = k_{15} \oplus k_{11} \\ s_8 = k_{13} & s_9 = k_{12} \oplus k_8 \oplus k_4 \oplus k_0 & s_{10} = k_{15} \oplus k_7 & s_{11} = k_{14} \oplus k_{10} \\ s_{12} = k_{12} & s_{13} = k_{15} \oplus k_{11} \oplus k_7 \oplus k_3 & s_{14} = k_{14} \oplus k_6 & s_{15} = k_{13} \oplus k_9 \end{array} \tag{1}$$

After defining s' with the same transformation from k', we can verify that:

$$\begin{aligned} s'_0 &= k'_{15} = k_{15} \oplus k_{11} \oplus k_7 \oplus k_3 \oplus S(k_{12}) &&= s_{13} \oplus S(s_{12}) \\ s'_1 &= k'_{14} \oplus k'_{10} \oplus k'_6 \oplus k'_2 = k_{14} \oplus k_6 &&= s_{14} \\ s'_2 &= k'_{13} \oplus k'_5 = k_{13} \oplus k_9 &&= s_{15} \\ s'_3 &= k'_{12} \oplus k'_8 = k_{12} &&= s_{12} \\ s'_4 &= k'_{14} = k_{14} \oplus k_{10} \oplus k_6 \oplus k_2 \oplus S(k_{15}) &&= s_1 \oplus S(s_0) \\ s'_5 &= k'_{13} \oplus k'_9 \oplus k'_5 \oplus k'_1 = k_{13} \oplus k_5 &&= s_2 \\ s'_6 &= k'_{12} \oplus k'_4 = k_{12} \oplus k_8 &&= s_3 \\ s'_7 &= k'_{15} \oplus k'_{11} = k_{15} &&= s_0 \\ s'_8 &= k'_{13} = k_{13} \oplus k_9 \oplus k_5 \oplus k_1 \oplus S(k_{14}) &&= s_5 \oplus S(s_4) \\ s'_9 &= k'_{12} \oplus k'_8 \oplus k'_4 \oplus k'_0 = k_{12} \oplus k_4 &&= s_6 \\ s'_{10} &= k'_{15} \oplus k'_7 = k_{15} \oplus k_{11} &&= s_7 \\ s'_{11} &= k'_{14} \oplus k'_{10} = k_{14} &&= s_4 \\ s'_{12} &= k'_{12} = k_{12} \oplus k_8 \oplus k_4 \oplus k_0 \oplus S(k_{13}) \oplus c_i &&= s_9 \oplus S(s_8) \oplus c_i \\ s'_{13} &= k'_{15} \oplus k'_{11} \oplus k'_7 \oplus k'_3 = k_{15} \oplus k_7 &&= s_{10} \\ s'_{14} &= k'_{14} \oplus k'_6 = k_{14} \oplus k_{10} &&= s_{11} \\ s'_{15} &= k'_{13} \oplus k'_9 = k_{13} &&= s_8 \end{aligned} \tag{2}$$

This is represented by Fig. 4. In the rest of this paper we use the notation k_i^r to denote byte i of the round-r subkey, and s_i^r to denote bytes of the alternative representation at round r, where the relations between k_i^r and s_i^r follow (1).

To further simplify the description, we write the output as

$$(s'_4, s'_5, s'_6, s'_7, \quad s'_8, s'_9, s'_{10}, s'_{11}, \quad s'_{12}, s'_{13}, s'_{14}, s'_{15}, \quad s'_0, s'_1, s'_2, s'_3).$$

This corresponds to "untwisting" the rotation of the 4-byte blocks, so that each block of 4 output bytes depends on the same 4 input bytes. This results in our alternate representation of the AES-128 key schedule:

1. We first apply the linear transformation A to the state, corresponding to the change of variable above.
2. Then the rounds of the key schedule are seen as the concatenation of 4 functions each acting on 32-bit words (4 bytes), as seen in Fig. 5.
3. In order to extract the subkey of round r, another linear transformation $C_{r \bmod 4}$ is applied to the state, depending of the round number modulo 4. C_i is defined as $C_i = A^{-1} \times \mathsf{SR}^i$, with SR the matrix corresponding to rotation of 4 bytes to the right (see below). In particular $C_0 = A^{-1}$.

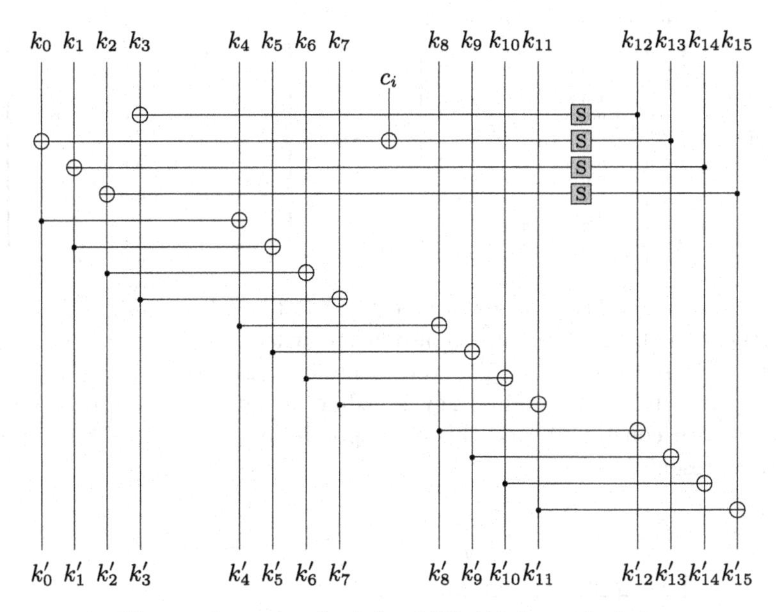

Fig. 3. One round of the AES-128 key schedule.

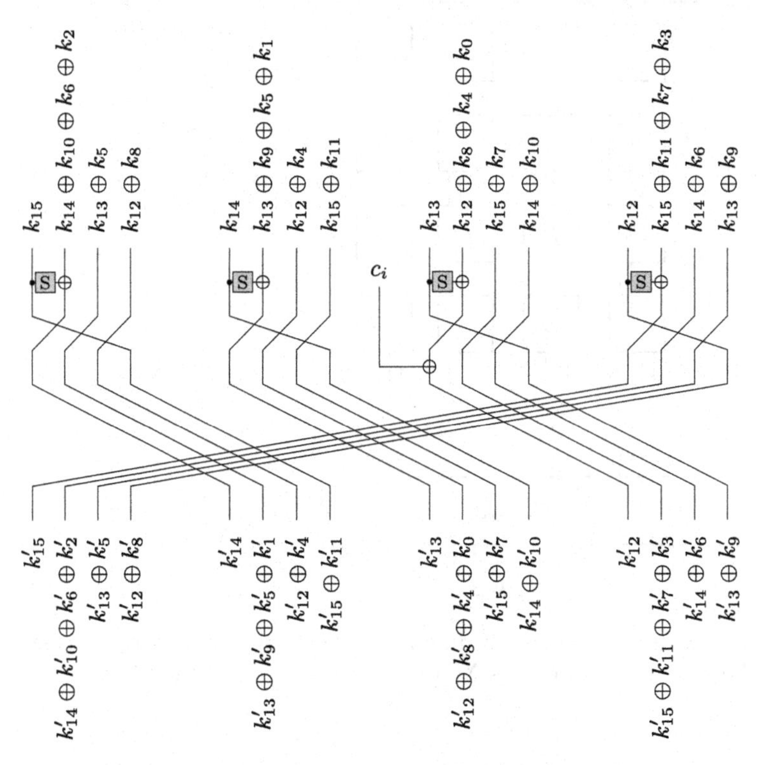

Fig. 4. One round of the AES-128 key schedule (alternative representation).

$$A = \begin{pmatrix}
0&0&0&0&0&0&0&0&0&0&0&0&0&0&0&1\\
0&0&1&0&0&0&1&0&0&0&1&0&0&0&1&0\\
0&0&0&0&0&1&0&0&0&0&0&0&0&1&0&0\\
0&0&0&0&0&0&0&0&1&0&0&0&1&0&0&0\\
0&0&0&0&0&0&0&0&0&0&0&0&0&0&1&0\\
0&1&0&0&0&1&0&0&0&1&0&0&0&1&0&0\\
0&0&0&0&1&0&0&0&0&0&0&0&1&0&0&0\\
0&0&0&0&0&0&0&0&0&0&0&1&0&0&0&1\\
0&0&0&0&0&0&0&0&0&0&0&0&0&1&0&0\\
1&0&0&0&1&0&0&0&1&0&0&0&1&0&0&0\\
0&0&0&0&0&0&0&1&0&0&0&0&0&0&0&1\\
0&0&0&0&0&0&0&0&0&0&1&0&0&0&1&0\\
0&0&0&0&0&0&0&0&0&0&0&0&1&0&0&0\\
0&0&0&1&0&0&0&1&0&0&0&1&0&0&0&1\\
0&0&0&0&0&0&1&0&0&0&0&0&0&0&1&0\\
0&0&0&0&0&0&0&0&0&1&0&0&0&1&0&0
\end{pmatrix}
\qquad
\mathsf{SR} = \begin{pmatrix}
0&0&0&0&0&0&0&0&0&0&0&0&1&0&0&0\\
0&0&0&0&0&0&0&0&0&0&0&0&0&1&0&0\\
0&0&0&0&0&0&0&0&0&0&0&0&0&0&1&0\\
0&0&0&0&0&0&0&0&0&0&0&0&0&0&0&1\\
1&0&0&0&0&0&0&0&0&0&0&0&0&0&0&0\\
0&1&0&0&0&0&0&0&0&0&0&0&0&0&0&0\\
0&0&1&0&0&0&0&0&0&0&0&0&0&0&0&0\\
0&0&0&1&0&0&0&0&0&0&0&0&0&0&0&0\\
0&0&0&0&1&0&0&0&0&0&0&0&0&0&0&0\\
0&0&0&0&0&1&0&0&0&0&0&0&0&0&0&0\\
0&0&0&0&0&0&1&0&0&0&0&0&0&0&0&0\\
0&0&0&0&0&0&0&1&0&0&0&0&0&0&0&0\\
0&0&0&0&0&0&0&0&1&0&0&0&0&0&0&0\\
0&0&0&0&0&0&0&0&0&1&0&0&0&0&0&0\\
0&0&0&0&0&0&0&0&0&0&1&0&0&0&0&0\\
0&0&0&0&0&0&0&0&0&0&0&1&0&0&0&0
\end{pmatrix}$$

In this new representation, there are clearly 4 independent chunks each acting on 4 bytes, and the subkeys are reconstructed with linear combinations of the alternative key schedule state. This representation also preserves the symmetry of the key schedule: the original key schedule is invariant by rotation of the columns (up to constants), and this corresponds to a rotation of four bytes in the new representation.

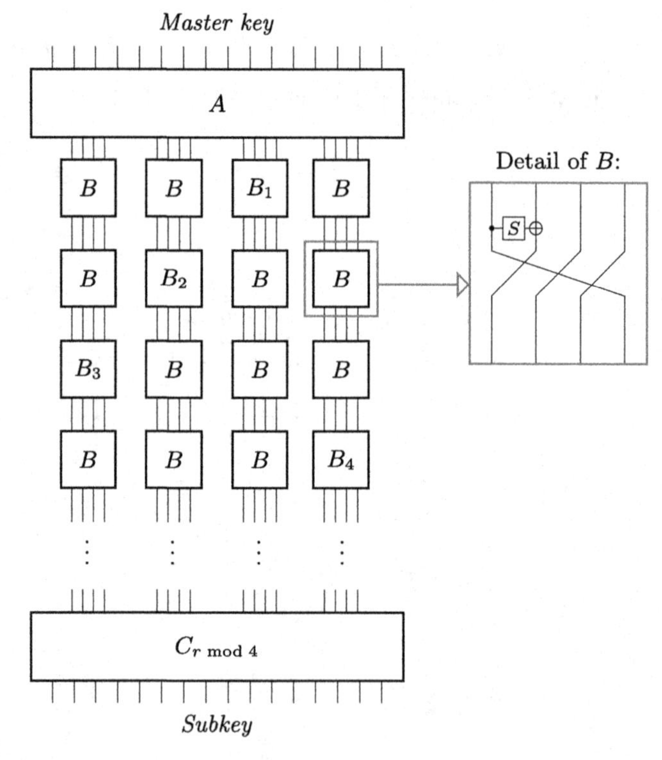

Fig. 5. r rounds of the key schedule in the new representation. B_i is similar to B but the round constant c_i is XORed to the output of the Sbox.

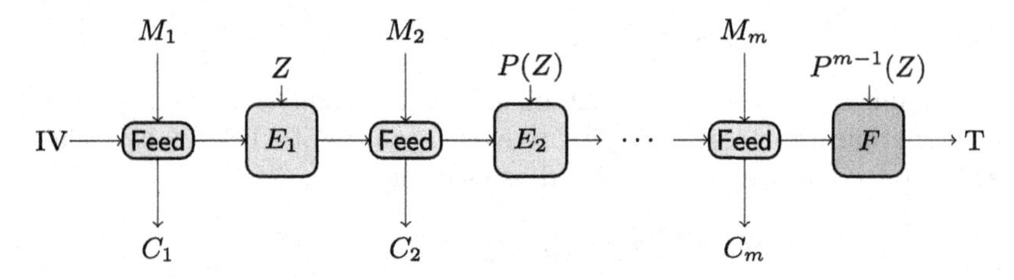

Fig. 6. Simplified scheme of mixFeed encryption.

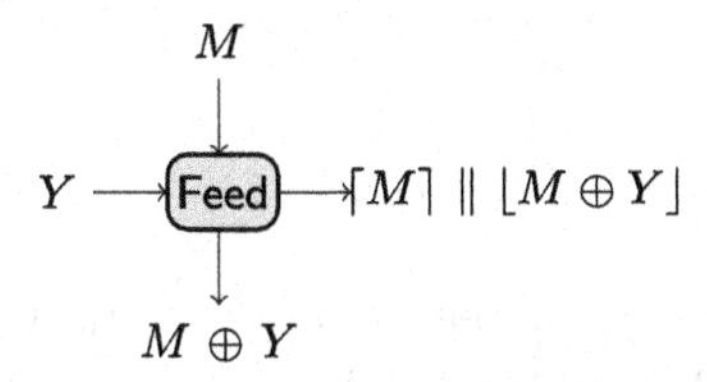

Fig. 7. Function Feed with a full message block.

3 Application to mixFeed

The AEAD scheme mixFeed [14] is a second-round candidate in the NIST Lightweight Standardization Process, submitted by Chakraborty and Nandi, and based on the AES block cipher. It is a rate-1 feedback-based mode inspired by COFB. For each message block, a Feed function is used to compute the ciphertext and the block cipher input from the previous internal state, and the internal state is replaced by the block cipher output. In COFB, there is a need for an extra state variable, to make each Feed function different. In order to reduce the state size, mixFeed instead makes each block cipher call different, applying a permutation P to the key between each block. For optimal efficiency, the permutation P just corresponds to eleven round of the AES key schedule, so that the subkeys for all the AES calls just correspond to running the AES key schedule indefinitely.

In [29], Khairallah observed that some keys generate short cycles when iterating the P permutation, and he built a forgery attack for keys in short cycles. In this work, we show that the new representation of the key schedule explains the existence of these short cycles, and we characterize the keys belonging to such cycles. This shows that the permutation P cannot be considered as a random permutation.

3.1 Description of mixFeed

For simplicity, we only describe a simplified mixFeed without associated data; the full description of mixFeed can be found in [14].

Notations: We use M and C to denote the plaintext and ciphertext. For the sake of simplicity, we assume that M is made of m 128-bit blocks.

The following functions are used in mixFeed:

- E: a modified version of AES-128 including MixColumns in the last round;
- P: the permutation corresponding to eleven rounds of AES-128 key schedule;
- Feed: the feedback function defined as (see Fig. 7):

$$\begin{aligned}\mathsf{Feed}(Y, M) &= (X, C) \\ &= (\lceil M \rceil \| \lfloor M \oplus Y \rfloor, M \oplus Y),\end{aligned}$$

where $\lceil D \rceil$ represent the 64 most significant bits of D, and $\lfloor D \rfloor$ the 64 least significant bits.

The computations are as follow (see Fig. 6):

Initialization of the state. An initial value IV $= Y_0$ and a internal key Z are computed from the nonce N and the key K.

Encryption and authentication. For i from 1 to m, the Feed function is applied to the current state Y_{i-1} and message block M_i. Feed returns the ciphertext block C_i, and a new state X_i which is then encrypted under the key $P^{i-1}(Z)$ using E to obtain Y_i. At the end of this step, a finalization function computes the tag from the final state and the internal key $P^{m-1}(Z)$, we denote as F the composition of the cipher call of last round and the finalization function.

3.2 Short Cycles of P

In [29], Khairallah found 20 keys belonging to small cycles of P, and observed that all of them have the same cycle length[1]: 14018661024. He deduced a forgery attack, assuming that the subkey falls in one of those cycles, but did not further analyse the probability of having such a subkey. Later the designers of mixFeed published a security proof for the scheme [15], under the assumption that the number of keys in a short cycle is sufficiently small. More precisely, they wrote:

Assumption 1 ([15]). *For any $K \in \{0,1\}^n$ chosen uniformly at random, probability that K has a period at most ℓ is at most $\ell/2^{n/2}$.*

The 20 keys identified by Khairallah do not contradict this assumption, but if there are many such keys the assumption does not hold, and mixFeed can be broken by a forgery attack. We now provide a theoretical explanation of the observation of Khairallah, and a full characterization of the cycles of P. We find that a random key is in a cycle of length smaller than 2^{34} with probability 0.44; this contradicts the assumption made in [15], and allows a practical forgery attack.

[1] Khairallah actually reported the length as 1133759136, probably because of a 32-bit overflow.

Analysis of the structure of P. Using our new representation, the 11-round key schedule P consists of:

- The linear transformation A
- 4 parallel 32-bit functions that we denote $f_1 \| f_2 \| f_3 \| f_4$, with

$$f_1 = B_{11} \circ B \circ B \circ B \circ B_7 \circ B \circ B \circ B \circ B_3 \circ B \circ B$$
$$f_2 = B \circ B_{10} \circ B \circ B \circ B \circ B_6 \circ B \circ B \circ B \circ B_2 \circ B$$
$$f_3 = B \circ B \circ B_9 \circ B \circ B \circ B \circ B_5 \circ B \circ B \circ B \circ B_1$$
$$f_4 = B \circ B \circ B \circ B_8 \circ B \circ B \circ B \circ B_4 \circ B \circ B \circ B$$

  (the functions differ only by the round constants)
- The linear transformation $C_3 = A^{-1} \times \mathsf{SR}^{-1}$

To simplify the analysis, we consider the cycle structure of $\widetilde{P} = A \circ P \circ A^{-1}$, which is the same as the cycle structure of P:

$$\widetilde{P} : (a, b, c, d) \mapsto (f_2(b), f_3(c), f_4(d), f_1(a))$$

To further simplify the analysis, we consider the cycle structure of $\widetilde{P}^4$, which is closely related to the cycle structure of $\widetilde{P}$. A cycle of $\widetilde{P}^4$ of length ℓ corresponds to a cycle of $\widetilde{P}$, of length ℓ, 2ℓ or 4ℓ. Conversely a cycle of $\widetilde{P}$ of length ℓ corresponds to one or several cycles of $\widetilde{P}^4$, of length ℓ, $\ell/2$ or $\ell/4$ (depending on the divisibility of ℓ). Analyzing $\widetilde{P}^4$ is easier because it can be decomposed into 4 parallel functions, cancelling the left rotation induced by SR^{-1}:

$$\widetilde{P}^4 : (a, b, c, d) \mapsto (\phi_1(a), \phi_2(b), \phi_3(c), \phi_4(d))$$
$$\phi_1(a) = f_2 \circ f_3 \circ f_4 \circ f_1(a)$$
$$\phi_2(b) = f_3 \circ f_4 \circ f_1 \circ f_2(b)$$
$$\phi_3(c) = f_4 \circ f_1 \circ f_2 \circ f_3(c)$$
$$\phi_4(d) = f_1 \circ f_2 \circ f_3 \circ f_4(d)$$

If (a, b, c, d) is in a cycle of length ℓ of $\widetilde{P}^4$, we have $\widetilde{P}^{4\ell}(a, b, c, d) = (a, b, c, d)$, that is to say:

$$\phi_1^\ell(a) = a \qquad \phi_2^\ell(b) = b \qquad \phi_3^\ell(c) = c \qquad \phi_4^\ell(d) = d$$

In particular, a, b, c and d must be in cycles of ϕ_1, ϕ_2, ϕ_3, ϕ_4 (respectively) of length dividing ℓ. Conversely, if a, b, c, d are in small cycles of the corresponding ϕ_i, then (a, b, c, d) is in a cycle of $\widetilde{P}^4$ of length the lowest common multiple of the small cycle lengths.

Moreover, due to the structure of the ϕ_i functions, all of them have the same cycle structure. This implies that $\widetilde{P}$ has a large number of small cycles. Indeed, if we consider a cycle of ϕ_i of length ℓ, and elements a, b, c, d in the corresponding cycles, (a, b, c, d) is in a cycle of P^4 of length ℓ. There are ℓ^4 choices of a, b, c, d, which correspond to ℓ^3 different cycles of P. If we assume that ϕ_i behaves like a random 32-bit permutation, we expect that the largest cycle has length about 2^{31}, which gives around 2^{93} cycles of $\widetilde{P}^4$ of length $\approx 2^{31}$, and around 2^{93} cycles of $\widetilde{P}$ of length $\approx 2^{33}$.

Cycle analysis of 11-round AES-128 key schedule. In order to identify the small cycles of the permutation P, we start by analyzing the cycle structure of the 32-bit function $\phi_1 = f_2 \circ f_3 \circ f_4 \circ f_1$: it can be decomposed into cycles of lengths 3504665256, 255703222, 219107352, 174977807, 99678312, 13792740, 8820469, 7619847, 5442633, 4214934, 459548, 444656, 14977, 14559, 5165, 4347, 1091, 317, 27, 6, 5 (3 cycles), 4 (2 cycles), 2 (3 cycles), and 1 (2 fixed points). In particular, the largest cycle has length $\ell = 3504665256$. Consequently, with probability $(3504665256 \times 2^{-32})^4 \approx 0.44$, we have a, b, c and d in a cycle of length ℓ, resulting in a cycle of length ℓ for $\widetilde{P}^4$, and a cycle of length at most $4\ell = 14018661024$ for $\widetilde{P}$ and P. This explains the observation of Khairallah [29], and clearly contradicts the assumption of [15].

More generally, when a, b, c, d belong to a cycle of length ℓ_i, the corresponding cycle for $\widetilde{P}^4$ is of length $\ell = \text{lcm}(\ell_1, \ell_2, \ell_3, \ell_4)$, and we can compute the associated probability. In most cases, a cycle of length ℓ of $\widetilde{P}^4$ corresponds to a cycle of $\widetilde{P}$ of length 4ℓ. However, the cycle of $\widetilde{P}$ is of length ℓ when $\widetilde{P}^\ell(a, b, c, d) = (a, b, c, d)$, and of length 2ℓ when $\widetilde{P}^{2\ell}(a, b, c, d) = (a, b, c, d)$ (this can only be the case with odd ℓ, by definition of ℓ). This is unlikely for short cycles, but as an example we can construct a fixed-point for $\widetilde{P}$ and P from a fixed-point of ϕ_1:

- $a =$ `7e be d1 92`
- $b =$ `de d4 b7 cc` $= f_3 \circ f_4 \circ f_1(a)$
- $c =$ `9f 95 88 26` $= f_4 \circ f_1(a)$
- $d =$ `d4 b9 79 91` $= f_1(a)$

Since $f_2 \circ f_3 \circ f_4 \circ f_1(a) = a$, we have $\widetilde{P}(a, b, c, d) = (f_2(b), f_3(c), f_4(d), f_1(a)) = (a, b, c, d)$. Since $\widetilde{P} = A \circ P \circ A^{-1}$, the corresponding key in the original representation is:

$$A^{-1} \times \begin{pmatrix} a \\ b \\ c \\ d \end{pmatrix} = \begin{pmatrix} \texttt{64 0b 3f 83 63 4e a7 f6 46 0e f8 b2 d4 9f de 7e} \end{pmatrix}^\top$$

This results in a fixed point of P.

We can generalize this construction for all odd cycle lengths ℓ. We choose w an element of a cycle of length ℓ, and then we can build an element which belongs to a cycle of length ℓ for the permutation P:

- if $\ell = 1 \bmod 4$:

$$\begin{aligned} a &= w \\ b &= f_3 \circ f_4 \circ f_1 \circ \ldots \circ f_1(w), && \text{with } 3\ell \text{ terms } f_i \\ c &= f_4 \circ f_1 \circ f_2 \circ \ldots \circ f_1(w), && \text{with } 2\ell \text{ terms } f_i \\ d &= f_1 \circ f_2 \circ f_3 \circ \ldots \circ f_1(w), && \text{with } \ell \text{ terms } f_i \end{aligned}$$

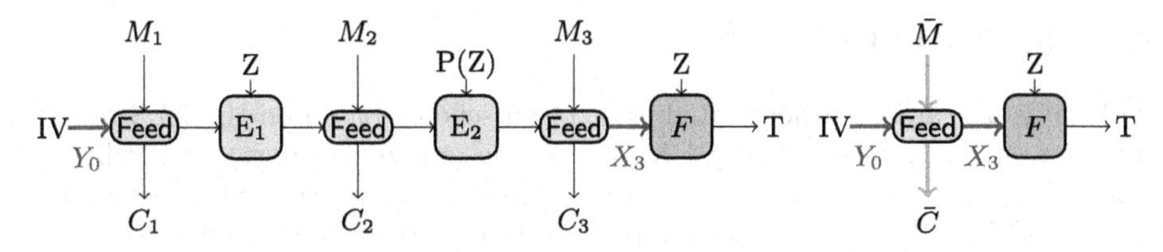

Fig. 8. Forgery attack when Z belongs to a cycle of length 2.

- if $\ell = 3 \bmod 4$:

$$\begin{aligned} a &= w \\ b &= f_3 \circ f_4 \circ f_1 \circ \ldots \circ f_1(w), && \text{with } \ell \text{ terms } f_i \\ c &= f_4 \circ f_1 \circ f_2 \circ \ldots \circ f_1(w), && \text{with } 2\ell \text{ terms } f_i \\ d &= f_1 \circ f_2 \circ f_3 \circ \ldots \circ f_1(w), && \text{with } 3\ell \text{ terms } f_i \end{aligned}$$

3.3 Forgery Attack Against mixFeed

Khairallah [29] proposed a forgery attack assuming that Z belongs to a cycle of length ℓ, considering a message M made of m blocks, with $m > \ell$ (Fig. 8):

1. Encrypt the message M to obtain the ciphertext C and tag T.
2. Compute Y_0 using M_1 and C_1 and $X_{\ell+1}$ using $M_{\ell+1}$ and $C_{\ell+1}$.
3. Compute $\bar{M}$ and $\bar{C}$ such that $(X_{\ell+1}, \bar{C}) = \mathsf{Feed}(Y_0, \bar{M})$.
4. The T tag will also authenticate the new ciphertext $C' = \bar{C}||C_{\ell+2}|| \cdots ||C_m$.

The computations required for the forge are negligible with only a few XORs to invert the Feed function. Therefore the complexity of the attack is just the encryption of a message with at least $(\ell+1)$ blocks, with ℓ the length of the cycle. As explained above, the probability of success is approximately 0.44, using $\ell = 14018661024$. When the forgery fails, we can repeat the attack with a different nonce, because the internal key Z depends on the nonce; for each master key K, the attack works on 44% of the nonces.

We have verified this attack using the reference implementation provided by the designers. We take a message of $\ell + 1 = 14018661025$ blocks of 16 bytes (220 Gbytes[2]), choose a random key and nonce, and encrypt the message with mixFeed. We modify the ciphertext according to the previous explanation, and we check if the new ciphertext is accepted. We obtained 41% of success over 100 attempts. This result is close to the expected 44% success rate, and confirms our analysis.

[2] Note that there is no need to store the plaintext or ciphertext in memory if we have access to an online implementation of mixFeed.

4 Application to ALE

ALE [8] is an earlier authenticated encryption scheme based on the AES round function, strongly inspired by LEX [4] (for the encryption part) and Pelican-MAC [16] (for the authentication part). Attacks have already been presented against ALE [30,34] but the new representation of the key schedule gives new types of attacks, based on previous attacks against LEX [11,20].

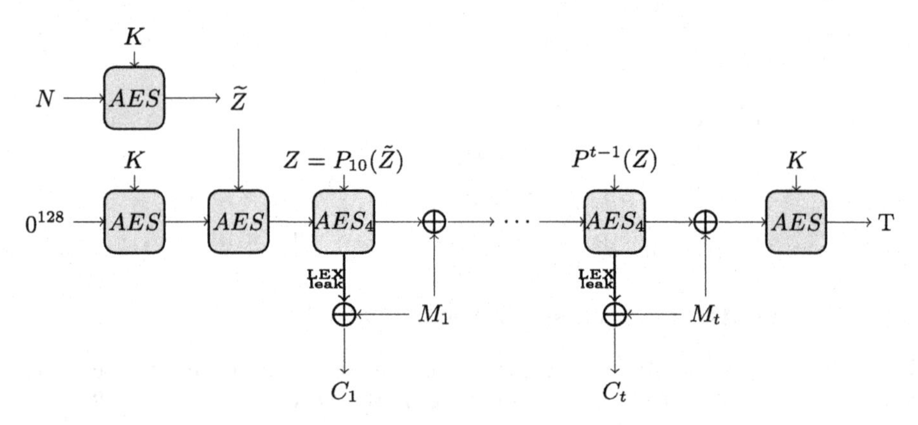

Fig. 9. Authenticated encryption with ALE (simplified).

4.1 Description of ALE

For the sake of simplicity, we will consider ALE without associated data, and we only consider blocks of 16 bytes for the plaintext (to ignore the padding). ALE maintains a state composed of an internal state and an internal key, and operates with 3 steps (cf Fig. 9). As for mixFeed, the internal key is updated with iterative applications of a permutation P corresponding to AES key schedule rounds. In the case of ALE, P corresponds to 5 rounds of key schedule rather than 11, but we have again many short cycles because 5 is also an odd number.

Initialization. The state is initialized from the key K and a nonce N, using a session key $\widetilde{Z} = E_K(N)$. The internal state is initialized to $IV = E_{\widetilde{Z}}(E_K(0))$, and the internal key is initialized to $P_{10}(Z)$, where P_{10} correspond to 11 rounds of AES key schedule.

Message processing phase. For each block of message, the internal state is encrypted with 4-round AES, and the internal key is updated by five rounds of AES key schedule. During the encryption, four bytes are leaked in each AES round according to the LEX specification (bytes 0, 2, 8, 10 for odd rounds, and bytes 4, 6, 12 and 14 for even rounds), and used as keystream to encrypt the message. Then the message block is xored to the current internal state, following the Pelican-MAC construction.

Finalization. Finally, the internal state is encrypted with the full AES using the master key K to generate the tag T.

Rekeying. The designers of ALE require that the master key is changed after processing 2^{48} bits (*i.e.* 2^{41} blocks).

Previous results. ALE was designed to thwart attacks against LEX [11,20] that use a pair of partially-colliding internal states to recover the key. Indeed, each AES call uses a different key, which prevents those attacks. Other attacks have been proposed against LEX, based on differential trails between two message injections [30,34]. We compare the previous attacks in Table 1. To make the results comparable, we assume that attacks with a low success rate are repeated until they succeed. For attacks using more than 2^{41} blocks of data, the master key will be rotated.

4.2 Internal Key Recovery

We describe a new attack against ALE, based on previous analysis of LEX. The key update of ALE was supposed to avoid these attacks, but since the update function has small cycles, there is a large probability that the key state is repeated, which makes the attack possible.

We analyze cycles of P in the same way as for mixFeed: four iterations of the 5-round key schedule are equivalent to the application in parallel of four 32-bit functions. The study of one of these functions gives us information about the cycle structure of the permutation P. The 32-bit function has a cycle of length $\ell = 4010800805 \approx 2^{31.9}$; therefore the permutation P admits many cycles of length $4 \times \ell \approx 2^{33.9}$ which are reached with probability $(\ell \times 2^{-32})^4 \approx 0.76$.

Previous attacks against LEX [11,12,20] are based on the search for a pair of internal states that partially collides, with two identical columns. This pattern can occur in odd or even round: we use columns 0 and 2 for odd rounds, and columns 1 and 3 for even rounds. The partial collision occurs with probability 2^{-64}, and 32 bits of the colliding state can be directly observed, due to the leak extractions. A candidate pair can be tested with complexity 2^{64} [12, Section 7.1], using the leak extraction of rounds before and after the collision; if it actually corresponds to a partial collision this reveals the internal state and key.

In the case of ALE, we perform a chosen plaintext attack: we choose a message M of 2^{41} blocks (the maximum length allowed by the ALE specification) which admits cycles of length $4 \times \ell$. With probability 0.76, the key cycles after $4 \times \ell \approx 2^{33.9}$ iterations of the permutation P. When this happens, we can split the message into $2^{33.9}$ sets of $2^{7.1}$ blocks encrypted under the same key. In each set we can construct $2^{13.2}$ pairs. In total, from one message M of 2^{41} blocks, we get on average $0.76 \times 2^{13.2} \times 2^{33.9} \approx 2^{46.7}$ pairs encrypted with the same key.

Unfortunately, the attack against LEX uses five consecutive AES rounds, but in ALE, the subkeys used in five consecutive rounds do not follow the exact AES key schedule. It is not possible to apply exactly the same attack on ALE, but we can use the tool developed by Bouillaguet, Derbez, and Fouque [10,12] in order to find an attack in this setting. This tool found an attack that can test a candidate pair with time complexity 2^{72}, and a memory requirement of 2^{72}, for two different positions of the partial collision:

- when the collision occurs in round 4, the attack uses the leak of rounds 1, 2, 3, 4 and of round 1 of the next 4-round AES.
- when the collision occurs in round 1, the attack uses the leak of rounds 1 and 2, and of rounds 2, 3, 4 of the previous 4-round AES.

Starting with $2^{16.3}$ messages of length 2^{48} (encrypted under different master keys) we obtain $2^{16.3} \times 2^{13.2} \times 2^{33.9} \approx 2^{63.4}$ pairs, such that each pair uses the same key with probability 0.76. Each pair can be used twice, assuming a collision at round 1 or at round 4, so we have in total $2^{64.4}$ pairs to consider, and we expect one of them to actually collide ($0.76 \times 2^{64.4} \approx 2^{64}$). After filtering on 32 bits, we have $2^{32.4}$ candidate pairs to analyse, so that the time complexity is $2^{32.4} \times 2^{72} = 2^{104.4}$, and the data complexity is $2^{16.3} \times 2^{41} = 2^{57.3}$.

This attack recovers the internal state, and we can compute backwards the initial state $E_K(0)$ and the session key $\widetilde{Z} = E_K(N)$. We can also generate almost universal forgeries: when $E_K(0)$ and $\widetilde{Z}$ are known we can compute the internal state and ciphertext corresponding to an arbitrary message, and we can match the value of the final internal state (and hence the tag) by choosing one block of message or associated data appropriately.

5 Application to Impossible Differential Attacks

In 1999, Biham, Biryukov and Shamir introduced Impossible Differential attacks: a new cryptanalysis technique that they applied to Skipjack [3]. This attack is based on the existence of an impossible differential, *i.e.* a differential occurring with probability 0. If a key guess leads to this differential, then it can be deduced that this guess was wrong. This allows to eliminate key candidates and thus to obtain an attack faster than exhaustive search. Impossible differentials have been applied to various cryptosystems, including reduced versions of AES [2,13,33].

The framework described in [13] is composed of two parts: firstly, combinations of bytes from the first and last subkeys are shown impossible, and secondly, the master keys associated to the remaining candidates are reconstructed and tested. When reconstructing the master key, previous attacks only exploit the subkeys bytes in the first rounds, guess the missing bytes, and evaluate the key schedule to check the bytes in the last subkeys. Our results significantly improve this part, by combining information from the first and the last subkeys. Indeed, the new representation shows that some bytes of a given subkey depend on fewer than 128 bits of information of another subkey, even if the subkeys are separated by many rounds. The complexity of the attack is a trade-off between the first and second parts. After improving the second part we obtain slightly better trade-offs. The improvement is limited because a small increase of the data complexity (corresponding to the cost of the part) leads to a large reduction in the number of remaining candidates (corresponding to the complexity of the second part).

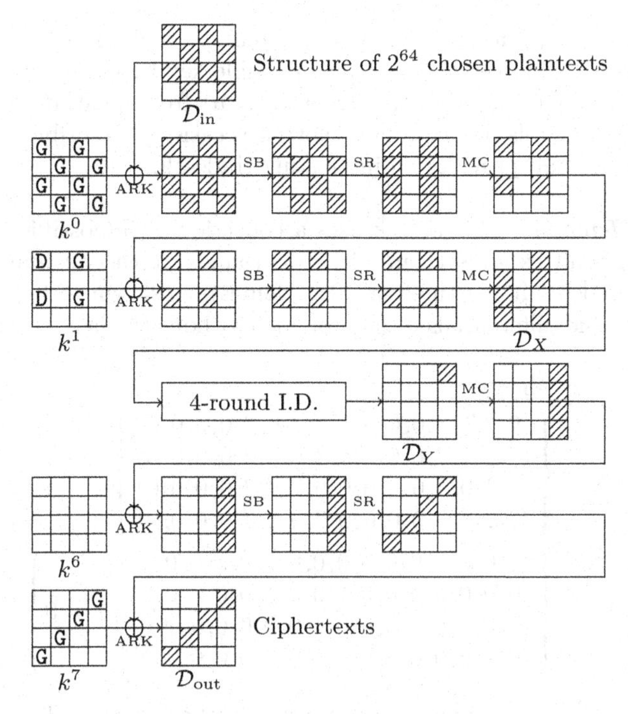

Fig. 10. 7-round impossible differential attack of [33] (figure adapted from [28]). Key bytes marked G and D are respectively guessed, and deduced from guessed bytes.

5.1 The AES Round Function

The AES state is represented as a 4×4-byte array, and the round function iterates the following operations:

- SubBytes applies an Sbox on each byte of the state;
- ShiftRows shifts by the left the second row of the state by 1 cell, the third row by 2 cells, and the last row by 3 cells;
- MixColumns multiplies each column of the state by an MDS matrix;
- AddRoundKey xors the state with the round key.

Sbox property. During this attack, we will use a well-known property for a n-bit to m-bit Sbox: given an input and an output difference, there is on average 2^{n-m} possible values. For the AES Sbox, $n = m = 8$, so in average one value is expected. We pre-compute those values, and refer to that table as the DDT.

5.2 Previous Results

The best impossible differential attacks against AES-128 are variants of an attack from Mala, Dakhilalian, Rijmen and Modarres-Hashemi [33]. Several trade-off are proposed in [13] with four output differentials and using a technique to reduce the memory by iterating over the possible key bytes values, rather that iterating over the data pairs. In this work, we start from a variant with a single output differential explained in detail below; it is easier to describe than variants considered in [13] and provides an interesting trade-off.

Impossible differential. This attack uses a collection of impossible differentials over 4 rounds, and extends them with two rounds at the beginning and one round at the end (omitting the final MixColumns), as shown in Fig. 10. We use a set of impossible differentials over 4-rounds (without the last MixColumns):

$$\mathcal{D}_X \nrightarrow \mathcal{D}_Y$$

$$\mathcal{D}_X = \left\{\begin{matrix} (0,?,?,?,0,0,0,0,?,?,0,?,0,0,0,0) \\ (?,0,?,?,0,0,0,0,?,?,?,0,0,0,0,0) \\ (?,?,0,?,0,0,0,0,0,?,?,?,0,0,0,0) \\ (?,?,?,0,0,0,0,0,?,0,?,?,0,0,0,0) \end{matrix}\right\}$$

$$\mathcal{D}_Y = \left\{\begin{matrix} (0,0,0,0,0,0,0,0,0,0,0,0,x,0,0,0) \\ (0,0,0,0,0,0,0,0,0,0,0,0,0,x,0,0) \\ (0,0,0,0,0,0,0,0,0,0,0,0,0,0,x,0) \\ (0,0,0,0,0,0,0,0,0,0,0,0,0,0,0,x) \end{matrix} \mid x \neq 0 \right\}$$

We assume to be given a pair of plaintexts and the corresponding ciphertexts such that the plaintext difference is in a set $\mathcal{D}_{\text{in}}$ corresponding to two active diagonals, and the ciphertext difference is in a set $\mathcal{D}_{\text{out}}$ corresponding to one active anti-diagonal:

$$\mathcal{D}_{\text{in}} = \{(?,0,?,0,0,?,0,?,?,0,?,0,0,?,0,?)\}$$
$$\mathcal{D}_{\text{out}} = \{(0,0,0,?,0,0,?,0,0,?,0,0,?,0,0,0)\}$$

After guessing the values of the key bytes $k^0_{\langle 0,2,5,7,8,10,13,15\rangle}$, $k^1_{\langle 8,10\rangle}$, $k^7_{\langle 3,6,9,12\rangle}$, we can deduce that some values result in differences in $\mathcal{D}_X$ and $\mathcal{D}_Y$. Since this transition holds with probability 0, we can discard those key candidates. Eventually with a large number N of pairs of plaintexts, we eliminate most of the key candidates, and we can verify the remaining candidates exhaustively. We now detail how to perform this attack efficiently, following Algorithm 1.

Pre-computation. After the MixColumns of the first round, in column 1 and 3, we want non-zero differences only in the first and the third bytes. There are 2^{16} possible differences; by inverting the linear operations MixColumns and ShiftRows, we obtain 2^{16} possible differences for the diagonal (bytes $\langle 0,5,10,15\rangle$ and $\langle 2,7,8,13\rangle$ respectively) after the SubBytes of the first round. We store these 2^{16} differences in the table T_1. Similarly, we build a table T_2 with the 2^{10} possible differences before the SubBytes of the last round by propagating the 2^{10} differences in $\mathcal{D}_Y$.

Construction of pairs. We start with $2^{37+\epsilon}$ structures of 2^{64} plaintexts such that all the plaintexts in a structure are identical in bytes 1, 3, 5, 7, 9, 11, 13, and 15. For each set, we construct $\binom{2^{64}}{2} \approx 2^{127}$ pairs. We identify the pairs with a ciphertext difference in $\mathcal{D}_{out}$ and store them in a list L_1; we expect to have $N = 2^{127} \times 2^{-96} \times 2^{37+\epsilon} = 2^{68+\epsilon}$ pairs.

Step 1. First, we identify plaintext/ciphertext pairs and values of $k^0_{\langle 0,5,10,15\rangle}$ that result in a zero difference in bytes 1 and 3 after the first MixColumns. To this end, we sort the list L_1 according to the plaintext difference and value in bytes 0, 5, 10 and 15. We obtain 2^{64} sublists of approximatively $2^{4+\epsilon}$ pairs. From now on, all the steps are repeated for all guesses of the key bytes $k^0_{\langle 0,5,10,15\rangle}$. For each possible difference δ in bytes 0, 5, 10 and 15 before SubBytes, we confront the difference with each of the possible differences after SubBytes in T_1. Then, using the DDT of the AES Sbox, we extract the input values of the SubBytes operation of the first round, corresponding to this input and output difference. Since the key $k^0_{\langle 0,5,10,15\rangle}$ has been guessed, we can deduce the value of the plaintext in bytes 0, 5, 10 and 15, and locate the right sublist of L_1 with $2^{4+\epsilon}$ pairs that follow this part of the trail for this key guess. We store those pairs in a list L_2; after iterating over δ and T_1 we have on average $2^{32+16+4+\epsilon} = 2^{52+\epsilon}$ pairs in L_2.

Step 2. During this step, we filter data pairs and values of $k^0_{\langle 2,7,8,13\rangle}$ leading to a zero difference in bytes 13 and 15 after the first MixColumns. To do this, we consider each pair of L_2, and iterate over the possible differences after SubBytes in bytes 2, 7, 8, 13, stored in T_1. Since we have the input and output differences of those Sboxes, we retrieve the corresponding values from the DDT. By xoring these values with the plaintext, we obtain the associated key bytes $k^0_{\langle 2,7,8,13\rangle}$ and we add this pair to a list indexed by the key bytes, $L_3[k^0_{\langle 2,7,8,13\rangle}]$.

The following steps are repeated for each value of $k^0_{\langle 2,7,8,13\rangle}$; we have a list $L_3[k^0_{\langle 2,7,8,13\rangle}]$ of $2^{52+\epsilon+16-32} = 2^{36+\epsilon}$ plaintext pairs that satisfy the required difference after the first round.

Step 3. During this step, we associate each pair of $L_3[k^0_{\langle 2,7,8,13\rangle}]$ to the key bytes k^1_8 and k^1_{10} such that difference after the MixColumns of round 2 is in $\mathcal{D}_X$. We recall that at this point, the bytes $k^0_{\langle 0,2,5,7,8,10,13,15\rangle}$ have already been guessed. Following the AES-128 key schedule, we can easily deduce bytes k^1_0 and k^1_2. For each pair of $L_3[k^0_{\langle 2,7,8,13\rangle}]$, we compute the values of the first and the third column of both plaintexts after the MixColumns of the first round. Using $k^1_{\langle 0,2\rangle}$ We can also compute the values of both states on bytes 0 and 2 after AddRoundKey and SubBytes in the second round, corresponding to bytes 0 and 10 after ShiftRows. Looking at the MixColumns operations in columns 1 and 3 in the second round, we know the difference in 3 input bytes (2 zeros given by the differential trail, and value just recovered) and one output byte (a zero given by the differences in $\mathcal{D}_X$). Therefore we can recover the full input and output difference in those columns by solving a linear system (the solution is unique because of the MDS property). By inverting the ShiftRows operation, we recover the difference after the SubBytes operation of the second round in bytes 8 and 10. The difference before this operation is also known, therefore we recover the values of bytes 8 and 10 before SubBytes, and deduce the value of $k^1_{\langle 8,10 \rangle}$ by xoring the value at the end of the first round. We have to repeat this deduction four time, because we have four different positions of the zero differences in $\mathcal{D}_X$. Each pair of $L_3[k^0_{\langle 2,7,8,13 \rangle}]$ suggests on average four candidates for $k^1_{\langle 8,10 \rangle}$, and we store the pairs in a list indexed by the key bytes, $L_4[k^1_{\langle 8,10 \rangle}]$.

The next steps are repeated for each value of $k^1_{\langle 8,10 \rangle}$, using the list $L_4[k^1_{\langle 8,10 \rangle}]$ with on average $2^{36+\epsilon+2-16} = 2^{22+\epsilon}$ pairs leading to a difference in $\mathcal{D}_X$.

Step 4. This step determines the key candidates $k^7_{\langle 3,6,9,12 \rangle}$ that are ruled out with the available data, for each $k^0_{\langle 0,2,5,7,8,10,13,15 \rangle}, k^1_{\langle 8,10 \rangle}$. For this purpose, we use a list L_5 of 2^{32} bits to mark impossible key candidates $k^7_{\langle 3,6,9,12 \rangle}$. For each pair of $L_4[k^1_{\langle 8,10 \rangle}]$, we consider all the differences at the end of the sixth round that correspond to a difference in $\mathcal{D}_Y$, stored in T_2. From the differences before and after the last SubBytes, we compute the value of the output of SBox in bytes 3, 6, 9 and 12 using the DDT. Then, using the ciphertext values, we recover the bytes $k^7_{\langle 3,6,9,12 \rangle}$ and mark this value in the list L_5.

On average we mark $2^{22+\epsilon+10} = 2^{32+\epsilon}$ keys as impossible, so that each key remains possible with probability $P = (1 - 2^{-32})^{2^{32+\epsilon}} \approx e^{-2^\epsilon}$.

Step 5. Finally, we reconstruct the master keys corresponding to the candidates $k^0_{\langle 0,2,5,7,8,10,13,15 \rangle}, k^1_{\langle 8,10 \rangle}, k^7_{\langle 3,6,9,12 \rangle}$ not marked as impossible. Following [13,33], knowing $k^0_{\langle 0,2,5,7,8,10,13,15 \rangle}$ and $k^1_{\langle 8,10 \rangle}$ is equivalent to knowing $k^0_{\langle 0,2,4,5,6,7,8,10,13,15 \rangle}$, but it is hard to combine this with information about the last round. Therefore, for each of the $2^{112} \times P$ candidates, we just consider the 10 known bytes of k^0, do an exhaustive search for the 6 missing bytes and recompute k^7 to see if it matches the candidate. This requires $2^{112} \times P \times 2^{48} = 2^{160} \times P$ evaluations of the key schedule. We verify the $2^{160} \times P \times 2^{-32} = 2^{128} \times P$ remaining candidates with a know plaintext/ciphertext pair, for a cost of $2^{128} \times P$ encryptions.

Complexity. There are three dominant terms in the complexity of the attack. First we need to make $2^{101+\epsilon}$ calls to the encryption oracle. Then, the generation of key candidates (steps 1 to 4) is dominated by step 4. This step is done 2^{80} times (for each guess of $k^0_{\langle 0,2,5,7,8,10,13,15 \rangle}$ and $k^1_{\langle 8,10 \rangle}$) and during this step we go through the whole list $L_4[k^1_{\langle 8,10 \rangle}]$, containing $2^{22+\epsilon}$ pairs. For each pair and for each of the 2^{10} differences in T_2, we use 4 times the DDT. In order to express this complexity using one encryption as the unit, we follow the common practice of counting the number of table look-up. A 7 round AES encryption, requires 20×7 table lookups (including the Sboxes in the key schedule), therefore the cost of 4 DDT lookups is similar to $4/140 = 1/35$ encryptions. In total, the complexity of Step 4 is $2^{80} \times 2^{22+\epsilon} \times 2^{10}/35$. Finally step 5 requires the equivalent

Algorithm 1. Construction of possible key candidates (Steps 1 to 4)

Require: Tables T_1, T_2 and a list L_1 of $2^{68+\epsilon}$ pairs satisfying $\mathcal{D}_{in}$ and $\mathcal{D}_{out}$.
  Sort L_1 according to the plaintext difference and value in bytes 0, 5, 10 and 15.
  Let $L_1[\delta][x]$ be the sub-list with difference δ and value x in those bytes.
  for all $k^0_{\langle 0,5,10,15\rangle}$ **do**
    $L_2 \leftarrow \varnothing$
    for all 32-bits difference δ **do**
      for all difference θ in T_1 **do** ▷ bytes $\langle 0,5,10,15\rangle$
        Compute value(s) $x_{\langle 0,5,10,15\rangle}$ before first SubBytes from DDT.
        Add all pairs of $L_1[\delta][x_{\langle 0,5,10,15\rangle} \oplus k^0_{\langle 0,5,10,15\rangle}]$ to L_2.
    $L_3 \leftarrow [\varnothing$, **for all** $k^0_{\langle 2,7,8,13\rangle}]$
    for all pairs $((p,p'),(c,c'))$ in L_2 **do**
      for all difference θ in T_1 **do** ▷ bytes $\langle 2,7,8,13\rangle$
        Compute value(s) $x_{\langle 2,7,8,13\rangle}$ before first SubBytes from DDT.
        Add pair to $L_3[x_{\langle 2,7,8,13\rangle} \oplus p_{\langle 2,7,8,13\rangle}]$.
    for all $k^0_{\langle 2,7,8,13\rangle}$ **do**
      $L_4 \leftarrow [\varnothing$, **for all** $k^1_{\langle 8,10\rangle}]$
      Compute $k^1_{\langle 0,2\rangle}$ using the AES key schedule.
      for i in $\{0,1,2,3\}$ **do**
        for all pairs in $L_3[k^0_{\langle 2,7,8,13\rangle}]$ **do**
          Deduce $k^1_{\langle 8,10\rangle}$, assuming that diagonal i is inactive at end of round 2.
          Add pair to $L_4[k^1_{\langle 8,10\rangle}]$.
      for all $k^1_{\langle 8,10\rangle}$ **do**
        $L_5 \leftarrow$ [**True, for all** $k^7_{\langle 3,6,9,12\rangle}]$
        for all pairs $((p,p'),(c,c'))$ in $L_4[k^1_{\langle 8,10\rangle}]$ **do**
          for all difference θ in T_2 **do** ▷ bytes $\langle 12,13,14,15\rangle$
            Compute value(s) $x_{\langle 15,14,13,12\rangle}$ after last SubBytes from DDT.
            $L_5[x_{\langle 15,14,13,12\rangle} \oplus c_{\langle 3,6,9,12\rangle}] \leftarrow$ **False**.
        for all $k^7_{\langle 3,6,9,12\rangle}$ **do**
          if $L_5[k^7_{\langle 3,6,9,12\rangle}]$ **then**
            Check key candidate $k^0_{\langle 0,2,5,7,8,10,13,15\rangle}, k^1_{\langle 8,10\rangle}, k^7_{\langle 3,6,9,12\rangle}$.

of $e^{-2^\epsilon} \cdot 2^{160}/5 + e^{-2^\epsilon} \cdot 2^{128}$ encryptions, because the cost of the key schedule compared to an encryption[3] is $4/20 = 1/5$. In total, the time complexity is:

$$T = 2^{101+\epsilon} + 2^{112+\epsilon}/35 + e^{-2^\epsilon} \cdot (2^{160}/5 + 2^{128})$$

The best time complexity is obtained by taking $\epsilon = 5.1$, leading to a time complexity of $2^{112.1}$, a data complexity of $2^{106.1}$ chosen plaintexts, and a memory complexity of $N = 2^{73.1}$ words.

[3] This ratio is given as $2^{-3.6} \approx 1/12$ in [13], but we don't see how to achieve this result. In any case the impact on the total complexity is negligible because it is compensated by a very small change of ϵ.

Variant with multiple differentials. Boura, Lallemand, Naya-Plasencia and Suder describe [13] in a variant of this attack using multiple output differentials. More precisely, instead of using a fixed column for $\mathcal{D}_Y$ and a fixed anti-diagonal for $\mathcal{D}_{\text{out}}$, they consider the four possible columns for $\mathcal{D}_Y$ and the four corresponding anti-diagonal for $\mathcal{D}_{\text{out}}$. The attacks is essentially the same, but there are two important differences.

To construct the pairs, they start from only $2^{35+\epsilon}$ structures of 2^{64} plaintexts, but they obtain $2^{68+\epsilon}$ pairs matching $\mathcal{D}_{in}$ and $\mathcal{D}_{out}$ when considering the four anti-diagonal in $\mathcal{D}_{out}$. Steps 1 to 3 of the attack are the same a given above, but in step 4 each pair can give information about different bytes of k^7, depending on which anti-diagonal is active in the ciphertext. For each choice of $k^0_{\langle 0,2,5,7,8,10,13,15\rangle}, k^1_{\langle 8,10\rangle}$, they build a list of possible values for each anti-diagonal of k^7, and each key value remains possible with probability $e^{-2^{\epsilon-2}}$ because one fourth of the data correspond to each diagonal. Finally, in step 5, they merge the 4 lists, for a cost of $2^{80} \times (e^{-2^{\epsilon-2}} \cdot 2^{32})^4 = e^{-2^{\epsilon}} \cdot 2^{208}$.

The total time complexity of this variant is:

$$T = 2^{99+\epsilon} + 2^{112+\epsilon}/35 + e^{-2^{\epsilon}} \cdot (2^{208}/5 + 2^{128})$$

The best time complexity is obtained by taking $\epsilon = 6.1$, leading to a time complexity of 2^{113}, a data complexity of $2^{105.1}$ chosen plaintexts, and a memory complexity of $N = 2^{74.1}$ words.

This attack is listed with a time complexity of $2^{106.88}$ with $\epsilon = 6$ in [13], but this seems to be a mistake. There are not enough details of this attack in [13] to verify where their attack would differ from our understanding, but we don't see how to avoid having $2^{112+\epsilon}$ iterations at step 4, when we are eliminating 112-bit keys. Applying the generic formula (7) from the same paper also gives a term $2^{112+\epsilon}/35$ in the complexity (written as $2^{k_A+k_B}\frac{N}{2^{c_{\text{in}}+c_{\text{out}}}} \cdot C'_E$ in [13]).

Variant with state-test technique. In [13], the authors describe in details a variant using four output differentials and the state-test technique. This allows them to reduce by one byte the number of key bytes to be guessed, but they must use smaller structures, and this increases the data complexity.

The attack requires $N = 2^{68+\epsilon}$ chosen plaintexts, with a time complexity of:

$$T = 2^{107+\epsilon} + 2^{104+\epsilon}/35 + e^{-2^{\epsilon}} \cdot (2^{200}/5 + 2^{128})$$

The optimal time complexity[4] is 2^{113} with $\epsilon = 6$.

5.3 Our Improvement

We now explain how to improve the first attack using properties of the key schedule. We keep steps 1 to 4 as given in Algorithm 1, but we improve the reconstruction of the master key from bytes of the first and last round keys (Step 5). With

[4] In [13] they report the complexity as $2^{113.1}$ with $\epsilon = 6.1$.

this improvement, generating the key candidates is actually cheaper than verifying them with a known plaintext/ciphertext pair. We use the following property of the key schedule, in order to guess the missing key bytes of k^0 iteratively, and to efficiently verify whether they match the known bytes of k^7.

Proposition 1. *Let k_i^r a byte of an AES-128 subkey. If the byte is in the last column ($12 \leq i < 16$), then it depends on only 32 bits of information of the master key. If the byte is in the second or third column ($4 \leq i < 12$), then it depends on only 64 bits of information of the master key.*

Proof. Bytes in the last column correspond to basis vectors in the new representation, following Eq. (1) (for instance $k_{12}^r = s_{12}^r$). Therefore they depend only on one 32-bit chunk at any given round (k_{12}^7 can be computed from $s_{\langle 0,1,2,3\rangle}^0$).

Bytes in the second column correspond to the sum of two basis vector in the new representation (for instance $k_6^r = s_{14}^r \oplus s_4^r$). Since the two elements do not belong to the same chunk, the byte depends on two 32-bit chunks at any given round (k_6^7 can be computed from $s_{\langle 0,1,2,3,8,9,10,11\rangle}^0$).

Similarly, bytes in the third column correspond to the sum of two basis vector in the new representation (for instance $k_9^r = s_{15}^r \oplus s_8^r$). Therefore they depend only on two 32-bit chunks at any given round (k_9^7 can be computed from $s_{\langle 0,1,2,3,12,13,14,15\rangle}^0$).

Bytes in the first column correspond to the sum of four basis vector from four different chunks, therefore they depend on the full state in general (for instance $k_3^r = s_{13}^r \oplus s_{10}^r \oplus s_7^r \oplus s_0^r$). □

Initially we are given the values of $k_{\langle 0,2,4,5,6,7,8,10,13,15\rangle}^0$ and $k_{\langle 3,6,9,12\rangle}^7$. According to the property above, k_{12}^7 can be computed from k_{15}^0, $k_{14}^0 \oplus k_{10}^0 \oplus k_6^0 \oplus k_2^0$, $k_{13}^0 \oplus k_5^0$, $k_{12}^0 \oplus k_8^0$, k_{14}^0, and k_6^7 can be computed from k_{15}^0, $k_{14}^0 \oplus k_{10}^0 \oplus k_6^0 \oplus k_2^0$, $k_{13}^0 \oplus k_5^0$, $k_{12}^0 \oplus k_8^0$, k_{13}^0, $k_{12}^0 \oplus k_8^0 \oplus k_4^0 \oplus k_0^0$, $k_{15}^0 \oplus k_7^0$, $k_{14}^0 \oplus k_{10}^0$. Therefore we can verify their value after guessing $k_{\langle 12,14\rangle}^0$.

At this point two chunks are completely known: $s_{\langle 0,1,2,3\rangle}^0$ and $s_{\langle 8,9,10,11\rangle}^0$ or equivalently $s_{\langle 12,13,14,15\rangle}^7$ and $s_{\langle 4,5,6,7\rangle}^7$. In particular, we can deduce the value of $k_{13}^7 = s_8^7 = s_{15}^7 \oplus k_9^7$, which can also be computed from $s_{\langle 12,13,14,15\rangle}^0$, *i.e.* from k_{12}^0, $k_{15}^0 \oplus k_{11}^0 \oplus k_7^0 \oplus k_3^0$, $k_{14}^0 \oplus k_6^0$, $k_{13}^0 \oplus k_9^0$. Therefore, we only need to guess $k_{11}^0 \oplus k_3^0$ and k_9^0 to verify k_{13}^7.

Finally, we focus of the remaining 32-bit chunk, corresponding to $s_{\langle 4,5,6,7\rangle}^0$ and $s_{\langle 0,1,2,3\rangle}^7$. We already have the value of $s_4^0 = k_{14}^0$ and $s_6^0 = k_{12}^0 \oplus k_4^0$, and we can compute $s_0^7 = s_{10}^7 \oplus s_{13}^7 \oplus s_7^7 \oplus k_3^7$. Using a pre-computed table, we recover the 2^8 values of the chunk corresponding to those constraints.

Algorithm 2 describes the full process. The cost of this step is $e^{-2^\epsilon} \times 2^{128}/5$, where 1/5 is the cost of computing the key schedule compared to a full encryption. Finally the total time complexity of our attack is:

$$T = 2^{101+\epsilon} + 2^{112+\epsilon}/35 + e^{-2^\epsilon} \cdot (2^{128}/5 + 2^{128})$$

Algorithm 2. Improved version of the key candidate checking (Step 5)

Require: A key candidate $k^0_{\langle 0,2,5,7,8,10,13,15\rangle}, k^1_{\langle 8,10\rangle}, k^7_{\langle 3,6,9,12\rangle}$.
  for all $k^0_{\langle 12,14\rangle}$ **do**
    Compute $s^7_{\langle 12,13,14,15\rangle}$ from $s^0_{\langle 0,1,2,3\rangle}$
    if $k^7_{12} = s^7_{12}$ **then**
      Compute $s^7_{\langle 4,5,6,7\rangle}$ from $s^0_{\langle 8,9,10,11\rangle}$
      if $k^7_6 = s^7_4 \oplus s^7_{14}$ **then**
        $T \leftarrow [\varnothing$, **for all** $k^7_{15}]$
        for all $k^0_{11}, k^0_1 \oplus k^0_9$ **do**
          Compute $s^7_{\langle 0,1,2,3\rangle}$ from $s^0_{\langle 4,5,6,7\rangle}$
          Add $(k^0_{11}, k^0_1 \oplus k^0_9)$ to $T[s^7_0]$
        for all $k^0_9, k^0_3 \oplus k^0_{11}$ **do**
          Compute $s^7_{\langle 8,9,10,11\rangle}$ from $s^0_{\langle 12,13,14,15\rangle}$
          if $k^7_9 = s^7_8 \oplus s^7_{15}$ **then**
            for all $(k^0_{11}, k^0_1 \oplus k^0_9)$ in $T[s^7_{13} \oplus s^7_{10} \oplus s^7_7 \oplus k^7_3]$ **do**
              Check the master key k^0 with a pair (p, c).

The best time complexity is obtained by taking $\epsilon = 3.9$ leading to a time complexity of $2^{110.9}$, a data complexity of $2^{104.9}$ chosen plaintext, and a memory complexity of $2^{71.9}$ words.

We remark that the improvement is only applicable when the last MixColumns is omitted. In general, it does not affect the complexity of attacks, because removing the last MixColumns defines an equivalent cipher up to a modification of the key schedule. However, when attacks exploit relations between the subkeys, the relations are simpler if the last MixColumns is omitted [22].

6 New Representations of the AES-192 and AES-256 Key Schedules

The same techniques can also be applied to other variants of AES: we apply the algorithm of Leander, Minaud and Rønjom [32] to extract invariant subspaces of the key schedule, and we use a change of variables corresponding to the subspaces to obtain a simplified representation.

AES-192. We find two invariant subpaces of dimension 12, and obtain a simplified representation with 2 independent chunks each acting on 12 bytes, as shown in Fig. 11.

AES-256. We find four invariant subpaces of dimension 8, and obtain a simplified representation with 4 independent chunks each acting on 8 bytes, as shown in Fig. 12.

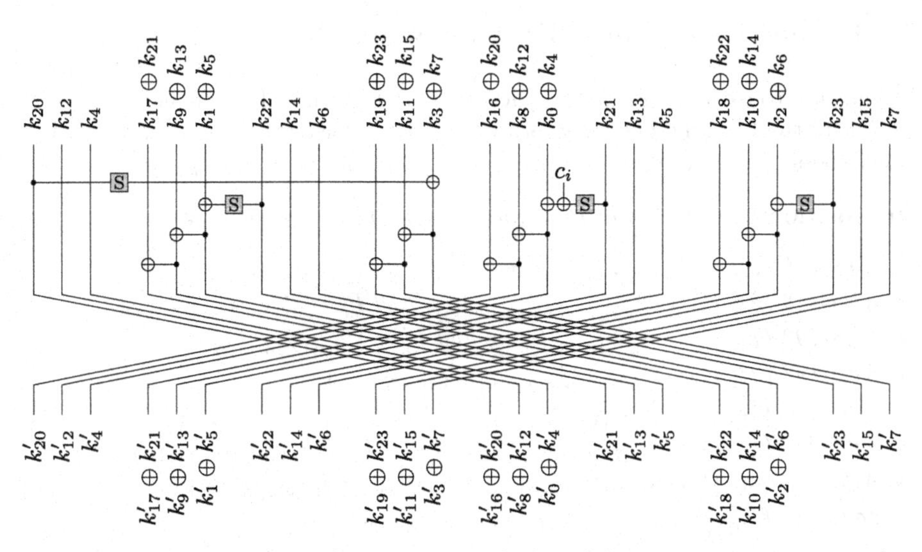

Fig. 11. One round of the AES-192 key schedule (alternative representation).

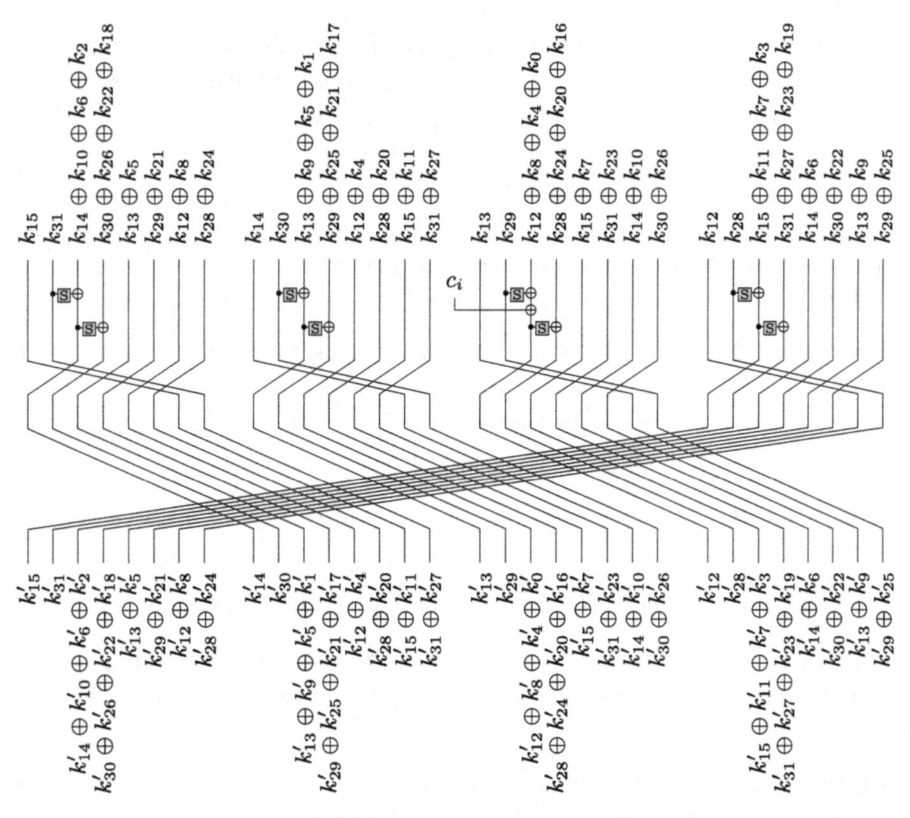

Fig. 12. One round of the AES-256 key schedule (alternative representation).

7 Properties on the AES Key Schedule

In addition to explaining the presence of short length cycles, our new representations of the key schedule also permits us to demonstrate some properties. For conciseness, we use the notation $k^r_{i,j_1 \oplus j_2,\ldots}$ to denote $k^r_i, k^r_{j_1} \oplus k^r_{j_2}, \ldots$

Proposition 2. *Let P_r and P'_r defined in one of the following ways:*

- *AES-128 (1): $P_r = k^r_{\langle 5,7,13,15\rangle}$, and $P'_r = k^r_{\langle 4,6,12,14\rangle}$*
- *AES-128 (2): $P_r = k^r_{\langle 0\oplus 4,2\oplus 6,8\oplus 12,10\oplus 14\rangle}$, and $P'_r = k^r_{\langle 1\oplus 5,3\oplus 7,9\oplus 13,11\oplus 15\rangle}$*
- *AES-192 (1): $P_r = k^r_{\langle 5,7,13,15,21,23\rangle}$, and $P'_r = k^r_{\langle 4,6,12,14,20,22\rangle}$*
- *AES-192 (2): $P_r = k^r_{\langle 0\oplus 4,2\oplus 6,8\oplus 12,10\oplus 14,16\oplus 20,18\oplus 22\rangle}$, and $P'_r = k^r_{\langle 1\oplus 5,3\oplus 7,9\oplus 13,11\oplus 15,17\oplus 21,19\oplus 23\rangle}$*
- *AES-256 (1): $P_r = k^r_{\langle 5,7,13,15,21,23,29,31\rangle}$, and $P'_r = k^r_{\langle 4,6,12,14,20,22,28,30\rangle}$*
- *AES-256 (2): $P_r = k^r_{\langle 0\oplus 4,2\oplus 6,8\oplus 12,10\oplus 14,16\oplus 20,18\oplus 22,24\oplus 28,26\oplus 30\rangle}$, and $P'_r = k^r_{\langle 1\oplus 5,3\oplus 7,9\oplus 13,11\oplus 15,17\oplus 21,19\oplus 23,25\oplus 29,27\oplus 31\rangle}$*

If there exists an r_0 such as P_{r_0} and $P'_{r_0\pm 1}$ are known, then for all $i \in \mathbb{Z}$, the bytes P_{r_0+2i} and P'_{r_0+2i+1} are known (and they are easily computable).

Proof. The AES-128 (1) case is considered here, the other cases are demonstrated in the same way. Knowing $k^r_{\langle 5,7,13,15\rangle}$ and $k^{r+1}_{\langle 4,6,12,14\rangle}$ is equivalent to knowing two chunks of the state: $s^r_{\langle 0,1,2,3\rangle}$ and $s^r_{\langle 8,9,10,11\rangle}$. This can be verified using Eq. (2). The knowledge of these 2 chunks allows us to extract the value of the bytes in position $k_{\langle 5,7,13,15\rangle}$ or $k_{\langle 4,6,12,14\rangle}$ at any round. □

This byte position of this proposition is represented in Fig. 13. This proposition is a generalization of the observations made for AES-128 by Dunkelman and Keller:

Observation 3 ([21]). *For each $0 \le i \le 3$, the subkeys of AES satisfy the relations:*

$$k_{r+2}(i,0) \oplus k_{r+2}(i,2) = k_r(i,2).$$

$$k_{r+2}(i,1) \oplus k_{r+2}(i,3) = k_r(i,3).$$

Observation 4 ([21]). *For each $0 \le i \le 3$, the subkeys of AES satisfy the relation:*

$$k_{r+2}(i,1) \oplus SB(k_{r+1}((i+1)\ mod\ 4,3)) \oplus RCON_{r+2}(i) = k_r(i,1).$$

Another property can also be demonstrated on the AES-128 key schedule, using the value of one byte of the last column per round over 4 consecutive rounds:

Proposition 3. *If there exists $r \in \mathbb{N}$ and $i \in \{0,1,2,3\}$ such that the bytes $k^r_{15-i}, k^{r+1}_{15-(i+1)\%4}, k^{r+2}_{15-(i+2)\%4}, k^{r+3}_{15-(i+3)\%4}$ are known, then for all $j \in \mathbb{Z}$, the value of the byte $k^{r+j}_{15-(i+j\%4)}$ is known.*

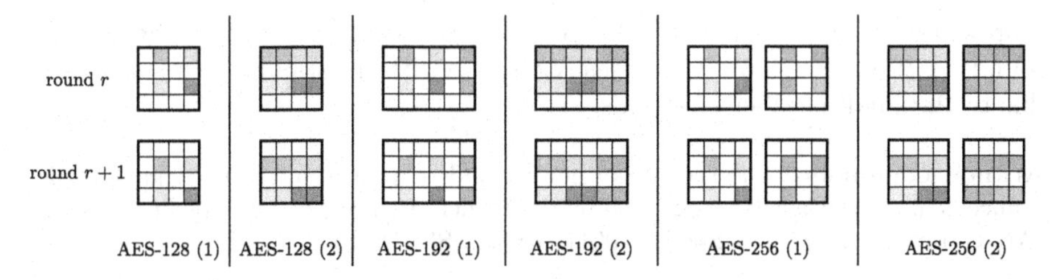

Fig. 13. Representation of the position of the bytes of the proposition. In variants (2), only the XOR of the two bytes of the same color must be known.

Proof. Knowing the bytes $k^r_{15-i}, k^{r+1}_{15-(i+1)\%4}, k^{r+2}_{15-(i+2)\%4}, k^{r+3}_{15-(i+3)\%4}$ is equivalent to knowing one chunk of the state in our representation: $s^r_{\langle 4i,4i+1,4i+2,4i+3\rangle}$. Given that $\forall r \in \mathbb{N}$, $s^r_{4i} = k^r_{15-i}$, we can calculate a byte of the last column at any round because we have the knowledge of a chunk in our new representation. □

The property can also be generalized when bytes at the correct position are known in non-consecutive rounds.

8 Conclusion

Alternative representations of the AES data operations have been used in several previous works; in particular, the super-box property [26] of Gilbert and Peyrin is an alternative representation of two AES rounds that led to several improved cryptanalysis results on AES-based schemes. Gilbert has later shown a more general untwisted representation of the AES data path, resulting in the first known-key attack against the full AES-128 [25].

In this work we use techniques from invariant subspace attacks to discover an equivalent representation of the AES key schedule, and we derive new cryptanalysis results, based on two main observations. First, iterating an odd number of key schedule rounds defines a permutation with short cycles. This undermine the security of AES-based schemes using iterations of the key schedule as a type of tweak to make each encryption call different. More generally, the AES key schedule cannot and should not be considered as a random permutation, even after a large number of rounds. Second, the alternative representation makes it easier to combine information from the first subkeys and from the last subkeys, improving previous key recovery attacks. This topic has been studied before and many attacks use key schedule relations to reduce the complexity (in particular, we can mention the *key bridging* notion of Dunkelman, Keller and Shamir [23,24]). However our alternative representation shows non-linear relations that have not been exploited before. In particular, we show that bytes in the last column of an AES-128 subkey depend on only 32 bits of information from the master key.

We expect that this alternative representation can open the way to further results exploiting properties of the AES key schedule. For instance, the new representation can be used to characterize keys that stay symmetric for two rounds, as used in [27], but this is easily be done with the standard representation due to the small number of rounds.

Acknowledgement. The second author is funded by a grant from Région Ile-de-France. This work was also supported by the French Agence Nationale de la Recherche (ANR), under grant ANR-20-CE48-0017 (project SELECT).

References

1. Advanced Encryption Standard (AES). National Institute of Standards and Technology (NIST), FIPS PUB 197, U.S. Department of Commerce, November 2001
2. Bahrak, B., Aref, M.R.: Impossible differential attack on seven-round AES-128. IET Inf. Secur. **2**(2), 28–32 (2008). https://doi.org/10.1049/iet-ifs:20070078
3. Biham, E., Biryukov, A., Shamir, A.: Cryptanalysis of skipjack reduced to 31 rounds using impossible differentials. In: Stern, J. (ed.) EUROCRYPT 1999. LNCS, vol. 1592, pp. 12–23. Springer, Heidelberg (1999). https://doi.org/10.1007/3-540-48910-X_2
4. Biryukov, A.: The design of a stream cipher LEX. In: Biham, E., Youssef, A.M. (eds.) SAC 2006. LNCS, vol. 4356, pp. 67–75. Springer, Heidelberg (2007). https://doi.org/10.1007/978-3-540-74462-7_6
5. Biryukov, A., Dunkelman, O., Keller, N., Khovratovich, D., Shamir, A.: Key recovery attacks of practical complexity on AES-256 variants with up to 10 rounds. In: Gilbert, H. (ed.) EUROCRYPT 2010. LNCS, vol. 6110, pp. 299–319. Springer, Heidelberg (2010). https://doi.org/10.1007/978-3-642-13190-5_15
6. Biryukov, A., Khovratovich, D.: Related-key cryptanalysis of the full AES-192 and AES-256. In: Matsui, M. (ed.) ASIACRYPT 2009. LNCS, vol. 5912, pp. 1–18. Springer, Heidelberg (2009). https://doi.org/10.1007/978-3-642-10366-7_1
7. Biryukov, A., Khovratovich, D., Nikolić, I.: Distinguisher and related-key attack on the full AES-256. In: Halevi, S. (ed.) CRYPTO 2009. LNCS, vol. 5677, pp. 231–249. Springer, Heidelberg (2009). https://doi.org/10.1007/978-3-642-03356-8_14
8. Bogdanov, A., Mendel, F., Regazzoni, F., Rijmen, V., Tischhauser, E.: ALE: AES-based lightweight authenticated encryption. In: Moriai, S. (ed.) FSE 2013. LNCS, vol. 8424, pp. 447–466. Springer, Heidelberg (2014). https://doi.org/10.1007/978-3-662-43933-3_23
9. Bonnetain, X., Naya-Plasencia, M., Schrottenloher, A.: Quantum security analysis of AES. IACR Trans. Symm. Cryptol. **2019**(2), 55–93 (2019)
10. Bouillaguet, C., Derbez, P.: AES attacks finder (2011). https://github.com/cbouilla/AES-attacks-finder
11. Bouillaguet, C., Derbez, P., Fouque, P.-A.: Automatic search of attacks on round-reduced AES and applications. In: Rogaway, P. (ed.) CRYPTO 2011. LNCS, vol. 6841, pp. 169–187. Springer, Heidelberg (2011). https://doi.org/10.1007/978-3-642-22792-9_10
12. Bouillaguet, C., Derbez, P., Fouque, P.A.: Automatic search of attacks on round-reduced AES and applications. Cryptology ePrint Archive, Report 2012/069 (2012). http://eprint.iacr.org/2012/069
13. Boura, C., Lallemand, V., Naya-Plasencia, M., Suder, V.: Making the impossible possible. J. Cryptol. **31**(1), 101–133 (2018)

14. Chakraborty, B., Nandi, M.: mixFeed. Submission to the NIST Lightweight Cryptography standardization process (2019). https://csrc.nist.gov/CSRC/media/Projects/lightweight-cryptography/documents/round-2/spec-doc-rnd2/mixFeed-spec-round2.pdf
15. Chakraborty, B., Nandi, M.: Security proof of mixFeed (2019). https://csrc.nist.gov/CSRC/media/Events/lightweight-cryptography-workshop-2019/documents/papers/security-proof-of-mixfeed-lwc2019.pdf
16. Daemen, J., Rijmen, V.: The Pelican MAC function 2.0. Cryptology ePrint Archive, Report 2005/088 (2005). http://eprint.iacr.org/2005/088
17. Daemen, J., Rijmen, V.: The design of Rijndael: AES - the advanced encryption standard (2013)
18. Demirci, H., Selçuk, A.A.: A meet-in-the-middle attack on 8-round AES. In: Nyberg, K. (ed.) FSE 2008. LNCS, vol. 5086, pp. 116–126. Springer, Heidelberg (2008). https://doi.org/10.1007/978-3-540-71039-4_7
19. Derbez, P., Fouque, P.-A., Jean, J.: Improved key recovery attacks on reduced-round AES in the single-key setting. In: Johansson, T., Nguyen, P.Q. (eds.) EUROCRYPT 2013. LNCS, vol. 7881, pp. 371–387. Springer, Heidelberg (2013). https://doi.org/10.1007/978-3-642-38348-9_23
20. Dunkelman, O., Keller, N.: A new attack on the LEX stream cipher. In: Pieprzyk, J. (ed.) ASIACRYPT 2008. LNCS, vol. 5350, pp. 539–556. Springer, Heidelberg (2008). https://doi.org/10.1007/978-3-540-89255-7_33
21. Dunkelman, O., Keller, N.: Treatment of the initial value in time-memory-data tradeoff attacks on stream ciphers. Inf. Process. Lett. **107**, 133–137 (2008)
22. Dunkelman, O., Keller, N.: The effects of the omission of last round's mixcolumns on AES. Inf. Process. Lett. **110**(8–9), 304–308 (2010)
23. Dunkelman, O., Keller, N., Shamir, A.: Improved single-key attacks on 8-round AES-192 and AES-256. In: Abe, M. (ed.) ASIACRYPT 2010. LNCS, vol. 6477, pp. 158–176. Springer, Heidelberg (2010). https://doi.org/10.1007/978-3-642-17373-8_10
24. Dunkelman, O., Keller, N., Shamir, A.: Improved single-key attacks on 8-round AES-192 and AES-256. J. Cryptol. **28**(3), 397–422 (2015)
25. Gilbert, H.: A simplified representation of AES. In: Sarkar, P., Iwata, T. (eds.) ASIACRYPT 2014, Part I. LNCS, vol. 8873, pp. 200–222. Springer, Heidelberg (2014). https://doi.org/10.1007/978-3-662-45611-8_11
26. Gilbert, H., Peyrin, T.: Super-Sbox cryptanalysis: improved attacks for AES-like permutations. In: Hong, S., Iwata, T. (eds.) FSE 2010. LNCS, vol. 6147, pp. 365–383. Springer, Heidelberg (2010). https://doi.org/10.1007/978-3-642-13858-4_21
27. Grassi, L., Leander, G., Rechberger, C., Tezcan, C., Wiemer, F.: Weak-key subspace trails and applications to AES. In: Jacobson Jr., M.J., Dunkelman, O., O'Flynn, C. (eds.) SAC 2020. LNCS. Springer, Heidelberg (2019)
28. Jean, J.: TikZ for Cryptographers (2016). https://www.iacr.org/authors/tikz/
29. Khairallah, M.: Weak keys in the rekeying paradigm: application to COMET and mixFeed. IACR Trans. Symm. Cryptol. **2019**(4), 272–289 (2019)
30. Khovratovich, D., Rechberger, C.: The LOCAL attack: cryptanalysis of the authenticated encryption scheme ALE. In: Lange, T., Lauter, K., Lisoněk, P. (eds.) SAC 2013. LNCS, vol. 8282, pp. 174–184. Springer, Heidelberg (2014). https://doi.org/10.1007/978-3-662-43414-7_9
31. Leander, G., Abdelraheem, M.A., AlKhzaimi, H., Zenner, E.: A cryptanalysis of PRINTCIPHER: the invariant subspace attack. In: Rogaway, P. (ed.) CRYPTO 2011. LNCS, vol. 6841, pp. 206–221. Springer, Heidelberg (2011). https://doi.org/10.1007/978-3-642-22792-9_12

32. Leander, G., Minaud, B., Rønjom, S.: A generic approach to invariant subspace attacks: cryptanalysis of robin, iSCREAM and Zorro. In: Oswald, E., Fischlin, M. (eds.) EUROCRYPT 2015, Part I. LNCS, vol. 9056, pp. 254–283. Springer, Heidelberg (2015). https://doi.org/10.1007/978-3-662-46800-5_11
33. Mala, H., Dakhilalian, M., Rijmen, V., Modarres-Hashemi, M.: Improved impossible differential cryptanalysis of 7-round AES-128. In: Gong, G., Gupta, K.C. (eds.) INDOCRYPT 2010. LNCS, vol. 6498, pp. 282–291. Springer, Heidelberg (2010). https://doi.org/10.1007/978-3-642-17401-8_20
34. Wu, S., Wu, H., Huang, T., Wang, M., Wu, W.: Leaked-state-forgery attack against the authenticated encryption algorithm ALE. In: Sako, K., Sarkar, P. (eds.) ASIACRYPT 2013. LNCS, vol. 8269, pp. 377–404. Springer, Heidelberg (2013). https://doi.org/10.1007/978-3-642-42033-7_20

Public-Key Cryptography

Analysing the HPKE Standard

Joël Alwen[1(✉)], Bruno Blanchet[2], Eduard Hauck[3], Eike Kiltz[3], Benjamin Lipp[2], and Doreen Riepel[3]

[1] Wickr, New York, USA
jalwen@wickr.com
[2] Inria Paris, Paris, France
{bruno.blanchet,benjamin.lipp}@inria.fr
[3] Ruhr-Universität Bochum, Bochum, Germany
{eduard.hauck,eike.kiltz,doreen.riepel}@rub.de

Abstract. The *Hybrid Public Key Encryption* (HPKE) scheme is an emerging standard currently under consideration by the Crypto Forum Research Group (CFRG) of the IETF as a candidate for formal approval. Of the four modes of HPKE, we analyse the authenticated mode $\mathsf{HPKE}_{\mathsf{Auth}}$ in its single-shot encryption form as it contains what is, arguably, the most novel part of HPKE.

$\mathsf{HPKE}_{\mathsf{Auth}}$'s intended application domain is captured by a new primitive which we call Authenticated Public Key Encryption (APKE). We provide syntax and security definitions for APKE schemes, as well as for the related Authenticated Key Encapsulation Mechanisms (AKEMs). We prove security of the AKEM scheme $\mathsf{DH\text{-}AKEM}$ underlying $\mathsf{HPKE}_{\mathsf{Auth}}$ based on the Gap Diffie-Hellman assumption and provide general AKEM/DEM composition theorems with which to argue about $\mathsf{HPKE}_{\mathsf{Auth}}$'s security. To this end, we also formally analyse $\mathsf{HPKE}_{\mathsf{Auth}}$'s key schedule and key derivation functions. To increase confidence in our results we use the automatic theorem proving tool CryptoVerif. All our bounds are quantitative and we discuss their practical implications for $\mathsf{HPKE}_{\mathsf{Auth}}$.

As an independent contribution we propose the new framework of *nominal groups* that allows us to capture abstract syntactical and security properties of practical elliptic curves, including the Curve25519 and Curve448 based groups (which do not constitute cyclic groups).

Keywords: Public-key encryption · Authentication · Signcryption · Key encapsulation mechanisms

1 Introduction

An effort is currently underway by the Crypto Forum Research Group (CFRG) to agree upon a new open standard for public key encryption [5]. The standard will be called *Hybrid Public Key Encryption* (HPKE) and it is, in particular, expected to be used as a building block by the Internet Engineering Task Force (IETF) in at least two further upcoming standardized security protocols [4,30]. The primary source for HPKE is an RFC [5] (currently on draft 8) which lays out the details of the construction and provides some rough intuition for its security properties.

A. Canteaut and F.-X. Standaert (Eds.): EUROCRYPT 2021, LNCS 12696, pp. 87–116, 2021.
https://doi.org/10.1007/978-3-030-77870-5_4

At first glance the HPKE standard might be thought of as a "public key encryption" scheme in the spirit of the KEM/DEM paradigm [15]. That is, it combines a Key Encapsulation Mechanism (KEM) and an Authenticated Encryption with Associated Data (AEAD) acting as a Data Encapsulation Mechanism (DEM) according to the KEM/DEM paradigm. However, upon closer inspection HPKE turns out to be more complex than this perfunctory description implies.

First, HPKE actually consists of 2 different KEM/DEM constructions. Moreover, each construction can be instantiated with a pre-shared key (PSK) known to both sender and receiver, which is used in the key schedule to derive the DEM key. In total this gives rise to 4 different *modes* for HPKE. The *basic* mode $\mathsf{HPKE}_{\mathsf{Base}}$ makes use of a standard (say IND-CCA-secure) KEM to obtain a "message privacy and integrity" only mode. This mode can be extended to $\mathsf{HPKE}_{\mathsf{PSK}}$ to support authentication of the sender via a PSK.

The remaining 2 HPKE modes make use of a different KEM/DEM construction built from a rather non-standard KEM variant which we call an *Authenticated KEM* (AKEM). Roughly speaking, an AKEM can be thought of the KEM analogue of signcryption [31]. In particular, sender and receiver both have their own public/private keys. Each party requires their own private and the other party's public key to perform en/decryption. The HPKE RFC constructs an AKEM based on a generic Diffie-Hellman group. It goes on to fix concrete instantiations of such groups using either the P-256, P-384, or P-521 NIST curves [28] or the Curve25519 or Curve448 curves [25]. The AKEM-based HPKE modes also intend to authenticate the sender to the receiver. Just as in the KEM-based case, the AKEM/DEM construction can be instantiated in modes either with or without a PSK. We refer to the AKEM/DEM-based mode without a PSK as the *authenticated mode* and, for reasons described below, it is the main focus of this work. The corresponding HPKE scheme is called $\mathsf{HPKE}_{\mathsf{Auth}}$.

Orthogonal to the choice of mode in use, HPKE also provides a so called single-shot and a multi-shot API. The single-shot API can be thought of as pairing a single instance of the DEM with a KEM ciphertext while the multi-shot API establishes a key schedule allowing a single KEM to be used to derive keys for an entire sequence of DEMs. Finally, HPKE also supports exporting keys from the key schedule for use by arbitrary higher-level applications.

APPLICATIONS. As an open standard of the IETF, we believe HPKE to be an interesting topic of study in its own right. Indeed, HPKE is already slated for use in at least two upcoming protocols; the Messaging Layer Security (MLS) [4] secure group messaging protocol and the Encrypted Server Name Indication (ESNI) extension for TLS 1.3 [30]. Both look to be well-served by the single-shot API as they require a single DEM to be produced (at the same time as the KEM) and the combined KEM/DEM ciphertext to be sent as one packet.

More interestingly, at least for MLS, authenticating the sender of an HPKE ciphertext (based on their public keys) is clearly also a useful property. (For the ESNI application things are less clear.[1])

In a bit more detail, MLS is already equipped with a notion of a PKI involving public keys bound to long-term identities of parties (as described in [29]). To invite a new member to an existing MLS protocol session the inviter must send an HPKE ciphertext to the new member. In line with MLS's strong authentication goals, the new member is expected to be able to cryptographically validate the (supposed) identity of the sender of such ciphertexts.

Currently, MLS calls for the HPKE ciphertext to be produced using HPKE's basic mode $\mathsf{HPKE}_{\mathsf{Base}}$ and the resulting ciphertext to be signed by the inviter using a digital signature scheme (either ECDSA or EdDSA). However, an alternative approach to achieve the same ends could be to directly use HPKE in its authenticated mode $\mathsf{HPKE}_{\mathsf{Auth}}$. This would save on at least 2 modular exponentiations as well as result in packets containing 2 fewer group elements. Reducing computational and communication complexity has been a central focus of the MLS design process as such costs are considered the main hurdles to achieving the MLS's stated goal of supporting extremely large groups. Unfortunately, in our analysis, we discovered that $\mathsf{HPKE}_{\mathsf{Auth}}$ does not authenticate the sender when the receiver's secret key leaked, a key compromise impersonation (KCI) attack (Sect. 4.4). MLS aims to provide strong security in the face of state leakage (which includes KCI attacks), so switching from $\mathsf{HPKE}_{\mathsf{Base}}$ and signatures to $\mathsf{HPKE}_{\mathsf{Auth}}$ would result in a significant security downgrade.

$\mathsf{HPKE}_{\mathsf{Auth}}$ could also be a replacement for the public-key authenticated encryption originally implemented by the NaCl cryptographic library. $\mathsf{HPKE}_{\mathsf{Auth}}$ is safer than the NaCl implementation because, in $\mathsf{HPKE}_{\mathsf{Auth}}$, the shared secret is bound to the intended sender and recipient public keys.

1.1 Our Contributions

So far, there has been no formal analysis of the HPKE standard. Unfortunately, due to its many modes, options and features a complete analysis of HPKE from scratch seems rather too ambitious for a single work such as this one. Thus, we are forced to choose our scope more carefully. The basic mode $\mathsf{HPKE}_{\mathsf{Base}}$ (especially using the single-shot API) seems to be a quite standard construction. Therefore, and in light of the above discussion around MLS, we have opted to focus on the more novel authenticated mode in its single-shot API form $\mathsf{HPKE}_{\mathsf{Auth}}$. To this end we make the following contributions.

AUTHENTICATED KEM AND PKE. We begin, in Sect. 4, by introducing *Authenticated Key Encapsulation Mechanisms* (AKEM) and *Authenticated Public Key*

[1] The ESNI RFC calls for a client initiating a TLS connection to send an HPKE ciphertext to the server. Although not as common, TLS can also be used in settings with bi-directional authentication. In particular, clients can use certificates binding their identities to their public key to authenticate themselves to the server. Unfortunately, it is unclear how the server would know, a priori, which public key to use for the client when attempting to decrypt the HPKE ciphertext.

Table 1. Security properties needed to prove Outsider-Auth, Outsider-CCA, and Insider-CCA security of APKE obtained by the AKEM/DEM construction.

	AKEM			AEAD	
	Outsider-Auth	Outsider-CCA	Insider-CCA	INT-CTXT	IND-CPA
$\mathsf{Outsider\text{-}Auth}_{\mathsf{APKE}}$	X	X		X	
$\mathsf{Outsider\text{-}CCA}_{\mathsf{APKE}}$		X		X	X
$\mathsf{Insider\text{-}CCA}_{\mathsf{APKE}}$			X	X	X

Encryption (APKE) schemes, where the syntax of APKE matches that of the single-shot authenticated mode of $\mathsf{HPKE}_{\mathsf{Auth}}$. In terms of security, we define (multi-user) security notions capturing both authenticity and (2 types of) privacy for an AKEM and an APKE. In a bit more detail, both for authenticity and for privacy we consider so called weaker *outsider* and stronger *insider* variants. Intuitively, outsider notions model settings where the adversary is an outside observer. Conversely, insider notions model settings where the adversary is somehow directly involved; in particular, even selecting some of the secrets used to produce target ciphertexts. A bit more formally, we call an honestly generated key pair *secure* if the secret key was not (explicitly) leaked to the adversary and *leaked* if it was. A key pair is called *bad* if it was sampled arbitrarily by the adversary. A scheme is outsider-secure if target ciphertexts are secure when produced using secure key pairs. Meanwhile, insider security holds even if one secure *and one bad key pair* are used. For example, insider privacy (Insider-CCA) for AKEM requires that an encapsulated key remains indistinguishable from random despite the encapsulating ciphertext being produced using bad sender keys (but secure receiver keys). Similarly, insider authenticity (Insider-Auth) requires that an adversary cannot produce a valid ciphertext for bad receiver keys as long as the sender keys are secure. In particular, insider authenticity implies (but is strictly stronger than) Key Compromise Impersonation (KCI) security as KCI security only requires authenticity for leaked (but not bad) receiver keys.

Moreover, as an independent contribution we show that for each security notion of an AKEM a (significantly simpler) single-user and single-challenge-query version already implies security for its (more complex but practically relevant) multi-user version. In particular, this provides an easier target for future work on AKEMs, e.g. when building a post-quantum variant of $\mathsf{HPKE}_{\mathsf{Auth}}$.

AKEM/DEM: FROM AKEM TO APKE. Next we turn to the AKEM/DEM construction used in the HPKE standard. We prove a set of composition results each showing a different type of security for the single-shot AKEM/DEM construction depending on which properties the underlying AKEM guarantees. Each of these results also assumes standard security properties for the AEAD (namely IND-CPA and INT-CTXT) and for the key schedule KS (namely pseudo-randomness). In particular, these results are proven in the standard model. Somewhat to our surprise, it turns out that the APKE obtained by the AKEM/DEM construction does not provide insider authenticity (and so, nor does $\mathsf{HPKE}_{\mathsf{Auth}}$ itself). Indeed, we give an attack in Sect. 4.4.

Table 1 summarises the AKEM and AEAD properties we use to prove each of the remaining 3 types of security for the AKEM/DEM APKE construction.

THE $\mathsf{HPKE}_{\mathsf{Auth}}$ SCHEME. In Sect. 5 we analyse the generic $\mathsf{HPKE}_{\mathsf{Auth}}$ scheme proposed in the RFC. $\mathsf{HPKE}_{\mathsf{Auth}}$ is an instantiation of the AKEM/DEM paradigm discussed above.

Thus, we first analyse DH-AKEM, the particular AKEM underlying $\mathsf{HPKE}_{\mathsf{Auth}}$. The RFC builds DH-AKEM from a key-derivation function KDF and an underlying generic Diffie-Hellman group. As one of our main results we show that DH-AKEM provides authenticity and privacy based on the Gap Diffie-Hellman assumption over the underlying group. To show this we model KDF as a random oracle.

Next we consider $\mathsf{HPKE}_{\mathsf{Auth}}$'s key schedule and prove it to be pseudo-random based on pseudo-randomness of its building blocks, the functions Extract and Expand. Similarly, we argue why DH-AKEM's key derivation function KDF can be modelled as a random oracle. Finally, by applying our results about the AKEM/DEM paradigm from the previous sections, we obtain security proofs capturing the privacy and authenticity of $\mathsf{HPKE}_{\mathsf{Auth}}$ as an APKE. Our presentation ends with concrete bounds of $\mathsf{HPKE}_{\mathsf{Auth}}$'s security and their interpretation.

PRACTICE-ORIENTED CRYPTOGRAPHY. Due to the very applied nature of HPKE we have taken care to maximise the practical relevance of our results. All security properties we analyse for $\mathsf{HPKE}_{\mathsf{Auth}}$ are defined directly for a multi-user setting. Further, to help practitioners set sound parameters for their HPKE applications, our results are stated in terms of very fine-grained exact (as opposed to asymptotic) terms. That is, the security loss for each result is bounded as an explicit function of various parameters such as the numbers of key pairs, queries, etc.

Finally, instead of relying on a generic prime-order group to state our underlying security assumptions, we ultimately reduce security to assumptions on each of the concrete elliptic-curve-based instantiations. For the P-256, P-384, and P-521 curves, this is relatively straightforward. However, for Curve25519 and Curve448, this is a less than trivial step as those groups (and their associated Diffie-Hellman functions X25519 and X448) depart significantly from the standard generic group abstraction. To this end we introduce the new abstraction of *nominal groups* which allows us to argue about correctness and security of our schemes over all above-mentioned elliptic curve groups, including Curve25519 and Curve448. (We believe this abstraction has applications well beyond its use in this work.) Ultimately, this approach results in both an additional security loss and the explicit consideration of (potential) new attacks not present for generic groups. In particular, both Curve25519 and Curve448 exhibit similar (but different) idiosyncrasies such as having non-equal but functionally equivalent curve points as well as self-reducibility with non-zero error probability, all of which we take into account in our reductions to the respective underlying assumption.

1.2 Proof Techniques

The results in this work have been demonstrated using a combination of traditional "pen-and-paper" techniques and the automated theorem proving tool CryptoVerif [13], which was already used to verify important practical protocols such as TLS 1.3 [12], Signal [22], and WireGuard [27]. CryptoVerif produces game-based proofs: it starts from an initial game provided by the user, which represents the protocol or scheme to prove; it transforms this game step by step using a predefined set of game transformations, until it reaches a game on which the desired security properties can easily be proved from the form of the game. The game transformations are guaranteed to produced computationally indistinguishable games, and either rely on a proof by reduction to a computational assumption or are syntactic transformations (e.g. replace a variable with its value). Using CryptoVerif to prove statements can result in greater confidence in their correctness, especially when the proofs require deriving (otherwise quite tedious) exact bounds on the security loss and/or reasoning about relatively complicated, e.g. multi-instance, security games.

However, CryptoVerif also has its limitations. Fortunately, these can be readily overcome using traditional techniques. The language used to define security statements in CryptoVerif is rather unconventional in the context of cryptography, not to mention (necessarily) very formal and detailed. Together this can make it quite challenging to build an intuitive understanding for a given notion (e.g. to verify that it captures the desired setting). To circumvent this, we present each of our security definitions using the more well-known language of game-based security. Next we map these to corresponding CryptoVerif definitions. Thus, the intuition can be built upon a game-based notion and it remains only to verify the *functional equivalence* of the CryptoVerif instantiation.

CryptoVerif was designed with multi-instance security in mind and so relies on more unconventional multi-instance number theoretic assumptions. However, the simpler a definition (say, for a KEM) the easier it is to demonstrate for a given construction. Similarly, in cryptography we tend to prefer simpler, static, not to mention well-known, number theoretic assumptions so as to build more confidence in them. Consequently, we have augmented the automated proofs with further pen-and-paper proofs reducing multi-instance security notions and assumptions to simpler (and more conventional) single-instance versions.

1.3 Related Work

Hybrid cryptography (of which the AKEM/DEM construction in this work is an example) is a widely used technique for constructing practically efficient asymmetric primitives. In particular, there exist several hybrid PKE-based concrete standards predating HPKE, mostly based on the DHIES scheme of [1] defined over a generic (discrete log) group. When the group is instantiated using elliptic curves the result is often referred to as ECIES (much like the Diffie-Hellman scheme over an elliptic curve group is referred to as ECDH). A description and comparison of the most important such standards can be found in [20]. However, per the HPKE RFC, "All these existing schemes have problems, e.g., because they rely on outdated primitives, lack proofs of IND-CCA2 security, or fail to provide test vectors." Moreover, to the best of our knowledge, none of these standards provide a means for authenticating senders.

The APKE primitive we analyse in this paper can be viewed as a flavour of signcryption [31]; a family of primitives intended to efficiently combine signatures and public key encryption. Signcryption literature is substantial and we refer to the textbook [18] for an extensive exposition thereof. We highlight some chapters of particular relevance. Chapters 2 and 3 cover 2-party and multi-party security notions, respectively; both for insider and outsider variants. Chapter 4 of [18] contains several (Gap)-Diffie-Hellman-based signcryption constructions. Finally, Chapter 7 covers some AKEM security notions and constructions (aka. "signcryption KEM") as well as hybrid signcryption constructions such as the outsider-secure one of [17] and insider-secure one of [16]. In contrast to our work, almost all security notions in [18] forbid honest parties from reusing the same key pair for both sending and receiving (even if sender and receiver keys have identical distribution).[2] Nor is it clear that a scheme satisfying a "key-separated" security notion could be converted into an equally efficient scheme supporting key reuse. The naïve transformation (embedding a sender and receiver key pair into a single reusable key pair) would double key sizes. However, an HPKE public key consists of a *single* group element which can be used simultaneously as a sender and receiver public key.

Recently, Bellare and Stepanovs analysed the signcryption scheme underlying the iMessage secure messaging protocol [9]. Although their security notions allow for key reuse as in our work, they fall outside the outsider/insider taxonomy common in signcryption literature. Instead, they capture an intermediary variant more akin to KCI security.

A detailed model of Curve25519 [25] in CryptoVerif was already presented in [27]; such a model was needed for the proof of the WireGuard protocol. In this paper, we present a more generic model that allows us to deal not only with Curve25519 but also with prime order groups such as NIST curves [28] in a single model. Moreover, we handle rerandomisation of curve elements, which was not taken into account in [27].

A very preliminary version of this work analyses HPKE as a single protocol, not in a modular KEM/DEM setting [26]. The proven theorems are less strong than the ones in this work, e.g. the adversary cannot choose secret keys but only compromise them. However, the analysis covers the single-shot encryption form of all four modes including the secret export API.

2 Preliminaries

Sets and Algorithms. We write $h \stackrel{\$}{\leftarrow} \mathcal{S}$ to denote that the variable h is uniformly sampled from the finite set $\mathcal{S}$. For integers $N, M \in \mathbb{N}$, we define $[N, M] := \{N, N+1, \ldots, M\}$ (which is the empty set for $M < N$), $[N] := [1, N]$ and $[N]_0 := [0, N]$. The statistical distance between two random variables U and V having a common domain $\mathcal{U}$ is defined as $\Delta[U, V] = \sum_{u \in \mathcal{U}} |\Pr[U = u] - \Pr[V = u]|$. The notation $[\![B]\!]$, where B is a boolean statement, evaluates to 1 if the statement is true and 0 otherwise.

[2] The only exception we are aware of are the security notions used to analyse 2 bilinear-pairing-based schemes in Sections 5.5 and 5.6 of [18].

We use uppercase letters $\mathcal{A}, \mathcal{B}$ to denote algorithms. Unless otherwise stated, algorithms are probabilistic, and we write $(y_1, \ldots) \xleftarrow{\$} \mathcal{A}(x_1, \ldots)$ to denote that $\mathcal{A}$ returns $(y_1, \ldots)$ when run on input $(x_1, \ldots)$. We write $\mathcal{A}^{\mathcal{B}}$ to denote that $\mathcal{A}$ has oracle access to $\mathcal{B}$ during its execution. For a randomised algorithm $\mathcal{A}$, we use the notation $y \in \mathcal{A}(x)$ to denote that y is a possible output of $\mathcal{A}$ on input x. We denote the running time of an algorithm $\mathcal{A}$ by $t_{\mathcal{A}}$.

SECURITY GAMES. We use standard code-based security games [8]. A *game* $\mathbf{G}$ is a probability experiment in which an adversary $\mathcal{A}$ interacts with an implicit challenger that answers oracle queries issued by $\mathcal{A}$. The game $\mathbf{G}$ has one *main procedure* and an arbitrary amount of additional *oracle procedures* which describe how these oracle queries are answered. We denote the (binary) output b of game $\mathbf{G}$ between a challenger and an adversary $\mathcal{A}$ as $\mathbf{G}^{\mathcal{A}} \Rightarrow b$. $\mathcal{A}$ is said to *win* $\mathbf{G}$ if $\mathbf{G}^{\mathcal{A}} \Rightarrow 1$. Unless otherwise stated, the randomness in the probability term $\Pr[\mathbf{G}^{\mathcal{A}} \Rightarrow 1]$ is over all the random coins in game $\mathbf{G}$.

3 Elliptic Curves

In this section we introduce the elliptic curves relevant for the HPKE standard, P-256, P-384, P-521 [28], Curve25519 and Curve448 [25], together with relevant security assumptions.

3.1 Nominal Groups

We first define *nominal groups*, a general abstract model of elliptic curves, and then show how we instantiate it for each of the above-mentioned curves.

Definition 1. *A nominal group $\mathcal{N} = (\mathcal{G}, g, p, \mathcal{E}_H, \mathsf{exp})$ consists of an efficiently recognizable finite set of elements $\mathcal{G}$ (also called "group elements"), a base element $g \in \mathcal{G}$, a prime p, a finite set of honest exponents $\mathcal{E}_H \subset \mathbb{Z}$, and an efficiently computable exponentiation function $\mathsf{exp} : \mathcal{G} \times \mathbb{Z} \to \mathcal{G}$, where we write X^y for $\mathsf{exp}(X, y)$. The exponentiation function is required to have the following properties:*

(1) $(X^y)^z = X^{yz}$ for all $X \in \mathcal{G}$, $y, z \in \mathbb{Z}$
(2) $g^{x+py} = g^x$ for all $x, y \in \mathbb{Z}$.

We remark that even though $\mathcal{G}$ is called the set of (group) elements, it is not required to form a group.

For a nominal group $\mathcal{N} = (\mathcal{G}, g, p, \mathcal{E}_H, \mathsf{exp})$ we let G_H be the distribution of honestly generated elements, that is, the distribution of g^x with $x \xleftarrow{\$} \mathcal{E}_H$. Let G_U be the distribution of g^x with $x \xleftarrow{\$} [1, p-1]$. Depending on the choice of $\mathcal{E}_H$, these distributions may differ. We define the two statistical parameters

$$\Delta_{\mathcal{N}} := \Delta[\mathrm{G}_H, \mathrm{G}_U], \quad \text{and} \quad P_{\mathcal{N}} = \max_{Y \in \mathcal{G}} \Pr_{x \xleftarrow{\$} \mathcal{E}_H} [Y = g^x] .$$

We summarise the expected security level and the concrete upper bounds for $\Delta_{\mathcal{N}}$ and $P_{\mathcal{N}}$ in Table 2 of Sect. 5.3 and compute them below.

Prime-Order Groups. The simplest example of a nominal group is when $\mathcal{G} = \mathbb{G}$ is a prime-order group with generator g, exp is defined via the usual scalar multiplication on $\mathbb{G}$, and $\mathcal{E}_H = [1, p-1]$. The two distributions G_H and G_U are identical, so $\Delta_{\mathcal{N}} = 0$. Since all elements have the same probability, we have $P_{\mathcal{N}} = 1/(p-1)$. The NIST curves P-256, P-384, and P-521 [28] are examples of prime-order groups.

Curve25519 and Curve448. We now show that Curve25519 and Curve448 [25] can also be seen as nominal groups. They are elliptic curves defined by equations of the form $Y^2 = X^3 + AX^2 + X$ in the field $\mathbb{F}_q$ for a large prime q. The curve points are represented only by their X coordinate. When $X^3 + AX^2 + X$ is a square Y^2, X represents the curve point (X, Y) or $(X, -Y)$. When X^3+AX^2+X is not a square, X does not represent a point on the curve, but on its quadratic twist. The curve is a group of cardinal kp and the twist is a group of cardinal $k'p'$, where p and p' are large primes and k and k' are small integers. For Curve25519, $q = 2^{255} - 19$, $k = 8$, $k' = 4$, $p = 2^{252} + \delta$, $p' = 2^{253} - 9 - 2\delta$ with $0 < \delta < 2^{125}$. For Curve448, $q = 2^{448} - 2^{224} - 1$, $k = k' = 4$, $p = 2^{446} - 2^{223} - \delta$, $p' = 2^{446} + \delta$ with $0 < \delta < 2^{220}$. The base point Q_0 is an element of the curve, of order p, which generates a subgroup $\mathbb{G}_s$ of the curve. The set of elements $\mathcal{G}$ is the set of bitstrings of 32 bytes for Curve25519, of 56 bytes for Curve448.

The exponentiation function is specified as follows, using [11, Theorem 2.1]: We consider the elliptic curve $E(\mathbb{F}_{q^2})$ defined by the equation $Y^2 = X^3+AX^2+X$ in a quadratic extension $\mathbb{F}_{q^2}$ of $\mathbb{F}_q$. We define $X_0 : E(\mathbb{F}_{q^2}) \to \mathbb{F}_{q^2}$ by $X_0(\infty) = 0$ and $X_0(X, Y) = X$. For $X \in \mathbb{F}_q$ and y an integer, we define $y \cdot X \in \mathbb{F}_q$ as $y \cdot X = X_0(yQ_X)$, where $Q_X \in E(\mathbb{F}_{q^2})$ is any of the two elements satisfying $X_0(Q_X) = X$. (It is not hard to verify that this mapping is well-defined.) Elements in $\mathcal{G}$ are mapped to elements of $\mathbb{F}_q$ by the function $\mathsf{decode\_pk} : \mathcal{G} \to \mathbb{F}_q$ and conversely, elements of $\mathbb{F}_q$ are mapped to the group elements by the function $\mathsf{encode\_pk} : \mathbb{F}_q \to \mathcal{G}$, such that $\mathsf{decode\_pk} \circ \mathsf{encode\_pk}$ is the identity. (For Curve25519 we have $\mathsf{decode\_pk}(X) = (X \bmod 2^{255}) \bmod q$, for Curve448 $\mathsf{decode\_pk}(X) = X \bmod q$, and $\mathsf{encode\_pk}(X)$ is the representation of X as an element of $\{0, \dots, q-1\}$.) Finally, $X^y = \mathsf{encode\_pk}(y \cdot \mathsf{decode\_pk}(X))$.

As required by Definition 1, we have $(X^y)^z = X^{yz}$. Indeed,

$$\begin{aligned}(X^y)^z &= \mathsf{encode\_pk}(z \cdot \mathsf{decode\_pk}(\mathsf{encode\_pk}(y \cdot \mathsf{decode\_pk}(X)))) \\ &= \mathsf{encode\_pk}(z \cdot y \cdot \mathsf{decode\_pk}(X)) \\ &= \mathsf{encode\_pk}(yz \cdot \mathsf{decode\_pk}(X)) = X^{yz}.\end{aligned}$$

The base element is $g = \mathsf{encode\_pk}(X_0(Q_0))$. It is easy to check that $g^{x+py} = g^x$, since Q_0 is an element of order p. The honest exponents are chosen uniformly in the set $\mathcal{E}_H = \{kn \mid n \in [M, N]\}$. For Curve25519, $M = 2^{251}$, $N = 2^{252} - 1$. For Curve448, $M = 2^{445}$, $N = 2^{446} - 1$.

Our exponentiation function is closely related to the function X25519 (resp. X448 for Curve448) as defined in [25], namely $\mathsf{X25519}(y, X) = X^{\mathsf{clamp}(y)}$, where $\mathsf{clamp}(y)$ sets and resets some bits in the bitstring y to make sure that $\mathsf{clamp}(y) \in \mathcal{E}_H$. Instead of clamping secret keys together with exponentiation, we clamp them when we generate them, hence we generate honest secret keys in $\mathcal{E}_H$.

The proof of the following Lemma 1 is in the long version [3].

Lemma 1. *For Curve25519, $\Delta_{\mathcal{N}} < 2^{-125}$ and $P_{\mathcal{N}} = 2^{-250}$, and for Curve448, $\Delta_{\mathcal{N}} < 2^{-220}$ and $P_{\mathcal{N}} = 2^{-444}$.*

3.2 Diffie-Hellman Assumptions

Let us first recall the Gap Diffie-Hellman and Square Gap Diffie-Hellman assumptions. We adapt them to the setting of a nominal group $\mathcal{N} = (\mathcal{G}, g, p, \mathcal{E}_H, \mathsf{exp})$ of the previous section, by allowing elements in $\mathcal{G}$ as arguments of the Diffie-Hellman decision oracle. Moreover, we still choose secret keys in $[1, p-1]$, not in $\mathcal{E}_H$, as it guarantees that the secret key p, or equivalently 0, is never chosen, which helps in the following theorems.

Definition 2 (Gap Diffie-Hellman (GDH) Problem). *We define the advantage function of an adversary $\mathcal{A}$ against the Gap Diffie-Hellman problem over nominal group $\mathcal{N}$ as*

$$\mathsf{Adv}^{\mathsf{GDH}}_{\mathcal{A},\mathcal{N}} := \Pr_{x,y \xleftarrow{\$} [1,p-1]} [Z = g^{xy} \mid Z \xleftarrow{\$} \mathcal{A}^{\mathrm{DH}}(g^x, g^y)]$$

where DH *is a decision oracle that on input $(g^{\hat{x}}, Y, Z)$, with $Y, Z \in \mathcal{G}$, returns* 1 *iff $Y^{\hat{x}} = Z$ and* 0 *otherwise.*

Definition 3 (Square Gap Diffie-Hellman (sqGDH) Problem). *We define the advantage function of an adversary $\mathcal{A}$ against the Square Gap Diffie-Hellman problem over nominal group $\mathcal{N}$ as*

$$\mathsf{Adv}^{\mathsf{sqGDH}}_{\mathcal{A},\mathcal{N}} := \Pr_{x \xleftarrow{\$} [1,p-1]} \left[Z = g^{x^2} \mid Z \xleftarrow{\$} \mathcal{A}^{\mathrm{DH}}(g^x)\right]$$

where DH *is a decision oracle that on input $(g^{\hat{x}}, Y, Z)$, with $Y, Z \in \mathcal{G}$, returns* 1 *iff $Y^{\hat{x}} = Z$ and* 0 *otherwise.*

CryptoVerif cannot use cryptographic assumptions directly in this form: it requires assumptions to be formulated as computational indistinguishability axioms between a left game G_ℓ and a right game G_r. In order to use such assumptions, it automatically recognizes when a game corresponds to an adversary interacting with G_ℓ, and it replaces G_ℓ with G_r in that game. Moreover, CryptoVerif requires the games G_ℓ and G_r to be formulated in a multi-key setting. That allows CryptoVerif to apply the assumption directly in case the scheme is used with several keys, without having to do a hybrid argument itself. (CryptoVerif infers the multi-key assumption automatically from a single-key assumption only in very simple cases.) Therefore, we reformulate the Gap Diffie-Hellman assumption to satisfy these requirements, and prove that our formulation is implied by the standard assumption.

We also take into account at this point that secret keys are actually chosen in $\mathcal{E}_H$ rather than in $[1, p-1]$.

Definition 4 (Left-or-Right (n, m)-Gap Diffie-Hellman Problem). *We define the advantage function of an adversary $\mathcal{A}$ against the left-or-right (n, m)-Gap Diffie-Hellman problem over nominal group $\mathcal{N}$ as*

$$\mathsf{Adv}^{\mathsf{LoR}\text{-}(n,m)\text{-}\mathsf{GDH}}_{\mathcal{A},\mathcal{N}} := \Bigg| \Pr_{\substack{\forall i \in [n]:\, x_i \xleftarrow{\$} \mathcal{E}_H \\ \forall j \in [m]:\, y_j \xleftarrow{\$} \mathcal{E}_H}} \left[\mathcal{A}^{\mathrm{DH}_\ell, \mathrm{DH}_0}(g^{x_1}, \ldots, g^{x_n}, g^{y_1}, \ldots, g^{y_m}) \Rightarrow 1\right] - \Pr_{\substack{\forall i \in [n]:\, x_i \xleftarrow{\$} \mathcal{E}_H \\ \forall j \in [m]:\, y_j \xleftarrow{\$} \mathcal{E}_H}} \left[\mathcal{A}^{\mathrm{DH}_r, \mathrm{DH}_0}(g^{x_1}, \ldots, g^{x_n}, g^{y_1}, \ldots, g^{y_m}) \Rightarrow 1\right] \Bigg| ,$$

where DH_0 *is a decision oracle that on input* $(g^{\hat{x}}, Y, Z)$ *returns* 1 *iff* $Y^{\hat{x}} = Z$ *and* 0 *otherwise;* DH_ℓ *is a decision oracle that on input* (i, j, Z) *for* $i \in [n], j \in [m]$ *returns* 1 *iff* $Z = g^{x_i y_j}$ *and* 0 *otherwise; and* DH_r *is an oracle that on input* (i, j, Z) *for* $i \in [n], j \in [m]$ *always returns* 0.

Definition 5 (Left-or-Right n-Square Gap Diffie-Hellman Problem). *We define the advantage function of an adversary $\mathcal{A}$ against the left-or-right n-Square Gap Diffie-Hellman problem over nominal group $\mathcal{N}$ as*

$$\mathsf{Adv}^{\mathsf{LoR}\text{-}n\text{-}\mathsf{sqGDH}}_{\mathcal{A},\mathcal{N}} := \Bigg| \Pr_{\forall i \in [n]:\, x_i \xleftarrow{\$} \mathcal{E}_H} \left[\mathcal{A}^{\mathrm{DH}_\ell, \mathrm{DH}_0}(g^{x_1} \ldots, g^{x_n}) \Rightarrow 1\right] - \Pr_{\forall i \in [n]:\, x_i \xleftarrow{\$} \mathcal{E}_H} \left[\mathcal{A}^{\mathrm{DH}_r, \mathrm{DH}_0}(g^{x_1}, \ldots, g^{x_n}) \Rightarrow 1\right] \Bigg| ,$$

where DH_0 *is a decision oracle that on input* $(g^{\hat{x}}, Y, Z)$ *returns* 1 *iff* $Y^{\hat{x}} = Z$ *and* 0 *otherwise;* DH_ℓ *is a decision oracle that on input* (i, j, Z) *for* $i, j \in [n]$ *returns* 1 *iff* $Z = g^{x_i x_j}$ *and* 0 *otherwise; and* DH_r *is an oracle that on input* (i, j, Z) *for* $i, j \in [n]$ *always returns* 0.

The proofs of Theorems 1 and 2 are in the long version [3].

Theorem 1 (GDH $\Rightarrow$ LoR-(n, m)-GDH). *For any adversary $\mathcal{A}$ against* LoR-(n, m)-GDH, *there exists an adversary $\mathcal{B}$ against* GDH *such that*

$$\mathsf{Adv}^{\mathsf{LoR}\text{-}(n,m)\text{-}\mathsf{GDH}}_{\mathcal{A},\mathcal{N}} \leq \mathsf{Adv}^{\mathsf{GDH}}_{\mathcal{B},\mathcal{N}} + (n + m)\Delta_{\mathcal{N}} ,$$

$\mathcal{B}$ queries the DH *oracle as many times as $\mathcal{A}$ queries* DH_0, DH_ℓ, *or* DH_r, *and* $t_{\mathcal{B}} \approx t_{\mathcal{A}}$.

Theorem 2 (sqGDH $\Rightarrow$ LoR-n-sqGDH). *For any adversary $\mathcal{A}$ against* LoR-n-sqGDH, *there exists an adversary $\mathcal{B}$ against* sqGDH *such that*

$$\mathsf{Adv}^{\mathsf{LoR}\text{-}n\text{-}\mathsf{sqGDH}}_{\mathcal{A},\mathcal{N}} \leq \mathsf{Adv}^{\mathsf{sqGDH}}_{\mathcal{B},\mathcal{N}} + n\Delta_{\mathcal{N}} ,$$

$\mathcal{B}$ queries the DH *oracle as many times as $\mathcal{A}$ queries* DH_0, DH_ℓ, *or* DH_r, *and* $t_{\mathcal{B}} \approx t_{\mathcal{A}}$.

In these theorems, the terms in $\Delta_{\mathcal{N}} = \Delta[\mathrm{G}_H, \mathrm{G}_U]$ come from the rerandomisation of keys, which yields keys distributed according to G_U, while the adversary expects keys distributed according to G_H. (Choosing secret keys in $\mathcal{E}_H$ in Definitions 2 and 3 would not avoid this term.)

Implementation in CryptoVerif. Definitions in this style for many cryptographic primitives are included in a standard library of cryptographic assumptions in CryptoVerif. As a matter of fact, this library includes a more general variant of the Gap Diffie-Hellman assumption, with corruption oracles and with a decision oracle $\mathrm{DH}(g, X, Y, Z)$, which allows the adversary to choose g. In this paper, we use the definition above as it is sufficient for our proofs.

4 Authenticated Key Encapsulation and Public Key Encryption

In Sect. 4.1, we introduce notation and security notions for an authenticated key encapsulation mechanism (AKEM), namely Outsider-CCA, Insider-CCA and Outsider-Auth. In Sect. 4.2, we introduce notation and security notions for authenticated public key encryption (APKE) which follow the ideas of the notions defined for AKEM. Additionally, we define Insider-Auth security.

In Sect. 4.3, we show how to construct an APKE scheme which achieves Outsider-CCA, Insider-CCA and Outsider-Auth, from an AKEM, a pseudo-random function (PRF), and a nonce-based authenticated encryption with associated data (AEAD) scheme. For Insider-Auth, we give a concrete attack in Sect. 4.4.

4.1 Authenticated Key Encapsulation Mechanism

Definition 6 (AKEM). *An authenticated key encapsulation mechanism* AKEM *consists of three algorithms:*

- Gen *outputs a key pair* (sk, pk)*, where* pk *defines a key space* $\mathcal{K}$.
- AuthEncap *takes as input a (sender) secret key* sk *and a (receiver) public key* pk*, and outputs an encapsulation* c *and a shared secret* $K \in \mathcal{K}$.
- *Deterministic* AuthDecap *takes as input a (receiver) secret key* sk*, a (sender) public key* pk*, and an encapsulation* c*, and outputs a shared key* $K \in \mathcal{K}$.

We require that for all $(sk_1, pk_1) \in \mathsf{Gen}$, $(sk_2, pk_2) \in \mathsf{Gen}$,

$$\Pr_{(c,K) \xleftarrow{\$} \mathsf{AuthEncap}(sk_1, pk_2)} [\mathsf{AuthDecap}(sk_2, pk_1, c) = K] = 1 .$$

The two sets of secret and public keys, $\mathcal{SK}$ and $\mathcal{PK}$, are defined via the support of the Gen algorithm as $\mathcal{SK} := \{sk \mid (sk, pk) \in \mathsf{Gen}\}$ and $\mathcal{PK} := \{pk \mid (sk, pk) \in \mathsf{Gen}\}$. We assume that there exists a projection function $\mu : \mathcal{SK} \to \mathcal{PK}$, such that for all $(sk, pk) \in \mathsf{Gen}$ it holds that $\mu(sk) = pk$. Note that such a function exists without loss of generality by defining sk to be the randomness rnd used in the key generation.

Finally, the key collision probability P_{AKEM} of AKEM is defined as

$$P_{\mathsf{AKEM}} := \max_{pk \in \mathcal{PK}} \Pr_{(sk',pk') \xleftarrow{\$} \mathsf{Gen}} [pk = pk'] .$$

Privacy. We define the games (n, q_e, q_d)-Outsider-CCA$_\ell$ and (n, q_e, q_d)-Outsider-CCA$_r$ in Listing 1 and the games (n, q_e, q_d, q_c)-Insider-CCA$_\ell$ and (n, q_e, q_d, q_c)-Insider-CCA$_r$ in Listing 2. The games follow the left-or-right style, as CryptoVerif requires this for assumptions, and we use these notions as assumptions in the composition theorems. In the long version [3, Appendix B], we compare the code-based game syntax with the CryptoVerif syntax for Outsider-CCA.

In all games, we generate key pairs for n users and run the adversary on the public keys. In the Outsider-CCA games, the adversary has access to oracles AEncap and ADecap. AEncap takes as input an index specifying a sender, as well as an arbitrary public key specifying a receiver, and returns a ciphertext and a KEM key. In the left game Outsider-CCA$_\ell$, AEncap always returns the real KEM key. In the right game Outsider-CCA$_r$, it outputs a uniformly random key if the receiver public key was generated by the experiment. This models the adversary as an outsider and ensures that target ciphertexts from an honest sender to an honest receiver are secure, i. e. do not leak any information about the shared key. Queries to ADecap, where the adversary specifies an index for a receiver public key, an arbitrary sender public key and a ciphertext, output a KEM key. In the Outsider-CCA$_r$ game, the output is kept consistent with the output of AEncap.

In the Insider-CCA games, there is an additional challenge oracle Chall. The adversary gives an index specifying the receiver and the secret key of the sender, thus taking the role of an insider. Chall will then output the real KEM key in the Insider-CCA$_\ell$ game, and a uniformly random key in the Insider-CCA$_r$ game. Thus, even if the target ciphertext was produced with a bad sender secret key (and honest receiver public key), the KEM key should be indistinguishable from a random key. AEncap will always output the real key and the output of ADecap is kept consistent with challenges.

In all games, the adversary makes at most q_e queries to oracle AEncap and at most q_d queries to oracle ADecap. In the Insider-CCA experiment, it can additionally make at most q_c queries to oracle Chall. We define the advantage of an adversary $\mathcal{A}$ as

$$\begin{aligned}
\mathsf{Adv}^{(n,q_e,q_d)\text{-Outsider-CCA}}_{\mathcal{A},\mathsf{AKEM}} &:= \Big| \Pr[(n, q_e, q_d)\text{-Outsider-CCA}_\ell(\mathcal{A}) \Rightarrow 1] \\
&\quad - \Pr[(n, q_e, q_d)\text{-Outsider-CCA}_r(\mathcal{A}) \Rightarrow 1] \Big| , \\
\mathsf{Adv}^{(n,q_e,q_d,q_c)\text{-Insider-CCA}}_{\mathcal{A},\mathsf{AKEM}} &:= \Big| \Pr[(n, q_e, q_d, q_c)\text{-Insider-CCA}_\ell(\mathcal{A}) \Rightarrow 1] \\
&\quad - \Pr[(n, q_e, q_d, q_c)\text{-Insider-CCA}_r(\mathcal{A}) \Rightarrow 1] \Big| .
\end{aligned}$$

Authenticity. Furthermore, we define the games (n, q_e, q_d)-Outsider-Auth$_\ell$ and (n, q_e, q_d)-Outsider-Auth$_r$ in Listing 3.

Listing 1: Games (n, q_e, q_d)-Outsider-CCA$_\ell$ and (n, q_e, q_d)-Outsider-CCA$_r$ for AKEM. Adversary $\mathcal{A}$ makes at most q_e queries to AENCAP and at most q_d queries to ADECAP.

```
(n, q_e, q_d)-Outsider-CCA_ℓ and [(n, q_e, q_d)-Outsider-CCA_r]
01 for i ∈ [n]
02     (sk_i, pk_i) <-$ Gen
03 E <- ∅
04 b <-$ A^{AENCAP,ADECAP}(pk_1, ..., pk_n)
05 return b

Oracle AENCAP(i ∈ [n], pk)
06 (c, K) <-$ AuthEncap(sk_i, pk)
[07 if pk ∈ {pk_1, ..., pk_n}
[08     K <-$ K
[09     E <- E ∪ {(pk_i, pk, c, K)}
10 return (c, K)

Oracle ADECAP(j ∈ [n], pk, c)
[11 if ∃K : (pk, pk_j, c, K) ∈ E
[12     return K
13 K <- AuthDecap(sk_j, pk, c)
14 return K
```

Listing 2: Games (n, q_e, q_d, q_c)-Insider-CCA$_\ell$ and (n, q_e, q_d, q_c)-Insider-CCA$_r$ for AKEM. Adversary $\mathcal{A}$ makes at most q_e queries to AENCAP, at most q_d queries to ADECAP and at most q_c queries to CHALL.

```
(n, q_e, q_d, q_c)-Insider-CCA_ℓ and [(n, q_e, q_d, q_c)-Insider-CCA_r]
01 for i ∈ [n]
02     (sk_i, pk_i) <-$ Gen
03 E <- ∅
04 b <-$ A^{AENCAP,ADECAP,CHALL}(pk_1, ..., pk_n)
05 return b

Oracle CHALL(j ∈ [n], sk)
06 (c, K) <-$ AuthEncap(sk, pk_j)
[07 K <-$ K
[08 E <- E ∪ {(μ(sk), pk_j, c, K)}
09 return (c, K)

Oracle AENCAP(i ∈ [n], pk)
10 (c, K) <-$ AuthEncap(sk_i, pk)
11 return (c, K)

Oracle ADECAP(j ∈ [n], pk, c)
[12 if ∃K : (pk, pk_j, c, K) ∈ E
[13     return K
14 K <- AuthDecap(sk_j, pk, c)
15 return K
```

Listing 3: Games (n, q_e, q_d)-Outsider-Auth$_\ell$ and (n, q_e, q_d)-Outsider-Auth$_r$ for AKEM. Adversary $\mathcal{A}$ makes at most q_e queries to AENCAP and at most q_d queries to ADECAP.

```
(n, q_e, q_d)-Outsider-Auth_ℓ and [(n, q_e, q_d)-Outsider-Auth_r]
01 for i ∈ [n]
02     (sk_i, pk_i) <-$ Gen
03 E <- ∅
04 b <-$ A^{AENCAP,ADECAP}(pk_1, ..., pk_n)
05 return b

Oracle AENCAP(i ∈ [n], pk)
06 (c, K) <-$ AuthEncap(sk_i, pk)
07 E <- E ∪ {(pk_i, pk, c, K)}
08 return (c, K)

Oracle ADECAP(j ∈ [n], pk, c)
[09 if ∃K : (pk, pk_j, c, K) ∈ E
[10     return K
11 K <- AuthDecap(sk_j, pk, c)
[12 if pk ∈ {pk_1, ..., pk_n} and K ≠ ⊥
[13     K <-$ K
[14     E <- E ∪ {(pk, pk_j, c, K)}
15 return K
```

The adversary has access to oracles AEncap and ADecap. AEncap will always output the real KEM key. ADecap will output the real key in game $\mathsf{Outsider\text{-}Auth}_\ell$. In the $\mathsf{Outsider\text{-}Auth}_r$ game, the adversary (acting as an outsider) will receive a uniformly random key if the receiver public key was generated by the experiment. Thus, the adversary should not be able to distinguish the real KEM key from a random key for two honest users, even if it can come up with the target ciphertext.

The adversary makes at most q_e queries to oracle AEncap and at most q_d queries to oracle ADecap. We define the advantage of an adversary $\mathcal{A}$ as

$$\begin{aligned}\mathsf{Adv}^{(n,q_e,q_d)\text{-}\mathsf{Outsider\text{-}Auth}}_{\mathcal{A},\mathsf{AKEM}} := \big| &\Pr[(n,q_e,q_d)\text{-}\mathsf{Outsider\text{-}Auth}_\ell(\mathcal{A}) \Rightarrow 1] \\ &- \Pr[(n,q_e,q_d)\text{-}\mathsf{Outsider\text{-}Auth}_r(\mathcal{A}) \Rightarrow 1]\big| .\end{aligned}$$

In the long version [3, Appendix A], we provide simpler single-user or 2-user versions of these properties, and show that they non-tightly imply the definitions above. These results could be useful to simplify the proof for new AKEMs that could be added to HPKE, such as post-quantum AKEMs. However, because the reduction is not tight, a direct proof of multi-user security may yield better probability bounds. This is the case for our proof of DH-AKEM in Sect. 5.1.

4.2 Authenticated Public Key Encryption

Definition 7 (APKE). *An authenticated public key encryption scheme* APKE *consists of the following three algorithms:*

- Gen *outputs a key pair* (sk, pk).
- AuthEnc *takes as input a (sender) secret key* sk*, a (receiver) public key* pk*, a message* m*, associated data* aad*, a bitstring* $info$*, and outputs a ciphertext* c.
- *Deterministic* AuthDec *takes as input a (receiver) secret key* sk*, a (sender) public key* pk*, a ciphertext* c*, associated data* aad *and a bitstring* $info$*, and outputs a message* m.

We require that for all messages $m \in \{0,1\}^*, aad \in \{0,1\}^*, info \in \{0,1\}^*$,

$$\Pr_{\substack{(sk_S,pk_S)\xleftarrow{\$}\mathsf{Gen}\\(sk_R,pk_R)\xleftarrow{\$}\mathsf{Gen}}}\begin{bmatrix}c \leftarrow \mathsf{AuthEnc}(sk_S, pk_R, m, aad, info),\\ \mathsf{AuthDec}(sk_R, pk_S, c, aad, info) = m\end{bmatrix} = 1 .$$

Privacy. We define the games (n, q_e, q_d, q_c)-Outsider-CCA and (n, q_e, q_d, q_c)-Insider-CCA in Listing 4, which follow ideas similar to the games for outsider and insider-secure AKEM. The security notions for APKE use the common style where challenge queries are with respect to a random bit b. In particular, the additional challenge oracle Chall will encrypt either message m_0 or m_1 provided by the adversary, depending on b. Oracles AEnc and ADec will always encrypt and decrypt honestly (except for challenge ciphertexts).

Listing 4: Games (n, q_e, q_d, q_c)-Outsider-CCA and (n, q_e, q_d, q_c)-Insider-CCA for APKE, where (n, q_e, q_d, q_c)-Outsider-CCA uses oracle CHALL in the dashed box and (n, q_e, q_d, q_c)-Insider-CCA uses oracle CHALL in the solid box. Adversary $\mathcal{A}$ makes at most q_e queries to AENC, at most q_d queries to ADEC and at most q_c queries to CHALL.

```
(n, q_e, q_d, q_c)-Outsider-CCA and
(n, q_e, q_d, q_c)-Insider-CCA
01 for i ∈ [n]
02     (sk_i, pk_i) <-$ Gen
03 E <- ∅
04 b <-$ {0,1}
05 b' <-$ A^{AEnc,ADec,Chall}(pk_1, ..., pk_n)
06 return [[b = b']]

Oracle ADec(j ∈ [n], pk, c, aad, info)
07 if (pk, pk_j, c, aad, info) ∈ E
08     return ⊥
09 m <- AuthDec(sk_j, pk, c, aad, info)
10 return m

Oracle AEnc(i ∈ [n], pk, m, aad, info)
11 c <-$ AuthEnc(sk_i, pk, m, aad, info)
12 return c

Oracle Chall(i ∈ [n], j ∈ [n], m_0, m_1, aad, info)   [dashed box]
13 if |m_0| ≠ |m_1| return ⊥
14 c <-$ AuthEnc(sk_i, pk_j, m_b, aad, info)
15 E <- E ∪ {(pk_i, pk_j, c, aad, info)}
16 return c

Oracle Chall(j ∈ [n], sk, m_0, m_1, aad, info)   [solid box]
17 if |m_0| ≠ |m_1| return ⊥
18 c <-$ AuthEnc(sk, pk_j, m_b, aad, info)
19 E <- E ∪ {(μ(sk), pk_j, c, aad, info)}
20 return c
```

Listing 5: Games (n, q_e, q_d)-Outsider-Auth and (n, q_e, q_d)-Insider-Auth for APKE. Adversary $\mathcal{A}$ makes at most q_e queries to AENC and at most q_d queries to ADEC.

```
(n, q_e, q_d)-Outsider-Auth
01 for i ∈ [n]
02     (sk_i, pk_i) <-$ Gen
03 E <- ∅
04 (i*, j*, c*, aad*, info*) <-$ A^{AEnc,ADec}(pk_1, ..., pk_n)
05 return [[(pk_{i*}, pk_{j*}, c*, aad*, info*) ∉ E
       and AuthDec(sk_{j*}, pk_{i*}, c*, aad*, info*) ≠ ⊥]]

(n, q_e, q_d)-Insider-Auth
06 for i ∈ [n]
07     (sk_i, pk_i) <-$ Gen
08 E <- ∅
09 (i*, sk, c*, aad*, info*) <-$ A^{AEnc,ADec}(pk_1, ..., pk_n)
10 return [[(pk_{i*}, μ(sk), c*, aad*, info*) ∉ E
       and AuthDec(sk, pk_{i*}, c*, aad*, info*) ≠ ⊥]]

Oracle AEnc(i ∈ [n], pk, m, aad, info)
11 c <-$ AuthEnc(sk_i, pk, m, aad, info)
12 E <- E ∪ {(pk_i, pk, c, aad, info)}
13 return c

Oracle ADec(j ∈ [n], pk, c, aad, info)
14 m <- AuthDec(sk_j, pk, c, aad, info)
15 return m
```

Listing 6: Authenticated PKE scheme APKE[AKEM, KS, AEAD] construction from AKEM, KS and AEAD, where APKE.Gen = AKEM.Gen.

```
AuthEnc(sk, pk, m, aad, info)
01 (c_1, K) <-$ AuthEncap(sk, pk)
02 (k, nonce) <- KS(K, info)
03 c_2 <- AEAD.Enc(k, m, aad, nonce)
04 return (c_1, c_2)

AuthDec(sk, pk, (c_1, c_2), aad, info)
05 K <- AuthDecap(sk, pk, c_1)
06 (k, nonce) <- KS(K, info)
07 m <- AEAD.Dec(k, c_2, aad, nonce)
08 return m
```

In these games, the adversary $\mathcal{A}$ makes at most q_e queries to oracle AEnc, at most q_d queries to oracle ADec, and at most q_c queries to oracle Chall. The advantage of $\mathcal{A}$ is

$$\mathsf{Adv}^{(n,q_e,q_d,q_c)\text{-Outsider-CCA}}_{\mathcal{A},\mathsf{APKE}} := \left|\Pr[(n,q_e,q_d,q_c)\text{-}\mathsf{Outsider\text{-}CCA}(\mathcal{A}) \Rightarrow 1] - \frac{1}{2}\right| ,$$

$$\mathsf{Adv}^{(n,q_e,q_d,q_c)\text{-Insider-CCA}}_{\mathcal{A},\mathsf{APKE}} := \left|\Pr[(n,q_e,q_d,q_c)\text{-}\mathsf{Insider\text{-}CCA}(\mathcal{A}) \Rightarrow 1] - \frac{1}{2}\right| .$$

Authenticity. Furthermore, we define the games (n, q_e, q_d)-Outsider-Auth and (n, q_e, q_d)-Insider-Auth in Listing 5. The adversary has access to an encryption and decryption oracle and has to come up with a new tuple of ciphertext, associated data and info for any honest receiver secret key (Outsider-Auth) or any (possibly leaked or bad) receiver secret key (Insider-Auth), provided that the sender public key is honest.

In these games, adversary $\mathcal{A}$ makes at most q_e queries to oracle AEnc and at most q_d queries to oracle ADec. The advantage of $\mathcal{A}$ is defined as

$$\mathsf{Adv}^{(n,q_e,q_d)\text{-Outsider-Auth}}_{\mathcal{A},\mathsf{APKE}} := \Pr[(n,q_e,q_d)\text{-}\mathsf{Outsider\text{-}Auth}(\mathcal{A}) \Rightarrow 1] ,$$

$$\mathsf{Adv}^{(n,q_e,q_d)\text{-Insider-Auth}}_{\mathcal{A},\mathsf{APKE}} := \Pr[(n,q_e,q_d)\text{-}\mathsf{Insider\text{-}Auth}(\mathcal{A}) \Rightarrow 1] .$$

4.3 From AKEM to APKE

In this section we define and analyse a general transformation that models HPKE's way of constructing APKE from an AKEM (c.f. Definition 6) and an AEAD (c.f. [3, Section 3]). It also uses a so-called *key schedule* KS which we model as a keyed function $\mathsf{KS} : \mathcal{K} \times \{0,1\}^* \rightarrow \{0,1\}^*$, where $\mathcal{K}$ matches the AKEM's key space. KS outputs an AEAD key k and an initialisation vector *nonce* (called *base nonce* in the RFC) from which the AEAD's nonces are computed. (The key schedule defined in the HPKE standard also outputs an additional key called *exporter secret* that can be used to derive keys for use by arbitrary higher-level applications. This export API is not part of the single-shot encryption API that we are analysing, and thus we omit it in our definitions.) Listing 6 gives the formal specification of APKE built from AKEM, KS and AEAD.

We observe that in the single-shot encryption API, every AEAD key k is used to produce exactly one ciphertext, and thus is only used with one nonce. In HPKE, messages are counted with a sequence number s starting at 0 and the nonce for a message is computed by $nonce \oplus s$. For the single-shot encryption API this means that the nonce is equal to the initialisation vector *nonce*. At the same time, this means that *nonce* is by definition unique.

We now give theorems stating the (n, q_e, q_d, q_c)-Outsider-CCA, (n, q_e, q_d)-Outsider-Auth and (n, q_e, q_d, q_c)-Insider-CCA security of APKE[AKEM, KS, AEAD] defined in Listing 6. Theorems 3 to 5 are proven using CryptoVerif version 2.04. This version includes an improvement in the computation of probability bounds that allows us to express these bounds as functions of the total numbers of queries to the AEnc, ADec, and Chall oracles instead of the number of users and the numbers of queries per user. The CryptoVerif input files are given in hpke.auth.outsider-cca.ocv, hpke.auth.insider-cca.ocv, and hpke.auth.outsider-auth.ocv [2]. These proofs are fairly straightforward. As an example, we prefer explaining the proof of Theorem 7 later, which is more interesting. In Sect. 4.4, we show that APKE[AKEM, KS, AEAD] cannot achieve Insider-Auth security.

As detailed in the long version [3, Section 3], we define a multi-key PRF security experiment (n_k, q_{PRF})-PRF with n_k keys, in which the adversary makes at most q_{PRF} queries for each key. We also define multi-key IND-CPA and INT-CTXT security experiments for the AEAD: n_k-IND-CPA and (n_k, q_d)-INT-CTXT, with n_k keys, in which the adversary makes at most one encryption query for each key and, for the INT-CTXT experiment, at most q_d decryption queries in total. In these experiments, the nonces of the AEAD are chosen randomly.

Theorem 3 (AKEM Outsider-CCA + KS PRF + AEAD IND-CPA + AEAD INT-CTXT ⇒ APKE Outsider-CCA). *For any (n, q_e, q_d, q_c)-Outsider-CCA adversary $\mathcal{A}$ against APKE[AKEM, KS, AEAD], there exist an $(n, q_e + q_c, q_d)$-Outsider-CCA adversary $\mathcal{B}$ against AKEM, an $(q_c, q_c + q_d)$-PRF adversary $\mathcal{C}$ against KS, an q_c-IND-CPA adversary $\mathcal{D}_1$ against AEAD and an (q_c, q_d)-INT-CTXT adversary $\mathcal{D}_2$ against AEAD such that $t_{\mathcal{B}} \approx t_{\mathcal{A}}$, $t_{\mathcal{C}} \approx t_{\mathcal{A}}$, $t_{\mathcal{D}_1} \approx t_{\mathcal{A}}$, $t_{\mathcal{D}_2} \approx t_{\mathcal{A}}$, and*

$$\begin{aligned}\mathsf{Adv}^{(n,q_e,q_d,q_c)\text{-}\mathsf{Outsider\text{-}CCA}}_{\mathcal{A},\mathsf{APKE[AKEM,KS,AEAD]}} \leq\ & 2 \cdot \mathsf{Adv}^{(n,q_e+q_c,q_d)\text{-}\mathsf{Outsider\text{-}CCA}}_{\mathcal{B},\mathsf{AKEM}} + 2 \cdot \mathsf{Adv}^{(q_c,q_c+q_d)\text{-}\mathsf{PRF}}_{\mathcal{C},\mathsf{KS}} \\ & + 2 \cdot \mathsf{Adv}^{q_c\text{-}\mathsf{IND\text{-}CPA}}_{\mathcal{D}_1,\mathsf{AEAD}} + 2 \cdot \mathsf{Adv}^{(q_c,q_d)\text{-}\mathsf{INT\text{-}CTXT}}_{\mathcal{D}_2,\mathsf{AEAD}} \\ & + 6n^2 \cdot P_{\mathsf{AKEM}}\ .\end{aligned}$$

Theorem 4 (AKEM Insider-CCA + KS PRF + AEAD IND-CPA + AEAD INT-CTXT ⇒ APKE Insider-CCA). *For any (n, q_e, q_d, q_c)-Insider-CCA adversary $\mathcal{A}$ against APKE[AKEM, KS, AEAD], there exist an (n, q_e, q_d, q_c)-Insider-CCA adversary $\mathcal{B}$ against AKEM, an $(q_c, q_c + q_d)$-PRF adversary $\mathcal{C}$ against KS, an q_c-IND-CPA adversary $\mathcal{D}_1$ against AEAD and an (q_c, q_d)-INT-CTXT adversary $\mathcal{D}_2$ against AEAD such that $t_{\mathcal{B}} \approx t_{\mathcal{A}}$, $t_{\mathcal{C}} \approx t_{\mathcal{A}}$, $t_{\mathcal{D}_1} \approx t_{\mathcal{A}}$, $t_{\mathcal{D}_2} \approx t_{\mathcal{A}}$, and*

$$\begin{aligned}\mathsf{Adv}^{(n,q_e,q_d,q_c)\text{-}\mathsf{Insider\text{-}CCA}}_{\mathcal{A},\mathsf{APKE[AKEM,KS,AEAD]}} \leq\ & 2 \cdot \mathsf{Adv}^{(n,q_e,q_d,q_c)\text{-}\mathsf{Insider\text{-}CCA}}_{\mathcal{B},\mathsf{AKEM}} + 2 \cdot \mathsf{Adv}^{(q_c,q_c+q_d)\text{-}\mathsf{PRF}}_{\mathcal{C},\mathsf{KS}} \\ & + 2 \cdot \mathsf{Adv}^{q_c\text{-}\mathsf{IND\text{-}CPA}}_{\mathcal{D}_1,\mathsf{AEAD}} + 2 \cdot \mathsf{Adv}^{(q_c,q_d)\text{-}\mathsf{INT\text{-}CTXT}}_{\mathcal{D}_2,\mathsf{AEAD}} \\ & + 6n^2 \cdot P_{\mathsf{AKEM}}\ .\end{aligned}$$

Theorem 5 (AKEM Outsider-CCA + AKEM Outsider-Auth + KS PRF + AEAD INT-CTXT ⇒ APKE Outsider-Auth). *For any (n, q_e, q_d)-Outsider-Auth adversary $\mathcal{A}$ against APKE[AKEM, KS, AEAD], there exist an $(n, q_e, q_d + 1)$-Outsider-CCA adversary $\mathcal{B}_1$ against AKEM, an $(n, q_e, q_d + 1)$-Outsider-Auth adversary $\mathcal{B}_2$ against AKEM, an $(q_e + q_d + 1, q_e + 2q_d + 1)$-PRF adversary $\mathcal{C}$ against KS, and an $(q_e + 3q_d + 3, 4q_d + 1)$-INT-CTXT adversary $\mathcal{D}$ against AEAD such that $t_{\mathcal{B}_1} \approx t_{\mathcal{A}}$, $t_{\mathcal{B}_2} \approx t_{\mathcal{A}}$, $t_{\mathcal{C}} \approx t_{\mathcal{A}}$, $t_{\mathcal{D}} \approx t_{\mathcal{A}}$, and*

$$\mathsf{Adv}^{(n,q_e,q_d)\text{-Outsider-Auth}}_{\mathcal{A},\mathsf{APKE[AKEM,KS,AEAD]}} \leq \mathsf{Adv}^{(n,q_e,q_d+1)\text{-Outsider-CCA}}_{\mathcal{B}_1,\mathsf{AKEM}} + \mathsf{Adv}^{(n,q_e,q_d+1)\text{-Outsider-Auth}}_{\mathcal{B}_2,\mathsf{AKEM}}$$
$$+ \mathsf{Adv}^{(q_e+q_d+1,q_e+2q_d+1)\text{-PRF}}_{\mathcal{C},\mathsf{KS}}$$
$$+ \mathsf{Adv}^{(q_e+3q_d+3,4q_d+1)\text{-INT-CTXT}}_{\mathcal{D},\mathsf{AEAD}} + n(q_e + 13n) \cdot P_{\mathsf{AKEM}} .$$

4.4 Infeasibility of Insider-Auth Security

For any AKEM, KS, and AEAD, the construction APKE[AKEM, KS, AEAD] given in Listing 6 is not (n, q_e, q_d)-Insider-Auth secure. The inherent reason for this construction to be vulnerable against this attack is that the KEM ciphertext does not depend on the message. Thus, the KEM ciphertext can be reused and the DEM ciphertext can be exchanged by the encryption of any other message.

Theorem 6. *There exists an efficient adversary $\mathcal{A}$ against (n, q_e, q_d)-*Insider-Auth *security of* APKE[AKEM, KS, AEAD] *such that*

$$\mathsf{Adv}^{(n,q_e,q_d)\text{-Insider-Auth}}_{\mathcal{A},\mathsf{APKE[AKEM,KS,AEAD]}} = 1 .$$

Proof. We construct adversary $\mathcal{A}$ in Listing 7. It takes as input n public keys and has oracle access to AEnc and ADec. It first generates a key pair (sk^*, pk^*) and queries the AEnc oracle on any index i^*, receiver public key pk^*, an arbitrary message m_1, as well as arbitrary associated data *aad* and string *info*.

Listing 7: Adversary $\mathcal{A}$ against (n, q_e, q_d)-Insider-Auth as defined in Listing 5, of APKE[AKEM, KS, AEAD].

```
Adversary A^{AEnc,ADec}(pk_1, ..., pk_n)
01 (sk*, pk*) ← AKEM.Gen
02 i* := 1; m_1 := aad := info := 1
03 (c_1, c_2) ← AEnc(i*, pk*, m_1, aad, info)
04 K ← AuthDecap(sk*, pk_{i*}, c_1)
05 (k, nonce) ← KS(K, info)
06 m_2 := 2
07 c'_2 ← AEAD.Enc(k, m_2, aad, nonce)
08 return (i*, sk*, (c_1, c'_2), aad, info)
```

The challenger computes $(c_1, K) \xleftarrow{\$} \mathsf{AuthEncap}(sk_{i^*}, pk^*)$, $(k, nonce) \leftarrow \mathsf{KS}(K, info)$ and $c_2 \leftarrow \mathsf{AEAD.Enc}(k, m_1, aad, nonce)$, and returns (c_1, c_2) to $\mathcal{A}$.

Since $\mathcal{A}$ knows the secret key sk^*, it is able to compute the underlying KEM key K using AuthDecap. Next, it computes $(k, nonce)$ and thus retrieves the key k used in the AEAD scheme. Finally, $\mathcal{A}$ encrypts any other message m_2 to ciphertext c'_2 and replaces the AEAD ciphertext c_2 with the new ciphertext. Since $(c_1, c_2) \neq (c_1, c'_2)$, the latter constitutes a valid forgery in the (n, q_e, q_d)-Insider-Auth security experiment. □

Listing 8: $\mathsf{DH\text{-}AKEM}[\mathcal{N}, \mathsf{KDF}] = (\mathsf{Gen}, \mathsf{AuthEncap}, \mathsf{AuthDecap})$ as defined in the RFC [5], constructed from a nominal group $\mathcal{N}$ and key derivation function $\mathsf{KDF} : \{0,1\}^* \rightarrow \mathcal{K}$, with $\mathcal{K} = \{0,1\}^N$.

```
Gen
01 sk <-$ E_H
02 pk <- g^sk
03 return (sk, pk)

ExtractAndExpand(dh, context)
04 IKM <- "HPKE-v1" || suite_id ||
          "eae_prk" || dh
05 info <- Encode(N) || "HPKE-v1" ||
          suite_id || "shared_secret" ||
          context
06 return KDF("", IKM, info)

AuthEncap(sk ∈ E_H, pk ∈ G)
07 (esk, epk) <-$ Gen
08 context <- (epk, pk, g^sk)
09 dh <- (pk^esk, pk^sk)
10 K <- ExtractAndExpand(dh, context)
11 return (epk, K)

AuthDecap(sk ∈ E_H, pk ∈ G, epk ∈ G)
12 context <- (epk, g^sk, pk)
13 dh <- (epk^sk, pk^sk)
14 return ExtractAndExpand(dh, context)
```

5 The HPKE Standard

In Sect. 5.1, we show how to construct HPKE's abstract AKEM construction DH-AKEM from a nominal group $\mathcal{N}$ and a key derivation function KDF. In Sect. 5.2, we define and analyse HPKE's specific key schedule $\mathsf{KS}_{\mathsf{Auth}}$ and key derivation function HKDF_N. Finally, in Sect. 5.3 we put everything together and obtain the HPKE standard in Auth mode from all previous sections.

5.1 HPKE's AKEM Construction DH-AKEM

In this section we present the RFC's instantiation of the AKEM definition, and prove that it satisfies the security notions defined earlier. Listing 8 shows the formal definition of $\mathsf{DH\text{-}AKEM}[\mathcal{N}, \mathsf{KDF}]$ relative to a nominal group $\mathcal{N}$ (c.f. Definition 1) and a key derivation function $\mathsf{KDF} : \{0,1\}^* \rightarrow \mathcal{K}$, where $\mathcal{K}$ is the key space. (The RFC uses a key space $\mathcal{K}$, consisting of bitstrings of length N, which corresponds to `Nsecret` in the RFC.) The construction also depends on the fixed-size protocol constants `"HPKE-v1"` and $suite_{id}$, where $suite_{id}$ identifies the KEM in use: it is a string `"KEM"` plus a two-byte identifier of the KEM algorithm. The bitstring $\mathsf{Encode}(N)$ is the two-byte encoding of the length N expressed in bytes. Correctness follows by property (1) of Definition 1. We make the implicit convention that AuthEncap and AuthDecap return reject ($\perp$) if their inputs are not of the right data type as specified in Listing 8.

We continue with statements about the (n, q_e, q_d)-Outsider-CCA, (n, q_e, q_d, q_c)-Insider-CCA, and (n, q_e, q_d)-Outsider-Auth security of $\mathsf{DH\text{-}AKEM}[\mathcal{N}, \mathsf{KDF}]$, modelling KDF as a random oracle. The proofs are written with CryptoVerif version 2.04; the input files are dhkem.auth.outsider-cca-lr.ocv, dhkem.auth.insider-cca-lr.ocv, and dhkem.auth.outsider-auth-lr.ocv [2]. We sketch the proof of one of the three theorems as an example, to help understand CryptoVerif's approach.

Our results hold for any nominal group, which covers the three NIST curves allowed by the RFC, as well as for the other two allowed curves, Curve25519 and Curve448. The bounds given in Theorems 7 to 9 depend on the probabilities $\Delta_{\mathcal{N}}$ and $P_{\mathcal{N}}$, which can be instantiated for these five different curves using the values indicated in Table 2 on Page 27.

At the end of this section, we sketch the attack against the Insider-Auth security.

Theorem 7 (**Outsider-CCA security of DH-AKEM**). *Under the* GDH *assumption in* $\mathcal{N}$ *and modelling* KDF *as a random oracle,* DH-AKEM[$\mathcal{N}$, KDF] *is* Outsider-CCA *secure. In particular, for any adversary* $\mathcal{A}$ *against* (n, q_e, q_d)-Outsider-CCA *security of* DH-AKEM[$\mathcal{N}$, KDF] *that issues at most* q_h *queries to the random oracle* KDF, *there exists an adversary* $\mathcal{B}$ *against* GDH *such that*

$$\mathsf{Adv}^{(n,q_e,q_d)\text{-Outsider-CCA}}_{\mathcal{A},\mathsf{DH\text{-}AKEM}[\mathcal{N},\mathsf{KDF}]} \leq \mathsf{Adv}^{\mathsf{GDH}}_{\mathcal{B},\mathcal{N}} + (n + q_e) \cdot \Delta_{\mathcal{N}} + (q_e q_d + 2nq_e + 7q_e^2 + 13n^2) \cdot P_{\mathcal{N}}$$

$\mathcal{B}$ *issues* $nq_e + nq_d + 2q_d q_h + 3nq_h$ *queries to the* DH *oracle, and* $t_{\mathcal{B}} \approx t_{\mathcal{A}}$.

Proof. This proof is mechanized using the tool CryptoVerif. We give to the tool the assumptions that $\mathcal{N}$ is a nominal group that satisfies the GDH assumption, formalized by Definition 4, and that KDF is a random oracle. We also give the definition of DH-AKEM, and ask it to show that the games (n, q_e, q_d)-Outsider-CCA$_\ell$ and (n, q_e, q_d)-Outsider-CCA$_r$ are computationally indistinguishable. In the particular case of DH-AKEM, these two games include an additional oracle: the random oracle KDF. The theorem, the initial game definitions, and the proof indications are available in the file dhkem.auth.outsider-cca-lr.ocv [2].

The proof proceeds by transforming the game (n, q_e, q_d)-Outsider-CCA$_\ell$ by several steps into a game G_{final} and the game (n, q_e, q_d)-Outsider-CCA$_r$ into the same game G_{final}. Since all transformation steps performed by CryptoVerif are designed to preserve computational indistinguishability, we obtain that (n, q_e, q_d)-Outsider-CCA$_\ell$ and (n, q_e, q_d)-Outsider-CCA$_r$ are computationally indistinguishable. We guide the transformations with the following main steps.

Starting from (n, q_e, q_d)-Outsider-CCA$_\ell$, in the oracle AEncap, we first distinguish whether the provided public key pk is honest, by testing whether $pk = pk_i$ for some i (a test that appears in (n, q_e, q_d)-Outsider-CCA$_r$). We rename some variables to give them different names when $pk \in \{pk_1, \dots, pk_n\}$ and when $pk \notin \{pk_1, \dots, pk_n\}$, to facilitate future game transformations. In the oracle ADecap, we test whether $\exists K : (pk, pk_j, c, K) \in \mathcal{E}$, which corresponds to a test done in (n, q_e, q_d)-Outsider-CCA$_r$. Furthermore, when this test succeeds, we replace the result normally returned by ADecap, AuthDecap(sk_j, pk, c) with the key K found in $\mathcal{E}$. CryptoVerif shows that this replacement does not modify the result, which corresponds to the correctness of DH-AKEM. In the random oracle, we distinguish whether the argument received from the adversary has a format that matches the one used by DH-AKEM or not. Only when the format matches, this argument may coincide with a call to the hash oracle made from DH-AKEM.

Next, we apply the random oracle assumption. Each call to the random oracle is replaced with the following test: if the argument is equal to the argument of a previous call, we return the previous result; otherwise, we return a fresh random value. Finally, we apply the GDH assumption, which allows us to show that some comparisons between Diffie-Hellman values are false. In particular, CryptoVerif shows that the arguments of calls to the random oracle coming from AENCAP with $pk \in \{pk_1, \dots, pk_n\}$ cannot coincide with arguments of other calls. Hence, they return a fresh random key, as in (n, q_e, q_d)-$\mathsf{Outsider\text{-}CCA}_r$.

Starting from (n, q_e, q_d)-$\mathsf{Outsider\text{-}CCA}_r$, in the random oracle, we distinguish whether the argument received from the adversary has a format that matches the one used by DH-AKEM or not. Next, we apply the random oracle assumption, as we did on the left-hand side.

The transformed games obtained respectively from (n, q_e, q_d)-$\mathsf{Outsider\text{-}CCA}_\ell$ and from (n, q_e, q_d)-$\mathsf{Outsider\text{-}CCA}_r$ are then equal, which concludes the proof.

CryptoVerif computes the bound on the probability of distinguishing the games (n, q_e, q_d)-$\mathsf{Outsider\text{-}CCA}_\ell$ and (n, q_e, q_d)-$\mathsf{Outsider\text{-}CCA}_r$ by adding bounds computed at each transformation step. During this proof, CryptoVerif automatically eliminates unlikely collisions, in particular between public Diffie-Hellman keys. By default, CryptoVerif eliminates these collisions aggressively, even when that is not required for the proof to succeed, which results in a large probability bound. To avoid that, we guide the tool by giving estimates for n, $q_e^{per\ user}$, $q_d^{per\ user}$, q_h, $P_\mathcal{N}$, where $q_e^{per\ user}$ and $q_d^{per\ user}$ are the number of AENCAP and ADECAP queries respectively, per user. We also give a maximum probability for which we allow eliminating collisions. Our estimates are such that we allow eliminating collisions of probability $P_\mathcal{N}$ times a cubic factor in n, $q_e^{per\ user}$, and $q_d^{per\ user}$, but do not allow eliminating collisions with more than a cubic factor in n, $q_e^{per\ user}$, and $q_d^{per\ user}$, nor collisions that involve q_h. These estimates are used only to decide whether to eliminate collisions. The obtained probability formula is then valid even if the actual numbers do not match the given estimates.

The probability formula computed by CryptoVerif involves both the total numbers of queries q_e, q_d and the number of queries per user $q_e^{per\ user}$, $q_d^{per\ user}$. For simplicity, we upper bound $q_e^{per\ user}$ by q_e and $q_d^{per\ user}$ by q_d, yielding the formula given in the theorem. □

Theorem 8 (Insider-CCA **security of** DH-AKEM)**.** *Under the* GDH *assumption in* $\mathcal{N}$ *and modelling* KDF *as a random oracle,* DH-AKEM[$\mathcal{N}$, KDF] *is* Insider-CCA *secure. In particular, for any* (n, q_e, q_d, q_c)-Insider-CCA *adversary* $\mathcal{A}$ *against* DH-AKEM$_\mathcal{N}$ *that issues at most* q_h *queries to the random oracle, there exists an adversary* $\mathcal{B}$ *against* GDH *such that*

$$\mathsf{Adv}^{(n,q_e,q_d,q_c)\text{-}\mathsf{Insider\text{-}CCA}}_{\mathcal{A},\mathsf{DH\text{-}AKEM}[\mathcal{N},\mathsf{KDF}]} \leq \mathsf{Adv}^{\mathsf{GDH}}_{\mathcal{B},\mathcal{N}} + (n + q_c) \cdot \Delta_\mathcal{N} + (2q_e q_d + q_c q_d + q_c q_e + 2nq_e + 7q_e^2 + 2q_c^2 + 17n^2) \cdot P_\mathcal{N}$$

$\mathcal{B}$ *makes* $nq_e + 2q_c q_e + 2q_d q_h + 3nq_h$ *queries to the* DH *oracle, and* $t_\mathcal{B} \approx t_\mathcal{A}$.

Theorem 9 (Outsider-Auth **security of** DH-AKEM)**.** *Under the* sqGDH *assumption in* $\mathcal{N}$ *and modelling* KDF *as a random oracle,* DH-AKEM[$\mathcal{N}$, KDF] *is*

$\mathsf{Outsider\text{-}Auth}$ *secure. In particular, for any* (n, q_e, q_d)-$\mathsf{Outsider\text{-}Auth}$ *adversary* $\mathcal{A}$ *against* $\mathsf{DH\text{-}AKEM}_{\mathcal{N}}$ *that issues at most* q_h *queries to the random oracle, there exists an adversary* $\mathcal{B}$ *against* sqGDH *such that*

$$\mathsf{Adv}^{(n,q_e,q_d)\text{-}\mathsf{Outsider\text{-}Auth}}_{\mathcal{A},\mathsf{DH\text{-}AKEM}[\mathcal{N},\mathsf{KDF}]} \leq 2\mathsf{Adv}^{\mathsf{sqGDH}}_{\mathcal{B},\mathcal{N}} + 2(n+q_e)\cdot \Delta_{\mathcal{N}} + (q_e q_d + 4nq_d + 12q_e^2 + 4nq_e + 20n^2)\cdot P_{\mathcal{N}}$$

$\mathcal{B}$ *issues* $nq_e + nq_d + 4q_d q_h + 3nq_h$ *queries to the* DH *oracle, and* $t_{\mathcal{B}} \approx t_{\mathcal{A}}$.

Infeasibility of $\mathsf{Insider\text{-}Auth}$ security. As for APKE, we could define an Insider-Auth security notion for AKEM, which precludes forgeries even when the receiver key pair is dishonest, provided the sender key pair is honest. However, the $\mathsf{DH\text{-}AKEM}$ construction does not even achieve KCI security, a relaxation of Insider-Auth security only precluding forgeries for leaked, but still honestly generated, receiver key pairs. Indeed, in $\mathsf{DH\text{-}AKEM}$, knowledge of an arbitrary receiver secret key is already sufficient to compute the Diffie-Hellman shared key for any sender public key. Thus, in a KCI attack, an adversary that learns a target receiver's keys can trivially produce a KEM ciphertext and corresponding encapsulated key for any target sender public key.

5.2 HPKE's Key Schedule and Key Derivation Function

HPKE's key schedule $\mathsf{KS}_{\mathsf{Auth}}$ and key derivation function HKDF_{N} are both instantiated via the functions $\mathsf{Extract}$ and Expand which are defined below. We proceed to prove a theorem that $\mathsf{KS}_{\mathsf{Auth}}$ is a PRF, as needed for the composition results presented in Theorems 3 to 5. Then, we argue why HKDF_{N} can be modelled as a random oracle, as assumed by Theorems 7 to 9 on $\mathsf{DH\text{-}AKEM}$. Finally, we indicate how the entire $\mathsf{HPKE}_{\mathsf{Auth}}$ scheme is assembled from the individual building blocks presented in the previous sections.

$\mathsf{Extract}$ and Expand. The RFC defines two functions $\mathsf{Extract}$ and Expand as follows.

- $\mathsf{Extract}(salt, IKM)$ is a function keyed by a bitstring *salt*, with input keying material *IKM* as parameter, and returns a bitstring of fixed length N_h bits.
- $\mathsf{Expand}(PRK, info, L)$ is a function keyed by *PRK*, with an arbitrary bitstring *info* and a length L as parameters, and returns a bitstring of length L.

In Theorem 10, we assume that $\mathsf{Extract}$ and Expand are PRFs with the first parameter being the PRF key. HPKE instantiates $\mathsf{Extract}$ and Expand with HMAC-SHA-2, for which the PRF assumption is justified by [6,7]. (Generally, HPKE's instantiation of Expand uses HMAC iteratively to achieve the variable output length L. However, all values L used in HPKE are less or equal than the output length of one HMAC call.) We also assume that $\mathsf{Extract}$ is collision resistant, provided its keys are not larger than blocks of SHA-2, which is needed to avoid that the keys be hashed before computing HMAC, and true in HPKE. This property is immediate from the collision resistance of SHA-2, studied in [21].

Listing 9: The key schedule $\mathsf{KS}_{\mathsf{Auth}}$ used in $\mathsf{HPKE}_{\mathsf{Auth}}$ [5].

```
KS_Auth(k_PRF, info)
01 return KeySchedule(k_PRF, 0x02, info, "", "")

KeySchedule(k_PRF, mode, info, psk, psk_id)
02 context ← mode ||
            LabeledExtract("", "psk_id_hash", psk_id) ||
            LabeledExtract("", "info_hash"   , info)
03 secret ← LabeledExtract(k_PRF, "secret", psk)
04 k ← LabeledExpand(secret, "key", context, N_k)
05 nonce ← LabeledExpand(secret, "base_nonce", context, N_n)
06 return (k, nonce)

LabeledExtract(salt, label, IKM')
07 return Extract(salt, "HPKE-v1" || suite_id || label || IKM')

LabeledExpand(PRK, label, context, L)
08 return Expand(PRK, Encode(L) || "HPKE-v1" || suite_id || label || context, L)
```

KEY SCHEDULE. The key schedule $\mathsf{KS}_{\mathsf{Auth}}$ serves as a bridging step between the AKEM and the AEAD of APKE. The computations done by $\mathsf{KS}_{\mathsf{Auth}}$ are as indicated in Listing 9. The function KeySchedule used internally is the common key schedule function that the RFC defines for all modes. In $\mathsf{HPKE}_{\mathsf{Auth}}$, the *mode* parameter is set to the constant one-byte value 0x02 identifying the mode Auth. Similarly, mode Auth does not use a pre-shared key, so the *psk* parameter is always set to the empty string "", and the value *psk_id* that is identifying which pre-shared key is used, is equally set to "". The RFC defines LabeledExtract and LabeledExpand as wrappers around Extract and Expand, for domain separation and context binding. The value $suite_{id}$ is a 10-byte string identifying the ciphersuite, composed as a concatenation of the string "HPKE", and two-byte identifiers of the KEM, the KDF, and the AEAD algorithm in use. The bitstring $\mathsf{Encode}(L)$ is the two-byte encoding of the length L expressed in bytes. The values N_k and N_n indicate the length of the AEAD key and nonce.

The composition results established by Theorems 3 to 5 assume that $\mathsf{KS}_{\mathsf{Auth}}$ is a PRF. The following theorem proves this property for $\mathsf{HPKE}_{\mathsf{Auth}}$'s instantiation of $\mathsf{KS}_{\mathsf{Auth}}$.

Theorem 10 (Extract CR + Extract PRF + Expand PRF $\Rightarrow$ $\mathsf{KS}_{\mathsf{Auth}}$ PRF). *Assuming that* Extract *is a collision-resistant hash function for calls with the labels "psk_id_hash" and "info_hash", that* Extract *is a* PRF *for calls with the label "secret", and that* Expand *is a* PRF, *it follows that* $\mathsf{KS}_{\mathsf{Auth}}$ *is a* PRF.

In particular, for any (n_k, q_{PRF})-PRF *adversary* $\mathcal{A}$ *against* $\mathsf{KS}_{\mathsf{Auth}}$, *there exist an adversary* $\mathcal{B}$ *against the collision resistance of* Extract, *a* (n_k, n_k)-PRF *adversary* $\mathcal{C}_1$ *against* Extract, *and a* $(n_k, 2q_{\mathsf{PRF}})$-PRF *adversary* $\mathcal{C}_2$ *against* Expand *such that* $t_{\mathcal{B}} \approx t_{\mathcal{A}}$, $t_{\mathcal{C}_1} \approx t_{\mathcal{A}}$, $t_{\mathcal{C}_2} \approx t_{\mathcal{A}}$, *and*

$$\mathsf{Adv}^{(n_k, q_{\mathsf{PRF}})\text{-}\mathsf{PRF}}_{\mathcal{A}, \mathsf{KS}_{\mathsf{Auth}}} \leq \mathsf{Adv}^{\mathsf{CR}}_{\mathcal{B}, \mathsf{Extract}} + \mathsf{Adv}^{(n_k, n_k)\text{-}\mathsf{PRF}}_{\mathcal{C}_1, \mathsf{Extract}} + \mathsf{Adv}^{(n_k, 2q_{\mathsf{PRF}})\text{-}\mathsf{PRF}}_{\mathcal{C}_2, \mathsf{Expand}} .$$

This theorem is proven by CryptoVerif in keyschedule.auth.prf.ocv [2].

THE KEY DERIVATION FUNCTION KDF IN $\mathsf{DH\text{-}AKEM}$. The AKEM instantiation $\mathsf{DH\text{-}AKEM}$ as we defined it in Listing 8 uses a function KDF to derive the KEM shared secret. In $\mathsf{HPKE}_{\mathsf{Auth}}$, this function is instantiated by HKDF_N, as defined in Listing 10, using the above-defined $\mathsf{Extract}$ and Expand internally. The output length N corresponds to `Nsecret` in the RFC.

In the analysis of the key schedule presented above, we assume that $\mathsf{Extract}$ and Expand are pseudo-random functions. However, this assumption would not be sufficient to prove the security of $\mathsf{DH\text{-}AKEM}$: the random oracle model is required. The simplest choice is to assume that the whole key derivation function $\mathsf{KDF} = \mathsf{HKDF}_N$ is a random oracle, as we do in Theorems 7 to 9. (Alternatively, we could probably rely on some variant of the PRF-ODH assumption [14]. While in principle the PRF-ODH assumption is weaker than the random oracle model, Brendel et al. [14] show that it is implausible to instantiate the PRF-ODH assumption without a random oracle, so that would not make a major difference.) The invocations of $\mathsf{Extract}$ and Expand in $\mathsf{DH\text{-}AKEM}$ and $\mathsf{KS}_{\mathsf{Auth}}$ use different labels for domain separation, so choosing different assumptions is sound. Next, we further justify the random oracle assumption for HKDF_N.

As mentioned at the beginning of Sect. 5, HPKE instantiates $\mathsf{Extract}$ and Expand with HMAC [23], which makes HKDF_N exactly the widely-used HKDF key derivation function [24]. HPKE specifies SHA-2 as the hash function underlying HMAC. Lemma 6 in [27] shows that HKDF is indifferentiable from a random oracle under the following assumptions[3]: (1) HMAC is indifferentiable from a random oracle. For HMAC-SHA-2, this is justified by Theorem 4.4 in [19] assuming the compression function underlying SHA-2 is a random oracle. The theorem's restriction on HMAC's key size is fulfilled, because $\mathsf{DH\text{-}AKEM}$ uses either the empty string, or a bitstring of hash output length as key. (2) Values of IKM do not collide with values of $info \,||\, \texttt{0x01}$. This is guaranteed by the prefix `"HPKE-v1"` of IKM, which is used as a prefix for $info$ as well, but shifted by two characters, because the two-byte encoding of the length N comes before it. The shared secret lengths `Nsecret` specified in the RFC correspond exactly to the output length of the hash function; this means there is only one internal call to Expand, and thus we do not need to consider collisions of IKM with the input to later HMAC calls.

5.3 HPKE's APKE Scheme $\mathsf{HPKE}_{\mathsf{Auth}}$

Let $\mathsf{HPKE}_{\mathsf{Auth}} := \mathsf{APKE}[\mathsf{DH\text{-}AKEM}[\mathcal{N}, \mathsf{HKDF}_N], \mathsf{KS}_{\mathsf{Auth}}, \mathsf{AEAD}]$ be the APKE construction obtained by applying the black-box AKEM/DEM composition of Listing 6 to the $\mathsf{DH\text{-}AKEM}[\mathcal{N}, \mathsf{HKDF}_N]$ authenticated KEM (Listing 8), where $\mathcal{N}$ is a nominal group. For the key schedule of $\mathsf{HPKE}_{\mathsf{Auth}}$ we use $\mathsf{KS}_{\mathsf{Auth}}$ of Listing 9 and for the key derivation function we use HKDF_N of Listing 10. For both

[3] The exact probability bound is indicated in Lemma 8 of that paper's full version.

Listing 10: Function $\mathsf{HKDF}_{\mathcal{N}}[\mathsf{Extract}, \mathsf{Expand}]$ as used in $\mathsf{HPKE}_{\mathsf{Auth}}$.

```
HKDF_N(salt, IKM, info)
01 PRK ← Extract(salt, IKM)
02 return Expand(PRK, info, N)
```

$\mathsf{KS}_{\mathsf{Auth}}$ and $\mathsf{HKDF}_{\mathcal{N}}$ we implement the Extract and Expand functions using HMAC (as described in the HPKE specification). Finally, we instantiate HMAC using one of the SHA2 family of hash functions. (Which one depends on the target bit security of $\mathsf{HPKE}_{\mathsf{Auth}}$, as we discuss below.)

The AKEM/DEM composition Theorems 3 to 5, together with Theorem 10 on the key schedule $\mathsf{KS}_{\mathsf{Auth}}$, and Theorems 7 to 9 on DH-AKEM's security, and $P_{\mathsf{DH\text{-}AKEM}} = P_{\mathcal{N}}$ provide the following concrete security bounds for $\mathsf{HPKE}_{\mathsf{Auth}}$. For simplicity, we ignore all constants and set $q := q_e + q_d + q_c$.

$$\begin{aligned}
\mathsf{Adv}_{\mathcal{A},\mathsf{HPKE}_{\mathsf{Auth}}}^{(n,q_e,q_d,q_c)\text{-}\mathsf{Outsider\text{-}CCA}} &\leq \mathsf{Adv}_{\mathcal{B}_1,\mathcal{N}}^{\mathsf{GDH}} + (n+q)^2 \cdot P_{\mathcal{N}} + (n+q) \cdot \Delta_{\mathcal{N}} \\
&+ \mathsf{Adv}_{\mathcal{C},\mathsf{KS}_{\mathsf{Auth}}}^{(q,q)\text{-}\mathsf{PRF}} + \mathsf{Adv}_{\mathcal{D}_1,\mathsf{AEAD}}^{q\text{-}\mathsf{IND\text{-}CPA}} + \mathsf{Adv}_{\mathcal{D}_2,\mathsf{AEAD}}^{(q,q)\text{-}\mathsf{INT\text{-}CTXT}} \\
\mathsf{Adv}_{\mathcal{A},\mathsf{HPKE}_{\mathsf{Auth}}}^{(n,q_e,q_d)\text{-}\mathsf{Outsider\text{-}Auth}} &\leq \mathsf{Adv}_{\mathcal{B}_1,\mathcal{N}}^{\mathsf{GDH}} + \mathsf{Adv}_{\mathcal{B}_2,\mathcal{N}}^{\mathsf{sqGDH}} + (n+q)^2 \cdot P_{\mathcal{N}} + (n+q) \cdot \Delta_{\mathcal{N}} \\
&+ \mathsf{Adv}_{\mathcal{C},\mathsf{KS}_{\mathsf{Auth}}}^{(q,q)\text{-}\mathsf{PRF}} + \mathsf{Adv}_{\mathcal{D}_1,\mathsf{AEAD}}^{(q,q)\text{-}\mathsf{INT\text{-}CTXT}} \quad .
\end{aligned}$$

The bound for Insider-CCA is the same as the one for Outsider-CCA. In all bounds, we have $\mathsf{Adv}_{\mathcal{C},\mathsf{KS}_{\mathsf{Auth}}}^{(q,q)\text{-}\mathsf{PRF}} \leq \mathsf{Adv}_{\mathcal{C}_1,\mathsf{Extract}}^{\mathsf{CR}} + \mathsf{Adv}_{\mathcal{C}_2,\mathsf{Extract}}^{(q,q)\text{-}\mathsf{PRF}} + \mathsf{Adv}_{\mathcal{C}_3,\mathsf{Expand}}^{(q,q)\text{-}\mathsf{PRF}}$. Moreover, the adversaries $\mathcal{B}_1, \mathcal{B}_2, \mathcal{C}, \mathcal{D}_1, \mathcal{D}_2$ have (roughly) the same running time as $\mathcal{A}$.

Parameter Choices of $\mathsf{HPKE}_{\mathsf{Auth}}$. To obtain a concrete instance of $\mathsf{HPKE}_{\mathsf{Auth}}$, the HPKE standard allows different choices of nominal groups $\mathcal{N}$ that lead to different bounds on the statistical parameters $P_{\mathcal{N}}$ and $\Delta_{\mathcal{N}}$. The standard also fixes the length N of the KEM keyspace, c.f. Table 2. Even though lengths are expressed in bytes in the RFC and the implementation, we express them in bits in this section as this is more convenient to discuss the number of bits of security.

All concrete instances of $\mathsf{HPKE}_{\mathsf{Auth}}$ proposed by the HPKE standard build Extract and Expand from HMAC which, in turn, uses a hash function. HPKE proposes several concrete hash functions (all in the SHA2 family). For our security bounds, the relevant consequence of choosing a particular hash function is the resulting key length N_h of Expand when used as a PRF, c.f. Table 3.

Finally, to instantiate $\mathsf{HPKE}_{\mathsf{Auth}}$, we must also specify the AEAD scheme. HPKE allows for several choices which affect the AEAD key length N_k, nonces length N_n, and tag length N_t, c.f. Table 4.

Discussion. We say that an instance of $\mathsf{HPKE}_{\mathsf{Auth}}$ achieves κ *bits of security* if the success ratio $\mathsf{Adv}_{\mathcal{A},\mathsf{HPKE}_{\mathsf{Auth}}}/t_{\mathcal{A}}$ is upper bounded by $2^{-\kappa}$ for any adversary $\mathcal{A}$ with runtime $t_{\mathcal{A}} \leq 2^{\kappa}$. In particular, we say that a term ε *has* κ *bits of security* if $\varepsilon/t_{\mathcal{A}} \leq 2^{-\kappa}$. We discuss the implications of our results for the bit security of the various instances of $\mathsf{HPKE}_{\mathsf{Auth}}$ proposed by the standard.

Table 2. Parameters of $\mathsf{DH\text{-}AKEM}[\mathcal{N}, \mathsf{HKDF}_{\mathcal{N}}]$ depending on the choice of the nominal group $\mathcal{N}$.

	P-256	P-384	P-521	Curve25519	Curve448
Security level $\kappa_{\mathcal{N}}$ (bits)	128	192	256	128	224
$P_{\mathcal{N}} \leq$	2^{-255}	2^{-383}	2^{-520}	2^{-250}	2^{-444}
$\Delta_{\mathcal{N}} \leq$	0	0	0	2^{-125}	2^{-220}
KEM keyspace N (bits)	256	384	512	256	512

Table 3. Choices of HMAC and the PRF key lengths of Expand, instantiated with HMAC.

	HMAC-SHA256	HMAC-SHA384	HMAC-SHA512
PRF key length N_h of Expand (bits)	256	384	512

Table 4. Choices of the AEAD scheme and their parameters.

	AES-128-GCM	AES-256-GCM	ChaCha20-Poly1305
AEAD key length N_k (bits)	128	256	256
AEAD nonces length N_n (bits)	96	96	96
AEAD tag length N_t (bits)	128	128	128

The runtime $t_{\mathcal{A}}$ of any adversary $\mathcal{A}$ in an APKE security game is lower-bounded by $n + q$, since the adversary needs n steps to parse the n public keys and additional q steps to make the oracle queries. We assume that $t_{\mathcal{A}} \leq 2^{\kappa}$, where κ is the target security level.

We now estimate the security level supported by each term in $\mathsf{Adv}_{\mathcal{A},\mathsf{HPKE}_{\mathsf{Auth}}}$.

- **Term** $\mathsf{Adv}^{\mathsf{GDH}}_{\mathcal{B}_1,\mathcal{N}}$. Nominal groups $\mathcal{N}$ proposed for use by the HPKE standard were designed to provide $\kappa_{\mathcal{N}}$ bits of security (c.f. Table 2). That is, we assume that $\mathsf{Adv}^{\mathsf{GDH}}_{\mathcal{B}_1,\mathcal{N}}/t_{\mathcal{B}_1} \leq 2^{-\kappa_{\mathcal{N}}}$. Since $t_{\mathcal{A}} \approx t_{\mathcal{B}_1}$, we conclude that this term has $\kappa_{\mathcal{N}}$ bits of security. The same arguments hold for $\mathsf{Adv}^{\mathsf{sqGDH}}_{\mathcal{B}_2,\mathcal{N}}$.
- **Term** $(n+q)^2 \cdot P_{\mathcal{N}}$. Let us show that this term also has $\kappa_{\mathcal{N}}$ bits of security. We have $n + q \leq t_{\mathcal{A}}$. Thus, it suffices to show that $(n+q) \cdot P_{\mathcal{N}} \leq 2^{-\kappa_{\mathcal{N}}}$. Since $t_{\mathcal{A}} \leq 2^{\kappa_{\mathcal{N}}}$, we get that $(n+q) \leq 2^{\kappa_{\mathcal{N}}}$. The statement now follows as, according to Table 2, $P_{\mathcal{N}} \lesssim 2^{-2\kappa_{\mathcal{N}}}$.
- **Term** $(n+q) \cdot \Delta_{\mathcal{N}}$. Let us show that this term also has $\kappa_{\mathcal{N}}$ bits of security. For all NIST curves, we have $\Delta_{\mathcal{N}} = 0$ trivially implying the statement. In contrast, for Curve25519 and Curve448, $\Delta_{\mathcal{N}} \lesssim 2^{-\kappa_{\mathcal{N}}}$, so $(n+q) \cdot \Delta_{\mathcal{N}} \approx (n+q)2^{-\kappa_{\mathcal{N}}}$. As $n + q \leq t_{\mathcal{A}}$, the statement also holds for these curves.
- **Term** $\mathsf{Adv}^{\mathsf{CR}}_{\mathcal{C}_1,\mathsf{Extract}}$. The output length N_h of the concrete hash functions are listed in Table 3. Since the generic bound on collision resistance is $t^2_{\mathcal{C}_1}/2^{N_h}$, this term has $N_h/2$ bits of security.

- **Term** $\mathsf{Adv}^{(q,q)\text{-}\mathsf{PRF}}_{\mathcal{C}_3,\mathsf{Expand}}$. The PRF key lengths N_h of Expand are specified in Table 3. Modelling the PRF as a random oracle, we have $\mathsf{Adv}^{(q,q)\text{-}\mathsf{PRF}}_{\mathcal{C}_3,\mathsf{Expand}} \leq q^2/2^{N_h}$. So this term also has $N_h/2$ bits of security.
- **Term** $\mathsf{Adv}^{(q,q)\text{-}\mathsf{PRF}}_{\mathcal{C}_2,\mathsf{Extract}}$. The PRF key length N of $\mathsf{Extract}$ is specified in Table 2. By the same argument as for the previous term, this term has $N/2$ bits of security. Since $N/2 \geq \kappa_{\mathcal{N}}$ by Table 2, this term has $\kappa_{\mathcal{N}}$ bits of security.
- **Terms** $\mathsf{Adv}^{q\text{-}\mathsf{IND}\text{-}\mathsf{CPA}}_{\mathcal{D}_1,\mathsf{AEAD}} + \mathsf{Adv}^{(q,q)\text{-}\mathsf{INT}\text{-}\mathsf{CTXT}}_{\mathcal{D}_2,\mathsf{AEAD}}$. The terms refer to the multi-key security of the AEAD schemes (c.f. [3, Section 3]), studied for instance in [10]. However, the current results are not sufficient to guarantee the expected security level, such as 128 bits for AES-128-GCM. We recommend further research to study the exact bounds of the terms instantiated with the AEAD schemes from Table 4. In any case a simple key/nonce-collision attack has success probability $\mathsf{Adv}^{q\text{-}\mathsf{IND}\text{-}\mathsf{CPA}}_{\mathcal{D}_1,\mathsf{AEAD}} = q^2/2^{N_k+N_n}$, where N_k is the AEAD key length and N_n is the nonce length. A simple computation shows that this term has at most N_k bits of security (assuming $q \leq 2^{N_n}$). Moreover, a simple attack against $\mathsf{INT}\text{-}\mathsf{CTXT}$ by guessing the authentication tag has success probability $\mathsf{Adv}^{(q,q)\text{-}\mathsf{INT}\text{-}\mathsf{CTXT}}_{\mathcal{D}_2,\mathsf{AEAD}} = q/2^{N_t}$, where N_t is the length of the authentication tag. Hence, this term has at most N_t bits of security. Assuming these attacks also serve as an upper bound, these terms would have $\min(N_k, N_t)$ bits of security if $q \leq 2^{N_n}$. Since for all AEAD schemes of Table 4, we have $N_t = 128$ bits, that limits the security level of HPKE to 128 bits.

To sum up, the analysis above suggests that HPKE has about $\kappa = \min(\kappa_{\mathcal{N}}, N_h/2, N_k, N_t)$ bits of security, under the assumption that $t_{\mathcal{A}} \leq 2^{\kappa}$ and $q \leq 2^{N_n}$. Since the tag length of the AEAD is $N_t = 128$ bits, we obtain $\kappa = 128$ bits; a greater security level could be obtained by using AEADs with longer tags. More research on the multi-key security of AEAD schemes is still needed to confirm this analysis.

Acknowledgements. The authors would like to thank the HPKE RFC co-authors Richard Barnes, Karthikeyan Bhargavan, and Christopher Wood for fruitful discussions during the preparation of this paper.

Bruno Blanchet was supported by ANR TECAP (decision number ANR-17-CE39-0004-03). Eduard Hauck was supported by the DFG SPP 1736 Big Data. Eike Kiltz was supported by the BMBF iBlockchain project, the EU H2020 PROMETHEUS project 780701, the DFG SPP 1736 Big Data, and the DFG Cluster of Excellence 2092 CASA. Benjamin Lipp was supported by ERC CIRCUS (grant agreement n° 683032) and ANR TECAP (decision number ANR-17-CE39-0004-03). Doreen Riepel was supported by the Cluster of Excellence 2092 CASA.

References

1. Abdalla, M., Bellare, M., Rogaway, P.: The oracle Diffie-Hellman assumptions and an analysis of DHIES. In: Naccache, D. (ed.) CT-RSA 2001. LNCS, vol. 2020, pp. 143–158. Springer, Heidelberg (2001). https://doi.org/10.1007/3-540-45353-9_12

2. Alwen, J., Blanchet, B., Hauck, E., Kiltz, E., Lipp, B., Riepel, D.: Analysing the HPKE standard - supplementary material. https://doi.org/10.5281/zenodo.4297811
3. Alwen, J., Blanchet, B., Hauck, E., Kiltz, E., Lipp, B., Riepel, D.: Analysing the HPKE standard. Cryptology ePrint Archive, Report 2020/1499 (2020). https://eprint.iacr.org/2020/1499
4. Barnes, R.L., Beurdouche, B., Millican, J., Omara, E., Cohn-Gordon, K., Robert, R.: The Messaging Layer Security (MLS) Protocol. Internet-Draft draft-ietf-mls-protocol-09, IETF Secretariat, March 2020. https://tools.ietf.org/html/draft-ietf-mls-protocol-09
5. Barnes, R.L., Bhargavan, K., Lipp, B., Wood, C.A.: Hybrid Public Key Encryption. Internet-Draft draft-irtf-cfrg-hpke-08, IETF Secretariat, October 2020. https://tools.ietf.org/html/draft-irtf-cfrg-hpke-08
6. Bellare, M.: New proofs for NMAC and HMAC: security without collision resistance. J. Cryptol. **28**(4), 844–878 (2015)
7. Bellare, M., Canetti, R., Krawczyk, H.: Keying hash functions for message authentication. In: Koblitz, N. (ed.) CRYPTO 1996. LNCS, vol. 1109, pp. 1–15. Springer, Heidelberg (1996). https://doi.org/10.1007/3-540-68697-5_1
8. Bellare, M., Rogaway, P.: Code-based game-playing proofs and the security of triple encryption. Cryptology ePrint Archive, Report 2004/331 (2004). http://eprint.iacr.org/2004/331
9. Bellare, M., Stepanovs, I.: Security under message-derived keys: signcryption in iMessage. In: Canteaut, A., Ishai, Y. (eds.) EUROCRYPT 2020, Part III. LNCS, vol. 12107, pp. 507–537. Springer, Cham (2020). https://doi.org/10.1007/978-3-030-45727-3_17
10. Bellare, M., Tackmann, B.: The multi-user security of authenticated encryption: AES-GCM in TLS 1.3. In: Robshaw, M., Katz, J. (eds.) CRYPTO 2016, Part I. LNCS, vol. 9814, pp. 247–276. Springer, Heidelberg (2016). https://doi.org/10.1007/978-3-662-53018-4_10
11. Bernstein, D.J.: Curve25519: new Diffie-Hellman speed records. In: Yung, M., Dodis, Y., Kiayias, A., Malkin, T. (eds.) PKC 2006. LNCS, vol. 3958, pp. 207–228. Springer, Heidelberg (2006). https://doi.org/10.1007/11745853_14
12. Bhargavan, K., Blanchet, B., Kobeissi, N.: Verified models and reference implementations for the TLS 1.3 standard candidate. In: 2017 IEEE Symposium on Security and Privacy, pp. 483–502. IEEE Computer Society Press, May 2017
13. Blanchet, B.: A computationally sound mechanized prover for security protocols. IEEE Trans. Dependable Secure Comput. **5**(4), 193–207 (2008)
14. Brendel, J., Fischlin, M., Günther, F., Janson, C.: PRF-ODH: relations, instantiations, and impossibility results. In: Katz, J., Shacham, H. (eds.) CRYPTO 2017, Part III. LNCS, vol. 10403, pp. 651–681. Springer, Cham (2017). https://doi.org/10.1007/978-3-319-63697-9_22
15. Cramer, R., Shoup, V.: Design and analysis of practical public-key encryption schemes secure against adaptive chosen ciphertext attack. SIAM J. Comput. **33**(1), 167–226 (2003)
16. Dent, A.W.: Hybrid signcryption schemes with insidersecurity. In: Boyd, C., González Nieto, J.M. (eds.) ACISP 2005. LNCS, vol. 3574, pp. 253–266. Springer, Heidelberg (2005). https://doi.org/10.1007/11506157_22
17. Dent, A.W.: Hybrid signcryption schemes with outsider security. In: Zhou, J., Lopez, J., Deng, R.H., Bao, F. (eds.) ISC 2005. LNCS, vol. 3650, pp. 203–217. Springer, Heidelberg (2005). https://doi.org/10.1007/11556992_15

18. Dent, A.W., Zheng, Y. (eds.): Practical Signcryption. Information Security and Cryptography. Springer, HeidelbergHeidelberg (2010). https://doi.org/10.1007/978-3-540-89411-7
19. Dodis, Y., Ristenpart, T., Steinberger, J., Tessaro, S.: To hash or not to hash again? (In)differentiability results for H^2 and HMAC. Cryptology ePrint Archive, Report 2013/382 (2013). http://eprint.iacr.org/2013/382
20. Gayoso Martínez, V., Alvarez, F., Hernandez Encinas, L., Sánchez Ávila, C.: A comparison of the standardized versions of ECIES. In: 2010 6th International Conference on Information Assurance and Security, IAS 2010, August 2010
21. Gilbert, H., Handschuh, H.: Security analysis of SHA-256 and sisters. In: Matsui, M., Zuccherato, R.J. (eds.) SAC 2003. LNCS, vol. 3006, pp. 175–193. Springer, Heidelberg (2004). https://doi.org/10.1007/978-3-540-24654-1_13
22. Kobeissi, N., Bhargavan, K., Blanchet, B.: Automated verification for secure messaging protocols and their implementations: a symbolic and computational approach. In: 2nd IEEE European Symposium on Security and Privacy, pp. 435–450. IEEE, April 2017
23. Krawczyk, H., Bellare, M., Canetti, R.: HMAC: Keyed-hashing for message authentication. RFC 2104, RFC Editor, February 1997. https://www.rfc-editor.org/rfc/rfc2104.html
24. Krawczyk, H., Eronen, P.: HMAC-based extract-and-expand key derivation function (HKDF). RFC 5869, RFC Editor, May 2010. https://www.rfc-editor.org/rfc/rfc5869.html
25. Langley, A., Hamburg, M., Turner, S.: Elliptic curves for security. RFC 7748, RFC Editor, January 2016. https://www.rfc-editor.org/rfc/rfc7748.html
26. Lipp, B.: An analysis of hybrid public key encryption. Cryptology ePrint Archive, Report 2020/243 (2020). https://eprint.iacr.org/2020/243
27. Lipp, B., Blanchet, B., Bhargavan, K.: A mechanised cryptographic proof of the WireGuard virtual private network protocol. In: 4th IEEE European Symposium on Security and Privacy, Stockholm, Sweden, pp. 231–246. IEEE Computer Society, June 2019. https://hal.inria.fr/hal-02100345
28. National Institute of Standards and Technology: Digital Signature Standard (DSS). FIPS Publication 186-4, July 2013. https://doi.org/10.6028/nist.fips.186-4
29. Omara, E., Beurdouche, B., Rescorla, E., Inguva, S., Kwon, A., Duric, A.: The Messaging Layer Security (MLS) Architecture. Internet-Draft draft-ietf-mls-architecture-05, IETF Secretariat, July 2020. https://tools.ietf.org/html/draft-ietf-mls-architecture-05
30. Rescorla, E., Oku, K., Sullivan, N., Wood, C.A.: TLS Encrypted Client Hello. Internet-Draft draft-ietf-tls-esni-07, IETF Secretariat, June 2020. https://tools.ietf.org/html/draft-ietf-tls-esni-07
31. Zheng, Y.: Digital signcryption or how to achieve cost(signature & encryption) $\ll$ cost(signature) + cost(encryption). In: Kaliski, B.S. (ed.) CRYPTO 1997. LNCS, vol. 1294, pp. 165–179. Springer, Heidelberg (1997). https://doi.org/10.1007/BFb0052234

Tightly-Secure Authenticated Key Exchange, Revisited

Tibor Jager[1(✉)], Eike Kiltz[2], Doreen Riepel[2], and Sven Schäge[2]

[1] Bergische Universität Wuppertal, Wuppertal, Germany
tibor.jager@uni-wuppertal.de
[2] Ruhr-Universität Bochum, Bochum, Germany
{eike.kiltz,doreen.riepel,sven.schaege}@rub.de

Abstract. We introduce new tightly-secure authenticated key exchange (AKE) protocols that are extremely efficient, yet have only a *constant* security loss and can be instantiated in the random oracle model both from the standard DDH assumption and a subgroup assumption over RSA groups. These protocols can be deployed with optimal parameters, independent of the number of users or sessions, without the need to compensate a security loss with increased parameters and thus decreased computational efficiency.

We use the standard "Single-Bit-Guess" AKE security (with forward secrecy and state corruption) requiring all challenge keys to be simultaneously pseudo-random. In contrast, most previous papers on tightly secure AKE protocols (Bader et al., TCC 2015; Gjøsteen and Jager, CRYPTO 2018; Liu et al., ASIACRYPT 2020) concentrated on a non-standard "Multi-Bit-Guess" AKE security which is known not to compose tightly with symmetric primitives to build a secure communication channel.

Our key technical contribution is a new generic approach to construct tightly-secure AKE protocols based on non-committing key encapsulation mechanisms. The resulting DDH-based protocols are considerably more efficient than all previous constructions.

Keywords: Authenticated key exchange · Tightness · Non-committing encryption · Forward security

1 Introduction

Authenticated Key Exchange (AKE) is a fundamental cryptographic primitive with immense practical importance. The goal is to securely establish a session key between two parties in a network where an adversary can read, send, modify or delete messages and may also corrupt selected parties and sessions.

TIGHTNESS OF AKE. When proving a cryptographic scheme secure, one commonly describes a security reduction which transforms an adversary $\mathcal{A}$ that breaks the cryptographic scheme into an adversary $\mathcal{B}$ that solves some underlying complexity assumption. For instance, if $\mathcal{A}$ has advantage ϵ in breaking the

A. Canteaut and F.-X. Standaert (Eds.): EUROCRYPT 2021, LNCS 12696, pp. 117–146, 2021.
https://doi.org/10.1007/978-3-030-77870-5_5

scheme and $\mathcal{B}$ solves the problem with advantage $\epsilon' = \epsilon/L$, then L is called the reduction's security loss. If L is constant (and in particular independent of the number of $\mathcal{A}$'s oracle queries) and additionally the running times of $\mathcal{A}$ and $\mathcal{B}$ are roughly identical, then we say the reduction is *tight*. Especially when choosing protocol-specific system parameters, the tightness of a security proof plays an important role. In the security model for AKE the attacker can actively control all messages sent between the involved parties and is additionally allowed to reveal secret information such as a long-term secret key (by corrupting a party), or a session key. The adversary breaks security if it is able to distinguish non-revealed session keys from random.

MULTI-CHALLENGE SECURITY DEFINITIONS. The standard and well established security notion in the context of multiple challenges [3,10,18,20] is "Single-Bit Guess" (SBG) security. The blueprint of a SBG security experiment is as follows. First, the experiment picks a secret random bit $b \in \{0, 1\}$. Next, the adversary is allowed to make multiple (up to, say, T) challenge queries. On each challenge query, the experiment returns a "real key" if $b = 0$, and an independent "random key" if $b = 1$. The adversary wins if it can guess the challenge bit b with a probability better than $1/2$.

In AKE protocols, challenge queries are usually called test queries and non-revealed session keys can be accessed by making multiple calls to a TEST oracle. If $b = 0$, a query to TEST returns the real challenge key; if $b = 1$, a query to TEST returns an independent random challenge key. This notation of multi-challenge SBG security for AKE was first formalized in 2019 by Cohn-Gordon et al. [10]. By conditioning on bit b, SBG security is known to be tightly equivalent to (single-bit) "Real-Or-Random" (ROR) security, where the adversary has to distinguish a real game (where all challenge keys output by TEST are real) from a random game (where all challenge keys are random). Using the above equivalence, SBG security precisely captures the intuition that *all challenge keys* are simultaneously pseudo-random.

Surprisingly, in the first publication on tightly secure AKE protocols in 2015, Bader et al. [1] defined a different and non-standard "Multi-Bit-Guess" (MBG) AKE security notion. In MBG security, the experiment picks multiple independent challenge bits $b_1, \ldots, b_T$ and, on the i-th TEST query, it returns a real challenge key if $b_i = 0$ and a random challenge key if $b_i = 1$. That is, each of the T challenge keys depends on an independent challenge bit b_i. The adversary wins if it can guess correctly one of the T challenge bits b_{i^*} with a probability better than $1/2$. We are not aware of any meaningful multi-bit ROR security game that is tightly equivalent to MBG security.[1] This makes it difficult to provide a good intuition of what MBG security tries to model.

[1] If one tries to apply a similar conditioning argument as in the single-bit case, MBG can be shown equivalent to a ROR-type security experiment where in the real game ($b_{i^*} = 0$) the i^*-th challenge key output by TEST is real and in the random game ($b_{i^*} = 1$) it is random. However, the remaining $T - 1$ keys still depends on the random bits b_i ($i \neq i^*$): the i-th challenge key is real if $b_i = 0$ and it is random if $b_i = 1$. Hence, about one half of the challenge keys is expected to be real (the ones

CHOOSING A MEANINGFUL SECURITY MODEL FOR AKE. SBG and MBG security are asymptotically equivalent but only imply each other with a security loss of T, the total number of TEST queries. Hence, when considering tightness, one has to carefully choose a meaningful security model.

First off, as already pointed out, SBG security is the standard and well established security notion in the context of multiple challenges [3,10,18,20]. Cohn-Gordon et al. [10, Section 3] already pointed out that, in the AKE setting, SBG security tightly composes with symmetric primitives, whereas MBG security doesn't. Let us elaborate. AKE is not intended to be used as a stand-alone primitive. Rather, it is naturally composed with symmetric primitives to establish a secure channel [7,24], for example to encrypt (e.g., using AES) a message with the session key. Since SBG security is tightly equivalent to ROR security, it offers precisely the right security interface to switch *all challenge keys at once* from real to random. This step allows to infer the privacy of the encrypted messages from the security properties of the symmetric primitive. MBG security, on the other hand, does not have a meaningful ROR-style security, which makes it difficult to argue about the privacy of the encrypted messages without relying on a hybrid argument. In summary, in the context of tightness of AKE protocols, SBG security is a meaningful notion whereas MBG isn't.

PREVIOUS RESULTS. Previous work on tight AKE protocols by Gjøsteen and Jager [21] and Liu et al. [32] exclusively concentrated on the MBG model by Bader et al. [1]. We now give a brief overview of existing AKE protocols in the context of tight SBG security.

- At CRYPTO 2019, Cohn-Gordon et al. [10] presented highly efficient two message AKE protocols with implicit authentication, in the style of HMQV [26] and similar protocols. Their schemes achieve a loss of $O(N)$ in the SBG security model with weak forward secrecy, where N is the number of users. They also extend the impossibility results from [2] to show that a loss of $O(N)$ is unavoidable for many natural protocols (including HMQV [26], NAXOS [28], Kudla-Paterson [27], KEA+ [29], and more) with respect to typical cryptographic security proofs (so-called simple reductions). Furthermore, since their protocol does not feature explicit authentication, a well-known impossibility result applies [6,26,34] and their protocol cannot achieve full forward security.
- Diemert and Jager [16] and independently Davis and Günther [15] considered the three message TLS 1.3 handshake AKE protocol with explicit authentication. Its design follows the standard "1×KEM+2×SIG" (aka. signed Diffie-Hellman) AKE approach [9,14–16,21,32]. TLS 1.3, when instantiated with standardized signatures (e.g., RSA-PSS, RSA-PKCS #1 v1.5, ECDSA, or EdDSA), has rather non-tight SBG security with full forward security. But when instantiated with tightly secure signatures in the multi-user setting with adaptive corruptions [1], then SBG security of TLS 1.3 actually becomes tight. Since the TLS 1.3 protocol contains two signatures, the inefficiency of cur-

with $b_i = 0$) whereas the other half is random, and the adversary does not have any information on them.

rently known tightly secure signature schemes [1,21] makes the resulting TLS instantiation very impractical.

1.1 The Difficulty of Constructing Tightly Secure AKE

Security models for authenticated key exchange are extremely complex, as they consider very strong adversaries that may modify, drop, or inject messages. Furthermore, usually an adversary may adaptively corrupt users' long-term secrets via CORRUPT-queries, session keys via REVEAL-queries, and sometimes even ephemeral states of sessions via REV-STATE-queries. Security is formalized with multiple TEST queries, where the adversary specifies a session, receives back a real key or a random key, and has to distinguish these. This complexity makes achieving tight security challenging, particularly because all the following difficulties must be tackled simultaneously.

THE "COMMITMENT" PROBLEM. As explained in more detail in [21], this problem is the reason why nearly all security proofs of classical key exchange protocols have a quadratic security loss. Essentially, the problem is that most AKE protocols have security proofs where a reduction can only extract a solution to a computationally hard problem if an instance of the problem is embedded into the protocol messages of the TESTed sessions, but at the same time the reduction is not able to answer REVEAL queries for such sessions. The standard way to resolve this is to let the reduction guess the TESTed session, and to embed an instance of a computationally hard problem only there. However, this incurs a significant security loss. A tight reduction has to be able to respond to *both* TEST and REVEAL queries for *every* session.

THE PROBLEM OF LONG-TERM KEY REVEALS. A CORRUPT query in typical AKE security models enables the adversary to obtain the long-term key of certain users. If we want to avoid a security loss that results from guessing corrupted and non-corrupted parties, then we must be able to construct a reduction that "knows" valid-looking long-term keys for all users throughout the security experiment. However, this is a major difficulty, for instance, in protocols where the long-term keys are key pairs for a digital signature scheme. The difficulty is that in the security proof we would have to describe a reduction that is able to extract a solution to a computationally hard problem from a forged signature, even though it "knows" the signing key and thus is able to compute a valid signature itself. Hence, in order to obtain a tightly-secure AKE protocol, one needs to devise a way such that a reduction always knows all secret keys, yet is able to argue that an adversary is, e.g., not able to forge signatures.

In order to resolve this issue, previous works [1,21] constructed signature schemes based on non-interactive OR-proof systems, which enable a reduction to "know" one out of two signing keys. It is argued that the adversary will forge a signature with respect to the other, unknown key with sufficiently high probability. However, these signature schemes are much less efficient than classical ones, and thus impose a performance penalty on the protocols.

THE PROBLEM OF EPHEMERAL STATE REVEALS. Yet another difficulty arises when the security model allows ephemeral state reveals. Previous works on tightly-secure AKE did not consider this very strong security notion at all, therefore we face (and solve) this problem for the first time. From a high-level perspective, the issue is similar to the long-term key reveal problem, except that ephemeral states are considered. In order to achieve tightness, the reduction must be able to output valid-looking states for all sessions. Note that this includes even TESTed sessions, where ephemeral states may be revealed when parties are not corrupted.

1.2 Main Contributions

Summarizing the previous paragraphs, we can formulate the following natural questions related to tightly secure AKE:

Q1: Do there exist implicitly authenticated two-message AKEs with tight SBG security, state reveals, and weak forward security?

Q2: Do there exist explicitly authenticated two-message AKEs with tight SBG security, state reveals, and full forward security, with *one single* signature?

In this work, we answer the two questions to the positive. Following [4,10], we consider SBG security, allowing adaptive corruptions of long-term secrets, adaptive reveals of session keys, and multiple adaptive TEST queries. Our model also captures (weak and full) forward security (FS), and prevents key-compromise impersonation and reflection attacks. In comparison to prior work on tightly-secure key exchange [1,10,15,16,21], we consider a model which additionally allows to reveal some internal state information.

OUR DDH-BASED AKE PROTOCOLS. Our two protocols instantiated from DDH are given in Fig. 1. $\mathsf{AKE}_{\mathsf{wFS,DDH}}$ is an implicitly-authenticated two-message protocol $\mathsf{AKE}_{\mathsf{wFS,DDH}}$ in the sense of [26]. It requires the exchange of only five group elements in total, and thus is the first efficient implicitly-authenticated protocol with weak FS that achieves full tightness.

Our second protocol $\mathsf{AKE}_{\mathsf{FS,DDH}}$ achieves full FS. Instead of using the standard "1×KEM+2×SIG" approach, it replaces one of the signatures with a more efficient MAC and an additional KEM ciphertext, which yields a "2×KEM+1×SIG+1×MAC" construction. When instantiated at "128-bit security" with the most efficient tightly-secure signatures of [21],[2] the communication complexity is 448 bytes, again with ephemeral state reveals. In comparison, the previously most efficient tightly and fully forward-secure protocol with SBG security TLS^* (which is TLS 1.3 instantiated with the tightly-secure signature of [21]) requires three messages, the transmission of 704 bytes and does not allow state reveals. See Fig. 2 for a comparison of our protocols with previous works. Note that the communication bottleneck in all full FS protocols is the number of signatures. For completeness the figure also list previous protocols with tight MBG security [21,32].

[2] The signatures of [21] consist of 2 group elements, 4 elements in $\mathbb{Z}_p$ and 2 hashes in $\{0,1\}^{\kappa}$. At "128-bit security" this corresponds to 256 bytes per signature.

Alice $((a_1, a_2, \mathsf{sigk}), (A := g_1^{a_1} g_2^{a_2}, \mathsf{vk}))$		**Bob** $((b_1, b_2), B := g_1^{b_1} g_2^{b_2})$
$(x_1, x_2) \xleftarrow{\$} \mathbb{Z}_p^2;\ X := g_1^{x_1} g_2^{x_2}$		
$s \xleftarrow{\$} \mathbb{Z}_p$		
$\sigma \leftarrow \mathsf{Sign}(\mathsf{sigk}, u)$	$u = (X, g_1^s, g_2^s),\ \sigma$ $\longrightarrow$	if $\mathsf{Vrfy}(\mathsf{vk}, u, \sigma) \neq 1$ abort
		$t \xleftarrow{\$} \mathbb{Z}_p,\ K := \mathsf{H}(\text{context}, K_A, K_B, K_X)$
if $\pi \neq \mathsf{F}(K_B, u, \sigma, v)$ abort	$\longleftarrow$ $v = (g_1^t, g_2^t),\ \pi$	$\pi := \mathsf{F}(K_B, u, \sigma, v)$
$K := \mathsf{H}(\text{context}, K_A, K_B, K_X)$		$K := \mathsf{H}(\text{context}, K_X)$
$K := \mathsf{H}(\text{context}, K_X)$		

$$\begin{aligned}
\mathsf{H}_A(g_1^t, g_2^t, (g_1^t)^{a_1}(g_2^t)^{a_2}) &= K_A = \mathsf{H}_A(g_1^t, g_2^t, A^t)\\
\mathsf{H}_B(g_1^s, g_2^s, B^s) &= K_B = \mathsf{H}_B(g_1^s, g_2^s, (g_1^s)^{b_1}(g_2^s)^{b_2})\\
\mathsf{H}_X(g_1^t, g_2^t, (g_1^t)^{x_1}(g_2^t)^{x_2}) &= K_X = \mathsf{H}_X(g_1^t, g_2^t, X^t)
\end{aligned}$$

Fig. 1. The two message protocols $\mathsf{AKE}_{\mathsf{wFS,DDH}}$ (without the gray boxes) and $\mathsf{AKE}_{\mathsf{FS,DDH}}$ (including the gray boxes), where K is the resulting session key. We define context := $(A, B, X, \mathsf{vk}, g_1^s, g_2^s, g_1^t, g_2^t, \sigma, \pi)$. $\mathsf{H}, \mathsf{H}_A, \mathsf{H}_B, \mathsf{H}_X$ and F are hash functions.

Generic constructions of AKE from NCKE. Our main technical tool is a new approach to achieve a tight reduction for authenticated key exchange protocols. Our starting point is an extension of (receiver) non-committing encryption (NCE) [8,33] to *non-committing key encapsulation (NCKE) in the multi-user setting with corruptions.* We construct an NCKE scheme in the random oracle model from any smooth projective hash proof system (HPS) [11]. If the HPS' subset membership problem (SMP) is hard in the multi-instance setting, then the NCKE scheme is also tightly secure in our multi-user setting. We provide two such HPS, one from the DDH assumption, and another one from a subgroup assumption over groups of unknown order. The construction allows us to address the commitment problem described above.

We give a generic construction of an implicitly authenticated two-message AKE protocol $\mathsf{AKE}_{\mathsf{wFS}}$ with weak forward security from any NCKE scheme, whose security is tightly based on the multi-user security of the underlying NCKE scheme. Furthermore, we give a generic construction of an explicitly authenticated two-message AKE protocol $\mathsf{AKE}_{\mathsf{FS}}$ with perfect forward security by adding a tightly-secure signature scheme and a message authentication code (MAC) to our first construction, see Fig. 3. Thus, we require only a single signature which is particularly useful for tightly-secure key exchange, because known constructions of suitable tightly-secure signature schemes [1,21] have relatively large signatures and replacing one signature with a MAC significantly improves the computational efficiency and communication complexity of the protocol.[3]

All these generic constructions leverage NCKE in order to resolve the technical difficulties in constructing tightly-secure AKE protocols described before.

[3] [31] showed how to generically avoid signatures in forward-secure AKE protocols, but at the cost of additional messages.

Protocol	**Comm.** $(\mathbb{G}, \{0,1\}^\kappa, \mathsf{Sig})$	**Bytes**	**#Msg.**	**Assump.**	**Auth.**	**Model**	**State Reveal**	**Sec. Loss**
Protocols with full forward security								
TLS* [16,15]	(2, 4, 2)	704	3	Strong-DH + DDH	expl.	SBG	no	$O(1)$
GJ [21]	(2, 1, 2)	608	3	DDH	expl.	MBG	no	$O(1)$
LLGW [32]	(3, 0, 2)	608	2	DDH	expl.	MBG	no	$O(1)$
$\mathsf{AKE}_{\mathsf{FS,DDH}}$ (Fig. 1)	(5, 1, 1)	448	2	DDH	expl.	SBG	yes	$O(1)$
Protocols with weak forward security								
HMQV [26]	(2, 0, 0)	64	2	CDH	impl.	SBG	yes	$O(TN^2\ell^2)$
CCGJJ [10]	(2, 0, 0)	64	2	Strong-DH	impl.	SBG	no	$O(N)$
$\mathsf{CCGJJ}_{\mathsf{Twin}}$ [10]	(3, 0, 0)	96	2	CDH	impl.	SBG	no	$O(N)$
$\mathsf{AKE}_{\mathsf{wFS,DDH}}$ (Fig. 1)	(5, 0, 0)	160	2	DDH	impl.	SBG	yes	$O(1)$

Fig. 2. Comparison of AKE protocols over a group $\mathbb{G}$, where N refers to the number of parties, ℓ to the number of sessions per party and T is the number of test queries. TLS* refers to the TLS 1.3 handshake, instantiated with the tightly-secure signatures of [21]. The column **Comm.** counts the communication complexity of the protocols in terms of the number of group elements, hashes, and signatures. The column **Model** lists the AKE security model and distinguishes between multi-bit guessing (MBG) and the single-bit-guessing (SBG) security.

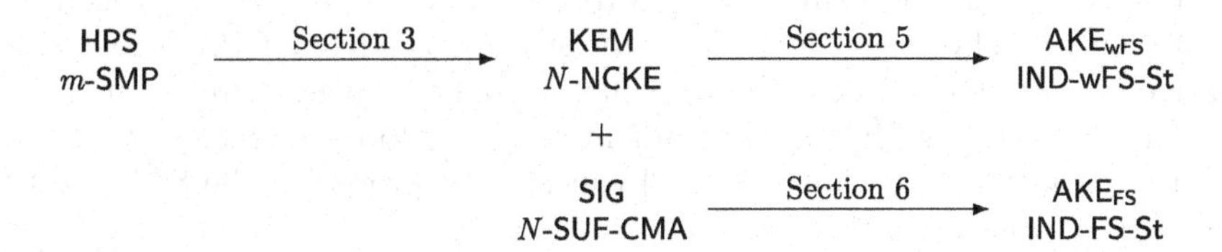

Fig. 3. Overview of our transformations, where N is the maximum number of users in the NCKE security game and in the SUF-CMA security game. The subset membership problem of HPS is m-fold for $m = N \cdot q$, where q is the maximum number of challenge queries in the NCKE security game.

Handling Ephemeral State Reveals. Our protocols are secure against ephemeral state reveals. We construct the first tightly-secure protocols to achieve this. Note that this requires us to deal with the situation that the reduction must "know" valid ephemeral states for *all* sessions, even tested sessions. To this end, we encrypt the state information with a symmetric long-term key. An adversary now needs to query both long-term secret key and ephemeral state to reveal the secret state information, similarly to the approach used in the NAXOS protocol [28]. While the idea of achieving security against ephemeral state reveals by relying on the security of long-term keys was used before [5,19,28,36], the approach to simply encrypt the state is new. It avoids the expensive re-computation of protocol messages required in prior generic approaches, which makes it particularly efficient. Also, previous work did not focus on tightness and it is unclear if a tight proof can be achieved in an even stronger security model which requires to reveal the randomness.

Our approach does not work generically, e.g., it cannot be applied to the protocols in [10,21], so we have to design our protocols such that they are compatible. This is due to the fact that in both works, the state is a secret DH exponent which is implicitly determined by rerandomizing the CDH (or DDH) challenge and then is embedded in multiple sessions. Thus, the reduction is able to extract the solution independently of which session is the test session, but it also does not know any of the secret exponents, which the adversary could reveal for non-test sessions.

1.3 Related Work and Open Problems

Concurrent and independent work of Liu et al. [32] also proposed a tightly secure 2-message AKE with full forward security. Compared to our protocols, they do not consider state reveal attacks and their proofs only hold in the MBG security model. Their AKE construction LLGW follows the well known $1\times$KEM$+2\times$SIG approach, meaning that even neglecting the issues with the MBG security model, it is still considerably less efficient than ours (c.f. Fig. 2). The main novelty of [32] is the new KEM security notion of (multi-bit) "IND-mCPA with adaptive reveals" that gives them the handle to prove tight MBG security. It is a natural question whether this KEM security notion can be adapted to a single-bit notion such that the resulting AKE protocol achieves tight SBG (rather than MBG) security. This is in particular interesting since IND-mCPA KEMs with adaptive reveals can be instantiated in the standard model, whereas our NCKE notion seem to inherently rely on random oracles. More concretely this raises the question whether (variants of) [32] can also be proved in the SBG model, without relying on random oracles.

2 Preliminaries

For an integer n, $[n]$ denotes the set $\{1, ..., n\}$. For a set S, $s \stackrel{\$}{\leftarrow} S$ denotes that s is sampled uniformly and independently at random from S. $y \leftarrow \mathcal{A}(x_1, x_2, ...)$ denotes that on input $x_1, x_2, ...$ the probabilistic algorithm $\mathcal{A}$ returns y. $\mathcal{A}^{\mathrm{O}}$ denotes that algorithm $\mathcal{A}$ has access to oracle O. We will use code-based games as introduced in [35]. An adversary is a probabilistic algorithm. $\Pr[G^{\mathcal{A}} \Rightarrow 1]$ denotes the probability that the final output $G^{\mathcal{A}}$ of game G running adversary $\mathcal{A}$ is 1.

3 Multi-receiver Non-committing Key Encapsulation

In this section, we introduce Multi-Receiver Non-Committing Key Encapsulation (NCKE). We will use this concept to resolve the "commitment problem" described in the introduction, which often makes proofs for multi-party protocols with adaptive corruptions non-tight, as for example AKE protocols.

SYNTAX. A key encapsulation mechanism $\mathsf{KEM} = (\mathsf{Gen}, \mathsf{Encaps}, \mathsf{Decaps})$ consists of three algorithms. The key generation algorithm Gen outputs a key pair $(\mathsf{pk}, \mathsf{sk})$, where pk is the public key and sk the secret key. The encapsulation algorithm inputs a public key pk and outputs a ciphertext c and a key K from the key space $\mathcal{K}$, where c is called an encapsulation of K. The deterministic decapsulation algorithm inputs the secret key sk and a ciphertext c and outputs K.

By μ we denote the *collision probability* of the key generation algorithm. In particular,

$$\Pr[(\mathsf{pk}, \mathsf{sk}) \leftarrow \mathsf{Gen}, (\mathsf{pk}', \mathsf{sk}') \leftarrow \mathsf{Gen} : \mathsf{pk} = \mathsf{pk}'] \leq 2^{-\mu} \ .$$

We denote the *min-entropy* of the encapsulation algorithm Encaps by $\gamma(\mathsf{pk}) := -\log \max_{c \in \mathcal{C}} \Pr[c = \mathsf{Encaps}(\mathsf{pk})]$. We say KEM is γ-spread if for all $(\mathsf{pk}, \mathsf{sk}) \leftarrow \mathsf{Gen} : \gamma(\mathsf{pk}) \geq \gamma$. This implies that for all $c \in \mathcal{C}$:

$$\Pr[c = \mathsf{Encaps}(\mathsf{pk})] \leq 2^{-\gamma} \ .$$

Security. Following [33], we introduce a security definition of Multi-Receiver Non-Committing Key Encapsulation (NCKE) for a key encapsulation mechanism KEM in the random oracle model, i. e., the KEM algorithms have access to a random oracle $\mathsf{H} : \{0,1\}^* \rightarrow \{0,1\}^\kappa$, indicated by $\mathsf{Encaps}^\mathsf{H}$. Our definition is relative to a simulator $\mathsf{Sim} = (\mathsf{SimGen}, \mathsf{SimEncaps}, \mathsf{SimHash})$. The simulated key generation algorithm SimGen generates a key pair $(\mathsf{pk}, \mathsf{sk})$. The simulated encapsulation algorithm $\mathsf{SimEncaps}$ takes both the public and private key and outputs a ciphertext c. The simulated hash algorithm $\mathsf{SimHash}$ inputs the key pair as well as three sets (used for bookkeeping) and deterministically computes a simulated hash value.

We define the two games $\mathsf{NCKE}_\mathsf{real}$ and $\mathsf{NCKE}_\mathsf{sim}$ in Fig. 4 where we consider N receivers each holding a key pair $(\mathsf{pk}_n, \mathsf{sk}_n)$. In the $\mathsf{NCKE}_\mathsf{real}$ game, the original Encaps algorithm is used. We give each user an individual hash function H_n such that keys are computed independently. (In general, this can be implemented by using the user's public key and identity as input to the hash function as well, where collisions have to be considered.) In the $\mathsf{NCKE}_\mathsf{sim}$ game, the $\mathsf{SimEncaps}$ algorithm is used to compute the ciphertexts. Keys are chosen uniformly at random. The adversary may also adaptively corrupt some receivers. We require that ciphertexts of corrupted receivers always decapsulate to the key output by Encaps, which is modeled by the $\mathsf{SimHash}$ algorithm. Therefore, if the receiver is corrupted, the algorithm takes sets $\mathcal{CK}$, $\mathcal{D}$ and $\mathcal{H}$, where the first one stores all challenge ciphertexts and keys output to the adversary, the second one stores all decapsulation queries and the third one stores all hash queries which have been issued so far. Thus, the $\mathsf{SimHash}$ algorithm can answer future queries based on everything that is known to the adversary. If the receiver is not corrupted, set $\mathcal{C}$ is used instead of $\mathcal{CK}$. This set stores only challenge ciphertexts and thus a hash value is computed independently of previous challenge keys.

The goal of an adversary $\mathcal{A}$ is to distinguish between the real KEM algorithms used in game $\mathsf{NCKE}_\mathsf{real}$ and the simulated algorithms used in game $\mathsf{NCKE}_\mathsf{sim}$. This is captured in Definition 1. Note that the non-committing property is due to the $\mathsf{SimHash}$ algorithm. In particular, the $\mathsf{SimHash}$ algorithm ensures that a (uniformly random) challenge key can be explained by the corresponding ciphertext generated by $\mathsf{SimEncaps}$ as soon as the receiver is corrupted.

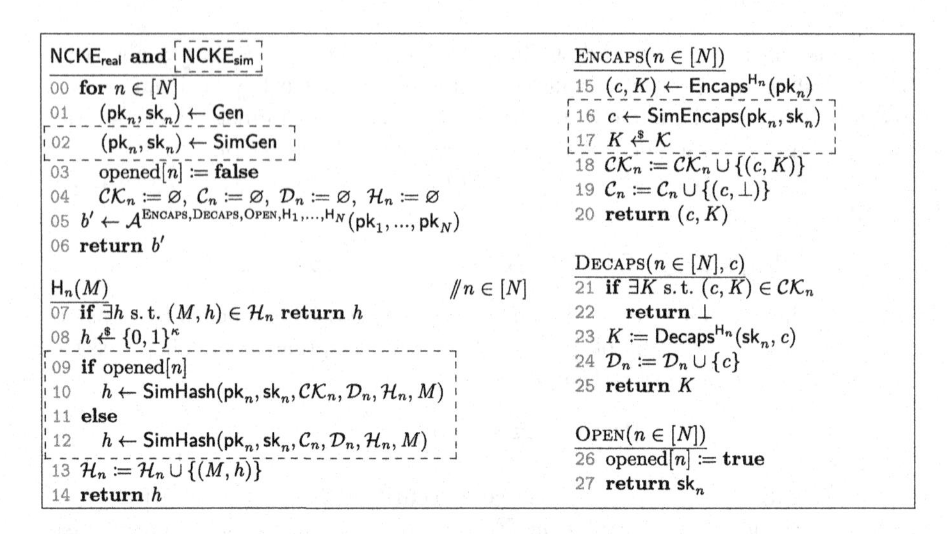

Fig. 4. Real and simulated game for N-receiver non-committing key encapsulation in the random oracle model.

Definition 1 (N-Receiver Non-Committing Key Encapsulation). *We define games* $\mathsf{NCKE}_{\mathsf{real}}$ *and* $\mathsf{NCKE}_{\mathsf{sim}}$ *as in Fig. 4, where* N *is the number of users. The simulator* $\mathsf{Sim} = (\mathsf{SimGen}, \mathsf{SimEncaps}, \mathsf{SimHash})$ *is defined relative to* KEM *and is used in* $\mathsf{NCKE}_{\mathsf{sim}}$*. The advantage of an adversary* $\mathcal{A}$ *against* KEM *and* Sim *is defined as*

$$\mathrm{Adv}^{N\text{-}\mathsf{NCKE}}_{\mathsf{KEM},\mathsf{Sim}}(\mathcal{A}) := \left|\Pr[\mathsf{NCKE}^{\mathcal{A}}_{\mathsf{real}} \Rightarrow 1] - \Pr[\mathsf{NCKE}^{\mathcal{A}}_{\mathsf{sim}} \Rightarrow 1]\right| \ .$$

When we write NCKE, we mean NCKE-CCA, where the adversary is allowed to access a decapsulation oracle. Sometimes we will explicitly write NCKE-CCA to differentiate from NCKE-CPA, where the adversary cannot issue decapsulation queries.

We stress that compared to the standard definition of non-committing encryption in the random oracle model (e.g., [33]), Definition 1 is for KEMs (rather than encryption), only considers receiver corruptions (rather than sender and receiver corruptions), and considers multiple receivers (rather than one single receiver).

Instantiations from Hash Proof Systems. We recall the definition of hash proof systems by Cramer and Shoup [11] and properties defined in [25].

Smooth Projective Hashing. Let $\mathcal{Y}$ and $\mathcal{Z}$ be sets and $\mathcal{X} \subset \mathcal{Y}$ a language. Let $\Lambda_{\mathsf{sk}} : \mathcal{Y} \to \mathcal{Z}$ be a hash function indexed with $\mathsf{sk} \in \mathcal{SK}$, where $\mathcal{SK}$ is a set. A hash function Λ_{sk} is projective if there exists a projection $\mu : \mathcal{SK} \to \mathcal{PK}$ such that $\mu(\mathsf{sk}) \in \mathcal{PK}$ defines the action of Λ_{sk} over $\mathcal{X}$. In particular, for every $c \in \mathcal{X}$, $Z = \Lambda_{\mathsf{sk}}(c)$ is uniquely determined by $\mu(\mathsf{sk})$ and c. However, there is no guarantee for $c \in \mathcal{Y} \setminus \mathcal{X}$ and it may not be possible to compute $\Lambda_{\mathrm{SK}}(c)$ from $\mu(\mathrm{SK})$ and C. A projective hash function is k-entropic if for all $c \in \mathcal{Y} \setminus \mathcal{X}$ it holds that $H_\infty(\Lambda_{\mathsf{sk}}(c) \mid \mathsf{pk}) \geq k$, where $\mathsf{pk} = \mu(\mathsf{sk})$ for $\mathsf{sk} \xleftarrow{\$} \mathcal{SK}$.

Gen(par)	Encaps^{H}(pk)	Decaps^{H}(sk, c)
00 $\mathsf{sk} \overset{\$}{\leftarrow} \mathcal{SK}$	03 $c \overset{\$}{\leftarrow} \mathcal{X}$ with witness r	06 $K := \mathsf{H}(c, \mathsf{Priv}(\mathsf{sk}, c))$
01 $\mathsf{pk} := \mu(\mathsf{sk})$	04 $K := \mathsf{H}(c, \mathsf{Pub}(\mathsf{pk}, c, r))$	07 **return** K
02 **return** $(\mathsf{pk}, \mathsf{sk})$	05 **return** (c, K)	

Fig. 5. Key encapsulation mechanism KEM = (Gen, Encaps, Decaps).

SimEncaps(pk, sk)	SimHash(pk, sk, $\mathcal{E}, \mathcal{D}, \mathcal{H}, M$)
00 $c \overset{\$}{\leftarrow} \mathcal{Y} \setminus \mathcal{X}$	02 $(c, Z) := M$
01 **return** c	03 **if** $\exists K$ s.t. $(c, K) \in \mathcal{E}$ **and** $\mathsf{Priv}(\mathsf{sk}, c) = Z$
	04 $\quad h := K$
	05 **else**
	06 $\quad h \overset{\$}{\leftarrow} \{0,1\}^\kappa$
	07 **return** h

Fig. 6. Simulator Sim = (SimGen, SimEncaps, SimHash) for KEM, where SimGen = Gen. List $\mathcal{E}$ is either $\mathcal{CK}$ or $\mathcal{C}$.

HASH PROOF SYSTEM. A hash proof system HPS = (Par, Priv, Pub) consists of three algorithms. The randomized algorithm Par generates parametrized instances of $\mathsf{par} = (group, \mathcal{Z}, \mathcal{Y}, \mathcal{X}, \mathcal{PK}, \mathcal{SK}, \Lambda_{(\cdot)} : \mathcal{Y} \to \mathcal{Z}, \mu : \mathcal{SK} \to \mathcal{PK})$, where *group* may contain additional structural parameters. The deterministic public evaluation algorithm Pub inputs the projection key $\mathsf{pk} = \mu(\mathsf{sk})$, $c \in \mathcal{X}$ and a witness r of the fact that $c \in \mathcal{X}$ and returns $Z = \Lambda_{\mathsf{sk}}(c)$. The deterministic private evaluation algorithm Priv takes $\mathsf{sk} \in \mathcal{SK}$ and returns $\Lambda_{\mathsf{sk}}(c)$ without knowing a witness. Furthermore, we assume that μ is efficiently computable and that there are efficient algorithms for sampling $c \in \mathcal{X}$ uniformly together with a witness r, sampling $c \in \mathcal{Y}$ uniformly and checking membership in $\mathcal{Y}$.

(m-FOLD) SUBSET MEMBERSHIP PROBLEM. We define the m-fold subset membership problem for HPS which requires to distinguish m ciphertexts uniformly drawn from $\mathcal{X}$ from m ciphertexts uniformly drawn from $\mathcal{Y} \setminus \mathcal{X}$. The advantage of an adversary $\mathcal{A}$ is defined as

$$\mathrm{Adv}_{\mathsf{HPS}}^{m-\mathsf{SM}}(\mathcal{A}) := |\Pr[\mathcal{A}(\mathcal{Y}, \mathcal{X}, c_1, ..., c_m) \Rightarrow 1] - \Pr[\mathcal{A}(\mathcal{Y}, \mathcal{X}, c_1', ..., c_m') \Rightarrow 1]| \ ,$$

where $c_1, ..., c_m \overset{\$}{\leftarrow} \mathcal{X}$ and $c_1', ..., c_m' \overset{\$}{\leftarrow} \mathcal{Y} \setminus \mathcal{X}$.

N-RECEIVER NCKE FROM HPS. We use a k-entropic hash proof system HPS = (Par, Pub, Priv) with m-fold subset membership problem and a random oracle $\mathsf{H} : \{0,1\}^* \to \{0,1\}^\kappa$ in order to construct a key encapsulation algorithm KEM and a simulator Sim as shown in Figs. 5 and 6. The encapsulation algorithm Encaps samples an element c from $\mathcal{X}$ and a witness r. It runs the public evaluation algorithm and computes the key K as $\mathsf{H}(c, \mathsf{Pub}(\mathsf{pk}, c, r))$. The decapsulation algorithm Decaps uses the result of the private evaluation algorithm Priv as input to H to compute K. Instead of sampling an element from $\mathcal{X}$, the SimEncaps algorithm samples an element c uniformly at random from$\mathcal{Y} \setminus \mathcal{X}$ and only returns c. The SimHash algorithm takes as input three sets $\mathcal{E}, \mathcal{D}, \mathcal{H}$, where $\mathcal{E} \in \{\mathcal{C}, \mathcal{CK}\}$, and the value $M = (c, Z)$ chosen by the adversary. If there exists a key K such that $(c, K) \in \mathcal{E}$ (note that for $\mathcal{E} = \mathcal{C}$ this will never be true) and the adversary's input to H satisfies $\mathsf{Priv}(\mathsf{sk}, c) = Z$, then the output value h is set to K.

Theorem 1 (**k-entropic HPS with $(N \cdot q_E)$-fold SMP $\Rightarrow$ N-NCKE**). *For any N-NCKE adversary $\mathcal{A}$ against KEM and Sim that issues at most q_E queries to* Encaps*, q_D queries to* Decaps *and at most q_{H} queries to each random oracle H_n for $n \in [N]$, there exists an adversary $\mathcal{B}$ against the $(N \cdot q_E)$-fold subset membership problem of* HPS *such that*

$$\mathrm{Adv}_{\mathsf{KEM},\mathsf{Sim}}^{N\text{-}\mathsf{NCKE}}(\mathcal{A}) \leq \mathrm{Adv}_{\mathsf{HPS}}^{(N \cdot q_E)\text{-}\mathsf{SM}}(\mathcal{B}) + \frac{N \cdot q_E \cdot q_{\mathsf{H}}}{2^k} + \frac{N \cdot q_E \cdot q_D}{|\mathcal{Y} \setminus \mathcal{X}|},$$

where HPS *is k-entropic, $\mathcal{Y}$ is the set of all ciphertexts and $\mathcal{X}$ is the set of valid ciphertexts.*

We will give an instantiation based on the DDH assumption in Sect. 7.1. For the proof of Theorem 1 and an instantiation based on the higher residuosity assumption, we refer to the full version [23].

4 Security Model for Two-Message Authenticated Key Exchange

A two-message key exchange protocol $\mathsf{AKE} = (\mathsf{Gen}_{\mathsf{AKE}}, \mathsf{Init}_{\mathrm{I}}, \mathsf{Der}_{\mathsf{R}}, \mathsf{Der}_{\mathrm{I}})$ consists of four algorithms which are executed interactively by two parties as shown in Fig. 7. We denote the party which initiates the session by P_i and the party which responds to the session by P_r. The key generation algorithm $\mathsf{Gen}_{\mathsf{AKE}}$ outputs a key pair $(\mathsf{pk}, \mathsf{sk})$ for one party. The initialization algorithm $\mathsf{Init}_{\mathrm{I}}$ inputs the initiator's long-term secret key sk_i and the responder's long-term public key pk_r and outputs a message I and a state st. The responder's derivation algorithm $\mathsf{Der}_{\mathsf{R}}$ takes as input the responder's long-term secret key sk_r, the initiator's long-term public key pk_i and a message I. It computes a message R and a session key K. The initiator's derivation algorithm $\mathsf{Der}_{\mathrm{I}}$ inputs the initiator's long-term secret key sk_i, the responder's long-term public key pk_r, a message R and a state st. It outputs a session key K. Note that in contrast to the initiating party P_i, the responding party P_r will not be required to save any (secret) state information besides the session key K. The session key can be derived immediately after receiving the initiator's message.

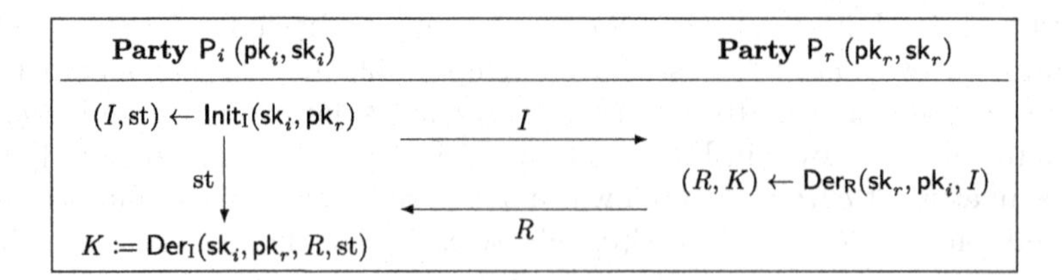

Fig. 7. Running a key exchange protocol between two parties.

Following [22], we define a game-based security model for authenticated key exchange using pseudocode. Our models for two different levels of security

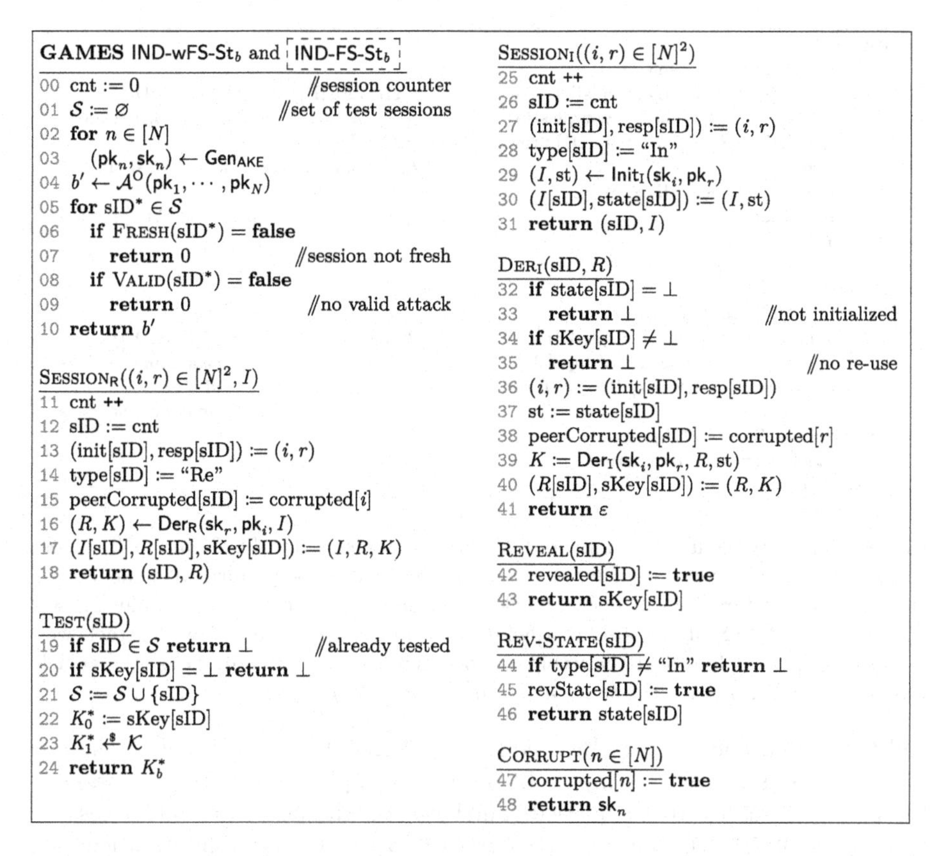

Fig. 8. Games IND-wFS-St$_b$ and IND-FS-St$_b$ for AKE, where $b \in \{0, 1\}$. $\mathcal{A}$ has access to oracles $\mathrm{O} := \{$SESSION$_\mathrm{I}$, SESSION$_\mathrm{R}$, DER$_\mathrm{I}$, REVEAL, REV-STATE, CORRUPT, TEST$\}$. Helper procedures FRESH and VALID are defined in Fig. 9. If there exists any test session which is neither fresh nor valid, the game will return 0.

are given in Fig. 8. We consider N parties $\mathsf{P}_1, ..., \mathsf{P}_N$ with long-term key pairs $(\mathsf{pk}_n, \mathsf{sk}_n)$, $n \in [N]$. Each session between two parties has a unique identification number sID and variables which are defined relative to sID:

- init[sID] $\in [N]$ denotes the initiator of the session.
- resp[sID] $\in [N]$ denotes the responder of the session.
- type[sID] $\in \{$"In", "Re"$\}$ denotes the session's view, i. e. whether the initiator or the responder computes the session key.
- I[sID] denotes the message that was computed by the initiator.
- R[sID] denotes the message that was computed by the responder.
- state[sID] denotes the state information that is stored by the initiator.
- sKey[sID] denotes the session key.

To establish a session between two parties, the adversary is given access to oracles SESSION$_\mathrm{I}$ and SESSION$_\mathrm{R}$, where the first one starts a session of type "In"

```
FRESH(sID*)
00 (i*, r*) := (init[sID*], resp[sID*])
01 𝔐(sID*) := {sID | (init[sID], resp[sID]) = (i*, r*) ∧ (I[sID], R[sID]) = (I[sID*], R[sID*])
                  ∧ type[sID] ≠ type[sID*]}                          //matching sessions
02 if revealed[sID*] or (∃sID ∈ 𝔐(sID*) : revealed[sID] = true)
03    return false                            //A trivially learned the test session's key
04 if ∃sID ∈ 𝔐(sID*) s.t. sID ∈ S
05    return false                                   //A also tested a matching session
06 return true

VALID(sID*)
07 (i*, r*) := (init[sID*], resp[sID*])
08 𝔐(sID*) := {sID | (init[sID], resp[sID]) = (i*, r*) ∧ (I[sID], R[sID]) = (I[sID*], R[sID*])
                  ∧ type[sID] ≠ type[sID*]}                          //matching sessions
09 𝔓(sID*) := {sID | I[sID] = I[sID*] ∧ type[sID] = "In" ∧ type[sID] ≠ type[sID*]}
                                                            //partially matching sessions
10 for attack ∈ Table 1 [Table 2]
11    if attack = true return true
12 return false
```

Fig. 9. Helper procedures FRESH and VALID for games IND-wFS-St and IND-FS-St defined in Fig. 8. Procedure FRESH checks if the adversary performed some trivial attack. In procedure VALID, each attack is evaluated by the set of variables shown in Table 1 (IND-wFS-St) or Table 2 (IND-FS-St) and checks if an allowed attack was performed. If the values of the variables are set as in the corresponding row, the attack was performed, i. e. attack = **true**, and thus the session is valid.

and the second one of type "Re". Following [26,28], these oracles also take the intended peer's identity as input. In order to complete the initiator's session, the oracle DER_I has to be queried. Furthermore, the adversary has access to oracles CORRUPT, REVEAL and REV-STATE to obtain secret information. As the responder can directly compute the key in a two-message protocol, we only require the initiator to store a state. The state contains information that is needed to compute the session key when the response is received, so it will consist of public and private information. We do not require to reveal the full randomness as in the eCK model [28]. A REV-STATE query may be issued at any time. We use the following boolean values to keep track of which queries the adversary made:

- corrupted[n] denotes whether the long-term secret key of party P_n was given to the adversary.
- revealed[sID] denotes whether the session key was given to the adversary.
- revState[sID] denotes whether the state information of that session was given to the adversary.
- peerCorrupted[sID] denotes whether the peer of the session was corrupted at the time the session key is computed, which is important for forward security.

The adversary can forward messages between sessions or modify them. By that, we can define the relationship between two sessions:

- **Matching Session**: Two sessions sID, sID′ *match* if the same parties are involved (init[sID] = init[sID′] and resp[sID] = resp[sID′]), the messages sent

and received are the same ($I[\text{sID}] = I[\text{sID}']$ and $R[\text{sID}] = R[\text{sID}']$) and they are of different types ($\text{type}[\text{sID}] \neq \text{type}[\text{sID}']$).
- **Partially Matching Session**: A session sID′ of type "In" is *partially matching* to session sID of type "Re" if the initial messages are the same ($I[\text{sID}] = I[\text{sID}']$).

Finally, the adversary is given access to oracle TEST which will return either the session key of the specified session or a uniformly random key. In our security models, we allow multiple test queries. We store test sessions in a set $\mathcal{S}$. In general, the adversary can disclose the complete interaction between two parties by querying the long-term secret keys, the state information and the session key. However, for each test session, we require that the adversary does not issue queries such that the session key can be trivially computed. We define the properties of freshness and validity which all test sessions have to satisfy:

- **Freshness**: A (test) session is called *fresh* if the session key was not revealed. Furthermore, if there exists a matching session, we require that this session's key is not revealed and that this session is not also a test session.
- **Validity**: A (test) session is called *valid* if it is fresh and the adversary performed any attack which is defined in the security model. We capture this with attack tables (cf. Tables 1 and 2). A description of how to read the tables is given below.

Attack Tables. All attacks are defined using variables to indicate which queries the adversary may (not) make. We consider three dimensions covering all possible combinations of reveal queries the adversary can make:

- whether the test session is on the initiator's ($\text{type}[\text{sID}^*] =$"In") or the responder's side ($\text{type}[\text{sID}^*] =$"Re"),
- all combinations of long-term secret key and state reveals (corrupted and revState variables), also taking into account when a corruption happened (peerCorrupted),
- whether the adversary acted passively (matching session), partially active (partially matching session) or actively (no matching session).

This yields a full table of 24 attacks, in particular capturing *key compromise impersonation* (KCI) and *maximal exposure* (MEX) attacks. An attack was performed if the variables are set to the corresponding values in the table. However, when considering two-message protocols, where the responder's side does not have a state, and we only consider *weak forward security*, some of the attacks are redundant. Thus, we obtain *distilled* tables. We exclude trivial attacks, e.g., the generic attack on two-message AKE protocols with state-reveals described in [30]. Therefore, the adversary is not allowed to obtain the state of a partially matching session. Also note that by definition, a partially matching session for a two-message protocol can only be of type "Re". Table 1 is the distilled table used for the IND-wFS-St security game and Table 2 is used for the IND-FS-St security game. Note that the numbering of attacks in the distilled tables is inherited from the full table given in the full version [23].

Table 1. Distilled table of attacks for wFS adversaries against two-message protocols. This table is obtained from the full table of attacks by using that responders do not have a state and that we are considering weak forward security. The numbering of attacks is inherited from the full table. An attack is regarded as an AND conjunction of variables with specified values as shown in the each line, where "–" means that this variable can take arbitrary value. **F** means "false" and "n/a" indicates that there is no state which can be revealed as no (partially) matching session exists.

$\mathcal{A}$ gets (Initiator, Responder)		corrupted[i^*]	corrupted[r^*]	type[sID*]	revState[sID*]	$\exists$sID $\in \mathfrak{M}$(sID*) : revState[sID]	$\lvert\mathfrak{M}$(sID*)$\rvert$	$\exists$sID $\in \mathfrak{P}$(sID*) : revState[sID]	$\lvert\mathfrak{P}$(sID*)$\rvert$
(0)	**multiple partially matching sessions**	–	–	–	–	–	–	–	> 1
(1∨2)	**(long-term, long-term)**	–	–	–	**F**	**F**	1	–	–
(7∨8)	**(state, long-term)**	**F**	–	–	–	–	1	–	–
(10)	**(long-term, long-term)**	–	–	"Re"	**F**	n/a	0	**F**	1
(16)	**(state, long-term)**	**F**	–	"Re"	**F**	n/a	0	–	1
(19)	**(state, state)**	**F**	**F**	"In"	–	n/a	0	n/a	0
(21)	**(long-term, state)**	–	**F**	"In"	**F**	n/a	0	n/a	0
(24)	**(state, long-term)**	**F**	–	"Re"	**F**	n/a	0	n/a	0

Table 2. Distilled table of attacks for full FS adversaries against two-message protocols. This table is obtained from the full table of attacks by removing redundant rows and using that responders do not have a state. The numbering of attacks is inherited from the full table. An attack is regarded as an AND conjunction of variables with specified values as shown in the each line, where "–" means that this variable can take arbitrary value. **F** means "false" and "n/a" indicates that there is no state which can be revealed as no (partially) matching session exists.

$\mathcal{A}$ gets (Initiator, Responder)		corrupted[i^*]	corrupted[r^*]	peerCorrupted[sID*]	type[sID*]	revState[sID*]	$\exists$sID $\in \mathfrak{M}$(sID*) : revState[sID]	$\lvert\mathfrak{M}$(sID*)$\rvert$	$\exists$sID $\in \mathfrak{P}$(sID*) : revState[sID]	$\lvert\mathfrak{P}$(sID*)$\rvert$
(0)	**multiple partially matching sessions**	–	–	–	–	–	–	–	–	> 1
(1∨2)	**(long-term, long-term)**	–	–	–	–	**F**	**F**	1	–	–
(7∨8)	**(state, long-term)**	**F**	–	–	–	–	–	1	–	–
(10)	**(long-term, long-term)**	–	–	**F**	"Re"	**F**	n/a	0	**F**	1
(16)	**(state, long-term)**	**F**	–	–	"Re"	**F**	n/a	0	–	1
(17)	**(long-term, long-term)**	–	–	**F**	"In"	**F**	n/a	0	n/a	0
(18)	**(long-term, long-term)**	–	–	**F**	"Re"	**F**	n/a	0	n/a	0
(23)	**(state, long-term)**	**F**	–	**F**	"In"	–	n/a	0	n/a	0

However, if the protocol does not use appropriate randomness, it should not be considered secure in our model. Thus, if the adversary is able to create more than one (partially) matching session to a test session, it may also run a trivial attack. We model this in row (0) of Tables 1 and 2.

Example. If the test session is an initiating session (type[sID*] ="In"), the state was not revealed (revState[sID*] = **false**) and there is a matching session ($|\mathfrak{M}(\mathrm{sID}^*)| = 1$), then row ($1 \vee 2$) will evaluate to true. In this scenario, the adversary is allowed to query both long-term secret keys.

For all test sessions, at least one attack has to evaluate to true. Then, the adversary wins if it distinguishes the session keys from uniformly random keys which it obtains through queries to the TEST oracle.

Definition 2 (Key Indistinguishability of AKE). *We define games* $\mathsf{IND\text{-}wFS\text{-}St}_b$ *and* $\mathsf{IND\text{-}FS\text{-}St}_b$ *for* $b \in \{0,1\}$ *as in Figs. 8 and 9. The advantage of an adversary* $\mathcal{A}$ *against* AKE *in these games is defined as*

$$\mathrm{Adv}_{\mathsf{AKE}}^{\mathsf{IND\text{-}wFS\text{-}St}}(\mathcal{A}) := \left|\Pr[\mathsf{IND\text{-}wFS\text{-}St}_1^{\mathcal{A}} \Rightarrow 1] - \Pr[\mathsf{IND\text{-}wFS\text{-}St}_0^{\mathcal{A}} \Rightarrow 1]\right| \quad \textit{and}$$

$$\mathrm{Adv}_{\mathsf{AKE}}^{\mathsf{IND\text{-}FS\text{-}St}}(\mathcal{A}) := \left|\Pr[\mathsf{IND\text{-}FS\text{-}St}_1^{\mathcal{A}} \Rightarrow 1] - \Pr[\mathsf{IND\text{-}FS\text{-}St}_0^{\mathcal{A}} \Rightarrow 1]\right| .$$

When proving the security of a protocol, the success probability for each attack strategy listed in the corresponding table will have to be analyzed, thus showing that independently of which queries the adversary makes, it cannot distinguish the session key from a uniformly random key.

4.1 Relation to Other Definitions

In this section, we will refer to the most widely used security definitions for authenticated key exchange protocols. In the first place, these include the CK model [9] and the stronger definition used for the HMQV protocol (CK+) in [26], the eCK model [28] and the strengthened version of [14], the definitions given in [24] and [1] which are both extensions of the BR model [4], and the definition of IND-Å security in [22]. In [12,13], Cremers showed that the CK, CK+ und eCK model are incomparable. Thus, we will not do a formal comparison of security models, but only point out similarities and differences between our definition and the definitions listed above.

PARTY CORRUPTION. We allow the adversary to corrupt a party which means that it will obtain that party's long-term secret key as in the eCK model and the models given in [1,22,24]. In contrast, a corrupt query in the CK and CK+ model will reveal all information in the memory of that party, i. e. long-term secrets and session-specific information.

STATE-REVEALS. Our model only allows state-reveal queries on initiating sessions because the initiator has to wait for the response to compute the session key. Thus, the state contains all that information that is needed to derive the session key as soon as the responder's message is received. The responder can directly compute the session key and does not have to store other information. The eCK model explicitly defines the state as the randomness that is used in the protocol. In the CK model, it is not clear which information is included in the state, but it is left to be specified by the AKE protocol itself. Other models such as [24], its extension given in [1] and the one used in [10] do not allow state-reveals at all. Here, we want to emphasize that in particular all previous work on tight AKE does not consider state reveals and we are the first ones to address this problem.

(WEAK) FORWARD SECURITY. Following Krawczyk [26], we specify two levels of forward security. IND-wFS-St models weak forward security, whereas IND-FS-St models full forward security. The first one is intended for 2-message protocols with implicit authentication, as those cannot achieve full forward security [26]. The second one is intended for protocols with explicit authentication. With those definitions, we capture the same properties as the most common security models given in [1,9,24,26,28], where some of them only define either weak or full forward security depending on whether they consider implicitly or explicitly authenticated protocols.

MATCHING SESSIONS AND PARTNERING. As most security models, ours use the concept of matching sessions to define a relation between two sessions. Following Cremer and Feltz [14], we additionally use the term of origin (or partially matching) sessions, which refers to a relaxation of the definition of matching sessions. The concept of origin sessions is used for full forward security, in particular we need this to handle the no-match attack described by Li and Schäge [30], where two sessions compute the same session key but do not have matching conversations. Recent works such as [10,21] take up the approach of origin sessions and oracle partnering based on session keys as additional requirement.

ON REGISTERING CORRUPT KEYS. Some security models for AKE allow the adversary also to *register* adversarially-generated keys, this holds in particular for previous works considering tightly-secure key exchange [1,10,21]. Technically this makes the security model strictly stronger, as one can easily construct contrived protocols that are insecure with adversarially-registered keys, but secure without.

However, in the actual security proofs in [1,10,21], adversarially-registered keys are treated no differently than corrupted keys. We chose to keep model, security proofs and notation as simple as possible (it is already complex enough, anyway), and thus omitted this query. However, it is straightforward to extend our model with it, and the proofs need not to be changed. Whenever the adversary registers a new key, it would immediately be marked as "corrupted" (just like in [1,10,21]). Apart from that, no additional changes to the proofs are required, since the proofs deal with all corrupted keys in the same way, regardless of their distribution or whether they are generated by the experiment or an external entity. We also do not require a proof of knowledge of the corresponding secret key for the registration, or a proof that the registered public key is valid in any sense.

5 AKE with Weak Forward Security

In this section, we show how to build an implicitly authenticated AKE protocol using the concept of non-committing key encapsulation.

In particular, from two key encapsulation mechanisms $\mathsf{KEM}_{\mathsf{CPA}} = (\mathsf{Gen}_{\mathsf{CPA}}, \mathsf{Encaps}_{\mathsf{CPA}}, \mathsf{Decaps}_{\mathsf{CPA}})$ and $\mathsf{KEM}_{\mathsf{CCA}} = (\mathsf{Gen}_{\mathsf{CCA}}, \mathsf{Encaps}_{\mathsf{CCA}}, \mathsf{Decaps}_{\mathsf{CCA}})$, we construct a two-message authenticated key exchange protocol $\mathsf{AKE}_{\mathsf{wFS}} = (\mathsf{Gen}_{\mathsf{AKE}}, \mathsf{Init}_{\mathrm{I}}, \mathsf{Der}_{\mathrm{R}}, \mathsf{Der}_{\mathrm{I}})$ as shown in Figs. 10 and 11. W.l.o.g. $\mathsf{KEM}_{\mathsf{CPA}}$, $\mathsf{KEM}_{\mathsf{CCA}}$, $\mathsf{AKE}_{\mathsf{wFS}}$ have identical key space $\mathcal{K}$. Each party holds a long-term key pair $(\mathsf{pk}, \mathsf{sk})$ for $\mathsf{KEM}_{\mathsf{CCA}}$ and a symmetric key k to encrypt the secret state information which has to be stored by the initiating party. State encryption protects against state attacks and is implemented using a symmetric encryption scheme defined as $\mathsf{E}_k(\mathrm{st}') := (IV, \mathsf{G}(k, IV) \oplus \mathrm{st}')$ for a random nonce IV. Here $\mathsf{G} : \{0,1\}^* \to \{0,1\}^d$ is a random oracle and d is an integer denoting the maximum bit length of the unencrypted state st'. The protocol uses an additional cryptographic hash function $\mathsf{H} : \{0,1\}^* \to \mathcal{K}$ to output the session key.

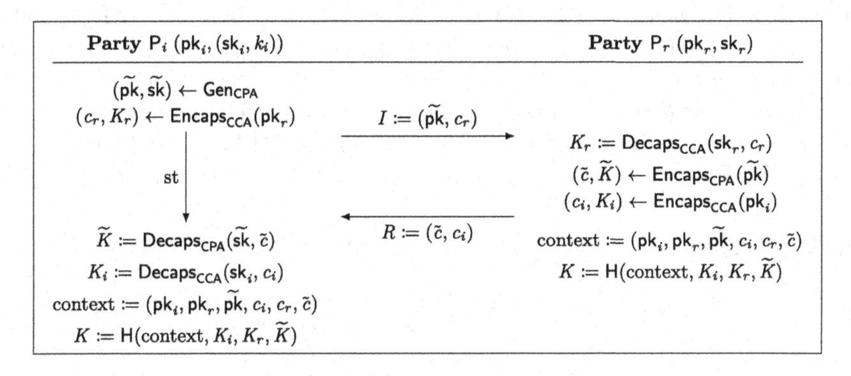

Fig. 10. Visualization: Running protocol $\mathsf{AKE}_{\mathsf{wFS}}$ between two parties.

The initiating party generates an ephemeral key pair for $\mathsf{KEM}_{\mathsf{CPA}}$, then runs the $\mathsf{Encaps}_{\mathsf{CCA}}$ algorithm on the peer's public key to output a ciphertext c_r and a key K_r and sends the ephemeral public key and c_r to the intended receiver. All values are stored temporarily and encrypted as described above, as they will later be needed to compute the session key. The responding party uses its secret key sk_r to compute key K_r from c_r. Next, it runs the $\mathsf{Encaps}_{\mathsf{CPA}}$ algorithm on the received ephemeral public key to compute a ciphertext $\tilde{c}$ and a key $\widetilde{K}$ and then the $\mathsf{Encaps}_{\mathsf{CCA}}$ algorithm on the initiator's public key to output c_i and K_i. It sends both ciphertexts to the initiating party and computes the session key evaluating the hash function H on all public context and the three shared keys K_r, K_i and $\widetilde{K}$. The initiator retrieves the secret state information and computes K_i and $\widetilde{K}$ from c_i and $\tilde{c}$. Now, it can also establish the session key.

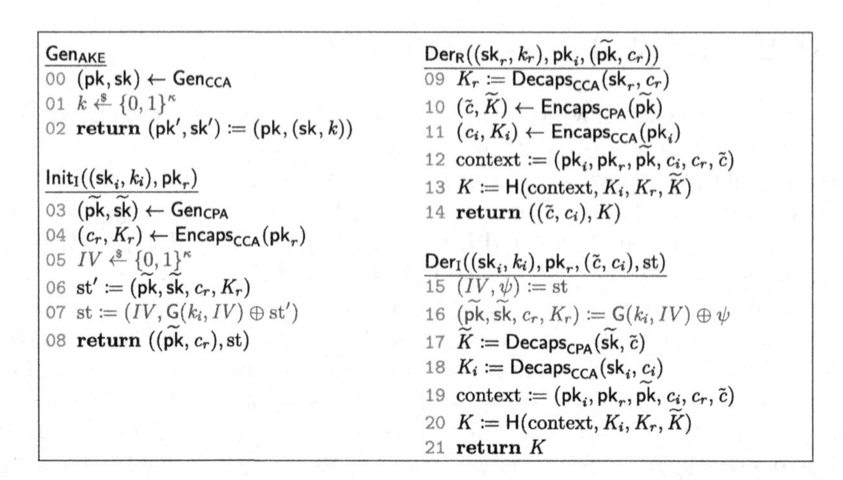

Fig. 11. Authenticated key exchange protocol $\mathsf{AKE_{wFS}}$ from $\mathsf{KEM_{CPA}}$ and $\mathsf{KEM_{CCA}}$. Lines written in purple color are only used to encrypt the state.

Theorem 2 ($\mathsf{KEM_{CPA}}$ NCKE-CPA + $\mathsf{KEM_{CCA}}$ NCKE-CCA $\overset{\mathrm{ROM}}{\Rightarrow}$ $\mathsf{AKE_{wFS}}$ IND-wFS-St). *For any* IND-wFS-St *adversary* $\mathcal{A}$ *against* $\mathsf{AKE_{wFS}}$ *with* N *parties that establishes at most* S *sessions and issues at most* T *queries to the test oracle* TEST, q_G *queries to random oracle* G *and at most* q_H *queries to random oracle* H, *there exists an* N-NCKE-CCA *adversary* $\mathcal{B}$ *against* $\mathsf{KEM_{CCA}}$ *and* $\mathsf{Sim_{CCA}}$ *and an* S-NCKE-CPA *adversary* $\mathcal{C}$ *against* $\mathsf{KEM_{CPA}}$ *and* $\mathsf{Sim_{CPA}}$ *such that*

$$\mathrm{Adv}^{\text{IND-wFS-St}}_{\mathsf{AKE_{wFS}}}(\mathcal{A}) \leq 2 \cdot \left(\mathrm{Adv}^{N\text{-NCKE-CCA}}_{\mathsf{KEM_{CCA}},\mathsf{Sim_{CCA}}}(\mathcal{B}) + \mathrm{Adv}^{S\text{-NCKE-CPA}}_{\mathsf{KEM_{CPA}},\mathsf{Sim_{CPA}}}(\mathcal{C})\right) + T \cdot \left(\frac{q_\mathsf{G}}{2^\kappa} + \frac{q_\mathsf{H}}{|\mathcal{K}|}\right)$$
$$+ N^2 \cdot \left(\frac{1}{2^{\mu_\mathsf{CCA}}} + \frac{1}{2^\kappa}\right) + S^2 \cdot \left(\frac{1}{2^{\mu_\mathsf{CPA}}} + \frac{1}{2^{\gamma_\mathsf{CCA}}} + \frac{1}{2^{\gamma_\mathsf{CPA}}} + \frac{1}{2^\kappa}\right) + 2S \cdot \frac{q_\mathsf{G}}{2^{2\kappa}},$$

where $\mathsf{Sim_{CCA}}$ *and* $\mathsf{Sim_{CPA}}$ *are the simulators from the* NCKE *experiments,* μ_CCA *and* μ_CPA *are the collision probability of the key generation algorithms* $\mathsf{Gen_{CCA}}$ *and* $\mathsf{Gen_{CPA}}$*,* γ_CCA *and* γ_CPA *are the spreadness parameters of the encapsulation algorithms* $\mathsf{Encaps_{CCA}}$ *and* $\mathsf{Encaps_{CPA}}$ *and* κ *is a security parameter. The running times of* $\mathcal{B}$ *and* $\mathcal{C}$ *consist essentially of the time required to execute the security experiment with the adversary once, plus a minor number of additional operations (including bookkeeping, lookups etc.).*

Proof (Sketch). Let $\mathcal{A}$ be an adversary against IND-wFS-St security of $\mathsf{AKE_{wFS}}$. For $b \in \{0,1\}$, game $G_{0,b}$ is the IND-wFS-St$_b$ game, where we additionally exclude that collisions between long-term key pairs, ephemeral key pairs, ciphertexts and nonces occur.

In game $G_{1,b}$, we replace the computations for $\mathsf{KEM_{CCA}}$ by the simulator $\mathsf{Sim_{CCA}}$, which allows to draw keys K_i and K_r uniformly at random. This change affects all sessions which makes the proof tight. If the adversary reveals a long-term key pair of any user, the property of receiver non-committing key encapsulation ensures that the correct keys K_i and K_r can be computed by the adversary.

Next, we want to replace the computations for $\mathsf{KEM}_{\mathsf{CPA}}$ by the simulator $\mathsf{Sim}_{\mathsf{CPA}}$, which allows to draw keys $\widetilde{K}$ uniformly at random. However, the ephemeral secret key $\widetilde{\mathsf{sk}}$ is part of the state and will not be available to the NCKE-CPA reduction in the first place. Thus, we introduce an intermediate game $G_{2,b}$ and do not compute the state when the session is initiated but only when the adversary queries the REV-STATE oracle. In game $G_{3,b}$, we can then use the simulator for $\mathsf{KEM}_{\mathsf{CPA}}$ and draw keys $\widetilde{K}$ uniformly at random, whenever the ephemeral public key $\widetilde{\mathsf{pk}}$ comes from the experiment (i.e. the adversary creates a partially matching session). Again, the non-committing property of $\mathsf{KEM}_{\mathsf{CPA}}$ ensures consistency in case the adversary reveals both the state of a session and the long-term key of the initiator, which reveals the ephemeral secret key $\widetilde{\mathsf{sk}}$.

Depending on whether there exists a (partially) matching session and which queries to REV-STATE and CORRUPT the adversary makes, we can argue that at least one key K_i, K_r or $\widetilde{K}$ in each test session is chosen uniformly at random and unknown to $\mathcal{A}$ and thus it cannot distinguish the session key from a uniformly random key in the last game $G_{4,b}$. □

The full proof of Theorem 2 can be found in the full version [23]. Note that the non-committing property is essential to embed random KEM keys in each session and thus to achieve tightness. This way, we only need to make a case distinction at the end and can argue that for all test sessions at least one KEM key is independent of the adversary's view no matter which queries it has made (provided it did not make a trivial attack). Relying on a weaker assumption requires to make a case distinction earlier in the proof and may involve guessing as in some cases it is not clear which KEM key will be revealed (through corruption and/or reveal or state reveal) at a later point in time.

6 AKE with Full Forward Security

We show how to build an explicitly authenticated AKE protocol using the concept of non-committing key encapsulation. As we also need a signature scheme, we will first give the relevant definitions.

6.1 Digital Signatures

A digital signature scheme $\mathsf{SIG} = (\mathsf{Gen}_{\mathsf{SIG}}, \mathsf{Sign}, \mathsf{Vrfy})$ consists of three algorithms. The key generation algorithm $\mathsf{Gen}_{\mathsf{SIG}}$ outputs a key pair $(\mathsf{vk}, \mathsf{sigk})$, where vk is the verification key and sigk the signing key. The signing algorithm Sign inputs a signing key sigk and a message m and outputs a signature σ. The deterministic verification algorithm Vrfy inputs the verification key vk, a message m and a signature σ and outputs 1 if σ is a valid signature for m, otherwise it outputs 0.

In Fig. 12, we define the security game N user Strong UnForgeability under Chosen Message Attacks with corruptions (N-SUF-CMA). The definition is similar to the one given in [1], except that we require *strong* unforgeability, i. e. the adversary may also find a new signature for a message it queried to the SIGN oracle before. The advantage of an adversary $\mathcal{A}$ is defined as

$$\mathrm{Adv}_{\mathsf{SIG}}^{N\text{-}\mathsf{SUF}\text{-}\mathsf{CMA}}(\mathcal{A}) := \Pr[N\text{-}\mathsf{SUF}\text{-}\mathsf{CMA}^{\mathcal{A}} \Rightarrow 1] \ .$$

```
GAME N-SUF-CMA
00 S^corr := ∅
01 for n ∈ [N]
02    (vk_n, sigk_n) ← Gen_SIG
03    S_n := ∅
04 (n*, m*, σ*) ← A^{SIGN,CORRUPT}(vk_1, ..., vk_N)
05 if Vrfy(vk_{n*}, m*, σ*) = 1 and n* ∉ S^corr
   and (m*, σ*) ∉ S_{n*}
06    return 1
07 else
08    return 0

SIGN(n ∈ [N], m)
09 σ ← Sign(sigk_n, m)
10 S_n := S_n ∪ {(m, σ)}
11 return σ

CORRUPT(n ∈ [N])
12 S^corr := S^corr ∪ {n}
13 return sigk_n
```

Fig. 12. Game N-SUF-CMA for SIG.

6.2 Transformation Using NCKE and a Signature Scheme

From two key encapsulation mechanisms $\mathsf{KEM}_{\mathsf{CPA}} = (\mathsf{Gen}_{\mathsf{CPA}}, \mathsf{Encaps}_{\mathsf{CPA}}, \mathsf{Decaps}_{\mathsf{CPA}})$ and $\mathsf{KEM}_{\mathsf{CCA}} = (\mathsf{Gen}_{\mathsf{CCA}}, \mathsf{Encaps}_{\mathsf{CCA}}, \mathsf{Decaps}_{\mathsf{CCA}})$ with key space $\mathcal{K}$ and a digital signature scheme $\mathsf{SIG} = (\mathsf{Gen}_{\mathsf{SIG}}, \mathsf{Sign}, \mathsf{Vrfy})$, we construct a two-message authenticated key exchange protocol $\mathsf{AKE}_{\mathsf{FS}} = (\mathsf{Gen}_{\mathsf{AKE}}, \mathsf{Init}_{\mathrm{I}}, \mathsf{Der}_{\mathsf{R}}, \mathsf{Der}_{\mathrm{I}})$ with key space $\mathcal{K}$ as shown in Figs. 13 and 14. Each party has a key pair $(\mathsf{vk}, \mathsf{sigk})$ for SIG, a key pair $(\mathsf{pk}, \mathsf{sk})$ for $\mathsf{KEM}_{\mathsf{CCA}}$ and a symmetric key k to encrypt the secret state information which has to be stored by the initiating party (cf. Sect. 5). The protocol uses additional cryptographic hash functions $\mathsf{F} : \{0,1\}^* \to \{0,1\}^\kappa$ to compute value π and $\mathsf{H} : \{0,1\}^* \to \mathcal{K}$ to output the session key.

The initiating party computes an ephemeral key pair for $\mathsf{KEM}_{\mathsf{CPA}}$, runs the $\mathsf{Encaps}_{\mathsf{CCA}}$ algorithm on the intended receiver's public key pk_r to obtain a ciphertext c_r and a key K_r and signs both the ephemeral public key and c_r, which are sent to the receiver along with the signature. The receiver verifies the signature and then runs the $\mathsf{Encaps}_{\mathsf{CPA}}$ algorithm on the ephemeral public key to output a ciphertext $\tilde{c}$ and a key $\widetilde{K}$. It computes K_r using its secret key sk_r. It then tags the received message together with $\tilde{c}$ and K_r by evaluating hash function F and sends the output together with $\tilde{c}$ to the initiator. The initiator retrieves K_r from the secret state and also evaluates F. If the output is the same, it computes $\widetilde{K}$ using the ephemeral secret key. The session key is computed evaluating hash function H on all public context and key $\widetilde{K}$. We establish the following theorem and give a proof sketch. The full proof can be found in the full version [23].

Theorem 3 ($\mathsf{KEM}_{\mathsf{CPA}}$ NCKE-CPA + $\mathsf{KEM}_{\mathsf{CCA}}$ NCKE-CCA + SIG N-SUF-CMA $\overset{\mathrm{ROM}}{\Rightarrow}$ $\mathsf{AKE}_{\mathsf{FS}}$ IND-FS-St). *For any* IND-FS-St *adversary* $\mathcal{A}$

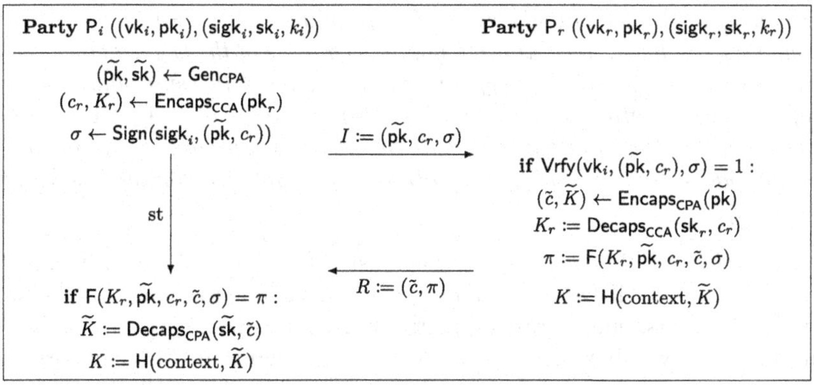

Party P_i $((\mathsf{vk}_i, \mathsf{pk}_i), (\mathsf{sigk}_i, \mathsf{sk}_i, k_i))$		Party P_r $((\mathsf{vk}_r, \mathsf{pk}_r), (\mathsf{sigk}_r, \mathsf{sk}_r, k_r))$
$(\widetilde{\mathsf{pk}}, \widetilde{\mathsf{sk}}) \leftarrow \mathsf{Gen}_{\mathsf{CPA}}$ $(c_r, K_r) \leftarrow \mathsf{Encaps}_{\mathsf{CCA}}(\mathsf{pk}_r)$ $\sigma \leftarrow \mathsf{Sign}(\mathsf{sigk}_i, (\widetilde{\mathsf{pk}}, c_r))$	$I := (\widetilde{\mathsf{pk}}, c_r, \sigma)$ $\longrightarrow$	
st $\downarrow$		if $\mathsf{Vrfy}(\mathsf{vk}_i, (\widetilde{\mathsf{pk}}, c_r), \sigma) = 1$: $(\tilde{c}, \widetilde{K}) \leftarrow \mathsf{Encaps}_{\mathsf{CPA}}(\widetilde{\mathsf{pk}})$ $K_r := \mathsf{Decaps}_{\mathsf{CCA}}(\mathsf{sk}_r, c_r)$ $\pi := \mathsf{F}(K_r, \widetilde{\mathsf{pk}}, c_r, \tilde{c}, \sigma)$
if $\mathsf{F}(K_r, \widetilde{\mathsf{pk}}, c_r, \tilde{c}, \sigma) = \pi$: $\widetilde{K} := \mathsf{Decaps}_{\mathsf{CPA}}(\widetilde{\mathsf{sk}}, \tilde{c})$ $K := \mathsf{H}(\text{context}, \widetilde{K})$	$\longleftarrow$ $R := (\tilde{c}, \pi)$	$K := \mathsf{H}(\text{context}, \widetilde{K})$

Fig. 13. Visualization: Running $\mathsf{AKE}_{\mathsf{FS}}$ between two parties, where K is the resulting session key and context $:= (\mathsf{vk}_i, \mathsf{pk}_i, \mathsf{vk}_r, \mathsf{pk}r, \widetilde{\mathsf{pk}}, c_r, \tilde{c}, \sigma, \pi)$

```
GenAKE
00 (vk, sigk) ← GenSIG
01 (pk, sk) ← GenCCA
02 k ←$ {0,1}^κ
03 return (pk', sk') :=
     ((vk, pk), (sigk, sk, k))

InitI((sigk_i, sk_i, k_i), (vk_r, pk_r))
04 (p̃k, s̃k) ← GenCPA
05 (c_r, K_r) ← EncapsCCA(pk_r)
06 σ ← Sign(sigk_i, (p̃k, c_r))
07 IV ←$ {0,1}^κ
08 st' := (p̃k, s̃k, c_r, K_r, σ)
09 st := (IV, G(k_i, IV) ⊕ st')
10 I := (p̃k, c_r, σ)
11 return (I, st)

DerR((sigk_r, sk_r, k_r), (vk_i, pk_i), (p̃k, c_r, σ))
12 if Vrfy(vk_i, (p̃k, c_r), σ) ≠ 1
13   return ⊥
14 (c̃, K̃) ← EncapsCPA(p̃k)
15 K_r := DecapsCCA(sk_r, c_r)
16 π := F(K_r, p̃k, c_r, c̃, σ)
17 context := (vk_i, pk_i, vk_r, pk_r, p̃k, c_r, c̃, σ, π)
18 K := H(context, K̃)
19 R := (c̃, π)
20 return (R, K)

DerI((sigk_i, sk_i, k_i), (vk_r, pk_r), (c̃, π), st)
21 (IV, ψ) := st
22 (p̃k, s̃k, c_r, K_r, σ) := G(k_i, IV) ⊕ ψ
23 if F(K_r, p̃k, c_r, c̃, σ) ≠ π
24   return ⊥
25 K̃ := DecapsCPA(s̃k, c̃)
26 context := (vk_i, pk_i, vk_r, pk_r, p̃k, c_r, c̃, σ, π)
27 K := H(context, K̃)
28 return K
```

Fig. 14. Authenticated key exchange protocol $\mathsf{AKE}_{\mathsf{FS}}$ from $\mathsf{KEM}_{\mathsf{CPA}}$, $\mathsf{KEM}_{\mathsf{CCA}}$ and SIG. Lines written in purple color are only used to encrypt the state.

against $\mathsf{AKE}_{\mathsf{FS}}$ *with* N *parties that establishes at most* S *sessions and issues at most* T *queries to test oracle* TEST, *at most* q_{H}, q_{G} *and* q_{F} *queries to random oracles* H, G *and* F, *there exists an* N-SUF-CMA *adversary* $\mathcal{B}$ *against* SIG, *an* S-NCKE-CPA *adversary* $\mathcal{C}$ *against* $\mathsf{KEM}_{\mathsf{CPA}}$ *and* $\mathsf{Sim}_{\mathsf{CPA}}$ *and an* N-NCKE-CCA *adversary* $\mathcal{D}$ *against* $\mathsf{KEM}_{\mathsf{CCA}}$ *and* $\mathsf{Sim}_{\mathsf{CCA}}$ *such that*

$$\begin{aligned}\mathrm{Adv}^{\mathsf{IND\text{-}FS\text{-}St}}_{\mathsf{AKE}_{\mathsf{FS}}}(\mathcal{A}) \leq\ & 2 \cdot \left(\mathrm{Adv}^{N\text{-}\mathsf{SUF\text{-}CMA}}_{\mathsf{SIG}}(\mathcal{B}) + \mathrm{Adv}^{S\text{-}\mathsf{NCKE\text{-}CPA}}_{\mathsf{KEM}_{\mathsf{CPA}},\mathsf{Sim}_{\mathsf{CPA}}}(\mathcal{C}) + \mathrm{Adv}^{N\text{-}\mathsf{NCKE\text{-}CCA}}_{\mathsf{KEM}_{\mathsf{CCA}},\mathsf{Sim}_{\mathsf{CCA}}}(\mathcal{D})\right) \\ & + T \cdot \left(\frac{q_{\mathsf{G}}}{2^{\kappa}} + \frac{q_{\mathsf{H}}}{|\mathcal{K}|}\right) + N^2 \cdot \left(\frac{1}{2^{\mu_{\mathsf{SIG}}}} + \frac{1}{2^{\mu_{\mathsf{CCA}}}} + \frac{1}{2^{\kappa}}\right) \\ & + S^2 \cdot \left(\frac{1}{2^{\mu_{\mathsf{CPA}}}} + \frac{1}{2^{\gamma_{\mathsf{CCA}}}} + \frac{1}{2^{\gamma_{\mathsf{CPA}}}} + \frac{1}{2^{\kappa}}\right) + 2S \cdot \frac{q_{\mathsf{G}}}{2^{2\kappa}} ,\end{aligned}$$

where $\mathsf{Sim}_{\mathsf{CPA}}$ *and* $\mathsf{Sim}_{\mathsf{CCA}}$ *are the simulators from the* NCKE-CPA *and* NCKE-CCA *experiment,* $\mu_{\mathsf{SIG}}, \mu_{\mathsf{CPA}}, \mu_{\mathsf{CCA}}$ *are collision probabilities of the key generation algorithms* $\mathsf{Gen}_{\mathsf{SIG}}$, $\mathsf{Gen}_{\mathsf{CPA}}$ *and* $\mathsf{Gen}_{\mathsf{CCA}}$ *and* $\gamma_{\mathsf{CPA}}, \gamma_{\mathsf{CCA}}$ *are the spreadness parameters of the encapsulation algorithms. The running times of* $\mathcal{B}$, $\mathcal{C}$ *and* $\mathcal{D}$ *consist essentially of the time required to execute the security experiment with the adversary once, plus a minor number of additional operations (including bookkeeping, lookups etc.).*

Proof (Sketch). Let $\mathcal{A}$ be an adversary against IND-FS-St security of $\mathsf{AKE}_{\mathsf{FS}}$. For $b \in \{0,1\}$, $G_{0,b}$ is the $\mathsf{IND\text{-}FS\text{-}St}_b$ game, where we exclude that collisions between long-term key pairs, ephemeral key pairs, ciphertexts and nonces occur.

In game $G_{1,b}$, we abort when $\mathcal{A}$ computes a valid signature for an uncorrupted user that was not output by the experiment, reducing to N-SUF-CMA security of the signature scheme.

In game $G_{2,b}$, we replace the computations for $\mathsf{KEM}_{\mathsf{CCA}}$ by the simulator $\mathsf{Sim}_{\mathsf{CCA}}$ in all sessions using the non-committing property of $\mathsf{KEM}_{\mathsf{CCA}}$, which allows to draw key K_r which serves as key for the MAC uniformly at random. Thus, the adversary cannot compute a valid MAC for an uncorrupted user.

In game $G_{3,b}$ (as in the proof of Theorem 2), we do not compute the state when the session is initiated but only when the adversary queries the REV-STATE oracle. After that, we can switch $\mathsf{KEM}_{\mathsf{CPA}}$ to the corresponding simulator $\mathsf{Sim}_{\mathsf{CPA}}$ in game $G_{4,b}$ and draw keys $\widetilde{K}$ uniformly at random, whenever the ephemeral public key $\widetilde{\mathsf{pk}}$ comes from the experiment (i.e. the adversary creates a partially matching session). As the adversary can only complete a (partially) matching sessions (otherwise it would have forged a signature or MAC), we can argue that $\widetilde{K}$ in each test session is chosen uniformly at random and unknown to $\mathcal{A}$ and thus he cannot distinguish the session key from a uniformly random key in the last game $G_{5,b}$. □

7 Concrete Instantiation of AKE Protocols

7.1 NCKE from the DDH Assumption

Let us first describe the hash proof system we will use. Therefore, let GGen be a group generation algorithm which takes the security parameter 1^κ as input and returns $(\mathbb{G}, p, g_1)$, where g_1 is a generator of the cyclic group $\mathbb{G}$ with prime order p. Define $group = (\mathbb{G}, p, g_1, g_2)$, where $g_2 = g_1^w$ for $w \xleftarrow{\$} \mathbb{Z}_p$. Define $\mathcal{Y} = \mathbb{Z}_p^2$ and $\mathcal{X} = \{(g_1^r, g_2^r) : r \in \mathbb{Z}_p\}$. A value r is a witness that $(c_1, c_2) \in \mathcal{X}$. Define $\mathcal{SK} = \mathbb{Z}_p^2$, $\mathcal{PK} = \mathbb{Z}_p$ and $\mathcal{Z} = \mathbb{Z}_p$. For $\mathsf{sk} = (x_1, x_2) \in \mathbb{Z}_p^2$, define $\mu(\mathsf{sk}) = X = g_1^{x_1} g_2^{x_2}$. This defines the output of the parameter generation algorithm Par.

For $(c_1, c_2) \in \mathcal{Y}$ define $\Lambda_{\mathsf{sk}}(c_1, c_2) := Z = (c_1^{x_1} c_2^{x_2})$. This defines the private evaluation algorithm $\mathsf{Priv}(\mathsf{sk}, (c_1, c_2))$. Given $\mathsf{pk} = \mu(\mathsf{sk}) = X$, $(c_1, c_2) \in \mathcal{X}$ and a witness $r \in \mathbb{Z}_p$ such that $(c_1, c_2) = (g_1^r, g_2^r)$, the public evaluation algorithm $\mathsf{Pub}(\mathsf{pk}, (c_1, c_2), r)$ computes $Z = \Lambda_{\mathsf{sk}}(c_1, c_2)$ as $Z = X^r$.

We define $\mathsf{KEM}_{\mathsf{DDH}} = (\mathsf{Gen}_{\mathsf{DDH}}, \mathsf{Encaps}_{\mathsf{DDH}}, \mathsf{Decaps}_{\mathsf{DDH}})$ with global parameters $\mathsf{par} := (\mathbb{G}, p, g_1, g_2)$ as shown in Fig. 15.

$\mathsf{Gen}_{\mathsf{DDH}}(\mathsf{par})$	$\mathsf{Encaps}^{\mathsf{H}}_{\mathsf{DDH}}(\mathsf{pk}, m)$	$\mathsf{Decaps}^{\mathsf{H}}_{\mathsf{DDH}}(\mathsf{sk}, (c_1, c_2))$
00 $(x_1, x_2) \xleftarrow{\$} \mathbb{Z}_p^2$	03 $r \xleftarrow{\$} \mathbb{Z}_p$	07 $K := \mathsf{H}(c_1, c_2, c_1^{x_1} c_2^{x_2})$
01 $X := g_1^{x_1} g_2^{x_2}$	04 $(c_1, c_2) := (g_1^r, g_2^r)$	08 **return** K
02 **return** ($\mathsf{pk} := X$,	05 $K := \mathsf{H}(c_1, c_2, X^r)$	
$\mathsf{sk} := (x_1, x_2)$)	06 **return** $((c_1, c_2), K)$	

Fig. 15. Key encapsulation mechanism $\mathsf{KEM}_{\mathsf{DDH}} = (\mathsf{Gen}_{\mathsf{DDH}}, \mathsf{Encaps}_{\mathsf{DDH}}, \mathsf{Decaps}_{\mathsf{DDH}})$.

Definition 3 (m-fold DDH Problem). *Let* GGen *be a PPT algorithm that on input 1^κ outputs a cyclic group $\mathbb{G}$ of prime order $2^{k-1} \leq p \leq 2^k$ with generator g_1. Furthermore let $g_2 = g_1^\omega$ for $\omega \xleftarrow{\$} \mathbb{Z}_p$. The m-DDH problem requires to distinguish m DDH tuples from m uniformly random tuples:*

$$\begin{aligned}\mathrm{Adv}^{m\text{-}\mathsf{DDH}}_{\mathsf{GGen}}(\mathcal{A}) := \Big|&\Pr[\mathcal{A}(\mathbb{G}, p, g_1, g_2, (g_1^{r_i}, g_2^{r_i})_{i\in[m]}) \Rightarrow 1] \\ &- \Pr[\mathcal{A}(\mathbb{G}, p, g_1, g_2, (g_1^{r_i}, g_2^{r'_i})_{i\in[m]}) \Rightarrow 1]\Big| ,\end{aligned}$$

where probability is taken over $(\mathbb{G}, p, g) \leftarrow \mathsf{GGen}$, $r_i, r'_i \xleftarrow{\$} \mathbb{Z}_p$ for $i \in [m]$, as well as the coin tosses of $\mathcal{A}$.

Lemma 1 (Random self-reducibility of DDH [17]). *For any adversary $\mathcal{C}$ against the m-fold DDH problem, there exists an adversary $\mathcal{B}$ against the DDH problem with roughly the same running time such that*

$$\mathrm{Adv}^{m\text{-}\mathsf{DDH}}_{\mathsf{GGen}}(\mathcal{C}) \leq \mathrm{Adv}^{\mathsf{DDH}}_{\mathsf{GGen}}(\mathcal{B}) + \frac{1}{p-1} .$$

The following theorem establishes that the construction given in Fig. 15 is an N-receiver non-committing encapsulation mechanism under the DDH assumption.

Theorem 4. *Under the* DDH *assumption and in the random oracle model,* $\mathsf{KEM}_{\mathsf{DDH}}$ *is an N-receiver non-committing key encapsulation mechanism. In particular, for any N-NCKE-CCA adversary $\mathcal{A}$ against* $\mathsf{KEM}_{\mathsf{DDH}}$ *and* $\mathsf{Sim}_{\mathsf{DDH}}$ *that issues at most q_E queries per user to* ENCAPS, *q_D queries to* DECAPS *and at most q_{H} queries to each random oracle H_n, $n \in [N]$, there exists an adversary $\mathcal{B}$ against* DDH *with roughly the same running time such that*

$$\mathrm{Adv}^{N\text{-}\mathsf{NCKE}\text{-}\mathsf{CCA}}_{\mathsf{KEM}_{\mathsf{DDH}},\mathsf{Sim}_{\mathsf{DDH}}}(\mathcal{A}) \leq \mathrm{Adv}^{\mathsf{DDH}}_{\mathsf{GGen}}(\mathcal{B}) + \frac{N \cdot q_E \cdot (q_{\mathsf{H}} + q_D + 1)}{p} + \frac{1}{p-1} ,$$

where $\mathsf{Sim}_{\mathsf{DDH}}$ *is the simulator defined relative to* $\mathsf{KEM}_{\mathsf{DDH}}$.

Proof. We apply Theorem 1 and analyze the entropy of the underlying HPS. The key space $\mathcal{Z}$ is $\mathbb{Z}_p$. For $\mathsf{sk} = (x_1, x_2) \xleftarrow{\$} \mathbb{Z}_p^2$, $\mathsf{pk} = \mu(\mathsf{sk}) = g_1^{x_1} g_2^{x_2}$ and $Z = \mathsf{Priv}(\mathsf{sk}, (c_1, c_2)) = c_1^{x_1} c_2^{x_2}$, where $(c_1, c_2) = (g_1^r, g_2^{r'})$ and $(r, r') \xleftarrow{\$} \mathbb{Z}_p^2$, we have

$$\begin{pmatrix} \log_{g_1} \mathsf{pk} \\ \log_{g_1} Z \end{pmatrix} = M \begin{pmatrix} x_1 \\ x_2 \end{pmatrix}, \text{ where } M = \begin{pmatrix} 1 & w \\ r & wr' \end{pmatrix} .$$

If $r \neq r'$, then $\det M = w(r' - r) \neq 0$, which implies that pk and Z are random and independent group elements as long as x_1, x_2 are unknown. Thus, for all $Z' \in \mathcal{Z}$, holds that $\Pr[Z = Z'] = 1/p$. In Definition 3, all values r_i and r_i' are drawn uniformly at random from $\mathbb{Z}_p$. The probability that $r_i = r_i'$ for any $i \in [N \cdot q_E]$ is upper bounded by $N \cdot q_E/p$. Furthermore, the probability that a specific challenge ciphertext is issued to DECAPS before it is output by ENCAPS is at most q_D/p. It follows that

$$\mathrm{Adv}_{\mathsf{KEM,Sim}}^{N\text{-}\mathsf{NCKE\text{-}CCA}}(\mathcal{A}) \leq \mathrm{Adv}_{\mathsf{GGen}}^{m\text{-}\mathsf{DDH}}(\mathcal{B}) + \frac{N \cdot q_E}{p} + \frac{N \cdot q_E \cdot q_{\mathsf{H}}}{p} + \frac{N \cdot q_E \cdot q_D}{p} \ .$$

Now Theorem 4 follows directly from Lemma 1. □

7.2 Concrete Instantiation of AKE Protocols

We instantiate protocols $\mathsf{AKE}_{\mathsf{wFS}}$ (Sect. 5) and $\mathsf{AKE}_{\mathsf{FS}}$ (Sect. 6.2) with $\mathsf{KEM}_{\mathsf{DDH}}$ (Sect. 7.1) for both $\mathsf{KEM}_{\mathsf{CPA}}$ and $\mathsf{KEM}_{\mathsf{CCA}}$. We will not give a concrete instantiation of the signature scheme used in $\mathsf{AKE}_{\mathsf{FS}}$ at this point. The resulting protocols $\mathsf{AKE}_{\mathsf{wFS,DDH}}$ and $\mathsf{AKE}_{\mathsf{FS,DDH}}$ are shown in Fig. 1 in the introduction.

Note that for $\mathsf{AKE}_{\mathsf{wFS,DDH}}$ we can improve efficiency by sending only one ciphertext for both $\widetilde{\mathsf{pk}}$ and $\mathsf{pk}i$ in the second message, as $\mathsf{KEM}_{\mathsf{DDH}}$ is a multi-recipient KEM. We establish Theorem 5 and give a proof sketch.

Theorem 5 (IND-wFS-St security of $\mathsf{AKE}_{\mathsf{wFS,DDH}}$). *Under the DDH assumption, $\mathsf{AKE}_{\mathsf{wFS,DDH}}$ is IND-wFS-St secure in the random oracle model. In particular, for any IND-wFS-St adversary $\mathcal{A}$ against $\mathsf{AKE}_{\mathsf{wFS,DDH}}$ with N parties that establishes at most S sessions and issues at most T queries to the test oracle* TEST, *q_{G} queries to random oracle G, $q_{\widetilde{\mathsf{H}}}$, q_{H_n} queries to each random oracle $\widetilde{\mathsf{H}}_{\mathrm{sID}}$ and H_n and at most q_{H} queries to random oracle H, there exists an adversary $\mathcal{B}$ against DDH with roughly the same running time such that*

$$\mathrm{Adv}_{\mathsf{AKE}_{\mathsf{wFS,DDH}}}^{\mathsf{IND\text{-}wFS\text{-}St}}(\mathcal{A}) \leq 2 \cdot \mathrm{Adv}_{\mathsf{GGen}}^{\mathsf{DDH}}(\mathcal{B}) + T \cdot \frac{q_{\mathsf{G}} + q_{\mathsf{H}}}{2^\kappa} + (N+S)^2 \cdot \frac{1}{p} + N^2 \cdot \frac{1}{2^\kappa}$$
$$+ S^2 \cdot \left(\frac{2}{p} + \frac{1}{2^\kappa}\right) + 2S \cdot \left(\frac{q_{\mathsf{G}}}{2^{2\kappa}} + \frac{q_{\widetilde{\mathsf{H}}} + q_{\mathsf{H}_n} + 1}{p}\right) + \frac{2}{p-1} \ ,$$

where κ is a security parameter.

Due to the improved construction, we cannot apply Theorem 2 directly, but we give a proof sketch from the DDH assumption and show that the same technique as in the proofs of Theorems 2 and 4 can be used.

Proof. We proceed similar and consider collisions first. We assume that all key pairs generated by $\mathsf{Gen}_{\mathsf{DDH}}$ are different. Note that we also have to consider collisions between long-term and ephemeral public keys. It holds that

$$\Pr[x_1, x_2, x_1', x_2' \xleftarrow{\$} \mathbb{Z}_p : g_1^{x_1} g_2^{x_2} = g_1^{x_1'} g_2^{x_2'}] = 1/p \ .$$

Union bound yields $(N+S)^2/p$, as we have N long-term public keys and at most S ephemeral public keys. For ciphertexts $(c_1, c_2) \in \mathcal{C}$ output by the encapsulation algorithm $\mathsf{Encaps}_{\mathsf{DDH}}$, it holds that $\Pr[r \xleftarrow{\$} \mathbb{Z}_p : (c_1, c_2) = (g_1^r, g_2^r)] = 1/p$, which yields an upper bound for collisions of S^2/p, as there are at most S sessions with one ciphertext. We also assume that values IV are different in all sessions and keys k_n are different for all parties.

We use the secret keys to compute keys K_i, K_r and $\widetilde{K}$. Next, we replace all ciphertexts by uniformly random group elements at the same time, reducing to the S-fold DDH assumption and use the random self-reducibility property. In addition to that, we ensure that all ciphertexts are indeed invalid by adding S/p which is the probability that exponents are the same for any ciphertext.

Instead of the corresponding random oracles, we use internal hash functions $\widetilde{\mathsf{H}}'_{\mathrm{sID}}$ and H'_n for sID $\in [S]$ and $n \in [N]$ to compute keys K_i, K_r and $\widetilde{K}$, but patch the random oracles if the secret key is known to the adversary. As there are at most S challenge keys computed with a long-term key pair and at most S challenge keys computed with an ephemeral key pair, the difference can be upper bounded by $S \cdot q_{\mathsf{H}_n}/p + S \cdot q_{\widetilde{\mathsf{H}}}/p$ using a hybrid argument. Now we can replace K_i, K_r and $\widetilde{K}$ by uniformly random keys.

The rest of the proof is equal to the proof of Theorem 2. The size of the key space of $\mathsf{KEM}_{\mathsf{DDH}}$ is 2^κ and the bound follows by collecting all probabilities. $\square$

For protocol $\mathsf{AKE}_{\mathsf{FS,DDH}}$, we apply Theorem 3 to show IND-FS-St security. The collision probabilities for $\mathsf{KEM}_{\mathsf{DDH}}$ are already shown in the previous proof. Additionally, we need a strongly unforgeable signature scheme.

Theorem 6 (**IND-FS-St security of $\mathsf{AKE}_{\mathsf{FS,DDH}}$**). *For an N-SUF-CMA secure signature scheme SIG and under the DDH assumption, $\mathsf{AKE}_{\mathsf{FS,DDH}}$ is IND-FS-St secure in the random oracle model. In particular, for any IND-FS-St adversary $\mathcal{A}$ against $\mathsf{AKE}_{\mathsf{FS,DDH}}$ with N parties that establishes at most S sessions and issues at most T queries to the test oracle* TEST, *q_{G} queries to random oracle G, q_{F} queries to random oracle F, $q_{\widetilde{\mathsf{H}}}$, q_{H_n} queries to each random oracle $\widetilde{\mathsf{H}}_{\mathrm{sID}}$ and H_n and at most q_{H} queries to random oracle H, there exists an adversary $\mathcal{B}$ against DDH and an adversary $\mathcal{C}$ against N-SUF-CMA such that*

$$\begin{aligned}\mathrm{Adv}^{\mathsf{IND\text{-}FS\text{-}St}}_{\mathsf{AKE}_{\mathsf{FS,DDH}}}(\mathcal{A}) \leq\ & 4 \cdot \mathrm{Adv}^{\mathsf{DDH}}_{\mathsf{GGen}}(\mathcal{B}) + 2 \cdot \mathrm{Adv}^{N\text{-}\mathsf{SUF\text{-}CMA}}_{\mathsf{SIG}}(\mathcal{C}) + T \cdot \frac{q_{\mathsf{F}} + q_{\mathsf{G}} + q_{\mathsf{H}}}{2^\kappa} \\ & + N^2 \cdot \left(\frac{1}{2^{\mu_{\mathsf{SIG}}}} + \frac{1}{p} + \frac{1}{2^\kappa}\right) + S^2 \cdot \left(\frac{2q_{\widetilde{\mathsf{H}}} + 6}{p} + \frac{1}{2^\kappa}\right) \\ & + 2NS \cdot \frac{q_{\mathsf{H}_n} + 2}{p} + 2S \cdot \frac{q_{\mathsf{G}}}{2^{2\kappa}} + \frac{4}{p-1}\ ,\end{aligned}$$

where μ_{SIG} is the collision probability of the key generation algorithm $\mathsf{Gen}_{\mathsf{SIG}}$ and κ is a security parameter.

The signature scheme can be instantiated with the tight scheme based on the DDH and CDH assumption proposed by Gjøsteen and Jager in [21], which is also used in their authenticated key exchange protocol.

Acknowledgments. Tibor Jager was supported by the European Research Council (ERC) under the European Union's Horizon 2020 research and innovation programme, grant agreement 802823. Eike Kiltz was supported by the BMBF iBlockchain project, the EU H2020 PROMETHEUS project 780701, DFG SPP 1736 Big Data, and the DFG Cluster of Excellence 2092 CASA. Doreen Riepel was supported by the Deutsche Forschungsgemeinschaft (DFG) Cluster of Excellence 2092 CASA. Sven Schäge was supported by the German Federal Ministry of Education and Research (BMBF), Project DigiSeal (16KIS0695) and Huawei Technologies Düsseldorf, Project vHSM.

References

1. Bader, C., Hofheinz, D., Jager, T., Kiltz, E., Li, Y.: Tightly-secure authenticated key exchange. In: Dodis, Y., Nielsen, J.B. (eds.) TCC 2015, Part I. LNCS, vol. 9014, pp. 629–658. Springer, Heidelberg (2015). https://doi.org/10.1007/978-3-662-46494-6_26
2. Bader, C., Jager, T., Li, Y., Schäge, S.: On the impossibility of tight cryptographic reductions. In: Fischlin, M., Coron, J.-S. (eds.) EUROCRYPT 2016, Part II. LNCS, vol. 9666, pp. 273–304. Springer, Heidelberg (2016). https://doi.org/10.1007/978-3-662-49896-5_10
3. Bellare, M., Boldyreva, A., Micali, S.: Public-key encryption in a multi-user setting: security proofs and improvements. In: Preneel, B. (ed.) EUROCRYPT 2000. LNCS, vol. 1807, pp. 259–274. Springer, Heidelberg (2000). https://doi.org/10.1007/3-540-45539-6_18
4. Bellare, M., Rogaway, P.: Entity authentication and key distribution. In: Stinson, D.R. (ed.) CRYPTO 1993. LNCS, vol. 773, pp. 232–249. Springer, Heidelberg (1994). https://doi.org/10.1007/3-540-48329-2_21
5. Bergsma, F., Jager, T., Schwenk, J.: One-round key exchange with strong security: an efficient and generic construction in the standard model. In: Katz, J. (ed.) PKC 2015. LNCS, vol. 9020, pp. 477–494. Springer, Heidelberg (2015). https://doi.org/10.1007/978-3-662-46447-2_21
6. Boyd, C., Nieto, J.G.: On forward secrecy in one-round key exchange. In: Chen, L. (ed.) IMACC 2011. LNCS, vol. 7089, pp. 451–468. Springer, Heidelberg (2011). https://doi.org/10.1007/978-3-642-25516-8_27
7. Brzuska, C., Fischlin, M., Warinschi, B., Williams, S.C.: Composability of Bellare-Rogaway key exchange protocols. In: Chen, Y., Danezis, G., Shmatikov, V. (eds.) ACM CCS 2011, pp. 51–62. ACM Press (2011). https://doi.org/10.1145/2046707.2046716
8. Canetti, R., Feige, U., Goldreich, O., Naor, M.: Adaptively secure multi-party computation. In: 28th ACM STOC, pp. 639–648. ACM Press (1996). https://doi.org/10.1145/237814.238015
9. Canetti, R., Krawczyk, H.: Analysis of key-exchange protocols and their use for building secure channels. In: Pfitzmann, B. (ed.) EUROCRYPT 2001. LNCS, vol. 2045, pp. 453–474. Springer, Heidelberg (2001). https://doi.org/10.1007/3-540-44987-6_28
10. Cohn-Gordon, K., Cremers, C., Gjøsteen, K., Jacobsen, H., Jager, T.: Highly efficient key exchange protocols with optimal tightness. In: Boldyreva, A., Micciancio, D. (eds.) CRYPTO 2019, Part III. LNCS, vol. 11694, pp. 767–797. Springer, Cham (2019). https://doi.org/10.1007/978-3-030-26954-8_25

11. Cramer, R., Shoup, V.: Universal hash proofs and a paradigm for adaptive chosen ciphertext secure public-key encryption. In: Knudsen, L.R. (ed.) EUROCRYPT 2002. LNCS, vol. 2332, pp. 45–64. Springer, Heidelberg (2002). https://doi.org/10.1007/3-540-46035-7_4
12. Cremers, C.: Examining indistinguishability-based security models for key exchange protocols: the case of CK, CK-HMQV, and eCK. In: Cheung, B.S.N., Hui, L.C.K., Sandhu, R.S., Wong, D.S. (eds.) ASIACCS 11, pp. 80–91. ACM Press (2011)
13. Cremers, C.J.F.: Session-state Reveal is stronger than Ephemeral Key Reveal: attacking the NAXOS authenticated key exchange protocol. In: Abdalla, M., Pointcheval, D., Fouque, P.-A., Vergnaud, D. (eds.) ACNS 2009. LNCS, vol. 5536, pp. 20–33. Springer, Heidelberg (2009). https://doi.org/10.1007/978-3-642-01957-9_2
14. Cremers, C., Feltz, M.: Beyond eCK: perfect forward secrecy under actor compromise and ephemeral-key reveal. In: Foresti, S., Yung, M., Martinelli, F. (eds.) ESORICS 2012. LNCS, vol. 7459, pp. 734–751. Springer, Heidelberg (2012). https://doi.org/10.1007/978-3-642-33167-1_42
15. Davis, H., Günther, F.: Tighter proofs for the SIGMA and TLS 1.3 key exchange protocols. Cryptology ePrint Archive, Report 2020/1029 (2020). https://eprint.iacr.org/2020/1029
16. Diemert, D., Jager, T.: On the tight security of TLS 1.3: theoretically-sound cryptographic parameters for real-world deployments. Cryptology ePrint Archive, Report 2020/726 (2020). https://eprint.iacr.org/2020/726
17. Escala, A., Herold, G., Kiltz, E., Ràfols, C., Villar, J.: An algebraic framework for Diffie-Hellman assumptions. In: Canetti, R., Garay, J.A. (eds.) CRYPTO 2013, Part II. LNCS, vol. 8043, pp. 129–147. Springer, Heidelberg (2013). https://doi.org/10.1007/978-3-642-40084-1_8
18. Freire, E.S.V., Hofheinz, D., Kiltz, E., Paterson, K.G.: Non-interactive key exchange. In: Kurosawa, K., Hanaoka, G. (eds.) PKC 2013. LNCS, vol. 7778, pp. 254–271. Springer, Heidelberg (2013). https://doi.org/10.1007/978-3-642-36362-7_17
19. Fujioka, A., Suzuki, K., Xagawa, K., Yoneyama, K.: Strongly secure authenticated key exchange from factoring, codes, and lattices. In: Fischlin, M., Buchmann, J., Manulis, M. (eds.) PKC 2012. LNCS, vol. 7293, pp. 467–484. Springer, Heidelberg (2012). https://doi.org/10.1007/978-3-642-30057-8_28
20. Gay, R., Hofheinz, D., Kiltz, E., Wee, H.: Tightly CCA-secure encryption without pairings. In: Fischlin, M., Coron, J.-S. (eds.) EUROCRYPT 2016, Part I. LNCS, vol. 9665, pp. 1–27. Springer, Heidelberg (2016). https://doi.org/10.1007/978-3-662-49890-3_1
21. Gjøsteen, K., Jager, T.: Practical and tightly-secure digital signatures and authenticated key exchange. In: Shacham, H., Boldyreva, A. (eds.) CRYPTO 2018. LNCS, vol. 10992, pp. 95–125. Springer, Cham (2018). https://doi.org/10.1007/978-3-319-96881-0_4
22. Hövelmanns, K., Kiltz, E., Schäge, S., Unruh, D.: Generic authenticated key exchange in the quantum random oracle model. In: Kiayias, A., Kohlweiss, M., Wallden, P., Zikas, V. (eds.) PKC 2020, Part II. LNCS, vol. 12111, pp. 389–422. Springer, Cham (2020). https://doi.org/10.1007/978-3-030-45388-6_14
23. Jager, T., Kiltz, E., Riepel, D., Schäge, S.: Tightly-secure authenticated key exchange, revisited. Cryptology ePrint Archive, Report 2020/1279 (2020). https://eprint.iacr.org/2020/1279

24. Jager, T., Kohlar, F., Schäge, S., Schwenk, J.: On the security of TLS-DHE in the standard model. In: Safavi-Naini, R., Canetti, R. (eds.) CRYPTO 2012. LNCS, vol. 7417, pp. 273–293. Springer, Heidelberg (2012). https://doi.org/10.1007/978-3-642-32009-5_17
25. Kiltz, E., Pietrzak, K., Stam, M., Yung, M.: A new randomness extraction paradigm for hybrid encryption. In: Joux, A. (ed.) EUROCRYPT 2009. LNCS, vol. 5479, pp. 590–609. Springer, Heidelberg (2009). https://doi.org/10.1007/978-3-642-01001-9_34
26. Krawczyk, H.: HMQV: a high-performance secure Diffie-Hellman protocol. In: Shoup, V. (ed.) CRYPTO 2005. LNCS, vol. 3621, pp. 546–566. Springer, Heidelberg (2005). https://doi.org/10.1007/11535218_33
27. Kudla, C., Paterson, K.G.: Modular security proofs for key agreement protocols. In: Roy, B. (ed.) ASIACRYPT 2005. LNCS, vol. 3788, pp. 549–565. Springer, Heidelberg (2005). https://doi.org/10.1007/11593447_30
28. LaMacchia, B., Lauter, K., Mityagin, A.: Stronger security of authenticated key exchange. In: Susilo, W., Liu, J.K., Mu, Y. (eds.) ProvSec 2007. LNCS, vol. 4784, pp. 1–16. Springer, Heidelberg (2007). https://doi.org/10.1007/978-3-540-75670-5_1
29. Lauter, K., Mityagin, A.: Security analysis of KEA authenticated key exchange protocol. In: Yung, M., Dodis, Y., Kiayias, A., Malkin, T. (eds.) PKC 2006. LNCS, vol. 3958, pp. 378–394. Springer, Heidelberg (2006). https://doi.org/10.1007/11745853_25
30. Li, Y., Schäge, S.: No-match attacks and robust partnering definitions: defining trivial attacks for security protocols is not trivial. In: Thuraisingham, B.M., Evans, D., Malkin, T., Xu, D. (eds.) ACM CCS 2017, pp. 1343–1360. ACM Press (2017). https://doi.org/10.1145/3133956.3134006
31. Li, Y., Schäge, S., Yang, Z., Bader, C., Schwenk, J.: New modular compilers for authenticated key exchange. In: Boureanu, I., Owesarski, P., Vaudenay, S. (eds.) ACNS 2014. LNCS, vol. 8479, pp. 1–18. Springer, Cham (2014). https://doi.org/10.1007/978-3-319-07536-5_1
32. Liu, X., Liu, S., Gu, D., Weng, J.: Two-pass authenticated key exchange with explicit authentication and tight security. In: Moriai, S., Wang, H. (eds.) ASIACRYPT 2020, Part II. LNCS, vol. 12492, pp. 785–814. Springer, Cham (2020). https://doi.org/10.1007/978-3-030-64834-3_27
33. Nielsen, J.B.: Separating random oracle proofs from complexity theoretic proofs: the non-committing encryption case. In: Yung, M. (ed.) CRYPTO 2002. LNCS, vol. 2442, pp. 111–126. Springer, Heidelberg (2002). https://doi.org/10.1007/3-540-45708-9_8
34. Schäge, S.: TOPAS: 2-pass key exchange with full perfect forward secrecy and optimal communication complexity. In: Ray, I., Li, N., Kruegel, C. (eds.) ACM CCS 2015, pp. 1224–1235. ACM Press (2015). https://doi.org/10.1145/2810103.2813683
35. Shoup, V.: Sequences of games: a tool for taming complexity in security proofs. Cryptology ePrint Archive, Report 2004/332 (2004). http://eprint.iacr.org/2004/332
36. Yoneyama, K.: One-round authenticated key exchange with strong forward secrecy in the standard model against constrained adversary. In: Hanaoka, G., Yamauchi, T. (eds.) IWSEC 2012. LNCS, vol. 7631, pp. 69–86. Springer, Heidelberg (2012). https://doi.org/10.1007/978-3-642-34117-5_5

Aggregatable Distributed Key Generation

Kobi Gurkan[1(✉)], Philipp Jovanovic[2], Mary Maller[3], Sarah Meiklejohn[2], Gilad Stern[4], and Alin Tomescu[5]

[1] cLabs, Ethereum Foundation, Berlin, Germany
me@kobi.one
[2] University College London, London, UK
{p.jovanovic,s.meiklejohn}@ucl.ac.uk
[3] Ethereum Foundation, Berlin, Germany
mary.maller@ethereum.org
[4] Hebrew University, Jerusalem, Israel
gilad.stern@mail.huji.ac.il
[5] VMware Research, Jersey City, USA
alint@vmware.com

Abstract. In this paper, we introduce a distributed key generation (DKG) protocol with aggregatable and publicly-verifiable transcripts. Compared with prior *publicly-verifiable* approaches, our DKG reduces the size of the final transcript and the time to verify it from $\mathcal{O}(n^2)$ to $\mathcal{O}(n \log n)$, where n denotes the number of parties. As compared with prior non-publicly-verifiable approaches, our DKG leverages *gossip* rather than *all-to-all communication* to reduce verification and communication complexity. We also revisit existing DKG security definitions, which are quite strong, and propose new and natural relaxations. As a result, we can prove the security of our aggregatable DKG as well as that of several existing DKGs, including the popular Pedersen variant. We show that, under these new definitions, these existing DKGs can be used to yield secure threshold variants of popular cryptosystems such as El-Gamal encryption and BLS signatures. We also prove that our DKG can be securely combined with a new efficient verifiable unpredictable function (VUF), whose security we prove in the random oracle model. Finally, we experimentally evaluate our DKG and show that the per-party overheads scale linearly and are practical. For 64 parties, it takes 71 ms to share and 359 ms to verify the overall transcript, while for 8192 parties, it takes 8 s and 42.2 s respectively.

1 Introduction

System designers who strive to remove single points of failure often rely on tools provided by threshold cryptography [21,58] and secure multi-party computation [20,34]. In this paper, we study *distributed key generation (DKG)* [31,53], a method from threshold cryptography that often plays an essential role during the setup of distributed systems, including Byzantine consensus [6,64], time-stamping services [14,61], public randomness beacons [29,59], and data archive systems [45,63]. A DKG enables a set of parties to generate a keypair such that

A. Canteaut and F.-X. Standaert (Eds.): EUROCRYPT 2021, LNCS 12696, pp. 147–176, 2021.
https://doi.org/10.1007/978-3-030-77870-5_6

any sufficiently large subset can perform an action that requires the secret key while any smaller subset cannot. To achieve this, a DKG essentially turns each party into a dealer for a verifiable secret sharing (VSS) scheme [19,25,54]. This process yields a single collective public key, generated in a distributed manner, with each party keeping a share of the secret key for themselves.

Current DKGs [27,31,32,53] commonly require that all n parties broadcast $\mathcal{O}(n)$-sized messages that are then used by each party to verify the shares they received from their peers. This results in each party communicating $\mathcal{O}(n^2)$ sized messages via broadcast. While some DKGs have $\mathcal{O}(n)$ communication and verification per party, they rely on constant-sized polynomial commitment schemes that require trusted setup [42,43]. In this work, we show how to reduce the size of the final DKG transcript to $\mathcal{O}(n)$ by making the parties' contributions *aggregatable*. This enables us to relay (partial) transcripts in an efficient and resilient manner, *e.g.*, over gossip networks, ensuring that transcripts do not grow in size since aggregation can be done in a continuous manner. Aggregatability also enables us to refresh the transcript if and when shares get compromised.

Our DKG transcripts contain the information needed for parties to decrypt their secret shares. During aggregation it is therefore essential to ensure that only valid (partial) transcripts are aggregated. We achieve this by making our transcripts *publicly-verifiable* so that anybody receiving and aggregating transcripts can verify their correctness. Making the transcripts publicly-verifiable has several other advantages: It ensures that all parties can obtain their secret shares, even if they go offline momentarily, and also enables us to remove the "complaint rounds" that are used in previous DKGs to expose misbehaving parties. This improves overall latency, since fewer communication rounds are required, and reduces the protocol's complexity from an implementation perspective.

A consequence of our approach is that the DKG secret key and its shares are group elements rather than field elements. While this prevents us from using it for many well-known cryptosystems, we demonstrate its applicability by introducing a *verifiable unpredictable function (VUF)* [23,48] whose secret key is a group element, and prove its security in the random oracle model. Threshold VUFs are useful in the construction of verifiable random beacons, which themselves are invaluable in building proof-of-stake-based cryptocurrencies [33,44]. To the best of our knowledge, our VUF is the first that takes a group element as the secret key, and its performance is also reasonable: our VUF output consists of 6 source group elements and can be verified using 10 pairings. We also provide further optimizations enabling us to reduce the VUF contributions in the threshold protocol to just 2 source group elements that can be verified with 3 pairings.

We also revisit the definitions for DKGs in the hope of reducing complexities and inefficiencies. In particular, previous definitions [31] required *secrecy*, in the sense that the output of the DKG must be indistinguishable from random. While this notion has the benefit of making the DKG modular (one can replace key generation with a DKG in any context), it also is difficult to realise. Indeed, Gennaro et al. [31] demonstrated that the popular Pedersen DKG does not have secrecy, and introduced an alternative and considerably less efficient protocol to achieve secrecy. Additionally, no one-round DKG can achieve secrecy because a rushing adversary (an adversary that plays last) can always influence the final distribution. Instead, we look to prove that a DKG is *robust* (see Definition 6) and *security-preserving* (see Definition 8) in the sense that any adversary that breaks the security of a threshold version of the scheme (i.e., one using a DKG) also breaks the original security property.

Gennaro et al. [32] previously observed that the Pedersen DKG suffices to construct threshold Schnorr signatures [57]. Recently, Benhamouda et al. [10] found an attack on this approach when the adversary is *concurrent.* (Gennaro et al. had not considered concurrent adversaries.) Komlo and Goldberg [47] show that it is possible to avoid the attack but, in doing so, they lose robustness (*e.g.*, if a single party goes offline a signature will not be produced). This raises questions as to whether it is still okay to use the Pedersen DKG with respect to other signature schemes such as BLS. In this paper, we provide a positive answer in the form of a security proof that holds concurrently and does not rely on rewinding the adversary. Specifically, we show that the Pedersen DKG is security-preserving with respect to any *rekeyable* encryption scheme, signature scheme, or VUF scheme where the sharing algorithm is the same as encryption or signing (see Definition 5).

Our contributions. In Sect. 5, we construct an *aggregatable* and *publicly-verifiable* DKG. The aggregation can be completed by any party (there are no additional secrets) and can also be done incrementally. The cost of verifying our transcripts is $\mathcal{O}(n \log n)$ whereas prior approaches were $\mathcal{O}(n^2)$ [27]. If any user temporarily goes offline, they can still recover their secret shares. Dealing DKG shares takes $\mathcal{O}(n \log n)$ time and aggregation costs are $\mathcal{O}(n)$.

We prove security of our DKG using a natural definition (see Sect. 3.6), which roughly states that, if it is possible to break a cryptosystem's security game with a DKG swapped in, then it is possible to break that cryptosystem's original security game that did not use the DKG for key generation. We further demonstrate that, counter-intuitively, it is possible to prove that a DKG realises this definition without needing a separate proof for each cryptosystem. Indeed, we show that any encryption scheme, signature scheme, or VUF that are *rekeyable*, and where the sharing algorithm is the same as encryption or signing, can be securely instantiated using a *key-expressable* DKG (see Definition 7). This includes El-Gamal encryption, BLS signatures, a new VUF we introduce in Sect. 7, and, we suspect, many others.

We further demonstrate the applicability of our techniques by showing that all three of the Pedersen DKG [53], the Fouque-Stern DKG [27], and our aggregatable DKG are key-expressable and thus can be used securely with rekeyable encryption schemes, signature schemes, and VUFs whose decryption/ signing algorithms are the same as the algorithms to generate decryption/ signature shares. Our proof allows for rushing adversaries and holds concurrently (*i.e.*, with respect to an adversary that can open many sessions at the same time). We cannot cover Schnorr signatures, however, because their threshold variants do not appear to be rekeyable.

Our final contribution, in Sect. 8, is a Rust implementation of our aggregatable DKG to demonstrate its practicality by showing that its overheads are indeed linear. For example, the evaluation of our implementation shows that for 64/128/8192 nodes it takes 71 ms/137 ms/8,000 ms to share one secret and 359 ms/747 ms/42,600 ms to verify the corresponding transcript.

2 Related Work

Table 1. Complexities of prior DKG protocols with n parties, per party. In the "Broadcast" column, we count the number of broadcasts by size (either $O(n)$ or $O(1)$-sized). "P2P" means the total size of the messages sent over public and private communication channels (excludes broadcast messages). "PV" means publicly-verifiable. "Verifier local" indicates the per-party time spent verifying their shares from other parties, while "global" indicates the time to verify the final DKG transcript.

DKG	Broadcasts		P2P	PV	Complaints	Rounds		Prover	Verifier	
	$\mathcal{O}(n)$	$\mathcal{O}(1)$				Broadcast	Gossip		Local	Global
Pedersen	n	–	n	no	yes	3	–	$n \lg n$	n^2	–
Kate	–	n	n	no	yes	3	–	n^2	n	–
AMT	–	n	$n \lg n$	no	yes	3	–	$n \lg n$	$n \lg n$	–
Fouque-Stern	n	–	n^2	yes	no	1	–	$n \lg n$	n^2	n^2
Our work	$\lg n$	n	$n \lg^2 n$	yes	no	2	$\lg n$	$n \lg n$	$n \lg^2 n$	$n \lg n$

We provide an asymptotic overview of the state-of-the-art for DKGs in Table 2. Here we assume that the threshold t is linear in n. Our comparisons consider the optimistic case where there is no more than a constant number of complaints for protocols where these are relavant (recall in our protocol there are no complaints) (Table 1).

Pedersen introduced the first efficient DKG protocol for discrete log-based cryptosystems [54], building on top of Feldman's VSS [25]. Gennaro et al. [32] showed Pedersen's DKG does not generate uniformly distributed secrets, and proposed a protocol that does but at the cost of lower efficiency. They also fix problems with the complaint phase in Pedersen's DKG. Neji et al. [50] gave a more efficient protocol that ensures uniformity in Pedersen's DKG. Kate [42] reduced the broadcast overhead per DKG party from $\mathcal{O}(n)$ to $\mathcal{O}(1)$ using their constant-sized polynomial commitment scheme [43]. However, there scheme depends on a trusted setup algorithm, the costs of which are not considered in Table 2. Trusted setup algorithms have a round complexity of t, and each of these rounds require users to broadcast $\mathcal{O}(n)$ sized messages [15,37]. Unlike our protocol, all of these protocols rely on complaints rounds, are not publicly verifiable and have $\mathcal{O}(n^2)$ communication complexity.

Fouque and Stern present a one-round, publicly-verifiable DKG that uses only public channels. However, their final transcript size is $\mathcal{O}(n^2)$ whereas ours is $\mathcal{O}(n)$ because we can aggregate. Furthermore, their security proof does not allow for rushing adversaries. While they do not measure performance, their use of Paillier encryption [52] is likely to make their DKG slow and have high communication costs. Nonetheless, unlike our DKG, theirs has the advantage of outputting secrets that are field, rather than group, elements.

Other works tackle the DKG problem from different angles. Canetti et al.'s DKG [16] has adaptive security, while ours is secure only against static adversaries that fix the set of corrupted parties before the protocol starts. Canny and Sorkin [17] study DKG protocols with poly-logarithmic communication and computation cost per-party, but their protocol relies on a trusted dealer that permutes the parties before the protocol starts. Kate et al. [41,42] and Kokoris-Kogias et al. [46] study DKG protocols in the *asynchronous* setting, unlike our work and most previous work. Schindler et al. [56] use the Ethereum blockchain to instantiate the synchronous broadcast channel all DKG protocols mentioned so far assume, including ours. Tomescu et al. [60] lower the computational cost of dealing in Kate et al.'s DKG [43], at a logarithmic increase in communication. Lastly, several works implement and benchmark synchronous, statically-secure DKG protocols for discrete log-based cryptosystems [22,39,40,51,56].

Abe et al. [1] observed that any fully structure preserving signature scheme that depends solely on algebraic operations cannot be used as a VUF or VRF. Unlike our VUF (which is not algebraic), this rules out a number of structure preserving signatures from being candidates for building VUFs [2–4,62].

3 Definitions

3.1 Preliminaries

If x is a binary string then $|x|$ denotes its bit length. If S is a finite set then $|S|$ denotes its size and $x \xleftarrow{\$} S$ denotes sampling a member uniformly from S and assigning it to x. We use $\lambda \in \mathbb{N}$ to denote the security parameter and 1^λ to denote its unary representation. Algorithms are randomized unless explicitly noted otherwise. "PPT" stands for "probabilistic polynomial time." We use $\boldsymbol{y} \leftarrow A(\boldsymbol{x}; r)$ to denote running algorithm A on inputs $\boldsymbol{x}$ and randomness r and assigning its output to $\boldsymbol{y}$. We use $\boldsymbol{y} \xleftarrow{\$} A(\boldsymbol{x})$ to denote $y \leftarrow A(x;r)$ for uniformly random r. We use $[A(\boldsymbol{x})]$ to denote the set of values that have non-zero probability of being output by A on input $\boldsymbol{x}$. For two functions $f, g : \mathbb{N} \rightarrow [0,1]$, we use $f(\lambda) \approx g(\lambda)$ to denote $|f(\lambda) - g(\lambda)| = \lambda^{-\omega(1)}$. We use code-based games in security definitions and proofs [9]. A game $\mathsf{Sec}_{\mathcal{A}}(\lambda)$, played with respect to a security notion Sec and adversary $\mathcal{A}$, has a MAIN procedure whose output is the output of the game. The notation $\Pr[\mathsf{Sec}_{\mathcal{A}}(\lambda)]$ denotes the probability that this output is 1.

We formalize bilinear groups via a *bilinear group sampler*, which is an efficient *deterministic* algorithm GroupGen that given a security parameter 1^λ (represented in unary), outputs a tuple $\mathsf{bp} = (p, \mathbb{G}_1, \mathbb{G}_2, \mathbb{G}_T, e, g_1, \hat{h}_1)$ where $\mathbb{G}_1$, $\mathbb{G}_2$, $\mathbb{G}_T$ are groups with order divisible by the prime $p \in \mathbb{N}$, g_1 generates $\mathbb{G}_1$, $\hat{h}_1$ generates $\mathbb{G}_2$, and $e \colon \mathbb{G}_1 \times \mathbb{G}_2 \rightarrow \mathbb{G}_T$ is a (non-degenerate) bilinear map.

Galbraith et al. distinguish between three types of bilinear group samplers [28]. Type I groups have $\mathbb{G}_1 = \mathbb{G}_2$ and are known as *symmetric* bilinear groups. Types II and III are *asymmetric* bilinear groups, where $\mathbb{G}_1 \neq \mathbb{G}_2$. Type II groups have an efficiently computable homomorphism $\psi : \mathbb{G}_2 \rightarrow \mathbb{G}_1$, while Type III groups do not have an efficiently computable homomorphism in either direction. Certain assumptions are provably false with respect to certain group types (*e.g.*, SXDH only holds for Type III groups), and we work only with Type III groups.

3.2 Communication and Threat Models

In this section we discuss our communication and threat models.

Synchrony: We assume perfect synchrony. There is a strict time bound between rounds. All messages (honest and adversarial) within a round will be seen by all parties by the end of the round.

Communication channel: We assume the existence of a broadcast channel for sending messages. If a non-faulty party broadcasts a message then it will be seen by everyone by the end of the round. It is not possible to forge messages from non-faulty parties.

Adversarial threshold: We denote by t the adversarial threshold; i.e., the number of parties that the adversary can corrupt. The total number of parties is denoted by n. We set no specific bounds on the adversarial threshold because a rational adversary might prefer to attack the secrecy of the DKG over blocking the communication channels [5,30].

Assumptions on the adversary: Our security proofs are given with respect to static adversaries, meaning the adversary must state at the start of the security game all of the parties that it has corrupted. We allow the adversary to control the ordering of messages within a round, and in particular the adversary can wait to receive all messages within a round before it broadcasts its message (this is called a *rushing* adversary). The adversary can also choose not to participate at all.

Byzantine adversary: A byzantine adversary is a malicious entity that may differ arbitrarily from the protocol.

Crashed party: A crashed party is a party that has gone offline e.g. due to a faulty internet connection. After a party has crashed they will not send any more messages.

3.3 Assumptions

Our security proofs are provided in the random oracle model; i.e., there exists a simulator that can program the output of a hash function provided that their chosen outputs are indistinguishable from random.

We rely on the SXDH assumption [7,8], which is an extension of the DDH assumption to Type III bilinear groups. Informally, it states that given g_1^α and g_1^β it is hard to distinguish $g_1^{\alpha\beta}$ from random.

The BDH assumption is an extension of the CDH assumption to Type III bilinear groups [12]. Informally, it states that given $g_1^\alpha, g_1^\beta, \hat{h}_1^\gamma, \hat{h}_1^{\alpha\gamma}$ it is hard to compute $e(g_1, \hat{h}_1)^{\alpha\beta}$.

3.4 Verifiable Unpredictable Functions (VUFs)

A VUF allows a party with a secret key to compute a deterministic (keyed) function and prove to an external verifier that the result is correct. The notion is related to signatures, with the extra requirement that the output of the signer must be unique, even to a party that can choose the secret key. We have made the following changes to prior definitions [23] in order to better suit our setting: (1) we include a global setup algorithm to generate a common reference string; (2) we include a derive algorithm to map the prover's output onto the unique function output.

Definition 1 (Verifiable Unpredictable Function). *Let Π =* (VUF.Setup, VUF.Gen, VUF.Eval, VUF.Sign, VUF.Derive, VUF.Ver) *be the following set of efficient algorithms:*

$\mathsf{crs}_{\mathsf{vuf}} \leftarrow \mathsf{VUF.Setup}(1^\lambda)$: *a DPT algorithm that takes as input the security parameter and outputs a common reference string.*

$(\mathsf{pk}, \mathsf{sk}) \xleftarrow{\$} \mathsf{VUF.Gen}(\mathsf{crs}_{\mathsf{vuf}})$*: a PPT algorithm that takes as input a common reference string and returns a public key and a secret key.*

$\mathsf{out} \leftarrow \mathsf{VUF.Eval}(\mathsf{crs}_{\mathsf{vuf}}, \mathsf{sk}, m)$*: a DPT algorithm that takes as input a common reference string, secret key, and message $m \in \{0,1\}^\lambda$ and returns* $\mathsf{out} \in \{0,1\}^\lambda$.

$\sigma \xleftarrow{\$} \mathsf{VUF.Sign}(\mathsf{crs}_{\mathsf{vuf}}, \mathsf{sk}, m)$*: a PPT algorithm that takes as input a common reference string, secret key, and message, and returns a signature σ.*

$\mathsf{out} \leftarrow \mathsf{VUF.Derive}(\mathsf{crs}_{\mathsf{vuf}}, \mathsf{pk}, m, \sigma)$*: a DPT algorithm that takes as input a common reference string, public key, message and signature and returns* $\mathsf{out} \in \{0,1\}^\lambda$.

$0/1 \leftarrow \mathsf{VUF.Ver}(\mathsf{crs}_{\mathsf{vuf}}, \mathsf{pk}, m, \sigma)$*: a DPT algorithm that takes as input a common reference string, public key, message and signature and returns 1 to indicate acceptance and 0 to indicate rejection.*

We say that Π is a verifiable unpredictable function (VUF) *if it satisfies correctness, uniqueness, and unpredictability (defined below).*

A VUF is correct if an honest signer always convinces an honest verifier and always outputs a seed such that the derive function outputs the correct value.

Definition 2 (Correctness). *A VUF is correct if for all* $\lambda \in \mathbb{N}$ *and* $m \in \{0,1\}^\lambda$ *we have that*

$$\Pr\left[\begin{array}{l|l} \mathsf{crs}_{\mathsf{vuf}} \leftarrow \mathsf{Setup}(1^\lambda), & \mathsf{VUF.Derive}(\mathsf{crs}_{\mathsf{vuf}}, \mathsf{pk}, m, \sigma) = \\ (\mathsf{pk}, \mathsf{sk}) \xleftarrow{\$} \mathsf{VUF.Gen}(\mathsf{crs}_{\mathsf{vuf}}), & \mathsf{VUF.Eval}(\mathsf{crs}_{\mathsf{vuf}}, \mathsf{sk}, m) \\ \sigma \xleftarrow{\$} \mathsf{VUF.Sign}(\mathsf{crs}_{\mathsf{vuf}}, \mathsf{sk}, m) & \wedge\ \mathsf{VUF.Ver}(\mathsf{crs}_{\mathsf{vuf}}, \mathsf{pk}, m, \sigma) = 1 \end{array}\right] = 1.$$

A VUF is unique if an adversary (even one that chooses the secret key) cannot output a verifying signature such that the derive function outputs the wrong value.

Definition 3 (Uniqueness). *For a VUF* Π *and an adversary* $\mathcal{A}$, *let* $\mathsf{Adv}^{\mathsf{unique}}_{\mathcal{A}}(\lambda) = \Pr[\mathsf{Game}^{\mathsf{unique}}_{\mathcal{A}}(\lambda)]$, *where* $\mathsf{Game}^{\mathsf{unique}}_{\mathcal{A}}(\lambda)$ *is defined as follows:*

<u>MAIN $\mathsf{Game}^{\mathsf{unique}}_{\mathcal{A}}(\lambda)$</u>
$\mathsf{crs}_{\mathsf{vuf}} \leftarrow \mathsf{VUF.Setup}(1^\lambda)$
$(\mathsf{pk}, m, \sigma_1, \sigma_2) \xleftarrow{\$} \mathcal{A}(\mathsf{crs}_{\mathsf{vuf}})$
$y_1 \leftarrow \mathsf{VUF.Derive}(\mathsf{crs}_{\mathsf{vuf}}, \mathsf{pk}, m, \sigma_1)$
$y_2 \leftarrow \mathsf{VUF.Derive}(\mathsf{crs}_{\mathsf{vuf}}, \mathsf{pk}, m, \sigma_2)$
return $(y_1 \neq y_2)\ \wedge\ \mathsf{VUF.Ver}(\mathsf{crs}_{\mathsf{vuf}}, \mathsf{pk}, m, \sigma_1)\ \wedge\ \mathsf{VUF.Ver}(\mathsf{crs}_{\mathsf{vuf}}, \mathsf{pk}, m, \sigma_2)$

We say that Π *is unique if for all PPT adversaries* $\mathcal{A}$ *we have that* $\mathsf{Adv}^{\mathsf{unique}}_{\mathcal{A}}(\lambda) \leq \mathsf{negl}(\lambda)$.

Finally, a VUF is unpredictable if an adversary cannot predict the output of the function VUF.Eval on a message for which it has not seen any valid signatures.

Definition 4 (Unpredictability). *For a VUF* Π *and an adversary* $\mathcal{A}$, *let* $\mathsf{Adv}^{\mathsf{predict}}_{\mathcal{A}}(\lambda) = \Pr[\mathsf{Game}^{\mathsf{predict}}_{\mathcal{A}}(\lambda)]$ *where* $\mathsf{Game}^{\mathsf{predict}}_{\mathcal{A}}(\lambda)$ *is defined as follows:*

<u>MAIN $\mathsf{Game}^{\mathsf{predict}}_{\mathcal{A}}(\lambda)$</u>
$H \leftarrow \emptyset$
$\mathsf{crs}_{\mathsf{vuf}} \leftarrow \mathsf{VUF.Setup}(1^\lambda)$
$(\mathsf{pk}, \mathsf{sk}) \leftarrow \mathsf{VUF.Gen}(\mathsf{crs}_{\mathsf{vuf}})$
$(m, y) \xleftarrow{\$} \mathcal{A}^{\mathsf{VUF.Sign}(\mathsf{crs}_{\mathsf{vuf}}, \mathsf{sk}, \cdot)}(\mathsf{crs}_{\mathsf{vuf}}, \mathsf{pk})$
return $(\mathsf{VUF.Eval}(\mathsf{crs}_{\mathsf{vuf}}, \mathsf{sk}, m) = y)\ \wedge\ (m \notin H)$

<u>ORACLE $\mathcal{O}^{\mathsf{VUF.Sign}(\mathsf{crs}_{\mathsf{vuf}}, \mathsf{sk}, m)}$</u>
add m *to query set* H
return $\mathsf{VUF.Sign}(\mathsf{crs}_{\mathsf{vuf}}, \mathsf{sk}, m)$

We say that Π *is unpredictable if for all PPT adversaries* $\mathcal{A}$ *we have that* $\mathsf{Adv}^{\mathsf{predict}}_{\mathcal{A}}(\lambda) \leq \mathsf{negl}(\lambda)$.

3.5 Rekeyability

To show that existing cryptographic primitives can be instantiated with our DKG, and other DKGs in the literature, we rely on a property called *rekeyability*. Intuitively, rekeyability says that it is possible to transform an object (e.g., a ciphertext or signature) that was formed using one cryptographic key into an object formed with a related key. As one concrete example, in the BLS signature scheme, in which a signature on a message m is of the form $\sigma = H(m)^{\mathsf{sk}_1}$, it is possible to transform this into a signature under the key $\alpha\mathsf{sk}_1 + \mathsf{sk}_2$ by computing $\sigma^\alpha \cdot H(m)^{\mathsf{sk}_2}$. This means that BLS can be efficiently rekeyed with respect to the secret key. While this notion is related to the idea of re-randomizability [26, 35, 55], we are not aware of any formalizations in the literature and it may be of independent interest.

Definition 5 (Rekeyability). *For a public-key primitive $\Pi = (\mathsf{KeyGen}, \Pi_1, \ldots, \Pi_n)$ and functions $f_{\mathsf{k}}(\alpha, \mathsf{k}_1, \mathsf{k}_2)$ that outputs $\alpha\mathsf{k}_1 \oplus \mathsf{k}_2$ for some binary operator $\oplus$ (typically $+$ or $\times$), we define rekeyability as follows for all $\alpha \in \mathbb{N}$ and $(\mathsf{pk}_1, \mathsf{sk}_1), (\mathsf{pk}_2, \mathsf{sk}_2) \in [\mathsf{KeyGen}(1^\lambda)]$:*

- *We say that an algorithm Π_i is* rekeyable *with respect to the secret key if there exists an efficient function rekey_i such that*

$$\mathsf{rekey}_i(\alpha, \mathsf{pk}_1, \mathsf{sk}_2, x, \Pi_i(\mathsf{sk}_1, x; r)) = \Pi_i(f_{\mathsf{sk}}(\alpha, \mathsf{sk}_1, \mathsf{sk}_2), x; r)$$

  for all $x \in \mathsf{Domain}(\Pi_i)$ and randomness r. Likewise, we say that an algorithm Π_j is rekeyable *with respect to the public key if there exists an efficient function rekey_j such that*

$$\mathsf{rekey}_j(\alpha, \mathsf{pk}_1, \mathsf{sk}_2, \Pi_j(\mathsf{pk}_1, x; r)) = \Pi_j(f_{\mathsf{pk}}(\alpha, \mathsf{pk}_1, \mathsf{pk}_2), x; r)$$

  for all $x \in \mathsf{Domain}(\Pi_i)$ and randomness r.
- *We say that (Π_i, Π_j) is* rekeyable *with respect to the secret key if (1) Π_i is rekeyable with respect to the secret key and (2)*

$$\Pi_j(\mathsf{pk}_1, y) = \Pi_j(f_{\mathsf{pk}}(\alpha, \mathsf{pk}_1, \mathsf{pk}_2), \mathsf{rekey}_i(\alpha, \mathsf{pk}_1, \mathsf{sk}_2, y)).$$

  Likewise we say that (Π_i, Π_j) is rekeyable *with respect to the public key if (1) Π_i is rekeyable with respect to the public key and (2)*

$$\Pi_j(\mathsf{sk}_1, y) = \Pi_j(f_{\mathsf{sk}}(\alpha, \mathsf{sk}_1, \mathsf{sk}_2), \mathsf{rekey}_i(\alpha, \mathsf{pk}_1, \mathsf{sk}_2, y)).$$

For encryption, we would want that $(\mathsf{Encrypt}, \mathsf{Decrypt})$ is rekeyable with respect to the public key, meaning new key material can be folded into ciphertexts without affecting the ability to decrypt. For signing, we would want that $(\mathsf{Sign}, \mathsf{Verify})$ is rekeyable with respect to the secret key, meaning that if signatures verify then so do their rekeyed counterparts.

3.6 Distributed Key Generation (DKG)

We define a *distributed key generation* (DKG) as an interactive protocol that is used to generate a keypair $(\mathsf{pk}, \mathsf{sk})$. We define this as $(\mathsf{transcript}, \mathsf{pk}) \xleftarrow{\$} \mathtt{DKG}(I, n)$, where n is the number of participants in the DKG, I is the indices of the adversarial participants (so $|I| \leq t$), pk is the resulting public key, and $\mathsf{transcript}$ is some representation of the messages that have been exchanged.

We additionally consider an algorithm Reconstruct that, given transcript and the shares submitted by $t+1$ honest parties, outputs the secret key sk corresponding to pk. With this in place, we can define an omniscient interactive protocol $(\mathsf{transcript}, (\mathsf{pk}, \mathsf{sk}), \mathsf{state}_{\mathcal{A}}) \xleftarrow{\$} \mathtt{OmniDKG}(I, n)$ that is aware of the internal state of each participant and thus can output sk (by running the Reconstruct algorithm) and $\mathsf{state}_{\mathcal{A}}$; i.e., the internal state of the adversary.

The Reconstruct algorithm is useful not only in defining this extra interactive protocol, but also in defining a notion of *robustness* for DKGs (initially called correctness by Gennaro et al. [31]). We define this as follows:

Definition 6 (Robustness). *A DKG protocol is* robust *if the following properties hold:*

- *A DKG transcript* dkg *determines a public key* pk *that all honest parties agree on.*
- *There is an efficient algorithm* Reconstruct

$$\mathsf{sk} \leftarrow \mathsf{Reconstruct}(\mathsf{dkg}, \mathsf{sk}_1, \ldots, \mathsf{sk}_\ell) \textit{ for } t+1 \leq \ell \leq n$$

that takes as input a set of secret key shares where at least $t+1$ are from honest parties and verifies them against the public transcript produced by the DKG protocol. It outputs the unique value sk *such that* $\mathsf{pk} \leftarrow \mathsf{KeyGen}(1^\lambda; \mathsf{sk})$.

Beyond robustness, we also want a DKG to *preserve security* of the underlying primitive for which it is run. Previous related definitions of *secrecy* for DKGs required there to exist a simulator that could fix the output of the DKG; i.e., given an input y, could output $(\mathsf{transcript}, y)$ that the adversary could not distinguish from a real $(\mathsf{transcript}, \mathsf{pk})$ output by the DKG run with t adversarial participants. While general, this definition is strong and required previous constructions to have more rounds or constraints than would otherwise be necessary; *e.g.*, there seem to be significant barriers to satisfying this definition in any DKG where the adversary is allowed to go last, as they the know the entire transcript and can bias the final result.

In defining what it means for a DKG to preserve security, we first weaken this previous definition. Rather than require a simulator given pk_1 to have the DKG output exactly pk_1, we consider that it can instead fix the output public key to have a known relation with its input public key. In particular, a simulator given pk_1 can fix the output of the DKG to be pk, where the simulator knows $(\alpha, \mathsf{pk}_2, \mathsf{sk}_2)$ such that $\mathsf{pk} = f(\alpha, \mathsf{pk}_1, \mathsf{pk}_2)$ for $\alpha \neq 0$ and f as defined in the rekeyability definition (see Definition 5). We call this property *key expressability.*

Definition 7 (Key expressability). *For a simulator* Sim, *define as* $(\mathsf{transcript}, \mathsf{pk}, \alpha, \mathsf{pk}_2, \mathsf{sk}_2) \xleftarrow{\$} \mathtt{SimDKG}(\mathsf{Sim}, I, n)$ *a run of the DKG protocol in which all honest participants are controlled by* Sim, *which takes as input a public key* pk_1 *and has private outputs* α, pk_2, *and* sk_2. *We say that a DKG is* key-expressable *if there exists such a simulator* Sim *such that (1)* $(\mathsf{transcript}, \mathsf{pk})$ *is distributed identically to the output of* $\mathtt{DKG}(I, n)$, *(2)* $(\mathsf{pk}_2, \mathsf{sk}_2)$ *is a valid keypair, and (3)* $\mathsf{pk} = f(\alpha, \mathsf{pk}_1, \mathsf{pk}_2) = \alpha\mathsf{pk}_1 \oplus \mathsf{pk}_2$.

To now define a security-preserving DKG, we intuitively consider a DKG being run in the context of a security game. To keep our definition as general as possible, our only requirements are that (1) the security game contains a line of the form $(\mathsf{pk}, \mathsf{sk}) \xleftarrow{\$} \mathsf{KeyGen}(1^\lambda)$ (it also works if KeyGen takes a common reference string as additional input), and (2) pk is then later given as input to the adversary. We then say that the DKG preserves security if it is not possible for an adversary participating in the DKG to do better than it would have done in the original security game, in which it was given pk directly. Formally, we have the following definition.

Definition 8 (Security-preserving). *Define* Game *as any security game containing the line* $(\mathsf{pk}, \mathsf{sk}) \xleftarrow{\$} \mathsf{KeyGen}(1^\lambda)$, *denoted* $\mathsf{line}_{\mathsf{pk}}$, *and where* pk *is later input to an adversary* $\mathcal{A}$ *(in addition to other possible inputs). Define* $\mathsf{Game}'(\mathsf{line}, x)$, *parameterized by a starting line* line *and some value* x, *as* Game *but with* $\mathsf{line}_{\mathsf{pk}}$ *replaced by* line *and* $\mathcal{A}$ *given* x *as input rather than* pk. *It is clear that* $\mathsf{Game} = \mathsf{Game}'(\mathsf{line}_{\mathsf{pk}}, \mathsf{pk})$.

Define $\mathsf{line}_{\mathsf{dkg}}$ *as the line* $(\mathsf{transcript}, (\mathsf{pk}, \mathsf{sk}), \mathsf{state}_{\mathcal{A}}) \xleftarrow{\$} \mathtt{OmniDKG}(I, n)$, *and define* $\mathsf{DKG\text{-}Game} \leftarrow \mathsf{Game}'(\mathsf{line}_{\mathsf{dkg}}, \mathsf{state}_{\mathcal{A}})$. *We say the DKG* preserves security *for* Game *if*

$$\mathsf{Adv}_{\mathcal{A}}^{\mathsf{DKG\text{-}Game}}(\lambda) \leq \mathsf{Adv}_{\mathcal{A}}^{\mathsf{Game}}(\lambda) + \mathsf{negl}(\lambda)$$

for all PPT adversaries $\mathcal{A}$.

We do not view our requirements for the original security game as restrictive, given the number of security games that satisfy them. For signature unforgeability, for example, our definition says that an adversary that participates in the DKG, and can carry its state from that into the rest of the game (including all of the messages it saw), cannot achieve better advantage than when it is just given the public key (as in the standard EUF-CMA game).

While the relationship between key expressability and security-preserving DKGs is not obvious, we show in the full version of our paper [38] that it is typically the case that when key-expressable DKGs are used for rekeyable primitives, they preserve the security of that primitive's underlying security game.

4 Our Enhanced Scrape PVSS

A *secret sharing scheme* allows a *dealer* to deal out n secret *shares* so that any subset of $t + 1$ shares suffices to reconstruct the secret, but subsets of size $\leq t$ shares do not. A *publicly verifiable secret sharing (PVSS)* scheme is a secret sharing scheme in which any third party can verify that the dealer has behaved honestly. Importantly, PVSS obviates the need for a complaint round in VSS protocols, which simplifies designing PVSS-based DKG protocols [27]. Cascudo and David designed an elegant PVSS scheme called *Scrape* [18] with $O(n)$ verification costs. In this section, we describe a slightly-modified variant of Scrape that supports *aggregation* and uses Type III pairings and an additional element

$\hat{u}_2 \in \mathbb{G}_2$ that will help our DKG security proofs later on. We rely on Type III pairings, not only for efficiency, but also because the SXDH assumption does not hold in symmetric groups. We give a formal description in Fig. 1. In Sect. 5, we use this slightly-modified variant of Scrape to construct our DKG.

Common reference string (CRS). All parties use the same CRS consisting of (1) a bilinear group description bp (which fixes $g_1 \in \mathbb{G}_1$ and $\hat{h}_1 \in \mathbb{G}_2$), (2) a group element $\hat{u}_1 \in \mathbb{G}_2$ and (3) *encryption keys* $\mathsf{ek}_i \in \mathbb{G}_2$ for every party P_i with corresponding *decryption keys* $\mathsf{dk}_i \in \mathbb{F}$ known only to P_i such that $\mathsf{ek}_i = \hat{h}_1^{\mathsf{dk}_i}$.

Dealing. Scrape resembles other Shamir-based [58] secret sharing schemes. The Scrape dealer will share a secret $\hat{h}_1^{a_0} \in \mathbb{G}_2$, whose corresponding $a_0 \in \mathbb{F}$ the dealer knows. (This is different than other VSS schemes, which typically share a secret in $\mathbb{F}$ rather than in $\mathbb{G}_2$.) The dealer picks a random, degree-t polynomial $f(X) = (a_0, a_1, \ldots, a_t)$, where $f(0) = a_0$, and commits to it via Feldman [25] as $F_i = g_1^{a_i}, \forall i \in [0, t]$. Party P_i's share will be $\hat{h}_1^{f(\omega_i)}$. The dealer then computes Feldman commitments $A_i = g_1^{f(\omega_i)}$ and encryptions $\mathsf{ek}_i^{f(\omega_i)}$ of each share. (The term "encryption" here is slightly abused since these are not IND-CPA-secure ciphertexts.) The *PVSS transcript* will consist of the Feldman commitments to $f(X)$ and to the shares, plus the encryptions of the shares. Additionally, we augment the transcript with $\hat{u}_2 = \hat{u}_1^{a_0}$, which helps our DKG security proofs.

Verifying. Each party P_i can verify that the PVSS transcript is a correct sharing of $\hat{h}_1^{a_0}$. For this, P_i checks the Feldman commitments A_i to the shares $f(\omega_i)$ are consistent with the Feldman commitment to $f(X)$ via Lagrange interpolation in the exponent (see Fig. 1). Then, each P_i checks their encryption of $f(\omega_i)$ against A_i. Altogether, this guarantees that the encrypted shares are indeed the evaluations of the committed polynomial f.

Aggregating transcripts. One of our key contributions is an algorithm for aggregating two Scrape PVSS transcripts pvss_1 and pvss_2 for polynomials f_1 and f_2 into a single transcript for their sum $f_1 + f_2$. This is a key ingredient of our DKG from Sect. 5. Our aggregation leverages the homomorphism of Feldman commitments and of the encryption scheme. Indeed, suppose we have Feldman commitments to f_b consisting of $F_{b,i} = g_1^{a_{b,i}}, \forall i \in [0, t]$, where $a_{b,i}$'s are the coefficients of f_b, for $b \in \{1, 2\}$. Then, $F_i = F_{1,i} F_{2,i} = g^{a_{1,i}+a_{2,i}}, \forall i \in [0, t]$ will be a Feldman commitment to $f_1 + f_2$. Similarly, we can aggregate the share commitments $A_{b,i} = g^{f_b(\omega_i)}$ as $A_i = A_{1,i} A_{2,i} = g^{(f_1+f_2)(\omega_i)}, \forall i \in [n]$. Lastly, the encryptions $\mathsf{ek}_i^{f_b(\omega_i)}$ can be aggregated as $\mathsf{ek}_i^{(f_1+f_2)(\omega_i)} = \mathsf{ek}_i^{f_1(\omega_i)} \mathsf{ek}_i^{f_2(\omega_i)}$. We summarize this aggregation algorithm in Fig. 1.

Reconstructing the secret. At the end of the PVSS protocol, each party P_i can decrypt their share as $\hat{A}_i = \hat{Y}_i^{\mathsf{dk}_i^{-1}} = (\mathsf{ek}_i^{f(\omega_i)})^{\mathsf{dk}_i^{-1}} = \hat{h}_1^{f(\omega_i)}$. Recall that the degree t polynomial $f(X)$ encodes the secret $f(0) = a_0$. Thus, any set S of $\geq t + 1$ honest parties can reconstruct $\mathsf{sk} = \hat{h}_1^{f(0)}$ as follows:

$\mathsf{Scrape.Deal}(\mathsf{bp}, \mathbf{ek}, \hat{u}_1, a_0) \to \mathsf{pvss}$

$(a_1, \ldots, a_t) \xleftarrow{\$} \mathbb{F}^t$, $f(X) \leftarrow \sum_{i=0}^{t} a_i X^i$
$F_0, \ldots, F_t \leftarrow g_1^{a_0}, \ldots, g_1^{a_t}$
$\hat{u}_2 \leftarrow \hat{u}_1^{a_0}$
$A_1, \ldots, A_n \leftarrow g_1^{f(\omega_1)}, \ldots, g_1^{f(\omega_n)}$
$\hat{Y}_1 \ldots, \hat{Y}_n \leftarrow \mathsf{ek}_1^{f(\omega_1)}, \ldots, \mathsf{ek}_n^{f(\omega_n)}$
return $\boldsymbol{F}, \hat{u}_2, \boldsymbol{A}, \hat{\boldsymbol{Y}}$

$\mathsf{Scrape.Verify}(\mathsf{bp}, \mathbf{ek}, \hat{u}_1, \hat{u}_2, \mathsf{pvss}) \to 0/1$

$\boldsymbol{F}, \hat{u}_2, \boldsymbol{A}, \hat{\boldsymbol{Y}} \leftarrow \mathsf{parse}(\mathsf{pvss})$
$\alpha \xleftarrow{\$} \mathbb{F}$
check $\prod_{j=1}^{n} A_j^{\ell_j(\alpha)} = \prod_{j=0}^{t} F_j^{\alpha^j}$
check $e(F_0, \hat{u}_1) = e(g_1, \hat{u}_2)$
check $e(g_1, \hat{Y}_j) = e(A_j, \mathsf{ek}_j)$ for $1 \leq j \leq n$
return 1 if all checks pass, else return 0

$\mathsf{Scrape.Aggregate}(\mathsf{bp}, \mathsf{pvss}_1, \mathsf{pvss}_2) \to \mathsf{pvss}$

$((F_{1,0}, \ldots, F_{1,t}), \hat{u}_{1,2}, (A_{1,1}, \ldots, A_{1,n}), (\hat{Y}_{1,1}, \ldots, \hat{Y}_{1,n})) \leftarrow \mathsf{parse}(\mathsf{pvss}_1)$
$((F_{2,0}, \ldots, F_{2,t}), \hat{u}_{2,2}, (A_{2,1}, \ldots, A_{2,n}), (\hat{Y}_{2,1}, \ldots, \hat{Y}_{2,n})) \leftarrow \mathsf{parse}(\mathsf{pvss}_2)$

for $0 \leq i \leq t$:
  $F_i \leftarrow F_{1,i} F_{2,i}$

for $1 \leq i \leq n$:
  $A_i \leftarrow A_{1,i} A_{2,i}$, $\hat{Y}_i \leftarrow \hat{Y}_{1,i} \hat{Y}_{2,i}$

$\hat{u}_2 \leftarrow \hat{u}_{1,2} \hat{u}_{2,2}$
return $\boldsymbol{F}, \hat{u}_2, \boldsymbol{A}, \hat{\boldsymbol{Y}}$

Fig. 1. Dealing, verification and aggregation algorithms for the Scrape PVSS. Here, $\mathbf{ek}, \boldsymbol{F}, \boldsymbol{A}, \hat{\boldsymbol{Y}}$ denote vectors of ek_i's, F_i's, A_i's and $\hat{Y}_i$'s. The polynomial $\ell_j(X)$ denotes the Lagrange polynomial equal to 1 at ω_j and 0 at $\omega_i \neq \omega_j$. The ω_i's are public predetermined values which, for efficiency purposes, should be chosen as roots of unity of degree n. For more details, see the full version of our paper [38].

1. For each share $\hat{A}_i$ provided, check that $e(A_i, \hat{h}_1) = e(g_1, \hat{A}_i)$, where $A_i = g_1^{f(\omega_i)}$ is part of the PVSS transcript. If this check fails, or if P_i does not provide a share, then remove P_i from S.
2. Return, $\mathsf{sk} = \prod_{i \in S} \hat{A}_i^{\ell_{S,i}(0)}$ where $\ell_{S,i}(X)$ is a Lagrange polynomial equal to 0 at $\omega_j \in S$ for $i \neq j$, and 1 at ω_i.

5 Distributed Key Generation

In this section, we describe our distributed key generation (DKG) protocol for generating a key-pair $(\mathsf{pk}, \mathsf{sk})$ of the form

$$\mathsf{pk} = (g_1^a, \hat{u}_1^a) \in \mathbb{G}_1 \times \mathbb{G}_2 \quad \text{and} \quad \mathsf{sk} = \hat{h}_1^a \in \mathbb{G}_2, \text{ where } a \in \mathbb{F}$$

We often refer to $a \in \mathbb{F}$ as the *DKG secret.* All parties P_i use the same Scrape CRS (see Sect. 4) but augmented with *verification keys* vk_i (defined later).

At a high level, our DKG protocol resembles previous protocols based on verifiable secret sharing: each party P_i deals a secret $\hat{h}_1^{c_i}$ to all other parties using the Scrape PVSS from Sect. 4. Additionally, each party P_i includes a proof-of-knowledge of their secret c_i. At this point, each party P_j would have to verify the PVSS transcript of every other party P_i, resulting in $O(n^2)$ work. Then, the final secret would be $\mathsf{sk} = \hat{h}_1^a$ with $a = \sum_{i \in Q} c_i$, where Q is the set of all parties who dealt honestly (i.e., whose PVSS transcript verified). Note that since PVSS transcripts are publicly-verifiable, all parties P_i agree on Q and there is no need for a complaint round. We often refer to an honest party P_i as having *contributed* to the final secret key and to c_i as its *contribution*.

Gossip and aggregate. To avoid the $O(n^2)$ verification work per party, we leverage aggregation of Scrape PVSS transcripts. We observe that a party who verified several transcripts can aggregate them into a single one and forward it to another party, who can now verify this aggregated transcript faster. By carefully aggregating and *gossiping* transcripts in this manner, we decrease verification time per party from $O(n^2)$ to $O(n \log^2 n)$. One caveat is that, due to the randomized nature of gossiping, a party's contribution c_i might be incorporated multiple times, say w_i times, into the final secret $\mathsf{sk} = \hat{h}_1^a$. As a result, the final $a = \sum_{i \in Q} w_i c_i$, where w_i is called the *weight* of each c_i.

Signatures-of-knowledge of contributions. Similar to previous DKGs [32], our DKG requires each party P_i to prove knowledge of its contribution c_i to the final DKG secret. However, since our DKG transcripts must be publicly-verifiable, we also require each party to sign their contributions. We achieve both of these goals using a *signature-of-knowledge (SoK)*. Specifically, P_i signs $C_i = g_1^{c_i}$ using its secret key sk_i, with corresponding *verification key* $\mathsf{vk}_i = g_1^{\mathsf{sk}_i}$:

$$\sigma_i = (\sigma_{i,1}, \sigma_{i,2}) = (\mathsf{Hash}_{\mathbb{G}_2}(C_i)^{c_i}, \mathsf{Hash}_{\mathbb{G}_2}(\mathsf{vk}_i, C_i)^{\mathsf{sk}_i})$$

where $\mathsf{Hash}_{\mathbb{G}_2}$ is a hash function that maps to $\mathbb{G}_2$. Any verifier with vk_i can verify the signature-of-knowledge σ_i of c_i as:

$$e(C_i, \mathsf{Hash}_{\mathbb{G}_2}(C_i)) = e(g_1, \sigma_{i,1}) \;\wedge\; e(\mathsf{vk}_i, \mathsf{Hash}_{\mathbb{G}_2}(\mathsf{vk}_i, C_i)) = e(g_1, \sigma_{i,2})$$

Our signatures of knowledge are simulation-sound and thus cannot be compressed or combined. However, since they are constant-sized, this is not problematic. We refer to the signing algorithm as $\mathsf{SoK.Sign}(C_i, \mathsf{sk}_i, c_i) \to \sigma_i$ and the verification algorithm as $\mathsf{SoK.Verify}(\mathsf{vk}_i, C_i, \sigma_i) \to 0/1$.

DKG transcripts. To maintain their public-verifiability, aggregated PVSS transcripts must keep track of the weights w_i of each party's contribution c_i and of the σ_i's. This gives rise to a new notion of a *DKG transcript* defined as:

$$\mathsf{transcript} = ((C_1, \ldots, C_n), (w_1, \ldots, w_n), (\sigma_1, \ldots, \sigma_n), \mathsf{pvss}), \tag{1}$$

where $C_i = g_1^{c_i}$ is a commitment to the contribution c_i of party P_i, w_i is its weight, σ_i is the SoK of c_i and pvss is an (aggregated) PVSS transcript for secret $a = \sum_{i \in [n]} w_i c_i$.

Recall from Fig. 1 that pvss stores a Feldman commitment $\boldsymbol{F}$ to a polynomial $f(X)$ with $f(0) = a$ and that $F_0 = g_1^a$. In our protocol, each party

P_i initializes their DKG transcript by picking $c_i \xleftarrow{\$} \mathbb{F}$ and setting $\mathsf{pvss} \leftarrow \mathsf{Scrape.Deal}(\mathsf{bp}, \mathbf{ek}, \hat{u}_1, c_i)$, $C_i \leftarrow g_1^{c_i}$, $w_i \leftarrow 1$ and $\sigma_i \leftarrow \mathsf{SoK.Sign}(C_i, \mathsf{sk}_i, c_i)$. For $j \neq i$, P_i sets $C_j \leftarrow \perp$, $w_j \leftarrow 0$ and $\sigma_j \leftarrow \perp$. Importantly, in our protocol, each party will broadcast the C_i commitment to their contribution and gossip the rest of their DKG transcript to a subset of the other parties (we discuss this in more detail later on).

Verifying DKG transcripts. To verify the DKG transcript from Eq. (1), one first checks that its inner pvss transcript verifies. Second, for all *non-trivial contributions* with $w_i \neq 0$, one first checks if their signature of knowledge σ_i verifies. Finally, one checks that the contributions correctly combine to the commitment F_0 to the zero coefficient of $f(X)$ shared in pvss; i.e., that $C_1^{w_1} \cdots C_n^{w_n} = F_0$. If transcript passes these checks, then one can be sure that the players P_i which have $w_i \neq 0$ in transcript have contributed to its corresponding DKG secret. See Fig. 2 for a full description.

Aggregating DKG transcripts. Given two input DKG transcripts

$$(C_{b,1}, \ldots, C_{b,n}), (w_{b,1}, \ldots, w_{b,n}), (\sigma_{b,1}, \ldots, \sigma_{b,n}), \mathsf{pvss}_b, \text{ for } b \in \{1, 2\}$$

we can easily aggregate them into a single DKG transcript

$$(C_1, \ldots, C_n), (w_1, \ldots, w_n), (\sigma_1, \ldots, \sigma_n), \mathsf{pvss}$$

We first aggregate the pvss_b transcripts into pvss via Scrape.Aggregate (see Fig. 1). Second, we aggregate the weights, which are field elements, as $w_i = w_{1,i} + w_{2,i}, \forall i \in [n]$. Third, if P_i contributed in one of the input transcripts, then P_i's contribution should also be reflected in the aggregated transcript. In other words, for any $C_{b,i} \neq \perp$ and valid $\sigma_{b,i}$, we simply set $C_i = C_{b,i}$ and $\sigma_i = \sigma_{b,i}$. The choice of $C_{b,i}$ does not matter when they are both $\neq \perp$ since they were both obtained from the broadcast channel, so they must be equal. As a result, their corresponding $\sigma_{b,i}$'s will also be equal since our signatures of knowledge are unique.

Reconstructing the secret. As explained in the beginning of this section, the final key-pair will be $\mathsf{pk} = (g_1^{f(0)}, \hat{u}_2) = (g_1^{f(0)}, \hat{u}_1^{f(0)})$ and $\mathsf{sk} = \hat{h}_1^{f(0)}$. Since the final DKG transcript is just an augmented Scrape PVSS transcript, reconstruction of sk works as explained in Sect. 4.

5.1 A Gossip Protocol

In Step 4 of our DKG, we rely on a gossip protocol to communicate the $\mathcal{O}(n)$-sized DKG transcripts. By using gossip, we avoid both the need to broadcast these larger messages, which is expensive, and the need for a central aggregator. We detail our protocol in the full version of our paper [38], but provide some insight here into how it works.

We take an optimistic approach and provide robustness for up to $t_r < n/2 - \log n$ crashed parties but only up to $\log n$ Byzantine adversaries. We believe this

$\underline{\mathsf{DKG.Aggregate}(\mathsf{bp}, \mathsf{transcript}_1, \mathsf{transcript}_2) \to \mathsf{transcript}}$

$((C_{1,1}, \ldots, C_{1,n}), (w_{1,1}, \ldots, w_{1,n}), (\sigma_{1,1}, \ldots, \sigma_{1,n}), \mathsf{pvss}_1) \leftarrow \mathsf{parse}(\mathsf{transcript}_1)$
$((C_{2,1}, \ldots, C_{2,n}), (w_{2,1}, \ldots, w_{2,n}), (\sigma_{2,1}, \ldots, \sigma_{2,n}), \mathsf{pvss}_2) \leftarrow \mathsf{parse}(\mathsf{transcript}_2)$

for $1 \leq i \leq n$:
  $w_i \leftarrow w_{1,i} + w_{2,i}$
  if $\sigma_{1,i} \neq \bot$: $\sigma_i \leftarrow \sigma_{1,i}$, else: $\sigma_i \leftarrow \sigma_{2,i}$
  if $C_{1,i} \neq \bot$: $C_i \leftarrow C_{1,i}$, else: $C_i \leftarrow C_{2,i}$

$\mathsf{pvss} \leftarrow \mathsf{Scrape.Aggregate}(\mathsf{bp}, \mathsf{pvss}_1, \mathsf{pvss}_2)$
return $(C_1, \ldots, C_n), (w_1, \ldots, w_n), (\sigma_1, \ldots, \sigma_n), \mathsf{pvss}$

$\underline{\mathsf{DKG.Verify}(\mathsf{bp}, (\mathsf{ek}_i, \mathsf{vk}_i)_{i \in [n]}, \hat{u}_1, \mathsf{transcript}) \to 0/1}$

$((C_1, \ldots, C_n), (w_1, \ldots, w_n), (\sigma_1, \ldots, \sigma_n), \mathsf{pvss}) \leftarrow \mathsf{parse}(\mathsf{transcript})$
$((F_0, \ldots, F_t), \hat{u}_2, (A_1, \ldots, A_n), (\hat{Y}_1, \ldots, \hat{Y}_n)) \leftarrow \mathsf{parse}(\mathsf{pvss})$
check $\mathsf{Scrape.Verify}(\mathsf{bp}, (\mathsf{ek}_1, \ldots, \mathsf{ek}_n), \hat{u}_1, \hat{u}_2, \mathsf{pvss}) = 1$

for $1 \leq i \leq n$:
  if $w_i \neq 0$: check $\mathsf{SoK.Verify}(\mathsf{vk}_i, C_i, \sigma_i) = 1$

check $C_1^{w_1} \cdots C_n^{w_n} = F_0$
return 1 if all checks pass, else return 0

Fig. 2. Aggregation algorithm for the distributed key generation protocol.

approach is often reasonable in practice because if a Byzantine adversary attacks the robustness of a DKG, the only outcome is that the computation required to output the DKG is higher. Furthermore, Byzantine attacks on robustness are detectable, so any faulty party can be manually removed from the system. This is in contrast to an attack on the security preservation of the DKG, which could have far more serious consequences. If we want a security threshold of t_s, then we have to assume that t_s parties respond. A direct implication is that $n - t_r$ must be at least t_s, showing an inherent tradeoff between the security and robustness thresholds. In our scheme we can set t_s to be exactly equal to $n - t_r$.

The gossip protocol has each party send its currently aggregated DKG transcript to $\mathcal{O}(c \log n)$ parties in expectation in each round, and terminate when it has agreed on a *"full" transcript*; i.e., a valid transcript with at least $t_s + 1$ contributions. Here c is a small success parameter such that $c \geq 4$. However, deciding when to terminate is non-trivial, because the aggregated "full" transcripts may all be different. We thus still rely on broadcast to agree on which transcript to use, but our goal is to minimize the number of total broadcasts. We do this by having each party with a full transcript broadcast it with probability $2/n$ in a given round. We argue that this makes the protocol likely to terminate within $\mathcal{O}(c \log n)$ rounds. Parties agree to use the transcript whose public key has a binary representation with the smallest bit-count (but any other publicly-verifiable convention works too). In terms of complexity, our gossip protocol

Our aggregatable DKG protocol

Common reference string: Scrape CRS consiting of $\mathsf{bp} = (p, \mathbb{G}_1, \mathbb{G}_2, \mathbb{G}_T, e, g_1, \hat{h}_1)$, encryption and verficiation keys $(\mathsf{ek}_i, \mathsf{vk}_i)_{i \in [n]}$, nth roots of unity $(\omega_i)_{i \in [n]}$ in $\mathbb{F}$, random $\hat{u}_1 \in \mathbb{G}_2$ such that nobody knows $\log_{\hat{h}_1}(\hat{u}_1)$.

Party P_i's private input: Decryption key dk_i for ek_i and secret key sk_i for vk_i.

1. Each P_i picks random $c_i \in \mathbb{F}$, computes $C_i = g_1^{c_i}$ and broadcasts C_i.
2. Each P_i picks random polynomial $f_i(X) \in \mathbb{F}[X]$ of degree at most t

$$f_i(X) = a_{i,0} + a_{i,1}X + \cdots + a_{i,t}X^t$$

such that $a_{i,0} = c_i$. They compute $f_i(\omega_j)$ for $j \in [n]$. Each party *gossips* (see Section 5.1) their DKG transcript consisting of (1) $F_{i,k} = g_1^{a_{i,k}}$ for $k \in [0, t]$; (2) $\hat{u}_{i,2} = \hat{u}_1^{c_i}$; (3) a vector $\boldsymbol{w}_i$ such that $w_{i,j} = 1$ if $i = j$ and 0 otherwise; (4) $A_{i,j} = g_1^{f_i(\omega_j)}$ for $j \in [n]$; (5) $Y_{i,j} = \mathsf{ek}_j^{f_i(\omega_j)}$ for $j \in [n]$; and (6) a vector $\boldsymbol{\sigma}_i$ such that $\sigma_{i,j} = (\mathsf{Hash}_{\mathbb{G}_2}(C_i)^{c_i}, \mathsf{Hash}_{\mathbb{G}_2}(\mathsf{vk}_i, C_i)^{\mathsf{sk}_i})$ if $i = j$ and $\perp$ otherwise.
3. During the gossip phase, each P_i verifies the transcripts it receives using DKG.Verify (see Fig. 2). If two transcripts verify, it aggregates them using DKG.Aggregate (also in Fig. 2), and gossips the aggregated transcript. The aggregated transcripts contain a list of weights $(w_1, \ldots, w_n) \in \mathbb{F}^n$ indicating how many times each party has contributed to the current transcript. When a party receives a *"full" transcript* with $\geq t + 1$ non-zero weights, it broadcasts this as a candidate final transcript.
4. Parties terminate in the round where they first broadcast a "full" transcript. If several candidate "full" transcripts were broadcast, the one whose pk has the lowest bit count is chosen as the final one. The final $\mathsf{pk} = (\prod_{i=1}^n C_i^{w_i}, \prod_{i=1}^n \hat{u}_{i,2}^{w_i})$. Each party computes their secret key share as $Y_i^{\mathsf{dk}_i^{-1}}$ such that they can reconstruct.

Fig. 3. Our DKG with reconstruction threshold $t + 1$ run by parties $P_1, \ldots, P_n$.

requires $\mathcal{O}(cn^2 \log n)$ total words to be communicated in private messages and $\mathcal{O}(c \log^2 n)$ broadcasts.

5.2 Security Analysis

Robustness. Our DKG is robust in the sense that all honest parties agree on the final public key, and in the sense that any set S containing at least $t + 1$ honest parties can reconstruct the secret key.

Theorem 1 (DKG is robust). *The scheme in Fig. 3 is robust for any primitive with keys of the form* $\mathsf{pk} = (g_1^a, \hat{u}_1^a) \in \mathbb{G}_1 \times \mathbb{G}_2$.

Proof. First we show that all honest parties have the same value pk. By perfect synchrony we have that in each round all honest parties agree on a completing set of broadcasts. From the broadcast messages that complete and verify, one must have the most sparse binary decomposition. This message defines a public key pk that all parties agree on.

We show that reconstruction always succeeds on input of n shares where at least $t+1$ are input by non-faulty parties. First observe that if the DKG transcript verifies, then for some random value α we have that

$$A_1^{\ell_1(\alpha)} \cdots A_n^{\ell_n(\alpha)} = F_0 F_1^{\alpha} \cdots F_t^{\alpha^t}$$

By the Schwartz-Zippel Lemma this implies that with overwhelming probability

$$f(X) = f_0 + f_1 X + \cdots + f_t X^t = a_1 \ell_1(X) + \cdots + a_n \ell_n(X)$$

and $a_i = f(\omega_i)$. Second observe that $e(A_i, \hat{h}_1) = e(g_1, \hat{A}_i)$ if and only if $\hat{A}_i = \hat{h}_1^{f(\omega_i)}$. Where at least $t+1$ parties are honest the reconstruction algorithm receives at least $t+1$ verifying shares. With $t+1$ verifying shares the reconstruction algorithm always succeeds because f has degree t.

Security preserving. We now prove that our DKG satisfies key expressability; i.e., we construct a simulator that is able to fix the output to be a value $\alpha \mathsf{pk}_1 + \mathsf{pk}_2$, where pk_1 is given as input and $\alpha \neq 0$. This does not directly prove that the DKG preserves security, but in the full version of our paper we detail how combining a key-expressable DKG with rekeyable encryption schemes, signature schemes, and VUFs implies that the DKG also preserves security of these primitives. We cover these three due to their popularity (and our VUF construction in Sect. 7), but envisage that there are many other primitives that are rekeyable and thus similarly preserve their security when combined with key-expressable DKGs.

Theorem 2 (DKG). *The scheme in Fig. 3 is key-expressable as per Definition 7 in the random oracle model for any primitive with keys of the form* $\mathsf{pk} = (g_1^a, \hat{u}_1^a) \in \mathbb{G}_1 \times \mathbb{G}_2$ *and* $\mathsf{sk} = \hat{h}_1^a \in \mathbb{G}_2$.

Proof. We design an adversary $\mathcal{B}$ that takes as input pk_1 such that whenever the DKG outputs pk, $\mathcal{B}$ outputs $\alpha, \mathsf{pk}_2, \mathsf{sk}_2$ such that $\mathsf{pk} = \alpha \mathsf{pk}_1 + \mathsf{pk}_2$. Suppose $\mathcal{B}$ receives input $\mathsf{pk}_1 = (g_2, \hat{v}_2)$.

First $\mathcal{B}$ runs the DKG with $\mathcal{A}$. Let $\mathbb{I}_B \subset [1, n]$ be the set of corrupted (i.e. "bad") parties and $\mathbb{I}_G \subset [1, n]$ be the set of uncorrupted ("good") parties. For good parties P_k, $\mathcal{B}$ *simulates* the adversarial view of this party's output, so that public view $C_k, \hat{u}_{k,2}$ sent by P_k is equal to $(g_2^{a_k}, \hat{v}_2^{a_k})$.

In the course of this simulation, $\mathcal{B}$ answers $\mathcal{A}$'s queries to the oracle $\mathsf{Hash}_{\mathbb{G}_2}$ by selecting $r \xleftarrow{\$} \mathbb{F}$ at random, and returning $\hat{h}_1^r$.

In the registration round, when $\mathcal{A}$ queries $\mathcal{B}$ on the k-th honest value, $\mathcal{B}$ chooses $\mu_k, \kappa_k \xleftarrow{\$} \mathbb{F}$ randomly from the field and returns the public key $(\mathsf{ek}_k, \mathsf{vk}_k) = (\hat{u}_1^{\mu_k}, g_2^{\kappa_k})$.

In the broadcast round, $\mathcal{B}$ chooses $a_k \xleftarrow{\$} \mathbb{F}$ randomly for each honest party and computes $C_k = g_2^{a_k}$. It then samples $\chi_k, \psi_k \xleftarrow{\$} \mathbb{F}$ and programs $\mathsf{Hash}_{\mathbb{G}_2}$ to return $\hat{u}_1^{\chi_k}$ and $\hat{u}_1^{\psi_k}$ on input C_k and (vk_k, C_k) respectively. Finally it broadcasts C_k. With overwhelming probability, $\mathcal{A}$ is yet to query the randomised value C_k.

In the share creation round, when queried on P_k, $\mathcal{B}$ is required to output

$$(\boldsymbol{F}_k, \hat{u}_{k,2}, \hat{\sigma}_k, \boldsymbol{A}_k, \hat{\boldsymbol{Y}}_k)$$

that are indistinguishable from a valid output. Assume without loss of generality that $|\mathbb{I}_B| = t$. It then behaves as follows

1. Choose random $\bar{x}_{k,j} \xleftarrow{\$} \mathbb{F}$ for each $j \in \mathbb{I}_B$ and interpolate in the exponent to find $(F_{k,0}, \ldots, F_{k,t})$ such that $F_{k,i} = g_1^{c_i}$, where $\sum_{i=0}^{t} c_i X^i$ evaluates to $\bar{x}_{k,j}$ at ω_j for $j \in \mathbb{I}_B$ and $a_k \log_{g_1}(g_2)$ at 0. These c_i values are unknown to $\mathcal{B}$.
2. Set $\hat{u}_{k,2} = \hat{v}_2^{a_k}$.
3. Set $\sigma_k = (\hat{v}_2^{a_k \chi_k}, \hat{v}_2^{\kappa_k \psi_k})$.
4. To compute $A_{k,1}, \ldots, A_{k,n}$, set $A_{k,j} = \prod_{i=0}^{t} F_{k,i}^{\omega_j^i}$.
5. To compute $\hat{Y}_{k,j}$ for $j \in \mathbb{I}_B$, return $\mathsf{ek}_j^{\bar{x}_{k,j}}$. To compute $\hat{Y}_{k,j}$ for $j \in \mathbb{I}_G$, interpolate in the exponent to find $\hat{u}_1^{c_0}, \ldots, u_1^{c_t}$ for $c_0, \ldots, c_{t-1}$ as in Step 1 (recall that $\mathcal{B}$ knows $\hat{u}_1^{\log_{g_1}(g_2)}$). Return $\hat{Y}_{k,j} = \prod_{i=0}^{t} \hat{u}_1^{c_i \mu_j \omega_j^i}$.

This simulation is perfect. Indeed $c_0 = \log_{g_1}(C_k)$ and $c_1, \ldots, c_t$ are randomly distributed. We have that $\hat{u}_{k,2} = \hat{v}_2^{a_k(\nu+1)} = \hat{u}_1^{\log_{g_1}(C_k)}$. Also, $\sigma_{k,1} = \mathsf{Hash}_{\mathbb{G}_2}(C_k)^{\log_{g_1}(C_k)}$ and $\sigma_{k,2} = \mathsf{Hash}_{\mathbb{G}_2}(\mathsf{vk}_k, C_k)^{\log_{g_1}(\mathsf{vk}_k)}$. The values $A_{k,1}, \ldots, A_{k,n}$ are computed honestly and are the unique encryptions that satisfy the verifier.

Suppose that the DKG terminates with transcript $((C_1', \ldots, C_n'), (w_1, \ldots, w_n), (\sigma_1', \ldots, \sigma_n'), \mathsf{pvss})$. The public key is given by $C = C_1' \cdots C_n', \hat{u}_2 = \hat{u}_{1,2} \cdots \hat{u}_{n,2}$. For each adversarial contribution C_i', $\mathcal{B}$ looks up r such that $\mathsf{Hash}_{\mathbb{G}_2}(C_j') = \hat{h}_1^r$. Here, i can be any index, as $\mathcal{A}$ might have forged one of $\mathcal{B}$'s contributions. If the adversary has not queried $\mathsf{Hash}_{\mathbb{G}_2}$ on C' then the probability of them returning a verifying signature σ is negligible. To get the secret key share, $\mathcal{B}$ extracts $\hat{C}_i = \hat{\sigma}^{\frac{1}{r}}$ such that $\hat{C}_i = \hat{h}_1^{\log_{g_1}(C_i)}$.

If $\mathcal{A}$ has included at least one of $\mathcal{B}$'s contributions, then $\mathcal{B}$ computes $z = \sum_{k \in S} w_k$ for S the set of honest participants whose contribution is included in the transcript. Additionally, $\mathcal{B}$ computes $\mathsf{pk}_2 = (\prod_{i \notin S} C_i', \prod_{i \notin S} \hat{u}_{i,2})$ and $\mathsf{sk}_2 = \prod_{i \notin S} \hat{C}_i$. Then, we have that $\mathsf{pk} = \alpha \mathsf{pk}_1 + \mathsf{pk}_2$ for $\alpha \neq 0$ and sk_2 is a key for pk_2. Thus $\mathcal{B}$ returns (α, sk_2).

If $\mathcal{A}$ has not included any contributions from $\mathcal{B}$, then that $\mathcal{A}$ has forged a signature σ_k' with respect to some $\mathsf{vk}_k = g_2^{\kappa_k}$ and contribution C_k'. Using the oracle queries, $\mathcal{B}$ looks up r such that $\mathsf{Hash}_{\mathbb{G}_2}(\mathsf{vk}_k, C_k') = \hat{h}_1^r$. Since $\sigma_k' = (\sigma_{k,1}', \sigma_{k,2}')$ verifies, we have that $\sigma_{k,2}' = \hat{h}_1^{r\kappa_k \log_{g_1}(g_2)}$. Thus, $\mathcal{B}$ computes $\mathsf{sk}_1 = (\sigma_{k,2}')^{\frac{1}{r\kappa_k}}$. Additionally, $\mathcal{B}$ computes $\mathsf{pk}_2 = (g_2^{-1} \prod_i C_i', \hat{v}_2^{-1} \prod_i \hat{u}_{i,2})$ and $\mathsf{sk}_2 = \mathsf{sk}_1^{-1} \prod_i \hat{C}_i$. Then, we have that $\mathsf{pk} = \mathsf{pk}_1 + \mathsf{pk}_2$ and sk_2 is a key for pk_2 and $\mathcal{B}$ returns $(1, \mathsf{sk}_2)$.

6 Alternative DKGs Have Provable Security

In this section we demonstrate that two popular DKGs, the Pedersen DKG and the Fouque-Stern DKG, are also key-expressable. As a direct consequence, they can be used to securely instantiate a DKG for both El-Gamal encryption and BLS signatures, as we prove in the full version of our paper [38]. Our results generalise to other rekeyable constructions that have public keys in $\mathbb{G}$ and secret keys in $\mathbb{F}$. In addition to justifying the applicability of our security definitions and proof techniques, we hope this also fills a gap in the literature as we are unaware of other works that provide correct proofs for these DKGs.

6.1 Pedersen DKG from Feldman's VSS

We prove that key expressability holds for Pedersen's DKG provided the threshold of adversarial participants is less than $n/2$. It is our belief that this bound on the number of adversarial participants can be removed provided that one gives signatures of knowledge of the individual contributions. Pedersen's DKG can be seen as n parallel instantiations of the Feldman VSS [25]. We remind the reader that key expressability does not imply secrecy (invalidating the attack of Gennaro et al. [31]) but does allow us to prove the security preservation of certain rekeyable schemes. A proof of the following theorem is provided in the full version of our paper.

Theorem 3. *The Pedersen DKG is a key-expressable DKG against static adversaries with adversarial threshold $t < n/2$ for any scheme whose key generation outputs values $\mathsf{pk} = g_1^a \in \mathbb{G}_1$, $\mathsf{sk} = a \in \mathbb{F}$.*

6.2 The Fouque-Stern Publicly Verifiable DKG

We now show the key expressability of the publicly verifiable Fouque Stern DKG [27]. This DKG has the benefit of outputting field elements as secret keys, but the total communication and verification costs are of order $\mathcal{O}(n^2)$. Unlike Fouque and Stern's original argument, we allow for the existence of rushing adversaries. Indeed Fouque and Stern rely in their reduction on an honest party playing last. Instantiating such an assumption would require the use of a trusted third party and therefore negate the benefits of distributing the key generation.

A proof of the following theorem is provided in the full version of our paper.

Theorem 4. *The Fouque-Stern DKG is a key-expressable DKG in the random oracle model against static adversaries under the decisional composite residuosity assumption for any scheme whose key generation outputs values $\mathsf{pk} = g_1^a \in \mathbb{G}_1$, $\mathsf{sk} = a \in \mathbb{F}$.*

6.3 El-Gamal and BLS

In the full version of our paper, we observe that El-Gamal encryption and BLS signatures are both rekeyable (and both have field elements as secret keys). We thus obtain the following two corollaries:

Corollary 1. *The El-Gamal encryption scheme is IND-CPA-secure when instantiated with the Pedersen DKG or the Fouque-Stern DKG.*

Corollary 2. *The BLS signature scheme is EUF-CMA-secure when instantiated with the Pedersen DKG or the Fouque-Stern DKG.*

7 A Structure-Preserving VUF

In this section, we introduce a verifiable unpredictable function (VUF), secure in the random oracle model, that has group elements as the secret key. We can thus securely instantiate our VUF using our DKG.

As one application, VUFs can be used to create randomness beacons, where unlike in, *e.g.*, BLS multi-signatures [11], if a threshold of signers is reached, then the same signature is always produced. By hashing the outcome of this VUF with a random oracle we can obtain a verifiable random function (VRF). Abe et al. [1] proved that it is impossible to construct an algebraic VUF with a secret key as a group element. Since we are using a hash function, however, we are not fully algebraic and therefore sidestep this impossibility result.

7.1 Our Construction

Our VUF scheme is given in Fig. 4. The techniques were inspired by a combination of BLS signatures [13] and Escala-Groth NIZKs [24] (which are an improvement of Groth-Sahai proofs [36]). Unlike BLS signatures our secret keys are group elements and unlike Escala-Groth NIZKs our VUFs are non-malleable.

Given an input $m \in \mathbb{F}$ under public key $g_1^a, \hat{u}_1^a$ and secret key $\hat{h}_1^a$, the unique output given by $\mathsf{VUF.Eval}(\mathsf{sk}, m)$ is $e(\mathsf{Hash}_{\mathbb{G}_1}(m), \hat{h}_1^a)$. Given $g_1^a \in \mathbb{G}_1$ and $\mathsf{Hash}_{\mathbb{G}_1}(m) \in \mathbb{G}_1$, it is hard for an adversary to compute $e(\mathsf{Hash}_{\mathbb{G}_1}(m), \hat{h}_1)^a \in \mathbb{G}_T$. We formally prove in Theorem 5 and 6 that our VUF satisfies uniqueness (see Definition 3) and unpredictability (see Definition 4) under the SXDH and BDH assumptions.

$\mathsf{VUF.Setup}(\mathsf{bp}, \mathsf{Hash}_{\mathbb{G}_1})$	$\mathsf{VUF.Gen}(\mathsf{crs}_{\mathsf{vuf}})$
$\hat{u}_1, \hat{h}_2, \hat{h}_3, \hat{h}_4 \xleftarrow{\$} \mathbb{G}_2$ $\mathsf{crs}_{\mathsf{vuf}} \leftarrow (\mathsf{bp}, \mathsf{Hash}_{\mathbb{G}_1}, \hat{h}_2, \hat{h}_3, \hat{h}_4)$ return $\mathsf{crs}_{\mathsf{vuf}}$	$a \xleftarrow{\$} \mathbb{F}$, $\mathsf{pk} \leftarrow g_1^a, \hat{u}_1^a \in (\mathbb{G}_1 \times \mathbb{G}_2)$ $\mathsf{sk} \leftarrow \hat{h}_1^a \in \mathbb{G}_2$ return $(\mathsf{pk}, \mathsf{sk})$
$\mathsf{VUF.Eval}(\mathsf{crs}_{\mathsf{vuf}}, \mathsf{sk}, m)$	$\mathsf{VUF.Derive}(\mathsf{crs}_{\mathsf{vuf}}, \mathsf{pk}, m, \sigma)$
$Z \leftarrow \mathsf{Hash}_{\mathbb{G}_1}(m)$ return $e(Z, \mathsf{sk})$	$(\pi_1, \pi_2, \pi_3, \pi_4 \in \mathbb{G}_1^4, \hat{\pi}_1, \hat{\pi}_2 \in \mathbb{G}_2^2) \leftarrow \mathsf{parse}(\sigma)$ $Z \leftarrow \mathsf{Hash}_{\mathbb{G}_1}(m)$ return $e(Z, \hat{\pi}_2)e(\pi_2, \hat{h}_3)e(\pi_4, \hat{h}_4)$
$\mathsf{VUF.Sign}(\mathsf{crs}_{\mathsf{vuf}}, \mathsf{sk}, m)$	$\mathsf{VUF.Ver}(\mathsf{crs}_{\mathsf{vuf}}, \mathsf{pk}, m, \sigma)$
$Z \leftarrow \mathsf{Hash}_{\mathbb{G}_1}(m)$ $\alpha, \beta \xleftarrow{\$} \mathbb{F}$ $\pi_1, \pi_2, \pi_3, \pi_4 \leftarrow g_1^\alpha, Z^\alpha, g_1^\beta, Z^\beta$ $\hat{\pi}_1, \hat{\pi}_2 \leftarrow \hat{h}_1^{-\alpha}\hat{h}_2^{-\beta},\ \hat{h}_3^{-\alpha}\hat{h}_4^{-\beta} \cdot \mathsf{sk}$ return $(\pi_1, \pi_2, \pi_3, \pi_4, \hat{\pi}_1, \hat{\pi}_2)$	$(A, \hat{u}_2) \leftarrow \mathsf{parse}(\mathsf{pk})$ $(\pi_1, \pi_2, \pi_3, \pi_4 \in \mathbb{G}_1^4, \hat{\pi}_1, \hat{\pi}_2 \in \mathbb{G}_2^2) \leftarrow \mathsf{parse}(\sigma)$ $Z \leftarrow \mathsf{Hash}_{\mathbb{G}_1}(m)$ check $1 = e(g_1, \hat{\pi}_1)e(\pi_1, \hat{h}_1)e(\pi_3, \hat{h}_2)$ check $1 = e(Z, \hat{\pi}_1)e(\pi_2, \hat{h}_1)e(\pi_4, \hat{h}_2)$ check $e(A, \hat{h}_1) = e(g_1, \hat{\pi}_2)e(\pi_1, \hat{h}_3)e(\pi_3, \hat{h}_4)$ return 1 if all checks pass, else return 0

Fig. 4. Verifiable unpredictable function with group elements as the secret key.

Setup: The setup algorithm is a transparent algorithm that takes as input the bilinear group $\mathsf{bp} = (p, \mathbb{G}_1, \mathbb{G}_2, \mathbb{G}_T, e, g_1, \hat{h}_1)$ and returns four group elements in the second source group: $\hat{u}_1, \hat{h}_2, \hat{h}_3, \hat{h}_4 \in \mathbb{G}_2^4$.

KeyGen: The $\mathsf{VUF.Gen}$ algorithm takes as input the common reference string. It samples a random field element $a \xleftarrow{\$} \mathbb{F}$. The public key $\mathsf{pk} \in \mathbb{G}_1 \times \mathbb{G}_2$ and the secret key $\mathsf{sk} \in \mathbb{G}_2$ are given as $\mathsf{pk} = (g_1^a, \hat{u}_1^a)$ and $\mathsf{sk} = \hat{h}_1^a$.

Sign: The $\mathsf{VUF.Sign}$ algorithm first hashes the message m to obtain $Z \in \mathbb{G}_1$ as $Z = \mathsf{Hash}_{\mathbb{G}_1}(m)$. The signer generates a commitment to sk by sampling random elements $\alpha, \beta \in \mathbb{F}$ and computing

$$(\hat{\pi}_1, \hat{\pi}_2) = (\hat{h}_1^{-\alpha}\hat{h}_2^{-\beta}, \mathsf{sk} \cdot \hat{h}_3^{-\alpha}\hat{h}_4^{-\beta}).$$

If $\hat{h}_1, \hat{h}_2, \hat{h}_3, \hat{h}_4$ are randomly distributed, this commitment is perfectly hiding. However, if $\hat{h}_1, \hat{h}_2, \hat{h}_3, \hat{h}_4$ form an SXDH challenge, then there exists some ξ such that $\hat{h}_3 = \hat{h}_1^\xi$ and $\hat{h}_4 = \hat{h}_2^\xi$, meaning that the commitment forms an El-Gamal encryption of sk. In this case, we say that the commitment is perfectly binding.

Having generated $(\hat{\pi}_1, \hat{\pi}_2)$, the signer now generates $(\pi_1, \pi_2, \pi_3, \pi_4) \in \mathbb{G}_1^4$ such that

$$(\pi_1, \pi_2, \pi_3, \pi_4) = (g_1^\alpha,\ Z^\alpha,\ g_1^\beta,\ Z^\beta)$$

These signature elements have been designed such that the random blinders α, β are canceled out in the verifier's equations.

The signer returns the output $\sigma = (\pi_1, \pi_2, \pi_3, \pi_4, \hat{\pi}_1, \hat{\pi}_2)$.

Derive: The $\mathsf{VUF.Derive}$ computes $Z = \mathsf{Hash}_{\mathbb{G}_1}(m)$ and then returns

$$T = e(Z, \hat{\pi}_2)e(\pi_2, \hat{h}_3)e(\pi_4, \hat{h}_4)$$

as the unique and unpredictable component. If the signer is honest then $T = e(Z, \mathsf{sk}) = \mathsf{VUF.Eval}(\mathsf{crs}_{\mathsf{vuf}}, \mathsf{sk}, m)$.

Verify: The $\mathsf{VUF.Ver}$ algorithm parses the signature to check that $(\pi_1, \pi_2, \pi_3, \pi_4)$ is in $\mathbb{G}_1^4$, and $(\hat{\pi}_1, \hat{\pi}_2)$ is in $\mathbb{G}_2^2$. The verifier computes Z identically to the signer, *i.e.*, $Z = \mathsf{Hash}_{\mathbb{G}_1}(m)$. The verifier then checks that three pairing equations are satisfied in order to be convinced that there exist α, β such that

$$(\pi_2, \pi_4, \hat{\pi}_2) = (Z^\alpha, Z^\beta, \hat{h}_3^{-\alpha}\hat{h}_4^{-\beta} \cdot \mathsf{sk})$$

Specifically, they check that:

$$1 = e(g_1, \hat{\pi}_1)e(\pi_1, \hat{h}_1)e(\pi_3, \hat{h}_2) \quad (2)$$

$$1 = e(Z, \hat{\pi}_1)e(\pi_2, \hat{h}_1)e(\pi_4, \hat{h}_2) \quad (3)$$

$$e(\mathsf{pk}, \hat{h}_1) = e(g_1, \hat{\pi}_2)e(\pi_1, \hat{h}_3)e(\pi_3, \hat{h}_4) \quad (4)$$

They return 1 if all these checks pass and 0 otherwise.

Given a signature that satisfies these equations, an extractor that knows a trapdoor SXDH relation between the CRS elements can output a valid witness sk. However, there also exists a simulated CRS indistinguishable from random such that we can simulate signatures without knowing sk.

Threshold VUF Scheme We discuss how to transform our VUF into a threshold VUF. The individual VUF shares can be made shorter using an optimisation in the full version of our paper [38]. Suppose that there are n parties $P_1, \ldots, P_n$ and we want that any $t+1$ of them can jointly sign a message, but that t of them cannot. We use Shamir's secret sharing scheme and choose a degree t polynomial $f(X)$. Let $\omega_1, \ldots, \omega_n$ denote unique evaluation points and $\ell_{S,1}(X), \ldots, \ell_{S,t+1}(X)$ denote the Lagrange polynomials such that for all $\omega_j \in S$ we have that $\ell_{S,i}(\omega_j)$ is equal to 1 if $i = j$ and 0 otherwise.

The threshold setup algorithm runs identically to the non-threshold version to return $\mathsf{crs}_{\mathsf{vuf}}$. The key generation outputs a public key and n secret key shares of the form

$$\mathsf{pk} = (g_1^{f(0)}, \hat{u}_1^{f(0)}), \mathsf{sk}_1 = \hat{h}_1^{f(\omega_1)}, \ldots, \mathsf{sk}_n = \hat{h}_1^{f(\omega_n)}.$$

To compute their share of the threshold signature on m party P_i outputs

$$\sigma_i = (\pi_{i,1}, \pi_{i,2}, \pi_{i,3}, \pi_{i,4}, \hat{\pi}_{i,1}, \hat{\pi}_{i,2}) \xleftarrow{\$} \mathsf{VUF.Sign}(\mathsf{crs}_{\mathsf{vuf}}, \mathsf{sk}_i, m)$$

To aggregate t signature shares on m from parties $\{P_i\}_{i \in S}$ compute

$$\sigma = \left(\prod_{i \in S} \pi_{i,1}^{\ell_{S,i}(0)}, \prod_{i \in S} \pi_{i,2}^{\ell_{S,i}(0)}, \prod_{i \in S} \pi_{i,3}^{\ell_{S,i}(0)}, \prod_{i \in S} \pi_{i,4}^{\ell_{S,i}(0)}, \prod_{i \in S} \hat{\pi}_{i,1}^{\ell_{S,i}(0)}, \prod_{i \in S} \hat{\pi}_{i,2}^{\ell_{S,i}(0)} \right)$$

The verification and derive algorithms run identically to their non-threshold counterparts on the input $(\mathsf{crs}_{\mathsf{vuf}}, \mathsf{pk}, m, \sigma)$

We briefly show that σ is correct. Set $Z = \mathsf{Hash}_{\mathbb{G}_1}(m)$ and see that $\sigma = (\pi_1, \pi_2, \pi_3, \pi_4, \hat{\pi}_1, \hat{\pi}_2)$ is given by

$$\begin{aligned}
\pi_1 &= \textstyle\prod_{i \in S} \pi_{i,1}^{\ell_{S,i}(0)} = g_1^{\sum_{i \in S} \alpha_i \ell_{S,i}(0)} \\
\pi_2 &= \textstyle\prod_{i \in S} \pi_{i,2}^{\ell_{S,i}(0)} = Z^{\sum_{i \in S} \alpha_i \ell_{S,i}(0)} \\
\pi_3 &= \textstyle\prod_{i \in S} \pi_{i,3}^{\ell_{S,i}(0)} = g_1^{\sum_{i \in S} \beta_i \ell_{S,i}(0)} \\
\pi_4 &= \textstyle\prod_{i \in S} \pi_{i,4}^{\ell_{S,i}(0)} = Z^{\sum_{i \in S} \beta_i \ell_{S,i}(0)} \\
\hat{\pi}_1 &= \textstyle\prod_{i \in S} \hat{\pi}_{i,1}^{\ell_{S,i}(0)} = \hat{h}_1^{-\sum_{i \in S} \alpha_i \ell_{S,i}(0)} \hat{h}_2^{-\sum_{i \in S} \beta_i \ell_{S,i}(0)} \\
\hat{\pi}_2 &= \textstyle\prod_{i \in S} \hat{\pi}_{i,2}^{\ell_{S,i}(0)} = \hat{h}_3^{-\sum_{i \in S} \alpha_i \ell_{S,i}(0)} \hat{h}_4^{-\sum_{i \in S} \beta_i \ell_{S,i}(0)} \textstyle\prod_{i \in S} \mathsf{sk}_i^{\ell_{S,i}(0)} \\
&= \hat{h}_3^{-\sum_{i \in S} \alpha_i \ell_{S,i}(0)} \hat{h}_4^{-\sum_{i \in S} \beta_i \ell_{S,i}(0)} \hat{h}_1^{f(\omega_i) \ell_{S,i}(0)}
\end{aligned}$$

Since f has degree t we have that $f(\omega_i)\ell_{S,i}(0) = f(0)$. Denote $\alpha = \sum_{i \in S} \alpha_i \ell_{S,i}(0)$ and $\beta = \sum_{i \in S} \beta_i \ell_{S,i}(0)$ in the above equation to get that

$$(\pi_1, \pi_2, \pi_3, \pi_4, \hat{\pi}_1, \hat{\pi}_2) = (g_1^{\alpha}, Z^{\alpha}, g_1^{\beta}, Z^{\beta}, \hat{h}_1^{-\alpha} \hat{h}_2^{-\beta}, \hat{h}_3^{-\alpha} \hat{h}_4^{-\beta} \hat{h}_1^{f(0)}) \ .$$

Thus the threshold signature is distributed identically to the non-threshold counterpart and the verifier and deriver output 1 and $e(Z, \hat{h}_1)^{f(0)}$, respectively.

Aggregatable signature scheme. It is also possible to use our VUF to instantiate an aggregatable signature scheme with secret keys as group elements. For aggregating, one simply takes the product of the public key elements output by $\mathsf{VUF.Gen}$ and the signature elements output by $\mathsf{VUF.Sign}$. Similar to the BLS scheme, this aggregatable signature scheme would be susceptible to *rogue key attacks* [49]. It is thus important to provide simulation-extractable proofs of knowledge of secret keys as part of a public key infrastructure.

7.2 Security Analysis

To prove that our VUF is secure, we need to prove that it satisfies uniqueness and unpredictability.

Theorem 5. *The VUF in Fig. 4 satisfies uniqueness (Definition 3) under the* SXDH *assumption in the random oracle model.*

Theorem 6. *The VUF in Fig. 4 satisfies unpredictability (Definition 4) under the* SXDH *and the* BDH *assumption in the random oracle model.*

We provide formal proofs of these theorems in the full version of our paper. Intuitively, uniqueness relies on the fact that it would be statistically impossible to satisfy the verifiers equations for a wrong evaluation if the CRS was made up of an SXDH instance. Since VUF.Eval is deterministic there can only be one correct evaluation. Thus if an adversary could break uniqueness in the general case, then we could use them as a subroutine to determine SXDH instances from random.

Our unpredictability proof uses an adversary who predicts the VUF to compute a BDH output. To do this we embed one component of the BDH challenge into the public key being targeted, and the other into the adversaries random oracle queries. However, we also need to simulate responses to the adversaries signature requests, and to do this (after jumping to a hybrid game with a structured CRS) we need to program the oracle such that we know a discrete log. This could present a collision as the adversary may have already queried that point. To counteract, we take a random guess as to which oracle query the adversary will output their prediction for, and if we guess wrong we abort. Thus our reduction is not tight, but does provide us with a polynomial chance of success whenever the adversary succeeds.

After observing that our VUF is rekeyable, we prove the following corollary in the full version of our paper.

Corollary 3. *The VUF in Fig. 4 is unique and unpredictable when instantiated with the DKG in Fig. 3.*

8 Implementation

We implement our DKG and VUF and summarise the performance of our schemes in Tables 2 and 3. Our implementation is written in Rust on top of the `libzexe` library, which performs efficient finite field arithmetic, elliptic curve arithmetic, and finite field FFTs. We evaluate our DKG and VUF on a desktop machine with an $i7$-8700k CPU at 3.7 GHz and 32 GB of DDR4 RAM. We use the BLS12-381 curve. For hashing to groups, we use the try-and-reject method by instantiating a ChaCha20 RNG with a Blake2s hash of the input message, sampling field elements and checking if they are valid x-coordinates, deriving the corresponding point if so. Our implementation is not constant-time. Upon publication, we plan to release our implementation as open-source software.

We utilise a few optimization techniques throughout the implementation. First, when verifying multiple pairing equations, we instead compute a randomised check of a single pairing equation so as to amortise the cost of the final exponentiations. We then compute the pairing product efficiently using the underlying `libzexe` implementation. In the same vein, when verifying pairing equations where two pairings are computed with respect to the same source group element, we combine the two into a randomised check. For large multi-exponentiations we use the `libzexe` implementation of Pippenger's algorithm. For large polynomial evaluations we use FFTs. We additionally utilise batch normalization of projective points.

Table 2. The performance of our DKG, averaged across 10 samples of each operation. For n parties, we use a threshold of $t = 2n/3$.

Parties	DKG.Deal (ms)	Scrape.Verify (ms)	DKG.Verify (ms)	Transcript size (kB)
64	72	96	376	25
128	124	178	704	50
256	271	346	1305	99
8192	8000	9900	42600	3146

Table 3. The performance of our VUF (Sect. 7), our optimised VUF, and the BLS signature scheme. These numbers were averaged across four distinct runs, with 100 samples of each operation per run.

	Our VUF	Our optimised VUF	BLS [13]
Key prove (ms)	–	2.89	–
Public key (bytes)	48	336	96
Key verify (ms)	–	4.00	–
Sign (ms)	3.47	0.58	0.44
Signature size (bytes)	384	96	48
Verify (ms)	4.73	2.39	2.15
Derive (ms)	2.37	2.37	–

We evaluate our DKG with respect to 64, 128, 256, and 8192 parties. We see that the time taken to compute, verify, and aggregate a transcript all increase linearly in the number of parties. Verifying a transcript with 256 parties takes a little more than a second.

In addition to our VUF presented in Sect. 7, we also evaluate an optimised VUF that we present in the full version of our paper [38]. We compare the performance of our VUF and our optimised VUF with BLS [13], which is the state of the art in the random oracle model. We do not give the derivation time for BLS because this is the identity function. It can be seen that signing and verifying our optimised VUF is only fractionally more expensive than BLS, but that verifying our full VUF is approximately twice as expensive.

Acknowledgements. Thank you to Ittai Abraham for helpful discussions and feedback. Sarah Meiklejohn was supported in part by EPSRC Grant EP/N028104/1. Gilad Stern was supported by the HUJI Federmann Cyber Security Research Center in conjunction with the Israel National Cyber Directorate (INCD) in the Prime Minister's Office.

References

1. Abe, M., Camenisch, J., Dowsley, R., Dubovitskaya, M.: On the impossibility of structure-preserving deterministic primitives. J. Cryptol. **32**(1), 239–264 (2019)
2. Abe, M., Chase, M., David, B., Kohlweiss, M., Nishimaki, R., Ohkubo, M.: Constant-size structure-preserving signatures: generic constructions and simple assumptions. J. Cryptol. **29**(4), 833–878 (2016)
3. Abe, M., Groth, J., Kohlweiss, M., Ohkubo, M., Tibouchi, M.: Efficient fully structure-preserving signatures and shrinking commitments. J. Cryptol. **32**(3), 973–1025 (2019)
4. Abe, M., Groth, J., Ohkubo, M., Tibouchi, M.: Unified, Minimal and Selectively Randomizable Structure-Preserving Signatures. In: Lindell, Y. (ed.) TCC 2014. LNCS, vol. 8349, pp. 688–712. Springer, Heidelberg (2014). https://doi.org/10.1007/978-3-642-54242-8_29
5. Abraham, I., Dolev, D., Gonen, R., Halpern, J.Y.: Distributed computing meets game theory: robust mechanisms for rational secret sharing and multiparty computation. In: Proceedings of the Twenty-Fifth Annual ACM Symposium on Principles of Distributed Computing, PODC 2006, Denver, CO, USA, 23–26 July 2006, pp. 53–62 (2006)
6. Abraham, I., Malkhi, D., Spiegelman, A.: Asymptotically optimal validated asynchronous byzantine agreement. In: Proceedings of the 2019 ACM Symposium on Principles of Distributed Computing, PODC 2019 (2019). https://doi.org/10.1145/3293611.3331612
7. Ateniese, G., Camenisch, J., de Medeiros, B.: Untraceable RFID tags via insubvertible encryption. In: Proceedings of the 12th ACM Conference on Computer and Communications Security, CCS 2005, Alexandria, VA, USA, 7–11 November, 2005, pp. 92–101 (2005)
8. Ballard, L., Green, M., de Medeiros, B., Monrose, F.: Correlation-resistant storage via keyword-searchable encryption. IACR Cryptol. ePrint Arch. 417 (2005). http://eprint.iacr.org/2005/417
9. Bellare, M., Rogaway, P.: The Security of Triple Encryption and a Framework for Code-Based Game-Playing Proofs. In: Vaudenay, S. (ed.) EUROCRYPT 2006. LNCS, vol. 4004, pp. 409–426. Springer, Heidelberg (2006). https://doi.org/10.1007/11761679_25
10. Benhamouda, F., Lepoint, T., Orrù, M., Raykova, M.: On the (in)security of ROS. Cryptol. ePrint Arch. 945 (2020). https://eprint.iacr.org/2020/945
11. Boldyreva, A.: Threshold Signatures, Multisignatures and Blind Signatures Based on the Gap-Diffie-Hellman-Group Signature Scheme. In: Desmedt, Y.G. (ed.) PKC 2003. LNCS, vol. 2567, pp. 31–46. Springer, Heidelberg (2003). https://doi.org/10.1007/3-540-36288-6_3
12. Boneh, D., Boyen, X.: Efficient selective identity-based encryption without random oracles. J. Cryptol. **24**(4), 659–693 (2011)
13. Boneh, D., Lynn, B., Shacham, H.: Short Signatures from the Weil Pairing. In: Boyd, C. (ed.) ASIACRYPT 2001. LNCS, vol. 2248, pp. 514–532. Springer, Heidelberg (2001). https://doi.org/10.1007/3-540-45682-1_30
14. Bonnecaze, A., Trebuchet, P.: Threshold signature for distributed time stamping scheme. Ann. Telecommun. **62**, 1353–1364 (2007)
15. Bowe, S., Gabizon, A., Miers, I.: Scalable multi-party computation for zk-snark parameters in the random beacon model. IACR Cryptol. ePrint Arch. 1050 (2017). http://eprint.iacr.org/2017/1050

16. Canetti, R., Gennaro, R., Jarecki, S., Krawczyk, H., Rabin, T.: Adaptive Security for Threshold Cryptosystems. In: Wiener, M. (ed.) CRYPTO 1999. LNCS, vol. 1666, pp. 98–116. Springer, Heidelberg (1999). https://doi.org/10.1007/3-540-48405-1_7
17. Canny, J., Sorkin, S.: Practical Large-Scale Distributed Key Generation. In: Cachin, C., Camenisch, J.L. (eds.) EUROCRYPT 2004. LNCS, vol. 3027, pp. 138–152. Springer, Heidelberg (2004). https://doi.org/10.1007/978-3-540-24676-3_9
18. Cascudo, I., David, B.: SCRAPE: Scalable Randomness Attested by Public Entities. In: Gollmann, D., Miyaji, A., Kikuchi, H. (eds.) ACNS 2017. LNCS, vol. 10355, pp. 537–556. Springer, Cham (2017). https://doi.org/10.1007/978-3-319-61204-1_27
19. Chor, B., Goldwasser, S., Micali, S., Awerbuch, B.: Verifiable secret sharing and achieving simultaneity in the presence of faults. In: 26th Annual Symposium on Foundations of Computer Science (SFCS 1985), pp. 383–395 (1985). https://ieeexplore.ieee.org/document/4568164
20. De. Santis, A., Desmedt, Y., Frankel, Y., Yung, M.: How to share a function securely. Proceedings of the Twenty-Sixth Annual ACM Symposium on Theory of Computing, STOC **1994**, 522–533 (1994). https://doi.org/10.1145/195058.195405
21. Desmedt, Y., Frankel, Y.: Threshold cryptosystems. In: Brassard, G. (ed.) CRYPTO 1989. LNCS, vol. 435, pp. 307–315. Springer, New York (1990). https://doi.org/10.1007/0-387-34805-0_28
22. DFINITY: Distributed key generation in JS. https://github.com/dfinity/dkg
23. Dodis, Y., Yampolskiy, A.: A Verifiable Random Function with Short Proofs and Keys. In: Vaudenay, S. (ed.) PKC 2005. LNCS, vol. 3386, pp. 416–431. Springer, Heidelberg (2005). https://doi.org/10.1007/978-3-540-30580-4_28
24. Escala, A., Groth, J.: Fine-Tuning Groth-Sahai Proofs. In: Krawczyk, H. (ed.) PKC 2014. LNCS, vol. 8383, pp. 630–649. Springer, Heidelberg (2014). https://doi.org/10.1007/978-3-642-54631-0_36
25. Feldman, P.: A practical scheme for non-interactive verifiable secret sharing. In: Proceedings of the 28th Annual Symposium on Foundations of Computer Science, SFCS 1987, pp. 427–438. IEEE Computer Society (1987)
26. Fleischhacker, N., Krupp, J., Malavolta, G., Schneider, J., Schröder, D., Simkin, M.: Efficient Unlinkable Sanitizable Signatures from Signatures with Re-randomizable Keys. In: Cheng, C.-M., Chung, K.-M., Persiano, G., Yang, B.-Y. (eds.) PKC 2016. LNCS, vol. 9614, pp. 301–330. Springer, Heidelberg (2016). https://doi.org/10.1007/978-3-662-49384-7_12
27. Fouque, P.-A., Stern, J.: One Round Threshold Discrete-Log Key Generation without Private Channels. In: Kim, K. (ed.) PKC 2001. LNCS, vol. 1992, pp. 300–316. Springer, Heidelberg (2001). https://doi.org/10.1007/3-540-44586-2_22
28. Galbraith, S.D., Paterson, K.G., Smart, N.P.: Pairings for cryptographers. Discrete Appl. Math. **156**(16), 3113–3121 (2008). https://doi.org/10.1016/j.dam.2007.12.010
29. Galindo, D., Liu, J., Ordean, M., Wong, J.M.: Fully distributed verifiable random functions and their application to decentralised random beacons. Cryptol. ePrint Arch. 096 (2020). https://eprint.iacr.org/2020/096
30. Garay, J.A., Katz, J., Maurer, U., Tackmann, B., Zikas, V.: Rational protocol design: cryptography against incentive-driven adversaries. In: 54th Annual IEEE Symposium on Foundations of Computer Science, FOCS 2013, Berkeley, CA, USA, 26–29 October 2013, pp. 648–657 (2013)
31. Gennaro, R., Jarecki, S., Krawczyk, H., Rabin, T.: Secure distributed key generation for discrete-log based cryptosystems. J. Cryptol. **20**, 51–83 (2007)

32. Gennaro, R., Jarecki, S., Krawczyk, H., Rabin, T.: Secure Applications of Pedersen's Distributed Key Generation Protocol. In: Joye, M. (ed.) CT-RSA 2003. LNCS, vol. 2612, pp. 373–390. Springer, Heidelberg (2003). https://doi.org/10.1007/3-540-36563-X_26
33. Gilad, Y., Hemo, R., Micali, S., Vlachos, G., Zeldovich, N.: Algorand: scaling byzantine agreements for cryptocurrencies. In: Proceedings of the 26th Symposium on Operating Systems Principles, SOSP 2017 (2017). https://doi.org/10.1145/3132747.3132757
34. Goldreich, O., Micali, S., Wigderson, A.: How to play ANY mental game. In: Proceedings of the Nineteenth Annual ACM Symposium on Theory of Computing, STOC 1987, pp. 218–229. Association for Computing Machinery (1987). https://doi.org/10.1145/28395.28420
35. Groth, J.: Rerandomizable and Replayable Adaptive Chosen Ciphertext Attack Secure Cryptosystems. In: Naor, M. (ed.) TCC 2004. LNCS, vol. 2951, pp. 152–170. Springer, Heidelberg (2004). https://doi.org/10.1007/978-3-540-24638-1_9
36. Groth, J., Sahai, A.: Efficient noninteractive proof systems for bilinear groups. SIAM J. Comput. **41**(5), 1193–1232 (2012)
37. Groth, J., Kohlweiss, M., Maller, M., Meiklejohn, S., Miers, I.: Updatable and Universal Common Reference Strings with Applications to zk-SNARKs. In: Shacham, H., Boldyreva, A. (eds.) CRYPTO 2018. LNCS, vol. 10993, pp. 698–728. Springer, Cham (2018). https://doi.org/10.1007/978-3-319-96878-0_24
38. Gurkan, K., Jovanovic, P., Maller, M., Meiklejohn, S., Stern, G., Tomescu, A.: Aggregatable distributed key generation (2021). https://eprint.iacr.org/2021/005
39. Stamer, H.: Distributed privacy guard. https://www.nongnu.org/dkgpg/
40. GNOSIS: Distributed key generation. https://github.com/gnosis/dkg
41. Kate, A., Goldberg, I.: Distributed key generation for the internet. In: 29th IEEE International Conference on Distributed Computing Systems, pp. 119–128 (2009). https://ieeexplore.ieee.org/document/5158416
42. Kate, A.: Distributed key generation and its applications. PhD thesis, Waterloo, Ontario, Canada (2010)
43. Kate, A., Zaverucha, G.M., Goldberg, I.: Constant-Size Commitments to Polynomials and Their Applications. In: Abe, M. (ed.) ASIACRYPT 2010. LNCS, vol. 6477, pp. 177–194. Springer, Heidelberg (2010). https://doi.org/10.1007/978-3-642-17373-8_11
44. Kiayias, A., Russell, A., David, B., Oliynykov, R.: Ouroboros: A Provably Secure Proof-of-Stake Blockchain Protocol. In: Katz, J., Shacham, H. (eds.) CRYPTO 2017. LNCS, vol. 10401, pp. 357–388. Springer, Cham (2017). https://doi.org/10.1007/978-3-319-63688-7_12
45. Kokoris-Kogias, E., Alp, E.C., Gasser, L., Jovanovic, P., Syta, E., Ford, B.: Verifiable management of private data under byzantine failures. Cryptol. ePrint Arch. 209 (2018). https://eprint.iacr.org/2018/209
46. Kokoris-Kogias, E., Malkhi, D., Spiegelman, A.: Asynchronous distributed key generation for computationally-secure randomness, consensus, and threshold signatures. Cryptol. ePrint Arch., Report 2019/1015 (2019). https://eprint.iacr.org/2019/1015
47. Komlo, C., Goldberg, I.: FROST: flexible round-optimized Schnorr threshold signatures. IACR Cryptol. ePrint Arch. **2020**, 852 (2020)
48. Micali, S., Rabin, M., Vadhan, S.: Verifiable random functions. In: 40th Annual Symposium on Foundations of Computer Science, pp. 120–130, October 1999. https://ieeexplore.ieee.org/document/814584

49. Micali, S., Ohta, K., Reyzin, L.: Accountable-subgroup multisignatures: extended abstract. In: Proceedings of the 8th ACM Conference on Computer and Communications Security, CCS 2001, pp. 245–254 (2001). Association for Computing Machinery, New York. https://doi.org/10.1145/501983.502017
50. Neji, W., Blibech, K., Ben Rajeb, N.: Distributed key generation protocol with a new complaint management strategy. Secur. Commun. Netw. **9**(17), 4585–4595 (2016). https://doi.org/10.1002/sec.1651
51. Orbs Network: Orbs network: DKG for BLS threshold signature scheme on the EVM using solidity (2018). https://github.com/orbs-network/dkg-on-evm
52. Paillier, P.: Public-Key Cryptosystems Based on Composite Degree Residuosity Classes. In: Stern, J. (ed.) EUROCRYPT 1999. LNCS, vol. 1592, pp. 223–238. Springer, Heidelberg (1999). https://doi.org/10.1007/3-540-48910-X_16
53. Pedersen, T.P.: A Threshold Cryptosystem without a Trusted Party. In: Davies, D.W. (ed.) EUROCRYPT 1991. LNCS, vol. 547, pp. 522–526. Springer, Heidelberg (1991). https://doi.org/10.1007/3-540-46416-6_47
54. Pedersen, T.P.: Non-Interactive and Information-Theoretic Secure Verifiable Secret Sharing. In: Feigenbaum, J. (ed.) CRYPTO 1991. LNCS, vol. 576, pp. 129–140. Springer, Heidelberg (1992). https://doi.org/10.1007/3-540-46766-1_9
55. Prabhakaran, M., Rosulek, M.: Rerandomizable RCCA Encryption. In: Menezes, A. (ed.) CRYPTO 2007. LNCS, vol. 4622, pp. 517–534. Springer, Heidelberg (2007). https://doi.org/10.1007/978-3-540-74143-5_29
56. Schindler, P., Judmayer, A., Stifter, N., Weippl, E.: ETHDKG: distributed key generation with ethereum smart contracts. Cryptol. ePrint Arch., Report 2019/985 (2019). https://eprint.iacr.org/2019/985
57. Schnorr, C.P.: Efficient Identification and Signatures for Smart Cards. In: Brassard, G. (ed.) CRYPTO 1989. LNCS, vol. 435, pp. 239–252. Springer, New York (1990). https://doi.org/10.1007/0-387-34805-0_22
58. Shamir, A.: How to share a secret. Commun. ACM **22**(11), 612–613 (1979). https://doi.org/10.1145/359168.359176
59. Syta, E., et al.: Scalable bias-resistant distributed randomness. In: 38th IEEE Symposium on Security and Privacy, May 2017. https://www.ieee-security.org/TC/SP2017/papers/413.pdf
60. Tomescu, A., et al.: Towards scalable threshold cryptosystems. In: IEEE S&P 2020, May 2020
61. Tulone, D.: A scalable and intrusion-tolerant digital time-stamping system. In: 2006 IEEE International Conference on Communications, vol. 5, pp. 2357–2363 (2006). https://ieeexplore.ieee.org/abstract/document/4024517
62. Wang, Y., Zhang, Z., Matsuda, T., Hanaoka, G., Tanaka, K.: How to Obtain Fully Structure-Preserving (Automorphic) Signatures from Structure-Preserving Ones. In: Cheon, J.H., Takagi, T. (eds.) ASIACRYPT 2016. LNCS, vol. 10032, pp. 465–495. Springer, Heidelberg (2016). https://doi.org/10.1007/978-3-662-53890-6_16
63. Wong, T.M., Wang, C., Wing, J.M.: Verifiable secret redistribution for archive systems. In: First International IEEE Security in Storage Workshop, pp. 94–105 (2002). https://www.cs.cmu.edu/wing/publications/Wong-Winga02.pdf
64. Yin, M., Malkhi, D., Reiter, M.K., Gueta, G.G., Abraham, I.: HotStuff: BFT consensus with linearity and responsiveness. In: Proceedings of the 2019 ACM Symposium on Principles of Distributed Computing, PODC 2019 (2019). https://doi.org/10.1145/3293611.3331591

Decentralized Multi-authority ABE for DNFs from LWE

Pratish Datta[1(✉)], Ilan Komargodski[1,2], and Brent Waters[1,3]

[1] NTT Research, Sunnyvale, CA 94085, USA
pratish.datta@ntt-research.com
[2] Hebrew University of Jerusalem, 91904 Jerusalem, Israel
ilank@cs.huji.ac.il
[3] University of Texas at Austin, Austin, TX 78712, USA
bwaters@cs.utexas.edu

Abstract. We construct the first decentralized multi-authority attribute-based encryption (MA-ABE) scheme for a non-trivial class of access policies whose security is based (in the random oracle model) solely on the Learning With Errors (LWE) assumption. The supported access policies are ones described by DNF formulas. All previous constructions of MA-ABE schemes supporting any non-trivial class of access policies were proven secure (in the random oracle model) assuming various assumptions on bilinear maps.

In our system, any party can become an authority and there is no requirement for any global coordination other than the creation of an initial set of common reference parameters. A party can simply act as a standard ABE authority by creating a public key and issuing private keys to different users that reflect their attributes. A user can encrypt data in terms of any DNF formulas over attributes issued from any chosen set of authorities. Finally, our system does not require any central authority. In terms of efficiency, when instantiating the scheme with a global bound s on the size of access policies, the sizes of public keys, secret keys, and ciphertexts, all grow with s.

Technically, we develop new tools for building ciphertext-policy ABE (CP-ABE) schemes using LWE. Along the way, we construct the first provably secure CP-ABE scheme supporting access policies in NC^1 under the LWE assumption that avoids the generic universal-circuit-based key-policy to ciphertext-policy transformation. In particular, our construction relies on linear secret sharing schemes with new properties and in some sense is more similar to CP-ABE schemes that rely on bilinear maps. While our CP-ABE construction is not more efficient than existing ones, it is conceptually intriguing and further we show how to extend it to get the MA-ABE scheme described above.

1 Introduction

Attribute-based encryption (ABE) is a generalization of traditional public-key encryption [26] that offers fine-grained access control over encrypted data based on the credentials (or attributes) of the recipients. ABE comes in two avatars:

A. Canteaut and F.-X. Standaert (Eds.): EUROCRYPT 2021, LNCS 12696, pp. 177–209, 2021.
https://doi.org/10.1007/978-3-030-77870-5_7

ciphertext-policy and *key-policy*. In a ciphertext-policy ABE (CP-ABE), as the name suggests, ciphertexts are associated with access policies and keys are associated with attributes. In a key-policy ABE (KP-ABE), the roles of the attribute sets and the access policies are swapped, i.e., ciphertexts are associated with attributes and keys are associated with access policies. In both cases, decryption is possible *only when* the attributes satisfy the access policy.

Since its inception by Sahai and Waters, and Goyal et al. [37,54], ABE has become a fundamental cryptographic primitive with a long list of potential applications. Therefore, designing ABE schemes has received tremendous attention by the cryptographic community resulting in a long sequence of works achieving various trade-offs between expressiveness, efficiency, security, and underlying assumptions [5,8,11,13,15,18,19,23,24,27,31–33,36,40,41,43,50,56,58].

Most of the aforementioned works base their security on cryptographic assumptions related to bilinear maps. It is very natural to seek for constructions based on other assumptions. First, this is important from a conceptual perspective as not only more constructions increase our confidence in the existence of a scheme, but constructions using different assumptions often require new techniques which in turn improves our understanding of the primitive. Second, this is important in light of the known attacks on group-based constructions by quantum computers [55]. Within this general goal, we currently have a handful of ABE schemes (that go beyond Identity-Based Encryption) [4,5,15,16,18,19,33,34,56] which avoid bilinear maps as their underlying building blocks.

All of these works derive their security from the hardness of the *learning with errors* (LWE) problem, which is currently also believed to be hard against quantum computers [29,46,47,51,52]. However, one striking fact is that all existing LWE-based ABE schemes (mentioned above) are designed in the key-policy setting. To date, the natural dual problem of constructing CP-ABE schemes based on the LWE assumption is essentially completely open.

The only known way to realize an LWE-based CP-ABE scheme is to convert either of the circuit-based KP-ABE schemes of [15,19,33] into a CP-ABE scheme by using a universal circuit to represent an access policy as an attribute and an attribute set as a circuit. However, this transformation will inherently result with a CP-ABE for a restricted class of access policies and with parameters that are far from ideal. Concretely, for any polynomials s, d in the security parameter, it allows to construct a CP-ABE for access policies with circuits of size s and depth d. Moreover, the size of a ciphertext generated with respect to some access policy f will be $|f| \cdot \mathsf{poly}(\lambda, s, d)$ (no matter what KP-ABE we start off with). That is, even if an f being encrypted has a very small circuit, the CP-ABE ciphertext would scale with the *worst-case* bounds s, d.

Open Problem 1: *Improve (even modestly) upon the universal-circuit based* CP-ABE *construction described above while assuming only* LWE.

There have been few recent exciting attempts towards this problem [6–8,18]. The works of [6,8,18] attempt go all the way and construct a *succinct* CP-ABE, where there is no global size bound s and ciphertexts and keys are of size independent of s. The works [6,8], rely on LWE as well as on bilinear groups (either generic [8] or a particular knowledge assumption [6]). The work [18] lacks a security proof. Most recently, [7] constructed a CP-ABE scheme based on LWE that still requires a universal circuit size bound but the sizes of ciphertexts and keys are independent of it.

Multi-authority Attribute-Based Encryption: In an ABE scheme, keys can only be generated and issued by a central authority. A natural extension of this notion, introduced by Chase [21] and termed multi-authority ABE (MA-ABE), allows multiple parties to play the role of an authority. In an MA-ABE, there are multiple authorities which control different attributes and each of them can issue secret keys to users possessing attributes under their control without any interaction with the other authorities in the system. Specifically, given a ciphertext generated with respect to some access policy, a user possessing a set of attributes satisfying the access policy can decrypt the ciphertext by pulling the individual secret keys it obtained from the various authorities controlling those attributes. The security requires collusion resistance against unauthorized users with the important difference that now some of the attribute authorities may be corrupted and therefore may collude with the adversarial users.

To date, there are only a few works which have dealt with the problem of constructing MA-ABE schemes. After few initial attempts [21,22,44,48,49] that had various limitations, Lewko and Waters [42] were able to design a truly decentralized MA-ABE scheme in which any party can become an authority and there is no requirement for any global coordination other than the creation of an initial trusted setup. In their scheme, a party can simply act as an authority by publishing a public key of its own and issuing private keys to different users that reflect their attributes. Different authorities need not even be aware of each other and they can join the system at any point of time. There is also no bound on the number of attribute authorities that can ever come into play during the lifetime of the system. Their scheme supports access policies computable by NC^1 circuits and their security is proven in the random oracle model and further relies on assumptions on bilinear groups (similarly to all previous MA-ABE constructions). Rouselakis and Waters [53] provided further efficiency improvements over [42], albeit they rely, in addition to a random oracle, on a non-standard q-type assumption.

Open Problem 2: *Is there a truly decentralized* MA-ABE *for some non-trivial class of access policies assuming hardness of* LWE *(and in the random oracle model)?*

There has been few recent attempts at this problem as well [39,57]. Both constructions [39,57] assume a central authority which generates the public and secret keys for all the attribute authorities in the system. Thus all authorities that will ever exist in the system are forever fixed once setup is complete which runs counter to the truly decentralized spirit of [42]. Additionally, both schemes guarantee security only against a bounded collusion of parties. In fact, the scheme of Kim [39] is built in a new model, called the “OT model”, which is incapable of handling even bounded collusion.[1] In this sense, both constructions suffer from related limitations to the early MA-ABE constructions [21,22,44,48,49] describe above. The differences between the two constructions are that the scheme of Wang et al. [57] supports NC^1 access policies, while the scheme due to Kim [39] support arbitrary bounded depth circuits.

1.1 Our Contributions

In this paper, we make progress with respect to Open Problem 2, stated above. We construct a new MA-ABE scheme supporting an unbounded number of attribute authorities for access policies captured by DNF formulas. Our scheme is proven secure in the random oracle model and relies on the hardness of the LWE problem.

Theorem 1.1 (Informal): *There exist a decentralized* MA-ABE *scheme for access policies captured by* DNF *formulas under the* LWE *assumption. Our scheme is (statically) secure against an arbitrary collusion of parties in the random oracle model and assuming the* LWE *assumption with subexponential modulus-to-noise ratio.*

Similarly to [42,53], in our MA-ABE scheme, any party can become an authority at any point of time and there is no bound on the number of attribute authorities that can join the system or need for any global coordination other than the creation of an initial set of common reference parameters created during a trusted setup. We prove the security of our MA-ABE scheme in the static security model introduced by Rouselakis and Waters [53] where all of the ciphertexts, secret keys, and corruption queries must be issued by the adversary before the public key of any attribute authority is published.

Towards obtaining Theorem 1.1, we make conceptual contribution towards Open Problem 1. We present the first provably secure direct CP-ABE construction which avoids the generic universal-circuit-based key-policy to ciphertext-policy transformation. In particular, our approach deviates from all previous LWE-based expressive ABE constructions [5,15,18,19,33,34,56] that are in turn based on techniques inspired by fully homomorphic encryption [28,30]. In contrast, our CP-ABE is based on useful properties of linear secret sharing schemes

[1] All previous multi-authority ABE schemes were designed in the so called global identifier (GID) model where each user in the system is identified by a unique global identity string $\mathsf{GID} \in \{0,1\}^*$. The global identity of a user remains fixed for the entire lifetime of the system and users have no freedom to choose their global identities. Kim [39] introduced a drastically relaxed model, the so called "OT model", where each user can self-generate some key-request string and produce it to the attribute authorities while requesting secret keys. To briefly see why this model fails to guarantee collusion resistance, imagine that there are two users A who has attribute u and B who has attribute v. Suppose there is a ciphertext encrypting to the policy "u AND v". User A and B can collude to decrypt it. Morally, the issue is that user A can go with the authority for attribute u and produce a key with identity George. User B can then present the same identity to the authority for attribute v. Then they can combine their keys.

and can be viewed as the LWE analog of the CP-ABE scheme of Waters [58] which relies on the decisional bilinear Diffie-Hellman assumption.

Theorem 1.2 (Informal): *There exist a* CP-ABE *scheme supporting all access policies in* NC^1. *The scheme is selectively secure assuming the* LWE *assumption with subexponential modulus-to-noise ratio.*

Our CP-ABE scheme achieves the standard selective security where the adversary must disclose its ciphertext query before the master public key is published but is allowed to make secret key queries adaptively throughout the security experiment. Again, Theorem 1.2 does not improve upon previously known constructions in any parameter. It is in fact worse in several senses: it only supports NC^1 access policies, its efficiency is worse, and it requires the LWE assumption to hold with subexponential modulus-to-noise ratio. However, the new construction is interesting not only because we show how to generalize it to get the new MA-ABE scheme from Theorem 1.1, but also because we introduce a conceptually new approach and develop several interesting tools and proof techniques.

One highlight is that we distill a set of properties of linear secret sharing schemes (LSSS) which makes them compatible with LWE-based constructions. Specifically, we instantiate both of our CP-ABE and MA-ABE schemes with such LSSS schemes. In the security model of CP-ABE we are able to construct such a compatible LSSS for all NC^1 while in the (much harder) security model of MA-ABE we are only able to get such a scheme for DNFs. The properties are:

- **Small reconstruction coefficients:** The reconstruction coefficients of the LSSS must be small, say $\{0, 1\}$. This property of LSSS secret sharing schemes was recently formally defined by [14]. They observed that a well-known construction by Lewko and Waters [42] actually results with an LSSS with this property for all access structures in NC^1.
- **Linear independence for unauthorized rows:** This property says that rows of the share generating matrix that correspond to an unauthorized set of parties are linearly independent. Agrawal et al. [1] recently observed that the aforementioned construction by Lewko and Waters [42], when applied on DNF access structures, results with a share generating matrix that has this property as well.

Both of our constructions, the CP-ABE as well as the MA-ABE, are actually designed to work with any access structure that has an LSSS with the above two properties.

Theorem 1.3 (Informal): *Consider a class of access policies* $\mathbb{P}$ *that has an associated* LSSS *with the above two properties. Then, there exists a* CP-ABE *and an* MA-ABE *supporting access policies from the class* $\mathbb{P}$. *Both schemes are secure assuming the* LWE *assumption with subexponential modulus-to-noise ratio and the* MA-ABE *scheme also requires a random oracle.*

To obtain Theorem 1.2 we design a new (non-monotone) LSSS for all NC^1 that has the above two properties. This is summarized in the following theorem.

Theorem 1.4 (Informal): *There exists a non-monotone* LSSS *scheme for all* NC^1 *circuits satisfying the* small reconstruction coefficients *and* linear independence for unauthorized rows *properties.*

By non-monotone, we mean that an attribute and its negation are treated separately (both having corresponding shares) and it is implicitly assumed that the attacker will never see shares corresponding to both the positive and the negative instances of the same attribute. This can be enforced in case of CP-ABE due to its centralized nature and this when combined with Theorem 1.3 implies Theorem 1.2. However, in MA-ABE attackers can get hold of the master secret key of any attribute authority and generate secret keys corresponding to both the attribute under control and its negation, and so non-monotone LSSS does not seem to suffice. We therefore settle for the (monotone) LSSS scheme for DNFs to obtain Theorem 1.1 (see further discussion in Sect. 2.3 below and [25, Remark 6.1] in the full version).

Boyen's [16] **scheme:** In TCC 2013 Boyen [16] suggested a lattice-based KP-ABE scheme for NC^1. While being conceptually similar to analogous constructions from the bilinear-maps LSSS-based schemes, soon after the publication a flaw was found and a recent work of Agrawal et al. [1] shows an explicit attack. The attack of [1] is based on identifying a subset of attributes which correspond to rows of the policy matrix that non-trivially span the $\mathbf{0}$ vector (i.e., linearly dependent rows). To rescue Boyen's construction, Agrawal et al. [1] suggest to use an LSSS which has the linear independence of unauthorized rows property (they call it an *admissible LSSS*), however, they fail to obtain such a scheme for any class larger than DNFs. Our non-monotone LSSS scheme for NC^1 (Theorem 1.4) can be used to resurrect the KP-ABE scheme of Boyen [16]. Although this does not imply any new result (as other constructions of KP-ABE for all polynomial-size circuits have since been discovered [15,19,33]), we believe that this is an important conceptual contribution.

Paper Organization: In Sect. 2 we provide a high-level overview of our techniques. Prerequisites on lattices and LWE are provided in Sect. 3. In Sect. 4 we give our construction of the new non-monotone LSSS for all NC^1 with the linear independence property. In Sect. 5 we give the construction of our CP-ABE scheme and prove its correctness. The proof of security is provided in the full version [25]. In Sect. 6 we give the construction of our MA-ABE scheme. The proofs of correctness and security are again deferred to the full version [25]. We further omit the formal syntax and security definitions of CP-ABE and MA-ABE in this version. Those can be found in the full version [25].

2 Technical Overview

In this section we provide a high level overview of our main ideas and techniques. In a very high level, our CP-ABE construction is composed of two main conceptual ideas:

1. *A linear non-monotone secret sharing scheme with small reconstruction coefficients and a* linear independence *guarantee*: We design a new linear non-monotone secret sharing scheme for all access structures that can be described by a Boolean *formula*, namely NC^1 access structures. The new secret sharing scheme possesses two properties which turns out to be key for our correctness and security proof. The first property states that it is possible to reconstruct a shared secret using only coefficients that come from $\{0,1\}$. An LSSS with this property is called $\{0,1\}$-LSSS [14]. The second property, called the linear independence property, says that the shares held by any unauthorized set, not only are independent of the secret, but are also linearly independent among each other. We give an overview of the new construction in Sect. 2.1.
2. *An* LWE*-based direct construction of* CP-ABE: We show how to leverage any $\{0,1\}$-LSSS with the above extra property to get a CP-ABE scheme. Conceptually, to some extent the construction can be viewed as a "translation" of Waters' [58, Section 6] construction of a CP-ABE scheme under the Decisional Bilinear Diffie-Hellman (DBDH) Assumption into the LWE regime. However, since we are basing the construction of the LWE assumption, the details and implementation are completely different and much more involved. We will give an overview of this part in Sect. 2.2.

Combining the two parts, we obtain a CP-ABE scheme for all NC^1 assuming the LWE assumption. The CP-ABE scheme we design is already amenable for extension to the multi-authority setting. We briefly discuss the main idea in the extension to MA-ABE in Sect. 2.3.

2.1 The New Linear Secret Sharing Scheme

Our goal is to construct a linear secret sharing scheme with $\{0,1\}$ reconstruction coefficients where the shares of unauthorized parties are linearly independent. Recall first that an access structure f is a partition of the universe of possible subsets of n parties into two sets, one is called *authorized* and its complement is called *unauthorized*. The partition is monotone in the sense that if some subset of parties is unauthorized, one can make it authorized only by adding more parties to it. A secret sharing scheme is a method by which it is possible to "split" a given secret into "shares" and distributes them among parties so that authorized subsets would be able to jointly recover the secret while others would not. Linear secret sharing schemes (LSSS) [38] are a subset of all possible schemes where there is an additional structural guarantee about the reconstruction procedure: For an authorized subset of parties to reconstruct the secret, all that is needed is to compute a *linear* function over its shares.

Every linear secret sharing scheme can be described by a share generating matrix. This is a matrix $\boldsymbol{M} \in \mathbb{Z}_q^{\ell \times d}$ where each row is associated to some party. A set of parties is qualified if and only if when we restrict $\boldsymbol{M}$ to rows of this set, we get a subspace that spans the vector $(1,0,\ldots,0)$. For a secret $z \in \mathbb{Z}_q$, computing $\boldsymbol{M} \cdot \boldsymbol{v}^\top$, where $\boldsymbol{v} \in \mathbb{Z}_q^d$ is a vector whose first entry is z and the rest are uniformly random, gives a vector of ℓ shares of the secret z. Here, we need a more specialized share generating matrix with an additional property.

Specifically, we need that for any unauthorized set of parties, restricting $\boldsymbol{M}$ to those rows, results with a set of linearly independent vectors. We construct such a share generating matrix for access structure given as a Boolean formula.

To see the challenge, it is useful to recall the standard construction of a share generating matrix for Boolean formulas, as adapted from the secret sharing scheme of [12] by Lewko and Waters [42, Appendix G]. Given a Boolean formula, the share generating matrix is constructed by labeling the wires of the formula from the root to the leaves. The labels of the leaves will form the rows of the share generating matrix. We first label the root node of the tree with the vector (1) (a vector of length 1). Then, we go down the levels of the tree one by one, labeling each node with a vector determined by the vector assigned to its parent node. Throughout the process, we maintain a global counter variable c which is initialized to 1. Consider a gate g with output wire w whose label is $\boldsymbol{w}$ and two input wires u, v. If g is an OR gate, we associate with u the label $\boldsymbol{u} = \boldsymbol{w}$ and with v the label $\boldsymbol{v} = \boldsymbol{w}$ (and do not change c). If g is an AND gate, we associate with u the label $\boldsymbol{u} = \boldsymbol{w}\|1$ and associate with v the label $\boldsymbol{v} = \mathbf{0}\| - 1$, where $\mathbf{0}$ denoted a length c vector of 0s. We now increment the value of c by 1. Finally all vectors are padded with 0s in the end to the length of the longest one.

Let us mention that this scheme already has several appealing properties. First, the entries of the share generating matrix are from $\{-1, 0, 1\}$. Moreover, it is already a $\{0, 1\}$-LSSS, namely, when reconstructing a secret using the shares corresponding to an authorized set, the coefficients used are only from $\{0, 1\}$. Nevertheless, a property that we need yet the above construction does not satisfy is linear independence. Consider, for instance, the formula $(A \vee B) \wedge C$. Here, an adversary controlling A and B cannot recover the secret, yet the rows corresponding to A and B in the share generating matrix are identical and thereby linearly dependent. The more intuitive way to see the problem is that during the reconstruction process, since we are dealing with an OR gate, we can choose to continue "either from the left or from the right" and in both cases we will see the same computation. Nevertheless, it is not hard to verify that when considering only DNF formulas, this construction already results with linearly independent rows for unqualified sets.

We next describe our new secret sharing scheme and argue that the rows corresponding to any unauthorized set are linearly independent. We make our task a little bit easier by allowing every wire in the formula have two associated labels. (This is why our scheme is a non-monotone LSSS.) The first is for "satisfying" the wire, i.e., the 1-label, and the other is for not satisfying it, i.e., the 0-label. (Whereas above we only had a label for satisfying the wire and hence it is a monotone LSSS.) Our procedure is similar to the one above in the sense that it also labels wires from the root to the leaves and the leaf labels form the rows of the share generating matrix. Since we have two labels per wire, we first label the root node of the tree with the vector (1,0) and (0,1). Our global counter c is initialized to 2.

Consider a gate g with output wire w whose labels are $\boldsymbol{w}_1, \boldsymbol{w}_0$, and two input wires u, v. We associate with u the labels $\boldsymbol{u}_1, \boldsymbol{u}_0$ and with v the label $\boldsymbol{v}_1, \boldsymbol{v}_0$. If

g is an AND gate, we set

$$\boldsymbol{u}_1 = \boldsymbol{0}\|1, \quad \boldsymbol{u}_0 = \boldsymbol{w}_0, \quad \boldsymbol{v}_1 = \boldsymbol{w}_1\| - 1, \quad \boldsymbol{v}_0 = \boldsymbol{w}_0\| - 1$$

If g is an OR gate, we set

$$\boldsymbol{u}_1 = \boldsymbol{w}_1, \quad \boldsymbol{u}_0 = \boldsymbol{0}\|1, \quad \boldsymbol{v}_1 = \boldsymbol{w}_1\| - 1, \quad \boldsymbol{v}_0 = \boldsymbol{w}_0\| - 1$$

We increment the value of c by 1 and pad all vectors with 0s in the end to be of size c.

Correctness and security of the construction (which can be proven by induction) say that for every wire in the formula, if it can be successfully satisfied, then there is a linear combination to recover the 1-label of that wire but not the 0-label. Analogously, if it cannot be satisfied, then there is also a linear combination to recover the 0-label of that wire but not the 1-label. Also, it is not hard to verify that, as with the previous construction, the matrix contains only values from $\{-1, 0, 1\}$ and the reconstruction coefficients needed to recover the secret for an authorized set are from $\{0, 1\}$.

For the new linear independent property, let us focus for now on a single gate g and assume that it is an OR gate. Observe that $\boldsymbol{w}_1$ can only be reconstructed using either $\boldsymbol{u}_1$ or using $\boldsymbol{u}_0 + \boldsymbol{v}_1$. As opposed to the "attack" we suggested before, now to continue the computation in the reconstruction phase, there is only one valid way, depending on the available shares. To see this more precisely, one needs to consider the 4 possible cases: (1) u, v are satisfied, (2) u is satisfied but v is not, (3) u is not satisfied but v is, and (4) both u, v are unsatisfied. Checking each case separately one can get convinced that there is exactly one way to compute the corresponding label of the output wire. An analogous case analysis can be done also for the case where g is an AND gate. This idea can be generalized and formalized to show that the vectors held by an attacker who controls an unauthorized must be linearly independent.

2.2 The CP-ABE Scheme

Here we describe our CP-ABE scheme. This serves as a warm up for our full MA-ABE scheme and includes most of the technical ideas. We discuss briefly the additional technicalities that arise in the multi-authority setting in Sect. 2.3 below. Note that the problem of constructing CP-ABE schemes directly has traditionally been much more challenging compared to its KP-ABE counterpart. Let us highlight two challenges:

- The first challenge is of course to prevent collusion attacks by users, that is, to somehow "bind" the key components of a particular user corresponding to the various attributes it possesses so that those key components cannot be combined with the key components possessed by other users.
- The second and more serious challenge is (in the selective model) how to embed a complex access policy in a short number of parameters.

In order to prove selective security, the standard strategy is to follow a "partitioning" technique where the reduction algorithm sets up the master public key such that it knows all the secret keys that it needs to give out, yet it cannot give out secret keys that can trivially decrypt the challenge ciphertext. In the context of KP-ABE, the challenge ciphertext is associated with an attribute set and therefore the public parameters for each attribute can be simply treated differently depending whether it is in the challenge attribute set or not. In CP-ABE, the situation is much more complicated as ciphertexts are associated with access policies which essentially encode a huge (maybe exponential size) set of authorized subsets of attributes. Consequently, there is no simple "on or off" method of programming this information into the master public key. While techniques have eventually been developed to overcome this challenge in the bilinear map world, devising the LWE analogs has remained elusive. One of the main technical contributions of our paper is a method for directly embedding an LSSS access policy into the master public key within the LWE-based framework in our reduction.

For concreteness, in what follows we assume that the LSSS access policy used in our CP-ABE scheme was generated using our transformation described above. Moreover, we assume that there is a public bound $s_{\max}$ on the number of columns in the matrix (which translates to a bound on the size of the Boolean formula while using our Boolean formula LSSS transformations above). We further assume that the row labeling function is injective, i.e., each attribute corresponds to exactly one row. In the precise description of the scheme we use several different noise distributions with varying parameters. Some of them are used to realize the standard noise smudging technique at various steps of the security proof. In order to keep the exposition simple, we will ignore such noise smudging and just use a single noise distribution, denoted noise. By default, vectors are thought of as row vectors.

Setup: For each attribute u in the system, sample $\boldsymbol{A}_u \in \mathbb{Z}_q^{n \times m}$ together a trapdoor $\boldsymbol{T}_{\boldsymbol{A}_u}$, and another uniformly random matrix $\boldsymbol{H}_u \leftarrow \mathbb{Z}_q^{n \times m}$. Additionally sample $\boldsymbol{y} \leftarrow \mathbb{Z}_q^n$. Output

$$\mathsf{PK} = (\boldsymbol{y}, \{\boldsymbol{A}_u\}, \{\boldsymbol{H}_u\}), \qquad \mathsf{SK} = \{\boldsymbol{T}_{\boldsymbol{A}_u}\}$$

Key Generation for attribute set U: Let $\hat{\boldsymbol{t}} \leftarrow \mathsf{noise}^{m-1}$ and $\boldsymbol{t} = (1, \hat{\boldsymbol{t}}) \in \mathbb{Z}^m$. This vector $\boldsymbol{t}$ will intuitively serve as the linchpin that will tie together all the secret key components of a specific user. For each attribute $u \in U$, using $\boldsymbol{T}_{\boldsymbol{A}_u}$, sample a short vector $\tilde{\boldsymbol{k}}_u$ such that $\boldsymbol{A}_u \tilde{\boldsymbol{k}}_u^\top = \boldsymbol{H}_u \boldsymbol{t}^\top$ and output

$$\mathsf{SK} = (\{\tilde{\boldsymbol{k}}_u\}, \boldsymbol{t})$$

Encryption of $\mathsf{msg} \in \{0, 1\}$ given matrix $\boldsymbol{M}$: Assume that ρ is a function that maps between row indices of $\boldsymbol{M}$ and attributes, that is, $\rho(i)$ is the attribute associated with the ith row in $\boldsymbol{M}$. The procedure samples $\boldsymbol{s} \leftarrow \mathbb{Z}_q^n$ and

$v_2, \ldots, v_{s_{\max}} \leftarrow \mathbb{Z}_q^m$ and computes

$$c_i = sA_{\rho(i)} + \mathsf{noise}$$
$$\hat{c}_i = M_{i,1}(sy^\top, \overbrace{0, \ldots, 0}^{m-1}) + \left[\sum_{j \in \{2,\ldots,s_{\max}\}} M_{i,j} v_j \right] - sH_{\rho(i)} + \mathsf{noise}$$

and outputs the ciphertext

$$\mathsf{CT} = \left(\{c_i\}_{i \in [\ell]}, \{\hat{c}_i\}_{i \in [\ell]}, C = \mathsf{MSB}(sy^\top) \oplus \mathsf{msg} \right).$$

Decryption: Assume that the available attributes are qualified to decrypt. Let I be the set of row indices corresponding to the available attributes and let $\{w_i\}_{i \in I} \in \{0,1\} \subset \mathbb{Z}_q$ be the reconstruction coefficients. For each $i \in I$, let $\rho(i)$ be the attribute associated with the ith row. The procedure computes

$$K' = \sum_{i \in I} w_i \left(c_i \tilde{k}_{\rho(i)}^\top + \hat{c}_i t^\top \right)$$

and outputs

$$\mathsf{msg}' = C \oplus \mathsf{MSB}(K').$$

Correctness
Consider a ciphertext CT w.r.t some matrix M and a key for a set of attributes U that satisfies M. By construction it is enough to show that $\mathsf{MSB}(K') = \mathsf{MSB}(sy^\top)$ with all but negligible probability. Here, for simplicity, we shall ignore small noise-like terms. Expanding $\{c_i\}_{i \in I}$ and $\{\hat{c}_i\}_{i \in I}$, we get

$$K' \approx \sum_{i \in I} w_i s A_{\rho(i)} \tilde{k}_{\rho(i)}^\top + \sum_{i \in I} w_i M_{i,1}(sy^\top, 0, \ldots, 0) t^\top + \sum_{i \in I, j \in \{2,\ldots,s_{\max}\}} w_i M_{i,j} v_j t^\top - \sum_{i \in I} w_i s H_{\rho(i)} t^\top$$

First, observe that each $w_i \in \{0,1\}$ since the reconstruction coefficients in our secret sharing scheme are guaranteed to be Boolean.

Now, recall that for each $u \in U$, we have $A_u \tilde{k}_u^\top = H_u t^\top$. Therefore, for each $i \in I$, it holds that

$$A_{\rho(i)} \tilde{k}_{\rho(i)}^\top = H_{\rho(i)} t^\top.$$

Hence,

$$\begin{aligned}K' &\approx \cancel{\sum_{i\in I} w_i \boldsymbol{s}\boldsymbol{H}_{\rho(i)}\boldsymbol{t}^\top} + \sum_{i\in I} w_i M_{i,1}(\boldsymbol{s}\boldsymbol{y}^\top, 0, \ldots, 0)\boldsymbol{t}^\top \\ &+ \sum_{i\in I, j\in\{2,\ldots,s_{\max}\}} w_i M_{i,j}\boldsymbol{v}_j\boldsymbol{t}^\top - \cancel{\sum_{i\in I} w_i \boldsymbol{s}\boldsymbol{H}_{\rho(i)}\boldsymbol{t}^\top} \\ &= \sum_{i\in I} w_i M_{i,1}(\boldsymbol{s}\boldsymbol{y}^\top, 0, \ldots, 0)\boldsymbol{t}^\top + \sum_{i\in I, j\in\{2,\ldots,s_{\max}\}} w_i M_{i,j}\boldsymbol{v}_j\boldsymbol{t}^\top \\ &= \left(\sum_{i\in I} w_i M_{i,1}\right)(\boldsymbol{s}\boldsymbol{y}^\top, 0, \ldots, 0)\boldsymbol{t}^\top + \sum_{j\in\{2,\ldots,s_{\max}\}} \left(\sum_{i\in I} w_i M_{i,j}\right)\boldsymbol{v}_j\boldsymbol{t}^\top.\end{aligned}$$

Recall that we have $\sum_{i\in I} w_i M_{i,1} = 1$ while for $1 < j \leq s_{\max}$, it holds that $\sum_{i\in I} w_i M_{i,j} = 0$. Also, recall that $\boldsymbol{t} = (1, \hat{\boldsymbol{t}})$, and hence, $(\boldsymbol{s}\boldsymbol{y}^\top, 0, \ldots, 0)\boldsymbol{t}^\top = \boldsymbol{s}\boldsymbol{y}^\top$. Thus,

$$K' \approx \boldsymbol{s}\boldsymbol{y}^\top.$$

By choosing the noise magnitude carefully, we can make sure that $\mathsf{MSB}(K') = \mathsf{MSB}(\boldsymbol{s}\boldsymbol{y}^\top)$, except with negligible probability.

Security

As mentioned, we prove that our scheme is selectively secure, namely, we require the challenge LSSS policy $(\boldsymbol{M}, \rho)$ to be submitted by the adversary ahead of time before seeing the public parameters. The proof is obtained by a hybrid argument where we start off with the security game played with the real scheme as the first hybrid and end up with a hybrid where the game is played with a scheme where the challenge ciphertext is independent of the underlying message.

In more detail, in the last hybrid we want to get rid of the secret $\boldsymbol{s}$. Recall that $\boldsymbol{s}$ appears in two places: (1) $\boldsymbol{c}_i$ and (2) $\hat{\boldsymbol{c}}_i$. Intuitively, the term $\boldsymbol{c}_i$ looks like an LWE sample and indeed our goal is to use LWE to argue that $\boldsymbol{s}$ is hidden there. The challenge is that to use LWE we need to get rid of the trapdoor $\boldsymbol{T}_{\boldsymbol{A}_u}$ of $\boldsymbol{A}_u$ which is used in the key generation procedure to sample $\tilde{\boldsymbol{k}}_u$. For $\hat{\boldsymbol{c}}_i$, our high level approach is to program $\boldsymbol{H}_u$ in such a way that it will cancel the terms that depend on $\boldsymbol{s}$ in $\hat{\boldsymbol{c}}_i$. However, at the same time $\boldsymbol{H}_u$ is used in the sampling procedure of $\tilde{\boldsymbol{k}}_u$ as well, and so (1) and (2) are actually related and need to be handled together.

We program $\boldsymbol{H}_u$ as follows

$$\boldsymbol{H}_u = M_{\rho^{-1}(u),1}\left[\boldsymbol{y}^\top | \overbrace{\boldsymbol{0}^\top | \cdots | \boldsymbol{0}^\top}^{m-1}\right] + \sum_{j\in\{2,\ldots,s_{\max}\}} M_{\rho^{-1}(u),j}\boldsymbol{B}_j + \boldsymbol{A}_u\boldsymbol{R}_u,$$

where $\boldsymbol{R}_u, \boldsymbol{B}_2, \ldots, \boldsymbol{B}_{s_{\max}}$ are matrices of the appropriate sizes and sampled from some distributions which we shall skip for now. Here we crucially use the fact that the row labeling function ρ is injective to ensure that the above definition of $\boldsymbol{H}_u$ is unambiguous. One of the purposes of the $\boldsymbol{R}_u$ matrices is to make sure that the programmed $\boldsymbol{H}_u$ is indistinguishable from the original $\boldsymbol{H}_u$. We make use of an extended version of the leftover hash lemma, we call the "leftover hash lemma with trapdoors" (see Lemma 3.4 in the full version [25]), to guarantee this indistinguishability. This programming allows us to embed the challenge access policy into the master public key. Also notice that indeed the first term of $\boldsymbol{H}_u$ cancels out the dependence on $\boldsymbol{s}$ in $\hat{\boldsymbol{c}}_i$.

Let us go back to how the keys look like with this $\boldsymbol{H}_u$. Recall that we chose $\tilde{\boldsymbol{k}}_u$ such that $\boldsymbol{A}_u\tilde{\boldsymbol{k}}_u^\top = \boldsymbol{H}_u\boldsymbol{t}^\top$. Our goal is to sample $\tilde{\boldsymbol{k}}_u$ directly and not through the trapdoor $\boldsymbol{T}_{\boldsymbol{A}_u}$ of $\boldsymbol{A}_u$ so that we can eventually do away with $\boldsymbol{T}_{\boldsymbol{A}_u}$. To this end, we program $\boldsymbol{t}$ so that $\boldsymbol{H}_u\boldsymbol{t}^\top$ is completely random. Note that once $\boldsymbol{H}_u\boldsymbol{t}^\top$ becomes random, we would be able to directly sample $\tilde{\boldsymbol{k}}_u$ via the properties of lattice trapdoors. At a high level for this purpose, we use the $\boldsymbol{B}_j$ matrices, which we actually generate along with trapdoors. Observe that with our programming of the $\boldsymbol{H}_u$ matrices above, we have

$$\boldsymbol{H}_u\boldsymbol{t}^\top = M_{\rho^{-1}(u),1}\left[\boldsymbol{y}^\top|\overbrace{\boldsymbol{0}^\top|\cdots|\boldsymbol{0}^\top}^{m-1}\right]\boldsymbol{t}^\top + \boxed{\sum_{j\in\{2,\ldots,s_{\max}\}} M_{\rho^{-1}(u),j}\boldsymbol{B}_j\boldsymbol{t}^\top} + \boldsymbol{A}_u\boldsymbol{R}_u\boldsymbol{t}^\top.$$

Roughly, $\boldsymbol{H}_u\boldsymbol{t}^\top$ would become uniformly random if we can make the boxed part above uniformly random. We plan to do this by first sampling some uniformly random vector $\boldsymbol{z}_u$ and then solving for $\left\{\boldsymbol{B}_j\boldsymbol{t}^\top\right\}_{j\in\{2,\ldots,s_{\max}\}}$ such that $\sum_{j\in\{2,\ldots,s_{\max}\}} M_{\rho^{-1}(u),j}\left(\boldsymbol{B}_j\boldsymbol{t}^\top\right) = \boldsymbol{z}_u$. Note that once we have a solution for the above system of equations, we can use the trapdoor of the $\boldsymbol{B}_j$ matrices to sample an appropriate $\boldsymbol{t}$ and our goal will be accomplished. It is for solving the above system of linear equations that we use the fact that the corresponding rows of $\boldsymbol{M}$ are linearly independent and so the above system of linear equations is solvable.

2.3 The MA-ABE Scheme

The MA-ABE scheme is a generalization of the above scheme and we avoid repeating the scheme here. Instead, let us go over our main ideas to overcome the technical challenges that prevented getting a collusion resistant decentralized MA-ABE scheme from LWE before this work. First, it is important to understand that a main challenge in CP-ABE constructions is collusion resistance. The standard technique to achieve collusion resistance in the literature is to tie together the different key components representing the different attributes of a user with the help of fresh randomness specific to that user. Such randomization would make the different key components of a user compatible with each other, but not with the parts of a key issued to another user. This is relatively easy to implement in the single-authority setting since there is only one central authority who is responsible to generate secret keys for users.

In a multi-authority, we want to satisfy the simultaneous goals of autonomous key generation and collusion resistance. The requirement of autonomous key generation means that established techniques for key randomization cannot be applied since there is no one party to compile all the pieces together. Furthermore, in a decentralized MA-ABE system each component may come from a different authority, where such authorities have no coordination and are possibly not even aware of each other. In order to overcome the above challenge, we aim to adapt the high level design rationally of the previous bilinear-map-based decentralized MA-ABE schemes [42,53] to not rely on one key generation call to tie all key components together and instead use the output of a public hash function applied on the user's global identity, GID, as the randomness tying together multiple key components issued by different authorities. However, this means that the randomness responsible for tying together the different key components must be publicly computable, that is, even known to the attacker. Unfortunately, all the CP-ABE schemes realizable under LWE so far fail to satisfy this property.

Importantly, and deviating from previous approaches, we design our CP-ABE scheme carefully so as to have this property. Observe that in our CP-ABE scheme above, the vector $\boldsymbol{t}$ is the one that is used to bind together different key components. A main feature of our CP-ABE scheme is that this vector $\boldsymbol{t}$ is actually part of the output of the key generation procedure. In particular, as we show, the system remains secure *even* when $\boldsymbol{t}$ is public and known to the attacker.

The second challenge in making a CP-ABE scheme compatible for extension to the decentralized multi-authority setting is modularity. Very roughly speaking, the setup and key generation procedures should have the structure such that it should be possible to view their operations as well as their outputs, that is, the master public/secret key and the secret keys of the users as aggregates of individual modules each of which relates to exactly one of the attributes involved. This is important since in a decentralized MA-ABE system, authorities/attributes should be able to join the system at any point of time without requiring any prior coordination with a central authority or a system reset and there is no bound on the number of authorities/attributes that can ever come into existence. Any CP-ABE scheme obtained from an underlying KP-ABE scheme via the universal-circuit-based transformation inherently fails to achieve the above modularity property roughly because in such a system, the master key and the user keys all become associated with the descriptions of circuits rather than the attributes directly. Hence it is not surprising that no prior CP-ABE scheme realizable under LWE achieves the above modularity feature. In contrast, we design our CP-ABE scheme above in such a way that everything is modular and fits into the decentralized multi-authority setting.

As is the design, the proof strategy for our MA-ABE scheme is also somewhat similar to the proof of the CP-ABE scheme. Although, since we are in the multi-authority setting, notation and various technical details become much more involved. For instance, the application of the linear independence property becomes much more delicate. Ignoring notational differences, one additional step we need to make for our proof to go through, is to somehow make the ciphertext components corresponding to corrupted authorities independent of the secret. This is because in our security model, we allow the adversary to generate the master keys for the corrupted authorities. Hence the simulator cannot hope to program any of the $\boldsymbol{H}_u$ matrices corresponding to the corrupted authorities and thereby cancel the secret present inside those ciphertext components as was possible in the single-authority scheme above.

To solve this, we are inspired by a previous technique of Rouselakis and Waters [53] in the bilinear map world for handling the same problem and we adapt it for our setting. After applying the idea under their transformation we reach a hybrid world which is more similar to the CP-ABE one where we only need to deal with the ciphertext components corresponding to uncorrupted authorities. As an additional contribution, en route to adapting their lemma to our setting, we observe a non-trivial gap in their proof which we resolve (please refer to [25, Section 4.3] for more details).

Lastly, let us explain why the new secret sharing scheme from Sect. 2.1 (see also Theorem 1.4) does not apply here. Since our LSSS from Sect. 2.1 is non-monotone, the share generating matrix has rows for both the positive and negative instances of an attribute. Now, in case of an MA-ABE for non-monotone LSSS, an attacker which corrupts an authority can generate keys for both the positive and negative instances of the attribute controlled by the authority and thus can get hold of both the rows of the LSSS matrix associated with both instances of that attribute. Unfortunately, in our LSSS, the linear independence property only holds when the set of unauthorized rows of an LSSS matrix does not include both the positive and negative instances of a particular attribute simultaneously. (Note that this is not an issue for our CP-ABE scheme since there is only one central authority which remains uncorrupted throughout the system.) We currently do not know of any non-monotone LSSS which achieves the linear independence property even when a set of unauthorized rows include both instances of the same attribute. We therefore settle for an LSSS which only considers attributes in their positive form, that is, monotone LSSS, and still satisfies the linear independence property for unauthorized rows. We use the direct construction of Lewko and Waters [42] which was recently observed by Agrawal et al. [1] to satisfy the linear independence property for unauthorized rows when implemented for the class of DNF formulas.

3 Preliminaries

3.1 Notations

Throughout this paper we will denote the underlying security parameter by λ. A function $\mathsf{negl}\colon \mathbb{N} \to \mathbb{R}$ is *negligible* if it is asymptotically smaller than any inverse-polynomial function, namely, for every constant $c > 0$ there exists an integer N_c such that $\mathsf{negl}(\lambda) \leq \lambda^{-c}$ for all $\lambda > N_c$. We let $[n] = \{1, \ldots, n\}$.

Let PPT stand for probabilistic polynomial-time. For a distribution $\mathcal{X}$, we write $x \leftarrow \mathcal{X}$ to denote that x is sampled at random according to distribution $\mathcal{X}$. For a set X, we write $x \leftarrow X$ to denote that x is sampled according to the uniform distribution over the elements of X. We use bold lower case letters, such as $\boldsymbol{v}$, to denote vectors and upper-case, such as $\boldsymbol{M}$, for matrices. We assume all vectors, by default, are row vectors. The jth row of a matrix is denoted $\boldsymbol{M}_j$ and analogously for a set of row indices J, we denote $\boldsymbol{M}_J$ for the submatrix of $\boldsymbol{M}$ that consists of the rows $\boldsymbol{M}_j$ for all $j \in J$. For a vector $\boldsymbol{v}$, we let $\|\boldsymbol{v}\|$ denote its ℓ_2 norm and $\|\boldsymbol{v}\|_\infty$ denote its ℓ_∞ norm.

For an integer $q \geq 2$, we let $\mathbb{Z}_q$ denote the ring of integers modulo q. We represent $\mathbb{Z}_q$ as integers in the range $(-q/2, q/2]$.

Indistinguishability: Two sequences of random variables $\mathcal{X} = \{\mathcal{X}_\lambda\}_{\lambda \in \mathbb{N}}$ and $\mathcal{Y} = \{\mathcal{Y}_\lambda\}_{\lambda \in \mathbb{N}}$ are *computationally indistinguishable* if for any non-uniform PPT algorithm $\mathcal{A}$ there exists a negligible function $\mathsf{negl}(\cdot)$ such that $|\Pr[\mathcal{A}(1^\lambda, \mathcal{X}_\lambda) = 1] - \Pr[\mathcal{A}(1^\lambda, \mathcal{Y}_\lambda) = 1]| \leq \mathsf{negl}(\lambda)$ for all $\lambda \in \mathbb{N}$.

For two distributions $\mathcal{D}$ and $\mathcal{D}'$ over a discrete domain Ω, the statistical distance between $\mathcal{D}$ and $\mathcal{D}'$ is defined as $\mathsf{SD}(\mathcal{D}, \mathcal{D}') = (1/2) \cdot \sum_{\omega \in \Omega} |\mathcal{D}(\omega) - \mathcal{D}'(\omega)|$. A family of distributions $\mathcal{D} = \{\mathcal{D}_\lambda\}_{\lambda \in \mathbb{N}}$ and $\mathcal{D}' = \{\mathcal{D}'_\lambda\}_{\lambda \in \mathbb{N}}$, parameterized by security parameter λ, are said to be *statistically indistinguishable* if there is a negligible function $\mathsf{negl}(\cdot)$ such that $\mathsf{SD}(\mathcal{D}_\lambda, \mathcal{D}'_\lambda) \leq \mathsf{negl}(\lambda)$ for all $\lambda \in \mathbb{N}$.

Smudging: The following lemma says that adding large noise "smudges out" any small values. This lemma was originally proven in [10, Lemma 2.1] and we use a paraphrased version from [35, Lemma 2.1]. Let us first define the notion of a B-bounded distribution.

Definition 3.1 (B-Bounded): For a family of distributions $\mathcal{D} = \{\mathcal{D}_\lambda\}_{\lambda \in \mathbb{N}}$ over the integers and a bound $B = B(\lambda) > 0$, we say that $\mathcal{D}$ is *B-bounded* if for every $\lambda \in \mathbb{N}$ it holds that $\Pr_{x \leftarrow \mathcal{D}_\lambda}[|x| \leq B(\lambda)] = 1$.

Lemma 3.1 (Smudging Lemma): *Let $B_1 = B_1(\lambda)$ and $B_2 = B_2(\lambda)$ be positive and let $\mathcal{D} = \{\mathcal{D}_\lambda\}_\lambda$ be a B_1-bounded distribution family. Let $\mathcal{U} = \{\mathcal{U}_\lambda\}_\lambda$ be the uniform distribution over $[-B_2(\lambda), B_2(\lambda)]$. The family of distributions $\mathcal{D} + \mathcal{U}$ and $\mathcal{U}$ are statistically indistinguishable if there exists a negligible function* $\mathsf{negl}(\cdot)$ *such that for all $\lambda \in \mathbb{N}$ it holds that $B_1(\lambda)/B_2(\lambda) \leq \mathsf{negl}(\lambda)$.*

Leftover Hash Lemma: We recall the well known leftover hash lemma, stated in a convenient form for our needs (e.g., [2,52]).

Lemma 3.2 (Leftover Hash Lemma): *Let $n \colon \mathbb{N} \to \mathbb{N}$, $q \colon \mathbb{N} \to \mathbb{N}$, $m > (n+1)\log q + \omega(\log n)$, and $k = k(n)$ be some polynomial. Then, the following two distributions are statistically indistinguishable:*

$$\mathcal{D}_1 \equiv \left\{ (\boldsymbol{A}, \boldsymbol{AR}) \mid \boldsymbol{A} \leftarrow \mathbb{Z}_q^{n \times m}, \boldsymbol{R} \leftarrow \{-1, 1\}^{m \times k} \right\},$$

$$\mathcal{D}_2 \equiv \left\{ (\boldsymbol{A}, \boldsymbol{S}) \mid \boldsymbol{A} \leftarrow \mathbb{Z}_q^{n \times m}, \boldsymbol{S} \leftarrow \mathbb{Z}_q^{n \times k} \right\}.$$

3.2 Lattice and LWE Preliminaries

Here, we provide necessary background on lattices, the LWE assumption, and various useful tools that we use.

Lattices: An m-dimensional lattice $\mathcal{L}$ is a discrete additive subgroup of $\mathbb{R}^m$. Given positive integers n, m, q and a matrix $\boldsymbol{A} \in \mathbb{Z}_q^{n \times m}$, we let $\lambda_q^{\perp}(\boldsymbol{A})$ denote the lattice $\{\boldsymbol{x} \in \mathbb{Z}^m \mid \boldsymbol{A}\boldsymbol{x}^\top = \boldsymbol{0}^\top \bmod q\}$. For $\boldsymbol{u} \in \mathbb{Z}_q^n$, we let $\lambda_q^{\boldsymbol{u}}(\boldsymbol{A})$ denote the coset $\{\boldsymbol{x} \in \mathbb{Z}^m \mid \boldsymbol{A}\boldsymbol{x}^\top = \boldsymbol{u}^\top \bmod q\}$.

Discrete Gaussians: Let σ be any positive real number. The Gaussian distribution $\mathcal{D}_\sigma$ with parameter σ is defined by the probability distribution function $\rho_\sigma(\boldsymbol{x}) = \exp(-\pi\|x\|^2/\sigma^2)$. For any discrete set $\mathcal{L} \subseteq \mathbb{R}^m$, define $\rho_\sigma(\mathcal{L}) = \sum_{\boldsymbol{x}\in\mathcal{L}} \rho_\sigma(\boldsymbol{x})$. The discrete Gaussian distribution $\mathcal{D}_{\mathcal{L},\sigma}$ over $\mathcal{L}$ with parameter σ is defined by the probability distribution function $\rho_{\mathcal{L},\sigma}(\boldsymbol{x}) = \rho_\sigma(\boldsymbol{x})/\rho_\sigma(\mathcal{L})$.

The following lemma (e.g., [47, Lemma 4.4]) shows that if the parameter σ of a discrete Gaussian distribution is small, then any vector drawn from this distribution will be short (with high probability).

Lemma 3.3: *Let m, n, q be positive integers with $m > n$, $q > 2$. Let $\boldsymbol{A} \in \mathbb{Z}_q^{n \times m}$ be a matrix of dimensions $n \times m$, $\sigma = \tilde{\Omega}(n)$, and $\mathcal{L} = \lambda_q^{\perp}(\boldsymbol{A})$. Then, there is a negligible function* $\mathsf{negl}(\cdot)$ *such that*

$$\Pr_{\boldsymbol{x}\leftarrow\mathcal{D}_{\mathcal{L},\sigma}}\left[\|\boldsymbol{x}\| > \sqrt{m}\sigma\right] \leq \mathsf{negl}(n),$$

where $\|\boldsymbol{x}\|$ denotes the ℓ_2 norm of $\boldsymbol{x}$.

Truncated Discrete Gaussians: The truncated discrete Gaussian distribution over $\mathbb{Z}^m$ with parameter σ, denoted by $\widetilde{\mathcal{D}}_{\mathbb{Z}^m,\sigma}$, is the same as the discrete Gaussian distribution $\mathcal{D}_{\mathbb{Z}^m,\sigma}$ except that it outputs 0 whenever the ℓ_∞ norm exceeds $\sqrt{m}\sigma$. Note that, by definition, $\widetilde{\mathcal{D}}_{\mathbb{Z}^m,\sigma}$ is $\sqrt{m}\sigma$-bounded. Also, by Lemma 3.3 we get that $\widetilde{\mathcal{D}}_{\mathbb{Z}^m,\sigma}$ and $\mathcal{D}_{\mathbb{Z}^m,\sigma}$ are statistically indistinguishable.

3.2.1 Lattice Trapdoors

Lattices with trapdoors are lattices that are indistinguishable from randomly chosen lattices, but have certain "trapdoors" that allow efficient solutions to hard lattice problems. A trapdoor lattice sampler [9,29,45], denoted $\mathsf{LT} = (\mathsf{TrapGen}, \mathsf{SamplePre})$, consists of two algorithms with the following syntax and properties:

- $\mathsf{TrapGen}(1^n, 1^m, q) \mapsto (\boldsymbol{A}, T_{\boldsymbol{A}})$: The lattice generation algorithm is a randomized algorithm that takes as input the matrix dimensions n, m, modulus q, and outputs a matrix $\boldsymbol{A} \in \mathbb{Z}_q^{n \times m}$ together with a trapdoor $T_{\boldsymbol{A}}$.
- $\mathsf{SamplePre}(\boldsymbol{A}, T_{\boldsymbol{A}}, \sigma, \boldsymbol{u}) \mapsto \boldsymbol{s}$: The presampling algorithm takes as input a matrix $\boldsymbol{A}$, trapdoor $T_{\boldsymbol{A}}$, a vector $\boldsymbol{u} \in \mathbb{Z}_q^n$, and a parameter $\sigma \in \mathbb{R}$ (which determines the length of the output vectors). It outputs a vector $\boldsymbol{s} \in \mathbb{Z}_q^m$ such that $\boldsymbol{A} \cdot \boldsymbol{s}^\top = \boldsymbol{u}^\top$ and $\|\boldsymbol{s}\| \leq \sqrt{m} \cdot \sigma$.

Well-sampledness: Following Goyal et al. [35], we further require that the aforementioned sampling procedures output well-sampled elements. That is, the

1. The adversary $\mathcal{A}$ receives input 1^λ and sends $1^n, 1^m, 1^z$ such that $m > n \log q(\lambda) + \lambda$ to the challenger.
2. Upon receipt, the challenger first selects a random bit $b \leftarrow \{0,1\}$. Next, it samples $\{(\boldsymbol{A}_{i,0}, \boldsymbol{T}_{\boldsymbol{A}_{i,0}})\}_{i \in [z]} \leftarrow \mathsf{TrapGen}(1^n, 1^m, q)$ and $\{\boldsymbol{A}_{i,1}\}_{i \in [z]} \leftarrow \mathbb{Z}_q^{n \times m}$. It sends $\{\boldsymbol{A}_{i,b}\}_{i \in [z]}$ to $\mathcal{A}$.
3. Finally, $\mathcal{A}$ outputs its guess $b' \in \{0,1\}$. The experiment outputs 1 if and only if $b = b'$.

Fig. 1. $\mathsf{Exp}_{\mathsf{LT},\mathcal{A}}^{\mathsf{matrix},q}$

matrix outputted by TrapGen looks like a uniformly random matrix, and the preimage outputted by SamplePre with a uniformly random vector/matrix is indistinguishable from a vector/matrix with entries drawn from an appropriate Gaussian distribution. These two properties are summarized next.

Definition 3.2 (Well-Sampledness of Matrix): Fix any function $q \colon \mathbb{N} \to \mathbb{N}$. The procedure TrapGen is said to satisfy the *q-well-sampledness of matrix* property if for any PPT adversary $\mathcal{A}$, there exists a negligible function $\mathsf{negl}(\cdot)$ such that for all $\lambda \in \mathbb{N}$,

$$\mathsf{Adv}_{\mathsf{LT},\mathcal{A}}^{\mathsf{matrix},q}(\lambda) \triangleq \left| \Pr\left[\mathsf{Exp}_{\mathsf{LT},\mathcal{A}}^{\mathsf{matrix},q}(\lambda) = 1 \right] - 1/2 \right| \leq \mathsf{negl}(\lambda),$$

where $\mathsf{Exp}_{\mathsf{LT},\mathcal{A}}^{\mathsf{matrix},q}(\lambda)$ is defined in Fig. 1.

Definition 3.3 (Well-Sampledness of Preimage): Fix any function $q \colon \mathbb{N} \to \mathbb{N}$ and $\sigma \colon \mathbb{N} \to \mathbb{N}$. The procedure SamplePre is said to satisfy the (q, σ)*-well-sampledness* property if for any stateful PPT adversary $\mathcal{A}$, there exists a negligible function $\mathsf{negl}(\cdot)$ such that for all $\lambda \in \mathbb{N}$,

$$\mathsf{Adv}_{\mathsf{LT},\mathcal{A}}^{\mathsf{preimage},q,\sigma}(\lambda) \triangleq \left| \Pr\left[\mathsf{Exp}_{\mathsf{LT},\mathcal{A}}^{\mathsf{preimage},q,\sigma}(\lambda) = 1 \right] - 1/2 \right| \leq \mathsf{negl}(\lambda),$$

where $\mathsf{Exp}_{\mathsf{LT},\mathcal{A}}^{\mathsf{preimage},q,\sigma}$ is defined in Fig. 2.

Both the above properties are satisfied by the gadget-based trapdoor lattice sampler presented in [45].

Enhanced trapdoor sampling: Let $q \colon \mathbb{N} \to \mathbb{N}$, $\sigma \colon \mathbb{N} \to \mathbb{R}^+$ be functions and $\mathsf{LT} = (\mathsf{TrapGen}, \mathsf{SamplePre})$ be a trapdoor lattice sampler satisfying the q-well-sampledness of matrix and (q, σ)-well-sampledness of preimage properties. We describe enhanced trapdoor lattice sampling algorithms $\mathsf{EnLT} = (\mathsf{EnTrapGen}, \mathsf{EnSamplePre})$ due to Goyal et al. [35] (which are, in turn, reminiscent of the trapdoor extension algorithms of [3,20]).

1. The adversary $\mathcal{A}$ receives input 1^λ and sends $1^n, 1^m, 1^z$ such that $\sigma(\lambda) > \sqrt{n \cdot \log q(\lambda) \cdot \log m} + \lambda$ and $m > n \cdot \log q(\lambda) + \lambda$ to the challenger.
2. Upon receipt, the challenger first selects a random bit $b \leftarrow \{0,1\}$. Next, it samples $\{(\boldsymbol{A}_i, T_{\boldsymbol{A}_i})\}_{i \in [z]} \leftarrow \mathsf{TrapGen}(1^n, 1^m, q)$ and sends $\{\boldsymbol{A}_i\}_{i \in [z]}$ to $\mathcal{A}$.
3. Then, $\mathcal{A}$ makes a $\mathsf{poly}(\lambda)$ number of pre-image queries of the form $i \in [z]$ to the challenger and the challenger responds as follows:
   (a) It samples $\boldsymbol{w} \leftarrow \mathbb{Z}_q^n$, $\boldsymbol{u}_0 \leftarrow \mathsf{SamplePre}(\boldsymbol{A}_i, T_{\boldsymbol{A}_i}, \sigma, \boldsymbol{w})$, and $\boldsymbol{u}_1 \leftarrow \mathcal{D}^m_{\mathbb{Z},\sigma}$. It sends $\boldsymbol{u}_b$ to $\mathcal{A}$.
4. Finally, $\mathcal{A}$ outputs its guess $b' \in \{0,1\}$. The experiment outputs 1 if and only if $b = b'$.

Fig. 2. $\mathsf{Exp}^{\mathsf{preimage},q,\sigma}_{\mathsf{LT},\mathcal{A}}$

- $\mathsf{EnTrapGen}(1^n, 1^m, q) \mapsto (\boldsymbol{A}, T_{\boldsymbol{A}})$: The trapdoor generation algorithm generates two matrices $\boldsymbol{A}_1 \in \mathbb{Z}_q^{n \times \lceil m/2 \rceil}$ and $\boldsymbol{A}_2 \in \mathbb{Z}_q^{n \times \lfloor m/2 \rfloor}$ as $(\boldsymbol{A}_1, T_{\boldsymbol{A}_1}) \leftarrow \mathsf{TrapGen}(1^n, 1^{\lceil m/2 \rceil}, q)$, $(\boldsymbol{A}_2, T_{\boldsymbol{A}_2}) \leftarrow \mathsf{TrapGen}(1^n, 1^{\lfloor m/2 \rfloor}, q)$. It appends both matrices column-wise to obtain a larger matrix $\boldsymbol{A}$ as $\boldsymbol{A} = (\boldsymbol{A}_1 | \boldsymbol{A}_2)$ and sets the associated trapdoor $T_{\boldsymbol{A}}$ to be the combined trapdoor information $T_{\boldsymbol{A}} = (T_{\boldsymbol{A}_1}, T_{\boldsymbol{A}_2})$.
- $\mathsf{EnSamplePre}(\boldsymbol{A}, T_{\boldsymbol{A}}, \sigma, \boldsymbol{Z}) \mapsto \boldsymbol{S}$: The pre-image sampling algorithm takes as input a matrix $\boldsymbol{A} = (\boldsymbol{A}_1 | \boldsymbol{A}_2)$ with trapdoor $T_{\boldsymbol{A}} = (T_{\boldsymbol{A}_1}, T_{\boldsymbol{A}_2})$, a parameter $\sigma = \sigma(\lambda)$, and a matrix $\boldsymbol{Z} \in \mathbb{Z}_q^{n \times k}$. It chooses a uniformly random matrix $\boldsymbol{W} \leftarrow \mathbb{Z}_q^{n \times k}$ and sets $\boldsymbol{Y} = \boldsymbol{Z} - \boldsymbol{W}$. Next, it computes matrices $\boldsymbol{S}_1, \boldsymbol{S}_2 \in \mathbb{Z}^{\lceil m/2 \rceil \times k}$ as $\boldsymbol{S}_1 \leftarrow \mathsf{SamplePre}(\boldsymbol{A}_1, T_{\boldsymbol{A}_1}, \sigma, \boldsymbol{W})$ and $\boldsymbol{S}_2 \leftarrow \mathsf{SamplePre}(\boldsymbol{A}_2, T_{\boldsymbol{A}_2}, \sigma, \boldsymbol{Y})$. It computes the final output matrix $\boldsymbol{S} \in \mathbb{Z}^{m \times k}$ by column-wise appending matrices $\boldsymbol{S}_1$ and $\boldsymbol{S}_2$ as $\boldsymbol{S} = (\boldsymbol{S}_1 | \boldsymbol{S}_2)$.

The well-sampledness properties (Definition 3.2 and Definition 3.3) of EnLT are inherited from the same properties of the underlying LT [35, Section 7.3].

We show that the enhanced trapdoor sampling procedures EnLT satisfy another property (which as far as we know has not been used or formalized before). We refer this property as "leftover hash lemma with trapdoors". This property is crucial in the security proofs of our constructions. Recall that in the original leftover hash lemma (Lemma 3.2 above) the matrix $\boldsymbol{A} \in \mathbb{Z}_q^{n \times m}$ appearing in the two indistinguishable distributions $\mathcal{D}_1$ and $\mathcal{D}_2$ is sampled uniformly at random. The "leftover hash lemma with trapdoors" property of EnLT basically states that the leftover hash lemma holds even when the matrix $\boldsymbol{A} \in \mathbb{Z}_q^{n \times m}$ is generated by the $\mathsf{EnTrapGen}$ algorithm and is not uniformly random. (See the full version [25] for the formal description and proof of the lemma.)

3.2.2 Learning with Errors

Assumption 1 (Learning With Errors (LWE) [52]): For a security parameter $\lambda \in \mathbb{N}$, let $n\colon \mathbb{N} \to \mathbb{N}$, $q\colon \mathbb{N} \to \mathbb{N}$, and $\sigma\colon \mathbb{N} \to \mathbb{R}^+$ be functions of λ. The Learning with Errors (LWE) assumption $\mathsf{LWE}_{n,q,\sigma}$, parametrized by $n = n(\lambda), q = q(\lambda), \sigma = \sigma(\lambda)$, states that for any PPT adversary $\mathcal{A}$, there exists a negligible function $\mathsf{negl}(\cdot)$ such that for any $\lambda \in \mathbb{N}$,

$$\mathsf{Adv}_{\mathcal{A}}^{\mathsf{LWE}_{n,q,\sigma}}(\lambda) \triangleq \left|\mathsf{Pr}\left[1 \leftarrow \mathcal{A}^{\mathcal{O}_1^{s}(\cdot)}(1^\lambda) \mid s \leftarrow \mathbb{Z}_q^n\right] - \mathsf{Pr}\left[1 \leftarrow \mathcal{A}^{\mathcal{O}_2(\cdot)}(1^\lambda)\right]\right| \\ \leq \mathsf{negl}(\lambda),$$

where the oracles $\mathcal{O}_1^{s}(\cdot)$ and $\mathcal{O}_2(\cdot)$ are defined as follows: $\mathcal{O}_1^{s}(\cdot)$ has $s \in \mathbb{Z}_q^n$ hard-wired, and on each query it chooses $a \leftarrow \mathbb{Z}_q^n$, $e \leftarrow \mathcal{D}_{\mathbb{Z},\sigma}$ and outputs $(a, sa^\top + e \bmod q)$, and $\mathcal{O}_2(\cdot)$ on each query chooses $a \leftarrow \mathbb{Z}_q^n$, $u \leftarrow \mathbb{Z}_q$ and outputs (a, u).

Regev [52] showed that if there exists a PPT adversary that can break the LWE assumption, then there exists a PPT quantum algorithm that can solve some hard lattice problems in the worst case. Given the current state of the art of lattice problems [17,29,46,47,51,52], the LWE assumption is believed to be true for any polynomial $n(\cdot)$ and any functions $q(\cdot), \sigma(\cdot)$ such that for all $\lambda \in \mathbb{N}$, $n = n(\lambda), q = q(\lambda), \sigma = \sigma(\lambda)$ satisfy the following constraints:

$$2\sqrt{n} < \sigma < q < 2^n, \quad n \cdot q/\sigma < 2^{n^\epsilon}, \quad \text{and } 0 < \epsilon < 1/2$$

4 Linear Secret Sharing Schemes with Linear Independence

In this section, we first provide the necessary definitions and properties of linear secret sharing schemes. Then, we present a new linear secret sharing scheme for all non-monotone access structures realizable by NC^1 circuits. This new secret sharing scheme has some interesting properties which we crucially utilize while designing our CP-ABE scheme for all NC^1 circuits under the LWE assumption.

4.1 Background on Linear Secret Sharing Schemes

A secret sharing scheme consists of a dealer who holds a secret and a set of n parties. Informally, the dealer "splits" the secret into "shares" and distributes them among the parties. Subsets of parties which are "authorized" should be able to jointly recover the secret while others should not. The description of the set of authorized sets is called the *access structure*.

Definition 4.1 (Access Structures): An access structure on n parties associated with numbers in $[n]$ is a set $\mathbb{A} \subseteq 2^{[n]} \setminus \emptyset$ of non-empty subsets of parties. The sets in $\mathbb{A}$ are called the *authorized* sets and the sets not in $\mathbb{A}$ are called the *unauthorized* sets. An access structure is called *monotone* if $\forall B, C \in 2^{[n]}$ if $B \in \mathbb{A}$ and $B \subseteq C$, then $C \in \mathbb{A}$.

A secret sharing scheme for a monotone access structure $\mathbb{A}$ is a randomized algorithm that on input a secret z outputs n shares $\mathsf{sh}_1, \ldots, \mathsf{sh}_n$ such that for any $A \in \mathbb{A}$ the shares $\{\mathsf{sh}_i\}_{i \in A}$ determine z and other sets are independent of z (as random variables).

Non-monotone secret sharing: A natural generalization of the above notion that captures all access structures (rather than only monotone ones) is called *non-monotone* secret sharing. Concretely, a non-monotone secret sharing scheme for an access structure $\mathbb{A}$ is a randomized algorithm that on input a secret z outputs $2n$ shares viewed as n pairs $(\mathsf{sh}_{1,0}, \mathsf{sh}_{1,1}), \ldots, (\mathsf{sh}_{n,0}, \mathsf{sh}_{n,1})$ such that for any $A \in \mathbb{A}$ the shares $\{\mathsf{sh}_{i,1}\}_{i \in A} \cup \{\mathsf{sh}_{i,0}\}_{i \notin A}$ determine z and other sets are independent of z.

We will be interested in a subset of all (non-monotone) secret sharing schemes where the reconstruction procedure is a linear function of the shares [38]. These are known as *linear (non-monotone) secret sharing schemes.*

Definition 4.2 (Linear (non-monotone) secret sharing schemes): Let $q \in \mathbb{N}$ be a prime power and $[n]$ be a set of parties. A non-monotone secret-sharing scheme Π with domain of secrets $\mathbb{Z}_q$ realizing access structure $\mathbb{A}$ on parties $[n]$ is linear over $\mathbb{Z}_q$ if

1. Each share $\mathsf{sh}_{i,b}$ for $i \in [n]$ and $b \in \{0,1\}$ of a secret $z \in \mathbb{Z}_q$ forms a vector with entries in $\mathbb{Z}_q$.
2. There exists a matrix $\boldsymbol{M} \in \mathbb{Z}_q^{\ell \times d}$, called the share-generating matrix, and a function $\rho\colon [\ell] \to [2n]$, that labels the rows of $\boldsymbol{M}$ with a party index from $[n]$ or its corresponding negation, represented as another party index from $\{n+1, \ldots, 2n\}$, which satisfy the following: During the generation of the shares, we consider the vector $\boldsymbol{v} = (z, r_2, ..., r_d) \in \mathbb{Z}_q^d$. Then the vector of ℓ shares of the secret z according to Π is equal to $\mathbf{sh} = \boldsymbol{M} \cdot \boldsymbol{v}^\top \in \mathbb{Z}_q^{\ell \times 1}$. For $i \in [n]$ and $b \in \{0,1\}$, the share $\mathbf{sh}_{i,b}$ consists of all $\mathbf{sh}_j$ values for which $\rho(j) = n \cdot (1-b) + i$ (so the first n shares correspond to the "1 shares" and the last n shares correspond to the "0 shares").
   We will be referring to the pair $(\boldsymbol{M}, \rho)$ as the LSSS *policy* of the access structure $\mathbb{A}$.

It is well known that the above method of sharing a secret satisfies the desired correctness and security of a non-monotone secret sharing scheme as defined above (e.g., [38]). For an LSSS policy $(\boldsymbol{M}, \rho)$, where $\boldsymbol{M} \in \mathbb{Z}_q^{\ell \times d}$ and $\rho\colon [\ell] \to [2n]$, and a set of parties $S \subseteq [n]$, let $\widehat{S} = S \cup \{i \in \{n+1, \ldots, 2n\} \mid i - n \notin S\} \subset [2n]$. We denote $\boldsymbol{M}_{\widehat{S}}$ the submatrix of $\boldsymbol{M}$ that consists of all the rows of $\boldsymbol{M}$ that "belong" to $\widehat{S}$ according to ρ (i.e., rows j for which $\rho(j) \in \widehat{S}$). *Correctness* means that if $S \subseteq [n]$ is authorized, the vector $(1, \overbrace{0, \ldots, 0}^{d-1}) \in \mathbb{Z}_q^d$ is in the span of the rows of $\boldsymbol{M}_{\widehat{S}}$. *Security* means that if $S \subseteq [n]$ is unauthorized, the vector $(1, 0, \ldots, 0)$ is *not* in the span of the rows of $\boldsymbol{M}_{\widehat{S}}$. Also, in the unauthorized case, there exists a vector $\boldsymbol{d} \in \mathbb{Z}_q^d$, such that its first component $\boldsymbol{d}_1 = 1$ and $\boldsymbol{M}_{\widehat{S}} \boldsymbol{d}^\top = \mathbf{0}$, where $\mathbf{0}$ is the all 0 vector.

{0, 1}-LSSS: A special subset of all linear secret sharing schemes are ones where the reconstruction coefficients are always binary [14, Definition 4.13]. We call such LSSS a $\{0,1\}$-LSSS. This property of LSSS secret sharing schemes was recently formally defined by [14]. They observed that a well-known construction by Lewko and Waters [42] actually results with an LSSS with this property for all access structures in NC^1.

On sharing vectors: The above sharing and reconstruction methods directly extend to sharing a vector $\boldsymbol{z} \in \mathbb{Z}_q^m$ of dimension $m \in \mathbb{N}$ rather than just scalars.

4.2 Our Non-monotone LSSS for NC^1

We introduce a new non-monotone linear secret sharing scheme for all access structures that can be described by NC^1 circuits. The new scheme has some useful properties for us which we summarize next:

- The entries in the corresponding policy matrix are small, i.e., coming from $\{-1, 0, 1\}$.
- Reconstruction of the secret can be done by small coefficients, i.e., coming from $\{0, 1\}$.
- The rows of the corresponding policy matrix that correspond to an unauthorized set are *linearly independent.*

Remark 4.1: The well-known construction of Lewko and Waters [42] actually results with an LSSS with these properties for all access structures described by DNF formulas. This was recently observed by [1]. As opposed to our construction, this construction is a monotone LSSS, not a non-monotone one.

The construction: We are given an access structure $\mathbb{A}$ described by an NC^1 circuit. This circuit can be described by a Boolean formula of logarithmic depth that consists of (fan-in 2) AND, OR, and (fan-in 1) NOT gates. We further push the NOT gates to the leaves using De Morgan laws, and from now on we assume that internal nodes only constitute of OR and AND gates and leaves are labeled either by variables or their negations. In other words, we assume that we are given a monotone Boolean formula consisting only of AND and OR gates. We would like to highlight that even if we are starting off with a monotone Boolean formula, the LSSS secret sharing scheme we are going to construct would be a non-monotone one. More precisely, the algorithm associates with each input variable x_i of the monotone Boolean formula two vector shares $\mathbf{sh}_{i,0}$ and $\mathbf{sh}_{i,1}$. This is done in a recursive fashion starting from the root by associating with each internal wire w two labels $\boldsymbol{w}_1$ and $\boldsymbol{w}_0$ (and the labels of the leaves correspond to the shares). The labels of the root w are $\boldsymbol{w}_1 = (1, 0, \ldots, 0)$ and $\boldsymbol{w}_0 = (0, 1, 0, \ldots, 0)$, both of which are of dimension $\tilde{k} \triangleq k + 2$, where k is the number of gates in the formula. We maintain a global counter variable c which is initialized to 2 and is increased by one after labeling each gate. We shall traverse the tree from top (root) to bottom (leaves) and within a layer from left to right. Consider a gate whose output wire w labels are $\boldsymbol{w}_1$, $\boldsymbol{w}_0$ and denote its children wires, u and v, with corresponding labels (to be assigned) $\boldsymbol{u}_1, \boldsymbol{u}_0$ and $\boldsymbol{v}_1, \boldsymbol{v}_0$, respectively. The assignment is done as follows, depending on the type of the gate connecting u and v to w:

AND gate: $\boldsymbol{u}_1 = 0^c \| 1 \| 0^{\tilde{k}-c-1}, \quad \boldsymbol{u}_0 = \boldsymbol{w}_0, \quad \boldsymbol{v}_1 = \boldsymbol{w}_1 - \boldsymbol{u}_1, \quad \boldsymbol{v}_0 = \boldsymbol{w}_0 - \boldsymbol{u}_1.$
OR gate: $\boldsymbol{u}_1 = \boldsymbol{w}_1, \quad \boldsymbol{u}_0 = 0^c \| 1 \| 0^{\tilde{k}-c-1}, \quad \boldsymbol{v}_1 = \boldsymbol{w}_1 - \boldsymbol{u}_0, \quad \boldsymbol{v}_0 = \boldsymbol{w}_0 - \boldsymbol{u}_0.$

An example: Consider the monotone Boolean formula $(A \wedge B) \vee (C \wedge D)$. The 1-label of the root is $(1, 0, 0, 0, 0)$ and the 0-label is $(0, 1, 0, 0, 0)$. The 1-label of the left child of the OR gate is $(1, 0, 0, 0, 0)$ and the 0-label is $(0, 0, 1, 0, 0)$. The 1-label of the right child of the OR gate is $(1, 0, -1, 0, 0)$ and the 0-label is $(0, 1, -1, 0, 0)$. Therefore, the resulting policy is

$$\boldsymbol{M} = \begin{matrix} A_1 \\ A_0 \\ B_1 \\ B_0 \\ C_1 \\ C_0 \\ D_1 \\ D_0 \end{matrix} \begin{pmatrix} 0 & 0 & 0 & 1 & 0 \\ 0 & 0 & 1 & 0 & 0 \\ 1 & 0 & 0 & -1 & 0 \\ 0 & 0 & 1 & -1 & 0 \\ 0 & 0 & 0 & 0 & 1 \\ 0 & 1 & -1 & 0 & 0 \\ 1 & 0 & -1 & 0 & -1 \\ 0 & 1 & -1 & 0 & -1 \end{pmatrix}$$

The following lemma follows by induction on the number of gates in the formula. Recall that for $S \subseteq [n]$, we let $\widehat{S} = S \cup \{i \in \{n+1, \ldots, 2n\} \mid i - n \notin S\} \subset [2n]$ and let $\boldsymbol{M}_{\widehat{S}}$ be the submatrix that consists of all the rows of $\boldsymbol{M}$ that "belong" to $\widehat{S}$ according to ρ.

Lemma 4.1: *For any access structure $\mathbb{A}$ which is described by a Boolean formula, the above process for generating the matrix $\boldsymbol{M}$ results with*

1. *A non-monotone $\{0,1\}$-LSSS for $\mathbb{A}$, namely*
   (a) *For any authorized set of parties $S \subseteq [n]$, there is a linear combination of the rows of $\boldsymbol{M}_{\widehat{S}}$ that results with $(1, 0, \ldots, 0) \in \mathbb{Z}_q^d$. Moreover, the coefficients in this linear combination are from $\{0, 1\}$.*
   (b) *For any unauthorized set of parties $S \subseteq [n]$, no linear combination of the rows of $\boldsymbol{M}_{\widehat{S}}$ results in $(1, 0, \ldots, 0) \in \mathbb{Z}_q^d$. Also, there exists a vector $\boldsymbol{d} \in \mathbb{Z}_q^d$, such that its first component $\boldsymbol{d}_1 = 1$ and $\boldsymbol{M}_{\widehat{S}} \boldsymbol{d}^\top = \boldsymbol{0}$, where $\boldsymbol{0}$ is the all 0 vector.*
2. *For any unauthorized set of parties $S \subseteq [n]$, all of the rows of $\boldsymbol{M}_{\widehat{S}}$ are linearly independent.*

The proof of Lemma 4.1 can be found in the full version [25, Section 4.2].

5 Our Ciphertext-Policy ABE Scheme

In this section, we present our ciphertext-policy ABE (CP-ABE) scheme supporting access structures represented by NC^1 circuits. The scheme is associated with a fixed attribute universe $\mathbb{U}$ and we will use the transformation described in Sect. 4.2 to represent the access structures as non-monotone LSSS. More precisely, we only design a CP-ABE scheme for LSSS access policies $(\boldsymbol{M}, \rho)$ with properties stipulated in Lemma 4.1, that is, we construct a CP-ABE scheme for LSSS access policies $(\boldsymbol{M}, \rho)$ such that the entries of $\boldsymbol{M}$ come from $\{-1, 0, 1\}$ as well as reconstruction only involves coefficients coming from $\{0, 1\}$, and prove the scheme to be selectively secure under linear independence restriction (see [25, Definition 3.5] in the full version for the formal description of the security model). It then follows directly from Lemma 4.1, that our CP-ABE scheme actually achieves the standard notion of selective security (see [25, Definition 3.4] in the full version for the formal description of the security model) when implemented for the class of all access structures represented by NC^1 circuits. Further, we will assume that all LSSS access policies $(\boldsymbol{M}, \rho)$ used in our scheme correspond to matrices $\boldsymbol{M}$ with at most $s_{\max}$ columns and an injective row-labeling function ρ, i.e., an attribute is associated with at most one row of $\boldsymbol{M}$.[2] Since our Boolean formula to LSSS transformation from Sect. 4.2 generates a new column in the resulting LSSS matrix for each gate in the underlying Boolean formula, the bound $s_{\max}$ on the number of columns in our CP-ABE construction naturally translates to a bound on the circuit size of the supported NC^1 access policies at implementation. Also, in our scheme description below, we assume for simplicity of presentation that both the encryption and the decryption algorithms receive an access policy directly in its LSSS representation. However, we note that in the actual implementation, the encryption and decryption algorithms should instead take in the circuit representation of the access policy and deterministically compute its LSSS representation using our transformation algorithm from Sect. 4.2. This is because, without the circuit description of an access policy, the decryption algorithm may not be able to efficiently determine the $\{0, 1\}$ reconstruction coefficients needed for a successful decryption.

First, we provide the parameter constraints required by our correctness and security proof. Fix any $0 < \epsilon < 1/2$. For any $B \in \mathbb{N}$, let $\mathcal{U}_B$ denote the uniform distribution on $\mathbb{Z} \cap [-B, B]$, i.e., integers between $\pm B$. The Setup algorithm chooses parameters n, m, σ, q and noise distributions $\chi_{\mathsf{lwe}}, \chi_1, \chi_2, \chi_{\mathsf{big}}$, satisfying the following constraints:

- $n = \mathsf{poly}(\lambda)$, $\sigma < q$, $n \cdot q/\sigma < 2^{n^\epsilon}$, $\chi_{\mathsf{lwe}} = \widetilde{\mathcal{D}}_{\mathbb{Z},\sigma}$ (for LWE security)
- $m > 2 s_{\max} n \log q + \omega \log n + 2\lambda$ (for enhanced trapdoor sampling and LHL)
- $\sigma > \sqrt{s_{\max} n \log q \log m} + \lambda$ (for enhanced trapdoor sampling)
- $\chi_1 = \widetilde{\mathcal{D}}_{\mathbb{Z}^{m-1},\sigma}$, $\chi_2 = \widetilde{\mathcal{D}}_{\mathbb{Z}^m,\sigma}$ (for enhanced trapdoor sampling)
- $\chi_{\mathsf{big}} = \mathcal{U}_{\hat{B}}$, where $\hat{B} > (m^{3/2}\sigma + 1)2^\lambda$ (for smudging/security)
- $|\mathbb{U}| \cdot 3m^{3/2}\sigma\hat{B} < q/4$ (for correctness)

Now, we describe our CP-ABE construction.

[2] Note that following the simple encoding technique devised in [42,58], we can alleviate the injective restriction on the row labeling functions to allow an attribute to appear an a priori bounded number of times within the LSSS access policies.

Setup$(1^\lambda, s_{\max}, \mathbb{U})$**:** The setup algorithm takes in the security parameter λ encoded in unary, the maximum width $s_{\max} = s_{\max}(\lambda)$ of an LSSS matrix supported by the scheme, and the attribute universe $\mathbb{U}$ associated with the system. It first chooses an LWE modulus q, dimensions n, m, and also distributions $\chi_{\mathsf{lwe}}, \chi_1, \chi_2, \chi_{\mathsf{big}}$ as described above. Next, it chooses a vector $\boldsymbol{y} \leftarrow \mathbb{Z}_q^n$ and a sequence of matrices $\{\boldsymbol{H}_u\}_{u\in\mathbb{U}} \leftarrow \mathbb{Z}_q^{n\times m}$. Then, it samples pairs of matrices with trapdoors $\{(\boldsymbol{A}_u, T_{\boldsymbol{A}_u})\}_{u\in\mathbb{U}} \leftarrow \mathsf{EnTrapGen}(1^n, 1^m, q)$. Finally, it outputs

$$\mathsf{PK} = \left(n, m, q, \chi_{\mathsf{lwe}}, \chi_1, \chi_2, \chi_{\mathsf{big}}, \boldsymbol{y}, \{\boldsymbol{A}_u\}_{u\in\mathbb{U}}, \{\boldsymbol{H}_u\}_{u\in\mathbb{U}}\right), \qquad \mathsf{MSK} = \{T_{\boldsymbol{A}_u}\}_{u\in\mathbb{U}}.$$

KeyGen(MSK, U)**:** The key generation algorithm takes as input the master secret key MSK, and a set of attributes $U \subseteq \mathbb{U}$. It samples a vector $\hat{\boldsymbol{t}} \leftarrow \chi_1$ and sets the vector $\boldsymbol{t} = (1, \hat{\boldsymbol{t}}) \in \mathbb{Z}^m$. For each $u \in U$, it samples vectors $\hat{\boldsymbol{k}}_u \leftarrow \chi_{\mathsf{big}}^m$ and $\tilde{\boldsymbol{k}}_u \leftarrow \mathsf{EnSamplePre}(\boldsymbol{A}_u, T_{\boldsymbol{A}_u}, \sigma, \boldsymbol{t}\boldsymbol{H}_u^\top - \hat{\boldsymbol{k}}_u\boldsymbol{A}_u^\top)$, and sets $\boldsymbol{k}_u = \hat{\boldsymbol{k}}_u + \tilde{\boldsymbol{k}}_u$. Finally, it outputs

$$\mathsf{SK} = \left(\{\boldsymbol{k}_u\}_{u\in U}, \boldsymbol{t}\right).$$

Enc$(\mathsf{PK}, \mathsf{msg}, (\boldsymbol{M}, \rho))$**:** The encryption algorithm takes as input the public parameters PK, a message $\mathsf{msg} \in \{0,1\}$ to encrypt, and an LSSS access policy $(\boldsymbol{M}, \rho)$ generated by the transformation from Sect. 4.2, where $\boldsymbol{M} = (M_{i,j})_{\ell\times s_{\max}} \in \{-1,0,1\}^{\ell\times s_{\max}} \subset \mathbb{Z}_q^{\ell\times s_{\max}}$ (Lemma 4.1) and $\rho\colon [\ell] \to \mathbb{U}$. The function ρ associates rows of $\boldsymbol{M}$ to attributes in $\mathbb{U}$. We assume that ρ is an injective function. The procedure samples vectors $\boldsymbol{s} \leftarrow \mathbb{Z}_q^n$ and $\{\boldsymbol{v}_j\}_{j\in\{2,\ldots,s_{\max}\}} \leftarrow \mathbb{Z}_q^m$. It additionally samples vectors $\{\boldsymbol{e}_i\}_{i\in[\ell]} \leftarrow \chi_{\mathsf{lwe}}^m$ and $\{\hat{\boldsymbol{e}}_i\}_{i\in[\ell]} \leftarrow \chi_{\mathsf{big}}^m$. For each $i \in [\ell]$, it computes vectors $\boldsymbol{c}_i, \hat{\boldsymbol{c}}_i \in \mathbb{Z}_q^m$ as follows:

$$\boldsymbol{c}_i = \boldsymbol{s}\boldsymbol{A}_{\rho(i)} + \boldsymbol{e}_i$$

$$\hat{\boldsymbol{c}}_i = M_{i,1}(\boldsymbol{s}\boldsymbol{y}^\top, \overbrace{0,\ldots,0}^{m-1}) + \left[\sum_{j\in\{2,\ldots,s_{\max}\}} M_{i,j}\boldsymbol{v}_j\right] - \boldsymbol{s}\boldsymbol{H}_{\rho(i)} + \hat{\boldsymbol{e}}_i$$

and outputs

$$\mathsf{CT} = \left((\boldsymbol{M}, \rho), \{\boldsymbol{c}_i\}_{i\in[\ell]}, \{\hat{\boldsymbol{c}}_i\}_{i\in[\ell]}, C = \mathsf{MSB}(\boldsymbol{s}\boldsymbol{y}^\top) \oplus \mathsf{msg}\right).$$

Dec$(\mathsf{PK}, \mathsf{CT}, \mathsf{MSK})$**:** Decryption takes as input the public parameters PK, a ciphertext CT encrypting some message under some LSSS access policy $(\boldsymbol{M}, \rho)$ with the properties stipulated in Lemma 4.1, and the secret key SK corresponding to some subset of attributes $U \subseteq \mathbb{U}$. If $(1, 0, \ldots, 0)$ is *not* in the span of the rows of $\boldsymbol{M}$ associated with U, then decryption fails. Otherwise, let I be a set of row indices of the matrix $\boldsymbol{M}$ such that $\forall i \in I\colon \rho(i) \in U$ and let $\{w_i\}_{i\in I} \in \{0,1\} \subset \mathbb{Z}_q$ be scalars such that $\sum_{i\in I} w_i\boldsymbol{M}_i = (1, 0, \ldots, 0)$, where $\boldsymbol{M}_i$ is the ith row of $\boldsymbol{M}$.

Note that the existence of such scalars $\{w_i\}_{i \in I}$ and their efficient determination for the LSSS generated by the algorithm from Sect. 4.2 are guaranteed by Lemma 4.1. The procedure computes

$$K' = \sum_{i \in I} w_i \left(\boldsymbol{c}_i \boldsymbol{k}_{\rho(i)}^\top + \hat{\boldsymbol{c}}_i \boldsymbol{t}^\top \right)$$

and outputs

$$\mathsf{msg}' = C \oplus \mathsf{MSB}(K').$$

Correctness: We show that the scheme is correct. Consider a set of attributes $U \subseteq \mathbb{U}$ and any LSSS access policy $(\boldsymbol{M}, \rho)$ for which U constitute an authorized set. By construction,

$$K' = \sum_{i \in I} w_i \left(\boldsymbol{c}_i \boldsymbol{k}_{\rho(i)}^\top + \hat{\boldsymbol{c}}_i \boldsymbol{t}^\top \right).$$

Expanding $\{\boldsymbol{c}_i\}_{i \in I}$ and $\{\hat{\boldsymbol{c}}_i\}_{i \in I}$, we get

$$\begin{aligned} K' =& \sum_{i \in I} w_i \boldsymbol{s} \boldsymbol{A}_{\rho(i)} \boldsymbol{k}_{\rho(i)}^\top + \sum_{i \in I} w_i M_{i,1} (\boldsymbol{s}\boldsymbol{y}^\top, 0, \ldots, 0) \boldsymbol{t}^\top + \\ & \sum_{i \in I, j \in \{2, \ldots, s_{\max}\}} w_i M_{i,j} \boldsymbol{v}_j \boldsymbol{t}^\top - \sum_{i \in I} w_i \boldsymbol{s} \boldsymbol{H}_{\rho(i)} \boldsymbol{t}^\top + \sum_{i \in I} w_i \boldsymbol{e}_i \boldsymbol{k}_{\rho(i)}^\top + \sum_{i \in I} w_i \hat{\boldsymbol{e}}_i \boldsymbol{t}^\top. \end{aligned}$$

Recall that for each $u \in U$, we have $\boldsymbol{k}_u = \hat{\boldsymbol{k}}_u + \tilde{\boldsymbol{k}}_u$ and $\boldsymbol{A}_u \tilde{\boldsymbol{k}}_u^\top = \boldsymbol{H}_u \boldsymbol{t}^\top - \boldsymbol{A}_u \hat{\boldsymbol{k}}_u^\top$. Therefore, for each $i \in I$, it holds that

$$\boldsymbol{A}_{\rho(i)} \boldsymbol{k}_{\rho(i)}^\top = \boldsymbol{A}_{\rho(i)} \hat{\boldsymbol{k}}_{\rho(i)}^\top + \boldsymbol{A}_{\rho(i)} \tilde{\boldsymbol{k}}_{\rho(i)}^\top = \boldsymbol{H}_{\rho(i)} \boldsymbol{t}^\top.$$

Hence,

$$\begin{aligned} K' =& \cancel{\sum_{i \in I} w_i \boldsymbol{s} \boldsymbol{H}_{\rho(i)} \boldsymbol{t}^\top} + \sum_{i \in I} w_i M_{i,1} (\boldsymbol{s}\boldsymbol{y}^\top, 0, \ldots, 0) \boldsymbol{t}^\top + \\ & \sum_{i \in I, j \in \{2, \ldots, s_{\max}\}} w_i M_{i,j} \boldsymbol{v}_j \boldsymbol{t}^\top - \cancel{\sum_{i \in I} w_i \boldsymbol{s} \boldsymbol{H}_{\rho(i)} \boldsymbol{t}^\top} + \sum_{i \in I} w_i \boldsymbol{e}_i \boldsymbol{k}_{\rho(i)}^\top + \sum_{i \in I} w_i \hat{\boldsymbol{e}}_i \boldsymbol{t}^\top \\ =& \sum_{i \in I} w_i M_{i,1} (\boldsymbol{s}\boldsymbol{y}^\top, 0, \ldots, 0) \boldsymbol{t}^\top + \sum_{i \in I, j \in \{2, \ldots, s_{\max}\}} w_i M_{i,j} \boldsymbol{v}_j \boldsymbol{t}^\top + \sum_{i \in I} w_i \boldsymbol{e}_i \boldsymbol{k}_{\rho(i)}^\top \\ & + \sum_{i \in I} w_i \hat{\boldsymbol{e}}_i \boldsymbol{t}^\top \\ =& \left(\sum_{i \in I} w_i M_{i,1} \right) (\boldsymbol{s}\boldsymbol{y}^\top, 0, \ldots, 0) \boldsymbol{t}^\top + \sum_{j \in \{2, \ldots, s_{\max}\}} \left(\sum_{i \in I} w_i M_{i,j} \right) \boldsymbol{v}_j \boldsymbol{t}^\top \\ & + \sum_{i \in I} w_i \boldsymbol{e}_i \boldsymbol{k}_{\rho(i)}^\top + \sum_{i \in I} w_i \hat{\boldsymbol{e}}_i \boldsymbol{t}^\top. \end{aligned}$$

Recall that $\sum_{i \in I} w_i M_{i,1} = 1$. Also, for $1 < j \leq s_{\max}$, $\sum_{i \in I} w_i M_{i,j} = 0$. Additionally, $\boldsymbol{t} = (1, \hat{\boldsymbol{t}})$, and hence, $(\boldsymbol{s}\boldsymbol{y}^\top, 0, \ldots, 0)\boldsymbol{t}^\top = \boldsymbol{s}\boldsymbol{y}^\top$. Thus,

$$K' = \boldsymbol{s}\boldsymbol{y}^\top + \sum_{i \in I} w_i \boldsymbol{e}_i \boldsymbol{k}_{\rho(i)}^\top + \sum_{i \in I} w_i \hat{\boldsymbol{e}}_i \boldsymbol{t}^\top.$$

Correctness now follows since the last two terms are small and should not affect the MSB of $\boldsymbol{s}\boldsymbol{y}^\top$. To see this, we observe that the following inequalities hold except with negligible probability:

- $\|\boldsymbol{e}_i\| \leq \sqrt{m}\sigma$: This follows directly from Lemma 3.3 since each of the m coordinates of $\boldsymbol{e}_i$ comes from the truncated discrete Gaussian distribution $\widetilde{\mathcal{D}}_{\mathbb{Z},\sigma}$.
- $\|\hat{\boldsymbol{e}}_i\| \leq \sqrt{m}\hat{B}$: This holds since each of the m coordinates of $\hat{\boldsymbol{e}}_i$ comes from the uniform distribution over $\mathbb{Z} \cap [-\hat{B}, \hat{B}]$.
- $\|\boldsymbol{k}_{\rho(i)}\| \leq m\sigma + \sqrt{m}\hat{B}$: This holds since $\boldsymbol{k}_{\rho(i)} = \hat{\boldsymbol{k}}_{\rho(i)} + \tilde{\boldsymbol{k}}_{\rho(i)}$, where (1) $\|\hat{\boldsymbol{k}}_{\rho(i)}\| \leq \sqrt{m}\hat{B}$ since each of its m coordinates comes from the uniform distribution over $\mathbb{Z} \cap [-\hat{B}, \hat{B}]$ and (2) $\|\tilde{\boldsymbol{k}}_{\rho(i)}\| \leq m\sigma$ since it comes from a distribution that is statistically close to the truncated discrete Gaussian distribution $\widetilde{\mathcal{D}}_{\mathbb{Z}^m,\sigma}$.
- $\|\boldsymbol{t}\| < m\sigma$: This holds since $\boldsymbol{t} = (1, \hat{\boldsymbol{t}})$, where $\hat{\boldsymbol{t}}$ comes from a truncated discrete Gaussian distribution $\widetilde{\mathcal{D}}_{\mathbb{Z}^{m-1},\sigma}$.

Using the fact that the w_i's are in $\{0, 1\}$ (Lemma 4.1), we have that

$$\begin{aligned} \|\sum_{i \in I} w_i \boldsymbol{e}_i \boldsymbol{k}_{\rho(i)}^\top + \sum_{i \in I} w_i \hat{\boldsymbol{e}}_i \boldsymbol{t}^\top\| &< |\mathbb{U}| \left(m^{3/2}\sigma^2 + m\sigma\hat{B} + m^{3/2}\sigma\hat{B}\right) \\ &< |\mathbb{U}| \cdot 3m^{3/2}\sigma\hat{B} < q/4, \end{aligned}$$

where the last inequality is by the parameter setting as shown above. Thus, with all but negligible probability in λ, the MSB of $\boldsymbol{s}\boldsymbol{y}^\top$ is not affected by the above noise which is bounded by $q/4$ and therefore does not affect the most significant bit. Namely, $\mathsf{MSB}(K') = \mathsf{MSB}(\boldsymbol{s}\boldsymbol{y}^\top)$. This completes the proof of correctness.

6 Our Multi-authority ABE Scheme

In this section, we present our MA-ABE scheme for access structures represented by DNF formulas. The scheme is associated with a universe of global identifiers $\mathcal{GID} \subset \{0,1\}^*$, a universe of authority identifiers $\mathcal{AU}$, and we will use the Lewko-Waters [42] transformation to represent the DNF access policies as monotone LSSS. More precisely, we only design an MA-ABE scheme for LSSS access policies $(\boldsymbol{M}, \rho)$ with properties stipulated in Lemma 4.1, that is, we construct an MA-ABE scheme for LSSS access policies $(\boldsymbol{M}, \rho)$ such that the entries of $\boldsymbol{M}$ come from $\{-1, 0, 1\}$ as well as reconstruction only involves coefficients coming from $\{0, 1\}$, and prove the scheme to be statically secure under linear independence restriction (see [25, Definition 3.7] in the full version for the formal description of the security model). Thanks to the observation made by [1] as mentioned in Remark 4.1, our MA-ABE scheme actually achieves the standard notion of static security (see [25, Definition 3.6] in the full version for the formal description of the security model) when implemented for the class of all access structures represented by DNF formulas. We will assume each authority controls only one attribute in our scheme. However, it can be readily generalized to a scheme where each authority controls an a priori bounded number of attributes using standard techniques [42]. Further, we will assume that all access policies $(\boldsymbol{M}, \rho)$ used in our scheme correspond to a matrix $\boldsymbol{M}$ with at most $s_{\max}$ columns and an injective row-labeling function ρ, i.e., an authority/attribute is associated with at most one row of $\boldsymbol{M}$. Since the Lewko-Waters transformation [42] introduces a new column for the resulting LSSS matrix for each AND gate in the underlying formula, the bound in the number of columns of the LSSS matrices naturally translates to the number of AND gates of the supported DNF formulas at implementation. Similar to our CP-ABE scheme, in our scheme description below, we assume for simplicity of presentation that both the encryption and the decryption algorithms receive an access policy directly in its LSSS representation. However, we note that in the actual implementation, the encryption and decryption algorithms should instead take in the DNF representation of the access policy and deterministically compute its LSSS representation using the Lewko-Waters transformation algorithm [42].

First, we provide the parameter constraints required by our correctness and security proof. Fix any $0 < \epsilon < 1/2$. For any $B \in \mathbb{N}$, let $\mathcal{U}_B$ denote the uniform distribution on $\mathbb{Z} \cap [-B, B]$, i.e., integers between $\pm B$. The Setup algorithm chooses parameters n, m, σ, q and noise distributions $\chi_{\mathsf{lwe}}, \chi_1, \chi_2, \chi_{\mathsf{big}}$, satisfying the following constraints:

- $n = \mathsf{poly}(\lambda)$, $\sigma < q$, $n \cdot q/\sigma < 2^{n^\epsilon}$, $\chi_{\mathsf{lwe}} = \widetilde{\mathcal{D}}_{\mathbb{Z},\sigma}$ (for LWE security)
- $m > 2s_{\max} n \log q + \omega \log n + 2\lambda$ (for enhanced trapdoor sampling and LHL)
- $\sigma > \sqrt{s_{\max} n \log q \log m} + \lambda$ (for enhanced trapdoor sampling)
- $\chi_1 = \widetilde{\mathcal{D}}_{\mathbb{Z}^{m-1},\sigma}$, $\chi_2 = \widetilde{\mathcal{D}}_{\mathbb{Z}^m,\sigma}$ (for enhanced trapdoor sampling)
- $\chi_{\mathsf{big}} = \mathcal{U}_{\hat{B}}$, where $\hat{B} > m^{3/2}\sigma 2^\lambda$ (for smudging/security)
- $|\mathcal{AU}| (m^{3/2}\sigma^2 + 2m\hat{B}^2) < q/4$ (for correctness)

We will now describe our MA-ABE construction.

GlobalSetup$(1^\lambda, s_{\max})$: The global setup algorithm takes in the security parameter λ encoded in unary and the maximum width $s_{\max} = s_{\max}(\lambda)$ of an LSSS matrix supported by the scheme. It first chooses an LWE modulus q, dimensions n, m, and also distributions $\chi_{\mathsf{lwe}}, \chi_1, \chi_2, \chi_{\mathsf{big}}$ as described above. Next, it samples a vector $\boldsymbol{y} \leftarrow \mathbb{Z}_q^n$ and sets the matrix $\boldsymbol{B}_1 \in \mathbb{Z}_q^{n \times m}$ as $\boldsymbol{B}_1 = \left[\boldsymbol{y}^\top \| \overbrace{\boldsymbol{0}^\top \| \cdots \| \boldsymbol{0}^\top}^{m-1}\right]$, where each $\boldsymbol{0} \in \mathbb{Z}_q^n$. Furthermore, we assume a hash function $\mathsf{H}\colon \mathcal{GID} \to (\mathbb{Z} \cap [-\hat{B}, \hat{B}])^{m-1}$ mapping strings $\mathsf{GID} \in \mathcal{GID}$ to random $(m-1)$-dimensional vectors of integers in the interval $[-\hat{B}, \hat{B}]$. H will be modeled as a random oracle in the security proof. Finally, it outputs the hash function H and the global parameters

$$\mathsf{GP} = (n, m, q, s_{\max}, \chi_{\mathsf{lwe}}, \chi_1, \chi_2, \chi_{\mathsf{big}}, \boldsymbol{B}_1).$$

AuthSetup(GP, H, u): Given the global parameters GP, the hash function H, and an authority identifier $u \in \mathcal{AU}$, the algorithm generates a matrix-trapdoor pair $(\boldsymbol{A}_u, T_{\boldsymbol{A}_u}) \leftarrow \mathsf{EnTrapGen}(1^n, 1^m, q)$ such that $\boldsymbol{A}_u \in \mathbb{Z}_q^{n\times m}$, samples another matrix $\boldsymbol{H}_u \leftarrow \mathbb{Z}_q^{n\times m}$, and outputs the pair of public key and secret key for the authority u

$$\mathsf{PK}_u = (\boldsymbol{A}_u, \boldsymbol{H}_u), \quad \mathsf{MSK}_u = T_{\boldsymbol{A}_u}.$$

KeyGen(GP, H, GID, MSK$_u$): The key generation algorithm takes as input the global parameters GP, the hash function H, the user's global identifier GID, and the authority's secret key MSK_u. It first computes the vector $\boldsymbol{t}_{\mathsf{GID}} = (1, \mathsf{H}(\mathsf{GID})) \in \mathbb{Z}^m$. Next, it chooses a vector $\hat{\boldsymbol{k}}_{\mathsf{GID},u} \leftarrow \chi_{\mathsf{big}}^m$, samples a vector $\tilde{\boldsymbol{k}}_{\mathsf{GID},u} \leftarrow \mathsf{EnSamplePre}(\boldsymbol{A}_u, T_{\boldsymbol{A}_u}, \sigma, \boldsymbol{t}_{\mathsf{GID}}\boldsymbol{H}_u^\top - \hat{\boldsymbol{k}}_{\mathsf{GID},u}\boldsymbol{A}_u^\top)$, and outputs the secret key for the user GID as

$$\mathsf{SK}_{\mathsf{GID},u} = \hat{\boldsymbol{k}}_{\mathsf{GID},u} + \tilde{\boldsymbol{k}}_{\mathsf{GID},u}.$$

Enc(GP, H, msg, $(\boldsymbol{M}, \rho)$, {PK$_u$}): The encryption algorithm takes as input the global parameters GP, the hash function H, a message bit $\mathsf{msg} \in \{0,1\}$ to encrypt, an LSSS access policy $(\boldsymbol{M}, \rho)$ generated by the Lewko-Waters transformation [42], where $\boldsymbol{M} = (M_{i,j})_{\ell\times s_{\max}} \in \{-1,0,1\}^{\ell\times s_{\max}} \subset \mathbb{Z}_q^{\ell\times s_{\max}}$ (Lemma 4.1) and $\rho\colon [\ell] \to \mathcal{AU}$, and public keys of the relevant authorities $\{\mathsf{PK}_u\}$. The function ρ associates rows of $\boldsymbol{M}$ to authorities (recall that we assume that each authority controls a single attribute). We assume that ρ is an injective function. The procedure samples vectors $\boldsymbol{s} \leftarrow \mathbb{Z}_q^n$, $\{\boldsymbol{v}_j\}_{j\in\{2,\ldots,s_{\max}\}} \leftarrow \mathbb{Z}_q^m$, and $\{\boldsymbol{x}_i\}_{i\in[\ell]} \leftarrow \mathbb{Z}_q^n$. It additionally samples vectors $\{\boldsymbol{e}_i\}_{i\in[\ell]} \leftarrow \chi_{\mathsf{lwe}}^m$ and $\{\hat{\boldsymbol{e}}_i\}_{i\in[\ell]} \leftarrow \chi_{\mathsf{big}}^m$. For each $i \in [\ell]$, it computes vectors $\boldsymbol{c}_i, \hat{\boldsymbol{c}}_i \in \mathbb{Z}_q^m$ as follows:

$$\boldsymbol{c}_i = \boldsymbol{x}_i\boldsymbol{A}_{\rho(i)} + \boldsymbol{e}_i$$

$$\hat{\boldsymbol{c}}_i = M_{i,1}\boldsymbol{s}\boldsymbol{B}_1 + \left[\sum_{j\in\{2,\ldots,s_{\max}\}} M_{i,j}\boldsymbol{v}_j\right] - \boldsymbol{x}_i\boldsymbol{H}_{\rho(i)} + \hat{\boldsymbol{e}}_i$$

and outputs

$$\mathsf{CT} = \Big((\boldsymbol{M}, \rho), \{\boldsymbol{c}_i\}_{i\in[\ell]}, \{\hat{\boldsymbol{c}}_i\}_{i\in[\ell]}, C = \mathsf{MSB}(\boldsymbol{s}\boldsymbol{y}^\top) \oplus \mathsf{msg}\Big).$$

Dec(GP, H, CT, GID, {SK$_{\mathsf{GID},u}$}): Decryption takes as input the global parameters GP, the hash function H, a ciphertext CT generated with respect to an LSSS access policy $(\boldsymbol{M}, \rho)$ generated by the Lewko-Waters transformation [42], a user identity GID, and the secret keys $\{\mathsf{SK}_{\mathsf{GID},\rho(i)}\}_{i\in I}$ corresponding to a subset I of row indices of the access matrix $\boldsymbol{M}$ possessed by that user. If $(1, 0, \ldots, 0)$ is *not* in the span of the rows of $\boldsymbol{M}$ having indices in the set I, then decryption fails. Otherwise, let $\{w_i\}_{i\in I} \in \{0, 1\} \subset \mathbb{Z}_q$ be scalars such that $\sum_{i\in I} w_i \boldsymbol{M}_i = (1, 0, \ldots, 0)$, where $\boldsymbol{M}_i$ is the ith row of $\boldsymbol{M}$. The existence of such scalars $\{w_i\}_{i\in I}$ and their efficient determination are guaranteed by [1,14,42]. The algorithm computes the vector $\boldsymbol{t}_{\mathsf{GID}} = (1, \mathsf{H}(\mathsf{GID})) \in \mathbb{Z}^m$ followed by

$$K' = \sum_{i\in I} w_i \cdot \left(\boldsymbol{c}_i \mathsf{SK}_{\mathsf{GID},\rho(i)}^{\top} + \hat{\boldsymbol{c}}_i \boldsymbol{t}_{\mathsf{GID}}^{\top}\right),$$

and outputs

$$\mathsf{msg}' = C \oplus \mathsf{MSB}(K').$$

References

1. Agrawal, S., Biswas, R., Nishimaki, R., Xagawa, K., Xie, X., Yamada, S.: Cryptanalysis of Boyen's attribute-based encryption scheme in TCC 2013 (2020, private communication)
2. Agrawal, S., Boneh, D., Boyen, X.: Efficient lattice (H)IBE in the standard model. In: Gilbert, H. (ed.) EUROCRYPT 2010. LNCS, vol. 6110, pp. 553–572. Springer, Heidelberg (2010). https://doi.org/10.1007/978-3-642-13190-5_28
3. Agrawal, S., Boneh, D., Boyen, X.: Lattice basis delegation in fixed dimension and shorter-ciphertext hierarchical IBE. In: Rabin, T. (ed.) CRYPTO 2010. LNCS, vol. 6223, pp. 98–115. Springer, Heidelberg (2010). https://doi.org/10.1007/978-3-642-14623-7_6
4. Agrawal, S., Freeman, D.M., Vaikuntanathan, V.: Functional encryption for inner product predicates from learning with errors. In: Lee, D.H., Wang, X. (eds.) ASIACRYPT 2011. LNCS, vol. 7073, pp. 21–40. Springer, Heidelberg (2011). https://doi.org/10.1007/978-3-642-25385-0_2
5. Agrawal, S., Maitra, M., Yamada, S.: Attribute based encryption (and more) for nondeterministic finite automata from LWE. In: Boldyreva, A., Micciancio, D. (eds.) CRYPTO 2019. LNCS, vol. 11693, pp. 765–797. Springer, Cham (2019). https://doi.org/10.1007/978-3-030-26951-7_26
6. Agrawal, S., Wichs, D., Yamada, S.: Optimal broadcast encryption from LWE and pairings in the standard model. In: Pass, R., Pietrzak, K. (eds.) TCC 2020. LNCS, vol. 12550, pp. 149–178. Springer, Cham (2020). https://doi.org/10.1007/978-3-030-64375-1_6
7. Agrawal, S., Yamada, S.: CP-ABE for circuits (and more) in the symmetric key setting. In: Pass, R., Pietrzak, K. (eds.) TCC 2020. LNCS, vol. 12550, pp. 117–148. Springer, Cham (2020). https://doi.org/10.1007/978-3-030-64375-1_5
8. Agrawal, S., Yamada, S.: Optimal broadcast encryption from pairings and LWE. In: Canteaut, A., Ishai, Y. (eds.) EUROCRYPT 2020. LNCS, vol. 12105, pp. 13–43. Springer, Cham (2020). https://doi.org/10.1007/978-3-030-45721-1_2
9. Ajtai, M.: Generating hard instances of the short basis problem. In: Wiedermann, J., van Emde Boas, P., Nielsen, M. (eds.) ICALP 1999. LNCS, vol. 1644, pp. 1–9. Springer, Heidelberg (1999). https://doi.org/10.1007/3-540-48523-6_1

10. Asharov, G., Jain, A., Wichs, D.: Multiparty computation with low communication, computation and interaction via threshold FHE. IACR Cryptology ePrint Archive 2011/613 (2011)
11. Attrapadung, N.: Unbounded dynamic predicate compositions in attribute-based encryption. In: Ishai, Y., Rijmen, V. (eds.) EUROCRYPT 2019. LNCS, vol. 11476, pp. 34–67. Springer, Cham (2019). https://doi.org/10.1007/978-3-030-17653-2_2
12. Benaloh, J., Leichter, J.: Generalized secret sharing and monotone functions. In: Goldwasser, S. (ed.) CRYPTO 1988. LNCS, vol. 403, pp. 27–35. Springer, New York (1990). https://doi.org/10.1007/0-387-34799-2_3
13. Bethencourt, J., Sahai, A., Waters, B.: Ciphertext-policy attribute-based encryption. In: S&P 2007, pp. 321–334. IEEE (2007)
14. Boneh, D., et al.: Threshold cryptosystems from threshold fully homomorphic encryption. In: Shacham, H., Boldyreva, A. (eds.) CRYPTO 2018. LNCS, vol. 10991, pp. 565–596. Springer, Cham (2018). https://doi.org/10.1007/978-3-319-96884-1_19
15. Boneh, D., et al.: Fully key-homomorphic encryption, arithmetic circuit ABE and compact garbled circuits. In: Nguyen, P.Q., Oswald, E. (eds.) EUROCRYPT 2014. LNCS, vol. 8441, pp. 533–556. Springer, Heidelberg (2014). https://doi.org/10.1007/978-3-642-55220-5_30
16. Boyen, X.: Attribute-based functional encryption on lattices. In: Sahai, A. (ed.) TCC 2013. LNCS, vol. 7785, pp. 122–142. Springer, Heidelberg (2013). https://doi.org/10.1007/978-3-642-36594-2_8
17. Brakerski, Z., Langlois, A., Peikert, C., Regev, O., Stehlé, D.: Classical hardness of learning with errors. In: STOC 2013, pp. 575–584. ACM (2013)
18. Brakerski, Z., Vaikuntanathan, V.: Lattice-inspired broadcast encryption and succinct ciphertext-policy ABE. IACR Cryptology ePrint Archive 2020/191 (2020)
19. Brakerski, Z., Vaikuntanathan, V.: Circuit-ABE from LWE: unbounded attributes and semi-adaptive security. In: Robshaw, M., Katz, J. (eds.) CRYPTO 2016. LNCS, vol. 9816, pp. 363–384. Springer, Heidelberg (2016). https://doi.org/10.1007/978-3-662-53015-3_13
20. Cash, D., Hofheinz, D., Kiltz, E., Peikert, C.: Bonsai trees, or how to delegate a lattice basis. In: Gilbert, H. (ed.) EUROCRYPT 2010. LNCS, vol. 6110, pp. 523–552. Springer, Heidelberg (2010). https://doi.org/10.1007/978-3-642-13190-5_27
21. Chase, M.: Multi-authority attribute based encryption. In: Vadhan, S.P. (ed.) TCC 2007. LNCS, vol. 4392, pp. 515–534. Springer, Heidelberg (2007). https://doi.org/10.1007/978-3-540-70936-7_28
22. Chase, M., Chow, S.S.M.: Improving privacy and security in multi-authority attribute-based encryption. In: CCS 2009, pp. 121–130. ACM (2009)
23. Chen, J., Gay, R., Wee, H.: Improved dual system ABE in prime-order groups via predicate encodings. In: Oswald, E., Fischlin, M. (eds.) EUROCRYPT 2015. LNCS, vol. 9057, pp. 595–624. Springer, Heidelberg (2015). https://doi.org/10.1007/978-3-662-46803-6_20
24. Chen, J., Gong, J., Kowalczyk, L., Wee, H.: Unbounded ABE via bilinear entropy expansion, revisited. In: Nielsen, J.B., Rijmen, V. (eds.) EUROCRYPT 2018. LNCS, vol. 10820, pp. 503–534. Springer, Cham (2018). https://doi.org/10.1007/978-3-319-78381-9_19
25. Datta, P., Komargodski, I., Waters, B.: Decentralized multi-authority ABE for DNFs from LWE. IACR Cryptology ePrint Archive 2020/1386 (2020)
26. Diffie, W., Hellman, M.E.: New directions in cryptography. IEEE Trans. Inf. Theory **22**(6), 644–654 (1976)

27. Garg, S., Gentry, C., Halevi, S., Sahai, A., Waters, B.: Attribute-based encryption for circuits from multilinear maps. In: Canetti, R., Garay, J.A. (eds.) CRYPTO 2013. LNCS, vol. 8043, pp. 479–499. Springer, Heidelberg (2013). https://doi.org/10.1007/978-3-642-40084-1_27
28. Gentry, C., Gorbunov, S., Halevi, S.: Graph-induced multilinear maps from lattices. In: Dodis, Y., Nielsen, J.B. (eds.) TCC 2015. LNCS, vol. 9015, pp. 498–527. Springer, Heidelberg (2015). https://doi.org/10.1007/978-3-662-46497-7_20
29. Gentry, C., Peikert, C., Vaikuntanathan, V.: Trapdoors for hard lattices and new cryptographic constructions. In: STOC 2008, pp. 197–206. ACM (2008)
30. Gentry, C., Sahai, A., Waters, B.: Homomorphic encryption from learning with errors: conceptually-simpler, asymptotically-faster, attribute-based. In: Canetti, R., Garay, J.A. (eds.) CRYPTO 2013. LNCS, vol. 8042, pp. 75–92. Springer, Heidelberg (2013). https://doi.org/10.1007/978-3-642-40041-4_5
31. Gong, J., Waters, B., Wee, H.: ABE for DFA from k-Lin. In: Boldyreva, A., Micciancio, D. (eds.) CRYPTO 2019. LNCS, vol. 11693, pp. 732–764. Springer, Cham (2019). https://doi.org/10.1007/978-3-030-26951-7_25
32. Gong, J., Wee, H.: Adaptively secure ABE for DFA from k-Lin and more. In: Canteaut, A., Ishai, Y. (eds.) EUROCRYPT 2020. LNCS, vol. 12107, pp. 278–308. Springer, Cham (2020). https://doi.org/10.1007/978-3-030-45727-3_10
33. Gorbunov, S., Vaikuntanathan, V., Wee, H.: Attribute-based encryption for circuits. In: STOC 2013, pp. 545–554. ACM (2013)
34. Gorbunov, S., Vinayagamurthy, D.: Riding on asymmetry: efficient ABE for branching programs. In: Iwata, T., Cheon, J.H. (eds.) ASIACRYPT 2015. LNCS, vol. 9452, pp. 550–574. Springer, Heidelberg (2015). https://doi.org/10.1007/978-3-662-48797-6_23
35. Goyal, R., Koppula, V., Waters, B.: Collusion resistant traitor tracing from learning with errors. In: STOC 2018, pp. 660–670. ACM (2018)
36. Goyal, R., Koppula, V., Waters, B.: Lockable obfuscation. In: FOCS 2017, pp. 612–621. IEEE (2017)
37. Goyal, V., Pandey, O., Sahai, A., Waters, B.: Attribute-based encryption for fine-grained access control of encrypted data. In: CCS 2006, pp. 89–98. ACM (2006)
38. Karchmer, M., Wigderson, A.: On span programs. In: Structure in Complexity Theory Conference 1993, pp. 102–111. IEEE (1993)
39. Kim, S.: Multi-authority attribute-based encryption from LWE in the OT model. IACR Cryptology ePrint Archive 2019/280 (2019)
40. Kowalczyk, L., Wee, H.: Compact adaptively secure ABE for NC^1 from k-Lin. In: Ishai, Y., Rijmen, V. (eds.) EUROCRYPT 2019. LNCS, vol. 11476, pp. 3–33. Springer, Cham (2019). https://doi.org/10.1007/978-3-030-17653-2_1
41. Lewko, A., Okamoto, T., Sahai, A., Takashima, K., Waters, B.: Fully secure functional encryption: attribute-based encryption and (hierarchical) inner product encryption. In: Gilbert, H. (ed.) EUROCRYPT 2010. LNCS, vol. 6110, pp. 62–91. Springer, Heidelberg (2010). https://doi.org/10.1007/978-3-642-13190-5_4
42. Lewko, A., Waters, B.: Decentralizing attribute-based encryption. In: Paterson, K.G. (ed.) EUROCRYPT 2011. LNCS, vol. 6632, pp. 568–588. Springer, Heidelberg (2011). https://doi.org/10.1007/978-3-642-20465-4_31
43. Lewko, A., Waters, B.: Unbounded HIBE and attribute-based encryption. In: Paterson, K.G. (ed.) EUROCRYPT 2011. LNCS, vol. 6632, pp. 547–567. Springer, Heidelberg (2011). https://doi.org/10.1007/978-3-642-20465-4_30

44. Lin, H., Cao, Z., Liang, X., Shao, J.: Secure threshold multi authority attribute based encryption without a central authority. In: Chowdhury, D.R., Rijmen, V., Das, A. (eds.) INDOCRYPT 2008. LNCS, vol. 5365, pp. 426–436. Springer, Heidelberg (2008). https://doi.org/10.1007/978-3-540-89754-5_33
45. Micciancio, D., Peikert, C.: Trapdoors for lattices: simpler, tighter, faster, smaller. In: Pointcheval, D., Johansson, T. (eds.) EUROCRYPT 2012. LNCS, vol. 7237, pp. 700–718. Springer, Heidelberg (2012). https://doi.org/10.1007/978-3-642-29011-4_41
46. Micciancio, D., Peikert, C.: Hardness of SIS and LWE with small parameters. In: Canetti, R., Garay, J.A. (eds.) CRYPTO 2013. LNCS, vol. 8042, pp. 21–39. Springer, Heidelberg (2013). https://doi.org/10.1007/978-3-642-40041-4_2
47. Micciancio, D., Regev, O.: Worst-case to average-case reductions based on gaussian measures. SIAM J. Comput. **37**(1), 267–302 (2007)
48. Müller, S., Katzenbeisser, S., Eckert, C.: Distributed attribute-based encryption. In: Lee, P.J., Cheon, J.H. (eds.) ICISC 2008. LNCS, vol. 5461, pp. 20–36. Springer, Heidelberg (2009). https://doi.org/10.1007/978-3-642-00730-9_2
49. Müller, S., Katzenbeisser, S., Eckert, C.: On multi-authority ciphertext-policy attribute-based encryption. Bull. Korean Math. Soc. **46**, 803–819 (2009)
50. Ostrovsky, R., Sahai, A., Waters, B.: Attribute-based encryption with non-monotonic access structures. In: CCS 2007, pp. 195–203. ACM (2007)
51. Peikert, C.: Public-key cryptosystems from the worst-case shortest vector problem: extended abstract. In: STOC 2009, pp. 333–342. ACM (2009)
52. Regev, O.: On lattices, learning with errors, random linear codes, and cryptography. In: STOC 2005, pp. 84–93. ACM (2005)
53. Rouselakis, Y., Waters, B.: Efficient statically-secure large-universe multi-authority attribute-based encryption. In: Böhme, R., Okamoto, T. (eds.) FC 2015. LNCS, vol. 8975, pp. 315–332. Springer, Heidelberg (2015). https://doi.org/10.1007/978-3-662-47854-7_19
54. Sahai, A., Waters, B.: Fuzzy identity-based encryption. In: Cramer, R. (ed.) EUROCRYPT 2005. LNCS, vol. 3494, pp. 457–473. Springer, Heidelberg (2005). https://doi.org/10.1007/11426639_27
55. Shor, P.W.: Algorithms for quantum computation: discrete logarithms and factoring. In: FOCS 1994, pp. 124–134. IEEE (1994)
56. Tsabary, R.: Fully secure attribute-based encryption for t-CNF from LWE. In: Boldyreva, A., Micciancio, D. (eds.) CRYPTO 2019. LNCS, vol. 11692, pp. 62–85. Springer, Cham (2019). https://doi.org/10.1007/978-3-030-26948-7_3
57. Wang, Z., Fan, X., Liu, F.-H.: FE for inner products and its application to decentralized ABE. In: Lin, D., Sako, K. (eds.) PKC 2019. LNCS, vol. 11443, pp. 97–127. Springer, Cham (2019). https://doi.org/10.1007/978-3-030-17259-6_4
58. Waters, B.: Ciphertext-policy attribute-based encryption: an expressive, efficient, and provably secure realization. In: Catalano, D., Fazio, N., Gennaro, R., Nicolosi, A. (eds.) PKC 2011. LNCS, vol. 6571, pp. 53–70. Springer, Heidelberg (2011). https://doi.org/10.1007/978-3-642-19379-8_4

Isogenies

Compact, Efficient and UC-Secure Isogeny-Based Oblivious Transfer

Yi-Fu Lai[1(✉)], Steven D. Galbraith[1], and Cyprien Delpech de Saint Guilhem[2]

[1] University of Auckland, Auckland, New Zealand
ylai276@aucklanduni.ac.nz, s.galbraith@auckland.ac.nz
[2] imec-COSIC, KU Leuven, Leuven, Belgium
cyprien.delpechdesaintguilhem@kuleuven.be

Abstract. Oblivious transfer (OT) is an essential cryptographic tool that can serve as a building block for almost all secure multiparty functionalities. The strongest security notion against malicious adversaries is universal composability (UC-secure). An important goal is to have post-quantum OT protocols. One area of interest for post-quantum cryptography is isogeny-based crypto. Isogeny-based cryptography has some similarities to Diffie-Hellman, but lacks some algebraic properties that are needed for discrete-log-based OT protocols. Hence it is not always possible to directly adapt existing protocols to the isogeny setting.

We propose the first practical isogeny-based UC-secure oblivious transfer protocol in the presence of malicious adversaries. Our scheme uses the CSIDH framework and does not have an analogue in the Diffie-Hellman setting. The scheme consists of a constant number of isogeny computations. The underlying computational assumption is a problem that we call the computational reciprocal CSIDH problem, and that we prove polynomial-time equivalent to the computational CSIDH problem.

1 Introduction

Oblivious transfer (OT) was first introduced by Rabin [35] in 1981 to establish an exchange of secrets protocol based on the factoring problem. Say the sender has two messages, oblivious transfer allows the receiver to know one of them and keeps the sender oblivious to which message has been received. The unchosen message remains unknown to the receiver.

It has been shown that oblivious transfer is an important building block as a cryptographic tool. Oblivious transfer can be used to construct other cryptographic primitives [15,22,32]. Several oblivious transfer protocols based on Diffie-Hellman-related problems were proposed [3,4,13,30,34].

Oblivious transfer protocols exist for various hardness assumptions. However, cryptographic protocols based on problems subordinate to either the discrete logarithm problem or the factoring problem will suffer a polynomial-time quantum attack from Shor's algorithm [37]. Several post-quantum oblivious transfer schemes have been proposed, including Peikert et al.'s lattice-based OT [34],

A. Canteaut and F.-X. Standaert (Eds.): EUROCRYPT 2021, LNCS 12696, pp. 213–241, 2021.
https://doi.org/10.1007/978-3-030-77870-5_8

and code-based OTs [3,16,19]. Recently, some isogeny-based OTs have been proposed [2,17,38].

Concerning security of oblivious transfer, traditional security definitions aim at guaranteeing privacy for both parties, including one-sided simulation and the view-based definition for a two-message protocol [19,24,30]. These notions ensure privacy for both parties in a standalone setting. However, in real world deployment, protocols are always composed into an enormous and complex construction. To ensure security of the full system the leading oblivious transfer proposals [3,16,28,34] follow the real/ideal paradigm and universally-composable security (UC security) as defined by Canetti [9]. Impossibility results for some protocols have been given in [11].

The first isogeny-based cryptosystem was proposed by Couveignes [14], which included a key exchange scheme based on a hard homogeneous space. However, the paper was not published at that time. The approach was independently rediscovered by Rostovtsev and Stolbunov [36]. Then, Jao and De Feo proposed the Supersingular Isogeny Diffie Hellman (SIDH) [26]. Later, SIDH was transformed into the Supersingular Isogeny Key Encapsulation (SIKE) [25] which includes a public key encryption scheme and a key encapsulation mechanism and is now one of the third-round alternate candidates in the post-quantum cryptography standardization competition led by NIST [31]. Castryck et al. [12] devised an efficient implementation of the Couveignes/Rostovtsev-Stolbunov approach that they called commutative SIDH (CSIDH). CSIDH is conjectured to provide post-quantum security with smaller public keys than the candidates in the NIST competition [31]. In this work, we exploit the structure of CSIDH to construct our schemes.

In comparison with Diffie-Hellman-based protocols, due to the reduced number of algebraic operations available, it is arguably more challenging to develop isogeny-based cryptosystems achieving the desired notion. For example, neither the randomizing procedure $(g^s y^t, x^s z^t) \leftarrow RAND(g, x, y, z; s, t \leftarrow_\$ \mathbb{F}_p)$ used in [28,30,34], nor the fundamental one-trapdoor setup $pk_0 pk_1 = c$ where c is a public constant used in [3,30] can be realized in the isogeny-based setting with current techniques in an efficient way.

We review the aforementioned isogeny-based oblivious transfer proposals [2,17,38]. Their schemes can be viewed as "tweaks of two Diffie-Hellman key exchange agreements" or variants of the Diffie-Hellman problem as stated in [13]. As stated in [13], these schemes, including their scheme, cannot achieve full-simulatablility, even in the sense of the sequential composition (SC) theorem [8]. This is because a simulator cannot extract the input of the malicious adversary who delays the decryption.

To get a secure OT against malicious adversaries, an inefficient solution is embedding a zero-knowledge proof to force the adversary to follow the protocol specification, which is the idea of the GMW compiler [22]. But then using isogenies the ZK proof may require a polynomial number of isogeny computations [6,21]. Another solution is using the transformation provided by Döttling et al. [18]. The mechanism can transform a two-round semi-simulatable OT into one secure against malicious adversaries and keep the construction round-optimal (2-round). The cost is a polynomial number of executions of the original OT scheme. Chou and Orlandi [13] pointed out another potential solution that the receiver should show a proof of timely decryption while not leaking the secret input, which was realized by Barreto et al. in their framework in the updated version of [3].

On the aspect of security, the schemes [2,17,38] are all only proved secure under either semi-honest models or a non-simulation-based definition. In other words, the schemes all ensure nothing when executed within a larger environment against malicious adversaries. Regarding the underlying computational assumption, a reduction to a well-known one is a preferred choice over a reduction to a weaker variant. For instance, the scheme in [38], as stated in their work, relies on a non-standard computational assumption that does not hold in the CSIDH setting. In conclusion, before our work, a practical isogeny-based oblivious transfer protocol proved to be secure under an assumption equivalent to a well-known problem and with respect to a powerful security notion was missing from the literature.

1.1 Contributions

We present the first practical construction of a UC-secure isogeny-based oblivious transfer protocol in the presence of malicious adversaries, hence resolving all the issues discussed above. To achieve this we introduce a variety of novel techniques. The construction is not only compact with a constant number of isogeny computations (see Table 1) but also a robust scheme without compromising the hardness. Our schemes use a feature that is available for isogenies (the quadratic twist) that does not have an analogue in the DLP setting, but some of our other techniques are not limited to isogeny-based cryptography, see Sect. 6.

Firstly, we design a novel 1-out-of-2 oblivious transfer protocol by a small change to the Diffie-Hellman protocol to achieve a compact OT prototype with a trusted setup curve (or a public curve). Next, the 3-round protocol is transformed into a 2-round scheme through a new use of quadratic twists. The 2-round scheme is the most efficient isogeny-based OT scheme in the semi-honest model so far, see Sect. 5. Based on this modification, a secure mechanism can be established in which the receiver will demonstrate the "ability to decrypt" to the sender for one-sided simulation, which is based on a similar idea of [3,13] but with a different mechanism for group actions. Furthermore, we establish a trapdoor algorithm with a novel use of quadratic twists in the setup to accomplish the fully-simulatable construction. Finally, we introduce a new assumption, the reciprocal CSIDH problem, (Problem 5) that looks non-standard, but we prove equivalence to the computational CSIDH problem with a quantum reduction using a tool we call "self-reconciling" (Proposition 1).

As pointed out by Canetti et al. [11], it is impossible to achieve a UC-secure OT scheme without any trusted setup. Our construction is proposed in a hybrid model with two functionalities, see Table 2 for comparison with the related works concerning the hybrid models.

This paper is organized as follows. Section 2 briefly describes CSIDH, the related functionalities, our new assumptions, and recalls the simulation-based definition for two-party protocols. Section 3 constructs our oblivious transfer protocol. Section 4 gives security proofs against the semi-honest adversary and the malicious adversary. A comparison of our OT with the previous three isogeny-based OT protocols is given in Sect. 5. We conclude in Sect. 6. For comprehensibility, the content related to isogenies will frequently be accompanied or introduced by the counterparts in the Diffie-Hellman setting.

1.2 Related Work

There are three aforementioned isogeny-based OT protocols. All the adversary models are either semi-honest or non-simulation, which are both quite weak notions. While the semi-honest model cannot reflect vicious attackers in the real world, the non-simulation-based model cannot enjoy the composition theorems [8,9], see Sect. 5.

The first protocol was proposed by Barreto et al. [2] and used the common reference string (CRS) model along with the random oracle model. They revisited Chou and Orlandi's work [13] and proposed an SIDH-based OT. They exploited the properties of SIDH to mask one party's public points by randomly (up to the receiver's choice) adding shared selective points derived from the common reference string. However, the claim of security is false. It may not ensure privacy even in the semi-honest model.

The second protocol was proposed by de Saint Guilhem et al. [17]. They derived their two constructions from the Shamir-3-Pass key transport scheme and [13], respectively. Their framework is UC-secure against semi-honest adversaries based on a masking structure hard homogeneous space or on $\mathbb{F}_{p^2}$ supersingular isogenies.

The third protocol was proposed by Vitse [38]. It is derived from Wu, Zhang, and Wang's OT [39], based on a Diffie-Hellman-related problem. Their proposal naturally fits well in the general setting (including DH, SIDH, and CSIDH). They claimed UC-security in a semi-honest model and gave another game-based security definition (semantic security) for their OT protocol. They also proved the hardness of their special assumption in generic groups.

An independent and concurrent work of Alamati et al. [1] is concerned with giving a general framework for developing cryptographic primitives based on group actions such as CSIDH. As an application, they briefly present some OT schemes. Their paper is concerned with theoretical aspects, not practical ones. Hence, the efficiency is worse than using the GMW compiler [22] or using the transformation of [18].

2 Preliminaries

Notation

Let $\{X(a,n)\} =_c \{Y(a,n)\}$ denote computationally indistinguishable probabilistic ensembles X, Y, which means for any PPT non-uniform algorithm D there exists a negligible function f such that for all $a, n \in \mathbb{N}$ we have $|\Pr[D(X(a,n))] - \Pr[D(Y(a,n))]| \le f(n)$. Let $\{X(a,n)\} =_s \{Y(a,n)\}$ denote statistically indistinguishable probabilistic ensembles X, Y on the same set, which means the statistical distance between X and Y is negligible. The notation $a \leftarrow_{\$} S$ means a is uniformly generated from the set S. For simplicity, we often omit the security level parameter n but it is implicit in the indistinguishabiliy and the negligible function.

2.1 CSIDH

For a given prime p and an elliptic curve E defined over $\mathbb{F}_p$, $End_p(E)$ is the subring of the endomorphism ring $End(E)$ consisting the endomorphisms defined over $\mathbb{F}_p$.

Let $\mathcal{O}$ be an order in an imaginary quadratic field and $\pi \in \mathcal{O}$ an element of norm p. Define the set of isomorphism classes of elliptic curves $\mathcal{E}\ell\ell_p(\mathcal{O}, \pi)$ where E defined over $\mathbb{F}_p$, $End_p(E) = \mathcal{O}$, and π is the $\mathbb{F}_p$-Frobenius map of E. For any ideal $\mathfrak{a} \in \mathcal{O}$ and $E \in \mathcal{E}\ell\ell_p(\mathcal{O}, \pi)$, an action can be defined by $\mathfrak{a} * E = E'$ such that there exists an isogeny $\phi : E \to E'$ with $ker(\phi) = \cap_{\alpha \in \mathfrak{a}} \{P \in E(\bar{\mathbb{F}}_p) \mid \alpha(P) = 0\}$. The image curve of $\mathfrak{a} * E$ is well-defined up to $\mathbb{F}_p$-isomorphism. Moreover, the ideal class group $Cl(\mathcal{O})$ acts freely and transitively on $\mathcal{E}\ell\ell_p(\mathcal{O}, \pi)$.

Castryck et al. specified the prime to be $p = 4 \times \ell_1 \times ... \times \ell_n - 1$ where ℓ_i are small odd primes. In the case of $p = 3 \mod 8$, for any supersingular elliptic curve E defined over $\mathbb{F}_p$, the restricted endomorphism ring $End_p(E) = \mathbb{Z}\{\pi\} \cong \mathbb{Z}\{\sqrt{-p}\}$ if and only if E is $\mathbb{F}_p$-isomorphic to $E_A : y^2 = x^3 + Ax^2 + x$ for some unique $A \in \mathbb{F}_p$. The quadratic twist of a given elliptic curve $E : y^2 = f(x)$ is $E^t : dy^2 = f(x)$ where $d \in \mathbb{F}_p$ has Legendre symbol -1. When $p = 3 \mod 4$ let E_0 be such that $j(E_0) = 1728$, then E_0 and E_0^t are $\mathbb{F}_p$-isomorphic. The quadratic twist can be efficiently computed in the CSIDH setting [12]. Since the prime $p = 3 \mod 4$, $E' : -y^2 = x^3 + Ax^2 + x$ is the quadratic twist of $E_A : y^2 = x^3 + Ax^2 + x$ and E' is $\mathbb{F}_p$-isomorphic to E_{-A} by $(x, y) \mapsto (-x, y)$. Further, $(a * E_0)^t = a^{-1} * E_0$. Therefore, for any curve $E \in \mathcal{E}\ell\ell_p(\mathcal{O}, \pi)$, we have, by the transitivity of the action,

$$(a * E)^t = a^{-1} * E^t.$$

Throughout this paper, we concentrate on supersingular curves defined over $\mathbb{F}_p$. Denote the ideal class group $Cl(End_p(E))$ by Cl and the set of elliptic curves $\mathcal{E}\ell\ell_p(\mathcal{O}, \pi)$ by $\mathcal{E}$.

Uniform Sampling of Curves. In CSIDH, the method provided to sample elements of the class group Cl is heuristically assumed to be statistically close to uniform [12]. Here we make the same assumption and derive the following lemma when $p = 3 \mod 4$.

Lemma 1. *Given a curve $E \in \mathcal{E}$ and a distribution D on Cl, let $D * E$ be the distribution on $\mathcal{E}$ of $a * E$ for $a \leftarrow D$, and let $(D * E)^t$ be the distribution on $\mathcal{E}$*

of $(a * E)^t$ *for* $a \leftarrow D$. *If* D *is statistically indistinguishable from the uniform distribution on* Cl, *then* $D * E$ *and* $(D * E)^t$ *are statistically indistinguishable from the uniform distribution on* $\mathcal{E}$.

Proof. Let U be the uniform distribution on $\mathcal{E}$. Since Cl acts freely and transitively on $\mathcal{E}$, $D*E$ is statistically indistinguishable from U. Since taking quadratic twists is a transposition on $\mathcal{E}$, by taking a twist on both distributions, we have $D * E =_s U = U^t =_s (D * E)^t$.

CSIDH works by sampling ideal classes as $\prod_{i=1}^{n}(\ell_i, \pi - 1)^{e_i}$ where e_i are sampled from $[-B, B\ ] \cap \mathbb{Z}$ for a suitably chosen value B. Heuristically, increasing B means that sampling becomes closer to the uniform distribution on Cl. Beullens et al. [6] proposed an efficient instantiation of these sampling methods in CSI-FiSh which requires pre-processing to compute a lattice of relations in the class group. Implementations of the CSIDH scheme can be found in [5,12]. We refer to [29,33] for constant-time variants.

Computational Assumptions. The computational assumptions relevant to this work are defined as follows.

Problem 1. **(Computational CSIDH Problem)** Given curves E, $r * E$ and $s * E$ in $\mathcal{E}$ where $r, s \in Cl$, find $E' \in \mathcal{E}$ such that $E' = rs * E$.

Problem 2. Given curves $(E,\ s * E, r * E)$ in $\mathcal{E}$ where $r, s \in Cl$, find $E' \in \mathcal{E}$ such that $E' = s^{-1}r * E$.

The computational CSIDH problem is the main hardness assumption for [12]. Problem 2 is an equivalent problem. To see this, given an oracle O for Problem 1, one can obtain E' by taking $E' \leftarrow O(s * E, E, r * E)$ such that $E' = rs^{-1} * E$. Conversely, given an oracle O for Problem 2, one can obtain E' by taking $E' \leftarrow O(s * E, E, r * E)$ such that $E' = rs * E$.

The following two problems are the main underlying problems against semi-honest adversaries.

Problem 3. **(Computational Square CSIDH Problem)** Given curves E and $s * E$ in $\mathcal{E}$ where $s \in Cl$, find $E' \in \mathcal{E}$ such that $E' = s^2 * E$.

Problem 4. **(Computational Inverse CSIDH Problem)** Given curves E and $s * E$ in $\mathcal{E}$ where $s \in Cl$, find $E' \in \mathcal{E}$ such that $E' = s^{-1} * E$.

The equivalence between these two assumptions and a conditional reduction to the computational CSIDH problem were given in [20]. The condition for the second reduction is that the group order is given and odd. Therefore, we can say that there is quantum reduction [23,37] to the computational CSIDH problem when $p = 3 \mod 4$. In fact, there is also an efficient quantum reduction for the case of $p = 1 \mod 4$, see Appendix 1 of [27]. Note that the quantum computation is only to compute the group structure of Cl, and so can be considered as a precomputation; the remainder of the reduction is classical.

As Castryck et al. pointed out [12] both problems contain exceptional cases when E_0 takes part in the problems due to the symmetric structure. That is, $(a * E_0)^t = a^{-1} * E_0$, and so Problem 4 is easy in the special case $E = E_0$. The issue can be circumvented if the public curve is generated by a trusted third party.

Next, we will introduce the main underlying assumption for our UC-secure construction.

Problem 5. **(Reciprocal CSIDH Problem)** Given E in $\mathcal{E}$. Firstly, the adversary chooses and commits to $X \in \mathcal{E}$, then receives the challenge $s * E$ where $s \in Cl$. Then the adversary must compute a pair $(s * X, s^{-1} * X)$ with respect to the committed X.

Intuitively, the computational reciprocal CSIDH problem is a relaxed version of the square CSIDH problem or the inverse CSIDH problem. In particular, if one can solve the inverse CSIDH problem, then one can solve the reciprocal CSIDH problem by taking $X = E$ with $(s * X, s^{-1} * X) = (s * E, s^{-1} * E)$. Conversely, if an attacker knows the isogeny between X and E, or E^t, then this can be used to solve the *inverse CSIDH problem.* That is, if $X = r * E$, one can obtain $s^{-1} * E$ by computing $r^{-1} * (s^{-1} * X)$ with the given r. On the other hand, if $X = r * E^t$, one can obtain $s^{-1} * E$ by computing $r * (s * X)^t$ with the given r. However, note that the attacker is not required to know the isogeny between X and E or E^t in the problem.

The reciprocal CSIDH problem appears to be non-standard at first sight but, in fact, it is equivalent to the inverse CSIDH problem. Even though the problem provides additional freedom X for the attacker, yet notice that X is chosen prior to the challenge $s * E$. We show in the following reduction that the freedom to choose X can be handled. We call the reduction strategy self-reconciling.

Proposition 1. *The computational reciprocal CSIDH problem is equivalent to the computational inverse CSIDH problem.*

Proof. Given a challenge $(E, s * E)$ for the inverse CSIDH problem. Invoke the adversary for the reciprocal CSIDH with E. After receiving X from the adversary, send the challenge $t_1 s * E$ to the adversary where $t_1 \leftarrow_{\$} Cl$. Receive $(t_1 s * X, (t_1 s)^{-1} * X)$ from the adversary, then rewind the adversary to the time when it output X, and then send $t_2 s * X$ as the challenge with respect to committed X where $t_2 \leftarrow_{\$} Cl$. Receive (X_0, X_1) from the adversary. Output X_1.

Claim $(t_2) * X_1 = s^{-1} * E$. Write $X = b * E$ by the transitivity of the action, so $t_2 s * X = (t_2 s b) * E$. Then, since the second challenge is $t_2 s * X = (t_2 s b) * E$, we have $t_2 * X_1 = (sb)^{-1} * X = s^{-1} * E$. Precisely, if the adversary can solve the problem based on E with committed X with probability ϵ, then the adversary can be used to solve the inverse CSIDH problem based on E with probability ϵ^2.

In the proof Proposition 1, the reduction extracts the first entry of the first solution and the second entry of the second solution to obtain the solution for the inverse CSIDH problem. We can, therefore, conclude the following corollary.

Corollary 1. *In the experiment of Problem 5, after committing to the curve X, if the adversary can solve $(s*X, s'^{-1}*X)$ with respect to different given challenges $s*E$ and $s'*E$ then the adversary can be used to solve the computation inverse CSIDH problem.*

We end the subsection with the following relation.

$$\begin{aligned}\textbf{Computational CSIDH} &=_{quantum} \textbf{Computational Inverse CSIDH}\\ &=_{classical} \textbf{Computational Square CSIDH}\\ &=_{classical} \textbf{Computational Reciprocal CSIDH}\end{aligned}$$

Remark. The above results can all be extended to general (free and transitive) group actions and hard homogeneous spaces [14]. We leave the details to the reader.

2.2 Functionalities

In this subsection, we define the functionalities we need as well as the related security definitions.

A symmetric encryption scheme is a pair of algorithms (E, D) defined over message space $\mathcal{M}$ and ciphertext space $\mathcal{C}$ with key space $\mathcal{K}$.

Definition 1. *(non-committing encryption (NCE)) A symmetric encryption scheme (E, D) is said to be non-committing if there exists PPT algorithm B_1, B_2 such that for any PPT distinguisher $\mathcal{D}$, message $m \in \mathcal{M}$.*

$$|\Pr[\mathcal{D}(c, k) = 1] - \Pr[\mathcal{D}(c', k') = 1]| = negl(n),$$

where $k \leftarrow_{\$} \mathcal{K}, c = \mathsf{E}_k(m)$ and $c' \leftarrow_{\$} B_1(1^n), k' \leftarrow_{\$} B_2(c', m)$

Informally speaking, non-committing encryption allows a user to generate a dummy ciphertext indistinguishable from the real one by B_1 and later explain it with the assistance of B_2. The idea was introduced by Canetti et al. [10] with the one-time pad (OTP) as an instantiation. It was also used in some oblivious transfer constructions [3,13]. The non-committing proposition here is used to extract the input without rewinding in the simulation process.

$\mathcal{F}_{TSC}$**-Functionality of a trusted setup curve**
The functionality is o output an element of $\mathcal{E}$. It generates an ideal class $t \leftarrow_{\$} Cl$ and outputs the curve $t * E_0$.

The functionality of trusted setup curves $\mathcal{F}_{TSC}$ serves as a setup for generating a curve for the protocol. This setup hides the relation t between the public curve and the curve E_0. In practice, this can be replaced with a key exchange protocol [7]. That is, two parties do a key exchange first and obtain a curve such that the isogeny relation to E_0 remains unknown if the two parties do not share their ideal classes or collude.

$\mathcal{F}_{RO}$-Functionality of Random Oracle

The functionality is a function with the domain $\mathcal{E}$ and the codomain $\mathcal{K}$. It keeps a list L of pairs in $\mathcal{E} \times \mathcal{K}$ where the initial state is empty. It works as follows:

1. Upon receiving a query $C \in \mathcal{E}$, check whether (C, k') for some $k' \in \mathcal{K}$. If so, set $k = k'$; if not, generate $k \leftarrow_{\$} \mathcal{K}$ and store the pair (C, k) in the list L.
2. Output k.

The functionality of a random oracle $\mathcal{F}_{RO}$ internally contains an initially empty list. Upon receiving the query from the domain, it will check whether it is a repetition. If so, return the value assigned before; otherwise, it randomly assigns a value from the codomain, stores the pair, and returns the value. Formally speaking, an input of a random oracle can be any binary string. For simplicity, we restrict the domain to $\mathcal{E}$. This can be easily and compatibly extend to $\{0,1\}^*$, since supersingularity can be efficiently verified [12].

We briefly define the security terms. Let $output^{\pi}(x, y)$ denote the outputs of two parties with the inputs x, y respectively after the execution of π, and $view_i^{\pi}$ consist of the input, the internal random tape and all received messages of the ith participant after the execution of π. Let $IDEAL_{\mathcal{F},\mathcal{S},\mathcal{Z}}$ and $HYBRID^{\mathcal{G}}_{\pi,\mathcal{A},\mathcal{Z}}$ denote the ideal execution ensemble and the hybrid ensemble, respectively. A detailed explanation can be found in [27]. We refer [9,24] for more thorough descriptions for the security notions against semi-honest adversaries and the malicious adversaries, respectively.

Definition 2. *(security OT against semi-honest adversary). We say a protocol π securely (privately) computes $\mathcal{F}_{OT}$ in the presence of static semi-honest adversaries if there exists probabilistic polynomial-time algorithms S_1, S_2 such that*

$$output^{\pi}(x, y) = \mathcal{F}_{OT}(x, y)$$

$$\{S_1(x, (F_{OT}(x, y))_1)\}_{x,y} =_c \{view_1^{\pi}(x, y)\}_{x,y}$$

and

$$\{S_2(y, (F_{OT}(x, y))_2)\}_{x,y} =_c \{view_2^{\pi}(x, y)\}_{x,y}.$$

Definition 3. *(UC-realize). A protocol π is said to UC-realize an ideal functionality $\mathcal{F}$ in the presence of malicious adversaries and static corruption in the hybrid model with functionality $\mathcal{G}$ if for any adversary $\mathcal{A}$ there exists a simulator $\mathcal{S}$ such that for every interactive distinguisher environment $\mathcal{Z}$ we have*

$$IDEAL_{\mathcal{F},\mathcal{S},\mathcal{Z}} =_c HYBRID^{\mathcal{G}}_{\pi,\mathcal{A},\mathcal{Z}}.$$

3 Our Proposal

This section first presents the idea behind our tweaked key exchange by introducing the core of Chou and Orlandi's OT scheme [13]; we then derive a novel compact protocol as a prototype. Following this, we compress the three-round scheme to an optimal two rounds by using the quadratic twist technique. Finally, building on the round-optimal structure, we add a "proof of decryption" mechanism, which requires an extra round, in order to achieve security against malicious adversaries.

3.1 Passively Secure Schemes

Tweaked Key Exchange. Figure 1 presents the Chou–Orlandi OT scheme [13] which is based on Diffie–Hellman key exchange. In Diffie–Hellman, the sender and the receiver first share their public "keys", g^s and g^r, with each other, after which both of them can secretly obtain a shared secret g^{rs}. To adapt this for the purpose of OT, the receiver can use the second round to obfuscate his secret bit i. In the third round, the sender can communicate an encryption of the two OT messages by deriving two keys, one which cancels out the obfuscation, and one which does not. Because of this key derivation, the receiver can then only decrypt the message corresponding to his input bit.

Sender		**Receiver**
Input: (M_0, M_1)		Input: $i \in \{0,1\}$
Output: N/A		Output: M_i
$s \leftarrow_{\$} \mathbb{Z}$		$r \leftarrow_{\$} \mathbb{Z}_p^*$
	$A = g^s$ $\longrightarrow$	
		if $i = 0 : B = g^r$
		if $i = 1 : B = Ag^r$
	$\longleftarrow$ B	
$k_0 = H(B^s)$		$k_i = H(A^r)$
$k_1 = H((B/A)^s)$		
	$c_0 \leftarrow \mathsf{E}_{k_0}(M_0)$ $c_1 \leftarrow \mathsf{E}_{k_1}(M_1)$ $\longrightarrow$	
		$M_i = \mathsf{D}_{k_i}(c_i)$

Fig. 1. Chou and Orlandi's OT scheme in a nutshell [13]

We can view the isogeny-based oblivious transfer constructions of previous works in the same way. In Barreto et al.'s work [2], the shared secret between the sender and the receiver is the j-invariant of the isomorphic elliptic curves $\phi_{B'}\phi_A(E)$ and $\phi_{A'}\phi_B(E)$ [2]. Here, the receiver hides his input bit by masking his $p_3^{e_3}$-torsion subgroup public basis by a pair of special $p_3^{e_3}$-torsion points $U, V \in \phi_B(E)$; the sender then requires the same pair of points U, V to remove the noise. A coin-flipping mechanism is then used to guarantee that both parties obtain the same points U, V.

Proposals by de Saint Guilhem et al. and Vitse rely on a similar idea to use a fixed key from the key exchange to decrypt the chosen ciphertext [17,38]. In the first OT construction of [17], two public curves are required as a trusted setup, which serve the same role as two fixed keys from the perspective of key exchange. In [38], one more $p_2^{e_2}$-torsion subgroup generated by the sender is required to obtain two fixed keys.

Our Three-Round Protocol. We present our three-round protocol in Fig. 2 using the notation of the CSIDH setting. In this work we approach the change from key exchange to OT with a different strategy. The essence is that the sender and the receiver can exponentiate by both s and by s^{-1}, and by both r and r^{-1} respectively.

Upon receiving g^s from the sender, the receiver computes both g^r and g^{sr}, and sends one of them to the sender depending on its choice bit. The sender then exponentiates it by both s and by s^{-1} as the encryption keys, which is like doing the key exchange as Problem 1 and 2. One can verify that the shared secret in each case is g^{rs} and g^r, resp.

The other encryption keys are $g^{rs^{-1}}$ and g^{rs^2}, resp. They are intractable to the honest-but-curious receiver due to the hardness of the inverse and square CSIDH problems, respectively. Furthermore, the receiver's input bit remains unknown since the sender only knows either g^r or g^{sr}.

Note that in this isogeny-based setting, it is necessary that the relation between the shared public curve $E \in \mathcal{E}$ and a fixed base curve E_0 remains unknown. Should the receiver know that $E = t * E_0$, then he can always input $i = 0$ and compute the other key as $t^2r^2 * (rs * E)^t = t^2r^2 * (trs * E_0)^t = trs^{-1} * E_0 = rs^{-1} * E$.

Our Two-Round Protocol. To address the drawbacks of our three-round protocol, we observe that the quadratic twist provides additional flexibility for the curve computations.

To first break the dependency of C on A, we let the receiver compute $C = (r * E)^t$ in the case $i = 1$, instead of $r * A$. Lemma 1 guarantees that this still statistically hides i. Now that C is independent of A, the receiver can send his message first, reducing the protocol to only two rounds. Furthermore, this removes the hypothetical attack of a malicious receiver choosing C in response to A and enables a direct reduction to the computational CSIDH problem.

We then note that the sender's second encryption curve can be computed as $(s*C^t)^t$, instead of $s^{-1}*C$, in the three-round version. Here again we can simplify by letting the sender compute the second curve as $s * C^t$, without the additional

Trusted Setup: random $E \in \mathcal{E}$		
Sender		**Receiver**
Input: (M_0, M_1)		Input: $i \in \{0,1\}$
Output: N/A		Output: M_i
$s \leftarrow_{\$} Cl$		$r \leftarrow_{\$} Cl$
	$A = s * E$ $\longrightarrow$	
		if $i = 0 : C = r * E$
		if $i = 1 : C = r * A$
	C $\longleftarrow$	
$k_0 = H(s * C)$		$k_i = H(r * (s^{1-i} * E))$
$k_1 = H(s^{-1} * C)$		
	$c_0 \leftarrow \mathsf{E}_{k_0}(M_0)$	
	$c_1 \leftarrow \mathsf{E}_{k_1}(M_1)$ $\longrightarrow$	
		$M_i = \mathsf{D}_{k_i}(c_i)$

Fig. 2. Our three-round OT protocol.

twisting operation. This then results in a simplification for key computation too: for $i = 0$, the encryption curve is $s * (r * E) = r * A$, and for $i = 1$ it is $s * ((r * E)^t)^t = r * A$; thus we return to the idea of using a single Diffie–Hellman key by way of using the twist operation. The modified two-round protocol is described in Fig. 3. We give a formal security proof in Sect. 4.1.

In this simplified variant the number of isogeny computations remains the same as in the three-round variant. We note that taking quadratic twists is an efficient operation via field negation.

3.2 The Full Construction Against Malicious Adversaries

The full protocol is shown in Fig. 4 below. To be secure against malicious adversaries who may deviate from the specification, both parties will do a simple verification of the received elements. In the CSIDH setting, both parties will check whether the curve is supersingular, which can be done efficiently, as shown in [12].

Protocol. (CSIDH-based OT). *Let* (E, D) *be a symmetric encryption scheme with message space* $\mathcal{M}$ *and ciphertext space* $\mathcal{C}$. *Let* $H : \mathcal{E} \to \mathcal{K}$ *be modeled as a random oracle* $\mathcal{F}_{RO}$ *that serves as the key derivation function from the group* $\mathcal{E}$ *to the key space* $\mathcal{K}$ *for the symmetric encryption scheme.*

Trusted Setup: $E \in \mathcal{E}$		
Sender		**Receiver**
Input: (M_0, M_1)		Input: $i \in \{0,1\}$
Output: $\perp$		Output: M_i
$s \leftarrow_{\$} Cl$		$r \leftarrow_{\$} Cl$
$A = s * E$		if $i = 0$: $C = r * E$
		if $i = 1$: $C = (r * E)^t$
	$\xleftarrow{\quad C \quad}$	
$k_0 = H(s * C)$		
$k_1 = H(s * C^t)$		
	$A, c_0 \leftarrow \mathsf{E}_{k_0}(M_0)$, $c_1 \leftarrow \mathsf{E}_{k_1}(M_1)$ $\longrightarrow$	
		$k_i = H(r * A)$
		$M_i = \mathsf{D}_{k_i}(c_i)$

Fig. 3. The core of our two-round OT scheme. No analogue exists in the Diffie–Hellman setting due to the use of the quadratic twist.

- ***Trusted Setup:*** *Let $E = t * E_0$ where $t \leftarrow_{\$} Cl$ is unknown.*
- ***Input:*** *As input, the sender $\mathcal{S}$ takes two messages M_0, M_1 of the same length; the receiver $\mathcal{R}$ takes a bit $i \in \{0,1\}$.*
- ***Procedure:***
  1. *$\mathcal{S}$ samples independent ideals $s_0, s_1 \leftarrow_{\$} Cl$, a random string $str \leftarrow_{\$} \{0,1\}^n$ and computes $A_0 = s_0 * E$, $A_1 = s_1 * E$.*
  2. *$\mathcal{R}$ generates an ideal $r \leftarrow_{\$} Cl$ and computes $C = r * E$; if $i = 1$, overwrites $C = C^t$; and sends C to $\mathcal{S}$.*
  3. *$\mathcal{S}$ checks whether $C \in \mathcal{E}$. If not, $\mathcal{S}$ aborts and outputs **abort**$_2$. Otherwise, $\mathcal{S}$ computes four keys $k_{j,0} = H(s_j * C)$ and $k_{j,1} = H(s_j * C^t)$ for $j \in \{0,1\}$. Then, $\mathcal{S}$ computes four ciphertexts $c_{0,j} \leftarrow \mathsf{E}_{k_{0,j}}(M_j)$ and $c_{1,j} \leftarrow \mathsf{E}_{k_{1,j}}(s_1 \parallel str)$ for $j \in \{0,1\}$. $\mathcal{S}$ sends $(A_0, A_1, c_{0,0}, c_{0,1}, c_{1,0}, c_{1,1})$ to $\mathcal{R}$.*
  4. *$\mathcal{R}$ runs the proof of ability to decrypt first. $\mathcal{R}$ checks whether $A_1 \in \mathcal{E}$. If not, $\mathcal{R}$ aborts and outputs **abort**$_1$. Otherwise, $\mathcal{R}$ computes $k'_{1,i} = H(r * A_1)$ and $(s'_1 \parallel str') \leftarrow \mathsf{D}_{k'_{1,i}}(c_{1,i})$. Verify whether $s'_1 * (r * E) = r * A_1$. If not, output **abort**$_1$. Otherwise, continue.*
  5. *$\mathcal{R}$ computes $k'_{1,1-i} = H(s'_1 * (r * E)^t)$. Verify whether $\mathsf{D}_{k'_{1,1-i}}(c_{1,i-i}) = (s'_1 \parallel str')$. If not, output **abort**$_1$. Otherwise, continue.*
  6. *$\mathcal{R}$ verifies $A_0 \in \mathcal{E}$. If not, $\mathcal{R}$ aborts and outputs **abort**$_1$. Otherwise, compute the decryption key $k'_{0,1} = H(r * A_0)$ and output $M_i \leftarrow \mathsf{D}_{k'_{0,i}}(c_{0,i})$. And send str' to $\mathcal{S}$.*

7. *$\mathcal{S}$ checks whether $str = str'$. If not, $\mathcal{S}$ aborts and outputs* $\mathbf{abort}_2$. *Otherwise, $\mathcal{S}$ accepts and outputs $\perp$.*

Intuitively, to simulate a sender controlled by an adversary, we have to show that the receiver's message's distribution with input $i = 0$ and that with input $i = 1$ are indistinguishable. Asides from that, the simulator needs to extract the real input of the message pair since the adversary can replace the original input. Lemma 1 assures the first requirement. The second condition is attained by controlling the functionality $\mathcal{F}_{TSC}$. As a result, the simulator can decrypt two ciphertexts by using the trapdoor of $\mathcal{F}_{TSC}$ and extract the real input of the sender.

To simulate a receiver corrupted by an adversary, the simulator should extract the adversary's input by observing the hash queries. In order to extract the input, the receiver should demonstrate the ability to decrypt. The reason to do this is that the corrupted receiver who skips all hash queries makes the input intractable to the simulator. The additional proof of ability to decrypt mechanism forces the adversary either to abort or to prove its ability to decrypt by querying the hash function. Here the sender will send another curve $s' * E$ distinct from $s * E$ for transferring messages. The sender encrypts the ideal s' and a concatenated random string with key pair derived from $s' * E$. The receiver decrypts one ciphertext with X, and the other ciphertext serves as a verification of the equality of encrypted messages. By requiring this together with Corollary 1, the mechanism enables the simulator to extract the input by observing the random oracle queries. Furthermore, since the simulator can only obtain one real message from the trusted third party (corresponding to the extracted input i), the simulator must forge the other ciphertext via the non-committing encryption scheme. The difference between the unchosen ciphertexts is not noticeable unless the environment machine knows the corresponding decryption key. In this case, the environment machine contains a pair of curves which is exactly the solution for the reciprocal CSIDH problem. See Sect. 4 for more details.

4 Security Analysis

In this section, we prove the security of our two schemes from Sects. 3.1 and 3.2 against semi-honest and malicious adversaries respectively.

4.1 Semi-honest Security

Eavesdropper. An eavesdropper receives all the communications of parties and does not intervene in the execution. We assume that such an adversary knows the parties' inputs while the simulator tasked with simulating an indistinguishable transcript is given nothing. The reason for this assumption is to match the definition of UC-security [9] where the environment machine decides the inputs. In fact, security against such eavesdroppers corresponds exactly to the honest-honest case discussed in the proof below.

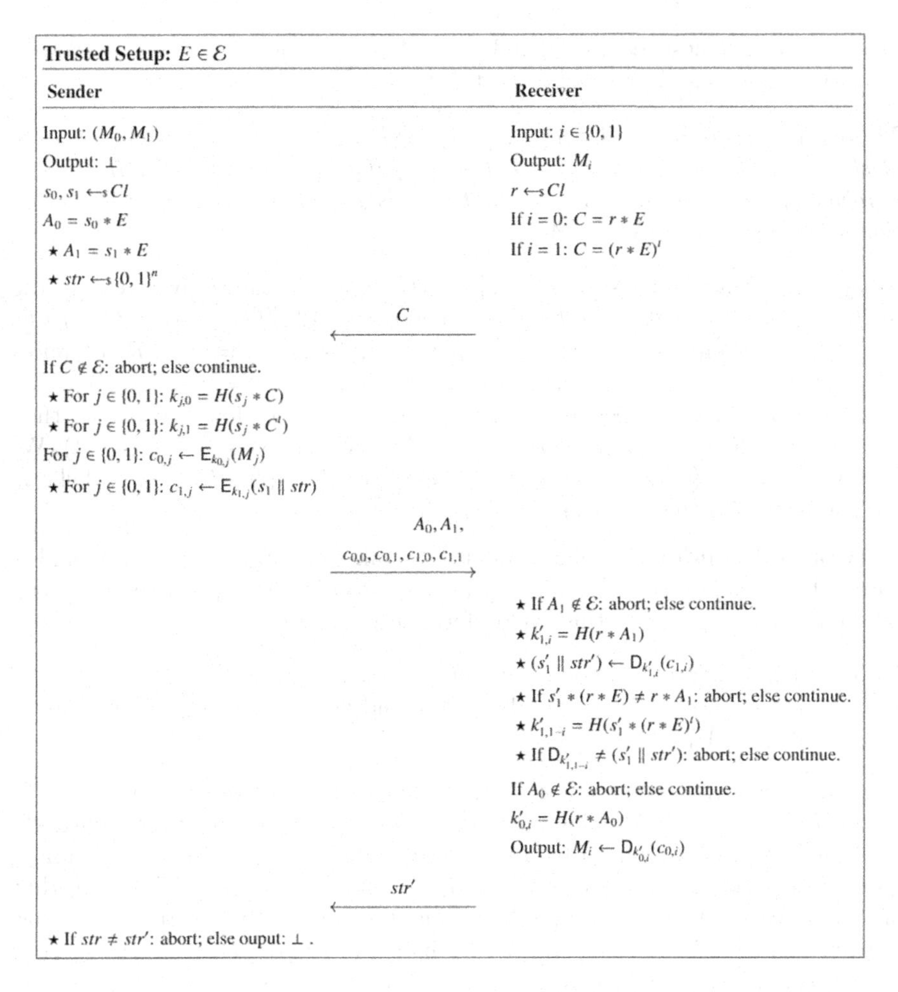

Fig. 4. Our CSIDH-based oblivious transfer protocol. For the sake of readability, we label the steps related to the process of "proof of ability to decrypt" with $\star$.

Semi-Honest Adversary. A static semi-honest adversary can choose to corrupt either, both or neither of the parties and will follow the protocol specification. We will prove that such adversary cannot obtain any information from the transcript of our two-round protocol (Fig. 3) assuming that the computational inverse CSIDH problem is hard.

We remark that it is not meaningless to design two different protocols for different security levels. As security against semi-honest adversaries is easier to achieve, it is better to use a simpler and more efficient protocol when only such guarantees are required. This then implies that it is not necessary to prove the semi-honest security of our second protocol since the first provides a simpler secure variant. We highlight the fact that some maliciously secure protocols fail to also be semi-honest secure [24] and stress that we do not claim the semi-honest security of our second protocol of Sect. 3.2.

Theorem 1. *The protocol π of Fig. 3 securely computes $\mathcal{F}_{OT}$ in the presence of static semi-honest adversaries if the computational inverse CSIDH problem (Problem 4) is infeasible, assuming that $H(\cdot)$ is a random oracle and the encryption scheme* (E, D) *is IND-CPA.*

Proof. (**Correctness**). Let $i \in \{0, 1\}$ be the input of the receiver $\mathcal{R}$. Say the sender $\mathcal{S}$ generates ideal $s \in Cl$ and $\mathcal{R}$ generates $r \in Cl$. If $i = 0$, then $C = r * E$. $\mathcal{S}$ computes the encryption key k_0 as $H(s*C)$, and sends $A = s*E$. $\mathcal{R}$ computes $k_0' = H(r * A)$ as the decryption key; as we have $r * A = r * (s * E) = s * (r * E) = s * C$, we indeed have $k_0' = k_0$. On the other hand, if $i = 1$, then $C = (r * E)^t$. $\mathcal{S}$ computes $k_1 = H(s * C^t)$ while $\mathcal{R}$ computes $k_1' = H(r * A)$. We have $s * C^t = s * ((r * E)^t)^t = s \cdot r * E = r * A$ which implies $k_1' = k_1$ and shows the correctness of the protocol.

(**Corrupt sender $\mathcal{S}^*$**) The simulator S_1 takes as input $(M_0, M_1, \perp)$ and is required to simulate the view $view_1^{\pi}(M_0, M_1, i) = (M_0, M_1, rp, C)$ where rp is a random tape. To generate this, S_1 performs these steps:

1. Uniformly generate a random tape rp for $\mathcal{S}^*$.
2. Generate $r' \leftarrow_{\$} Cl$ acting as an honest $\mathcal{R}$ and using a private random tape.
3. Output $(M_0, M_1, rp, C' = r' * E)$.

In a real execution, the curve C sent by the honest receiver is either $r*E$ if $i = 0$, or $(r * E)^t$ if $i = 1$. In the first case, the transcript output by S_1 is identically distributed to that produced by a real execution. In the second case, Lemma 1 gives us that the distribution of C' produced by S_1 is statistically close to that of C produced by the real receiver. Thus, any polynomial-time distinguisher that is given a tuple (M_0, M_1, i) is not able to distinguish $\{S_1((M_0, M_1), \perp)\}_{(M_0,M_1),i}$ from $\{view_1^{\pi}(M_0, M_1, i)\}_{M_0,M_1,i}$.

(**Corrupt receiver $\mathcal{R}^*$**) The simulator S_2 takes as input (i, M_i) and is required to simulate the view $view_2^{\pi}(M_0, M_1, i) = (i, rp, A, c_0, c_1)$ where rp is a random tape. To generate this, S_2 performs these steps:

1. Choose a uniform generated random tape rp for $\mathcal{R}^*$.
2. Generate $s' \leftarrow_{\$} Cl$ acting as an honest $\mathcal{S}$ and using a private random tape, and generate $r' \leftarrow_{\$} Cl$ using rp. Compute the curve C as $r' * E$ or $(r' * E)^t$ depending on i.
3. Compute the decryption keys k_i', k_{1-i}' honestly using s' and C. Replace k_{1-i}' with $\widetilde{k'} \leftarrow_{\$} \mathcal{K}$
4. Compute ciphertexts $c_i = \mathsf{E}_{k_i'}(M_i)$ and $c_{1-i} = \mathsf{E}_{\widetilde{k'}}(\widetilde{M})$ where $\widetilde{M}$ is a string of the same length as M_i sampled at random from the message space $\mathcal{M}$.
5. Output $(i, rp, s' * E, c_0, c_1)$.

We claim that if there exists a successful PPT distinguisher between the simulated view and the real view, then reductions can be made to solve the computational problems (Problem 3 or the equivalent Problem 4) or to break the IND-CPA security of the encryption scheme.

To show this, we build a series of hybrid views. Let $\mathcal{H}_0$ be the view of the real adversary, and $\mathcal{H}_2$ be the view generated by S_2 (i.e., $\{view_2^\pi(M_0, M_1, i)\}_{(M_0,M_1),i}$ and $\{S_2((M_0, M_1), \perp)\}_{(M_0,M_1),i}$, resp). Let the intermediate $\mathcal{H}_1$ be the view produced by running a real execution and replacing the encryption key k_{1-i} with a random $\widetilde{k} \leftarrow\!\!\$\, \mathcal{K}$. The difference between $\mathcal{H}_1$ and $\mathcal{H}_2$ is then that the real message M_{1-i} is replaced with a random one $\widetilde{M} \leftarrow\!\!\$\, \mathcal{M}$.

Hybrid 1. We first claim $\mathcal{H}_0 =_c \mathcal{H}_1$ if the computational inverse CSIDH problem (Problem 4) is hard. To offer an intuition: let E_{1-i} denote the curve from which the replaced key k_{1-i} is derived. When $i = 0$, we have $E_{1-i} = s * C^t = s * (r * E)^t = r^{-1} * (s^{-1} * E)^t$; and when $i = 1$, we have $E_{1-i} = s * C = s * (r * E)^t = r^{-1} * (s^{-1} * E)^t$ as well. In both cases we see that the hard-to-compute curve contains $s^{-1} * E$ which we use to reduce a successful distinguisher to the computational inverse CSIDH problem (Problem 4).

Let $\mathcal{Z}$ be an environment that can successfully distinguish between $\mathcal{H}_0$ and $\mathcal{H}_1$, then a solver $\mathcal{B}$ for Problem 4 with the assistance of $\mathcal{Z}$ runs as follows:

1. Receive challenge $(E', s' * E')$ from Problem 4, where $s' \in Cl$ is unknown.
2. Set E' to be the public curve used by the protocol π and set $s' * E'$ as the curve A sent to the receiver.
3. Randomly generate random tape rp for the receiver, use it to sample r, and compute C according to i.
4. While running, simulate the random oracle by assigning a random value from $\mathcal{K}$ whenever a new query is made and recording a list of past queries during the execution.
5. When deriving the real encryption key k_i, compute it as $r * (s' * E')$ (since s' from the challenge is unknown).
6. Replace the other encryption key k_{1-i} with $\widetilde{k} \leftarrow\!\!\$\, \mathcal{K}$ to simulate the output of $\mathcal{H}_1$; abort if $\widetilde{k}$ already appears on the list of answers to random oracle queries.
7. Invoke the distinguisher $\mathcal{Z}$ with the produced output of $\mathcal{H}_1$.
8. When $\mathcal{Z}$ terminates, randomly select a curve $\widetilde{E}$ in the list of past queries of the simulated random oracle and return $(r * \widetilde{E})^t$ as the computational inverse CSIDH solution.

Note that, if $\mathcal{B}$ does not abort, the only difference between $\mathcal{H}_0$ and $\mathcal{H}_1$ is the key for M_{i-1}, thus a distinguisher $\mathcal{Z}$ which does not query this key must have a zero advantage.

Let **A** denote the event that $\mathcal{B}$ aborts when sampling the replacement key. Denoting by q_H the maximum number of queries made to H during the reduction, we have that $\Pr[\mathbf{A}] \leq \frac{q_H}{|\mathcal{K}|}$. Also let **E** denote the event that the targeted curve $E'_{1-i} = r^{-1} * (s^{-1} * E')^t$ is present on the query list. We see that the reduction $\mathcal{B}$ wins with probability $1/q_H$ when $\mathbf{E}$ happens, and we can then write:

$$\begin{aligned}\mathbf{Adv}_{\mathcal{B}}^{\text{Problem 4}} = \Pr[\mathcal{B} \text{ wins}] &= \Pr[\mathcal{B} \text{ wins} \mid \neg\mathbf{A}] \cdot \Pr[\neg\mathbf{A}] + \Pr[\mathcal{B} \text{ wins} \mid \mathbf{A}] \cdot \Pr[\mathbf{A}] \\ &\geq \Pr[\mathcal{B} \text{ wins} \mid \neg\mathbf{A}] \cdot (1 - \Pr[\mathbf{A}]) \\ &\geq \Pr[\mathcal{B} \text{ wins} \mid \neg\mathbf{A}] \cdot \left(1 - \frac{q_H}{|\mathcal{K}|}\right) \\ \Leftrightarrow \frac{1}{1 - \frac{q_H}{|\mathcal{K}|}} \cdot \Pr[\mathcal{B} \text{ wins}] &\geq \Pr[\mathcal{B} \text{ wins} \mid \neg\mathbf{A}] = \frac{1}{q_H} \cdot \Pr[\mathbf{E}] \end{aligned} \tag{1}$$

Looking an arbitrary distinguisher $\mathcal{Z}$, we then have

$$\begin{aligned}|\Pr[\mathcal{Z}(\mathcal{H}_0) = 1] - \Pr[\mathcal{Z}(\mathcal{H}_1) = 1]| = &|\Pr[\mathcal{Z}(\mathcal{H}_0) = 1|\mathbf{E}] \cdot \Pr[\mathbf{E}] \\ &- \Pr[\mathcal{Z}(\mathcal{H}_1) = 1|\mathbf{E}] \cdot \Pr[\mathbf{E}] \\ &+ \Pr[\mathcal{Z}(\mathcal{H}_0) = 1|\neg\mathbf{E}] \cdot \Pr[\neg\mathbf{E}] \\ &- \Pr[\mathcal{Z}(\mathcal{H}_1) = 1|\neg\mathbf{E}] \cdot \Pr[\neg\mathbf{E}]| \\ \leq &\Pr[\mathbf{E}] \end{aligned} \tag{2}$$

since $|\Pr[\mathcal{Z}(\mathcal{H}_0) = 1|\neg\mathbf{E}] - \Pr[\mathcal{Z}(\mathcal{H}_1) = 1|\neg\mathbf{E}]| = 0$ and $|\Pr[\mathcal{Z}(\mathcal{H}_0) = 1|\mathbf{E}] - \Pr[\mathcal{Z}(\mathcal{H}_1) = 1|\mathbf{E}]| \leq 1$ by definition. By combining (1) and (2) we see that if $\mathcal{Z}$ distinguishes the two views with non-negligible advantage ϵ, then $\mathcal{B}$ successfully solves Problem 4 with probability at least $\epsilon \cdot (1 - \frac{q_H}{|\mathcal{K}|})/q_H$ which is non-negligible if $q_H = \mathsf{poly}(n)$ and $1/|\mathcal{K}| = \mathsf{negl}(n)$. This contradicts the assumption that Problem 4 is intractable and therefore implies that $\mathcal{H}_0$ and $\mathcal{H}_1$ are computationally indistinguishable to any PPT environment $\mathcal{Z}$.

Hybrid 2. We now claim $\mathcal{H}_1 =_c \mathcal{H}_2$ for any PPT distinguisher if the encryption scheme (E, D) is IND-CPA secure. The only difference is the encryption $\mathsf{E}_{\widetilde{k}}(M_{1-i})$ in $\mathcal{H}_1$ and the encryption $\mathsf{E}_{\widetilde{k}}(\widetilde{M})$ in $\mathcal{H}_2$, where $\widetilde{k}$ is uniformly sampled from $\mathcal{K}$. A successful distinguisher $\mathcal{Z}$ between the two distributions can be reduced to an adversary against the IND-CPA security of (E, D) in a straightforward manner. As this reduction is common in the literature, we only include a sketch here.

The IND-CPA adversary $\mathcal{B}$ has access to a left-right encryption oracle which uses a secret key randomly sampled from $\mathcal{K}$ to encrypt either the left or the right input; this hidden key plays the role of $\widetilde{k}$ in the generation of the view given to $\mathcal{Z}$. After setting up and executing the protocol honestly, $\mathcal{B}$ uses the left-right oracle to encrypt either M_{1-i} or a random $\widetilde{M}$ as the ciphertext c_{1-i}; depending on the hidden bit (left or right) of the oracle, the view $view_{\mathcal{B}}$ generated by $\mathcal{B}$ for $\mathcal{Z}$ is distributed identically to either $\mathcal{H}_1$ or $\mathcal{H}_2$. After the distinguisher terminates, the reduction returns its output as the guess of the oracle's hidden bit. Labelling the oracle's hidden bit as b, we then have

$$\begin{aligned}\mathsf{Adv}_{\mathcal{B},(\mathsf{E},\mathsf{D})}^{\text{IND-CPA}} &= |\Pr[\mathcal{B} = 1 \mid \mathsf{b} = 0] - \Pr[\mathcal{B} = 1 \mid \mathsf{b} = 1]| \\ &= |\Pr[\mathcal{Z}(view_{\mathcal{B}}) = 1 \mid \mathsf{b} = 0] - \Pr[\mathcal{Z}(view_{\mathcal{B}}) = 1 \mid \mathsf{b} = 1]| \\ &= |\Pr[\mathcal{Z}(\mathcal{H}_1) = 1] - \Pr[\mathcal{Z}(\mathcal{H}_2) = 1]|\end{aligned}$$

which immediately shows that if $\mathcal{Z}$ is successful with non-negligible advantage, then so is $\mathcal{B}$ which contradicts the assumption that (E, D) is IND-CPA secure.

(**Honest sender and honest receiver**) We now claim that there exists a PPT simulator that can generate a transcript tuple, without knowledge of the parties' inputs, which is indistinguishable from the view of an eavesdropper $\mathcal{Z}$ that knows the parties' inputs (but not their random tapes). This simulator is constructed from the following sequence:

1. $\mathcal{S}_0$ knows the real inputs (M_0, M_1) and i of the parties; by sampling random tapes and acting honestly, it produces a perfect simulation.
2. $\mathcal{S}_1$ always uses $i = 0$; by Lemma 1 and the argument made in the case of a corrupt sender, the output of $\mathcal{S}_1$ is either identically distributed or statistically indistinguishable from the output of $\mathcal{S}_0$.
3. $\mathcal{S}_2$ replaces k_1 with a randomly sampled key; as above, this is computationally indistinguishable from the output of $\mathcal{S}_1$ assuming that Problem 4 is intractable.
4. $\mathcal{S}_3$ replaces M_1 with a randomly sampled message; as above, this is computationally indistinguishable from the output of $\mathcal{S}_2$ assuming that the encryption scheme is IND-CPA secure.
5. $\mathcal{S}_4$ always uses $i = 1$; as above, the output of $\mathcal{S}_4$ is statistically indistinguishable from the output of $\mathcal{S}_3$.
6. $\mathcal{S}_5$ and $\mathcal{S}_6$ respectively first replace k_0 and then M_0 with random values; as above, these changes are computationally indistinguishable assuming the hardness of Problem 4 and the IND-CPA security of the encryption scheme.

Finally, we observe that the last simulator $\mathcal{S}_6$ does not use any of the real inputs to produce a random transcript. By the sequence above, this simulation is indistinguishable from the transcript of a real execution.

(**Corrupt sender and corrupt receiver**) In this case, the simulator knows the inputs of both corrupt parties; as for $\mathcal{S}_0$ in the previous case, it can generate a perfect simulation of the views of the parties.

The four cases considered above cover all possible corruption strategies; this thus completes the proof that the protocol π securely computes $\mathcal{F}_{OT}$.

4.2 Malicious Adversary

Malicious Adversary. A malicious adversary with static corruptions can corrupt either, both or neither of the parties prior to the execution. The environment machine decides the initial inputs of all parties. The adversary will be in charge of the corrupted party or parties, and decide all messages to be sent. In particular, the adversary can replace the inputs of the participants from the environment machine and deviate from the protocol specification. We will prove that the construction in Fig. 4 UC-realizes the functionality $\mathcal{F}_{OT}$ in the presence of malicious adversaries with static corruptions.

Theorem 2. *The protocol π of Fig. 4, where the encryption scheme* (E, D) *is non-committing, securely UC-realizes the functionality $\mathcal{F}_{OT}$ in the hybrid model with the functionality $\mathcal{F}_{RO}$ and a trusted setup $\mathcal{F}_{TSC}$ in the presence of malicious adversaries and static corruption if the computational reciprocal CSIDH problem is infeasible.*

Proof. **(Honest Sender and Honest Receiver)** We start with the honest sender and the honest receiver. The goal is to show that the execution of π is indistinguishable from the ideal functionality when the parties follow the specification.

By following the same process as the honest-sender-and-honest-receiver case in Theorem 1, we can construct the simulator that simulates the first-half messages. By continuing the process of $\mathcal{S}_1$ or $\mathcal{S}_4$, the simulator can simulate the second-half messages $A_1, c_{1,0}$ and $c_{1,1}$ by generating s_1 and str. Since the second-half part requires no inputs from either the sender or the receiver, it produces a perfect simulation. Therefore, the simulator outputs a transcript indistinguishable from the one of a real execution.

(Corrupted Sender and Corrupted Receiver) When two parties are corrupted, the simulator can invoke the adversary with the input $(x = (M_0, M_1), y = i, z)$ given by the environment $\mathcal{Z}$ to run the whole execution. The simulator outputs whatever the adversary outputs for both parties to produce a perfect simulation.

(Honest Sender and Corrupted Receiver) Let $\mathcal{A}$ be the malicious adversary controlling the receiver. In order to emulate the adversary, the simulator needs to extract the input of the adversary, and send it to the trusted party in the ideal execution. Say the environment $\mathcal{Z}$ generates input $(x = (M_0, M_1), y = i, z)$ and gives (y, z) to the simulator. The simulator $\mathcal{S}_2$ passes any query from $\mathcal{Z}$ to $\mathcal{A}$ and returns the output of $\mathcal{A}$. The simulator $\mathcal{S}_2$ with auxiliary input (y, z) proceeds the protocol execution with the adversary as follows:

1. The simulator $\mathcal{S}_2$ emulates a random oracle $\mathcal{F}_{RO}$ by keeping a list L in $\mathcal{E} \times \mathcal{K}$ that records each past query. It initializes the random oracle with an empty list L. If the simulator receives a query on $E' \in \mathcal{E}$, the simulator checks whether $(E', k') \in L$ for some $k' \in \mathcal{K}$. If not, generate $k' \leftarrow_{\$} \mathcal{K}$ and add the entry (E', k') to the list L. Finally, $\mathcal{S}_2$ returns k' to emulate the random oracle.
2. Generate the public curve $E = t * E_0$ by sampling $t \leftarrow_{\$} Cl$ to simulate $\mathcal{F}_{TSC}$. Invoke the adversary $\mathcal{A}$ with the input (y, z) and E.
3. Receive a curve X, the first message, from the adversary. Check whether $X \in \mathcal{E}$, if not, end the session by outputting $\mathbf{abort}_2$ to the trusted party in the ideal execution. Otherwise, continue.
4. Activate the algorithm B_1 of the non-committing encryption scheme. Generate $c_{0,0}, c_{0,1}$ with B_1, $s_0, s_1 \leftarrow_{\$} Cl$ and $str \leftarrow_{\$} \{0,1\}^n$. Compute A_0, A_1 and $c_{1,0}, c_{1,1}$ as the honest sender. Send $(A_0, A_1, c_{0,0}, c_{0,1}, c_{1,0}, c_{1,1})$ to the receiver.

5. After Step 4, the simulator starts to do an additional process for any hash query of a curve $E' \in \mathcal{E}$. Firstly, check whether $E' = s_j * X$ or $s_j * X^t$ for any $j \in \{0, 1\}$ (any one out of four). If not, then skip this step and process the query in a standard way as Step 1. Else, check whether both $s_0 * X$ and $s_0 * X^t$ (i.e., the other decryption key) have been queried. If so, then abort the session by outputting **abort**$_2$. Else, check whether E' is listed in the past queries $(E', k') \in L$. If so, then skip this step and return k'. Else, send the ideal message i to $\mathcal{F}_{OT}$ in the ideal execution where $i = 0$ for the case $E' = s_j * X$ or $i = 1$ for the case $E' = s_j * X^t$, which is the extraction procedure. After obtaining M_i from $\mathcal{F}_{OT}$, generate the decryption key $k' \leftarrow B_2(c_{0,i}, M_i)$ and store $(s_0 * X, k')$ for the case $i = 0$ or $(s_0 * X^t, k')$ for the case $i = 1$ in the list, which is the response for the case $j = 0$. For the case $j = 1$, process the hash query in a standard way as Step 1.
6. After receiving str', the third message, from the adversary, verify $str = str'$. If not, end the session by outputting **abort**$_2$. Otherwise, continue.
7. After the outputs of the adversary, if none of $s_0 * X$, $s_0 * X^t$, $s_1 * X$, and $s_1 * X^t$ are in the list L, then end the session by outputting **abort**$_2$. Otherwise, the simulator outputs whatever the adversary outputs.

We claim $\{HYBRID^{\mathcal{F}_{RO}, \mathcal{F}_{TSC}}_{\pi, \mathcal{A}(z), 2}(x, y)\}_{x,y,z} =_c \{IDEAL_{\mathcal{F}_{OT}, \mathcal{S}_2(z), 2}(x, y)\}_{x,y,z}$. In comparison with the real execution, the abort in Step 5 implies the solution to the reciprocal CSIDH problem $(E, s * E)$ lies in the list L, which contradicts the assumption. The other abort in Step 7, together with the result of Step 6, implies the adversary decrypts the ciphertext $c_{1,j}$ without the knowledge of the key. If this occurs with non-negligible probability, then it contradicts the non-committing assumption since the real ciphertext can be decrypted without the key, while the dummy ciphertext cannot be (because it can be generated before the plaintext by B_1).

Other differences caused by the simulator are the ciphertexts for the receiver. The ciphertext $c_{0,i}$ in the pair $(c_{0,0}, c_{0,1})$ is indistinguishable from the one in the real execution due to the non-committing encryption scheme. The only suspicious part is $c_{0,i-1}$, which is a dummy ciphertext generated by the algorithm B_1 of the encryption scheme. The counterpart in the real execution is the encrypted message $\mathsf{E}_{k_{1-i}}(M_{1-i})$ where k_{1-i} is either $H(s * X)$ or $H(s^{-1} * X)$.

Similar to the previous proof, the distinguisher (the environment machine) can only succeed with negligible advantage only without the knowledge of k_{1-i}. Precisely, let $\mathbf{E}$ denote the event that the targeted curves $s * X, s * X^t$ are both queried where $(s * X^t)^t = s^{-1} * X$. We have that $|\Pr[\mathcal{Z}(\mathcal{H}_0) = 1] - \Pr[\mathcal{Z}(\mathcal{H}_1) = 1]|$ is not greater than $\Pr[E] + |\Pr[\mathcal{Z}(\mathcal{H}_0) = 1 \mid \neg E] - \Pr[\mathcal{Z}(\mathcal{H}_1) = 1 \mid \neg E]|$.

Claim that $|\Pr[\mathcal{Z}(\mathcal{H}_0) = 1 \mid \neg \mathbf{E}] - \Pr[\mathcal{Z}(\mathcal{H}_1) = 1 \mid \neg \mathbf{E}]|$ is negligible if the encryption scheme is non-committing. Given the non-committing challenge (c, k), a solver runs as follows:

1. Randomly generate $j \in \{0, 1\}$.
2. Run as the simulator $\mathcal{S}_2$ with the environment machine except in Step 4 that assign value c to the variable $c_{0,j}$

3. Say the simulation in Step 2 extracts i from the input of the receiver. If $i \neq j$, then abort and restart the session.
4. If the environment machine judges the machine as the ideal machine, then output 1. Otherwise, output 0.

If $\mathcal{Z}$ succeeds with non-negligible advantage $p(n)$ without the knowledge of the key, then the reduction can win the non-committing challenge with non-negligible advantage $p(n)/2$ where the loss is caused by the guess in Step 2.

Since $|\Pr[\mathcal{Z}(\mathcal{H}_0) = 1 \mid \neg\mathbf{E}] - \Pr[\mathcal{Z}(\mathcal{H}_1) = 1 \mid \neg\mathbf{E}]|$ is negligible, we have

$$|\Pr[\mathcal{Z}(\mathcal{H}_0) = 1] - \Pr[\mathcal{Z}(\mathcal{H}_1) = 1]| \leq \Pr[\mathbf{E}] + negl(n).$$

Therefore, if the distinguisher can succeed with non-negligible advantage, then the solution for the reciprocal CSIDH problem (Problem 5) is in the list of the hash queries with non-negligible probability. Let the challenge of the reciprocal CSIDH problem start with E. A solver $\mathcal{B}$ for the problem runs as follows:

1. Run as the simulator $\mathcal{S}_2$ with the environment machine except for the changes in Step 4 and 5, and an extraction in Step 3. The solver $\mathcal{B}$ commits to the curve X obtained in Step 3 in the reciprocal CSIDH experiment.
2. Say $\mathcal{B}$ receives $s * E$ from the challenge. Then, in Step 4, assign $s * E$ to the variable A_0.
3. In Step 5, guess $i \in \{0,1\}$ and obtain the decryption key k_i via $B_2(c_{0,i}, M_i)$. Randomly pick a curve X_1 of $\mathcal{F}_{RO}$ queries, and assign k_i to it. (Due to the unknown element s, the solver needs to guess here.)
4. After the simulation, if the environment machine judges the machine as the ideal machine, then randomly pick a curve X_2 in the hash query list, and output (X_1, X_2). Otherwise, restart the challenge.

If the environment machine can win with non-negligible advantage $p(n)$ with q hash queries, then the solver $\mathcal{B}$ can win the reciprocal CSIDH challenge with non-negligible advantage $p(n)/(2q^2)$ where the loss is caused by the guesses in Step 3 and 4. To sum up, if the encryption scheme is non-committing, and the reciprocal CSIDH problem is hard, then the simulator $\mathcal{S}_2$ indistinguishably simulates the adversary.

We remark that the simulator $\mathcal{S}_2$ correctly extracts the input of the adversary in Step 5. According to Corollary 1, if the simulator extracts the wrong input, then the adversary can also be used to solve the inverse CSIDH problem.

(Corrupted Sender and Honest Receiver) Let $\mathcal{A}$ be a malicious adversary controlling the sender. In order to emulate the adversary, the simulator needs to extract the input of the adversary, and send it to the trusted party in the ideal execution. The input here is the message pair which the honest receiver will read. Say the environment machine $\mathcal{Z}$ generates input $(x = (M_0, M_1), y = i, z)$ and gives (x, z) to the simulator. The simulator $\mathcal{S}_1$ with input (x, z) proceeds as follows:

1. Firstly, the simulator $\mathcal{S}_1$ emulates a random oracle $\mathcal{F}_{RO}$ by keeping a list L in $\mathcal{E} \times \mathcal{K}$ that records every past query. It initializes the random oracle with an empty list L. Whenever it receives a query on $E' \in \mathcal{E}$, the simulator checks whether $(E', k') \in L$ for some $k' \in \mathcal{K}$. If not, it generates $k' \leftarrow_\$ \mathcal{K}$ and adds the entry (E', k') to the list L. Finally, $\mathcal{S}_1$ returns k' to emulate the random oracle.
2. Generate the public curve $E = t * E_0$ by sampling $t \leftarrow_\$ Cl$ to simulate $\mathcal{F}_{TSC}$. Invoke the adversary $\mathcal{A}$ with the input (x, z) and E. Keep t as the trapdoor secret.
3. Generate $r \leftarrow_\$ Cl$,, and compute $C = r * E$. Send C to the adversary, and act as the procedure of an honest receiver with the input $i = 0$ throughout the remaining execution. (Note that the simulator does not know the input of the receiver here.)
4. If the adversary aborts, then send $\mathbf{abort}_1$ to $\mathcal{F}_{OT}$ and finish the session. Otherwise, assume the execution is not aborted. Say it receives $(A_0, A_1, c_{0,0}, c_{0,1}, c_{1,0}, c_{1,1})$ from the adversary. Compute $k_0 = H(r * A_0)$, $k_1 = H((tr * (t^{-1} * A_0)^t)^t)$, and $m_j = \mathsf{D}_{k_j}(c_{0,j})$ for $j \in \{0, 1\}$.
5. Send (m_0, m_1) to the trusted third party in the ideal execution, output whatever the adversary outputs to complete the simulation. (Note that (M_0, M_1) and (m_0, m_1) are not necessary the same since the adversary can change the original input.)

Claim $\{HYBRID^{\mathcal{F}_{RO}, \mathcal{F}_{TSC}}_{\pi, \mathcal{A}(z), 1}(x, y)\}_{x,y,z} =_c \{IDEAL_{\mathcal{F}_{OT}, \mathcal{S}(z), 1}(x, y)\}_{x,y,z}$. In contrast to the real execution, there are two differences here. Firstly, the simulator possesses the trapdoor t of the public curve. The process is identical to $\mathcal{F}_{TSC}$, and the simulator acts as an honest receiver throughout the process. Hence, this difference is unnoticeable to the adversary.

The other difference is the receiver the simulator plays always uses input $i = 0$. By Lemma 1, the distribution of the first message (C) in the protocol as $i = 0$ is indistinguishable to that generated as $i = 1$. Hence, it suffices to show the correctness of the extraction in Step 4.

If an honest receiver sends C to the sender with the input $i = 0$, then the decryption key is $k_0 = H(r * A_0)$. The message the receiver will obtain is $D_{k_0}(c_{0,0}) = m_0$. Besides, if an honest receiver sends C to the sender with the input $i = 1$, then the private ideal is equivalent to $r^{-1}t^{-2}$ since $(r^{-1}t^{-2} * E)^t = (r^{-1}t^{-1} * E_0)^t = (r^{-1} * E^t)^t = (r * E) = C$. Hence, the receiver will decrypt $c_{0,1}$ with $H(r^{-1}t^{-2} * A_0)$. Due to $k_1 = H((tr * (t^{-1} * A_0)^t)^t) = H((tr)^{-1}t^{-1} * A_0) = H(r^{-1}t^{-2} * A_0)$, the receiver will therefore get the message $m_1 = \mathsf{D}_{k_1}(c_{0,1})$. That is, the simulator correctly extracts the input of the adversary. Hence, the real execution is indistinguishable from the ideal execution.

Remark. In the formal description of [9], the environment machine and the adversary (simulator) starts with z, and the inputs of the parties are given through further instruction messages. Regarding readability and simplicity, we combine them into a single statement here without undermining the effectiveness of the proof.

5 Comparison

5.1 Efficiency

Table 1 illustrates a comparison between our oblivious transfer protocols with [1,2,17,38] in terms of efficiency, including the number of curves in the domain parameters or generated by a trusted party, the number of curves in the public keys for the sender and the receiver, the total number of isogeny computations for the sender and the receiver, and the number of rounds, respectively. Among the isogeny-based OTs, our 2-round OT proposal is the most efficient with respect to every criteria against semi-honest adversaries. It only takes an additional round and two isogeny computations for each participant to achieve UC-secure against static malicious adversaries.

Table 1. Comparison between isogeny-based OTs on efficiency where n is the security parameter. We give the costs for both our 2-round protocol from Fig. 3 and the full construction from Fig. 4.

Proposal	DP	PK_S	PK_R	$\# \; Iso_S$	$\# \; Iso_R$	$\# \; rounds$	Others
[2]	1	1	1	3	2	3	SIDH-based
[17] I	2	1	1	3	2	2	
[17] II	1	3	1	5	2	3	
[38]	1	2	1	4	2	3	Insecure in CSIDH
[1] I	4	$2n$	2	$4n$	$n+2$	2	Group-action-based
[1] II	1	$2n$	5	$4n$	$n+5$	2	Single Bit Transfer
This paper (Fig. 3)	1	1	1	3	2	2	CSIDH-based
This paper (Fig. 4)	1	1	1	5	4	3	CSIDH-based

In [2], they used some properties of SIDH. The receiver randomly subtracts two selected points $U, V \in E_B$ to the points $(\phi_B(P_A), \phi_B(Q_A))$ to produce public points $(\hat{G_A}, \hat{H_A})$ with respect to the secret bit i. The sender adds the same points jU, jV to the received points for $j \in \{0, 1\}$ to produce two decryption keys. The additional mechanism allows the receiver and the sender to generate the same points U, V. As stated in their work, randomly generated $U, V \in E_B$ may reveal the secret bit to an honest-but-curious sender by checking the equality of Weil pairings $e(P_A, Q_A)^{l_A^{e_A}}$, $e(\hat{G_A}, \hat{H_A})$, and $(\hat{G_A} + \lambda U, \hat{H_A} + \lambda V)$ for $\lambda \in \mathbb{Z}$. On the other hand, it is also possible that the honest-but-curious receiver gets the isomorphic curves. In order to prevent these, the U, V are generated through a delicate process.

The two frameworks of [17] includes DH, SIDH, and CSIDH settings. The first construction is a two-message oblivious transfer and requires one more curve in the trusted setup phase.

The paper [38] showed a construction based on exponentiation-only Diffie-Hellman. The construction can fit in the DH, SIDH, and CISDH settings. But, as stated in their work, it will be totally insecure in the CSIDH setting against a malicious receiver. Specifically, their two-inverse problem is given curves $(E, a * E, b * E)$ to find some curve tuple $(X, a^{-1} * X, b^{-1} * X)$ where X is isogenous to E. This can be done in the CSIDH setting by taking quadratic twists of $(E, a * E, b * E)$.

In [1], both constructions are based on the decisional group action problem (the decisional CSIDH problem for instance). If the number of isogeny computations in the encryption (and decryption) algorithm is $\ell = \omega(\log(n))$, then the statistical distance between a pair of ciphertexts is $\Delta = n^{-\omega(1)}$. In particular, the parameter ℓ here is taken to be n so that the distance is less than 2^{-n}.

5.2 Security

Table 2. Comparison between previous isogeny OTs and our constructions. The models include the random oracle model (ROM), the common reference string model (CRS) and trusted setup curves (TSC).

	Adversary Model	Security Definition	Model
[2]	≤Semi-honest*	Simulatable*	ROM+CRS
[17] I	Semi-honest	UC-realize	ROM+TSC
[17] II	Semi-honest	UC-realize	ROM+TSC
[38]	Malicious	Semantic	Plain
[1] I	Malicious	UC-realize	CRS
[1] II	Malicious	SSP	Plain
This paper (Fig. 3)	Semi-honest	UC-realize	ROM+TSC
This paper (Fig. 4)	Malicious	UC-realize	ROM+TSC

On the issue of security, a comparison is shown in Table 2. In [2], the claim is incorrect. Firstly, the adapted definition is Definition 2.6.1 of [24] that guarantees the privacy in the presence of *malicious* adversaries for a *two-round* oblivious transfer protocol while the scheme in [2] is *three-round*. Except for the misuse of the definition, the view-based simulation proof is incomplete even against semi-honest adversaries. The evidence is the further algebraic analysis appended after the proof. The context manifests that the protocol might still leak information even both the sender and the receiver follow the protocol specification. In other words, the proof is incomplete even against semi-honest adversaries.

In [17], the schemes are universally composable secure in the semi-honest model. In [38], they proposed a security definition called the semantic security of oblivious transfer, which guarantees indistinguishability for the sender within the distinct executions. The scheme is under a weak decisional problem which, in the SIDH setting, is easier than the decisional SIDH problem.

Section 4.2 and 4.3 of [1] present two OT constructions. Through using group actions and developing new tools, the first one is derived from a dual-mode public key encryption based on the Diffie-Hellman setting of [34]. The second construction is a plain model OT, which is *statistically sender-private*. The notion ensures computational indistinguishability privacy for the receiver and statistical indistinguishability privacy for the sender. The schemes' main drawback is efficiency since both of them are bit-transferring and require a $poly(n)$ number of isogeny computations.

Remark. One can also show that the construction of Fig. 3 is a *private oblivious transfer* (Definition 2.6.1 in [24]) ensuring privacy for both parties in the malicious model. Since the proof would sidetrack the goal of this work, we leave this to the reader.

6 Conclusion

In this paper, we present the first practical UC-secure isogeny-based oblivious transfer protocol in the presence of static corruptions and malicious adversaries. The construction is simple and compact, and the number of isogeny computations is constant. Moreover, the scheme shares the same hardness as the CSIDH key agreement scheme.

To achieve this outcome, we developed six techniques in this work. In the beginning, the communication bandwidth is reduced through mixing the key-exchange-type problem and an equivalent variant. Next, by utilizing a new use of quadratic twists, we not only compress the number of rounds of the protocol but also fortify the hardness of the underlying assumption (achieving the self-reconciling property). By combining the self-reconciling proposition and proof of ability to decrypt at the cost of one extra round, the simulator is able to extract the input of the receiver to achieve one-sided simulation. Furthermore, for the purpose of extracting the input of the sender, we set up trapdoors for the protocol via a new use of quadratic twists to get a fully-simulatable construction. Finally, we develop a new computational assumption as well as the inverse and square variants and prove equivalence to the standard CSIDH assumption with quantum reductions.

We remark that these techniques are not exclusive to isogeny-based cryptography except for the use of quadratic twists. We envisage that these techniques can serve as potential cryptographic tools in future work.

Acknowledgments. We sincerely thank the anonymous reviewers of EUROCRYPT 2021 for their patience and valuable comments that helped to substantially improve the presentation of this work. We are also grateful to Wouter Castryck for sharing his knowledge of isogenies and Yehuda Lindell for sharing his knowledge of MPC. This research is partially funded by the Ministry for Business, Innvovation and Employment in New Zealand.

This work was supported in part by ERC Advanced Grant ERC-2015-AdG-IMPaCT and by CyberSecurity Research Flanders with reference number VR20192203.

Any opinions, findings and conclusions or recommendations expressed in this material are those of the author(s) and do not necessarily reflect the views of the ERC or of Cyber Security Research Flanders.

References

1. Alamati, N., De Feo, L., Montgomery, H., Patranabis, S.: Cryptographic group actions and applications. In: Moriai, S., Wang, H. (eds.) ASIACRYPT 2020. LNCS, vol. 12492, pp. 411–439. Springer, Cham (2020). https://doi.org/10.1007/978-3-030-64834-3_14
2. Barreto, P., Oliveira, G., Benits, W., Nascimento, A.: Supersingular isogeny oblivious transfer. Cryptology ePrint Archive, report 2018/459 (2018). https://eprint.iacr.org/2018/459
3. Barreto, P.S., David, B., Dowsley, R., Morozov, K., Nascimento, A.C.: A framework for efficient adaptively secure composable oblivious transfer in the ROM, arXiv preprint arXiv:1710.08256 (2017)
4. Bellare, M., Micali, S.: Non-interactive oblivious transfer and applications. In: Brassard, G. (ed.) CRYPTO 1989. LNCS, vol. 435, pp. 547–557. Springer, New York (1990). https://doi.org/10.1007/0-387-34805-0_48
5. Bernstein, D., de Feo, L., Leroux, A., Smith, B.: Faster computation of isogenies of large prime degree, arXiv preprint arXiv:2003.10118 (2020)
6. Beullens, W., Kleinjung, T., Vercauteren, F.: CSI-FiSh: efficient isogeny based signatures through class group computations. In: Galbraith, S.D., Moriai, S. (eds.) ASIACRYPT 2019. LNCS, vol. 11921, pp. 227–247. Springer, Cham (2019). https://doi.org/10.1007/978-3-030-34578-5_9
7. Burdges, J., Feo, L.D. Delay encryption. Cryptology ePrint Archive, report 2020/638 (2020). https://eprint.iacr.org/2020/638
8. Canetti, R.: Security and composition of multiparty cryptographic protocols. J. Cryptol. **13**, 143–202 (2000)
9. Canetti, R.: Universally composable security: a new paradigm for cryptographic protocols, in Proceedings 42nd IEEE Symposium on Foundations of Computer Science, pp. 136–145. IEEE (2001)
10. Canetti, R., Feige, U., Goldreich, O., Naor, M.: Adaptively secure multi-party computation. In: Proceedings of the Twenty-Eighth Annual ACM Symposium on Theory of Computing, pp. 639–648 (1996)
11. Canetti, R., Kushilevitz, E., Lindell, Y.: On the limitations of universally composable two-party computation without set-up assumptions. In: Biham, E. (ed.) EUROCRYPT 2003. LNCS, vol. 2656, pp. 68–86. Springer, Heidelberg (2003). https://doi.org/10.1007/3-540-39200-9_5
12. Castryck, W., Lange, T., Martindale, C., Panny, L., Renes, J.: CSIDH: an efficient post-quantum commutative group action. In: Peyrin, T., Galbraith, S. (eds.) ASIACRYPT 2018. LNCS, vol. 11274, pp. 395–427. Springer, Cham (2018). https://doi.org/10.1007/978-3-030-03332-3_15
13. Chou, T., Orlandi, C.: The simplest protocol for oblivious transfer. In: Lauter, K., Rodríguez-Henríquez, F. (eds.) LATINCRYPT 2015. LNCS, vol. 9230, pp. 40–58. Springer, Cham (2015). https://doi.org/10.1007/978-3-319-22174-8_3
14. Couveignes, J.M.: Hard homogeneous spaces. 1997, IACR Cryptology ePrint Archive, 2006, p. 291 (2006)

15. Crépeau, C., van de Graaf, J., Tapp, A.: Committed oblivious transfer and private multi-party computation. In: Coppersmith, D. (ed.) CRYPTO 1995. LNCS, vol. 963, pp. 110–123. Springer, Heidelberg (1995). https://doi.org/10.1007/3-540-44750-4_9
16. David, B.M., Nascimento, A.C.A., Müller-Quade, J.: Universally composable oblivious transfer from lossy encryption and the McEliece assumptions. In: Smith, A. (ed.) ICITS 2012. LNCS, vol. 7412, pp. 80–99. Springer, Heidelberg (2012). https://doi.org/10.1007/978-3-642-32284-6_5
17. de Saint Guilhem, C., Orsini, E., Petit, C., Smart, N.P.: Secure oblivious transfer from semi-commutative masking. IACR Cryptology ePrint Archive, 2018, p. 648 (2018)
18. Döttling, N., Garg, S., Hajiabadi, M., Masny, D., Wichs, D.: Two-round oblivious transfer from CDH or LPN. In: Canteaut, A., Ishai, Y. (eds.) EUROCRYPT 2020. LNCS, vol. 12106, pp. 768–797. Springer, Cham (2020). https://doi.org/10.1007/978-3-030-45724-2_26
19. Dowsley, R., van de Graaf, J., Müller-Quade, J., Nascimento, A.C.A.: Oblivious transfer based on the McEliece assumptions. In: Safavi-Naini, R. (ed.) ICITS 2008. LNCS, vol. 5155, pp. 107–117. Springer, Heidelberg (2008). https://doi.org/10.1007/978-3-540-85093-9_11
20. Felderhoff, J.: Hard homogeneous spaces and commutative supersingular isogeny based diffie-hellman, internship report, LIX, Ecole polytechnique, ENS de Lyon, August 2019
21. De Feo, L., Galbraith, S.D.: SeaSign: compact isogeny signatures from class group actions. In: Ishai, Y., Rijmen, V. (eds.) EUROCRYPT 2019. LNCS, vol. 11478, pp. 759–789. Springer, Cham (2019). https://doi.org/10.1007/978-3-030-17659-4_26
22. Goldreich, O., Micali, S., Wigderson, A.: How to play any mental game. In: Proceedings of the Nineteenth ACM Symposium on Theory of Computing, STOC, pp. 218–229. ACM (1987)
23. Hallgren, S.: Fast quantum algorithms for computing the unit group and class group of a number field. In: Proceedings of the 37th Annual ACM Symposium on Theory of Computing, Baltimore, MD, USA, May 22–24, 2005, pp. 468–474 (2005)
24. Hazay, C., Lindell, Y.: Efficient Secure Two-Party Protocols. ISC, Springer, Heidelberg (2010). https://doi.org/10.1007/978-3-642-14303-8
25. Jao, D., et al.: Sike: supersingular isogeny key encapsulation (2017). https://sike.org/
26. Jao, D., De Feo, L.: Towards quantum-resistant cryptosystems from supersingular elliptic curve isogenies. In: Yang, B.-Y. (ed.) PQCrypto 2011. LNCS, vol. 7071, pp. 19–34. Springer, Heidelberg (2011). https://doi.org/10.1007/978-3-642-25405-5_2
27. Lai, Y.-F., Galbraith, S.D., de Saint Guilhem, C.D.: Compact, efficient and UC-secure isogeny-based oblivious transfer. Cryptology ePrint Archive, report 2020/1012 (2020). https://eprint.iacr.org/2020/1012
28. Lindell, A.Y.: Efficient fully-simulatable oblivious transfer. In: Malkin, T. (ed.) CT-RSA 2008. LNCS, vol. 4964, pp. 52–70. Springer, Heidelberg (2008). https://doi.org/10.1007/978-3-540-79263-5_4
29. Meyer, M., Campos, F., Reith, S.: On Lions and elligators: an efficient constant-time implementation of CSIDH. In: Ding, J., Steinwandt, R. (eds.) PQCrypto 2019. LNCS, vol. 11505, pp. 307–325. Springer, Cham (2019). https://doi.org/10.1007/978-3-030-25510-7_17
30. Naor, M., Pinkas, B.: Efficient oblivious transfer protocols. In: Proceedings of the Twelfth Annual ACM-SIAM Symposium on Discrete Algorithms, Society for Industrial and Applied Mathematics, pp. 448–457 (2001)

31. NIST: National institute of standards and technology (2020). https://csrc.nist.gov/Projects/post-quantum-cryptography/round-3-submissions
32. Oded, G.: Foundations of cryptography: Volume 2, basic applications (2009)
33. Onuki, H., Aikawa, Y., Yamazaki, T., Takagi, T.: A constant-time algorithm of CSIDH keeping two points. IEICE Trans. Fundam. Electron. Commun. Comput. Sci. **103**, 1174–1182 (2020)
34. Peikert, C., Vaikuntanathan, V., Waters, B.: A framework for efficient and composable oblivious transfer. In: Wagner, D. (ed.) CRYPTO 2008. LNCS, vol. 5157, pp. 554–571. Springer, Heidelberg (2008). https://doi.org/10.1007/978-3-540-85174-5_31
35. Rabin, M.O.: How to exchange secrets with oblivious transfer, Technical report TR-81, p. 187. Harvard University, Aiken Computation Lab (1981)
36. Rostovtsev, A., Stolbunov, A.: Public-key cryptosystem based on isogenies. IACR Cryptology ePrint Archive, 2006, p. 145 (2006)
37. Shor, P.W.: Polynomial-time algorithms for prime factorization and discrete logarithms on a quantum computer. SIAM Rev. **41**, 303–332 (1999)
38. Vitse, V.: Simple oblivious transfer protocols compatible with supersingular isogenies. In: Buchmann, J., Nitaj, A., Rachidi, T. (eds.) AFRICACRYPT 2019. LNCS, vol. 11627, pp. 56–78. Springer, Cham (2019). https://doi.org/10.1007/978-3-030-23696-0_4
39. Wu, Q.-H., Zhang, J.-H., Wang, Y.-M.: Practical t-out-n oblivious transfer and its applications. In: Qing, S., Gollmann, D., Zhou, J. (eds.) ICICS 2003. LNCS, vol. 2836, pp. 226–237. Springer, Heidelberg (2003). https://doi.org/10.1007/978-3-540-39927-8_21

One-Way Functions and Malleability Oracles: Hidden Shift Attacks on Isogeny-Based Protocols

Péter Kutas[1(✉)], Simon-Philipp Merz[2], Christophe Petit[1,3], and Charlotte Weitkämper[1]

[1] University of Birmingham, Birmingham, UK
P.Kutas@bham.ac.uk
[2] Royal Holloway, University of London, London, UK
[3] Université libre de Bruxelles, Brussels, Belgium

Abstract. Supersingular isogeny Diffie-Hellman key exchange (SIDH) is a post-quantum protocol based on the presumed hardness of computing an isogeny between two supersingular elliptic curves given some additional torsion point information. Unlike other isogeny-based protocols, SIDH has been widely believed to be immune to subexponential quantum attacks because of the non-commutative structure of the endomorphism rings of supersingular curves.

We contradict this commonly believed misconception in this paper. More precisely, we highlight the existence of an abelian group action on the SIDH key space, and we show that for sufficiently *unbalanced* and *overstretched* SIDH parameters, this action can be efficiently computed (heuristically) using the torsion point information revealed in the protocol. This reduces the underlying hardness assumption to a hidden shift problem instance which can be solved in quantum subexponential time.

We formulate our attack in a new framework allowing the inversion of one-way functions in quantum subexponential time provided a malleability oracle with respect to some commutative group action. This framework unifies our new attack with earlier subexponential quantum attacks on isogeny-based protocols, and it may be of further interest for cryptanalysis.

1 Introduction

The hardness of solving mathematical problems such as integer factorization or the computation of discrete logarithms in finite fields and elliptic curve groups guarantees the security of most currently deployed cryptographic protocols. However, these classical problems can be solved efficiently using quantum algorithms. Quantum computers with sufficient processing power to threaten cryptographic primitives currently in use do presumably not yet exist, but progress towards their realization is being made. The possibility of large scale quantum computers and the need for long-term security in some applications necessitate the development of quantum-secure cryptographic algorithms.

A. Canteaut and F.-X. Standaert (Eds.): EUROCRYPT 2021, LNCS 12696, pp. 242–271, 2021.
https://doi.org/10.1007/978-3-030-77870-5_9

Different approaches to attain quantum resistance are based on problems in lattices, codes, multivariate polynomials over finite fields, and elliptic curve isogenies. Within the field of post-quantum cryptography, isogeny-based cryptography is a relatively new area which is particularly interesting due to the small key sizes required. The main problem underlying this branch of post-quantum cryptography is to find an isogeny $\varphi : E_1 \to E_2$ between two given isogenous elliptic curves E_1 and E_2 over some finite field $\mathbb{F}_q$.

An early isogeny-based cryptographic system utilizing *ordinary* elliptic curves was proposed by Couveignes but at first only circulated privately [7]. Meanwhile, the first construction using *supersingular* curves was a hash function developed by Charles, Lauter and Goren [4]. Later, Rostovtsev and Stolbunov independently rediscovered and further developed Couveignes' construction [27]. In 2010, Childs, Jao and Soukharev [5] showed how to break this scheme in quantum subexponential time using a reduction to an instance of abelian hidden shift problem. While this attack is tolerable for sufficiently large parameters, the main drawback of the Couveignes-Rostovtsev-Stolbunov (CRS) construction is its unacceptable lack of speed. Adapting the CRS scheme to supersingular elliptic curves, Castryck et al. managed to eliminate most of the performance issues allowing for larger practical parameters when introducing CSIDH [3]. While it is known that CSIDH can be attacked in quantum subexponential time, there have been several works on establishing its concrete security levels [2,22].

The attack due to Childs, Jao and Soukharev crucially relies on the commutativity of the ideal class groups acting on the endomorphism rings of the relevant elliptic curves over $\mathbb{F}_q$. This motivated Jao and De Feo [14] to consider the full isogeny graph of supersingular elliptic curves, whose endomorphism rings are maximal orders in a quaternion algebra (in particular, the endomorphism rings are non-commutative). The result of their work, the *Supersingular Isogeny Diffie-Hellman* (SIDH) key agreement scheme, underlies the SIKE submission to NIST's post-quantum standardization process [1,13].

The hard problem SIDH is based on is to find an isogeny between two isogenous curves, further given the images of certain torsion points under this isogeny. The best known way to break SIDH with balanced parameters on both classical and a quantum computers is a claw-finding approach on the isogeny graph [15] which does not use any torsion point information. Yet, the supply of this additional public information has fueled cryptanalytic research. It has been shown that the torsion point information can be used in *active* attacks [10] or when parameters are sufficiently overstretched [19,23]. However, a widespread misconception amongst cryptographers assumes that due to SIDH's non-commutative nature there is no quantum attack reducing the SIDH problem to an abelian hidden shift problem. In particular, many believe that no reasonable variant of Childs-Jao-Soukharev's attack applies in the supersingular case [14, p. 18, Sect. 5].

Our Contributions. We provide a new quantum attack on overstretched SIDH which uses a reduction of the underlying computational problem to an injective abelian hidden shift problem. This can be solved in quantum subexponential time and thus disproves the common misbelief mentioned above.

Let $\varphi : E_0 \to E_0/K$ be a secret isogeny that an attacker wishes to recover. As in SIDH, let E_0, E_0/K, $\deg(\varphi)$, and some torsion point images under the secret isogeny be known publicly. The idea underlying our cryptanalysis is to construct an abelian group G of E_0-endomorphisms acting freely and transitively on certain cyclic subgroups of E_0. These subgroups are kernels of $\deg(\varphi)$-isogenies, and therefore they can be mapped to supersingular elliptic curves $\deg(\varphi)$-isogenous to E_0. The group action of G can then be understood as an action on the curves. Forcing the endomorphisms in G to be of a certain degree, the public torsion point information allows an adversary to compute the action on E_0/K efficiently under some heuristics. Finally, solving an abelian hidden shift problem of two functions mapping G to a set of curves $\deg(\varphi)$-isogenous to E_0 containing E_0/K enables an attacker to recover K and therefore φ. We stress that this is a novel way of exploiting torsion point information.

While this attack does not threaten SIDH with balanced parameter sets as originally proposed by Jao and De Feo [14] and used in SIKE [13], it shows that an attack using a hidden shift algorithm is possible despite SIDH's non-commutative nature.

We describe our new attack as a special instance in a more general setting. This allows us to unify our new cryptanalysis with other quantum attacks on isogeny-based schemes such as the one due to Childs, Jao and Soukharev [5] constructing isogenies between ordinary curves, or a similar application of quantum hidden shift algorithms to CSIDH [2,3,22].

This framework might be of interest beyond isogeny-based cryptography. To define one of the key properties required, we introduce the notion of a *malleability oracle* for a function with respect to some group action. Under some additional assumptions, access to this oracle is sufficient to compute preimages of the function via solving a hidden shift problem.

Outline. In Sect. 2, we provide an overview of the notations we use, we recall mathematical background for isogeny-based cryptography, and we review quantum algorithms used in our attack. In Sect. 3, we present our general framework, namely sufficient conditions for computing preimages of one-way functions via reduction to a hidden shift problem, and then present our new attack on overstretched SIDH in Sect. 4. In Sect. 5, we additionally instantiate our general framework with the attack of Childs, Jao and Soukharev and its generalization to CSIDH. We conclude the paper in Sect. 6 with a discussion of potential improvements and future work.

2 Preliminaries

In this section, we introduce terminology and notation, and we recall relevant background on isogeny-based protocols and quantum algorithms.

2.1 Terminology

We call a function $\mu : \mathbb{N} \to \mathbb{R}$ *negligible* if for every positive integer c there exists an integer N_c such that $|\mu(x)| < \frac{1}{x^c}$ for every $x > N_c$. We call an algorithm *efficient* if the execution time is bounded by a polynomial in the security parameter of the underlying cryptographic scheme. Given any function, by having *oracle access* to this function we mean that it is feasible to evaluate the function at any possible element in an efficient way. In particular, we assume that the oracle acts like a black box such that one query with an element from the domain outputs the corresponding value of the function.

Further, we call a function $f : \{0,1\}^* \to \{0,1\}^*$ *one-way*, if f can be computed by a polynomial-time algorithm, but for all polynomial-time randomized algorithms F, all positive integers c and all sufficiently large $n = \text{length}(x)$, $\Pr[f(F(f(x))) = f(x)] < n^{-c}$, where the probability is taken over the choice of x from the discrete uniform distribution on $\{0,1\}^n$, and the randomness of F.

2.2 Mathematical Background on Isogenies

For more complete introductions to elliptic curves and to isogeny-based cryptography we refer to Silverman [29] and De Feo [8], respectively.

Let $\mathbb{F}_q$ be a finite field of characteristic p. In the following we assume $p \geq 3$ and therefore an elliptic curve E over $\mathbb{F}_q$ can be defined by its short Weierstrass form

$$E(\mathbb{F}_q) = \{(x, y) \in \mathbb{F}_q^2 \mid y^2 = x^3 + Ax + B\} \cup \{\mathcal{O}_E\}$$

where $A, B \in \mathbb{F}_q$ and $\mathcal{O}_E$ is the point $(X : Y : Z) = (0 : 1 : 0)$ on the associated projective curve $Y^2Z = X^3 + AXZ^2 + BZ^3$. The set of points on an elliptic curve is an abelian group under the "chord and tangent rule" with $\mathcal{O}_E$ being the identity element. The *j-invariant* of an elliptic curve is $j(E) = 1728\frac{4A^3}{4A^3+27B^2}$ and there is an isomorphism of curves $f : E_0 \to E_1$ if and only if $j(E_0) = j(E_1)$.

Given two elliptic curves E_0 and E_1 over a finite field $\mathbb{F}_q$, an *isogeny* is a non-constant rational map $\phi : E_0 \to E_1$ defined over $\mathbb{F}_q$ which is also a group homomorphism from $E_0(\mathbb{F}_q)$ to $E_1(\mathbb{F}_q)$. Two curves are called *isogenous* if there exists an isogeny between them. The *degree* of an isogeny ϕ is its degree as a rational map. For separable isogenies, the degree is also equal to the number of elements in the kernel of ϕ. Note that we will always consider the separable case in the following.

Since an isogeny defines a group homomorphism $E_0 \to E_1$, its kernel is a subgroup of E_0. Conversely, any subgroup $S \subset E_0$ determines a (separable) isogeny $\phi : E_0 \to E_1$ with $\ker \phi = S$ and $E_1 = E_0/S$.

An *endomorphism* of an elliptic curve E defined over $\mathbb{F}_q$ is an isogeny defined over an extension of $\mathbb{F}_q$ mapping E onto itself. The set of endomorphisms of E together with the zero map forms a ring under pointwise addition and function composition. This ring is the *endomorphism ring* of E, denoted $\text{End}(E)$, and it is isomorphic either to an order in a quaternion algebra and E is called *supersingular*, or to an order in an imaginary quadratic field and E is referred to as an *ordinary* curve [29].

Let d be a positive integer. Throughout the paper, we say a supersingular elliptic curve E is *at distance* d from E_0 if there exists a separable isogeny ϕ with cyclic kernel of degree d from E_0 to E.

For any isogeny $\phi : E_0 \to E_1$, there exists another isogeny $\hat{\phi}$, called the *dual isogeny*, satisfying $\phi \circ \hat{\phi} = \hat{\phi} \circ \phi = [\deg(\phi)]$, where $[\cdot]$ denotes scalar multiplication. Therefore, the property of being isogenous is an equivalence relation on the set of isomorphism classes of elliptic curves defined over $\mathbb{F}_q$.

2.3 Hard Homogeneous Spaces and CSIDH

Recall the notion of Couveignes' *hard homogeneous spaces* (HHS) [7], a finite commutative group action for which some operations are easy to compute and others are hard.

Instances of Couveignes' hard homogeneous spaces can be constructed using elliptic curve isogenies and have been the basis of one branch of isogeny-based cryptography which uses the group action we will describe in the following.

Denote the set of all isomorphism classes over $\overline{\mathbb{F}_q}$ of isogenous curves with n points and endomorphism ring $\mathcal{O}$ by $\mathrm{Ell}_{q,n}(\mathcal{O})$, and represent the isomorphism class of a curve E in $\mathrm{Ell}_{q,n}(\mathcal{O})$ by the j-invariant $j(E)$. Any isogeny $\varphi : E \to E_{\mathfrak{b}}$ between curves having the same endomorphism ring in $\mathrm{Ell}_{q,n}(\mathcal{O})$ is determined by E and $\ker \varphi$ up to isomorphism. This kernel corresponds to an ideal $[\mathfrak{b}]$ in $\mathcal{O}$. Recall that the ideal class group of $\mathcal{O}$, $\mathrm{Cl}(\mathcal{O})$, is the quotient group of the abelian group of fractional $\mathcal{O}$-ideals under ideal multiplication and all principal fractional $\mathcal{O}$-ideals. Since principal ideals in $\mathcal{O}$ correspond to isomorphisms, ideals that are equivalent in $\mathrm{Cl}(\mathcal{O})$ induce the same isogeny up to isomorphism. Hence, we have a well-defined group action

$$\begin{aligned} \cdot : \mathrm{Cl}(\mathcal{O}) \times \mathrm{Ell}_{q,n}(\mathcal{O}) &\to \mathrm{Ell}_{q,n}(\mathcal{O}), \\ ([\mathfrak{b}], j(E)) &\mapsto j(E_{\mathfrak{b}}), \end{aligned}$$

which is free and transitive ([32], Thm. 4.5, and erratum Thm. 4.5 of [28]).

Given two elliptic curves E_0, E_1 in $\mathrm{Ell}_{q,n}(\mathcal{O})$ up to isomorphism, it is in general assumed to be hard to find an isogeny $\varphi : E_0 \to E_1$.

A similar construction can be performed with endomorphism rings of supersingular curves. This occurrence of hard homogenous spaces is used for the *Commutative SIDH* (CSIDH) protocol [3] proposed for post-quantum non-interactive key exchange. Since the endomorphism rings of such curves are orders in a quaternion algebra, they are non-commutative and hence yield a group action with less desirable properties than in the construction for ordinary curves. Therefore, Castryck et al. suggest restricting the endomorphism ring to the subring of $\mathbb{F}_p$-rational endomorphisms which is an order in an imaginary quadratic field, and as such commutative. Again, the ideal class group of this order $\mathcal{O}$ acts on $\mathrm{Ell}_p(\mathcal{O})$, the set of all isomorphism classes of supersingular isogenous curves over $\mathbb{F}_p$ with $\mathbb{F}_p$-rational endomorphism ring (isomorphic to) $\mathcal{O}$.

Given that the set $\mathrm{Ell}_p(\mathcal{O})$ is non-empty, the group action is free and transitive (see [3], Thm. 7, summarizing results from [28,32]), and can be used to perform a Diffie-Hellman-type key exchange. Note that CSIDH is strictly speaking not an instance of a HHS as it is not possible to compute the group action efficiently for *all* group elements.

There have been multiple proposals to attack concrete parameter suggestions for CSIDH with quantum algorithms. Peikert [22] uses Kuperberg's collimation sieve algorithm to solve the hidden shift instance with quantum accessible classical memory and subexponential quantum time, a strategy independently also explored by Bonnetain-Schrottenloher [2].

2.4 SIDH

We recall the *Supersingular Isogeny Diffie-Hellman* (SIDH) protocol which was introduced by Jao and De Feo in [14] and forms the basis of *Supersingular Isogeny Key Encapsulation* (SIKE) [13] which has been submitted to NIST's post-quantum competition.

Fix some supersingular elliptic curve E_0 over a field $\mathbb{F}_{p^2}$, where p is a prime, and let N_1 and N_2 be two smooth integers coprime to p with $(N_1, N_2) = 1$. Further choose some points $P_A, Q_A, P_B, Q_B \in E_0$ such that P_A and Q_A generate the N_1-torsion of E_0, $E_0[N_1]$, and similarly, $\langle P_B, Q_B \rangle = E_0[N_2]$. Then the protocol is as follows:

1. Alice chooses a random cyclic subgroup of $E_0[N_1]$ generated by a point of the form $A = P_A + [x_A]Q_A$ and Bob chooses some random cyclic subgroup of $E_0[N_2]$ generated by $B = P_B + [x_B]Q_B$.
2. Alice then computes her secret isogeny $\varphi_A : E_0 \to E_0/\langle A \rangle$ and Bob computes his secret isogeny $\varphi_B : E_0 \to E_0/\langle B \rangle$.
3. Alice sends the curve $E_A := E_0/\langle A \rangle$ and the two points $\varphi_A(P_B), \varphi_A(Q_B)$ to Bob while Bob sends $\big(E_B := E_0/\langle B \rangle, \varphi_B(P_A), \varphi_B(Q_A)\big)$ to Alice.
4. Alice and Bob both compute the shared secret curve $E_{AB} := E_0/\langle A, B \rangle$ using the given torsion information, $E_{AB} = E_B/\langle \varphi_B(A) \rangle = E_A/\langle \varphi_A(B) \rangle$.

For SIDH, one chooses the prime p of the form $p = N_1 N_2 f - 1$ with N_1 and N_2 being powers of 2 and 3, respectively. As the above protocol is vulnerable to adaptive attacks (see e.g., [10]), SIKE applies a variant of the Fujisaki-Okamoto transformation due to Hofheinz, Hövelmanns and Kiltz [12] to standard SIDH. To ensure that both Alice and Bob enjoy the same level of security, the recommended parameter sets for SIDH and SIKE suggest balanced parameters, i.e., $N_1 \approx N_2$.

The active attack on standard SIDH presented by Galbraith et al. [10] utilizes the additional information on torsion points to recover a secret key through multiple executions of the protocol with malformed messages. Further, the given torsion point information is exploited in Petit's passive attack [23] on a non-standard variant of SIDH with unbalanced and comparatively large torsion parameters. The requirements on unbalancedness and size of parameters have recently been improved upon by Kutas et al. [19] who additionally show that, even with balanced parameters, there exist certain primes which facilitate an effective-torsion point attack on SIDH.

For our quantum attack to work, we will need to relax the balancedness condition of standard SIDH and require one of N_1 and N_2 to be larger than the other by a certain factor. In particular, we need $N_1 N_2 \gg p$ which prohibits choosing p

as suggested by Jao-De Feo. We call this variant of SIDH *overstretched*. Note that this variant of SIDH is still polynomial time as long as N_1 and N_2 are smooth numbers, albeit much slower in practice than with the suggested parameters.

SIDH is believed to be immune to subexponential quantum attacks [1,13,14]. In particular, it has been claimed and widely accepted that no reasonable variant of Childs et al.'s attack [5] exists for SIDH [14, p.18, Sect. 5]. Yet, we will show in this paper how to reduce SIDH with overstretched parameters to an abelian hidden shift problem.

2.5 Quantum Algorithms to Solve Hidden Shift Problems

First, we recall what is meant when two functions are said to be shifts of each other, or equivalently that these two functions *hide a shift*.

Definition 2.1. *Let* $F_0, F_1 \colon G \to X$ *be two functions defined on some group* G*, such that there exists some* $s \in G$ *satisfying* $F_0(g) = F_1(g \cdot s)$ *for all* $g \in G$*. The* hidden shift problem *is to find* s *given oracle access to the functions* F_0 *and* F_1*.*

Multiple approaches utilizing quantum computation have been proposed to solve the hidden shift problem. Some of these works have considered different group structures as well as variations on the promise. We summarize some quantum algorithms solving the injective abelian hidden shift problem, i.e., where the functions F_i are injective functions and G is abelian.

The first quantum subexponential algorithm is due to Kuperberg [17] and reduces the hidden shift problem to the hidden subgroup problem in the dihedral group $D_G \simeq C_2 \ltimes G$, i.e., to finding a subgroup of D_G such that a function obtained from combining the input functions of the hidden shift problem is constant exactly on its cosets. It requires quantum subexponential time, namely $2^{\mathcal{O}(\sqrt{\log |G|})}$ quantum queries, for a finite abelian group G. A modification of this method proposed by Regev [26] reduces the memory required by Kuperberg's approach (from super-polynomial to polynomial) while keeping the running time quantum subexponential. Another, slightly faster algorithm, the collimation sieve, using polynomial quantum space was proposed later by Kuperberg [18]. In this variant, parameter trade-offs between classical and quantum running time and quantumly accessible memory are possible.

These algorithms to solve the hidden shift problem when G is abelian generally begin by producing some random quantum states, each with an associated classical "label", by evaluating the group action on a uniform superposition over the group G. For this generation of states, oracle access to the two functions F_0 and F_1 is needed. Then, the hidden shift s is extracted bitwise through performing measurements on specific quantum states (i.e., ones with desirable labels) which are generated from the random states via some sieve algorithm.

3 Malleability Oracles and Hidden Shift Attacks

In this section, we introduce the notion of a *malleability oracle* for a one-way function. Under some conditions, such an oracle allows the computation of preimages of given elements in quantum subexponential time by reduction to the hidden shift problem.

3.1 Malleability Oracles

Recall the definition of a free and transitive group action.

Definition 3.1. *Let G be a group with neutral element e, and let $\mathcal{I}$ be a set. A* (left) group action $\star$ *of G on $\mathcal{I}$ is a function*

$$\star\colon G \times \mathcal{I} \to \mathcal{I}, \quad (g, x) \mapsto g \star x,$$

that satisfies $e \star x = x$, and $gh \star x = g \star (h \star x)$ for all $x \in \mathcal{I}$ and $g, h \in G$.

The group action is called transitive *if and only if $\mathcal{I}$ is non-empty and for every pair of elements $x, y \in \mathcal{I}$ there exists $g \in G$ such that $g \star x = y$. The group action is called* free *if and only if $g \star x = x$ implies $g = e$.*

Next, we define an oracle capturing the main premise required for our strategy to compute preimages of one-way functions.

Definition 3.2. *Let $f : \mathcal{I} \to \mathfrak{O}$ be an injective (one-way) function and let $\star$ be the action of a group G on $\mathcal{I}$. A* malleability oracle *for G at $o := f(i)$ provides the value of $f(g \star i)$ for any input $g \in G$, i.e., the malleability oracle evaluates the map*

$$g \mapsto f(g \star i).$$

We call the function f malleable, *if a malleability oracle is available at every $o \in f(\mathcal{I})$.*

In Sect. 4 we show how a polynomial-time malleability oracle can be constructed in the context of SIDH with overstretched parameters, and in Sect. 5 we describe other contexts where such an oracle arises naturally.

For the remainder of the paper, we will denote the action of a group element $g \in G$ on a set element $i \in \mathcal{I}$ by $g \cdot i$.

3.2 Reduction to Hidden Shift Problem

Given a malleability oracle at $o = f(i)$, computing a preimage of o reduces to a hidden shift problem in the following case.

Theorem 3.3. *Let $f : \mathcal{I} \to \mathfrak{O}$ be an injective (one-way) function and let G be a group acting transitively on $\mathcal{I}$. Given a malleability oracle for G at $o := f(i)$, the preimage of o can be computed by solving a hidden shift problem.*

Proof. Let k be an arbitrary but fixed element in $\mathcal{I}$ and define

$$F_k \ : \ G \to \mathfrak{O} \ , \ \theta \mapsto f(\theta \cdot k).$$

Since f is an injective function, $i = f^{-1}(o)$ is unique and thus F_i is well-defined. Moreover, the malleability oracle allows us to evaluate the function F_i on any $\theta \in G$, as $F_i(\theta) = f(\theta \cdot i)$.

Fix some arbitrary $j \in \mathcal{I}$. Since we know j, we can evaluate F_j on any group element θ by evaluating $f(\theta \cdot j)$ via simply computing the group action. Due to the transitivity of the group action of G, there exists $\sigma \in G$ such that $i = \sigma \cdot j$. Since for all $\theta \in G$

$$F_i(\theta) = f(\theta \cdot i) = f(\theta\sigma \cdot j) = F_j(\theta\sigma),$$

the functions F_j and F_i are shifts of each other. Hence, solving the hidden shift problem for F_i and F_j allows us to recover σ, and thus to compute $i = \sigma \cdot j$. □

The following corollary will be used in our attack on overstretched SIDH.

Corollary 3.4. *Let $f : \mathcal{I} \to \mathfrak{O}$ be an injective (one-way) function and let G be a finitely generated abelian group acting freely and transitively on $\mathcal{I}$. Given a malleability oracle for G at $o := f(i)$, the preimage of o can be computed in quantum subexponential time.*

Proof. To obtain a hidden shift instance solvable by subexponential quantum algorithm such as Kuperberg's, we only have to show that for every $k \in \mathcal{I}$ the function $F_k(\theta) = f(\theta \cdot k)$ is injective. Then the claim follows from Theorem 3.3 and the discussion in Sect. 2.5.

Suppose that $F_k(g) = f(g \cdot k) = f(h \cdot k) = F_k(h)$ for some $g, h \in G$. Since f is injective and the group action is free, this implies $g = h$. □

4 Attack on Overstretched SIDH Instances in Quantum Subexponential Time

Despite the non-commutative nature of SIDH, we show in this section that one can find an abelian group action on its private key space. Moreover for sufficiently *overstretched* SIDH parameters, the torsion point information revealed in the protocol allows us to build a malleability oracle for this group action. This gives rise to an attack using quantum subexponential hidden shift algorithms as outlined in Sect. 3.2.

This section is organized as follows: We first sketch our approach to exploit the torsion point information in Sect. 4.1. We then solve some technical issues in Sects. 4.2, 4.2 and 4.4. These issues require small tweaks to our general approach, and we summarize the resulting algorithm in Sect. 4.5. Finally in Sect. 4.6, we present a hybrid approach to combine guessing part of the secret and computing the remaining part using our new attack; this allows us to slightly extend the attack to further parameter sets.

Throughout this section, we use the following notation. Let $p \equiv 3 \pmod 4$ be prime, let E_0 be the supersingular elliptic curve with j-invariant 1728 defined over $\mathbb{F}_p$, given by the equation $y^2 = x^3 + x$, and let $\mathcal{O}_0 = \mathrm{End}(E_0)$ be its endomorphism ring. Note that $\mathcal{O}_0$ is well-known. More precisely, it is the $\mathbb{Z}$-module generated by $1, \iota, \frac{1+\pi}{2}$ and $\frac{\iota+\iota\pi}{2}$, where ι denotes the non-trivial automorphism of E_0, $(x, y) \mapsto (-x, iy)$, and π is the Frobenius endomorphism, $(x, y) \mapsto (x^p, y^p)$.

Remark 4.1. The attack we describe can be expanded to other curves that are close to E_0, such as the curve used in the updated parameters of SIKE for the second round of NIST's post-quantum standardization effort [1], by computing the isogeny to E_0 and translating the problem to there.

4.1 Overview of the Attack

Let $\mathcal{I}$ be the set of cyclic N_1-order subgroups of E_0, and let $\mathfrak{O}$ be the set of j-invariants of all supersingular curves that are N_1-isogenous to E_0. Let f be the function sending any element of $\mathcal{I}$ to the j-invariant of the codomain of its corresponding isogeny, i.e.,

$$f : \mathcal{I} \to \mathfrak{O}, \quad K \mapsto j(E_0/K). \tag{1}$$

The function f can be efficiently computed on any input using Vélu's formulae [30], provided N_1 is sufficiently smooth and that the N_1-torsion is defined over a sufficiently small extension field of $\mathbb{F}_p$. In SIDH, the latter is achieved by choosing $N_1 | p - 1$, but this is true more generally for sufficiently powersmooth N_1.

On the other hand, inverting f amounts to finding an isogeny of degree N_1 from E_0 to a curve in a given isomorphism class, or equivalently to finding the subgroup of E_0 defining this isogeny. The conjectured hardness of this problem is at the heart of isogeny-based cryptography.

In the SIDH protocol, additional torsion point information is transmitted publicly as part of the exchange, and thus also given to adversaries. For the security proof it is assumed that a variant of the following problem with $N_1 \approx N_2$ is hard [14].

Problem 4.2. *Let p be a large prime, let N_1 and N_2 be two smooth coprime integers such that $E_0[N_1]$ and $E_0[N_2]$ can be represented efficiently, let $K \in \mathcal{I}$ be a cyclic subgroup of order N_1 of E_0 chosen uniformly at random, and let $\varphi : E_0 \to E_0/K$. Given the supersingular elliptic curves E_0 and E_0/K together with the restriction of φ to $E_0[N_2]$, compute K.*

Our attack exploits the information provided by the restriction of the secret isogeny to $E_0[N_2]$ to construct a malleability oracle for f at the (unknown) secret. Following the framework outlined in Sect. 3, this gives rise to an attack on *overstretched* SIDH.

Let G be a subgroup of $(\mathcal{O}_0/N_1\mathcal{O}_0)^*$. Then G induces a group action on $\mathcal{I}$ given by

$$G \times \mathcal{I} \to \mathcal{I}\ , \ (\theta, K) \mapsto \theta(K).$$

Indeed, the degree of any non-trivial representative θ is coprime to N_1 and thus preserves the order of any generator of K.

Note that the full group $(\mathcal{O}_0/N_1\mathcal{O}_0)^*$ is not abelian. Our attack will require an abelian subgroup G acting on $\mathcal{I}$ such that G acts freely and transitively on the orbit of an isogeny kernel of an isogeny $E_0 \to E_0/K$ under this group action, as well as one element in this orbit. This leads to the following task.

Task 4.3. *Let $K \in \mathcal{I}$ be any cyclic subgroup of E_0 of order N_1 chosen uniformly at random and let $\varphi : E_0 \to E_A := E_0/K$. Compute an element $L \in \mathcal{I}$ and an abelian subgroup G of $(\mathcal{O}_0/N_1\mathcal{O}_0)^*$ such that G acts freely and transitively on the orbit $G \cdot L$, f is injective on $G \cdot L$ and $j(E_A)$ is contained in $f(G \cdot L) \subset \mathfrak{D}$.*

We solve this task in Sect. 4.2. More precisely, we find three subsets of $\mathcal{I}$ restricted to which f is injective, and we give abelian groups that induce the required action on these subsets. Furthermore, the image of f restricted to one of these three subsets of $\mathcal{I}$ will always contain $j(E_0/K)$.

In order to apply our general framework from Sect. 3, it remains to construct a malleability oracle for f at $j(E_0/K)$ for any secret $K \in \mathcal{I}$. To construct this oracle, we use both the torsion point information provided in the SIDH protocol and a solution to the following task.

Task 4.4. *Given an endomorphism $\theta \in G$ of degree coprime to N_1 and an integer N_2 coprime to N_1, compute an endomorphism θ' of degree N_2 such that θ and θ' induce the same action on the set $\mathcal{I}$ of cyclic subgroups of $E_0[N_1]$ of order N_1.*

In Appendix C of the full version of this paper [20], we give a direct solution to a variation of this task when using sufficiently overstretched and unbalanced parameters, i.e. $N_2 > p^2 N_1^4$. However, in Sect. 4.3 we show that it suffices to lift elements of πG, where π is the Frobenius map. A solution to Task 4.4 for these elements requiring only $N_2 > pN_1^4$ is described in Sect. 4.4.

The following lemma results from the coprimality of deg(θ) and N_1 and is depicted in Fig. 1.

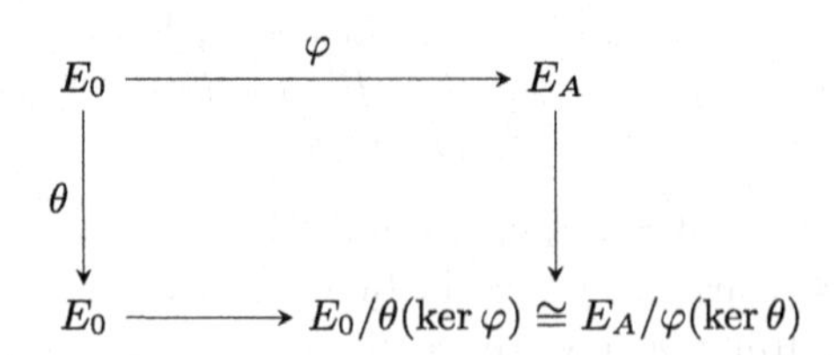

Fig. 1. The isogenies φ and the endomorphism θ are of coprime degrees.

Lemma 4.5. *Let $\varphi : E_0 \to E_A$ be an isogeny of degree N_1 and let $\theta \in End(E_0)$ be of degree coprime to N_1. Then $E_A/\varphi(\ker\theta)$ is isomorphic to $E_0/\theta(\ker\varphi)$.*

Let N_3 be the degree of θ. We cannot compute the curve $E_0/\theta(\ker\varphi)$ in general without the knowledge of the isogeny φ or its action on the N_3-torsion. However, we can compute the curve if we find an endomorphism θ' of degree N_3' such that θ and θ' have the same action on the N_1-torsion and $\varphi|_{E_0[N_3']}$ is known. This is the motivation behind Task 4.4, as we know the action of φ on the N_2-torsion in Problem 4.2. A solution to this task yields a malleability oracle for f with respect to the previously described group action of G on $\mathcal{I}$ in the SIDH setting.

We outline the construction of the malleability oracle in Algorithm 1. Correctness will follow from the proof of Proposition 4.26 given a suitable choice of the acting group G which we will discuss in Subsect. 4.2.

Algorithm 1: Computation of $f(\theta(K))$, given $f(K)$ and $\theta \in G$

Let $\varphi : E_0 \to E_A := E_0/K$ be an isogeny of degree N_1, let N_2 be coprime to N_1 and $G \subset (\mathcal{O}_0/N_1\mathcal{O}_0)^*$ one of the abelian groups as in Task 4.3 that acts freely and transitively on K.

Input: E_0, $f(K) = j(E_A)$, $\varphi|_{E_0[N_2]}$ and $\theta \in G$.

Output: $f(\theta(K)) = j(E_0/\theta(K))$.

1 Compute endomorphism θ' of degree N_2 having the same action as θ on cyclic N_1-order subgroups of $E_0[N_1]$ as provided by a solution to Task 4.4;
2 Determine $\varphi(\ker\theta')$, using the knowledge of φ on $E_0[N_2]$;
3 Compute $j(E_A/\varphi(\ker\theta')) = j(E_0/\theta(K))$;
4 **return** $f(\theta(K)) = j(E_0/\theta(K))$

For parameters that allow us to construct a malleability oracle, we can then solve Problem 4.2 underlying SIDH-like protocols via a reduction to an injective abelian hidden shift problem using the framework introduced in Sect. 3.2.

Informal result 4.6. *Suppose the parameters allow the efficient solution of Task 4.4, then Problem 4.2 can be solved in quantum subexponential time.*

We use the remainder of this section to prove this result formally under certain assumptions. To this end, we first give solutions to Task 4.3 and, for some parameters, to a variant of Task 4.4. More precisely, we show in Sect. 4.3 that it is sufficient to lift elements from πG instead of G. For this case, we then give a more efficient lifting procedure requiring unbalanced and overstretched parameters. We construct a malleability oracle using the torsion point information provided in SIDH and a subroutine solving our variant of Task 4.4. Apart from some technical details that we will address in the following, the informal result follows from Corollary 3.4. An overview of the attack is depicted in Algorithm 2.

4.2 A Free and Transitive Group Action

Recall that E_0 is the supersingular curve with j-invariant 1728, given by the equation $y^2 = x^3 + x$. In this section we provide a solution to Task 4.3. For simplicity, we treat N_1 as a power of 2, but the results generalize to any power

Algorithm 2: Solving SIDH's underlying hardness assumption via an abelian hidden shift problem

Let $\varphi : E_0 \to E_0/K$ be an N_1-isogeny and $N_2 \in \mathbb{Z}$ such that $\gcd(N_1, N_2) = 1$.
Input: E_0, E_0/K, $\varphi(E_0[N_2])$.
Output: Isogeny $E_0 \to E$, where $j(E) = j(E_0/K)$.

1. Compute an abelian group $G \subset (\mathcal{O}_0/N_1\mathcal{O}_0)^*$ acting freely and transitively on the orbit $G(K)$ and some $J \in G(K) \subset \mathcal{I}$;
2. Define $F_K : G \to \mathfrak{D}, g \mapsto f(g(K))$ and $F_J : G \to \mathfrak{D}, g \mapsto f(g(J))$;
3. Compute injective abelian hidden shift $\theta \in G$ of F_K and F_J, i.e., $\theta \in G$ such that $F_K(g) = F_J(\theta g)$ for all $g \in G$, using a quantum algorithm such as Kuperberg's. To this end, one evaluates F_K using Algorithm 1 and F_J using the knowledge of J;
4. **return** Isogeny $E_0 \to E_0/\theta(J)$

of a small prime. A generalization to powers of 3 is sketched in Appendix B of the full version of this paper [20].

We provide the solution by identifying three subsets of $\mathcal{I}$ that are orbits under a free and transitive action of abelian subgroups of $(\mathcal{O}_0/N_1\mathcal{O}_0)^*$. More precisely, let $P \in E_0$ such that $\langle P, \iota(P)\rangle = E_0[N_1]$, where ι denotes the automorphism $(x, y) \mapsto (-x, iy)$ of E_0. Let $Q := P + \iota(P)$ and define the following three subsets of $\mathcal{I}$.

$$\mathcal{I}_1 := \{\langle P + [\alpha]\iota(P)\rangle \mid \alpha \text{ even } \}$$

$$\mathcal{I}_2 := \left\{\langle Q + \alpha\iota(Q)\rangle \mid \alpha \text{ even and } \alpha \in \left[0, \frac{N_1}{2} - 1\right]\right\}$$

$$\mathcal{I}_3 := \left\{\langle Q + \alpha\iota(Q)\rangle \mid \alpha \text{ even and } \alpha \in \left[\frac{N_1}{2}, N_1 - 1\right]\right\}$$

Recall the function f defined in (1), mapping cyclic subgroups of $E_0[N_1]$ of order N_1 to j-invariants of curves at distance N_1 from E_0,

$$f : \mathcal{I} \to \mathfrak{D}, \quad K \mapsto j(E_0/K).$$

We will show that restricting the function f to any of the subsets $\mathcal{I}_1$, $\mathcal{I}_2$, or $\mathcal{I}_3$ yields an injective function and we will prove that $f(\cup_i \mathcal{I}_i) = f(\mathcal{I})$. Furthermore, we will see that

$$G_0 := \{a + b\iota \mid a \text{ odd}, b \text{ even } \}/N_1\mathcal{O}_0^*$$

acts transitively on $\mathcal{I}_1$. In order to ensure that the action is free, we identify two endomorphisms $a + b\iota$ and $a' + b'\iota$ in G_0 if there exists an odd $\lambda \in \mathbb{Z}/N_1\mathbb{Z}$ such that $a \equiv \lambda a' \pmod{N_1}$ and $b \equiv \lambda b' \pmod{N_1}$. We denote the resulting group by G.

In order to define free and transitive group actions on $\mathcal{I}_2$, and $\mathcal{I}_3$ we define similarly to G_0

$$H_0 := \{a + b\iota \mid a \text{ odd}, b \text{ even } \}/(N_1/2)\mathcal{O}_0^*.$$

Again, we identify two endomorphisms $a+b\iota$ and $a'+b'\iota$ in H_0 if there exists an odd $\lambda \in \mathbb{Z}/(N_1/2)\mathbb{Z}$ such that $a \equiv \lambda a' \pmod{N_1/2}$ and $b \equiv \lambda b' \pmod{N_1/2}$, we obtain a group H. The group H will act freely and transitively on $\mathcal{I}_2$ and $\mathcal{I}_3$.

Hence, one of these three options will always be a solution to Task 4.3.

The map f is based on the well-known correspondence between $\mathcal{I}$ and curves at distance N_1 from E_0. However, this correspondence is not necessarily one-to-one. In particular, if E_0 has a non-scalar endomorphism of degree N_1^2, then that endomorphism can be decomposed as $\hat{\tau}_1 \circ \tau_2$, where τ_1 and τ_2 are non-isomorphic isogenies of degree N_1 from E_0 to the same curve E. For small enough N_1, the following lemma shows that two kernels correspond to the same curve if and only if they are linked by the automorphism ι.

Lemma 4.7. *Suppose that $N_1^2 < \frac{p+1}{4}$. Then the only endomorphisms of degree N_1^2 of E_0 are $[N_1]$ and $[N_1] \cdot \iota$, where $\iota : E_0 \to E_0, (x,y) \mapsto (-x, iy)$ is the non-trivial automorphism.*

Proof. Due to the condition $N_1^2 < \frac{p+1}{4}$, an endomorphism θ of degree N_1^2 lies in $\mathbb{Z}[\iota]$. Let $\theta = a + b\iota$ for some $a, b \in \mathbb{Z}$. Then the degree of θ is a^2+b^2. Now we have to prove that the only ways to decompose N_1^2 as a sum of two squares are trivial, i.e., $N_1^2 = N_1^2 + 0^2 = 0^2 + N_1^2$.

Let $N_1 = 2^k$, and we prove the statement by induction on k. For $k=1$ the statement is trivial. Suppose that $k > 1$ and that $N_1^2 = a^2 + b^2$. Then a and b cannot both be odd as N_1^2 is divisible by four. If they were both even, then dividing by four yields a decomposition of $(N_1/2)^2 = (a/2)^2 + (b/2)^2$. By the induction hypothesis, this decomposition is trivial implying that N_1^2 can also only be decomposed in a trivial way. □

Corollary 4.8. *Suppose that $N_1^2 < \frac{p+1}{4}$. Let ϕ and ϕ' be two isogenies of degree N_1 from E_0 to a curve E. Then either $\ker\phi = \ker\phi'$ or $\ker\phi = \iota(\ker\phi')$.*

Proof. Consider the endomorphism $\tau = \hat{\phi}' \circ \phi$ of E_0. The degree of τ is N_1^2, so $\tau = [N_1]$ or $\tau = [N_1] \cdot \iota$ by Lemma 4.7. In the former case, the isogenies ϕ and ϕ' are identical by the uniqueness of the dual. In the latter case, we have $\ker\phi = \iota(\ker\phi')$. □

Thus, an element in the image of f has precisely one preimage if the kernel of the corresponding isogeny is fixed by the automorphism ι.

Identifying an Abelian Group with $\mathcal{I}_1$: Now, we will give the free and transitive group action on $\mathcal{I}_1$ and show that f restricted to $\mathcal{I}_1$ is injective.

Let P be a point such that $\{P, \iota(P)\}$ is a basis of $E_0[N_1]$ and recall

$$\mathcal{I}_1 := \{\langle P + [\alpha]\iota(P)\rangle \mid \alpha \text{ even }\}.$$

We show that the restriction of f to $\mathcal{I}_1$ is injective.

Proposition 4.9. *Let $j(E_0) = 1728$ and suppose that $N_1^2 < \frac{p+1}{4}$. The restriction of f to $\mathcal{I}_1$ is injective.*

Proof. We apply Corollary 4.8 to show that the codomains of isogenies with kernel in $\mathcal{I}_1$ are pairwise non-isomorphic curves. It is clear that $P + \alpha\iota(P)$ and $P + \alpha'\iota(P)$ are not scalar multiples of each other if $\alpha \neq \alpha'$ as $P, \iota(P)$ generate $E_0[N_1]$. It remains to show that for any even α, α', the points $P + \alpha\iota(P)$ and $-\alpha' P + \iota(P)$ are not scalar multiples of each other. Suppose there exists an odd λ such that

$$P + \alpha\iota(P) = \lambda(-\alpha' P + \iota(P)).$$

Note that we can restrict to odd λs as the order of both points is N_1. Since $\{P, \iota(P)\}$ is a basis of the N_1-torsion, this implies that $1 \equiv -\lambda\alpha' \pmod{N_1}$. Since α' is even this is a contradiction concluding the proof. □

Clearly, $f(\mathcal{I}_1)$ does not include all elliptic curves at distance N_1 from E_0, i.e., all curves in $f(\mathcal{I})$. Every curve at distance N_1 from E_0 is of the form $E_0/\langle P + \alpha\iota(P)\rangle$ for some $\alpha \in \mathbb{Z}/N_1\mathbb{Z}$, which follows from the observation that the curves $E_0/\langle \beta_1 P + \beta_2\iota(P)\rangle$ and $E_0/\langle -\beta_2 P + \beta_1\iota(P)\rangle$ are isomorphic since their kernels are linked by ι. We first restrict ourselves to define a free and transitive group action on $\mathcal{I}_1$ and define the free and transitive group action on the kernels corresponding to the remaining curves later.

Recall that E_0 is a curve with well-known endomorphism ring, and we are interested in the endomorphisms that are of degree coprime to N_1. While there are infinitely many such endomorphisms, we are only concerned with their action on $E_0[N_1]$, i.e., we are looking at the group $(\mathcal{O}_0/N_1\mathcal{O}_0)^*$ which is isomorphic to $GL_2(\mathbb{Z}/N_1\mathbb{Z})$ [31, p. 676]. Furthermore, we are only concerned with the action of the endomorphisms on $\mathcal{I}$, i.e., on cyclic subgroups of $E_0[N_1]$ of order N_1, and we can therefore identify even more endomorphisms with each other by the following lemma.

Lemma 4.10. *Let (a, b, c, d) and (a', b', c', d') be the coefficients of θ and θ' with respect to some $\mathbb{Z}$-basis of the endomorphism ring $\mathcal{O}_0$ of E_0, and let $\mathcal{I}$ be the set of cyclic N_1-order subgroups of $E_0[N_1]$. Then $\theta(K) = \theta'(K)$ for every $K \in \mathcal{I}$ if and only if there exists some $\lambda \in (\mathbb{Z}/N_1\mathbb{Z})^*$ such that*

$$(a, b, c, d) \equiv \lambda(a', b', c', d') \pmod{N_1}.$$

Proof. Considering the respective restrictions to $E_0[N_1]$, two endomorphisms are equal if they lie in the same class in $(\mathcal{O}_0/N_1\mathcal{O}_0)^*$. Moreover, let θ_1, θ_2 be two endomorphisms such that $\theta_1 = [\lambda]\theta_2$ for some integer λ, and let P be an element of order N_1. Since scalar multiplication commutes with any endomorphism, it is easy to see that $\theta_1(P)$ and $\theta_2(P)$ generate the same subgroup in $E_0[N_1]$ if and only if λ is coprime to N_1. □

Now, we are ready to give a solution to Task 4.3 if $K \in \mathcal{I}_1$.

Proposition 4.11. *Let G be the group of equivalence classes of elements*

$$\{a + b\iota \mid a \text{ odd, } b \text{ even }\} \subset \mathbb{Z}[\iota]/N_1\mathcal{O}_0^* \subset (\mathcal{O}_0/N_1\mathcal{O}_0)^*,$$

where we identify two elements if and only if they differ by multiplication by an odd scalar modulo N_1. Then G is an abelian group, and it acts freely and transitively on $\mathcal{I}_1$.

Proof. It is easy to see that the endomorphisms in $\mathbb{Z}[\iota]$ of degree coprime to N_1 form an abelian subgroup of $\mathcal{O}_0$. Using any basis for $E_0[N_1]$ of the form $\{P, \iota(P)\}$, we can write the elements of this subgroup as matrices of the form $\begin{pmatrix} a & b \\ -b & a \end{pmatrix}$, where a is odd and b is even. By identifying two endomorphisms $a_1+b_1\iota$ and $a_2 + b_2\iota$ if there exists an integer λ coprime to N_1 and an endomorphism δ such that $a_1 - \lambda a_2 + (b_1 - \lambda b_2) = N_1\delta$, which is possible by Lemma 4.10, we obtain G. As G is closed under multiplication and reduction modulo N_1, it is a subgroup of an abelian group and therefore abelian itself. Note that G contains all equivalence classes under Lemma 4.10 of endomorphisms of the form $a + b\iota$ for even b, independently of the chosen basis.

To examine the orbit of an element in $\mathcal{I}$, which is a cyclic N_1-order subgroup of $E_0[N_1]$, under the action of G, it is sufficient to look at the orbit of a generator of this cyclic group in $\mathcal{I}$. We consider the orbit of P which has coordinates $(1, 0)$ with respect to our basis under the group action of G. The image of $(1, 0)$ under an element $\begin{pmatrix} 1 & b \\ -b & 1 \end{pmatrix}$ is $(1, b)$. Inspecting the cyclic subgroups of E_0 these points generate, we get $G \cdot \langle P \rangle = \mathcal{I}_1$. □

Free and Transitive Group Action on $\mathcal{I}_2$ and $\mathcal{I}_3$: So far we have defined a free and transitive group action on $\mathcal{I}_1$ and thus for the curves in $f(\mathcal{I}_1)$. However, when the secret kernel is generated by $P+\alpha\iota(P)$ with α odd, the curve $E_0/\langle P+\alpha\iota(P)\rangle$ is not contained in $f(\mathcal{I}_1)$. This is the case we handle next.

One can show that the action of the previously defined group G acting on curves at distance N_1 from E_0 considered via f has three orbits (see Appendix A of the full version [20] for details). We have already seen that $f(\mathcal{I}_1)$ is one orbit, but the odd-α cases will split into two orbits. Clearly, G cannot be free and transitive on both orbits, since the size of the orbits is smaller than the cardinality of the group. We avoid this issue by choosing a different (but related) group of cardinality $N_1/4$ acting on the curves corresponding to an odd α.

Lemma 4.12. *Let $Q := P + \iota(P)$ and define*

$$\mathcal{I}_2 := \left\{ \langle Q + \alpha\iota(Q) \rangle \mid \alpha \text{ even and } \alpha \in \left[0, \frac{N_1}{2} - 1\right] \right\}$$
$$\mathcal{I}_3 := \left\{ \langle Q + \alpha\iota(Q) \rangle \mid \alpha \text{ even and } \alpha \in \left[\frac{N_1}{2}, N_1 - 1\right] \right\}.$$

The restrictions $f_{|\mathcal{I}_2}$ and $f_{|\mathcal{I}_3}$ of f to $\mathcal{I}_2$ and $\mathcal{I}_3$ are injective.

Proof. We show that two distinct isogenies with kernel both in $\mathcal{I}_2$ (or both in $\mathcal{I}_3$) map to two non-isomorphic curves. Let α, α' be such that $\langle Q + \alpha\iota(Q) \rangle$ and

$\langle Q + \alpha' \iota(Q) \rangle$ are both in $\mathcal{I}_2$, or $\mathcal{I}_3$, respectively. Suppose there exists an odd λ such that

$$Q + \alpha\iota(Q) = \lambda(Q + \alpha'\iota(Q)).$$

This means $1 - \lambda \equiv 0 \pmod{N_1/2}$ and $\alpha - \lambda\alpha' \equiv 0 \pmod{N_1/2}$ which implies $\alpha \equiv \alpha' \pmod{N_1/2}$. We are left to show that $Q + \alpha\iota(Q)$ is never an odd multiple of $-\alpha Q + \iota(Q)$. Suppose there exists an odd λ such that

$$Q + \alpha\iota(Q) = \lambda(-\alpha' Q + \iota(Q)).$$

This implies $1 + \alpha'\lambda \equiv \alpha - \lambda \equiv 0 \pmod{N_1/2}$, which is a contradiction, since $\alpha - \lambda \equiv 0 \pmod{N_1/2}$ implies that λ is even while $1 + \alpha'\lambda \equiv 0 \pmod{N_1/2}$ implies that λ is odd. Therefore, the curves $E_0/\langle Q + \alpha\iota(Q) \rangle$ and $E_0/\langle Q + \alpha'\iota(Q) \rangle$ are pairwise non-isomorphic. □

Finally, we give a free and transitive group action on $\mathcal{I}_2$ and $\mathcal{I}_3$. We start by defining the acting group.

We identify two endomorphisms $a + b\iota$ and $a' + b'\iota$ if there exists an odd $\lambda \in \mathbb{Z}/(N_1/2)\mathbb{Z}$ such that $a \equiv \lambda a' \pmod{N_1/2}$ and $b \equiv \lambda b' \pmod{N_1/2}$ and we call the resulting group H_0. Let H be the subgroup of H_0 containing elements with even b.

Proposition 4.13. *H acts freely and transitively on $\mathcal{I}_2$ and $\mathcal{I}_3$.*

Proof. It is enough to show that H acts transitively on $\mathcal{I}_2$ and $\mathcal{I}_3$ because $H, \mathcal{I}_2$ and $\mathcal{I}_3$ have the same cardinality. We show that the orbit $H \cdot \langle Q \rangle$ contains every element in $\mathcal{I}_2$. This follows immediately from $(1 + \alpha\iota)Q = Q + \alpha\iota(Q)$. Similarly, H acts transitively on $\mathcal{I}_3$ as

$$(1+\alpha\iota)(Q+N_1\iota(Q)/2) = (1-\alpha N_1/2)Q+(\alpha+N_1/2)\iota(Q) = Q+(\alpha+N_1/2)\iota(Q),$$

where $(\alpha N_1/2)Q = 0$ as α is even. □

What remains to be shown is that every curve $E_0/\langle P + \alpha\iota(P) \rangle$ with odd α has a j-invariant contained in $f(\mathcal{I}_2)$ or $f(\mathcal{I}_3)$.

Proposition 4.14. *Let α be an odd integer. Then $f(\langle P + \alpha\iota(P) \rangle)$ is contained in $f(\mathcal{I}_2)$ or $f(\mathcal{I}_3)$.*

Proof. Observe that

$$P + \alpha\iota(P) = \frac{1+\alpha}{2}(P + \iota(P)) + \frac{\alpha - 1}{2}(-P + \iota(P)) = \frac{1+\alpha}{2}Q + \frac{\alpha-1}{2}\iota(Q).$$

The sum of $\frac{1+\alpha}{2}$ and $\frac{\alpha-1}{2}$ is odd and therefore one of the fractions is even while the other one is odd. If $\frac{\alpha-1}{2}$ is even, then it is clear that the curve is contained in $f(\mathcal{I}_2)$ or $f(\mathcal{I}_3)$. In the case where $\frac{1+\alpha}{2}$ is even, $E_0/\langle \frac{1+\alpha}{2}Q + \frac{\alpha-1}{2}\iota(Q) \rangle$ is isomorphic to $E_0/\langle \frac{1-\alpha}{2}Q + \frac{\alpha+1}{2}\iota(Q) \rangle$ (because their kernels are related by ι) and thus the curve is contained in $f(\mathcal{I}_2)$ or $f(\mathcal{I}_3)$. □

In this subsection, we have identified three subsets of $\mathcal{I}$, restricted to which f is injective. Moreover, we have seen that the union $\cup_{i=1}^{3} f(\mathcal{I}_i)$ contains the j-invariants of all curves at distance N_1 from E_0. Finally, we gave an abelian subgroup of $(\mathcal{O}_0/N_1\mathcal{O}_0)^*$ for each of these subsets of $\mathcal{I}$ that acts freely and transitively on it. Thus, we solve Task 4.3 as long as one determines or guesses which of the three $f(\mathcal{I}_i)$ contains $j(E_0/K)$.

4.3 Using the Frobenius Map

In the previous subsection, we described how to choose suitable abelian subgroups of $(\mathcal{O}_0/N_1\mathcal{O}_0)^*$ in order to solve Task 4.3 after guessing whether $j(E_0/K)$ is a j-invariant in $f(\mathcal{I}_1)$, $f(\mathcal{I}_2)$, or $f(\mathcal{I}_3)$.

The elements of the acting groups chosen as described in the previous section can be trivially lifted to $\mathbb{Z}[\iota] := \mathbb{Q}[\iota] \cap \mathcal{O}_0$. In Appendix C of the full version [20] we show how these representatives can be lifted directly to elements of norm N_2 or eN_2, where e is a small positive integer, whenever the SIDH parameters N_1 and N_2 are sufficiently overstretched and unbalanced with $N_2 > p^2 N_1^4$. For these parameters, this solves a variation of Task 4.4.

In this section we reduce the required unbalancedness partially by proving that we can lift elements from $\pi\mathbb{Z}[\iota]$ instead. Assuming that $N_2 > pN_1^4$, we will show in Subsect. 4.4 how an endomorphisms from $\pi\mathbb{Z}[\iota]$ can be lifted efficiently to another endomorphism of norm N_2 or eN_2, for some small integer e, inducing the same action on $\mathcal{I}$. Note that it is not possible to choose a group generated by an element in $\pi\mathbb{Z}[\iota]$ to solve Task 4.3 directly, acting freely and transitively on a large number of N_1-isogeny kernels, as such an element has multiplicative order at most 4.

As before, let $\varphi : E_0 \to E_0/K$ denote the secret N_1-isogeny we want to compute. Recall that to run our attack we need to be able to compute $E_0/\theta(K)$ for every θ in the groups G acting on $\mathcal{I}_1$, and H acting on $\mathcal{I}_2$ and $\mathcal{I}_2$. We have seen that we can represent θ as an element in $\mathbb{Z}[\iota]$.

Let π denote the Frobenius map. Assuming that we can lift $\pi\theta$ to an endomorphism of degree N_2 inducing the same action on $\mathcal{I}$, we can compute $E_0/\pi\theta(K)$ using knowledge of $\varphi(E_0[N_2])$ as described in Sect. 4.1. Now let $B := \theta(K)$. Given $E_0/\pi(B)$, we can compute E_0/B using the Frobenius map as follows.

Lemma 4.15. *Let E be an elliptic curve defined over $\mathbb{F}_p$, π the Frobenius map and let $B \subset E$ be a cyclic subgroup. $E/\pi(B)$ is isomorphic to the image of the Frobenius map of E/B.*

Proof. Let ϕ_1 be the isogeny with kernel B and ϕ_2 the isogeny with kernel $\pi(B)$. The isogeny ϕ_1 is separable and its kernel is contained in the kernel of $\phi_2 \circ \pi$. Then, there exists a unique isogeny $\psi : E/B \to E/\pi(B)$ satisfying $\phi_2 \circ \pi = \psi \circ \phi_1$ (see [29, Corollary III. 4.11.]), i.e., the following diagram commutes.

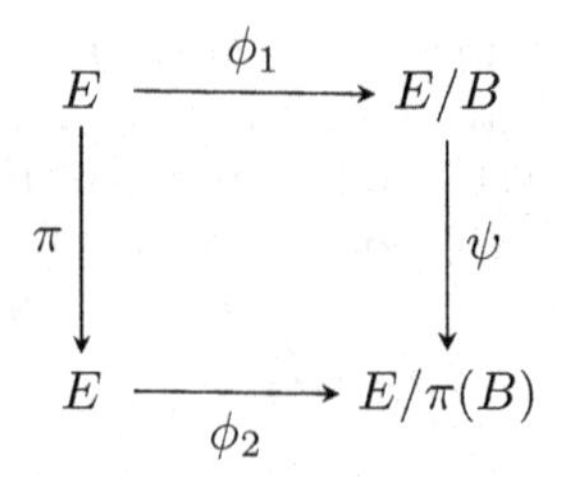

The degree of a composition of isogenies is the product of its factors which implies $\deg(\psi) = p$. Furthermore, ψ is not separable as the Frobenius map is not. As ψ can be decomposed as a composition of the Frobenius map and a separable isogeny (see [29, Corollary II.2.12.]), $\deg(\psi) = p$ implies that ψ must be a composition of Frobenius and an automorphism. Hence, E_0/B and $E_0/\pi(B)$ are linked by the Frobenius map. □

By Lemma 4.15 we can compute $E_0/\theta(K)$ by first computing $E_0/\pi\theta(K)$ and then applying the Frobenius map. This gives rise to the following strategy when constructing the malleability oracle.

Assume we want to compute $E_0/\theta(K)$ for some $\theta \in \mathbb{Z}[\iota]$ and unknown K, given the image of the N_2-torsion of the isogeny $\varphi : E_0 \to E_0/K$. Using the lifting algorithm of Subsect. 4.4, we compute an endomorphism θ' of degree N_2 or eN_2 for a small e that induces the same action on $\mathcal{I}$ as $\pi\theta$. As described previously, the torsion point information allows us to compute $E_0/\theta'(K) = E_0/\pi\theta(K)$. By Lemma 4.15, applying the Frobenius map yields $E_0/\pi\theta'(K) = E_0/\theta(K)$.

4.4 Lifting $\theta \in \pi\mathbb{Z}[\iota]$ to an Element of Norm eN_2

In this subsection we give an efficient algorithm to lift endomorphisms from $\pi\mathbb{Z}[\iota] = \pi(\mathbb{Q}[\iota] \cap \mathrm{End}(E_0))$ to another endomorphism of $E_0/\mathbb{F}_p$ of degree N_2 or eN_2 that induces the same action on $\mathcal{I}$, whenever $N_2 > pN_1^4$. Here, e is the smallest positive integer such that $eN_2/p(c_0^2 + d_0^2)$ is a quadratic residue modulo $2N_1$, where $\pi(c_0 + d_0\iota) \in \pi\mathbb{Z}[\iota]$ is the endomorphism we want to lift.

This will solve the following task, which is a variant of Task 4.4, efficiently.

Task 4.16. *Let N_1, N_2 be coprime integers such that $N_2 > pN_1^4$, let $\theta := \pi(c_0 + d_0\iota) \in \pi\mathbb{Z}[\iota]$ be an E_0-endomorphism of degree coprime to N_1 and let e denote the smallest positive integer such that $eN_2/p(c_0^2 + d_0^2) \pmod{2N_1}$ is a quadratic residue. Compute an endomorphism θ' of degree N_2 or eN_2 such that $\theta(K) = \theta'(K)$ for all $K \in \mathcal{I}$.*

We have discussed in Sect. 4.3 that we can lift $\pi(c_0 + d_0\iota)$ instead of $c_0 + d_0\iota$. Therefore, this task solves Task 4.4 up to the following two relaxations. First, we require N_2 to be sufficiently large and unbalanced compared to N_1. Second, we allow θ' to be either of degree N_2 or eN_2 for some small positive integer e.

We have implemented the lifting algorithm of this section in MAGMA and made it publicly available[1].

[1] https://github.com/SimonMerz/lifting-for-malleability-oracles..

Remark 4.17. If N_1 were a prime, e could be chosen as the smallest quadratic non-residue modulo N_1. However, in our case N_1 is a composite number. Thus, the product of two quadratic non-residues might not be a quadratic residue if there are multiple cosets of the subgroup of quadratic residues in the group of units modulo $2N_1$.

We are primarily interested in the case where N_1 is a prime power ℓ^n. By Hensel's lemma, being a quadratic residue modulo ℓ^n is equivalent to being a quadratic residue modulo ℓ, if ℓ is odd, and equivalent to being a quadratic residue modulo 8, if $\ell = 2$.

Consequently, there is one coset of the quadratic residues in the group of units of $2N_1$ if ℓ is an odd prime. Therefore, e can be chosen to be the smallest quadratic non-residue modulo ℓ. For example, if N_1 is a power of 3 one can choose $e = 2$.

If $\ell = 2$, then there are three cosets of the quadratic residues in the group of units, i.e., the ones that contain $3, 5$, and 7 respectively. Consequently, e can always be chosen to be one of $3, 5$, or 7 in this case.

In case N_1 has distinct prime factors, for $eN_2/p(c_0^2 + d_0^2)$ to be a quadratic residue it has to be a quadratic residue modulo the largest prime power dividing $2N_1$ for each distinct prime factor. If the number of cosets grows, so do the possibilities for e und thus the size of the smallest e that is guaranteed to work.

We now describe an algorithm to solve Task 4.16. By Lemma 4.10, it suffices to solve the following task, which is similar to the problem solved at the core of the KLPT algorithm [16].

Task 4.18. *Given $\theta = a_0 + b_0\iota + (c_0 + d_0\iota)\pi$, find $\theta' = a_1 + b_1\iota + (c_1 + d_1\iota)\pi$ of degree N_2 or eN_2 with coefficients $(a_1, b_1, c_1, d_1) \equiv \lambda(a_0, b_0, c_0, d_0) \pmod{N_1}$ for some scalar $\lambda \in (\mathbb{Z}/N_1\mathbb{Z})^*$.*

In the following, we provide a solution to this task. Let

$$\theta' = \lambda a_0 + N_1 a_1 + \iota(\lambda b_0 + N_1 b_1) + (\lambda c_0 + N_1 c_1 + \iota(\lambda d_0 + N_1 d_1))\pi.$$

As $\mathrm{Norm}(x + y\iota) = x^2 + y^2$, its norm equals

$$\mathrm{Norm}(\theta') = (\lambda a_0+N_1 a_1)^2+(\lambda b_0+N_1 b_1)^2+p\big((\lambda c_0+N_1 c_1)^2+(\lambda d_0+N_1 d_1)^2\big). \quad (2)$$

Since $\theta \in \pi\mathbb{Z}[\iota]$ implies $a_0 = b_0 = 0$, Eq. (2) simplifies to

$$\mathrm{Norm}(\theta') = N_1^2(a_1^2 + b_1^2) + p\big((\lambda c_0 + N_1 c_1)^2 + (\lambda d_0 + N_1 d_1)^2\big). \quad (3)$$

Set e to be the smallest positive integer such that $eN_2/(p(c_0^2+d_0^2))$ is a quadratic residue modulo $2N_1$.

The goal is to compute θ' such that $\mathrm{Norm}(\theta') = eN_2$. Considering Eq. (3) modulo N_1, we obtain

$$eN_2 \equiv \lambda^2 p(c_0^2 + d_0^2) \pmod{N_1}. \quad (4)$$

Since $eN_2/p(c_0^2+d_0^2)$ is a quadratic residue modulo $2N_1$ by the choice of e, there exists a solution for λ in Eq. (4) modulo $2N_1$. Compute any such solution, and lift it to the integers in $[1, 2N_1 - 1]$. Note that we do not lose generality by the lift as any other lift of λ corresponds to a change in c_1, d_1 instead.

For fixed c_0, d_0 and λ, this gives an affine relation between c_1 and d_1 modulo N_1, i.e.,

$$c_0c_1 + d_0d_1 \equiv \frac{\mathrm{Norm}(\theta') - \lambda^2 p(c_0^2 + d_0^2)}{2\lambda p N_1} \pmod{N_1}. \tag{5}$$

Finally, one is left with the problem of representing an integer r as the sum of two squares, namely to find a solution (a_1, b_1) for

$$a_1^2 + b_1^2 = r := \frac{\mathrm{Norm}(\theta') - p((\lambda c_0 + N_1c_1)^2 + (\lambda d_0 + N_1d_1)^2)}{N_1^2} \tag{6}$$

where λ, c_0 and d_0 are fixed, and c_1, d_1 satisfy an affine equation modulo N_1.

As Petit and Smith pointed out at Mathcrypt 2018, the solution space to Eq. (5) is a translated lattice modulo N_1 [24]. More precisely, we know that c_0 or d_0 is coprime to N_1. Without loss of generality, let d_0 be coprime to N_1. Furthermore, let C denote the right hand side of Eq. (5). Then, (c_1, d_1) lies in the lattice

$$\langle (c_0/d_0, -1), (N_1, 0) \rangle + (C/d_0, 0). \tag{7}$$

Clearly, r from Eq. (6) can only be represented as a sum of two squares, if it is positive. This happens when the parameters N_1 and N_2 are sufficiently overstretched and unbalanced. To find a solution, one computes close vectors (c_1, d_1) to the target vector $(-\lambda c_0/N_1, -\lambda d_0/N_1)$ in the translated lattice.

Given the factorisation of r as defined in Eq. (6), Cornacchia's algorithm [6] can then efficiently solve for a_1, b_1 or determine that no such solution exists. If no solution exists, a different vector (c_1, d_1) is chosen.

Remark 4.19. Cornacchia's algorithm requires the factorization of r. This can be done in classical subexponential time or in quantum polynomial time. To avoid such computations, we apply Cornacchia's algorithm only when r is a prime and otherwise sample another close vector from the lattice.

Assuming the values of r behave like random values around pN_1^3 for the close vectors, one expects to choose $\log(pN_1^3)$ different vectors (c_1, d_1) before finding a solution for a_1, b_1 with Cornacchia's algorithm. If we do not apply Cornacchia's algorithm unless r is prime, we expect furthermore to sample roughly $\log(pN_1^3)$ values for (c_1, d_1) until r is prime.

The volume of the translated lattice is N_1. Thus, for a generic lattice for which the Gaussian heuristic holds we expect to find a lattice point at distance N_1 from $(\lambda c_0/N_1, \lambda d_0/N_1)$. Furthermore, we can use the Hermite constant for 2-dimensional lattices to trivially bound the distance between this lattice point and the next $2\log(pN_1^3)$ closest lattice points by $\frac{8}{3}\log(pN_1^3)\sqrt{N_1}$. Thus, heuristically r is positive for the expected number of vectors (c_1, d_1) that we need to sample, whenever $eN_2 > pN_1^3 + 8/3\log(pN_1^3)\sqrt{N_1^3}$.

Remark 4.20. Note that for specific lattices, the Gaussian heuristic might be violated. In the worst case, we can only expect to find a lattice point at distance N_1^2 from $(\lambda c_0/N_1, \lambda d_0/N_1)$ and overall solutions require roughly $eN_2 > pN_1^4$.

It is easy to see that a solution for (a_1, b_1, c_1, d_1) as computed with the routine described above satisfies Eq. (3). The full algorithm is summarized in Algorithm 3 and an implementation in MAGMA is available[4].

Algorithm 3: Lift element from $\pi\mathbb{Z}[\iota]$ to quaternion of norm N_2 or eN_2

Input: $\theta = \pi(c_0 + d_0\iota) \in \mathrm{End}(E_0)$, and parameters p, ε, N_1, N_2
Output: $\theta' = N_1a_1 + N_1b_1\iota + (\lambda c_0 + N_1c_1)\pi + (\lambda d_0 + N_1d_1)\iota\pi$ satisfying $\mathrm{Norm}(\theta') = N_2$ or eN_2 with probability $1 - \varepsilon$ and $\perp$ otherwise

1 $e \leftarrow$ least positive integer s.t. $eN_2/p(c_0^2 + d_0^2) \pmod{2N_1}$ is a quadratic residue;
2 Compute λ in $eN_2 \equiv \lambda^2 p(c_0^2 + d_0^2) \pmod{2N_1}$;
3 Compute affine relation $c_0c_1 + d_0d_1 \equiv C \pmod{N_1}$;
4 Define translated lattice L containing all (c_1, d_1) satisfying the affine relation;
5 $B \leftarrow \log(\varepsilon)\log(pN_1^3)/\log(1 - \log^{-1}(pN_1^3))$;
6 **for** $m = 1, \ldots, B$ **do**
7 　Compute next closest vector (c_1, d_1) to $(-\lambda c_0/N_1, -\lambda d_0/N_1)$ in L;
8 　$r \leftarrow \frac{\mathrm{Norm}(\theta') - p((\lambda c_0 + N_1c_1)^2 + (\lambda d_0 + N_1d_1)^2)}{N_1^2}$;
9 　**if** r prime **then**
10 　　Use Cornacchia's algorithm to find a_1, b_1 such that $a_1^2 + b_1^2 = r$ or determine that no solution exists;
11 　**if** solution found **then**
12 　　**return** $\theta' = N_1a_1 + N_1b_1\iota + (\lambda c_0 + N_1c_1)\pi + (\lambda d_0 + N_1d_1)\iota\pi$;
13 **return** $\perp$

An examination of Algorithm 3 shows that it aborts after a fixed number of trials for pairs (c_1, d_1), which leads to the following result.

Lemma 4.21. *Algorithm 3 always terminates and is correct if it returns a solution.*

We conclude this section by investigating the heuristic probability of the lifting algorithm returning a solution or aborting unsuccessfully, as well as its complexity.

Lemma 4.22. *Let $0 < \varepsilon < 1$. Assume r in Line 8 of Algorithm 3 behaves like a random value around pN_1^3. Then we expect Algorithm 3 heuristically to return a correct lift with probability $1 - \varepsilon$ and an error $\perp$ otherwise.*

Proof. If r in Line 8 of Algorithm 3 behaves like a random value around pN_1^3, we expect it to be prime with probability roughly $1/\log(pN_1^3)$ and Cornacchia's algorithm to provide a solution with probability approximately $1/(\log(pN_1^3))$ due to Landau [21] and Ramanujan [25]. Iterating over B short vectors (c_1, d_1) of the lattice as defined in Step 6 of Algorithm 3, we therefore expect our algorithm to return $\perp$ with probability

$$\left(1 - \frac{1}{\log(pN_1^3)}\right)^{B/\log(pN_1^3)}.$$

Hence, iterating over $B \geq \log(\varepsilon)\log(pN_1^3)/\log(1-\log^{-1}(pN_1^3))$ as in Algorithm 3, we fail to find a solution with probability less than ε heuristically. □

Remark 4.23. In Algorithm 2 the lifting of endomorphisms is used for every element of the acting group G or H with cardinality $N_1/2$ and $N_1/4$, respectively. Since we expect the lifting algorithm to fail heuristically with probability ε for every single group element and the functions in Algorithm 2 are only exact shifts of each other when it does not fail a single time, we need to choose ε sufficiently small. Assuming independence between the different executions of the lifting algorithm, we expect to find two functions satisfying the promise of a hidden shift with probability $(1-\varepsilon)^{N_1/2} \approx 1-\varepsilon N_1/2$ by first order Taylor approximation. Thus, choosing $\varepsilon < \frac{1}{N_1}$ we expect our lifting to work with probability roughly $\frac{1}{2}$ on all endomorphisms of G and similarly $\varepsilon < \frac{2}{N_1}$ for the elements in H. By the previous lemma, the lifting remains polynomial in $\log(N_1)$ and $\log(p)$ for any such ε. Choosing ε smaller allows us to heuristically achieve a larger success probability of the algorithm. The worst-case complexity of the lifting increases linearly in $|\log(\varepsilon)|$.

Lemma 4.24. *Let $0 < \varepsilon < 1$. Algorithm 3 runs in time polynomial in $\log p$, $\log N_1$, and $|\log(\varepsilon)|$.*

Proof. The worst-case runtime of the algorithm stems from sampling B (as defined in Algorithm 3, Line 5) potential values of (c_1, d_1) from a lattice of dimension 2. In each iteration one needs to run a primality test, and apply Cornacchia's algorithm to a prime of size polynomial in p and N_1. □

The main drawback of our lifting algorithm is the requirement of approximately $N_2 > pN_1^3$ in case the Gaussian heuristic is satisfied for the lattice defined in Eq. (7), and roughly $N_2 > pN_1^4$ otherwise (see Remark 4.20). This bound might be partially caused by inefficiencies in the lifting algorithm. However, the following remark discusses why we can a priori not expect to find a lifting algorithm for balanced parameters.

Remark 4.25. A randomly chosen non-homogeneous quadratic equation in two variables has in general no solution. Similarly, for arbitrary endomorphisms and any N_1, N_2, we would not expect to find an endomorphism $a_1 + b_1\iota \in \mathbb{Z}[\iota]$ (in the variables a_1, b_1) inducing the same action on $\mathcal{I}$ of degree N_2. Yet, as soon as we lift an endomorphism θ to an endomorphism $\theta' = N_1(a_1 + b_1\iota + c_1\pi) + \lambda\theta$, the degree of the lift will be of degree larger than pN_1^2.

4.5 Algorithm Summary

We begin the summary of our attack by proving that a solution to Task 4.4 allows us to construct a malleability oracle for f.

Proposition 4.26. *Let $f_{|\mathcal{I}'} : \mathcal{I}' \to \mathfrak{O}$ be the function defined in (1) restricted to a domain $\mathcal{I}'$ so it is injective, let G be an abelian subgroup of $(\mathcal{O}_0/N_1\mathcal{O}_0)^*$ acting freely and transitively on $\mathcal{I}'$ and let $\varphi : E_0 \to E_0/K$, where $K \in \mathcal{I}'$ is chosen uniformly at random and unknown. Suppose the public parameters allow us to solve Task 4.4 for endomorphisms in G efficiently. Given $\varphi|_{E_0[N_2]}$, we then have a polynomial-time malleability oracle for G at $f_{|\mathcal{I}'}(K)$.*

Proof. We need to show that there exists an efficient algorithm that, on input $f(K) = f_{|\mathcal{I}'}(K) = j(E_0/K)$ and $\theta \in G$, computes $f(\theta(K))$. Let φ be the isogeny corresponding to the cyclic subgroup $K \subset E_0$ of order N_1.

The endomorphism θ has degree N_2 coprime to N_1 and using the efficient solution to Task 4.4, we can compute some θ' of degree N_2 such that it has the same action on the N_1-torsion as θ. Therefore, $f(\theta(K)) = E_0/\theta(K) = E_0/\theta'(K)$ up to isomorphism. By Lemma 4.5, this equals $(E_0/K)/\varphi(\ker\theta')$. Since $\ker\theta'$ lies in $E_0[N_2]$, we can compute its image under φ and therefore we can calculate $f(\theta(K)) = (E_0/K)/\varphi(\ker\theta')$ efficiently. □

Proposition 4.26 calls for solutions to the Tasks 4.3 and 4.4. In Sects. 4.2 and 4.4 we presented solutions to *variants* of these tasks. We use the remainder of this section to summarize the impact of these variations on the success of our approach.

Restricting the function $f : \mathcal{I} \to \mathfrak{O}$ to a subset $\mathcal{I}'$ such that $f_{|\mathcal{I}'}$ is injective and its image contains $j(E_0/K)$ for the K one aspires to recover requires information on the secret we do not posses. However, we gave three subsets $\mathcal{I}_1$, $\mathcal{I}_2$, $\mathcal{I}_3$ of $\mathcal{I}$ in Sect. 4.2 such that f restricted to any of these subsets is injective. The images of these sets under f partition all curves at distance N_1 from E_0 up to isomorphism, i.e., one of the three subsets will yield the desired result. Moreover, we provided abelian subgroups of $\mathbb{Q}[\iota] \cap (\mathcal{O}_0/N_1\mathcal{O}_0)^*$ acting freely and transitively on $\mathcal{I}_1$, $\mathcal{I}_2$, and $\mathcal{I}_3$.

We then supply an algorithm to solve Task 4.16, a variant of Task 4.4 when N_1 and N_2 are sufficiently unbalanced, lifting endomorphisms from $\pi\mathbb{Z}[\iota]$ to ones with the same action on $\mathcal{I}$ of degree N_2 or eN_2. Here, e is a small integer depending on the parameters p, N_1, N_2 and the endomorphism. As a consequence, to use the torsion point information of $E_0[eN_2]$ under the secret isogeny given the image of $E_0[N_2]$, we need to guess the action on $E_0[e]$. Furthermore, we lift all endomorphisms in the acting group and thus we need to guess the action on $E_0[E]$, where E is the least common multiple of all e appearing for the different lifts. In Remark 4.17 we discuss which e might appear depending on the factorisation of N_1. For example, E is 2 if N_1 is a power of 3, or $\mathrm{lcm}(3, 5, 7)$ if N_1 is a power of 2. Guessing the action of the secret isogeny on $E_0[E]$ takes $O(E^3)$ trials. Finally, for efficiency reasons we lift endomorphisms from $\pi\mathbb{Z}[\iota]$, whereas the elements in the abelian groups acting on $\mathcal{I}_1$, $\mathcal{I}_2$, and $\mathcal{I}_3$ have representatives in $\mathbb{Z}[\iota]$. In Sect. 4.3 we showed that this is no restriction via the computation of an action of the Frobenius map.

For each combination of guesses of $E_0[E]$ under the secret isogeny and whether f maps the secret K into $f(\mathcal{I}_1)$, $f(\mathcal{I}_2)$ or $f(\mathcal{I}_3)$, we can use a subexponential quantum algorithm such as Kuperberg's [18] to compute the hidden shift for the functions F_K and F_J as defined in Algorithm 2 and verify the output of the algorithm. Both functions are injective and therefore the verification can be achieved by computing both functions on a single element and its shift respectively. Once the premise of a hidden shift is satisfied, Kuperberg's algorithm [18] recovers the (correct) solution to the injective abelian hidden shift problem. Thus, we recover the secret isogeny as described in Sect. 4. We can summarize our result as follows.

Theorem 4.27. *Let $N_2 > pN_1^4$. Under the heuristics used for the lifting of endomorphisms in Sect. 4.4, the SIDH problem can be solved in quantum subexponential time via a reduction to the injective abelian hidden shift problem.*

During this section, we have made some restrictions to simplify the presentation of our cryptanalysis. We assumed the starting curve E_0 to be a supersingular curve with j-invariant 1728. However, the attack also applies to other curves with known endomorphism rings that are close to E_0. In Sect. 4.2, we described the required group action on $\mathcal{I}$ under the further assumption that N_1 is a power of 2, which can be generalized to powers of small primes. A sketch for powers of 3 can be found in [20, Appendix B]. Finally, we assumed that $N_1^2 < \frac{p+1}{4}$ in Lemma 4.7. However, to run our attack we can slightly ease this restriction. Namely, if $N_1^2 > \frac{p+1}{4}$, then we choose a divisor N_1' of N_1 such that $N_1'^2 < \frac{p+1}{4}$ and run the attack with N_1' instead. This will reveal the N_1'-part of the isogeny and then we can guess the remaining part. For sufficiently small $\frac{N_1}{N_1'}$, this is only a minor inefficiency.

4.6 Hybrid Attacks on Overstretched SIDH

In this section, we examine to what extent partial knowledge of the secret, i.e., knowledge of the most or least significant bits, renders the attack more efficient. Moreover, we describe how the attack can be adapted to some further parameters that are not quite sufficiently unbalanced. The idea is to apply exhaustive search to recover parts of the secret isogeny until the remaining part of the isogeny is of such small degree that the attack outlined in this paper can be used to recover the remaining part.

We start with the case where the most significant bits of the secret are leaked or correctly guessed. These bits correspond to the last steps of the secret isogeny in the isogeny graph. Assume N_1 is a power of a prime ℓ. If the most significant k digits of the secret with respect to their representation in base ℓ are leaked or guessed correctly, the partial isogeny which remains to be recovered is of degree N_1/ℓ^k and we can run our attack as soon as N_1/ℓ^k fulfills the unbalancedness criterion $N_2 > p(N_1/\ell^k)^4$.

The case where the least significant digits are known or guessed requires a little more work. For simplicity of our exposition we assume again that N_1 is a power of 2 as in Sect. 4.2, but the results generalize to powers of small primes.

Lemma 4.28. *Let G be the group of Proposition 4.11, and let $G' \subset G$ be the subset of the form $\{a + b\iota \mid a \text{ odd}, b \text{ divisible by } 2^k\}$ where we identify two endomorphisms with each other if they differ by multiplication by an odd scalar modulo N_1. Then G' is an abelian subgroup of G.*

Proof. Since G is abelian, it suffices to show that G' is a subgroup. Consider $(a + b\iota)(a' + b'\iota) = (aa' - bb') + (ab' + a'b)\iota$. It is easy to see that $aa' - bb'$ is odd and $ab' + a'b$ is divisible by 2^k if $a + b\iota$ and $a' + b'\iota$ are in G'. □

Assume the least significant k bits of the secret, or equivalently the first k steps of the secret isogeny, are known. Kernels of isogenies of degree $N_1 > 2^k$ that share the same first k steps lie in the same 2^k-torsion subgroup and are therefore congruent modulo 2^k.

Recall the subsets of $\mathcal{I}$ introduced in Sect. 4.2.

Proposition 4.29. *Let $\mathcal{I}'$ be any subsets of $\mathcal{I}_1 := \{\langle P + [\alpha]Q\rangle \text{ with } 2|\alpha\}$ containing all those cyclic subgroups where the αs are congruent modulo 2^k. The group G' of Lemma 4.28 acts freely and transitively on any $\mathcal{I}'$.*

Proof. First, we need to show that $G' \times \mathcal{I}' \to \mathcal{I}'$ is well-defined. Let $(1 + b\iota)$ be a representative of some element in G' and let $P + k\iota(P)$, for some $k \in \mathbb{Z}$, be the kernel of an isogeny leading to a curve in $\mathcal{I}'$. We have

$$(1 + b\iota)(P + k\iota(P)) = P + k\iota(P) + b(\iota(P) - kP) \equiv P + k\iota(P) \pmod{b}$$

and as b is divisible by 2^k, $P + k\iota(P) \in \mathcal{I}'$ implies $(1 + b\iota)(P + k\iota(P)) \in \mathcal{I}'$. That the action is free and transitive follows from Proposition 4.11 and a counting argument as $|G|/|G'| = 2^{k-1} = |\mathcal{I}_1|/|\mathcal{I}'|$. □

Similarly, we can take subsets of $\mathcal{I}_2$ and $\mathcal{I}_3$ and restrict the acting group.

This gives rise to an attack strategy when N_2 is not large enough. Guessing k bits of the secret before applying the attack on the remaining part allows an attacker to reduce the requirements on the parameters to $N_2 > p(N_1/2^k)^4$. This is the same as when guessing the last bits of the secret.

Given such a partial isogeny, one computes the correct equivalence class $\mathcal{I}'$ from the kernel of the known part of the isogeny. Moreover, one needs to compute the lifts of elements of G' to endomorphisms of norm N_2 or eN_2. Computing the action of G' on the set $\mathcal{I}'$ allows one to test for the hidden shift property. Once it is satisfied, the secret can be recovered by solving an injective abelian hidden shift problem. Otherwise, one can make another guess on the k bits of the secret.

Apart from reducing the requirements on the unbalancedness, guessing part of the isogeny reduces the number of elements one needs to lift and the size of the hidden shift instance. Depending on the concrete parameter sets provided, one may combine exhaustive search and the attack presented in this paper to recover secrets more efficiently.

5 Childs-Jao-Soukharev's Attack on HHS

We begin by providing more detail on how the algorithm proposed by Childs, Jao and Soukharev [5] succeeds to construct an isogeny between two given ordinary elliptic curves in quantum subexponential time. The provided strategy can further be applied to CSIDH [3].

Recall the free and transitive group action from Sect. 2.3 of the class group on the set of isogenous ordinary curves with the same endomorphism ring. The hard problem is to find an isogeny between two isogenous ordinary elliptic curves with the same endomorphism ring, i.e., reversing this group action. Childs-Jao-Soukharev provide an algorithm that constructs such an isogeny in quantum subexponential time [5] using a reduction to the hidden shift problem.

We summarize the core idea as another instance of our framework using malleability oracles. Let $\mathcal{I} := \mathrm{Cl}(\mathcal{O})$ and $\mathfrak{O} := \mathrm{Ell}_{q,n}(\mathcal{O})$. We can look at the group action defined in Sect. 2.3 as a one-way function

$$f : \mathcal{I} \to \mathfrak{O}\ ,\ [x] \mapsto [x] \cdot j(E_0).$$

Note that the class group $\mathrm{Cl}(\mathcal{O})$ acts on itself and therefore f has a malleability oracle with respect to the class group readily available everywhere on the image, i.e., f is malleable with respect to this group action.

Finding an isogeny φ is now equivalent to finding the ideal class $[\mathfrak{b}]$ in $\mathrm{Cl}(\mathcal{O})$ containing the ideal corresponding to the kernel of φ, i.e., we would like to compute the preimage of f at $j(E_1) = [\mathfrak{b}] \cdot j(E_0)$.

Childs-Jao-Soukharev observed that the functions $F_i : \mathrm{Cl}(\mathcal{O}) \to \mathrm{Ell}_{q,n}(\mathcal{O})$, $[x] \mapsto [x] \cdot j(E_i)$ for $i = 0, 1$ are shifts of each other. Moreover, they are injective functions since the action of the class group on $\mathrm{Ell}_{q,n}(\mathcal{O})$ is free and transitive. The injective abelian hidden shift problem can be solved in quantum subexponential time, which allows one to recover $[\mathfrak{b}]$ and therefore an isogeny $\varphi : E_0 \to E_1$.

Analogously to the case for ordinary curves, the group action in CSIDH utilizing supersingular curves can be attacked this way. Recall that CSIDH uses the $\mathbb{F}_p$-rational endomorphism ring of the fixed starting curve E_0, $\mathcal{O}$. In the Diffie-Hellman-type key exchange, recovering a party's secret key constitutes of computing their secret ideal class $[\mathfrak{b}] \in \mathrm{Cl}(\mathcal{O})$ which satisfies $[\mathfrak{b}] \cdot E_0 = E_B$ for the party's public curve E_B. Through defining functions $F_0, F_1 : \mathrm{Cl}(\mathcal{O}) \to \mathrm{Ell}_p(\mathcal{O})$ by $F_0([x]) = [x] \cdot E_0$ and $F_1([x]) = [x] \cdot E_B$, it is possible to reduce finding Bob's secret key $[\mathfrak{b}]$ to an instance of the injective hidden shift problem: We have $F_1([x]) = F_0([x] \cdot [\mathfrak{b}])$ for any ideal class $[x] \in \mathrm{Cl}(\mathcal{O})$, where the functions are both injective due to the group action being free and transitive.

6 Conclusion and Further Work

In this paper, we constructed an abelian group action on the key space of the inherently non-commutative SIDH. Having this group action in place allows us to construct a heuristic malleability oracle using the torsion point information provided in SIDH when overstretched and sufficiently unbalanced parameters are being used. This contradicts the commonly believed misconception that no such group action exists in the branch of isogeny-based cryptography where one considers the full isogeny graph of supersingular elliptic curves. We embedded our attack in a more general framework that also captures other quantum attacks on schemes in isogeny-based cryptography.

The attack does *not* apply to balanced parameters as specified in the original SIDH proposal [14] or the NIST post-quantum candidate SIKE [13]. Furthermore, the unbalancedness condition between N_1 and N_2 is stronger than required by the attack from [19]. Interestingly, the obstruction to attack SIDH with balanced parameters in our case does not seem to be directly related to the hindrances in other attacks on unbalanced SIDH exploiting torsion point information [19, 23] but to limitations of the KLPT algorithm [16] and the ones described in Remark 4.25 instead. Improvements to the lifting subroutine included in the KLPT algorithm would not only partially decrease the required unbalancedness of SIDH parameters in this work, but also improve various isogeny-based schemes such as Galbraith-Petit-Silva's signatures [11] and SQISign [9].

Future work will extend the given quantum algorithm to more general group actions of quadratic orders that embed optimally into the (known) endomorphism ring of the starting curve. Hereby, the starting curve does not necessarily need to be of j-invariant 1728. Furthermore, we will generalize the approach to higher genus generalizations of SIDH. Finally, providing applications of this work to areas beyond isogeny-based cryptography is left for future investigation.

It remains an open problem to improve the framework further and to give conditions on the malleability oracle that are sufficient to invert one-way functions in quantum polynomial time.

Acknowledgement. We thank Lorenz Panny for helpful comments on a previous version of this paper and the anonymous reviewers of Eurocrypt2021 for their work and useful feedback. The work of Péter Kutas and Christophe Petit was supported by EPSRC grant EP/S01361X/1. Simon-Philipp Merz was supported by EPSRC grant EP/P009301/1.

References

1. Azarderakhsh, R., et al.: Supersingular isogeny key encapsulation. Updated parameters for round 2 of NIST Post-Quantum Standardization project (2019)
2. Bonnetain, X., Schrottenloher, A.: Quantum security analysis of CSIDH. In: Canteaut, A., Ishai, Y. (eds.) EUROCRYPT 2020. LNCS, vol. 12106, pp. 493–522. Springer, Cham (2020). https://doi.org/10.1007/978-3-030-45724-2_17
3. Castryck, W., Lange, T., Martindale, C., Panny, L., Renes, J.: CSIDH: an efficient post-quantum commutative group action. In: Peyrin, T., Galbraith, S. (eds.) ASIACRYPT 2018. LNCS, vol. 11274, pp. 395–427. Springer, Cham (2018). https://doi.org/10.1007/978-3-030-03332-3_15
4. Charles, D.X., Lauter, K.E., Goren, E.Z.: Cryptographic hash functions from expander graphs. J. Cryptol. **22**(1), 93–113 (2009)

5. Childs, A., Jao, D., Soukharev, V.: Constructing elliptic curve isogenies in quantum subexponential time. J. Math. Cryptol. **8**(1), 1–29 (2014)
6. Cornacchia, G.: Su di un metodo per la risoluzione in numeri interi dell'equazione $\sum_{h=0}^{n} c_h x^{n-h} y^h = p$. Giornale di Matematiche di Battaglini **46**, 33–90 (1908)
7. Couveignes, J.-M.: Hard homogeneous spaces. IACR Cryptology ePrint Archive, 2006:291 (1999)
8. Feo, L.D.: Mathematics of isogeny based cryptography. arXiv preprint: 1711.04062 (2017)
9. De Feo, L., Kohel, D., Leroux, A., Petit, C., Wesolowski, B.: SQISign: compact post-quantum signatures from quaternions and isogenies. IACR Cryptology ePrint Archive, 2020:1240 (2020)
10. Galbraith, S.D., Petit, C., Shani, B., Ti, Y.B.: On the security of supersingular isogeny cryptosystems. In: Cheon, J.H., Takagi, T. (eds.) ASIACRYPT 2016. LNCS, vol. 10031, pp. 63–91. Springer, Heidelberg (2016). https://doi.org/10.1007/978-3-662-53887-6_3
11. Galbraith, S.D., Petit, C., Silva, J.: Identification protocols and signature schemes based on supersingular isogeny problems. J. Cryptol. **33**(1), 130–175 (2020)
12. Hofheinz, D., Hövelmanns, K., Kiltz, E.: A modular analysis of the Fujisaki-Okamoto transformation. IACR Cryptology ePrint Archive, 2017:604 (2017)
13. Jao, D., et al.: SIKE: Supersingular isogeny key encapsulation (2017). http://sike.org/
14. Jao, D., De Feo, L.: Towards quantum-resistant cryptosystems from supersingular elliptic curve isogenies. In: Yang, B.-Y. (ed.) PQCrypto 2011. LNCS, vol. 7071, pp. 19–34. Springer, Heidelberg (2011). https://doi.org/10.1007/978-3-642-25405-5_2
15. Jaques, S., Schanck, J.M.: Quantum cryptanalysis in the RAM model: claw-finding attacks on SIKE. In: Boldyreva, A., Micciancio, D. (eds.) CRYPTO 2019. LNCS, vol. 11692, pp. 32–61. Springer, Cham (2019). https://doi.org/10.1007/978-3-030-26948-7_2
16. Kohel, D., Lauter, K., Petit, C., Tignol, J.-P.: On the quaternion ℓ-isogeny path problem. LMS J. Comput. Math. **17**(A), 418–432 (2014)
17. Kuperberg, G.: A subexponential-time quantum algorithm for the dihedral hidden subgroup problem. SIAM J. Comput. **35**(1), 170–188 (2005)
18. Kuperberg, G.: Another subexponential-time quantum algorithm for the dihedral hidden subgroup problem. arXiv preprint:1112.3333 (2011)
19. Kutas, P., Martindale, C., Panny, L., Petit, C., Stange, K.E.: Weak instances of SIDH variants under improved torsion-point attacks. IACR Cryptology ePrint Archive, 2020:633 (2020)
20. Kutas, P., Merz, S.-P., Petit, C., Weitkämper, C.: One-way functions and malleability oracles: Hidden shift attacks on isogeny-based protocols. IACR Cryptology ePrint Archive, 2021:282 (2021)
21. Landau, E.: Über die Einteilung der positiven ganzen Zahlen in vier Klassen nach der Mindestzahl der zu ihrer additiven Zusammensetzung erforderlichen Quadrate (1909)
22. Peikert, C.: He gives C-sieves on the CSIDH. In: Canteaut, A., Ishai, Y. (eds.) EUROCRYPT 2020. LNCS, vol. 12106, pp. 463–492. Springer, Cham (2020). https://doi.org/10.1007/978-3-030-45724-2_16
23. Petit, C.: Faster algorithms for isogeny problems using torsion point images. In: Takagi, T., Peyrin, T. (eds.) ASIACRYPT 2017. LNCS, vol. 10625, pp. 330–353. Springer, Cham (2017). https://doi.org/10.1007/978-3-319-70697-9_12
24. Petit, C., Smith, S.: An improvement to the quaternion analogue of the l-isogeny problem. Presentation at MathCrypt (2018)

25. Ramanujan, S.: First letter to G.H. Hardy (1913)
26. Regev, O.: A subexponential time algorithm for the dihedral hidden subgroup problem with polynomial space. arXiv preprint:0406151 (2004)
27. Rostovtsev, A., Stolbunov, A.: Public-key cryptosystem based on isogenies. IACR Cryptology ePrint Archive, 2006:145 (2006)
28. Schoof, R.: Nonsingular plane cubic curves over finite fields. J. Comb. Theory, Ser. A **46**(2), 183–211 (1987)
29. Silverman, J.H.: The Arithmetic of Elliptic Curves. GTM, vol. 106. Springer, New York (2009). https://doi.org/10.1007/978-0-387-09494-6
30. Vélu, J.: Isogénies entre courbes elliptiques. CR Acad. Sci. Paris, Séries A **273**, 305–347 (1971)
31. Voight, J.: Quaternion algebras. Preprint (2018)
32. Waterhouse, W.C.: Abelian varieties over finite fields. In: Annales scientifiques de l'École Normale Supérieure, vol. 2, pp. 521–560 (1969)

Sieving for Twin Smooth Integers with Solutions to the Prouhet-Tarry-Escott Problem

Craig Costello[1(✉)], Michael Meyer[2,3], and Michael Naehrig[1]

[1] Microsoft Research, Redmond, WA, USA
{craigco,mnaehrig}@microsoft.com
[2] University of Applied Sciences Wiesbaden, Wiesbaden, Germany
[3] University of Würzburg, Würzburg, Germany
michael@random-oracles.org

Abstract. We give a sieving algorithm for finding pairs of consecutive smooth numbers that utilizes solutions to the Prouhet-Tarry-Escott (PTE) problem. Any such solution induces two degree-n polynomials, $a(x)$ and $b(x)$, that differ by a constant integer C and completely split into linear factors in $\mathbb{Z}[x]$. It follows that for any $\ell \in \mathbb{Z}$ such that $a(\ell) \equiv b(\ell) \equiv 0 \bmod C$, the two integers $a(\ell)/C$ and $b(\ell)/C$ differ by 1 and necessarily contain n factors of roughly the same size. For a fixed smoothness bound B, restricting the search to pairs of integers that are parameterized in this way increases the probability that they are B-smooth. Our algorithm combines a simple sieve with parametrizations given by a collection of solutions to the PTE problem.

The motivation for finding large *twin smooth* integers lies in their application to compact isogeny-based post-quantum protocols. The recent key exchange scheme B-SIDH and the recent digital signature scheme SQISign both require large primes that lie between two smooth integers; finding such a prime can be seen as a special case of finding twin smooth integers under the additional stipulation that their sum is a prime p.

When searching for cryptographic parameters with $2^{240} \leq p < 2^{256}$, an implementation of our sieve found primes p where $p+1$ and $p-1$ are 2^{15}-smooth; the smoothest prior parameters had a similar sized prime for which $p-1$ and $p+1$ were 2^{19}-smooth. In targeting higher security levels, our sieve found a 376-bit prime lying between two 2^{21}-smooth integers, a 384-bit prime lying between two 2^{22}-smooth integers, and a 512-bit prime lying between two 2^{28}-smooth integers. Our analysis shows that using previously known methods to find high-security instances subject to these smoothness bounds is computationally infeasible.

Keywords: Post-quantum cryptography · Isogeny-based cryptography · Prouhet-Tarry-Escott problem · Twin smooth integers · B-SIDH · SQISign

A. Canteaut and F.-X. Standaert (Eds.): EUROCRYPT 2021, LNCS 12696, pp. 272–301, 2021.
https://doi.org/10.1007/978-3-030-77870-5_10

1 Introduction

We study the problem of finding *twin smooth* integers, i.e. finding two consecutive large integers, m and $m+1$, whose product is as smooth as possible. Though the literature on the role of smooth numbers in computational number theory and cryptography is vast (see for example the surveys by Pomerance [20] and Granville [12]), the problem of finding consecutive smooth integers of cryptographic size has only been motivated very recently: optimal instantiations of the key exchange scheme B-SIDH [8] and the digital signature scheme SQISign [10] require a large prime that lies between two smooth integers, and this is a special case of the twin smooth problem in which $2m+1$ is prime.

This paper presents a sieving algorithm for finding twin smooth integers that improves on the methods used in [8] and [10]. The high-level idea is to use two monic polynomials of degree n that split in $\mathbb{Z}[x]$ and that differ by a constant, i.e.

$$a(x) = \prod_{i=1}^{n}(x - a_i) \quad \text{and} \quad b(x) = \prod_{i=1}^{n}(x - b_i), \quad \text{where} \quad a(x) - b(x) = C \tag{1}$$

for $C \in \mathbb{Z}$. Whenever $\ell \in \mathbb{Z}$ such that $a(\ell) \equiv b(\ell) \equiv 0 \bmod C$, it follows that the integers $a(\ell)/C$ and $b(\ell)/C$ differ by 1.

Assume that $|\ell| \gg |a_i|$ and $|\ell| \gg |b_i|$ for $1 \leq i \leq n$, and fix a smoothness bound B. Rather than directly searching for two consecutive B-smooth integers m and $m+1$, roughly of size N, the search instead becomes one of finding a value of ℓ such that the $2n$ (not necessarily distinct) integers

$$\ell - a_1, \ \ldots, \ \ell - a_n, \ \ell - b_1, \ \ldots, \ \ell - b_n, \tag{2}$$

each of size roughly $N^{1/n}$, are B-smooth. For $n > 1$, and under rather mild heuristics, the probability of finding twin smooth integers in this fashion is significantly greater than the searches used in [8] and [10]. Put another way, the same computational resources are likely to succeed in finding twin smooth integers subject to an appreciably smaller smoothness bound.

To search for $\ell \approx N^{1/n}$ such that the $2n$ integers in (2) are B-smooth, we adopt the simple sieve of Eratosthenes as described by Crandall and Pomerance [9, §3.2.5]; this identifies all of the B-smooth numbers in an arbitrary interval. If w is the largest difference among the $2n$ integers in $\{a_i\} \cup \{b_i\}$, then a sliding window of size $|w|$ can be used to scan the given interval for simultaneous smoothness among the integers in (2). This approach has a number of benefits. Firstly, smooth numbers in a given interval can be recognized once-and-for-all, meaning we can combine arbitrarily many solutions to (1) into one scan of the interval. Secondly, different processors can scan disjoint intervals in parallel, and each of the interval sizes can be tailored to the available memory of the processor. Finally, the simple sieve we use to identify the smooth numbers in an interval (which is the bottleneck of the overall procedure) is open to a range of modifications and improvements – see Sect. 7.

The approach in this paper hinges on being able to find solutions to (1). Such solutions are related to a classic problem in Diophantine Analysis.

1.1 The Prouhet-Tarry-Escott Problem

The Prouhet-Tarry-Escott (PTE) problem of size n and degree k asks to find two distinct multisets of integers $\{a_1, \ldots, a_n\}$ and $\{b_1, \ldots, b_n\}$ for which

$$\begin{aligned} a_1 + \cdots + a_n &= b_1 + \cdots + b_n, \\ a_1^2 + \cdots + a_n^2 &= b_1^2 + \cdots + b_n^2, \\ &\vdots \\ a_1^k + \cdots + a_n^k &= b_1^k + \cdots + b_n^k. \end{aligned}$$

The most interesting case is $k = n - 1$, which is maximal (see Sect. 3), and such *ideal* solutions immediately satisfy (1). For example, when $n = 4$, the sets $\{0, 4, 7, 11\}$ and $\{1, 2, 9, 10\}$ are such that

$$\begin{aligned} 0 + 4 + 7 + 11 &= 1 + 2 + 9 + 10 &= 22, \\ 0^2 + 4^2 + 7^2 + 11^2 &= 1^2 + 2^2 + 9^2 + 10^2 &= 186, \\ 0^3 + 4^3 + 7^3 + 11^3 &= 1^3 + 2^3 + 9^3 + 10^3 &= 1738, \end{aligned}$$

from which it follows (see Proposition 1) that

$$a(x) = x(x-4)(x-7)(x-11) \text{ and } b(x) = (x-1)(x-2)(x-9)(x-10)$$

differ by a constant $C \in \mathbb{Z}$. Indeed, $a(x) - b(x) = -180$.

Origins of the PTE problem are found in the 18th century works of Euler and Goldbach, and it remains an active area of investigation [5–7]. In 1935, Wright [28] conjectured that ideal solutions to the PTE problem should exist for all n, but at present this conjecture is open: for $n = 11$ and for $n \geq 13$, no ideal solutions to the PTE problem have been found, see [5, p. 94] and [7, p. 73]. However, Borwein states that "heuristic arguments suggest that Wright's conjecture should be false. [...] It is intriguing, however, that ideal solutions exist for as many n as they do" [5, p. 87].

The PTE solutions that *are* known for $n \in \{2, 3, 4, 5, 6, 7, 8, 9, 10, 12\}$ are a nice fit for our purposes. If we were to fix a smoothness bound, B, and then search for the largest pair of consecutive B-smooth integers we could find, having PTE solutions for n as large as possible would be helpful. But for our cryptographic applications (see Sect. 1.3), we will instead fix a target range for our twin smooth integers to match a given security level, and then aim to find the smoothest twins within that range. In this case, the degree n of $a(x)$ and $b(x)$ cannot be too large, since a larger n means fewer $\ell \in \mathbb{Z}$ to search over. Ideally, n needs to be large enough such that the splitting of $a(x)$ and $b(x)$ into n linear factors helps with the smoothness probability, but small enough so that we still have ample $\ell \in \mathbb{Z}$ to find $a(\ell)$ and $b(\ell)$ such that

(i) $a(\ell) \equiv b(\ell) \equiv 0 \bmod C$,
(ii) $(m, m+1) = (b(\ell)/C, a(\ell)/C)$ are B-smooth, and (if desired)
(iii) $2m + 1$ is prime.

It turns out that those $n \leq 12$ for which PTE solutions are known are the *sweet spot* for our target applications, where $2^{240} \leq m \leq 2^{512}$.

1.2 Prior Methods of Finding Twin Smooth Integers

After defining *twin smooth integers* for concreteness, we recall previous methods used to find large twin smooth integers.

Definition 1 (Twin smooth integers). *For a given $B > 1$, we call $(m, m+1)$ with $m \in \mathbb{Z}$ a pair of* twin B-smooth integers *or* B-smooth twins *if $m \cdot (m+1)$ contains no prime factor larger than B.*

As Lehmer notes in [18], consecutive pairs of smooth integers have occurred in 18th century works and have been mentioned by Gauss in the context of computing logarithms of integers.

Hildebrand [13, Corollary 2] has shown that there are infinitely many pairs of consecutive smooth integers $(m, m+1)$, however this result notably holds for a smoothness bound that depends on m. More precisely, there are infinitely many such pairs of m^ϵ-smooth integers for any fixed $\epsilon > 0$. An analogous result holds for tuples of k consecutive smooth integers (for any k), as shown by Balog and Wooley [1].

For a fixed, constant smoothness bound B, the picture is different. A theorem by Størmer [25] states that there are only a finite number of such pairs. We begin with some historical results which show that deterministically computing the largest pair of consecutive B-smooth integers requires a number of operations that is exponential in the number of primes up to B.

Solving Pell Equations. Fix B, let $\{2, 3, \ldots q\}$ be the set of primes up to B with cardinality $\pi(B)$, and suppose that m and $m+1$ are both B-smooth. Let $x = 2m+1$, so that $x-1$ and $x+1$ are also B-smooth, and let D be the squarefree part of the product $(x-1)(x+1)$, so that $x^2 - 1 = Dy^2$ for some $y \in \mathbb{Z}$. Since the product $(x-1)(x+1)$ is B-smooth, it follows that Dy^2 is B-smooth, which (since D is squarefree) means that

$$D = 2^{\alpha_2} \cdot 3^{\alpha_3} \cdot \cdots \cdot q^{\alpha_q}$$

with $\alpha_i \in \{0, 1\}$ for $i = 2, 3, \ldots, q$. For each of the $2^{\pi(B)}$ squarefree possibilities for D, an effective theorem of Størmer [25] (and further work by Lehmer [18]) reverses the above argument and proposes to solve the $2^{\pi(B)}$ Pell equations

$$x^2 - Dy^2 = 1,$$

finding *all* of the solutions for which y is B-smooth, and in doing so finding the complete set of B-smooth consecutive integers m and $m+1$.

Ideally, this process could be used to deterministically find optimally smooth consecutive integers at any size, by increasing B until the largest pair of twin smooths is large enough. For example, the largest pair of twin smooth integers with $B = 3$ is $(8, 9)$, the largest pair of twin smooth integers with $B = 5$ is $(80, 81)$, and the largest pair of twin smooth integers with $B = 7$ is $(4374, 4375)$. Unfortunately, solving $2^{\pi(B)}$ Pell equations becomes infeasible before the size of m grows large enough to meet our requirements.

For $B = 113$, [8] reports that the largest twins $(m, m+1)$ found upon solving all 2^{30} Pell equations have $m = 19316158377073923834000 \approx 2^{74}$, and the largest twins found among the set when adding the requirement that $2m+1$ is prime have $m = 75954150056060186624 \approx 2^{66}$.

The Extended Euclidean Algorithm. One naïve way of searching for twin smooth integers is to compute B-smooth numbers m until either $m-1$ or $m+1$ also turns out to be B-smooth. A much better method, which was used in [4,8,10], is to instead choose two coprime B-smooth numbers α and β that are both of size roughly the square root of the targets m and $m+1$. Since α and β are coprime, Euclid's extended GCD algorithm outputs two integers (s, t) such that $\alpha s + \beta t = 1$ with $|s| < |\beta/2|$ and $|t| < |\alpha/2|$. We can then take $\{m, m+1\} = \{|\alpha s|, |\beta t|\}$, and the probability of m and $m+1$ being B-smooth is now the probability that $s \cdot t$ is B-smooth. The key observation here is that the product $s \cdot t$ with $s \approx t$ is much more likely to be B-smooth than a random integer of similar size. In Sect. 2 we will develop methods and heuristics that allow us to closely approximate these probabilities.

Searching with $m = x^n - 1$. The method from [8] that proved most effective in finding twin smooth integers with $2^{240} \le m \le 2^{256}$ is by searching with $(m, m+1) = (x^n - 1, x^n)$ for various n, where the best instances were found with $n = 4$ and $n = 6$. Our approach can be seen as an extension of this method, where the crucial difference is that for $n > 2$ the polynomial $x^n - 1$ does not split in $\mathbb{Z}[x]$, and the presence of higher degree terms significantly hampers the probability that values of $\ell^n - 1 \in \mathbb{Z}$ are smooth. For example, with $n = 6$ we have $m = (x^2 - x + 1)(x^2 + x + 1)(x - 1)(x + 1)$ and, assuming $B \ll \ell$, the probability that integer values of this product are B-smooth is far less than if it was instead a product of six monic, linear terms. On the other hand, the probability that $m+1$ is B-smooth for a given ℓ is the probability that ℓ itself is B-smooth, which works in favor of the *non-split* method. However, as we shall see in the sections that follow, this is not enough to counteract the presence of the higher degree terms. Furthermore, several of the PTE solutions we will be using also benefit from repeated factors.

1.3 Cryptographic Applications of Twin Smooth Integers

The field of supersingular isogeny-based cryptography continues to gain increased popularity in large part due to the conjectured quantum-hardness of variants of the *supersingular isogeny problem.* In its most general form, this problem asks to find a secret isogeny $\phi \colon E \to E'$ between two given supersingular elliptic curves $E/\mathbb{F}_{p^2}$ and $E'/\mathbb{F}_{p^2}$.

The most famous isogeny-based cryptosystems are Jao and De Feo's SIDH key exchange protocol [15] and its actively secure incarnation SIKE [14], which recently advanced to the third round of the NIST post-quantum standardization effort [26]. On the one hand, SIKE offers the advantage of having the smallest public key and ciphertext sizes of all of the key encapsulation schemes under consideration, but on the other, its performance is currently around an order of magnitude slower than its code- and lattice-based counterparts.

Two supersingular isogeny-based schemes have recently emerged that require a new type of instantiation. Rather than defining primes p for which either $p-1$ *or* $p+1$ is smooth (as in SIDH/SIKE), the key exchange scheme B-SIDH [8] and the digital signature scheme SQISign [10] instead require primes for which (large factors of) both $p-1$ *and* $p+1$ are smooth. As both of those papers discuss, finding primes that lie between two smooth integers is not an easy task, but the practical incentive to do this is again related to the compactness of these schemes: B-SIDH's public keys are even smaller than the analogous SIDH/SIKE compressed public keys, and the sum of the SQISign public key and signature sizes is significantly smaller than those of all of the remaining NIST signature candidates.

In both B-SIDH and SQISign, the overall efficiency of the protocol is closely tied to the smoothness of $p-1$ and $p+1$. Roughly speaking, any prime ℓ appearing in the factorizations of these two integers implies that an ℓ-isogeny needs to be computed somewhere in the protocol. Such ℓ-isogenies have traditionally been computed in $O(\ell)$ field operations using Vélu's formulas [27], but recent work by Bernstein, De Feo, Leroux, and Smith [4] improved the asymptotic complexity to $\widetilde{O}(\sqrt{\ell})$ by clever use of a baby-step giant-step algorithm. Nevertheless, the large ℓ-isogenies that are required in these protocols dominate the runtime, and the best instantiations of both schemes will use large primes p lying between two integers that are as smooth as possible.

In this paper we will view the search for such primes as one that imposes an additional stipulation on the more general problem of finding twin smooth integers: cryptographically useful instances of the twin smooth integers $(m, m+1)$ are those where the sum $2m+1$ is a prime, p.

Security analyses of B-SIDH and SQISign suggest that it is possible to relax the requirements and to tolerate *cofactors* that divide either or both of $p-1$ and $p+1$ and have prime factors somewhat larger than the target smoothness bound, such that (the size of) any primes dividing these cofactors have no impact on the efficiency. For simplicity and concreteness, we will focus our analysis on the pure problem of finding twin smooth integers that disallows any primes larger than our smoothness bound, but we will oftentimes point out the modifications and relaxations that account for cofactors; this is discussed in Sect. 7.

The heuristic analysis summarized in Table 3 predicts that sieving with PTE solutions finds twin smooth integers $(m, m+1)$ that are smoother than one expects to find using the same computational resources and the prior methods described in Sect. 1.2. Indeed, in Sect. 6 we present a number of examples we found with our sieve whose largest prime divisors are several bits smaller than the largest prime divisors in instantiations found in the literature. In reference to Table 3, we briefly sketch some intuition on how these smoother examples translate into practical speedups. For example, the best prior instantiation of a prime p with $2^{240} \leq p < 2^{256}$ found that $(p-1)$ and $(p+1)$ are simultaneously 2^{19}-smooth, whereas our sieve found a similarly sized p subject to a smoothness bound of 2^{15}. Given the current (square root) complexity of state-of-the-art ℓ-isogeny computations, this suggests that the most expensive isogeny computed in our example will be roughly 4 times faster than that of the prior example.

The source code for our sieving algorithm is publicly available at

https://github.com/microsoft/twin-smooth-integers.

This code can be used by implementers to find their own instantiations; in particular, the code is intended to be general and users should be able to tailor it to their own requirements, e.g., to allow for different requirements, cofactors, or to target other security levels.

Roadmap. First time readers may benefit from jumping straight to Sect. 5, where all the theory developed in Sects. 2–4 is put into action by way of a full worked example. Section 2 gathers some results that allow us to approximate the smoothness probabilities of both integers and integer-valued polynomials. Section 3 starts by making the connection between our method of finding twin smooth integers and the PTE problem, before going into the theory of the PTE problem and showing how to generate infinitely many solutions for certain degrees. Section 4 describes our sieving algorithm. Section 6 presents some of the best examples found with our sieve and compares them with the previous examples in the literature. Section 7 discusses a number of possible modifications and improvements to the sieve.

2 Smoothness Probabilities

In this section we recall some well-known results concerning smoothness probabilities that will be used to analyse various approaches throughout the paper: Sect. 2.1 shows how to approximate the probability that $m \gg B$ is B-smooth using the Dickman–de Bruijn function; Sect. 2.2 shows how to approximate the probability that integer values of a polynomial $f(x) \in \mathbb{Z}[x]$ are B-smooth.

2.1 Smoothness Probabilities for Large N

Recall that an integer is said to be B-smooth if it does not have any prime factor exceeding B. Let

$$\Psi(N, B) = \#\{1 \leq m \leq N \colon m \text{ is } B\text{-smooth}\}$$

be the number of positive B-smooth integers. For each real number $u > 0$, Dickman's theorem [9, Theorem 1.4.9] states that there is a real number $\rho(u) > 0$ such that

$$\frac{\Psi(N, N^{1/u})}{N} \sim \rho(u) \text{ as } N \to \infty. \tag{3}$$

Dickman described $\rho(u)$ as the unique continuous function on $[0, \infty)$ that satisfies $\rho(u) = 1$ for $0 \leq u \leq 1$, and $\rho'(u) = -\frac{\rho(u-1)}{u}$ for $u > 1$. For $1 \leq u \leq 2$, $\rho(u) = 1 - \ln(u)$, but for $u > 2$ there is no known closed form for $\rho(u)$. Nevertheless, it is easy to evaluate $\rho(u)$ (up to any specified precision) for a given value of u, and popular computer algebra packages (like Magma and Sage) have this function built in.

In this paper we will be using (3) to approximate the probability that certain large numbers are smooth. For example, with $N = 2^{128}$ and $u = 8$, the value $\rho(8) \approx 2^{-25}$ approximates the probability that a 128-bit number is 2^{16}-smooth. With u fixed, this approximation becomes better as N tends towards infinity. Using $\rho(u)$ as the *smoothness probability* assumes the heuristic that $N^{1/u}$-smooth numbers are uniformly distributed in $[1, N]$.

While there are methods to more precisely estimate $\Psi(N, B)$, see e.g. [24] and [2], we are content with the simple approximation given by ρ. Using a basic sieve to identify smooth integers, we have counted all B-smooth integers up to $N = 2^{43}$ for B up to 2^{16} and compared their numbers with those predicted by the Dickman–de Bruijn function. Except for the lower end of the studied interval and for very small smoothness bounds, we have found the approximation by ρ to be sufficiently close to the actual values.

2.2 Smoothness Heuristics for Polynomials

For a polynomial $f(x) \in \mathbb{Z}[x]$, define

$$\Psi_f(N, B) = \#\{1 \leq m \leq N \colon f(m) \text{ is } B\text{-smooth}\}.$$

Throughout the paper we will use the following conjecture (see [19, Eq. 1.4] and [12, Eq. 1.20]) as a heuristic to estimate the probability that $f(N)$ is $N^{1/u}$-smooth.

Heuristic 1. *Suppose that the polynomial $f(x) \in \mathbb{Z}[x]$ has distinct irreducible factors over $\mathbb{Z}[x]$ of degrees $d_1, d_2, \ldots d_k \geq 1$, respectively, and fix $u > 0$. Then*

$$\frac{\Psi_f(N, N^{1/u})}{N} \sim \rho(d_1 u) \ldots \rho(d_k u) \tag{4}$$

as $N \to \infty$.

With $B = N^{1/u}$, Heuristic 1 says that for $m \leq N$, the probability of $f(m)$ being B-smooth is the product of the probabilities of each of its factors being B-smooth (these are computed via (3)). Martin proved this conjecture for a certain range of u [19, Theorem 1.1] that does not apply in our case. Heuristic 1 inherently assumes that the smoothness probabilities of each of the factors are independent of one another; here, the roots of our split polynomials all lie in relatively short intervals, and thus are not uniformly distributed in, say, $[1, N]$. For example, with $f(m) = \prod_{1 \leq i \leq d}(m - f_i) \in \mathbb{Z}$, any prime q that divides $m - f_1$ only divides $m - f_i$ for some $1 < i \leq d$ if $q \mid f_i - f_1$, which in particular means that any prime which is larger than the interval size can divide at most one of the (unique) $m - f_i$. Nevertheless, our experiments have shown Heuristic 1 to be a very accurate approximation for our purposes; we simply use it as a means to approximate how many values of $m \in \mathbb{Z}$ need to be searched before we can expect to start finding twin smooth integers, and to draw comparisons between approaches for various target sizes.

3 Split Polynomials that Differ by a Constant

Henceforth we will use $a(x)$ and $b(x)$ to denote two polynomials of degree $n > 1$ in $\mathbb{Z}[x]$ that differ by an integer constant $C \in \mathbb{Z}$, i.e. $a(x) - b(x) = C$. Moreover, unless otherwise stated, both a and b are assumed to split into linear factors over $\mathbb{Z}$, i.e.

$$a(x) = \prod_{1 \leq i \leq n} (x - a_i) \quad \text{and} \quad b(x) = \prod_{1 \leq i \leq n} (x - b_i),$$

where the a_i and b_i (which are not necessarily distinct) are all in $\mathbb{Z}$.

The core idea of this paper is to search for twin smooth integers by searching over $\ell \in \mathbb{Z}$ such that

$$a(\ell) \equiv b(\ell) \equiv 0 \bmod C.$$

Then, the two polynomials $a_C(x) := a(x)/C$ and $b_C(x) := b(x)/C \in \mathbb{Q}[x]$ evaluate to integer values $a_C(\ell)$ and $b_C(\ell)$ at ℓ, and moreover

$$a_C(\ell) = b_C(\ell) + 1.$$

Since a and b split into n linear factors over $\mathbb{Z}$, $a_C(\ell)$ and $b_C(\ell)$ necessarily contain n integer factors of approximately the same size. In Sect. 4.4 we approximate the probability that $a_C(\ell)$ and $b_C(\ell)$ are B-smooth, and show that these probabilities are favorable (in the ranges of practical interest) compared to the previously known methods of searching for large twin smooths.

3.1 The Prouhet-Tarry-Escott Problem

For degrees $n \leq 3$, infinite families of split polynomials $a(x)$ and $b(x)$ with $a(x) - b(x) = C \in \mathbb{Z}$ can be constructed by solving the system that arises from equating all but the constant coefficients. Although there are n equations in $2n$ unknowns, for $n > 3$ this process becomes unwieldy; the equations are nonlinear and we are seeking solutions that assume values in $\mathbb{Z}$. Moreover, relaxing the monic requirement (which permits $4n$ unknowns) and allowing for solutions in $\mathbb{Q}$ does not seem to help beyond $n > 3$. Fortunately, finding these pairs of polynomials is closely connected to the computational hardness of solving the PTE problem of size n.

Definition 2 (The Prouhet-Tarry-Escott problem). *The Prouhet-Tarry-Escott (PTE) problem of size n and degree k asks to find distinct multisets of integers $\mathcal{A} = \{a_1, \ldots, a_n\}$ and $\mathcal{B} = \{b_1, \ldots, b_n\}$, such that*

$$\sum_{i=1}^{n} a_i^j = \sum_{i=1}^{n} b_i^j$$

for $j = 1 \ldots k$. We abbreviate solutions to this problem by writing $[a_1, \ldots, a_n] =_k [b_1, \ldots, b_n]$ or $\mathcal{A} =_k \mathcal{B}$.

A classic result that links PTE solutions to polynomials is the following [6, Proposition 1].

Proposition 1. *The following are equivalent:*

$$\sum_{i=1}^{n} a_i^j = \sum_{i=1}^{n} b_i^j \quad \text{for} \quad j = 1, \ldots, k. \tag{5}$$

$$\deg\left(\prod_{i=1}^{n}(x - a_i) - \prod_{i=1}^{n}(x - b_i)\right) \leq n - (k+1). \tag{6}$$

Proposition 1 implies that for any PTE solution of size n and degree $k = n-1$, the polynomials $a(x) = \prod_{i=1}^{n}(x-a_i)$ and $b(x) = \prod_{i=1}^{n}(x-b_i)$ differ by a constant. For a given n, this choice for k is the maximal possible choice [6, Proposition 2], hence the respective solutions are called *ideal solutions*. Ideal solutions are known for $n \leq 10$ and $n = 12$, but it remains unclear if there are ideal solutions for other sizes [7]. Unless stated otherwise, henceforth we will only speak of PTE solutions that are ideal solutions.

As we will see later, the most useful PTE solutions for our purposes are those for which the constant C is as small as possible. We now recall some useful results from the literature concerning the constants that can arise from PTE solutions.

Definition 3 (Fundamental constant C_n). *Let n be a positive integer, and write $C_{n,\mathcal{A},\mathcal{B}}$ for the associated constant of an ideal PTE solution $\mathcal{A} =_{n-1} \mathcal{B}$ of size n. Then we define*

$$C_n = \gcd\{C_{n,\mathcal{A},\mathcal{B}} \mid \mathcal{A} =_{n-1} \mathcal{B}\}$$

as the fundamental constant associated to ideal PTE solutions of size n.

A result by Kleiman [17] gives a lower bound on the fundamental constant.

Proposition 2. *Let n be a positive integer. Then $(n-1)! \mid C_n$.*

For concrete choices of n, more divisibility results are presented by Rees and Smyth [21], and Caley [7]. These results form sharper bounds for C_n, and thus for constants arising from any given PTE solution. Upper bounds for C_n can be

Table 1. Divisibility results for the PTE problem

n	Lower bound for C_n	Upper bound for C_n
2	1	1
3	2^2	2^2
4	$2^2 \cdot 3^2$	$2^2 \cdot 3^2$
5	$2^4 \cdot 3^2 \cdot 5 \cdot 7$	$2^4 \cdot 3^2 \cdot 5 \cdot 7$
6	$2^5 \cdot 3^2 \cdot 5^2$	$2^5 \cdot 3^2 \cdot 5^2$
7	$2^6 \cdot 3^3 \cdot 5^2 \cdot 7 \cdot 11$	$2^6 \cdot 3^3 \cdot 5^2 \cdot 7 \cdot 11$
8	$2^4 \cdot 3^3 \cdot 5^2 \cdot 7^2 \cdot 11 \cdot 13$	$2^8 \cdot 3^3 \cdot 5^2 \cdot 7^2 \cdot 11 \cdot 13$
9	$2^7 \cdot 3^3 \cdot 5^2 \cdot 7^2 \cdot 11 \cdot 13$	$2^9 \cdot 3^4 \cdot 5^2 \cdot 7^2 \cdot 11 \cdot 13 \cdot 17 \cdot 23 \cdot 29$
10	$2^7 \cdot 3^4 \cdot 5^2 \cdot 7^2 \cdot 13 \cdot 17$	$2^{11} \cdot 3^6 \cdot 5^2 \cdot 7^2 \cdot 11 \cdot 13 \cdot 17 \cdot 23 \cdot 37$
11	$2^8 \cdot 3^4 \cdot 5^3 \cdot 7^2 \cdot 11 \cdot 13 \cdot 17 \cdot 19$	none known
12	$2^8 \cdot 3^4 \cdot 5^3 \cdot 7^2 \cdot 11^2 \cdot 17 \cdot 19$	$2^{12} \cdot 3^8 \cdot 5^3 \cdot 7^2 \cdot 11^2 \cdot 13^2 \cdot 17 \cdot 19 \cdot 23 \cdot 29 \cdot 31$

directly computed by taking the GCD of all known solutions of size n. This is detailed in [7], where for example it is known that for $n = 9$ we have

$$2^7 \cdot 3^3 \cdot 5^2 \cdot 7^2 \cdot 11 \cdot 13 \mid C_9 \text{ and } C_9 \mid 2^9 \cdot 3^4 \cdot 5^2 \cdot 7^2 \cdot 11 \cdot 13 \cdot 17 \cdot 23 \cdot 29.$$

Table 1 is an updated version of [7, Table 3.2], and gives an overview of the bounds for the fundamental constants C_n. These results give estimates for the optimal choices of solutions for our searches. In particular, choosing solutions with associated constants close to the upper bound for C_n yields the best preconditions for finding twin smooth integers.

For our application of finding twin smooth integers, it may seem unnecessarily restrictive to only make use of PTE solutions, yielding monic polynomials a and b with integer roots. However, it can be proven that *all* polynomials that are split over $\mathbb{Q}$ and that differ by a constant arise from PTE solutions. In order to prove this, we make use of the following result ([6, Lemma 1], [7, Proposition 2.1.2]).

Proposition 3. *Let* $[a_1, \ldots, a_n] =_k [b_1, \ldots, b_n]$ *with associated constant* C *and* M, K *arbitrary integers with* $M \neq 0$. *Define a linear transform* $h(x) = Mx + K$ *and let* $a'_i = h(a_i)$ *and* $b'_i = h(b_i)$ *for* $i = 1, \ldots, n$. *Then* $[a'_1, \ldots, a'_n] =_k [b'_1, \ldots, b'_n]$, *and the associated constant is* $C' = C \cdot M^n$.

Two such solutions that are connected through a linear transform are called *equivalent*. Note that Proposition 3 also holds for the PTE problem over rational numbers instead of integers, i.e. for $a_i, b_i \in \mathbb{Q}$ for $1 \le i \le n$.

Corollary 1. *Let* $a(x)$ *and* $b(x)$ *be polynomials of degree* n *with rational roots* $\mathcal{A} = \{a_1, \ldots, a_n\}$ *and* $\mathcal{B} = \{b_1, \ldots, b_n\}$, *such that* $a(x) - b(x) = C \in \mathbb{Q}$. *Then* $\mathcal{A} =_{n-1} \mathcal{B}$ *for the PTE problem over* $\mathbb{Q}$, *and there is an equivalent solution* $\mathcal{A}' =_{n-1} \mathcal{B}'$ *to the PTE problem over* $\mathbb{Z}$.

Proof. Since $\deg(a(x) - b(x)) = 0$, Proposition 1 implies that $\mathcal{A} =_{n-1} \mathcal{B}$. Let $M \in \mathbb{Z}$ be a common denominator of $a_1, \ldots, a_n, b_1, \ldots, b_n$ and define the linear transform $h(x) = Mx$. Let $a'_i = h(a_i)$ and $b'_i = h(b_i)$ for $i = 1, \ldots, n$. Then $\mathcal{A}' = \{a'_1, \ldots, a'_n\}$ and $\mathcal{B}' = \{b'_1, \ldots, b'_n\}$ consist of integers, and by Proposition 3, $\mathcal{A}' =_{n-1} \mathcal{B}'$ is a solution for the PTE problem over $\mathbb{Z}$. □

Corollary 1 allows us to focus entirely on integer PTE solutions without imposing any further restrictions. For our search for smooth values of the polynomials, Proposition 3 further implies that we only have to search with one polynomial per equivalence class.

Corollary 2. *Let $\mathcal{A} = \{a_1, \ldots, a_n\}$, $\mathcal{B} = \{b_1, \ldots, b_n\}$, and $\mathcal{A}' = \{a'_1, \ldots, a'_n\}$, $\mathcal{B}' = \{b'_1, \ldots, b'_n\}$ be equivalent ideal PTE solutions. Let $a(x)$, $b(x)$, and $a'(x)$, $b'(x)$ be the respective polynomials such that $a(x) - b(x) = C \in \mathbb{Z}$ resp. $a'(x) - b'(x) = C' \in \mathbb{Z}$, and $h(x)$ be the associated linear transform. Then for given $x_{\min}$ and $x_{\max}$, $a_C(x)$ and $b_C(x)$ take on the same integer values for $x \in I = [x_{\min}, x_{\max}]$ as $a'_{C'}(x)$ and $b'_{C'}(x)$ for $x \in h(I)$.*

In order to efficiently identify equivalent solutions, we make use of Proposition 3 to define a representation of equivalence classes, which we call the *normalized form* of a class of solutions.

Definition 4 (Normalized form of PTE solutions). *A normalized form of a given PTE solution is a solution such that $a_1 \leq a_2 \leq \cdots \leq a_n$, $b_1 \leq b_2 \leq \cdots \leq b_n$, $0 = a_1 < b_1$, and $\gcd(a_1, \ldots, a_n, b_1, \ldots, b_n) = 1$.*

Another classification of solutions, which is of importance for our searches, is the distinction between *symmetric* and *non-symmetric* solutions [5].

Definition 5 (Symmetric PTE solutions). *For n even, an even ideal symmetric solution to the PTE problem is of the form*

$$[\pm a_1, \pm a_2, \ldots, \pm a_{n/2}] =_{n-1} [\pm b_1, \pm b_2, \ldots, \pm b_{n/2}].$$

For n odd, an odd ideal symmetric solution to the PTE problem is of the form

$$[a_1, a_2, \ldots, a_n] =_{n-1} [-a_1, -a_2, \ldots, -a_n].$$

It can immediately be seen that the normalized form of a symmetric solution is unique, but no longer has the form satisfying Definition 5. However, we will still be calling these solutions symmetric, since they are symmetric with respect to the integer K (instead of symmetric with respect to 0, as in the classic formulation of Definition 5), where $h(x) = Mx + K$ is the linear transform connecting these solutions. Thus, we define solutions as non-symmetric if and only if their equivalence class does not contain a symmetric solution according to Definition 5.

Note that in the special case of non-symmetric solutions, the normalized form is not unique. In particular, if $[a_1, \ldots, a_n] =_{n-1} [b_1, \ldots, b_n]$ is a non-symmetric normalized solution, then so is the solution arising from the linear transform $h(x) = Mx + K$, where $M = -1$ and $K = \max\{a_n, b_n\}$. In this case, we take the solution with minimal b_1 to represent the normalized solution, and refer to the second normalized solution as the *flipped solution.*

Finally, in Sect. 4.4 we will see that PTE solutions with repeated factors have higher probabilities (than those without repeated factors) of finding twin smooth integers. The following result [7, Theorem 2.1.3] shows that repeated factors can only occur with multiplicity at most 2.

Proposition 4 (Interlacing). *Let $\mathcal{A} = \{a_1, \ldots, a_n\}$ and $\mathcal{B} = \{b_1, \ldots, b_n\}$ be an ideal PTE solution, where $a_1 \leq a_2 \leq \cdots \leq a_n$ and $b_1 \leq b_2 \leq \cdots \leq b_n$, and w.l.o.g., we assume that $a_1 < b_1$. Then, $a_1 \neq b_j$ for all j. If n is odd, we have*

$$a_1 < b_1 \leq b_2 < a_2 \leq a_3 < \cdots < a_{n-1} \leq a_n < b_n,$$

and if n is even, then

$$a_1 < b_1 \leq b_2 < a_2 \leq a_3 < \cdots < a_{n-2} \leq a_{n-1} < b_{n-1} \leq b_n < a_n.$$

3.2 PTE Solutions

An important prerequisite for searching for twin smooth integers is a large number of normalized ideal PTE solutions with relatively small associated constants. To this end, we briefly review solutions from the literature as well as methods to construct ideal solutions. Henceforth, we will refer to normalized ideal PTE solutions only as PTE solutions.

A database of Shuwen collects several PTE solutions, both symmetric and non-symmetric [22]. In particular, special solutions, such as the smallest solutions with respect to the associated constants, and the first solutions found for each size, are presented there.

Apart from this, several methods for generating PTE solutions have been found. Parametric solutions are known for $n \in \{2, 3, 4, 5, 6, 7, 8, 10, 12\}$, and these can be used to generate infinitely many symmetric solutions [7]. However, the number of solutions with small associated constants is limited. For $n = 9$, only two non-equivalent solutions are known.

For $n \in \{5, 6, 7, 8\}$, we implemented the methods from [5] to generate as many symmetric solutions with small associated constants as possible. For $n = 10$ and $n = 12$, there are parametric symmetric solutions due to Smyth [23] and Choudhry and Wróblewski [29], resp., both following an earlier method from Letac [11]. In both methods, the two parameters that form solutions come from a quadratic equation in two variables. This equation can be transformed into an elliptic curve equation, and thus finding suitable parameters is equivalent to finding rational points on this elliptic curve. In [7, Section 6], Caley implements these methods by adding multiples of a non-torsion point, P, to the eight known

Table 2. Number of PTE solutions up to an upper bound for the constants. $C_{\min,n}$ denotes the smallest constant known for each degree.

n	$\lceil \log_2(C_{\min,n}) \rceil$	Bitlength of upper bound	# of solutions
5	13	50	49
6	14	50	2438
7	33	60	8
8	31	60	51
9	52	60	2
10	73	100	1
12	76	100	1

torsion points.[1] However, it is evident from the underlying transforms that PTE solutions with small constants can only arise from rational elliptic curve points with small denominators in their coordinates. Caley's approach thus proves to be non-optimal for our aims, as the denominators in the coordinates of $[i]P$ become too large already for very small i, resulting in PTE solutions with huge constants. We implemented these methods with the curves and transforms from [7], but deviated from Caley's approach by first searching for non-torsion points with integer coordinates, resp. coordinates with very small denominators. We then followed Caley's algorithm and computed small multiples of these points and their sums with torsion points. Despite finding many PTE solutions, none of them proved to have an associated constant close to the upper bound for C_{10} resp. C_{12}. Further, taking the GCD of all found solutions, we did not succeed in reducing the known upper bounds for C_{10} resp. C_{12}.

For each size n, we identified an upper bound for constants that permit acceptable success probabilities for our searches, and collected as many solutions as possible up to this value. Table 2 reports on the numbers of solutions we found, including solutions from [22].

4 Sieving with PTE Solutions

Our sieving algorithm consists of two phases. The first phase identifies the B-smooth numbers in a given interval (Sect. 4.1). The second phase then scans the interval using either a single PTE solution (Sect. 4.2) or the combination of many PTE solutions (Sect. 4.3).

4.1 Identifying Smooth Numbers in an Interval

We follow the exposition of Crandall and Pomerance [9, §3.2.5] and adopt the simple sieve of Eratosthenes to identify the B-smooth integers in an interval

[1] The elliptic curves that arise for $n = 10$ and $n = 12$ have Mordell-Weil-groups $\mathbb{Z}/4\mathbb{Z} \times \mathbb{Z}/2\mathbb{Z} \times \mathbb{Z}$ resp. $\mathbb{Z}/4\mathbb{Z} \times \mathbb{Z}/2\mathbb{Z} \times \mathbb{Z} \times \mathbb{Z}$. Thus there are eight torsion points in each case, and the non-torsion groups are generated by one resp. two non-torsion points.

$[L, R)$. We set up an array of $R - L$ integers corresponding to the integers $L, L + 1, \ldots, R - 1$, and initialize each entry with 1. For all primes with $p < B$, we identify the smallest non-negative $i \in \mathbb{Z}$, for which $L + i \equiv 0 \bmod p$, and multiply the array elements at positions $i + jp$ by p for all $j \in \mathbb{Z}$ such that $L \leq i + jp < R$. Additionally, for all primes with $p < \sqrt{R}$, we have to identify the maximal exponent e such that $p^e < R$, and analogously perform sieving steps with the relevant prime powers, where further multiplications by p take place. After this process is finished, the B-smooth integers in the interval are precisely those for which the number at position i is $L + i$. Subsequently, we transform this array of integers into a bitstring, where a '1' indicates a B-smooth number, while a '0' represents a non-smooth number.

This simple approach allows for several optimizations and modifications, some of which are discussed further in Sect. 7.

4.2 Searching with a Single PTE Solution

Assume that we are searching with a normalized ideal PTE solution of size n, writing $a(x) = \prod_{i=1}^{n}(x - a_i)$ and $b(x) = \prod_{i=1}^{n}(x - b_i)$, together with $C \in \mathbb{Z}$ such that $a(x) - b(x) = C$. We will assume $C > 0$, since $a(x)$ and $b(x)$ can otherwise swap roles accordingly, and as usual we write $a_C(x) = a(x)/C$ and $b_C(x) = b(x)/C$ as the two polynomials in $\mathbb{Q}[x]$.

We are searching for ℓ such that $m + 1 = a_C(\ell)$ and $m = b_C(\ell)$ are both B-smooth and of a given size, and thus the size of the constant C affects the size of the ℓ we should search over. Moreover, we only wish to search over the values of ℓ for which $a_C(\ell)$ and $b_C(\ell)$ are integers, and we determine this set of residues (modulo C) as follows. If $C = \prod p_i^{e_i}$ is the prime factorization of the constant, then for each prime-power factor we determine all residues $r_i \bmod p_i^{e_i}$ for which $a(r_i) \equiv b(r_i) \equiv 0 \bmod p_i^{e_i}$ (note that it is sufficient to check that one of $a(r_i)$ or $b(r_i)$ is a multiple of $p_i^{e_i}$). We then use the Chinese Remainder Theorem (CRT) to reconstruct the full set of residues $\{r \bmod C\}$ for which $a(r) \equiv b(r) \equiv 0 \bmod C$. Depending on the size of the constant, the full list of suitable residues may be rather large; if not, they can be stored in a lookup table, but if so, only the smaller sets (i.e. the $\{r_i\}$ corresponding to $p_i^{e_i}$) need to be stored. We can then either loop over the suitable residues by constructing them on the fly using the CRT, or we can check whether a candidate ℓ is a suitable residue by reducing it modulo each of the $p_i^{e_i}$.

It is worth pointing out that when searching for cryptographic parameters with a single PTE solution, the condition that $2m + 1$ is prime can be used to discard the residues $\{\tilde{r} \bmod C\}$ for which $2b_C(r) + 1$ can never be prime if $r \equiv \tilde{r} \bmod C$. In a very rare number of cases, the polynomial $2b_C(x) + 1 = 2/C \cdot (b(x) + C/2)$ in $\mathbb{Q}[x]$ is such that $(b(x) + C/2)$ is reducible in $\mathbb{Z}[x]$, in which case the PTE solution can be completely discarded. For example, this happens for *both* of the PTE solutions with $n = 9$.

Recall from Sect. 3 that the constants of the PTE solutions are (for our purposes) always B-smooth. When processing an interval $[L, R)$, the problem therefore reduces to finding $\ell \in [L, R)$ such that all of the factors of $a(\ell)$ and

$b(\ell)$ are marked as B-smooth. For the PTE solution in use, these factors are given by $\ell_i = \ell - i$, where $i \in \{a_1, \ldots, a_n, b_1, \ldots, b_n\}$. Note that since $a_1 = 0$ for our normalized representation, we have $\ell = \ell_0$. Starting with ℓ at the left end of the interval requires some care since for a given ℓ, we need to be able to check for the smoothness of all ℓ_i. Hence, to be able to cover the full space when processing consecutive intervals, we have to run the first phase of the sieve for a slightly larger interval, namely $[L - w, R)$ (overlapping to the left with the previous interval), where $w = \max\{a_n, b_n\}$. This allows us to process $\ell \in [L, R)$ such that ℓ_w will cover $[L - w, R - w)$.

In the second phase of the sieve we advance ℓ through all of the elements in the bitstring marked '1', each time checking the bits corresponding to the remaining ℓ_i, i.e. $i \in \{a_2, \ldots, a_n, b_1, \ldots, b_n\}$. If, at any time, we see that any of the ℓ_i corresponds to a '0', we advance ℓ such that it is aligned with the next '1' and repeat the process until all of the ℓ_i correspond to a '1'. At this point, we can then check whether ℓ is a suitable residue modulo C as above; if not, ℓ is again advanced to the next set bit, but if so, we have found twin smooth integers, and it is here that we can optionally check whether their sum is prime.

We note that when using a single PTE solution, the algorithm could be modified to sieve in arithmetic progressions given by the suitable residues modulo C. We leave the exploration of whether this can be more efficient than the above approach for future work.

In the case of a large interval $[L, R)$, the memory requirements can be significantly reduced by dividing $[L, R)$ into several subintervals, which can be processed separately. The only downside is that a naïve implementation of the first phase processes certain intervals twice due to the overlap of length w. This can be easily mitigated by copying the last w entries of the previous interval at each step. However, due to both the large (sub)intervals used in our implementation and the relatively small w's that arise in PTE solutions, the impact of this overlap is negligible in practice, so the naïve approach can be taken without a noticeable performance penalty.

Parallelization. Our implementation parallelizes the sieve in a straightforward way by assigning processors distinct subintervals of $[L, R)$, e.g. according to their own memory/performance capabilities. However, if many processors have rapid access to the same memory, then it may be faster for some resources being devoted to identifing smooth numbers in the next interval while the remaining resources sieve the current interval.

Negative Input Values. Until now we have only considered positive input values $\ell \in [L, R)$, but our approach also permits negative inputs to the polynomials $a(x)$ and $b(x)$. For example, for even n, this gives another pair of integers that could potentially be smooth. At first glance, this seems to imply that each time ℓ is advanced, we must also check the values $\ell_i' = \ell + i$ with $i \in \{a_1, \ldots, a_n, b_1, \ldots, b_n\}$ for smoothness. Moreover, it seems that the overlap of size w for each search interval must also be added to both sides. We note, however, that if the PTE solution in use is symmetric (see Definition 5), then the values ℓ_i' are the same as the values $(\ell+w)_i$, and thus are naturally checked by our previous algorithm at position $\ell+w$. This is not the case for general non-symmetric solutions, but for those non-symmetric solutions that are normalized (see Definition 4), we can instead search with positive inputs to the *flipped* solution arising from the linear transform $h(x) = -x + w$, which is especially beneficial when searching with many solutions simultaneously.

4.3 Searching with Many PTE Solutions

One of the main benefits of our sieve is that it can combine many PTE solutions into the same search and rapidly process them together. Many PTE solutions tend to share at least one non-zero element in common, and if checking this element returns a '0', all such solutions can be discarded at once. In what follows we describe a method to arrange the set of PTE solutions in a *tree*, such that (on average) a minimal number of checks is used to check the full set of solutions. Note that computing this tree is a one-time precomputation that is performed at initialization.

Suppose we have t solutions, written as $[a_{i,1},\ldots,a_{i,n}] =_{n-1} [b_{i,1},\ldots,b_{i,n}]$ for $1 \le i \le t$. Noting that $a_{i,1} = 0$ for all i, write $S_i = \{a_{i,2},\ldots,a_{i,n},b_{i,1},\ldots,b_{i,n}\}$, i.e. S_i is the set of distinct non-zero integers in the i-th PTE solution. Now, as in the single solution sieve above, suppose we have advanced ℓ to a set bit at some stage of our sieving algorithm. Rather than checking each of the PTE solutions individually, we would like to share any checks that are common to multiple PTE solutions. The key observation is that we are highly unlikely[2] to have a PTE solution whose elements all correspond to '1', so in combining many PTE solutions we would ultimately like to minimize the number of checks required before we can rule all of them out and move ℓ to the next set bit.

In looking for the minimum number of checks whose failures rule out all PTE solutions, we are looking for a set H of minimal cardinality such that $H \cap S_i \neq \{\emptyset\}$ for $1 \le i \le t$, i.e. the smallest-sized set that shares at least one element with each of the PTE solutions. Finding this set is an instance of the *hitting set problem*; this problem is NP-complete in general, but for the sizes of the problem in this paper, a good approximation is given by the greedy algorithm [16]. We start by looking for the element that occurs most among all of the S_i, call this g_1; we then look for the element that occurs most among the S_i that do not contain g_1, call this g_2; we then look for the element that occurs most among those S_i that do not contain g_1 or g_2, and continue in this way until we have $H = \{g_1, g_2, \ldots, g_h\}$ such that every S_i contains at least one of the g_j, for $1 \le i \le t$ and $1 \le j \le h$. This process naturally partitions the PTE solutions to fall under h different *branches*. For each PTE solution in a given branch, the corresponding element of the hitting set is removed and the process is repeated recursively until there is no common element between the remaining solutions,

[2] We assume that the smoothness bound is aggressive enough to make the smooth integers sparse.

at which point they become *leaves*. In Sect. 5.2 we give a toy example with 20 PTE solutions that produces the tree in Fig. 2. In this example the first hitting set is $\{1, 2\}$; if a search was to use these 20 solutions, then most of the time only two checks will be required before ℓ can be advanced to the next set bit.

At a high level, our multi-solution sieve then runs the same way as the single solution sieve in Sect. 4.2, except that we must traverse our tree each time ℓ is advanced. We do this by checking all of the elements of a the hitting set, and we only enter the branch corresponding to a given element if the associated check finds a '1' (an example sequence of checks is included in Sect. 5.2). This is repeated recursively until we either encounter a leaf, where we simply check the remaining elements sequentially, or until all of the elements in the hitting set at the current level of the tree return a '0', at which point we can move up to the branch above and continue. As mentioned above, in practice the most common scenario is that all of the elements in the highest hitting set correspond to a '0', and the number of checks performed in order to rule out the full set of PTE solutions is minimal. Note that checking the divisibility of $a(\ell)$ and $b(\ell)$ by the constant C is, in practice, best left until the point where a match is found. Since solutions have different constants and different sets of suitable relations, it is not useful to incorporate modular relations into the sieving step of the multi-solution algorithm.

The efficiency of checking all PTE solutions simultaneously is therefore heavily dependent on the size of the first hitting set. In cases where we have many PTE solutions (see Sect. 3.2), the first hitting set can be used to decide which PTE solutions to search with. If a pre-existing set of PTE solutions has a hitting set H, then including any additional solutions that share at least one element with H incurs nearly no performance cost.

4.4 Success Probabilities

In Table 3 we use Heuristic 1 to draw comparisons between our method of finding twin smooth integers and the prior methods discussed in Sect. 1.2. The entries in the table are the approximate smoothness bounds that should be used to give success probabilities of 2^{-20}, 2^{-30}, 2^{-40} and 2^{-50}. The term *success probability* is used to estimate how large a search space needs to be covered before we can expect to find twin smooth integers; these probabilities are computed directly via (1.2). For example (refer to the bold element in the last row of the table), using *one* PTE solution with $n = 8$ and a smoothness bound of $B \approx 2^{26.9}$, we can expect to find a pair of twin smooth numbers in $[1, N] = [1, 2^{384}]$ after searching roughly 2^{20} inputs $\ell \in [1, N^{1/n}] = [1, 2^{48}]$, for which $a_C(\ell)$ and $b_C(\ell)$ are integers.[3] To find similarly sized twin smooth integers using the XGCD approach, we would have to search roughly 2^{20} elements with a smoothness bound of $B \approx 2^{41.5}$, or 2^{30} elements with a smoothness bound of $B \approx 2^{32.8}$; on

[3] The total number of inputs required for this (including the ones which lead to non-integer polynomial values) depends on the PTE solution and associated constant in use, and can easily be computed via the CRT approach described before.

Table 3. Table of smoothness bounds and success probabilities for known methods and our method. All numbers are given as base-2 logarithms. Further explanation in text.

method	N														
	256					384					512				
	n	probability				n	probability				n	probability			
		−50	−40	−30	−20		−50	−40	−30	−20		−50	−40	−30	−20
naïve	–	20.2	23.4	28.4	36.7	–	30.2	35.2	42.6	55.1	–	40.3	46.9	56.7	73.4
XGCD	–	15.9	18.4	21.9	27.7	–	23.9	27.5	32.8	41.5	–	31.9	36.7	43.7	55.3
$2x^n - 1$	4	15.6	17.8	20.8	25.8	6	19.9	22.6	26.4	32.3	6	26.6	30.1	35.2	43.1
	6	13.3	15.1	17.6	21.6	8	20.4	23.2	27.2	33.8	12	22.0	24.9	28.9	35.2
	8	13.6	15.5	18.2	22.5	10	20.3	23.1	27.2	33.8	16	25.8	29.3	34.6	43.5
	9	15.4	17.7	21.0	26.4	12	16.5	18.7	21.7	26.4	18	23.3	26.3	30.9	38.4
	10	13.5	15.4	18.2	22.5	16	19.3	22.0	25.9	32.7	20	23.2	26.3	31.0	38.5
	12	11.0	12.4	14.5	17.6	18	17.4	19.8	23.1	28.8	24	20.2	22.9	26.7	32.8
PTE	3	20.4	23.0	26.6	32.2	3	30.6	34.5	39.9	48.4	4*	30.6	34.5	39.9	48.4
	3*	16.2	18.4	21.6	26.6	3*	24.3	27.7	32.4	39.9	5	31.9	25.6	40.6	48.2
	4	17.8	20.0	22.9	27.5	4	26.7	29.9	34.4	41.2	6	29.1	32.2	36.6	43.0
	4*	15.3	17.2	20.0	24.2	4*	22.9	25.8	29.9	36.3	6*	25.2	28.2	32.2	38.5
	5	16.0	17.8	20.3	24.1	5	24.0	26.7	30.4	36.1	7	26.8	29.6	33.5	39.0
	6	14.5	16.1	18.3	21.5	6	21.8	24.2	27.5	32.3	8	24.9	27.5	30.9	35.8
	6*	12.6	14.1	16.1	19.3	6*	18.9	21.1	24.2	28.9	9	23.3	25.7	28.7	33.2
	7	13.4	14.8	16.7	19.5	7	20.1	22.2	25.1	29.3	10	22.0	24.1	26.8	31.1
	8	12.5	13.7	15.4	17.9	8	18.7	20.6	23.2	**26.9**	12	19.8	21.5	23.9	27.5

the other hand, if we were using XGCD with the same $B \approx 2^{26.9}$ as the PTE solution, we should expect to have to search a space larger han 2^{40} before finding twin smooths.

We stress that Table 3 is merely intended as a rough guide to the size of the smoothness bounds we should use in a given search, and similarly to provide an approximate comparison between the methods. As mentioned in Sect. 2, Heuristic 1 makes the rather strong assumption that the elements in our PTE solutions are uniform in $[1, N^{1/n}]$, and using the Dickman–de Bruijn function is a rather crude blanket treatment of the concrete combinations of B, N and n of interest to us. Moreover, the best version of our sieve (like the one used in Sect. 6) combines hundreds of PTE solutions into one search, and extending a theoretical analysis to cover such a collection of solutions is unnecessary. We point out that the application of Heuristic 1 to our scenario further assumes that the denominator C gets absorbed by the different factors uniformly. In other words, we assume that after canceling the denominator, all factors of $a_C(\ell)$ and $b_C(\ell)$ roughly have the same size. Although this is not true in general, our experiments and the smoothness of C (see Sect. 3.1) suggest this to be a good approximation for the average case.

The elements of the table that are faded out correspond to instances where the size of the possible search space is not large enough to expect to find solutions with the given probability. Moreover, Table 3 does not incoporate the additional probabilities associated with the twin smooth integers having a prime sum. Searches for cryptographic parameters typically need to find several twin smooth integers before finding a pair with a prime sum, so our search spaces tend to be a little larger than Table 3 suggests.[4] We chose 2^{-20} as the largest success probability in the table under the assumption that any search for twin smooth integers will cover a space of size at least 2^{20}.

A number of rows in the lower section of the table are marked (*) to indicate that these are PTE solutions with repeated factors. Viewing Heuristic 1, we see that these solutions find twin smooth integers with a higher probability than those PTE solutions without repeated factors, which is why they show a lower smoothness bound (for a fixed probability). PTE solutions with repeated factors are only known for $n \in \{3, 4, 6\}$.

5 A Worked Example

We now give concrete examples found with the sieve described in Sect. 4, referring back to the theory developed in Sect. 3 where applicable. We first illustrate a simple search that uses a single PTE solution, and then move to combining many PTE solutions into the same sieve.

5.1 Searching with a Single PTE Solution

Suppose we are searching for twin smooth integers $(m, m+1)$ with $2^{240} \leq m < 2^{256}$. Table 3 suggests that the best chances of success are with $n \in \{6, 7, 8\}$, and in particular with the $n = 6$ solutions that have repeated factors. Since the search spaces using polynomials of degree $n = 7$ and $n = 8$ are rather confined when targeting $m < 2^{256}$ (see Table 3), for this example we use a PTE solution of size $n = 6$ containing repeated factors, namely

$$[1, 1, 8, 8, 15, 15] =_5 [0, 3, 5, 11, 13, 16], \tag{7}$$

which corresponds to the polynomials

$$a(x) = (x-1)^2(x-8)^2(x-15)^2, \quad b(x) = x(x-3)(x-5)(x-11)(x-13)(x-16).$$

Proposition 1 induces that $a(x)$ and $b(x)$ differ by an integer constant, which in this case is

$$C = a(x) - b(x) = 14400 = 2^6 3^2 5^2.$$

Observe that Proposition 2 guaranteed that C was a multiple of $(n-1)! = 5!$.

[4] It is beyond the scope of this work to make any statements about the probability of a prime sum, except to say that in practice we observe that twin smooth sums have a much higher probability of being prime than a random number of the same size.

Given that $2^{13} < C < 2^{14}$, searching for m with $2^{240} \leq m < 2^{256}$ means searching for values ℓ such that $a(\ell)$ and $b(\ell)$ lie between 2^{254} and 2^{269}, so that $a_C(\ell)$ and $b_C(\ell)$ are then of the right size. Since $a(x)$ and $b(x)$ have degree 6, this means searching with $2^{42} \leq \ell < 2^{45}$.

Recall from Sect. 4 that our sieving algorithm alternates between two main phases. The first is independent of the PTE solution(s) we are searching with, and simply involves identifying all smooth numbers in a given interval (see Sect. 4.1). In this example, we chose interval sizes of $2^{20} = 1048576$, so at the conclusion of this first phase, we have a bitstring of length 1048576 to search over: a '1' in this string means the number associated with its index is B-smooth, while a '0' indicates that it is not.

With $B \approx 2^{16.1}$, Table 3 suggests that searching with the PTE solution in (7) will find twin smooth integers for roughly 1 in every 2^{30} values of ℓ that are tried. Thus, we set $B = 2^{16}$ and started the search at $\ell = 2^{42}$. With this ℓ and B, the Dickman–de Bruijn function tells us that we can expect the proportion of B-smooth numbers to be close to $\rho(42/16) \approx 0.103$.

At the top of Fig. 1, we give 30 bits of an interval (found after sieving for some time) that correspond to $\ell = 5170314186700 + t$, for $t \in \{30, 31, \ldots 59\}$. Here 11 of the 30 bits are 1, so the proportion of B-smooth numbers in this small interval is exceptionally high; indeed, these are the kinds of substrings we are sieving for, in hope that our PTE solution aligns favorably to find 1's in all of the required places. Viewing (7), we write $\ell_i = \ell - i$ for $i \in \{0, 1, 3, 5, 8, 11, 13, 15, 16\}$. As depicted in Fig. 1, each step in the second phase starts by finding the next smooth number (i.e. the next '1' in the string), advancing $\ell = \ell_0$ to align there before sequentially checking from ℓ_1 through to ℓ_{16}. If, at any stage, one of the ℓ_i is aligned with a '0', we advance ℓ to the next '1' in the string and repeat the procedure. Once we have finished processing a full interval (of size 2^{20} in this case), we advance to the next interval by first computing the string that identifies all B-smooth numbers, then processing the interval by aligning ℓ_0 with the next set bit, and checking the remaining ℓ_i.

In Fig. 1 we see that when $\ell_0 = 5170314186747$, the next bit checked reveals that ℓ_1 corresponds to a '0', so this position is immediately discarded and we advance to the next set bit taking $\ell_0 = 5170314186750$. Again, ℓ_1 discovers a '0', so ℓ_0 advances to 5170314186752, and then to 5170314186754 (both of these also have ℓ_1 aligned with '0'). Advancing to $\ell_0 = 5170314186755$, we see that the remaining ℓ_i correspond to set bits and are thus all smooth, namely

$$\ell_0 = 5 \cdot 29 \cdot 31 \cdot 211 \cdot 557 \cdot 9787, \qquad \ell_1 = 2 \cdot 71 \cdot 919 \cdot 1237 \cdot 32029,$$
$$\ell_3 = 2^{12} \cdot 11^2 \cdot 13 \cdot 277 \cdot 2897, \qquad \ell_5 = 2 \cdot 3 \cdot 5^3 \cdot 181 \cdot 4783 \cdot 7963,$$
$$\ell_8 = 3^2 \cdot 23 \cdot 41 \cdot 83 \cdot 1117 \cdot 6571, \qquad \ell_{11} = 2^3 \cdot 3 \cdot 7^2 \cdot 17 \cdot 43 \cdot 191 \cdot 31489,$$
$$\ell_{13} = 2 \cdot 103 \cdot 1093 \cdot 2663 \cdot 8623, \qquad \ell_{15} = 2^2 \cdot 5 \cdot 1163 \cdot 11927 \cdot 18637,$$
$$\ell_{16} = 13 \cdot 53 \cdot 113 \cdot 3347 \cdot 19841.$$

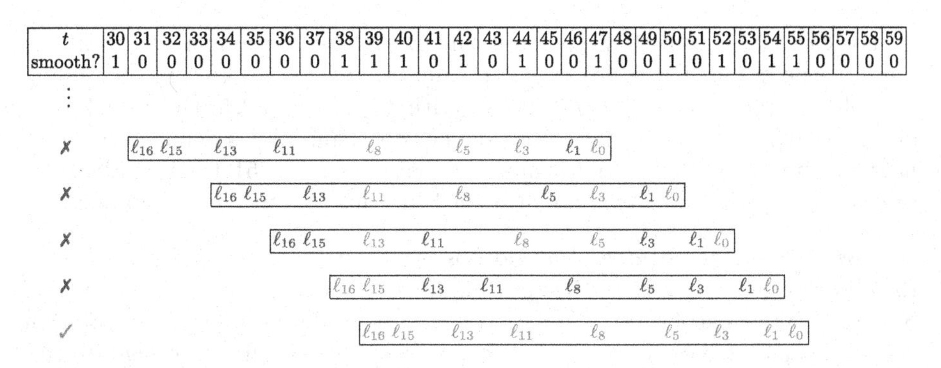

t	30	31	32	33	34	35	36	37	38	39	40	41	42	43	44	45	46	47	48	49	50	51	52	53	54	55	56	57	58	59
smooth?	1	0	0	0	0	0	0	0	1	1	1	0	1	0	1	0	0	1	0	0	1	0	1	0	1	1	0	0	0	0

Fig. 1. Sieving with the PTE solution $[1, 1, 8, 8, 15, 15] =_5 [0, 3, 5, 11, 13, 16]$ across the subinterval $\ell = 5170314186700 + t$ for $t \in \{30, 31, \ldots 59\}$. Further explanation in text.

The PTE solution (7) translates into the twin-smooth numbers

$$(m, m+1) = \left(\frac{\ell_0 \ell_3 \ell_5 \ell_{11} \ell_{13} \ell_{16}}{C}, \frac{(\ell_1 \ell_8 \ell_{15})^2}{C} \right).$$

In this case their sum is a prime p, which lies between the B-smooth numbers $2m$ and $2(m+1)$, namely

$$p = 2m + 1 = 2653194648913198538763028808847267222102564753030025033104122760223436801.$$

Remark 1. When searching with a single solution, in practice we only want to search over the $\ell \in \mathbb{Z}$ for which $a(\ell) \equiv b(\ell) = 0 \bmod C$. As described in Sect. 3, we use the CRT to find these ℓ by first working modulo each of the prime power factors of C. In this case we find

- 40 residues $r_1 \in [0, 2^6)$ such that $a(\ell) \equiv b(\ell) \equiv 0 \bmod 2^6$ iff $\ell \equiv r_1 \bmod 2^6$;
- 9 residues $r_2 \in [0, 3^2)$ such that $a(\ell) \equiv b(\ell) \equiv 0 \bmod 3^2$ iff $\ell \equiv r_2 \bmod 3^2$;
- 15 residues $r_3 \in [0, 5^2)$ such that $a(\ell) \equiv b(\ell) \equiv 0 \bmod 5^2$ iff $\ell \equiv r_3 \bmod 5^2$.

Here we see that $a(\ell) \equiv b(\ell) \equiv 0 \bmod 3^2$ for all $\ell \in \mathbb{Z}$ (this can be seen immediately by looking at the expression for $a(x)$ above), so we can ignore the factor of 3^2 and work with the effective denominator $C' = 2^6 5^2 = 1600$. Of the 1600 possible residues in $[0, 2^6 3^5)$, we only search over the $40 \cdot 15 = 600$ values of ℓ that will produce $a(\ell) \equiv b(\ell) \equiv 0 \bmod C'$. In this case the list of residues is small enough that we can simply store them once and for all and avoid recomputing them on the fly with the CRT at runtime. However, many of the PTE solutions we use have much larger denominators and a much smaller proportion of residues to be searched over, and in these cases storing residues modulo each prime power and then using the CRT on the fly is much faster than looking up the full set of residues (modulo C) in one huge table.

For ease of exposition, we ignored this in the above example. Returning to Fig. 1, we point out that none of the four values that were checked prior to finding the solution (i.e. $\ell = 5170314186700 + t$ with $t \in \{47, 50, 52, 54\}$) are such that $a(\ell) \equiv b(\ell) \equiv 0 \bmod C$. In fact, none of the other smooth ℓ's depicted in Fig. 1 have this property; the previous smooth ℓ that does is $\ell = 5170314186728$, so in practice we would have advanced straight from this ℓ to the successful one.

Remark 2. Since the degree of a and b is even, negative values for ℓ will lead to valid positive twin smooth integers and possibly a corresponding prime sum. Negative values can be taken into account by considering the flipped solution (as defined at the end of Sect. 3.1). Because the solution considered here is symmetric, any pattern corresponding to a negative value also occurs for a positive value.

5.2 Sieving with Many PTE Solutions

We now turn to illustrating the full sieving algorithm that combines many PTE solutions into one search. The degree 6 sieves we used in practice combined hundreds of PTE solutions into one search (see Table 2), but for ease of exposition we will illustrate using the first 20 solutions (ordered by the size of the constant). These range from the solution S_1, which has $C = 14400 = 2^6 \cdot 3^2 \cdot 5^2$, to S_{20}, which has $C = 13305600 = 2^8 \cdot 3^3 \cdot 5^2 \cdot 7 \cdot 11$. These solutions are listed below.

$S_1 : [0, 3, 5, 11, 13, 16] =_5 [1, 1, 8, 8, 15, 15];$ $S_2 : [0, 5, 6, 16, 17, 22] =_5 [1, 2, 10, 12, 20, 21],$

$S_3 : [0, 4, 9, 17, 22, 26] =_5 [1, 2, 12, 14, 24, 25],$ $S_4 : [0, 7, 7, 21, 21, 28] =_5 [1, 3, 12, 16, 25, 27],$

$S_5 : [0, 7, 8, 22, 23, 30] =_5 [2, 2, 15, 15, 28, 28],$ $S_6 : [0, 5, 13, 23, 31, 36] =_5 [1, 3, 16, 20, 33, 35],$

$S_7 : [0, 8, 9, 25, 26, 34] =_5 [1, 4, 14, 20, 30, 33],$ $S_8 : [0, 7, 11, 25, 29, 36] =_5 [1, 4, 15, 21, 32, 35],$

$S_9 : [0, 9, 11, 29, 31, 40] =_5 [1, 5, 16, 24, 35, 39],$ $S_{10} : [0, 8, 11, 27, 30, 38] =_5 [2, 3, 18, 20, 35, 36],$

$S_{11} : [0, 5, 16, 26, 37, 42] =_5 [2, 2, 21, 21, 40, 40],$ $S_{12} : [0, 6, 17, 29, 40, 46] =_5 [1, 4, 20, 26, 42, 45],$

$S_{13} : [0, 7, 14, 28, 35, 42] =_5 [2, 3, 20, 22, 39, 40],$ $S_{14} : [0, 10, 13, 33, 36, 46] =_5 [1, 6, 18, 28, 40, 45],$

$S_{15} : [0, 9, 17, 34, 36, 46] =_5 [1, 6, 24, 25, 42, 44],$ $S_{16} : [0, 9, 14, 32, 37, 46] =_5 [2, 4, 21, 25, 42, 44],$

$S_{17} : [0, 9, 16, 34, 41, 50] =_5 [1, 6, 20, 30, 44, 49],$ $S_{18} : [0, 11, 15, 37, 41, 52] =_5 [1, 7, 20, 32, 45, 51],$

$S_{19} : [0, 7, 21, 35, 49, 56] =_5 [1, 5, 24, 32, 51, 55],$ $S_{20} : [0, 12, 13, 37, 38, 50] =_5 [2, 5, 22, 28, 45, 48].$

In regards to Remark 1, recall from Sect. 4 that each PTE solution has a different constant C and thus a different set of residues. In general these residues are incompatible with one another, so we choose to ignore them until the sieve identifies candidate pairs (ℓ, S_i), at which point we only mark the pair as a solution if the corresponding polynomials have $a(\ell) \equiv b(\ell) \equiv 0 \bmod C$.

Now, recall from Sect. 4 that our sieving tree is built by recursively identifying *hitting sets* among the set of solutions, and then removing the corresponding element in the hitting set from each solution. The first hitting set is (always) $\{0\}$, which is the root of our tree. After removing 0 from all of the solutions, we see that the next hitting set is $\{1, 2\}$; some PTE solutions contain both 1 and 2, but 1 appears in more solutions than 2 does, so the solutions S_2 and S_3 occur in the branches that fall beneath 1 in the tree. Repeating this process produces the tree in Fig. 2. Note that this is a precomputation that is done once-and-for-all before the sieve begins.

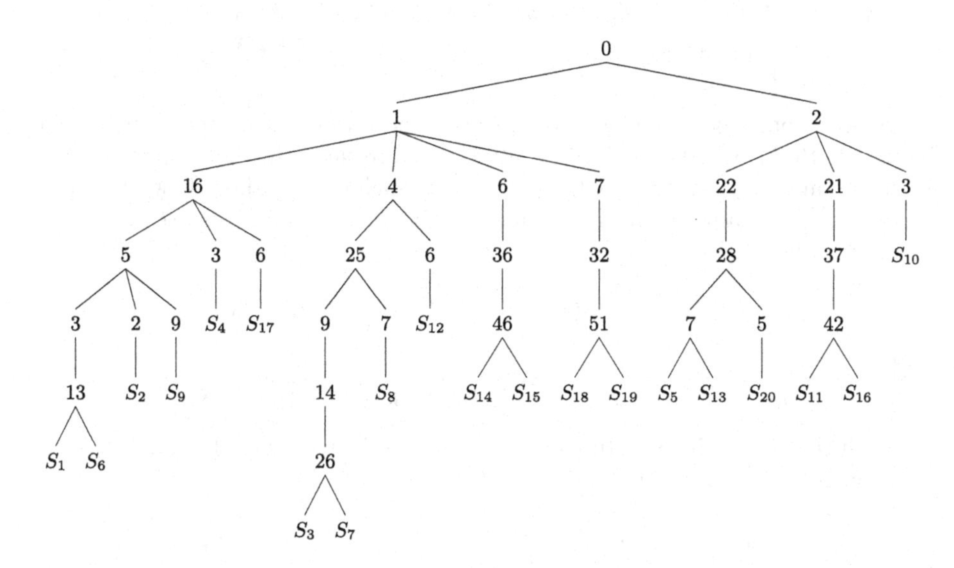

Fig. 2. A sieving tree for 20 example PTE solutions. Further explanation in text.

Again we target $2^{240} \leq m < 2^{256}$ by searching with $2^{42} \leq \ell < 2^{45}$, set our smoothness bound as $B = 2^{16}$, and alternate between identifying the B-smooth numbers in intervals of size $2^{20} = 1048576$, processing each interval by advancing through all of the set bits (smooth numbers) within it. Write $\ell_i = \ell - i$ as before. Here the hitting set has only two elements, so given that the probability of smoothness is roughly $\rho(42/16) \approx 0.103$, most of the time we will only need to check two neighboring bits (ℓ_1 and ℓ_2) before discarding each candidate ℓ.

Viewing Fig. 2, we traverse the tree by moving down the levels and processing each subsequent hitting set from left to right. If, at any stage, we find a smooth number, we immediately move down a level and process the numbers branching beneath it. We are only permitted to move up a level and continue to the right once the entire hitting set at a given level has been checked. Finally, if at any stage we arrive at a leaf and find that all of the remaining numbers are smooth, we then identify this solution as a candidate. At this stage we check whether $a(\ell) \equiv b(\ell) \equiv 0 \bmod C$, in which case we have found twin smooth integers, and then optionally check whether their sum is a prime, in which case we have found cryptographically suitable parameters.

After some time, our sieve advances to the B-smooth number

$$\ell_0 = 5435932476400 = 2^4 \cdot 5^2 \cdot 199 \cdot 4817 \cdot 14177.$$

In this case the subsequent set of ordered checks made in traversing the tree in Fig. 2 are given below (we use ✓ to indicate that ℓ_i is B-smooth, ✗

otherwise). Checking the entire leaf marked S_{17} is combined into Check 5 for brevity; the remaining values here are ℓ_i with $i \in \{9, 20, 30, 34, 41, 44, 49, 50\}$.

Check 1. ℓ_1 ✓ *Check 2.* ℓ_{16} ✓ *Check 3.* ℓ_5 ✗ *Check 4.* ℓ_3 ✗
Check 5. S_{17} ✓ *Check 6.* ℓ_4 ✗ *Check 7.* ℓ_6 ✓ *Check 8.* ℓ_{36} ✗
Check 9. ℓ_7 ✗ *Check 10.* ℓ_2 ✗

At the conclusion of Check 5, we now know that all of the elements in S_{17}: $[0, 9, 16, 34, 41, 50] =_5 [1, 6, 20, 30, 44, 49]$ are smooth, and thus we have found a candidate solution. Checks 6–10 are included to show how the sieve continues. It remains to check whether $\ell = 5435932476400$ gives $a(\ell) \equiv b(\ell) \equiv 0 \bmod C$, when

$$a(x) = x(x-9)(x-16)(x-34)(x-41)(x-50)$$

and

$$b(x) = (x-1)(x-6)(x-20)(x-30)(x-44)(x-49).$$

are such that $C = 7761600$. In this case we do find that $a(\ell) \equiv 0 \bmod C$ (which is sufficient), so we know that

$$m = \ell_0 \ell_9 \ell_{16} \ell_{34} \ell_{41} \ell_{50} / C \qquad \text{and} \qquad m+1 = \ell_1 \ell_6 \ell_{20} \ell_{30} \ell_{44} \ell_{49} / C$$

are both B-smooth integers. Indeed, factoring reveals that

$$\begin{aligned} m &= 2^5 \cdot 3^4 \cdot 5^2 \cdot 109 \cdot 173 \cdot 199 \cdot 233 \cdot 571 \cdot 677 \cdot 743 \cdot 1303 \cdot 2351 \cdot 2729 \\ &\quad \cdot 3191 \cdot 4817 \cdot 12071 \cdot 12119 \cdot 14177 \cdot 16979 \cdot 30389 \cdot 37159 \cdot 39979, \text{ and} \\ m+1 &= 13 \cdot 17 \cdot 23 \cdot 31 \cdot 61 \cdot 103 \cdot 263 \cdot 643 \cdot 1153 \cdot 1429 \cdot 1889 \cdot 2213 \cdot 3359 \\ &\quad \cdot 5869 \cdot 7951 \cdot 9281 \cdot 18307 \cdot 28163 \cdot 34807 \cdot 41077 \cdot 41851 \cdot 64231. \end{aligned}$$

In this case $2m+1$ is the product of two large primes, so a sieve for cryptographic parameters would continue by advancing to the next smooth ℓ_0 in the interval.

6 Cryptographic Examples of Twin Smooth Integers

We implemented the sieve including the tree structure for searching with multiple PTE solutions in Python 3 and used it to run our experiments. The first phase of the algorithm, i.e. the sieve that identifies smooth numbers was written in C and called from the python code, which resulted in a significant speedup. The code takes as input the left and right bounds of a desired interval to be searched, a size for the sub-intervals that are processed by the sieve at a time, as well as a smoothness bound and a list of PTE solutions. It then computes the PTE solution search tree and starts the sieve as described in Sects. 4 and 5. Another input is a desired number of threads, between which the interval is divided and then run on the available processors in a multi-processing fashion.

After examining the PTE solution counts in Table 2 and the smoothness probabilities in Table 3, we chose to launch a sieve with 520 PTE solutions of size $n = 6$ that searched $\ell \in [2^{40}, 2^{45}]$ with a smoothness bound of $B = 2^{16}$ and intervals of size 2^{20}. The 520 solutions are all the ones we found that have a constant of at most 38 bits. The first hitting set of the PTE solutions had cardinality 13, and the Dickman–de Bruijn function estimates that the proportion of B-smooth numbers in our interval is $\rho(45/16) \approx 0.0715$. The search ran on 128 logical processors (Intel Xeon CPU E5-2450L @1.8 GHz) for just over a week before the entire interval was scanned.

Table 4 reports one of the cryptographic primes that was found with our sieve for each bitlength between 240 and 257 (excluding 253, 254 and 256, for which no primes were found), and compares it to the primes found with prior methods in the literature. For the primes found using PTE solutions, we give the search parameter ℓ together with the corresponding PTE solution, which is one of

$$\begin{aligned} S_1^6&: [0,3,5,11,13,16] =_5 [1,1,8,8,15,15],\\ S_2^6&: [0,7,8,22,23,30] =_5 [2,2,15,15,28,28],\\ S_3^6&: [0,7,33,47,73,80] =_5 [3,3,40,40,77,77],\\ S_4^6&: [0,5,16,26,37,42] =_5 [2,2,21,21,40,40]. \end{aligned}$$

For each prime we report the smoothness bound B, which is the largest prime divisor of $(p-1)(p+1)$, together with its bitlength. In the case of the 241- and 250-bit primes, we see that $B < 2^{15}$. The smallest prior B corresponding to primes of around this size was the 19-bit $B = 486839$ from [8]. Referring back to Table 3, we see that a search through an interval of this size should find a few twin smooth integers with $B < 2^{15}$, but finding enough twin smooths with $B < 2^{14}$ to hope for a prime sum among them may have been out of the question.

To check whether $n = 6$ produces the smoothest twins of this size (as Table 3 predicts), we ran similar sieves using the 8 PTE solutions with $n = 7$ and the 51 PTE solutions with $n = 8$ with $B = 2^{18}$, and in both cases we covered the full range of possible inputs that would produce a $p < 2^{256}$. Despite finding a handful of twin smooth integers with $B < 2^{17}$, the search spaces were not large enough to find any primes among them.

Table 4 also reports three cryptographic primes that target higher security levels. When searching for $p \approx 2^{384}$, the PTE solutions with $n = 6$ again proved to produce the smoothest twins; the 376- and 384-bit primes reported correspond to twin smooths with $B = 2^{21}$ and $B = 2^{22}$, respectively. When searching for $p \approx 2^{512}$, the PTE solution

$$\begin{aligned} S_1^{12}&: [0,11,24,65,90,129,173,212,237,278,291,302]\\ &=_{11} [3,5,30,57,104,116,186,198,245,272,297,299] \end{aligned}$$

with $n = 12$ found the reported 512-bit prime, which lies between two integers that are both 2^{29}-smooth.

Table 4. A comparison between some of the best instances found with our sieve and the best instances from the literature. Further explanation in text.

Method	where		p (bits)	B	$\lceil \log_2 B \rceil$
XGCD	[4, App. A]		256	6548911	**23**
$p = 2x^n - 1$	[8, Ex. 5]		247	652357	**20**
	[8, Ex. 6]		237	709153	**20**
	[8, Ex. 7]		247	745309897	**30**
	[8, Ex. 8]		250	486839	**19**
PTE sieve	19798693013832	S_3^6	240	54503	**16**
	5170314186755	S_1^6	241	32039	**15**
	11434786499430	S_2^6	242	62653	**16**
	6387061913711	S_1^6	243	56711	**16**
	32519458118257	S_3^6	244	64591	**16**
	16232865719280	S_2^6	245	49711	**16**
	8812545447095	S_1^6	246	40151	**16**
	20173246926702	S_2^6	247	40289	**16**
	22687888853658	S_2^6	248	59497	**16**
	13061439823095	S_2^6	249	38119	**16**
	36144284257450	S_4^6	250	32191	**15**
	16189037375263	S_2^6	251	65029	**16**
	17545941442175	S_1^6	252	35291	**16**
	27071078665441	S_1^6	255	52069	**16**
	32554839816383	S_1^6	257	42979	**16**
	7493998973665338152 0	S_4^6	376	1604719	**21**
	74939982689644756283	S_1^6	384	3726773	**22**
	510796126391672	S_1^{12}	512	238733063	**28**

7 Relaxations and Modifications

There are numerous ways to modify our sieving approach for performance reasons, or to relax the search conditions in order to precisely match the security requirements imposed by B-SIDH or SQISign.

Approximate Sieves. There are several sieving optimizations discussed in [9, §3.2.5–3.3] that can be applied to the sieving phase of our algorithm. For large scale searches, it could be preferred to sacrifice the exactness of the sieve we implemented for more performant approximate sieves. For example, the smallest primes are the most expensive to sieve with due to the large number of multiplications. Thus, an approximate sieve can choose to skip these small primes (but still include the larger prime powers) and choose to tag numbers as being B-smooth as soon as the result is close enough to the expected number. This requires to choose an error bound, which also determines if and how many false positive and false negative results are going to occur.

A standard approach for sieving algorithms is discussed by Crandall and Pomerance [9, §3.2.5]. This approach replaces multiplications by additions in Eratosthenes-like sieves, by choosing to represent numbers as their (base-2) logarithms. Moreover, sieves can use approximate logarithms, i.e. round these logarithms to nearby integers and tolerate errors in the logarithms; for example, if we choose to tolerate errors up to $\log B$, then we are guaranteed that factors that are unaccounted for in the approximation are also less than the smoothness bound [9, p. 124]. Rather than accumulating products, we are then accumulating sums of relatively small integers. This approach is used in our C implementation and for the ranges targeted here, allows the accumulated approximate logarithms to be stored in a single byte.

Recall from Sect. 4.2 that when a single PTE solution is used we are only interested to sieve the subset of integers for which $a(\ell) \equiv b(\ell) \equiv 0 \bmod C$. In this case it may be preferable to employ Bernstein's batch smoothness algorithm [3]; this can be used to gain a better overall complexity (per element) when sieving through an arbitrary set.

Lastly, we point out that the set of primes used in the factor base can be tailored to our needs. For example, if future research reveals that certain types of prime isogeny degrees are favored over others (i.e. when invoking the $\tilde{O}(\sqrt{\ell})$ algorithm from [4]), then it may be preferable to increase the bound B and only include those primes in our sieve.

Non-smooth Cofactors Vs. Fully Smooth Numbers. The security analyses of B-SIDH or SQISign suggest that both systems can tolerate a non-smooth cofactor in either or both of $p-1$ and $p+1$. In these cases, relaxing conditions in the second part of our sieve to allow non-smooth cofactors is straightforward. When searching with PTE solutions of size n, we could e.g. only require $n-1$ of the factors on each side to be B-smooth. The naïve way to do this when traversing the tree would be to incorporate a counter that only allows branches to be discarded when two non-smooth numbers have been discovered, but this approach makes things unnecessarily complicated and significantly slower, e.g. it no longer suffices to start the sieving procedure at each '1' in the interval, since ℓ_0 is now allowed to be non-smooth.

A much better approach can be taken by simply creating many relaxed PTE solutions from the original solution $\mathcal{A} =_{n-1} \mathcal{B}$, and including them in the solution tree. For example, if the security analysis corresponding to a search with $n=6$ suggests we only need 5 smooth factors from each side of the PTE solution, then the solution $[0, 7, 11, 25, 29, 36] =_5 [1, 4, 15, 21, 32, 35]$ can be modified into 36 relaxed solutions, each of which corresponds from wiping out one number from $\mathcal{A}$ and one number from $\mathcal{B}$; these new solutions only include 10 distinct elements. By building a tree from these solutions and running the same algorithm as in Sect. 4, we are effectively allowing for one of the factors of the original solution to be non-smooth. The only minor modification required appears when 0 is wiped out from a solution, in which case we have to shift all elements such that the new solution contains 0, by the means of Proposition 3. We reiterate that all of these modifications are a one-time precomputation before the sieve begins. In the case of the PTE solutions with repeated factors, e.g. $[0, 3, 5, 11, 13, 16] =_5 [1, 1, 8, 8, 15, 15]$, we may not be able to tolerate a non-smooth cofactor that would arise from removing any of 1, 8 or 15 from the PTE solution. On the other hand, if the security analysis does permit such a cofactor (which appears to be the case for SQISign), then our relaxed solutions would either remove one of the repeated numbers from $\mathcal{B}$, or two of the numbers from $\mathcal{A}$; the latter would have a better success probability, but (assuming the hitting set remains unchanged) our tree approach would not pay any noticeable overhead by including all such relaxations.

Acknowledgments. We thank Patrick Longa for his help with implementing the smoothness sieve in C, and Fabio Campos for running and overseeing some of our experiments.

References

1. Balog, A., Wooley, T.: On strings of consecutive integers with no large prime factors. J. Austral. Math. Soc. (Series A) **64**, 266–276 (1998)
2. Bernstein, D.J.: Arbitrarily tight bounds on the distribution of smooth integers. In: Proceedings of the Millennial Conference on Number Theory, pp. 49–66 (2002)
3. Bernstein, D.J.: How to find smooth parts of integers (2004). http://cr.yp.to/papers.html#smoothparts
4. Bernstein, D.J., De Feo, L., Leroux, A., Smith, B.: Faster computation of isogenies of large prime degree. In: ANTS-XIV: Fourteenth Algorithmic Number Theory Symposium (2020). https://eprint.iacr.org/2020/341
5. Borwein, P.: The Prouhet-Tarry-Escott problem. In: Computational Excursions in Analysis and Number Theory, pp. 85–95. Springer, New York (2002). https://doi.org/10.1007/978-0-387-21652-2_11
6. Borwein, P., Ingalls, C.: The Prouhet-Tarry-Escott problem revisited. http://www.cecm.sfu.ca/personal/pborwein/PAPERS/P98.pdf
7. Caley, T.: The Prouhet-Tarry-Escott problem. Ph.D. thesis, University of Waterloo (2012)
8. Costello, C.: B-SIDH: supersingular isogeny Diffie-Hellman using twisted torsion. In: Moriai, S., Wang, H. (eds.) ASIACRYPT 2020. LNCS, vol. 12492, pp. 440–463. Springer, Cham (2020). https://doi.org/10.1007/978-3-030-64834-3_15
9. Crandall, R., Pomerance, C.B.: Prime Numbers: A Computational Perspective, vol. 182. Springer, New York (2006). https://doi.org/10.1007/0-387-28979-8
10. De Feo, L., Kohel, D., Leroux, A., Petit, C., Wesolowski, B.: SQISign: compact post-quantum signatures from quaternions and isogenies. In: Moriai, S., Wang, H. (eds.) ASIACRYPT 2020. LNCS, vol. 12491, pp. 64–93. Springer, Cham (2020). https://doi.org/10.1007/978-3-030-64837-4_3
11. Gloden, A.: Mehrgradige Gleichungen. Noordhoff (1944)
12. Granville, A.: Smooth numbers: computational number theory and beyond. Algorithmic Num. Theory Latt. Number Fields, Curves Cryptogr. **44**, 267–323 (2008)

13. Hildebrand, A.: On a conjecture of Balog. Proc. Am. Math. Soc. **95**(4), 517–523 (1985)
14. Jao, D., et al.: SIKE: Supersingular Isogeny Key Encapsulation (2017). Manuscript sike.org/
15. Jao, D., De Feo, L.: Towards quantum-resistant cryptosystems from supersingular elliptic curve isogenies. In: PQCrypto, pp. 19–34 (2011)
16. Karp, R.M.: Reducibility among combinatorial problems. In: Jünger, M., et al. (eds.) 50 Years of Integer Programming 1958–2008, pp. 219–241. Springer, Heidelberg (2010). https://doi.org/10.1007/978-3-540-68279-0_8
17. Kleiman, H.: A note on the Tarry-Escott problem. J. Reine Angew. Math. **278**(279), 48–51 (1975)
18. Lehmer, D.H.: On a problem of Störmer. Illinois J. Math. **8**(1), 57–79 (1964)
19. Martin, G.: An asymptotic formula for the number of smooth values of a polynomial. J. Number Theory **93**, 108–182 (2002)
20. Pomerance, C.: The Role of Smooth Numbers in Number Theoretic Algorithms. In: Chatterji, S.D. (ed.) Proceedings of the International Congress of Mathematicians, pp. 411–422. Birkhäuser Basel, Basel (1995). https://doi.org/10.1007/978-3-0348-9078-6_34
21. Rees, E., Smyth, C.: On the constant in the Tarry-Escott problem. In: Langevin, M., Waldschmidt, M. (eds.) Cinquante Ans de Polynômes Fifty Years of Polynomials. LNM, vol. 1415, pp. 196–208. Springer, Heidelberg (1990). https://doi.org/10.1007/BFb0084888
22. Shuwen, C.: The Prouhet-Tarry-Escott Problem. http://eslpower.org/TarryPrb.htm
23. Smyth, C.J.: Ideal 9th-order multigrades and Letac's elliptic curve. Math. Comput. **57**(196), 817–823 (1991)
24. Sorenson, J.P.: A fast algorithm for approximately counting smooth numbers. In: Bosma, W. (ed.) ANTS 2000. LNCS, vol. 1838, pp. 539–549. Springer, Heidelberg (2000). https://doi.org/10.1007/10722028_36
25. Størmer, C.: Quelques théorèmes sur l'équation de Pell $x^2 - dy^2 = \pm 1$ et leurs applications. Christiania Videnskabens Selskabs Skrifter, Math. Nat. Kl, (2), 48 (1897)
26. The National Institute of Standards and Technology (NIST): Submission requirements and evaluation criteria for the post-quantum cryptography standardization process, December 2016. https://csrc.nist.gov/CSRC/media/Projects/Post-Quantum-Cryptography/documents/call-for-proposals-final-dec-2016.pdf
27. Vélu, J.: Isogénies entre courbes elliptiques. C.R. Acad. Sc. Paris, Série A., **271**, 238–241 (1971)
28. Wright, E.: On Tarry's problem (I). Quart. J. Math. **1**, 261–267 (1935)
29. Wróblewski, J., Choudhry, A.: Ideal solutions of the Tarry-Escott problem of degree eleven with applications to sums of thirteenth powers. Hardy-Ramanujan J., 31 2008

Delay Encryption

Jeffrey Burdges[1(✉)] and Luca De Feo[2]

[1] Web 3 Foundation, Zug, Switzerland
burdges@gnunet.org
[2] IBM Research Europe, Zürich, Switzerland
eurocrypt21@defeo.lu

Abstract. We introduce a new primitive named Delay Encryption, and give an efficient instantiation based on isogenies of supersingular curves and pairings. Delay Encryption is related to Time-lock Puzzles and Verifiable Delay Functions, and can be roughly described as "time-lock identity based encryption". It has several applications in distributed protocols, such as sealed bid Vickrey auctions and electronic voting.

We give an instantiation of Delay Encryption by modifying Boneh and Frankiln's IBE scheme, where we replace the master secret key by a long chain of isogenies, as in the isogeny VDF of De Feo, Masson, Petit and Sanso. Similarly to the isogeny-based VDF, our Delay Encryption requires a trusted setup before parameters can be safely used; our trusted setup is identical to that of the VDF, thus the same parameters can be generated once and shared for many executions of both protocols, with possibly different delay parameters.

We also discuss several topics around delay protocols based on isogenies that were left untreated by De Feo *et al.*, namely: distributed trusted setup, watermarking, and implementation issues.

Keywords: Delay functions · Isogenies · Pairings · Supersingular elliptic curves

1 Introduction

The first appearance of *delay cryptography* was in Rivest, Shamir and Wagner's [29] *Time-lock Puzzle*, an encryption primitive where the holder of a trapdoor can encrypt (or decrypt) "fast", but where anyone not knowing the trapdoor can only decrypt (or encrypt) "slowly".

Recently, a revival of delay cryptography has been promoted by research on blockchains, in particular thanks to the introduction of *Verifiable Delay Functions (VDF)* [4]: deterministic functions f that can only be evaluated "sequentially" and "slowly", but such that verifying that $y = f(x)$ is "fast".

After their definition, VDFs quickly gained attention, prompting two independent solutions in the space of a few weeks [27,34]. Both proposals are based on repeated squaring in groups of unknown order, and are similar in spirit to Rivest *et al.*'s Time-lock Puzzle, however they use no trapdoor.

One year later, another VDF, based on a different algebraic structure, was proposed by De Feo, Masson, Petit and Sanso [17]. This VDF uses chains of

A. Canteaut and F.-X. Standaert (Eds.): EUROCRYPT 2021, LNCS 12696, pp. 302–326, 2021.
https://doi.org/10.1007/978-3-030-77870-5_11

supersingular isogenies as "sequential slow" functions, and pairings for efficient verification. Interestingly, it is not known how to build a Time-lock Puzzle from isogenies; in this work we introduce a new primitive, in some respects more powerful than Time-lock Puzzles, that we are able to instantiate from isogenies.

Limitations of Time-lock Puzzles. Time-lock Puzzles allow one to "encrypt to the future", i.e., to create a puzzle π that encapsulates a message m for a set amount of time T. They have the following two properties:

- Puzzle generation is efficient: there exists an algorithm which, on input the message m and the *delay* T, generates π in time much less than T.
- Puzzle solving is predictably slow and sequential: on input π, the message m can be recovered by a circuit of depth approximately T, and no circuit of depth less than T can recover m reliably.

Time-lock Puzzles can be used to remove trusted parties from protocols, replacing them with a time-consuming puzzle solving. Prototypical applications are auctions and electronic voting, we will use auctions as a motivating example.

In a highest bidder auction, the easy solution in presence of a trusted authority is to encrypt bids to the authority, who then decrypts all the bids and selects the winner. Lacking a trusted authority, the standard solution is to divide the auction in two phases: in the *bidding phase* all bidders commit to their bids using a commitment; in the *tallying phase* bidders open their commitments, and the highest bidder wins. However, this design has one flaw in contexts where it is required that all bidders reveal their bids at the end of the auction. For example, in *Vickrey auctions*, the highest bidder wins the auction, but only pays the price of the second highest bid. If at the end of the auction some bidders refuse to open their commitment, the result of the auction may be invalid.

Time-lock Puzzles solve this problem: by having bidders encapsulate their bid in a Time-lock Puzzle, it is guaranteed that all bids can be decrypted in the tallying phase. However this solution becomes very expensive in large auctions, because one puzzle per bidder must be solved: if several thousands of bidders participate, the tallyers must strike a balance between running thousands of puzzle solving computations in parallel, and having a tallying phase that is thousands of times longer than the bidding phase. Since Time-lock Puzzles use trapdoors for puzzle generation, a potential mitigation is to have the bidders reveal their trapdoors in the tallying phase, thus speeding up decryption; however this does not help in presence of a large number of uncollaborative bidders.

An elegant solution introduced in [25] is to use *Homomorphic Time-lock Puzzles (HTLP)*, i.e., Time-lock Puzzles where the puzzles can be efficiently combined homomorphically. Using these, the tallyers can efficiently evaluate the desired tallying circuit on the unopen puzzles, and then run only a single slow puzzle-solving algorithm. Unfortunately, the only efficient HTLPs introduced in [25] are simply homomorphic (either additively or multiplicatively), and they are thus only useful for voting; fully homomorphic TLPs, which are necessary for auctions, are only known from Fully Homomorphic Encryption [9] or from Indistinguishability Obfuscation [25], and are thus unpractical.

On top of that, it can be argued that Time-lock Puzzles are not the appropriate primitive to solve the problem: why do the tallyers need to run one of two different algorithms to open the puzzles? Are trapdoors really necessary? In this work, we introduce a new primitive, *Delay Encryption*, that arguably solves the problem more straightforwardly and elegantly.

Delay Encryption. Delay Encryption is related to both Time-lock Puzzles and VDFs, however it does not seem to be subsumed by either. It can be viewed as a time-lock analogue of Identity Based Encryption, where the derivation of individual private keys is sequential and slow.

Instead of senders and receivers, Delay Encryption has a concept of *sessions*. A session is defined by a *session identifier*, which must be a hard to predict string. When a session identifier id is issued, anyone knowing id can *encrypt to the session for* id; decryption is however unfeasible without a *session key*, which is to be derived from id. The defining feature of Delay Encryption is *extraction*: the process of deriving a session key from a session identifier. Extraction must be a sequential and slow operation, designed to run in time T and no less for almost any id.

Since there are no secrets in Delay Encryption, anyone can run extraction. It is thus important that session identifiers are hard to predict, and thrown away after the first use, otherwise an attacker may precompute session keys and immediately decrypt any ciphertext to the sessions.

Delay Encryption is different from known Time-lock Puzzles in that it has no trapdoor, and from VDFs in that it provides a fast encryption, rather than just a fast verification. It has similar applications to Homomorphic Time-lock Puzzles, it is however more efficient and solves many problems more straightforwardly.

Applications of Delay Encryption. We already mentioned the two main applications of Time-lock Puzzles. We review here how Delay Encryption offers better solutions.

Vickrey auctions. Sealed bid auctions are easily implemented with standard commitments: in the bidding phase each bidder commits to its bid; later, in the tallying phase each bidder opens their commitment. However this solution is problematic when some bidders may refuse to open their commitments.

Delay Encryption provides a very natural solution: at the beginning of the auction an *auction identifier* is selected using some unpredictable and unbiased randomness, e.g., coming from a randomness beacon. After the auction identifier is published, all bidders encrypt to the auction as senders of a Delay Encryption scheme. In the meantime, anyone can start computing the auction key using the extraction functionality. When the *auction key* associated with the auction identifier is known, anyone in possession of it can decrypt all bids and determine the winner.

Electronic voting. In electronic voting it is often required that the partial tally of an election stays unknown until the end, to avoid influencing the outcome.

Delay Encryption again solves the problem elegantly: once the *election identifier* is published, all voters can cast their ballot by encrypting to it. Only after the election key is published, anyone can verify the outcome by decrypting the ballots.

Of course this idea can be combined with classical techniques for anonymity, integrity, etc.

In both applications it is evident that the session/auction/election identifier must be unpredictable and unbiased: if it is not, someone may start computing the session key before anyone else can, and thus break the delay property. Fortunately, this requirement is easily satisfied by using randomness beacons, which, conveniently, can be implemented using VDFs.

Contributions. Our main contribution is the introduction of Delay Encryption: we formally define the primitive and its security, then argue about its naturalness by relating it to other well known primitives such as IBE and VDFs.

Building on Boneh and Franklin's IBE scheme [6], and on a framework introduced in [17] for VDFs, we give an instantiation of Delay Encryption from isogeny walks in graphs of pairing friendly supersingular elliptic curves. We prove the security of our instantiation using a new assumption, related to both the Bilinear Diffie-Hellman assumption typical of pairing based protocols, and the Isogeny Shortcut assumption used for isogeny based VDFs.

Additionally, we cover some topics related to isogeny-based delay functions which apply to both our Delay Encryption and to VDFs, which were left untreated by [17]:

1. We show how to realize the trusted setup needed in all isogeny-based delay protocols in a distributed manner, and propose an efficient implementation based on a new zero-knowledge proof of isogeny knowledge—whose security we are only able to prove heuristically using a non-falsifiable assumption.
2. We show how to claim "ownership" of a delay function evaluation (aka extraction, in the Delay Encryption jargon), by attaching a "watermark" to the result of the evaluation. Watermarking can be used in distributed consensus protocols to reward the party who bears the load of evaluating the delay function.
3. We provide new elliptic curve representations and isogeny formulas optimized for the operations occurring in isogeny based delay functions. Based on these, we estimate the length of the isogeny walk needed to achieve a certain delay, and the size of the associated public parameters.

Plan. Delay Encryption is defined in Sect. 2, and our instantiation is given in Sect. 3. Each of the following sections discusses one topic related to both Delay Encryption and VDFs based on isogenies: Sect. 4 discusses the trusted setup, Sect. 5 covers watermarking, Sect. 6 introduces the new isogeny formulas and makes some practical considerations.

2 Definitions

Our definition of Delay Encryption uses an API similar to a Key Encapsulation Mechanism: it consists of four algorithms—Setup, Extract, Encaps and Decaps—with the following interface.

$\mathsf{Setup}(\lambda, T) \to (\mathsf{ek}, \mathsf{pk})$. Takes a *security parameter* λ, a *delay parameter* T, and produces public parameters consisting of an *extraction key* ek and an *encryption key* pk. Setup must run in time $\mathrm{poly}(\lambda, T)$; the encryption key pk must have size $\mathrm{poly}(\lambda)$, but the evaluation key ek is allowed to have size $\mathrm{poly}(\lambda, T)$.

$\mathsf{Extract}(\mathsf{ek}, \mathsf{id}) \to \mathsf{idk}$. Takes the extraction key ek and a *session identifier* $\mathsf{id} \in \{0,1\}^*$, and outputs a *session key* idk. Extract is expected to run in time *exactly* T, see below.

$\mathsf{Encaps}(\mathsf{pk}, \mathsf{id}) \to (c, k)$. Takes the encryption key pk and a *session identifier* $\mathsf{id} \in \{0,1\}^*$, and outputs a *ciphertext* $c \in \mathcal{C}$ and a *key* $k \in \mathcal{K}$. Encaps must run in time $\mathrm{poly}(\lambda)$.

$\mathsf{Decaps}(\mathsf{pk}, \mathsf{id}, \mathsf{idk}, c) \to k$. Takes the encryption key pk, a *session identifier* id, a *session key* idk, a ciphertext $c \in \mathcal{C}$, and outputs a key $k \in \mathcal{K}$. Decaps must run in time $\mathrm{poly}(\lambda)$.

When Encaps and Decaps are combined with a symmetric encryption scheme keyed by k, they become the encryption and decryption routines of a hybrid encryption scheme, which can then be used as in the applications described previously. Alternatively we could have used a PKE-like API directly, however we prefer the KEM one as it is closer to known instantiations.

A Delay Encryption scheme is correct if for any $(\mathsf{ek}, \mathsf{pk}) = \mathsf{Setup}(\lambda, T)$ and any id

$$\mathsf{idk} = \mathsf{Extract}(\mathsf{ek}, \mathsf{id}) \ \wedge\ (c, k) = \mathsf{Encaps}(\mathsf{pk}, \mathsf{id}) \ \Rightarrow\ \mathsf{Decaps}(\mathsf{pk}, \mathsf{id}, \mathsf{idk}, c) = k.$$

The security of Delay Encryption is defined similarly to that of public key encryption schemes, and in particular of identity-based ones; however one additional property is required of Extract: that for a randomly selected identifier id, the probability that any algorithm outputs idk in time less than T is negligible. We now give the formal definition.

The security game. It is apparent from the definitions that Delay Encryption has no secrets: after public parameters $(\mathsf{ek}, \mathsf{pk})$ are generated, anyone can run any of the algorithms. Thus, the usual notion of indistinguishability will only be defined with respect to the delay parameter T: no adversary is able to distinguish a key k from a random string in time $T - o(T)$, but anyone can in time T. Properly defining what is meant by "time" requires fixing a computation model. Here we follow the usual convention from VDFs, and assume a model of parallel computation: in this context, "time T" may mean T steps of a parallel Turing machine, or an arithmetic circuit of depth T. Crucially, we do not bound the amount of parallelism of the Turing machine, or the breadth of the circuit, i.e., we focus on *sequential delay* functions.

We consider the following Δ-IND-CPA game. Note that the game involves no oracles, owing to the fact that the scheme has no secrets.

Precomputation. The adversary receives $(\mathsf{ek}, \mathsf{pk})$ as input, and outputs an algorithm $\mathcal{D}$.

Challenge. The challenger selects a random id and computes a key encapsulation $(c, k_0) \leftarrow \mathsf{Encaps}(\mathsf{pk}, \mathsf{id})$. It then picks a uniformly random $k_1 \in \mathcal{K}$, and a random bit $b \in \{0, 1\}$. Finally, it outputs (c, k_b, id).

Guess. Algorithm $\mathcal{D}$ is run on input (c, k_b, id). The adversary wins if $\mathcal{D}$ terminates in time less than Δ, and the output is such that $\mathcal{D}(c, k_b, \mathsf{id}) = b$.

We stress that the game is intrinsically non-adaptive, in the sense that no computation is "free" after the adversary has seen the challenge.

We say a Delay Encryption scheme is *Δ-Delay Indistinguishable under Chosen Plaintext Attacks* if any adversary running the precomputation in time $\mathrm{poly}(\lambda, T)$ has negligible advantage in winning the game. Obviously, the interesting schemes are those where $\Delta = T - o(T)$.

Remark 1. Although it would be possible to define an analogue of chosen ciphertext security for Delay Encryption, by giving algorithm $\mathcal{D}$ access to a decryption oracle for id, it is not clear what kind of real world attacks this security notion could model. Indeed, an instantaneous decryption oracle for id would go against the idea that the session key idk needed for decryption is not known to anyone before time T.

Similarly, one could imagine giving $\mathcal{D}$ access to an extraction oracle, to allow it instantaneous adaptive extraction queries after the challenge (note that in the precomputation phase the adversary is free to run polynomially many non-adaptive extractions). However it is not clear what component of a real world system could provide such instantaneous extraction in practice, since extraction is a public (and slow) operation.

2.1 Relationship with Other Primitives

Delay Encryption and Identity Based Encryption. Although there is no formal relationship between Identity Based Encryption (IBE) and Delay Encryption, the similarity is evident.

Recall that an IBE scheme is a public key encryption with three parties: a dealer who possesses a master private/public key pair, a receiver who has an *identity* that acts as its public key (e.g., its email address), and a sender who wants to send a message to the receiver. In IBE, the dealer runs an *extraction* on the identity to generate the receiver's secret key. The sender encrypts messages to the receiver using both the identity and the master public key. The receiver decrypts using the master public key and the private key provided by the dealer.

Delay Encryption follows the same blueprint, but has no secrets: there is no master key anymore, but only a set of public parameters $(\mathsf{ek}, \mathsf{pk})$. Receiver identities become session identifiers id: public but unpredictable. The dealer is replaced by the public functionality $\mathsf{Extract}(\mathsf{ek}, \mathsf{id})$: sequential and slow. Senders encrypt messages to the sessions by using pk and id. After extraction has produced idk, anyone can decrypt messages "sent" to id by using idk.

The similarity with IBE is not fortuitous. Indeed, the instantiation we present next is obtained from Boneh and Franklin's IBE scheme [6], by replacing the master secret with a public, slow to evaluate, isogeny. This is analogous to the way De Feo *et al.*'s VDF [17] is obtained from the Boneh–Lynn–Shacham signature scheme [7].

The similarity with IBE will be mirrored both in the reductions we discuss next, and in the security proof of our instantiation.

Delay Encryption and Verifiable Delay Functions. Boneh and Franklin attribute to Naor the observation that IBE implies signatures. The construction is straightforward: messages are encoded to identities; to sign a message id, simply output the derived private key idk associated to it. To verify a signature $(\mathsf{id}, \mathsf{idk})$: run encapsulation to id obtaining a random (c, k), decapsulate $(\mathsf{id}, \mathsf{idk}, c)$ to obtain k', and accept the signature if $k = k'$. The signature scheme is existentially unforgeable if the IBE scheme is indistinguishable under chosen ciphertext attacks.

Precisely the same construction shows that Delay Encryption implies (sequential) Proof of Work. Furthermore, if we define *extraction soundness* as the property that adversaries have negligible chance of finding $\mathsf{idk} \neq \mathsf{idk}'$ such that

$$\mathsf{Decaps}(\mathsf{pk}, \mathsf{id}, \mathsf{idk}, c) = \mathsf{Decaps}(\mathsf{pk}, \mathsf{id}, \mathsf{idk}', c),$$

then we see that extraction sound Delay Encryption implies Verifiable Delay Functions. It is easily verified that the derived VDF is Δ-sequential if the Delay Encryption scheme is Δ-IND-CPA.

The signature scheme derived from Boneh and Franklin's IBE is equivalent to the Boneh–Lynn–Shacham scheme. Unsurprisingly, the instantiation of Delay Encryption that we give in the next section is extraction sound, and the derived VDF is equivalent to De Feo *et al.*'s VDF.

Delay Encryption and Time-lock Puzzles. Both Delay Encryption and Time-lock Puzzles permit a form of *encryption to the future*: encrypt a message now, so that it can only be decrypted at a set time in the future. There is no formal definition of Time-lock Puzzles commonly agreed upon in the literature, it is thus difficult to say what they exactly are and how they compare to Delay Encryption.

Bitansky *et al.* [3] define Time-lock Puzzles as two algorithms

- $\mathsf{Gen}(\lambda, T, s) \to Z$ that takes as input a delay parameter T and a solution $s \in \{0,1\}^\lambda$, and outputs a puzzle Z;
- $\mathsf{Solve}(Z) \to s$ that takes as input a puzzle Z and outputs the solution s;

under the constraints that Gen runs in time $\mathrm{poly}(\lambda, \log T)$ and that no algorithm computes s from Z in parallel time significantly less than T.

One might be tempted to construct a Time-lock Puzzle from Delay Encryption by defining Gen as follows:

1. Compute $(\mathsf{ek}, \mathsf{pk}) \leftarrow \mathsf{Setup}(\lambda, T)$;
2. Sample a random $\mathsf{id} \in \{0,1\}^\lambda$;
3. Compute $(c, k) \leftarrow \mathsf{Encaps}(\mathsf{pk}, \mathsf{id})$;
4. Compute $m = E_k(s)$;
5. Return $(\mathsf{ek}, \mathsf{pk}, \mathsf{id}, c, m)$;

where E_k is a symmetric encryption scheme. Then Solve is naturally defined as

1. Compute $\mathsf{idk} \leftarrow \mathsf{Extract}(\mathsf{ek}, \mathsf{id})$;
2. Compute $k \leftarrow \mathsf{Decaps}(\mathsf{pk}, \mathsf{id}, \mathsf{idk}, c)$;
3. Return $s = D_k(m)$;

where D_k is the decryption routine associated to E_k.

However this fails to define a Time-lock Puzzle in the sense above, because Setup can take time $\mathrm{poly}(\lambda, T)$ instead of $\mathrm{poly}(\lambda, \log T)$. If we take Setup out of Gen, though, we obtain something very similar to what Bitansky *et al.* call Time-lock Puzzles *with pre-processing*, albeit slightly weaker.[1]

We see no technical obstruction to having Setup run in time $\mathrm{poly}(\lambda, \log T)$, and thus being a strictly stronger primitive than Time-lock Puzzles. However our instantiation does not satisfy this stronger notion of Delay Encryption, and, lacking any other candidate, we prefer to keep our definitions steeping in reality.

To summarize, Delay Encryption is a natural analogue of Identity Based Encryption in the world of time delay cryptography. It requires Proofs of Work to exist, and a mild strengthening of it (which we are able to instantiate) implies Verifiable Delay Functions. It also implies a weak form of Time-lock Puzzles, and a strengthening of it (of which we know no instantiation) implies standard Time-lock Puzzles. At the same time, no dependency is known between Time-lock Puzzles and Verifiable Delay Functions, indicating that Delay Encryption is possibly a stronger primitive than both.

3 Delay Encryption from Isogenies and Pairings

We instantiate Delay Encryption from the same framework De Feo, Masson, Petit and Sanso used to instantiate Verifiable Delay Functions [17]. We briefly recall it here for completeness.

An elliptic curve E over a finite field $\mathbb{F}_{p^n}$ is said to be supersingular if the trace of its Frobenius endomorphism is divisible by p, i.e., if $\#E(\mathbb{F}_{p^n}) = 1 \mod p$. Over the algebraic closure of $\mathbb{F}_p$, there is only a finite number of isomorphism classes of supersingular curves, and every class contains a curve defined over $\mathbb{F}_{p^2}$.

We will only work with supersingular curves $E/\mathbb{F}_{p^2}$ whose group of $\mathbb{F}_{p^2}$-rational points is isomorphic to $(\mathbb{Z}/(p+1)\mathbb{Z})^2$. For these curves, if N is a divisor of $p+1$, we will denote by $E[N]$ the subgroup of $\mathbb{F}_{p^2}$-rational points of N-torsion, which is isomorphic to $(\mathbb{Z}/N\mathbb{Z})^2$. We will write $E[N]^\circ$ for the subset of points

[1] Bitansky *et al.* require pre-processing to run in sequential time $T \cdot \mathrm{poly}(\lambda)$, but parallel time only $\mathrm{poly}(\lambda, \log T)$.

in $E[N]$ of order exactly N; when N is prime, this is simply a shorthand for $E[N] \setminus \{\mathcal{O}\}$.

However, among these curves some will be curves $E/\mathbb{F}_p$ defined over $\mathbb{F}_p$, seen as curves over $\mathbb{F}_{p^2}$ (in algebraic jargon, with scalars extended from $\mathbb{F}_p$ to $\mathbb{F}_{p^2}$). For this special case, if N is an odd divisor of $p+1$, the $\mathbb{F}_{p^2}$-rational torsion subgroup $E[N]$ contains two distinguished subgroups: the subgroup $E[N] \cap E(\mathbb{F}_p)$ of points of order N defined over $\mathbb{F}_p$, which we also denote by $E[(N, \pi - 1)]$; and the subgroup of points of order N not in $E(\mathbb{F}_p)$, but with x-coordinate in $\mathbb{F}_p$, which we denote by $E[(N, \pi + 1)]$. Again, we write $E[(N, \pi - 1)]^\circ$ and $E[(N, \pi + 1)]^\circ$ for the subsets of points of order exactly N.

An isogeny is a group morphism of elliptic curves with finite kernel. In particular, isogenies preserve supersingularity. Isogenies can be represented by ratios of polynomials, and, like polynomials, have a *degree*. Isogenies of degree ℓ are also called ℓ-isogenies; the degree is multiplicative with respect to composition, thus $\deg \phi \circ \psi = \deg \phi \cdot \deg \psi$. The degree is an important invariant of isogenies, roughly measuring the amount of information needed to represent them.

An isogeny graph is a graph whose vertices are isomorphism classes of elliptic curves, and whose edges are isogenies, under some restrictions. Isogeny-based cryptography mainly uses two types of isogeny graphs:

- The *full supersingular graph* of (the algebraic closure of) $\mathbb{F}_p$, whose vertices are all isomorphism classes of supersingular curves over $\mathbb{F}_{p^2}$, and whose edges are all isogenies of a prime degree ℓ; typically $\ell = 2, 3$.
- The $\mathbb{F}_p$*-restricted supersingular graph*, or *supersingular CM graph* of $\mathbb{F}_p$, whose vertices are all $\mathbb{F}_p$-isomorphism classes of supersingular curves over $\mathbb{F}_p$, and whose edges are ℓ-isogenies for all primes ℓ up to some bound; typically $\ell \lessapprox \lambda \log \lambda$, where λ is the security parameter.

Any ℓ-isogeny $\phi : E \to E'$ has a unique *dual* ℓ-isogeny $\hat{\phi} : E' \to E$ such that

$$e'_N(\phi(P), Q) = e_N(P, \hat{\phi}(Q)), \tag{1}$$

for any integer N and any points $P \in E[N]$, $Q \in E'[N]$, where e_N is the Weil pairing on E, and e'_N the one on E'. The same equation, with the same $\hat{\phi}$, also holds for any other known pairing, such as the Tate and Ate pairings.

The framework of De Feo *et al.* uses chains of small degree isogenies as delay functions, and the pairing Eq. (1) as an efficient means to verify the computation. Formally, they propose two related instantiations of VDF, following the same pattern: they both use the same base field $\mathbb{F}_p$, where p is a prime of the form $p+1 = N \cdot f$ with N prime, chosen so that discrete logarithms in the group of N-th roots of unity in $\mathbb{F}_{p^2}$ (the target group G_T of the pairing) are hard (i.e., $N \approx 2^{2\lambda}$ and $p \sim 2^{\lambda^3}$). They have a common trusted setup, independent of the delay parameter, and the usual functionalities of a VDF:

Trusted setup selects a random supersingular elliptic curve E over $\mathbb{F}_p$.
Setup takes as input p, N, E, a delay parameter T, and performs a walk in an ℓ-isogeny graph to produce a degree ℓ^T isogeny $\phi : E \to E'$.

It also computes a point $P \in E(\mathbb{F}_p)$ of order N. It outputs $\phi, E', P, \phi(P)$.

Evaluation takes as input a random point $Q \in E'[N]$ and outputs $\hat{\phi}(Q)$.

Verification uses Eq. (1) to check that the value output by evaluation is $\hat{\phi}(Q)$ as claimed.

The two variants only differ in the way the isogeny walk is set up, and in minor details of the verification; these differences will be irrelevant to us.

The delay property of this VDF rests, roughly speaking, on the assumption that a chain of T isogenies of small prime degree ℓ cannot be computed more efficiently than by going through each of the isogenies one at a time, sequentially. The case $\ell = 2$ is very similar to repeated squaring in groups of unknown order as used by other VDFs [27,34] and Time-lock Puzzles [29]: in groups, one iterates T times the function $x \mapsto x^2$, a polynomial of degree 2; in isogeny graphs, one iterates rational fractions of degree 2. See Sect. 6 for more details.

It is important to remark that both setup and evaluation in these VDFs are "slow" algorithms, indeed both need to evaluate an isogeny chain (either ϕ, or $\hat{\phi}$) at one input point of order N; this is in stark contrast with VDFs based on groups of unknown order, where the setup, and thus its complexity, is independent of the delay parameter T.

3.1 Instantiation

The isogeny-based VDF of De Feo *et al.* can be understood as a modification on the Boneh–Lynn–Shacham [7] signature scheme, where the secret key is replaced by a long chain of isogenies: signing becomes a "slow" operation and thus realizes the evaluation function, whereas verification stays efficient.

Similarly, we obtain a Delay Encryption scheme by modifying the IBE scheme of Boneh and Franklin [6]: the master secret is replaced by a long chain of isogenies, while session identifiers play the role of identities, so that producing the decryption key for a given identity becomes a slow operation.

Concretely, setup is identical to that of the VDF. A prime of the form $p = 4 \cdot N \cdot f - 1$ is fixed according to the security parameter, then setup is actually split into two algorithms: a $\mathsf{TrustedSetup}$ independent of the delay parameter T and reusable for arbitrarily many untrusted setups, and a Setup which depends on T.

$\mathsf{TrustedSetup}(\lambda)$. Generate a nearly uniformly random supersingular curve $E/\mathbb{F}_p$ by starting from the curve $y^2 = x^3 + x$ and performing a random walk in the $\mathbb{F}_p$-restricted supersingular graph. Output E.

$\mathsf{Setup}(E, T)$.

1. Perform an ℓ-isogeny walk $\phi : E \to E'$ of length T;
2. Select a random point $P \in E(\mathbb{F}_p)$ of order N, and compute $\phi(P)$;
3. Output $\mathsf{ek} := (E', \phi)$ and $\mathsf{pk} := (E', P, \phi(P))$.

We stress that known homomorphic Time-lock Puzzles [25] also require a one-shot trusted setup. Furthermore, unlike constructions based on RSA groups, there is no evidence that trusted setup is unavoidable for isogeny-based delay functions, and indeed removing this trusted setup is an active area of research [12,24].

The isogeny chain ϕ in Setup can be generated by any of the two methods proposed by De Feo *et al.*, the difference will be immaterial for Delay Encryption; as discussed in [17], a (deterministic) walk limited to curves and isogenies defined over $\mathbb{F}_p$ will be more efficient, however a generic (pseudorandom) walk over $\mathbb{F}_{p^2}$ will offer some partial protection against quantum attacks.

Before defining the other routines, we need two hash functions. The first, $H_1 : \{0,1\}^\lambda \to E'[N]^\circ$, will be used to hash session identifiers to points of order N in $E'/\mathbb{F}_{p^2}$ (although the curve E' may be defined over $\mathbb{F}_p$). The second, $H_2 : \mathbb{F}_{p^2} \to \{0,1\}^\lambda$, will be a key derivation function.

$\mathsf{Extract}(E, E', \phi, \mathsf{id})$.
1. Let $Q = H_1(\mathsf{id})$;
2. Output $\hat{\phi}(Q)$.

$\mathsf{Encaps}(E, E', P, \phi(P), \mathsf{id})$.
1. Select a uniformly random $r \in \mathbb{Z}/N\mathbb{Z}$;
2. Let $Q = H_1(\mathsf{id})$;
3. Let $k = e'_N(\phi(P), Q)^r$;
4. Output $(rP, H_2(k))$.

$\mathsf{Decaps}(E, E', \hat{\phi}(Q), rP)$.
1. Let $k = e_N(rP, \hat{\phi}(Q))$.
2. Output $H_2(k)$.

Correctness of the scheme follows immediately from Eq. (1) and the bilinearity of the pairing.

Remark 2. Notice that two hashed identities Q, Q' such that $Q - Q' \in \langle P \rangle$ are equivalent for encapsulation and decapsulation purposes, and thus an adversary only needs to compute the image of one of them under $\hat{\phi}$. However, if we model H_1 as a random oracle, the probability of two identities colliding remains negligible (about $1/N$).

Alternatively, if E' is defined over $\mathbb{F}_p$, one can restrict the image of H_1 to $E'[(N, \pi + 1)]$, like in [17].

3.2 Security

To prove security of their VDF schemes, De Feo *et al.* defined the following *isogeny shortcut game*:

Precomputation. The adversary receives N, p, E, E', ϕ, and outputs an algorithm $\mathcal{S}$ (in time $\mathrm{poly}(\lambda, T)$).

Challenge. The challenger outputs a uniformly random $Q \in E'[N]$.

Guess. The algorithm $\mathcal{S}$ is run on input Q. The adversary wins if $\mathcal{S}$ terminates in time less than Δ, and $\mathcal{S}(Q) = \hat{\phi}(Q)$.

However, it is clear that the Δ-hardness of this game is insufficient to prove Δ-IND-CPA security of our Delay Encryption scheme. Indeed, while the hardness of the isogeny shortcut obviously guarantees that the output of Extract cannot be computed in time less than Δ, there is at least one other way to decapsulate a ciphertext rP, which consists in evaluating $\phi(rP)$ and computing $k = e'_N(\phi(rP), Q)$. Computing $\phi(rP)$ is expected to be at least as "slow" as computing $\hat{\phi}(Q)$, however this fact is not captured by the isogeny shortcut game.

Instead, we define a new security assumption, analogous to the Bilinear Diffie-Hellman assumption used in standard pairing-based protocols. The *bilinear isogeny shortcut game* is defined as follows:

Precomputation. The adversary receives p, N, E, E', ϕ, and outputs an algorithm $\mathcal{S}$.
Challenge. The challenger outputs uniformly random $R \in E[(N, \pi - 1)]$ and $Q \in E'[N]$.
Guess. Algorithm $\mathcal{S}$ is run on (R, Q). The adversary wins if $\mathcal{S}$ outputs $\mathcal{S}(R, Q) = e'_N(\phi(R), Q) = e_N(R, \hat{\phi}(Q))$.

We say that the bilinear isogeny shortcut game is Δ-hard if no adversary running the precomputation in time $\text{poly}(\lambda, T)$ produces an algorithm $\mathcal{S}$ that wins in time less than Δ with non-negligible probability. The reduction to Δ-IND-CPA of our Delay Encryption scheme closely follows the proof of security of Boneh and Franklin's IBE scheme.

Theorem 1. *The Delay Encryption scheme presented above is Δ-IND-CPA secure, assuming the Δ'-hardness of the bilinear isogeny shortcut game, with $\Delta \in \Delta' - o(\Delta')$, when H_1 and H_2 are modeled as random oracles.*

Concretely, suppose there is a Δ-IND-CPA adversary $\mathcal{A}$ with advantage ϵ and complexity $\text{poly}(\lambda, T)$*, making q queries to H_2 (including the queries made by the sub-algorithm $\mathcal{D}$). Then there is a* $\text{poly}(\lambda, T)$ *algorithm $\mathcal{B}$ that wins the bilinear isogeny shortcut game with probability at least $2\epsilon/q$ and delay $\Delta' = \Delta + q \cdot \text{poly}(\lambda)$.*

Proof. In the precomputation phase, when $\mathcal{B}$ receives the parameters p, N, E, E', ϕ, it draws a random $P \in E(\mathbb{F}_p)$ of order N, and evaluates $\phi(P)$. It then passes $p, N, E, E', \phi, P, \phi(P)$ to $\mathcal{A}$ for its own precomputation phase. Whenever $\mathcal{A}$ makes calls to H_1 or H_2, algorithm $\mathcal{B}$ checks whether the input has already been requested, in which case it responds with the same answer previously given, otherwise it responds with a uniformly sampled output and records the query.

When $\mathcal{A}$ requests its challenge, $\mathcal{B}$ does the same, receiving $R \in E[(N, \pi - 1)]$ and $Q \in E'[N]$. If R or Q is the point at infinity, it outputs 1 and terminates. Otherwise it draws a random string $s \in \{0,1\}^\lambda$, a random $\mathsf{id} \in \{0,1\}^\lambda$ that was not already queried to H_1, it programs the random oracle so that $H_1(\mathsf{id}) = Q$, and challenges $\mathcal{A}$ with the tuple (R, s, id).

During the guessing phase, whenever $\mathcal{A}$ (actually, $\mathcal{D}$) makes a call to H_1 or H_2, algorithm $\mathcal{B}$ (actually, $\mathcal{S}$) responds as before. Finally, when $\mathcal{D}$ outputs its guess, $\mathcal{S}$ simply returns a random entry among those that were queried to H_2.

Let $\mathcal{H}$ be the event that $\mathcal{A}$ (or $\mathcal{D}$) queries H_2 on input $e_N(R, \hat{\phi}(Q))$. We prove that $\Pr(\mathcal{H}) \geq 2\epsilon$, which immediately gives the claim of the theorem. To this end, we first prove that $\Pr(\mathcal{H})$ in the simulation is equal to $\Pr(\mathcal{H})$ in the real attack; then we prove that $\Pr(\mathcal{H}) \geq 2\epsilon$ in the real attack.

To prove the first claim, it suffices to show that the simulation is indistinguishable from the real world for $\mathcal{A}$. Indeed, public parameters are distributed identically to a Delay Encryption scheme, and the point R that is part of the challenge is necessarily a multiple of P, since $E[(N, \pi - 1)]$ is cyclic. The proof that the two probabilities are equal, then proceeds as in [6, Lemma 4.3, Claim 1].

The proof that $\Pr(\mathcal{H}) \geq 2\epsilon$ is identical to [6, Lemma 4.3, Claim 2]. This proves the part of the statement on the winning probability of $\mathcal{B}$.

If Algorithm $\mathcal{D}$ runs in time less than Δ, algorithm $\mathcal{S}$ runs in the same time, plus the time necessary for drawing the random string s and for answering queries to H_2. Depending on the computational model, a lookup in the table for H_2 can take anywhere from $O(1)$ (e.g., RAM model) to $O(q)$ (e.g., Turing machine). We err on the safe side, and estimate that $\mathcal{S}$ runs in time less than $\Delta + q \cdot \mathrm{poly}(\lambda)$.

3.3 Known Attacks

We now shift our attention to attacks. As discussed in [17], there are three types of known attacks: *shortcut* attacks, discrete logarithm attacks, and attacks on the computation.

Parameters for a Delay Encryption scheme must be chosen so that all known attacks have exponential difficulty in the security parameter λ. Given that (total) attacks successfully compute decapsulation in exponential time in λ, it is evident that the delay parameter T must grow at most subexponentially in λ.

Shortcut attacks aim at computing a shorter path $\psi : E \to E'$ in the isogeny graph from the knowledge of $\phi : E \to E'$. The name should not be confused with the isogeny shortcut game described above, as shortcut attacks are only one of the possible ways to beat the game.

De Feo *et al.* show that shortcut attacks are possible when the endomorphism ring of at least one of E or E' is known. Indeed, in this case, the isogeny ϕ can be translated to an ideal class in the endomorphism ring, then smoothing techniques similar to [22] let us convert the ideal to one of smaller norm, and finally to an isogeny $\psi : E \to E'$ of smaller degree.

The only way out of these attacks is to select the starting curve E as a uniformly random supersingular curve over $\mathbb{F}_p$, then no efficient algorithm is known to compute $\mathrm{End}(E)$, nor $\mathrm{End}(E')$. Unfortunately, the only way we currently know to sample nearly uniformly in the supersingular class over $\mathbb{F}_p$, is, paraphrasing [20], to choose the endomorphism ring first and then compute E given $\mathrm{End}(E)$.

Thus, the solution put forth in [17] is to generate the starting curve E via a trusted setup that first selects $\mathrm{End}(E)$, and then outputs E and throws away the information about its endomorphism ring. We stress that, given a random supersingular curve E, computing $\mathrm{End}(E)$ is a well known hard problem, upon which almost most of isogeny-based cryptography is founded. We explain in the next section how to mitigate the inconvenience of having a trusted setup, using a distributed protocol.

As stressed in [17], there is no evidence that "hashing" in the supersingular class, i.e., sampling nearly uniformly without gaining knowledge of the endomorphism ring, should be a hard problem. But there is no evidence it should be easy either, and several attempts have failed already [12,24].

Another possibility hinted at in [17] would be to generate ordinary pairing friendly curves with large isogeny class, as the shortcut attack is then thwarted by the difficulty of computing the order of the class group of the endomorphism ring. However finding such curves possibly seems an even harder problem than hashing to the supersingular class.

Discrete logarithm attacks compute $\hat{\phi}(Q)$ by directly solving the pairing Eq. (1). In our case, we can even directly attack the key encapsulation. Indeed, knowing rP, we obtain r through a discrete logarithm, and then compute $k = e'_N(\phi(P), Q)^r$.

Thanks to the efficiently computable pairing, the discrete logarithm can actually be solved in $\mathbb{F}_{p^2}$, which justifies taking p, N large enough to resist finite field discrete logarithm computations. Obviously, this also shows that our scheme is easily broken by quantum computers. See [17], however, for a discussion of how a setup with pseudo-random walks over $\mathbb{F}_{p^2}$ resists quantum attacks in a world where quantum computers are available, but much slower than classical ones.

Attacks on the computation do not seek to deviate from the description of the protocol, but simply try to speed up Extract beyond the way officially prescribed by the scheme. In this sort of attacks, the adversary may be given more resources than the legitimate user: for example, it may be allowed a very large precomputation, or it may dispose of an unbounded amount of parallelism, or it may have access to an architecture not available to the user (e.g., a quantum computer).

These attacks are the most challenging to analyze, because standard complexity-theoretical techniques are of little help here. On some level, this goal is unachievable: given a sufficiently abstract computational model, and a sufficiently powerful adversary, any scheme is broken. For example, an adversary may precompute all possible pairs $(Q, \hat{\phi}(Q))$ and store them in a $O(1)$-accessible RAM, then extraction amounts to a table lookup. However, such an adversary with exponential precomputation, exponential storage, and constant time RAM is easily dismissed as unreasonable. More subtle trade-offs between precomputation, storage and efficiency can be obtained, like for example RNS-based techniques to attack group-based VDFs [1]. However the real impact of these theoretical algorithms has yet to be determined.

In practice, a pragmatic approach to address attacks on the computation is to massively invest in highly specialized hardware development to evaluate the "sequential and slow" function quickly, and then produce the best designs at scale, so that they are available to anyone who wants to run the extraction. This is the philosophy of the competitions organized by Ethereum [33] and Chia [21], targeting, respectively, the RSA based VDF and the class group based VDF.

We explore this topic more in detail in Sect. 6.

4 Distributed Trusted Setup

Trusted setup is an obvious annoyance to distributed protocols. A way to mitigate this negative impact is to distribute trust over several participants, ensuring through a multi-party computation that, if at least one participant is honest, then the setup can be trusted.

Ethereum is notoriously investing in the RSA-based VDF with Wesolowski's proof [33,34], which is known to require a trusted setup. To generate parameters, the Ethereum network will need to run a distributed RSA modulus generation, for which all available techniques essentially trace back to the work of Boneh and Franklin [5].

Distributed RSA modulus generation is notoriously a difficult task: the cost is relatively high, scales badly with the number of participants, and the attempts at optimizing it have repeatedly led to subtle and powerful attacks [31,32]. Worse still, specialized hardware for the delay function must be designed specifically for the generated modulus, which means that little design can be done prior to the distributed generation, and that if the distributed generation is then found to be rigged, a new round of distributed-generation-then-design is needed.

On the contrary, distributed parameter generation for our Delay Encryption candidate, or for the isogeny based VDF, is extremely easy. The participants start from a well known supersingular curve with known endomorphism ring, e.g., $E_0 : y^2 = x^3 - x$, and repeat, each at its own turn, the following steps:

1. Participant i checks all zero-knowledge proofs published by participants that preceded them;
2. They perform a pseudorandom isogeny walk $\psi_i : E_{i-1} \to E_i$ of length $c \log(p)$ in the $\mathbb{F}_p$-restricted supersingular graph;
3. They publish E_i, and a zero-knowledge proof that they know an isogeny $\psi : E_{i-1} \to E_i$.

The constant c is to be determined as a function of the expansion properties of the isogeny graph, and is meant to be large enough to ensure nearly uniform mixing of the walk. In practice, this constant is usually small, say $c < 10$, implying that each participant needs to evaluate a few thousands isogenies, a computation that is expected to take in the order of seconds [11].

The setup is clearly secure as long as at least one participant is honest. Indeed it is well known that computing a path from E_i to E_0 is equivalent to computing the endomorphism ring of E_i [18,22], and, since E_i is nearly uniformly distributed in the supersingular graph, the dishonest participants have no advantage in solving this problem compared to a generic attacker.

This distributed computation scales linearly with the number of participants, each participant needing to check the proofs of the previous ones. It can be left running for a long period of time, allowing many participants to contribute trust without any need for prior registration. More importantly, it is *updatable*, meaning that after the distributed generation is complete, the final curve E can be used as the starting point for a new distributed trusted setup. This way the trusted setup can be updated regularly, building upon the trust accumulated in previous distributed generations.

Compared with the trusted setup for RSA, the outcome of the setup is much less critical for the design of hardware. Indeed, the primes p, N can be publicly chosen in advance, and hardware can be designed for them before the trusted setup is performed. The trusted curve E only impacts the first few steps of the "slow" isogeny walk $\phi : E \to E'$ generated by the untrusted setup, and can easily be integrated in the hardware design at a later stage.

4.1 Proofs of Isogeny Knowledge

We take a closer look at the last step each participant takes in the trusted setup: the proof of isogeny knowledge. Ignoring zero-knowledge temporarily, Eq. (1) already provides a proof of knowledge of a non-trivial relation between E_{i-1} and E_i. Let F be a deterministic function which takes as input a pair of curves E_i, E_j and outputs a pair of points in $E_i[(N, \pi - 1)]^\circ \times E_j[(N, \pi + 1)]^\circ$. Also let e_N^i denote the Weil pairing on E_i. The proof proceeds as follows

1. Map (E_{i-1}, E_i) to a pair of points $(P, Q) \leftarrow F(E_{i-1}, E_i)$;
2. Choose a random $r \in (\mathbb{Z}/N\mathbb{Z})^\times$,
3. Publish $(R, S) \leftarrow (r\psi_i(P), r\hat{\psi}_i(Q))$.

Then verification simply consists of:

1. Compute $(P, Q) \leftarrow F(E_{i-1}, E_i)$,
2. Check that $R \in E_i[(N, \pi - 1)]^\circ$ and $S \in E_{i-1}[(N, \pi + 1)]^\circ$;
3. Check that $e_N^i(R, Q) = e_N^{i-1}(P, S)$.

This proof is compact, requiring only four elements of $\mathbb{F}_p$, and efficient because computing $\psi_i(P), \hat{\psi}_i(Q)$ only adds a small overhead to the computation of ψ_i, and verification takes essentially two pairing computations. Note that the restriction in step 2 of the verification implies that for any R there exists a unique S satisfying the equation in step 3, and *vice versa*.

Remark 3. While we believe that an adversary not knowing an isogeny from E_{i-1} to E_i has a negligible probability of convincing a verifier in the protocol above, it is not clear what kind of knowledge is *exactly* proved by it. Ideally, we would like to prove that, given an algorithm that passes verification with non negligible probability, one can extract a description of some isogeny $\psi' : E_{i-1} \to E_i$.

However, no such algorithm is currently known. Related problems have been studied in the context of cryptanalyses of SIDH, under the name of "torsion point attacks" [23,26,30], however these algorithms crucially rely on the knowledge of the endomorphism ring of E_{i-1}, something we cannot exploit here.

The only way out is apparently to define a non-falsifiable "knowledge of isogeny" assumption, which would tautologically state that the protocol above is indeed a proof of knowledge of an isogeny. We defer investigation of this type of assumptions to future work.

As stated, the proof above is clearly not zero-knowledge, because the values $r\psi_i(P)$ and $r\hat{\psi}_i(Q)$ reveal a considerable amount of additional information on ψ_i. To turn the proof into a zero-knowledge one, we use Pedersen commitments to mask $r\psi_i(P)$ and $r\hat{\psi}_i(Q)$, then prove their knowledge using standard Schnorr-like proofs of knowledge.

Let F' be a function with same domain and range as F (possibly $F' = F$), and let H be a cryptographic hash function with values in $\mathbb{Z}/N\mathbb{Z}$.

We compute $P', Q' \leftarrow F'(E_i, E_{i-1})$ and choose $x, y \in (\mathbb{Z}/N\mathbb{Z})^\times$ secret, then we publish a NIZK proof for (X, Y) satisfying

$$e_N^i(X, Q)e_N^{i-1}(P, Q')^y = e_N^{i-1}(P, Y)e_N^i(P', Q)^x.$$

More precisely, we publish:

- two Pedersen commitments $X = xP' + r\psi_i(P) \in E_i[(N, \pi - 1)]$ and $Y = yQ' + r\hat{\psi}_i(Q) \in E_{i-1}[(N, \pi + 1)]$,
- two "public keys" $Y' = e_N^{i-1}(P, Q')^y$ and $X' = e_N^i(P', Q)^x$, and
- a Schnorr-like proof of knowledge (c, s_x, s_y) for x of X' and y of Y', where:
  - $s_x = k - xc$ and $s_y = k - yc$ for a randomly chosen $k \in (\mathbb{Z}/N\mathbb{Z})^\times$, and
  - $c = H(e_N^{i-1}(P, Q') \| e_N^i(P', Q) \| X' \| Y' \| e_N^i(P', Q)^k \| e_N^{i-1}(P, Q')^k)$.

At this point, our verifier now checks

- that $X \in E_i[(N, \pi - 1)]$ and $Y \in E_{i-1}[(N, \pi + 1)]$,
- that $e_N^i(X, Q)Y' = e_N^{i-1}(P, Y)X'$,
- non-triviality $X' \neq e_N^i(X, Q)$ and $Y' \neq e_N^{i-1}(P, Y)$ of the commitments, and
- the proofs of knowledge s_x, s_y using

$$c = H(e_N^{i-1}(P, Q') \| e_N^i(P', Q) \| X' \| Y' \| (X')^c e_N^i(P', Q)^{s_x} \| (Y')^c e_N^{i-1}(P, Q')^{s_y}).$$

In this, we ask verifiers to compute four pairings, which only doubles the verifier time.

The following lemma shows that this is a NIZK proof for the same statement that was proven in the simple protocol revealing $r\psi_i(P)$ and $r\hat{\psi}_i(Q)$, and thus it is a NIZK proof of isogeny knowledge, if we accept the non-falsifiable assumption mentioned in Remark 3.

Lemma 1. *Let E_{i-1}, E_i be a pair of isogenous elliptic curves, let $(P, Q) = F(E_{i-1}, E_i)$. The protocol above is a NIZK proof of knowledge of a pair (R, S) of points such that $e_N^i(R, Q) = e_N^{i-1}(P, S) \neq 1$, assuming CDH in the target group $G_T \subset \mathbb{F}_{p^2}$, and modeling H as a random oracle.*

Proof. To simulate the proof, it is enough to choose X, Y, and X' at random, set $Y' = X' e_N^{i-1}(P, Y)/e_N^i(X, Q)$, then use the simulator of the Schnorr proof to simulate knowledge of the discrete logarithms of X' and Y'.

To extract (R, S) from a prover, we start by using the Schnorr extractor to get x and y. Then, by hypothesis

$$e_N^i(X - xP', Q) = \frac{e_N^i(X, Q)}{X'} = \frac{e_N^{i-1}(P, Y)}{Y'} = e_N^{i-1}(P, Y - yQ') \neq 1,$$

thus $R = X - xP'$ and $S = Y - yQ$ is the solution we were looking for.

Since the Schnorr proof is proven secure under CDH in the ROM, the same hypotheses carry over.

For completeness, we also mention some other tools with which one might prove knowledge of this isogeny in zero knowledge, although none seem to be competitive with the technique above.

First, there exists a rapidly expanding SNARK toolbox from which one could perform $\mathbb{F}_p$ arithmetic inside the SNARK to check the verification of the second and third conditions directly. As instantiating the delay function imposes restrictions on p, one cannot necessarily select p using the Cocks-Pinch method to provide a pairing friendly elliptic curve with group order p, like in [8]. There are optimisations for arithmetic in arbitrary $\mathbb{F}_p$ however, especially using polynomial commitments, like in [19].

Second, there are well known post-quantum isogeny-based proofs:

SIDH-style proofs [15] are very inconvenient, because they require primes of a specific form, and severely limit the length of pseudo-random walks. On top of that, they are very inefficient, and do not have perfect zero-knowledge.

SeaSign-style proofs [14] have sizes in the hundreds of kilobytes, and their generation and verification are extremely slow (dozens of hours). Note that several of the optimizations used for signatures, including the class group order precomputation of CSI-FiSh [2], are not available in this context. More research on the optimization of SeaSign-style proofs for this specific context would be welcome.

SQISign-style proofs [16] are not compatible with our setting, because they require knowledge of the endomorphism rings.

5 Watermarking

When delay functions are used in distributed consensus protocols, it is common to want to reward participants who spend resources to evaluate the function. For example, in the auction application the participants who compute the session key may receive a percentage on the sale to compensate for the cost of running the extraction routine.

This raises the question of how to prove that some participant did run the computation, rather than simply steal the public output from someone else.

In the context of VDFs based on groups of unknown order, Wesolowski [34] introduced the concept of *proof watermarking*. The output of these VDFs consists of two parts: a delay function output and a proof. Wesolowski's idea is to attach to the proof a watermark based on the identity of the evaluator, so that anyone verifying the output immediately associates it to the legitimate participant. Since producing the proof costs at least as much as evaluating the delay function, a usurper who would like to claim an output for themself would need to do an amount of work comparable to legitimately evaluating the delay function, which strongly reduces the incentive for usurpation.

In the context of isogeny based VDFs, or of extraction in Delay Encryption, proof watermarking is a meaningless concept, because there is simply no proof to watermark. Nevertheless, it is possible to produce a watermark next to the output of the delay function, giving evidence that the owner of the watermark spent an amount of effort comparable to legitimately computing the output. The idea is to publish a *mid-point* update on the progress of the evaluation, and attach this mid-point to the identity of the evaluator.

Concretely, given parameters $\phi : E \to E'$ and $(P, \phi(P))$, the isogeny walk is split into two halves of equal size $\phi_1 : E \to E_{\text{mid}}$ and $\phi_2 : E_{\text{mid}} \to E'$ so that $\phi = \phi_2 \circ \phi_1$, and $\phi_1(P)$ is added to the public parameters. Each evaluator then generates a secret key $s \in \mathbb{Z}/N\mathbb{Z}$ and a public key $S = s\phi(P)$. When evaluating $\hat{\phi} = \hat{\phi}_1 \circ \hat{\phi}_2$ at a point $Q \in E'[N]$, the evaluator:

1. Computes $Q_{\text{mid}} = \hat{\phi}_2(Q)$,
2. Computes and publishes sQ_{mid},
3. Finishes off the computation by computing $\hat{\phi}(Q) = \hat{\phi}_1(Q_{\text{mid}})$.

A watermark can then be verified by checking that

$$e_N^{\text{mid}}(\phi_1(P), sQ_{\text{mid}}) = e'_N(S, Q).$$

Interestingly, this proof is *blind*, meaning that it can be verified even before the work is finished.

Given $\hat{\phi}(Q)$, a usurper wanting to claim the computation for themselves would need to either start from Q and compute $\hat{\phi}_2(Q)$, or start from $\hat{\phi}(Q)$ and compute $\frac{\phi_1(\hat{\phi}(Q))}{\deg \phi_1}$. Either way, they would perform at least half as much work as if they had legitimately evaluated the function.

This scheme is easily generalized to several equally spaced mid-points along the isogeny evaluation chain: if the isogeny is split into n pieces of equal size, a usurper would need to do at least $(n-1)/n$ times as much work as a legitimate evaluator, thus linearly decreasing the incentive for usurpation.

It is possible, nevertheless, for a usurper to target a specific evaluator, by generating a random $u \in \mathbb{Z}/N\mathbb{Z}$, and choosing $us\phi_1(P)$ as public key. Then, any proof sQ_{mid} for the legitimate evaluator is easily transformed to a proof usQ_{mid} for the usurper. This attack is easily countered by having all evaluators publish a zero-knowledge proof of knowledge of their secret exponent s, along with their public key $s\phi_1(P)$.

6 Challenges in Implementing Isogeny-Based Delay Functions

For a delay function to be useful, there need to be convincing arguments as to why the evaluation cannot be performed considerably better than with the legitimate algorithm.

In this sense, repeated squaring modulo an RSA modulus is especially appealing: modular arithmetic has been studied for a long time, and we are reasonably confident that we know all useful algorithms and hardware in this respect; and the repeated application of the function $x \mapsto x^2$ is so simple that one may hope no better algorithm exists (see [1], though).

Repeated squaring in class groups, already, raises more skepticism, as the arithmetic of class groups is a much less studied area. This clearly had an impact on Ethereum's choice to go with RSA-based VDFs, despite class group based ones not needing a trusted setup.

For isogeny based delay functions, we argue that the degree of assurance seems to be nearly as good as for RSA based ones, although more research is certainly needed. To support this claim, we give here more details on the way the evaluation of $\hat{\phi}$ is performed, that were omitted by [17].

For a start, we must choose a prime degree ℓ. Intuitively, the smaller, the better, thus we shall fix $\ell = 2$, although $\ell = 3$ also deserves to be studied. A 2-isogeny is represented by rational maps of degree 2, thus we expect one isogeny evaluation to require at least one multiplication modulo p. Our goal is to get as close as possible to this lower bound, by choosing the best representation for the elliptic curves, their points, and their isogenies.

It is customary in isogeny based cryptography to use curves in Montgomery form, and projective points in $(X : Z)$ coordinates, as these give the best formulas for arithmetic operations and isogenies [13,28]. Montgomery curves satisfy the equation

$$E : y^2 = x^3 + Ax^2 + x,$$

in particular they have a point of order two in $(0, 0)$, and two other points of order two with x-coordinates α and $1/\alpha$, where α is a root of the polynomial x^2+Ax+1, and possibly lives in $\mathbb{F}_{p^2}$. These three points define the three possible isogenies of degree 2 starting from E. The Montgomery form is almost unique, there being only six possible choices for the A coefficient for a given isomorphism class, corresponding to the three possible choices for the point to send in $(0, 0)$ (each taken twice).

In our case, all three points (in projective coordinates) $(0 : 1)$, $(\alpha : 1)$ and $(1 : \alpha)$, are defined over $\mathbb{F}_p$, we thus choose to distinguish one additional point by writing the curves as

$$E_\alpha : y^2 = x(x - \alpha)(x - 1/\alpha),$$

with $\alpha \neq 0, \pm 1$. We call this a *semi-Montgomery form*; although it is technically equivalent to the Montgomery form, 2-isogeny formulas are expressed in it more easily. Recovering the Montgomery form is easy via $A = -\alpha - 1/\alpha$.

Using the formula of Renes [28], we readily get the isogeny with kernel generated by $(\alpha : 1)$ as

$$\phi_\alpha(x, y) = \left(x\frac{x\alpha - 1}{x - \alpha}, \dots\right), \tag{2}$$

and its image curve is the Montgomery curve defined by $A = 2 - 4\alpha^2$. By comparing with the multiplication-by-2 map on E_α, we obtain the dual map to ϕ_α as

$$\hat{\phi}_\alpha(x, y) = \left(\frac{(x+1)^2}{4\alpha x}, \dots\right). \tag{3}$$

It is clear from this formula that the kernel of $\hat{\phi}_\alpha$ is generated by $(0, 0)$.

This formula is especially interesting, as we verify that its projective version in $(X : Z)$ coordinates only requires 2 multiplications and 1 squaring:

$$\hat{\phi}_\alpha(X : Z) = \left((X + Z)^2 : 4\alpha XZ\right), \tag{4}$$

and the squaring can be performed in parallel with one multiplication. The analogous formulas for $\phi_{1/\alpha}$ are readily obtained by replacing $\alpha \to 1/\alpha$ in the previous ones, and moving around projective coefficients to minimize work.

But, if we want to chain 2-isogenies, we need a way to compute the semi-Montgomery form of the image curve. For the given $A = 4\alpha^2 - 2$, direct calculation shows that the two possible choices are

$$\alpha' = 2\alpha\left(\alpha \pm \sqrt{\alpha^2 - 1}\right) - 1 = \left(\alpha \pm \sqrt{\alpha^2 - 1}\right)^2. \tag{5}$$

As we know that $(0, 0)$ generates the dual isogeny to ϕ_α, neither choice of α' will define a backtracking walk. Interestingly, Castryck and Decru [10] show that when $p = 7 \mod 8$, if $\alpha \in \mathbb{F}_p$, if ϕ_α is a horizontal isogeny (see definition in [10]), and if α' is defined as

$$\alpha' = \left(\alpha + \sqrt{\alpha^2 - 1}\right)^2$$

where $\sqrt{\alpha^2 - 1}$ denotes the principal square root, then $\alpha' \in \mathbb{F}_p$ and $\phi_{\alpha'}$ is horizontal too. This gives a very simple algorithm to perform a non-backtracking 2-isogeny walk staying in the $\mathbb{F}_p$-restricted isogeny graph, i.e., a walk on the 2-*crater*. Alternatively, if a pseudo-random walk in the full supersingular graph is wanted, one simply takes a random square root of $\alpha^2 - 1$.

Using these formulas, the isogeny walk $\phi : E \to E'$ is simply represented by the list of coefficients α encountered, and the evaluation of $\hat{\phi}$ using Formula (4) costs 2 multiplications and 1 parallel squaring per isogeny.

Implementation challenges. Following the recommendations of [17], for a 128-bits security level we need to choose a prime p of around 1500 bits, which is comparable to the 2048-bits RSA arithmetic targeted by Ethereum, although possibly open to optimizations for special primes.

In software, the latency of multiplication modulo such a prime is today around 1 µs. The winner of the Ethereum FPGA competition [33] achieved a latency of 25 ns for 2048-bits RSA arithmetic. Assuming a pessimistic baseline of 50 ns for one 2-isogeny evaluation, for a target delay of 1 h we need an isogeny walk of length $\approx 7 \cdot 10^{10}$. That represents as many coefficients α to store, each occupying $\approx$1500 bits, i.e., $\approx$12 TiB of storage!

We stress that only the evaluation key ek requires such large storage, and thus only evaluators (running extraction in Delay Encryption, or evaluating a VDF) need to store it. However any FPGA or hardware design for the evaluation of isogeny-based delay functions must take this constraint into account, and provide fast storage with throughputs of the order of several GiB/s.

At present, we do not know any configuration that pushes these 2-isogeny computations into being memory bandwidth bound. In fact, computational adversaries only begin encountering current DRAM and CPU bus limits when going an order of magnitude faster than the hypothetical high speeds above.

An isogeny-based VDF could dramatically reduce storage requirements by doing repeated shorter evaluations, and simply hashing each output to be the input for the next evaluation. We sacrifice verifier time by doing so, but verifiers remain fast since they still only compute two pairings per evaluation. We caution however that this trick does not apply to Delay Encryption.

In [17], De Feo *et al.* describe an alternative implementation that divides the required storage by a factor of 1244, at the cost of slowing down evaluation by a factor of at least $\log_2(1244)$. While this trade-off could be acceptable in some non-fully distributed applications, it seems incompatible with applications where evaluators want to get to the result as quickly as possible, e.g., when several evaluators are competing to compute the output.

It would be very interesting to find compact representations of very long isogeny chains which do not come at the expense of efficiently evaluating them.

Optimality. Formula (4) is, intuitively, almost optimal, as we expect that a 2-isogeny in projective $(X : Z)$ coordinates should require at least 2 multiplications. And indeed we know of at least one case where a 2-isogeny can be evaluated with 2 parallel multiplications: the isogeny of kernel $(0 : 1)$ is given by

$$\phi_0(x, y) = \left(\frac{(x-1)^2}{x}, \dots\right), \tag{6}$$

or, in projective coordinates,

$$\phi_0(X : Z) = \big((X - Z)^2 : XZ\big), \tag{7}$$

which only requires one parallel multiplication and squaring.

We tried to construct elliptic curve models and isogeny formulas that could evaluate 2-isogeny chains using only 2 parallel multiplications per step, however any formula we could find had a coefficient similar to α intervene in it, and thus bring the cost up by at least one multiplication.

Intuitively, this is expected: there are exponentially many isogeny walks, and the coefficients α must necessarily intervene in the formulas to distinguish between them. However this is far from being a proof. Even proving a lower bound of 2 *parallel* multiplications seems hard.

It would be interesting to prove that any 2-isogeny chain needs at least 2 *sequential* multiplications for evaluation, or alternatively find a better way to represent and evaluate isogeny chains.

7 Conclusion

We introduced a new time delay primitive, named Delay Encryption, related to Time-lock Puzzles and Verifiable Delay Functions. Delay Encryption has some interesting applications such as sealed-bid auctions and electronic voting. We gave an instantiation of Delay Encryption using isogenies of supersingular curves and pairings, and discussed several related topics that also apply to the VDF of De Feo, Masson, Petit and Sanso.

Several interesting questions are raised by our work, such as, for example, clarifying the relationship between Delay Encryption, Verifiable Delay Functions and Time-lock Puzzles.

Like the isogeny-based VDF, our Delay Encryption requires a trusted setup. We described an efficient way to perform a distributed trusted setup, however the associated zero-knowledge property relies on a non-falsifiable assumption which requires more scrutiny.

The implementation of delay functions from isogenies presents several practical challenges, such as needing very large storage for the public parameters. On top of that, it is not evident how to prove the optimality of isogeny formulas used for evaluating the delay function. While we gave here extremely efficient formulas, these seem to be at least one multiplication more expensive than the theoretical optimum. More research on the arithmetic of elliptic curves best adapted to work with extremely long chains of isogenies is needed.

Finally, we invite the community to look for more constructions of Delay Encryption, in particular quantum-resistant ones.

References

1. Bernstein, D.J., Sorenson, J.: Modular exponentiation via the explicit Chinese remainder theorem. Math. Comput. **76**, 443–454 (2007). https://doi.org/10.1090/S0025-5718-06-01849-7
2. Beullens, W., Kleinjung, T., Vercauteren, F.: CSI-FiSh: efficient isogeny based signatures through class group computations. In: Galbraith, S.D., Moriai, S. (eds.) ASIACRYPT 2019. LNCS, vol. 11921, pp. 227–247. Springer, Cham (2019). https://doi.org/10.1007/978-3-030-34578-5_9
3. Bitansky, N., Goldwasser, S., Jain, A., Paneth, O., Vaikuntanathan, V., Waters, B.: Time-lock puzzles from randomized encodings. In: Proceedings of the 2016 ACM Conference on Innovations in Theoretical Computer Science, ITCS 2016, New York, NY, USA, pp. 345–356. Association for Computing Machinery (2016). https://doi.org/10.1145/2840728.2840745

4. Boneh, D., Bonneau, J., Bünz, B., Fisch, B.: Verifiable delay functions. In: Shacham, H., Boldyreva, A. (eds.) CRYPTO 2018. LNCS, vol. 10991, pp. 757–788. Springer, Cham (2018). https://doi.org/10.1007/978-3-319-96884-1_25
5. Boneh, D., Franklin, M.: Efficient generation of shared RSA keys. In: Kaliski, B.S. (ed.) CRYPTO 1997. LNCS, vol. 1294, pp. 425–439. Springer, Heidelberg (1997). https://doi.org/10.1007/BFb0052253
6. Boneh, D., Franklin, M.: Identity-based encryption from the Weil pairing. SIAM J. Comput. **32**(3), 586–615 (2003). https://doi.org/10.1137/S0097539701398521
7. Boneh, D., Lynn, B., Shacham, H.: Short signatures from the Weil pairing. J. Cryptol. **17**(4), 297–319 (2004). https://doi.org/10.1007/s00145-004-0314-9
8. Bowe, S., Chiesa, A., Green, M., Miers, I., Mishra, P., Wu, H.: ZEXE: enabling decentralized private computation. In: 2020 IEEE Symposium on Security and Privacy (SP), pp. 947–964 (2020). https://doi.org/10.1109/SP40000.2020.00050
9. Brakerski, Z., Döttling, N., Garg, S., Malavolta, G.: Leveraging linear decryption: rate-1 fully-homomorphic encryption and time-lock puzzles. In: Hofheinz, D., Rosen, A. (eds.) TCC 2019. LNCS, vol. 11892, pp. 407–437. Springer, Cham (2019). https://doi.org/10.1007/978-3-030-36033-7_16
10. Castryck, W., Decru, T.: CSIDH on the surface. In: Ding, J., Tillich, J.-P. (eds.) PQCrypto 2020. LNCS, vol. 12100, pp. 111–129. Springer, Cham (2020). https://doi.org/10.1007/978-3-030-44223-1_7
11. Castryck, W., Lange, T., Martindale, C., Panny, L., Renes, J.: CSIDH: an efficient post-quantum commutative group action. In: Peyrin, T., Galbraith, S. (eds.) ASIACRYPT 2018. LNCS, vol. 11274, pp. 395–427. Springer, Cham (2018). https://doi.org/10.1007/978-3-030-03332-3_15
12. Castryck, W., Panny, L., Vercauteren, F.: Rational isogenies from irrational endomorphisms. In: Canteaut, A., Ishai, Y. (eds.) EUROCRYPT 2020. LNCS, vol. 12106, pp. 523–548. Springer, Cham (2020). https://doi.org/10.1007/978-3-030-45724-2_18
13. Costello, C., Longa, P., Naehrig, M.: Efficient algorithms for supersingular isogeny Diffie-Hellman. In: Robshaw, M., Katz, J. (eds.) CRYPTO 2016. LNCS, vol. 9814, pp. 572–601. Springer, Heidelberg (2016). https://doi.org/10.1007/978-3-662-53018-4_21
14. De Feo, L., Galbraith, S.D.: SeaSign: compact isogeny signatures from class group actions. In: Ishai, Y., Rijmen, V. (eds.) EUROCRYPT 2019. LNCS, vol. 11478, pp. 759–789. Springer, Cham (2019). https://doi.org/10.1007/978-3-030-17659-4_26
15. De Feo, L., Jao, D., Plût, J.: Towards quantum-resistant cryptosystems from supersingular elliptic curve isogenies. J. Math. Cryptol. **8**(3), 209–247 (2014). https://doi.org/10.1007/978-3-642-25405-5_2
16. De Feo, L., Kohel, D., Leroux, A., Petit, C., Wesolowski, B.: SQISign: compact post-quantum signatures from quaternions and isogenies. In: Moriai, S., Wang, H. (eds.) ASIACRYPT 2020, Part I. LNCS, vol. 12491, pp. 64–93. Springer, Cham (2020). https://doi.org/10.1007/978-3-030-64837-4_3
17. De Feo, L., Masson, S., Petit, C., Sanso, A.: Verifiable delay functions from supersingular isogenies and pairings. In: Galbraith, S.D., Moriai, S. (eds.) ASIACRYPT 2019. LNCS, vol. 11921, pp. 248–277. Springer, Cham (2019). https://doi.org/10.1007/978-3-030-34578-5_10
18. Eisenträger, K., Hallgren, S., Lauter, K., Morrison, T., Petit, C.: Supersingular isogeny graphs and endomorphism rings: reductions and solutions. In: Nielsen, J.B., Rijmen, V. (eds.) EUROCRYPT 2018. LNCS, vol. 10822, pp. 329–368. Springer, Cham (2018). https://doi.org/10.1007/978-3-319-78372-7_11

19. Gabizon, A., Williamson, Z.J.: plookup: a simplified polynomial protocol for lookup tables. Cryptology ePrint Archive, Report 2020/315 (2020). https://eprint.iacr.org/2020/315
20. Galbraith, S.D., Vercauteren, F.: Computational problems in supersingular elliptic curve isogenies. Quantum Inf. Process. **17**(10), 1–22 (2018). https://doi.org/10.1007/s11128-018-2023-6
21. Howard, M., Cohen, B.: Chia network announces 2nd VDF competition with $100,000 in total prize money (2019). https://www.chia.net/2019/04/04/chia-network-announces-second-vdf-competition-with-in-total-prize-money.en.html
22. Kohel, D.R., Lauter, K., Petit, C., Tignol, J.P.: On the quaternion-isogeny path problem. LMS J. Comput. Math. **17**(A), 418–432 (2014)
23. Kutas, P., Martindale, C., Panny, L., Petit, C., Stange, K.E.: Weak instances of SIDH variants under improved torsion-point attacks. Cryptology ePrint Archive, Report 2020/633 (2020). https://eprint.iacr.org/2020/633
24. Love, J., Boneh, D.: Supersingular curves with small noninteger endomorphisms. Open Book Series **4**(1), 7–22 (2020)
25. Malavolta, G., Thyagarajan, S.A.K.: Homomorphic time-lock puzzles and applications. In: Boldyreva, A., Micciancio, D. (eds.) CRYPTO 2019, Part I. LNCS, vol. 11692, pp. 620–649. Springer, Cham (2019). https://doi.org/10.1007/978-3-030-26948-7_22
26. Petit, C.: Faster algorithms for isogeny problems using torsion point images. In: Takagi, T., Peyrin, T. (eds.) ASIACRYPT 2017. LNCS, vol. 10625, pp. 330–353. Springer, Cham (2017). https://doi.org/10.1007/978-3-319-70697-9_12
27. Pietrzak, K.: Simple verifiable delay functions. In: Blum, A. (ed.) 10th Innovations in Theoretical Computer Science Conference (ITCS 2019). Leibniz International Proceedings in Informatics (LIPIcs), vol. 124, pp. 60:1–60:15. Schloss Dagstuhl-Leibniz-Zentrum fuer Informatik, Dagstuhl, Germany (2018). https://doi.org/10.4230/LIPIcs.ITCS.2019.60
28. Renes, J.: Computing isogenies between montgomery curves using the action of (0, 0). In: Lange, T., Steinwandt, R. (eds.) PQCrypto 2018. LNCS, vol. 10786, pp. 229–247. Springer, Cham (2018). https://doi.org/10.1007/978-3-319-79063-3_11
29. Rivest, R.L., Shamir, A., Wagner, D.A.: Time-lock puzzles and timed-release crypto. Technical report, Cambridge, MA, USA (1996). https://people.csail.mit.edu/rivest/pubs/RSW96.pdf
30. Delpech de Saint Guilhem, C., Kutas, P., Petit, C., Silva, J.: SÉTA: supersingular encryption from torsion attacks. Cryptology ePrint Archive, Report 2019/1291 (2019). https://eprint.iacr.org/2019/1291
31. Shlomovits, O.: Diogenes Octopus: Playing red team for Eth2.0 VDF, June 2020. https://medium.com/zengo/dac3f2e3cc7b
32. Shlomovits, O.: DogByte attack: playing red team for Eth2.0 VDF, August 2020. https://medium.com/zengo/ea2b9b2152af
33. VDF Alliance: VDF FPGA competition (2019). https://supranational.atlassian.net/wiki/spaces/VA/pages/36569208/FPGA+Competition
34. Wesolowski, B.: Efficient verifiable delay functions. In: Ishai, Y., Rijmen, V. (eds.) EUROCRYPT 2019, Part III. LNCS, vol. 11478, pp. 379–407. Springer, Cham (2019). https://doi.org/10.1007/978-3-030-17659-4_13

Post-Quantum Cryptography

The Nested Subset Differential Attack

A Practical Direct Attack Against LUOV Which Forges a Signature Within 210 Minutes

Jintai Ding[1], Joshua Deaton[2(✉)], Vishakha[2], and Bo-Yin Yang[3]

[1] Tsinghua University, Beijing, China
[2] University of Cincinnati, Cincinnati, OH, USA
{deatonju,sharmav4}@mail.uc.edu
[3] Academia Sinica, Taipei, Taiwan
byyangat@iis.sinica.edu.tw

Abstract. In 2017, Ward Beullens *et al.* submitted Lifted Unbalanced Oil and Vinegar [3], which is a modification to the Unbalanced Oil and Vinegar Scheme by Patarin. Previously, Ding *et al.* proposed the Subfield Differential Attack [22] which prompted a change of parameters by the authors of LUOV for the second round of the NIST post quantum standardization competition [4].

In this paper we propose a modification to the Subfield Differential Attack called the Nested Subset Differential Attack which fully breaks half of the parameter sets put forward. We also show by experimentation that this attack is practically possible to do in under 210 min for the level I security parameters and not just a theoretical attack. The Nested Subset Differential attack is a large improvement of the Subfield differential attack which can be used in real world circumstances. Moreover, we will only use what is called the "lifted" structure of LUOV, and our attack can be thought as a development of solving "lifted" quadratic systems.

1 Introduction

1.1 Signature Schemes, Post-quantum Cryptography and the NIST Post Quantum Standardization

Signature schemes allow one to digitally sign a document. These were first theoretically proposed by Whitfield Diffie and Martin Hellman using public key cryptography in [12]. The first and still most commonly used scheme is that of RSA made by Rivest, Shamir, and Adleman [35]. As technology and long distance communication become increasingly more a part of everyone's life, it becomes vital that one can verify who sent them a message and sign off on any message they intend to send. However, quantum computers utilizing Shor's algorithm threaten the security of the RSA scheme and many others now in use [37]. With the recent progress of building quantum computers, post-quantum cryptography able to resist quantum attacks has become a central research topic [1,7,8,30]. In 2016, NIST put out a call for proposals for post-quantum cryptosystems for

A. Canteaut and F.-X. Standaert (Eds.): EUROCRYPT 2021, LNCS 12696, pp. 329–347, 2021.
https://doi.org/10.1007/978-3-030-77870-5_12

standardization. These cryptosystems, though using classical computing in their operations, would resist quantum attacks [31]. We are currently in the third round of the "competition," with many different types of schemes being proposed. Multivariate cryptography is one family of post-quantum cryptosystems which is promising to resist quantum attacks [13,16].

1.2 Multivariate Cryptography

Public key encryption and signature schemes rely on a trapdoor function, one which is very difficult to invert except if one has special knowledge about the specific function. Multivariate cryptography bases its trapdoors on the difficulty of solving a random system of m polynomials in n variables over a finite field. For efficiency these polynomials are generally of degree 2. This has been proven to be NP hard [25], and thus is a good candidate for a public key cryptosystem. Moreover, working over these finite fields is often more efficient than older number-theory based methods like RSA. The difficulty lies in the fact that, as these systems must be invertible for the user and thus require a trapdoor, they are not truly random and must have a specific form which undermines the supposed NP hardness of solving them. Generally their special form is hidden by composition by invertible affine maps. Though there are interesting and practical multivariate encryption schemes [17,38,39], multivariate schemes are better known for simple and efficient signature scheme.

The first real breakthrough for multivariate cryptography was the MI or C^* cryptosystem proposed by Matsumoto and Imai in 1988 [29]. Their insight was to use the correspondence ψ between a n dimensional vector space k^n over a finite field k and a n degree extension K over k. They constructed their univariate trapdoor function $\mathscr{F} : K \to K$ over the large field which they were able to solve due to its special shape, and then composed it with two invertible affine maps $\mathscr{S}, \mathscr{T} : k^n \to k^n$ hiding its structure. Their public key is then $\mathscr{P} = \mathscr{S} \circ \psi \circ \mathscr{F} \circ \psi^{-1} \circ \mathscr{T}$. Though broken today, the MI cryptosystem is the inspiration for all "big field" schemes which have their trapdoor over a larger field. But the attack against MI is the inspiration for what are called oil and vinegar schemes, which LUOV is a extension of. The Linearization Equation Attack was developed by Patarin [32]. To be brief, Patarin discovered that plain-text/cipher-text pairs $(\mathbf{x}, \mathbf{y})$ will satisfy equations (called the linearization equations) of the form

$$\sum \alpha_{ij} x_i y_j + \sum \beta_i x_i + \sum \gamma_i y_i + \delta = 0$$

Collecting enough such pairs and plugging them into the above equation produces linear equations in the α_{ij}'s, β_i's, γ_i's, and δ which then can be solved for. Then for any cipher-text $\mathbf{y}$, its corresponding plain-text $\mathbf{x}$ will satisfy the linear equations found by plugging in $\mathbf{y}$ into the linearization equations. This will either solve for the $\mathbf{x}$ directly if enough linear equations were found or at least massively increase the efficiency of other direct attacks of solving for $\mathbf{x}$. So a quadratic problem becomes linear and thus easy to solve.

1.3 Oil and Vinegar Schemes

Inspired by the Linearization Equation Attack, Patarin introduced the Oil and Vinegar scheme [33]. The key idea is to reduce the problem of solving a quadratic system of equations into solving a linear system by separating the variables into two types, the vinegar variables which can be guessed for and the oil variables which will be solved for. Let $\mathbb{F}$ be a (generally small) finite field, m and v be two integers, and $n = m + v$. The central map $\mathscr{F} : \mathbb{F}^n \to \mathbb{F}^m$ is a quadratic map whose components $f_1, \ldots, f_m$ are in the form

$$f_k(X) = \sum_{i=1}^{v} \sum_{j=i}^{n} \alpha_{i,j,k} x_i x_j + \sum_{i=1}^{n} \beta_{i,k} x_i + \gamma_k$$

where each coefficient is in $\mathbb{F}$. Here the set of variables $V = \{x_1, \ldots, x_v\}$ are called the vinegar variables, and the set $O = \{x_{v+1}, \ldots, x_n\}$ are the oil variables. While the vinegar variables are allowed to be multiplied to any other variables, there are no oil times oil terms. Hence, if we guess for the vinegar variables we are left with a system of m linear equations in m variables. This has a high probability of being invertible (and one can always guess again for the vinegar variables if it is not). By composing with an affine transformation $\mathscr{T} : \mathbb{F}^n \to \mathbb{F}^n$ one gets the trapdoor function $\mathscr{P} = \mathscr{F} \circ \mathscr{T}$. This is indeed a trapdoor as by composing with $\mathscr{T}$, the oil and vinegar shape of the polynomials is lost and they appear just to be random. Thus for a oil and vinegar system the public key is $\mathscr{P}$ and the private key is $(\mathscr{F}, \mathscr{T})$. To sign a document Y, one first computes $\mathscr{F}^{-1}(Y) = Z$ by guessing the vinegar variables until $\mathscr{F}$ is an invertible linear system. Then one computes $\mathscr{T}^{-1}(Z) = W$. One verifies that W is a signature for Y by noting that $\mathscr{P}(W) = Y$.

Patarin originally proposed that the number of oil variables would equal the number of vinegar variables. Hence the original scheme is now called Balanced Oil and Vinegar. However, Balanced Oil Vinegar was broken by Kipnis and Shamir using the method of invariant subspaces [27]. This attack, however, is thwarted by making the number of vinegar variables sufficiently greater than the number of oil variables. Generally this is between 2 and 4 times as many vinegar variables to oil variables. Thus modern oil and vinegar schemes are called Unbalanced Oil and Vinegar (UOV) The other major attack using the structure of UOV is the Oil and Vinegar Reconciliation attack proposed by Ding *et al.* However, with appropriate parameters this attack can be avoided as well [20]. UOV remains unbroken to this day, and offers competitive signing and verifying times compared to other signatures schemes. Its main flaw is its rather large key size. Thus there have been many modifications to UOV designed to reduce the key size. One, due to Petzoldt, is to use a pseudo-random number generator to generate large portions of the key from a smaller seed which is easier to store [34]. Other schemes use the basic mathematical structure of UOV, but modify it in a way to increase efficiency. However, any changes can generate weakness for the system as can be seen from the first round contender of the NIST competition HIMQ-3 [36] which was broken by the Singularity Attack from

Ding *et al.* [21]. Two of the nine signature schemes left in the second round of the competition are also based on UOV. Rainbow, originally proposed in 2005, gains efficiency by forming multiple UOV layers where the oil variables in the previous layers are the vinegar variables in the latter layers [20]. The other scheme first proposed in [3] is Lifted Unbalanced Oil and Vinegar (LUOV) whose core idea is to reduce its key size by selecting all the coefficients of its polynomials from $\mathbb{F}_2 = \{0, 1\}$. However, LUOV signs its messages in some extension field $\mathbb{F}_{2^r}$. LUOV was attacked by Ding *et al.* using the Subfield Differential Attack (SDA) in [22]. SDA uses the lifted form of the polynomials to always work in a smaller field and thus increase efficiency of direct attacks (those which try to solve the quadratic system outright) against LUOV. The authors of LUOV have amended their parameters in order to prevent SDA. However, in this paper we will show that LUOV is still vulnerable to a modified form of SDA which we will call the Nested Subset Differential Attack (NSDA).

1.4 Lifted Unbalanced Oil and Vinegar (LUOV)

The LUOV, proposed in [3], is a UOV scheme with three main modifications. Let $\mathbb{F}_{2^r}$ be an extension of $\mathbb{F}_2$, m and v be positive integers, and $n = m + v$. The central maps $\mathscr{F} : \mathbb{F}_{2^r}^n \to \mathbb{F}_{2^r}^m$ is a system of quadratic maps $\mathscr{F}(X) = (F^{(1)}(X), \ldots, F^{(m)}(X))$ whose components are in oil and vinegar form

$$F^{(k)}(X) = \sum_{i=1}^{v} \sum_{j=i}^{n} \alpha_{i,j,k} x_i x_j + \sum_{i=1}^{n} \beta_{i,k} x_i + \gamma_k.$$

The first modification is that each $F^{(k)}$ is "lifted," meaning that the coefficients are taken from the prime field $\mathbb{F}_2$. Messages are still taken over the extension field, hence the name Lifted Unbalanced Oil Vinegar. The second modification is that the affine map $\mathscr{T}$ has the easier to store and computationally faster to sign form

$$\begin{bmatrix} \mathbf{1}_v & \mathbf{T} \\ \mathbf{0} & \mathbf{1}_o \end{bmatrix}.$$

This was first proposed by Czypek [11]. This does not affect security as for any given UOV private key $(\mathscr{F}', \mathscr{T}')$ there is highly likely an equivalent private key $(\mathscr{F}, \mathscr{T})$ where $\mathscr{T}$ is of the form above [41]. The third modification is that LUOV uses Petzdolt's method of generating the keys from a PRNG instead of storing them directly [34].

1.5 Our Contributions

In this paper we will first present the original SDA and then NSDA which is a modified version of the SDA attack which will defeat fully half of the new parameter sets used by LUOV. These parameters will fall well short of their targeted NIST security levels. We will also document an attack against one of these parameters sets which we were able to perform in under 210 min. Our attack does not rely on the oil and vinegar structure of LUOV, and can be seen as a way to solve "lifted" polynomial equations in general.

2 A Lemma on Random Maps

For both the Subfield Differential Attack and the Nested Subset Differential Attack we will require a short lemma on random maps which, under the assumption that quadratic systems of polynomials act like random maps, will allow us to say when it is possible to forge signatures.

Lemma 1. *Let A and B be two finite sets and $\mathcal{Q} : A \to B$ be a random map. For each $b \in B$, the probability that $\mathcal{Q}^{-1}(b)$ is non-empty is approximately $1 - e^{-|A|/|B|}$.*

Proof. As the output of each element of A is independent, it is elementary that the probability for there to be at least one $a \in A$ such that $\mathcal{Q}(a) = b$ is

$$1 - \Pr(\mathcal{Q}(\alpha) \neq b, \forall \alpha \in A) = 1 - \prod_{\alpha \in A} \Pr(\mathcal{Q}(\alpha) \neq b)$$

$$= 1 - \left(1 - \frac{1}{|B|}\right)^{|A|} = 1 - \left(1 - \frac{1}{|B|}\right)^{|B|\frac{|A|}{|B|}}.$$

Using $\lim_{n\to\infty} \left(1 - \frac{1}{n}\right)^n = e^{-1}$, we achieve the desired result.

3 The Subfield Differential Attack

3.1 Transforming the Public Key into Better Form

In this section we recall the Subfield Differential Attack proposed in [22]. Let $\mathcal{P} : \mathbb{F}_{2^r}^n \to \mathbb{F}_{2^r}^m$ be a LUOV public key. Let $X = (x_1, \ldots, x_n) \in \mathbb{F}_{2^r}^n$ be an indeterminate point. Then

$$\mathcal{P}(X) = \begin{cases} P^{(1)}(X) = \sum_{i=1}^{n}\sum_{j=i}^{n} \alpha_{i,j,1} x_i x_j + \sum_{i=1}^{n} \beta_{i,1} x_i + \gamma_1 \\ P^{(2)}(X) = \sum_{i=1}^{n}\sum_{j=i}^{n} \alpha_{i,j,2} x_i x_j + \sum_{i=1}^{n} \beta_{i,2} x_i + \gamma_2 \\ \vdots \\ P^{(m)}(X) = \sum_{i=1}^{n}\sum_{j=i}^{n} \alpha_{i,j,m} x_i x_j + \sum_{i=1}^{n} \beta_{i,m} x_i + \gamma_m \end{cases}$$

where for each i, j, k we have $\alpha_{i,j,k}, \beta_{i,k}, \gamma_k \in \mathbb{F}_2$. Due to this special structure we are able to transform $\mathcal{P}$ to be over a subfield of $\mathbb{F}_{2^r}$ which, depending on the parameters, will allow us to forge signatures.

First we recall for every positive integer d which divides r we may represent $\mathbb{F}_{2^r}$ as a quotient ring

$$\mathbb{F}_{2^r} \cong \mathbb{F}_{2^d}[t]/\langle g(t) \rangle$$

where $g(t)$ is a irreducible degree $s = r/d$ polynomial. For details see [28]. Let $\overline{X} = (\overline{x}_1, \ldots, \overline{x}_n) \in \mathbb{F}_{2^d}^n$ be an indeterminate point and $X' = (x'_1, \ldots, x'_n) \in \mathbb{F}_{2^r}^n$ be a random fixed point. So $\tilde{\mathscr{P}}(\overline{X}) := \mathscr{P}(\overline{X} + X') : \mathbb{F}_{2^d}^n \to \mathbb{F}_{2^r}^m$. Further this map is of a special form. Examining the kth component of $\tilde{\mathscr{P}}(\overline{X})$

$$\tilde{P}^{(k)}(\overline{X}) = \sum_{i=1}^{n}\sum_{j=i}^{n} \alpha_{i,j,k}(\overline{x}_i + x'_i)(\overline{x}_j + x'_j) + \sum_{i=1}^{n} \beta_{i,k}(\overline{x}_i + x'_i) + \gamma_k.$$

Expanding the above and separating the quadratic terms leads to

$$\begin{aligned}\tilde{P}^{(k)}(\overline{X}) =& \sum_{i=1}^{n}\sum_{j=i}^{n} \alpha_{i,j,k}(x'_i\overline{x}_i + x'_j\overline{x}_j + x'_i x'_j) \\ &+ \sum_{i=1}^{n} \beta_{i,k}(\overline{x}_i + x'_i) + \gamma_k + \sum_{i=1}^{n}\sum_{j=i}^{n} \alpha_{i,j,k}\overline{x}_i\overline{x}_j.\end{aligned}$$

We see that, due to $\alpha_{i,j,k} \in \mathbb{F}_2$, the coefficients of the quadratic terms $\overline{x}_i\overline{x}_j$ are all in the prime field. However, as the x'_i are random elements from $\mathbb{F}_{2^r}$, the coefficients of the linear $\overline{x}_i$ terms will contain all the powers of t up to $s - 1$. This means that, by grouping by the various powers of t, we may rewrite $\tilde{\mathscr{P}}(\overline{X})$ as

$$\tilde{\mathscr{P}}(\overline{X}) = \begin{cases} \tilde{P}^{(1)}(\overline{X}) = Q_1(\overline{X}) + \sum_{i=1}^{s-1} L_{i,1}(\overline{X})t^i \\ \tilde{P}^{(2)}(\overline{X}) = Q_2(\overline{X}) + \sum_{i=1}^{s-1} L_{i,2}(\overline{X})t^i \\ \quad\vdots \\ \tilde{P}^{(m)}(\overline{X}) = Q_m(\overline{X}) + \sum_{i=1}^{s-1} L_{i,m}(\overline{X})t^i \end{cases}$$

3.2 Forging a Signature

Now suppose we wanted to forge a signature for a message Y. First decompose Y into the sum of vectors

$$Y = Y_0 + Y_1 t + \cdots + Y_{s-1}t^{s-1}$$

where for each i, $Y_i = (y_{i,1}, \ldots, y_{i,m}) \in \mathbb{F}_{2^d}^m$.

First one finds the solution space S for the system of linear equations

$$A = \left\{L_{i,j}(\overline{X}) = y_{i,j} : 1 \leq i \leq s-1, 1 \leq j \leq m\right\}.$$

As A is essentially a random system of linear equations, it will have a high probability to be full rank $(s-1)m$ (or n if $(s-1)m \geq n$). So the dimension of S will be

$$\dim(S) = \max\{n - (s-1)m, 0\}.$$

Next, one tries to solve the system of m quadratic equations

$$B = \left\{Q_i(\overline{X}) = y_{0,i} : 1 \leq i \leq m, \overline{X} \in S\right\}.$$

If S is of large enough dimension, which depends on the choice of d, n, and m, The solution $\overline{X}$ to B yields $\tilde{P}(X) = Y$ which implies that $\mathscr{P}(\overline{X} + X') = Y$. Hence $\overline{X} + X'$ is the signature we seek. As the most costly step is solving the m quadratic equations of B over $\mathbb{F}_{2^d}$, we always choose d to be as small as possible for the S to likely have a solution according to Lemma 1 where in this case the domain is S and then range is $\mathbb{F}_{2^d}$. Generally, the domain will be much larger than the range for the attack and in this case we can assume that the probability for success on the first try is 1, or the domain is smaller and then the attack will fail as we almost never expect a solution to exist.

4 Nested Subset Differential Attack

4.1 The Change of Parameters for LUOV

In response to the Subfield Differential Attack, the authors of LUOV proposed the size of the extension r should be made prime so that the only subfield will be the prime field $\mathbb{F}_2$ [4]. They claim that given their new parameters, $\mathbb{F}_2^n$ will be far too small for a signature to exist for any given differential with any probability. The new parameters are in Table 1. We note that they are for different NIST security levels than before.

Table 1. The new parameter sets for LUOV

Name	Security level	(r, m, v, n)
LUOV-7-57-197	I	$(7, 57, 197, 254)$
LUOV-7-83-283	III	$(7, 83, 283, 366)$
LUOV-7-110-374	V	$(7, 110, 374, 484)$
LUOV-47-42-182	I	$(47, 42, 182, 224)$
LUOV-61-60-261	III	$(61, 60, 261, 321)$
LUOV-79-76-341	V	$(79, 76, 341, 417)$

Indeed, by Lemma 1 the Subfield Differential Attack will not work without modification, but it is the claim of this paper that such a modification, which we will call the Nested Subset Differential Attack (NSDA), is indeed possible for the three cases for which $r = 7$. In fact for the level I security level the complexity will be brought into the range where the attack is not theoretical but possible in practice in under 210 min as we will later show. This is due to the special construction of lifted polynomials given by the following lemma.

4.2 A Lemma on Lifted Polynomials

Lemma 2. *Let*

$$\tilde{f}(X) = \sum_{i=1}^{n}\sum_{j=i}^{n} \alpha_{i,j}x_i x_j + \sum_{i=1}^{n} \beta_i x_i + \gamma$$

be a lifted polynomial and $A_0, A_1, \cdots, A_{\ell-1} \in \mathbb{F}_2^n$ *with*

$$A_i = (a_{i,1}, \cdots, a_{i,n}).$$

Set $\mathbf{A} = A_0 + A_1 t + A_2 t^2 + \cdots + A_{\ell-1}t^{\ell-1}$. *We have that for* $\tilde{f}\left(\mathbf{A} + Xt^{\ell}\right)$ *all the quadratic terms are coefficients of* $t^{2\ell}$, *the linear terms are coefficients of* $t^{\ell}, t^{\ell+1}, \cdots, t^{2\ell-1}$, *and the coefficients of* t^h *depends only on* $\alpha_{i,j}, \beta_i$, *and* A_k *for* $k \leq h$ *and* X *for* $h \geq \ell$.

Proof. This follows from the following calculation and the fact that for each $i, j \in \{1, \ldots, n\}$, $\alpha_{i,j}, \beta_i \in \mathbb{F}_2$.

$$\begin{aligned}
f(\mathbf{A} + Xt^{\ell}) &= \sum_{i=1}^{n}\sum_{j=i}^{n} \alpha_{i,j} \left(\sum_{k=0}^{\ell-1} a_{k,i}t^k + x_i t^{\ell}\right)\left(\sum_{k=0}^{\ell-1} a_{k,j}t^k + x_j t^{\ell}\right) \\
&\quad + \sum_{i=1}^{n} \beta_i \left(\sum_{k=0}^{\ell-1} a_{k,i}t^k + x_i t^{\ell}\right) + \gamma \\
&= \sum_{i=1}^{n}\sum_{j=i}^{n} \alpha_{i,j} \left(x_i x_j t^{2\ell} + x_i \sum_{k=0}^{\ell-1} a_{k,j}t^{k+\ell} + x_j \sum_{k=0}^{\ell-1} a_{k,i}t^{k+\ell}\right) \\
&\quad + \sum_{i=1}^{n} \beta_i x_i t^{\ell} + \sum_{i=1}^{n}\sum_{j=i}^{n} \alpha_{i,j} \sum_{h=0}^{2\ell-2} \left(\sum_{\substack{0 \leq k,k' \leq \ell \\ k+k'=h}} a_{k,i}a_{k',j}t^h \right) \\
&\quad + \sum_{i=1}^{n} \beta_i \left(\sum_{k=0}^{\ell} a_{k,i}t^k\right) + \gamma.
\end{aligned}$$

4.3 s-Truncation

It will also be convenient later to define the concept of s-truncation for an element of the extension field. For $0 \leq s \leq r-1$, we define the s-truncation of a element

$$a = \sum_{i=0}^{r-1} a_i t^i \quad \text{to be} \quad \overline{a}^s = \sum_{i=0}^{s} a_i t^i.$$

Similarly for a polynomial

$$f(\overline{X}) = \sum_{i=1}^{n}\sum_{j=i}^{n} a_{i,j}\overline{x_i x_j} + \sum_{i=1}^{n} b_i \overline{x_i} + c$$

we define the s-truncation to be term by term

$$\overline{f}^s(\overline{X}) = \sum_{i=1}^{n}\sum_{j=i}^{n} \overline{a_{i,j}}^s \overline{x_i x_j} + \sum_{i=1}^{n} \overline{b_i}^s \overline{x_i} + \overline{c}^s.$$

Finally, for a system of polynomials

$$\mathscr{G}(\overline{X}) = \left(g_1(\overline{X}), g_2(\overline{X}), \ldots, g_m(\overline{X})\right)$$

we define the s-truncation to by truncating each polynomial individually

$$\overline{\mathscr{G}}^s(\overline{X}) = \left(\overline{g_1}^s(\overline{X}), \overline{g_2}^s(\overline{X}), \ldots, \overline{g_m}^s(\overline{X})\right).$$

4.4 The Attack

Let $P : \mathbb{F}_{2^r}^n \to \mathbb{F}_{2^r}^m$ be a LUOV public key with $r = 7$ and suppose we want to forge a signature for a message $Y \in \mathbb{F}_{2^r}^m$. We will denote by $\overline{X} = (\overline{x}_1, \ldots, \overline{x}_n)$ an indeterminate in $\mathbb{F}_2^n$ and decompose the message Y into the sum of vectors

$$Y = Y_0 + Y_1 t + \cdots + Y_{r-1} t^{r-1}$$

where for each i, $Y_i = (y_{i,1}, \ldots, y_{i,m}) \in \mathbb{F}_2^m$.

Consider the set of polynomials in $\mathbb{F}_2[t]/\langle g(t)\rangle$ which are truncated to the third power

$$E := \left\{\overline{a}^3 : a \in \mathbb{F}_{2^r}\right\}.$$

Table 2 calculates the probability that there will exist a signature for Y in E^n for the relevant parameters using Lemma 1. In this case the domain is E^n which has a size of 2^{4n} and the range is $\mathbb{F}_{2^7}^m$ which has a size of 2^{7m}. So in each case the probability of success is $1 - \exp(-2^{4n}/2^{7m})$.

Table 2. Probability that a signature exists in E^n

Name	Probability
LUOV-7-57-197	$1 - \exp(-2^{617})$
LUOV-7-83-283	$1 - \exp(-2^{883})$
LUOV-7-110-374	$1 - \exp(-2^{2366})$

We thus see that it is very likely that we need to only consider signatures from E^n when we attempt to forge. Similar to SDA's usage of the differential

X' to transform the direct attack into solving equations over a subfield, we do not need to look over all of E^n at once but can instead construct a signature piece by piece using differentials. However, instead of choosing the differentials randomly, we will instead solve for them in such a manner that will eventually construct a signature. For our attack to be efficient, we will want to always solve no more than m quadratic equations over $\mathbb{F}_2$ with at least as many variables as equations. This can be done in four steps using Lemma 2.

First we see that

$$\overline{\mathscr{P}}^0(\overline{X}) = \begin{cases} Q_{0,1}(\overline{X}) \\ Q_{0,2}(\overline{X}) \\ \vdots \\ Q_{0,m}(\overline{X}) \end{cases}$$

where each $Q_{0,i}(\overline{X})$ is a quadratic polynomial over $\mathbb{F}_2$. So we may solve the system of m equations in n variables $\overline{\mathscr{P}}^0(\overline{X}) = Y_0$ using a direct attack method like exhaustive search [6], a variant of XL (eXtended Linerization) [10], or a Gröbner Basis method like F4 [24]. We will forestall discussion of which algorithm to use until Sect. 4.6. Let us call the solution we found A_0.

For the second step, let us examine $\overline{\mathscr{P}}^1(A_0 + \overline{X}t)$. By the definition of s-truncation, this will be a system of polynomials of degree at most 1 in t. Following from Lemma 2, the coefficients of the t^1 terms will be linear in the variables $\overline{X}$. Furthermore, the coefficients of the t^0 terms will depend only on A_0. As $\overline{\mathscr{P}}^0(A_0) = Y_0$, we see that

$$\overline{\mathscr{P}}^1(A_0 + \overline{X}t) = \begin{cases} y_{0,1} + L_{1,1}(\overline{X})t \\ y_{0,2} + L_{1,2}(\overline{X})t \\ \vdots \\ y_{0,m} + L_{1,m}(\overline{X})t \end{cases}$$

where each $L_{1,i}(\overline{X})$ is a linear polynomial over $\mathbb{F}_2$ in the variables $\overline{X}$. Now find a solution A_1 to the system of linear equations

$$\{L_{1.i}(\overline{X}) = y_{1,i} : 1 \leq i \leq m\}.$$

Then we have $\overline{\mathscr{P}}^1(A_0 + A_1t) = Y_0 + Y_1t$.

For the third step, examine $\overline{\mathscr{P}}^2(A_0 + A_1t + \overline{X}t^2)$. Again the s-truncation will make this a system of polynomials of degree 2 in t. Lemma 2 states that the coefficients of the t^2 terms will be linear in the variables $\overline{X}$. The coefficients of the t^0 terms will depend only on A_0, and the coefficients of the t^1 will depend only on A_0 and A_1. But by construction of A_0 and A_1 we see that

$$\overline{\mathscr{P}}^2(A_0 + A_1 t + \overline{X}t^2) = \begin{cases} y_{0,1} + y_{1,1}t + L_{2,1}(\overline{X})t^2 \\ y_{0,2} + y_{1,2}t + L_{2,2}(\overline{X})t^2 \\ \vdots \\ y_{0,m} + y_{1,m}t + L_{2,m}(\overline{X})t^2 \end{cases}$$

where each $L_{2,i}(\overline{X})$ is a linear polynomial over $\mathbb{F}_2$ in the variables $\overline{X}$. Again find a solution A_2 to the system of linear equations

$$\left\{L_{2.i}(\overline{X}) = y_{2,i} : 1 \leq i \leq m\right\}.$$

Then we have $\overline{\mathscr{P}}^2(A_0 + A_1 t + A_2 t^2) = Y_0 + Y_1 t + Y_2 t^2$.

As a final step, we drop the need for s-truncation and look at $\mathscr{P}(A_0 + A_1 t + A_2 t^2 + \overline{X}t^3)$. We note that this will be a system of polynomials of degree 6 in t, the highest degree for polynomials in $\mathbb{F}_2[t]/\langle g(t)\rangle$ as $r = 7$. Further, by Lemma 2, only the coefficients of the t^6 terms will be quadratic in $\overline{X}$. The coefficients of the t^3, t^4 and t^5 terms will be linear in $\overline{X}$. Finally, the coefficients of the t^0, t^1, t^2 terms depend only on A_0, A_0 and A_1, and A_0 A_1 and A_2 respectively. Let $\mathbf{A} = A_0 + A_1 t + A_2 t^2$. By construction of A_0, A_1, and A_2 we see that

$$\mathscr{P}(\mathbf{A} + \overline{X}t^3) = \begin{cases} y_{0,1} + y_{1,1}t + y_{2,1}t^2 + L_{3,1}(\overline{X})t^3 + L_{4,1}(\overline{X})t^4 \\ \qquad + L_{5,1}(\overline{X})t^5 + Q_{6,1}(\overline{X})t^6 \\ y_{0,2} + y_{1,2}t + y_{2,2}t^2 + L_{3,2}(\overline{X})t^3 + L_{4,2}(\overline{X})t^4 \\ \qquad + L_{5,2}(\overline{X})t^5 + Q_{6,2}(\overline{X})t^6 \\ \vdots \\ y_{0,m} + y_{1,m}t + y_{2,m}t^2 + L_{3,m}(\overline{X})t^3 + L_{4,m}(\overline{X})t^4 \\ \qquad + L_{5,m}(\overline{X})t^5 + Q_{6,m}(\overline{X})t^6 \end{cases}$$

Now we proceed largely in the same manner as the last step in the SDA attack. Find the solution space S for the system of linear equations

$$A = \left\{L_{i,j}(\overline{X}) = y_{i,j} : 3 \leq i \leq 5, 1 \leq j \leq m\right\}.$$

As A will most likely be full rank $3m$, the dimension of S will have high probability of being $n - 3m$. Thus, the system of m quadratic equations

$$B = \left\{Q_{6,j}(\overline{X}) = y_{6,j} : 1 \leq j \leq m, \overline{X} \in S\right\}$$

has a high probability of having a solution given the parameter sets of LUOV which we record in Table 3. Again, we used Lemma 1 with the domain being S which has size 2^{n-3m}, and the range being $\mathbb{F}_2^m$ which has size 2^m. So the probability of success is $1 - \exp(-2^{n-4m})$.

Table 3. Probability of success for NSDA

Name	Probability
LUOV-7-57-197	$1-\exp(-2^{26})$
LUOV-7-83-283	$1-\exp(-2^{34})$
LUOV-7-110-374	$1-\exp(-2^{344})$

Find a solution A_3 to B. Then we see that

$$\mathscr{P}(A_0 + A_1t + A_2t^2 + A_3t^3) = Y$$

and thus $\sigma = A_0 + A_1t + A_2t^2 + A_3t^3$ is a forged signature for Y.

Note that in each case we assumed that it was possible to find the solutions A_i for the various systems. The last quadratic system is when this is most unlikely, and still we see that the odds are overwhelmingly in our favor for the parameter sets we attacked for the solutions to exist assuming that polynomial systems act as random maps. Thus, we may ignore the potential that a solution does not exist in our attack for any step, and even if that were the case one can always go back a previous step for a different solution and try again.

For the different parameter sets this is no longer so. They use a larger value for r, which means that the number of linear equations to solve along side the final quadratic system also increases to the point where we no longer expect a final solution to exist. This bring into question when LUOV is safe from SDA and NSDA, which depends on the relationship between n, m, r, and any factors d of r, but is still competitive. This is beyond the scope of this paper, and further work will need to be done to see the exact value of the lifting modification.

4.5 Hiding the Signature

It might be argued that signatures that come from E^n are in a very special shape and thus can be rejected as obviously forged. However, it is possible to hide the shape of the signatures generated from the NSDA attack. Due to the special shape of the lifted polynomials, it is possible to know about preimages of a more generic form which are connected to the preimages we can find. Let $\mathscr{P}$ be a LUOV public key so that

$$\mathscr{P}(X) = \begin{cases} P^{(1)}(X) = \sum_{i=1}^{n}\sum_{j=i}^{n} \alpha_{i,j,1}x_ix_j + \sum_{i=1}^{n} \beta_{i,1}x_i + \gamma_1 \\ P^{(2)}(X) = \sum_{i=1}^{n}\sum_{j=i}^{n} \alpha_{i,j,2}x_ix_j + \sum_{i=1}^{n} \beta_{i,2}x_i + \gamma_2 \\ \\ P^{(m)}(X) = \sum_{i=1}^{n}\sum_{j=i}^{n} \alpha_{i,j,m}x_ix_j + \sum_{i=1}^{n} \beta_{i,m}x_i + \gamma_m \end{cases}$$

Suppose we wanted to forge a signature for a message $Y = (y_1, \ldots, y_m) \in \mathbb{F}_{2^r}^m$. As we are in a finite field of characteristic 2, we may take square roots of any element. For some natural number N, define a vector $Z = (z_1, \ldots, z_m) = Y^{1/2^N}$ by which we mean that, for each i, $z_i = y_i^{1/2^N}$ the 2^Nth root of y_i. Now let $X = (x_1, \ldots, x_n) \in E^n$ be a signature for Z so that $\mathscr{P}(X) = Z$. Define $X^{2^N} = (x_i^{2^N}, \ldots, x_n^{2^N})$. Let us recall the freshman's dream.

Theorem 1 (Freshman's Dream). *If $\mathbb{F}$ is a field of characteristic p then for any natural number N and elements $x, y \in \mathbb{F}$ we have $(x+y)^{p^N} = x^{p^N} + x^{p^N}$.*

Then examining the kth component of $\mathscr{P}(X^{2^N})$ we see that due to the freshman's dream and the fact that the coefficients of $\mathscr{P}$ are in $\mathbb{F}_2$

$$\begin{aligned} P^{(k)}(X^{2^N}) &= \sum_{i=1}^{n}\sum_{j=i}^{n} \alpha_{i,j,k} x_i^{2^N} x_j^{2^N} + \sum_{i=1}^{n} \beta_{i,k} x_i^{2^N} + \gamma_k \\ &= \left(\sum_{i=1}^{n}\sum_{j=i}^{n} \alpha_{i,j,k} x_i x_j + \sum_{i=1}^{n} \beta_{i,k} x_i + \gamma_1 \right)^{2^N} \\ &= z_k^{2^N} = y_k. \end{aligned}$$

As the elements of X are degree three polynomials in $\mathbb{F}_2[t]/\langle g(t) \rangle$, X^{2^N}'s elements will appear to be generic degree six polynomials. Now, the signature can still be seen by checking the 2^Nth roots for each N less than r, but this procedure still masks the forged signature from against lazy implementations of the verification process.

4.6 Complexity

The complexity of our attack is determined by solving the two quadratic systems of m equations over $\mathbb{F}_2$. The overhead from solving the linear systems we may ignore as the size of the linear systems is always not much larger than the quadratic systems, and linear systems are much more efficient to solve.

Let us take a system $\mathscr{P} = \left(P^{(1)}(X), \ldots, P^{(m)}(X)\right)$ of m quadratic equations in n variables over $\mathbb{F}_2$ and attempt to find a solution. The best method in our case given the small field size and the limited number of variables we will have is exhaustive search. In our practical experiment on LUOV-7-57-197, we used a variant of the "forcepq_fpga" algorithm [5,6], so this algorithm is how we will estimate the complexity of solving the system. We will give a brief sketch of the main idea here.

We will denote the solution set of the first k equations as

$$Z_\ell = \{A \in \mathbb{F}_2 \mid P^{(i)}(A) = 0, 1 \leq i \leq \ell\}.$$

For some well chosen ℓ, the algorithm first utilizes Grey-code and partial derivatives to find Z_ℓ by solving the first ℓ equations individually. We begin by ordering the elements of $\mathbb{F}_2$ according to a Grey-code order $A_1, A_2, \ldots, A_{2^n}$. This means that for an element $A_s \in \mathbb{F}_2^n$, A_{s+1} will only have one component different than A_s. The authors of [5] noticed that, as we are working under $\mathbb{F}_2$ and if A_{s+1} differs from A_s only at the ith component

$$P^{(k)}(A_{s+1}) = P^{(k)}(A_s) + \frac{\partial P^{(k)}}{\partial x_i}(A_{s+1}).$$

As the partial derivative is one degree smaller, it is more efficient to evaluate. It was also found that this trick can be used recursively for evaluating the first partial derivatives utilizing the second partial derivatives.

Notice, though, that Z_ℓ is no longer in Gray-code order as it is essentially a random subset of $\mathbb{F}_2^n$. Thus, it is not possible to fully utilize the Gray-code method to compute $Z_{\ell+1}$ from Z_ℓ. One would have to add multiple evaluations of different partial derivatives, one for each change in component, when selecting the next element of Z_ℓ. This was only found to be twice as efficient as simply evaluating the original equations in view of finding Z_m at the very end.

It was estimated in [5] that the number of bit operations for finding all the solutions would be $\log_2(n)2^{n+2}$ for a determined system ($n = m$) with an optimal value of $\ell = 1 + \log_2(n)$. We will use this estimate on determined systems as for the cases we consider we will have more variables than equations. As we only need one solution we can randomly assign values until the system is either determined or only slightly underdetermined ($n > m$) if we want a solution on the first attempt. In our experiment we guessed for all but $m+2$ of the variables to assure a solution first try, so we will do likewise in our estimate.

We will note though that if n is multiple times the size of m, we can first use the method of Thomae and Wolf [40], which is an improvement of the work of the Kipnis, Patarin, and Goubin [26], to reduce the number of variables *and* equations. While we will not go into the details of the method in this paper, the core idea is to make the random system act as if that is was at least partly an oil and vinegar system. By this we mean we attempt to find some linear transformation of the variables $\mathcal{S}$ such that $\mathcal{P} \circ \mathcal{S}$ has a set of vinegar variables V and a set of oil variables O. The result is part of the resulting system is linear in the oil variables after fixing the vinegar variables. As we are in characteristic 2, square terms act linearly. Thus, we search for $\mathcal{S}$ to set each $O \times O$ coefficient $\alpha_{i,j,k} = 0$ for $i \neq j$. Thomae and Wolf showed that this process can be done solving a relatively small system of linear equations. The statement of their result is as follows.

Theorem 2 (Thomae and Wolf). *By a linear change of variables, the complexity of solving an under-determined quadratic system of m equations and $n = \omega m$ variables can be reduced to solving a determined quadratic system of $m - \lfloor \omega \rfloor + 1$ equations. Furthermore, provided $\lfloor \omega \rfloor | m$ the complexity can be further reduced to the complexity of solving a determined quadratic system of $m - \lfloor \omega \rfloor$ equations [40].*

In Table 4 we compute the complexity for solving the final quadratic system. This will be the most complex part of the attack as we had to first solve a linear system. We will have approximately $n - 3m$ variables and m equations. We note that as $(n - 3m)/m < 2$ in each case, Thomae and Wolf's method will not apply. We will guess all but $m + 2$ variables and estimate the complexity as $\log_2(m + 2)2^{m+4}$.

Table 4. Complexity in terms of number of bit operations

Name	$\log_2$ NSDA's complexity (NIST Requirement)
LUOV-7-57-197	61 (143)
LUOV-7-83-283	89 (207)
LUOV-7-110-374	116 (272)

As the classical $\log_2$ classical gate operations for NIST security level I is 143, III is 207, and V is 272 [31], we see that LUOV falls short in every category for these parameters. Moreover, the actual complexity for NSDA is possible in practice as we show with experimental results in Sect. 4.7.

Before we continue, we will mention that if the subfield over which we solved had been larger, or if the number of variables to guess for had been too great, then exhaustive search would not be the optimal choice for the solver for the quadratic systems. Generally, after applying the method of Thomae and Wolf, either XL [10] with the Block Wiedemann Algorithm [9] or the F4 algorithm by Faugère [24] is the preferred choice for such systems using a hybrid method [2] (meaning guessing a certain number of variables before applying the mentioned algorithms). The complexity of both algorithms relies on solving/reducing very large, sparse Macaulay matrices. Roughly, the highest degree found in XL is denoted by D_0 (called the operating degree), and the highest degree in F4 is D_{reg} (called the degree of regularity [14,15,19,23]). Yeh *et al.* [42] have shown that for the resulting overdetermined systems after using the hybrid method, $0 \leq D_0 - D_{reg} \leq 1$ and often $D_0 = D_{reg}$. So the matrices are roughly the same size, but XL is sparser and is thus the preferred method to use. Please see [42] for full details.

4.7 Experimental Results

We have performed practical experiments on the LUOV parameter set LUOV-7-57-197.

For the hardware, we used a field-programmable gates array cluster from Sci-engines, a "Rivyera S6-LX150" with 64 Xilinx Spartan 6 LX150 FPGAs chips. The LX150 were so named because each contains nearly "150,000 gate equivalent units". They were driven on 8 PCI express cards in a chassis containing a Supermicro motherboard, an Intel Xeon(R) CPU (E3-1230 V2). When new in 2012, the machine cost 55,000 EUR. Although not directly comparable, a machine with current FPGAs costing the same 55,000 EUR today will probably have at least 2× as much computing power and cost less in electricity.

We use a variant of the "forcepq_fpga" algorithm from the paper [6], using the input format of the Fukuoka MQ Challenge. We processed the early parts of our LUOV attack using the computer algebra system Magma and output the resulting system in this format, which is basically binary quadratic systems with zero-one coefficients lined up in graded reverse lexicographic order.

The "forcemp_fpga" implementation allows us to test 2^{10} input vectors per cycle (at 200 MHz) per FPGA chip. In general this lets us solve a 48 × 48 MQ system in a maximum of slightly less than 23 min using one single chip, or find a solution to $n \times m$ quadratic equations, where $n \geq m$, in $2^{m-48} \times 23$ min. We could accelerate this somewhat if we can implement a variation of the Joux-Vitse algorithm.

For a 55-equation system, using all 64 FPGAs, the maximum is 46 min. In general it is a little shorter. The expectation is half of that or 23 min. For a 57-equation system, it is 4 times that, hence about 3 h, expectation is about half of that or 92 min. When we solved the 59-variable, 57-equation system in practice, the run ended after 105 min. This, like all our runs in this experiment, happened to be slightly unlucky.

As there are two quadratic systems to solve, we can forge a signature in under 210 min.

5 Inapplicability to Non-lifted Schemes

Before we conclude, lets discuss why NSDA or any similar attack does not work on UOV [33], Rainbow [18], or any other multivariate scheme which does not use the lifting modification. In these schemes, though some coefficients in the central map are forced to be 0 (like the oil × oil coefficients in UOV and Rainbow) to allow efficient pre-image finding, most of the coefficients in the central maps are taken randomly from a finite field $\mathbb{F}_q$. Thus, in the public key $\mathscr{P} : \mathbb{F}_q^n \to \mathbb{F}_q^m$ all of the coefficients are seemingly random elements of $\mathbb{F}_q$. This makes any differential we add seemingly mixed randomly.

To be explicit, Let us assume that $\mathbb{F}_q$ contains a subfield $\mathbb{F}_{q'}$ so that $\mathbb{F}_q \cong \mathbb{F}_{q'}[t]/\langle g(t)\rangle$ where $\deg(g) = s$. We will assume that $\mathbb{F}_{q'}$ is to small to find pre-images in. Let $\overline{X} = (\overline{x}_1, \ldots, \overline{x}_n)$ be an indeterminate point in $\mathbb{F}_{q'}^n$, $f(t) \in \mathbb{F}_{q'}[t]$ (say $f(t) = t$ like in NSDA), and $A = (a_1, \ldots, a_n)$ be a fixed point (whether in a special form like in NSDA or not). Let $\tilde{\mathscr{P}}(\overline{X}) := \mathscr{P}(A + \overline{X} f(t))$. Similar to the SDA section we find that in the kth component of $\tilde{\mathscr{P}}$

$$\begin{aligned}\tilde{P}^{(k)}(\overline{X}) &= \sum_{i=1}^{n}\sum_{j=i}^{n} \alpha_{i,j,k}(a_j\overline{x}_i f(t) + a_i\overline{x}_j f(t) + a_i a_j) \\ &\quad + \sum_{i=1}^{n} \beta_{i,k}(\overline{x}_i f(t) + a_i) + \gamma_k + \sum_{i=1}^{n}\sum_{j=i}^{n} \alpha_{i,j,k}\overline{x}_i\overline{x}_j f(t)^2.\end{aligned}$$

Note that there are no restrictions on the coefficients, $\alpha_{i,j,k}, \beta_{i,k}$ and γ_k as they are random elements from $\mathbb{F}_{q^r}$. The quadratic terms' coefficients will contain powers of t from t^0 to t^{s-1}. Hence, we are trading one random quadratic system $\mathscr{P}$ which $\mathbb{F}_{q'}^n$ is too small to find pre-images in for another equally random quadratic system $\tilde{\mathscr{P}}$ which $\mathbb{F}_{q'}^n$ is still too small. So, NSDA is inapplicable to non-lifted systems.

6 Conclusion

We have proposed a modified version of the Subfield Differential Attack called Nested Subset Differential Attack which fully breaks half the parameters set forward by the round 2 version of Lifted Unbalanced Oil and Vinegar. We reduced attacking these parameters sets to the problem of solving quadratic equations over the prime field $\mathbb{F}_2$. This makes our attack effective enough to be performed practically. As our attack did not use the Unbalanced Oil and Vinegar Structure of LUOV, it can be seen as a method of solving lifted quadratic systems in general. We feel that more research into solving these types of quadratic systems using the NSDA attack is needed. We also performed experimental attacks on actual LUOV parameters and were able to forge a signature in under 210 min.

References

1. Bernstein, D.J.: Introduction to post-quantum cryptography. In: Bernstein, D.J., Buchmann, J., Dahmen, E. (eds.) Post-Quantum Cryptography. Springer, Heidelberg (2009). https://doi.org/10.1007/978-3-540-88702-7_1
2. Bettale, L., Faugère, J.-C., Perret, L.: Hybrid approach for solving multivariate systems over finite fields. J. Math. Cryptol. **3**(3), 177–197 (2009)
3. Beullens, W., Preneel, B.: Field lifting for smaller UOV public keys. In: Patra, A., Smart, N.P. (eds.) INDOCRYPT 2017. LNCS, vol. 10698, pp. 227–246. Springer, Cham (2017). https://doi.org/10.1007/978-3-319-71667-1_12
4. Beullens, W., Preneel, B., Szepieniec, A., Vercauteren, F.: Luov signature scheme proposal for NIST PQC project (round 2 version) (2019)
5. Bouillaguet, C., et al.: Fast exhaustive search for polynomial systems in $\mathbb{F}_2$. In: Mangard, S., Standaert, F.-X. (eds.) CHES 2010. LNCS, vol. 6225, pp. 203–218. Springer, Heidelberg (2010). https://doi.org/10.1007/978-3-642-15031-9_14

6. Bouillaguet, C., Cheng, C.-M., Chou, T., Niederhagen, R., Yang, B.-Y.: Fast exhaustive search for quadratic systems in $\mathbb{F}_2$ on FPGAs. In: Lange, T., Lauter, K., Lisoněk, P. (eds.) SAC 2013. LNCS, vol. 8282, pp. 205–222. Springer, Heidelberg (2014). https://doi.org/10.1007/978-3-662-43414-7_11
7. Buchmann, J., Ding, J. (eds.): Post-Quantum Cryptography. Springer, Heidelberg (2008). https://doi.org/10.1007/978-3-540-88702-7
8. Campagna, M., et al.: ETSI whitepaper: Quantum safe cryptography and security. Technical report, ETSI (2015)
9. Coppersmith, D.: Solving homogeneous linear equations over GF(2) via block Wiedemann algorithm. Math. Comput. **62**(205), 333–350 (1994)
10. Courtois, N., Klimov, A., Patarin, J., Shamir, A.: Efficient algorithms for solving overdefined systems of multivariate polynomial equations. In: Preneel, B. (ed.) EUROCRYPT 2000. LNCS, vol. 1807, pp. 392–407. Springer, Heidelberg (2000). https://doi.org/10.1007/3-540-45539-6_27
11. Czypek, P.: Implementing multivariate quadratic public key signature schemes on embedded devices. Ph.D. thesis, Citeseer (2012)
12. Diffie, W., Hellman, M.: New directions in cryptography. IEEE Trans. Inf. Theory **22**(6), 644–654 (1976)
13. Ding, J., Gower, J.E., Schmidt, D.: Multivariate Public Key Cryptosystems. Advances in Information Security, vol. 25. Springer, Cham (2006). https://doi.org/10.1007/978-0-387-36946-4
14. Ding, J., Hodges, T.J.: Inverting HFE systems is quasi-polynomial for all fields. In: Rogaway, P. (ed.) CRYPTO 2011. LNCS, vol. 6841, pp. 724–742. Springer, Heidelberg (2011). https://doi.org/10.1007/978-3-642-22792-9_41
15. Ding, J., Kleinjung, T.: Degree of regularity for HFE-. IACR Cryptology ePrint Archive 2011, 570 (2011)
16. Ding, J., Petzoldt, A.: Current state of multivariate cryptography. IEEE Secur. Privacy **15**(4), 28–36 (2017)
17. Ding, J., Petzoldt, A., Wang, L.: The cubic simple matrix encryption scheme. In: Mosca, M. (ed.) PQCrypto 2014. LNCS, vol. 8772, pp. 76–87. Springer, Cham (2014). https://doi.org/10.1007/978-3-319-11659-4_5
18. Ding, J., Schmidt, D.: Rainbow, a new multivariable polynomial signature scheme. In: Ioannidis, J., Keromytis, A., Yung, M. (eds.) ACNS 2005. LNCS, vol. 3531, pp. 164–175. Springer, Heidelberg (2005). https://doi.org/10.1007/11496137_12
19. Ding, J., Yang, B.-Y.: Degree of regularity for HFEv and HFEv-. In: Gaborit, P. (ed.) PQCrypto 2013. LNCS, vol. 7932, pp. 52–66. Springer, Heidelberg (2013). https://doi.org/10.1007/978-3-642-38616-9_4
20. Ding, J., Yang, B.-Y., Chen, C.-H.O., Chen, M.-S., Cheng, C.-M.: New differential-algebraic attacks and reparametrization of rainbow. In: Bellovin, S.M., Gennaro, R., Keromytis, A., Yung, M. (eds.) ACNS 2008. LNCS, vol. 5037, pp. 242–257. Springer, Heidelberg (2008). https://doi.org/10.1007/978-3-540-68914-0_15
21. Ding, J., Zhang, Z., Deaton, J.: The singularity attack to the multivariate signature scheme HiMQ-3. Advances in Mathematics of Communications (2019)
22. Ding, J., Zhang, Z., Deaton, J., Schmidt, K., Vishakha, F.: New attacks on lifted unbalanced oil vinegar. In: The 2nd NIST PQC Standardization Conference (2019)
23. Dubois, V., Gama, N.: The degree of regularity of HFE systems. In: Abe, M. (ed.) ASIACRYPT 2010. LNCS, vol. 6477, pp. 557–576. Springer, Heidelberg (2010). https://doi.org/10.1007/978-3-642-17373-8_32
24. Faugère, J.-C.: A new efficient algorithm for computing Gröbner bases (F4). J. Pure Appl. Algebra **139**(1–3), 61–88 (1999)

25. Johnson, D.S., Garey, M.R.: Computers and Intractability: A Guide to the Theory of NP-Completeness. WH Freeman, New York (1979)
26. Kipnis, A., Patarin, J., Goubin, L.: Unbalanced oil and vinegar signature schemes. In: Stern, J. (ed.) EUROCRYPT 1999. LNCS, vol. 1592, pp. 206–222. Springer, Heidelberg (1999). https://doi.org/10.1007/3-540-48910-X_15
27. Kipnis, A., Shamir, A.: Cryptanalysis of the oil and vinegar signature scheme. In: Krawczyk, H. (ed.) CRYPTO 1998. LNCS, vol. 1462, pp. 257–266. Springer, Heidelberg (1998). https://doi.org/10.1007/BFb0055733
28. Lidl, R., Niederreiter, H.: Finite Fields, vol. 20. Cambridge University Press, Cambridge (1997)
29. Matsumoto, T., Imai, H.: Public quadratic polynomial-tuples for efficient signature-verification and message-encryption. In: Barstow, D., et al. (eds.) EUROCRYPT 1988. LNCS, vol. 330, pp. 419–453. Springer, Heidelberg (1988). https://doi.org/10.1007/3-540-45961-8_39
30. National Institute of Standards and Technology: Workshop on cybersecurity in a post-quantum world. Technical report, National Institute of Standards and Technology (2015)
31. National Institute of Standards and Technology: Submission requirements and evaluation criteria for the post-quantum cryptography standardization process. Technical report, National Institute of Standards and Technology (2017)
32. Patarin, J.: Cryptanalysis of the Matsumoto and Imai public key scheme of Eurocrypt'88. In: Coppersmith, D. (ed.) CRYPTO 1995. LNCS, vol. 963, pp. 248–261. Springer, Heidelberg (1995). https://doi.org/10.1007/3-540-44750-4_20
33. Patarin, J.: The oil and vinegar algorithm for signatures. In: Dagstuhl Workshop on Cryptography (1997)
34. Petzoldt, A., Bulygin, S.: Linear recurring sequences for the UOV key generation revisited. In: Kwon, T., Lee, M.-K., Kwon, D. (eds.) ICISC 2012. LNCS, vol. 7839, pp. 441–455. Springer, Heidelberg (2013). https://doi.org/10.1007/978-3-642-37682-5_31
35. Rivest, R.L., Shamir, A., Adleman, L.: A method for obtaining digital signatures and public-key cryptosystems. Commun. ACM **21**(2), 120–126 (1978)
36. Shim, K.-A., Park, C.-M., Kim, T.: HiMQ-3: A high speed signature scheme based on multivariate quadratic equations, NIST submission (2017). Internet: https://csrc.nist.gov/Projects/Post-Quantum-Cryptography/Round-1-Submissions
37. Shor, P.W.: Polynomial-time algorithms for prime factorization and discrete logarithms on a quantum computer. SIAM Rev. **41**(2), 303–332 (1999)
38. Tao, C., Diene, A., Tang, S., Ding, J.: Simple matrix scheme for encryption. In: Gaborit, P. (ed.) PQCrypto 2013. LNCS, vol. 7932, pp. 231–242. Springer, Heidelberg (2013). https://doi.org/10.1007/978-3-642-38616-9_16
39. Tao, C., Xiang, H., Petzoldt, A., Ding, J.: Simple matrix - a multivariate public key cryptosystem (MPKC) for encryption. Finite Fields Appl. **35**, 352–368 (2015)
40. Thomae, E., Wolf, C.: Solving underdetermined systems of multivariate quadratic equations revisited. In: Fischlin, M., Buchmann, J., Manulis, M. (eds.) PKC 2012. LNCS, vol. 7293, pp. 156–171. Springer, Heidelberg (2012). https://doi.org/10.1007/978-3-642-30057-8_10
41. Wolf, C., Preneel, B.: Equivalent keys in multivariate quadratic public key systems. J. Math. Cryptol. **4**(4), 375–415 (2011)
42. Yeh, J.Y.-C., Cheng, C.-M., Yang, B.-Y.: Operating degrees for XL vs. F_4/F_5 for generic $\mathcal{MQ}$ with number of equations linear in that of variables. In: Fischlin, M., Katzenbeisser, S. (eds.) Number Theory and Cryptography. LNCS, vol. 8260, pp. 19–33. Springer, Heidelberg (2013). https://doi.org/10.1007/978-3-642-42001-6_3

Improved Cryptanalysis of UOV and Rainbow

Ward Beullens(✉)

imec-COSIC, KU Leuven, Leuven, Belgium
ward.beullens@esat.kuleuven.be

Abstract. The contributions of this paper are twofold. First, we simplify the description of the Unbalanced Oil and Vinegar scheme (UOV) and its Rainbow variant, which makes it easier to understand the scheme and the existing attacks. We hope that this will make UOV and Rainbow more approachable for cryptanalysts. Second, we give two new attacks against the UOV and Rainbow signature schemes; the intersection attack that applies to both UOV and Rainbow and the rectangular MinRank attack that applies only to Rainbow. Our attacks are more powerful than existing attacks. In particular, we estimate that compared to previously known attacks, our new attacks reduce the cost of a key recovery by a factor of 2^{17}, 2^{53}, and 2^{73} for the parameter sets submitted to the second round of the NIST PQC standardization project targeting the security levels I, III, and V respectively. For the third round parameters, the cost is reduced by a factor of 2^{20}, 2^{40}, and 2^{55} respectively. This means all these parameter sets fall short of the security requirements set out by NIST.

1 Introduction

The Oil and Vinegar scheme and its Rainbow variant are two of the oldest and most studied signature schemes in multivariate cryptography. The Oil and Vinegar scheme was proposed by Patarin in 1997 [17]. Soon thereafter, Kipnis and Shamir discovered that the original choice of parameters was weak and could be broken in polynomial time [15]. However, it is possible to pick parameters differently, such that the scheme resists the Kipnis-Shamir attack. This variant is called the Unbalanced Oil and Vinegar scheme (UOV), and has withstood all cryptanalysis since 1999 [14].

The rainbow signature scheme can be seen as multiple layers of UOV stacked on top of each other. This was proposed by Ding and Schmidt in 2005 [9]. The design philosophy is that by iterating the UOV construction, the Kipnis-Shamir attack becomes less powerful, which enables the use of more efficient parameters. However, the additional complexity opened up more attack strategies, such as the

This work was supported by CyberSecurity Research Flanders with reference number VR20192203, and by the Research Council KU Leuven grant C14/18/067 on Cryptanalysis of post-quantum cryptography. Ward Beullens is funded by FWO SB fellowship 1S95620N.

A. Canteaut and F.-X. Standaert (Eds.): EUROCRYPT 2021, LNCS 12696, pp. 348–373, 2021.
https://doi.org/10.1007/978-3-030-77870-5_13

MinRank attack, the Billet-Gilbert attack [4], and the Rainbow Band Separation attack [10]. Even though our understanding of the complexity of these attacks has been improving over the last decade, there have been no new attacks since 2008.

Multivariate cryptography is believed to resist attacks from adversaries with access to large scale quantum computers, which is why there has been renewed interest in this field of research during recent years. Seven out of the nineteen signature schemes that were submitted to the NIST post-quantum cryptography standardization project were multivariate signature schemes. From those seven schemes, four were allowed to proceed to the second round [3,5,8,18], and only the Rainbow submission was selected as a finalist. The UOV scheme was not submitted to the NIST PQC project.

Contributions. As a first contribution, we simplify the description of the UOV and Rainbow schemes. Traditionally, the public key is a multivariate quadratic map $\mathcal{P}$, and the secret key is a factorization $\mathcal{P} = \mathcal{S} \circ \mathcal{F} \circ \mathcal{T}$ where $\mathcal{S}$ and $\mathcal{T}$ are invertible linear maps, and $\mathcal{F}$ is a so-called central map. Our description avoids the use of a central map and only talks about properties of $\mathcal{P}$ instead. This new perspective makes it easier to understand the scheme and the existing attacks.

Secondly, we introduce two new key-recovery attacks: the intersection attack and the rectangular MinRank attack. The intersection attack relies on the idea behind the Kipnis-Shamir attack and applies to both the UOV scheme and the Rainbow scheme. The rectangular MinRank attack reduces key recovery to an instance of the MinRank problem. In this problem the task is, given a number of matrices, to find a linear combination of these matrices with exceptionally low rank. When Ding and Schmidt designed the Rainbow scheme in 2005 they were already aware that Rainbow was susceptible to MinRank attacks. However, our new attack shows that there was another instance of the MinRank problem lurking in the Rainbow public keys that went undiscovered until now. We call our attack the rectangular MinRank attack because unlike previous attacks, the matrices in the new MinRank instance are rectangular instead of square.

Roadmap. After giving some necessary background in Sect. 2, we introduce our simplified description of the Oil and Vinegar scheme and the existing attacks in Sect. 3. In Sect. 4 we introduce our intersection attack on UOV. In Sect. 5 we give a simplified description of the Rainbow scheme, and we review the existing attacks. The following Sects. 6 and 7 introduce the intersection attack for Rainbow and the rectangular MinRank attack respectively. We conclude in Sect. 8 with an overview of our attack complexities and new parameter sets for UOV and Rainbow.

2 Preliminaries

2.1 Notation

For a vector space $V \subset K^n$ over a field K, we define its orthogonal complement $V^\perp$ as the space of vectors that are orthogonal to all the vectors in V, i.e. $V^\perp = \{\mathbf{w} | \langle \mathbf{w}, \mathbf{v} \rangle = 0, \forall \mathbf{v} \in V\}$. For a linear subspace $W \subset V$, we denote by V/W the quotient space of V by W. This is the vector space whose elements are the cosets of W in V:

$$V/W = \{\overline{\mathbf{x}} := \mathbf{x} + W \,|\, \mathbf{x} \in V\} \,.$$

Let $\mathbf{x} = x_1, \cdots, x_{n_x}$ and $\mathbf{y} = y_1, \cdots, y_{n_y}$ be two groups of variables in $\mathbb{F}_q$. We denote by $\mathcal{M}(a,b)$ the number of monomial functions of degree a in the $\mathbf{x}$ variables and degree b in the $\mathbf{y}$ variables. We denote by $\overline{\mathcal{M}}(a,b)$ the number of monomial functions of degree at most a in $\mathbf{x}$ and at most b in $\mathbf{y}$. If a and b are lower than q we have

$$\mathcal{M}(a,b) = \binom{a+n_x-1}{a}\binom{b+n_y-1}{b} \text{ and } \overline{\mathcal{M}}(a,b) = \binom{a+n_x}{a}\binom{b+n_y}{b}$$

2.2 Multivariate Quadratic Maps

The central object in Multivariate Quadratic cryptography is the multivariate quadratic map. A multivariate quadratic map $\mathcal{P}$ with m components and n variables is a sequence $p_1(\mathbf{x}), \cdots, p_m(\mathbf{x})$ of m multivariate quadratic polynomials in n variables $\mathbf{x} = (x_1, \cdots, x_n)$, with coefficients in a finite field $\mathbb{F}_q$.

To evaluate the map $\mathcal{P}$ at a value $\mathbf{a} \in \mathbb{F}_q^n$, we simply evaluate each of its component polynomials in $\mathbf{a}$ to get a vector $\mathbf{b} = (b_1 = p_1(\mathbf{a}), \cdots, b_m = p_m(\mathbf{a}))$ of m output elements. We denote this by $\mathcal{P}(\mathbf{a}) = \mathbf{b}$.

MQ problem. The main source of computational hardness for multivariate cryptosystems is the Multivariate Quadratic (MQ) problem. Given a multivariate quadratic map $\mathcal{P} : \mathbb{F}_q^n \to \mathbb{F}_q^m$, and given a target $\mathbf{t} \in \mathbb{F}_q^m$, the MQ problem asks to find a solution $\mathbf{s}$ such that $\mathcal{P}(\mathbf{s}) = \mathbf{t}$. This problem is NP-hard, and it is believed to be exponentially hard on average, even for quantum adversaries. Currently, the best algorithms to solve instances of this problem (for cryptographically relevant parameters) are algorithms such as F_4/F_5 or XL that use a Gröbner-basis-like approach [6,11].

Polar forms. For a multivariate quadratic polynomial $p(\mathbf{x})$, we can define its *polar form*

$$p'(\mathbf{x},\mathbf{y}) := p(\mathbf{x}+\mathbf{y}) - p(\mathbf{x}) - p(\mathbf{y}) + p(0) \,.$$

Similarly, for a multivariate quadratic map $\mathcal{P}(\mathbf{x}) = p_1(\mathbf{x}), \cdots, p_m(\mathbf{x})$, we define its polar form as $\mathcal{P}'(\mathbf{x},\mathbf{y}) = p_1'(\mathbf{x},\mathbf{y}), \cdots, p_m'(\mathbf{x},\mathbf{y})$. This polar form will allow us to simplify the description of the UOV and Rainbow schemes, and will play a major role in the attacks on UOV and Rainbow. The multivariate quadratic maps of interest in this paper are homogenous, so we will often omit the $\mathcal{P}(0)$ term.

Theorem 1. *Given a multivariate quadratic map $\mathcal{P}(\mathbf{x}) : \mathbb{F}_q^n \to \mathbb{F}_q^m$, its polar form $\mathcal{P}'(\mathbf{x},\mathbf{y}) : \mathbb{F}_q^n \times \mathbb{F}_q^n \to \mathbb{F}_q^m$ is a symmetric and bilinear map.*

Proof. We can write $p(\mathbf{x}) = \mathbf{x}^\top Q\mathbf{x} + \mathbf{v}\cdot\mathbf{x} + c$, where Q is an upper triangular matrix that contains the coefficients of the quadratic terms of p, where $\mathbf{v}$ contains the coefficients of the linear terms of $p(\mathbf{x})$, and where c is the constant term of $p(\mathbf{x})$. Then we have

$$\begin{aligned} p'(\mathbf{x},\mathbf{y}) &:= p(\mathbf{x}+\mathbf{y}) - p(\mathbf{x}) - p(\mathbf{y}) + p(0) \\ &= (\mathbf{x}+\mathbf{y})^\top Q(\mathbf{x}+\mathbf{y}) - \mathbf{y}^\top Q\mathbf{y} - \mathbf{x}^\top Q\mathbf{x} + \mathbf{v}\cdot(\mathbf{x}+\mathbf{y}) - \mathbf{v}\cdot\mathbf{x} - \mathbf{v}\cdot\mathbf{y} \\ &= \mathbf{x}^\top Q\mathbf{y} + \mathbf{y}^\top Q\mathbf{x} \\ &= \mathbf{x}^\top (Q+Q^\top)\mathbf{y}\,. \end{aligned}$$

□

2.3 Solving MinRank with Support Minors Modeling

The MinRank problem asks, given k matrices $L_1,\cdots,L_k$ with n rows and m columns and a target rank r, to find coefficients $y_i \in \mathbb{F}_q$ for i from 1 to k, not all zero, such that the linear combination $\sum_{i=1}^k y_i L_i$ has rank at most r.

Recently, Bardet *et al.* introduced the Support Minors Modeling algorithm for solving this problem [1]. Let $\mathbf{y} \in \mathbb{F}_q^k$ be a solution, and let C be a matrix whose rows form a basis for the rowspan of $L_\mathbf{y} = \sum_{i=1}^k y_i L_i$. For each subset $S \subset \{1,\cdots,m\}$ of size $|S| = r$, let c_S be the determinant of the r-by-r submatrix of C whose columns are the columns of C with index in S.

The Support Minors Modeling approach considers for each $j \in \{1,\cdots,n\}$ the matrix

$$C_j = \begin{pmatrix} r_j \\ C \end{pmatrix},$$

where r_j is the j-th row of $L_\mathbf{y}$. Then the rank of C_j is at most r, which implies that all its $(r+1)$-by-$(r+1)$ minors vanish. Using cofactor expansion on the first row, each minor gives a bilinear equation in the y_i variables and the c_S variables. The Support Minors Modeling algorithm then uses the XL algorithm to find a solution to this system of $n\binom{m}{r+1}$ bilinear equations.

Analysis. The attack constructs the Macaulay matrix M_b at bi-degree $(b,1)$, a large sparse matrix, whose columns correspond to the monomials of degree b in the y_i variables, and of degree 1 in the c_S variables. So at degree $(b,1)$, the matrix has $\mathcal{M}(b,1)$ columns. The rows of the matrix contain the degree $(b,1)$ polynomials of the form $\mu(\mathbf{y})\cdot f(\mathbf{y},\mathbf{c})$, where $\mu(\mathbf{y})$ is a monomial of degree $b-1$, and $f(\mathbf{y}, \mathbf{c})$ is one of the bilinear equations of the Support Minors Modeling system. The goal of the attack is then to use the Wiedemann algorithm to find a non-trivial solution to the linear system $M_b\mathbf{x} = 0$, so that $\mathbf{x}$ reveals a solution to the MinRank problem. This approach works if the rank of M_b is $\mathcal{M}(b, 1) - 1$, so that there is only a one-dimensional solution space that corresponds to the unique (up to a scalar) solution of the MinRank problem.

Bardet *et al.* calculate that whenever $b < r + 2$, the rank of the Macaulay matrix is

$$R_{k,n,m,r}(b) = \sum_{i=1}^{b} (-1)^{i+1} \binom{m}{r+i} \binom{n+i-1}{i} \binom{k+b-i-1}{b-i}, \tag{1}$$

unless $R_{k,n,m,r}(b') > \mathcal{M}(b', 1) - 1$ for some $b' \leq b$, in which case the rank is equal to $\mathcal{M}(b, 1) - 1$. This allows to calculate for which b the attack will succeed.

If b_{min} is the smallest integer for which the attack will succeed, then solving the XL system with the Wiedemann algorithm requires

$$3\mathcal{M}(b_{min}, 1)^2(r + 1)k$$

field multiplications. Bardet *et al.* found that it is often advantageous to ignore a number of columns of the L_i matrices and only consider the first m' columns of the matrices, for some optimal value of m' in the range $[r + 1, m]$. For more details on the Support Minors Modeling algorithm, we refer to [1].

3 The UOV Signature Scheme

The Oil and Vinegar signature scheme, introduced in 1997 by Patarin [17], is based on an elegant MQ-based trapdoor function. The trapdoor function is a multivariate quadratic map $\mathcal{P} : \mathbb{F}_q^n \to \mathbb{F}_q^m$ for which it is assumed that finding preimages (i.e. solving the MQ problem) is hard. However, if one knows some extra information (called the trapdoor), then it is easy to find preimages for any arbitrary output. Originally, Patarin proposed to use the system with $n = 2m$. This parameter choice was cryptanalysed by Kipnis and Shamir [15], which is why current proposals use $n > 2m$. This is known as the Unbalanced Oil and Vinegar (UOV) signature scheme. The conservative recommendation is to use $n = 3m$ or even $n = 4m$, but more aggressive and (more efficient) parameter sets have been proposed that use $n \approx 2.35m$ [7].

The UOV signature scheme is created from the UOV trapdoor function with the Full Domain Hash approach: The public key is the trapdoor function $\mathcal{P} : \mathbb{F}_q^n \to \mathbb{F}_q^m$, the secret key contains the trapdoor information, and a signature on a message M is simply an input $\mathbf{s}$ such that $\mathcal{P}(\mathbf{s}) = \mathcal{H}(M||\mathsf{salt})$, where $\mathcal{H}$ is a cryptographic hash function that outputs elements in the range of $\mathcal{P}$ and where salt is a fixed-length bit string chosen uniformly at random for every signature. Therefore, to understand the UOV signature scheme, we only need to understand how the UOV trapdoor function works.

3.1 UOV Trapdoor Function

The UOV trapdoor function is a multivariate quadratic map $\mathcal{P} : \mathbb{F}_q^n \to \mathbb{F}_q^m$ that vanishes on a secret linear subspace $O \subset \mathbb{F}_q^n$ of dimension $\dim(O) = m$, i.e.

$$\mathcal{P}(\mathbf{o}) = 0 \quad \text{for all } \mathbf{o} \in O\,.$$

The trapdoor information is nothing more than a description of O. To generate the trapdoor function one first picks the subspace O uniformly at random and then one picks $\mathcal{P}$ uniformly at random from the set of multivariate quadratic maps with m components in n variables that vanish on O. Note that on top of the q^m "artificial" zeros in the subspace O, we expect roughly q^{n-m} "natural" zeros that do not lie in O.

Given a target $\mathbf{t} \in \mathbb{F}_q^m$, how do we use this trapdoor to find $\mathbf{x} \in \mathbb{F}_q^n$ such that $\mathcal{P}(\mathbf{x}) = \mathbf{t}$? To do this, one picks a vector $\mathbf{v} \in \mathbb{F}_q^n$ and solves the system $\mathcal{P}(\mathbf{v} + \mathbf{o}) = \mathbf{t}$ for a vector $\mathbf{o} \in O$. This can simply be done by solving a linear system for $\mathbf{o}$, because

$$\mathcal{P}(\mathbf{v}+\mathbf{o}) = \underbrace{\mathcal{P}(\mathbf{v})}_{\text{fixed by choice of } \mathbf{v}} + \underbrace{\mathcal{P}(\mathbf{o})}_{=0} + \underbrace{\mathcal{P}'(\mathbf{v},\mathbf{o})}_{\text{linear function of } \mathbf{o}} = \mathbf{t}\,.$$

With probability roughly $1 - 1/q$ over the choice of $\mathbf{v}$ the linear map $\mathcal{P}'(\mathbf{v}, \cdot)$ will be non-singular, in which case the linear system $\mathcal{P}(\mathbf{v} + \mathbf{o}) = \mathbf{t}$ has a unique solution. If this is not the case, one can simply pick a new value for $\mathbf{v}$ and try again.

3.2 Traditional Description of UOV

Traditionally, the UOV signature is described as follows: The secret key is a pair $(\mathcal{F}, \mathcal{T})$, where $\mathcal{T} \in GL(n, q)$ is a random invertible linear map, and $\mathcal{F} : \mathbb{F}_q^n \to \mathbb{F}_q^m$ is the so-called central map, whose components $f_1, \cdots, f_m$ are chosen uniformly at random of the form

$$f_i(\mathbf{x}) = \sum_{i=1}^{n} \sum_{j=i}^{n-m} \alpha_{i,j} x_i x_j\,.$$

Note that the second sum only runs from i to $n - m$. So all the terms have at least one variable in $x_1, \cdots, x_{n-m}$.

The public key that corresponds to $(\mathcal{F}, \mathcal{T})$ is the multivariate quadratic map $\mathcal{P} = \mathcal{F} \circ \mathcal{T}$. To sign a message M, the strategy is to first solve for $\mathbf{s}' \in \mathbb{F}_q^n$ such that $\mathcal{F}(\mathbf{s}') = \mathcal{H}(M||\mathsf{salt})$, and then the final signature is $\mathbf{s} = \mathcal{T}^{-1}(\mathbf{s}')$, such that $\mathcal{P}(\mathbf{s}) = \mathcal{F}(\mathbf{s}') = \mathcal{H}(M||\mathsf{salt})$.

The description in Sect. 3.1 is just a slightly different way of thinking about the same scheme. In particular, the distribution of public keys for this signature scheme is the same: The central map $\mathcal{F}$ is chosen uniformly from the set of maps that vanish on the m-dimensional space of vectors O' that consists of all the vectors whose first $n-m$ entries are zero, i.e. $O' = \{\mathbf{v} \mid v_i = 0 \text{ for all } i \leq n-m\}$. After composing with $\mathcal{T}$, we get a public key $\mathcal{P} = \mathcal{F} \circ \mathcal{T}$ that vanishes on some secret linear subspace $O = \mathcal{T}^{-1}(O')$.

3.3 Attacks on UOV

A straightforward approach to attack the UOV signature scheme is to completely ignore the existence of the oil subspace and directly try to solve the system $\mathcal{P}(\mathbf{x}) = \mathcal{H}(M||\mathsf{salt})$ to produce a signature for the message M. This can be done with a Gröbner basis-like approach such as XL or F_4/F_5 [6,11]. This is called a direct attack.

More interestingly, the attacker can first try to find the oil space O. After O is found, the attacker can sign any message as if he was a legitimate signer. Two attacks in the literature take this approach.

Reconciliation attack. The reconciliation attack was developed by Ding *et al.* as a stepping stone towards the Rainbow Band Separation (RBS) attack on Rainbow [10]. As an attack on UOV, the reconciliation attack is not very useful, since it never outperforms a direct attack on UOV for properly chosen parameters. Nevertheless, we describe the attack here, since it can also be seen as a precursor to our intersection attack of Sect. 4.

The attack tries to find a vector $\mathbf{o} \in O$ by solving the system $\mathcal{P}(\mathbf{o}) = 0$. We know that $\dim(O) = m$, so if we impose m affine constraints on the entries of $\mathbf{o}$, we still expect a unique solution $\mathbf{o} \in O$.

If $n - m \leq m$, then we expect $\mathcal{P}(\mathbf{o}) = 0$ to have a unique solution after fixing m entries of $\mathbf{o}$. This is a system of m equations in fewer than m variables, so solving this system is more efficient than a direct attack.

If $n - m > m$ then $\mathcal{P}(\mathbf{o}) = 0$ will have a lot of solutions, only one of which corresponds to an $\mathbf{o} \in O$. Enumerating all the solutions is too costly, and the attack will not outperform a direct attack. We can try to solve the following system to find multiple vectors $\mathbf{o}_1, \cdots, \mathbf{o}_k$ in O simultaneously:

$$\begin{cases} \mathcal{P}(\mathbf{o}_i) = 0 & \forall i \in \{1, \cdots, k\} \\ \mathcal{P}'(\mathbf{o}_i, \mathbf{o}_j) = 0 & \forall i < j \in \{1, \cdots, k\} \end{cases} .$$

However, this increases the number of variables that appear in the system, and therefore the attack will usually not outperform a direct attack.

Once a first vector in O is found, finding subsequent vectors is much easier. If $\mathbf{o}$ is the first vector that we found, then a second vector $\mathbf{o}' \in O$ will satisfy

$$\begin{cases} \mathcal{P}(\mathbf{o}') = 0 \\ \mathcal{P}'(\mathbf{o}, \mathbf{o}') = 0 \end{cases},$$

which means we get m linear equations on $\mathbf{o}'$ for free. Therefore, the complexity of the attack is dominated by the complexity of finding the first vector in O.

Kipnis-Shamir attack. Historically, the first attack on the OV signature scheme was given by Kipnis and Shamir [15]. The basic version of this attack works when $n = 2m$, which was the case for the parameter sets initially proposed by Patarin.

Attack if $\mathbf{n = 2m}$. The attack looks at the m components of $\mathcal{P}'(\mathbf{x}, \mathbf{y})$. Each component $p_i'(\mathbf{x}, \mathbf{y}) = p_i(\mathbf{x} + \mathbf{y}) - p_i(\mathbf{x}) - p_i(\mathbf{y})$, defines a matrix M_i such that $p_i'(\mathbf{x}, \mathbf{y}) = \mathbf{x}^\top M_i \mathbf{y}$. Kipnis and Shamir observed the following useful property of M_i.

Lemma 2. *For each $i \in \{1, \cdots, m\}$, we have that $M_i O \subset O^\perp$. That is, each M_i sends O into its own orthogonal complement $O^\perp$.*

Proof. For any $\mathbf{o}_1, \mathbf{o}_2 \in O$ we need to prove that $\langle \mathbf{o}_2, M_i \mathbf{o}_1 \rangle = 0$. This follows from the assumption that p_i vanishes on O:

$$\langle \mathbf{o}_2, M_i \mathbf{o}_1 \rangle = \mathbf{o}_2^\top M_i \mathbf{o}_1 = p_i'(\mathbf{o}_1, \mathbf{o}_2) = p_i(\mathbf{o}_1 + \mathbf{o}_2) - p_i(\mathbf{o}_1) - p_i(\mathbf{o}_2) = 0\,. \quad \square$$

If $n = 2m$, then $\dim(O^\perp) = n - m = m$, so if M_i is nonsingular (which happens with high probability[1]), then Lemma 2 turns into an equality $M_i O = O^\perp$. This means that for any pair of invertible M_i, M_j, we have that $M_j^{-1} M_i O = O$, i.e. that O is an invariant subspace of $M_j^{-1} M_i$. It turns out that finding a common invariant subspace of a large number of linear maps can be done in polynomial time, so this gives an efficient algorithm for finding O. For more details we refer to [15].

Remark 3. Note that, as a map from $\mathbb{F}_q^n$ to itself, M_i implicitly depends on a choice of basis for $\mathbb{F}_q^n$. A more natural approach would be to define M_i as a map from $\mathbb{F}_q^n$ to its dual $\mathbb{F}_q^{n\vee}$ given by $\mathbf{x} \mapsto p_i'(\mathbf{x}, \cdot)$. Lemma 2 would then say

[1] In fields of characteristic 2 and in case n is odd, the M_i are never invertible, because M_i is skew-symmetric and with zeros on the diagonal and therefore has even rank. (Recall that $M_i = Q_i + Q_i^\perp$ as in the proof of Theorem 1.) To avoid this case we can always set one of the variables to zero. This has the effect of reducing n by one (which gets us back to the case where n is even), and it also reduces the dimension of O by one, which makes the attack slightly less powerful. Since this trick is always possible, we will assume that n is even in the remainder of the paper.

$M_iO \subset O^0$, where $O^0 \subset \mathbb{F}_q^{n\vee}$ is the annihilator of O. We chose not to take this approach to avoid the dual vector space and annihilators, which some readers might not be familiar with.

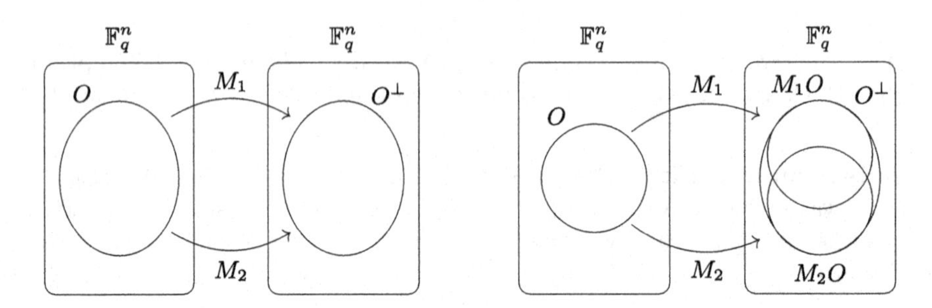

Fig. 1. Behavior of O under M_1 and M_2, in case $n = 2m$ (on the left) and $2m < n < 3m$ (on the right).

Attack if n > 2m. If $n > 2m$, then it is still the case that M_i sends O into $O^\perp$, but because $\dim(O^\perp) = n - m > m = \dim(O)$ the equality $M_iO = M_jO$ may no longer hold. Therefore, $M_i^{-1}M_j$ is no longer guaranteed to have O as an invariant subspace and the basic attack fails. However, even though in general $M_iO \neq M_jO$, they still have an unusually large intersection (see Fig. 1): M_iO and M_jO are both subspaces of $O^\perp$, so their intersection has dimension at least $\dim(M_iO) + \dim(M_jO) - \dim(O^\perp) = 3m - n$. Kipnis *et al.* [14] realized that this means that vectors in O are more likely to be eigenvectors of $M_j^{-1}M_i$.

Heuristically, for $\mathbf{x} \in O$, the probability that it gets mapped by M_i to some point in the intersection $M_iO \cap M_jO$ is approximately

$$\frac{|M_iO \cap M_jO|}{|M_iO|} = q^{2m-n}.$$

If this happens, then the probability that M_j^{-1} maps $M_i\mathbf{x}$ back to a multiple of $\mathbf{x}$ is expected to be $(q-1)/|O| \approx q^{1-m}$. Therefore, we can estimate that the probability that a vector in O is an eigenvector of $M_j^{-1}M_i$ is approximately q^{1+m-n}, and the expected number of eigenvectors in O is therefore q^{1+2m-n}.

The same analysis holds when you replace M_i and M_j by arbitrary invertible linear combinations of the M_i. The attacker can repeatedly compute the eigenvectors of $F^{-1}G$, where F and G are random invertible linear combinations of the M_i. After q^{n-2m} attempts he can expect to find a vector in O (he can verify whether a given eigenvector $\mathbf{x}$ is in O by checking that $\mathcal{P}(\mathbf{x}) = 0$). The complexity of the attack is $\tilde{O}(q^{n-2m})$, so the attack runs in polynomial time if $n = 2m$, but quickly becomes infeasible for unbalanced instances of the OV construction[2]. For more details on the attack, we refer to [14].

4 Intersection Attack on UOV

In this section, we introduce a new attack that uses the ideas behind the Kipnis-Shamir attack, in combination with a system-solving approach such as in the reconciliation attack. We first describe a basic version of the attack that works as long as $n < 3m$. Then we also give a more efficient version of the attack that works if $n < 2.5m$.

4.1 Attack if $n < 3m$

Like in the Kipnis-Shamir attack, we consider for each $i \in \{1, \cdots, m\}$ the matrix M_i such that $p_i'(\mathbf{x}, \mathbf{y}) = \mathbf{x}^\top M_i \mathbf{y}$, and we choose two indices $i, j \in \{1, \cdots, m\}$ such that M_i and M_j are invertible matrices. The goal of our attack is to find a vector $\mathbf{x}$ in the intersection $M_i O \cap M_j O$. Recall from Sect. 3.3 that this intersection has dimension at least $3m - n$, so non-trivial solutions exist if $n < 3m$.

If $\mathbf{x}$ is in the intersection $M_i O \cap M_j O$, then both $M_i^{-1}\mathbf{x}$ and $M_j^{-1}\mathbf{x}$ are in O. Therefore, $\mathbf{x}$ is a solution to the following system of quadratic equations

$$\begin{cases} \mathcal{P}(M_i^{-1}\mathbf{x}) = 0 \\ \mathcal{P}(M_j^{-1}\mathbf{x}) = 0 \\ \mathcal{P}'(M_i^{-1}\mathbf{x}, M_j^{-1}\mathbf{x}) = 0 \end{cases} . \tag{2}$$

Since there is a $3m - n$ dimensional subspace of solutions, we can impose $3m - n$ affine constraints on $\mathbf{x}$, so that we expect a unique solution. The attack is then to simply use the XL algorithm to find a solution to this system of $3m$ quadratic equations in $n - (3m - n) = 2n - 3m$ variables.

Once $\mathbf{x}$ is found, we know 2 vectors $M_i^{-1}\mathbf{x}$ and $M_j^{-1}\mathbf{x}$ in O, and the remaining vectors in O can be found more easily with the approach described in Sect. 3.3.

4.2 Attack When $n < 2.5m$

If n is small enough compared to m we can make the attack more efficient by solving for an $\mathbf{x}$ in the intersection of more than 2 subspaces $M_i O$ at the same time. Suppose $n < \frac{2k-1}{k-1}m$ for an integer $k \geq 1$, and let $L_1, \cdots, L_k$ be k randomly chosen invertible linear combinations of the M_i, then the intersection $L_1 O \cap \cdots \cap L_k O$ will have dimension at least $km - (k-1)(n-m) > 0$, which

[2] The $\tilde{O}$-notation ignores polynomial factors.

means there is a nonzero $\mathbf{x}$ such that $L_i^{-1}\mathbf{x} \in O$ for all i from 1 to k. We can then solve the following system of equations:

$$\begin{cases} \mathcal{P}(L_i^{-1}\mathbf{x}) = 0\,, & \forall i \in \{1, \cdots, k\} \\ \mathcal{P}'(L_i^{-1}\mathbf{x}, L_j^{-1}\mathbf{x}) = 0\,, & \forall i < j \in \{1, \cdots, k\} \end{cases} \tag{3}$$

We expect to find a unique solution after imposing $km - (k-1)(n-m)$ linear conditions on $\mathbf{x}$ to random values, so the complexity of the attack is dominated by the complexity of solving a system of $\binom{k+1}{2}m$ quadratic equations in $nk - (2k-1)m$ variables.

Remark 4. Note that in the case $n = 2m$ the requirement $n < \frac{2k-1}{k-1}m$ is satisfied for every $k > 1$. If we pick $k \approx \sqrt{m}$, then we have more than $\binom{m+1}{2}$ equations in m variables, which means we can linearize the system and solve it with Gaussian elimination in polynomial time. This is not surprising, because Kipnis and Shamir have already shown that UOV can be broken in polynomial time if $n = 2m$.

4.3 Complexity Analysis of the Attack

We noticed that the equations of system (3) are not linearly independent: even though there are $\binom{k+1}{2}m$ equations they only span a subspace of dimension $\binom{k+1}{2}m - 2\binom{k}{2}$. This is because if we have $L_i = \sum_{l=1}^{m} \alpha_{il} M_i$, for all i from 1 to k, then for all $1 \leq i < j \leq k$ we have

$$\begin{aligned} \sum_{l=1}^{m} \alpha_{il} \mathcal{P}_l'(L_i^{-1}\mathbf{x}, L_j^{-1}\mathbf{x}) &= \sum_{l=1}^{m} \alpha_{il} (L_i^{-1}\mathbf{x})^{\perp} M_l L_j^{-1}\mathbf{x}) \\ &= (L_i^{-1}\mathbf{x})^{\perp} L_i L_j^{-1}\mathbf{x}) \\ &= \mathbf{x}^{\perp} L_j^{-1}\mathbf{x} = \sum_{l=1}^{m} \alpha_{jl} \mathcal{P}_l(L_j^{-1}\mathbf{x}) \end{aligned}$$

Similarly, we have

$$\sum_{l=1}^{m} \alpha_{jl} \mathcal{P}_l'(L_i^{-1}\mathbf{x}, L_j^{-1}\mathbf{x}) = \mathbf{x}^{\perp} L_i^{-1}\mathbf{x} = \sum_{l=1}^{m} \alpha_{il} \mathcal{P}_l(L_i^{-1}\mathbf{x})\,,$$

so for each choice of $0 \leq i < j \leq k$ there are two linear dependencies between the equations of system (3). This explains why they only span a subspace of dimension $\binom{k+1}{2}m - 2\binom{k}{2}$.

Our experiments show that, after removing the $2\binom{k}{2}$ redundant equations, the systems (2) and (3) behave like random systems of $M = \binom{k+1}{2}m - 2\binom{k}{2}$ quadratic equations in $N = nk - (2k-1)m$ variables. For some small UOV systems, we computed the ranks of the Macaulay matrices at various degrees, and we found that they exactly match the ranks of generic systems (see Table 1).

That is, at degree d, the rank is equal to the coefficient of t^d in the power series expansion of

$$\frac{1-(1-t^2)^M}{(1-t)^{N+1}},$$

assuming that this coefficient does not exceed the number of columns of the Macaulay matrix.

We can use the standard methodology for estimating the complexity of system solving with an XL Wiedemann approach as

$$3\binom{N+d_{reg}}{d_{reg}}^2\binom{N+2}{2}$$

field multiplications, where the degree of regularity d_{reg} is the first d such that the coefficient of t^d in

$$\frac{(1-t^2)^M}{(1-t)^{N+1}}$$

is non-positive [2,8].

Table 1. The rank and the number of columns of the Macaulay matrices for the system of equations of the intersection attack. The rank at degree d always matches the coefficients of t^d the corresponding generating function, except if the coefficient is larger or equal to the number of columns. In this case (marked by boldface in the table) the rank equals the number of columns minus 1, and the XL system can be solved at that degree d.

Parameters				Macaulay matrix at degree d				Generating function
n	m	k		$d=2$	$d=3$	$d=4$	$d=5$	
8	4	2	Rank	10	**34**			$\frac{1-(1-t^2)^{10}}{(1-t)^5}$
			#Columns	15	35			
10	4	2	Rank	10	90	405	1245	$\frac{1-(1-t^2)^{10}}{(1-t)^9}$
			#Columns	45	165	495	1287	
12	5	2	Rank	13	130	673	**2001**	$\frac{1-(1-t^2)^{13}}{(1-t)^{10}}$
			#Columns	55	220	715	2002	
12	5	3	Rank	24	288	**1364**		$\frac{1-(1-t^2)^{24}}{(1-t)^{12}}$
			#Columns	78	364	1365		
14	6	2	Rank	16	176	936	**3002**	$\frac{1-(1-t^2)^{16}}{(1-t)^{11}}$
			#Columns	66	286	1001	3003	
14	6	3	Rank	30	390	**1819**		$\frac{1-(1-t^2)^{30}}{(1-t)^{13}}$
			#Columns	91	455	1820		

Concrete costs. To demonstrate that the new attack is more efficient than existing attacks, we apply it to the UOV parameters proposed by Czypek *et al.* [7]. They proposed to use $q = 256, n = 103, m = 44$, targeting 128 bits of security. More precisely, they estimate that the direct attack requires 2^{130} field multiplications and that the Kipnis-Shamir attack requires 2^{136} multiplications.

Their parameter choice satisfies $n < 2.5m$, so we can use the more efficient version of the attack with $k = 3$ (i.e. where we solve for $\mathbf{x}$ in the intersection of 3 subspaces of the form $M_i O$). This results in a system of $M = \binom{3+1}{2}m - 2\binom{3}{2} = 258$ equations in $N = nk - (2k-1)m = 89$ variables. The complexity of finding a solution is 2^{95} multiplications ($d_{reg} = 9$), which is lower than the claimed security level of 2^{128} multiplications.

In general, it seems that the new attack only outperforms a direct forgery attack, if $n < 2.5$. The usual recommendation in the literature is to use $n = 3m$ or even $n = 4m$, so these parameters are not affected by the new attack. In contrast, the example above shows that more aggressive parameters (which are tempting because they are much more efficient and previously no attacks were known) are no longer secure.

5 The Rainbow Signature Scheme

The Rainbow signature scheme is a variant of the UOV signature scheme proposed in 2004 by Ding and Schmidt [9]. The Rainbow trapdoor function is a multivariate quadratic map $\mathcal{P} : \mathbb{F}_q^n \to \mathbb{F}_q^m$. The trapdoor consists of a sequence of nested subspaces $\mathbb{F}_q^n \supset O_1 \supset \cdots \supset O_l$ of the input space, and a sequence of nested subspaces $\mathbb{F}_q^m \supset W_1 \supset \cdots \supset W_l = \{0\}$ of the output space, with $\dim O_1 = m$, and $\dim O_i = \dim W_{i-1}$ for $i > 1$ and such that the following hold:

1. $\mathcal{P}(\mathbf{x}) \in W_i$ for all $\mathbf{x} \in O_i$, and
2. $\mathcal{P}'(\mathbf{x}, \mathbf{y}) \in W_{i-1}$ for all $\mathbf{x} \in \mathbb{F}_q^n$, all $\mathbf{y} \in O_i$.

Rainbow with one layer (i.e. $l = 1$) is nothing more than UOV. In the rest of the paper, we focus on Rainbow with two layers (i.e. $l = 2$), because this results in the most efficient schemes and because this covers all the parameter sets submitted to the NIST PQC standardization project. In this case, there are 3 secret subspaces: O_1, O_2 and W (see Fig. 2). An instantiation of Rainbow is then described by 4 parameters:

- q: the size of the finite field
- n: the number of variables
- m: the number of equations in the public key, also the dimension of O_1.
- o_2: the dimension of O_2, also the dimension of W.

Given the trapdoor information (i.e. O_1, O_2 and W), a solution $\mathbf{s}$ to $\mathcal{P}(\mathbf{s}) = \mathbf{t}$ can be found with an efficient 2-step algorithm.

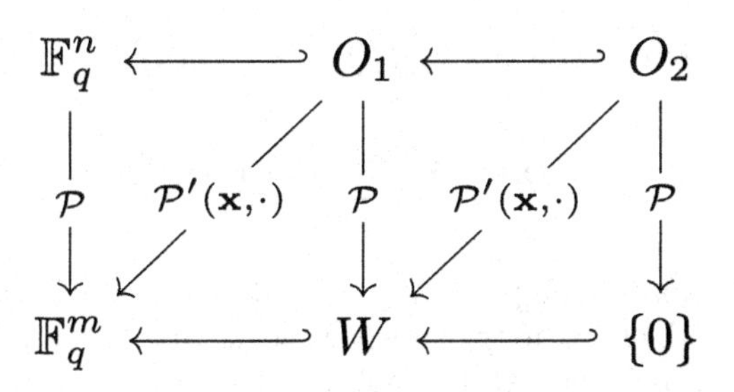

Fig. 2. The structure of a Rainbow public key with 2 layers. The polar form $\mathcal{P}'(\mathbf{x},\cdot)$ maps O_2 to W for every $\mathbf{x} \in \mathbb{F}_q^n$.

1. In the first step, pick $\mathbf{v} \in \mathbb{F}_q^n$ uniformly at random, and solve for $\overline{\mathbf{o}}_1 \in O_1/O_2$, such that $\mathcal{P}(\mathbf{v}+\overline{\mathbf{o}}_1) + W = \mathbf{t} + W$. This can be rewritten as

$$\underbrace{\mathcal{P}(\mathbf{v})}_{\text{fixed by choice of } \mathbf{v}} + \underbrace{\mathcal{P}(\overline{\mathbf{o}}_1)}_{\in W} + \underbrace{\mathcal{P}'(\mathbf{v},\overline{\mathbf{o}}_1)}_{\text{linear in } \mathbf{o}_1} + W = \mathbf{t} + W.$$

   This is a system of linear equations in the quotient space $\mathbb{F}_q^m/W$, so we can efficiently sample a solution with Gaussian elimination. Note that the system has $m - \dim W$ constraints and $m - \dim W$ degrees of freedom, so we expect there to be a unique solution (mod O_2) with probability approximately $1-1/q$. If there is no unique solution we pick a new value of $\mathbf{v}$ and start over.
2. In the second step, we solve for $\mathbf{o}_2 \in O_2$, such that $\mathcal{P}(\mathbf{v}+\mathbf{o}_1+\mathbf{o}_2) = \mathbf{t}$. Writing it as

$$\underbrace{\mathcal{P}(\mathbf{v}+\mathbf{o}_1) - \mathbf{t}}_{\text{fixed},\in W} + \underbrace{\mathcal{P}(\mathbf{o}_2)}_{=0} + \underbrace{\mathcal{P}'(\mathbf{v}+\mathbf{o}_1,\mathbf{o}_2)}_{\text{linear in } \mathbf{o}_2,\in W} = 0\,,$$

   we see that this is a system of $\dim W$ linear equations (because all the values are in W) in $\dim W$ variables, so we expect to find a unique solution with Gaussian elimination with probability $1-1/q$. If no unique solution exists we return to step 1 with a new guess of $\mathbf{v}$.

Remark 5. If we put $W = \mathbb{F}_q^m$ and $O_1 = O_2$, or if we put $O_2 = \{0\}$ and $W = \{0\}$ then we get back the original UOV construction.

5.1 Traditional Description of Rainbow

Traditionally, a Rainbow public key is generated as $\mathcal{P} = \mathcal{S} \circ \mathcal{F} \circ \mathcal{T}$, where $\mathcal{S} \in GL(m,q)$ and $\mathcal{T} \in GL(n,q)$ are uniformly random invertible linear maps, and where $\mathcal{F}(\mathbf{x}) = f_1(\mathbf{x}), \cdots, f_m(\mathbf{x})$ is the so-called central map, whose first o_1 components $f_1(\mathbf{x}), \ldots, f_{o_1}(\mathbf{x})$ are of the form

$$f_i(\mathbf{x}) = \sum_{j=1}^{n-o_1} \sum_{k=1}^{n-m} \alpha_{ijk} x_j x_k\,,$$

and whose remaining components $f_{o_1+1}(\mathbf{x}), \cdots, f_m(\mathbf{x})$ are of the form

$$f_i(\mathbf{x}) = \sum_{j=1}^{n} \sum_{k=1}^{n-o_1} \alpha_{ijk} x_j x_k .$$

Let O_1' be the subspace of $\mathbb{F}_q^n$ consisting of all the vectors whose first $n-m$ entries are zeros, and let O_2' be the subspace consisting of the vectors whose first $n-o_2$ entries are zero. Then all the polynomials in the central map vanish on O_2', and the first o_1 polynomials also vanish on O_1'. In other words, $\mathcal{F}(O_2') = 0$ and $\mathcal{F}(O_1') \subset W'$, where W' is the subspace of $\mathbb{F}_q^m$ consisting of the vectors whose first o_1 entries are zero. Moreover, $\mathcal{F}'(\mathbf{x}, \mathbf{y}) \in W'$ for any $\mathbf{x} \in \mathbb{F}_q^n$ and any $\mathbf{y} \in O_2$. Therefore, the central map $\mathcal{F}$ satisfies the diagram in Fig. 2 with the publicly known subspaces O_1', O_2' and W' taking the roles of O_1, O_2 and W. This means that after composing $\mathcal{F}$ with secret random linear maps $\mathcal{S}$ and $\mathcal{T}$ we obtain a public key $\mathcal{P} = \mathcal{S} \circ \mathcal{F} \circ \mathcal{T}$ that satisfies the diagram in Fig. 2 for uniformy random secret subspaces $O_1 = \mathcal{T}^{-1} O_1'$, $O_2 = \mathcal{T}^{-1} O_2'$ and $W = \mathcal{S}^{-1} W'$.

5.2 Rainbow NIST PQC Parameter Sets

In this paper, we focus on the Rainbow parameter sets that were proposed to the second round and the finals of the NIST PQC standardization project [8]. These parameter sets and the corresponding key and signature sizes are displayed in Table 2.

Table 2. The Rainbow parameter sets that were submitted to the second round and the finals of the NIST PQC standardization project.

Parameter set		Parameters				$\lvert\mathsf{pk}\rvert$	$\lvert\mathsf{sk}\rvert$	$\lvert\mathsf{sig}\rvert$
		q	n	m	o_2	(kB)	(kB)	(Bytes)
Second round	Ia	16	96	64	32	149	93	64
	IIIc	256	140	72	36	710	511	156
	Vc	256	188	96	48	1705	1227	204
Finals	Ia	16	100	64	32	157	101	66
	IIIc	256	148	80	48	861	611	164
	Vc	256	196	100	64	1885	1376	212

5.3 Attacks on Rainbow

A straightforward method to forge a signature is to simply try to find a solution $\mathbf{s}$ to the system $\mathcal{P}(\mathbf{s}) = \mathcal{H}(M||\mathsf{salt})$. This is called a direct attack. More interesting attacks try to exploit the hidden structure of the Rainbow trapdoor.

OV attack. The OV attack of Kipnis and Shamir to find the subspace O in the OV construction can be used against Rainbow to find O_2. The complexity of the attack is $\tilde{O}(q^{n-2o_2})$.

When O_2 is found, it is easy to find W, because

$$\{\mathcal{P}'(\mathbf{x},\mathbf{y}) \mid \mathbf{x} \in \mathbb{F}_q^n, \mathbf{y} \in O_2\} \subset W\,,$$

and with overwhelming probability this will be an equality. Once W is found, we have reduced the problem to a small UOV instance with parameters $n' = n - o_2$ and $m' = m - o_2$, so the Kipnis and Shamir attack can be used again to find O_1, with complexity $\tilde{O}(q^{n'-2m'}) = \tilde{O}(q^{n+o_2-2m})$, which is negligible compared to the complexity of the first step.

MinRank/HighRank attack. For all $i \in \{1, \cdots, m\}$, we define $M_i \in \mathbb{F}_q^{n\times n}$ like we did in the description of the OV attack. For $\mathbf{v} \in \mathbb{F}_q^m$ we define the linear combination $M_\mathbf{v} := \sum_{i=0}^m v_i M_i$. Then it follows that $\langle \mathbf{v}, \mathcal{P}'(\mathbf{x},\mathbf{y})\rangle = \mathbf{x}^\top M_\mathbf{v}\mathbf{y}$. The second property of the Rainbow public key says that if $\mathbf{v} \in W^\perp$, then $\langle \mathbf{v}, \mathcal{P}'(\mathbf{x},\mathbf{y})\rangle = \mathbf{x}^\top M_\mathbf{v}\mathbf{y} = 0$ for all values of $\mathbf{x}$ and all $\mathbf{y} \in O_2$. This implies that O_2 is in the kernel of $M_\mathbf{v}$, so $M_\mathbf{v}$ has an exceptionally small rank of at most $n - \dim O_2$.

The MinRank attack attempts to exploit this property to find a vector in $W^\perp$. The problem is, given the M_i for $i \in \{1, \cdots, m\}$, to find a linear combination of these maps that has rank $n - \dim O_2$. This can be done with 2 strategies:

Guessing strategy [13]. Repeatedly pick $\mathbf{v} \in \mathbb{F}_q^m$. With probability q^{-o_2}, we have $\mathbf{v} \in W^\perp$. To check if a guess is correct, we simply check if the rank of $M_\mathbf{v}$ is at most $n - \dim O_2$. The complexity of the attack is $\tilde{O}(q^{o_2})$. There is a more efficient version of this attack by Billet and Gilbert, that runs in time $\tilde{O}(q^{2n-3m+o_2+1})$ [4].

Algebraic strategy. One expresses $\mathrm{rank}(M_\mathbf{v}) \leq n - \dim O_2$ as a system of multivariate polynomial equations in the entries of $\mathbf{v}$ and uses an algorithm such as XL to find a solution. There exist several methods to translate the rank condition into a system of polynomial equations, such as the Kipnis-Shamir modeling, and Minors modeling [12,16]. Recently, a more efficient approach by Bardet *et al.* called "Support Minors Modeling" drastically improved the efficiency of this attack (see Sect. 2.3 and [1]). The algebraic approach is asymptotically more efficient than the guessing strategy.

As soon as a single vector $\mathbf{v} \in W^\perp$ is found, the attacker knows O_2, because it is the kernel of $M_\mathbf{v}$. Then, once O_2 is known he can finish the key recovery attack as described in the previous section on the UOV attack.

Rainbow band separation attack. This attack, proposed by Ding *et al.* [10], tries to simultaneously find a vector $\mathbf{o} \in O_2$, and a vector $\mathbf{v} \in W^\perp$. This gives rise to the following system of equations

$$\begin{cases} \mathcal{P}(\mathbf{o}) = 0 \\ \langle \mathbf{v}, \mathcal{P}'(\mathbf{o},\mathbf{x})\rangle = 0\,, \quad \forall \mathbf{x} \in \mathbb{F}_q^n \end{cases} . \tag{4}$$

To get a unique solution, we can impose o_2 linear relations on the entries of $\mathbf{o}$ and $m - o_2$ linear relations on the entries of $\mathbf{v}$. This results in a system with $n - o_2$ variables for $\mathbf{o}$ and o_2 variables for $\mathbf{v}$, which makes a total of n variables.

It looks like we get q^n bilinear equations (one for each choice of $\mathbf{x} \in \mathbb{F}_q^n$), but these equations are obviously not independent. Extend $\mathbf{o}$ to a basis $\mathbf{x}_1 = \mathbf{o}, \mathbf{x}_2, \cdots, \mathbf{x}_n$ for $\mathbb{F}_q^n$ (since we fixed some entries of $\mathbf{o}$, we can pick the $\mathbf{x}_i$ with $i > 1$ without having to know the precise value of $\mathbf{o}$). We can rewrite system (4) as

$$\begin{cases} \mathcal{P}(\mathbf{o}) = 0 \\ \langle \mathbf{v}, \mathcal{P}'(\mathbf{o}, \mathbf{x}_i) \rangle = 0\,, \quad \forall i \in \{1, \cdots, n\} \end{cases} . \tag{5}$$

Note that the first bilinear equation is $\langle \mathbf{v}, \mathcal{P}'(\mathbf{o}, \mathbf{o}) \rangle = 0$, which is equivalent to $\langle \mathbf{v}, \mathcal{P}(2\mathbf{o}) - 2\mathcal{P}(\mathbf{o}) \rangle = \langle \mathbf{v}, 2\mathcal{P}(\mathbf{o}) \rangle = 0$, (recall that $\mathcal{P}$ is homogenous), so this equation is already implied by the $\mathcal{P}(\mathbf{o}) = 0$ equations. This leaves us with a system of m quadratic equations in $\mathbf{o}$, and $n-1$ bi-linear equations in the entries of $\mathbf{o}$ and $\mathbf{v}$. The complexity of this attack is studied in detail in [19], where they introduce a variant of the XL algorithm that exploits the bi-homogenous structure of the system.

6 Intersection Attack on Rainbow

In this section we introduce a new key-recovery attack against the Rainbow signature scheme that is similar to our intersection attack on UOV from Sect. 4. Let k be such that $n < \frac{2k-1}{k-1} o_2$, and pick invertible matrices $L_1, \cdots, L_k$ from the span of the M_i. Our goal is to find a vector $\mathbf{x}$ in the intersection

$$\mathbf{x} \in \bigcap_{i=1}^{k} L_i O_2 \,.$$

This intersection has dimension at least $ko_2 - (k-1)(n - o_2) > 0$, so non-zero vectors in the intersection exist. We could try to find $\mathbf{x}$ by solving the system (3). However, similar to the RBS attack, we can improve the efficiency of the attack by simultaneously looking for a vector $\mathbf{v} \in W^{\perp}$. Let $\mathbf{e}_1, \cdots, \mathbf{e}_n$ be a basis for $\mathbb{F}_q^n$, where all the entries of $\mathbf{e}_i$ are zero, except the i-th entry which equals 1. Then we get the following system of quadratic equations:

$$\begin{cases} \mathcal{P}(L_i^{-1}\mathbf{x}) = 0\,, & \forall i \in \{1, \cdots, k\} \\ \mathcal{P}'(L_i^{-1}\mathbf{x}, L_j^{-1}\mathbf{x}) = 0\,, & \forall i < j \in \{1, \cdots, k\} \\ \langle \mathbf{v}, \mathcal{P}'(L_i^{-1}\mathbf{x}, \mathbf{e}_j) \rangle = 0\,, & \forall i \in \{1, \cdots, k\} \text{ and } \forall j \in \{1, \cdots, n\} \end{cases} . \tag{6}$$

If we impose $ko_2 - (k-1)(n - o_2)$ affine constraints on the entries of $\mathbf{x}$, and $m - o_2$ affine constraints on the entries of $\mathbf{v}$ we expect to have a unique solution.

It looks like we get $\binom{k+1}{2} m$ quadratic equations in the $\mathbf{x}$ variables and kn equations that are linear in the $\mathbf{x}$ variables and the $\mathbf{v}$ variables. However, the quadratic equations are the same set of equations as in the Intersection attack on UOV, so we know that they give only $\binom{k+1}{2}m - 2\binom{k}{2}$ linearly independent equations. We can then use the Bilinear XL variant of Smith-Tone and Perlner [19] to find the unique solution to the system of equations.

Remark 6. If we put $k = 1$ then we recover the Rainbow Band Separation attack (see Sect. 5.3), so our attack can be seen as a generalization of the RBS attack. However, note that previous works have assumed that only $n - 1$ out of the n bilinear equations are useful. We find that this is not quite correct. Even though there is a syzygy at degree $(2, 1)$ (which we will discuss later) it is still useful to consider all n bilinear equations.

6.1 Extending to $n \geq 3o_2$

If $n \geq 3o_2$, then we expect there to be no non-trivial intersection, so the attack is not guaranteed to succeed with $k = 2$. However, if we model L_1O_2 and L_2O_2 as uniformly random subspaces of $O_2^{\perp}$, then the probability that they intersect non-trivially is approximately q^{-n+3o_2-1}. Therefore, we can expect the attack to succeed after q^{n-3o_2+1} guesses for (L_1, L_2).

6.2 Complexity Analysis of the Attack

The system of equations (6) is clearly not generic, since the first $\binom{k+1}{2}m$ equations only contain the entries of $\mathbf{x}$ as variables, and the remaining $k(n-k)$ equations are bi-linear in the entries of $\mathbf{x}$ and $\mathbf{v}$. This is the same structure as the systems that appear in the RBS attack (Sect. 5.3). Smith-Tone and Perlner investigated the complexity of solving such systems, and they proposed a variant of the XL algorithm that exploits the bi-homogeneous structure of the system [19]. Their algorithm works for systems of polynomial equations in $n_x + n_y$ variables, where m_x equations are quadratic in the first n_x variables, and m_{xy} equations are bi-linear in the first n_x and last n_y variables respectively. Under a maximal rank assumption, their XL variant terminates at bi-degree (A, B) if the coefficient corresponding to $t^a s^b$ in

$$\frac{(1-t^2)^{m_x}(1-ts)^{m_{xy}}}{(1-t)^{n_x+1}(1-s)^{n_y+1}} \tag{7}$$

is non-positive for some a, b with $a \leq A$ and $b \leq B$. If this is the case, an upper bound for the number of multiplications in the attack is given by

$$3\overline{\mathcal{M}}(A,B)^2\binom{n_x+2}{2}, \tag{8}$$

where $\overline{\mathcal{M}}(A, B)$ is the number of monomials with bi-degree bounded by (A, B).

The maximal rank assumption is not valid for small instances of Rainbow, because there are k^2 non-trivial syzygies: For each $(i,j) \in \{1,\cdots,k\}^2$ we have

$$\begin{aligned}\langle \mathbf{v}, \mathcal{P}'(L_i^{-1}\mathbf{x}, L_j^{-1}\mathbf{x})\rangle &= \sum_{l=1}^{m} v_l \cdot \mathcal{P}'_l(L_i^{-1}\mathbf{x}, L_j^{-1}\mathbf{x}) \\ &= \sum_{t=1}^{m} \langle \mathbf{v}, \mathcal{P}'(L_i^{-1}\mathbf{x}, \mathbf{e}_t)\rangle \cdot (L_j^{-1}\mathbf{x})_t\,,\end{aligned}$$

which gives a non-trivial syzygy for the system (6) at bi-degree $(2,1)$.

Since adding an equation with bi-degree (a,b) to the polynomial system corresponds to an extra factor $(1-t^a s^b)$ in the generating function (7), it seems natural that a syzygy at degree (a,b) results in a factor $(1-t^a s^b)^{-1}$. We therefore conjecture that the generating function for the system (6) is

$$\frac{(1-t^2)^{m_x}(1-ts)^{m_{xy}}(1-t^2s)^{-k^2}}{(1-t)^{n_x+1}(1-s)^{n_y+1}} \tag{9}$$

where

$$\begin{aligned} n_x &= \min(nk-(2k-1)o_2, n-1), & n_y &= o_2, \\ m_x &= \binom{k+1}{2}m - 2\binom{k}{2}, \quad \text{and} & m_{xy} &= kn\,. \end{aligned}$$

We experimentally verified that this generating function exactly predicts the ranks of the Macaulay matrices for small instances of Rainbow (see Table 4). That is, we found that the rank of the Macaulay matrix at bi-degree (A,B) equals $\overline{\mathcal{M}}(A,B)$ minus the coefficient of $t^A s^B$ in (9), unless one of the coefficient of $t^a s^b$ with $a \le A$ and $b \le B$ is non-positive, in which case the rank is $\overline{\mathcal{M}}(A,B)-1$, and the bilinear XL algorithm will succeed at bi-degree (A,B).

Under our assumption, we can estimate the cost of our attack by iterating over all minimal bi-degrees (A,B) for which the attack will succeed (i.e. for which the coefficient of $t^A s^B$ in the generating function is non-positive), and picking the bi-degree (A,B) that minimizes the cost (8).

6.3 Application to Rainbow NIST Submissions

We now estimate the complexity of our attack on the Rainbow parameter sets that were submitted to the NIST PQC project. For all the proposed parameter sets we have $n \ge 3o_2$, which means the basic attack will need to be repeated multiple times before we expect to recover the secret key. For the Ia parameter set on the second-round submission, we have $n = 3m$, and for all the parameter sets of the final round submission we have $n = 3m+4$. In these cases, we need to repeat the attack q and q^5 times respectively. For the IIIc and Vc parameter set of the second-round submission, n is much larger than $3m$, so the attack is very inefficient in these cases.

Table 3 reports the estimated gate count of our attack. To convert from the number of multiplications to the gate count, we use the model that is standard in the MQ literature; each multiplication costs $2(\log_2(q)^2 + \log_2(q))$ gates. We see that our attack outperforms the best known attacks for 4 out of the 6 proposed parameter sets. The improvement is the largest for the Ia parameter set of the first round and the Vc parameter set of the finals, where we improve on existing attacks by almost 20 bits.

Table 3. The estimated gate count of our Intersection attack on Rainbow compared to the best known attacks (taken from [19] for the second round parameters and the Rainbow NIST submission for the finals parameters).

Parameter set		Attack parameters						New attack	Known attacks
		n_x	n_y	m_x	m_{xy}	Guesses	(A, B)		
Second round	Ia	95	32	190	192	q^1	(10, 1)	<u>123</u>	140
	IIIc	139	36	214	280	q^{33}	(6, 9)	412	<u>204</u>
	Vc	187	48	286	376	q^{45}	(6, 15)	548	<u>264</u>
Finals	Ia	99	32	190	200	q^5	(7, 4)	<u>140</u>	147
	IIIc	147	48	238	296	q^5	(10, 6)	<u>213</u>	217
	Vc	195	64	298	392	q^5	(10, 12)	<u>262</u>	281

7 The Rectangular MinRank Attack

In this section we introduce a new MinRank attack that exploits the property that for $\mathbf{y} \in O_2$, we have that $\mathcal{P}'(\mathbf{x}, \mathbf{y}) \in W$ for all $\mathbf{x} \in \mathbb{F}_q^n$. Let $\mathbf{e}_1, \cdots, \mathbf{e}_n$ be the basis for $\mathbb{F}_q^n$ where $\mathbf{e}_i$ is a vector whose entries are zero, except for the i-th entry which equals one. For a vector $\mathbf{x} \in \mathbb{F}_q^n$, we define the matrix

$$L_{\mathbf{x}} = \begin{pmatrix} \mathcal{P}'(\mathbf{e}_1, \mathbf{x}) \\ \cdots \\ \mathcal{P}'(\mathbf{e}_n, \mathbf{x}) \end{pmatrix} .$$

If $\mathbf{y} \in O_2$, then all the rows of $L_{\mathbf{y}}$ are in W, which implies that the matrix has rank at most $\dim W = o_2$. Moreover, it follows from the bilinearity of $\mathcal{P}'$ that

$$L_{\mathbf{y}} = \sum_{i=1}^{n} y_i L_{\mathbf{e}_1} .$$

Since the $L_{\mathbf{e}_i}$ matrices are public information, it follows that finding $\mathbf{y} \in O$ reduces to an instance of a rectangular MinRank problem; if an attacker can find a linear combination $\sum_{i=1}^{n} L_{\mathbf{e}_i} y_i$ with rank at most o_2, then we can assume that $\mathbf{y}$ is in O_2. If we set $o_2 - 1$ entries of $\mathbf{y}$ to zero, we still expect a non-trivial solution, so it suffices to look for a linear combinations of only the matrices $L_{\mathbf{e}_1}$ up to $L_{\mathbf{e}_{n-o_2+1}}$. Note that this MinRank instance is fundamentally different from the one that was already known in the literature (see Table 5).

Table 4. The rank and the number of columns of the Macaulay matrices for the system of equations of the intersection attack. The rank at degree (A, B) always matches the coefficient of $t^A s^B$ in $\frac{1-(1-t^2)^{m_x}(1-ts)^{m_{xy}}(1-t^2 s)^{-k^2}}{(1-t)^{n_x}(1-s)^{n_y}}$, except if the coefficient is larger or equal to the number of columns. In this case (marked by boldface in the table) the rank equals the number of columns minus 1, and the XL system can be solved at bi-degree (A, B).

Parameters					Macaulay matrix at bi-degree (A, B)							
n	m	o_2	k		$(2,0)$	$(1,1)$	$(3,0)$	$(2,1)$	$(1,2)$	$(3,1)$	$(2,2)$	$(1,3)$
8	6	3	2	Rank	16	12	**119**	**143**	64			**159**
				Cols	36	32	120	144	80			160
10	6	3	1	Rank	6	10	48	103	40	**479**	331	100
				Cols	36	32	120	144	80	480	360	160
12	8	4	1	Rank	8	12	72	147	60	795	589	180
				Cols	45	45	165	225	135	825	675	315
12	8	4	2	Rank	22	24	264	**389**	120			360
				Cols	78	60	364	390	180			420
14	10	5	1	Rank	10	14	100	199	84	1220	953	294
				Cols	55	60	220	330	210	1320	1155	650
14	10	5	2	Rank	28	28	392	556	168	**3359**	**2204**	588
				Cols	105	84	560	630	294	3360	2205	784

Table 5. Comparison of the new MinRank instance with the known instance of the MinRank problem.

	Known instance of MinRank problem	New instance of MinRank problem
Size of matrices	n-by-n	n-by-m
Number of matrices	$o_2 + 1$	$n - o_2 + 1$
Rank of linear combination	m	o_2
Solution	Vector in $W^{\perp}$	vector in O_2

We can use generic algorithms to solve this instance of the MinRank problem, such as the guessing strategy, or the algebraic methods of Sect. 5.3. However, in our case we can do slightly better because we have more information about $\mathbf{y}$; on top of knowing that $L_{\mathbf{y}}$ has low rank, we also know that $\mathcal{P}(\mathbf{y}) = 0$. Note that the variables y_i already appear in the system of equations that model the rank condition $\text{rank}(L_{\mathbf{y}}) \leq o_2$. Therefore, we can add the equations $\mathcal{P}(\mathbf{y}) = 0$ to the system without having to introduce additional variables. This will make the attack slightly more efficient.

7.1 Complexity Analysis

We first estimate the complexity of solving the pure MinRank problem with the support minors modeling approach of Sect. 2.3, without using the additional equations $\mathcal{P}(\mathbf{y}) = 0$. From experiments it seems that in case we are working in a field of odd characteristic, the MinRank instance behaves like a generic instance of the MinRank problem, so we can use the methodology of Bardet *et al.* to estimate the complexity of a random MinRank instance with $n - o_2 + 1$ matrices of size n-by-m with target rank o_2 (see Sect. 2.3). However, in case of a field with characteristic 2 (which includes all the Rainbow parameters submitted to NIST), there are some syzygies that do not appear in the case of random MinRank instances. This stems from the fact that, in characteristic 2, we have

$$\mathcal{P}'(\mathbf{y}, \mathbf{y}) = \mathcal{P}(2\mathbf{y}) - \mathcal{P}(\mathbf{y}) - \mathcal{P}(\mathbf{y}) = 2\mathcal{P}(\mathbf{y}) = 0\,,$$

so the $(r+1)$-by-$r+1$ minors of

$$\binom{\mathcal{P}'(\mathbf{y}, \mathbf{y})}{C} = \sum_{i=0}^{n} y_i \binom{\mathcal{P}'(\mathbf{e}_i, \mathbf{y})}{C}$$

all vanish, which gives $\binom{m}{r+1}$ non-trivial linear relation between the equations at degree $(2, 1)$. It is possible to carefully count how many linearly independent equations we have at each degree (b, i), with an analysis similar to the analysis of Bardet *et al.* [1].

However, to simplify the analysis, we can side-step the syzygies by ignoring one of the rows of the $L_1, \cdots, L_{n-o_2+1}$ matrices; since all the syzygies use all the rows of the L_i, the syzygies do not occur anymore if we omit a row from all the L_i matrices. Experimentally, we find that after removing a row, the instance behaves exactly like a random instance of the MinRank problem with $n - o_1 + 1$ matrices of size $(n-1)$-by-m and with rank o_2. We can therefore use the methodology of Bardet *et al.* to estimate the complexity of the attack (see Sect. 2.3). The first half of Table 6 reports on the estimated complexities for the Rainbow parameter sets that were submitted to the second round and the finals of the NIST PQC standardization project.

The attack using $\mathcal{P}(\mathbf{y}) = 0$. We use the notation of Sect. 2.3, where M_b is the Macaulay matrix for the Support Minors Modelling system at bi-degree $(b, 1)$ (omitting one row of the L_i matrices, as discussed earlier), and where $\mathcal{M}(b, 1)$ is the number of monomials of degree b in the y_i variables and of degree 1 in the c_S variables. Let M_b^+ be the Macaulay matrix of the SMM system after appending the $\mathcal{P}(\mathbf{y}) = 0$ equations. We want to figure out the minimal value of b, for which the rank of M_b^+ is equal to $\mathcal{M}(b, 1) - 1$, because in that case the system $M_b^+\mathbf{x} = 0$ will have a one-dimensional solution space that corresponds to the solutions of the MinRank problem.

Table 6. The optimal attack parameters of the new MinRank attack, and the corresponding gate complexity for the Rainbow parameter sets submitted to the second round and the finals of the NIST PQC standardization project.

Parameter set		Plain MinRank			MinRank and $\mathcal{P}(\mathbf{y}) = 0$		
		m'	b	$\log_2$ gates	m'	b	$\log_2$ gates
Second round	Ia	51	2	131	40	6	124
	IIIc	59	2	153	52	4	151
	Vc	80	2	197	74	3	191
Finals	Ia	51	2	131	44	4	127
	IIIc	72	3	184	68	4	177
	Vc	95	4	235	87	6	226

Bardet *et al.* already computed the rank of M_b, so we only need to figure out how much the rank increases by including the $\mathcal{P}(\mathbf{y}) = 0$ equations. Let $G(t)$ be a generating function for the dimension of the kernel of M_b, and $G^+(t)$ a generating function for the dimension of the kernel of M_b^+. Note that, even though we do not have a nice expression for $G(t)$, we can compute its coefficients from the expression of Bardet *et al.* for the rank of M_b, because the coefficient corresponding to t^b in $G(t)$ is $\mathcal{M}(b,1) - \text{rank}(M_b)$. Under some genericity assumptions we have that $G^+(t) = (1-t^2)^m G(t)$, from which we can get the rank of M_b^+.

Experimentally, we found for all the instances of Rainbow we could check, that this predicts the rank of M_b^+ exactly (see Table 7).

To estimate the complexity of the attack, we compute the first few terms of $G(t)$ until we encounter the first non-positive coefficient. If the first non-positive coefficient corresponds to t^b, then we assume the bilinear XL algorithm will work at bi-degree $(b,1)$ and we can upper bound its cost as

$$3\mathcal{M}(b_{min},1)^2\, W$$

multiplications, where $W = \max((o_2+1)(n-o_2+1), \binom{(n-o_2+3)}{2})$ is the maximal weight of the equations in the system. We found that, as already observed by Bardet *et al.* , it is helpful to consider only the first m' columns of the matrices $L_{\mathbf{e}_i}$. For each value of $m' \in [o_2+1, m]$ we estimate the attack cost, and we pick the value of m' that results in the smallest cost. The optimal attack parameters (m', b) and the corresponding costs (in terms of gate count) are reported in Table 6. We see that adding the $\mathcal{P}(\mathbf{y}) = 0$ equations to the Support minors modeling system reduces the attack complexity by a modest factor between 2^2 and 2^9 for the NIST parameter sets.

Table 7. The rank and the number of columns of the Macaulay matrices for the system of equations of the rectangular MinRank attack. The rank at bi-degree $(b, 1)$ always matches the predicted values, except if the prediction is larger or equal to the number of columns. In this case (marked by boldface in the table) the rank equals the number of columns minus 1, and the XL system can be solved at bi-degree $(b, 1)$.

Parameters					Macaulay matrix at bi-degree $(b,1)$			
n	m	o_2	m'		$b=1$	$b=2$	$b=3$	$b=4$
9	6	3	5	Rank	40	244	**839**	
				Rank with $\mathcal{P}(\mathbf{y})=0$	40	**279**		
				Number of columns	70	280	840	
12	8	4	6	Rank	66	528	2376	**7424**
				Rank with $\mathcal{P}(\mathbf{y})=0$	66	648	**2474**	
				Number of columns	135	675	2475	7425
15	10	5	6	Rank	14	154	924	4004
				Rank with $\mathcal{P}(\mathbf{y})=0$	14	214	1444	**6005**
				Number of columns	66	396	1716	6006
18	12	6	8	Rank	136	1615	10387	
				Rank with $\mathcal{P}(\mathbf{y})=0$	136	1951	**12739**	
				Number of columns	364	2548	12740	

8 Conclusion

This paper offers a new perspective on the UOV and Rainbow signature schemes that avoids the use of a central map. This makes it easier to understand the existing attacks on these schemes, and allowed us to discover some new, more powerful, attacks. We hope that our simpler perspective will encourage more researchers to scrutinize the UOV and Rainbow signature schemes.

We introduce two new attacks: the intersection attack, which applies to both the UOV and the Rainbow signature schemes, and the rectangular MinRank Attack that applies only to the Rainbow scheme. Although methods for solving systems of multivariate quadratic equations (and our understanding of their complexity) have been improving over the last decades, the intersection attack is the first improvement in the cryptanalysis of UOV that is specific to the structure of the UOV public keys since 1999. Similarly, even though our understanding of the complexity of attacks on Rainbow has been improving (recent examples are [19] and [1]), there had not been any fundamentally new attacks on Rainbow since 2008.

Table 8. An overview of the estimated gate counts of our attacks versus known attacks and the target security level for the six Rainbow parameter sets submitted to the second round and the finals of the NIST PQC standardization project. The complexities of the known attacks are taken from [19] for the second round parameters and the Rainbow NIST submission for the finals parameters. The security target is taken from the NIST PQC call for proposals.

Parameter set		Intersection attack	New MinRank attack	Known attacks	Security target
Second round	Ia	<u>123</u>	124	140	143
	IIIc	412	<u>151</u>	204	207
	Vc	548	<u>191</u>	264	272
Finals	Ia	140	<u>127</u>	147	143
	IIIc	213	<u>177</u>	217	207
	Vc	262	<u>226</u>	281	272

New parameters for UOV and Rainbow. Both of our attacks reduce the security level of the Rainbow NIST submission below the requirements set out by NIST (see Table 8). However, our attacks are still exponential, and Rainbow can be saved by increasing the parameter sizes by a relatively small amount. For example, using $q = 16, n = 109, m = 68, o_2 = 36$ would presumably reach NIST security level I and would result in a signature size of 71 Bytes (a 10 % increase) an key size of roughly 203 KB (an increase of 25 %). Alternatively, one could use the UOV scheme with $q = 64, n = 118, m = 47$, which results in 89 Byte signatures and a key size of 242 Kilobytes. It seems questionable whether the small performance advantage of Rainbow over UOV is worth the additional complexity. We leave a more carefully optimized parameter choice for UOV and Rainbow for future work.

Acknowledgments. I would like to thank Bo-Yin Yang and Jintai Ding for providing helpful feedback on an earlier version of this paper.

References

1. Bardet, M., et al.: Improvements of algebraic attacks for solving the rank decoding and MinRank problems. In: Moriai, S., Wang, H. (eds.) ASIACRYPT 2020. LNCS, vol. 12491, pp. 507–536. Springer, Cham (2020). https://doi.org/10.1007/978-3-030-64837-4_17
2. Bardet, M., Faugere, J.-C., Salvy, B., Yang, B.-Y.: Asymptotic behaviour of the degree of regularity of semi-regular polynomial systems. In: Proceedings of MEGA, Eighth International Symposium on Effective Methods in Algebraic Geometry, vol. 5 (2005)
3. Beullens, W., Preneel, B., Szepieniec, A., Vercauteren, F.: LUOV. Technical report, National Institute of Standards and Technology (2019). https://csrc.nist.gov/projects/post-quantum-cryptography/round-2-submissions

4. Billet, O., Gilbert, H.: Cryptanalysis of rainbow. In: De Prisco, R., Yung, M. (eds.) SCN 2006. LNCS, vol. 4116, pp. 336–347. Springer, Heidelberg (2006). https://doi.org/10.1007/11832072_23
5. Casanova, A., Faugère, J.-C., Macario-Rat, G., Patarin, J., Perret, L., Ryckeghem, J. GeMSS. Technical report, National Institute of Standards and Technology (2019). https://csrc.nist.gov/projects/post-quantum-cryptography/round-2-submissions
6. Courtois, N., Klimov, A., Patarin, J., Shamir, A.: Efficient algorithms for solving overdefined systems of multivariate polynomial equations. In: Preneel, B. (ed.) EUROCRYPT 2000. LNCS, vol. 1807, pp. 392–407. Springer, Heidelberg (2000). https://doi.org/10.1007/3-540-45539-6_27
7. Czypek, P., Heyse, S., Thomae, E.: Efficient implementations of MQPKS on constrained devices. In: Prouff, E., Schaumont, P. (eds.) CHES 2012. LNCS, vol. 7428, pp. 374–389. Springer, Heidelberg (2012). https://doi.org/10.1007/978-3-642-33027-8_22
8. Ding, J., Chen, M.-S., Petzoldt, A., Schmidt, D., Yang, B.-Y.: Rainbow. Technical report, National Institute of Standards and Technology (2019). https://csrc.nist.gov/projects/post-quantum-cryptography/round-2-submissions
9. Ding, J., Schmidt, D.: Rainbow, a new multivariable polynomial signature scheme. In: Ioannidis, J., Keromytis, A., Yung, M. (eds.) ACNS 2005. LNCS, vol. 3531, pp. 164–175. Springer, Heidelberg (2005). https://doi.org/10.1007/11496137_12
10. Ding, J., Yang, B.-Y., Chen, C.-H.O., Chen, M.-S., Cheng, C.-M.: New differential-algebraic attacks and reparametrization of rainbow. In: Bellovin, S.M., Gennaro, R., Keromytis, A., Yung, M. (eds.) ACNS 2008. LNCS, vol. 5037, pp. 242–257. Springer, Heidelberg (2008). https://doi.org/10.1007/978-3-540-68914-0_15
11. Faugère, J.C.: A new efficient algorithm for computing Gröbner bases without reduction to zero (F_5). In: Proceedings of the 2002 International Symposium on Symbolic and Algebraic Computation, pp. 75–83 (2002)
12. Faugère, J.-C., El Din, M.S., Spaenlehauer, P.-J.: Computing loci of rank defects of linear matrices using Gröbner bases and applications to cryptology. In: Proceedings of the 2010 International Symposium on Symbolic and Algebraic Computation, pp. 257–264 (2010)
13. Goubin, L., Courtois, N.T.: Cryptanalysis of the TTM cryptosystem. In: Okamoto, T. (ed.) ASIACRYPT 2000. LNCS, vol. 1976, pp. 44–57. Springer, Heidelberg (2000). https://doi.org/10.1007/3-540-44448-3_4
14. Kipnis, A., Patarin, J., Goubin, L.: Unbalanced oil and vinegar signature schemes. In: Stern, J. (ed.) EUROCRYPT 1999. LNCS, vol. 1592, pp. 206–222. Springer, Heidelberg (1999). https://doi.org/10.1007/3-540-48910-X_15
15. Kipnis, A., Shamir, A.: Cryptanalysis of the oil and vinegar signature scheme. In: Krawczyk, H. (ed.) CRYPTO 1998. LNCS, vol. 1462, pp. 257–266. Springer, Heidelberg (1998). https://doi.org/10.1007/BFb0055733
16. Kipnis, A., Shamir, A.: Cryptanalysis of the HFE public key cryptosystem by relinearization. In: Wiener, M. (ed.) CRYPTO 1999. LNCS, vol. 1666, pp. 19–30. Springer, Heidelberg (1999). https://doi.org/10.1007/3-540-48405-1_2
17. Patarin, J.: The oil and vinegar signature scheme. In: Dagstuhl Workshop on Cryptography, September 1997
18. Samardjiska, S., Chen, M.-S., Hülsing, A., Rijneveld, J., Schwabe, P.: MQDSS. Technical report, National Institute of Standards and Technology (2019). https://csrc.nist.gov/projects/post-quantum-cryptography/round-2-submissions
19. Smith-Tone, D., Perlner, R.: Rainbow band separation is better than we thought. Technical report, Cryptology ePrint Archive preprint (2020)

Cryptanalytic Applications of the Polynomial Method for Solving Multivariate Equation Systems over GF(2)

Itai Dinur(✉)

Department of Computer Science, Ben-Gurion University, Be'er Sheva, Israel
dinuri@cs.bgu.ac.il

Abstract. At SODA 2017 Lokshtanov et al. presented the first worst-case algorithms with exponential speedup over exhaustive search for solving polynomial equation systems of degree d in n variables over finite fields. These algorithms were based on the polynomial method in circuit complexity which is a technique for proving circuit lower bounds that has recently been applied in algorithm design. Subsequent works further improved the asymptotic complexity of polynomial method-based algorithms for solving equations over the field $\mathbb{F}_2$. However, the asymptotic complexity formulas of these algorithms hide significant low-order terms, and hence they outperform exhaustive search only for very large values of n.

In this paper, we devise a concretely efficient polynomial method-based algorithm for solving multivariate equation systems over $\mathbb{F}_2$. We analyze our algorithm's performance for solving random equation systems, and bound its complexity by about $n^2 \cdot 2^{0.815n}$ bit operations for $d = 2$ and $n^2 \cdot 2^{(1-1/2.7d)n}$ for any $d \geq 2$.

We apply our algorithm in cryptanalysis of recently proposed instances of the Picnic signature scheme (an alternate third-round candidate in NIST's post-quantum standardization project) that are based on the security of the LowMC block cipher. Consequently, we show that 2 out of 3 new instances do not achieve their claimed security level. As a secondary application, we also improve the best-known preimage attacks on several round-reduced variants of the Keccak hash function.

Our algorithm combines various techniques used in previous polynomial method-based algorithms with new optimizations, some of which exploit randomness assumptions about the system of equations. In its cryptanalytic application to Picnic, we demonstrate how to further optimize the algorithm for solving structured equation systems that are constructed from specific cryptosystems.

1 Introduction

The security of many cryptographic schemes is based on the conjectured hardness of solving systems of polynomial equations over a finite field. This problem is known to be NP-hard even for systems of quadratic equations over $\mathbb{F}_2$.

A. Canteaut and F.-X. Standaert (Eds.): EUROCRYPT 2021, LNCS 12696, pp. 374–403, 2021.
https://doi.org/10.1007/978-3-030-77870-5_14

The input to the problem consists of m polynomials in n variables over a finite field $\mathbb{F}$, denoted by $E = \{P_j(x_1, \ldots, x_n)\}_{j=1}^{m}$, where each polynomial is given as a sum of monomials. The algebraic degree of each polynomial is bounded by a small constant d. The goal is to find a solution of the system, namely $\hat{x} = (\hat{x}_1, \ldots, \hat{x}_n) \in \mathbb{F}^n$ such that $P_j(\hat{x}) = 0$ for every $j \in \{1, \ldots, m\}$, or to determine that a solution does not exist.[1]

In this paper, we will be interested in the concrete (rather than asymptotic) complexity of algorithms for solving polynomial systems over the field $\mathbb{F}_2$ and in their applications to cryptanalysis.

1.1 Previous Algorithms for Solving Polynomial Equation Systems

The problem of solving polynomial equation systems over finite fields is widely studied. We give a brief overview of the main algorithms that were applied in cryptanalysis.

Classical techniques developed to solve polynomial systems attempt to find a reduced representation of the ideal generated by the polynomials in the form of a Gröbner basis (e.g., the F4 [17] and F5 [18] algorithms). These methods have had success in solving some very structured polynomial systems that arise from certain cryptosystems (e.g., see [19]), but it is difficult to estimate their complexity in solving arbitrary systems. Related methods such as XL [11] and its variants typically work well only for largely over-defined systems in which $m \gg n$.

In [3] Bardet et al. analyzed the problem of solving quadratic equations over $\mathbb{F}_2$ and devised an algorithm that combines exhaustive search and sparse linear algebra. The authors estimated the asymptotic complexity of their randomized algorithm for $m = n$ by $O(2^{0.792n})$, under some algebraic assumptions that were empirically found to hold for random systems. However, due to a large overhead hidden in the asymptotic formula, the authors expect their algorithm to beat exhaustive search only when the number of variables is at least 200. Another algorithm based on a different hybrid approach was published by Joux and Vitse, who gave experimental evidence that it outperforms in practice previous algorithms for a wide range of parameters [22]. Analyzing the complexity of the algorithm is non-trivial, but according to the recent work of Duarte [16], the algorithm for solving quadratic systems over $\mathbb{F}_2$ with $m = n + 1$ does not beat existing algorithms (such as the one of [3]) asymptotically.

Finally, we mention the work of Bouillaguet et al. [8], which devised an optimized exhaustive search algorithm for solving polynomial systems of degree d over $\mathbb{F}_2$ whose complexity is $2d \log n \cdot 2^n$ bit operations.

1.2 The Polynomial Method

In [27] Lokshtanov et al. presented the first worst-case algorithms for solving polynomial equations over finite fields that have exponential speedup over

[1] We denote an assignment to the formal variable vector x in the polynomial $P_j(x)$ by $\hat{x}$ and the value of $P_j(x)$ on this assignment by $P_j(\hat{x})$.

exhaustive search. These algorithms were based on a technique known as the *polynomial method*. It was borrowed from circuit complexity [4] and recently applied in algorithm design (see [35] for a survey). The randomized algorithm of Lokshtanov et al. for solving equations over $\mathbb{F}_2$ has runtime of $O(2^{0.8765n})$ for quadratic systems and $O(2^{(1-1/(5d))n})$ in general. Following [27], Björklund, Kaski and Williams [7] reduced the complexity of these algorithms to $O(2^{0.804n})$ for $d = 2$ and $O(2^{(1-1/2.7d)n})$ in general. More recently, these complexities were further improved in [12] by Dinur to $O(2^{0.6943n})$ for $d = 2$ and $O(2^{(1-1/2d)n})$ for $d > 2$.

Although these recent algorithms have better asymptotic complexity than exhaustive search, a close examination reveals that their concrete (non-asymptotic) complexity is above 2^n for parameter ranges that are relevant to cryptography.

1.3 Our Results

We introduce the polynomial method for solving multivariate equation systems over $\mathbb{F}_2$ as a tool in cryptanalysis. For this purpose, we devise a concretely efficient algorithm for solving such systems.

Our algorithm is relatively simple and its analysis assumes the degree d polynomials are selected uniformly at random. Up to small constants, we bound the complexity of our algorithm by $n^2 \cdot 2^{0.815n}$ bit operations for $d = 2$, and $n^2 \cdot 2^{(1-1/2.7d)n}$ for $d \geq 2$. The analysis we present here is heuristic, but is it formally established in the full version of this paper.

In a straightforward implementation of our algorithm, its memory complexity is significant and only about n times lower than its time complexity. In fact, this is the case for all previous polynomial method-based algorithms. We address this issue by presenting a memory-optimized variant of the algorithm which maintains roughly the same time complexity, but whose memory complexity is reduced to about $n^2 \cdot 2^{0.63n}$ bits for $d = 2$ and $n^2 \cdot 2^{(1-1/1.35d)n}$ in general.[2]

Potential fast implementation. Even after the substantial reduction in memory complexity, it remains high and would present a challenge for obtaining a fast practical implementation of the algorithm. To address this challenge, future works may present an additional reduction in memory complexity or utilize time-memory tradeoffs. Taking this optimistic viewpoint, our work may be viewed as a step towards a practically efficient implementation of a polynomial method-based algorithm for solving multivariate equation systems over $\mathbb{F}_2$.

We stress, however, that the main goal of the paper is to give a good *analytical* concrete estimate of the complexity of polynomial method algorithms for problem sizes that are too large to be solved in practice. Consequently, it can be used in the security analysis of cryptosystems and serve as a starting point for additional optimizations.

[2] Asymptotically, the polynomial factor in the memory complexity formula is between n^2 and n^3, but it is close to n^2 for relevant parameters.

Asymptotic complexity. Our algorithm can actually be viewed as a concretely efficient variant of the algorithm of [12]. The only reason that [12] seems to have better asymptotic complexity is that it uses a self-reduction to a smaller multivariate system. Essentially, each recursive call reduces the exponent in the complexity formula, where the gain diminishes with the number of recursive calls. On the other hand, each such call increases the lower order terms.

An estimated calculation suggests that a self-reduction is profitable for $d = 2$ starting from about $n = 100$. Beyond $n = 200$ for $d = 2$ the advantage in concrete complexity is made substantial using several recursive calls. On the other hand, for $d = 4$ the reduction seems profitable only beyond $n = 200$. We chose not to augment our algorithm with any self-reduction and simply replace it with exhaustive search for two reasons. First, as described below, the applications we present in this paper require solving multivariate systems with $d > 2$, for which the benefit of the self-reduction seems marginal for relevant parameters. Second, this self-reduction significantly complicates the concrete analysis, whereas we aim for simplicity. Yet, this estimation suggests that the full potential of the algorithm is still to be discovered.

Cryptanalytic applications. We estimate the concrete complexities of our algorithm for solving quadratic systems in 80, 128 and 256 variables by $2^{77}, 2^{117}$ and 2^{223} bit operations, respectively. In terms of cryptanalysis, the main targets of our algorithm for $d = 2$ are multivariate public-key cryptosystems (e.g., HFE by Patarin [30] and UOV by Kipnis, Patarin and Goubin [24]), whose security is directly based on the hardness of solving quadratic systems. However, recent multivariate cryptosystems such as GeMSS [9] (an alternate third-round candidate signature scheme in NIST's post-quantum standardization project [29]) were designed with a sufficiently large security margin and resist our attack. Nevertheless, the security margin for some of these cryptosystems seems to be reduced by our algorithm.

Interestingly, the main application of our algorithm is for solving multivariate systems of degree $d > 2$ which have generally received less attention in the literature compared to quadratic systems. In particular, we apply it to cryptanalyze recently proposed instances [23] of the Picnic signature scheme [10] (an alternate third-round candidate in NIST's post-quantum standardization project) that is based on the security of the LowMC block cipher [1].

We focus on the three Picnic instances where the LowMC block cipher has a full Sbox layer and 4 internal rounds. These instances have claimed security levels of $S \in \{128, 192, 255\}$ bits. The best-known attacks on these instances were recently published by Banik et al. [2], but they are only applicable to weakened variants where the number of LowMC rounds is reduced from 4 to 2. On the other hand, our attacks on the full 4-round instances with $S \in \{128, 192, 255\}$ have complexities of $2^{130}, 2^{188}$ and 2^{245} bit operations, respectively. Thus, 2 out of the 3 instances do not achieve their claimed security level, while the security of the instance with $S = 128$ is somewhat marginal. When optimized for time complexity, the attacks for $S \in \{128, 192, 255\}$ require about 2^{112}, 2^{164} and 2^{219} bits of memory, respectively. However, there is no consensus among researchers on a model that takes memory complexity into account and the formal security claims of the Picnic (and LowMC) designers only involve time complexity.[3]

The authors of [23] also proposed conservative instantiations where the number of LowMC rounds is increased by 1. The attack complexities for these instances with $S \in \{128, 192, 255\}$ bits become $2^{133}, 2^{192}$ and 2^{251}, respectively. Hence the instance with $S = 255$ still does not achieve the desired security level, while security remains very marginal for the strengthened instance with $S = 192$.

We also analyze round-reduced variants of the Keccak hash function [5], which was selected by NIST in 2015 as the SHA-3 standard. In particular, we describe the best-known preimage attacks on 4 rounds of Keccak-k for all $k \in \{224, 256, 384, 512\}$. These have complexities of $2^{217}, 2^{246}, 2^{374}$ and 2^{502} bit operations, respectively. We further describe the first collision attack on 4-round Keccak-512 that is (slightly) faster than the birthday bound. We consider the cryptanalysis of round-reduced Keccak as a secondary application since our attacks are very far from threatening Keccak's security.

Complexity evaluation. It is important to emphasize that the complexities of our attacks are measured in *bit operations*. On the other hand, the complexity of exhaustive search for the cryptanalytic problems we consider on a space of size 2^n is larger than 2^n bit operations. Hence the improvement we obtain over exhaustive search is more significant than it may first appear.

In particular, the encryption algorithms of the LowMC instances we cryptanalyze (that are used in Picnic) have complexities of at least 2^{17} bit operations. However, evaluating an attack in terms of the complexity of the LowMC encryption algorithm is misleading, as naive exhaustive search is not the most efficient generic attack on the 4-round LowMC instances. Indeed, breaking these instances is easily reduced to solving a multivariate system in (about) n variables with $d = 4$, for which the best-known generic attack is the optimized exhaustive search algorithm of Bouillaguet et al. [8], whose complexity is about $2d \log n \cdot 2^n = 8 \log n \cdot 2^n$ bit operations (also see [15] for an alternative algorithm). Overall, in terms of bit operations, our algorithm is more efficient than the one of [8] by a factor which is between 32 and $2^{16} = 65536$ (depending on the LowMC instance considered).

1.4 Comparison to Previous Works

The analysis of our algorithm for random equation systems over $\mathbb{F}_2$ is simple. In contrast, the analysis of previous cryptanalytic algorithms that beat brute force for random equation systems over $\mathbb{F}_2$ (particularly, the one by Bardet et al. [3]) is based on heuristic algebraic assumptions that are difficult to analyze.

The algorithm of [3] (applied to systems with $m = n$) may seem to have a slightly better asymptotic complexity than ours, but this is a misleading comparison, since (as noted above) our algorithm can be extended to have better

[3] The Picnic designers have confirmed our findings and plan to update the parameter sets accordingly.

asymptotic complexity. In terms of concrete complexity for relevant parameters, [3] only beats exhaustive search for $d = 2$ beyond $n = 200$, whereas the complexity of our algorithm for $d = 2, n = 200$ is about 2^{177}.

The algorithm of Joux and Vitse [22] performs well in practice. However, its concrete complexity has not been established analytically and it is not clear how to use it in the security analysis of cryptosystems.

Finally, previous polynomial method-based algorithms were only analyzed asymptotically. While it is difficult to calculate their exact concrete complexity, our optimizations reduce complexity by many orders of magnitude for relevant parameters.

1.5 Technical Contribution

Let $E = \{P_j(x_1, \ldots, x_n)\}_{j=1}^m$ be a polynomial system of degree d over $\mathbb{F}_2$. Algorithms based on the polynomial method consider the polynomial

$$F(x) = (1 + P_1(x))(1 + P_2(x)) \ldots (1 + P_m(x))$$

(operations are over $\mathbb{F}_2$). Note that $F(\hat{x}) = 1$ if and only if $\hat{x}$ is a solution of E. However, the degree of $F(x)$ can be as high as $d \cdot m$ and it generally contains too many monomials to manipulate efficiently. It is thus replaced by a *probabilistic polynomial* $\tilde{F}(x)$ with a lower degree that agrees with $F(x)$ on most assignments. Taking advantage of the low degree of $\tilde{F}(x)$ by using fast polynomial interpolation and evaluation algorithms allows to solve E faster than brute force.

Our main algorithm includes various concrete optimizations and simplifications to previous polynomial method-based algorithms. These are described in detail in Sect. 3. For example, we reduce the number of polynomials which need to be interpolated and evaluated and show how to jointly interpolate several polynomials with improved amortized complexity.

Then, we show how to reduce the memory complexity of the algorithm by an exponential factor with essentially no penalty in time complexity. The optimization is based on a memory-reduced variant of the Möbius transform over $\mathbb{F}_2$ which is a fast polynomial interpolation and evaluation algorithm. This variant allows to evaluate a low-degree polynomial on its entire domain with memory complexity proportional to the memory required to store the input polynomial itself (and time complexity proportional to the domain size). Although the Möbius transform is widely used and the variant we describe is simple, it seems not to be well-known. The way this variant is used in our algorithm is, however, slightly more involved.

As an additional technical contribution, we show how to optimize our algorithm for solving *structured* equation systems that are constructed from specific cryptosystems. In particular, we observe that in cryptographic settings, a probabilistic polynomial can be replaced with a deterministic construction of a polynomial that preserves the structure of the polynomials of E. We show that in some cases (e.g., in the analysis of Picnic in Sect. 5.2) this alternative polynomial has reduced degree, optimizing the attack. We view this optimization as one of the main contributions of this paper, as it may serve as a starting point for future works in cryptanalysis.

Paper structure. Next, we describe some preliminaries. We overview our algorithm in Sect. 3 and give its details in Sect. 4. Applications are described in Sect. 5.

2 Preliminaries

2.1 Boolean Algebra

Given a finite ordered set S, denote by $|S|$ its size and by $S[i]$ its i'th element. For a positive integer n, let $[n] = \{1, 2, \ldots, n\}$.

It is well-known that any Boolean function $F : \mathbb{F}_2^n \to \mathbb{F}_2$ can be uniquely described as a multilinear polynomial, whose algebraic normal form (ANF) is given by $F(x_1, \ldots, x_n) = \sum_{u \in \{0,1\}^n} \alpha_u(F) M_u(x)$, where $\alpha_u(F) \in \{0,1\}$ is the coefficient of the monomial $M_u(x) = \prod_{i=1}^n x_i^{u_i}$ (operations are over $\mathbb{F}_2$).

We denote by $\mathrm{HW}(x)$ the Hamming weight of a vector $x \in \{0,1\}^n$. The algebraic degree of a function F is defined as $\max\{\mathrm{HW}(u) \mid \alpha_u(F) \neq 0\}$. Let W_w^n be the set $\{x \in \{0,1\}^n \mid \mathrm{HW}(x) \leq w\}$. Thus, a function F of degree $d \leq n$ can be described using $|W_d^n| = \sum_{i=0}^d \binom{n}{i}$ coefficients. We simplify notation by denoting $\binom{n}{\downarrow w} = \sum_{i=0}^w \binom{n}{i}$.

For $i \in \{1, \ldots, n\}$ and $b \in \{0,1\}$, define the function $F_{x_i \leftarrow b} : \mathbb{F}_2^{n-1} \to \mathbb{F}_2$ by

$$F_{x_i \leftarrow b}(x_1, \ldots, x_{i-1}, x_{i+1}, \ldots, x_n) = F(x_1, \ldots, x_{i-1}, b, x_{i+1}, \ldots, x_n).$$

Interpolation. Any ANF coefficient $\alpha_u(F)$ can be interpolated by summing (over $\mathbb{F}_2$) over $2^{\mathrm{HW}(u)}$ evaluations of F: for $u \in \{0,1\}^n$, define $I_u = \{i \in \{1, \ldots, n\} \mid u_i = 1\}$ and let $S_u = \{x \in \{0,1\}^n \mid I_x \subseteq I_u\}$. Then,

$$\alpha_u(F) = \sum_{\hat{x} \in S_u} F(\hat{x}). \tag{1}$$

Indeed, among all monomials only $M_u(\hat{x})$ attains a value of 1 an odd number of times in the expression $\sum_{\hat{x} \in S_u} F(\hat{x}) = \sum_{\hat{x} \in S_u} \sum_{v \in \{0,1\}^n} \alpha_v(F) M_v(\hat{x})$.

Proposition 2.1. *Let $F : \mathbb{F}_2^n \to \mathbb{F}_2$ be a Boolean function. For some $1 \leq n_1 \leq n$, partition its n variables into two sets $y_1, \ldots, y_{n-n_1}, z_1, \ldots, z_{n_1}$. Given the* ANF *of F, write it as $F(y, z) = (z_1 \ldots z_{n_1}) F_1(y) + F_2(y, z)$ by factoring out all the monomials that are multiplied with $z_1 \ldots z_{n_1}$. Then, $F_1(y) = \sum_{\hat{z} \in \{0,1\}^{n_1}} F(y, \hat{z})$.*

The proposition follows from (1) by considering the polynomial $F_1(y)$ as the symbolic coefficient of the monomial $z_1 \ldots z_{n_1}$. Observe that if $F(y, z)$ is of degree d then $F_1(y)$ is of degree at most $\max(d - n_1, 0)$.

Remark 2.1. Proposition 2.1 is also at the basis of cube attacks [14]. However, in cube attacks, the z variables are public bits controlled by the attacker (e.g., plaintext bits) while the y variables are secret key bits. In our case, we will apply Proposition 2.1 in a setting where all variables are secret key bits.

2.2 Model of Computation

We estimate the complexity of a straight-line implementation of our algorithm by counting the number of bit operations (e.g., AND, OR, XOR) on pairs of bits. This ignores bookkeeping operations such as moving a bit from one position to another (which merely requires renaming of variables in straight-line programs).

2.3 Basic Algorithms

We describe the basic algorithms that our main algorithm uses as sub-procedures.

Möbius transform. Given the truth table of an arbitrary function F (as a bit vector of 2^n entries), the ANF of F can be represented as a bit vector of 2^n entries, corresponding to its 2^n coefficients. This ANF representation can be computed from the truth table of F via the *Möbius transform* over $\mathbb{F}_2^n$.

A fast algorithm for computing this transform is based on the decomposition

$$F(x_1, \ldots, x_n) = x_1 \cdot F_1(x_2, \ldots, x_n) + F_2(x_2, \ldots, x_n). \tag{2}$$

Thus, one recursively computes the ANF of $F_1(x_2, \ldots, x_n)$ and $F_2(x_2, \ldots, x_n)$. Given the evaluations of F, for every $(\hat{x}_2, \ldots, \hat{x}_n) \in \{0,1\}^{n-1}$,

$$F_2(\hat{x}_2, \ldots, \hat{x}_n) = F(0, \hat{x}_2, \ldots, \hat{x}_n)$$

and

$$F_1(\hat{x}_2, \ldots, \hat{x}_n) = F(0, \hat{x}_2, \ldots, \hat{x}_n) + F(1, \hat{x}_2, \ldots, \hat{x}_n).$$

Therefore, computing the evaluations of F_1 requires 2^{n-1} bit operations. Denoting the time complexity by $T(n)$, we have $T(n) = 2T(n-1) + 2^{n-1}$, and hence $T(n) \leq n \cdot 2^{n-1} < n \cdot 2^n$.

By (1), a function F of degree bounded by $d \leq n$ can be interpolated from its evaluations on the set W_d^n. Adapting the Möbius transform for such a function F using the decomposition above gives an algorithm with complexity $T(n,d) \leq T(n-1,d) + T(n-1,d-1) + \binom{n-1}{\downarrow d}$ and $T(n,n) \leq n \cdot 2^n$. It can be shown by induction that $T(n,d) \leq n \cdot \binom{n}{\downarrow d}$ bit operations.

The Möbius transform over $\mathbb{F}_2^n$ coincides with its inverse which corresponds to evaluating the ANF representation of F (i.e., computing its truth table). More details about the Möbius transform over $\mathbb{F}_2^n$ and its applications in crypography can be found in [21, p.285].

Memory complexity. A standard in-place implementation of the Möbius transform performs n iterations on its input vector, where in each iteration, half of the array entries are XORed to the other half. This requires 2^n bits of memory.

Fast exhaustive search for polynomial systems over $\mathbb{F}_2$ [8]. At CHES 2010 Bouillaguet et al. presented an optimized exhaustive search algorithm for enumerating over all solutions of a polynomial system over $\mathbb{F}_2$. For a polynomial system of degree d with n variables, the complexity of their algorithm is $2d \cdot \log n \cdot 2^n$. The algorithm also requires a preprocessing phase that has complexity of n^{2d}, which is negligible when d is much smaller than n. We note that the analysis of the algorithm makes some randomness assumptions about the polynomial system, and requires that the expected number of solutions to a system with m equations over n variables is about 2^{n-m}.

In this paper we will use this algorithm to find solutions inside sets of the special form $W_w^{n-n_1} \times \{0,1\}^{n_1}$ for some values of w and n_1 in time

$$2d \cdot \log n \cdot |W_w^{n-n_1} \times \{0,1\}^{n_1}| = 2d \cdot \log n \cdot 2^{n_1} \cdot \binom{n-n_1}{\downarrow w}.$$

For an arbitrary value of n_1, obtaining this complexity may not be trivial because the algorithm of [8] iterates over the search space using a Gray code, and hence it cannot be implemented on the set of low Hamming weight vectors $W_w^{n-n_1}$ (for $w < n - n_1$) in a straightforward manner.

On the other hand, the least significant bits of the vectors in the set $W_w^{n-n_1} \times \{0,1\}^{n_1}$ can be traversed using a standard Gray code, paying a penalty only every 2^{n_1} iterations. Since we can naively enumerate any set of n-bit vectors by flipping at most n bits at a time, we can conservatively estimate the multiplicative penalty by about n. Thus, the amortized penalty over [8] is about $2^{-n_1} \cdot n$. In our setting, $2^{n_1} \gg n$, and the overhead is negligible.

Remark 2.2. Asymptotically, we will set $n_1 = \Theta(n)$, hence $2^{n_1} = \omega(n)$. Concretely, the advantage of the algorithm over exhaustive search will be roughly $\frac{2^{n_1}}{n^2}$. Thus, we may assume that $2^{n_1} \gg n$ holds without loss of generality, as otherwise, the algorithm would not obtain any advantage over exhaustive search.

2.4 Probabilistic Polynomials

Previous algorithms for solving multivariate equation systems based on the polynomial method (starting with [27]) make use of probabilistic polynomials. In particular, these works use the following construction (credited to Razborov [31] and Smolensky [32]). Given m polynomial equations of degree d in the n Boolean variables $x_1, \ldots, x_n$, $E = \{P_j(x)\}_{j=1}^m$, consider the polynomial $F(x) = (1 + P_1(x))(1 + P_2(x)) \ldots (1 + P_m(x))$. Note that $\hat{x}$ is a solution of E if and only if $F(\hat{x}) = 1$. Thus, we call the polynomial F the *identifying polynomial* of E.

The degree of $F(x)$ is generally too high and we work with a probabilistic polynomial with a lower degree, defined as follows. Let $\ell < m$ be a parameter. Pick a uniformly random matrix of full rank ℓ, $A \in \mathbb{F}_2^{\ell \times m}$ and define ℓ degree d polynomials as

$$R_i(x) = \sum_{j=1}^{m} A_{i,j} \cdot P_j(x). \quad (3)$$

We note that previous works [7,12,27] did not restrict the rank of A. In our case, this restriction will slightly simplify the analysis. Let

$$\tilde{F}(x) = (1 + R_1(x))(1 + R_2(x)) \ldots (1 + R_\ell(x)) \quad (4)$$

be the identifying polynomial of the system $\tilde{E} = \{R_i(x)\}_{i=1}^{\ell}$. Note that the degree of $\tilde{F}(x)$ (denoted by $d_{\tilde{F}}$) is at most $d \cdot \ell$.

Proposition 2.2. *For any* $\hat{x} \in \{0,1\}^n$, *if* $F(\hat{x}) = 1$, *then* $\tilde{F}(\hat{x}) = 1$. *Otherwise,* $F(\hat{x}) = 0$ *and then* $\Pr[\tilde{F}(\hat{x}) = 0] \geq 1 - 2^{-\ell}$.

Proof. If $F(\hat{x}) = 1$, then $P_j(\hat{x}) = 0$ for all $j \in [m]$ and therefore $R_i(\hat{x}) = 0$ for all $i \in [\ell]$. Hence, $\tilde{F}(\hat{x}) = 1$.

Otherwise, $F(\hat{x}) = 0$. Let $v \in \mathbb{F}_2^m$ be a vector such that $v_j = P_j(\hat{x})$ and $u \in \mathbb{F}_2^\ell$, a vector such that $u_i = R_i(\hat{x})$. Note that $u = A \cdot v$. Since $F(\hat{x}) = 0$, there exists $j \in [m]$ such that $P_j(\hat{x}) = 1$ and thus $v \neq 0$. On the other hand, if $\tilde{F}(\hat{x}) = 1$, then $R_i(\hat{x}) = 0$ for all $i \in [\ell]$, implying that $u = 0$. Therefore, v is a non-zero vector in the kernel of A. Since A is a uniform matrix of full rank ℓ, any fixed non-zero vector (including v) belongs to its kernel with probability $2^{-\ell} - 2^{-m} < 2^{-\ell}$. ∎

2.5 Previous Polynomial Method Algorithms for Solving Equation Systems over $\mathbb{F}_2$

In this section we give a short description of the previous polynomial method-based algorithms of [7,12]. We focus on the parts which are most relevant to this work.

The Björklund et al. algorithm [7]. In the algorithm of [7], the search problem of finding a solution to the system E is reduced to the parity-counting problem, where the goal is to compute the parity of the number of solutions.

The first step reduces the search problem to the problem of deciding whether a solution exists. The reduction iteratively fixes one variable of the solution at a time using $\Theta(n)$ calls to the decision algorithm. Then, the decision problem is reduced to the parity-counting problem. This reduction uses the Valiant-Vazirani affine hashing [34], adding random linear equations to the system with the goal of *isolating* some solution of E (assuming a solution exists), such that it is the only solution to the new system. In this case, the output of the parity-counting algorithm is 1. The number of linear equations to add that ensures isolation with high probability depends on the logarithm of the number of solutions to E, which is unknown. Hence, the algorithm exhausts all its possible n values.

We now overview the Björklund et al. parity-counting algorithm. Algebraically, the parity of solutions of $E = \{P_j(x)\}_{j=1}^m$ is computed by $\sum_{\hat{x} \in \{0,1\}^n} F(\hat{x})$, where $F(x) = (1 + P_1(x)) \ldots (1 + P_m(x))$. This sum (parity) is computed in parts by partitioning the n variables into 2 sets $y = y_1, \ldots, y_{n-n_1}$ and $z = z_1, \ldots z_{n_1}$, where $n_1 < n$ is a parameter. Thus, $\sum_{\hat{x} \in \{0,1\}^n} F(\hat{x}) = \sum_{\hat{y} \in \{0,1\}^{n-n_1}} \sum_{\hat{z} \in \{0,1\}^{n_1}} F(\hat{y}, \hat{z})$.

Replace the polynomial $F(y,z)$ with the probabilistic polynomial $\tilde{F}(y,z) = (1 + R_1(y,z)) \dots (1 + R_\ell(y,z))$ for $\ell = n_1 + 2$, similarly to (4). By Proposition 2.2 and a union bound over all $\hat{z} \in \{0,1\}^{n_1}$, for each $\hat{y} \in \{0,1\}^{n-n_1}$, $\Pr\left[\sum_{\hat{z}\in\{0,1\}^{n_1}} \tilde{F}(\hat{y},\hat{z}) = \sum_{\hat{z}\in\{0,1\}^{n_1}} F(\hat{y},\hat{z})\right] \geq 1 - 2^{n_1-\ell} = \frac{3}{4}$. In order to compute the sum for each $\hat{y} \in \{0,1\}^{n-n_1}$ efficiently, let $G(y) = \sum_{\hat{z}\in\{0,1\}^{n_1}} \tilde{F}(y,\hat{z})$. It follows from Proposition 2.1 that the degree of $G(y)$ is at most $d_G = d \cdot \ell - n_1$. Proceed by interpolating $G(y)$ by first computing its values on the set $W^{n-n_1}_{d_G}$. For this purpose, find all solutions of the equation system $\{R_i(y,z)\}_{i=1}^{\ell}$ in $W^{n-n_1}_{d_G} \times \{0,1\}^{n_1}$ via brute force, giving $\tilde{F}(\hat{y},\hat{z})$ for $(\hat{y},\hat{z}) \in W^{n-n_1}_{d_G} \times \{0,1\}^{n_1}$. Then, compute $G(\hat{y}) = \sum_{\hat{z}\in\{0,1\}^{n_1}} \tilde{F}(\hat{y},\hat{z})$ for $\hat{y} \in W^{n-n_1}_{d_G}$. Given these values of $G(y)$, interpolate it using the Möbius transform.

Next, evaluate $G(y)$ on all $\hat{y} \in \{0,1\}^{n-n_1}$ (using the Möbius transform) to obtain the partial parity of each part $\sum_{\hat{z}\in\{0,1\}^{n_1}} \tilde{F}(\hat{y},\hat{z})$. However, each partial parity is correct only with probability $\frac{3}{4}$. Thus, repeat the above steps with $t = 48n + 1$ independent probabilistic polynomials. Obtain t suggestions for each part and take their majority to obtain the true partial parity, except with exponentially small probability. Assuming that all partial parities are computed correctly, return $\sum_{\hat{x}\in\{0,1\}^n} F(\hat{x}) = \sum_{\hat{y}\in\{0,1\}^{n-n_1}} \sum_{\hat{z}\in\{0,1\}^{n_1}} F(\hat{y},\hat{z})$.

Optimizing complexity by self-reduction. Ignoring low-order (but concretely substantial) terms, the complexity of the above algorithm is dominated by brute force on sets $W^{n-n_1}_{d_G} \times \{0,1\}^{n_1}$ and polynomial evaluations on the sets $\{0,1\}^{n-n_1}$. The complexity is thus

$$O^*(\tbinom{n-n_1}{\downarrow d_G} \cdot 2^{n_1} + 2^{n-n_1}) \tag{5}$$

(O^* hides polynomial factors in n). The parameter n_1 can be set to balance these terms and optimize complexity.

In [7], the asymptotic complexity is optimized. Instead of brute force, for each $\hat{y} \in W^{n-n_1}_{d_G}$, compute $G(\hat{y})$ by a self-reduction to a parity-counting problem with input $\{R_i(\hat{y},z)\}_{i=1}^{\ell}$, which is a system with n_1 variables $z_1, \dots, z_{n_1}$.

Dinur's algorithm for enumerating solutions [12]. The main result of the followup paper [12] is an asymptotically more efficient algorithm. Essentially, it defines a multiple parity-counting problem (which computes many parities at once), and shows that all the $\binom{n-n_1}{\downarrow d_G}$ parities returned by recursive calls in [7] can be computed more efficiently by a single recursive call to a multiple parity-counting problem. As noted in Sect. 1.3, we do not use this reduction.

Instead, we focus on the secondary result of [12], which is an algorithm for enumerating *all* solutions of a polynomial system. In our context, we will use a related technique to eliminate the initial reduction of [7] from solving E to parity-counting (which has a high concrete overhead) and replace it with a more direct way of recovering solutions from parity computations.

Isolating solutions. The first observation is that we can isolate many solutions of E at once using a variable partition $x = (y, z) = (y_1, \ldots, y_{n-n_1}, z_1, \ldots, z_{n_1})$.

Definition 2.1 (Isolated solutions). *A solution $\hat{x} = (\hat{y}, \hat{z})$ to $E = \{P_j(y, z)\}_{j=1}^{m}$ is called isolated (with respect to the variable partition (y, z)), if for any $\hat{z}' \neq \hat{z}$, $(\hat{y}, \hat{z}')$ is not a solution to E.*

The goal is to "evenly spread" solutions across the different $\hat{y}$ values. Thus, for many solutions $(\hat{y}, \hat{z})$, there is no additional solution that shares the same $\hat{y}$ value (namely, $(\hat{y}, \hat{z})$ is isolated). This is made possible by a random linear change of variables applied to the polynomials of E and a careful choice of n_1.

Enumerating isolated solutions. The second observation is that *all* solutions isolated by the variable partition (y, z) can be recovered bit-by-bit by computing $n_1 + 1$ sums (parities) for each $y \in \{0,1\}^{n-n_1}$. Let

$$V_0(y) = \sum_{\hat{z} \in \{0,1\}^{n_1}} F(y, \hat{z}), \text{and let } V_i(y) = \sum_{\hat{z} \in \{0,1\}^{n_1 - 1}} F_{z_i \leftarrow 0}(y, \hat{z})$$

for $i \in \{1, \ldots, n_1\}$, where $F(y, z) = (1 + P_1(y, z)) \ldots (1 + P_m(y, z))$.

Proposition 2.3. *Assume that $(\hat{y}, \hat{z})$ is an isolated solution of E with respect to (y, z). Then, $V_0(\hat{y}) = 1$ and $V_i(\hat{y}) = \hat{z}_i + 1$ for all $i \in \{1, \ldots, n_1\}$.*

As a result, in order to recover $(\hat{y}, \hat{z})$ (assuming $V_0(\hat{y}) = 1$), it is sufficient to compute $V_i(\hat{y})$ for each $i \in \{1, \ldots, n_1\}$. The formal polynomials $V_i(y)$ cannot be directly interpolated (they are derived from F and hence of high-degree). In [12], these sums are computed using its first algorithm.

Proof. First, since F is the identifying polynomial of E, then $V_0(\hat{y})$ counts the parity of solutions of the system $\{P_j(\hat{y}, z)\}_{i=1}^{m}$ (in which y is fixed to $\hat{y}$). Therefore, if $(\hat{y}, \hat{z})$ is an isolated solution of E with respect to (y, z), then $V_0(\hat{y}) = 1$.

Next, if $\hat{z}_i = 0$, then the assignment $(\hat{y}, \hat{z})$ continues to be an isolated solution of E with respect to (y, z) after setting $z_i = 0$ and hence $V_i(\hat{y}) = 1$. Otherwise, $\hat{z}_i = 1$, and E has no solutions after setting $y = \hat{y}$ and $z_i = 0$, implying that $V_i(\hat{y}) = 0$. In both cases, $V_i(\hat{y}) = \hat{z}_i + 1$ for all $i \in \{1, \ldots, n_1\}$ as required. ∎

Testing solutions. For each $\hat{y} \in \{0,1\}^{n-n_1}$, the algorithm computes $n_1 + 1$ sums. Those with $V_0(\hat{y}) = 1$ suggest some solution $(\hat{y}, \hat{z})$. The suggestion is correct if $(\hat{y}, \hat{z})$ is an isolated solution. Otherwise, the suggestion may be a "false alarm". Consequently, all suggested solutions are tested on E.

3 Overview of the New Algorithm

The starting point of our algorithm is the solution enumeration algorithm of [12]. We first notice that in order to find a solution to E, it is, in fact, sufficient to enumerate isolated solutions of $\tilde{E}$, (as they form a superset of the solution set of E) and test each one on E. This has a significant advantage in terms of concrete complexity, as detailed below.

We now describe how we isolate solutions of $\tilde{E}$ with respect to a variable partition $x = (y, z)$ according to a parameter n_1 and then overview our algorithm that outputs all isolated solutions of $\tilde{E}$ and tests them.

Isolating solutions. We use the following proposition.

Proposition 3.1. *Let E and $\tilde{E}$ be polynomials systems with identifying polynomials $F(x)$ and $\tilde{F}(x)$, respectively. For $n_1 = \ell - 1$, define a variable partition $x = (y, z) = (y_1, \ldots, y_{n-n_1}, z_1, \ldots, z_{n_1})$. Assume that $(\hat{y}, \hat{z})$ is an isolated solution of E. Then,* $\Pr[(\hat{y}, \hat{z})$ *is an isolated solution of* $\tilde{E}] \geq 1 - 2^{n_1 - \ell} = \frac{1}{2}$.

Proof. The proposition follows by Proposition 2.2 by a union bound over the set $\{(\hat{y}, \hat{z}') \mid \hat{z}' \neq \hat{z}\}$, whose size is $2^{n_1} - 1$. ■

Hence, assuming that E has an isolated solution, setting $\ell = n_1 + 1$ ensures that this solution is also isolated in $\tilde{E}$ with probability at least $\frac{1}{2}$. Consequently, the algorithm has to be repeated a few times (with independent probabilistic polynomials) until it is output.

We also need to argue that E has an isolated solution with high probability. The (y, z) variable partition groups a solution to E together with $2^{n_1} - 1$ different assignments. In a cryptographic setting, we may assume that each such assignment satisfies E (containing m equations) with probability 2^{-m}. Thus, a solution to E is isolated with probability at least $1 - 2^{n_1 - m}$, which is typically very close to 1, as in our case we set $n_1 < n/5$ (to optimize complexity) and we usually have $m \gg n/5$.

Enumerating isolated solutions. Similarly to [12], isolated solutions are recovered bit-by-bit by computing n_1 sums, but we enumerate isolated solutions of $\tilde{E}$ rather than E. Define the polynomials

$$U_0(y) = \sum_{\hat{z} \in \{0,1\}^{n_1}} \tilde{F}(y, \hat{z}), \text{and } U_i(y) = \sum_{\hat{z} \in \{0,1\}^{n_1 - 1}} \tilde{F}_{z_i \leftarrow 0}(y, \hat{z})$$

for $i \in \{1, \ldots, n_1\}$.

Proposition 3.2. *Assume that $(\hat{y}, \hat{z})$ is an isolated solution of $\tilde{E}$ with respect to (y, z). Then, $U_0(\hat{y}) = 1$ and $U_i(\hat{y}) = \hat{z}_i + 1$ for all $i \in \{1, \ldots, n_1\}$.*

Proof. The proposition follows from Proposition 2.3, applied with $\tilde{E}$ and $\tilde{F}$, instead of E and F. ■

Exploiting the low degree of $\tilde{F}$, our algorithm interpolates all $n_1 + 1$ polynomials $U_i(y)$ for $i \in \{0, \ldots, n_1\}$ and then evaluates each one on all $\hat{y} \in \{0,1\}^{n-n_1}$ to recover isolated solutions. [4] We optimize the interpolation of the polynomials $U_i(y)$, exploiting the following proposition.

[4] We never explicitly interpolate the probabilistic polynomial $\tilde{F}$ itself, but only the polynomials $U_i(y)$ derived from it.

Proposition 3.3. *Let $\tilde{E} = \{R_i(y,z)\}_{i=1}^{\ell}$ and let $\tilde{F}$ by its identifying polynomial. Denote by $d_{\tilde{F}}$ the degree of $\tilde{F}$ and let $w = d_{\tilde{F}} - n_1$. The polynomial $U_0(y)$ can be interpolated from the solutions of $\tilde{E}$ on the set $W_w^{n-n_1} \times \{0,1\}^{n_1}$, while $U_i(y)$ for $i \in \{1,\dots,n_1\}$ can be interpolated from the solutions of $\tilde{E}$ on the set $W_{w+1}^{n-n_1} \times \{0,1\}^{i-1} \times \{0\} \times \{0,1\}^{n_1-i}$. Hence, all the n_1+1 polynomials can be interpolated from the solutions of $\tilde{E}$ on the set $W_{w+1}^{n-n_1} \times \{0,1\}^{n_1}$.*

Proposition 3.3 shows that the domains of the system $\tilde{E}$ solved for interpolating $U_i(y)$ for $i \in \{1,\dots,n_1\}$ overlap. Instead of naively solving $\tilde{E}$ on n_1+1 overlapping domains, we solve $\tilde{E}$ on one (slightly bigger) domain. Specifically, we use the exhaustive search algorithm of [8] for this purpose. We note that the analysis of [8] requires randomness assumptions about the input system, yet the other optimizations described here for enumerating isolation solutions do not.

Proof. By Proposition 2.1, the algebraic degree of $U_0(y) = \sum_{\hat{z} \in \{0,1\}^{n_1}} \tilde{F}(y,\hat{z})$ is at most $d_{\tilde{F}} - n_1 = w$, while the algebraic degree of each $U_i(y)$ for $i \in \{1,\dots,n_1\}$ is at most $d_{\tilde{F}} - n_1 + 1 = w+1$.

Therefore, $U_0(y)$ can be interpolated from its values on the set $W_w^{n-n_1}$, where the computation of each such value requires 2^{n_1} evaluations of $\tilde{F}(y,z)$. Thus, $U_0(y)$ can be interpolated from the values of $\tilde{F}(y,z)$ on the set $W_w^{n-n_1} \times \{0,1\}^{n_1}$. Similarly, $U_i(y)$ for $i \in \{1,\dots,n_1\}$ can be interpolated given the values of $\tilde{F}(y,z)$ on the set $W_{w+1}^{n-n_1} \times \{0,1\}^{i-1} \times \{0\} \times \{0,1\}^{n_1-i}$. The proposition follows since $\tilde{F}$ is the identifying polynomial of $\tilde{E}$, and hence $\tilde{F}(\hat{y},\hat{z}) = 1$ if and only if $(\hat{y},\hat{z})$ is a solution to $\tilde{E}$. ■

Testing solutions. Similarly to [12], for each $\hat{y} \in \{0,1\}^{n-n_1}$, the algorithm computes n_1+1 sums. In our case, those with $U_0(\hat{y}) = 1$ suggest some solution $(\hat{y},\hat{z})$ to $\tilde{E}$ (and hence to E), and we need to test them. However, these tests make expensive evaluations of polynomials, which generally require about $\binom{n}{d}$ bit operations. This may lead to a large overhead, particularly for $d > 2$. In order to reduce this overhead, we repeat the algorithm a small number of times (using independent probabilistic polynomials) and test only candidate solutions that are output more than once. This is an additional concrete optimization over the second algorithm of [12], and makes use of assumptions about the input system to argue that it is unlikely for an incorrect candidate solution to be suggested more than once.

Comparison to the previous works [7,12]. Our algorithm differs from previous works in each of the three elements mentioned above.

First, unlike the worst-case setting of [7,12], isolating solutions in a cryptographic setting is essentially trivial. In particular, there is no need for a random change of variables, and the parameter n_1 will simply be chosen to optimize the complexity. In addition, the procedure of testing solutions is more efficient than the one of [12] as explained above.

Technically, a more interesting modification is that we enumerate isolated solutions of $\tilde{E}$ instead of E as in [12]. As a result, we only need to compute sums of the form $\sum_{\hat{z}\in\{0,1\}^{n_1}} \tilde{F}(\hat{y},\hat{z})$. This is a significant concrete optimization, as accurate sums of the form $\sum_{\hat{z}\in\{0,1\}^{n_1}} F(\hat{y},\hat{z})$ are too expensive to compute directly due to the high degree of F. In previous algorithms of [7,12], such sums are computed by majority voting across $48 \cdot n + 1$ evaluations of different polynomials derived from E, which have to be interpolated and evaluated. Consequently, the complexity of our algorithm is reduced by a factor of $\Omega(n)$ while additional savings are obtained via Proposition 3.3. Thus, our algorithm eliminates majority voting altogether and uses probabilistic polynomials in a different and a more direct way to solve E.

The asymptotical complexity of the algorithm is determined by two terms, similarly to (5). It could be improved using the techniques of [12], essentially by recursively solving the multiple parity-counting problem on $\tilde{E}$ for sets of the form $W_w^{n-n_1} \times \{0,1\}^{n_1}$ (rather than applying brute force). Yet, these recursive calls have a significant concrete overhead (they require working with many more probabilistic polynomials) and we do not use them, as noted in Sect. 1.3.

4 Details and Analysis of the New Algorithm

The pseudo-code of our main algorithm in given in Algorithm 1. It uses procedures 1 and 2. We now describe it in detail and then analyze its complexity.

Details of Algorithm 1. The main loop of Algorithm 1 runs until we find a solution of E. In each iteration, we define a new probabilistic set of equations $\tilde{E}$ from E and call Procedure 1 to output all candidate solutions of $\tilde{E}$. The output of Procedure 1 is a 2-dimensional $2^{n-n_1} \times (n_1+1)$ array that contains for each $\hat{y} \in \{0,1\}^{n-n_1}$, the evaluations $U_0(\hat{y})$ and $U_i(\hat{y})+1$ for $i \in \{1,\ldots,n_1\}$. Hence, by Proposition 3.2, assuming that $(\hat{y},\hat{z})$ is an isolated solution of $\tilde{E}$, then $U_0(\hat{y}) = 1$ and $U_i(\hat{y}) + 1 = \hat{z}_i$ for $i \in \{1,\ldots,n_1\}$.

We store candidate solutions of $\tilde{E}$ in an array and check whether a potential solution has been output before (for a previous probabilistic set of equations). Such a potential solution is tested against the full system E.

Details of Procedures 1 **and** 2. Procedures 1 and 2 output the potential solutions of a given system $\tilde{E}$ by interpolating the polynomials $U_i(y)$ for $i \in \{0,\ldots,n_1\}$ and evaluating them on all $\hat{y} \in \{0,1\}^{n-n_1}$.

Recall from Proposition 3.3 that $U_i(y)$ for $i \in \{0,\ldots,n_1\}$ can be interpolated by solving $\tilde{E}$ on the set $W_{w+1}^{n-n_1} \times \{0,1\}^{n_1}$, where $w = d_{\tilde{F}} - n_1$ and $d_{\tilde{F}}$ is the degree of $\tilde{F}$. In procedure 2, these solutions are output by the fast exhaustive search algorithm of [8].

We denote by L the number of solutions and store them in memory. Next, we need to compute the values of each $U_i(y)$ on the sets described in Proposition 3.3 (with the aim of interpolating it).[5] These values are computed by summing the

[5] In practice, we do not need to store all the L solutions in memory at once, but we can interleave the exhaustive search with the computation of the $U_i(y)$ values.

evaluations of $\tilde{F}(y,z)$ on the corresponding subset of $\hat{z} \in \{0,1\}^{n_1}$. The values of all the polynomials $U_i(y)$ for $i \in \{0,\ldots,n_1\}$ are computed in parallel by iterating over the solutions of the system. For each solution, we calculate its contribution to each of the relevant polynomials. Having calculated the required values of each of the polynomials, we return them.

Then, in Procedure 1, we interpolate $U_i(y)$ for $i \in \{0,\ldots,n_1\}$ using the Möbius transform. Finally, we evaluate all the n_1+1 polynomials on the full range $\hat{y} \in \{0,1\}^{n-n_1}$ using the Möbius transform and output the potential solutions.

Parameters: $n_1, d_{\tilde{F}}$
Initialization: $\ell \leftarrow n_1 + 1, w \leftarrow d_{\tilde{F}} - n_1$
1: $PotentialSolutionsList[0\ldots] \leftarrow$ NewList()
2: **for all** $k = 0, 1, \ldots$ **do**
3: Pick a uniformly random matrix of full rank ℓ, $A^{(k)} \in \mathbb{F}_2^{\ell\times m}$. Compute $\tilde{E}^{(k)} = \{R_i^{(k)}(x)\}_{i=1}^{\ell} = \{\sum_{j=1}^{m} A_{i,j}^{(k)} \cdot P_j(x)\}_{i=1}^{\ell}$
4: $CurrPotentialSolutions \leftarrow$ OutputPotentialSolutions($\{R_i^{(k)}(x)\}_{i=1}^{\ell}, n_1, w$)
5: $PotentialSolutionsList[k] \leftarrow CurrPotentialSolutions$
6: **for all** $\hat{y} \in \{0,1\}^{n-n_1}$ **do**
7: **if** $CurrPotentialSolutions[\hat{y}][0] = 1$

▷test whether $U_0(\hat{y}) = 1$, i.e., entry is valid

then
8: **for all** $k_1 \in \{0,\ldots,k-1\}$ **do**
9: **if** $CurrPotentialSolutions[\hat{y}] = PotentialSolutionsList[k_1][\hat{y}]$

▷check whether solution appears twice

then
10: $sol \leftarrow \hat{y} \| CurrPotentialSolutions[\hat{y}][1\ldots n_1]$

▷concatenate n_1 least significant bits

11: **if** TestSolution($\{P_j(x)\}_{j=1}^{m}, sol$) = TRUE **then**
12: **return** sol
13: **break** ▷continue with next $\hat{y}$

Algorithm 1: Solve($\{P_j(x)\}_{j=1}^{m}$)

4.1 Time Complexity Analysis

We now analyze the expected time complexity of the algorithm, denoted by $T = T_{n_1,d_{\tilde{F}}}(n,m,d)$, in terms of bit operations. For this purpose we define the following notation:

- $\mathcal{E}$ is the event that E has an isolated solution.
- N_k is the number main loop iterations of Algorithm 1.
- T_1 is the average complexity of an iteration of Algorithm 1, not including the complexity of testing solutions.

1: $(V, ZV[1 \ldots n_1]) \leftarrow \text{ComputeUValues}(\{R_i(y,z)\}_{i=1}^{\ell}, n_1, w)$

▷obtain values for interpolating each $U_i(y)$

2: Interpolate $U_0(y)$: apply Möbius transform to $V[1 \ldots |W_w^{n-n_1}|]$
3: **for all** $i \in \{1, \ldots, n_1\}$ **do**
4: Interpolate $U_i(y)$: apply Möbius transform to $ZV[i][1 \ldots |W_{w+1}^{n-n_1}|]$
5: $Evals[0 \ldots n_1][0 \ldots 2^{n-n_1} - 1] \leftarrow \vec{0}$ ▷init evaluation array
6: **for all** $i \in \{0, 1, \ldots, n_1\}$ **do**
7: Evaluate $U_i(y)$ on $\{0,1\}^{n-n_1}$ by Möbius transform. Store result in $Evals[i][0 \ldots 2^{n-n_1} - 1]$
8: $Out[0 \ldots 2^{n-n_1} - 1][0 \ldots n_1] \leftarrow \vec{0}$ ▷init output
9: **for all** $\hat{y} \in \{0,1\}^{n-n_1}$ **do**
10: **if** $Evals[0][\hat{y}] = 1$ **then**
11: $Out[\hat{y}][0] \leftarrow 1$ ▷indicate $U_0(\hat{y}) = 1$, i.e., entry is valid
12: **for all** $i \in \{1, \ldots, n_1\}$ **do**
13: $Out[\hat{y}][i] \leftarrow Evals[i][\hat{y}] + 1$

▷copy potential solution by flipping evaluation bit

14: **return** Out

Procedure 1: OutputPotentialSolutions$(\{R_i(x)\}_{i=1}^{\ell}, n_1, w)$

- N_s is the total number of solutions tested by Algorithm 1.
- T_s is the average complexity of testing a solution.

We have

$$T \leq N_k \cdot T_1 + N_s \cdot T_s. \tag{6}$$

Probabilistic setting. We assume that the polynomials of E are chosen independently and uniformly at random from all degree d polynomials, conditioned on having a pre-fixed solution chosen initially (e.g., a cryptographic key). A formal analysis of Algorithm 1 is given in the full version of the paper.

Below, we give a simple heuristical analysis, assuming each assignment $\hat{x}$ that is not the pre-fixed solution satisfies any polynomial equation in E with probability 1/2 independently of the other assignments and equations.

Theorem 4.1 (heuristic). *For a random equation system, the success probability of Algorithm 1 is at least* $1 - 2^{n_1 - m}$. *Given that* $m \geq 2 \cdot (n_1 + 1) + 2$ *and* $T_s \ll n_1 \cdot n \cdot 2^{n_1}$, *ignoring negligible factors, its expected running time is at most*

$$4\left(2d \cdot \log n \cdot 2^{n_1} \cdot \binom{n-n_1}{\downarrow d_{\tilde{F}} - n_1 + 1} + n_1 \cdot n \cdot 2^{n-n_1}\right) \leq \tag{7}$$

$$4\left(2d \cdot \log n \cdot 2^{n_1} \cdot \binom{n-n_1}{\downarrow n_1 \cdot (d-1) + d + 1} + n_1 \cdot n \cdot 2^{n-n_1}\right). \tag{8}$$

1: $Sols[1 \ldots L] \leftarrow$ BruteForceSystem($\{R_i(y,z)\}_{i=1}^{\ell}, n-n_1, w+1$)

▷brute force on space $W_{w+1}^{n-n_1} \times \{0,1\}^{n_1}$

2: $V[1 \ldots |W_w^{n-n_1}|] \leftarrow \vec{0}$, $ZV[1 \ldots n_1][1 \ldots |W_{w+1}^{n-n_1}|] \leftarrow \vec{0}$ ▷init values for each $U_i(y)$
3: **for all** $(\hat{y}, \hat{z}) \in Sols[1 \ldots L]$ **do**
4: **if** $\mathrm{HW}(\hat{y}) \leq w$

▷values of HW more than w do not contribute to $U_0(y)$

then
5: $index \leftarrow$ IndexOf($\hat{y}, n-n_1, w$) ▷get index of $\hat{y}$ in $W_w^{n-n_1}$
6: $V[index] = V[index] + 1$ ▷sum is over $\mathbb{F}_2$
7: **for all** $i \in \{1, \ldots, n_1\}$ **do**
8: **if** $\hat{z}_i = 0$

▷only values with $z_i = 0$ contribute to $U_i(y)$ for $i > 1$

then
9: $index \leftarrow$ IndexOf($\hat{y}, n-n_1, w+1$)
10: $ZV[i][index] = ZV[i][index] + 1$
11: **return** $V, ZV[1 \ldots n_1]$

Procedure 2: ComputeUValues($\{R_i(y,z)\}_{i=1}^{\ell}, n_1, w$)

The "negligible factors" ignored are negligible both asymptotically and for relevant concrete parameter choices. Several terms will be neglected based on assumptions that n_1 is sufficiently large (such as $2^{n_1} \gg n$). We have already used and justified such assumptions (see Remark 2.2).

The total complexity of Algorithm 1. The complexity formula (8) establishes a tradeoff between two terms. First, the term

$$2d \cdot \log n \cdot 2^{n_1} \cdot \binom{n-n_1}{\downarrow w+1} \tag{9}$$

accounts for the brute force on the space $W_{w+1}^{n-n_1} \times \{0,1\}^{n_1}$ in Procedure 2 (based on the analysis of Sect. 2.3 for random systems). The second term accounts for the evaluation of the polynomials $U_i(y)$ on $\{0,1\}^{n-n_1}$ in Procedure 1 and is

$$(n_1+1) \cdot (n-n_1) \cdot 2^{n-n_1} \leq n_1 \cdot n \cdot 2^{n-n_1} \tag{10}$$

(given that $n_1^2 + n_1 \geq n$). The free parameter n_1 is set to balance these terms and optimize the complexity. Assuming the terms are equal, based on the second term, the gain in complexity over 2^n bit operations is roughly $\frac{2^{n_1}}{8 \cdot n_1 \cdot n}$. In the full version of the paper we estimate the total complexity of Algorithm 1 by about $n^2 \cdot 2^{0.815n}$ bit operations for $d = 2$ and $n^2 \cdot 2^{(1-1/2.7d)n}$ in general. This is obtained by setting $n_1 \approx \frac{n}{2.7d}$. Table 1 (at the end of this section) shows that $n^2 \cdot 2^{(1-1/2.7d)n}$ slightly overestimates the complexity for relevant parameters.

Next, we establish Theorem 4.1 and argue that the condition $T_s \ll n_1 \cdot n \cdot 2^{n_1}$ holds both asymptotically and for relevant concrete parameter choices.

Success probability analysis. To analyze $\mathcal{E}$ (the event that E has an isolated solution), examine a specific solution to E. It is placed in a group with $2^{n_1} - 1$ additional potential solutions. We thus estimate the probability that another solution exists in this group by $2^{n_1 - m}$, and $\mathcal{E}$ holds with probability $1 - 2^{n_1 - m}$.

Time complexity analysis. The algorithm succeeds once a solution to E is isolated twice. Given that $\mathcal{E}$ holds, by Proposition 3.1, every iteration isolates a particular solution with probability at least $\frac{1}{2}$. Hence the expected number of iterations is at most $N_k \leq 2 \cdot 2 = 4$.

We analyze the most expensive operations of procedures 2 and 1, showing that the terms (9) and (10) indeed dominate. We then analyze the cost of testing solutions.

Procedure 2. As noted above, using the analysis of Sect. 2.3, the brute force complexity is given by (9). In addition, we estimate

$$L = 2^{-\ell} \cdot |W_{w+1}^{n-n_1} \times \{0,1\}^{n_1}| = \frac{1}{2} \cdot \binom{n-n_1}{\downarrow w+1}, \tag{11}$$

as the equation systems we solve by brute force have $\ell = n_1 + 1$ equations.

The complexity of computing the values of the arrays V and ZV is slightly more[6] than $(n_1+1) \cdot L$, which is negligible compared to (9), given that $2d \cdot \log n \cdot 2^{n_1} \gg \frac{1}{2}(n+1)$.

Procedure 1. The complexity of interpolating $(U_0(y), U_1(y), \ldots, U_{n_1}(y))$ from their evaluations is $n \cdot \binom{n-n_1}{\downarrow w} + n_1 \cdot n \cdot \binom{n-n_1}{\downarrow w+1} < (n_1+1) \cdot n \cdot \binom{n-n_1}{\downarrow w+1}$, which is negligible compared to (9) given that $2d \cdot \log n \cdot 2^{n_1} \gg (n_1+1) \cdot n$.

The complexity of evaluating these polynomials on $\{0,1\}^{n-n_1}$ is given in (10).

Estimating N_s. Fix an iteration pair $0 \leq i < j < N_k$. Given that we have at least $m \geq 2 \cdot (n_1+1) + 2 = 2 \cdot \ell + 2$ equations, the 2ℓ rows of $A^{(i)}$ and $A^{(j)}$ are linearly independent vectors over $\{0,1\}^m$ with high probability[7] in which case $\tilde{E}^{(i)}$ and $\tilde{E}^{(j)}$ are independent equation systems with ℓ equations. Hence, restricting these systems to $\hat{y} \in \{0,1\}^{n-n_1}$, the pair suggests the same n_1-bit solution suffix $\hat{z}$ with probability 2^{-n_1}. Considering all $\hat{y} \in \{0,1\}^{n-n_1}$, the expected number of suggested solutions is about 2^{n-2n_1}. As the number of iteration pairs is small and many systems restricted to $\hat{y}$ do no suggest any solution since $U_0(\hat{y}) = 0$, we estimate

$$N_s = 2^{n-2n_1}. \tag{12}$$

[6] We note that the operations of the IndexOf functions can be implemented with small overhead because solutions are output by the brute force algorithm in fixed order.

[7] Even if the rows of $A^{(i)}$ and $A^{(j)}$ have a few linear dependencies, it does not substantially affect the analysis.

Since (9) and (10) dominate T_1, up to now we estimate

$$T \leq N_k \cdot T_1 + N_s \cdot T_s \lessapprox 4 \cdot (2d \cdot \log n \cdot 2^{n_1} \cdot \tbinom{n-n_1}{\downarrow w+1} + n_1 \cdot n \cdot 2^{n-n_1}) + 2^{n-2n_1} \cdot T_s.$$

In order to establish Theorem 4.1, we need to show that

$$T_s \ll n_1 \cdot n \cdot 2^{n_1}, \tag{13}$$

so the final term that corresponds to the total complexity of testing candidate solutions may be neglected compared to the term $4 \cdot n_1 \cdot n \cdot 2^{n-n_1}$.

Testing solutions. Naively testing a candidate solution requires evaluating two polynomials in E on average, which has complexity of about $2 \cdot \binom{n}{\downarrow d}$ bit operations. Asymptotically, for constant d, this complexity is negligible compared to $2^{n_1} = 2^{\Omega(n)}$, hence (13) holds.

Concretely, for $d = 2$, since $n \ll 2^{n_1}$, then $n^2 \ll n_1 \cdot n \cdot 2^{n_1}$ and (13) holds. However, when $d > 2$ is relatively large compared to n, then (13) may no longer hold for relevant parameter choices (e.g., for $d = 4$ and $n = 128$).

On the other hand, we can reduce this complexity such that it becomes negligible for relevant parameter choices by tweaking Algorithm 1. The main idea is to test the potential solutions in batches, reducing the amortized complexity as described in the full version of this paper.

In practice, E is constructed from a cryptosystem, and for relatively large d one may simply test candidate solutions by directly evaluating the cryptosystem.

Experimental validation. The most important probabilistic quantities analyzed are L (11) and N_s (12) (the bound on the expected value of N_k is rigourous). As detailed in the full version of this paper, we experimentally calculated their values (for random equation systems with $m = n$), and conclude that our estimate of L is accurate, while the estimate of N_s is somewhat conservative.

4.2 Optimizing Memory Complexity

The expected memory complexity of the algorithm is about $4 \cdot (n_1+1) \cdot 2^{n-n_1}$ bits, dominated by storing the potential solutions output by the different executions of Procedure 1. A simple way to obtain a time-memory tradeoff is to guess several bits of x and repeat the algorithm for each guess. However, we can improve the memory complexity with essentially no penalty by making use of a memory-efficient implementation of the Möbius transform.

Memory-efficient Möbius transform. We deal with the problem of evaluating a polynomial $F(x_1, \ldots, x_n)$ of degree d on the space $\{0,1\}^n$ using the Möbius transform. Assume that d is not too large and the polynomial is represented by a bit array of size $\binom{n}{\downarrow d} \ll 2^n$. Moreover, assume that the application does not need to store the evaluation of the polynomial on the full space, but can work even if the space is partitioned into smaller spaces on which the polynomial is evaluated on the fly. Then, the memory complexity can be reduced as follows. Instead of allocating an array of size 2^n, we work directly with the recursive formula (2), $F(x_1, \ldots, x_n) = x_1 \cdot F_1(x_2, \ldots, x_n) + F_2(x_2, \ldots, x_n)$, by first evaluating $F_2(x_2, \ldots, x_n)$ (i.e., $F(x)$ for $x_1 = 0$), and then calculating and evaluating $F_1(x_2, \ldots, x_n) + F_2(x_2, \ldots, x_n)$ (i.e., $F(x)$ for $x_1 = 1$).

This algorithm does not work in-place, but only keeps in memory the recursion stack. The memory complexity is bounded by the formulas $M(n, d) = M(n-1, d) + \binom{n}{\downarrow d}$ and $M(n, n) = 2^n$. Thus, the total memory complexity is less than $n \cdot \binom{n}{\downarrow d}$. The time complexity in bit operations is bounded by $\binom{n}{\downarrow d} + 2 \cdot \binom{n-1}{\downarrow d} + \ldots + 2^{n-d-1} \cdot \binom{d+1}{\downarrow d} + 2^{n-d} \cdot d \cdot 2^d$.

Remark 4.1. A more precise evaluation reveals that the total number of bit operations is about $d \cdot 2^n$. The exhaustive search algorithm of [8] for enumerating all zeroes of a polynomial of degree d also requires $d \cdot 2^n$ bit operations. It would be interesting to investigate whether the recursive Möbius transform can compete with [8] in practice. We note that it was already observed in [15] that the Möbius transform on degree d polynomials requires $d \cdot 2^n$ bit operations, but the algorithm used a standard implementation with memory complexity of 2^n.

In our context, we will exploit the lower complexity of the top level recursive calls to further reduce the memory complexity, while keeping the time complexity below $n \cdot 2^n$. Specifically, for a parameter $k \approx n - \log \binom{n}{\downarrow d}$, we perform the top k levels of the recursion independently without saving the recursion stack (i.e., we recursively evaluate the input polynomial on all 2^k values of $x_1, \ldots, x_k$ independently). At the bottom levels, we switch to the in-place implementation of the Möbius transform to evaluate the polynomial on all values of $x_{k+1}, \ldots, x_n$. In order to perform the independent evaluations, we only need to allocate two additional arrays, one for the input to the recursive call and one for its output. The roles of these arrays are interchanged on every recursive call. The memory required for each such additional array is bounded by $\binom{n}{\downarrow d}$. The memory required for the in-place Möbius transform is $2^{n-k} \approx \binom{n}{\downarrow d}$. Therefore, the in-place transform does not require additional memory and the total memory complexity is bounded by $3 \cdot \binom{n}{\downarrow d}$. The time complexity of the procedure is bounded by $2^k \cdot \left(\binom{n}{\downarrow d} + \ldots + \binom{n-k}{\downarrow d}\right) + (n-k) \cdot 2^n < n \cdot 2^n$.

We note that the algorithm of [8] could also be used for the same purpose. However, it requires a preprocessing phase of complexity n^{2d}. On the other hand, we will use the Möbius transform variant with relatively large d (e.g., $d = n/3$), for which such preprocessing is too expensive.

Improving the memory complexity of algorithm 1. In order to reduce the memory complexity we first interpolate the polynomials $(U_0(y), \ldots, U_{n_1}(y))$ for several executions (e.g., 4 or a bit more) of Procedure 1 in advance. Using the recursive version of the Möbius transform (as described in Sect. 2.3), the additional memory required is negligible.

The main idea that allows to save memory is to interleave the tasks of evaluating all the polynomials (in parallel) with testing solutions that are suggested at least twice. The parallel evaluation is performed using the memory-optimized implementation of the Möbius transform. This reduces the memory complexity to about 3 times the memory required to store all the polynomials. In fact, for the purpose of testing solutions, we only need to keep the evaluations of each such polynomial on a space proportional to its size, used by the in-place transform. Thus, when sequentially calculating several transforms, we reuse one of the two additional allocated arrays. In total, we require 2 times the memory used for storing all the polynomials, namely

$$8 \cdot (n_1 + 1) \cdot \binom{n-n_1}{\downarrow d_{\tilde{F}} - n_1 + 1}. \tag{14}$$

Choosing n_1 that minimizes time complexity (balancing the two terms of (7)), gives $\binom{n-n_1}{\downarrow d_{\tilde{F}} - n_1 + 1} \approx \frac{n_1 \cdot n}{2d \cdot \log n} \cdot 2^{n-2n_1}$ and total memory complexity of about $\frac{4 \cdot n_1 \cdot (n_1+1) \cdot n}{d \cdot \log n} \cdot 2^{n-2n_1}$ bits. Compared to Algorithm 1, this saves a multiplicative factor of about $\frac{d \cdot \log n}{n_1 \cdot n} \cdot 2^{n_1}$. Since $n_1 < \frac{n}{5}$, $\frac{8 \cdot n_1 \cdot (n_1+1) \cdot n}{2d \cdot \log n} \approx n^2$ for relevant concrete parameters.

Asymptotically, the memory complexity is $O(n^3 \cdot 2^{n-2n_1})$ bits. A choice of n_1 that minimizes time complexity gives $O\left(n^3 \cdot 2^{0.63n}\right)$ bits for $d = 2$ and $O\left(n^3 \cdot 2^{(1-1/1.35d)n}\right)$ in general.

Concrete parameters. In Table 1 we give concrete complexity estimates for interesting parameter sets after optimizing the free parameter n_1 of (8).

Table 1. Concrete complexity of (the memory-optimized variant of) Algorithm 1

Variables n	Degree d	Internal parameter n_1	Complexity (bit operations)	Memory (bits)	Exhaustive search [8] ($2d \log n \cdot 2^n$)
80	2	16	2^{77}	2^{60}	2^{84}
128	2	25	2^{117}	2^{91}	2^{133}
128	4	12	2^{129}	2^{112}	2^{134}
192	2	37	2^{170}	2^{132}	2^{197}
192	4	18	2^{188}	2^{164}	2^{198}
256	2	49	2^{223}	2^{173}	2^{261}
256	4	25	2^{246}	2^{219}	2^{262}

5 Cryptanalytic Applications

In this section we describe cryptanalytic applications of Algorithm 1. Our main application is in cryptanalysis of Picnic (and LowMC) variants and our secondary application is in cryptanalysis of round-reduced Keccak. We begin by describing the general optimization method we use in cryptanalysis of Picnic.

5.1 Deterministic Replacement of Probabilistic Polynomials

In cryptanalytic applications, E may have some properties that depend on the underlying cryptosystem and could be ruined in $\tilde{E}$ that mixes the equations of E. We observe that in cryptographic settings, we may replace the randomized construction of $\tilde{E}$ (and $\tilde{F}$) by a deterministic construction which simply takes subsets of equations from E, thus preserving the properties of E.

Essentially, the randomness of the probabilistic constructions is already "embedded" in E itself. For example, we still (heuristically) expect a variant of Proposition 3.1 to hold: if $\hat{x}$ is a solution to E, it is also a solution to $\tilde{E}$. On the other hand, since $\tilde{E}$ is a system with $\ell = n_1 + 1$ equations in n_1 variables, we expect it not to have an additional solution with high probability.

In order for the number of tested candidate solutions to remain small, we require the different equations systems $\tilde{E}$ analyzed to be roughly independent. Specifically, the intersections of the equations subsets taken for different "probabilistic" equations systems $\tilde{E}$ should be empty (or small).

5.2 Picnic and LowMC

The Picnic signature scheme [10] is an alternate third-round candidate in NIST's post-quantum standardization project [29]. It uses a zero-knowledge protocol in order to non-interactively prove knowledge of a preimage x to a public value y under a one-way function f, where y is part of the public key and x is the secret signing key. The one-way function is implemented using a block cipher, where the secret signing key is the block cipher's key, while the public key consists of a randomly chosen plaintext and the corresponding ciphertext (the encryption of the plaintext with the secret key). Thus a key-recovery attack on Picnic reduces to finding the block cipher's secret key from one plaintext-ciphertext pair.

Picnic uses the LowMC block cipher family, proposed at EUROCRYPT 2015 by Albrecht et al. [1]. It is optimized for practical instantiations of multi-party computation, fully homomorphic encryption, and zero-knowledge proofs, in which non-linear operations are typically much more expensive than linear ones. Consequently, LowMC uses a relatively small number of multiplications.

LowMC is an SP-network built using several rounds, where in each round, a round-key is added to the state, followed by an application of a linear layer and a non-linear layer (operations are over $\mathbb{F}_2$). Finally, an additional round-key is added to the state. Importantly, the key schedule of LowMC is linear.

Each non-linear layer of LowMC consists of identical Sboxes $\mathcal{S} : \{0,1\}^3 \to \{0,1\}^3$ of algebraic degree 2. The algebraic normal form of an Sbox is

$$\mathcal{S}(a_1, a_2, a_3) = (a_1 + a_2 a_3, a_1 + a_2 + a_1 a_3, a_1 + a_2 + a_3 + a_1 a_2). \tag{15}$$

We note that the inverse Sbox also has algebraic degree of 2.

In this paper, we focus on LowMC instances that were recently integrated into Picnic variants [33]. These instances have internal state and key sizes of 129, 192 and 255, claiming security levels of 128, 192 and 255 bits, respectively and have a full non-linear layer (as opposed to other instances of LowMC that have a partial non-linear layer). All of these instances have 4 rounds, although in a recent publication by some of the designers [23] additional instances with 5 rounds were proposed in order to provide a larger security margin.

Attacks on LowMC instances. As noted above, we analyze LowMC instances with a full Sbox layer given only a single plaintext-ciphertext pair. The best-known attacks on such instances were recently published by Banik et al. [2] where the authors analyzed instances reduced to 2 rounds. Their techniques are based on linearization and it is not clear how to extend them beyond 2 rounds without exceeding the complexity of (optimized) exhaustive search [8].

We focus on the full 4 and 5-round instances. We fix an arbitrary plaintext-ciphertext pair to a LowMC instance with key and internal state size of n bits. We denote the unknown key by $x = (x_1, \ldots, x_n)$.

Even round number. We begin by considering LowMC instances with an even number of rounds r. We focus on an arbitrary state bit b_i (for some $i \in \{1, \ldots, n\}$) after $r/2$ rounds. Starting from the plaintext, we symbolically evaluate the encryption process and express b_i as a polynomial $p_i(x)$. Similarly, we symbolically evaluate the decryption process starting from the ciphertext and express b_i as a polynomial $q_i(x)$. This standard meet-in-the-middle approach gives rise to the equation $p_i(x) + q_i(x) = 0$. As the algebraic degree of the LowMC round and its inverse is 2 and the key schedule is linear, the algebraic degree of both $p_i(x)$ and $q_i(x)$ is at most $2^{r/2}$. Repeating this process n times for all intermediate state bits, we obtain an equation systems with $m = n$ equations. For the small values of r we consider, the complexity of calculating the equation system is negligible. We can now apply Algorithm 1 and solve for the secret key.

Odd round number. For an odd number of rounds r, the approach above gives rise to equations of degree at least $2^{(r+1)/2}$, as in general, any intermediate state bit has algebraic degree of at least $2^{(r+1)/2}$ from either the encryption or the decryption side. If we apply Algorithm 1 in a straightforward manner, its complexity compared to an attack on an even number of $r - 1$ rounds would increase substantially. We now show that by a better choice of the equation system and careful analysis we can reduce the algorithm's complexity.

Consider the first 3 intermediate state bits b_1, b_2, b_3 that are outputs of the first Sbox in round $(r+1)/2$. From the decryption side, we can express them as polynomials of degree $2^{(r-1)/2}$, denoted by $q_1(x), q_2(x), q_3(x)$, respectively. From the encryption side, based on (15), we can express these bits as functions of the bits a_1, a_2, a_3 that are inputs the first Sbox in round $(r+1)/2$,

$$(b_1, b_2, b_3) = \mathcal{S}(a_1, a_2, a_3) = (a_1 + a_2a_3, a_1 + a_2 + a_1a_3, a_1 + a_2 + a_3 + a_1a_2).$$

From the encryption side, we can express each of a_1, a_2, a_3 as a polynomial of degree $2^{(r-1)/2}$ in the key. Equating each bit to its evaluation from the decryption side, we obtain the 3 equations

$$
\begin{aligned}
q_1(x) + a_1(x) + a_2(x)a_3(x) &= 0, \\
q_2(x) + a_1(x) + a_2(x) + a_1(x)a_3(x) &= 0, \\
q_3(x) + a_1(x) + a_2(x) + a_3(x) + a_1(x)a_2(x) &= 0.
\end{aligned}
$$

Each polynomial appearing in these equations is of algebraic degree $2^{(r-1)/2}$.

Recall that the main complexity formula (7) heavily depends on the value of $d_{\tilde{F}}$. This value is upper bounded by $d \cdot \ell$, but it could be lower if we choose $\tilde{F}$ more carefully. Indeed, we will construct the "probabilistic polynomials" deterministically using (non-overlapping) subsets of the original equation system E. The probabilistic analysis is essentially unchanged, as suggested in Sect. 5.1.

Specifically, in this case, each equation is of degree $2^{(r-1)/2} + 2^{(r-1)/2} = 2^{(r+1)/2}$ due to the multiplication of the a_i's. However, if we multiply the 3 polynomials for the purpose of calculating $\tilde{F}$ (as in (4), but with R_i's replaced by the equations above), the term $a_1(x)a_2(x)a_3(x)$ can only be multiplied with at most one of the $q_i(x)$'s and therefore the degree of multiplication of all the equations is at most $4 \cdot 2^{(r-1)/2} = 2^{(r+3)/2}$, rather than the trivial upper bound of $6 \cdot 2^{(r-1)/2}$, reducing the degree by a factor of $\frac{1}{3}$. For example, for $r = 3$, we get a bound of 8 instead of the general upper bound of 12, whereas for $r = 5$, we get a bound of 16 instead of the general upper bound of 24.

We proceed to collect n equations as before. However, in Algorithm 1, for some integer ℓ' we analyze equations that are computed as above using ℓ' triplets that are outputs of the Sbox layer of round $(r+1)/2$. We now have $\ell = 3\ell'$ equations. The total degree of $\tilde{F}$ in (4) is upper bounded by $d_{\tilde{F}} \leq \ell' \cdot 2^{(r+3)/2} = \ell/3 \cdot 2^{(r+3)/2}$. We choose $n_1 = \ell - 1$ as before. Revisiting the complexity analysis formula of (7), we obtain

$$
4\left(2 \cdot 2^{(r+1)/2} \cdot \log n \cdot 2^{n_1} \cdot \binom{n-n_1}{\downarrow \frac{\ell}{3} \cdot 2^{(r+3)/2} - n_1 + 1} + n_1 \cdot n \cdot 2^{n-n_1}\right). \tag{16}
$$

If we take ℓ which is not a multiple of 3, then $d_{\tilde{F}}$ increases more sharply due to the last Sbox. Specifically, if $\ell \bmod 3 = 1$, then $d_{\tilde{F}} \leq \frac{\ell-1}{3} \cdot 2^{(r+3)/2} + 2^{(r+1)/2}$, whereas if $\ell \bmod 3 = 2$, then $d_{\tilde{F}} \leq \frac{\ell-2}{3} \cdot 2^{(r+3)/2} + 3 \cdot 2^{(r-1)/2}$. Standard approaches to deal with the middle Sbox layer (e.g., linearization) result in higher complexity.

Results. Table 2 summarizes our attacks on instances of LowMC used in recent Picnic variants (and additional 5-round instances). Solutions can be tested simply by evaluating the LowMC encryption process, whose most expensive procedures are the evaluations of the linear layers, each consisting of a multiplication of an n-bit state with an $n \times n$ matrix. Naively, this has complexity of $2n^2$ bit operations. It can be checked that (13) holds for the parameters of Table 2.

5.3 Keccak

Keccak is a family of cryptographic functions, designed by Bertoni et al. in 2008 [5]. We focus on the Keccak hash function family, selected by NIST in 2015 as the SHA-3 standard. It is built using the sponge construction (cf. [6]) using a

Table 2. Attacks on LowMC instances with 4 rounds (used in Picnic) and 5 rounds

Security level S	Key length n	Rounds r	Internal parameters $(n_1, d, d_{\tilde{F}})$	Attack complexity (bit operations)
128	129	4	$(12, 4, 52)$	2^{130}
192	192	4	$(18, 4, 76)$	2^{188}
192	192	5	$(14, 8, 80)$	2^{192}
255	255	4	$(25, 4, 104)$	2^{245}
255	255	5	$(18, 8, 104)$	2^{251}

permutation that operates on a 1600-bit state. The permutation consists of 24 rounds, where each round consists of an application of a non-linear layer, followed by linear operations over $\mathbb{F}_2$. Importantly, the non-linear layer is of algebraic degree 2. We analyze the 4 basic Keccak variants which are parameterized by the output size of k bits and denoted by Keccak-k for $k \in \{224, 256, 384, 512\}$.

Preimage attacks of round-reduced Keccak. We consider messages of length that is smaller than the *rate* of the hash function (so that the output is produced after a single invocation of the permutation). We start by representing the message (preimage) bits as symbolic variables. We then linearize the first round of Keccak by setting some linear constraints on the variables, such that each state bit in the second round of the permutation is a linear polynomial in these variables. Using the linearization technique of [13,20] for Keccak-k (that selects variables that keep the *column parities* constant), this leaves more than 224 and 256 free variables for Keccak-224 and Keccak-256, respectively, and 256 and 128 free variables for Keccak-384 and Keccak-512, respectively (for the SHA-3 versions, the number is slightly smaller).

For Keccak-384 and Keccak-512, we can further partially invert the final non-linear layer (applied to the first 5×64 Sboxes) on the target image to obtain its input values. For Keccak-224 and Keccak-256, not all these 5×64 output bits are fixed by the image. However, we can obtain 192 and 256 linear relations among the input bits of the final Sbox layer, for Keccak-224 and Keccak-256 respectively (e.g., see [20]). Having peeled off 2 out of the 4 non-linear layers, we obtain equations of degree $2^2 = 4$ and solve for the preimage using Algorithm 1.

Results. We begin by considering preimage attacks on 4 rounds of Keccak-224 and Keccak-256. These instances were recently analyzed in by Li and Sun [25], who devised attacks with claimed complexities of 2^{207} and 2^{239}, on Keccak-224 and Keccak-256, respectively. However, the analysis of these attacks ignores the complexity of solving (numerous) linear equation systems over $\mathbb{F}_2$ with hundreds of variables, each requiring many thousands of bit operations.

For Keccak-224 and Keccak-256 we have sufficiently many free variables to obtain systems with 224 and 256 variables (respectively), which we assume to have a solution. Consequently, our preimage attacks have complexities of 2^{217}

and 2^{246} bit operations by choosing $n_1 = 22$ and $n_1 = 25$, respectively. These are lower complexities than those obtained in [25] in terms of bit operations.

For Keccak-384, the number of free variables is only 256, so we need to solve systems of degree 4 with 256 variables an expected number of $2^{384-256} = 2^{128}$ times (with different initial linear constraints on the variables) to obtain a solution. This requires time $2^{246+128} = 2^{374}$ (by choosing $n_1 = 25$). In terms of bit operations, this improves upon the recent result of [26], which estimated the attack complexity by 2^{375} evaluations of the 4-round Keccak-384 function.

For Keccak-512, we do not linearize the first round, but directly solve a system of degree 8 in 512 variables. This requires 2^{502} bit operations (by choosing $n_1 = 26$) and improves the previous attack [28] that requires 2^{506} Keccak calls.

Collision attacks on round-reduced Keccak. The problem of finding a collision can be formulated as a non-linear equation system (where the variables are the bits of two colliding messages). However, the complexity of solving such a system is unlikely to be more efficient than a generic birthday attack on the k-bit hash function which takes time $2^{k/2}$. A better idea is to directly speed up the generic birthday attack. In order to do so, for a parameter ℓ, we fix ℓ bits of the output to an arbitrary value v and try to find $2^{(k-\ell)/2}$ messages whose output value on the ℓ bits is equal to v. With high probability, these messages contain a pair whose outputs also agree of the remaining $k - \ell$ bits that we have not fixed, and therefore constitute a colliding pair. In order to find $2^{(k-\ell)/2}$ such messages, we apply Procedure 1.

Suppose we run Procedure 1 with n variables after fixing $\ell = n_1 + 1$ output bits. We expect the output to contain about $\frac{1}{2} \cdot 2^{n-n_1}$ isolated solutions (messages) that satisfy the n_1 constraints. We evaluate the hash function on each output message, and test whether it is indeed a solution. We store all the true solutions and sort them, trying to find a collision among them. If $\frac{1}{2} \cdot 2^{n-n_1} < 2^{(k-\ell)/2}$ (e.g., we lack degrees of freedom and are forced to choose a small value of n), we repeat the procedure several times (with a different set of message variables), while storing all the produced messages until a collision is found.

Since we test all outputs of Procedure 1, the complexity of this attack also directly depends on the number of bit operations required to evaluate the hash function on some message. We denote this number by τ.

We apply the framework to 4-round Keccak-512, as there are no published attacks better than the birthday bound on this variant. Assuming $\tau = 2^{13}$ (hence the complexity of the birthday attack is $2^{256+13} = 2^{269}$ bit operations), we set $d = 4$ and $n = 128$ (unlike the preimage attack, we linearize the first round) and choose $n_1 = \ell - 1 = 12$. Calculation reveals that the complexity is about 2^{263} bit operations, which is roughly 64 times faster than the birthday attack.[8] If we assume a larger value of τ, the complexity of the attack will increase, but also its relative advantage compared to the birthday attack.

[8] The complexity of sorting the $2^{(512-13)/2} = 2^{249.5}$ images is estimated to be smaller than 2^{262} bit operations.

Acknowledgements. The author was supported by the Israeli Science Foundation through grant No. 573/16 and grant No. 1903/20, and by the European Research Council under the ERC starting grant agreement No. 757731 (LightCrypt).

References

1. Albrecht, M.R., Rechberger, C., Schneider, T., Tiessen, T., Zohner, M.: Ciphers for MPC and FHE. In: Oswald, E., Fischlin, M. (eds.) EUROCRYPT 2015. Part I. LNCS, vol. 9056, pp. 430–454. Springer, Heidelberg (2015). https://doi.org/10.1007/978-3-662-46800-5_17
2. Banik, S., Barooti, K., Durak, F.B., Vaudenay, S.: Cryptanalysis of LowMC instances using single plaintext/ciphertext pair. IACR Trans. Symmetric Cryptol. **2020**(4), 130–146 (2020). https://tosc.iacr.org/index.php/ToSC/article/view/8751
3. Bardet, M., Faugère, J., Salvy, B., Spaenlehauer, P.: On the complexity of solving quadratic Boolean systems. J. Complex. **29**(1), 53–75 (2013)
4. Beigel, R.: The polynomial method in circuit complexity. In: Proceedings of the Eigth Annual Structure in Complexity Theory Conference, San Diego, CA, USA, 18–21 May 1993, pp. 82–95. IEEE Computer Society (1993)
5. Bertoni, G., Daemen, J., Peeters, M., Assche, G.V.: The Keccak reference. https://keccak.team/files/Keccak-reference-3.0.pdf
6. Bertoni, G., Daemen, J., Peeters, M., Van Assche, G.: On the indifferentiability of the sponge construction. In: Smart, N.P. (ed.) EUROCRYPT 2008. LNCS, vol. 4965, pp. 181–197. Springer, Heidelberg (2008). https://doi.org/10.1007/978-3-540-78967-3_11
7. Björklund, A., Kaski, P., Williams, R.: Solving systems of polynomial equations over GF(2) by a parity-counting self-reduction. In: Baier, C., Chatzigiannakis, I., Flocchini, P., Leonardi, S. (eds.) 46th International Colloquium on Automata, Languages, and Programming, ICALP 2019, Patras, Greece, 9–12 July 2019. LIPIcs, vol. 132, pp. 26:1–26:13. Schloss Dagstuhl - Leibniz-Zentrum für Informatik (2019)
8. Bouillaguet, C., et al.: Fast exhaustive search for polynomial systems in $\mathbb{F}_2$. In: Mangard, S., Standaert, F.-X. (eds.) CHES 2010. LNCS, vol. 6225, pp. 203–218. Springer, Heidelberg (2010). https://doi.org/10.1007/978-3-642-15031-9_14
9. Casanova, A., Faugère, J.C., Macario-Rat, G., Patarin, J., Perret, L., Ryckeghem, J.: GeMSS: A Great Multivariate Short Signature. Submission to NIST (2017). https://www-polsys.lip6.fr/Links/NIST/GeMSS.html
10. Chase, M., et al.: Post-quantum zero-knowledge and signatures from symmetric-key primitives. In: Thuraisingham, B.M., Evans, D., Malkin, T., Xu, D. (eds.) Proceedings of the 2017 ACM SIGSAC Conference on Computer and Communications Security, CCS 2017, Dallas, TX, USA, 30 October–03 November 2017, pp. 1825–1842. ACM (2017)
11. Courtois, N., Klimov, A., Patarin, J., Shamir, A.: Efficient algorithms for solving overdefined systems of multivariate polynomial equations. In: Preneel, B. (ed.) EUROCRYPT 2000. LNCS, vol. 1807, pp. 392–407. Springer, Heidelberg (2000). https://doi.org/10.1007/3-540-45539-6_27
12. Dinur, I.: Improved algorithms for solving polynomial systems over GF(2) by multiple parity-counting. In: Marx, D. (ed.) Proceedings of the 2021 ACM-SIAM Symposium on Discrete Algorithms, SODA 2021, 10–13 January 2021, pp. 2550–2564. SIAM (2021)

13. Dinur, I., Morawiecki, P., Pieprzyk, J., Srebrny, M., Straus, M.: Cube attacks and cube-attack-like cryptanalysis on the round-reduced Keccak sponge function. In: Oswald, E., Fischlin, M. (eds.) EUROCRYPT 2015. Part I. LNCS, vol. 9056, pp. 733–761. Springer, Heidelberg (2015). https://doi.org/10.1007/978-3-662-46800-5_28
14. Dinur, I., Shamir, A.: Cube attacks on tweakable black box polynomials. In: Joux, A. (ed.) EUROCRYPT 2009. LNCS, vol. 5479, pp. 278–299. Springer, Heidelberg (2009). https://doi.org/10.1007/978-3-642-01001-9_16
15. Dinur, I., Shamir, A.: An improved algebraic attack on Hamsi-256. In: Joux, A. (ed.) FSE 2011. LNCS, vol. 6733, pp. 88–106. Springer, Heidelberg (2011). https://doi.org/10.1007/978-3-642-21702-9_6
16. Duarte, J.D.: On the Complexity of the Crossbred Algorithm. IACR Cryptology ePrint Archive 2020, 1058 (2020). https://eprint.iacr.org/2020/1058
17. Faugère, J.C.: A new efficient algorithm for computing Gröbner bases (F4). J. Pure Appl. Algebra **139**(1–3), 61–88 (1999)
18. Faugère, J.C.: A new efficient algorithm for computing Gröbner bases without reduction to zero (F5). In: Proceedings of the 2002 International Symposium on Symbolic and Algebraic Computation. ISSAC 2002, pp. 75–83. Association for Computing Machinery, New York (2002)
19. Faugère, J.-C., Joux, A.: Algebraic cryptanalysis of Hidden Field Equation (HFE) cryptosystems using Gröbner bases. In: Boneh, D. (ed.) CRYPTO 2003. LNCS, vol. 2729, pp. 44–60. Springer, Heidelberg (2003). https://doi.org/10.1007/978-3-540-45146-4_3
20. Guo, J., Liu, M., Song, L.: Linear structures: applications to cryptanalysis of round-reduced Keccak. In: Cheon, J.H., Takagi, T. (eds.) ASIACRYPT 2016. Part I. LNCS, vol. 10031, pp. 249–274. Springer, Heidelberg (2016). https://doi.org/10.1007/978-3-662-53887-6_9
21. Joux, A.: Algorithmic Cryptanalysis, 1st edn, pp. 285–286. Chapman & Hall/CRC, Boca Raton (2009)
22. Joux, A., Vitse, V.: A crossbred algorithm for solving Boolean polynomial systems. In: Kaczorowski, J., Pieprzyk, J., Pomykała, J. (eds.) NuTMiC 2017. LNCS, vol. 10737, pp. 3–21. Springer, Cham (2018). https://doi.org/10.1007/978-3-319-76620-1_1
23. Kales, D., Zaverucha, G.: Improving the performance of the picnic signature scheme. IACR Trans. Cryptogr. Hardw. Embed. Syst. **2020**(4), 154–188 (2020)
24. Kipnis, A., Patarin, J., Goubin, L.: Unbalanced oil and vinegar signature schemes. In: Stern, J. (ed.) EUROCRYPT 1999. LNCS, vol. 1592, pp. 206–222. Springer, Heidelberg (1999). https://doi.org/10.1007/3-540-48910-X_15
25. Li, T., Sun, Y.: Preimage attacks on round-reduced Keccak-224/256 via an allocating approach. In: Ishai, Y., Rijmen, V. (eds.) EUROCRYPT 2019. Part III. LNCS, vol. 11478, pp. 556–584. Springer, Cham (2019). https://doi.org/10.1007/978-3-030-17659-4_19
26. Liu, F., Isobe, T., Meier, W., Yang, Z.: Algebraic attacks on round-reduced Keccak/Xoodoo. IACR Cryptology ePrint Archive 2020, 346 (2020). https://eprint.iacr.org/2020/346
27. Lokshtanov, D., Paturi, R., Tamaki, S., Williams, R.R., Yu, H.: Beating brute force for systems of polynomial equations over finite fields. In: Klein, P.N. (ed.) Proceedings of the Twenty-Eighth Annual ACM-SIAM Symposium on Discrete Algorithms, SODA 2017, Barcelona, Spain, Hotel Porta Fira, 16–19 January, pp. 2190–2202. SIAM (2017)

28. Morawiecki, P., Pieprzyk, J., Srebrny, M.: Rotational cryptanalysis of round-reduced KECCAK. In: Moriai, S. (ed.) FSE 2013. LNCS, vol. 8424, pp. 241–262. Springer, Heidelberg (2014). https://doi.org/10.1007/978-3-662-43933-3_13
29. NIST's Post-Quantum Cryptography Project. https://csrc.nist.gov/Projects/Post-Quantum-Cryptography
30. Patarin, J.: Hidden Fields Equations (HFE) and Isomorphisms of Polynomials (IP): two new families of asymmetric algorithms. In: Maurer, U.M. (ed.) EUROCRYPT 1996. LNCS, vol. 1070, pp. 33–48. Springer, Heidelberg (1996). https://doi.org/10.1007/3-540-68339-9_4
31. Razborov, A.A.: Lower bounds on the size of bounded-depth networks over a complete basis with logical addition. Math. Notes Acad. Sci. USSR **41**(4), 333–338 (1987). https://doi.org/10.1007/BF01137685
32. Smolensky, R.: Algebraic methods in the theory of lower bounds for Boolean circuit complexity. In: Aho, A.V. (ed.) 1987 Proceedings of the 19th Annual ACM Symposium on Theory of Computing, pp. 77–82. ACM, New York (1987)
33. The Picnic Design Team: The Picnic Signature Algorithm Specification. Version 3.0, April 2020. https://microsoft.github.io/Picnic/
34. Valiant, L.G., Vazirani, V.V.: NP is as easy as detecting unique solutions. Theoret. Comput. Sci. **47**(3), 85–93 (1986)
35. Williams, R.R.: The polynomial method in circuit complexity applied to algorithm design (invited talk). In: Raman, V., Suresh, S.P. (eds.) 34th International Conference on Foundation of Software Technology and Theoretical Computer Science, FSTTCS 2014, New Delhi, India, 15–17 December 2014. LIPIcs, vol. 29, pp. 47–60. Schloss Dagstuhl - Leibniz-Zentrum für Informatik (2014)

Round-Optimal Blind Signatures in the Plain Model from Classical and Quantum Standard Assumptions

Shuichi Katsumata[1(✉)], Ryo Nishimaki[2], Shota Yamada[1], and Takashi Yamakawa[2]

[1] AIST, Tokyo, Japan
{shuichi.katsumata,yamada-shota}@aist.go.jp
[2] NTT Secure Platform Laboratories, Tokyo, Japan
{ryo.nishimaki.zk,takashi.yamakawa.ga}@hco.ntt.co.jp

Abstract. Blind signatures, introduced by Chaum (Crypto'82), allows a user to obtain a signature on a message without revealing the message itself to the signer. Thus far, all existing constructions of round-optimal blind signatures are known to require one of the following: a trusted setup, an interactive assumption, or complexity leveraging. This state-of-the-affair is somewhat justified by the few known impossibility results on constructions of round-optimal blind signatures in the plain model (i.e., without trusted setup) from standard assumptions. However, since all of these impossibility results only hold *under some conditions*, fully (dis)proving the existence of such round-optimal blind signatures has remained open.

In this work, we provide an affirmative answer to this problem and construct the first round-optimal blind signature scheme in the plain model from standard polynomial-time assumptions. Our construction is based on various standard cryptographic primitives and also on new primitives that we introduce in this work, all of which are instantiable from *classical and post-quantum* standard polynomial-time assumptions. The main building block of our scheme is a new primitive called a blind-signature-conforming zero-knowledge (ZK) argument system. The distinguishing feature is that the ZK property holds by using a quantum polynomial-time simulator against non-uniform classical polynomial-time adversaries. Syntactically one can view this as a delayed-input three-move ZK argument with a reusable first message, and we believe it would be of independent interest.

1 Introduction

1.1 Background

Blind signatures enable users to obtain a signature without revealing a message to be signed to a signer. More precisely, a blind signature scheme is a two-party computation between a signer and a user. The signer has a pair of keys called verification-key and signing-key, and the user takes as input a message and the

A. Canteaut and F.-X. Standaert (Eds.): EUROCRYPT 2021, LNCS 12696, pp. 404–434, 2021.
https://doi.org/10.1007/978-3-030-77870-5_15

verification-key. They interact with each other, and the user obtains a signature for the message after the interaction. There are two security requirements on blind signatures: (1) users cannot forge a signature for a new message (unforgeability), and (2) the signer cannot obtain information about the signed messages (blindness).

Chaum introduced the notion of blind signatures and provided a concrete instantiation, while also showing an application to e-cash systems [Cha82]. After its invention, blind signatures have been used as a crucial building block for various other privacy-preserving crypto-systems such as e-voting [FOO93, Cha88], anonymous credential [CL01], and direct anonymous attestation [BCC04].

Round-complexity. One of the main performance measures for blind signatures is round-complexity. A round-optimal blind signature is a blind signature with only 2-moves[1], where the user and signer sends one message to each other. We focus on round-optimal blind signatures in this study since a high round-complexity is one of the main bottlenecks in cryptographic systems. Another advantage is that round-optimal blind signatures are automatically secure in the concurrent setting [Lin08, HKKL07].

Round-optimal scheme in the plain model from standard assumptions. From a theoretical point of view, using less and weaker assumptions is much better. However, all existing round-optimal blind signature schemes require either (1) a trusted setup [Fis06, AO12, AFG+16, BFPV11, BPV12, MSF10, SC12, Bol03, BNPS02], (2) an interactive assumption [FHS15, FHKS16, Gha17, BNPS02, Bol03], or (3) complexity leveraging [GRS+11, GG14]. We briefly discuss each item. In the trusted setup model, if an authority set a backdoor, we can no longer guarantee any security. Interactive assumptions are non-standard compared to standard non-interactive ones since an adversary can interact with the challenger.[2] Complexity leveraging uses a gap between the computational power of an adversary and the reduction algorithm in security proofs. To create this gap, we require super-polynomial-time assumptions[3] and large parameters, which hurt the overall efficiency. In fact, there are a few impossibility results on constructing round-optimal blind signatures in the plain model (i.e., without any trusted setup) from standard assumptions *under some conditions* [Lin08, FS10, Pas11]. So far, constructing a round-optimal blind signature scheme in the plain model from standard polynomial-time assumptions has proven to be elusive.

Thus, a natural and long-standing open question is the following:

Can we achieve a round-optimal blind signature scheme in the plain model from standard polynomial-time assumptions?

---

[1] We count one move when an entity sends information to the other entity.

[2] An adversary may have the flexibility to choose a problem instance or obtain auxiliary information related to a problem instance.

[3] A super-polynomial-time assumption means that a hard problem cannot be broken even by *super-polynomial-time* adversaries. This is stronger than a standard polynomial-time assumption, where adversaries are restricted to run in polynomial-time.

We affirmatively answer this open question in this study. Hereafter, we call blind signatures that satisfy all the above conditions as a blind signature with desired properties.

1.2 Our Result

We present a round-optimal blind signature scheme with desired properties. Our construction relies on various standard cryptographic primitives such as oblivious transfer and also new primitives that we introduce in this study, all of which are instantiable from *classical and quantum* standard polynomial-time assumptions.[4]

Our construction is based on the idea by Kalai and Khurana [KK19] that we can replace complexity leveraging with classical and quantum assumptions. However, our technique is not a simple application of their idea. There are several technical hurdles to avoid complexity leveraging in blind signatures, even if we use classical and quantum assumptions. We provide further details in Sect. 1.3.

The main building block of our scheme is a blind-signature-conforming zero-knowledge (ZK) argument system, which we introduce in this study. It is a 2-move ZK argument system in the reusable public key model where the ZK property holds by using a quantum polynomial-time simulator against non-uniform classical polynomial-time adversaries, and parties have access to a reusable public key (possibly maliciously) generated by a prover. We construct a blind-signature-conforming ZK argument for any NP language from standard classical and quantum assumptions. We give an overview of our technique in Sect. 1.3.

Although our scheme satisfies desirable features in the theoretical sense, it is not quite practical since we rely on general cryptographic tools such as garbled circuits. We believe our scheme opens the possibility of practical round-optimal blind signatures with the desired properties. We leave this question as an open problem.

1.3 Technical Overview

Here, we provide an overview of our construction.

Blind signature scheme by Garg et al. Our starting point is the blind signature scheme by Garg, Rao, Sahai, Schröder, and Unruh [GRS+11]. Their scheme is round-optimal and in the plain model, but the security proof requires complexity leveraging. Our goal is to remove the complexity leveraging and base the security on classical and quantum polynomial assumptions.

Here, we recall their construction. In their protocol, a signer publishes a verification key of a digital signature scheme as its public key and keeps the

[4] The learning with errors (LWE) assumption against quantum polynomial time adversaries and one of the following assumptions against (non-uniform) classical polynomial time adversaries: quadratic residuosity (QR), decisional composite residuosity (DCR), symmetric external Diffie-Hellman (SXDH) over pairing group, or decisional linear (DLIN) over pairing groups.

corresponding signing key secret. To blindly sign on a message, the signer and the user run secure function evaluation (SFE) protocol where the signer plays the role of the sender and the user plays the role of the receiver. In more detail, the user's input is the message to be signed, and the signer holds a circuit corresponding to the signing algorithm of the digital signature scheme where the signing key is hardwired. At the end of the protocol, only the user receives the output signature. To prevent malicious behaviors of the signer, such as using arbitrarily chosen randomness for the signing algorithm to break the blindness, they make the signing algorithm deterministic by using a PRF and include the perfectly binding commitment of the signing key into the public key. Furthermore, they have the signer prove that it honestly follows the SFE protocol using a zero-knowledge argument system.

The blindness of the protocol follows from the receiver's security of the SFE and from the fact that the signer cannot deviate from the honest execution of the protocol due to the soundness of the zero-knowledge argument system and the binding property of the commitment scheme. On the other hand, the unforgeability follows from the combination of the zero-knowledge property of the zero-knowledge argument system, the sender's security of the SFE protocol, and the unforgeability of the digital signature scheme. The former two properties intuitively imply that the user cannot obtain anything beyond the signatures corresponding to the messages it chooses. The final property implies that it cannot forge a new signature. While intuitively correct, there are two problems with this approach. The first problem is with the reduction algorithm that reduces the unforgeability of the blind signature scheme to that of the underlying digital signature scheme. The reduction algorithm has to simulate the signer and extract the message to be signed from the first message of the user. However, this should not be possible because of the receiver's security of the SFE. The second problem is that we need a 2-move zero-knowledge argument system to obtain round-optimal blind signatures. However, it is known that a 2-move zero-knowledge argument system is impossible [GO94].

To resolve these problems, they assume super-polynomial security for the underlying (plain) signature scheme and allow the corresponding reduction algorithm to run in super-polynomial time. Then, the first issue can be resolved by letting the reduction algorithm *break* the receiver's security of the SFE scheme and extract the message to be signed using its super polynomial power. Furthermore, allowing the reduction algorithm to run in super-polynomial time also enables them to sidestep the impossibility result mentioned above. They use a 2-move zero-knowledge argument system with a super-polynomial time simulator by Pass [Pas03] and run the super-polynomial time simulator in the reduction algorithm for unforgeability.[5] This also resolves the second issue above.

Our first step towards the goal is to replace the super-polynomial time reduction algorithm in their security proof with a quantum-polynomial time (QPT)

[5] Though Garg et al. [GRS+11] does not explicitly state that they use the zero-knowledge argument of [Pas03], we observe that their construction can be viewed in this way.

algorithm, which is inspired by Kalai and Khurana [KK19]. To make this work, we replace primitives with super-polynomial security with quantumly secure ones and the primitives broken by the super-polynomial time algorithm with quantumly insecure and classically secure ones. However, simple replacement of the underlying primitives does not work, because their security proof uses complexity leveraging twice, which requires three levels of security for the underlying primitives, while the combination of classical and quantum polynomial hardness can offer only two levels of security.[6] In particular, the above idea necessitates 2-move zero-knowledge arguments with QPT simulation, which cannot be obtained by a simple modification of the construction by Pass [Pas03]. As we elaborate in the following, we relax the notion of zero-knowledge argument system so that it still implies blind signatures and provides a construction that satisfies the notion by adding many modifications to the original zero-knowledge argument system by Pass [Pas03].

Zero-knowledge argument system by pass. To see the problem more closely, we review the zero-knowledge argument system by Pass [Pas03], which is used in the construction of round-optimal blind signatures by Garg et al. [GRS+11]. Their starting point is ZAP for NP languages [DN00, DN07]. Recall that ZAP is a 2-move public coin witness indistinguishable proof system without setup, where the first message can be reused. To make it zero-knowledge, they use the "OR-proof trick" by Feige, Lapidot, and Shamir [FLS90, FLS99]. This technique converts a witness indistinguishable proof into a zero-knowledge proof in the context of non-interactive proof systems by adding a trapdoor branch for the relation to be proven so that the zero-knowledge simulator can use the branch. In more detail, the protocol proceeds as follows.

1. In the first round of the protocol, the verifier sends the first round message r_{zap} of the ZAP system along with a random image $z = f(y)$ of a one-way permutation (OWP) $f : \{0,1\}^{\ell} \to \{0,1\}^{\ell}$.[7]
2. Given the message, the prover who proves $x \in \mathcal{L}$, where $\mathcal{L}$ is some NP language specified by a relation R, proceeds as follows. It first commits the string 0^{ℓ} by a non-interactive commitment with perfect biding to obtain $\mathsf{com} = \mathsf{Com}(0^{\ell}; r_{\mathsf{com}})$ using randomness r_{com}. It then proves that there is witness $(w', y', r'_{\mathsf{com}})$ such that

$$\Big((x, w') \in R\Big) \vee \Big(\mathsf{com} = \mathsf{Com}(y'; r'_{\mathsf{com}}) \wedge f(y') = z\Big)$$

[6] A reader might consider starting from the blind signature scheme by Garg and Gupta [GG14] instead since their security proof uses complexity leveraging only once. However, their construction may not be compatible with our idea of using quantum simulation since it is heavily dependent on a specific structure of the Groth-Sahai proofs [GS08], which is quantumly insecure.

[7] Though one-way functions with efficiently decidable images suffice, we use OWP in this overview for simplicity. In our construction, we rely on a slightly generalized notion of *hard problem generators* which we introduce in Sect. 3.1.

by the proving algorithm of the ZAP system to obtain a proof π_{zap} and sends $\pi = (\mathsf{com}, \pi_{\mathsf{zap}})$ to the verifier. Note that in the honest execution, the prover sets $(w', y', r'_{\mathsf{com}}) = (w, \bot, \bot)$.

3. Given the proof from the prover, the verifier parses $\pi \to (\mathsf{com}, \pi_{\mathsf{zap}})$ and verifies the proof π_{zap} for the above statement by the verification algorithm of the ZAP system.

We then discuss the security of the system. Let us start with the zero-knowledge property. As mentioned, the simulator will run in super-polynomial time, say T. Given (r_{zap}, z), the simulator uses its super-polynomial power to invert the permutation to compute $y = f^{-1}(z)$. It then computes a commitment $\mathsf{com} = \mathsf{Com}(y; r_{\mathsf{com}})$ and uses the witness $(w', y', r'_{\mathsf{com}}) = (\bot, y, r_{\mathsf{com}})$ to generate a proof. Due to the witness indistinguishability of the underlying ZAP and the hiding property of the commitment, the simulated proof is indistinguishable from the real one. The proof for soundness is a bit more complicated. Let us assume an adversary that can generate an accepting proof for a false statement $x \notin \mathcal{L}$. By the statistical soundness of ZAP and by the fact that $x \notin \mathcal{L}$, the output (com^*, π^*) of the successful adversary should satisfy the trapdoor branch of the relation. Namely, com^* should be a commitment of $y = f^{-1}(z)$. Intuitively, this contradicts the one-wayness of f, and thus the system is sound since generating such a commitment seems to require the knowledge of y. However, to turn this intuition into a formal argument, we have to construct a reduction algorithm (i.e., inverter for the OWP) that outputs $y = f^{-1}(z)$ in the clear, instead of the commitment of y. To do so, they turn to complexity leveraging. Namely, they consider an inversion algorithm that runs in super-polynomial time, say T', and have the algorithm extract y from com^* using its super-polynomial power. If we assume f is hard to invert in time T' and the commitment is broken in time T', we can derive the contradiction as desired.

We observe that the two super-polynomial functions T and T' should satisfy $T \gg T'$, since f should be invertible in time T for the zero-knowledge simulator to work, while f should be hard to invert in time T' for the above reduction to make sense. This seems to be incompatible with our approach of replacing T-time simulator with QPT simulator, since this requires hardness that lies between QPT hardness and classical polynomial hardness to replace T'-time secure primitives with something. However, we do not know how to do this without turning to complexity leveraging.

Replacing the commitment with encryption. As we observed above, the main technical hurdle to our goal is that there is no efficient way to extract the message from the commitment for the reduction algorithm that inverts the OWP. However, extraction should not be possible efficiently, since otherwise the commitment cannot be hiding and thus harms the zero-knowledge property. To satisfy these contradicting requirements, we switch to the non-uniform setting and use the standard trick of leveraging the gap between the information available for algorithms in the real-world and non-uniform reduction algorithms. As observed by Garg et al. [GRS+11], non-uniform algorithms can be regarded as two-stage algorithms. The pre-computation phase of the algorithm takes the security parameter as input and computes an advice string of polynomial length using *unbounded computational power*. Then, the online phase of the algorithm takes the problem instance along with the advice string as input and tries to solve the problem in polynomial time. In our context, the non-uniform reduction algorithm will use this advice string to efficiently extract the message from the commitment. On the other hand, this advice string is not available for the real world algorithms and hence does not harm the hiding property of the commitment.

To implement this idea, we replace the commitment with public key encryption (PKE) and change the protocol so that the prover encrypts 0^ℓ using a public key pk_P chosen by itself, instead of computing a commitment of 0^ℓ. The advice string in our context is the secret key corresponding to pk_P. Using the secret key, one can efficiently decrypt the ciphertext and extract the message as desired. Subtle yet, the important point is that the prover should choose the public key pk_P *before* the protocol is run and use the same public key for every invocation of the protocol. Then, the non-uniform reduction algorithm can find the secret key corresponding to pk_P in the pre-computation phase using its unbounded computational power, since pk_P is chosen before the problem instance $z = f(y)$ of the OWP is chosen. This is not possible if the prover chooses a fresh public key for every encryption because the problem instance z and the public key are chosen at the same time in this case. It is not possible to off-load the task of finding the secret key to the pre-computation phase.

In fact, with the above modification, the argument system is no longer in the plain model, since we allow the prover to choose a long-term public key. However, since the syntax of round optimal blind signatures allows the signer to have a long-term public parameter, this modification does not affect the application to blind signatures.

Dealing with maliciously generated public keys. While the above idea may seem to work at first sight, there is still an issue. The problem is that a malicious prover may choose an ill-formed public key for the PKE, for which there are no corresponding secret keys. In this case, we may not be able to extract the message from the ciphertext even with unbounded computational power. We should consider this kind of attack since a malicious signer against blind signatures may maliciously choose a public key. A simple countermeasure against this attack would be to use a PKE scheme such that one can efficiently decide whether the public key is honestly generated or not and have the verifier reject provers with ill-formed public keys. However, we cannot adopt this simple solution because we do not know how to instantiate such a PKE. In particular, we require the PKE to have security against QPT adversaries in addition to the above property due to a technical reason,[8] but there are no known PKE schemes satisfying these properties simultaneously.

---

[8] We need security against QPT adversaries for the PKE scheme because its security is used to prove zero-knowledge property, where the simulator is a QPT algorithm. Recall that the simulator needs quantum power to invert the OWP. One may try to show that non-uniform security instead of quantum security is enough for the PKE by using the pre-computation trick we mentioned. However, this does not seem possible because the inversion should be done *after* the public key is chosen.

To resolve the issue, we further change the protocol. Our first attempt is to let the verifier choose a public key pk_V of PKE and have the prover encrypt 0^ℓ under pk_V in addition to the long-term public key pk_P. Furthermore, we have the prover prove that it has valid witness w for x or it encrypts y under pk_P and pk_V. With this change, the reduction algorithm can extract the message from the ciphertext corresponding to pk_V even if pk_P is maliciously generated since pk_V is under the control of the reduction algorithm and honestly generated. However, this modification harms the zero-knowledge property. In particular, since the verifier has secret key corresponding to pk_V, it can know whether the proof is generated from the honest execution of the protocol or not by simply decrypting the ciphertext.

The reason why the above idea fails is that we allow too much flexibility for the verifier in the sense that it can choose a public key that enables the extraction even if the prover behaves honestly. What we really need is a mechanism where the verifier can extract the message only when the prover cheats. For this purpose, we use lossy encryption. Recall that lossy encryption [PVW08, BHY09] is an extension of PKE where we have an additional lossy key generation algorithm. While the normal key generation algorithm outputs a public key and secret key, the lossy key generation algorithm only outputs a public key. For lossy encryption, we require the lossiness property, which stipulates that the ciphertext generated under the lossy key does not carry any information of the message. As for security, we require that the lossy key and the normal key are indistinguishable. We then would like to change the protocol so that the verifier is restricted to choose the lossy public key in the honest execution of the protocol and can choose normal public key that allows the extraction only when the prover chooses an ill-formed public key. To restrict the behavior of the verifier, we have the verifier prove the following statement:

$$(\mathsf{pk}_V \text{ is chosen from the lossy key generation}) \vee (\mathsf{pk}_P \text{ is an ill-formed public key}) . \quad (1)$$

The former branch of the statement is used in the honest execution and the latter is for simulation. The proof is generated by running another instance of the ZAP system, where the roles of the prover and the verifier are swapped. To avoid increasing the round of the overall protocol, we put the first round message of the ZAP system into the public parameter of the prover and have the verifier generate the proof with respect to it and send the proof along with pk_V to the prover in the first round. Note that it is not clear how to prove the above statement by the ZAP system, since it is not necessarily in NP. In particular, we do not know of a general way of providing an NP witness for proving the ill-formedness of a public key. We skip this issue and simply assume that it is possible for the time being. We will get back to the issue at the end of the overview. The protocol now proceeds as follows.

1. The prover runs the key generation algorithm of the PKE to obtain a public key pk_P and chooses the first message r'_{zap} of the ZAP system. It then sets the long-term public parameter as $\mathsf{pp} = (\mathsf{pk}_P, r'_{\mathsf{zap}})$.

2. In the first round of the protocol, the verifier chooses the first round message r_{zap} of the ZAP system and a random image $z = f(y)$ of the OWP $f : \{0,1\}^{\ell} \to \{0,1\}^{\ell}$. It then runs lossy key generation of the lossy encryption to obtain a public key pk_V. It then proves statement (1) with respect to r'_{zap} by using the randomness for the lossy key generation as a witness to obtain a proof π'_{zap}. Finally, it sends $(r_{\mathsf{zap}}, \mathsf{pk}_V, \pi'_{\mathsf{zap}})$ to the prover.
3. Given the message, the prover verifies π'_{zap} for statement (1) and aborts the protocol if it is not valid. Otherwise, it encrypts the string 0^{ℓ} under pk_P and pk_V to obtain $\mathsf{ct}_P = \mathsf{PKE.Enc}_{\mathsf{pk}_P}(0^{\ell}; r_P)$ and $\mathsf{ct}_V = \mathsf{LE.Enc}_{\mathsf{pk}_V}(0^{\ell}; r_V)$, where LE stands for "lossy encryption". It then proves that there is a witness (w', y', r'_P, r'_V) such that
$$\Big((x, w') \in R\Big) \vee \Big(\mathsf{ct}_P = \mathsf{PKE.Enc}_{\mathsf{pk}_P}(y'; r'_P) \wedge \mathsf{ct}_V = \mathsf{LE.Enc}_{\mathsf{pk}_V}(y'; r'_V) \wedge f(y') = z\Big) \quad (2)$$
with respect to r_{zap} to obtain a proof π_{zap}. Note that in the honest execution, the prover sets $(w', y', r'_P, r'_V) = (w, \perp, \perp, \perp)$. It then sends $\pi = (\mathsf{ct}_P, \mathsf{ct}_V, \pi_{\mathsf{zap}})$ to the verifier.
4. Given the proof from the prover, the verifier parses $\pi \to (\mathsf{ct}_P, \mathsf{ct}_V, \pi_{\mathsf{zap}})$ and verifies the proof π_{zap} with respect to statement (2).

First attempt of the security proof. We now try to prove the security of the scheme. We first prove the zero-knowledge property with a QPT simulator. To do so, we start from the real game where a malicious verifier interacts with an honest prover and gradually change the prover into a zero-knowledge simulator through game hops. In the first step, we change the prover to be a quantum algorithm, which inverts the OWP to recover $y = f^{-1}(z)$ from the first round message by the verifier. We then change the game so that the prover encrypts y instead of 0^{ℓ} when it generates the ciphertext ct_P. Due to the security of PKE against QPT adversaries, this game is indistinguishable from the real game. In the next step, we replace the ciphertext ct_V with the encryption of y instead of 0^{ℓ}. We show that this game is indistinguishable from the previous game by combining the soundness of the ZAP system and the lossiness of the lossy encryption. Without loss of generality, we can assume that the prover does not abort the interaction, since otherwise the malicious verifier cannot obtain any information. However, if the prover does not reject the malicious verifier, this means that statement (1) holds by the soundness of the ZAP. Since pk_P is honestly chosen, pk_V should be a lossy key. Then, by the lossiness of the lossy encryption, we conclude that ct_V does not carry any information about the message, and the change does not alter the distribution of ct_V. Finally, we change the game so that the prover uses the latter branch of statement (2) to generate π_{zap}. Due to the witness indistinguishability of the ZAP, this game is indistinguishable from the previous game. Notice that the prover in the final game does not use the witness w for the statement x to generate the proof and thus constitutes a zero-knowledge simulator.

We then proceed to the proof of the soundness. The proof will be by case analysis. In both cases, we construct a non-uniform reduction algorithm that inverts the OWP. First, we consider the case where the malicious prover chooses honestly generated pk_P. In this case, the reduction algorithm receives pk_P from the malicious prover and finds the corresponding secret key sk_P in the pre-computation phase using its unbounded computational power. Then, in the online phase, it receives the problem instance $z = f(y)$ of the OWP and embeds it into the first-round message from the verifier to the prover. If the malicious prover manages to generate an accepting proof for $x \notin \mathcal{L}$, this should satisfy the trapdoor branch of statement (2) by the soundness of the ZAP. In particular, ct_P should be an encryption of $y = f^{-1}(z)$ under the public key pk_P and thus the reduction algorithm can successfully extract y from ct_P by using sk_P.

We next consider the other case where pk_P is ill-formed. In this case, we need a game hop. In the first step, we change the verifier to be a non-uniform algorithm and have it compute the NP witness for the ill-formedness of pk_P. Then, the verifier generates the proof using the latter branch of statement (1). This game is indistinguishable from the previous game by the witness indistinguishability of the ZAP. In the next step, we change the game so that the verifier generates pk_V by the normal key generation algorithm rather than the lossy key generation algorithm. This game is indistinguishable from the previous game by the security of the lossy encryption. Note that this game hop is possible because the verifier no longer needs the witness that proves pk_V is generated from the lossy key generation due to the change introduced in the previous game. We are now ready to construct the inverter for OWP. Similarly to the case where pk_P is honestly generated, the soundness of the ZAP implies that an accepting proof for $x \notin \mathcal{L}$ satisfies the latter branch of statement (1). This time, the inverter extracts y from ct_V, which is possible because pk_V is now changed to be a normal public key rather than a lossy one.

While the above proof sketch is almost correct, there is still a subtle issue. In particular, the proof of the soundness for the case of ill-formed pk_P is not correct. The problem is that we cannot prove that the winning probability of the malicious prover is changed only negligibly through the game changes because we cannot construct a corresponding reduction algorithm that establishes this. For example, we try to construct a reduction algorithm that breaks the witness indistinguishability of the ZAP by assuming a malicious prover whose success probability in the second game is non-negligibly different from that in the first game. A natural way to do so is to let the reduction algorithm output 1 only when the malicious prover successfully breaks the soundness of our argument system. However, this is not possible since the reduction algorithm cannot efficiently decide whether the output (x^*, π^*) of the malicious prover violates the soundness or not. In particular, even if the malicious prover outputs an accepting pair of a statement x^* and a proof π^*, x^* may be in $\mathcal{L}$ and the reduction algorithm cannot detect it, since $\mathcal{L}$ may be hard to decide language. To address this problem, we further change the protocol.

Making the winning condition efficiently checkable. As we observed above, the only reason why the winning condition is not efficiently checkable is that the language $\mathcal{L}$ is not efficiently decidable in general. To resolve the problem, we change the protocol so that the prover explicitly includes an encrypted version of witness w in the proof. In more details, we change the protocol so that we add a public key $\widehat{\mathsf{pk}}_P$ of another instance of PKE to the public parameter of the prover and change the prover so that it outputs $\widehat{\mathsf{ct}}_P = \mathsf{PKE.Enc}_{\widehat{\mathsf{pk}}_P}(w; \widehat{r}_P)$ along with ct_P and ct_V and proves that there is a witness $(w', \widehat{r}'_{\mathsf{KeyGen}}, \widehat{r}'_P, y', r'_P, r'_V)$ such that

$$\Big((x, w') \in R \wedge \big(\widehat{\mathsf{pk}}_P \text{ is generated by} \mathsf{PKE.KeyGen}(1^\kappa; \widehat{r}'_{\mathsf{KeyGen}})\big) \wedge \widehat{\mathsf{ct}}_P = \mathsf{PKE.Enc}_{\widehat{\mathsf{pk}}_P}(w'; \widehat{r}'_P)\Big)$$
$$\bigvee \Big(\mathsf{ct}_P = \mathsf{PKE.Enc}_{\mathsf{pk}_P}(y'; r'_P) \wedge \mathsf{ct}_V = \mathsf{LE.Enc}_{\mathsf{pk}_V}(y'; r'_V) \wedge f(y') = z\Big), \qquad (3)$$

where the former branch is used in the honest execution of the protocol and the latter is for the simulation and is not changed from the previous construction. Note that to prove the former branch, the prover needs randomness $\widehat{r}_{\mathsf{KeyGen}}$ used in the key generation of $\widehat{\mathsf{pk}}_P$ and thus it has to keep the randomness as a secret parameter. This needs to change the syntax of the zero-knowledge argument system again. However, it does not affect the application to blind signatures, since the syntax of the latter allows the prover to have a secret key.

We then explain how the above change helps. In our proof for the soundness, we relax the winning condition so that the adversary is said to semi-win the game if it outputs an accepting proof $\pi^* = (\mathsf{ct}^*_P, \mathsf{ct}^*_V, \widehat{\mathsf{ct}}^*_P, \pi^*_{\mathsf{zap}})$ for x^* and $\widehat{\mathsf{pk}}_P$ is not in the range of the key generation algorithm or $\widehat{\mathsf{ct}}^*_P$ is not an encryption of a witness w^* such that $R(x^*, w^*) = 1$. We observe that to check this modified winning condition, it is unnecessary to perform the membership test of the language $\mathcal{L}$. The modified winning condition is efficiently checkable for the non-uniform reduction algorithm as follows. It first checks whether $\widehat{\mathsf{pk}}_P$ is honestly generated or not in the pre-computation phase and find the corresponding secret key by brute-force search if it is so. Then, in the online phase, it decrypts the ciphertext $\widehat{\mathsf{ct}}^*_P$ using the secret key to see if the decryption result w^* satisfies $R(x^*, w^*) = 1$ or not. We note that since we relaxed the winning condition, the adversary is regarded as (semi-)winning the game even when it outputs an accepting proof for $x^* \in \mathcal{L}$ if it chooses ill-formed $\widehat{\mathsf{pk}}_P$ or $\widehat{\mathsf{ct}}^*_P$ that does not encrypt the witness for x^*. However, these events happen only with negligible probability and thus can be ignored, since these events imply that the soundness of the ZAP is violated.

Certifying invalid public keys. Now, the only remaining problem is how to prove the statement that pk_P is an ill-formed public key. We show that it is possible to provide an NP witness for this statement if we use Regev's PKE scheme [Reg05, Reg09]. In Regev's PKE scheme, a public key consists of description of a basis of a lattice L and a vector $\mathbf{v}$. The secret key is the vector in L closest to $\mathbf{v}$. For an honestly generated public key, the distance $\mathsf{dist}(L, \mathbf{v})$ between L and $\mathbf{v}$ is close, while for a maliciously generated key, the distance may be far. Therefore, our goal is to provide a proof that $\mathbf{v}$ is far from L. For this purpose, we use the result by Aharonov and Regev [AR04, AR05], who showed that a language consisting of a pair of a lattice and a vector whose distance is far constitutes an NP language. The subtle point is that their proof is for "gap language" in the sense that they cannot give an NP witness for the pair of a lattice and a vector whose distance is neither far enough nor close enough. Translated to our setting, this means that a malicious prover in our zero-knowledge argument system may choose a public key that is not in the support of the honest key generation algorithm without being caught, if the lattice and the vector are not very much far. We show that we can still define a secret key for such a public key that enables the extraction of the message from the ciphertext, which is sufficient for our purpose.

2 Preliminaries

Notation. For a positive integer n, $[n]$ denotes a set $\{1, ..., n\}$. For a bit string x, $|x|$ denotes its bit-length. For a set S, we write $s \xleftarrow{\$} S$ to denote the operation of sampling a random s from the uniform distribution over S. For a (probabilistic classical or quantum) algorithm $\mathcal{A}$, we write $y \xleftarrow{\$} \mathcal{A}(x)$ to mean that we run $\mathcal{A}$ on input x and the output is y. For a probabilistic classical algorithm $\mathcal{A}$, we write $\mathcal{A}(x; r)$ to mean the output of $\mathcal{A}$ on input x and randomness r. Moreover, by a slight abuse of notation, we write $y \xleftarrow{\$} \mathcal{A}(x; r)$ to mean that we uniformly pick r from the randomness space of $\mathcal{A}$ and then set $y := \mathcal{A}(x; r)$. For a probabilistic classical algorithm $\mathcal{A}$ that takes as input x and randomness r, "$y \in \mathcal{A}(x)$" means $\Pr_r[y' = y : y' \leftarrow \mathcal{A}(x; r)] > 0$. We use PPT and QPT to mean (classical) probabilistic polynomial time and quantum polynomial time.

A convention on non-uniform adversaries. When we consider the security of cryptographic primitives against non-uniform classical adversaries, we say that an adversary $\mathcal{A} = (\mathcal{A}_0, \mathcal{A}_1)$ is non-uniform PPT if $\mathcal{A}_0$ is a (possibly randomized) unbounded-time algorithm that takes as input the security parameter 1^κ and outputs a string of length $\mathsf{poly}(\kappa)$ and $\mathcal{A}_1$ is PPT. Typically, $\mathcal{A}_0$ and $\mathcal{A}_1$ can be understood respectively as a "pre-computation phase" that outputs a non-uniform advice and an "online phase" that takes as input the advice and a problem instance and outputs a solution. We note that the randomness of $\mathcal{A}$ does not increase the computational power of $\mathcal{A}$ since $\mathcal{A}_0$ can find the best randomness by using its unbounded computational power. We allow $\mathcal{A}$ to be randomized just for convenience for describing the reductions.

Definitions of standard cryptographic primitives, including non-interactive commitment, public key encryption, lossy encryption, ZAP, and digital signatures, can be found in the full version.

2.1 Secure Function Evaluation

A secure function evaluation (SFE) is a 2-move protocol between a sender who holds a (classical) circuit C and a receiver who holds x, where the goal is for the receiver to compute $C(x)$ without revealing the inputs to each other. Specifically, SFE consists of PPT algorithms $\Pi_{\mathsf{SFE}} = (\mathsf{Receiver}, \mathsf{Sender}, \mathsf{Derive})$ with the following syntax:

$\mathsf{Receiver}(1^\kappa, x) \to (\mathsf{sfe}_1, \mathsf{sfest})$: This is an algorithm supposed to be run by a receiver that takes the security parameter 1^κ and x as input and outputs a first message sfe_1 and a receiver's state sfest.

$\mathsf{Sender}(1^\kappa, \mathsf{sfe}_1, C) \to \mathsf{sfe}_2$: This is an algorithm supposed to be run by a sender that takes the security parameter 1^κ, a first message sfe_1 sent from a receiver and a description of a classical circuit C as input and outputs a second message sfe_2.

$\mathsf{Derive}(\mathsf{sfest}, \mathsf{sfe}_2) \to y$: This is an algorithm supposed to be run by a receiver that takes a receiver's state sfest and a second message sfe_2 as input and outputs a string y.

Correctness. For any $\kappa \in \mathbb{N}$, C, and x, we have

$$\Pr[\mathsf{Derive}(\mathsf{sfest}, \mathsf{sfe}_2) = C(x) : (\mathsf{sfe}_1, \mathsf{sfest}) \xleftarrow{\$} \mathsf{Receiver}(1^\kappa, x), \mathsf{sfe}_2 \xleftarrow{\$} \mathsf{Sender}(1^\kappa, \mathsf{sfe}_1, C)] = 1.$$

Security requirements are essentially the same as those in [GRS+11] except that we require the extraction algorithm to run in QPT instead of classical super-polynomial time. Specifically, we require the following two security notions.

Receiver's security against non-uniform PPT adversary. For any pair of inputs (x_0, x_1) and non-uniform PPT adversary $\mathcal{A} = (\mathcal{A}_0, \mathcal{A}_1)$, we have

$$\left| \Pr\left[\mathcal{A}_1(\mathsf{st}, \mathsf{sfe}_1) = 1 : \begin{array}{l} \mathsf{st} \xleftarrow{\$} \mathcal{A}_0(1^\kappa) \\ (\mathsf{sfe}_1, \mathsf{sfest}) \xleftarrow{\$} \mathsf{Receiver}(1^\kappa, x_0) \end{array} \right] - \Pr\left[\mathcal{A}_1(\mathsf{st}, \mathsf{sfe}_1) = 1 : \begin{array}{l} \mathsf{st} \xleftarrow{\$} \mathcal{A}_0(1^\kappa) \\ (\mathsf{sfe}_1, \mathsf{sfest}) \xleftarrow{\$} \mathsf{Receiver}(1^\kappa, x_1) \end{array} \right] \right| \leq \mathsf{negl}(\kappa).$$

Quantum-extraction sender's security against QPT adversary. There exists a QPT algorithm SFEExt and a PPT algorithm SFESim that satisfy the following: For any QPT adversary $\mathcal{A} = (\mathcal{A}_0, \mathcal{A}_1)$, we have

$$\left| \Pr\left[\mathcal{A}_1(\mathsf{st}_\mathcal{A}, \mathsf{sfe}_2) = 1 : \begin{array}{l} (\mathsf{sfe}_1, C, \mathsf{st}_\mathcal{A}) \xleftarrow{\$} \mathcal{A}_0(1^\kappa), \\ \mathsf{sfe}_2 \xleftarrow{\$} \mathsf{Sender}(1^\kappa, \mathsf{sfe}_1, C) \end{array} \right] - \Pr\left[\mathcal{A}_1(\mathsf{st}_\mathcal{A}, \mathsf{sfe}_2) = 1 : \begin{array}{l} (\mathsf{sfe}_1, C, \mathsf{st}_\mathcal{A}) \xleftarrow{\$} \mathcal{A}_0(1^\kappa), \\ x \xleftarrow{\$} \mathsf{SFEExt}(\mathsf{sfe}_1), \\ \mathsf{sfe}_2 \xleftarrow{\$} \mathsf{SFESim}(1^\kappa, \mathsf{sfe}_1, C(x)) \end{array} \right] \right| \leq \mathsf{negl}(\kappa).$$

An SFE protocol that satisfies these security notions can be constructed based on either of the DDH, QR, or decisional composite residuosity (DCR) assumptions against non-uniform PPT adversaries and LWE assumption against QPT adversaries. Namely, we can construct it based on Yao's 2PC protocol instantiated with secure garbled circuit against quantum adversaries (which can be instantiated based on OWF against quantum adversaries) and non-uniform classical-receiver-secure but quantumly receiver-insecure and statistically sender-private OT (which can be instantiated based on the non-uniform PPT hardness of DDH [NP01], QR, or DCR [HK12]). See the full version for details.

2.2 Blind Signatures

Here, we give a definition of blind signatures. For simplicity, we give a definition focusing on round-optimal blind signatures. A round-optimal blind signature scheme with a message space $\mathcal{M}$ consists of PPT algorithms $(\mathsf{BSGen}, \mathcal{U}_1, \mathcal{S}_2, \mathcal{U}_{\mathsf{der}}, \mathsf{BSVerify})$.

$\mathsf{BSGen}(1^\kappa) \to (\mathsf{pk}, \mathsf{sk})$: The key generation algorithm takes as input the security parameter 1^κ and outputs a public key pk and a signing key sk.

$\mathcal{U}_1(\mathsf{pk}, m) \to (\mu, \mathsf{st}_\mathcal{U})$: This is the user's first message generation algorithm that takes as input a public key pk and a message $m \in \mathcal{M}$ and outputs a first message μ and a state $\mathsf{st}_\mathcal{U}$.

$\mathcal{S}_2(\mathsf{sk}, \mu) \to \rho$: This is the signer's second message generation algorithm that takes as input a signing key sk and a first message μ as input and outputs a second message ρ.

$\mathcal{U}_{\mathsf{der}}(\mathsf{st}_\mathcal{U}, \rho) \to \sigma$: This is the user's signature derivation algorithm that takes as input a state $\mathsf{st}_\mathcal{U}$ and a second message ρ as input and outputs a signature σ.

$\mathsf{BSVerify}(\mathsf{pk}, m, \sigma) \to \top$ or $\bot$: This is a deterministic verification algorithm that takes as input a public key pk, a message $m \in \mathcal{M}$, and a signature σ, and outputs $\top$ to indicate acceptance or $\bot$ to indicate rejection.

Correctness. For any $\kappa \in \mathbb{N}$, $m \in \mathcal{M}$,

$$\Pr\left[\mathsf{BSVerify}(\mathsf{pk}, m, \sigma) = \bot : \begin{array}{l} (\mathsf{pk}, \mathsf{sk}) \xleftarrow{\$} \mathsf{BSGen}(1^\kappa) \\ (\mu, \mathsf{st}_\mathcal{U}) \xleftarrow{\$} \mathcal{U}_1(\mathsf{pk}, m) \\ \rho \xleftarrow{\$} \mathcal{S}_2(\mathsf{sk}, \mu) \\ \sigma \xleftarrow{\$} \mathcal{U}_{\mathsf{der}}(\mathsf{st}_\mathcal{U}, \rho) \end{array}\right] = \mathsf{negl}(\kappa).$$

Unforgeability against PPT adversary. For any $q = \mathsf{poly}(\kappa)$ and PPT adversary $\mathcal{A}$ that makes at most q queries, we have

$$\Pr\left[\begin{array}{l} \mathsf{BSVerify}(\mathsf{pk}, m_i, \sigma_i) = \top \text{ for all } i \in [q+1] \\ \wedge \{m_i\}_{i \in [q+1]} \text{ is pairwise distinct} \end{array} : \begin{array}{l} (\mathsf{pk}, \mathsf{sk}) \xleftarrow{\$} \mathsf{BSGen}(1^\kappa) \\ \{(m_i, \sigma_i)\}_{i \in [q+1]} \xleftarrow{\$} \mathcal{A}^{\mathcal{S}_2(\mathsf{sk}, \cdot)}(\mathsf{pk}) \end{array}\right] = \mathsf{negl}(\kappa)$$

where we say that $\{m_i\}_{i \in [q+1]}$ is pairwise distinct if we have $m_i \neq m_j$ for all $i \neq j$.

Blindness against PPT adversary. For defining blindness, we consider the following game between an adversary $\mathcal{A}$ and a challenger.

Setup. $\mathcal{A}$ is given as input the security parameter 1^κ, and sends a public key pk and a pair of messages (m_0, m_1) to the challenger.

First Message. The challenger generates $(\mu_b, \mathsf{st}_{\mathcal{U},b}) \xleftarrow{\$} \mathcal{U}_1(\mathsf{pk}, m_b)$ for each $b \in \{0,1\}$, picks $\mathsf{coin} \xleftarrow{\$} \{0,1\}$, and gives $(\mu_{\mathsf{coin}}, \mu_{1-\mathsf{coin}})$ to $\mathcal{A}$.

Second Message. The adversary sends $(\rho_{\mathsf{coin}}, \rho_{1-\mathsf{coin}})$ to the challenger.

Signature Derivation. The challenger generates $\sigma_b \xleftarrow{\$} \mathcal{U}_{\mathsf{der}}(\mathsf{st}_{\mathcal{U},b}, \rho_b)$ for each $b \in \{0,1\}$. If $\sigma_0 = \bot$ or $\sigma_1 = \bot$, then the challenger gives $(\bot, \bot)$ to $\mathcal{A}$. Otherwise, it gives (σ_0, σ_1) to $\mathcal{A}$.

Guess. $\mathcal{A}$ outputs its guess coin'

We say that $\mathcal{A}$ wins if $\mathsf{coin} = \mathsf{coin}'$. We say that a blind signature scheme satisfies blindness if for any PPT adversary $\mathcal{A}$, we have

$$\left|\Pr[\mathcal{A} \text{ wins}] - \frac{1}{2}\right| = \mathsf{negl}(\kappa).$$

Remark 2.1. In a definition of blindness for general (not necessarily round-optimal) blind signatures, $\mathcal{A}$ can schedule interactions with two sessions of a user in an arbitrary order. However, as observed in [GRS+11], the order can be fixed as above without loss of generality when we consider round-optimal schemes.

Remark 2.2. The above definition only requires security against uniform PPT adversaries. We can achieve security against non-uniform PPT adversaries if we assume all assumptions used in this paper hold against non-uniform adversaries. We primarily consider security against uniform adversaries to clarify which assumptions should hold against non-uniform adversaries even if our goal is to prove security against uniform PPT adversaries.

3 Preparations

In this section, we introduce two new primitives used in our construction of blind-signature-conforming zero-knowledge argument in Sect. 4.

3.1 Classical-Hard Quantum-Solvable Hard Problem Generator

A hard problem generator consists of algorithms $\Pi_{\mathsf{HPG}} = (\mathsf{ProbGen}, \mathsf{VerProb}, \mathsf{Solve}, \mathsf{VerSol})$.

$\mathsf{ProbGen}(1^\kappa) \to \mathsf{prob}$: The problem generation algorithm is a PPT algorithm that is given the security parameter 1^κ as input and outputs a problem $\mathsf{prob} \in \{0,1\}^*$.

$\mathsf{VerProb}(1^\kappa, \mathsf{prob}) \to \top$ or $\bot$: The problem verification algorithm is a deterministic classical polynomial-time algorithm that is given the security parameter 1^κ and a problem prob and returns $\top$ if it accepts and $\bot$ if it rejects.

$\mathsf{Solve}(\mathsf{prob}) \to \mathsf{sol}$: The solving algorithm is a QPT algorithm that is given a problem prob and returns a solution sol.

$\mathsf{VerSol}(\mathsf{prob}, \mathsf{sol}) \to \top$ or $\bot$: The solution verification algorithm is a deterministic classical polynomial-time algorithm that is given an problem prob and a solution sol, and returns $\top$ if it accepts and $\bot$ if it rejects.

We say that Π_{HPG} is *non-uniform-classical-hard quantum-solvable* if it satisfies the following properties.

Quantum Solvability. For any $\mathsf{prob} \in \{0,1\}^*$ such that $\mathsf{VerProb}(1^\kappa, \mathsf{prob}) = \top$, we have

$$\Pr[\mathsf{VerSol}(\mathsf{prob}, \mathsf{sol}) = \bot : \mathsf{sol} \xleftarrow{\$} \mathsf{Solve}(\mathsf{prob})] = \mathsf{negl}(\kappa).$$

Validity of Honestly Generated Problem. For all $\kappa \in \mathbb{N}$, we have

$$\Pr[\mathsf{VerProb}(1^\kappa, \mathsf{prob}) = \top : \mathsf{prob} \xleftarrow{\$} \mathsf{ProbGen}(1^\kappa)] = 1.$$

Non-Uniform Classical Hardness. For any non-uniform PPT adversary $\mathcal{A} = (\mathcal{A}_0, \mathcal{A}_1)$, we have

$$\Pr[\mathsf{VerSol}(\mathsf{prob}, \mathsf{sol}) = \top : \mathsf{st} \xleftarrow{\$} \mathcal{A}_0(1^\kappa), \mathsf{prob} \xleftarrow{\$} \mathsf{ProbGen}(1^\kappa), \mathsf{sol} \xleftarrow{\$} \mathcal{A}_1(\mathsf{st}, \mathsf{prob})] = \mathsf{negl}(\kappa).$$

Remark 3.1. HPG can be trivially constructed based on any OWF with an efficiently recognizable range that is uninvertible by non-uniform PPT adversaries and invertible in QPT by considering an image of the function as prob and its preimage as sol. The efficient recognizability of the range is needed since otherwise we cannot implement $\mathsf{VerProb}$ that verifies the existence of a solution. Such an OWF with an efficiently recognizable range can be constructed from the RSA assumption or the discrete logarithm assumption over $\mathbb{Z}_p$ for a prime p of a special form as shown by Goldreich, Levin, and Nisan [GLN11]. (Indeed, their construction is length-preserving and injective and thus any bit-string is in the range of the function.) On the other hand, to the best of our knowledge, there is no known construction of such an OWF from the hardness of factoring or DL over more general groups. This is why we introduce the notion of classical-hard quantum-solvable HPG, which can be seen as a relaxed notion of a OWF with an efficiently recognizable range that is secure against non-uniform classical adversaries and invertible in QPT.

Lemma 3.1. *Assuming the non-uniform classical hardness of factoring or discrete logarithm over an efficiently recognizable cyclic group, there exists classical-hard quantum-solvable hard problem generator.*

This is an easy consequence of Shor's algorithm [Sho94] that solves factoring and discrete logarithm in QPT. A full proof can be found in the full version.

3.2 Public Key Encryption with Invalid Key Certifiability

We introduce a new notion for PKE which we call *invalid key certifiability.* Roughly speaking, it requires that for any (malformed) encryption key $\mathsf{ek}_{\mathsf{ikc}}$, there exists a witness for the invalidness of $\mathsf{ek}_{\mathsf{ikc}}$ or otherwise there must exist a corresponding decryption key that can decrypt ciphertexts under $\mathsf{ek}_{\mathsf{ikc}}$.

More precisely, a PKE scheme $\Pi_{\mathsf{IKC}} = (\mathsf{IKC.KeyGen}, \mathsf{IKC.Enc}, \mathsf{IKC.Dec})$ has *invalid key certifiability* if it additionally has a deterministic classical polynomial-time algorithm $\mathsf{IKC.InvalidVerf}$ with the following syntax and properties:

$\mathsf{IKC.InvalidVerf}(1^\kappa, \mathsf{ek_{ikc}}, \mathsf{wit_{invalid}}) \rightarrow \top$ or $\bot$: This algorithm takes the security parameter 1^κ, an encryption key $\mathsf{ek_{ikc}}$ and a witness $\mathsf{wit_{invalid}} \in \{0,1\}^\ell$ as input where $\ell(\kappa) = \mathsf{poly}(\kappa)$ is a parameter fixed by the scheme, and outputs $\top$ or $\bot$.

We require the following two properties:

1. For any $\kappa \in \mathbb{N}$ and $(\mathsf{ek_{ikc}}, \mathsf{dk_{ikc}}) \xleftarrow{\$} \mathsf{IKC.KeyGen}(1^\kappa)$, there does not exist $\mathsf{wit_{invalid}} \in \{0,1\}^\ell$ such that $\mathsf{IKC.InvalidVerf}(1^\kappa, \mathsf{ek_{ikc}}, \mathsf{wit_{invalid}}) = \top$.
2. For any $\kappa \in \mathbb{N}$ and (possibly malformed) $\mathsf{ek_{ikc}}$, if there does not exist $\mathsf{wit_{invalid}} \in \{0,1\}^\ell$ such that $\mathsf{IKC.InvalidVerf}(1^\kappa, \mathsf{ek_{ikc}}, \mathsf{wit_{invalid}}) = \top$, then there exists $\mathsf{dk_{ikc}}$ such that for any m, we have

$$\Pr[\mathsf{IKC.Dec}(\mathsf{dk_{ikc}}, \mathsf{IKC.Enc}(\mathsf{ek_{ikc}}, m)) = m] = 1.$$

We call such $\mathsf{dk_{ikc}}$ a corresponding decryption key to $\mathsf{ek_{ikc}}$. We say that $\mathsf{ek_{ikc}}$ is undecryptable if there does not exist a corresponding decryption key to $\mathsf{ek_{ikc}}$.

Remark 3.2. Remark that we do not require the converse of Item 2, i.e., we do not require that "if there exists $\mathsf{wit_{invalid}} \in \{0,1\}^\ell$ such that $\mathsf{IKC.InvalidVerf}(1^\kappa, \mathsf{ek_{ikc}}, \mathsf{wit_{invalid}}) = \top$, then $\mathsf{ek_{ikc}}$ is undecryptable". That is, even if $\mathsf{ek_{ikc}}$ has a corresponding decryption key, it may also have a witness for the invalidness.

Remark 3.3. All dense PKE schemes, in which any string can be a valid encryption key, satisfy invalid key certifiability since all bit strings can be a valid encryption key that has a corresponding decryption key. However, a PKE scheme with invalid key certifiability may not be dense. We note that there is no known candidate of a dense PKE scheme against quantum adversaries.

Lemma 3.2. *There exists a PKE scheme that satisfies the CPA security against QPT adversaries and invalid key certifiability under the quantum hardness of LWE problem.*

The construction is almost identical to the Regev's PKE scheme [Reg09] (modulo some tweak in the parameter). To show the invalid key certifiability property, we rely on the result that the (approximated) gap closest vector (GapCVP) problem lies in **NP** $\cap$ **CoNP** [AR05]. In particular, $\mathsf{wit_{invalid}}$ will be a witness to a NO instance of the GapCVP problem. Then, Item 1 follows since a valid public key of Regev's PKE scheme can be seen as an YES instance to the GapCVP problem and there will exist no witness to prove otherwise (i.e., $\mathsf{wit_{invalid}}$ does not exist). On the other hand, to show Item 2, we rely on the fact that if the public key is *not* a NO instance to the GapCVP problem, then it is still a public key that admits a "good enough" decryption key (i.e., a short vector slightly larger than an honestly generated one). We refer the full details to the full version.

4 Blind-Signature-Conforming Zero-Knowledge Argument

In this section, we define blind-signature-conforming zero-knowledge arguments that are sufficient to construct round-optimal blind signatures and construct it based on standard assumptions. Roughly speaking, a blind-signature-conforming zero-knowledge argument is an interactive argument protocol that satisfies the following properties:

1. publicly verifiable[9] and 2-move with reusable setup by the prover,[10]
2. adaptive soundness with untrusted setup against classical prover, and
3. reusable quantum-simulation zero-knowledge against classical verifier.

4.1 Definition

Let $\mathcal{L}$ be an NP language and $\mathcal{R}$ be the corresponding relation. A blind-signature-conforming zero-knowledge argument for $\mathcal{L}$ has the following syntax:

$\mathsf{Setup}(1^\kappa) \to (\mathsf{pp}, \mathsf{sp})$: This is a setup algorithm (supposed to be run by a prover) that takes as input the security parameter 1^κ and outputs a public parameter pp and a secret parameter sp.

$\mathcal{V}_1(\mathsf{pp}) \to \mathsf{ch}$: This is the verifier's first message generation algorithm that takes as input a public parameter pp and outputs a first message ch referred to as a *challenge*.

$\mathcal{P}_2(\mathsf{sp}, \mathsf{ch}, x, w) \to \mathsf{resp}$: This is the prover's second message generation algorithm that takes as input a secret parameter sp, a challenge ch, a statement x, and a witness w, and outputs a second message resp referred to as a *response*.

$\mathcal{V}_{\mathsf{out}}(\mathsf{pp}, \mathsf{ch}, x, \mathsf{resp}) \to \top$ or $\bot$: This is the verification algorithm that takes a public parameter pp, a challenge ch, a statement x, and a response resp, and outputs $\top$ to indicate acceptance or $\bot$ to indicate rejection.

It should satisfy the following properties:

Completeness. For any $(x, w) \in \mathcal{R}$, we have

$$\Pr[\mathcal{V}_{\mathsf{out}}(\mathsf{pp}, \mathsf{ch}, x, \mathsf{resp}) = \top : (\mathsf{pp}, \mathsf{sp}) \xleftarrow{\$} \mathsf{Setup}(1^\kappa), \mathsf{ch} \xleftarrow{\$} \mathcal{V}_1(\mathsf{pp}), \mathsf{resp} \xleftarrow{\$} \mathcal{P}_2(\mathsf{sp}, \mathsf{ch}, x, w)] = 1.$$

Adaptive soundness with untrusted setup against non-uniform PPT adversary. For any non-uniform PPT cheating prover $\mathcal{P}^* = (\mathcal{P}^*_{\mathsf{Setup}}, \mathcal{P}^*_2)$, we have

$$\Pr\left[\begin{array}{l} \mathcal{V}_{\mathsf{out}}(\mathsf{pp}, \mathsf{ch}, x^*, \mathsf{resp}) = \top \\ \quad \wedge\ x^* \notin \mathcal{L} \end{array} : \begin{array}{l} (\mathsf{pp}, \mathsf{st}_{\mathcal{P}^*}) \xleftarrow{\$} \mathcal{P}^*_{\mathsf{Setup}}(1^\kappa), \\ \mathsf{ch} \xleftarrow{\$} \mathcal{V}_1(\mathsf{pp}), \\ (x^*, \mathsf{resp}) \xleftarrow{\$} \mathcal{P}^*_2(\mathsf{st}_{\mathcal{P}^*}, \mathsf{ch}) \end{array}\right] \le \mathsf{negl}(\kappa).$$

---

[9] Actually, the public verifiability is not needed in the construction of our blind signatures. We only require this because our construction satisfies this.

[10] We can also view it as a three-move protocol by considering the setup as the prover's first message. However, since the first message is reusable, we view the protocol as a two-move protocol with reusable setup.

Reusable quantum-simulation zero-knowledge against PPT adversary. Roughly speaking, we require that there exists a QPT simulator that simulates a view of a PPT cheating verifier that interacts with an honest prover even if the setup is reused many times.

More precisely, there exists a QPT simulator $\mathcal{S}$ such that for any PPT adversary $\mathcal{A}$, we have

$$\left|\Pr\left[\mathcal{A}^{\mathcal{O}_{\mathsf{real}}}(\mathsf{pp}) = 1 : (\mathsf{pp}, \mathsf{sp}) \xleftarrow{\$} \mathsf{Setup}(1^{\kappa})\right] - \Pr\left[\mathcal{A}^{\mathcal{O}_{\mathsf{sim}}}(\mathsf{pp}) = 1 : (\mathsf{pp}, \mathsf{sp}) \xleftarrow{\$} \mathsf{Setup}(1^{\kappa})\right]\right| \leq \mathsf{negl}(\kappa)$$

where oracles $\mathcal{O}_{\mathsf{real}}$ and $\mathcal{O}_{\mathsf{sim}}$ are defined as follows:

$\mathcal{O}_{\mathsf{real}}(\mathsf{ch}, x, w)$	$\mathcal{O}_{\mathsf{sim}}(\mathsf{ch}, x, w)$
If $(x, w) \in \mathcal{R}$	If $(x, w) \in \mathcal{R}$
Return $\mathsf{resp} \xleftarrow{\$} \mathcal{P}_2(\mathsf{sp}, \mathsf{ch}, x, w)$	Return $\mathsf{resp} \xleftarrow{\$} \mathcal{S}(\mathsf{pp}, \mathsf{ch}, x, 1^{\lvert w \rvert})$
Else	Else
Return $\perp$	Return $\perp$

4.2 Construction

Let $\mathcal{L}$ be an NP language and $\mathcal{R}$ be its corresponding relation (i.e., $x \in \mathcal{L}$ if and only if there exists w such that $(x, w) \in \mathcal{R}$). We construct a blind-signature-conforming zero-knowledge argument for $\mathcal{L}$ based on the following building blocks.

- A PKE scheme $\Pi_{\mathsf{PKE}} = (\mathsf{PKE.KeyGen}, \mathsf{PKE.Enc}, \mathsf{PKE.Dec})$ that is CPA secure against QPT adversaries.
- A PKE scheme with invalid key certifiability $\Pi_{\mathsf{IKC}} = (\mathsf{IKC.KeyGen}, \mathsf{IKC.Enc}, \mathsf{IKC.Dec}, \mathsf{IKC.InvalidVerf})$ that is CPA secure against QPT adversaries.
- A lossy PKE scheme $\Pi_{\mathsf{LE}} = (\mathsf{LE.InjGen}, \mathsf{LE.LossyGen}, \mathsf{LE.Enc}, \mathsf{LE.Dec})$ that satisfies key indistinguishability against non-uniform PPT adversaries.
- A classical-hard quantum-solvable hard problem generator $\Pi_{\mathsf{HPG}} = (\mathsf{ProbGen}, \mathsf{VerProb}, \mathsf{Solve}, \mathsf{VerSol})$.
- A ZAP system $\Pi_{\mathsf{zap}} = (\mathsf{ZAP.Prove}, \mathsf{ZAP.Verify})$ for the NP language $\widetilde{\mathcal{L}} = \widetilde{\mathcal{L}}_1 \cup \widetilde{\mathcal{L}}_2$ that satisfies completeness, adaptive statistical soundness, and adaptive computational witness indistinguishability against non-uniform PPT adversaries where languages $\widetilde{\mathcal{L}}_1$ and $\widetilde{\mathcal{L}}_2$ are defined as follows.
  1. $(x, \mathsf{ek}_{\mathsf{pke}}, \mathsf{ek}_{\mathsf{ikc}}, \mathsf{ek}_{\mathsf{le}}, \mathsf{prob}, \mathsf{ct}_{\mathsf{pke}}, \mathsf{ct}_{\mathsf{ikc}}, \mathsf{ct}_{\mathsf{le}}) \in \widetilde{\mathcal{L}}_1$ if there exists $(w, \mathsf{dk}_{\mathsf{pke}}, r_{\mathsf{pke\text{-}gen}}, r_{\mathsf{pke\text{-}enc}})$ such that

$$(x, w) \in \mathcal{L},$$

$$(\mathsf{ek}_{\mathsf{pke}}, \mathsf{dk}_{\mathsf{pke}}) = \mathsf{PKE.KeyGen}(1^{\kappa}; r_{\mathsf{pke\text{-}gen}}),$$

$$\mathsf{ct}_{\mathsf{pke}} = \mathsf{PKE.Enc}(\mathsf{ek}_{\mathsf{pke}}, w; r_{\mathsf{pke\text{-}enc}}).$$

2. $(x, \mathsf{ek}_{\mathsf{pke}}, \mathsf{ek}_{\mathsf{ikc}}, \mathsf{ek}_{\mathsf{le}}, \mathsf{prob}, \mathsf{ct}_{\mathsf{pke}}, \mathsf{ct}_{\mathsf{ikc}}, \mathsf{ct}_{\mathsf{le}}) \in \widetilde{\mathcal{L}}_2$ if there exists $(\mathsf{sol}, r_{\mathsf{ikc\text{-}enc}}, r_{\mathsf{le\text{-}enc}})$ such that

$$\mathsf{VerSol}(\mathsf{prob}, \mathsf{sol}) = \top,$$

$$\mathsf{ct}_{\mathsf{ikc}} = \mathsf{IKC.Enc}(\mathsf{ek}_{\mathsf{ikc}}, \mathsf{sol}; r_{\mathsf{ikc\text{-}enc}}),$$

$$\mathsf{ct}_{\mathsf{le}} = \mathsf{LE.Enc}(\mathsf{ek}_{\mathsf{le}}, \mathsf{sol}; r_{\mathsf{le\text{-}enc}}).$$

- A ZAP system $\Pi'_{\mathsf{zap}} = (\mathsf{ZAP.Prove}', \mathsf{ZAP.Verify}')$ for the NP language $\widetilde{\mathcal{L}}' = \widetilde{\mathcal{L}}'_1 \cup \widetilde{\mathcal{L}}'_2$ that satisfies completeness, adaptive statistical soundness, and adaptive computational witness indistinguishability against non-uniform PPT adversaries where languages $\widetilde{\mathcal{L}}'_1$ and $\widetilde{\mathcal{L}}'_2$ are defined as follows.
  1. $(\mathsf{ek}_{\mathsf{ikc}}, \mathsf{ek}_{\mathsf{le}}) \in \widetilde{\mathcal{L}}'_1$ if there exists $r_{\mathsf{le\text{-}gen}}$ such that $\mathsf{ek}_{\mathsf{le}} = \mathsf{LE.LossyGen}(1^\kappa; r_{\mathsf{le\text{-}gen}})$.
  2. $(\mathsf{ek}_{\mathsf{ikc}}, \mathsf{ek}_{\mathsf{le}}) \in \widetilde{\mathcal{L}}'_2$ if there exists $\mathsf{wit}_{\mathsf{invalid}}$ such that $\mathsf{IKC.InvalidVerf}(\mathsf{ek}_{\mathsf{ikc}}, \mathsf{wit}_{\mathsf{invalid}}) = \top$.

We assume that the first message spaces of Π_{zap} and Π'_{zap} are $\{0,1\}^\ell$, which can be assumed without loss of generality by taking ℓ as an arbitrarily large polynomial in κ. Then our blind-signature-conforming zero-knowledge argument $(\mathsf{Setup}, \mathcal{V}_1, \mathcal{P}_2, \mathcal{V}_{\mathsf{out}})$ is described as follows:

$\mathsf{Setup}(1^\kappa)$: The setup algorithm is given the security parameter 1^κ, and works as follows.
  1. Generate $(\mathsf{ek}_{\mathsf{pke}}, \mathsf{dk}_{\mathsf{pke}}) := \mathsf{PKE.KeyGen}(1^\kappa; r_{\mathsf{pke\text{-}gen}})$.
  2. Generate $(\mathsf{ek}_{\mathsf{ikc}}, \mathsf{dk}_{\mathsf{ikc}}) \xleftarrow{\$} \mathsf{IKC.KeyGen}(1^\kappa)$.
  3. Generate $r'_{\mathsf{zap}} \xleftarrow{\$} \{0,1\}^\ell$
  4. Output $\mathsf{pp} := (\mathsf{ek}_{\mathsf{pke}}, \mathsf{ek}_{\mathsf{ikc}}, r'_{\mathsf{zap}})$ and $\mathsf{sp} := (\mathsf{ek}_{\mathsf{pke}}, \mathsf{ek}_{\mathsf{ikc}}, r'_{\mathsf{zap}}, \mathsf{dk}_{\mathsf{pke}}, r_{\mathsf{pke\text{-}gen}})$.

$\mathcal{V}_1(\mathsf{pp})$: The verifier is given a public parameter $\mathsf{pp} = (\mathsf{ek}_{\mathsf{pke}}, \mathsf{ek}_{\mathsf{ikc}}, r'_{\mathsf{zap}})$, and works as follows.
  1. Generate $r_{\mathsf{zap}} \xleftarrow{\$} \{0,1\}^\ell$.
  2. Generate $\mathsf{prob} \xleftarrow{\$} \mathsf{ProbGen}(1^\kappa)$.
  3. Generate $\mathsf{ek}_{\mathsf{le}} \xleftarrow{\$} \mathsf{LE.LossyGen}(1^\kappa; r_{\mathsf{le\text{-}gen}})$.
  4. Generate $\pi'_{\mathsf{zap}} \xleftarrow{\$} \mathsf{ZAP.Prove}'(r'_{\mathsf{zap}}, (\mathsf{ek}_{\mathsf{ikc}}, \mathsf{ek}_{\mathsf{le}}), r_{\mathsf{le\text{-}gen}})$.
  5. Output $\mathsf{ch} := (r_{\mathsf{zap}}, \mathsf{prob}, \mathsf{ek}_{\mathsf{le}}, \pi'_{\mathsf{zap}})$.

$\mathcal{P}_2(\mathsf{sp}, \mathsf{ch}, x, w)$: The prover is given a secret parameter $\mathsf{sp} := (\mathsf{ek}_{\mathsf{pke}}, \mathsf{ek}_{\mathsf{ikc}}, r'_{\mathsf{zap}}, \mathsf{dk}_{\mathsf{pke}}, r_{\mathsf{pke\text{-}gen}})$, a challenge $\mathsf{ch} = (r_{\mathsf{zap}}, \mathsf{prob}, \mathsf{ek}_{\mathsf{le}}, \pi'_{\mathsf{zap}})$, a statement x, and a witness w, and works as follows.
  1. Immediately abort and output $\bot$ if $\mathsf{VerProb}(1^\kappa, \mathsf{prob}) = \bot$ or $\mathsf{ZAP.Verify}'(r'_{\mathsf{zap}}, (\mathsf{ek}_{\mathsf{ikc}}, \mathsf{ek}_{\mathsf{le}}), \pi'_{\mathsf{zap}}) = \bot$.
  2. Generate $\mathsf{ct}_{\mathsf{ikc}} \xleftarrow{\$} \mathsf{IKC.Enc}(\mathsf{ek}_{\mathsf{ikc}}, 0^{|\mathsf{sol}|})$ and $\mathsf{ct}_{\mathsf{le}} \xleftarrow{\$} \mathsf{LE.Enc}(\mathsf{ek}_{\mathsf{le}}, 0^{|\mathsf{sol}|})$.
  3. Generate $\mathsf{ct}_{\mathsf{pke}} \xleftarrow{\$} \mathsf{PKE.Enc}(\mathsf{ek}_{\mathsf{pke}}, w; r_{\mathsf{pke\text{-}enc}})$.
  4. Generate $\pi_{\mathsf{zap}} \xleftarrow{\$} \mathsf{ZAP.Prove}(r_{\mathsf{zap}}, (x, \mathsf{ek}_{\mathsf{pke}}, \mathsf{ek}_{\mathsf{ikc}}, \mathsf{ek}_{\mathsf{le}}, \mathsf{prob}, \mathsf{ct}_{\mathsf{pke}}, \mathsf{ct}_{\mathsf{ikc}}, \mathsf{ct}_{\mathsf{le}}), (w, \mathsf{dk}_{\mathsf{pke}}, r_{\mathsf{pke\text{-}gen}}, r_{\mathsf{pke\text{-}enc}}))$.
  5. Output $\mathsf{resp} := (\mathsf{ct}_{\mathsf{pke}}, \mathsf{ct}_{\mathsf{ikc}}, \mathsf{ct}_{\mathsf{le}}, \pi_{\mathsf{zap}})$.

$\mathcal{V}_{\text{out}}(\mathsf{pp}, \mathsf{ch}, x, \mathsf{resp})$: The verifier is given a public parameter $\mathsf{pp} = (\mathsf{ek}_{\mathsf{pke}}, \mathsf{ek}_{\mathsf{ikc}}, r'_{\mathsf{zap}})$, a challenge $\mathsf{ch} = (r_{\mathsf{zap}}, \mathsf{prob}, \mathsf{ek}_{\mathsf{le}}, \pi'_{\mathsf{zap}})$, a statement x, and a response $\mathsf{resp} = (\mathsf{ct}_{\mathsf{pke}}, \mathsf{ct}_{\mathsf{ikc}}, \mathsf{ct}_{\mathsf{le}}, \pi_{\mathsf{zap}})$, and works as follows.
1. Output $\mathsf{ZAP.Verify}(r_{\mathsf{zap}}, (x, \mathsf{ek}_{\mathsf{pke}}, \mathsf{ek}_{\mathsf{ikc}}, \mathsf{ek}_{\mathsf{le}}, \mathsf{prob}, \mathsf{ct}_{\mathsf{pke}}, \mathsf{ct}_{\mathsf{ikc}}, \mathsf{ct}_{\mathsf{le}}), \pi_{\mathsf{zap}})$.

The correctness of the scheme immediately follows from the correctness of Π_{zap} and Π'_{zap} and the validity of an honestly generated instance of Π_{HPG}.

4.3 Security

Here, we only give a proof sketch. A full proof can be found in the full version.

Adaptive soundness with untrusted setup. Consider an interaction between an honest verifier and a cheating prover $\mathcal{P}^*$ (that may maliciously generate pp). When $\mathcal{P}^*$ succeeds in breaking soundness, we have $x \notin \mathcal{L}$, which implies $(x, \mathsf{ek}_{\mathsf{pke}}, \mathsf{ek}_{\mathsf{ikc}}, \mathsf{ek}_{\mathsf{le}}, \mathsf{prob}, \mathsf{ct}_{\mathsf{pke}}, \mathsf{ct}_{\mathsf{ikc}}, \mathsf{ct}_{\mathsf{le}}) \notin \widetilde{\mathcal{L}}_1$. On the other hand, by soundness of Π_{zap}, if the verifier accepts, then we have $(x, \mathsf{ek}_{\mathsf{pke}}, \mathsf{ek}_{\mathsf{ikc}}, \mathsf{ek}_{\mathsf{le}}, \mathsf{prob}, \mathsf{ct}_{\mathsf{pke}}, \mathsf{ct}_{\mathsf{ikc}}, \mathsf{ct}_{\mathsf{le}}) \in \widetilde{\mathcal{L}} = \widetilde{\mathcal{L}}_1 \cup \widetilde{\mathcal{L}}_2$ with overwhelming probability. Therefore, if $\mathcal{P}^*$ wins with non-negligible probability, we have $(x, \mathsf{ek}_{\mathsf{pke}}, \mathsf{ek}_{\mathsf{ikc}}, \mathsf{ek}_{\mathsf{le}}, \mathsf{prob}, \mathsf{ct}_{\mathsf{pke}}, \mathsf{ct}_{\mathsf{ikc}}, \mathsf{ct}_{\mathsf{le}}) \in \widetilde{\mathcal{L}}_2$. We assume that this happens and construct a reduction algorithm that breaks non-uniform PPT hardness of Π_{HPG}. We consider the following two cases:

1. When $\mathsf{ek}_{\mathsf{ikc}}$ is decryptable (i.e., there is a corresponding decryption key $\mathsf{dk}_{\mathsf{ikc}}$ to $\mathsf{ek}_{\mathsf{ikc}}$): In this case, we can construct a reduction algorithm that finds $\mathsf{dk}_{\mathsf{ikc}}$ by brute-force and then extracts sol by decrypting $\mathsf{ct}_{\mathsf{ikc}}$ to break Π_{HPG}. We note that the brute-force search can be done *before* getting a problem instance prob, and thus non-uniform PPT hardness suffices.
2. When $\mathsf{ek}_{\mathsf{ikc}}$ is undecryptable: In this case, we consider several hybrids. In the first hybrid, π'_{zap} is generated by using a witness $\mathsf{wit}_{\mathsf{invalid}}$ instead of $r_{\mathsf{le\text{-}gen}}$. We note that such a witness $\mathsf{wit}_{\mathsf{invalid}}$ of invalidness of $\mathsf{ek}_{\mathsf{ikc}}$ must exist when $\mathsf{ek}_{\mathsf{ikc}}$ is undecryptable by the second property of invalid key certifiability. By witness indistinguishability of π'_{zap}, this only negligibly changes the cheating prover's winning probability.[11] In the next hybrid, $\mathsf{ek}_{\mathsf{le}}$ is generated in the injective mode instead of lossy mode. By key indistinguishability of Π_{LE}, this only negligibly changes the cheating prover's winning probability. At this point, a reduction algorithm can generate $\mathsf{ek}_{\mathsf{le}}$ in the injective mode with its corresponding decryption key, and thus it can extract sol by decrypting $\mathsf{ct}_{\mathsf{le}}$ to break Π_{HPG}. Similarly to the previous case, the non-uniform PPT hardness suffices even though the reduction algorithm runs a brute-force algorithm to find $\mathsf{wit}_{\mathsf{invalid}}$ since this can be done *before* getting a problem instance prob.

This contradicts non-uniform PPT hardness of $\Pi_{\Pi_{\mathsf{HPG}}}$. Therefore, the cheating prover's winning probability is negligible, and thus soundness holds.

[11] Strictly speaking, since the event that the cheating prover wins is not efficiently checkable, a more careful analysis is needed.

Reusable quantum-simulation zero-knowledge. A QPT simulator $\mathcal{S}$ is described as follows:

$\mathcal{S}(\mathsf{pp}, \mathsf{ch}, x, 1^{|w|})$: $\mathcal{S}$ is given $\mathsf{pp} = (\mathsf{ek}_{\mathsf{pke}}, \mathsf{ek}_{\mathsf{ikc}}, r'_{\mathsf{zap}})$, $\mathsf{ch} = (r_{\mathsf{zap}}, \mathsf{prob}, \mathsf{ek}_{\mathsf{le}}, \pi'_{\mathsf{zap}})$, a statement x, and a witness length $1^{|w|}$ as input, and works as follows.

1. Return $\bot$ if $\mathsf{VerProb}(1^\kappa, \mathsf{prob}) = \bot$ or $\mathsf{ZAP.Verify}'(r'_{\mathsf{zap}}, (\mathsf{ek}_{\mathsf{ikc}}, \mathsf{ek}_{\mathsf{le}}), \pi'_{\mathsf{zap}}) = \bot$.
2. Generate $\mathsf{sol} \xleftarrow{\$} \mathsf{Solve}(\mathsf{prob})$ (by using a QPT computation). If $\mathsf{VerSol}(\mathsf{prob}, \mathsf{sol}) = \bot$, immediately return $\bot$ and halt. Otherwise, generate $\mathsf{ct}_{\mathsf{ikc}} \xleftarrow{\$} \mathsf{Enc}(\mathsf{sol}; r_{\mathsf{ikc\text{-}enc}})$ and $\mathsf{ct}_{\mathsf{le}} \xleftarrow{\$} \mathsf{LE.Enc}(\mathsf{ek}_{\mathsf{le}}, \mathsf{sol}; r_{\mathsf{le\text{-}enc}})$.
3. Generate $\mathsf{ct}_{\mathsf{pke}} \xleftarrow{\$} \mathsf{PKE.Enc}(\mathsf{ek}_{\mathsf{pke}}, 0^{|w|})$.
4. Generate $\pi_{\mathsf{zap}} \xleftarrow{\$} \mathsf{ZAP.Prove}(r_{\mathsf{zap}}, (x, \mathsf{ek}_{\mathsf{pke}}, \mathsf{ek}_{\mathsf{ikc}}, \mathsf{ek}_{\mathsf{le}}, \mathsf{prob}, \mathsf{ct}_{\mathsf{pke}}, \mathsf{ct}_{\mathsf{ikc}}, \mathsf{ct}_{\mathsf{le}}), (\mathsf{sol}, r_{\mathsf{ikc\text{-}enc}}, r_{\mathsf{le\text{-}enc}}))$.
5. Return $\mathsf{resp} = (\mathsf{ct}_{\mathsf{pke}}, \mathsf{ct}_{\mathsf{ikc}}, \mathsf{ct}_{\mathsf{le}}, \pi_{\mathsf{zap}})$.

A response simulated by $\mathcal{S}$ is different from the real one in the following ways:

1. $\mathsf{ct}_{\mathsf{ikc}}$ is an encryption of sol instead of $0^{|\mathsf{sol}|}$, and
2. $\mathsf{ct}_{\mathsf{le}}$ is an encryption of sol instead of $0^{|\mathsf{sol}|}$, and
3. π_{zap} is generated by using a witness of $\widetilde{\mathcal{L}}_2$ instead of $\widetilde{\mathcal{L}}_1$, and
4. $\mathsf{ct}_{\mathsf{pke}}$ is an encryption of $0^{|w|}$ instead of w.

Roughly, the first difference is indistinguishable by the CPA security of Π_{IKC} against QPT adversaries. The second difference is indistinguishable due to the following reasons. (1) If $\mathsf{ek}_{\mathsf{le}}$ is a lossy key, encryptions of sol and $0^{|\mathsf{sol}|}$ are statistically indistinguishable. (2) If $\mathsf{ek}_{\mathsf{le}}$ is not a lossy key, we have $\mathsf{ZAP.Verify}'(r'_{\mathsf{zap}}, (\mathsf{ek}_{\mathsf{ikc}}, \mathsf{ek}_{\mathsf{le}}), \pi'_{\mathsf{zap}}) = \bot$ with overwhelming probability by the soundness of Π'_{ZAP} noting that $\mathsf{ek}_{\mathsf{ikc}}$ is honestly generated. In this case, $\mathsf{ct}_{\mathsf{le}}$ is not given to the adversary. The third difference is indistinguishable by the witness indistinguishability of Π_{ZAP}. The fourth difference is indistinguishable by the CPA security of Π_{PKE} against QPT adversaries after finishing the modification 3. We would be able to turn this intuition into a formal proof in a straightforward manner if we assumed witness indistinguishability against *quantum* adversaries. However, since we only assume witness indistinguishability against non-uniform *classical* adversaries, we have to be careful about the order of game hops.[12] Namely, if we first make the modifications 1 and 2 for all queries, then we cannot make the modification 3 since the game involves quantum computations in every query. To circumvent this issue, we make the modifications 1, 2, and 3 for each query one-by-one similarly to [GRS+11]. In this way, we can ensure that all quantum computations can be done in pre-computation stage when making the modification 3 for each query, and the proof goes through even with witness indistinguishability against non-uniform PPT adversaries.

[12] Note that there is no known ZAP with witness indistinguishability against QPT adversaries based on (quantum) polynomial hardness of standard assumptions.

5 Round-Optimal Blind Signatures

In this section, we construct round-optimal blind signatures.

5.1 Construction

Building blocks. We construct a round-optimal blind signature scheme based on the following building blocks.

- $\Pi_{\mathsf{Sig}} = (\mathsf{SigGen}, \mathsf{Sign}, \mathsf{SigVerify})$ is a digital signature scheme that is EUF-CMA against QPT adversaries. We assume that Sign is deterministic. This can be assumed without loss of generality by derandomizing the signing algorithm by using a quantumly secure PRF (which is only required to be secure against QPT adversaries that just make classical queries).
- $\Pi_{\mathsf{SFE}} = (\mathsf{Receiver}, \mathsf{Sender}, \mathsf{Derive})$ is an SFE protocol that satisfies receiver's security against non-uniform PPT adversaries and quantum-extraction sender's security against QPT adversaries.
- Com is a perfectly-binding non-interactive commitment with computational hiding against QPT adversaries.
- $\Pi_{\mathsf{ZK}} = (\mathsf{Setup}, \mathcal{V}_1, \mathcal{P}_2, \mathcal{V}_{\mathsf{out}})$ is blind-signature-conforming zero-knowledge arguments for a language $\mathcal{L}$, which is defined as follows: We have $(\mathsf{com}, \mathsf{sfe}_1, \mathsf{sfe}_2) \in \mathcal{L}$ if there exists $(\mathsf{ssk}, r_{\mathsf{com}}, r_{\mathsf{sfe}})$ such that

$$\mathsf{com} = \mathsf{Com}(\mathsf{ssk}; r_{\mathsf{com}})$$
$$\mathsf{sfe}_2 = \mathsf{Sender}(1^\kappa, \mathsf{sfe}_1, \mathsf{Sign}(\mathsf{ssk}, \cdot); r_{\mathsf{sfe}})$$

Construction. Our construction of a round-optimal blind signature scheme $\Pi_{\mathsf{BS}} = (\mathsf{BSGen}, \mathcal{U}_1, \mathcal{S}_2, \mathcal{U}_{\mathsf{der}}, \mathsf{BSVerify})$ is described as follows.

$\mathsf{BSGen}(1^\kappa)$: The key generation algorithm takes the security parameter 1^κ as input, and works as follows:
1. Generate $(\mathsf{svk}, \mathsf{ssk}) \xleftarrow{\$} \mathsf{SigGen}(1^\kappa)$.
2. Generate $\mathsf{com} \xleftarrow{\$} \mathsf{Com}(\mathsf{ssk}; r_{\mathsf{com}})$.
3. Generate $(\mathsf{pp}, \mathsf{sp}) \xleftarrow{\$} \mathsf{Setup}(1^\kappa)$.
4. Output a public key $\mathsf{pk} := (\mathsf{svk}, \mathsf{com}, \mathsf{pp})$ and a signing key $\mathsf{sk} := (\mathsf{ssk}, r_{\mathsf{com}}, \mathsf{sp})$.

$\mathcal{U}_1(\mathsf{pk}, m)$: The user's first message generation algorithm takes as input a public key $\mathsf{pk} = (\mathsf{svk}, \mathsf{com}, \mathsf{pp})$ and a message m, and works as follows:
1. Generate $(\mathsf{sfe}_1, \mathsf{sfest}) \xleftarrow{\$} \mathsf{Receiver}(1^\kappa, m)$.
2. Generate $\mathsf{ch} \xleftarrow{\$} \mathcal{V}_1(\mathsf{pp})$.
3. Output a first message $\mu := (\mathsf{sfe}_1, \mathsf{ch})$ and a state $\mathsf{st}_{\mathcal{U}} := \mathsf{sfest}$.

$\mathcal{S}_2(\mathsf{sk}, \mu)$: The signer's second message generation algorithm takes as input a signing key $\mathsf{sk} = (\mathsf{ssk}, r_{\mathsf{com}}, \mathsf{sp})$. and a first message $\mu = (\mathsf{sfe}_1, \mathsf{ch})$ and works as follows:
1. Generate $\mathsf{sfe}_2 \xleftarrow{\$} \mathsf{Sender}(1^\kappa, \mathsf{sfe}_1, \mathsf{Sign}(\mathsf{ssk}, \cdot); r_{\mathsf{sfe}})$.
2. Generate $\mathsf{resp} \xleftarrow{\$} \mathcal{P}_2(\mathsf{sp}, \mathsf{ch}, (\mathsf{com}, \mathsf{sfe}_1, \mathsf{sfe}_2), (\mathsf{ssk}, r_{\mathsf{com}}, r_{\mathsf{sfe}}))$.
3. Output a second message $\rho := (\mathsf{sfe}_2, \mathsf{resp})$.

$\mathcal{U}_{\mathsf{der}}(\mathsf{st}_{\mathcal{U}}, \rho)$: The user's signature derivation algorithm takes as input a state $\mathsf{st}_{\mathcal{U}} = \mathsf{sfest}$ and a second message $\rho = (\mathsf{sfe}_2, \mathsf{resp})$ as input, and works as follows:
1. Output $\bot$ if $\mathcal{V}_{\mathsf{out}}(\mathsf{pp}, \mathsf{ch}, (\mathsf{com}, \mathsf{sfe}_1, \mathsf{sfe}_2), \mathsf{resp}) = \bot$
2. Otherwise generate $\sigma \xleftarrow{\$} \mathsf{Derive}(\mathsf{sfest}, \mathsf{sfe}_2)$ and output a signature σ.

$\mathsf{BSVerify}(\mathsf{pk}, m, \sigma)$: The verification algorithm takes as input a public key $\mathsf{pk} = (\mathsf{svk}, \mathsf{com}, \mathsf{pp})$, a message m, and a signature σ as input, and outputs $\mathsf{SigVerify}(\mathsf{svk}, m, \sigma)$.

The correctness of the scheme immediately follows from the correctness of Π_{Sig}, Π_{ZK}, and Π_{SFE}.

5.2 Security

In this section, we give security proofs for the above scheme.

Unforgeability.

Theorem 5.1. *If Π_{Sig} satisfies unforgeability against QPT adversaries,* Com *satisfies computational hiding against QPT adversaries, Π_{SFE} satisfies quantum-extraction sender's security against QPT adversaries, and Π_{ZK} satisfies reusable quantum-simulation zero-knowledge against classical adversaries, then Π_{BS} satisfies unforgeability against classical adversaries.*

Proof. We consider the following sequence of games between a PPT adversary $\mathcal{A}$ and a challenger. We denote by E_i the event that Game i returns 1.

Game 1: This is the original unforgeability game. That is, this game proceeds as follows.
1. The challenger generates $(\mathsf{ssk}, \mathsf{svk}) \xleftarrow{\$} \mathsf{SigGen}(1^\kappa)$, $\mathsf{com} \xleftarrow{\$} \mathsf{Com}(\mathsf{ssk}; r_{\mathsf{com}})$, and $(\mathsf{pp}, \mathsf{sp}) \xleftarrow{\$} \mathsf{Setup}(1^\kappa)$, and defines a public key $\mathsf{pk} := (\mathsf{svk}, \mathsf{com}, \mathsf{pp})$ and a signing key $\mathsf{sk} := (\mathsf{ssk}, r_{\mathsf{com}}, \mathsf{sp})$, and sends pk to $\mathcal{A}$.
2. $\mathcal{A}$ can make arbitrarily many signing queries. When it makes a signing query $\mu = (\mathsf{sfe}_1, \mathsf{ch})$, the challenger generates $\mathsf{sfe}_2 \xleftarrow{\$} \mathsf{Sender}(1^\kappa, \mathsf{sfe}_1, \mathsf{Sign}(\mathsf{ssk}, \cdot); r_{\mathsf{sfe}})$ and $\mathsf{resp} \xleftarrow{\$} \mathcal{P}_2(\mathsf{sp}, \mathsf{ch}, (\mathsf{com}, \mathsf{sfe}_1, \mathsf{sfe}_2), (\mathsf{ssk}, r_{\mathsf{com}}, r_{\mathsf{sfe}}))$, and returns $\rho := (\mathsf{sfe}_2, \mathsf{resp})$.
3. Finally, $\mathcal{A}$ returns $\{(m_i, \sigma_i)\}_{i \in [q+1]}$ where q is the number of signing queries made by $\mathcal{A}$.

The game returns 1 if and only if $\mathcal{A}$ wins, i.e., $\{m_i\}_{i \in [q+1]}$ is pairwise distinct and $\mathsf{SigVerify}(\mathsf{svk}, m_i, \sigma_i) = \top$ for all $i \in [q+1]$. Our goal is to prove $\Pr[\mathsf{E}_1] = \mathsf{negl}(\kappa)$.

Game 2: This game is identical to the previous one except that resp is generated as $\mathsf{resp} \xleftarrow{\$} \mathcal{S}(\mathsf{pp}, \mathsf{ch}, (\mathsf{com}, \mathsf{sfe}_1, \mathsf{sfe}_2), 1^{|w|})$ when responding to each signing query where $\mathcal{S}$ is the simulator of Π_{ZK} and $|w|$ denotes the bit-length of $(\mathsf{ssk}, r_{\mathsf{com}}, r_{\mathsf{sfe}})$.

By a straightforward reduction to reusable quantum-simulation zero-knowledge property of Π_{ZK}, we have $|\Pr[\mathsf{E}_2] - \Pr[\mathsf{E}_1]| = \mathsf{negl}(\kappa)$.

Game 3: This game is identical to the previous one except that sfe_2 is generated as $m \xleftarrow{\$} \mathsf{SFEExt}(\mathsf{sfe}_1)$ and $\mathsf{sfe}_2 \xleftarrow{\$} \mathsf{SFESim}(1^\kappa, \mathsf{sfe}_1, \mathsf{Sign}(\mathsf{ssk}, m))$ when responding to each signing query.
Noting that r_{sfe} is no longer used for generating resp due to the modification made in Game 2, a straightforward reduction to quantum-extraction sender's security of Π_{SFE} gives $|\Pr[\mathsf{E}_3] - \Pr[\mathsf{E}_2]| = \mathsf{negl}(\kappa)$. We note that the reduction works even though these games involve QPT computations (for $\mathcal{S}$ and SFEExt) since we assume quantum-extraction sender's security against *quantum* adversaries.

Game 4: In this game, the challenger generates com as $\mathsf{com} \xleftarrow{\$} \mathsf{Com}(0^{|\mathsf{ssk}|})$.
Noting that r_{com} is no longer used for generating resp due to the modification made in Game 2, a straightforward reduction to computational hiding of Π_{SFE} gives $|\Pr[\mathsf{E}_4] - \Pr[\mathsf{E}_3]| = \mathsf{negl}(\kappa)$. We note that the reduction works even though these games involve QPT computations (for $\mathcal{S}$ and SFEExt) since we assume computational hiding against *quantum* adversaries.

What is left is to prove $\Pr[\mathsf{E}_4] = \mathsf{negl}(\kappa)$. We show this by considering the following QPT adversary $\mathcal{B}$ against unforgeability of Π_{Sig}.

$\mathcal{B}^{\mathsf{Sign}(\mathsf{ssk},\cdot)}(\mathsf{svk})$: It generates $\mathsf{com} \xleftarrow{\$} \mathsf{Com}(0^{|\mathsf{ssk}|})$ and $(\mathsf{pp}, \mathsf{sp}) \xleftarrow{\$} \mathsf{Setup}(1^\kappa)$ and gives a public key $\mathsf{pk} := (\mathsf{svk}, \mathsf{com}, \mathsf{pp})$ to $\mathcal{A}$. When $\mathcal{A}$ makes a signing query $\mu = (\mathsf{sfe}_1, \mathsf{ch})$, $\mathcal{B}$ computes $m \xleftarrow{\$} \mathsf{SFEExt}(\mathsf{sfe}_1)$ and queries m to its own signing oracle to obtain $\sigma = \mathsf{Sign}(\mathsf{ssk}, m)$. Then $\mathcal{B}$ generates $\mathsf{sfe}_2 \xleftarrow{\$} \mathsf{SFESim}(1^\kappa, \mathsf{sfe}_1, \sigma)$ and $\mathsf{resp} \xleftarrow{\$} \mathcal{S}(\mathsf{pp}, \mathsf{ch}, (\mathsf{com}, \mathsf{sfe}_1, \mathsf{sfe}_2), 1^{|w|})$, and returns $\rho := (\mathsf{sfe}_2, \mathsf{resp})$ to $\mathcal{A}$ as a response from the signing oracle. Let $\{(m_i, \sigma_i)\}_{i \in [q+1]}$ be $\mathcal{A}$'s final output. $\mathcal{B}$ finds $i^* \in [q+1]$ such that it has not queried m_{i^*} to its own signing oracle and $\mathsf{SigVerify}(\mathsf{svk}, m_{i^*}, \sigma_{i^*}) = \top$, and outputs (m_{i^*}, σ_{i^*}). If there does not exist such i^*, $\mathcal{B}$ aborts.

It is easy to see that $\mathcal{B}$ perfectly simulates the environment of Game 4 to $\mathcal{A}$, and when $\mathcal{A}$ wins, $\mathcal{B}$ also wins (i.e., it succeeds in outputting (m_{i^*}, σ_{i^*}) such that $\mathsf{SigVerify}(\mathsf{svk}, m_{i^*}, \sigma_{i^*}) = \top$ and $\mathcal{B}$ has not queried m_{i^*}). Therefore, by unforgeability of Π_{Sig}, we have $\Pr[\mathsf{E}_4] = \mathsf{negl}(\kappa)$. This completes the proof of Theorem 5.1. □

Blindness

Theorem 5.2. *If* Com *satisfies perfect binding,* Π_{SFE} *satisfies receiver's security against non-uniform PPT adversaries, and* Π_{ZK} *satisfies adaptive soundness with untrusted setup against non-uniform PPT adversaries, then* Π_{BS} *satisfies blindness against PPT adversaries.*

Proof. We consider the following sequence of games between a PPT adversary $\mathcal{A}$ against the blindness and a challenger. We denote by E_i the event that Game i returns 1.

Game 1: This is the original blindness game. That is, this game proceeds as follows:

1. $\mathcal{A}$ is given as input the security parameter 1^κ, and sends a public key $\mathsf{pk} = (\mathsf{svk}, \mathsf{com}, \mathsf{pp})$ and a pair (m_0, m_1) of messages to the challenger.
2. The challenger generates $(\mathsf{sfe}_{1,b}, \mathsf{sfest}_b) \xleftarrow{\$} \mathsf{Receiver}(1^\kappa, m_b)$ and $\mathsf{ch}_b \xleftarrow{\$} \mathcal{V}_1(\mathsf{pp})$ and defines $\mu_b := (\mathsf{sfe}_{1,b}, \mathsf{ch}_b)$ and $\mathsf{st}_{\mathcal{U},b} := \mathsf{sfest}_b$ for each $b \in \{0,1\}$, picks $\mathsf{coin} \xleftarrow{\$} \{0,1\}$, and sends $(\mu_{\mathsf{coin}}, \mu_{1-\mathsf{coin}})$ to $\mathcal{A}$.
3. $\mathcal{A}$ sends $(\rho_{\mathsf{coin}} = (\mathsf{sfe}_{2,\mathsf{coin}}, \mathsf{resp}_{\mathsf{coin}}), \rho_{1-\mathsf{coin}} = (\mathsf{sfe}_{2,1-\mathsf{coin}}, \mathsf{resp}_{1-\mathsf{coin}}))$ to the challenger.
4. The challenger gives $(\bot, \bot)$ to $\mathcal{A}$ if $\mathcal{V}_{\mathsf{out}}(\mathsf{pp}, \mathsf{ch}_b, (\mathsf{com}, \mathsf{sfe}_{1,b}, \mathsf{sfe}_{2,b}), \mathsf{resp}) = \bot$ for either of $b \in \{0,1\}$. Otherwise it generates $\sigma_b \xleftarrow{\$} \mathsf{Derive}(\mathsf{sfest}_b, \mathsf{sfe}_{2,b})$ for each $b \in \{0,1\}$ and gives (σ_0, σ_1) to $\mathcal{A}$.
5. $\mathcal{A}$ outputs its guess coin'.

This game returns 1 if and only if $\mathsf{coin} = \mathsf{coin}'$. Our goal is to prove $\left|\Pr[\mathsf{E}_1] - \frac{1}{2}\right| = \mathsf{negl}(\kappa)$.

Game 2: This game is identical to the previous game except that we insert Step 1.5 between Step 1 and 2 and Step 4 is replaced with Step 4′ described below: (Differences of Step 4' from Step 4 are marked by red underlines.)

1.5.: The challenger finds $(\mathsf{ssk}, r_{\mathsf{com}})$ such that $\mathsf{com} = \mathsf{Com}(\mathsf{ssk}; r_{\mathsf{com}})$ by a brute-force search. If such $(\mathsf{ssk}, r_{\mathsf{com}})$ does not exist, it sets $(\mathsf{ssk}, r_{\mathsf{com}}) := (\bot, \bot)$.

4′.: The challenger gives $(\bot, \bot)$ to $\mathcal{A}$ if $\mathcal{V}_{\mathsf{out}}(\mathsf{pp}, \mathsf{ch}_b, (\mathsf{com}, \mathsf{sfe}_{1,b}, \mathsf{sfe}_{2,b}), \mathsf{resp}) = \bot$ for either of $b \in \{0,1\}$ or $(\mathsf{ssk}, r_{\mathsf{com}}) = (\bot, \bot)$. Otherwise it generates $\sigma_b := \mathsf{Sign}(\mathsf{ssk}, m_b)$ for each $b \in \{0,1\}$ and gives (σ_0, σ_1) to $\mathcal{A}$.

In Lemma 5.1, we prove $|\Pr[\mathsf{E}_2] - \Pr[\mathsf{E}_1]| = \mathsf{negl}(\kappa)$.

Game 3: This game is identical to the previous game except that $\mathsf{sfe}_{1,b}$ is generated as $(\mathsf{sfe}_{1,b}, \mathsf{sfest}_b) \xleftarrow{\$} \mathsf{Receiver}(1^\kappa, m_0)$ for both $b \in \{0,1\}$.

In Lemma 5.2, we prove $|\Pr[\mathsf{E}_3] - \Pr[\mathsf{E}_2]| = \mathsf{negl}(\kappa)$.

Game 4: This game is identical to the previous game except that the challenger gives (μ_0, μ_1) to $\mathcal{A}$ instead of $(\mu_{\mathsf{coin}}, \mu_{1-\mathsf{coin}})$ in Step 2.

Since the distributions of μ_0 and μ_1 are identical, we have $\Pr[\mathsf{E}_4] = \Pr[\mathsf{E}_3]$. Moreover, since no information on coin is given to $\mathcal{A}$ in this game, we have $\Pr[\mathsf{E}_4] = \frac{1}{2}$.

What is left is to prove the following lemmata.

Lemma 5.1. *If* Com *satisfies perfect binding and* Π_{ZK} *satisfies adaptive soundness with untrusted setup against non-uniform PPT adversaries, then we have* $|\Pr[\mathsf{E}_2] - \Pr[\mathsf{E}_1]| = \mathsf{negl}(\kappa)$.

Proof. For each $b \in \{0,1\}$, we define Bad_b as an event that we have $\mathcal{V}_{\mathsf{out}}(\mathsf{pp}, \mathsf{ch}_b, (\mathsf{com}, \mathsf{sfe}_{1,b}, \mathsf{sfe}_{2,b}), \mathsf{resp}_b) = \top$ and

1. there does not exist $(\mathsf{ssk}, r_{\mathsf{com}})$ such that $\mathsf{com} = \mathsf{Com}(\mathsf{ssk}; r_{\mathsf{com}})$, or
2. there exists $(\mathsf{ssk}, r_{\mathsf{com}})$ such that $\mathsf{com} = \mathsf{Com}(\mathsf{ssk}; r_{\mathsf{com}})$ and $\mathsf{Derive}(\mathsf{sfest}_b, \mathsf{sfe}_{2,b}) \neq \mathsf{Sign}(\mathsf{ssk}, m_b)$.

Game 2 and Game 1 may be different only if Bad_0 or Bad_1 occurs. Therefore, it suffices to prove $\Pr[\mathsf{Bad}_b] = \mathsf{negl}(\kappa)$ for each $b \in \{0,1\}$. First, we prove $\Pr[\mathsf{Bad}_0] = \mathsf{negl}(\kappa)$ by considering a non-uniform PPT cheating prover $\mathcal{P}^* = (\mathcal{P}^*_{\mathsf{Setup}}, \mathcal{P}^*_2)$ against adaptive soundness with adaptive setup of Π_{ZK} as described below:

$\mathcal{P}^*_{\mathsf{Setup}}(1^\kappa)$: It runs the first stage of $\mathcal{A}(1^\kappa)$ to obtain $\mathsf{pk} = (\mathsf{svk}, \mathsf{com}, \mathsf{pp})$ and (m_0, m_1). It finds $(\mathsf{ssk}, r_{\mathsf{com}})$ such that $\mathsf{com} = \mathsf{Com}(\mathsf{ssk}; r_{\mathsf{com}})$ by a brute-force search. If such $(\mathsf{ssk}, r_{\mathsf{com}})$ does not exist, it sets $(\mathsf{ssk}, r_{\mathsf{com}}) := (\bot, \bot)$. It outputs pp and $\mathsf{st}_{\mathcal{P}^*} := (\mathsf{pk}, m_0, m_1, \mathsf{ssk}, r_{\mathsf{com}}, \mathsf{st}_{\mathcal{A}})$ where $\mathsf{st}_{\mathcal{A}}$ denotes the snapshot of $\mathcal{A}$ at this point.

$\mathcal{P}^*_2(\mathsf{st}_{\mathcal{P}^*}, \mathsf{ch})$: It parses $(\mathsf{pk}, m_0, m_1, \mathsf{ssk}, r_{\mathsf{com}}, \mathsf{st}_{\mathcal{A}}) \leftarrow \mathsf{st}_{\mathcal{P}^*}$, generates $(\mathsf{sfe}_{1,b}, \mathsf{sfest}_b) \xleftarrow{\$} \mathsf{Receiver}(1^\kappa, m_b)$ for each $b \in \{0,1\}$ and $\mathsf{ch}_1 \xleftarrow{\$} \mathcal{V}_1(\mathsf{pp})$, sets $\mathsf{ch}_0 := \mathsf{ch}$, and defines $\mu_b := (\mathsf{sfe}_{1,b}, \mathsf{ch}_b)$ for each $b \in \{0,1\}$, picks $\mathsf{coin} \xleftarrow{\$} \{0,1\}$, and sends $(\mu_{\mathsf{coin}}, \mu_{1-\mathsf{coin}})$ to $\mathcal{A}$ to run the second stage of $\mathcal{A}$ to obtain $(\rho_{\mathsf{coin}} = (\mathsf{sfe}_{2,\mathsf{coin}}, \mathsf{resp}_{\mathsf{coin}}), \rho_{1-\mathsf{coin}} = (\mathsf{sfe}_{2,1-\mathsf{coin}}, \mathsf{resp}_{1-\mathsf{coin}}))$. If Bad_0 occurs, then $\mathcal{P}^*_2$ outputs $(\mathsf{com}, \mathsf{sfe}_{1,0}, \mathsf{sfe}_{2,0})$ and resp_0.

We can see that $\mathcal{P}^*$ perfectly simulates Game 1 for $\mathcal{A}$ until the second stage of $\mathcal{A}$. Moreover, if Bad_0 occurs, we have $\mathcal{V}_{\mathsf{out}}(\mathsf{pp}, \mathsf{ch}_0, (\mathsf{com}, \mathsf{sfe}_{1,0}, \mathsf{sfe}_{2,0}), \mathsf{resp}_0) = \top$ and $(\mathsf{com}, \mathsf{sfe}_{1,0}, \mathsf{sfe}_{2,0}) \notin \mathcal{L}$ noting that Com is perfectly binding. Therefore, by the adaptive soundness with untrusted setup of Π_{ZK}, we have $\Pr[\mathsf{Bad}_0] = \mathsf{negl}(\kappa)$. We can prove $\Pr[\mathsf{Bad}_1] = \mathsf{negl}(\kappa)$ analogously. This completes a proof of Lemma 5.1. □

Lemma 5.2. *If Π_{SFE} satisfies receiver's security against non-uniform PPT adversaries, then we have $|\Pr[\mathsf{E}_3] - \Pr[\mathsf{E}_2]| = \mathsf{negl}(\kappa)$.*

Proof. We prove this by considering a non-uniform PPT cheating adversary $\mathcal{B} = (\mathcal{B}_0, \mathcal{B}_1)$ against receiver's security of Π_{SFE} as described below:

$\mathcal{B}_0(1^\kappa)$: It runs the first stage of $\mathcal{A}(1^\kappa)$ to obtain $\mathsf{pk} = (\mathsf{svk}, \mathsf{com}, \mathsf{pp})$ and (m_0, m_1). It finds $(\mathsf{ssk}, r_{\mathsf{com}})$ such that $\mathsf{com} = \mathsf{Com}(\mathsf{ssk}; r_{\mathsf{com}})$ by a brute-force search. If such $(\mathsf{ssk}, r_{\mathsf{com}})$ does not exist, it sets $(\mathsf{ssk}, r_{\mathsf{com}}) := (\bot, \bot)$. It outputs (m_0, m_1) and $\mathsf{st}_{\mathcal{P}^*} := (\mathsf{pk}, m_0, m_1, \mathsf{ssk}, r_{\mathsf{com}}, \mathsf{st}_{\mathcal{A}})$ where $\mathsf{st}_{\mathcal{A}}$ denotes the snapshot of $\mathcal{A}$ at this point.

$\mathcal{B}_1(\mathsf{st}_{\mathcal{B}}, \mathsf{sfe}_1)$: It sets $\mathsf{sfe}_{1,1} := \mathsf{sfe}_1$, generates $(\mathsf{sfe}_{1,0}, \mathsf{sfest}_0) \xleftarrow{\$} \mathsf{Receiver}(1^\kappa, m_0)$ and $\mathsf{ch}_b \xleftarrow{\$} \mathcal{V}_1(\mathsf{pp})$ for $b \in \{0,1\}$, defines $\mu_b := (\mathsf{sfe}_{1,b}, \mathsf{ch}_b)$ for each $b \in \{0,1\}$, picks $\mathsf{coin} \xleftarrow{\$} \{0,1\}$, and sends $(\mu_{\mathsf{coin}}, \mu_{1-\mathsf{coin}})$ to $\mathcal{A}$ to run the second stage of $\mathcal{A}$ to obtain $(\rho_{\mathsf{coin}} = (\mathsf{sfe}_{2,\mathsf{coin}}, \mathsf{resp}_{\mathsf{coin}}), \rho_{1-\mathsf{coin}} = (\mathsf{sfe}_{2,1-\mathsf{coin}}, \mathsf{resp}_{1-\mathsf{coin}}))$. Then $\mathcal{B}_1$ gives $(\bot, \bot)$ to $\mathcal{A}$ if $\mathcal{V}_{\mathsf{out}}(\mathsf{pp}, \mathsf{ch}_b, (\mathsf{com}, \mathsf{sfe}_{1,b}, \mathsf{sfe}_{2,b}), \mathsf{resp}) = \bot$ for either of $b \in \{0,1\}$ or $(\mathsf{ssk}, r_{\mathsf{com}}) = (\bot, \bot)$. Otherwise it generates $\sigma_b := \mathsf{Sign}(\mathsf{ssk}, m_b)$ for each $b \in \{0,1\}$ and gives (σ_0, σ_1) to $\mathcal{A}$. Let coin' be $\mathcal{A}$'s final output. $\mathcal{B}_1$ outputs 1 if $\mathsf{coin} = \mathsf{coin}'$.

Clearly, $\mathcal{B}$ perfectly simulates Game 3 (resp. Game 2) if sfe_1 is generated as $(\mathsf{sfe}_1, \mathsf{sfest}) \xleftarrow{\$} \mathsf{Receiver}(1^\kappa, m_0)$ (resp. $(\mathsf{sfe}_1, \mathsf{sfest}) \xleftarrow{\$} \mathsf{Receiver}(1^\kappa, m_1)$). Therefore, by receiver's security of Π_{SFE}, we have $|\Pr[\mathsf{E}_3] - \Pr[\mathsf{E}_2]| = \mathsf{negl}(\kappa)$. □

Combining the above, Theorem 5.2 is proven. □

Acknowledgement. We thank anonymous reviewers of Eurocrypt 2021 for their helpful comments. The first and third authors were supported by JST CREST Grant Number JPMJCR19F6. The third author is also supported by JSPS KAKENHI Grant Number 19H01109.

References

[AFG+16] Abe, M., Fuchsbauer, G., Groth, J., Haralambiev, K., Ohkubo, M.: Structure-preserving signatures and commitments to group elements. J. Cryptol. **29**(2), 363–421 (2016). https://doi.org/10.1007/s00145-014-9196-7

[AO12] Abe, M., Ohkubo, M.: A framework for universally composable non-committing blind signatures. Int. J. Appl. Cryptogr. **2**(3), 229–249 (2012)

[AR04] Aharonov, D., Regev, O.: Lattice problems in NP cap coNP. In: 45th FOCS, pp. 362–371. IEEE Computer Society Press, October 2004

[AR05] Aharonov, D., Regev, O.: Lattice problems in NP cap coNP. J. ACM **52**(5), 749–765 (2005)

[BCC04] Brickell, E.F., Camenisch, J., Chen, L.: Direct anonymous attestation. In: Atluri, V., Pfitzmann, B., McDaniel, P. (eds.) ACM CCS 2004, pp. 132–145. ACM Press, October 2004

[BFPV11] Blazy, O., Fuchsbauer, G., Pointcheval, D., Vergnaud, D.: Signatures on randomizable ciphertexts. In: Catalano, D., Fazio, N., Gennaro, R., Nicolosi, A. (eds.) PKC 2011. LNCS, vol. 6571, pp. 403–422. Springer, Heidelberg (2011). https://doi.org/10.1007/978-3-642-19379-8_25

[BHY09] Bellare, M., Hofheinz, D., Yilek, S.: Possibility and impossibility results for encryption and commitment secure under selective opening. In: Joux, A. (ed.) EUROCRYPT 2009. LNCS, vol. 5479, pp. 1–35. Springer, Heidelberg (2009). https://doi.org/10.1007/978-3-642-01001-9_1

[BNPS02] Bellare, M., Namprempre, C., Pointcheval, D., Semanko, M.: The power of RSA inversion oracles and the security of Chaum's RSA-based blind signature scheme. In: Syverson, P.F. (ed.) FC 2001. LNCS, vol. 2339, pp. 319–338. Springer, Heidelberg (2002). https://doi.org/10.1007/3-540-46088-8_25

[Bol03] Boldyreva, A.: Threshold signatures, multisignatures and blind signatures based on the gap-Diffie-Hellman-group signature scheme. In: Desmedt, Y.G. (ed.) PKC 2003. LNCS, vol. 2567, pp. 31–46. Springer, Heidelberg (2003). https://doi.org/10.1007/3-540-36288-6_3

[BPV12] Blazy, O., Pointcheval, D., Vergnaud, D.: Compact round-optimal partially-blind signatures. In: Visconti, I., De Prisco, R. (eds.) SCN 2012. LNCS, vol. 7485, pp. 95–112. Springer, Heidelberg (2012). https://doi.org/10.1007/978-3-642-32928-9_6

[Cha82] Chaum, D.: Blind signatures for untraceable payments. In: Chaum, D., Rivest, R.L., Sherman, A.T. (eds.) CRYPTO'82, pp. 199–203. Plenum Press, New York (1982)

[Cha88] Chaum, D.: Elections with unconditionally-secret ballots and disruption equivalent to breaking RSA. In: Barstow, D., et al. (eds.) EUROCRYPT 1988. LNCS, vol. 330, pp. 177–182. Springer, Heidelberg (1988). https://doi.org/10.1007/3-540-45961-8_15

[CL01] Camenisch, J., Lysyanskaya, A.: An efficient system for non-transferable anonymous credentials with optional anonymity revocation. In: Pfitzmann, B. (ed.) EUROCRYPT 2001. LNCS, vol. 2045, pp. 93–118. Springer, Heidelberg (2001). https://doi.org/10.1007/3-540-44987-6_7

[DN00] Dwork, C., Naor, M.: Zaps and their applications. In: 41st FOCS, pp. 283–293. IEEE Computer Society Press, November 2000

[DN07] Dwork, C., Naor, M.: Zaps and their applications. SIAM J. Comput. **36**(6), 1513–1543 (2007)

[FHKS16] Fuchsbauer, G., Hanser, C., Kamath, C., Slamanig, D.: Practical round-optimal blind signatures in the standard model from weaker assumptions. In: Zikas, V., De Prisco, R. (eds.) SCN 2016. LNCS, vol. 9841, pp. 391–408. Springer, Cham (2016). https://doi.org/10.1007/978-3-319-44618-9_21

[FHS15] Fuchsbauer, G., Hanser, C., Slamanig, D.: Practical round-optimal blind signatures in the standard model. In: Gennaro, R., Robshaw, M. (eds.) CRYPTO 2015. Part II. LNCS, vol. 9216, pp. 233–253. Springer, Heidelberg (2015). https://doi.org/10.1007/978-3-662-48000-7_12

[Fis06] Fischlin, M.: Round-optimal composable blind signatures in the common reference string model. In: Dwork, C. (ed.) CRYPTO 2006. LNCS, vol. 4117, pp. 60–77. Springer, Heidelberg (2006). https://doi.org/10.1007/11818175_4

[FLS90] Feige, U., Lapidot, D., Shamir, A.: Multiple non-interactive zero knowledge proofs based on a single random string (extended abstract). In: 31st FOCS, pp. 308–317. IEEE Computer Society Press, October 1990

[FLS99] Feige, U., Lapidot, D., Shamir, A.: Multiple noninteractive zero knowledge proofs under general assumptions. SIAM J. Comput. **29**(1), 1–28 (1999)

[FOO93] Fujioka, A., Okamoto, T., Ohta, K.: A practical secret voting scheme for large scale elections. In: Seberry, J., Zheng, Y. (eds.) AUSCRYPT 1992. LNCS, vol. 718, pp. 244–251. Springer, Heidelberg (1993). https://doi.org/10.1007/3-540-57220-1_66

[FS10] Fischlin, M., Schröder, D.: On the impossibility of three-move blind signature schemes. In: Gilbert, H. (ed.) EUROCRYPT 2010. LNCS, vol. 6110, pp. 197–215. Springer, Heidelberg (2010). https://doi.org/10.1007/978-3-642-13190-5_10

[GG14] Garg, S., Gupta, D.: Efficient round optimal blind signatures. In: Nguyen, P.Q., Oswald, E. (eds.) EUROCRYPT 2014. LNCS, vol. 8441, pp. 477–495. Springer, Heidelberg (2014). https://doi.org/10.1007/978-3-642-55220-5_27

[Gha17] Ghadafi, E.: Efficient round-optimal blind signatures in the standard model. In: Kiayias, A. (ed.) FC 2017. LNCS, vol. 10322, pp. 455–473. Springer, Cham (2017). https://doi.org/10.1007/978-3-319-70972-7_26

[GLN11] Goldreich, O., Levin, L.A., Nisan, N.: On constructing 1-1 one-way functions. In: Goldreich, O. (ed.) Studies in Complexity and Cryptography. Miscellanea on the Interplay between Randomness and Computation. LNCS, vol. 6650, pp. 13–25. Springer, Heidelberg (2011). https://doi.org/10.1007/978-3-642-22670-0_3

[GO94] Goldreich, O., Oren, Y.: Definitions and properties of zero-knowledge proof systems. J. Cryptol. **7**(1), 1–32 (1994). https://doi.org/10.1007/BF00195207

[GRS+11] Garg, S., Rao, V., Sahai, A., Schröder, D., Unruh, D.: Round optimal blind signatures. In: Rogaway, P. (ed.) CRYPTO 2011. LNCS, vol. 6841, pp. 630–648. Springer, Heidelberg (2011). https://doi.org/10.1007/978-3-642-22792-9_36

[GS08] Groth, J., Sahai, A.: Efficient non-interactive proof systems for bilinear groups. In: Smart, N.P. (ed.) EUROCRYPT 2008. LNCS, vol. 4965, pp. 415–432. Springer, Heidelberg (2008). https://doi.org/10.1007/978-3-540-78967-3_24

[HK12] Halevi, S., Kalai, Y.T.: Smooth projective hashing and two-message oblivious transfer. J. Cryptol. **25**(1), 158–193 (2010). https://doi.org/10.1007/s00145-010-9092-8

[HKKL07] Hazay, C., Katz, J., Koo, C.-Y., Lindell, Y.: Concurrently-secure blind signatures without random oracles or setup assumptions. In: Vadhan, S.P. (ed.) TCC 2007. LNCS, vol. 4392, pp. 323–341. Springer, Heidelberg (2007). https://doi.org/10.1007/978-3-540-70936-7_18

[KK19] Kalai, Y.T., Khurana, D.: Non-interactive non-malleability from quantum supremacy. In: Boldyreva, A., Micciancio, D. (eds.) CRYPTO 2019. Part III. LNCS, vol. 11694, pp. 552–582. Springer, Cham (2019). https://doi.org/10.1007/978-3-030-26954-8_18

[Lin08] Lindell, Y.: Lower bounds and impossibility results for concurrent self composition. J. Cryptol. **21**(2), 200–249 (2007). https://doi.org/10.1007/s00145-007-9015-5

[MSF10] Meiklejohn, S., Shacham, H., Freeman, D.M.: Limitations on transformations from composite-order to prime-order groups: the case of round-optimal blind signatures. In: Abe, M. (ed.) ASIACRYPT 2010. LNCS, vol. 6477, pp. 519–538. Springer, Heidelberg (2010). https://doi.org/10.1007/978-3-642-17373-8_30

[NP01] Naor, M., Pinkas, B.: Efficient oblivious transfer protocols. In: Rao Kosaraju, S. (ed.) 12th SODA, pp. 448–457. ACM-SIAM, January 2001

[Pas03] Pass, R.: Simulation in quasi-polynomial time, and its application to protocol composition. In: Biham, E. (ed.) EUROCRYPT 2003. LNCS, vol. 2656, pp. 160–176. Springer, Heidelberg (2003). https://doi.org/10.1007/3-540-39200-9_10

[Pas11] Pass, R.: Limits of provable security from standard assumptions. In: Fortnow, L., Vadhan, S.P. (eds.) 43rd ACM STOC, pp. 109–118. ACM Press, June 2011

[PVW08] Peikert, C., Vaikuntanathan, V., Waters, B.: A framework for efficient and composable oblivious transfer. In: Wagner, D. (ed.) CRYPTO 2008. LNCS, vol. 5157, pp. 554–571. Springer, Heidelberg (2008). https://doi.org/10.1007/978-3-540-85174-5_31

[Reg05] Regev, O.: On lattices, learning with errors, random linear codes, and cryptography. In: Gabow, H.N., Fagin, R. (eds.) 37th ACM STOC, pp. 84–93. ACM Press, May 2005

[Reg09] Regev, O.: On lattices, learning with errors, random linear codes, and cryptography. J. ACM **56**(6), 34:1–34:40 (2009)

[SC12] Seo, J.H., Cheon, J.H.: Beyond the limitation of prime-order bilinear groups, and round optimal blind signatures. In: Cramer, R. (ed.) TCC 2012. LNCS, vol. 7194, pp. 133–150. Springer, Heidelberg (2012). https://doi.org/10.1007/978-3-642-28914-9_8

[Sho94] Shor, P.W.: Algorithms for quantum computation: discrete logarithms and factoring. In: 35th FOCS, pp. 124–134. IEEE Computer Society Press, November 1994

Post-Quantum Multi-Party Computation

Amit Agarwal[1(✉)], James Bartusek[2], Vipul Goyal[3], Dakshita Khurana[1], and Giulio Malavolta[4]

[1] University of Illinois Urbana-Champaign, Urbana, USA
{amita2,dakshita}@illinois.edu
[2] UC Berkeley, Berkeley, USA
[3] NTT Research and CMU, Pittsburgh, USA
vipul@cmu.edu
[4] Max Planck Institute for Security and Privacy, Bochum, Germany

Abstract. We initiate the study of multi-party computation for classical functionalities in the plain model, with security against malicious quantum adversaries. We observe that existing techniques readily give a polynomial-round protocol, but our main result is a construction of *constant-round* post-quantum multi-party computation. We assume mildly super-polynomial quantum hardness of learning with errors (LWE), and quantum polynomial hardness of an LWE-based circular security assumption. Along the way, we develop the following cryptographic primitives that may be of independent interest:

- A spooky encryption scheme for relations computable by quantum circuits, from the quantum hardness of (a circular variant of) the LWE problem. This immediately yields the first quantum multi-key fully-homomorphic encryption scheme with classical keys.
- A constant-round post-quantum non-malleable commitment scheme, from the mildly super-polynomial quantum hardness of LWE.

To prove the security of our protocol, we develop a new straight-line non-black-box simulation technique against parallel sessions that does not clone the adversary's state. This technique may also be relevant to the classical setting.

1 Introduction

Secure multi-party computation (MPC) allows a set of parties to compute a joint function of their inputs, revealing only the output of the function while keeping their inputs private. General secure MPC, initiated in works such as [6,14,33,68],

A. Agarwal and D. Khurana—This material is based on work supported in part by DARPA under Contract No. HR001120C0024. Any opinions, findings and conclusions or recommendations expressed in this material are those of the author(s) and do not necessarily reflect the views of the United States Government or DARPA.
V. Goyal—Supported in part by the DARPA SIEVE program, NSF award 1916939, a gift from Ripple, a DoE NETL award, a JP Morgan Faculty Fellowship, a PNC center for financial services innovation award, and a Cylab seed funding award.

A. Canteaut and F.-X. Standaert (Eds.): EUROCRYPT 2021, LNCS 12696, pp. 435–464, 2021.
https://doi.org/10.1007/978-3-030-77870-5_16

has played a central role in modern theoretical cryptography. The last few years have seen tremendous research optimizing MPC in various ways, enabling a plethora of practical applications that include joint computations on distributed medical data, privacy-preserving machine learning, e-voting, distributed key management, among others. The looming threat of quantum computers naturally motivates the problem of constructing protocols with *provable security against quantum adversaries.*

After Watrous' breakthrough work on zero-knowledge against quantum adversaries [64], the works of [20,41,51] considered variants of quantum-secure computation protocols, in the *two*-party setting. Very recently, Bitansky and Shmueli [10] obtained the first *constant-round* classical zero-knowledge arguments with security against quantum adversaries. Their techniques (and those of [1] in a concurrent work) are based on the recent non-black-box simulation technique of [8], who constructed two-message *classically-secure* weak zero-knowledge in the plain model. Unfortunately, it is unclear whether these protocols compose under parallel repetition. As a result, they become largely inapplicable to the constant-round multi-party setting.

There has also been substantial effort in constructing protocols for securely computing quantum circuits [23,25,26] (see Sect. 2.6 for further discussion). However, to the best of our knowledge, generic multi-party computation protocols with classical communication and security against quantum adversaries have only been studied in models with *trusted pre-processing or setup.* To make things even worse, [23] construct a maliciously-secure multi-party protocol for computing quantum ciruits, assuming the existence of a maliciously-secure post-quantum classical MPC protocol. This means that the only available implementations of such a building block require trusted pre-processing or a common reference string.

Post-Quantum MPC. In this work we initiate the study of MPC protocols that allow classical parties to securely compute general classical functionalities, and where security is guaranteed against *malicious quantum adversaries.* Our focus is on MPC in the *plain model*: Fully classical participants interact with each other with no access to trusted/pre-processed parameters or a common reference string. Multi-party protocols achieving security in these settings do not seem to have been previously analyzed in *any* number of rounds.

We stress that the challenges of proving post-quantum security of MPC protocols stretch far beyond the appropriate instantiations of the cryptographic building blocks (e.g. avoiding factoring- or discrete logarithm-based cryptosystems): Because quantum information behaves very differently from classical information, designing post-quantum protocols often requires new techniques to achieve provable security. As an example, a common strategy to prove classical security of MPC protocols is to define a simulator that can extract the inputs of the corrupted parties by "rewinding" them, i.e. taking a snapshot of the state of the adversary and split the protocol execution in multiple branches. However, when the adversary is a quantum machine, this technique becomes largely inapplicable since the no-cloning theorem (one of the fundamental principles of quantum mechanics) prevents us from creating two copies of an arbitrary quantum state.

One of our key contributions is a new *parallel no-cloning non-black-box simulation technique* that extends the work of [10], to achieve security against multiple parallel quantum verifiers.

1.1 Our Results

We begin by summarizing our main result: Classical multi-party computation with security against quantum circuits in the plain model. Here, parties communicate classically via authenticated point-to-point channels as well as broadcast channels, where everyone can send messages in the same round. In each round, all parties simultaneously exchange messages. The network is assumed to be synchronous with rushing adversaries, i.e. adversaries may generate their messages for any round after observing the messages of all honest parties in that round, but before observing the messages of honest parties in the next round. The (quantum) adversary may corrupt upto all but one of the participants. In this model, we obtain the following main result.

Theorem 1 (Informal). *Assuming mildly super-polynomial quantum hardness of LWE and AFS-spooky encryption for relations computable by polynomial-size quantum circuits, there exists a constant-round classical MPC protocol (in the plain model) maliciously secure against quantum polynomial-time adversaries.*

In more detail, our protocol is secure against any adversary $\mathsf{A} = \{\mathsf{A}_\lambda, \rho_\lambda\}_\lambda$, where each A_λ is the (classical) description of a polynomial-size quantum circuit and ρ_λ is some (possibly inefficiently computable) non-uniform quantum advice. Beyond being interesting in its own right, our plain-model protocol may serve as a useful stepping stone to obtaining interesting protocols for securely computing quantum circuits in the plain model, as evidenced by the work of [23]. This protocol is constructed in Sects. 8 and 9 in the full version.

By "mildly" super-polynomial quantum hardness of LWE, we mean to assume that there exists a constant $c \in \mathbb{N}$, such that for large enough security parameter $\lambda \in \mathbb{N}$, no quantum polynomial time algorithm can distinguish LWE samples from uniform with advantage better than $\mathsf{negl}(\lambda^{\mathsf{ilog}(c,\lambda)})$, where $\mathsf{ilog}(c, \lambda)$ denotes the c-times iterated logarithm $\log\log\cdots_{c \text{ times}}(\lambda)$. We note that this is weaker than assuming the quasi-polynomial quantum hardness of LWE, i.e. the assumption that quantum polynomial-time adversaries cannot distinguish LWE samples from uniform with advantage better than $2^{-(\log\lambda)^c}$ for some constant $c > 1$.

A central technical ingredient of our work is an additive function sharing (AFS) spooky encryption scheme [21] for relations computable by quantum circuits. An AFS-spooky encryption scheme has a publicly-computable algorithm that, on input a set of ciphertexts $\mathsf{Enc}(\mathsf{pk}_1, m_1), \ldots, \mathsf{Enc}(\mathsf{pk}_n, m_n)$ encrypted under *independently sampled* public keys and a (possibly quantum) circuit C, computes a new set of ciphertexts

$$\mathsf{Enc}(\mathsf{pk}_1, y_1), \ldots, \mathsf{Enc}(\mathsf{pk}_n, y_n) \ s.t. \bigoplus_{i=1}^{n} y_i = C(m_1, \ldots, m_n).$$

In Sect. 4 in the full version we show how to construct AFS-spooky encryption for relations computable by quantum circuits, under an LWE-based circular security assumption. We refer the reader to Sect. 4.4 in the full version for the exact circular security assumption we need, which is similar to the one used in [52]. As a corollary, this immediately yields the first multi-key fully-homomorphic encryption [50] for quantum circuits with classical key generation and classical encryption of classical messages.

Theorem 2 (Informal). *Under an appropriate LWE-based circular security assumption, there exists an AFS-spooky encryption scheme for relations computable by polynomial-size quantum circuits with classical key generation and classical encryption of classical messages.*

Along the way to proving our main theorem, we construct and rely on constant-round zero-knowledge arguments against parallel quantum verifiers, and constant-round extractable commitments against parallel quantum committers. Parallel extractable commitments and zero-knowledge are formally constructed and analyzed in Sects. 5 and 6 in the full version, respectively. We only show the construction of parallel extractable commitments in Sect. 3 in this paper. We point out that we do not obtain protocols that compose under *unbounded* parallel repetition. Instead we build a bounded variant (that we also refer to as multi-verifier zero-knowledge and multi-committer extractable commitments) that suffices for our applications.

Theorem 3 (Informal). *Assuming the quantum polynomial hardness of LWE and the existence of AFS-spooky encryption for relations computable by polynomial-size quantum circuits, there exists:*

- *A constant-round classical argument for NP that is computational-zero-knowledge against parallel quantum polynomial-size verifiers.*
- *A constant-round classical commitment that is extractable against parallel quantum polynomial-size committers.*

In addition, we initiate the study of post-quantum non-malleable commitments. Specifically, we construct and rely on constant-round post-quantum non-malleable commitments based on the super-polynomial hardness assumption described above. The formal construction and analysis can be found in Sect. 7 in the full version.

Theorem 4 (Informal). *Assuming the mildly super-polynomial quantum hardness of LWE and the existence of fully-homomorphic encryption for quantum circuits, there exists a constant-round non-malleable commitment scheme secure against quantum polynomial-size adversaries.*

We also obtain quantum-secure non-malleable commitments in $O(\mathsf{ilog}(c, \lambda))$ rounds for any constant $c \in \mathbb{N}$ based on any quantum-secure extractable commitment. In particular, plugging in these commitments instead of our constant round non-malleable commitments gives an $O(\mathsf{ilog}(c, \lambda))$ round quantum-secure MPC from any quantum AFS-spooky encryption scheme.

2 Technical Overview

2.1 Background

Our starting point is any constant-round post-quantum MPC protocol maliciously secure in the programmable common reference string (CRS) model. Such a protocol can be obtained, for example, based on the semi-maliciously secure MPC protocols of [2,53] in the CRS model. Specifically, assuming the existence of post-quantum zero-knowledge in the CRS model (that can be obtained based on the quantum hardness of LWE [60]) and the quantum hardness of LWE, these works obtain multi-party computation for classical circuits in the CRS model with the following property: There exists an ideal-world simulator that *programs the CRS*, interacts in a straight-line, black-box manner with any quantum adversary corrupting an arbitrary subset of the players, and outputs a view that is indistinguishable from the real view of the adversary, including the output of honest parties.

Thus, a natural approach to achieving post-quantum MPC in the plain model is to then securely implement a multi-party functionality that generates the aforementioned CRS. Specifically, we would like a set of n parties to jointly execute a *coin-flipping protocol.* Such a protocol outputs a uniformly random string that may then be used to implement post-quantum secure MPC according to [2,53]. The programmability requirement on the CRS roughly translates to ensuring that for any quantum adversary, there exists a simulator that on input a random string s, can force the output of the coin-flipping protocol to be equal to s. A protocol satisfying this property is often referred to as a *fully-simulatable* multi-party coin-flipping protocol.

Post-Quantum Multi-Party Coin-Flipping. Existing constant-round protocols [36,65] for multi-party coin-flipping against classical adversaries make use of the following template. Each participant first commits to a uniformly random string using an appropriate perfectly binding commitment.[1] In a later phase, all participants reveal the values they committed to, without actually revealing the randomness used for commitment. Additionally, each participant proves (in zero-knowledge) to every other participant that they opened to the same value that they originally committed to. If all zero-knowledge arguments verify, the protocol output is computed as the sum of the openings of all participants. To highlight challenges to construct constant-round protocols, we elaborate on this template and outline a simple polynomial-round coin tossing protocol. Readers familiar with this template for multi-party coin-tossing may skip to the next page.

A Simple Protocol in Polynomially Many Rounds. In order to motivate the challenges involved in constructing a post-quantum *constant-round* multiparty coin tossing protocol, we first outline a simple protocol that requires *polynomially many* rounds, and follows from ideas in existing work. Our starting point is the polynomial-round post-quantum zero-knowledge protocol due to Watrous [64].

[1] We actually require this commitment to also satisfy a property called *non-malleability*, which we discuss later in this section.

Ideas developed in [10] can be used almost immediately to convert this to a post-quantum extractable commitment scheme, assuming polynomial hardness of LWE (or, more generally, any post-quantum oblivious transfer). For completeness, we outline how this is done in Appendix A in the full version. Next, it is possible to use the resulting post-quantum secure extractable commitment to obtain post-quantum multiparty fully-simulatable coin flipping, that admits a straight-line simulator in the dishonest majority setting. The protocol requires rounds that grow linearly with the number of parties and polynomially with the security parameter, as described in Fig. 1.

n-Party Coin tossing

Common input: 1^λ, 1^n.

1. For each $i \in [n]$, party P_i samples $r_i \leftarrow \{0,1\}^\lambda$.
2. Sequentially, for every $i \in [n]$, $j \in [n] \setminus \{i\}$ parties P_i, P_j execute a post-quantum extractable commitment where P_i commits to r_i and P_j is the receiver.
3. P_i broadcasts r_i.
4. For every $i \in [n]$, $j \in [n] \setminus \{i\}$ parties P_i, P_j *sequentially* execute a post-quantum ZK protocol where P_i is the prover and P_j is the verifier. P_i proves to P_j (in zero-knowledge) that the value committed via the extractable commitment (in Step 2) is consistent with the value broadcasted (in Step 3).
5. If all the proofs where P_i is verifier are accepting, P_i outputs $\bigoplus_{i=1}^n r_i$.

Fig. 1. Multiparty coin tossing

Recall that the simulator Sim of any coin-flipping protocol will obtain a uniformly random string r^* from the ideal functionality, and must force this value as the output. The Sim for the protocol in Fig. 1 samples r_i uniformly at random on behalf of each honest party P_i, and commits to r_i in Step 2 following honest sender strategy. At the same time, Sim runs Ext to (sequentially) extract the value committed by every corrupted party in Step 2. This allows the simulator to compute $\bigoplus_{i \in \mathbb{M}} r_i$, where $\mathbb{M}$ denotes the set of corrupted parties. In Step 3, the simulator broadcasts values r_i' on behalf of honest parties such that $\bigoplus_{i \in [n] \setminus \mathbb{M}} r_i' = \bigoplus_{i \in \mathbb{M}} r_i \oplus r^*$. Finally, it invokes the simulator of the ZK protocol to produce proofs on behalf of honest parties. It is easy to see that the output would indeed end up being the intended output r^*.

Notice that replacing Watrous' polynomial-round ZK protocol with the constant-round ZK of [1,10] only decreases the rounds to linear in the number of parties. To decrease the number of rounds to constant, it is clear that one would need to find a way to execute the commitment sessions (Step 2) and ZK sessions (Step 4) in parallel. While the recent work of Bitansky and Shmueli [10] builds constant-round post-quantum zero-knowledge, their protocol and its guarantees turn out to be insufficient for the parallel setting. In this setting, a single prover would typically need to interact in parallel with $(n-1)$ different verifiers, a subset or all of which may be adversarial. It should be possible for a simulator to *simultaneously* simulate the view of multiple parallel verifiers. In addition, the argument should continue to satisfy soundness, even if a subset of verifiers colludes with a (cheating) prover.

Post-Quantum Parallel Zero-Knowledge. We overcome this barrier by building the first constant-round zero-knowledge argument secure against *parallel quantum verifiers* from quantum polynomial hardness of an LWE-based circular security assumption. This improves upon the work of [1,10] who provided arguments with provable security only against a single quantum verifier. Very roughly, the approach in [1,10] relies on a modification of the [8] homomorphic trapdoors paradigm. We do not assume familiarity with the details of this protocol or paradigm, and will in fact discuss a (variant of) this in the next subsection. For now, we simply point out that in this paradigm, the verifier generates an initial FHE ciphertext and public key, as well as some additional information to enable simulation. The simulator *homomorphically evaluates* the verifier's (quantum) circuit over the initial FHE ciphertext and then uses the result of this evaluation to recover secrets that will enable simulation.

However, when a prover interacts with several verifiers at once, each verifier will generate its own FHE ciphertexts. In a nutshell, in the parallel setting the simulator can no longer perform individual homomorphic evaluations corresponding to each verifier, due to no-cloning. To address this issue, we develop a novel **parallel no-cloning** simulation strategy. This strategy relies on a novel technique that enables the simulator to *peel away* secret keys of this FHE scheme layer-by-layer. An overview of this technique can be found in Sect. 2.2.

Our technique also crucially relies on a strong variant of quantum fully-homomorphic encryption that allows for homomorphic operations under multiple keys at once. The encryption scheme that we use is a quantum generalization of the notion of additive function sharing (AFS) *spooky encryption* [21]. As a contribution of independent interest, we build the first AFS-spooky encryption (that also implies multi-key FHE) for quantum circuits from a circular variant of the LWE assumption. We give an overview of our construction in Sect. 2.3.

Post-Quantum Non-malleable Commitments. Our construction of zero-knowledge against parallel quantum verifiers gives rise to a coin-flipping protocol that is secure as long as at least one participant is honest, and all committed strings are independent of each other. However, ensuring such independence is not straightforward, even in the classical setting. In fact, upon seeing an honest party's commitment string c, a malicious, rushing adversary may be able to produce a string c' that commits to a related message. This is known as a malleability attack, and can be prevented by relying on *non-malleable commitments.* In this work, we devise the first post-quantum non-malleable commitments based on slightly superpolynomial hardness of LWE. An overview of our construction can be found in Sect. 2.4.

Finally, we discuss how to combine all these primitives to build our desired coin-tossing protocol, and a few additional subtleties that come up in the process, in Sect. 2.5.

2.2 A New Parallel No-Cloning Non-Black-Box Simulation Technique

In the following we give a high-level overview of our constant-round zero-knowledge protocol secure against parallel quantum verifiers. In favor of a simpler exposition, we first describe a *parallel extractable commitment* protocol. A parallel extractable commitment is a commitment where a single receiver interacts in parallel with multiple committers, each committing to its own independent message. The main challenge in this setting is to simulate the view of an adversary corrupting several of these committers, while *simultaneously* recovering all committed messages. Once we build a parallel extractable commitment, obtaining a parallel zero-knowledge protocol becomes a simple exercise (that we discuss towards the end of this overview).

Throughout the following overview we only consider adversaries that are (i) *non-aborting*, i.e. they never interrupt the execution of the protocol, and (ii) *explainable*, i.e. their messages always lie in the support of honestly generated messages, though they can select their random coins and inputs arbitrarily. We further simplify our overview by only considering (iii) *classical* adversaries, while being mindful to avoid any kind of state cloning during extraction. In the end of this overview we discuss how to remove these simplifications.

Cryptographic Building Blocks. Before delving into the description of our protocol, we introduce the technical tools needed for our construction. A fully-homomorphic encryption (FHE) scheme [29] allows one to compute any function (in its circuit representation) over some encrypted message $\mathsf{Enc}(\mathsf{pk}, m)$, without the need to decrypt it first. We say that an FHE is multi-key [50] if it supports the homomorphic evaluation of circuits even over messages encrypted under *independently sampled* public keys:

$$\{\mathsf{Enc}(\mathsf{pk}_i, m_i)\}_{i\in[n]} \xrightarrow{\mathsf{Eval}((\mathsf{pk}_1,\ldots,\mathsf{pk}_n),C,\cdot)} \mathsf{Enc}((\mathsf{pk}_1,\ldots,\mathsf{pk}_n), C(m_1,\ldots,m_n)).$$

Clearly, decrypting the resulting ciphertext should require the knowledge of all of the corresponding secret keys $(\mathsf{sk}_1,\ldots,\mathsf{sk}_n)$. Other than semantic security, we require that the scheme is compact, in the sense that the size of the evaluated ciphertext is proportional to $|C(m_1,\ldots,m_n)|$ (and possibly the number of parties n) but does not otherwise depend on the size of C.

The second tool that we use is compute and compare obfuscation [35,66]. A compute and compare program $\mathbf{CC}[f, u, z]$ program is defined by a function f, a lock value u, and an output z. On input a string x, the program returns z if and only if $f(x) = u$. The obfuscator Obf is guaranteed to return an obfuscated program $\widetilde{\mathbf{CC}}$ that is indistinguishable from a program that rejects any input, as long as u has sufficient entropy conditioned on f and z. Finally, we use a conditional disclosure of secret (CDS)[2] scheme. Recall that this is an interactive protocol parametrized by an NP relation $\mathcal{R}$ where both the sender and the receiver share a statement x and in addition, the sender has a secret message m. At the end of the interaction, the receiver obtains m if and only if it knows a valid witness w such that $\mathcal{R}(x, w) = 1$.

A Strawman Solution. We now describe a naive extension of the [1,10] approach to the parallel setting (where a receiver interacts with multiple committers), and highlight its pitfalls. We do not assume familiarity with [1,10]. To commit to messages $(m_1, \ldots, m_n)$, the committers and the receiver engage in the following protocol.

- Each committer samples a key pair of a multi-key FHE scheme $(\mathsf{pk}_i, \mathsf{sk}_i)$, a uniform trapdoor td_i, and a uniform lock value lk_i, and sends to the receiver:
  1. A commitment $\mathsf{c}_i = \mathsf{Com}(\mathsf{td}_i)$.
  2. An FHE encryption $\mathsf{Enc}(\mathsf{pk}_i, \mathsf{td}_i)$.
  3. An obfuscation $\widetilde{\mathbf{CC}}_i$ of the program $\mathbf{CC}[\mathsf{Dec}(\mathsf{sk}_i, \cdot), \mathsf{lk}_i, (\mathsf{sk}_i, m_i)]$.
- The receiver engages each committer in a (parallel) execution of a CDS protocol where the i'th committer sends lk_i if the receiver correctly guesses a valid preimage of c_i.

At a high level, the fact that the protocol hides the message m_i is ensured by the following argument. Since the receiver cannot invert c_i, it cannot guess td_i and therefore the CDS protocol will return 0. This in turn means that the lock lk_i is hidden from the receiver, and consequently that the obfuscated program is indistinguishable from a null program. This is, of course, an informal explanation, and we refer the reader to [1,8,10] for a formal security analysis.

We now turn to the description of the extractor. The high-level strategy is the following: Upon receiving the first message from all committers, the extractor uses the FHE encryption $\mathsf{Enc}(\mathsf{pk}_i, \mathsf{td}_i)$ and the code of the adversary to run the CDS protocol homomorphically (on input td_i) to recover an FHE encryption of lk_i. Then the extractor feeds it as an input to the obfuscated program $\widetilde{\mathbf{CC}}_i$, which returns (sk_i, m_i).

Unfortunately this approach has a major limitation: It implicitly assumes that each corrupted party is a local algorithm. In other words, we are assuming that the adversary consists of individual subroutines (one per corrupted party), which may not necessarily be the case. As an example, if the adversary were to somehow implement a strategy where corrupted machines do not respond until *all* receiver messages have been delivered, then the above homomorphic evaluation would get stuck and return no output. It is also worth mentioning that what makes the problem challenging is our inability to clone the state of the adversary. If we were allowed to clone its state, then we could extract messages one by one, by running a separate thread under each FHE key.

[2] In the body of the paper we actually resort to a slightly stronger tool, namely a secure function evaluation protocol with statistical circuit privacy.

Multi-Key Evaluation. A natural solution to circumvent the above issue is to rely on multi-key FHE evaluation. Using this additional property, the extractor can turn the ciphertexts $\mathsf{Enc}(\mathsf{pk}_1, \mathsf{td}_1), \ldots, \mathsf{Enc}(\mathsf{pk}_n, \mathsf{td}_n)$ into a single encryption

$$\mathsf{Enc}((\mathsf{pk}_1, \ldots, \mathsf{pk}_n), (\mathsf{td}_1, \ldots, \mathsf{td}_n))$$

under the hood of all public keys $(\mathsf{pk}_1, \ldots, \mathsf{pk}_n)$. Given this information, the extractor can homomorphically evaluate all instances of the CDS protocol at once, using the code of the adversary, no matter how intricate. This procedure allows the extractor to obtain the encryption of each lock value $\mathsf{Enc}((\mathsf{pk}_1, \ldots, \mathsf{pk}_n), \mathsf{lk}_i)$. In the single committer setting, we could then feed this into the corresponding obfuscated program and call it a day.

However, in the parallel setting, even given multi-key FHE, it is unclear how to proceed. If the compute and compare program $\widetilde{\mathbf{CC}}_i$ tried to decrypt such a ciphertext, it would obtain (at best) an encryption under the remaining public keys. Glossing over the fact that the structure of single-key and multi-key ciphertexts might be incompatible, it is unlikely that

$$\mathsf{Dec}(\mathsf{sk}_i, \mathsf{Enc}((\mathsf{pk}_1, \ldots, \mathsf{pk}_n), \mathsf{lk}_i)) = \mathsf{lk}_i$$

which is what we would need to trigger the compute and compare program. The general problem here is that each compute and compare program cannot encode information about other secret keys, thus making it infeasible to decrypt multi-key ciphertexts. One approach to resolve this issue would be to ask all committers to jointly obfuscate a compute and compare program that encodes all secret keys at once. However, this seems to require a general-purpose MPC protocol, which is what we are trying to build in the first place. Therefore, we outline a different approach by imagining a special kind of multi-key fully homomorphic encryption scheme.

A spooky encryption[3] scheme [21] is an FHE scheme that supports a special *spooky evaluation* algorithm, that generates no-signaling correlations among independently encrypted messages. We will restrict attention to a sub-class of no-signaling relations called *additive function sharing* (AFS) relations, and we will call the scheme AFS-spooky. More concretely, on input a circuit C and n independently generated ciphertexts (under independently generated public keys), the algorithm $\mathsf{Spooky.Eval}$ produces

$$\{\mathsf{Enc}(\mathsf{pk}_i, m_i)\}_{i \in [n]} \xrightarrow{\mathsf{Spooky.Eval}((\mathsf{pk}_1, \ldots, \mathsf{pk}_n), C, \cdot)} \{\mathsf{Enc}(\mathsf{pk}_i, y_i)\}_{i \in [n]} \ s.t. \ \bigoplus_{i=1}^{n} y_i = C(m_1, \ldots, m_n).$$

It is not hard to see that AFS-spooky encryption is a special case of multi-key FHE where multi-key ciphertexts have the following structure

$$\mathsf{Enc}((\mathsf{pk}_1, \ldots, \mathsf{pk}_n), m) = \{\mathsf{Enc}(\mathsf{pk}_i, y_i)\}_{i \in [n]} \ s.t. \ \bigoplus_{i=1}^{n} y_i = m.$$

[3] As a historical remark, while the name is inspired by Einstein's quote "spooky action at a distance" referring to entangled quantum states, the concept of spooky encryption (as defined in [21]) is entirely classical.

This additional structure is going to be our main leverage for constructing an efficient extractor.

The Extractor. Going back to our extractor, our next technical insight is to look for a mechanism to *peel away* encryption layers one by one from an AFS-spooky (multi-key) ciphertext. Our extractor will achieve this via careful *homomorphic* evaluation of the independently generated programs $(\widetilde{\mathbf{CC}}_1, \ldots, \widetilde{\mathbf{CC}}_n)$, as described below.

- First, homomorphically execute the code of the adversary using the AFS-spooky scheme to obtain

$$\mathsf{ct}_1 = \mathsf{Enc}((\mathsf{pk}_1, \ldots, \mathsf{pk}_n), \mathsf{lk}_1), \ldots, \mathsf{ct}_n = \mathsf{Enc}((\mathsf{pk}_1, \ldots, \mathsf{pk}_n), \mathsf{lk}_n),$$

  as described above.
- Parse ct_n as a collection of individual ciphertexts

$$\mathsf{Enc}((\mathsf{pk}_1, \ldots, \mathsf{pk}_n), \mathsf{lk}_n) = \{\mathsf{Enc}(\mathsf{pk}_i, y_i)\}_{i \in [n]} = \{\mathsf{Enc}(\mathsf{pk}_i, y_i)\}_{i \in [n-1]} \cup \underbrace{\{\mathsf{Enc}(\mathsf{pk}_n, y_n)\}}_{\tilde{\mathsf{ct}}_n}.$$

  Note that we can interpret the first $n-1$ elements as an AFS-spooky ciphertext encrypted under $(\mathsf{pk}_1, \ldots, \mathsf{pk}_{n-1})$:

$$\tilde{\mathsf{ct}} = \{\mathsf{Enc}(\mathsf{pk}_i, y_i)\}_{i \in [n-1]} = \mathsf{Enc}\left((\mathsf{pk}_1, \ldots, \mathsf{pk}_{n-1}), \bigoplus_{i=1}^{n-1} y_i\right) = \mathsf{Enc}\left((\mathsf{pk}_1, \ldots, \mathsf{pk}_{n-1}), \tilde{y}\right)$$

  where $\tilde{y} = \bigoplus_{i=1}^{n-1} y_i$.
- Let Γ be the following function

$$\Gamma(\zeta) : \mathsf{Spooky.Eval}(\mathsf{pk}_n, \zeta \oplus \cdot, \tilde{\mathsf{ct}}_n)$$

  which homomorphically computes the XOR of ζ with the plaintext of $\tilde{\mathsf{ct}}_n$. Compute the following nested AFS-spooky correlation

$$\begin{aligned}\widehat{\mathsf{ct}} &= \mathsf{Spooky.Eval}((\mathsf{pk}_1, \ldots, \mathsf{pk}_{n-1}), \Gamma, \tilde{\mathsf{ct}}) \\ &= \mathsf{Enc}\left((\mathsf{pk}_1, \ldots, \mathsf{pk}_{n-1}), \mathsf{Spooky.Eval}(\mathsf{pk}_n, \tilde{y} \oplus \cdot, \tilde{\mathsf{ct}}_n)\right) \quad (1) \\ &= \mathsf{Enc}\left((\mathsf{pk}_1, \ldots, \mathsf{pk}_{n-1}), \mathsf{Enc}\left(\mathsf{pk}_n, \bigoplus_{i=1}^{n} y_i\right)\right) \quad (2) \\ &= \mathsf{Enc}\left((\mathsf{pk}_1, \ldots, \mathsf{pk}_{n-1}), \mathsf{Enc}(\mathsf{pk}_n, \mathsf{lk}_n)\right) \quad (3)\end{aligned}$$

  by interpreting $\tilde{\mathsf{ct}}_n$ as a single key ciphertext. Here (1) follows by substituting Γ, and (2) follows by correctness of the AFS-spooky evaluation.
- Run the obfuscated compute and compare program homomorphically to obtain an encryption of sk_n and m_n under $(\mathsf{pk}_1, \ldots, \mathsf{pk}_{n-1})$

$$\begin{aligned}\mathsf{Spooky.Eval}\left((\mathsf{pk}_1, \ldots, \mathsf{pk}_{n-1}), \widetilde{\mathbf{CC}}_n, \widehat{\mathsf{ct}}\right) &= \mathsf{Enc}\left((\mathsf{pk}_1, \ldots, \mathsf{pk}_{n-1}), \widetilde{\mathbf{CC}}_n\left(\mathsf{Enc}(\mathsf{pk}_n, \mathsf{lk}_n)\right)\right) \\ &= \mathsf{Enc}\left((\mathsf{pk}_1, \ldots, \mathsf{pk}_{n-1}), (\mathsf{sk}_n, m_n)\right).\end{aligned}$$

- Using the encryption of sk_n under $(\mathsf{pk}_1, \ldots, \mathsf{pk}_{n-1})$, update the initial ciphertexts $(\mathsf{ct}_1, \ldots, \mathsf{ct}_{n-1})$ by homomorphically decrypting their last component and adding the resulting string. This allows the extractor to obtain

$$\mathsf{Enc}((\mathsf{pk}_1, \ldots, \mathsf{pk}_{n-1}), \mathsf{lk}_1), \ldots, \mathsf{Enc}((\mathsf{pk}_1, \ldots, \mathsf{pk}_{n-1}), \mathsf{lk}_{n-1}).$$

- Recursively apply the procedure described above until $\mathsf{Enc}(\mathsf{pk}_1, \mathsf{lk}_1)$ is recovered, then feed this ciphertext as an input to $\widetilde{\mathbf{CC}}_1$ to obtain (sk_1, m_1) in the clear. Iteratively recover $(\mathsf{sk}_2, \ldots, \mathsf{sk}_n)$ by decrypting the corresponding ciphertexts. At this point the extractor knows all secret keys and can decrypt the transcript of the interaction together with the committed messages.

To summarize, this extractor will isolate single-key ciphertexts (albeit in a nested form) by relying on AFS-spooky encryption. These ciphertexts by design will be compatible with compute and compare programs. In turn, evaluating the program *under the encryption* allows us to *escape* from the newly introduced layer. Repeating this procedure recursively eventually leads to a complete recovery of the plaintexts.

We stress that, although the extraction algorithm repeats the nesting operation n times, the additional encryption layer introduced in each iteration is immediately peeled off by executing the obfuscated compute and compare program. Thus the above procedure runs in (strict) polynomial time for *any polynomial* number of parties n.

Parallel Zero Knowledge. The above outline is deliberately simplified and ignores some subtle issues that arise during the analysis of the protocol. As an example, we need to ensure that the adversary is not able to *maul* the commitment of the trapdoor into a CDS encryption to be used in the CDS protocol. This issue also arose in [10], and we follow their approach of using non-uniformity in a reduction to the semantic security of the quantum FHE scheme. [10] also present the technical tools needed to lift the protocol to the setting of malicious and possibly aborting adversaries (as opposed to explainable), and we roughly follow their approach. However, it is worth pointing out that [10] directly construct a zero-knowledge argument, without first constructing and analyzing a stand-alone extractable commitment. Since we use a parallel extractable commitment as a building block in the our coin-flipping protocol, we analyze the above as a stand-alone commitment, which requires a few modifications to the protocol and proof techniques. More discussion about this can be found in Sect. 3.

Now, we describe how to obtain parallel zero-knowledge (i.e. zero-knowledge against multiple verifiers) from parallel extractable commitment. This is accomplished in a routine manner by enhancing a standard Σ protocol with a stage where each verifier commits to its Σ protocol challenge using a parallel extractable commitment. Using the extractor, the simulator can obtain the challenges ahead of time and can therefore simulate the rest of the transcript, without the need to perform state cloning.

It remains to argue that our extraction strategy does not break down in the presence of quantum adversaries. Observe that the only step that involves the execution of a quantum circuit is the AFS-spooky evaluation of the CDS protocol, under the hood of $(\mathsf{pk}_1, \ldots, \mathsf{pk}_n)$. Assuming that we can construct AFS-spooky encryption for relations computable by quantum circuits (which we show in Sect. 2.3), the remainder of the extraction algorithm only depends on the encryptions of $(\mathsf{lk}_1, \ldots, \mathsf{lk}_n)$, which are classical strings. Once the extractor recovers all the secret keys, it can decrypt the (possibly quantum) state of the adversary resulting from the homomorphic evaluation of the CDS, and resume the protocol execution, without the need to clone the adversary's state.

2.3 Quantum AFS-Spooky Encryption

We now turn to the construction of AFS-spooky encryption for relations computable by quantum circuits. The main technical contribution of this section is a construction of multi-key fully-homomorphic encryption for quantum circuits with classical key generation and classical encryption of classical messages. Such schemes were already known in the *single*-key setting, due to [11,52].

Background. At a very high level, these single-key schemes follow a paradigm introduced by Broadbent and Jeffery [13], which makes use of the quantum one-time pad (QOTP). The QOTP is a method of perfectly encrypting arbitrary quantum states with a key that consists of only classical bits. [13] suggest to encrypt a quantum state with a quantum one-time pad (QOTP), and then encrypt the classical bits that comprise the QOTP using a classical fully-homomorphic encryption scheme. One can then apply quantum gates to the encrypted quantum state, and update the classical encryption of the one-time pad appropriately. A key feature of this encryption procedure is that while an encryption of a quantum state necessarily must be a quantum state, an encryption of classical information does not necessarily have to include a quantum state. Indeed, one can simply give a classical one-time pad encryption of the data, along with a classical fully-homomorphic encryption of the pad.

However, the original schemes presented by Broadbent and Jeffery [13] and subsequent work [24] based on their paradigm left much to be desired. In particular, they required even a classical encryptor to supply quantum "gadgets" encoding their secret key. These gadgets were then used to evaluate a particular non-Clifford gate over encrypted data.[4] The main innovation in the work of [52] was to remove the need for quantum gadgets, instead showing how to evaluate an appropriate non-Clifford gate using just *classical* information supplied by the encryptor.

Encrypted CNOT Operation. In more detail, evaluating a non-Clifford gate on a ciphertext $(\mathsf{ct}, |\phi\rangle)$, where ct is an FHE encryption of a QOTP key and $|\phi\rangle$ is a quantum state encrypted under the QOTP key, involves an operation (referred to as encrypted CNOT) that somehow must "teleport" the bits encrypted in ct

[4] We also remark here that [34] presented a *multi*-key scheme based on this paradigm, but with the same drawbacks. Note that compactness and classical encryption are crucial in our setting, as per the discussion in the previous section.

into the state $|\phi\rangle$. [52] gave a method for doing this, as long as the ciphertext ct is encrypted under a scheme with some particular properties. Roughly, the scheme must support a "natural" XOR homomorphic operation, it must be circuit private with respect to this homomorphism, and perhaps most stringently, there must exist some trapdoor that can be used to recover the message and the *randomness* used to produce any ciphertext.

[52] observed that the dual-Regev encryption scheme [30] (with large enough modulus-to-noise ratio) does in fact satisfy these properties, as long as one generates the public key matrix $\mathbf{A}$ along with a trapdoor. However, recall that ct was supposed to be encrypted under a fully-homomorphic encryption scheme. [52] resolves this by observing that ciphertexts encrypted under the dual variant of the [31] fully-homomorphic encryption scheme actually already contain a dual-Regev ciphertext. In particular, a dual-GSW ciphertext encrypting a bit μ is a matrix $\mathbf{M} = \mathbf{AS} + \mathbf{E} + \mu\mathbf{G}$, where $\mathbf{G}$ is the gadget matrix. The final column of $\mathbf{M}$ is $\mathbf{As} + \mathbf{e} + \mu[0, \ldots, 0, q/2]^\top$, which is exactly a dual-Regev ciphertext encrypting μ under public key $\mathbf{A}$. Note that, crucially, if the dual-GSW public key $\mathbf{A}$ is drawn with a trapdoor, then this trapdoor also functions as a trapdoor for the dual-Regev ciphertext. Thus, an evaulator can indeed perform the encrypted CNOT operation on any ciphertext $(\mathsf{ct}, |\phi\rangle)$, by first extracting a dual-Regev ciphertext ct' from ct and then proceeding.

Challenges in the Multi-Key Setting. Now, it is natural to ask whether this approach readily extends to the multi-key setting. Namely, does there exist a multi-key FHE scheme where any (multi-key) ciphertext contains within it a dual-Regev ciphertext *with a corresponding trapdoor*? Unfortunately, this appears to be much less straightforward than in the single-key setting, for the following reason. Observe that (dual) GSW homomorphic operations over ciphertexts $\mathbf{M}_i = \mathbf{AS}_i + \mathbf{E}_i + \mu_i\mathbf{G}$ always maintain the same $\mathbf{A}$ matrix, while updating $\mathbf{S}_i$, $\mathbf{E}_i$, and μ_i. Thus, a trapdoor for $\mathbf{A}$ naturally functions as a trapdoor for the dual-Regev ciphertext that consitutes the last column of $\mathbf{M}_i$. However, LWE-based multi-key FHE schemes from the literature [12,18,53,59] include a *ciphertext expansion* procedure, which allows an evaluator, given public keys $\mathsf{pk}_1, \ldots, \mathsf{pk}_n$, and a ciphertext ct encrypted under some pk_i, to convert ct into a ciphertext $\hat{\mathsf{ct}}$ encrypted under all keys $\mathsf{pk}_1, \ldots, \mathsf{pk}_n$. Now, even if these public keys are indeed matrices $\mathbf{A}_1, \ldots, \mathbf{A}_n$ drawn with trapdoors $\tau_1, \ldots, \tau_n$, it is unclear how to combine $\tau_1, \ldots, \tau_n$ to produce a trapdoor $\hat{\tau}$ for the "expanded" ciphertext. Indeed, the expanded ciphertext generally can no longer be written as some $\mathbf{AS} + \mathbf{E} + \mu\mathbf{G}$, since the expansion procedure constructs a highly structured matrix that includes components from the ciphertexts $\mathsf{ct}_1, \ldots, \mathsf{ct}_n$, as well as auxiliary encryptions of the randomness used to produce the ciphertexts (see e.g. [53]).

A Solution Based on Key-Switching. Thus, we take a different approach. Rather than attempting to tweak known ciphertext expansion procedures to also support "trapdoor expansion", we rely on the notion of key-switching, which is a method of taking a ciphertext encrypted under one scheme and converting it into a ciphertext encrypted under another scheme. The observation, roughly, is that we do not need to explicitly maintain a trapdoor for the multi-key FHE scheme, as long as it is possible to convert a multi-key FHE ciphertext into a dual-Regev ciphertext that *does* explicitly have a trapdoor. In fact, we will consider a natural multi-key generalization of dual-Regev, as described below. Key switching is possible as long as the second scheme has sufficient homomorphic properties, namely, it can support homomorphic evaluation of the *decryption circuit* of the first scheme.

Fortunately, the dual-Regev scheme is already linearly homomorphic, and many known classical multi-key FHE schemes [12,18,53,59] support *nearly linear decryption*, which means that decrypting a ciphertext simply consists of applying a linear function (derived from the secret key) and then rounding. Thus, as long as the evaluator has the secret key of the multi-key FHE ciphertext encrypted under a dual-Regev public key with a trapdoor, they can first key-switch the multi-key FHE ciphertext ct into a dual-Regev ciphertext ct', and then proceed with the encrypted CNOT operation.

It remains to show how an evaluator may have access to such a dual-Regev encryption. Since we are still in the multi-key setting, we will need a ciphertext and corresponding trapdoor expansion procedure for dual-Regev. However, we show that such a procedure is much easier to come by when the scheme only needs to support *linear* homomorphism (as is the case for the dual-Regev scheme) rather than *full* homomorphism. Each party can draw its own dual-Regev public key $\mathbf{A}_i$ along with a trapdoor τ_i, and encrypt its multi-key FHE secret key under $\mathbf{A}_i$ to produce a ciphertext ct_i. The evaluator can then treat the block-diagonal matrix $\hat{\mathbf{A}} = \mathsf{diag}(\mathbf{A}_1, \ldots, \mathbf{A}_n)$ as an "expanded" public key.[5] Now, the message and randomness used to generate a ciphertext encrypted under $\hat{\mathbf{A}}$ may be recovered by applying τ_1 to the first set of entries of the ciphertext, applying τ_2 to the second set of entries and so on. This observation, combined with an appropriate expansion procedure for the ciphertexts ct_i, allows an evaluator to convert any multi-key FHE ciphertext into a multi-key dual-Regev ciphertext with trapdoor. Given a classical multi-key FHE scheme with nearly linear decryption, this suffices to build multi-key quantum FHE with classical key generation and encryption.

Distributed Setup. We showed above how to convert any classical multi-key FHE scheme into a quantum multi-key FHE scheme, as long as the classical scheme has nearly linear decryption. However, most LWE-based classical multi-key FHE schemes operate in the common random string (CRS) model, which assumes that all parties have access to a common source of randomness, generated by a trusted party. Thinking back to our application to parallel extractable commitments, it is clear that this will not suffice, since we have no CRS a priori, and a receiver that generates a CRS maliciously may be able to break hiding of the scheme. Thus, we rely on the multi-key FHE scheme of [12], where instead of assuming a CRS, the parties participate in a distributed setup procedure. In particular, each party (and in our application, each committer) generates some

[5] Actually this expansion should be done slightly more carefully, see Sect. 4.4 in the full version for details.

public parameters pp_i, which are then combined publicly to produce a single set of public parameters pp, which can be used by anyone to generate their own public key/secret key pair.

This form of distributed setup indeed suffices to prove the hiding of our parallel commitment, so it remains to show that our approach, combined with [12], yields a quantum multi-key FHE scheme with distributed setup. First, the [12] scheme does indeed enjoy nearly linear decryption, so plugging it into our compiler described above gives a functional quantum multi-key FHE scheme. Next, we need to confirm that our compiler does not destroy the distributed setup property. This follows since each party draws its own dual-Regev public key with trapdoor without relying on any CRS, or even any public parameters.

Quantum AFS-Spooky Encryption. Finally, we show, via another application of key-switching, how to construct a quantum AFS-spooky encryption scheme (with distributed setup). Recall that we only require "spooky" interactions to hold over classical ciphertexts. That is, for any *quantum* circuit C with classical outputs, given ciphertexts $\mathsf{ct}_1, \ldots, \mathsf{ct}_n$ encrypting $|\phi_1\rangle, \ldots, |\phi_n\rangle$ respectively under public keys $\mathsf{pk}_1, \ldots, \mathsf{pk}_n$, an evaluator can produce ciphertexts $\mathsf{ct}'_1, \ldots, \mathsf{ct}'_n$ where ct'_i encrypts y_i under pk_i, and such that $\bigoplus_{i=1}^{n} y_i = C(|\phi_1\rangle, \ldots, |\phi_n\rangle)$.

Now, using our quantum multi-key FHE scheme, it is possible to compute a single (multi-key) ciphertext $\hat{\mathsf{ct}}$ that encrypts $C(|\phi_1\rangle, \ldots, |\phi_n\rangle)$ under all public keys $\mathsf{pk}_1, \ldots, \mathsf{pk}_n$. Then, if each party additionally drew a key pair $(\mathsf{pk}'_i, \mathsf{sk}'_i)$ for a classical AFS-spooky encryption scheme, and released $\tilde{\mathsf{ct}}_1, \ldots, \tilde{\mathsf{ct}}_n$, where $\tilde{\mathsf{ct}}_i = \mathsf{Enc}(\mathsf{pk}'_i, \mathsf{sk}_i)$ encrypts the i-th party's quantum multi-key FHE secret key under their AFS-spooky encryption public key, then the evaluator can homomorphically evaluate the quantum multi-key FHE decryption circuit (which is classical for classical ciphertexts) with $\hat{\mathsf{ct}}$ hardcoded, where $\hat{\mathsf{ct}}$ is the multi-key ciphertext defined at the beginning of this paragraph. This circuit on input $\tilde{\mathsf{ct}}_1, \ldots, \tilde{\mathsf{ct}}_n$ produces the desired output $\mathsf{ct}'_1, \ldots, \mathsf{ct}'_n$. Finally, note that the classical AFS-spooky encryption scheme must also have distributed setup, and we show (see Sect. 4.5 in the full version) that one can derive a distributed-setup AFS-spooky encryption scheme from [12] using standard techniques [21].

2.4 Post-Quantum Non-malleable Commitments

In this section, we describe how to obtain constant-round post-quantum non-malleable commitments under the assumption that there exists a natural number $c > 0$ such that quantum polynomial-time adversaries cannot distinguish LWE samples from uniform with advantage better than $\lambda^{-\mathsf{ilog}(c,\lambda)}$, where $\mathsf{ilog}(c, \lambda) = \log\log \cdots_{c \text{ times}} \log(\lambda)$ and λ denotes the security parameter.

We will focus on perfectly binding and computationally hiding constant-round interactive commitments. Loosely speaking, a commitment scheme is said to be non-malleable if no adversary (also called a man-in-the-middle), when participating as a receiver in an execution of an honest commitment $\mathsf{Com}(m)$, can at the same time generate a commitment $\mathsf{Com}(m')$, such that the message m' is related to the original message m. This is equivalent (assuming the existence of one-way functions with security against quantum adversaries) to a tag-based notion where the commit algorithm obtains as an additional input a tag in $\{0,1\}^\lambda$, and the adversary is restricted to using a tag, or identity, that is different from the tag used to generate its input commitment. We will rely on tag-based definitions throughout this paper. We will also only focus on the *synchronous setting*, where the commitments proceed in rounds, and the man-in-the-middle sends its own message for a specific round before obtaining an honest party's message for the next round.

Before describing our ideas, we briefly discuss existing work on *classically-secure* non-malleable commitments. Unfortunately, existing constructions of constant-round non-malleable commitments against classical adversaries from standard polynomial hardness assumptions [4,16,17,36–40,44,46–48,54,56–58, 65] either rely on rewinding, or use Barak's non-black-box simulation technique, both of which require the reduction to perform state cloning. As such, known techniques fail to prove quantum security of these constructions.

We now discuss our techniques for constructing post-quantum non-malleable commitments. Just like several classical approaches, we will proceed in two steps.

- We will obtain simple "base" commitment schemes for very small tag/identity spaces from slightly superpolynomial hardness assumptions.
- Then assuming polynomial hardness of LWE against quantum adversaries, and making use of constant-round post-quantum zero-knowledge arguments, we will convert non-malleable commitments for a small tag space into commitments for a larger tag space, while only incurring a constant round overhead.

For the base schemes, there are known classical constructions [58] that assume hardness of LWE against 2^{λ^δ}-size adversaries, where λ denotes the security parameter and $0 < \delta < 1$ is a constant. We observe that these constructions can be proven secure in the quantum setting, resulting in schemes that are suitable for tag spaces of $O(\log \log \lambda)$ tags.

Tag Amplification. Since an MPC protocol could be executed among up to $\mathsf{poly}(\lambda)$ parties where $\mathsf{poly}(\cdot)$ is an arbitrary polynomial, we end up requiring non-malleable commitments suitable for tag spaces of $\mathsf{poly}(\lambda)$. This is obtained by combining classical tools for amplifying tag spaces [22] with constant round post-quantum zero-knowledge protocols. Our tag amplification protocol, on input a scheme with tag space $2t$, outputs a scheme with tag space 2^t, for any $t \leq \mathsf{poly}(\lambda)$. This follows mostly along the lines of existing classical protocols, and as such we do not discuss the protocol in detail here. Our protocol can be found in Sect. 7.3 in the full version.

Base Schemes from $\lambda^{-\mathsf{ilog}(c,\lambda)}$ Hardness. Returning to the question of constructing appropriate base schemes, we also improve the assumption from 2^{λ^δ}-quantum hardness of LWE (that follows based on [58]) to the mildly superpolynomial hardness assumption discussed at the beginning of this subsection.

Recall that we will only need to assume that there exists an (explicit) natural number $c > 0$ such that quantum polynomial time adversaries cannot distinguish LWE samples from uniform with advantage better than $\mathsf{negl}(\lambda^{\mathsf{ilog}(c,\lambda)})$ where $\mathsf{ilog}(c, \lambda) = \log\log \cdots_{c \text{ times}} \log(\lambda)$. Our base scheme will only be suitable for identities in $\mathsf{ilog}(c+1, \lambda)$, where $c > 0$ is a natural number, independent of λ. We will then repeatedly apply the tag amplification process referred to above to boost the tag space to 2^λ, by adding only a constant number of rounds.

To build our base scheme, we take inspiration from the classically secure non-malleable commitments of Khurana and Sahai [45]. However, beyond considering quantum as opposed to classical adversaries, our protocol and analysis will have the following notable differences from [45]:

- The work of [45] relies on sub-exponential hardness (i.e. 2^{λ^δ} security), which is stronger than the type of superpolynomial hardness we assume. This is primarily because [45] were restricted to two rounds, but we can improve parameters while allowing for a larger constant number of rounds.
- [45] build a reduction that rewinds an adversary to the beginning of the protocol, and executes the adversary several times, repeatedly sampling the adversary's initial state. This may be undesirable in the quantum setting.[6] On the other hand, we have a simpler fully straight-line reduction that only needs to run the adversary once.

Specifically, following [45], we will establish *an erasure channel* between the committer and receiver that transmits the committed message to the receiver with probability ϵ. To ensure that the commitment satisfies hiding, ϵ is chosen to be a value that is negligible in λ. At the same time, the exact value of ϵ is determined by the identity (tag) of the committer. Recall that $\mathsf{tag} \in [1, \mathsf{ilog}(c+1, \lambda)]$. We will set $\epsilon = \eta^{-\mathsf{tag}}$ where $\eta = \lambda^{\mathsf{ilog}(c+1,\lambda)}$ is a superpolynomial function of λ.

Next, for simplicity, we restrict ourselves to a case where the adversary's tag (which we denote by tag') is smaller than that of the honest party (which we denote by tag). In this case, the adversary's committed message is transmitted with probability $\epsilon' = \eta^{-\mathsf{tag}'}$, whereas the honest committer's message is transmitted with probability only $\epsilon = \eta^{-\mathsf{tag}}$, which is smaller than ϵ'.

We set this up so that the transcript of an execution transmits the adversary's message with probability ϵ' (over the randomness of the honest receiver), and on the other hand, an honestly committed message will remain hidden except with probability $\epsilon < \epsilon'$ (over the randomness of the honest committer). This gap in the probability of extraction will help us argue non-malleability, using a proof strategy that bears resemblance to the proof technique in [9] (who relied on stronger assumptions to achieve such a gap in the non-interactive setting).

We point out one subtlety in our proof that does not appear in [9]. We must rule out a man-in-the-middle adversary that on the one hand, does not commit to a related message if its message was successfully transmitted, but on the other hand,

---

[6] In particular this state may not always be efficiently sampleable, in which case it would be difficult to build an efficient reduction.

can successfully perform a mauling attack if its message was not transmitted. To rule out such an adversary, just like [45], we will design our erasure channel so that the adversary cannot distinguish transcripts where his committed message was transmitted from those where it wasn't.

Finally, our erasure channel can be cryptographically established in a manner similar to prior work [3,42,45] via an indistinguishability-based variant of two-party secure function evaluation, that can be based on quantum hardness of LWE. Specifically, we would like to ensure that the SFE error is (significantly) smaller than the transmission probabilities of our erasure channels: therefore, we will set parameters so that SFE error is $\lambda^{-\mathsf{ilog}(c,\lambda)}$. We refer the reader to Sect. 7 in the full version for additional details about our construction.

On Super-Constant Rounds from Polynomial Hardness. We also observe that for any $t(\lambda) \leq \mathsf{poly}(\lambda)$, non-malleable commitments for tag space of size $t(\lambda)$ can be obtained in $O(t(\lambda))$ rounds based on any extractable commitment using ideas from [15,22], where only one party speaks in every round. These admit a straight-line reduction, and can be observed to be quantum-secure. As such, based on quantum polynomial hardness of LWE and quantum FHE, we can obtain a base protocol for $O(\log\log\ldots_{c \text{ times}} \log\lambda)$ tags requiring $O(\log\log\ldots_{c \text{ times}} \log\lambda)$ rounds, for any constant $c \in \mathbb{N}$. Applying our tag-amplification compiler to this base protocol makes it possible to increase the tag space to 2^λ while only adding a constant number of rounds. Therefore, this technique gives $O(\log\log\ldots_{c \text{ times}} \log\lambda)$ round non-malleable commitments for exponentially large tags from quantum polynomial hardness. It also yields constant round non-malleable commitments for a constant number of tags from polynomial hardness.

2.5 Putting Things Together

Finally, we show how to combine the primitives described above to obtain a constant-round coin-flipping protocol that supports straight-line simulation. As we saw above, in the setting of multi-verifier zero-knowledge, simultaneously simulating the view of multiple parties without rewinding can be quite challenging, so a careful protocol and proof is needed.

Recall the outline presented at the beginning of this section, where each party first commits to a uniformly random string, then broadcasts the committed message, and finally proves in ZK that the message broadcasted is equal to the previously committed message. If all proofs verify, then the common output is the XOR of all broadcasted strings. Recall also that the coin-tossing protocol should be *fully-simulatable.* This means that a simulator should be able to force the common output to be a particular uniformly drawn string given to it as input.

It turns out that in order to somehow force a particular output, the simulator should be able to *simultaneously extract in advance* all the messages that adversarial parties committed to. In particular, we require commitments where a simulator can extract from multiple committers committing in parallel. Here, we will rely on our parallel extractable commitment described above. Note that we will also need to simulate the subsequent zero-knowledge arguments given by the malicious parties in parallel, and thus we instantiate these with our parallel zero-knowledge argument described above. However, an issue remains. What if an adversary could somehow *maul* an honest party's commitment to a related message and then broadcast that commitment as their own? This could bias the final outcome away from uniformly random.

Thus, we need to introduce some form of non-malleability into the protocol. Indeed, we will add another step at the beginning where each party commits to its message c_i and some randomness r_i using our post-quantum many-to-one non-malleable commitment.[7] Each party will then commit to c_i again with our extractable commitment, using randomness r_i. Finally, each party proves in zero-knowledge that the previous commitments were consistent.

This protocol can be proven to be fully simulatable. Intuitively, even though the simulator changes the behavior of honest players in order to extract from the adversary's commitments and then later force the appropriate output, the initial non-malleable commitments given by the adversary must not change in a meaningful way, due the the guarantee of non-malleability. However, additional subtleties arise in the proof of security. In particular, during the hybrids the simulator will first have to simulate the honest party zero-knowledge arguments, before changing the honest party commitments in earlier stages. However, when changing an honest party's commitment, we need to rely on non-malleability to ensure that the malicious party commitments will not also change in a non-trivial way. Here, we use a proof technique that essentially invokes soundness of the adversary's zero-knowledge arguments at an earlier hybrid but allows us to nevertheless rely on non-malleable commitments to enforce that the adversary behaves consistently in all future hybrids. More discussion and a formal analysis can be found in Sect. 8 in the full version.

2.6 Related Work

Classical secure multi-party computation was introduced and shown to be achievable in the two-party setting by [67] and in the multi-party setting by [33]. Since these seminal works, there has been considerable interest in reducing the round complexity of classical protocols. In the setting of malicious security against a dishonest majority, [49] gave the first *constant*-round protocol for two-party computation, and [43] gave the first constant-round protocol for multi-party computation. Since then, there has been a long line of work improving on the exact round complexity and assumptions necessary for classical multi-party computation (see e.g. [27,55]).

Post-quantum Classical Protocols. The above works generally focus on security against *classical* polynomial-time adversaries. Another line of work, most relevant to the present work, has considered the more general goal of proving the

[7] Above we described a construction of one-to-one non-malleable commitment, though a hybrid argument [48] shows that one-to-one implies many-to-one.

security of classical protocols against arbitrary *quantum* polynomial-time adversaries.

This study was initiated by van de Graaf [63], who observed that the useful rewinding technique often used to prove zero-knowledge in the classical setting may be problematic in the quantum setting. In a breakthrough work, Watrous [64] showed that several well-known classical zero-knowledge protocols are in fact zero-knowledge against quantum verifiers, via a careful rewinding argument. However, these protocols require a polynomial number of rounds to achieve negligible security against quantum attackers. Later, Unruh [62] developed a more powerful rewinding technique that suffices to construct classical zero-knowledge *proofs of knowledge* secure against quantum adversaries, though still in a polynomial number of rounds. In a recent work, [10] managed to construct a constant-round post-quantum zero-knowledge protocol, under assumptions similar to those required to obtain classical fully-homomorphic encryption. In another recent work, [1] constructed a constant-round protocol that is zero-knowledge against quantum verifiers under the quantum LWE assumption, though soundness holds against only classical provers.

There has also been some work on the more general question of post-quantum secure computation. In particular, [20] used the techniques developed in [64] to build a two-party coin-flipping protocol, and [41,51] constructed general two-party computation secure against quantum adversaries, in a polynomial number of rounds. More recently, [10] gave a *constant*-round two-party coin-flipping protocol, with full simulation of one party. However, prior to this work, nothing was known in the most general setting of post-quantum multi-party computation (in the plain model).

Finally, we remark that post-quantum classical protocols do exist in the literature, as long as some form of trusted setup is available. For example, the two-round protocol of [53] from LWE is in the programmable common random string model, and was shown to be semi-maliciously secure via straight-line simulation. Thus, applying the semi-malicious to malicious compiler of [2] instantiated with a NIZK from LWE [60] gives a post-quantum maliciously secure protocol in the common random string model from the quantum hardness of LWE. Another example is the maliciously secure OT-based two-round protocol of [7,28] instantiated with maliciously-secure oblivious transfer from LWE [61].

Quantum Protocols. Yet another line of work focuses on protocols for securely computing *quantum* circuits. General multi-party quantum computation was shown to be achievable in the information-theoretic setting (with honest majority) in the works of [5,19]. In the computational setting, [25] gave a two-party protocol secure against a quantum analogue of semi-honest adversaries, and [26] extended security of two-party quantum computation to the malicious setting. In a recent work [23] constructed a maliciously secure multi-party protocol for computing quantum circuits, assuming the existence of a maliciously secure post-quantum classical MPC protocol. We remark that all of the above protocols operate in a polynomial number of rounds.

3 Quantum-Secure Multi-Committer Extractable Commitment

In this section, we follow the outline presented in Sect. 2.2 to construct a commitment scheme that allows for simultaneous extraction from multiple parallel committers. The protocol is somewhat more involved than the high-level description given earlier, so we briefly highlight the differences.

First, the committer is instructed to (non-interactively) commit to its message and trapdoor at the very beginning of the protocol. We use these commitments to take advantage of non-uniformity in the reductions between hybrids in the extractability proof. In particular, hybrids that come before the step where the simulator goes "under the hood" of the FHE may still need access to the trapdoor and commitment, and this can be given to any reduction via non-uniform advice consisting of each committer's first message and corresponding openings.

Next, the CDS described earlier is replaced with a function-hiding secure function evaluation (SFE) protocol. In order to rule out the malleability attack mentioned in Sect. 2.2, where a malicious receiver mauls the AFS-spooky encryption of the committer's trapdoor into an SFE encryption of the trapdoor, we do the following. The first message sent by the receiver to each committer C_i will actually be a commitment to some key k_i of a generic secret-key encryption scheme. After C_i sends its AFS-spooky encryption ciphertext and compute and compare obfuscation, the receiver prepares and sends a secret-key encryption of an arbitrary message. Then, the receiver's input to the SFE consists of the opening to its earlier commitment k_i, and the SFE checks if the secret-key encryption sent by the receiver is actually an encryption of the committer's trapdoor under secret key k_i. If so, it returns the lock and otherwise it returns $\perp$. This setup ensures that a malicious receiver cannot maul the AFS-spooky encryption of the committer's trapdoor, for the following reason. If it could, then a non-uniform reduction to the semantic security of AFS-spooky encryption may obtain the receiver's committed k_i as advice and decrypt the receiver's secret-key encryption to obtain the trapdoor. Of course, this assumes the receiver actually acted explainably in sending a valid commitment at the beginning of the protocol, and this is ensured by the opening check performed under the SFE. We note that this mechanism is somewhat different than what was presented in [10], as they directly build a zero-knowledge argument (i.e. without first constructing a stand-alone extractable commitment) and are able to take advantage of witness indistinguishability to enforce explainable behavior.

Compliant Distinguishers. Finally, we discuss the issue of *committer* explainability. Recall from the high-level overview that a simulator is able to extract from a committer by homomorphically evaluating its code on an AFS-spooky encryption ciphertext *generated by the committer*. Thus, if the committer acts arbitrarily maliciously and does not return a well-formed ciphertext, the extraction may completely fail. Again, [10] address this issue by only analyzing their commitment within the context of a larger zero-knowledge argument protocol, and having the verifier prove to the prover using a witness indistinguishable proof that it performed the commitment explainably.

Thus, without adding zero-knowledge and performing [32]-style analysis to handle non-explainable and aborting committers, we will only obtain extractability against explainable committers. However, since we will be using this protocol inside larger protocols where participants are not assumed to be acting explainably, restricting the class of committers we consider in our definition is problematic. We instead consider arbitrary committers but restrict the class of *distinguishers* (who are supposed to decide whether they received the view of a committer interacting in the real protocol or the view of a committer interacting with the extractor) to those that always output 0 on input a non-explainable transcript. In other words, any advantage these distinguishers may have must be coming from their behavior on input explainable views. Even though checking whether a particular view is explainable or not is not efficient, it turns out that this definition lends itself quite nicely to composition, since one can use witness indistinguishability/zero-knowledge to construct provably compliant distinguishers between hybrids for the larger protocols.

For completeness, and because post-quantum multi-committer extractable commitments may be of independent interest, we also show in Appendix D in the full version how to add zero-knowledge within the extractable commitment protocol itself to obtain security against arbitrary committers.

3.1 Definition

Definition 1 (Quantum-Secure Multi-Committer Extractable Commitment). *A quantum-secure multi-committer extractable commitment scheme is a pair* (C, R) *of classical PPT interactive Turing machines. In the commit phase,* R *interacts with* n *copies* $\{\mathsf{C}_i\}_{i\in[n]}$ *of* C *(who do not interact with each other) on common input* 1^λ *and* 1^n*, with each* C_i *additionally taking a private input* $m_i \in \{0,1\}^*$*. This produces a transcript* τ*, which may be parsed as a set of* n *transcripts* $\{\tau_i\}_{i\in[n]}$*, one for each set of messages exchanged between* R *and* C_i*. In the decommitment phase, each* C_i *outputs* m_i *along with its random coins* r_i*, and* R *on input* $(1^\lambda, \tau_i, m_i, r_i)$ *either accepts or rejects. The scheme should satisfy the following properties.*

- **Perfect Correctness:** *For any* $\lambda, n \in \mathbb{N}, i \in [n]$*,*

$$\Pr[\mathsf{R}(1^\lambda, \tau_i, m_i, r_i) = 1 \mid \{\tau_i\}_{i\in[n]} \leftarrow \langle \mathsf{R}, \mathsf{C}_1(m_1; r_1), \ldots, \mathsf{C}_n(m_n; r_n)\rangle(1^\lambda, 1^n)] = 1.$$

- **Perfect Binding:** *For any* $\lambda \in \mathbb{N}$ *and string* $\tau \in \{0,1\}^*$*, there does not exist* (m, r) *and* (m', r') *with* $m \neq m'$ *such that* $\mathsf{R}(1^\lambda, \tau, m, r) = \mathsf{R}(1^\lambda, \tau, m', r') = 1$.
- **Quantum Computational Hiding:** *For any non-uniform quantum polynomial-size receiver* $\mathsf{R}^* = \{\mathsf{R}^*_\lambda, \rho_\lambda\}_{\lambda\in\mathbb{N}}$*, any polynomial* $\ell(\cdot)$*, and any sequence of sets of strings* $\{m^{(0)}_{\lambda,1}, \ldots, m^{(0)}_{\lambda,n}\}_{\lambda,n\in\mathbb{N}}$*,* $\{m^{(1)}_{\lambda,1}, \ldots, m^{(1)}_{\lambda,n}\}_{\lambda,n\in\mathbb{N}}$ *where each* $|m^{(b)}_{\lambda,i}| = \ell(\lambda)$*,*

$$\{\mathsf{VIEW}_{\mathsf{R}^*_\lambda}(\langle \mathsf{R}^*_\lambda(\rho_\lambda), \mathsf{C}_1(m^{(0)}_{\lambda,1}), \ldots, \mathsf{C}_n(m^{(0)}_{\lambda,n})\rangle(1^\lambda, 1^n))\}_{\lambda,n\in\mathbb{N}}$$
$$\approx_c \{\mathsf{VIEW}_{\mathsf{R}^*_\lambda}(\langle \mathsf{R}^*_\lambda(\rho_\lambda), \mathsf{C}_1(m^{(1)}_{\lambda,1}), \ldots, \mathsf{C}_n(m^{(1)}_{\lambda,n})\rangle(1^\lambda, 1^n))\}_{\lambda,n\in\mathbb{N}}.$$

The extractability property will require the following two definitions. First, for any adversary $\mathsf{C}^* = \{\mathsf{C}^*_\lambda, \rho_\lambda\}_{\lambda\in\mathbb{N}}$ *representing a subset* $I \subseteq [n]$ *of* n *committers, any honest party messages* $\{m_i\}_{i\notin I}$*, and any security parameter* $\lambda \in \mathbb{N}$*, define* $\mathsf{VIEW}^{\mathsf{msg}}_{\mathsf{C}^*_\lambda}(\langle \mathsf{R}, \mathsf{C}^*_\lambda(\rho_\lambda), \{\mathsf{C}_i(m_i)\}_{i\notin I}\rangle(1^\lambda, 1^n))$ *to consist of the following.*

1. *The view of* C^*_λ *on interaction with the honest receiver* R *and set* $\{\mathsf{C}_i(m_i)\}_{i\notin I}$ *of honest parties; this view includes a set of transcripts* $\{\tau_i\}_{i\in I}$ *and a state* st.
2. *A set of strings* $\{m_i\}_{i\in I}$*, where each* m_i *is defined relative to* τ_i *as follows. If there exists* m'_i, r_i *such that* $\mathsf{R}(1^\lambda, \tau_i, m'_i, r_i) = 1$*, then* $m_i = m'_i$*, otherwise,* $m_i = \bot$.

Next, we consider distinguishers $\mathsf{D} = \{\mathsf{D}_\lambda, \sigma_\lambda\}_{\lambda\in\mathbb{N}}$ *that take as input a sample* $(\{\tau_i\}_{i\in I}, \mathsf{st}, \{m_i\}_{i\in I})$ *from the distribution just described. We say that* D *is* compliant *if whenever* $\{\tau_i\}_{i\in I}$ *is not an explainable transcript with respect to the set* I*,* D *outputs 0 with overwhelming probability (over the randomness of* D*).*

- **Multi-Committer Extractability:** *There exists a quantum expected-polynomial-time extractor* Ext *such that for any* compliant *non-uniform polynomial-size quantum distinguisher* $\mathsf{D} = \{\mathsf{D}_\lambda, \sigma_\lambda\}_{\lambda\in\mathbb{N}}$*, there exists a negligible function* $\mu(\cdot)$*, such that for all adversaries* $\mathsf{C}^* = \{\mathsf{C}^*_\lambda, \rho_\lambda\}_{\lambda\in\mathbb{N}}$ *representing a subset of* n *committers, namely,* $\{\mathsf{C}_i\}_{i\in I}$ *for some set* $I \subseteq [n]$*, the following holds for all polynomial-size sequences of inputs* $\{\{m_{i,\lambda}\}_{i\notin I}\}_{\lambda\in\mathbb{N}}$ *and* $\lambda \in \mathbb{N}$.

$$\Big|\Pr[\mathsf{D}_\lambda(\mathsf{VIEW}^{\mathsf{msg}}_{\mathsf{C}^*_\lambda}(\langle \mathsf{R}, \mathsf{C}^*_\lambda(\rho_\lambda), \{\mathsf{C}_i(m_{i,\lambda})\}_{i\notin I}\rangle(1^\lambda, 1^n)), \sigma_\lambda) = 1]$$
$$- \Pr[\mathsf{D}_\lambda(\mathsf{Ext}(1^\lambda, 1^n, I, \mathsf{C}^*_\lambda, \rho_\lambda), \sigma_\lambda) = 1]\Big| \leq \mu(\lambda).$$

Remark 1. Observe that the above definition of quantum computational hiding does not consider potentially malicious committers that interact in the protocol to try to gain information about commitments made by other committers. This is without loss of generality, since all communication occurs between R and some C_i. In particular, no messages are sent between any C_i and C_j.

3.2 Construction

Ingredients: All of the following are assumed to be quantum-secure, and the construction is presented in Protocol Fig. 2.

- A non-interactive perfectly-binding commitment Com.
- A secret-key encryption scheme $(\mathsf{Enc}, \mathsf{Dec})$.[8]

[8] We use the syntax that for key k, a ciphertext of message m is computed as $\mathsf{ct} \leftarrow \mathsf{Enc}(k, m)$ and decrypted as $m := \mathsf{Dec}(k, \mathsf{ct})$.

- A compute-and-compare obfuscator Obf.
- A quantum AFS-spooky encryption scheme with distributed setup ($\mathsf{Spooky.Setup}$, $\mathsf{Spooky.KeyGen}$, $\mathsf{Spooky.Enc}$, $\mathsf{Spooky.QEnc}$, $\mathsf{Spooky.Eval}$, $\mathsf{Spooky.Dec}$, $\mathsf{Spooky.QDec}$).
- A two-message function-hiding secure function evaluation scheme ($\mathsf{SFE.Gen}$, $\mathsf{SFE.Enc}$, $\mathsf{SFE.Eval}$, $\mathsf{SFE.Dec}$).

Protocol 2

Common input: $1^\lambda, 1^n$.
C_i's additional input: A string m_i.

1. Each C_i computes $\mathsf{td}_i \leftarrow U_\lambda$ and sends $\mathsf{c}_i^{(\mathsf{msg})} \leftarrow \mathsf{Com}(1^\lambda, m_i)$, $\mathsf{c}_i^{(\mathsf{td})} \leftarrow \mathsf{Com}(1^\lambda, \mathsf{td}_i)$ to R.
2. For each $i \in [n]$, R computes $k_i, r_i \leftarrow U_\lambda$ and sends $\mathsf{c}_i^{(\mathsf{key})} := \mathsf{Com}(1^\lambda, k_i; r_i)$ to C_i.
3. Each C_i computes and sends $\mathsf{pp}_i \leftarrow \mathsf{Spooky.Setup}(1^\lambda)$ to R.
4. R defines $\mathsf{pp} := \{\mathsf{pp}_i\}_{i\in[n]}$, and sends pp to each C_i. Each C_i checks that the pp_i it received matches the pp_i it sent in Step 3, and if not, it aborts.
5. Each C_i computes
   - $\mathsf{lk}_i \leftarrow U_\lambda$,
   - $(\mathsf{pk}_i, \mathsf{sk}_i) \leftarrow \mathsf{Spooky.KeyGen}(1^\lambda, \mathsf{pp})$,
   - $\mathsf{ct}_i \leftarrow \mathsf{Spooky.Enc}(\mathsf{pk}_i, \mathsf{td}_i)$,
   - and $\widetilde{\mathbf{CC}}_i \leftarrow \mathsf{Obf}\left(\mathbf{CC}[\mathsf{Spooky.Dec}(\mathsf{sk}_i, \cdot), \mathsf{lk}_i, (\mathsf{sk}_i, m_i)]\right)$,

   and sends $(\mathsf{pk}_i, \mathsf{ct}_i, \widetilde{\mathbf{CC}}_i)$ to R.
6. For each $i \in [n]$, R computes $\mathsf{ct}_i^{(\mathsf{td})} \leftarrow \mathsf{Enc}(k_i, 0^\lambda)$, $\mathsf{dk}_i \leftarrow \mathsf{SFE.Gen}(1^\lambda)$, and $\mathsf{ct}_i^{(\mathsf{SFE})} \leftarrow \mathsf{SFE.Enc}(\mathsf{dk}_i, (k_i, r_i))$ and sends $(\mathsf{ct}_i^{(\mathsf{td})}, \mathsf{ct}_i^{(\mathsf{SFE})})$ to C_i.
7. Define the circuit $\mathsf{C}[\mathsf{c}_i^{(\mathsf{key})}, \mathsf{ct}_i^{(\mathsf{td})}, \mathsf{td}_i, \mathsf{lk}_i](\cdot)$ to take as input (k_i, r_i), check if $\mathsf{c}_i^{(\mathsf{key})}$ opens to k_i with opening r_i and if $\mathsf{td}_i = \mathsf{Dec}(k_i, \mathsf{c}_i^{(\mathsf{td})})$, and if so output lk_i, and otherwise output $\perp$. Each C_i computes and sends $\widehat{\mathsf{ct}}_i^{(\mathsf{SFE})} \leftarrow \mathsf{SFE.Eval}(\mathsf{C}[\mathsf{c}_i^{(\mathsf{key})}, \mathsf{ct}_i^{(\mathsf{td})}, \mathsf{td}_i, \mathsf{lk}_i], \mathsf{ct}_i^{(\mathsf{SFE})})$.

Fig. 2. Constant round post-quantum multi-committer extractable commitment.

Analysis. We state the security of our scheme in the following and we refer the reader to the full version of this work for the proofs.

Lemma 1. *Protocol Fig. 2 is quantum computational hiding.*

Lemma 2. *Protocol Fig. 2 is multi-committer extractable.*

References

1. Ananth, P., Placa, R.L.L.: Secure quantum extraction protocols. Cryptology ePrint Archive, Report 2019/1323 (2019). https://eprint.iacr.org/2019/1323

2. Asharov, G., et al.: Multiparty computation with low communication, computation and interaction via threshold FHE. In: Pointcheval, D., Johansson, T. (eds.) EUROCRYPT 2012. LNCS, vol. 7237, pp. 483–501. Springer, Heidelberg (2012). https://doi.org/10.1007/978-3-642-29011-4_29
3. Badrinarayanan, S., Fernando, R., Jain, A., Khurana, D., Sahai, A.: Statistical ZAP arguments. In: Canteaut, A., Ishai, Y. (eds.) EUROCRYPT 2020, Part III. LNCS, vol. 12107, pp. 642–667. Springer, Cham (2020). https://doi.org/10.1007/978-3-030-45727-3_22
4. Barak, B.: Constant-round coin-tossing with a man in the middle or realizing the shared random string model. FOCS **2002**, 345–355 (2002)
5. Ben-Or, M., Crépeau, C., Gottesman, D., Hassidim, A., Smith, A.: Secure multiparty quantum computation with (only) a strict honest majority. In: 47th FOCS, pp. 249–260. IEEE Computer Society Press, Berkeley, CA, USA, 21–24 October 2006. https://doi.org/10.1109/FOCS.2006.68
6. Ben-Or, M., Goldwasser, S., Wigderson, A.: Completeness theorems for non-cryptographic fault-tolerant distributed computation (extended abstract). In: 20th ACM STOC, pp. 1–10. ACM Press, Chicago, IL, USA, 2–4 May 1988. https://doi.org/10.1145/62212.62213
7. Benhamouda, F., Lin, H.: k-round multiparty computation from k-round oblivious transfer via garbled interactive circuits. In: Nielsen, J.B., Rijmen, V. (eds.) EUROCRYPT 2018, Part II. LNCS, vol. 10821, pp. 500–532. Springer, Cham (2018). https://doi.org/10.1007/978-3-319-78375-8_17
8. Bitansky, N., Khurana, D., Paneth, O.: Weak zero-knowledge beyond the black-box barrier. In: Proceedings of the 51st Annual ACM SIGACT Symposium on Theory of Computing, STOC 2019, Phoenix, AZ, USA, 23–26 June 2019, pp. 1091–1102 (2019). https://doi.org/10.1145/3313276.3316382
9. Bitansky, N., Lin, H.: One-message zero knowledge and non-malleable commitments. In: Beimel, A., Dziembowski, S. (eds.) TCC 2018, Part I. LNCS, vol. 11239, pp. 209–234. Springer, Cham (2018). https://doi.org/10.1007/978-3-030-03807-6_8
10. Bitansky, N., Shmueli, O.: Post-quantum zero knowledge in constant rounds. In: STOC (2020)
11. Brakerski, Z.: Quantum FHE (Almost) as secure as classical. In: Shacham, H., Boldyreva, A. (eds.) CRYPTO 2018, Part III. LNCS, vol. 10993, pp. 67–95. Springer, Cham (2018). https://doi.org/10.1007/978-3-319-96878-0_3
12. Brakerski, Z., Halevi, S., Polychroniadou, A.: Four round secure computation without setup. In: Kalai, Y., Reyzin, L. (eds.) TCC 2017, Part I. LNCS, vol. 10677, pp. 645–677. Springer, Cham (2017). https://doi.org/10.1007/978-3-319-70500-2_22
13. Broadbent, A., Jeffery, S.: Quantum homomorphic encryption for circuits of low T-gate complexity. In: Gennaro, R., Robshaw, M. (eds.) CRYPTO 2015, Part II. LNCS, vol. 9216, pp. 609–629. Springer, Heidelberg (2015). https://doi.org/10.1007/978-3-662-48000-7_30
14. Chaum, D., Crépeau, C., Damgård, I.: Multiparty unconditionally secure protocols (abstract). In: Pomerance, C. (ed.) CRYPTO 1987. LNCS, vol. 293, pp. 462–462. Springer, Heidelberg (1988). https://doi.org/10.1007/3-540-48184-2_43
15. Chor, B., Rabin, M.: Achieving independence in logarithmic number of rounds, pp. 260–268 (1987). https://doi.org/10.1145/41840.41862
16. Ciampi, M., Ostrovsky, R., Siniscalchi, L., Visconti, I.: Concurrent non-malleable commitments (and more) in 3 rounds. In: Robshaw, M., Katz, J. (eds.) CRYPTO 2016, Part III. LNCS, vol. 9816, pp. 270–299. Springer, Heidelberg (2016). https://doi.org/10.1007/978-3-662-53015-3_10

17. Ciampi, M., Ostrovsky, R., Siniscalchi, L., Visconti, I.: Four-round concurrent non-malleable commitments from one-way functions. In: Katz, J., Shacham, H. (eds.) CRYPTO 2017, Part II. LNCS, vol. 10402, pp. 127–157. Springer, Cham (2017). https://doi.org/10.1007/978-3-319-63715-0_5
18. Clear, M., McGoldrick, C.: Multi-identity and multi-key leveled FHE from learning with errors. In: Gennaro, R., Robshaw, M. (eds.) CRYPTO 2015, Part II. LNCS, vol. 9216, pp. 630–656. Springer, Heidelberg (2015). https://doi.org/10.1007/978-3-662-48000-7_31
19. Crépeau, C., Gottesman, D., Smith, A.: Secure multi-party quantum computation. In: 34th ACM STOC, pp. 643–652. ACM Press, Montréal, Québec, Canada, 19–21 May 2002. https://doi.org/10.1145/509907.510000
20. Damgård, I., Lunemann, C.: Quantum-secure coin-flipping and applications. In: Matsui, M. (ed.) ASIACRYPT 2009. LNCS, vol. 5912, pp. 52–69. Springer, Heidelberg (2009). https://doi.org/10.1007/978-3-642-10366-7_4
21. Dodis, Y., Halevi, S., Rothblum, R.D., Wichs, D.: Spooky encryption and its applications. In: Robshaw, M., Katz, J. (eds.) CRYPTO 2016, Part III. LNCS, vol. 9816, pp. 93–122. Springer, Heidelberg (2016). https://doi.org/10.1007/978-3-662-53015-3_4
22. Dolev, D., Dwork, C., Naor, M.: Non-malleable cryptography (extended abstract). In: STOC 1991 (1991)
23. Dulek, Y., Grilo, A.B., Jeffery, S., Majenz, C., Schaffner, C.: Secure multi-party quantum computation with a dishonest majority. In: Canteaut, A., Ishai, Y. (eds.) EUROCRYPT 2020, Part III. LNCS, vol. 12107, pp. 729–758. Springer, Cham (2020). https://doi.org/10.1007/978-3-030-45727-3_25
24. Dulek, Y., Schaffner, C., Speelman, F.: Quantum homomorphic encryption for polynomial-sized circuits. In: Robshaw, M., Katz, J. (eds.) CRYPTO 2016, Part III. LNCS, vol. 9816, pp. 3–32. Springer, Heidelberg (2016). https://doi.org/10.1007/978-3-662-53015-3_1
25. Dupuis, F., Nielsen, J.B., Salvail, L.: Secure two-party quantum evaluation of unitaries against specious adversaries. In: Rabin, T. (ed.) CRYPTO 2010. LNCS, vol. 6223, pp. 685–706. Springer, Heidelberg (2010). https://doi.org/10.1007/978-3-642-14623-7_37
26. Dupuis, F., Nielsen, J.B., Salvail, L.: Actively secure two-party evaluation of any quantum operation. In: Safavi-Naini, R., Canetti, R. (eds.) CRYPTO 2012. LNCS, vol. 7417, pp. 794–811. Springer, Heidelberg (2012). https://doi.org/10.1007/978-3-642-32009-5_46
27. Garg, S., Mukherjee, P., Pandey, O., Polychroniadou, A.: The exact round complexity of secure computation. In: Fischlin, M., Coron, J.-S. (eds.) EUROCRYPT 2016, Part II. LNCS, vol. 9666, pp. 448–476. Springer, Heidelberg (2016). https://doi.org/10.1007/978-3-662-49896-5_16
28. Garg, S., Srinivasan, A.: Two-round multiparty secure computation from minimal assumptions. In: Nielsen, J.B., Rijmen, V. (eds.) EUROCRYPT 2018, Part II. LNCS, vol. 10821, pp. 468–499. Springer, Cham (2018). https://doi.org/10.1007/978-3-319-78375-8_16
29. Gentry, C.: Fully homomorphic encryption using ideal lattices. In: Mitzenmacher, M. (ed.) 41st ACM STOC, pp. 169–178. ACM Press, Bethesda, MD, USA, 31 May - 2 Jun 2009. https://doi.org/10.1145/1536414.1536440
30. Gentry, C., Peikert, C., Vaikuntanathan, V.: Trapdoors for hard lattices and new cryptographic constructions. In: Ladner, R.E., Dwork, C. (eds.) 40th ACM STOC. pp. 197–206. ACM Press, Victoria, BC, Canada, 17–20 May 2008. https://doi.org/10.1145/1374376.1374407

31. Gentry, C., Sahai, A., Waters, B.: Homomorphic encryption from learning with errors: conceptually-simpler, asymptotically-faster, attribute-based. In: Canetti, R., Garay, J.A. (eds.) CRYPTO 2013, Part I. LNCS, vol. 8042, pp. 75–92. Springer, Heidelberg (2013). https://doi.org/10.1007/978-3-642-40041-4_5
32. Goldreich, O., Kahan, A.: How to construct constant-round zero-knowledge proof systems for NP. J. Cryptology **9**(3), 167–190 (1996)
33. Goldreich, O., Micali, S., Wigderson, A.: How to play any mental game or Aa completeness theorem for protocols with honest majority. In: Aho, A. (ed.) 19th ACM STOC, pp. 218–229. ACM Press, New York City, NY, USA, 25–27 May 1987. https://doi.org/10.1145/28395.28420
34. Goyal, R.: Quantum multi-key homomorphic encryption for polynomial-sized circuits. Cryptology ePrint Archive, Report 2018/443 (2018). https://eprint.iacr.org/2018/443
35. Goyal, R., Koppula, V., Waters, B.: Lockable obfuscation. In: Umans, C. (ed.) 58th FOCS, pp. 612–621. IEEE Computer Society Press, Berkeley, CA, USA, 15–17 October 2017. https://doi.org/10.1109/FOCS.2017.62
36. Goyal, V.: Constant round non-malleable protocols using one way functions. In: Fortnow, L., Vadhan, S.P. (eds.) 43rd ACM STOC, pp. 695–704. ACM Press, San Jose, CA, USA, 6–8 Jun 2011. https://doi.org/10.1145/1993636.1993729
37. Goyal, V., Lee, C.K., Ostrovsky, R., Visconti, I.: Constructing non-malleable commitments: a black-box approach. In: FOCS (2012)
38. Goyal, V., Pandey, O., Richelson, S.: Textbook non-malleable commitments. In: STOC, pp. 1128–1141. ACM, New York, NY, USA (2016). https://doi.org/10.1145/2897518.2897657
39. Goyal, V., Richelson, S.: Non-malleable commitments using Goldreich-Levin list decoding. In: Zuckerman, D. (ed.) 60th IEEE Annual Symposium on Foundations of Computer Science, FOCS 2019, Baltimore, Maryland, USA, 9–12 November 2019, pp. 686–699. IEEE Computer Society (2019). https://doi.org/10.1109/FOCS.2019.00047
40. Goyal, V., Richelson, S., Rosen, A., Vald, M.: An algebraic approach to non-malleability. FOCS **2014**, 41–50 (2014). https://doi.org/10.1109/FOCS.2014.13
41. Hallgren, S., Smith, A., Song, F.: Classical cryptographic protocols in a quantum world. In: Rogaway, P. (ed.) CRYPTO 2011. LNCS, vol. 6841, pp. 411–428. Springer, Heidelberg (2011). https://doi.org/10.1007/978-3-642-22792-9_23
42. Kalai, Y.T., Khurana, D., Sahai, A.: Statistical witness indistinguishability (and more) in two messages. In: Nielsen, J.B., Rijmen, V. (eds.) EUROCRYPT 2018, Part III. LNCS, vol. 10822, pp. 34–65. Springer, Cham (2018). https://doi.org/10.1007/978-3-319-78372-7_2
43. Katz, J., Ostrovsky, R., Smith, A.: Round efficiency of multi-party computation with a dishonest majority. In: Biham, E. (ed.) EUROCRYPT 2003. LNCS, vol. 2656, pp. 578–595. Springer, Heidelberg (2003). https://doi.org/10.1007/3-540-39200-9_36
44. Khurana, D.: Round optimal concurrent non-malleability from polynomial hardness. In: Kalai, Y., Reyzin, L. (eds.) TCC 2017, Part II. LNCS, vol. 10678, pp. 139–171. Springer, Cham (2017). https://doi.org/10.1007/978-3-319-70503-3_5
45. Khurana, D., Sahai, A.: How to achieve non-malleability in one or two rounds. In: Umans, C. (ed.) 58th IEEE Annual Symposium on Foundations of Computer Science, FOCS 2017, Berkeley, CA, USA, 15–17 October 2017, pp. 564–575. IEEE Computer Society (2017). https://doi.org/10.1109/FOCS.2017.58
46. Lin, H., Pass, R.: Non-malleability amplification. In: Proceedings of the 41st Annual ACM Symposium on Theory of Computing, pp. 189–198. STOC 2009 (2009)

47. Lin, H., Pass, R.: Constant-round non-malleable commitments from any one-way function. In: Fortnow, L., Vadhan, S.P. (eds.) 43rd ACM STOC, pp. 705–714. ACM Press, San Jose, CA, USA, 6–8 Jun 2011. https://doi.org/10.1145/1993636.1993730
48. Lin, H., Pass, R., Venkitasubramaniam, M.: Concurrent non-malleable commitments from any one-way function. In: Canetti, R. (ed.) TCC 2008. LNCS, vol. 4948, pp. 571–588. Springer, Heidelberg (2008). https://doi.org/10.1007/978-3-540-78524-8_31
49. Lindell, Y.: Parallel coin-tossing and constant-round secure two-party computation. J. Cryptology **16**(3), 143–184 (2003). https://doi.org/10.1007/s00145-002-0143-7
50. López-Alt, A., Tromer, E., Vaikuntanathan, V.: On-the-fly multiparty computation on the cloud via multikey fully homomorphic encryption. In: Karloff, H.J., Pitassi, T. (eds.) 44th ACM STOC, pp. 1219–1234. ACM Press, New York, NY, USA, 19–22 May 2012. https://doi.org/10.1145/2213977.2214086
51. Lunemann, C., Nielsen, J.B.: Fully simulatable quantum-secure coin-flipping and applications. In: Nitaj, A., Pointcheval, D. (eds.) AFRICACRYPT 2011. LNCS, vol. 6737, pp. 21–40. Springer, Heidelberg (2011). https://doi.org/10.1007/978-3-642-21969-6_2
52. Mahadev, U.: Classical homomorphic encryption for quantum circuits. In: Thorup, M. (ed.) 59th FOCS, pp. 332–338. IEEE Computer Society Press, Paris, France, 7–9 October 2018). https://doi.org/10.1109/FOCS.2018.00039
53. Mukherjee, P., Wichs, D.: Two round multiparty computation via multi-key FHE. In: Fischlin, M., Coron, J.-S. (eds.) EUROCRYPT 2016, Part II. LNCS, vol. 9666, pp. 735–763. Springer, Heidelberg (2016). https://doi.org/10.1007/978-3-662-49896-5_26
54. Pandey, O., Pass, R., Vaikuntanathan, V.: Adaptive one-way functions and applications. In: Wagner, D. (ed.) CRYPTO 2008. LNCS, vol. 5157, pp. 57–74. Springer, Heidelberg (2008). https://doi.org/10.1007/978-3-540-85174-5_4
55. Pass, R.: Bounded-concurrent secure multi-party computation with a dishonest majority. In: Babai, L. (ed.) 36th ACM STOC, pp. 232–241. ACM Press, Chicago, IL, USA, 13–16 Jun 2004. https://doi.org/10.1145/1007352.1007393
56. Pass, R., Rosen, A.: Concurrent Non-Malleable Commitments. In: Proceedings of the 46th Annual IEEE Symposium on Foundations of ComputerScience, pp. 563–572. FOCS 2005 (2005)
57. Pass, R., Rosen, A.: New and improved constructions of nonmalleable cryptographic protocols. SIAM J. Comput. **38**(2), 702–752 (2008)
58. Pass, R., Wee, H.: Constant-round non-malleable commitments from subexponential one-way functions. In: Gilbert, H. (ed.) EUROCRYPT 2010. LNCS, vol. 6110, pp. 638–655. Springer, Heidelberg (2010). https://doi.org/10.1007/978-3-642-13190-5_32
59. Peikert, C., Shiehian, S.: Multi-key FHE from LWE, revisited. In: Hirt, M., Smith, A. (eds.) TCC 2016, Part II. LNCS, vol. 9986, pp. 217–238. Springer, Heidelberg (2016). https://doi.org/10.1007/978-3-662-53644-5_9
60. Peikert, C., Shiehian, S.: Noninteractive zero knowledge for NP from (plain) learning with errors. In: Boldyreva, A., Micciancio, D. (eds.) CRYPTO 2019, Part I. LNCS, vol. 11692, pp. 89–114. Springer, Cham (2019). https://doi.org/10.1007/978-3-030-26948-7_4
61. Peikert, C., Vaikuntanathan, V., Waters, B.: A framework for efficient and composable oblivious transfer. In: Wagner, D. (ed.) CRYPTO 2008. LNCS, vol. 5157, pp. 554–571. Springer, Heidelberg (2008). https://doi.org/10.1007/978-3-540-85174-5_31

62. Unruh, D.: Quantum proofs of knowledge. In: Pointcheval, D., Johansson, T. (eds.) EUROCRYPT 2012. LNCS, vol. 7237, pp. 135–152. Springer, Heidelberg (2012). https://doi.org/10.1007/978-3-642-29011-4_10
63. Van De Graaf, J.: Towards a Formal Definition of Security for Quantum Protocols. Ph.D. thesis, CAN (1998), aAINQ35648
64. Watrous, J.: Zero-knowledge against quantum attacks. SIAM J. Comput. **39**(1), 25–58 (2009). https://doi.org/10.1137/060670997
65. Wee, H.: Black-box, round-efficient secure computation via non-malleability amplification. In: 51st FOCS, pp. 531–540. IEEE Computer Society Press, Las Vegas, NV, USA, 23–26 October 2010. https://doi.org/10.1109/FOCS.2010.87
66. Wichs, D., Zirdelis, G.: Obfuscating compute-and-compare programs under LWE. In: Umans, C. (ed.) 58th FOCS, pp. 600–611. IEEE Computer Society Press, Berkeley, CA, USA, 15–17 October 2017. https://doi.org/10.1109/FOCS.2017.61
67. Yao, A.C.C.: Protocols for secure computations (extended abstract). In: 23rd FOCS, pp. 160–164. IEEE Computer Society Press, Chicago, Illinois, 3–5 November 1982. https://doi.org/10.1109/SFCS.1982.38
68. Yao, A.C.C.: How to generate and exchange secrets. In: FOCS (1986)

Lattices

A $2^{n/2}$-Time Algorithm for $\sqrt{n}$-SVP and $\sqrt{n}$-Hermite SVP, and an Improved Time-Approximation Tradeoff for (H)SVP

Divesh Aggarwal[1(✉)], Zeyong Li[1], and Noah Stephens-Davidowitz[2]

[1] National University of Singapore, Singapore, Singapore
dcsdiva@nus.edu.sg, li.zeyong@u.nus.edu
[2] Cornell University, Ithaca, USA

Abstract. We show a $2^{n/2+o(n)}$-time algorithm that, given as input a basis of a lattice $\mathcal{L} \subset \mathbb{R}^n$, finds a (non-zero) vector in whose length is at most $\widetilde{O}(\sqrt{n}) \cdot \min\{\lambda_1(\mathcal{L}), \det(\mathcal{L})^{1/n}\}$, where $\lambda_1(\mathcal{L})$ is the length of a shortest non-zero lattice vector and $\det(\mathcal{L})$ is the lattice determinant. Minkowski showed that $\lambda_1(\mathcal{L}) \le \sqrt{n} \det(\mathcal{L})^{1/n}$ and that there exist lattices with $\lambda_1(\mathcal{L}) \ge \Omega(\sqrt{n}) \cdot \det(\mathcal{L})^{1/n}$, so that our algorithm finds vectors that are as short as possible relative to the determinant (up to a polylogarithmic factor).

The main technical contribution behind this result is new analysis of (a simpler variant of) a $2^{n/2+o(n)}$-time algorithm from [ADRS15], which was only previously known to solve less useful problems. To achieve this, we rely crucially on the "reverse Minkowski theorem" (conjectured by Dadush [DR16] and proven by [RS17]), which can be thought of as a partial converse to the fact that $\lambda_1(\mathcal{L}) \le \sqrt{n} \det(\mathcal{L})^{1/n}$.

Previously, the fastest known algorithm for finding such a vector was the $2^{.802n+o(n)}$-time algorithm due to [LWXZ11], which actually found a non-zero lattice vector with length $O(1) \cdot \lambda_1(\mathcal{L})$. Though we do not show how to find lattice vectors with this length in time $2^{n/2+o(n)}$, we do show that our algorithm suffices for the most important application of such algorithms: basis reduction. In particular, we show a modified version of Gama and Nguyen's slide-reduction algorithm [GN08], which can be combined with the algorithm above to improve the time-length tradeoff for shortest-vector algorithms in nearly all regimes—including the regimes relevant to cryptography.

1 Introduction

A lattice $\mathcal{L} \subset \mathbb{R}^n$ is the set of integer linear combinations

$$\mathcal{L} := \mathcal{L}(\mathbf{B}) = \{z_1\mathbf{b}_1 + \cdots + z_n\mathbf{b}_n \; : \; z_i \in \mathbb{Z}\}$$

of linearly independent basis vectors $\mathbf{B} = (\mathbf{b}_1, \ldots, \mathbf{b}_n) \in \mathbb{R}^{n \times n}$. We define the length of a shortest non-zero vector in the lattice as $\lambda_1(\mathcal{L}) := \min_{\mathbf{x} \in \mathcal{L}_{\neq \mathbf{0}}} \|\mathbf{x}\|$. (Throughout this paper, $\|\cdot\|$ is the Euclidean norm.)

A. Canteaut and F.-X. Standaert (Eds.): EUROCRYPT 2021, LNCS 12696, pp. 467–497, 2021.
https://doi.org/10.1007/978-3-030-77870-5_17

The Shortest Vector Problem (SVP) is the computational search problem whose input is a (basis for a) lattice $\mathcal{L} \subseteq \mathbb{R}^n$, and the goal is to output a shortest non-zero vector $\mathbf{y} \in \mathcal{L}$ with $\|\mathbf{y}\| = \lambda_1(\mathcal{L})$. For $\delta \geq 1$, the δ-approximate variant of SVP (δ-SVP) is the problem of finding a non-zero vector $\mathbf{y} \in \mathcal{L}$ of length at most $\delta \cdot \lambda_1(\mathcal{L})$ given a basis of $\mathcal{L}$.

δ-SVP and its many relatives have found innumerable applications over the past forty years. More recently, many cryptographic constructions have been discovered whose security is based on the (worst-case) hardness of δ-SVP or closely related lattice problems. See [Pei16] for a survey. Such lattice-based cryptographic constructions are likely to be used in practice on massive scales (e.g., as part of the TLS protocol) in the not-too-distant future [NIS18], and it is therefore crucial that we understand this problem as well as we can.

For most applications, it suffices to solve δ-SVP for superconstant approximation factors. E.g., cryptanalysis typically requires $\delta = \mathrm{poly}(n)$. However, our best algorithms for δ-SVP work via (non-trivial) reductions to δ'-SVP for much smaller δ' *over lattices with smaller rank*, typically $\delta' = 1$ or $\delta' = O(1)$. E.g., one can reduce n^c-SVP with rank n to $O(1)$-SVP with rank $n/(c+1)$ for constant $c \geq 1$ [GN08, ALNS20]. Such reductions are called *basis reduction algorithms* [LLL82, Sch87, SE94].

Therefore, even if one is only interested in δ-approximate SVP for large approximation factors, algorithms for $O(1)$-SVP are still relevant. (We make little distinction between exact SVP and $O(1)$-SVP in the introduction. Indeed, many of the algorithm that we call $O(1)$-SVP algorithms actually solve exact SVP.)

1.1 Sieving for Constant-Factor-Approximate SVP

There is a very long line of work (e.g., [Kan83, AKS01, NV08, PS09, MV13, LWXZ11, WLW15, ADRS15, AS18, AUV19]) on this problem.

The fastest known algorithms for $O(1)$-SVP run in time $2^{O(n)}$. With one exception [MV13], all known algorithms with this running time are variants of *sieving algorithms*. These algorithms work by sampling $2^{O(n)}$ not-too-long lattice vectors $\mathbf{y}_1, \ldots, \mathbf{y}_M \in \mathcal{L}$ from some nice distribution over the input lattice $\mathcal{L}$, and performing some kind of sieving procedure to obtain $2^{O(n)}$ shorter vectors $\mathbf{x}_1, \ldots, \mathbf{x}_m \in \mathcal{L}$. They then perform the sieving procedure again on the $\mathbf{x}_k$, and repeat this process many times.

The most natural sieving procedure was originally studied by Ajtai, Kumar, and Sivakumar [AKS01]. This procedure simply takes $\mathbf{x}_k := \mathbf{y}_i - \mathbf{y}_j \in \mathcal{L}$, where i, j are chosen so that $\|\mathbf{y}_i - \mathbf{y}_j\| \leq (1-\varepsilon)\min_\ell \|\mathbf{y}_\ell\|$. In particular, the resulting sieving algorithm clearly finds progressively shorter lattice vectors at each step. So, it is trivial to show that this algorithm will eventually find a short lattice vector. Unfortunately (and maddeningly), it seems very difficult to say nearly anything else about the distribution of the vectors when this very simple sieving technique is used, and in particular, while we know that the vectors must be short, we do not know how to show that they are *non-zero*. [AKS01] used clever tricks to modify the above procedure into one for which they could prove correctness, and the current state-of-the-art is a $2^{0.802n}$-time algorithm for γ-SVP for a sufficiently large constant $\gamma > 1$ [LWXZ11, WLW15, AUV19].

Another line of research [NV08, Laa15, MW16, BDGL16, Duc18] focuses on improving the time complexity of practical SVP algorithms by introducing various experimentally verified heuristics. These heuristic algorithms are thus more directly relevant for cryptanalysis. The fastest known heuristic algorithm for solving SVP has time complexity $(3/2)^{(n/2)+o(n)}$, illustrating a large gap between provably correct and heuristic algorithms. (In this regard, this work contributes to the ultimate goal of closing this gap.)

In this work, we are more interested in the "sieving by averages" technique, introduced in [ADRS15] to obtain a $2^{n+o(n)}$-time algorithm for exact SVP. This sieving procedure takes $\mathbf{x}_k := (\mathbf{y}_i + \mathbf{y}_j)/2$ to be the average of two lattice vectors. Of course, $\mathcal{L}$ is not closed under taking averages, so one must choose i, j so that $(\mathbf{y}_i + \mathbf{y}_j)/2 \in \mathcal{L}$. This happens if and only if $\mathbf{y}_i, \mathbf{y}_j$ lie in the same *coset* of $2\mathcal{L}$, $\mathbf{y}_i = \mathbf{y}_j \bmod 2\mathcal{L}$. Equivalently, the coordinates of $\mathbf{y}_i$ and $\mathbf{y}_j$ in the input basis should have the same parities. So, these algorithms pair vectors according to their cosets (and ignore all other information about the vectors) and take their averages $\mathbf{x}_k = (\mathbf{y}_i + \mathbf{y}_j)/2$.

The analysis of these algorithms centers around the *discrete Gaussian distribution* $D_{\mathcal{L},s}$ over a lattice, given by

$$\Pr_{\mathbf{X} \sim D_{\mathcal{L},s}}[\mathbf{X} = \mathbf{y}] \propto e^{-\pi \|\mathbf{y}\|^2/s^2}$$

for a parameter $s > 0$ and any $\mathbf{y} \in \mathcal{L}$. When the starting vectors come from this distribution, we are able to say quite a bit about the distribution of the vectors at each step. (Intuitively, this is because this algorithm only uses algebraic properties of the vectors—their cosets—and entirely ignores the geometry.) In particular, [ADRS15] used a careful rejection sampling procedure to guarantee that the vectors at each step are distributed exactly as $D_{\mathcal{L},s}$ for some parameter $s > 0$. Specifically, in each step the parameter lowers by a factor of $\sqrt{2}$, which is exactly what one would expect, taking intuition from the continuous Gaussian. More closely related to this work is [AS18], which showed that this rejection sampling procedure is actually unnecessary.

In addition to the above, [ADRS15, Ste17] also present a $2^{n/2+o(n)}$-time algorithm that samples from $D_{\mathcal{L},s}$ as long as the parameter $s > 0$ is not too small. In particular, we need s to be "large enough that $D_{\mathcal{L},s}$ looks like a continuous Gaussian." This algorithm is similar to the $2^{n+o(n)}$-time algorithms in that it starts with independent discrete Gaussian vectors with some high parameter, and it gradually lowers the parameter using a rejection sampling procedure together with a procedure that takes the averages of pairs of vectors that lie in the same coset modulo some sublattice (with index $2^{n/2+o(n)}$). But, it fails for smaller parameters because the rejection sampling procedure that it uses must throw out too many vectors in this case. (In [Ste17], a different rejection sampling procedure is used that never throws away too many vectors, but it is not clear how to implement it in $2^{n/2+o(n)}$ time for small parameters $s < \sqrt{2}\eta_{1/2}(\mathcal{L})$.) It was left as an open question whether there is a suitable variant of this algorithm that works for small parameters, which would lead to an algorithm to solve SVP in $2^{n/2+o(n)}$ time. For example, perhaps we could show that the simple algorithm that solves SVP without doing any rejection sampling at all (similar to what was shown for the $2^{n+o(n)}$-time algorithm in [AS18]).

1.2 Hermite SVP

We will also be interested in a variant of SVP called Hermite SVP (HSVP). HSVP is defined in terms of the determinant $\det(\mathcal{L}) := |\det(\mathbf{B})|$ of a lattice $\mathcal{L}$ with basis $\mathbf{B}$. (Though a lattice can have many bases, one can check that $|\det(\mathbf{B})|$ is the same for all such bases, so that this quantity is well-defined.) Minkowski's celebrated theorem says that $\lambda_1(\mathcal{L}) \le O(\sqrt{n}) \cdot \det(\mathcal{L})^{1/n}$, and Hermite's constant $\gamma_n = \Theta(n)$ is the maximal value of $\lambda_1(\mathcal{L})^2 / \det(\mathcal{L})^{2/n}$. (Hermite SVP is of course named in honor of Hermite and his study of γ_n. It is often alternatively called Minkowski SVP.)

For $\delta \ge 1$, it is then natural to define δ-HSVP as the variant of SVP that asks for any non-zero lattice vector $\mathbf{x} \in \mathcal{L}$ such that $\|\mathbf{x}\| \le \delta \det(\mathcal{L})^{1/n}$. One typically takes $\delta \ge \sqrt{\gamma_n} \ge \Omega(\sqrt{n})$, in which case the problem is total. In particular, there is a trivial reduction from $\delta\sqrt{\gamma_n}$-HSVP to δ-SVP. (There is also a non-trivial reduction from δ^2-SVP to δ-HSVP for $\delta \ge \sqrt{\gamma_n}$ [Lov86].)

δ-HSVP is an important problem in its own right. In particular, the random lattices most often used in cryptography typically satisfy $\lambda_1(\mathcal{L}) \ge \Omega(\sqrt{n}) \cdot \det(\mathcal{L})^{1/n}$, so that for these lattices δ-HSVP is equivalent to $O(\delta/\sqrt{n})$-SVP. This fact is quite useful as the best known basis reduction algorithms [GN08, MW16, ALNS20] yield solutions to both δ_S-SVP and δ_H-HSVP with, e.g.,

$$\delta_H := \gamma_k^{\frac{n-1}{2(k-1)}} \approx k^{n/(2k)} \qquad \delta_S := \gamma_k^{\frac{n-k}{k-1}} \approx k^{n/k-1}\,, \tag{1}$$

when given access to an oracle for (exact) SVP in dimension $k \le n/2$. Notice that δ_H is significantly better than the approximation factor $\sqrt{\gamma_n}\delta_S \approx \sqrt{n}k^{n/k-1}$ that one obtains from the trivial reduction to δ_S-SVP. (Furthermore, the approximation factor δ_H in Eq. (1) is achieved even for $n/2 < k \le n$.)

In fact, it is easy to check that we will achieve the same value of δ_H if the reduction is instantiated with a $\sqrt{\gamma_k}$-HSVP oracle in dimension k, rather than an SVP oracle. More surprisingly, a careful reading of the proofs in [GN08, ALNS20] shows that a $\sqrt{\gamma_k}$-HSVP oracle is "almost sufficient" to even solve δ_S-SVP. (We make this statement a bit more precise below.)

1.3 Our Results

Our main contribution is a simplified version of the $2^{n/2+o(n)}$-time algorithm from [ADRS15] and a novel analysis of the algorithm that gives an approximation algorithm for both SVP and HSVP.

Theorem 1.1 (Informal, approximation algorithm for (H)SVP). *There is a $2^{n/2+o(n)}$-time algorithm that solves δ-SVP and δ-HSVP for $\delta \le \widetilde{O}(\sqrt{n})$.*

Notice that this algorithm almost achieves the best possible approximation factor δ for HSVP since there exists a family of lattices for which $\lambda_1(\mathcal{L}) \geq \Omega(\sqrt{n}\det(\mathcal{L})^{1/n})$ (i.e., $\gamma_n \geq \Omega(n)$). So, δ is optimal for HSVP up to a polylogarithmic factor.

As far as we know, this algorithm might actually solve exact or near-exact SVP, but we do not know how to prove this. However, by adapting the basis reduction algorithms of [GN08, ALNS20], we show that Theorem 1.1 is nearly as good (when combined with known results) as a $2^{k/2}$-time algorithm for exact SVP in k dimensions, in the sense that we can already nearly match Eq. (1) in time $2^{k/2+o(k)}$ with this.

In slightly more detail, basis reduction procedures break the input basis vectors $\mathbf{b}_1, \ldots, \mathbf{b}_n$ into blocks $\mathbf{b}_{i+1}, \ldots, \mathbf{b}_{i+k}$ of length k. They repeatedly call their oracle on (projections of) the lattices generated by these blocks and use the result to update the basis vectors. We observe that the procedures in [GN08, ALNS20] only need to use an SVP oracle on the last block $\mathbf{b}_{n-k+1}, \ldots, \mathbf{b}_n$. For all other blocks, an HSVP oracle suffices. Since we now have a faster algorithm for HSVP than we do for SVP, we make this last block a bit smaller than the others, so that we can solve (near-exact) SVP on the last block in time $2^{k/2+o(k)}$.

When we apply the $2^{0.802n}$-time algorithm for $O(1)$-SVP from [LWXZ11, WLW15, AUV19] to instantiate this idea, it yields the following result, which gives the fastest known algorithm for δ-SVP for all $\delta \gtrsim n^c$.

Theorem 1.2 (Informal). *There is a* $(2^{k/2+o(k)} \cdot \mathrm{poly}(n))$*-time algorithm that solves* δ_H^**-HSVP with*

$$\delta_H^* \approx k^{n/(2k)} ,$$

for $k \leq .99n$ *and*

$$\delta_S^* \approx k^{(n/k)-0.62} ,$$

for $k \leq n/1.63$.

Notice that Theorem 1.2 matches Eq. (1) with block size k exactly for δ_H, and up to a factor of $k^{0.37}$ for δ_S. This small loss in approximation factor comes from the fact that our last block is slightly smaller than the other blocks.

Together, Theorems 1.1 and 1.2 give the fastest proven running times for n^c-HSVP for all $c > 1/2$ and for n^c-SVP for all $c > 1$, as well as $c \in (1/2, 0.802)$. Table 1 summarizes the current state of the art.

1.4 Our Techniques

Summing vectors over a tower of lattices. Like the $2^{n/2+o(n)}$-time algorithm in [ADRS15], our algorithm for $\widetilde{O}(\sqrt{n})$-(H)SVP constructs a tower of lattices $\mathcal{L}_0 \supset \mathcal{L}_1 \supset \cdots \supset \mathcal{L}_\ell = \mathcal{L}$ such that for every $i \geq 1$, $2\mathcal{L}_{i-1} \subset \mathcal{L}_i$. The idea of using a tower of lattices was independently developed in [BGJ14] (see also [GINX16]) for heuristic algorithms. The index of $\mathcal{L}_i$ over $\mathcal{L}_{i-1}$ is 2^α for an integer $\alpha = n/2 + o(n)$, and $\ell = o(n)$. For the purpose of illustrating our ideas, we make a simplifying assumption here that $\ell\alpha$ is an integer multiple of n, and hence $\mathcal{L}_0 = \mathcal{L}/2^{\alpha\ell/n}$ is a scalar multiple of $\mathcal{L}$.

Table 1. Proven running times for solving (H)SVP. We mark results that do not use basis reduction with [*]. We omit $2^{o(n)}$ factors in the running time, and except in the first two rows, polylogarithmic factors in the approximation factor.

Problem	Approximation factor	Previous Best		This work
SVP	Exact	2^n [*]	[ADRS15]	—
	$O(1)$	$2^{0.802n}$ [*]	[WLW15]	—
	n^c for $c \in (0.5, 0.802]$	$2^{\frac{0.401n}{c}}$	[ALNS20]	$2^{\frac{n}{2}}$ [*]
	n^c for $c \in (0.802, 1]$	$2^{\frac{0.401n}{c}}$	[ALNS20]	—
	n^c for $c > 1$	$2^{\frac{0.802n}{c+1}}$	[ALNS20]	$2^{\frac{n}{2c+1.24}}$
HSVP	$\sqrt{n}$	$2^{0.802n}$ [*]	[WLW15]	$2^{\frac{n}{2}}$ [*]
	n^c for $c \geq 1$	$2^{\frac{0.401n}{c}}$	[ALNS20]	$2^{\frac{n}{4c}}$

And, as in [ADRS15], we start by sampling $\mathbf{X}_1, \ldots, \mathbf{X}_N \in \mathcal{L}_0$ for $N = 2^{\alpha+o(n)}$ from $D_{\mathcal{L}_0,s}$. This can be done efficiently using known techniques, as long as s is large relative to, e.g., the length of the shortest basis of $\mathcal{L}_0$ [GPV08, BLP13]. Since $\mathcal{L}_0 = \mathcal{L}/2^{\alpha\ell/n}$, the parameter s can still be significantly smaller than, e.g., $\lambda_1(\mathcal{L})$. In particular, we can essentially take $s \leq \mathrm{poly}(n)\lambda_1(\mathcal{L})/2^{\alpha\ell/n}$.

The algorithm then takes disjoint pairs of vectors that are in the same coset of $\mathcal{L}_0/\mathcal{L}_1$, and adds the pairs together. Since $2\mathcal{L}_0 \subset \mathcal{L}_1$, for any such pair $\mathbf{X}_i, \mathbf{X}_i$, $\mathbf{Y}_k = \mathbf{X}_i + \mathbf{X}_j$ is in $\mathcal{L}_1$. (This adding is analogous to the averaging procedure from [ADRS15, AS18] described above. In that case, $\mathcal{L}_1 = 2\mathcal{L}_0$, so that it is natural to divide vectors in $\mathcal{L}$ by two, while here adding seems more natural.) We thus obtain approximately $N/2$ vectors in $\mathcal{L}_1$ (up to the loss due to the vectors that could not be paired), and repeat this procedure many times, until finally we obtain vectors in $\mathcal{L}_\ell = \mathcal{L}$, each the sum of 2^ℓ of the original $\mathbf{X}_i$.

To prove correctness, we need to prove that with high probability some of these vectors will be *both* short and non-zero. It is actually relatively easy to show that the vectors are short—at least in expectation. To prove this, we first use the fact that the expected squared norm of the $\mathbf{X}_i$ is bounded by ns^2 (which is what one would expect from the continuous Gaussian distribution). And, the original $\mathbf{X}_i$ are distributed symmetrically, i.e., $\mathbf{X}_i$ is as likely to equal $-\mathbf{x}$ as it is to equal $\mathbf{x}$).

Furthermore, our pairing procedure is symmetric, i.e., if we were to replace $\mathbf{X}_i$ with $-\mathbf{X}_i$, the pairing procedure would behave identically. (This is true precisely because $2\mathcal{L}_0 \subset \mathcal{L}_1$—we are using the fact that $\mathbf{x} = -\mathbf{x} \bmod \mathcal{L}_1$ for any $\mathbf{x} \in \mathcal{L}_0$.) This implies that

$$\mathbb{E}[\langle \mathbf{X}_i, \mathbf{X}_j \rangle \mid E_{i,j}] = \mathbb{E}[\langle \mathbf{X}_i, -\mathbf{X}_j \rangle \mid E_{i,j}] = 0 ,$$

where $E_{i,j}$ is the event that $\mathbf{X}_i$ is paired with $\mathbf{X}_j$. Therefore, $\mathbb{E}[\|\mathbf{X}_i + \mathbf{X}_j\|^2 \mid E_{i,j}]$ is equal to

$$\mathbb{E}[\|\mathbf{X}_i\|^2 \mid E_{i,j}] + \mathbb{E}[\|\mathbf{X}_j\|^2 \mid E_{i,j}] + 2\,\mathbb{E}[\langle \mathbf{X}_i, \mathbf{X}_j \rangle \mid E_{i,j}] \approx 2\,\mathbb{E}[\|\mathbf{X}_i\|^2] .$$

The same argument works at every step of the algorithm. So, (if we ignore the subtle distinction between $\mathbb{E}[\|\mathbf{X}_i\|^2 \mid E_{i,j}]$ and $\mathbb{E}[\|\mathbf{X}_i\|^2]$), we see that our final vectors have expected squared norm

$$2^\ell \,\mathbb{E}[\|\mathbf{X}_i\|^2] \le 2^\ell n s^2 \le \mathrm{poly}(n) 2^{\ell(1-2\alpha n)} \cdot \lambda_1(\mathcal{L})^2 \,. \qquad (2)$$

By taking, e.g., $\alpha = n/2 + n/\log n < n + o(n)$ and $\ell = \log^2 n$, we see that we can make this expectation small relative to $\lambda_1(\mathcal{L})$.

The difficulty, then, is "only" to show that the distribution of the final vectors is not heavily concentrated on zero. Of course, we can't hope for this to be true if, e.g., the expectation in Eq. (2) is much smaller than $\lambda_1(\mathcal{L})^2$. And, as we will discuss below, if we choose α and ℓ so that this expectation is sufficiently large, then techniques from prior work can show that the probability of zero is low. Our challenge is therefore to bound the probability of zero for the largest choices of α and ℓ (and therefore the lowest expectation in Eq. (2)) that we can manage.

Gaussians over unknown sublattices. Peikert and Micciancio (building on prior work) showed what they called a "convolution theorem" for discrete Gaussians. Their theorem said that the sum of discrete Gaussian vectors is statistically close to a discrete Gaussian (with parameter increased by a factor of $\sqrt{2}$), provided that the parameter s is a bit larger than the *smoothing parameter* $\eta(\mathcal{L})$ of the lattice $\mathcal{L}$ [MP13]. This (extremely important) parameter $\eta(\mathcal{L})$, was introduced by Micciancio and Regev [MR07], and has a rather technical (and elegant) definition. (See Sect. 2.4.) Intuitively, $\eta(\mathcal{L})$ is such that for any $s > \eta(\mathcal{L})$, $D_{\mathcal{L},s}$ "looks like a continuous Gaussian distribution." E.g., for $s > \eta(\mathcal{L})$, the moments of the discrete Gaussian distribution are quite close to the moments of the continuous Gaussian distribution (with the same parameter).

In fact, [MP13] showed a convolution theorem for lattice *cosets*, not just lattices, i.e., the sum of a vector sampled from coset $D_{\mathcal{L}+\mathbf{t}_1,s}$ and a vector sampled from $D_{\mathcal{L}+\mathbf{t}_2,s}$ yields a vector with a distribution that is statistically close to $D_{\mathcal{L}+\mathbf{t}_1+\mathbf{t}_2,\sqrt{2}s}$. Since our algorithm sums vectors sampled from a discrete Gaussian over $\mathcal{L}_0$, conditioned on their cosets modulo $\mathcal{L}_1$, it is effectively summing discrete Gaussians over cosets of $\mathcal{L}_1$. So, as long as we stay above the smoothing parameter of $\mathcal{L}_1 \supset \mathcal{L}$, our vectors will be statistically close to discrete Gaussians, allowing us to easily bound the probability of zero.

However, [ADRS15] already showed how to use a variant of this algorithm to obtain samples from *exactly* the discrete Gaussian above smoothing. And, more generally, there is a long line of work that uses samples from the discrete Gaussian above smoothing to find "short vectors" from a lattice, but the length of these short vectors is always proportional to $\eta(\mathcal{L})$. The problem is that in general $\eta(\mathcal{L})$ can be arbitrarily larger than $\lambda_1(\mathcal{L})$ and $\det(\mathcal{L})^{1/n}$. (To see this, consider the two-dimensional lattice generated by $(T, 0), (0, 1/T)$ for large T, which has $\eta(\mathcal{L}) \approx T$, $\lambda_1(\mathcal{L}) = 1/T$ and $\det(\mathcal{L}) = 1$.) So, this seems useless for solving (H)SVP, instead yielding a solution to another variant of SVP called SIVP.[1]

Our solution is essentially to apply these ideas from [MP13] to an *unknown* sublattice $\mathcal{L}' \subseteq \mathcal{L}$. (Here, one should imagine a sublattice generated by fewer than n vectors. Jumping ahead a bit, the reader might consider the example $\mathcal{L}' = \mathbb{Z}\mathbf{v} = \{\mathbf{0}, \pm\mathbf{v}, \pm 2\mathbf{v}, \ldots,\}$ the rank-one sublattice generated by $\mathbf{v}$, shortest non-zero vector in the lattice.) Indeed, the discrete Gaussian over $\mathcal{L}$, $D_{\mathcal{L},s}$, can be viewed as a *mixture of discrete Gaussians* over $\mathcal{L}'$, $D_{\mathcal{L},s} = D_{\mathcal{L}'+\mathbf{C},s}$, where $\mathbf{C} \in \mathcal{L}/\mathcal{L}'$ is some random variable over cosets of $\mathcal{L}'$. (Put another way, one could obtain a sample from $D_{\mathcal{L},s}$ by first sampling a coset $\mathbf{C} \in \mathcal{L}/\mathcal{L}'$ from some appropriately chosen distribution and then sampling from $D_{\mathcal{L}'+\mathbf{C},s}$.)

The basic observation behind our analysis is that we can now apply (a suitable variant of) [MP13]'s convolution theorem in order to see that the sum of two mixtures of Gaussians over $\mathcal{L}'$, $\mathbf{X}_1, \mathbf{X}_2 \sim D_{\mathcal{L}'+\mathbf{C},s}$, yields a new mixture of Gaussians $D_{\mathcal{L}'+\mathbf{C}',\sqrt{2}s}$ for *some* $\mathbf{C}'$, provided that s is sufficiently large *relative to* $\eta(\mathcal{L}')$.

Ignoring *many* technical details, this shows that our algorithm can be used to output a distribution of the form $D_{\mathcal{L}'+\mathbf{C},s}$ for some random variable $\mathbf{C} \in \mathcal{L}/\mathcal{L}'$ provided that $s \gg \eta(\mathcal{L}')$. Crucially, we only need to consider $\mathcal{L}'$ in the analysis; the algorithm does not need to know what $\mathcal{L}'$ is for this to work. Furthermore, we do not care at all about the distribution of $\mathbf{C}$! We already know that our algorithm samples from a distribution that is short in expectation (by the argument above), so that the only thing we need from the distribution $D_{\mathcal{L}'+\mathbf{C},s}$ is that it is not zero too often. Indeed, when $\mathbf{C}$ is not the zero coset (i.e., $\mathbf{C} \notin \mathcal{L}'$), then $D_{\mathcal{L}'+\mathbf{C},s}$ is never zero, and when $\mathbf{C}$ is zero, then we get a sample from $D_{\mathcal{L}',s}$ for $s \gg \eta(\mathcal{L})$, in which case well-known techniques imply that we are unlikely to get zero.

Smooth sublattices. So, in order to prove that our algorithm finds short vectors, it remains to show that there exists some sublattice $\mathcal{L}' \subseteq \mathcal{L}$ with low smoothing parameter—a "smooth sublattice." In more detail, our algorithm will find a non-zero vector with length less than $\sqrt{n} \cdot \eta(\mathcal{L}')$ for any sublattice $\mathcal{L}'$. Indeed, as one might guess, taking $\mathcal{L}' = \mathbb{Z}\mathbf{v} = \{\mathbf{0}, \pm\mathbf{v}, \pm 2\mathbf{v}, \ldots,\}$ to be the lattice generated by a shortest non-zero vector $\mathbf{v}$, we have $\eta(\mathcal{L}') = \mathrm{polylog}(n)\|\mathbf{v}\| = \mathrm{polylog}(n)\lambda_1(\mathcal{L})$ (where the polylogarithmic factor arises because of "how smooth we need $\mathcal{L}'$ to be"). This immediately yields our $\widetilde{O}(\sqrt{n})$-SVP algorithm.

To solve $\widetilde{O}(\sqrt{n})$-HSVP, we must argue that every lattice has a sublattice $\mathcal{L}' \subseteq \mathcal{L}$ with $\eta(\mathcal{L}') \leq \mathrm{polylog}(n) \cdot \det(\mathcal{L})^{1/n}$. In fact, for very different reasons, Dadush conjectured *exactly* this statement (phrased slightly differently), calling

[1] It is not known how to use an SIVP oracle for basis reduction, which makes it significantly less useful than SVP. [MR07,MP13] and other works used these ideas to reduce SIVP to the problem of breaking a certain cryptosystem, in order to argue that the cryptosystem is secure. They were therefore primarily interested in SIVP as an example of a hard lattice problem, rather than as a problem that one might actually wish to solve.

it a "reverse Minkowski conjecture" [DR16]. (The reason for this name might not be clear in this context, but one can show that this is a partial converse to Minkowski's theorem.) Later, Regev and Stephens-Davidowitz proved the conjecture [RS17]. Our HSVP result then follows from this rather heavy hammer.

1.5 Open Questions and Directions for Future Work

We leave one obvious open question: Does our algorithm (or some variant) solve γ-SVP for a better approximation factor? It is clear that our current analysis cannot hope to do better than $\delta \approx \sqrt{n}$, but we see no fundamental reason why the algorithm cannot achieve, say, $\delta = \text{polylog}(n)$ or even $\delta = 1$! (Indeed, we have been trying to prove something like this for roughly five years.)

We think that even a negative answer to this question would also be interesting. In particular, it is not currently clear whether our algorithm is "fundamentally an HSVP algorithm." For example, if one could show that our algorithm fails to output vectors of length $\text{polylog}(n) \cdot \lambda_1(\mathcal{L})$ for some family of input lattices $\mathcal{L}$, then this would be rather surprising. Perhaps such a result could suggest a true algorithmic separation between the two problems.

2 Preliminaries

We write log for the base-two logarithm. We use the notation $a = 1 \pm \delta$ and $a = e^{\pm\delta}$ to denote the statements $1 - \delta \le a \le 1 + \delta$ and $e^{-\delta} \le a \le e^{\delta}$, respectively.

Definition 2.1. *We say that a distribution $\widehat{D}$ is δ-similar to another distribution D if for all $\mathbf{x}$ in the support of D, we have*

$$\Pr_{\mathbf{X}\sim\widehat{D}}[\mathbf{X} = \mathbf{x}] = e^{\pm\delta} \cdot \Pr_{\mathbf{X}\sim D}[\mathbf{X} = \mathbf{x}] .$$

2.1 Probability

The following inequality gives a concentration result for the values of (sub-)martingales that have bounded differences.

Lemma 2.2 ([AS04] Azuma's inequality, Chapter 7). *Let $X_0, X_1, \ldots$ be a set of random variables that form a discrete-time sub-martingale, i.e., for all $n \ge 0$,*

$$\mathbb{E}[X_{n+1} \mid X_1, \ldots, X_n] \ge X_n .$$

If for all $n \ge 0$, $|X_n - X_{n-1}| \le c$, then for all integers N and positive real t,

$$\Pr[X_N - X_0 \le -t] \le \exp\left(\frac{-t^2}{2Nc^2}\right) .$$

We will need the following corollary of the above inequality.

Corollary 2.3. *Let $\alpha \in (0,1)$, and let $Y_1, Y_2, Y_3, \ldots$ be random variables in $[0,1]$ such that for all $n \geq 0$*

$$\mathbb{E}[Y_{n+1}|Y_1, \ldots, Y_n] \geq \alpha \,.$$

Then, for all positive integers N and positive real t,

$$\Pr[\sum_{i=1}^{N} Y_i \leq N\alpha - t] \leq \exp\left(\frac{-t^2}{2N}\right) \,.$$

Proof. Let $X_0 = 0$, and for all $i \geq 1$,

$$X_i := X_{i-1} + Y_i - \alpha = \sum_{j=1}^{i} Y_i - i \cdot \alpha \,.$$

The statement then follows immediately from Lemma 2.2. □

2.2 Lattices

A lattice $\mathcal{L} \subset \mathbb{R}^n$ is the set of integer linear combinations

$$\mathcal{L} := \mathcal{L}(\mathbf{B}) = \{z_1\mathbf{b}_1 + \cdots + z_k\mathbf{b}_k \ : \ z_i \in \mathbb{Z}\}$$

of linearly independent basis vectors $\mathbf{B} = (\mathbf{b}_1, \ldots, \mathbf{b}_k) \in \mathbb{R}^{n\times k}$. We call k the *rank* of the lattice. Given a lattice $\mathcal{L}$, the basis is not unique. For any lattice $\mathcal{L}$, we use $\mathrm{rank}(\mathcal{L})$ to denote its rank. We use $\lambda_1(\mathcal{L})$ to denote the length of the shortest non-zero vector in $\mathcal{L}$, and more generally, for $1 \leq i \leq k$,

$$\lambda_i(\mathcal{L}) := \min\{r \ : \ \dim \mathrm{span}(\{\mathbf{y} \in \mathcal{L} \ : \ \|\mathbf{y}\| \leq r\}) \geq i\} \,.$$

For any lattice $\mathcal{L} \subset \mathbb{R}^n$, its dual lattice $\mathcal{L}^*$ is defined to be the set of vectors in the span of $\mathcal{L}$ that have integer inner products with all vectors in $\mathcal{L}$. More formally:

$$\mathcal{L}^* = \{\mathbf{x} \in \mathrm{span}(\mathcal{L}) : \forall \mathbf{y} \in \mathcal{L}, \langle \mathbf{x}, \mathbf{y}\rangle \in \mathbb{Z}\} \,.$$

We often assume without loss of generality that the lattice is full rank, i.e., that $n = k$, by identifying $\mathrm{span}(\mathcal{L})$ with $\mathbb{R}^k$. However, we do often work with sublattices $\mathcal{L}' \subseteq \mathcal{L}$ with $\mathrm{rank}(\mathcal{L}') < \mathrm{rank}(\mathcal{L})$.

For any sublattice $\mathcal{L}' \subseteq \mathcal{L}$, $\mathcal{L}/\mathcal{L}'$ denotes the set of cosets which are translations of $\mathcal{L}'$ by vectors in $\mathcal{L}$. In particular, any coset can be denoted as $\mathcal{L}' + \mathbf{c}$ for $\mathbf{c} \in \mathcal{L}$. When there is no ambiguity, we drop the $\mathcal{L}'$ and use $\mathbf{c}$ to denote a coset.

2.3 The Discrete Gaussian Distribution

For any parameter $s > 0$, we define Gaussian mass function $\rho_s : \mathbb{R}^n \to \mathbb{R}$ to be:

$$\rho_s(\mathbf{x}) = \exp\Big(-\frac{\pi\|\mathbf{x}\|^2}{s^2}\Big) \,,$$

and for any discrete set $A \subset \mathbb{R}^n$, its Gaussian mass is defined as $\rho_s(A) = \sum_{\mathbf{x}\in A} \rho_s(\mathbf{x})$.

For a lattice $\mathcal{L} \subset \mathbb{R}^n$, shift $\mathbf{t} \in \mathbb{R}^n$, and parameter $s > 0$, we have the following convenient formula for the Gaussian mass of the lattice coset $\mathcal{L} + \mathbf{t}$, which follows from the Poisson Summation Formula

$$\rho_s(\mathcal{L} + \mathbf{t}) = \frac{s^n}{\det(\mathcal{L})} \cdot \sum_{\mathbf{w}\in\mathcal{L}^*} \rho_{1/s}(\mathbf{w}) \cos(2\pi\langle \mathbf{w}, \mathbf{t}\rangle) \,. \tag{3}$$

In particular, for the special case $\mathbf{t} = \mathbf{0}$, we have $\rho_s(\mathcal{L}) = s^n \rho_{1/s}(\mathcal{L}^*)/\det(\mathcal{L})$.

Definition 2.4. *For a lattice $\mathcal{L} \subset \mathbb{R}^n$, $\mathbf{u} \in \mathbb{R}^n$, the discrete Gaussian distribution $\mathcal{D}_{\mathcal{L}+\mathbf{u},s}$ over $\mathcal{L} + \mathbf{u}$ with parameter $s > 0$ is defined as follows. For any $\mathbf{x} \in \mathcal{L} + \mathbf{u}$,*

$$\Pr_{\mathbf{X}\sim\mathcal{D}_{\mathcal{L}+\mathbf{u},s}}[\mathbf{X} = \mathbf{x}] = \frac{\rho_s(\mathbf{x})}{\rho_s(\mathcal{L} + \mathbf{u})} \,.$$

We will need the following result about the discrete Gaussian distribution.

Lemma 2.5 ([DRS14] Lemma 2.13). *For any lattice $\mathcal{L} \subset \mathbb{R}^n$, $s > 0$, $\mathbf{u} \subset \mathbb{R}^n$, and $t > \frac{1}{\sqrt{2\pi}}$,*

$$\Pr_{\mathbf{X}\sim\mathcal{D}_{\mathcal{L}+\mathbf{u},s}}(\|\mathbf{X}\| > ts\sqrt{n}) < \frac{\rho_s(\mathcal{L})}{\rho_s(\mathcal{L} + \mathbf{u})} \left(\sqrt{2\pi e t^2} \exp(-\pi t^2)\right)^n \,.$$

2.4 The Smoothing Parameter

Definition 2.6. *For a lattice $\mathcal{L} \subset \mathbb{R}^n$ and $\varepsilon > 0$, the smoothing parameter $\eta_\varepsilon(\mathcal{L})$ is defined as the unique value that satisfies $\rho_{1/\eta_\varepsilon(\mathcal{L})}(\mathcal{L}^*\backslash\{\mathbf{0}\}) = \varepsilon$.*

We will often use the basic fact that $\eta_\varepsilon(\alpha\mathcal{L}) = \alpha\eta_\varepsilon(\mathcal{L})$ for any $\alpha > 0$ and $\eta_\varepsilon(\mathcal{L}') \geq \eta_\varepsilon(\mathcal{L})$ for any *full-rank* sublattice $\mathcal{L}' \subseteq \mathcal{L}$.

Claim 2.7 ([MR07] Lemma 3.3). *For any $\varepsilon \in (0, 1/2)$, we have*

$$\eta_\varepsilon(\mathbb{Z}) \leq \sqrt{\log(1/\varepsilon)} \,.$$

We will need the following simple results, which follows immediately from Eq. (3).

Lemma 2.8 ([Reg09] Claim 3.8). *For any lattice $\mathcal{L}$, $s \geq \eta_\varepsilon(\mathcal{L})$, and any vectors $\mathbf{c}_1, \mathbf{c}_2$, we have that*

$$\frac{1-\varepsilon}{1+\varepsilon} \leq \frac{\rho_s(\mathcal{L} + \mathbf{c}_1)}{\rho_s(\mathcal{L} + \mathbf{c}_2)} \leq \frac{1+\varepsilon}{1-\varepsilon} \,.$$

Thus, for $\varepsilon < 1/3$,

$$e^{-3\varepsilon} \leq \frac{\rho_s(\mathcal{L} + \mathbf{c}_1)}{\rho_s(\mathcal{L} + \mathbf{c}_2)} \leq e^{3\varepsilon} \,.$$

We prove the following statement.

Theorem 2.9. *For any lattice $\mathcal{L} \subset \mathbb{R}^n$ with rank $k \geq 20$,*

$$\eta_{1/2}(\mathcal{L}) \geq \lambda_k(\mathcal{L})/\sqrt{k}\,.$$

Proof. If $\mathcal{L}$ is not a full-rank lattice, then we can project to a subspace given by the span of $\mathcal{L}$. So, without loss of generality, we assume that $\mathcal{L}$ is a full-rank lattice, i.e., $k = n$.

Suppose $\lambda_n(\mathcal{L}) > \sqrt{n}\eta_{1/2}(\mathcal{L})$. Then there exists a vector $\mathbf{u} \in \mathbb{R}^n$ such that $\mathrm{dist}(\mathbf{u}, \mathcal{L}) > \frac{1}{2}\sqrt{n}\eta_{1/2}(\mathcal{L})$. Then, using Lemma 2.5 with $t = 1/2$, $s = \eta_{1/2}(\mathcal{L})$, we have

$$\begin{aligned}
1 &= \Pr_{\mathbf{X}\sim\mathcal{D}_{\mathcal{L}+\mathbf{u},\eta_{1/2}(\mathcal{L})}}[\|\mathbf{X}\| > st\sqrt{n}] \\
&< \frac{\rho_s(\mathcal{L})}{\rho_s(\mathcal{L}+\mathbf{u})}\left(\sqrt{2\pi e t^2}\exp(-\pi t^2)\right)^n \\
&\leq \frac{1+1/2}{1-1/2}(\sqrt{\pi e/2}\cdot e^{-\pi/4})^n && \text{using Lemma 2.8} \\
&\leq 3\cdot(0.943)^n \\
&< 1 && \text{since } k = n \geq 20\,,
\end{aligned}$$

which is a contradiction. □

Claim 2.10. *For any lattice $\mathcal{L} \subset \mathbb{R}^n$ and any parameters $s \geq s' \geq \eta_{1/2}(\mathcal{L})$,*

$$\frac{\rho_s(\mathcal{L})}{\rho_{s'}(\mathcal{L})} \geq \frac{2s}{3s'}\,.$$

Proof. By the Poisson Summation Formula (Eq. (3)), we have

$$\rho_s(\mathcal{L}) = s^n\cdot\frac{\rho_{1/s}(\mathcal{L}^*)}{\det(\mathcal{L})} \geq s^n/\det(\mathcal{L})\,,$$

and similarly,

$$\rho_{s'}(\mathcal{L}) = (s')^n\cdot\frac{\rho_{1/s'}(\mathcal{L}^*)}{\det(\mathcal{L})} \leq 3(s')^n/(2\det(\mathcal{L}))\,,$$

since $\rho_{1/s'}(\mathcal{L}^*) \leq 3/2$ for $s' \geq \eta_{1/2}(\mathcal{L})$. Combining the two inequalities gives $\rho_s(\mathcal{L}) \geq 2(s/s')^n/3 \geq 2(s/s')/3$, as needed. □

Claim 2.11. *For any lattice $\mathcal{L} \subset \mathbb{R}^n$ and any $s > 0$,*

$$\mathop{\mathbb{E}}_{\mathbf{X}\sim D_{\mathcal{L},s}}[\|\mathbf{X}\|^2] \leq \frac{ns^2}{2\pi}\,.$$

Lemma 2.12. *For $s \geq \eta_\varepsilon(\mathcal{L})$, and any real factor $k \geq 1$, $ks \geq \eta_{\varepsilon^{k^2}}(\mathcal{L})$.*

Proof.

$$\begin{aligned}\sum_{\mathbf{w}\in\mathcal{L}^*\setminus\{\mathbf{0}\}} \rho_{1/(ks)}(\mathbf{w}) &= \sum_{\mathbf{w}\in\mathcal{L}^*\setminus\{\mathbf{0}\}} e^{-\pi\|\mathbf{w}\|k^2s^2} \\ &= \sum_{\mathbf{w}\in\mathcal{L}^*\setminus\{\mathbf{0}\}} \rho_{1/s}(\mathbf{w})^{k^2} \\ &\le \Big(\sum_{\mathbf{w}\in\mathcal{L}^*\setminus\{\mathbf{0}\}} \rho_{1/s}(\mathbf{w})\Big)^{k^2} \\ &\le \varepsilon^{k^2} .\end{aligned}$$

□

Corollary 2.13. *For any lattice $\mathcal{L} \subset \mathbb{R}^n$ and $\varepsilon \in (0, 1/2)$, $\eta_\varepsilon(\mathcal{L}) \le \sqrt{\log(1/\varepsilon)} \cdot \eta_{1/2}(\mathcal{L})$.*

Proof. Let $k = \sqrt{\log(1/\varepsilon)}$ and thus $(\frac{1}{2})^{k^2} = \varepsilon$. By Lemma 2.12, $k\eta_{1/2}(\mathcal{L}) \ge \eta_\varepsilon(\mathcal{L})$. □

We will need the following useful lemma concerning the convolution of two discrete Gaussian distributions. See [GMPW20] for a very general result of this form (and a list of similar results). Our lemma differs from those in [GMPW20] and elsewhere in that we are interested in a stronger notion of statistical closeness: point-wise multiplicative distance, rather than statistical distance. One can check that this stronger variant follows from the proofs in [GMPW20], but we give a separate proof for completeness.

Lemma 2.14. *For any lattice $\mathcal{L} \subset \mathbb{R}^n$, $\varepsilon \in (0, 1/3)$, parameter $s \ge \sqrt{2}\eta_\varepsilon(\mathcal{L})$, and shifts $\mathbf{t}_1, \mathbf{t}_2 \in \mathbb{R}^n$, let $\mathbf{X}_i \sim D_{\mathcal{L}+\mathbf{t}_i,s}$ be independent random variables. Then the distribution of $\mathbf{X}_1 + \mathbf{X}_2$ is 6ε-similar to $D_{\mathcal{L}+\mathbf{t}_1+\mathbf{t}_2,\sqrt{2}s}$.*

Proof. Let $\mathbf{y} \in \mathcal{L} + \mathbf{t}_1 + \mathbf{t}_2$. We have

$$\begin{aligned}\Pr[\mathbf{X}_1 + \mathbf{X}_2 = \mathbf{y}] &= \frac{1}{\rho_s(\mathcal{L}+\mathbf{t}_1)\rho_s(\mathcal{L}+\mathbf{t}_2)} \sum_{\mathbf{x}\in\mathcal{L}+\mathbf{t}_1} \exp(-\pi(\|\mathbf{x}\|^2 + \|\mathbf{y}-\mathbf{x}\|^2)/s^2) \\ &= \frac{1}{\rho_s(\mathcal{L}+\mathbf{t}_1)\rho_s(\mathcal{L}+\mathbf{t}_2)} \sum_{\mathbf{x}\in\mathcal{L}+\mathbf{t}_1} \exp(-\pi(\|\mathbf{y}\|^2/2 + \|2\mathbf{x}-\mathbf{y}\|^2/2)/s^2) \\ &= \frac{\rho_{\sqrt{2}s}(\mathbf{y})}{\rho_s(\mathcal{L}+\mathbf{t}_1)\rho_s(\mathcal{L}+\mathbf{t}_2)} \rho_{s/\sqrt{2}}(\mathcal{L}+\mathbf{t}_1-\mathbf{y}/2) \\ &= e^{\pm 3\varepsilon} \rho_{\sqrt{2}s}(\mathbf{y}) \cdot \frac{\rho_{s/\sqrt{2}}(\mathcal{L})}{\rho_s(\mathcal{L}+\mathbf{t}_1)\rho_s(\mathcal{L}+\mathbf{t}_2)} ,\end{aligned}$$

where the last step follows from Lemma 2.8. By applying this for all $\mathbf{y}' \in \mathcal{L} + \mathbf{t}_1 + \mathbf{t}_2$, we see that

$$\Pr[\mathbf{X}_1 + \mathbf{X}_2 = \mathbf{y}] = e^{\pm 3\varepsilon} \cdot \frac{\rho_{\sqrt{2}s}(\mathbf{y})}{\sum_{\mathbf{y}'\in\mathcal{L}+\mathbf{t}_1+\mathbf{t}_2} \chi_{\mathbf{y}'}\rho_{\sqrt{2}s}(\mathbf{y}')}$$

for some $\chi_{\mathbf{y}'} = e^{\pm 3\varepsilon}$. Therefore,

$$\Pr[\mathbf{X}_1 + \mathbf{X}_2 = \mathbf{y}] = e^{\pm 6\varepsilon} \cdot \frac{\rho_{\sqrt{2}s}(\mathbf{y})}{\rho_{\sqrt{2}s}(\mathcal{L} + \mathbf{t}_1 + \mathbf{t}_2)} ,$$

as needed. □

2.5 Lattice Problems

In this paper, we study the algorithms for the following lattice problems.

Definition 2.15 (r-HSVP). *For an approximation factor $r := r(n) \geq 1$, the r-Hermite Approximate Shortest Vector Problem (r-HSVP) is defined as follows: Given a basis $\mathbf{B}$ for a lattice $\mathcal{L} \subset \mathbb{R}^n$, the goal is to output a vector $\mathbf{x} \in \mathcal{L}\backslash\{\mathbf{0}\}$ with $\|\mathbf{x}\| \leq r \cdot \det(\mathcal{L})^{1/n}$.*

Definition 2.16 (r-SVP). *For an approximation factor $r := r(n) \geq 1$, the r-Shortest Vector Problem (r-SVP) is defined as follows: Given a basis $\mathbf{B}$ for a lattice $\mathcal{L} \subset \mathbb{R}^n$, the goal is to output a vector $\mathbf{x} \in \mathcal{L}\backslash\{\mathbf{0}\}$ with $\|\mathbf{x}\| \leq r \cdot \lambda_1(\mathcal{L})$.*

It will be convenient to define a generalized version of SVP, of which HSVP and SVP are special cases.

Definition 2.17 (η-GSVP). *For a function η mapping lattices to positive real numbers, the η-Generalized Shortest Vector Problem η-GSVP is defined as follows: Given a basis $\mathbf{B}$ for a lattice $\mathcal{L} \subset \mathbb{R}^n$ and a length bound $d \geq \eta(\mathcal{L})$, the goal is to output a vector $\mathbf{x} \in \mathcal{L}\backslash\{\mathbf{0}\}$ with $\|\mathbf{x}\| \leq d$.*

To recover r-SVP or r'-HSVP, we can take $\eta(\mathcal{L}) = r\lambda_1(\mathcal{L})$ or $\eta(\mathcal{L}) = r' \det(\mathcal{L})^{1/n}$ respectively. Below, we will set η to be a new parameter, which in particular will satisfy $\eta(\mathcal{L}) \leq \widetilde{O}(\sqrt{n}) \cdot \min\{\lambda_1(\mathcal{L}), \det(\mathcal{L})^{1/n}\}$.

2.6 Gram-Schmidt Orthogonalization

For any given basis $\mathbf{B} = (\mathbf{b}_1, \ldots, \mathbf{b}_n) \in \mathbb{R}^{m \times n}$, we define the sequence of projections $\pi_i := \pi_{\{\mathbf{b}_1,\ldots,\mathbf{b}_{i-1}\}^\perp}$ where $\pi_{W^\perp}$ refers to the orthogonal projection onto the subspace orthogonal to W. As in [GN08, ALNS20], we use $\mathbf{B}_{[i,j]}$ to denote the projected block $(\pi_i(\mathbf{b}_i), \pi_i(\mathbf{b}_{i+1}), \ldots, \pi_i(\mathbf{b}_j))$.

The Gram-Schmidt orthogonalization (GSO) $\mathbf{B}^* := (\mathbf{b}_1^*, \ldots, \mathbf{b}_n^*)$ of a basis $\mathbf{B}$ is as follows: for all $i \in [1, n], \mathbf{b}_i^* := \pi_i(\mathbf{b}_i) = \mathbf{b}_i - \sum_{j<i} \mu_{i,j} \mathbf{b}_j^*$, where $\mu_{i,j} = \langle \mathbf{b}_i, \mathbf{b}_j^* \rangle / \|\mathbf{b}_j^*\|^2$.

Theorem 2.18 ([GPV08] Lemma 3.1). *For any lattice $\mathcal{L} \subset \mathbb{R}^n$ with basis $\mathbf{B} := (\mathbf{b}_1, \ldots, \mathbf{b}_n)$ and any $\varepsilon \in (0, 1/2)$,*

$$\eta_\varepsilon(\mathcal{L}) \leq \sqrt{\log(n/\varepsilon)} \cdot \max_i \|\mathbf{b}_i^*\| .$$

For $\gamma \geq 1$, a basis is *γ-HKZ-reduced* if for all $i \in \{1, \ldots, n\}$, $\|\mathbf{b}_i^*\| \leq \gamma \cdot \lambda_1(\pi_i(\mathcal{L}))$.

We say that a basis $\mathbf{B}$ is *size-reduced* if it satisfies the following condition: for all $i \neq j$, $|\mu_{i,j}| \leq \frac{1}{2}$. A size-reduced basis $\mathbf{B}$ satisfies that $\|\mathbf{B}\| \leq \sqrt{n}\|\mathbf{B}^*\|$, where $\|\mathbf{B}\|$ is the length of the longest basis vector in $\mathbf{B}$. It is known that we can efficiently transform any basis into a size-reduced basis while maintaining the lattice generated by the basis $\mathcal{L}(\mathbf{B})$ as well as the GSO $\mathbf{B}^*$. We call such operation *size reduction*.

2.7 Some Lattice Algorithms

Theorem 2.19 ([LLL82]). *Given a basis $\mathbf{B} \in \mathbb{Q}^{n \times n}$, there is an algorithm that computes a vector $\mathbf{x} \in \mathcal{L}(\mathbf{B})$ of length at most $2^{n/2} \cdot \lambda_1(\mathcal{L}(\mathbf{B}))$ in polynomial time.*

We will prove a strictly stronger result than the theorem below in the sequel, but this weaker result will still prove useful.

Theorem 2.20 ([ADRS15, GN08]). *There is a $2^{r+o(r)} \cdot \mathrm{poly}(n)$-time algorithm that takes as input a (basis for a) lattice $\mathcal{L} \subset \mathbb{R}^n$ and $2 \leq r \leq n$ and outputs a γ-HKZ-reduced basis for $\mathcal{L}$, where $\gamma := r^{n/r}$.*

Theorem 2.21 ([BLP13]). *There is a probabilistic polynomial-time algorithm that takes as input a basis $\mathbf{B}$ for an n-dimensional lattice $\mathcal{L} \subset \mathbb{R}^n$, a parameter $s \geq \|\mathbf{B}^*\|\sqrt{10 \log n}$ and outputs a vector that is distributed as $\mathcal{D}_{\mathcal{L},s}$, where $\|\mathbf{B}^*\|$ is the length of the longest vector in the Gram-Schmidt orthogonalization of $\mathbf{B}$.*

2.8 Lattice Basis Reduction

LLL reduction. A basis $\mathbf{B} = (\mathbf{b}_1, \ldots, \mathbf{b}_n)$ is *ε-LLL-reduced* [LLL82] for $\varepsilon \in [0, 1]$ if it is a size-reduced basis and for $1 \leq i < n$, the projected block $\mathbf{B}_{[i,i+1]}$ satisfies Lovász's condition: $\|\mathbf{b}_i^*\|^2 \leq (1 + \varepsilon)\|\mu_{i,i-1}\mathbf{b}_{i-1}^* + \mathbf{b}_i^*\|^2$. For $\varepsilon \geq 1/\mathrm{poly}(n)$, an ε-LLL-reduced basis for any given lattice can be computed efficiently.

SVP reduction and its extensions. Let $\mathbf{B} = (\mathbf{b}_1, \ldots, \mathbf{b}_n)$ be a basis of a lattice $\mathcal{L}$ and $\delta \geq 1$ be approximation factors.

We say that $\mathbf{B}$ is *δ-SVP-reduced* if $\|\mathbf{b}_1\| \leq \delta \cdot \lambda_1(\mathcal{L})$. Similarly, we say that $\mathbf{B}$ is *δ-HSVP-reduced* if $\|\mathbf{b}_1\| \leq \delta \cdot \mathrm{vol}(\mathcal{L})^{1/n}$.

$\mathbf{B}$ is *δ-DHSVP-reduced* [GN08, ALNS20] (where D stands for dual) if the reversed dual basis $\mathbf{B}^{-s}$ is δ-HSVP-reduced and it implies that

$$\mathrm{vol}(\mathcal{L})^{1/n} \leq \delta \cdot \|\mathbf{b}_n^*\| \,.$$

Given a δ-(H)SVP oracle on lattices with rank at most n, we can efficiently compute a δ-(H)SVP-reduced basis or a δ-D(H)SVP-reduced basis for any rank n lattice $\mathcal{L} \subseteq \mathbb{Z}^m$. Furthermore, this also applies for a projected block of basis. More specifically, with access to a δ-(H)SVP oracle for lattices with rank at most

k, given any basis $\mathbf{B} = (\mathbf{b}_1, \ldots, \mathbf{b}_n) \in \mathbb{Z}^{m \times n}$ of $\mathcal{L}$ and an index $i \in [1, n-k+1]$, we can efficiently compute a size-reduced basis

$$\mathbf{C} = (\mathbf{b}_1, \ldots, \mathbf{b}_{i-1}, \mathbf{c}_i, \ldots, \mathbf{c}_{i+k-1}, \mathbf{b}_{i+k}, \ldots, \mathbf{b}_n)$$

such that $\mathbf{C}$ is a basis for $\mathcal{L}$ and the projected block $\mathbf{C}_{[i,i+k-1]}$ is δ-(H)SVP-reduced or δ-D(H)SVP reduced. Moreover, we note the following:

- If $\mathbf{C}_{[i,i+k-1]}$ is δ-(H)SVP-reduced, the procedures in [GN08, MW16] equipped with δ-(H)SVP-oracle ensure that $\|\mathbf{C}^*\| \le \|\mathbf{B}^*\|$;
- If $\mathbf{C}_{[i,i+k-1]}$ is δ-D(H)SVP-reduced, the inherent LLL reduction implies $\|\mathbf{C}^*\| \le 2^k \|\mathbf{B}^*\|$. Indeed, the GSO of $\mathbf{C}_{[i,i+k-1]}$ satisfies

$$\|(\mathbf{C}_{[i,i+k-1]})^*\| \le 2^{k/2} \lambda_k(\mathcal{L}(\mathbf{C}_{[i,i+k-1]}))$$

(by [LLL82, p. 518, Line 27]) and $\lambda_k(\mathcal{L}(\mathbf{C}_{[i,i+k-1]})) \le \sqrt{k}\|\mathbf{B}^*\|$. Here, $\lambda_k(\cdot)$ denotes the k-th minimum.

Therefore, with size reduction, performing $\mathrm{poly}(n, \log \|\mathbf{B}\|)$ many such operations will increase $\|\mathbf{B}^*\|$ and hence $\|\mathbf{B}\|$ by at most a factor of $2^{\mathrm{poly}(n, \log \|B\|)}$. If the number of operations is bounded by $\mathrm{poly}(n, \log \|\mathbf{B}\|)$, all intermediate steps and the total running time (excluding oracle queries) will be polynomial in the initial input size; Details can be found in e.g., [GN08, LN14]. Hence, we will focus on bounding the number of calls to such block reduction subprocedures when we analyze the running time of basis reduction algorithms.

Twin reduction. The following notion of twin reduction and the subsequent fact comes from [GN08, ALNS20].

A basis $\mathbf{B} = (\mathbf{b}_1, \ldots, \mathbf{b}_{d+1})$ is *δ-twin-reduced* if $\mathbf{B}_{[1,d]}$ is *δ-HSVP-reduced* and $\mathbf{B}_{[2,d+1]}$ is *δ-DHSVP-reduced.*

Fact 2.22. *If* $\mathbf{B} := (\mathbf{b}_1, \ldots, \mathbf{b}_{d+1}) \in \mathbb{R}^{m \times (d+1)}$ *is δ-twin-reduced, then*

$$\|\mathbf{b}_1\| \le \delta^{2d/(d-1)} \|\mathbf{b}^*_{d+1}\| \,. \tag{4}$$

2.9 The DBKZ Algorithm

We augment Micciancio and Walter's elegant DBKZ algorithm [MW16] with a δ_H-HSVP-oracle instead of an SVP-oracle since the SVP-oracle is used as a $\sqrt{\gamma_k}$-HSVP oracle everywhere in their algorithm. See [ALNS20] for a high-level sketch of the proof.

Theorem 2.23. *For integers $n > k \ge 2$, an approximation factor $1 \le \delta_H \le 2^k$, an input basis $\mathbf{B}_0 \in \mathbb{Z}^{m \times n}$ for a lattice $\mathcal{L} \subseteq \mathbb{Z}^m$, and $N := \lceil (2n^2/(k-1)^2) \cdot \log(n \log(5\|\mathbf{B}_0\|)/\varepsilon) \rceil$ for some $\varepsilon \in [2^{-\mathrm{poly}(n)}, 1]$, Algorithm 1 outputs a basis $\mathbf{B}$ of $\mathcal{L}$ in polynomial time (excluding oracle queries) such that*

$$\|\mathbf{b}_1\| \le (1+\varepsilon) \cdot (\delta_H)^{\frac{n-1}{(k-1)}} \mathrm{vol}(\mathcal{L})^{1/n} ,$$

by making $N \cdot (2n - 2k + 1) + 1$ calls to the δ_H-HSVP oracle for lattices with rank k.

Algorithm 1. The Micciancio-Walter DBKZ algorithm [MW16, Algorithm 1]

Input: A block size $k \geq 2$, number of tours N, a basis $\mathbf{B} = (\mathbf{b}_1, \cdots, \mathbf{b}_n) \in \mathbb{Z}^{m \times n}$, and access to a δ_H-HSVP oracle for lattices with rank k.
Output: A new basis of $\mathcal{L}(\mathbf{B})$.
1: **for** $\ell = 1$ **to** N **do**
2: **for** $i = 1$ **to** $n - k$ **do**
3: δ_H-HSVP-reduce $\mathbf{B}_{[i,i+k-1]}$.
4: **end for**
5: **for** $j = n - k + 1$ **to** 1 **do**
6: δ_H-DHSVP-reduce $\mathbf{B}_{[j,j+k-1]}$
7: **end for**
8: **end for**
9: δ_H-HSVP-reduce $\mathbf{B}_{[1,k]}$.
10: **return B**.

3 Smooth Sublattices and $\overline{\eta}_\varepsilon(\mathcal{L})$

The analysis of our algorithm relies on the existence of a *smooth sublattice* $\mathcal{L}' \subseteq \mathcal{L}$ of our input lattice $\mathcal{L} \subset \mathbb{R}^n$, i.e., a sublattice $\mathcal{L}'$ such that $\eta_\varepsilon(\mathcal{L}')$ is small (relative to, say, $\lambda_1(\mathcal{L})$ or $\det(\mathcal{L})^{1/n}$). To that end, for $\varepsilon > 0$ and a lattice $\mathcal{L} \subset \mathbb{R}^n$, we define

$$\overline{\eta}_\varepsilon(\mathcal{L}) := \min_{\mathcal{L}' \subseteq \mathcal{L}} \eta_\varepsilon(\mathcal{L}') ,$$

where the minimum is taken over all sublattices $\mathcal{L}' \subseteq \mathcal{L}$. (It is not hard to see that the minimum is in fact achieved. Notice that any minimizer $\mathcal{L}'$ must be a primitive sublattice, i.e., $\mathcal{L}' = \mathcal{L} \cap \operatorname{span}(\mathcal{L}')$.)

We will now prove that $\overline{\eta}_\varepsilon(\mathcal{L})$ is bounded both in terms of $\lambda_1(\mathcal{L})$ and $\det(\mathcal{L})$.

Lemma 3.1. *For any lattice $\mathcal{L} \subset \mathbb{R}^n$ and any $\varepsilon \in (0, 1/2)$,*

$$\lambda_1(\mathcal{L})/\sqrt{n} \leq \overline{\eta}_\varepsilon(\mathcal{L}) \leq \sqrt{\log(1/\varepsilon)} \cdot \min\{\lambda_1(\mathcal{L}), 10(\log n + 2)\det(\mathcal{L})^{1/n}\} .$$

The bounds in terms of $\lambda_1(\mathcal{L})$ are more-or-less trivial. The bound $\overline{\eta}_\varepsilon(\mathcal{L}) \lesssim \sqrt{\log(1/\varepsilon)\log n}\det(\mathcal{L})^{1/n}$ follows from the main result in [RS17] (originally conjectured by Dadush [DR16]), which is called a "reverse Minkowski theorem" and which we present below. (In fact, Lemma 3.1 is essentially equivalent to the main result in [RS17].)

Definition 3.2. *A lattice $\mathcal{L} \subset \mathbb{R}^n$ is a* ***stable*** *lattice if $\det(\mathcal{L}) = 1$ and $\det(\mathcal{L}') \geq 1$ for all lattices $\mathcal{L}' \subseteq \mathcal{L}$.*

Theorem 3.3 ([RS17]). *For any stable lattice $\mathcal{L} \subset \mathbb{R}^n$, $\eta_{1/2}(\mathcal{L}) \leq 10(\log n + 2)$.*

Proof of Lemma 3.1. The lower bound on $\overline{\eta}_\varepsilon(\mathcal{L})$ follows immediately from Theorem 2.9 together with the fact that $\lambda_1(\mathcal{L}) \leq \lambda_1(\mathcal{L}') \leq \lambda_n(\mathcal{L}')$ for any sublattice $\mathcal{L}' \subseteq \mathcal{L}$. The bound $\overline{\eta}_\varepsilon(\mathcal{L}) \leq \sqrt{\log(1/\varepsilon)} \cdot \lambda_1(\mathcal{L})$ is immediate from Claim 2.7 applied to the one-dimensional lattice $\mathbb{Z}\mathbf{v}$ generated by $\mathbf{v} \in \mathcal{L}$ with $\|\mathbf{v}\| = \lambda_1(\mathcal{L})$.

So, we only need to prove that $\overline{\eta}_{1/2}(\mathcal{L}) \leq 10(\log n + 2)\det(\mathcal{L})^{1/n}$. The result for all $\varepsilon \in (0, 1/2)$ then follows from Corollary 2.13.

We prove this by induction on n. The result is trivial for $n = 1$. (Indeed, for $n = 1$ we have $\det(\mathcal{L})^{1/n} = \lambda_1(\mathcal{L})$.) For $n > 1$, we first assume without loss of generality that $\det(\mathcal{L}) = 1$. If $\mathcal{L} \subset \mathbb{R}^n$ is stable, then the result follows immediately from Theorem 3.3. Otherwise, there exists a sublattice $\mathcal{L}' \subset \mathcal{L}$ such that $\det(\mathcal{L}') < 1$. Notice that $k := \mathrm{rank}(\mathcal{L}') < n$. Therefore, by the induction hypothesis, $\overline{\eta}_{1/2}(\mathcal{L}') \leq 10(\log k + 2)\det(\mathcal{L}')^{1/k} < 10(\log n + 2)$. The result then follows from the fact that $\overline{\eta}_\varepsilon(\mathcal{L}) \leq \overline{\eta}_\varepsilon(\mathcal{L}')$ for any sublattice $\mathcal{L}' \subseteq \mathcal{L}$. □

3.1 Sampling with Parameter $\mathrm{poly}(n) \cdot \overline{\eta}_\varepsilon(\mathcal{L})$

Lemma 3.4. *For any lattice $\mathcal{L} \subset \mathbb{R}^n$, $\gamma \geq 1$, $\varepsilon \in (0, 1/2)$, γ-HKZ-reduced basis $\mathbf{B} = (\mathbf{b}_1, \ldots, \mathbf{b}_n)$ of $\mathcal{L}$, $\varepsilon \in (0, 1/2)$, and index $i \in \{2, \ldots, n\}$ such that*

$$\|\mathbf{b}_i^*\| > \gamma\sqrt{n} \cdot \overline{\eta}_\varepsilon(\mathcal{L}) ,$$

we have $\overline{\eta}_\varepsilon(\mathcal{L}(\mathbf{b}_1, \ldots, \mathbf{b}_{i-1})) = \overline{\eta}_\varepsilon(\mathcal{L})$.

Proof. Suppose that $\mathcal{L}' \subseteq \mathcal{L}$ satisfies $\eta_\varepsilon(\mathcal{L}') = \overline{\eta}_\varepsilon(\mathcal{L}) < \|\mathbf{b}_i^*\|/(\gamma\sqrt{n})$ with $k := \mathrm{rank}(\mathcal{L}')$. We wish to show that $\mathcal{L}' \subseteq \mathcal{L}(\mathbf{b}_1, \ldots, \mathbf{b}_{i-1})$, or equivalently, that $\pi_i(\mathcal{L}') = \{\mathbf{0}\}$. Indeed, by Theorem 2.9, $\lambda_k(\mathcal{L}') \leq \sqrt{k} \cdot \eta_\varepsilon(\mathcal{L}') \leq \sqrt{n} \cdot \overline{\eta}_\varepsilon(\mathcal{L})$. In particular, there exist $\mathbf{v}_1, \ldots, \mathbf{v}_k \in \mathcal{L}'$ with $\mathrm{span}(\mathbf{v}_1, \ldots, \mathbf{v}_k) = \mathrm{span}(\mathcal{L}')$ and

$$\|\pi_i(\mathbf{v}_j)\| \leq \|\mathbf{v}_j\| \leq \lambda_k(\mathcal{L}') \leq \sqrt{n} \cdot \overline{\eta}_\varepsilon(\mathcal{L}) < \|\mathbf{b}_i^*\|/\gamma$$

for all $j \in \{1, \ldots, k\}$. Therefore, if $\pi_i(\mathbf{v}_j) \neq \mathbf{0}$. Then, $\pi_i(\mathbf{v}_j) \in \pi_i(\mathcal{L})$ is a non-zero vector with norm strictly less than $\|\mathbf{b}_i^*\|/\gamma$, which implies that $\lambda_1(\pi_i(\mathcal{L})) < \|\mathbf{b}_i^*\|/\gamma$, contradicting the assumption that $\mathbf{B}$ is a γ-HKZ basis. Therefore, $\pi_i(\mathbf{v}_j) = \mathbf{0}$ for all j, which implies that $\pi_i(\mathcal{L}') = \{\mathbf{0}\}$, i.e., $\mathcal{L}' \subseteq \mathcal{L}(\mathbf{b}_1, \ldots, \mathbf{b}_{i-1})$, as needed. □

Proposition 3.5. *There is a $(2^{r+o(r)} + M) \cdot \mathrm{poly}(n, \log M)$-time algorithm that takes as input a (basis for a) lattice $\mathcal{L} \subset \mathbb{R}^n$, $2 \leq r \leq n$, an integer $M \geq 1$, and a parameter*

$$s \geq r^{n/r}\sqrt{n \log n} \cdot \overline{\eta}_\varepsilon(\mathcal{L})$$

for some $\varepsilon \in (0, 1/2)$ and outputs a (basis for a) sublattice $\widehat{\mathcal{L}} \subseteq \mathcal{L}$ with $\overline{\eta}_\varepsilon(\widehat{\mathcal{L}}) = \overline{\eta}_\varepsilon(\mathcal{L})$ and $\mathbf{X}_1, \ldots, \mathbf{X}_M \in \widehat{\mathcal{L}}$ that are sampled independently from $D_{\widehat{\mathcal{L}},s}$.

Proof. The algorithm takes as input a (basis for a) lattice $\mathcal{L} \subset \mathbb{R}^n$, $2 \leq r \leq n$, $M \geq 1$, and a parameter $s > 0$ and behaves as follows. It first uses the procedure from Theorem 2.20 to compute a γ-HKZ reduced basis $\mathbf{b}_1, \ldots, \mathbf{b}_n$, where $\gamma := r^{n/r}$. Let $i \in \{1, \ldots, n\}$ be maximal such that $\|\mathbf{b}_j^*\| \leq s/\sqrt{\log n}$ for all $j \leq i$, and let $\widehat{\mathcal{L}} := \mathcal{L}(\mathbf{b}_1, \ldots, \mathbf{b}_i)$. (If no such i exists, the algorithm simply fails.) The algorithm then runs the procedure from Theorem 2.21 repeatedly to sample $\mathbf{X}_1, \ldots, \mathbf{X}_M \sim D_{\widehat{\mathcal{L}},s}$ and outputs $\widehat{\mathcal{L}}$ and $\mathbf{X}_1, \ldots, \mathbf{X}_M$.

The running time of the algorithm is clearly $(2^r + M) \cdot \mathrm{poly}(n, \log M)$. By Theorem 2.21, the $\mathbf{X}_i$ have the correct distribution. Notice that, if the algorithm fails, then

$$\|\mathbf{b}_1\| > s/\sqrt{\log n} \geq \gamma\sqrt{n} \cdot \overline{\eta}_\varepsilon(\mathcal{L}) .$$

Recalling that $\|\mathbf{b}_1\| \leq \gamma\lambda_1(\mathcal{L})$, it follows that $\sqrt{n}\overline{\eta}_\varepsilon(\mathcal{L}) < \lambda_1(\mathcal{L})$, which contradicts Lemma 3.1. So, the algorithm never fails (provided that the promise on s holds).

It remains to show that $\overline{\eta}_\varepsilon(\mathcal{L}) = \overline{\eta}_\varepsilon(\mathcal{L}(\mathbf{b}_1, \ldots, \mathbf{b}_i))$. If $i = n$, then this is trivial. Otherwise, $i \in \{1, \ldots, n-1\}$, and we have

$$\|\mathbf{b}_{i+1}^*\| > s/\sqrt{\log n} \geq \gamma\sqrt{n} \cdot \overline{\eta}_\varepsilon(\mathcal{L}) .$$

The result follows immediately from Lemma 3.4. □

4 An Approximation Algorithm for HSVP and SVP

In this section, we present our algorithm that solves $\widetilde{O}(\sqrt{n})$-HSVP and $\widetilde{O}(\sqrt{n})$-SVP in $2^{n/2+o(n)}$ time. More precisely, we provide a detailed analysis of a simple "pair-and-sum" algorithm, which will solve $O(\sqrt{n}) \cdot \overline{\eta}_\varepsilon(\mathcal{L})$-GSVP for $\varepsilon = 1/\mathrm{poly}(n)$. This in particular yields an algorithm that simultaneously solves $\widetilde{O}(\sqrt{n})$-SVP and $\widetilde{O}(\sqrt{n})$-HSVP.

4.1 Mixtures of Gaussians

We will be working with random variables $\mathbf{X}$ that are "mixtures" of discrete Gaussians, i.e., random variables that can be written as $D_{\mathcal{L}+\mathbf{C},s}$ for some lattice $\mathcal{L} \subset \mathbb{R}^n$, parameter $s > 0$, and random variable $\mathbf{C} \in \mathbb{R}^n$. In other words, $\mathbf{X}$ can be sampled by first sampling $\mathbf{C} \in \mathbb{R}^n$ from some arbitrary distribution and then sampling $\mathbf{X}$ from $D_{\mathcal{L}+\mathbf{C},s}$. E.g., the discrete Gaussian $D_{\mathcal{L},s}$ itself is such a distribution, as is the discrete Gaussian $D_{\widehat{\mathcal{L}},s}$ for any superlattice $\widehat{\mathcal{L}} \supseteq \mathcal{L}$. Indeed, in our applications, we will always have $\mathbf{C} \in \widehat{\mathcal{L}}$ for some superlattice $\widehat{\mathcal{L}} \supseteq \mathcal{L}$, and we will initialize our algorithm with samples from $D_{\widehat{\mathcal{L}},s}$.

Our formal definition below is a bit technical, since we must consider the joint distribution of many such random variables that are only δ-similar to these distributions and satisfy a certain independence property. In particular, we will work with $\mathbf{X}_1, \ldots, \mathbf{X}_M$ such that each $\mathbf{X}_i$ is δ-similar to $\mathbf{Y}_i \sim D_{\mathcal{L}+\mathbf{C}_i,s}$, where $\mathbf{C}_i$ is an arbitrary random variable (that might depend on the $\mathbf{X}_j$) but once $\mathbf{C}_i$ is fixed, $\mathbf{Y}_i$ is sampled from $D_{\mathcal{L}+\mathbf{C}_i,s}$ independently of everything else. Here and below, we adopt the convention that $\Pr[A \mid B] = 0$ whenever $\Pr[B] = 0$, i.e., all probabilities are zero when conditioned on events with probability zero.

Definition 4.1. *For (discrete) random variables $\mathbf{X}_1, \ldots, \mathbf{X}_m \in \mathbb{R}^n$ and $i \in \{1, \ldots, m\}$, let us define the tuple of random variables*

$$\mathbf{X}_{-i} := (\mathbf{X}_1, \ldots, \mathbf{X}_{i-1}, \mathbf{X}_{i+1}, \ldots, \mathbf{X}_m) \in \mathbb{R}^{(m-1)n} .$$

We say that $\mathbf{X}_1, \ldots, \mathbf{X}_m$ *are* δ*-similar to a mixture of independent Gaussians over* $\mathcal{L}$ *with parameter* $s > 0$ *if for any* $i \in \{1, \ldots, m\}$, $\mathbf{y} \in \mathbb{R}^n$, *and* $\mathbf{w} \in \mathbb{R}^{(m-1)n}$,

$$\Pr[\mathbf{X}_i = \mathbf{y} \mid \mathbf{X}_{-i} = \mathbf{w}] = e^{\pm\delta} \cdot \frac{\rho_s(\mathbf{y})}{\rho_s(\mathcal{L} + \mathbf{y})} \cdot \Pr[\mathbf{X}_i \in \mathcal{L} + \mathbf{y} \mid \mathbf{X}_{-i} = \mathbf{w}] .$$

Additionally we will need the distribution we obtain at every step to be symmetric about the origin as defined below.

Definition 4.2. *We say that a list of (discrete) random variables* $\mathbf{X}_1, \ldots, \mathbf{X}_m \in \mathbb{R}^n$ *is symmetric if for any* $i \in \{1, \ldots, m\}$, *any* $\mathbf{y} \in \mathbb{R}^n$, *and any* $\mathbf{w} \in \mathbb{R}^{(m-1)n}$,

$$\Pr[\mathbf{X}_i = \mathbf{y} \mid \mathbf{X}_{-i} = \mathbf{w}] = \Pr[\mathbf{X}_i = -\mathbf{y} \mid \mathbf{X}_{-i} = \mathbf{w}] .$$

We need the following simple lemma that bounds the probability of $\mathbf{X}$ being $\mathbf{0}$, where $\mathbf{X}$ is distributed as a mixture of discrete Gaussians over $\mathcal{L}$.

Lemma 4.3. *For any lattice* $\mathcal{L} \subset \mathbb{R}^n$, *let* $\mathbf{X}_1, \ldots, \mathbf{X}_m \in \mathcal{L}$ *be* δ*-similar to a mixture of independent Gaussians over* $\mathcal{L}$ *with parameter* $s \geq \beta\eta_{1/2}(\mathcal{L})$ *for some* $\beta > 1$. *Then, for any* i, *and any* $\mathbf{w} \in \mathbb{R}^{(m-1)n}$

$$\Pr[\mathbf{X}_i = \mathbf{0} \mid \mathbf{X}_{-i} = \mathbf{w}] \leq \frac{3e^\delta}{2\beta} .$$

Proof. Let $s' := \eta_{1/2}(\mathcal{L})$. We have that

$$\Pr[\mathbf{X}_i = \mathbf{0} \mid \mathbf{X}_{-i} = \mathbf{w}] \leq \Pr[\mathbf{X}_i = \mathbf{0} \mid \mathbf{X}_i \in \mathcal{L},\ \mathbf{X}_{-i} = \mathbf{w}] \leq \frac{e^\delta}{\rho_s(\mathcal{L})} \leq e^\delta \cdot \frac{\rho_{s'}(\mathcal{L})}{\rho_s(\mathcal{L})} .$$

The result then follows from Claim 2.10. □

The following corollary shows that a mixture of discrete Gaussians must contain a short non-zero vector in certain cases.

Corollary 4.4. *For any lattices* $\mathcal{L}' \subseteq \mathcal{L} \subset \mathbb{R}^n$, *parameter* $s \geq 10e^\delta\eta_{1/2}(\mathcal{L}')$, $m \geq 100$, *and random variables* $\mathbf{X}_1, \ldots, \mathbf{X}_m$ *that are* δ*-similar to mixtures of independent Gaussians over* $\mathcal{L}'$ *with parameter* s,

$$\Pr[\exists i \in [1, m] \text{ such that } 0 < \|\mathbf{X}_i\|^2 < 4T] \geq 1/10 ,$$

where $T := \frac{1}{m}\sum_{i=1}^m \mathbb{E}[\|\mathbf{X}_i\|^2]$.

Proof. By Markov's inequality, we have

$$\Pr\Big[\sum_{i=1}^m \|\mathbf{X}_i\|^2 \geq 2mT\Big] \leq \frac{1}{2} .$$

Hence, with probability at least $\frac{1}{2}$, we have $\sum_{i=1}^m \|\mathbf{X}_i\|^2 < 2mT$.

We next note that many of the $\mathbf{X}_i$ must be non-zero with high probability. Let $Y_1, \ldots, Y_m \in \{0, 1\}$ such that $Y_i = 0$ if and only if $\mathbf{X}_i = \mathbf{0}$. By Lemma 4.3,

$$\mathbb{E}[Y_i \mid Y_1 = y_1, \ldots, Y_{i-1} = y_{i-1}] \geq 4/5$$

for any $y_1, \ldots, y_{i-1} \in \{0, 1\}$. By Corollary 2.3, we have that

$$\Pr[Y_1 + \cdots + Y_m \leq 3m/5] \leq e^{-m/100} \leq 1/e \; .$$

Finally, by union bound, we see that with probability at least $1 - 1/e - 1/2 > 1/10$ the average squared norm will be at most $2T$ and more than half of the $\mathbf{X}_i$ will be non-zero. It follows from another application of Markov's inequality that at least one of the non-zero $\mathbf{X}_i$ must have squared norm less than $4T$. □

4.2 Summing Vectors

Our algorithm will start with vectors $\mathbf{X}_1, \ldots, \mathbf{X}_m \in \mathcal{L}_0$, where $\mathcal{L}_0 \subset \mathcal{L}$ is some very dense superlattice of the input lattice $\mathcal{L}$. It then takes sums $\mathbf{Y}_k = \mathbf{X}_i + \mathbf{X}_j$ of pairs of these in such a way that the resulting $\mathbf{Y}_k$ lie in some appropriate sublattice $\mathcal{L}_1 \subset \mathcal{L}_0$, i.e., $\mathbf{Y}_k \in \mathcal{L}_1$. It does this repeatedly, finding vectors in $\mathcal{L}_2, \mathcal{L}_3, \ldots, \mathcal{L}_\ell$ until finally it obtains vectors in $\mathcal{L}_\ell := \mathcal{L}$.

Here, we study a single step of this algorithm, as shown below.

Algorithm 2. One step of the algorithm.

Input: Lattices $\mathcal{L}_0, \mathcal{L}_1 \subset \mathbb{R}^n$ with $2\mathcal{L}_0 \subseteq \mathcal{L}_1 \subseteq \mathcal{L}_0$, and lattice vectors $\mathbf{X}_1, \ldots, \mathbf{X}_m \in \mathcal{L}_0$ with $m \geq 2|\mathcal{L}_0/\mathcal{L}_1|$.
Output: Lattice vectors $\mathbf{Y}_1, \ldots, \mathbf{Y}_M \in \mathcal{L}_1$, with $M := \lceil (m - |\mathcal{L}_0/\mathcal{L}_1|)/2 \rceil$.

1: Set $\mathsf{USED}_i :=$ **false** for $i = 1, \ldots, m$, $k = 1$, and $i = 1$.
2: **while** $k \leq M$ **do**
3:   **if not** USED_i **and** $(\exists j \in \{1, \ldots, m\} \setminus \{i\}$ such that $\mathbf{X}_j \equiv \mathbf{X}_i \bmod \mathcal{L}_1$ **and** $\mathsf{USED}_j =$ **false**$)$ **then**
4:     Let $j \neq i$ be minimal such that $\mathbf{X}_j \equiv \mathbf{X}_i \bmod \mathcal{L}_1$ **and** $\mathsf{USED}_j =$ **false**.
5:     Set $\mathbf{Y}_k = \mathbf{X}_i + \mathbf{X}_j$.
6:     Set $\mathsf{USED}_i = \mathsf{USED}_j =$ **true** and increment k.
7:   **end if**
8:   Increment i.
9: **end while**
10: **return** $\mathbf{Y}_1, \ldots, \mathbf{Y}_M$

Notice that Algorithm 2 can be implemented in time $m \cdot \mathrm{poly}(n, \log m)$. This can be done, e.g., by creating a table of the $\mathbf{X}_i$ sorted according to $\mathbf{X}_i \bmod \mathcal{L}_1$. Then, for each i, such a j can be found (if it exists) by performing binary search on the table. Furthermore, the algorithm is guaranteed to find $M = \lceil (m - |\mathcal{L}_0/\mathcal{L}_1|)/2 \rceil$ output vectors because at most $|\mathcal{L}_0/\mathcal{L}_1|$ of the input vectors can be unpaired.

The key property that we will need from Algorithm 2 is that for any (possibly unknown) sublattice $\mathcal{L}' \subseteq \mathcal{L}_1 \subseteq \mathcal{L}_0$, the algorithm maps mixtures of Gaussians over $\mathcal{L}'$ to mixtures of Gaussians over $\mathcal{L}'$, provided that the parameter s is larger than $\eta_\varepsilon(\mathcal{L}')$ by a factor of $\sqrt{2}$. In other words, as long as there exists some sublattice $\mathcal{L}' \subseteq \mathcal{L}_1$ such that $\eta_\varepsilon(\mathcal{L}') \lesssim s$, then the output of the algorithm will be a mixture of Gaussians. Indeed, this is more-or-less immediate from Lemma 2.14.

Lemma 4.5. *For any lattices $\mathcal{L}_0, \mathcal{L}_1, \mathcal{L}' \subset \mathbb{R}^n$ with $2\mathcal{L}_0 \subseteq \mathcal{L}_1 \subseteq \mathcal{L}_0$ and $\mathcal{L}' \subseteq \mathcal{L}_1$, $\varepsilon \in (0, 1/3)$, $\delta > 0$, and parameter $s \geq \sqrt{2}\eta_\varepsilon(\mathcal{L}')$, if the input vectors $\mathbf{X}_1, \ldots, \mathbf{X}_m \in \mathcal{L}_0$ are sampled from the distribution that is δ-similar to a mixture of independent Gaussians over $\mathcal{L}'$ with parameter s, then the output vectors $\mathbf{Y}_1, \ldots, \mathbf{Y}_M \in \mathcal{L}_1$ are $(2\delta + 3\varepsilon)$-similar to a mixture of independent Gaussians over $\mathcal{L}'$ with parameter $\sqrt{2}s$.*

Proof. For a list of cosets $\mathbf{d} := (\mathbf{c}_1, \ldots, \mathbf{c}_m) \in (\mathcal{L}_0/\mathcal{L}')^m$ such that $\Pr[\mathbf{X}_1 = \mathbf{c}_1 \bmod \mathcal{L}', \ldots, \mathbf{X}_m = \mathbf{c}_m \bmod \mathcal{L}']$ is non-zero, let $\mathbf{Y}_{\mathbf{d},1}, \ldots, \mathbf{Y}_{\mathbf{d},M}$ be the random variables obtained by taking $\mathbf{Y}_1, \ldots, \mathbf{Y}_M$ conditioned on $\mathbf{X}_i \equiv \mathbf{c}_i \bmod \mathcal{L}'$ for all i. We similarly define $\mathbf{X}_{\mathbf{d},i}$. Notice that $\mathbf{Y}_1, \ldots, \mathbf{Y}_M$ is a convex combination of random variables of the form $\mathbf{Y}_{\mathbf{d},1}, \ldots, \mathbf{Y}_{\mathbf{d},M}$, and that the property of being close to a mixture of independent Gaussians is preserved by taking convex combinations. Therefore, it suffices to prove the statement for $\mathbf{Y}_{\mathbf{d},1}, \ldots, \mathbf{Y}_{\mathbf{d},M}$ for all fixed $\mathbf{d}$.

To that end, fix $k \in \{1, \ldots, M\}$ and such a $\mathbf{d} \in (\mathcal{L}_0/\mathcal{L}')^m$. Notice that $\mathbf{X}_{\mathbf{d},i} \in \mathcal{L}' + \mathbf{c}_i \subseteq \mathcal{L}_1 + \mathbf{c}_i$. Therefore, there exist fixed i, j such that $\mathbf{Y}_{\mathbf{d},k} = \mathbf{X}_{\mathbf{d},i} + \mathbf{X}_{\mathbf{d},j}$. Furthermore, by assumption, for any $\mathbf{w} \in \mathcal{L}_0^{m-1}$ and $\mathbf{x} \in \mathcal{L}_0$,

$$\Pr[\mathbf{X}_{\mathbf{d},i} = \mathbf{x} \mid \mathbf{X}_{\mathbf{d},-i} = \mathbf{w}] = e^{\pm\delta} \frac{\rho_s(\mathbf{x})}{\rho_s(\mathcal{L}' + \mathbf{c}_i)} ,$$

and likewise for j. It follows from Lemma 2.14 that for any $\mathbf{y} \in \mathcal{L}_1$ and $\mathbf{z} \in \mathcal{L}_1^{M-1}$,

$$\Pr[\mathbf{X}_{\mathbf{d},i} + \mathbf{X}_{\mathbf{d}_j} = \mathbf{y} \mid \mathbf{Y}_{\mathbf{d},-k} = \mathbf{z}] = e^{\pm(2\delta+3\varepsilon)} \frac{\rho_{\sqrt{2}s}(\mathbf{y})}{\rho_{\sqrt{2}s}(\mathcal{L}' + \mathbf{c}_i + \mathbf{c}_j)} ,$$

as needed. □

Lemma 4.6. *For any lattices $\mathcal{L}_0, \mathcal{L}_1 \subset \mathbb{R}^n$ with $2\mathcal{L}_0 \subseteq \mathcal{L}_1 \subseteq \mathcal{L}_0$, if the input vectors $\mathbf{X}_1, \ldots, \mathbf{X}_m \in \mathcal{L}_0$ are sampled from a symmetric distribution, then the distribution of the output vectors $\mathbf{Y}_1, \ldots, \mathbf{Y}_M$ will also be symmetric. Furthermore,*

$$\sum \mathbb{E}[\|\mathbf{Y}_k\|^2] \leq \sum \mathbb{E}[\|\mathbf{X}_i\|^2] .$$

Proof. Let $\mathbf{d} = (\mathbf{c}_1, \ldots, \mathbf{c}_m) \in (\mathcal{L}_0/\mathcal{L}_1)^m$ be a list of cosets such that with non-zero probability we have $\mathbf{X}_1 \in \mathcal{L}_1 + \mathbf{c}_1, \ldots, \mathbf{X}_m \in \mathcal{L}_1 + \mathbf{c}_m$. Let $\mathbf{X}_{\mathbf{d},1}, \ldots, \mathbf{X}_{\mathbf{d},m}$ be the distribution obtained by sampling the $\mathbf{X}_i$ conditioned on this event, and let $\mathbf{Y}_{\mathbf{d},1}, \ldots, \mathbf{Y}_{\mathbf{d},M}$ be the corresponding output.

Notice that the distribution of $\mathbf{X}_{\mathbf{d},1}, \ldots, \mathbf{X}_{\mathbf{d},m}$ is also symmetric, since $\mathcal{L}_1 + \mathbf{c} = -(\mathcal{L}_1 + \mathbf{c})$ for any $\mathbf{c} \in \mathcal{L}_0/\mathcal{L}_1$. (Here, we have used the fact that $2\mathcal{L}_0 \subseteq \mathcal{L}_1 \subseteq \mathcal{L}_0$.)

And, for fixed $\mathbf{d}$ and $k \in \{1, \ldots, M\}$ there exist fixed (distinct) $i, j \in \{1, \ldots, m\}$ such that

$$\mathbf{Y}_{\mathbf{d},k} = \mathbf{X}_{\mathbf{d},i} + \mathbf{X}_{\mathbf{d},j} .$$

But, since the $\mathbf{X}_{\mathbf{d},1}, \ldots, \mathbf{X}_{\mathbf{d},m}$ are distributed symmetrically, we see immediately that for any $\mathbf{y} \in \mathcal{L}_1$ and $\mathbf{w} \in \mathcal{L}_1^{M-1}$,

$$\Pr[\mathbf{Y}_{\mathbf{d},k} = \mathbf{y} \mid \mathbf{Y}_{\mathbf{d},-k} = \mathbf{w}] = \Pr[\mathbf{Y}_{\mathbf{d},k} = -\mathbf{y} \mid \mathbf{Y}_{\mathbf{d},-k} = \mathbf{w}] .$$

In other words, the distribution of $\mathbf{Y}_{\mathbf{d},1}, \ldots, \mathbf{Y}_{\mathbf{d},M}$ is symmetric.

Furthermore, $\mathbb{E}[\|\mathbf{X}_{\mathbf{d},i} + \mathbf{X}_{\mathbf{d},j}\|^2]$ is equal to

$$\mathbb{E}[\|\mathbf{X}_{\mathbf{d},i}\|^2] + \mathbb{E}[\|\mathbf{X}_{\mathbf{d},j}\|^2] + 2\,\mathbb{E}[\langle \mathbf{X}_i, \mathbf{X}_j \rangle] = \mathbb{E}[\|\mathbf{X}_{\mathbf{d},i}\|^2] + \mathbb{E}[\|\mathbf{X}_{\mathbf{d},j}\|^2] ,$$

where in the last step we have used the symmetry of $\mathbf{X}_{\mathbf{d},1}, \ldots, \mathbf{X}_{\mathbf{d},m}$. Since the $\mathbf{Y}_{\mathbf{d},k}$ are sums of disjoint pairs of the $\mathbf{X}_{\mathbf{d},i}$, it follows immediately that

$$\sum_{k=1}^{M} \mathbb{E}[\|\mathbf{Y}_{\mathbf{d},k}\|^2] \le \sum_{i=1}^{m} \mathbb{E}[\|\mathbf{X}_{\mathbf{d},i}\|^2] .$$

The results for $\mathbf{X}_1, \ldots, \mathbf{X}_m, \mathbf{Y}_1, \ldots, \mathbf{Y}_M$ then follow immediately from the fact that this distribution can be written as a convex combination of the vectors $\mathbf{X}_{\mathbf{d},1}, \ldots, \mathbf{X}_{\mathbf{d},m}, \mathbf{Y}_{\mathbf{d},1}, \ldots, \mathbf{Y}_{\mathbf{d},M}$ for different coset lists $\mathbf{d} \in (\mathcal{L}_0/\mathcal{L}_1)^m$, since both symmetry and the inequality on expectations are preserved by convex combinations. □

4.3 A Tower of Lattices

We will repeatedly apply Algorithm 2 on a "tower" of lattices similar to [ADRS15]. We use (a slight modification of) the definition and construction of the tower of lattices from [ADRS15].

Definition 4.7 ([ADRS15]). *For an integer α satisfying $n/2 \le \alpha \le n$, we say that $(\mathcal{L}_0, \ldots, \mathcal{L}_\ell)$ is a tower of lattices in $\mathbb{R}^n$ of index 2^α if for all i we have $2\mathcal{L}_{i-1} \subseteq \mathcal{L}_i \subset \mathcal{L}_{i-1}, \mathcal{L}_i/2 \subseteq \mathcal{L}_{i-2}$, $|\mathcal{L}_{i-1}/\mathcal{L}_i| = 2^\alpha$, and $2^{\lceil i\alpha/n \rceil}\mathcal{L}_0 \subseteq \mathcal{L}_i \subseteq 2^{\lfloor i\alpha/n \rfloor}\mathcal{L}_0$ for all i.*

Theorem 4.8 ([ADRS15]). *There is a polynomial-time algorithm that takes as input integers $\ell \ge 1$ and $n/2 \le \alpha \le n$ as well as a lattice $\mathcal{L} \subseteq \mathbb{R}^n$ and outputs a tower of lattice $(\mathcal{L}_0, \ldots, \mathcal{L}_\ell)$ with $\mathcal{L}_\ell = \mathcal{L}$.*

Proof. We give the construction below. The desired properties are immediate from the construction. Let $\mathbf{b}_1, \ldots, \mathbf{b}_n$ be a basis of $\mathcal{L}$. The tower is then defined by "cyclically halving α coordinates", namely,

$$\begin{aligned}\mathcal{L}_\ell &= \mathcal{L}(\mathbf{b}_1, \ldots, \mathbf{b}_n),\\ \mathcal{L}_{\ell-1} &= \mathcal{L}(\mathbf{b}_1/2, \ldots, \mathbf{b}_\alpha/2, \mathbf{b}_{\alpha+1}, \ldots \mathbf{b}_n),\\ \mathcal{L}_{\ell-2} &= \mathcal{L}(\mathbf{b}_1/4, \ldots, \mathbf{b}_{2\alpha-n}/4, \mathbf{b}_{2\alpha-n+1}/2, \ldots \mathbf{b}_n/2),\end{aligned}$$

etc. The required properties can be easily verified. □

The following proposition shows that starting with discrete Gaussian samples from $\mathcal{L}_0$ and then repeatedly applying Algorithm 2 gives us a list of vectors in $\mathcal{L}_\ell$ that is close to a mixture of Gaussians, provided that there exists an appropriate "smooth sublattice" $\mathcal{L}' \subseteq \mathcal{L}_0$.

Proposition 4.9. *There is an algorithm that runs in $m \cdot \mathrm{poly}(n, \ell, \log m)$ time; takes as input a tower of lattices $(\mathcal{L}_0, \ldots, \mathcal{L}_\ell)$ in $\mathbb{R}^n$ of index 2^α, and vectors $\mathbf{X}_1, \ldots, \mathbf{X}_m \in \mathcal{L}_0$ with $m := 2^{\ell+\alpha+1}$; and outputs $\mathbf{Y}_1, \ldots, \mathbf{Y}_M \in \mathcal{L}_\ell$ with $M := 2^\alpha$ with the following properties. If the input vectors $\mathbf{X}_1, \ldots, \mathbf{X}_m$ are symmetric and 0-similar to a mixture of Gaussians over $\mathcal{L}' \subseteq \mathcal{L}_0$ with parameter $s > 10 \cdot 2^{(\alpha/n-1/2)\ell}\eta_\varepsilon(\mathcal{L}')$ for some (possibly unknown) sublattice $\mathcal{L}' \subseteq \mathcal{L}_0$ and $\varepsilon \in (0, 1/3)$; then the output distribution is $(10^\ell\varepsilon)$-similar to a mixture of independent Gaussians over $2^{\lceil \ell\alpha/n \rceil}\mathcal{L}' \subseteq \mathcal{L}_\ell$ with parameter $2^{\ell/2}s$, and*

$$\sum_{k=1}^{M} \mathbb{E}[\|\mathbf{Y}_k\|^2] \le \sum_{i=1}^{m} \mathbb{E}[\|\mathbf{X}_i\|^2] \, .$$

Proof. The algorithm simply applies Algorithm 2 repeatedly, first using the input vectors in $\mathcal{L}_0$ to obtain vectors in $\mathcal{L}_1$, then using these to obtain vectors in $\mathcal{L}_2$, etc., until eventually it obtains vectors $\mathbf{Y}_1, \ldots, \mathbf{Y}_M \in \mathcal{L}_\ell$. The running time is clearly $m \cdot \mathrm{poly}(n, \ell, \log m)$, as claimed.

By Lemma 4.6 and a simple induction argument, we see that every call to Algorithm 2 results in a symmetric distribution, and the sum of the expected squared norms is non-increasing after each step. In particular,

$$\sum_{k=1}^{M} \mathbb{E}[\|\mathbf{Y}_k\|^2] \le \sum_{i=1}^{m} \mathbb{E}[\|\mathbf{X}_i\|^2] \, ,$$

as needed.

We suppose for induction that the distribution of the output of the ith call to Algorithm 2 is $10^i\varepsilon$-similar to a mixture of independent Gaussians over $2^{\lceil i\alpha/n \rceil}\mathcal{L}' \subseteq 2^{\lceil i\alpha/n \rceil}\mathcal{L}_0 \subseteq \mathcal{L}_i$ with parameter $2^{i/2}s$ (which is true by assumption for $i = 0$). Then, this distribution is also $10^i\varepsilon$-similar to a mixture of independent Gaussians over $2^{\lceil (i+1)\alpha/n \rceil}\mathcal{L}' \subseteq 2^{\lceil i\alpha/n \rceil}\mathcal{L}'$ (since a mixture of Gaussians over a lattice is also a mixture of Gaussians over any sublattice). Furthermore, $\eta_\varepsilon(2^{\lceil (i+1)\alpha/n \rceil}\mathcal{L}') = 2^{\lceil (i+1)\alpha/n \rceil}\eta_\varepsilon(\mathcal{L}') < 2^{i/2}s/\sqrt{2}$. Therefore, we may apply Lemma 4.5 to conclude that the distribution of the output of the $(i+1)$st call to Algorithm 2 is $10^{i+1}\varepsilon$-similar to a mixture of independent Gaussians over $2^{\lceil (i+1)\alpha/n \rceil}\mathcal{L}' \subseteq \mathcal{L}_{i+1}$ with parameter $2^{(i+1)/2}s$. In particular, the final output vectors are $10^{\ell}\varepsilon$-similar to a mixture of independent Gaussians over $2^{\lceil \ell\alpha/n \rceil}\mathcal{L}'$, as needed. □

4.4 The Algorithm

Theorem 4.10. *For any $\varepsilon = \varepsilon(n) \in (0, n^{-200})$, there is a $2^{n/2+O(n\log(n)/\log(1/\varepsilon))+o(n)}$-time algorithm that solves $(100\sqrt{n}\overline{\eta}_\varepsilon)$-GSVP. In particular, if $\varepsilon = n^{-\omega(1)}$, then the running time is $2^{n/2+o(n)}$.*

Proof. The algorithm takes as input a (basis for a) lattice $\mathcal{L} \subset \mathbb{R}^n$ with $n \geq 50$ and behaves as follows. Without loss of generality, we may assume that $\varepsilon > 2^{-n}$ and that the algorithm has access to a parameter $s > 0$ with $50\overline{\eta}_\varepsilon(\mathcal{L}) \leq s \leq 100\overline{\eta}_\varepsilon(\mathcal{L})$. Let $\ell := \lfloor \log(1/\varepsilon)/\log(10) \rfloor - 1$ and $\alpha := \lceil n/2 + 100n\log n/\log(1/\varepsilon) \rceil$.

The algorithm first runs the procedure from Theorem 4.8 on input ℓ, α, and $\mathcal{L}$, receiving as output a tower of lattices $(\mathcal{L}_0, \ldots, \mathcal{L}_\ell)$ with $\mathcal{L}_\ell = \mathcal{L}$. The algorithm then runs the procedure from Proposition 3.5 on input $\mathcal{L}_0$, $r := n/5$, $m := 2^{\ell+\alpha+1}$, and parameter $s' := 2^{-\ell/2}s$, receiving as output a sublattice $\widehat{\mathcal{L}} \subseteq \mathcal{L}_0$, and vectors $\mathbf{X}_1, \ldots, \mathbf{X}_m \in \widehat{\mathcal{L}} \subseteq \mathcal{L}_0$. Finally, the algorithm runs the procedure from Proposition 4.9 on input $(\mathcal{L}_0, \ldots, \mathcal{L}_\ell)$ and $\mathbf{X}_1, \ldots, \mathbf{X}_m$, receiving as output $\mathbf{Y}_1, \ldots, \mathbf{Y}_M \in \mathcal{L}_\ell = \mathcal{L}$. It then simply outputs the shortest non-zero vector amongst the $\mathbf{Y}_i \in \mathcal{L}$. (If all of the $\mathbf{Y}_i$ are zero, the algorithm fails.)

The running time of the algorithm is clearly $(m + 2^{r+o(r)}) \cdot \mathrm{poly}(n, \ell, \log m) = 2^{n/2+O(n\log n/\log(1/\varepsilon))+o(n)}$. We first show that the promise $s' \geq r^{n/r}\sqrt{n\log n} \cdot \overline{\eta}_\varepsilon(\mathcal{L}_0)$ needed to apply Proposition 3.5 is satisfied. Indeed, by the definition of a tower of lattices, we have $\mathcal{L} \subseteq 2^{\lfloor \ell\alpha/n \rfloor}\mathcal{L}_0$, so that

$$s' \geq 50 \cdot 2^{-\ell/2} \cdot \overline{\eta}_\varepsilon(\mathcal{L}) \geq 50 \cdot 2^{\lfloor \ell\alpha/n \rfloor - \ell/2} \cdot \overline{\eta}_\varepsilon(\mathcal{L}_0) \geq r^{n/r}\sqrt{n\log n} \cdot \overline{\eta}_\varepsilon(\mathcal{L}_0) ,$$

as needed. Therefore, the procedure from Proposition 3.5 succeeds, i.e. we have $\overline{\eta}_\varepsilon(\widehat{\mathcal{L}}) = \overline{\eta}_\varepsilon(\mathcal{L}_0)$ and that the $\mathbf{X}_i$ are distributed as independent samples from $D_{\widehat{\mathcal{L}},s'}$.

In particular, let $\mathcal{L}' \subseteq \widehat{\mathcal{L}} \subseteq \mathcal{L}_0$ such that $\eta_\varepsilon(\mathcal{L}') = \overline{\eta}_\varepsilon(\widehat{\mathcal{L}}) = \overline{\eta}_\varepsilon(\mathcal{L}_0)$. Then, the distribution of $\mathbf{X}_1, \ldots, \mathbf{X}_m$ is symmetric and 0-similar to a mixture of Gaussians over $\mathcal{L}'$ with parameter $s' > 10 \cdot 2^{(\alpha/n-1/2)\ell}\eta_\varepsilon(\mathcal{L}')$. We may therefore apply Proposition 4.9 and see that the $\mathbf{Y}_1, \ldots, \mathbf{Y}_M \in \mathcal{L}$ are δ-similar to a mixture of independent Gaussians over $2^{\lceil \ell\alpha/n \rceil}\mathcal{L}'$ with parameter s and $\delta := 10^{\ell}\varepsilon \leq 1/10$. Furthermore,

$$\sum_{k=1}^{M} \mathbb{E}[\|\mathbf{Y}_k\|^2] \leq \sum_{i=1}^{m} \mathbb{E}[\|\mathbf{X}_i\|^2] \leq \frac{nm(s')^2}{2\pi} = 2^{-\ell} \cdot \frac{nms^2}{2\pi} ,$$

where the last inequality is Claim 2.11.

Finally, we notice that

$$s \geq 50\overline{\eta}_\varepsilon(\mathcal{L}) \geq 50 \cdot 2^{\lfloor \ell\alpha/n \rfloor}\overline{\eta}_\varepsilon(\mathcal{L}_0) = 50\eta_\varepsilon(2^{\lfloor \ell\alpha/n \rfloor}\mathcal{L}') \geq 25\eta_\varepsilon(2^{\lceil \ell\alpha/n \rceil}\mathcal{L}')$$
$$\geq 10e^\delta \eta_{1/2}((2^{\lceil \ell\alpha/n \rceil}\mathcal{L}') .$$

Therefore, we may apply Corollary 4.4 to $\mathbf{Y}_1, \ldots, \mathbf{Y}_M$ to conclude that with probability at least 1/10, there exists $k \in \{1, \ldots, M\}$ such that

$$0 < \|\mathbf{Y}_k\|^2 < \frac{4}{M} \cdot \sum_{i=1}^{M} \mathbb{E}[\|\mathbf{Y}_i\|^2] \leq 2^{-\ell} \cdot \frac{nms^2}{2\pi M} \leq ns^2 \leq 100^2 n\overline{\eta}_\varepsilon(\mathcal{L})^2 .$$

In other words, $\mathbf{Y}_k \in \mathcal{L}$ is a valid solution to $(100\sqrt{n}\overline{\eta}_\varepsilon)$-GSVP, as needed. □

Corollary 4.11. *There is a $2^{n/2+o(n)}$-time algorithm that solves γ-SVP for any $\gamma = \gamma(n) > \omega(\sqrt{n \log n})$.*

Proof. Theorem 4.10 gives an algorithm with the desired running time that finds a non-zero lattice vector with norm bounded by $100\sqrt{n}\overline{\eta}_\varepsilon(\mathcal{L})$ for

$$\varepsilon := 2^{-\gamma^2/(100^2 n)} < n^{-\omega(1)} .$$

The result follows from Lemma 3.1, which in particular tells us that

$$\overline{\eta}_\varepsilon(\mathcal{L}) \leq \sqrt{\log(1/\varepsilon)}\lambda_1(\mathcal{L}) \leq \gamma/(100\sqrt{n}) \cdot \lambda_1(\mathcal{L}) ,$$

as needed. □

Corollary 4.12. *There is a $2^{n/2+o(n)}$-time algorithm that solves γ-HSVP for any $\gamma = \gamma(n) > \omega(\sqrt{n \log^3 n})$.*

Proof. Theorem 4.10 gives an algorithm with the desired running time that finds a non-zero lattice vector with norm bounded by $100\sqrt{n}\overline{\eta}_\varepsilon(\mathcal{L})$ for

$$\varepsilon := 2^{-\gamma^2/(10^{10} n \log^2 n)} < n^{-\omega(1)} .$$

The result follows from Lemma 3.1, which in particular tells us that

$$\overline{\eta}_\varepsilon(\mathcal{L}) \leq 10\sqrt{\log(1/\varepsilon)}(\log n + 2)\det(\mathcal{L})^{1/n} \leq \gamma/(100\sqrt{n}) \cdot \det(\mathcal{L})^{1/n} ,$$

as needed (where we have assumed that n is sufficiently large). □

5 Approximate SVP via Basis Reduction

Basis reduction algorithms solve δ-(H)SVP in dimension n by making polynomially many calls to a δ'-SVP algorithm on lattices in dimension $k < n$. We will show in this section how to modify the basis reduction algorithm from [GN08, ALNS20] to prove Theorem 1.2.

5.1 Slide-Reduced Bases

Here, we introduce our notion of a reduced basis. This differs from prior work in that we allow the length ℓ of the last block to be not equal to k, and we use HSVP reduction where other works use SVP reduction. E.g., taking $\ell = k$ and replacing (D)HSVP reduction with (D)SVP reduction in Item 2 recovers the definition from [ALNS20]. (Taking $\ell = k$ and $q = 0$ and replacing all (D)HSVP reduction with (D)SVP reduction recovers the original definition in [GN08].)

Definition 5.1 (Slide reduction). *Let n, k, p, q, ℓ be integers such that $n = pk + q + \ell$ with $p \geq 1, k, \ell \geq 2$ and $0 \leq q \leq k-1$. Let $\delta_H \geq 1$ and $\delta_S \geq 1$. A basis $\mathbf{B} \in \mathbb{R}^{m \times n}$ is $(\delta_H, k, \delta_S, \ell)$-slide-reduced if it is size-reduced and satisfies the following four sets of constraints.*

1. *The block $\mathbf{B}_{[1,k+q+1]}$ is η-twin-reduced for $\eta := \delta_H^{\frac{k+q-1}{k-1}}$.*
2. *For all $i \in [1, p-1]$, the block $\mathbf{B}_{[ik+q+1,(i+1)k+q+1]}$ is δ_H-twin-reduced.*
3. *The block $\mathbf{B}_{[pk+q+1,n]}$ is δ_S-SVP-reduced.*

Theorem 5.2. *For any $\delta_H, \delta_S \geq 1, k \geq 2, \ell \geq 2$, if $\mathbf{B} \in \mathbb{R}^{n \times n}$ is a $(\delta_H, k, \delta_S, \ell)$-slide-reduced basis of a lattice $\mathcal{L}$ with $\lambda_1(\mathcal{L}(\mathbf{B}_{[1,n-\ell]})) > \lambda_1(\mathcal{L})$ then*

$$\|\mathbf{b}_1\| \leq \delta_S (\delta_H^2)^{\frac{n-\ell}{k-1}} \lambda_1(\mathcal{L}) .$$

Proof. By Fact 2.22, $\|\mathbf{b}_1\| \leq \eta^{\frac{2(k+q)}{k+q-1}} \|\mathbf{b}^*_{k+q+1}\| = \delta_H^{\frac{2(k+q)}{k-1}} \|\mathbf{b}^*_{k+q+1}\|$. Also, for all $i \in [1, p-1]$, $\|\mathbf{b}^*_{ik+q+1}\| \leq \delta_H^{\frac{2k}{k-1}} \|\mathbf{b}^*_{(i+1)k+q+1}\|$. All together we have:

$$\|\mathbf{b}_1\| \leq (\delta_H^2)^{\frac{k+q+(p-1)k}{k-1}} \|\mathbf{b}^*_{pk+q+1}\| = (\delta_H^2)^{\frac{n-\ell}{k-1}} \|\mathbf{b}^*_{pk+q+1}\|$$

Lastly, since $\lambda_1(\mathcal{L}(\mathbf{B}_{[1,n-\ell]})) > \lambda_1(\mathcal{L})$, $\|\mathbf{b}^*_{pk+q+1}\| \leq \delta_S \lambda_1(\mathcal{L}(\mathbf{B}_{[pk+q+1,n]})) \leq \delta_S \lambda_1(\mathcal{L})$. The result does follow. □

5.2 The Slide Reduction Algorithm

We show our algorithm for generating a slide-reduced basis. We stress that this is essentially the same algorithm as in [ALNS20] (which itself is a generalization of the algorithm in [GN08]) with a slight modification that allows the last block to have arbitrary length ℓ. Our proof for bounding the running time of the algorithm is therefore essentially identical to the proof in [GN08, ALNS20].

Theorem 5.3. *For $\varepsilon \in [1/\mathrm{poly}(n), 1]$, Algorithm 3 runs in polynomial time (excluding oracle calls), makes polynomially many calls to its δ_H-HSVP oracle and δ_S-SVP oracle, and outputs a $((1+\varepsilon)\delta_H, k, \delta_S, \ell)$-slide-reduced basis of the input lattice $\mathcal{L}$.*

Algorithm 3. Our slide-reduction algorithm

Input: Block size $k \geq 2$, slack $\varepsilon > 0$, approximation factor $\delta_H, \delta_S \geq 1$, basis $\mathbf{B} = (\mathbf{b}_1, \ldots, \mathbf{b}_n) \in \mathbb{Z}^{m \times n}$ of a lattice $\mathcal{L}$ of rank $n = pk + q + \ell$ for $0 \leq q \leq k-1$, and access to a rank k δ_H-HSVP oracle and a rank ℓ δ_S-SVP oracle.
Output: A $((1+\varepsilon)\delta_H, k, \delta_S, \ell)$-slide-reduced basis of $\mathcal{L}(\mathbf{B})$.

1: **while** $\mathrm{vol}(\mathbf{B}_{[1,ik+q]})^2$ is modified by the loop for some $i \in [1,p]$ **do**
2: $(1+\varepsilon)\eta$-HSVP-reduce $\mathbf{B}_{[1,k+q]}$ using Alg. 1 for $\eta := (\delta_H)^{\frac{k+q-1}{k-1}}$.
3: **for** $i = 1$ **to** $p-1$ **do**
4: δ_H-HSVP-reduce $\mathbf{B}_{[ik+q+1,(i+1)k+q]}$.
5: **end for**
6: δ_S-SVP-reduce $\mathbf{B}_{[pk+q+1,n]}$.
7: **if** $\mathbf{B}_{[2,k+q+1]}$ is not $(1+\varepsilon)\eta$-DHSVP-reduced **then**
8: $(1+\varepsilon)^{1/2}\eta$-DHSVP-reduce $\mathbf{B}_{[2,k+q+1]}$ using Alg. 1.
9: **end if**
10: **for** $i = 1$ **to** $p-1$ **do**
11: Find a new basis $\mathbf{C} := (\mathbf{b}_1, \ldots, \mathbf{b}_{ik+q+1}, \mathbf{c}_{ik+q+2}, \ldots, \mathbf{c}_{(i+1)k+q+1}, \mathbf{b}_{ik+q+2}, \ldots, \mathbf{b}_n)$ of $\mathcal{L}$ by δ_H-DHSVP-reducing $\mathbf{B}_{[ik+q+2,(i+1)k+q+1]}$.
12: **if** $(1+\varepsilon)\|\mathbf{b}^*_{(i+1)k+q+1}\| < \|\mathbf{c}^*_{(i+1)k+q+1}\|$ **then**
13: $\mathbf{B} \leftarrow \mathbf{C}$.
14: **end if**
15: **end for**
16: **end while**
17: **return** $\mathbf{B}$.

Proof. First, notice that if Algorithm 3 ever terminates, the output must be $((1+\varepsilon)\delta_H, k, \delta_S, \ell)$-slide-reduced basis. It remains to show that the algorithm terminates in polynomially many steps (excluding oracle calls).

Let $\mathbf{B}_0 \in \mathbb{Z}^{m \times n}$ be the input basis and let $\mathbf{B} \in \mathbb{Z}^{m \times n}$ denote the current basis during the execution of Algorithm 3. Following the analysis of basis reduction algorithms in [LLL82, GN08, LN14, ALNS20], we consider an integral potential

$$P(\mathbf{B}) := \prod_{i=1}^{p} \mathrm{vol}(\mathbf{B}_{[1,ik+q]})^2 \in \mathbb{Z}^+.$$

At the beginning of the algorithm, the potential satisfies $\log P(\mathbf{B}_0) \leq 2n^2 \cdot \log \|\mathbf{B}_0\|$. For each of the primal steps (i.e., Steps 2, 4 and 6), the lattice $\mathcal{L}(\mathbf{B}_{[1,ik+q]})$ for any $i \geq 1$ is unchanged. Hence $P(\mathbf{B})$ does not change. On the other hand, the dual steps (i.e., Steps 8 and 13) either leave $\mathrm{vol}(\mathbf{B}_{[1,ik+q]})$ unchanged for all i or decrease $P(\mathbf{B})$ by a multiplicative factor of at least $(1+\varepsilon)$.

Therefore, there are at most $\log P(\mathbf{B}_0)/\log(1+\varepsilon)$ updates on $P(\mathbf{B})$ by Algorithm 3. This directly implies that the algorithm makes at most $4pn^2 \log \|\mathbf{B}_0\|/\log(1+\varepsilon)$ calls to the HSVP oracle, the SVP oracle, and Algorithm 1.

We then conclude that Algorithm 3's running time is bounded by some polynomial in the size of input (excluding the running time of oracle calls). □

Corollary 5.4. *For any constant $c \geq 1$, there is a randomized algorithm that solves $(\mathrm{polylog}(n)n^c)$-SVP that runs in $2^{k/2+o(k)}$ time for $k := \frac{n-c}{c+5/(8.02)}$.*

Proof. Let $\ell = \frac{0.5k}{0.802}$ and run Algorithm 3, using the $O(\text{polylog}(n)\sqrt{n})$-HSVP algorithm from Corollary 4.12 and the $O(1)$-SVP algorithm from [LWXZ11] as oracles. We receive a $((1+\varepsilon)\text{polylog}(k)\sqrt{k}, k, O(1), \ell)$-slide-reduced basis $\mathbf{B}$ for any input lattice $\mathcal{L}$. Now consider two cases:

CASE 1: $\lambda_1(\mathcal{L}(\mathbf{B}_{[1,n-\ell]})) > \lambda_1(\mathcal{L})$: By Theorem 5.2, we conclude that

$$\|\mathbf{b}_1\| \leq \delta_S(\delta_H^2)^{\frac{n-\ell}{k-1}}\lambda_1(\mathcal{L}) \leq O(\text{polylog}(k)^c n^c)\lambda_1(\mathcal{L}) ,$$

as desired.

CASE 2: $\lambda_1(\mathcal{L}(\mathbf{B}_{[1,n-\ell]})) = \lambda_1(\mathcal{L})$: Then we repeat the algorithm on the lattice $\mathcal{L}(\mathbf{B}_{[1,n-\ell]})$ with lower dimension. This can happen at most n/ℓ times, introducing at most a polynomial factor in the running time.

For the running time, the algorithm from Corollary 4.12 runs in time $2^{0.5k+o(k)}$. The algorithm from [LWXZ11] runs in time $2^{0.802\ell+o(\ell)}$, which is the same as $2^{0.5k+o(k)}$, by our choice of ℓ. This completes the proof. □

References

[ADRS15] Aggarwal, D., Dadush, D., Regev, O., Stephens-Davidowitz, N.: Solving the shortest vector problem in 2^n time via discrete gaussian sampling. In: STOC (2015). http://arxiv.org/abs/1412.7994

[AKS01] Ajtai, M., Kumar, R., Sivakumar, D.: A sieve algorithm for the shortest lattice vector problem. In: STOC (2001)

[ALNS20] Aggarwal, D., Li, J., Nguyen, P.Q., Stephens-Davidowitz, N.: Slide reduction, revisited—filling the gaps in SVP approximation. In: Micciancio, D., Ristenpart, T. (eds.) CRYPTO 2020. LNCS, vol. 12171, pp. 274–295. Springer, Cham (2020). https://doi.org/10.1007/978-3-030-56880-1_10

[AS04] Alon, N., Spencer, J.H.: The Probabilistic Method. Wiley, Hoboken (2004)

[AS18] Aggarwal, D., Stephens-Davidowitz, N.: Just take the average! An embarrassingly simple 2^n-time algorithm for SVP (and CVP). In: SOSA (2018). http://arxiv.org/abs/1709.01535

[AUV19] Aggarwal, D., Ursu, B., Vaudenay, S.: Faster sieving algorithm for approximate SVP with constant approximation factors (2019). https://eprint.iacr.org/2019/1028

[BDGL16] Becker, A., Ducas, L., Gama, N., Laarhoven, T.: New directions in nearest neighbor searching with applications to lattice sieving. In: SODA (2016)

[BGJ14] Becker, A., Gama, N., Joux, A.: A sieve algorithm based on overlattices. LMS J. Comput. Math. **17**(A), 49–70 (2014)

[BLP13] Brakerski, Z., Langlois, A., Peikert, C., Regev, O., Stehlé, D.: Classical hardness of learning with errors. In: STOC (2013)

[DR16] Dadush, D., Regev, O.: Towards strong reverse Minkowski-type inequalities for lattices. In: FOCS (2016). http://arxiv.org/abs/1606.06913

[DRS14] Dadush, D., Regev, O., Stephens-Davidowitz, N.: On the closest vector problem with a distance guarantee. In: CCC (2014)

[Duc18] Ducas, L.: Shortest vector from lattice sieving: a few dimensions for free. In: Nielsen, J.B., Rijmen, V. (eds.) EUROCRYPT 2018. LNCS, vol. 10820, pp. 125–145. Springer, Cham (2018). https://doi.org/10.1007/978-3-319-78381-9_5

[GINX16] Gama, N., Izabachène, M., Nguyen, P.Q., Xie, X.: Structural lattice reduction: generalized worst-case to average-case reductions and homomorphic cryptosystems. In: Fischlin, M., Coron, J.-S. (eds.) EUROCRYPT 2016. LNCS, vol. 9666, pp. 528–558. Springer, Heidelberg (2016). https://doi.org/10.1007/978-3-662-49896-5_19

[GMPW20] Genise, N., Micciancio, D., Peikert, C., Walter, M.: Improved discrete Gaussian and subgaussian analysis for lattice cryptography. In: Kiayias, A., Kohlweiss, M., Wallden, P., Zikas, V. (eds.) PKC 2020. LNCS, vol. 12110, pp. 623–651. Springer, Cham (2020). https://doi.org/10.1007/978-3-030-45374-9_21

[GN08] Gama, N., Nguyen, P.Q.: Finding short lattice vectors within Mordell's inequality. In: STOC (2008)

[GPV08] Gentry, C., Peikert, C., Vaikuntanathan, V.: Trapdoors for hard lattices and new cryptographic constructions. In: STOC (2008). https://eprint.iacr.org/2007/432

[Kan83] Kannan, R.: Improved algorithms for integer programming and related lattice problems. In: STOC (1983)

[Laa15] Laarhoven, T.: Sieving for shortest vectors in lattices using angular locality-sensitive hashing. In: Gennaro, R., Robshaw, M. (eds.) CRYPTO 2015. LNCS, vol. 9215, pp. 3–22. Springer, Heidelberg (2015). https://doi.org/10.1007/978-3-662-47989-6_1

[LLL82] Lenstra, A.K., Lenstra Jr., H.W., Lovász, L.: Factoring polynomials with rational coefficients. Math. Ann. **261**(4), 515–534 (1982)

[LN14] Li, J., Nguyen, P.Q.: Approximating the densest sublattice from Rankin's inequality. LMS J. Comput. Math. **17**(A), 92–111 (2014)

[Lov86] Lovász, L.: An algorithmic theory of numbers, graphs and convexity. Society for Industrial and Applied Mathematics (1986)

[LWXZ11] Liu, M., Wang, X., Xu, G., Zheng, X.: Shortest lattice vectors in the presence of gaps (2011). http://eprint.iacr.org/2011/139

[MP13] Micciancio, D., Peikert, C.: Hardness of SIS and LWE with small parameters. In: Canetti, R., Garay, J.A. (eds.) CRYPTO 2013. LNCS, vol. 8042, pp. 21–39. Springer, Heidelberg (2013). https://doi.org/10.1007/978-3-642-40041-4_2

[MR07] Micciancio, D., Regev, O.: Worst-case to average-case reductions based on Gaussian measures. SIAM J. Comput. **37**(1), 267–302 (2007)

[MV13] Micciancio, D., Voulgaris, P.: A deterministic single exponential time algorithm for most lattice problems based on Voronoi cell computations. SIAM J. Comput. **42**(3), 1364–1391 (2013)

[MW16] Micciancio, D., Walter, M.: Practical, predictable lattice basis reduction. In: Eurocrypt (2016). http://eprint.iacr.org/2015/1123

[NIS18] Computer Security Division NIST. Post-quantum cryptography (2018). https://csrc.nist.gov/Projects/Post-Quantum-Cryptography

[NV08] Nguyen, P.Q., Vidick, T.: Sieve algorithms for the shortest vector problem are practical. J. Math. Cryptol. **2**(2), 181–207 (2008)

[Pei16] Peikert, C.: A decade of lattice cryptography. Found. Trends Theor. Comput. Sci. **10**(4), 283–424 (2016)

[PS09] Pujol, X., Stehlé, D.: Solving the Shortest Lattice Vector Problem in time $2^{2.465n}$ (2009). http://eprint.iacr.org/2009/605

[Reg09] Regev, O.: On lattices, learning with errors, random linear codes, and cryptography. J. ACM (JACM) **56**(6), 34 (2009)

[RS17] Regev, O., Stephens-Davidowitz, N.: A reverse Minkowski theorem. In: STOC (2017)

[Sch87] Schnorr, C.-P.: A hierarchy of polynomial time lattice basis reduction algorithms. Theor. Comput. Sci. **53**(23), 201–224 (1987)

[SE94] Schnorr, C.P., Euchner, M.: Lattice basis reduction: improved practical algorithms and solving subset sum problems. Math. Program. **66**(1), 181–199 (1994)

[Ste17] Stephens-Davidowitz, N.: On the Gaussian measure over lattices. Ph.d. thesis, New York University (2017)

[WLW15] Wei, W., Liu, M., Wang, X.: Finding shortest lattice vectors in the presence of gaps. In: Nyberg, K. (ed.) CT-RSA 2015. LNCS, vol. 9048, pp. 239–257. Springer, Cham (2015). https://doi.org/10.1007/978-3-319-16715-2_13

New Lattice Two-Stage Sampling Technique and Its Applications to Functional Encryption – Stronger Security and Smaller Ciphertexts

Qiqi Lai[1,2(✉)], Feng-Hao Liu[3], and Zhedong Wang[3]

[1] School of Computer Science, Shaanxi Normal University, Xi'an, China
[2] State Key Laboratory of Integrated Service Networks, Xidian University, Xi'an, China
laiqq@snnu.edu.cn
[3] Florida Atlantic University, Boca Raton, FL, USA
{fenghao.liu,wangz}@fau.edu

Abstract. This work proposes a new lattice two-stage sampling technique, generalizing the prior two-stage sampling method of Gentry, Peikert, and Vaikuntanathan (STOC '08). By using our new technique as a key building block, we can significantly improve security and efficiency of the current state of the arts of simulation-based functional encryption. Particularly, our functional encryption achieves (Q, poly) simulation-based semi-adaptive security that allows arbitrary pre- and post-challenge key queries, and has succinct ciphertexts with only an additive $O(Q)$ overhead.

Additionally, our two-stage sampling technique can derive new feasibilities of indistinguishability-based adaptively-secure IB-FE for inner products and semi-adaptively-secure AB-FE for inner products, breaking several technical limitations of the recent work by Abdalla, Catalano, Gay, and Ursu (Asiacrypt '20).

1 Introduction

Functional Encryption (FE) [13,35] is a powerful generalization of public-key encryption (PKE), allowing more fine-grained information disclosure to a secret key holder. FE with regular syntax can be described as follows – every secret key is associated with a function f (in some class $\mathcal{F}$), and the decryptor given such key (i.e., sk_f) and a ciphertext $\mathsf{Enc}(u)$ can only learn $f(u)$. During the past decade, there has been tremendous progress of FE for various function classes, e.g., [2,4–6,21,26,27] and more.

To facilitate presentation and comparisons with prior work, we consider the notion of FE with a more fine-grained syntax, which has been studied in the literature to capture various settings of FE [1,2,13,27]. Particularly, each message u consists of two parts, namely $u := (\mathsf{x}, \mu)$, where x is some index (or attribute)[1],

[1] We note that both the names "index" and "attribute" have been used interchangeably in the literature.

A. Canteaut and F.-X. Standaert (Eds.): EUROCRYPT 2021, LNCS 12696, pp. 498–527, 2021.
https://doi.org/10.1007/978-3-030-77870-5_18

and μ is some message. Additionally, each function f consists of two parts, namely, $f := (\mathsf{P}, g) \in \mathcal{P} \times \mathcal{G}$, where P is a predicate over the index, and g is a function over the message. The overall function acts as:

$$f(u) := \begin{cases} g(\mu) & \text{if } \mathsf{P}(\mathsf{x}) = 1 \\ \perp & \text{otherwise.} \end{cases}$$

When decrypting the ciphertext $\mathsf{ct}_u = \mathsf{Enc}(\mathsf{x}, \mu)$ by $\mathsf{sk}_f := \mathsf{sk}_{(\mathsf{P},g)}$, the decryptor can learn $g(\mu)$ if $\mathsf{P}(\mathsf{x}) = 1$, and $\perp$ otherwise. Under this syntax, we call a key $\mathsf{sk}_{f:=(\mathsf{P},g)}$ a 1-key with respect to an index x if $\mathsf{P}(\mathsf{x}) = 1$, or otherwise a 0-key. Intuitively, a 1-key is allowed to open the ciphertext, but a 0-key is not.

Even though FE with the fine-grained syntax is essentially equivalent to the regular syntax for sufficiently expressive function/predicate classes, it is more convenient to present our new results in this way. Moreover as noticed since [13], many advanced encryption schemes such as identity-based encryption, attribute-based encryption, predicate encryption can be captured naturally from this notion, by different predicate and function classes $\mathcal{P} \times \mathcal{G}$.

There are two important settings studied in the literature – FE with private or public index, according to whether the index x is revealed to the decryption algorithm. In what follows, we first discuss in more details about challenges of the state of the arts in both settings. Then we present our contributions and new techniques to break these barriers and advance the research frontiers.

FE with Private Index. In this setting, FE provides very strong privacy guarantee where only $g(\mu)$ can be learned given a 1-key $\mathsf{sk}_{\mathsf{P},g}$ and a $\mathsf{Enc}(\mathsf{x}, \mu)$ with $\mathsf{P}(\mathsf{x}) = 1$. It is worthwhile to point out that in this setting, realizing the class $\mathcal{P} \times \{I\}$ for the identity function I is already general enough, as it suffices to capture FE (of regular syntax) for the boolean circuit class $\mathcal{P}$. In particular, we can use $\mathsf{sk}_{\mathsf{P},I}$ and $\mathsf{Enc}(\mathsf{x}, \mu)$ to simulate the exact effect of sk_{P} and $\mathsf{Enc}(\mathsf{x})$ of the regular syntax FE. Therefore, following some prior work [2], this work just focuses on the function class $\mathcal{P} \times \{I\}$ for FE in the private index setting by default. We discuss this in more details in the full version of this paper.

To capture security, there have been notions of indistinguishable-based (IND) and simulation-based (SIM) definitions proposed and studied in the literature since [13]. As raised by [13], the IND-based security is inadequate (i.e., too weak) in the private index setting for certain functionalities and applications. Thus, it would be much desirable to achieve the stronger notion of SIM-based notion.

However, there are various settings that the SIM-based notion is too strong to be attained. For example, the work [13] showed that for very simple functionalities (identity-based encryption), the SIM-based security is impossible for multiple challenge ciphertexts, even given just one post-challenge key query. Additionally, the work [4] showed that for FE scheme with respect to the class of general functions, the ciphertext size must grow linearly with the number of pre-challenge key queries. Therefore it is impossible to achieve the notion (poly, poly)-SIM security (allowing an unbounded number of both 1 and 0-keys) for general functions.

Despite these lower bounds, the work [27] identified important feasible settings for SIM-based security, by proposing new constructions in the setting of single challenge ciphertext and bounded collusion. More specifically, [27] achieved (Q, Q)-adaptive-SIM FE for the family of polynomial-sized circuits under the minimal assumption of PKE. Their attained SIM notion is very strong – the challenge index can be adaptively chosen and the adversary is allowed to query both pre- and post-challenge key queries, up to some bounded Q times for both 1 and 0-keys. The ciphertexts however, are not succinct (i.e., dependent on the circuit description of the function), and their size grows with a multiplicative factor of $O(Q^4)$. Even though a recent work [10] improved the multiplicative factor to $O(Q)$, their ciphertexts are still not succinct, prohibiting other important applications, such as reusable garbled circuits [26]. Thus, improvements in this dimension would be very significant.

A subsequent work [26] constructed the first single-key *succinct* FE for bounded depth circuits, and showed that this suffices for reusable garbled circuits, solving a long-term open question in the field. However, their scheme [26] has drawbacks in the following two aspects. First, the single-key FE of [26] achieves a weaker notion of selective security and only allows one pre-challenge key query (either a 1 or 0-key). Second, even though the single-key FE of [26] can be bootstrapped to Q key FE using the compiler of [27], yet the resulting ciphertexts grows with $O(Q^4)$ multiplicatively.

Tackling these drawbacks, two almost concurrent work [2,6] advanced this direction of work significantly. Particularly, the work [6] constructed a single key succinct FE for NC1, and then showed another bootstrapping method (from NC1 to general circuits) that only induces an $O(Q^2)$ additive overhead, yet the resulting (offline-part) ciphertexts become no longer succinct. The other concurrent work [2] designed a new succinct single key FE that supports $(1, \mathsf{poly})$ queries for general circuits, and a new bootstrapping method that achieves (Q, poly)-SIM security with succinct ciphertexts and $O(Q^2)$ additive overhead. As a substantial milestone, [2] for the first time identified an important and useful[2] subclass of key queries (i.e., 0-keys), where SIM-based security is feasible beyond bounded collusion. Recently, the work [10] designed a simple yet very novel compiler that turns any bounded-collusion FE into one with ciphertext growth $O(Q)$ multiplicatively. This compiler improves the ciphertext size significantly, but does not improve the security over the original scheme.

Challenges. The attainable SIM-based security of [2] is however weaker than that of the work [27] in three aspects – (1) the challenge index needs to be semi-adaptive (the adversary commits to the challenge right after the master public-key); (2) the 1-key queries need to be made at one-shot right before the challenge ciphertext; (3) no more 1-key is allowed for post-challenge phase. How to bridge the gap between the two methods [2,27] is an important open question.

To measure how large the gap is, we first notice that the semi-adaptive attribute (i.e., aspect (1)) can be mitigated (though not completely satisfac-

[2] For example in IBE and ABE, 0-keys are useful for decrypting other ciphertexts with satisfying indices. They just cannot decrypt the specific (challenge) index.

tory) by the generic complexity leveraging argument as also pointed out by [26]. Particularly, by scaling up the ℓ in bit-security of the selective scheme, we can achieve adaptive security over ℓ-bit index. Even though theoretically this would require to assume sub-exponential security of the underlying hard problem, yet nevertheless in practice this assumption is usually in use, given the estimations of the best-known concrete attacks, e.g., the concrete LWE estimation [7].

On the other hand, how to tackle adaptiveness for pre-challenge and post-challenge key queries seems beyond the current techniques, as the length to describe all possible key queries requires $Q \cdot \mathsf{poly}(\lambda)$ bits for some unbounded polynomial, which is too large for the complexity leveraging argument. Thus, how to improve aspects (2) and (3) would require substantial new techniques. This work aims to solve these challenges with the following particular goal.

(**Main Goal 1:**) Design a succinct FE for general bounded depth circuits with (Q, poly)-SIM-based security[3], allowing arbitrary pre- and post-challenge queries for both 1 and 0-keys.

FE with Public Index. The public index setting does not require the scheme to hide the index, and for many scenarios in this setting the IND security notion would already be adequate, as pointed out by [13]. Even though FE with public index can be generically derived from FE with private index, much more efficient solutions are desired. For example, current instantiations of FE with private index either use heavy tools such as garbled circuits or fully homomorphic encryption, while the identity-based encryption [3] (as a special case of FE with public index) only requires simple lattice operations and thus can be much more efficient.

A recent work [1] studied the class $\mathsf{IB} \times \mathsf{IP}$, where the IB is the class of identity comparison predicates and IP is the class of inner products. Particularly, the work [1] showed that by connecting ABB [3] encoding for IB and ALS [5] encoding for IP, one can derive a simple FE for $\mathsf{IB} \times \mathsf{IP}$ from lattices. Albeit simple and efficient, the work [1] can only prove the selective security (over IB) for their lattice design in the standard model, even though the ABB and ALS encodings both achieve the adaptive security in their encryption settings. Moreover, note that their construction idea [1] naturally extends to the setting of $\mathsf{AB} \times \mathsf{IP}$ by connecting the AB encoding of [11] with ALS, where AB is the general attribute-based policy functions. However, their proof of security [1] even for the selective security would hit a subtle yet challenging technical barrier. Our second goal is to tackle these challenges.

(**Main Goal 2:**) Determine new proof strategy for the class of $\mathsf{IB} \times \mathsf{IP}$ and $\mathsf{AB} \times \mathsf{IP}$ in the public index setting.

1.1 Our Contributions

This work aims at the two main goals and makes three major contributions.

[3] We notice that $(\mathsf{poly}, \mathsf{poly})$ SIM-based security is not possible by the lower bound of [4]. Thus, (Q, poly) SIM-based security is the best we can hope for in this model.

Contribution 1. First we propose a new two-stage lattice two-stage sampling technique, generalizing the prior GPV type two-stage sampling [24]. Using this new sampling technique, we design a unified framework that handles major challenges in our two (seemingly different) main goals as we elaborate next. The crux of our design relies on adding smudging noise over secret keys, which is critical in the analysis and conceptually new, as all prior work (to our knowledge) only considered adding smudging noise over ciphertexts, e.g., [2].

Contribution 2. By using our new sampling technique, we improve the prior designs of [2] substantially as we elaborate below.

- Our first step is to achieve a $(1, \mathsf{poly})$ selectively secure (over the challenge index) *partially hiding predicate encryption* (PHPE), allowing general pre-challenge but no post-challenge key queries. Technically, our construction simply replaces the key generation algorithm in the very-selective PHPE of [2][4] by our new sampler. Our result at this step is already stronger than the work [2] in the following ways.

  1. We notice that our PHPE can achieve the adaptive security by the complexity leveraging argument directly, yet the very-selective PHPE of [2] cannot, as the description of the function for key queries is too large.
  2. The two schemes can be upgraded to semi-adaptive security over the challenging index without the complexity leveraging, yet the transformation for ours is much more efficient. Particularly, our upgrade only applies the very light-weight method of [17,30], whereas the very-selective PHPE of [2] requires to compose PHPE with another FE (ALS [5]). Moreover, our resulting scheme allows arbitrary pre-challenge key queries, whereas the resulting scheme of [2] still requires the adversary to commit to the 1-key query before making further 0-key queries.

- Our $(1, \mathsf{poly})$ PHPE can be turned into FE by using the transformation of [2, 29], resulting in a *succinct* single key $(1, \mathsf{poly})$ FE that allows arbitrary pre-challenge key queries as long as there is at most one 1-key. This suffices to construct the reusable garbled circuits [26]. We present a comparison of our single key succinct FE with prior work in Table 1.
- Our next step is to achieve a succinct (Q, poly) FE that allows arbitrary pre- and post-challenge queries. To achieve this, we slightly modify the transformation (from $(1, \mathsf{poly})$ PHPE to (Q, poly) PHPE) of [2] by using the technique of secret sharing and a new way of generating cover-free sets inspired by [10]. By applying our new transformation to our $(1, \mathsf{poly})$ PHPE, we derive a (Q, poly) PHPE that allows arbitrary pre- and post-key queries. Then the desired FE again follows from the transformation of [2,29].
  Importantly, our transformation inherits many nice properties in [2], e.g., the succinctness of the ciphertexts is preserved. Thus, our resulting FE has

[4] A very-selective scheme requires the adversary to commit to both the challenge index and function in the very beginning of the security experiment.

succinct ciphertexts, whose size grows *additively* with $O(Q)$, and are independent of the function/circuit size. Our result is better than the transformation of [10], which incurs a *multiplicative* $O(Q)$ blowup in the ciphertexts.

Table 1. Comparison of prior work of single key SIM-secure public-key FE.

Ref.	(1-key, 0-key)	(Pre, Post)-Challenge	Index	Succinct ct
[27]	$(a, b) : a + b = 1$	(✓, ✓)	AD	✗
[26]	$(a, b) : a + b = 1$	(✓, ✗)	Sel[†]	✓
[6]	$(a, b) : a + b = 1$	(✓, ✗)	AD	✓ for NC1
[2]	(1, poly)	(✗, ✗)*	SA[†]	✓
Ours	(1, poly)	(✓, ✗)	SA[†]	✓

(∗) The scheme requires the adversary to commit to the 1-key query right after seeing the master public key. Then the adversary is allowed to make further arbitrary 0-key queries in the pre- and post-challenge phases, but not any more 1-key query.
(†) The selective (Sel)/semi-adaptive (SA) security can be raised to adaptive security (AD) by the complexity leveraging argument, at the cost of scaling up the security parameters.

In summary, we achieve our *Main Goal* 1 for semi-adaptive security over the challenge index, and the full-fledged of the goal if we further apply the complexity leveraging argument. Additionally, our scheme for the first time achieves succinct ciphertexts with only $O(Q)$ additive overhead. We present a comparison of our (Q, poly) FE with prior work in Table 2.

Table 2. Comparison of other private index SIM-secure public-key FE.

Ref.	(1-key, 0-key)	(Pre, Post)-Challenge	Index	Succinct ct	Ciphertext size
[27]	(Q, Q)	(✓, ✓)	AD	✗	$\times\, O(Q^4)$
[26]+ [27]	(Q, Q)	(✓, ✗)	Sel[†]	✓	$\times\, O(Q^4)$
[6]	(Q, Q)	(✓, ✗)	AD	✓ for NC1	$+\, O(Q^2)$
[2]	(Q, poly)	(✗, ✗)*	SA[†]	✓	$+\, O(Q^2)$
[2]+ [10][‡]	(Q, poly)	(✗, ✗)*	SA[†]	✓	$\times\, O(Q)$
Ours	(Q, poly)	(✓, ✓)	SA[†]	✓	$+\, O(Q)$

(∗) The scheme requires the adversary to commit to all the Q 1-key queries (in one shot) right after seeing the master public key. Then the adversary is allowed to make further arbitrary 0-key queries in the pre- and post-challenge phases, but not any more 1-key query.
(†) Similar to Table 1.
(‡) The generic method in [10] can transform any bounded collusion FE scheme into one whose ciphertext size grows with $O(Q)$ multiplicatively.

Contribution 3. Finally, we identify that our new sampling technique is the key to break the technical barriers of the lattice-based analysis of [1]. Particularly, for the setting of public index, we construct new FE schemes for IB × IP and AB × IP. The crux is to replace the key generation algorithm of [1] by our new pre-sampler. The novelty of this contribution majorly comes from the proof techniques. In Table 3 we compare our schemes with [1].

Table 3. Comparison of public index IND-based construction.

Reference	IB-FEIP	AB-FEIP
[1]	$(1, \mathsf{poly})$-Sel	✗
Ours	$(1, \mathsf{poly})$-AD	(Q, poly)-SA

1.2 Technical Overview

We present an overview of our new techniques. We first describe our central technique – a new two-stage sampling method, and then show how it can be used to achieve our main goals, together with further new insights. Our two-stage sampling method can be understood without the context of FE, and might be useful in other applications. Thus we believe that this technique can be of general interests.

Two-stage Sampling Method. At a high level, we would like to sample the following two-stage distribution:

- In the first stage, a random matrix $\mathbf{A}$ and a random vector $\boldsymbol{u}$ are sampled;
- In the second stage, an arbitrary small-norm matrix $\mathbf{R}$ is first specified, and then a short vector $\boldsymbol{y}$ is sampled conditioned on $[\mathbf{A}|\mathbf{AR}]\boldsymbol{y} = \boldsymbol{u}$.
- The overall distribution consists of $(\mathbf{A}, \mathbf{AR}, \boldsymbol{u}, \boldsymbol{y})$.

In a series of lattice-based work [1–3,11,14,24,28,29], the proof framework requires to sample this distribution (or its slight variations) in two ways – with $\mathbf{A}$'s trapdoor and without $\mathbf{A}$'s trapdoor. On the one hand, given the trapdoor of $\mathbf{A}$, one can efficiently sample this distribution. On the other hand, without the trapdoor of $\mathbf{A}$, one can also sample the distribution by using the $\mathbf{G}$-trapdoor technique [33]. Particularly, if we have the $\mathbf{G}$ matrix [33] in the right, i.e., the matrix is of the form $[\mathbf{A}|\mathbf{AR} + \gamma \cdot \mathbf{G}]$ with $\gamma \neq 0$, then this sampling task can be solved easily by the sample-right technique [3,33]. However, our task (and the security proofs in this work) does not have $\mathbf{G}$ in the second matrix, and thus the prior technique cannot be applied to sample the required distribution.

Is this task even doable? To answer this question, we first consider a simpler case where there is no $\mathbf{R}$. Then we notice that this task is achievable via the classic GPV two-stage sampling technique: we first pre-sample $\boldsymbol{y}$, and set $\boldsymbol{u} = \mathbf{A}\boldsymbol{y}$. By setting parameters appropriately, the work [24] showed that the distributions $(\mathbf{A}, \boldsymbol{u}, \boldsymbol{y})$ generated in the two ways (with trapdoor and without trapdoor) are statistically indistinguishable. Moreover, this idea can be generalized to achieve a weaker version of our task where $\mathbf{R}$ is given in the first stage – we simply pre-sample $\boldsymbol{y}$, set $\boldsymbol{u} = [\mathbf{A}|\mathbf{AR}]\boldsymbol{y}$, and output $(\mathbf{A}, \mathbf{AR}, \boldsymbol{u}, \boldsymbol{y})$. In fact, this approach has been explored by prior work [2] in the context of functional encryption (more precisely PHPE). Due to the technical barrier that $\mathbf{R}$ must be given in the first stage, schemes using this approach achieve a weak notion of very selective PHPE, where the adversary needs to commit to the challenge index and 1-key query at the beginning. We will elaborate more on the connection of FE and PHPE later.

As we discuss above, the challenge comes from the fact that if $\mathbf{R}$ is only given in the second stage, the prior two-stage sampling method cannot generate $\boldsymbol{u}$ in a way that depends on $\mathbf{R}$. To tackle this, we aim to "eliminate" the effect of this matrix $\mathbf{R}$ in the two-stage sampling process. In particular, we observe that if the matrix $\mathbf{R}$ has a small norm, we can "smudged" its effect by using a distribution with some larger parameter. With this intention in mind, we propose the following new two-stage sampling method:

- In the first stage, generate a random $\mathbf{A}$, and *pre-sample* $\boldsymbol{x}$ from a discrete Gaussian for some larger parameter ρ. Set $\boldsymbol{u} = \mathbf{A}\boldsymbol{x}$.
- In the second stage when $\mathbf{R}$ is given, sample $\boldsymbol{z}$ from a discrete Gaussian with a smaller parameter s, and then output $\boldsymbol{y} = \begin{pmatrix} \boldsymbol{x} - \mathbf{R}\boldsymbol{z} \\ \boldsymbol{z} \end{pmatrix}$.
- The sampler outputs $(\mathbf{A}, \mathbf{AR}, \boldsymbol{u}, \boldsymbol{y})$ at the end.

Clearly the output $\boldsymbol{y}$ satisfies $[\mathbf{A}|\mathbf{AR}]\boldsymbol{y} = \boldsymbol{u}$. If $\rho \gg s\|\mathbf{R}\|$, then we can intuitively think that $\boldsymbol{x}$ smudges $\mathbf{R}\boldsymbol{z}$, so $\boldsymbol{y} = \begin{pmatrix} \boldsymbol{x} - \mathbf{R}\boldsymbol{z} \\ \boldsymbol{z} \end{pmatrix}$ behaves like $\boldsymbol{y}' = \begin{pmatrix} \boldsymbol{x}' \\ \boldsymbol{z} \end{pmatrix}$ such that $[\mathbf{A}|\mathbf{AR}]\boldsymbol{y}' = \boldsymbol{u}$. By formalizing this idea, this task is achieved.

Improving FE with Private Index. Our two-stage sampling method can significantly improve FE with private index of [2]. Before presenting our insights, we first briefly review the framework of [2].

At a high level, [2] constructed FE in the following steps:

(1a) Construct a $(1, \mathsf{poly})$ very-selective partially hiding predicate encryption (PHPE) where the adversary needs to commit to the challenge index and 1-key query at the beginning of the security experiment.

(1b) Upgrade the basic scheme to $(1, \mathsf{poly})$ semi-adaptive PHPE by composing the basic scheme with ALS-FE for inner products [5].

(2) Upgrade the $(1, \mathsf{poly})$ semi-adaptive PHPE to (Q, poly) semi-adaptive PHPE. Here the transformation preserves succinctness of ciphertexts and only incurs an additive blow up of $O(Q^2)$.

(3) Transform the (Q, poly) semi-adaptive PHPE to (Q, poly) semi-adaptive FE. This step follows from [29] and an additional technique of adding smudging noise over the ciphertexts.

We notice that Step (3) is generic, so it suffices to focus on improving PHPE in Steps (1a)–(2). To facilitate presentation of our new ideas, we next identify the following four limitations in the current framework.

- First, Steps (1a) and (1b) require the adversary to commit to his 1-key challenge query before asking further 0-key queries.
- Second, the step (1b) uses a composition of FE over another FE, which could be overly complicated and inefficient.
- Third, Step (3) does not support post-challenge 1-key queries.
- Fourth, Step (3) incurs an additive overhead of $O(Q^2)$, which is incomparable with the multiplicative $O(Q)$ overhead the recent work by [10].

Next, we present our new insights to break all these limitations! To describe how our techniques work, we start with a highly simplified description of the very selective PHPE of [2]: the master public key contains matrices $\mathbf{A}$, $\mathbf{B}_1, \ldots, \mathbf{B}_\ell$ for ℓ being the length of the index (private and public combined), and a matrix $\mathbf{P}$. Given a key query f, the key generation algorithm defines another related function C_f and computes $\mathbf{B}_{C_f}$ from $\mathbf{B}_1, \ldots, \mathbf{B}_\ell$ by the technique of key homomorphic evaluation [11]. Then the key generation algorithm samples $\mathsf{sk}_f := \mathbf{Y}$ such that $[\mathbf{A}|\mathbf{B}_{C_f}] \cdot \mathbf{Y} = \mathbf{P}$. Clearly, this sampling task can be easily performed if the trapdoor of $\mathbf{A}$ is given.

In the proof of security, the trapdoor of $\mathbf{A}$ is not given. Yet we can set $\mathbf{B}_i := \mathbf{A}{\cdot}\mathbf{R}_i{+}\mathsf{x}_i^*\mathbf{G}$ for challenge index $\mathsf{x}^* = (\mathsf{x}_1^*, \ldots, \mathsf{x}_\ell^*)$. (Note that here we do not need to distinguish public/private index to demonstrate our idea.) Then by the key homomorphic evaluation method, we have $[\mathbf{A}|\mathbf{B}_{C_f}] = [\mathbf{A}|\mathbf{A}\mathbf{R}_{C_f}{+}C_f(\mathsf{x}^*)\mathbf{G}]$. From the design of C_f, we have $C_f(\mathsf{x}^*) = 0$ if the key query f corresponds to a 1-key with respect to x^*, or otherwise $C_f(\mathsf{x}^*) \neq 0$ if the key query corresponds to a 0-key. Therefore in the security analysis, one can clearly answer any 0-key queries as the $\mathbf{G}$-trapdoor appears in the second matrix.

At this moment, the reader can already see that answering the 1-key query corresponds to the two-stage sampling as we describe above. In fact, the reason why [2] starts with the very selective notion comes from the fact that the prior technique requires $\mathbf{R}_{C_f}$ to be given in the first stage. This requires the adversary to commit to the challenge 1-key function f and the challenge index at the beginning of the security experiment.

Note that by using our new two-stage sampling method for the key generation algorithm, we are able to answer the 1-key query at any moment just before the challenge ciphertext. Therefore, we can achieve (1, poly) selective FE, allowing arbitrary pre-challenge key queries. Moreover by the very light-weight method of [17,30], the FE can be upgraded to semi-adaptive security[5]. This solves the first two limitations, giving an improved way to achieve (1a) + (1b) of [2].

To further break the third and fourth limitations, we first briefly overview the transformation in Step (2) of [2]. At a high level, besides $\mathbf{A}, \mathbf{B}_1, \ldots, \mathbf{B}_\ell$,

[5] The reason why [2] cannot apply the light-weight method is because its basic construction only achieves very selective security, whereas the technique of [17,30] can be applied to the selective security only over index.

the method generate additional matrices $\mathbf{P}_1, \ldots, \mathbf{P}_N$. The key generation would choose a small subset $\Delta \subseteq [N]$ of some fixed cardinality and generate $\mathsf{sk}_f := \mathbf{Y}$ such that $[\mathbf{A}|\mathbf{B}_{C_f}] \cdot \mathbf{Y} = \mathbf{P}_\Delta$, where $\mathbf{P}_\Delta = \sum_{i\in\Delta} \mathbf{P}_i$. To encrypt a message μ, the encryption algorithm just additionally generates $\boldsymbol{\beta}_{1,i} = \boldsymbol{s}^\top \mathbf{P}_i + \boldsymbol{e} + \frac{q}{2|\Delta|}\mu$ for all $i \in [N]$. The decryption algorithm can figure out $\boldsymbol{\beta}_{1,\Delta} = \sum_{i\in\Delta} \boldsymbol{\beta}_{1,i} = \boldsymbol{s}^\top \mathbf{P}_i + \boldsymbol{e}' + \frac{q}{2}\mu$, and the rest of the procedure is similar to the $(1, \mathsf{poly})$-PHPE. The work [2] requires that for Q randomly sampled sets $\Delta_1, \ldots, \Delta_Q$ in $[N]$, it is overwhelming that the sets are cover-free. By using the result of [27], this would require $N = O(Q^2)$. This explains why the transformation incurs an additive $O(Q^2)$ overhead.

To further reduce the parameter N, it suffices to generate cover-free sets more efficiently. We then construct a simple set sampler that only requires requires $N = O(Q)$, inspired by an implicit construction in the work [10]. We identify that this more efficient cover-freeness suffices for the rest of the proof.

Finally, we show how to handle post-challenge key queries if the message space is small, e.g., bit encryption. (Here we do not need to place a constraint on the index length.) Our idea is to share the plaintext $\mu \in \{0, 1\}$, more precisely, $\frac{q}{2}\mu$, into $\mu_1, \ldots, \mu_N$, such that any subset Δ with some fixed cardinality would recover the message, i.e., $\frac{q}{2}\mu = \sum_{i\in\Delta} \mu_i$. Then we generate ciphertexts $\boldsymbol{\beta}_{1,i} = \boldsymbol{s}^\top \mathbf{P}_i + \boldsymbol{e} + \mu_i$ for all $i \in [N]$. As a critical proof insight, we show that given all secret keys of the form $(\Delta, \mathbf{Y})$, one can only learn $\sum_{i\in\Delta} \mu_i = \frac{q}{2}\mu$ but nothing more. By using this fact, we can design a simulator, who generates simulated shares $\mu_1, \ldots, \mu_N$ and $2Q$ sets $\Delta_1, \ldots, \Delta_Q, \Delta'_1, \ldots, \Delta'_Q$ such that $\sum_{\Delta_i} \mu_i = q/2$, and $\sum_{\Delta'_i} \mu_i = 0$. Thus in the post-challenge stage, the simulator can answer a 1-key query by using either $\{\Delta_i\}$ or $\{\Delta'_i\}$ according to whether $\mu = 1$ or $\mu = 0$.

Notice that the core and useful properties of the above process are that: (1) the simulation of the ciphertext does not depend on the plaintext μ; (2) the post-challenge key simulation can consistently generate a key that opens the simulated ciphertext to either $\mu = 1$ or $\mu = 0$. By further taking fine care of the details, we are able to achieve (Q, poly)-PHPE that supports arbitrary key queries and has succinct ciphertext that grows additively with $O(Q)$. This solves the third and fourth challenges as above and improves Step (2) of [2]. Clearly, this PHPE can also be transformed into an FE, following Step (3) as [2].

Improving FE with Public Index. Interestingly, the lattice-based construction of FE with public index [1] faces exactly the same technical challenge as the very selective PHPE of [2]. Our new two-stage sampling method is the key missing link of [1] to achieve adaptive IB $\times$ IP and semi-adaptive AB $\times$ IP. We further elaborate on this setting in Sect. 6. The reader would immediately see the point even just with a glance at the construction.

1.3 Other Related Work

We notice that FE can be obtained from indistinguishable obfuscation ($i\mathcal{O}$) [21], achieving the notion of $(\mathsf{poly}, \mathsf{poly})$-IND adaptive security via [9]. Even though recently there has been substantial progress for instantiating $i\mathcal{O}$ [15,22,31], the derived FE (as is) cannot achieve the simulation-based security. This is because the $i\mathcal{O}$-based FE has ciphertext length independent of the number of collusion Q, and thus according to the lower bound of [4], the scheme cannot be SIM secure. Moreover as mentioned in [13,27], IND-based FE does not imply SIM-based FE. Therefore for the direction of SIM-based FE, our work would shed light on new methods and feasibilities beyond what can be implied from the recent progress on the direction of $i\mathcal{O}$ [15,22,31].

In [18], Canetti and Chen show that a single key SIM-secure private-key FE suffices to construct reusable garbled circuits. Compared with the reusable garbled circuits derived from our (Q, poly)-SA-SIM FE with $Q = 1$,[6], the construction in [18] achieves the stronger adaptive security with respect to index without the complexity leveraging argument, yet can only support either a pre- or post-challenge key query for a NC1 circuit, rather than a general circuit.

2 Preliminaries

2.1 Notations

In this paper, $\mathbb{N}$, $\mathbb{Z}$ and $\mathbb{R}$ denote the sets of natural numbers, integers and real numbers, respectively. We use λ to denote the security parameter, which is the implicit input for all algorithms in this paper. A function $f(\lambda) > 0$ is negligible and denoted by $\mathsf{negl}(\lambda)$ if for any $c > 0$ and sufficiently large λ, $f(\lambda) < 1/\lambda^c$. A probability is called overwhelming if it is $1 - \mathsf{negl}(\lambda)$. A column vector is denoted by a bold lower case letter (e.g., $\boldsymbol{x}$). A matrix is denoted by a bold upper case letter (e.g., $\mathbf{A}$), and its transposition is denoted by $\mathbf{A}^\top$.

For a set D, we denote by $u \xleftarrow{\$} D$ the operation of sampling a uniformly random element u from D, and denote $|u|$ as the bit length of u. For an integer $\ell \in \mathbb{N}$, we use U_ℓ to denote the uniform distribution over $\{0,1\}^\ell$. Given a randomized algorithm or function $f(\cdot)$, we use $y \leftarrow f(x)$ to denote y as the output of f and x as input. For a distribution X, we denote by $x \leftarrow X$ the operation of sampling a random x according to the distribution X. Given two different distributions X and Y over a countable domain D, we denote their statistical distance as $\mathsf{SD}(X, Y) = \frac{1}{2}\sum_{d \in D} |X(d) - Y(d)|$, and say that X and Y are $\mathsf{SD}(X, Y)$ close. Moreover, if $\mathsf{SD}(X, Y)$ is negligible in λ, we say that the two distributions are statistically close, which is always denoted by $X \overset{s}{\approx} Y$. If for any PPT algorithm $\mathcal{A}$ that $\left|\Pr[\mathcal{A}(1^\lambda, X) = 1] - \Pr[\mathcal{A}(1^\lambda, Y) = 1]\right|$ is negligible in λ, then we say that the two distributions are computationally indistinguishable, denoted by $X \overset{c}{\approx} Y$.

Matrix Norms. For a vector $\boldsymbol{x}$, its Euclidean norm (also known as the ℓ_2 norm) is defined as $\|\boldsymbol{x}\| = (\sum_i x_i^2)^{1/2}$. For a matrix $\mathbf{R}$, we denote its ith column vector as $\boldsymbol{r}_i$, and use $\widetilde{\mathbf{R}}$ to denote its Gram-Schmidt orthogonalization. In addition,

[6] Notice that the reusable garbled circuits following from our SIM-secure FE can achieve SA-SIM security, and support general circuits and any arbitrary pre- and post-challenge key query, for one query.

- $\|\mathbf{R}\|$ denotes the Euclidean norm of $\mathbf{R}$, i.e., $\|\mathbf{R}\| = \max_i \|\boldsymbol{r}_i\|$.
- $s_1(\mathbf{R})$ denotes the spectral norm of $\mathbf{R}$, i.e., $s_1(\mathbf{R}) = \sup_{\|\boldsymbol{x}\|=1} \|\mathbf{R}\boldsymbol{x}\|$, with $\boldsymbol{x} \in \mathbb{Z}^m$.

We know the facts on the above norms: $\|\widetilde{\mathbf{R}}\| \leq \|\mathbf{R}\| \leq s_1(\mathbf{R}) \leq \sqrt{k}\|\mathbf{R}\|$ and $s_1(\mathbf{R}|\mathbf{S}) \leq \sqrt{s_1(\mathbf{R})^2 + s_1(\mathbf{S})^2}$, where k denote the number of columns of $\mathbf{R}$. Besides, we have the following lemma for the bounding spectral norm.

Lemma 2.1 ([20]). *Let $\mathbf{X} \in \mathbb{R}^{n\times m}$ be a subgaussian random matrix with parameter s. There exists a universal constant $c \approx 1/\sqrt{2\pi}$ such that for any $t > 0$, we have $s_1(\mathbf{X}) \leq c \cdot s \cdot (\sqrt{m} + \sqrt{n} + t)$ except with probability at most $\frac{2}{e^{\pi t^2}}$.*

At the same time, we rely on the following useful lemma on cover-free for our security proof.

Lemma 2.2 (Cover-Freeness [27]). *Let $\Delta_1, \cdots, \Delta_Q \subseteq [N]$ be randomly chosen subsets of size v. Let $v(\kappa) = \Theta(\kappa)$ and $N(\kappa) = \Theta(vQ^2)$. Then for all $i \in [Q]$, we have $\Pr\left[\Delta_i \backslash \left(\bigcup_{j\neq i} \Delta_j\right) \neq \phi\right] = 1 - 2^{-\Omega(\kappa)}$, where the probability is over the random choice of subsets $\Delta_1, \cdots, \Delta_Q$.*

2.2 Gaussians on Lattices

Due to space limit, we defer well-known background on lattices to the full version of this paper. Here we just give some useful preliminaries of gaussians on lattices.

Let σ be any positive real number. The Gaussian distribution $\mathcal{D}_{\sigma,\mathbf{c}}$ with parameter σ and $\mathbf{c}$ is defined by probability distribution function $\rho_{\sigma,\mathbf{c}}(\boldsymbol{x}) = exp(-\pi\|\boldsymbol{x} - \mathbf{c}\|^2/\sigma^2)$. For any set $S \subseteq \mathbb{R}^m$, define $\rho_{\sigma,\mathbf{c}}(S) = \sum_{\boldsymbol{x}\in S} \rho_{\sigma,\mathbf{c}}(\boldsymbol{x})$. The discrete Gaussian distribution $\mathcal{D}_{S,\sigma,\mathbf{c}}$ over S with parameter σ and $\mathbf{c}$ is defined by the probability distribution function $\rho_{\sigma,\mathbf{c}}(\boldsymbol{x}) = \rho_{\sigma,\mathbf{c}}(\boldsymbol{x})/\rho_{\sigma,\mathbf{c}}(S)$ for all $\boldsymbol{x} \in S$.

In [34], Micciancio and Regev introduced a useful quantity called smoothing parameter.

Definition 2.3. *For any m-dimensional lattice Λ and positive real $\epsilon > 0$, the smoothing parameter $\eta_\epsilon(\Lambda)$ is the smallest real $s > 0$ such that $\rho_{1/s}(\Lambda^* \backslash \{0\}) \leq \epsilon$.*

Then, we have the following upper bound for the smoothing parameter.

Lemma 2.4 ([24]). *For any m-dimensional lattice Λ and real $\epsilon > 0$, we have $\eta_\epsilon(\Lambda) \leq \frac{\sqrt{\log(2m/(1+1/\epsilon))/\pi}}{\lambda_1^\infty(\Lambda^*)}$. Then for any $\omega(\sqrt{\log m})$ function, there is a negligible $\epsilon(m)$ for which $\eta_\epsilon(\Lambda) \leq \omega(\sqrt{\log m})/\lambda_1^\infty(\Lambda^*)$.*

Furthermore, we have the following useful facts from the literature.

Lemma 2.5 ([24] **and Full Version of** [32]). *Let n, m, q are integers such that $m > 2n \log q$. Then for all but an at most q^{-n} fraction of $\mathbf{A} \in \mathbb{Z}_q^{n\times m}$, we have $\lambda_1^\infty(\Lambda_q(\mathbf{A})) > q/4$.*

Furthermore, for such $\mathbf{A}$ and any function $\omega(\sqrt{\log m})$, there is a negligible function $\varepsilon(m)$ such that $\eta_\varepsilon(\Lambda_q^\perp(\mathbf{A})) \leq \omega(\sqrt{\log m})$.

Lemma 2.6 *Let n, m, q are integers such that $m > 2n \log q$, and $\mathbf{R} \in \mathbb{Z}_q^{m \times m}$ be arbitrary. Then for all but an at most q^{-n} fraction of $\mathbf{A} \in \mathbb{Z}_q^{n \times m}$, we have $\lambda_1^{\infty}(\Lambda_q(\mathbf{A}|\mathbf{AR})) > q/4$.*

Furthermore, for such $\mathbf{A}$ and any function $\omega(\sqrt{\log m})$, there is a negligible function $\varepsilon(m)$ such that $\eta_\varepsilon(\Lambda_q^{\perp}(\mathbf{A}|\mathbf{AR})) \leq \omega(\sqrt{\log m})$.

Due to space limit, we defer the proof of Lemma 2.6 to full version.

Lemma 2.7 **([24], Lemma 5.2).** *Assume the columns of $\mathbf{A}$ generate $\mathbb{Z}_q^n$, let $\epsilon \in (0, 1/2)$ and $r \geq \eta_\epsilon(\Lambda^{\perp}(\mathbf{A}))$. Then for $\mathbf{e} \leftarrow \mathcal{D}_{\mathbb{Z}^m, r}$, the distribution of $\boldsymbol{u} = \mathbf{A}^T\mathbf{e} \bmod q$ is within statistical distance 2ϵ of uniform over $\mathbb{Z}_q^n$.*

Furthermore, for any fixed $\boldsymbol{u} \in \mathbb{Z}_q^n$, let $\mathbf{t} \in \mathbb{Z}^m$ be an arbitrary solution to $\mathbf{At} = \boldsymbol{u} \bmod q$. Then the conditional distribution of $e \sim \mathcal{D}_{\mathbb{Z}^m, s}$ given $\mathbf{Ae} = \boldsymbol{u} \bmod q$ is exactly $\boldsymbol{t} + \mathcal{D}_{\Lambda^{\perp}, s, -\boldsymbol{t}}$.

Lemma 2.8 **([34], Lemma 4.4).** *For any m-dimensional lattice Λ, $\mathrm{c} \in \mathbf{R}^m$, real $\epsilon \in (0, 1)$ and $s \geq \eta_\epsilon(\Lambda)$, we have $\Pr_{\boldsymbol{x} \leftarrow \mathcal{D}_{\Lambda, s, \mathrm{c}}}[\|\boldsymbol{x} - \mathbf{c}\| > s\sqrt{m}] \leqslant \frac{1+\epsilon}{1-\epsilon} \cdot 2^{-m}$.*

Lemma 2.9 (Smudging Lemma). *Let $n \in \mathbb{N}$. For any real $\sigma \geq \omega(\sqrt{\log n})$, and any $\mathbf{c} \in \mathbb{Z}^n$, it holds $\mathsf{SD}(\mathcal{D}_{\mathbb{Z}^n, \sigma}, \mathcal{D}_{\mathbb{Z}^n, \sigma, \mathbf{c}}) \leq \|\mathbf{c}\|/\sigma$.*

Learning With Errors. The Learning with Errors problem, or LWE, is the problem of determining a secret vector over $\mathbb{Z}_q$ given a polynomial number of "noisy" inner products. The decision variant is to distinguish such samples from random. More formally, we define the problem as follows:

Definition 2.10 **([37]).** *Let $n \geq 1$ and $q \geq 2$ be integers, and let χ be a probability distribution on $\mathbb{Z}_q$. For $\boldsymbol{s} \in \mathbb{Z}_q^n$, let $A_{\boldsymbol{s},\chi}$ be the probability distribution on $\mathbb{Z}_q^n \times \mathbb{Z}_q$ obtained by choosing a vector $\boldsymbol{a} \in \mathbb{Z}_q^n$ uniformly at random, choosing $e \in \mathbb{Z}_q$ according to χ and outputting $(\boldsymbol{a}, \langle \boldsymbol{a}, \boldsymbol{s} \rangle + e)$.*

The decision $\mathsf{LWE}_{q,n,\chi}$ problem is: for uniformly random $\boldsymbol{s} \in \mathbb{Z}_q^N$, given a poly(n) number of samples that are either (all) from $A_{\boldsymbol{s},\chi}$ or (all) uniformly random in $\mathbb{Z}_q^n \times \mathbb{Z}_q$, output 0 if the former holds and 1 if the latter holds.

We say the decision-$\mathsf{LWE}_{q,n,\chi}$ problem is infeasible if for all polynomial-time algorithms $\mathcal{A}$, the probability that $\mathcal{A}$ solves the decision-$\mathsf{LWE}_{q,n,\chi}$ problem (over $\boldsymbol{s}$ and $\mathcal{A}$'s random coins) is negligibly close to $1/2$ as a function of n. The works of [16,36,37] show that the LWE assumption is as hard as (quantum or classical) solving GapSVP and SIVP under various parameter regimes.

2.3 Lattice Trapdoor and Gaussian Sampling

Gadget Matrix. We recall the "gadget matrix" $\mathbf{G}$ defined in [33]. The "gadget matrix" $\mathbf{G} = \mathbf{I}_n \otimes \boldsymbol{g}^\top \in \mathbb{Z}_q^{n \times n\lceil \log q \rceil}$ where $\boldsymbol{g}^\top = (1, 2, 4, ..., 2^{\lceil \log q \rceil - 1})$. We can also extend the column dimension to any $m \geq n\lceil \log q \rceil$ by padding $\mathbf{0}_{n \times m'}$ to the right for $m' = (m - n\lceil \log q \rceil)$, i.e., $\mathbf{G} = [\mathbf{I}_n \otimes \boldsymbol{g}^\top | \mathbf{0}_{n \times m'}] \in \mathbb{Z}_q^{n \times m}$.

Lemma 2.11 (Theorem 4.1, [33]). *Let $q \geq 2$ be any integer, and $n, m \geq 2$ be integers with $m \geq n\lceil \log q \rceil$. There is a full-rank (of columns) matrix $\mathbf{G} \in \mathbb{Z}_q^{n\times m}$ such that the lattice $\Lambda_q^{\perp}(\mathbf{G})$ has a publicly known trapdoor matrix $\mathbf{T_G} \in \mathbb{Z}^{n\times m}$ with $\|\widetilde{\mathbf{T}}_\mathbf{G}\| \leq \sqrt{5}$, where $\widetilde{\mathbf{T}}_\mathbf{G}$ is the Gram-Schmidt orthogonalization of $\mathbf{T_G}$.*

Theorem 2.12 (Trapdoor Generation [8,33]). *There is a probabilistic polynomial-time algorithm* $\mathsf{TrapGen}(1^n, q, m)$ *that for all $m \geq m_0 = m_0(n, q) = O(n \log q)$, outputs $(\mathbf{A}, \mathbf{T_A})$ such that $\mathbf{A} \in \mathbb{Z}_q^{n\times m}$ is within statistical distance 2^{-n} from uniform, and $\mathbf{T_A}$ is a basis for $\Lambda_q^{\perp}(\mathbf{A})$ satisfying $\|\mathbf{T_A}\| \leq O(n \log q)$ and $\|\widetilde{\mathbf{T}}_\mathbf{A}\| \leq O(\sqrt{n \log q})$, where $\widetilde{\mathbf{T}}_\mathbf{A}$ denotes the Gram-Schmidt orthogonalization of $\mathbf{T_A}$.*

Lemma 2.13 (SampleLeft [3,19]). *Let $q > 2$, $\mathbf{A}, \mathbf{B} \in \mathbb{Z}_q^{n\times m}$ be two full rank matrices with $m > n$, $\mathbf{T_A}$ be a trapdoor matrix for $\mathbf{A}$, a matrix $\mathbf{U} \in \mathbb{Z}_q^{n\times \ell}$ and $s \geq \|\widetilde{\mathbf{T}}_\mathbf{A}\| \cdot \omega(\sqrt{\log m})$. Then there exists a* PPT *algorithm* $\mathsf{SampleLeft}(\mathbf{A}, \mathbf{T_A}, \mathbf{B}, \mathbf{U}, s)$ *that outputs a matrix $\mathbf{X} \in \mathbb{Z}_q^{2m\times \ell}$, which is distributed statistically close to $D_{\Lambda_q^{\mathbf{U}}(\mathbf{A}|\mathbf{B}),s}$.*

Lemma 2.14 (SampleRight [33]). *Let $q > 2$, $\mathbf{A} \in \mathbb{Z}_q^{n\times m}$ be a full rank matrix with $m > n$, $\mathbf{R} \in \mathbb{Z}^{m\times m}$, $\mathbf{U} \in \mathbb{Z}_q^{n\times \ell}$, $y \in \mathbb{Z}_q$ with $y \neq 0$, and $s \geq \sqrt{5} \cdot s_1(\mathbf{R}) \cdot \omega(\sqrt{\log m})$. Then there exists a* PPT *algorithm* $\mathsf{SampleRight}(\mathbf{A}, \mathbf{R}, y, \mathbf{U}, s)$ *that outputs a matrix $\mathbf{X} \in \mathbb{Z}_q^{2m\times \ell}$, which is distributed statistically close to $D_{\Lambda_q^{\mathbf{U}}(\mathbf{A}|\mathbf{A}\cdot\mathbf{R}+y\mathbf{G}),s}$, where $\mathbf{G}$ is the gadget matrix.*

2.4 Partially Hiding Predicate Encryption

We recall the notation of partially hiding predicate encryption (PHPE) proposed by [29], which interpolates attribute-based encryption and predicate encryption. A Partially-Hiding Predicate Encryption scheme PHPE for a pair of private-public index spaces $\mathcal{X}, \mathcal{Y}$, a function class $\mathcal{F}$ mapping $\mathcal{X} \times \mathcal{Y}$ to $\{0, 1\}$, and a message space $\mathcal{M}$, consists of four algorithms
($\mathsf{PHPE.Setup}$, $\mathsf{PHPE.Enc}$, $\mathsf{PHPE.KeyGen}$, $\mathsf{PHPE.Dec}$):

$\mathsf{PHPE.Setup}(1^\lambda, \mathcal{X}, \mathcal{Y}, \mathcal{F}, \mathcal{M}) \rightarrow (\mathsf{PHPE.mpk}, \mathsf{PHPE.msk})$. The setup algorithm gets as input the security parameter λ and a description of $(\mathcal{X}, \mathcal{Y}, \mathcal{F}, \mathcal{M})$ and outputs the public parameter $\mathsf{PHPE.mpk}$, and the master key $\mathsf{PHPE.msk}$.

$\mathsf{PHPE.Enc}(\mathsf{PHPE.mpk}, (\boldsymbol{x}, \boldsymbol{y}), \mu) \rightarrow \mathsf{ct}_{\boldsymbol{y}}$. The encryption algorithm gets as input $\mathsf{PHPE.mpk}$, a pair of private-public indexes $(\boldsymbol{x}, \boldsymbol{y}) \in \mathcal{X} \times \mathcal{Y}$ and a message $\mu \in \mathcal{M}$. It outputs a ciphertext $\mathsf{ct}_{\boldsymbol{y}}$.

$\mathsf{PHPE.KeyGen}(\mathsf{PHPE.msk}, f) \rightarrow \mathsf{sk}_f$. The key generation algorithm gets as input $\mathsf{PHPE.msk}$ and a function $f \in \mathcal{F}$. It outputs a secret key sk_f.

$\mathsf{PHPE.Dec}((\mathsf{sk}_f, f), (\mathsf{ct}_{\boldsymbol{y}}, \boldsymbol{y})) \rightarrow \mu \vee \bot$. The decryption algorithm gets as input the secret key sk_f, a function f, and a ciphertext $\mathsf{ct}_{\boldsymbol{y}}$ and the public part $\boldsymbol{y}$ of the attribute vector. It outputs a message $\mu \in \mathcal{M}$ or $\bot$.

Correctness. We require that for all $(\mathsf{PHPE.mpk}, \mathsf{PHPE.msk}) \leftarrow \mathsf{PHPE.Setup}(1^\lambda, \mathcal{X}, \mathcal{Y}, \mathcal{F}, \mathcal{M})$, for all $(\boldsymbol{x}, \boldsymbol{y}, f) \in \mathcal{X} \times \mathcal{Y} \times \mathcal{F}$ and for all $\mu \in \mathcal{M}$,

- For 1-queries, i.e., $f(\boldsymbol{x}, \boldsymbol{y}) = 1$, $\Pr[\mathsf{PHPE.Dec}((\mathsf{sk}_f, f), (\mathsf{ct}_{\boldsymbol{y}}, \boldsymbol{y})) \neq \mu] \leq \mathsf{negl}(\lambda)$.
- For 0-queries, i.e., $f(\boldsymbol{x}, \boldsymbol{y}) = 0$, $\Pr[\mathsf{PHPE.Dec}((\mathsf{sk}_f, f), (\mathsf{ct}_{\boldsymbol{y}}, \boldsymbol{y})) \neq \perp] \leq \mathsf{negl}(\lambda)$.

Due to space limit, we defer the full security definition of PHPE to full version.

3 Definitions of Functional Encryption

We first present the syntax of functional encryption.

Definition 3.1 (Functional Encryption). *Let $\mathcal{F}$ be a family of functions, where each $f \in \mathcal{F}$ is defined as $f : \mathcal{U} \to \mathcal{Y}$. A functional encryption (*FE*) scheme for $\mathcal{F}$ consists of four algorithms as follows.*

- $\mathsf{Setup}(1^\lambda, \mathcal{F})$: *Given as input the security parameter λ and a description of the function family $\mathcal{F}$, the algorithm outputs a pair of master public key and master secret key* $(\mathsf{mpk}, \mathsf{msk})$. *In the following algorithms,* mpk *is implicitly assumed to be part of their inputs.*
- $\mathsf{KeyGen}(\mathsf{msk}, f \in \mathcal{F})$: *Given as input the master secret key* msk *and a function $f \in \mathcal{F}$, the algorithm outputs a description key sk_f.*
- $\mathsf{Enc}(\mathsf{mpk}, u \in \mathcal{U})$: *Given as input the master public key and a message $u \in \mathcal{U}$, the algorithm outputs a ciphertext* ct.
- $\mathsf{Dec}(\mathsf{sk}_f, \mathsf{ct})$: *Given as input the secret key sk_f and a ciphertext* ct, *the algorithm outputs a value $y \in \mathcal{Y}$ or $\perp$ if it fails.*

A functional encryption scheme is correct, if for all security parameter λ, any message $u \in \mathcal{U}$ and any function $f \in \mathcal{F}$, the decryption algorithm outputs the right outcome, i.e., $\Pr[\mathsf{Dec}(\mathsf{sk}_f, \mathsf{ct}_u) = f(u)] \geq 1 - \mathsf{negl}(\lambda)$, where the probability is taken over $(\mathsf{mpk}, \mathsf{msk}) \leftarrow \mathsf{Setup}(1^\lambda, \mathcal{F})$, $\mathsf{sk}_f \leftarrow \mathsf{KeyGen}(\mathsf{msk}, f)$, $\mathsf{ct}_u \leftarrow \mathsf{Enc}(u)$.

More Fine-Grained Syntax of FE. For FE with fine-grained syntax, each message u consists of two parts, namely $u := (\mathsf{x}, \mu)$, where $\mathsf{x} \in \mathcal{X}$ for some index (or attribute) space $\mathcal{X}$, and $\mu \in \mathcal{M}$ for message space $\mathcal{M}$. Additionally, each function f consists of two parts, namely, $f := (\mathsf{P}, g) \in \mathcal{P} \times \mathcal{G}$, where P is a predicate over the index space $\mathcal{X}$, and g is a function of the message space $\mathcal{M}$. The overall function acts as the following:

$$f(u) := \begin{cases} g(\mu) & \text{if } \mathsf{P}(\mathsf{x}) = 1 \\ \perp & \text{otherwise.} \end{cases} \tag{1}$$

Therefore, when decrypting the ciphertext $\mathsf{ct}_u = \mathsf{Enc}(\mathsf{mpk}, (\mathsf{x}, \mu))$ by $\mathsf{sk}_f = \mathsf{KeyGen}(\mathsf{msk}, (\mathsf{P}, g))$, the algorithm outputs $g(\mu)$ if $\mathsf{P}(\mathsf{x}) = 1$, and $\perp$ otherwise. Under this fine-grained syntax, we call a key $\mathsf{sk}_{f:=(\mathsf{P},g)}$ a 1-key with respect to an index x if $\mathsf{P}(\mathsf{x}) = 1$, or otherwise a 0-key. Intuitively, a 1-key is allowed to open the ciphertext, but a 0-key is not.

To differentiate the regular FE in Definition 3.1 and FE with the fine-grained syntax, we use different types of function classes, i.e., FE for $\mathcal{F}$ refers to the former and FE for $\mathcal{P} \times \mathcal{G}$ refers to the latter.

There are two important types of index studied in the literature – FE with private or public index, according to whether the index x is revealed to the decryption algorithm or not.

Our security notions simply follow from those in prior work [2,13,27]. It is important that for the simulation-based security, we can achieve a notion where any pre- and post-challenge key queries are allowed, while the prior work [2] requires the adversary to commit in one-shot to all the 1-key queries right after seeing the master public key. Due to space limit, we defer the detailed security notions of interests on these two cases and comparisons between the notions in related work to the full version of this paper.

4 Our New Two-Stage Sampling Method

In this section, we present our key technical contribution – a new two-stage sampling method. At a high level, we would like to sample the following two-stage distribution: (1) in the first stage, a random matrix $\mathbf{A}$ and a random vector $\boldsymbol{u}$ are sampled, and (2) in the second stage, an arbitrary small-norm matrix $\mathbf{R}$ is given, and then some short vector $\boldsymbol{y}$ is sampled conditioned on $[\mathbf{A}|\mathbf{AR}]\boldsymbol{y} = \boldsymbol{u}$. The distribution then outputs $(\mathbf{A}, \mathbf{AR}, \boldsymbol{u}, \boldsymbol{y})$.

For a simpler case where there is no $\mathbf{R}$, this task is achievable via the following GPV two-stage sampling technique:

Lemma 4.1 ([24]). *For any prime q, integers integer $n \geq 1$, $m \geq 2n \log q$, $s \geq \omega(\sqrt{\log m})$, the following two distributions are statistically indistinguishable:*

- $(\mathbf{A}, \boldsymbol{u}, \boldsymbol{y})$: $\mathbf{A} \xleftarrow{\$} \mathbb{Z}_q^{n\times m}$, $\boldsymbol{u} \xleftarrow{\$} \mathbb{Z}_q^n$, $\boldsymbol{y} \leftarrow \mathcal{D}_{\Lambda_q^{\boldsymbol{u}}(\mathbf{A}),s}$.
- $(\mathbf{A}, \boldsymbol{u}, \boldsymbol{y})$: $\mathbf{A} \xleftarrow{\$} \mathbb{Z}_q^{n\times m}$, $\boldsymbol{y} \leftarrow \mathcal{D}_{\mathbb{Z}^m,s}$, $\boldsymbol{u} = \mathbf{A}\boldsymbol{y} \bmod q$.

Intuitively, we can pre-sample a short vector $\boldsymbol{y}$ from an appropriate Gaussian distribution and then set $\boldsymbol{u} = \mathbf{A}\boldsymbol{y}$. By the indistinguishability as Lemma 4.1, we can sample the desired distribution with or without the trapdoor of $\mathbf{A}$ as desired.[7] Moreover, this idea can be generalized to achieve a weaker version of our task where $\mathbf{R}$ is given in the first stage. The generalized idea has been explored in the context of functional encryption (more precisely PHPE) by prior work [2], yet the technique however, would inherently require to know $\mathbf{R}$ in the first stage, resulting in a weak notion of very selective PHPE, where the adversary needs to commit to the challenge index and 1-key query at the beginning.

[7] To sample $\mathcal{D}_{\Lambda_q^{\boldsymbol{u}}(\mathbf{A}),s}$, the current sampling algorithm requires that $s > \|\widetilde{\mathbf{T}}_{\mathbf{A}}\|\omega(\sqrt{\log m})$. According to the best known (to our knowledge) trapdoor generation, the smallest s we can sample would be $\omega(\sqrt{n \log q} \cdot \sqrt{\log m})$, which is much larger than the required bound for Lemma 4.1.

To break this limitation, we design a new two-stage sampling method that uses smudging noise over keys. Below we first present the two-stage sampling method and then explain the idea behind it.

For any integers $m > n \geq 1, q \geq 2$, we consider the following two procedures:

Sampler-1$(\mathbf{R}, \rho, s)$: Given a matrix $\mathbf{R} \in \mathbb{Z}^{m \times m}$ and two values $\rho, s \in \mathbb{R}$ as input, this sampler conducts the following steps in two stages.

1. Stage 1: (without the need of $\mathbf{R}$)
   - Sample a random matrix $\mathbf{A} \xleftarrow{\$} \mathbb{Z}_q^{n \times m}$;
   - Sample a random vector $\boldsymbol{u} \xleftarrow{\$} \mathbb{Z}_q^n$;
2. Stage 2:
   - Sample a random $\boldsymbol{x} \leftarrow \mathcal{D}_{\mathbb{Z}^m, \rho}$;
   - Compute $\boldsymbol{z} = \boldsymbol{u} - \mathbf{A}\boldsymbol{x} \pmod q$;
   - Sample a vector $\boldsymbol{z}' = \begin{pmatrix} \boldsymbol{z}_1 \\ \boldsymbol{z}_2 \end{pmatrix} \leftarrow \mathcal{D}_{\Lambda_q^{\boldsymbol{z}}(\mathbf{A}|\mathbf{AR}), s}$, satisfying $(\mathbf{A}|\mathbf{AR}) \begin{pmatrix} \boldsymbol{z}_1 \\ \boldsymbol{z}_2 \end{pmatrix} = \boldsymbol{z} \pmod q$;
   - Set $\boldsymbol{y} = \begin{pmatrix} \boldsymbol{x} + \boldsymbol{z}_1 \\ \boldsymbol{z}_2 \end{pmatrix} \in \mathbb{Z}^{2m}$, satisfying $(\mathbf{A}|\mathbf{AR})\boldsymbol{y} = \boldsymbol{u} \pmod q$;
   - Output the tuple $(\mathbf{A}, \mathbf{AR}, \boldsymbol{y}, \boldsymbol{u})$.

The Sampler-1$(\mathbf{R}, \rho, s)$ can be implemented efficiently given the trapdoor $\mathbf{T}_\mathbf{A}$ of $\mathbf{A}$, using the SampleLeft algorithm as Lemma 2.13 (with larger parameters of s than the required bound in Lemma 4.1). Next we present another way to sample the distribution without the need of the trapdoor.

Sampler-2$(\mathbf{R}, \rho, s)$: Given a matrix $\mathbf{R} \in \mathbb{Z}^{m \times m}$ and two values $\rho, s \in \mathbb{R}$ as input, this sampler conducts the following steps in two stages.

1. Stage 1: (without the need of $\mathbf{R}$)
   - Sample a random matrix $\mathbf{A} \xleftarrow{\$} \mathbb{Z}_q^{n \times m}$;
   - Sample a random vector $\boldsymbol{x} \leftarrow \mathcal{D}_{\mathbb{Z}^m, \sqrt{\rho^2 + s^2}}$, and set $\boldsymbol{u} = \mathbf{A}\boldsymbol{x} \pmod q$;
2. Stage 2:
   - Sample a random vector $\boldsymbol{z}_2 \leftarrow \mathcal{D}_{\mathbb{Z}^m, s}$;
   - Compute a vector $\boldsymbol{y} = \begin{pmatrix} \boldsymbol{x} - \mathbf{R}\boldsymbol{z}_2 \\ \boldsymbol{z}_2 \end{pmatrix}$, satisfying $(\mathbf{A}|\mathbf{AR})\boldsymbol{y} = \boldsymbol{u} \pmod q$;
   - Output the tuple $(\mathbf{A}, \mathbf{AR}, \boldsymbol{y}, \boldsymbol{u})$.

In a nutshell, this algorithm first pre-samples a (larger) $\boldsymbol{x}$ and sets $\boldsymbol{u} = \mathbf{A}\boldsymbol{x}$, without knowing $\mathbf{R}$. In the second stage when $\mathbf{R}$ is given, it samples a smaller $\boldsymbol{z}_2$ and adjusts $\boldsymbol{y}$ accordingly. Intuitively, the larger $\boldsymbol{x}$ servers as the smudging noise that "overwrites" the effect of $\mathbf{R}\boldsymbol{z}_2$ as long as the norm of $\boldsymbol{x}$ is super-polynomially larger. This would hide the information of $\mathbf{R}$, which needs to be kept secret as required by the proof framework in prior work [2,3]. We formalize this intuition by the following theorem.

Theorem 4.2. *For integers $q \geq 2$, $n \geq 1$, sufficiently large $m = O(n \log q)$, any $\mathbf{R} \in \mathbb{Z}^{m \times m}$, $s > \omega(\sqrt{\log m})$, and $\rho \geq s\sqrt{m}\|\mathbf{R}\| \cdot \lambda^{\omega(1)}$, the output distributions $(\mathbf{A}, \mathbf{AR}, \boldsymbol{y}, \boldsymbol{u})$ of the above two procedures are statistically close.*

Proof. Our high-level proof idea is to introduce an additional two-stage sampling algorithm Sampler-3, and then prove it statistically indistinguishable from both Sampler-1 and Sampler-2. Below, we describe the algorithm Sampler-3$(\mathbf{R}, \rho, s)$.

Sampler-3$(\mathbf{R}, \rho, s)$: Given a matrix $\mathbf{R} \in \mathbb{Z}_q^{m \times m}$ and two values $\rho, s \in \mathbb{R}$ as input, this sampler conducts the following steps in two stages.

1. Stage 1: Sample a random matrix $\mathbf{A} \stackrel{\$}{\leftarrow} \mathbb{Z}_q^{n \times m}$;
2. Stage 2:
   - Sample two random vectors $\boldsymbol{x}' \leftarrow \mathcal{D}_{\mathbb{Z}^m, \sqrt{\rho^2+s^2}}$, $\boldsymbol{z}_2 \leftarrow \mathcal{D}_{\mathbb{Z}^m, s}$;
   - Compute $\boldsymbol{u} = (\mathbf{A}|\mathbf{AR}) \begin{pmatrix} \boldsymbol{x}' \\ \boldsymbol{z}_2 \end{pmatrix} \pmod{q}$, and denote $\boldsymbol{y} = \begin{pmatrix} \boldsymbol{x}' \\ \boldsymbol{z}_2 \end{pmatrix} \in \mathbb{Z}^{2m}$;
   - Output a tuple $(\mathbf{A}, \mathbf{AR}, \boldsymbol{y}, \boldsymbol{u})$.

Claim 4.3. *For the parameters in the statement of Theorem 4.2, the output distributions of* Sampler-1 *and* Sampler-3 *are statistically close.*

Proof. We first observe that in Sampler-3, the $\boldsymbol{x}'$ component can be decomposed into $\boldsymbol{x} + \boldsymbol{z}_1$ (within a negligible statistical distance), where $\boldsymbol{x} \leftarrow \mathcal{D}_{\mathbb{Z}^m, \rho}$ and $\boldsymbol{z}_1 \leftarrow \mathcal{D}_{\mathbb{Z}^m, s}$. The decomposition holds as we have $\rho > s > \eta_\varepsilon(\mathbb{Z}^m)$ for some $\varepsilon = \mathsf{negl}(\lambda)$.

Next, we prove a generalization of Lemma 4.1 that the following two distributions are statistically close:

- D_1: $\left(\mathbf{A}, \mathbf{AR}, \begin{pmatrix} \boldsymbol{z}_1 \\ \boldsymbol{z}_2 \end{pmatrix}, \boldsymbol{u}'\right)$: $\mathbf{A} \stackrel{\$}{\leftarrow} \mathbb{Z}_q^{n \times m}$, $\boldsymbol{u}' \stackrel{\$}{\leftarrow} \mathbb{Z}_q^n$, $\begin{pmatrix} \boldsymbol{z}_1 \\ \boldsymbol{z}_2 \end{pmatrix} \leftarrow \mathcal{D}_{\Lambda_q^{\boldsymbol{u}'}(\mathbf{A}|\mathbf{AR}), s}$.
- D_2: $\left(\mathbf{A}, \mathbf{AR}, \begin{pmatrix} \boldsymbol{z}_1 \\ \boldsymbol{z}_2 \end{pmatrix}, \boldsymbol{u}'\right)$: $\mathbf{A} \stackrel{\$}{\leftarrow} \mathbb{Z}_q^{n \times m}$, $\begin{pmatrix} \boldsymbol{z}_1 \\ \boldsymbol{z}_2 \end{pmatrix} \leftarrow \mathcal{D}_{\mathbb{Z}^{2m}, s}$,
  $\boldsymbol{u}' = (\mathbf{A}|\mathbf{AR}) \begin{pmatrix} \boldsymbol{z}_1 \\ \boldsymbol{z}_2 \end{pmatrix} \bmod q$.

This simply follows from Lemmas 2.6 and 2.7 – for all but q^{-n} fraction of $\mathbf{A}$, we have $\eta_\epsilon(\Lambda^\perp(\mathbf{A}|\mathbf{AR})) \leq \omega(\sqrt{\log m}) < s$; for such an $\mathbf{A}$, the distribution of $(\mathbf{A}|\mathbf{AR}) \begin{pmatrix} \boldsymbol{z}_1 \\ \boldsymbol{z}_2 \end{pmatrix}$ is uniformly random over $\mathbb{Z}_q^n$, and the conditional distribution of $\begin{pmatrix} \boldsymbol{z}_1 \\ \boldsymbol{z}_2 \end{pmatrix}$ given the constraint is $\mathcal{D}_{\Lambda_q^{\boldsymbol{u}}(\mathbf{A}|\mathbf{AR}), s}$. Thus, we conclude that D_1 and D_2 are statistically close.

The above indistinguishability implies directly that the following two distributions are as well statistically indistinguishable:

- D_1': $\left(\mathbf{A}, \mathbf{AR}, \begin{pmatrix} \boldsymbol{z}_1 + \boldsymbol{x}' \\ \boldsymbol{z}_2 \end{pmatrix}, \boldsymbol{u}' + \mathbf{A}\boldsymbol{x}'\right)$: $\boldsymbol{x}' \leftarrow D_{\mathbb{Z}^m, \rho}$; the other random variables are sampled the same way as D_1.

– D_2': $\left(\mathbf{A}, \mathbf{AR}, \begin{pmatrix} z_1 + x' \\ z_2 \end{pmatrix}, u' + \mathbf{A}x'\right)$: $x' \leftarrow D_{\mathbb{Z}^m,\rho}$; the other random variables are sampled the same way as D_2.

As one can apply the same randomized procedure F such that $D_1' = F(D_1)$ and $D_2' = F(D_2)$, we conclude that $\mathsf{SD}(D_1', D_2') \leq \mathsf{SD}(D_1, D_2) < \mathsf{negl}(\lambda)$.

Finally, by change of variable with $u = u' + \mathbf{A}x'$, we can easily see that the marginal distribution of u is still uniformly random in D_1', i.e., (u' serves as a one-time pad). Then it is not hard to see that D_1' is distributed identical as Sampler-1 and D_2' is distributed statistically close to Sampler-3. This concludes the proof of the claim. □

Claim 4.4. *For the parameters in the statement of Theorem 4.2, the output distributions of* Sampler-2 *and* Sampler-3 *are statistically close.*

Proof. We first observe that for both Sampler-2 and Sampler-3, the component u can be determined (deterministically) from the first three components $(\mathbf{A}, \mathbf{AR}, y)$. Therefore, it suffices for us just to prove statistical closeness for the first three components.

We next note that $\mathbf{A}$ is uniformly random and independent with the component y in both Sampler-2 and Sampler-3. Therefore, it remains to show that the distributions of y in these two algorithms are statistically close.

In Sampler-2, we have $y = \begin{pmatrix} x - \mathbf{R}z_2 \\ z_2 \end{pmatrix}$, and in Sampler-3 we have $y = \begin{pmatrix} x \\ z_2 \end{pmatrix}$. As $\rho \geq s\sqrt{m}\|\mathbf{R}\| \cdot \lambda^{\omega(1)}$, by the smudging lemma (i.e., Lemma 2.9) and the Gaussian tail bound (i.e., Lemma 2.8), these two distributions are statistically close. This concludes the proof of the claim. □

The proof of this theorem follows directly from the above two claims. □

5 Constructions of PHPE and FE with Private Index

In this section, we present three constructions of partially hiding predicate encryption scheme PHPE. Particularly, we first construct a basic $(1, \mathsf{poly})$-Sel-SIM secure PHPE in Sect. 5.1. Then, we upgrade our basic scheme to a (Q, poly)-Sel-SIM secure PHPE for any polynomially bounded Q and general key queries in Sect. 5.2. In Sect. 5.3, we show how to obtain a (Q, poly)-SA-SIM secure PHPE via a simple transformation. Finally, we present the construction of (Q, poly)-SIM-secure Functional Encryption with private input in Sect. 5.4.

Throughout the whole section, we will work on the function class $\mathcal{F}$ as described below. Before presenting the class, we first define three basic functions.

Definition 5.1. *Let $t \in \mathbb{N}, q \in \mathbb{N}$ and $t = t' \log q$. Define the function $\mathsf{PT} : \{0,1\}^t \to \mathbb{Z}_q^{t'}$ as: on input $x \in \{0,1\}^t$, first parse the vector x into a bit matrix $\{x'_{i,j}\}_{i \in [t'], j \in [\log q]}$. The function then computes $z = (z_1, \ldots, z_{t'})^\top$ as $z_i = \sum_{j \in [\log q]} x'_{i,j} \cdot 2^{j-1}$ for $i \in [t']$ and outputs $z \in \mathbb{Z}_q^{t'}$.*

Definition 5.2. *Let $t' \in \mathbb{N}$ be the dimension of vectors, q be some modulus, and $\gamma \in \mathbb{Z}_q$ be some parameter. Define $\mathsf{IP} : \mathbb{Z}_q^{t'} \times \mathbb{Z}_q^{t'} \to \mathbb{Z}_q$ be the inner product modulo q, and $\mathsf{IP}_\gamma : \mathbb{Z}_q^{t'} \times \mathbb{Z}_q^{t'} \to \{0,1\}$ be function such that $\mathsf{IP}_\gamma(\boldsymbol{x}, \boldsymbol{y}) = 1$ if and only if $\gamma = \mathsf{IP}(\boldsymbol{x}, \boldsymbol{y})$ for inputs $\boldsymbol{x}, \boldsymbol{y} \in \mathbb{Z}_q^{t'}$.*

Intuitively, PT acts as the "power-of-two" function that maps $\{0,1\}^t$ to $\mathbb{Z}_q^{t'}$, and IP_γ acts as the comparison function between the parameter γ and the inner product of the inputs.

Function Class $\mathcal{F}$. We consider functions of the following form. Any function in the class $\mathcal{F}$, namely $C : \{0,1\}^t \times \{0,1\}^\ell \to \{0,1\}$ can be described as $\widehat{C} \circ \mathsf{IP}_\gamma$, where $\widehat{C} : \{0,1\}^\ell \to \{0,1\}^{t'}$ is a boolean circuit of depth d, $t' \log q = t$, and $\gamma \in \mathbb{Z}_q$. More formally, for $\boldsymbol{x} \in \{0,1\}^t$ and $\boldsymbol{y} \in \{0,1\}^t$, the function is defined as

$$(\mathsf{IP}_\gamma \circ \widehat{C})(\boldsymbol{x}, \boldsymbol{y}) = \mathsf{IP}_\gamma \left(\mathsf{PT}(\boldsymbol{x}), \widehat{C}(\boldsymbol{y})\right).$$

Similarly, we define a relevant function $(\mathsf{IP} \circ \widehat{C}) : \{0,1\}^t \times \{0,1\}^\ell \to \mathbb{Z}_q$ as

$$(\mathsf{IP} \circ \widehat{C})(\boldsymbol{x}, \boldsymbol{y}) = \mathsf{IP} \left(\mathsf{PT}(\boldsymbol{x}), \widehat{C}(\boldsymbol{y})\right) = \langle \mathsf{PT}(\boldsymbol{x}), \widehat{C}(\boldsymbol{y})\rangle (\text{mod} q).$$

Notice that our formulation is slightly different from that of the prior work [2,29], which directly defined the input $\boldsymbol{x}$ in the domain $\mathbb{Z}_q^{t'}$. In full version, we show that this formulation can also achieve the same effect as the prior work [2,29] with a simple tweak. Thus, it is without loss of generality to define functions in this way. In fact, our modified formulation is for the need of the transformation (from selective-security to semi-adaptive security) in Sect. 5.3, which requires to work on a small input base, e.g., $\{0,1\}$. We notice that both our selective PHPE and the scheme of [2] require a super-polynomial q, so without the modification of the input space, the selective scheme would not be compatible with the transformation.

5.1 $(1, \mathsf{poly})$-Partially Hiding Predicate Encryption

Our basic construction of PHPE is essentially the same as that of Agrawal [2] (her basic construction), except that we adopt our new sampling algorithm in Sect. 4 for the key generation. Our scheme achieves $(1, \mathsf{poly})$-Sel-Sim security, whose formal definition is deferred to the full version of this paper due to the space limit, where one 1-key pre-challenge query is allowed. This is stronger than the $(1, \mathsf{poly})$-very-selective scheme of Agrawal [2], which requires the adversary to commit to both his challenge index and function of the 1-key query at the beginning of the experiment. Below we present the construction.

$\mathsf{PH.Setup}(1^\lambda, 1^t, 1^\ell, 1^d)$: Given as input the security parameter λ, the length of the private and public indices, t and ℓ respectively, and the depth of the circuit family d, the algorithm does the following steps:

1. Choose public parameters (q, ρ, s) as described in the following parameter setting paragraph.
2. Choose random matrices $\mathbf{A}_i \in \mathbb{Z}_q^{n\times m}$ for $i \in [\ell], \mathbf{B}_j \in \mathbb{Z}_q^{n\times m}$ for $j \in [t]$, and $\mathbf{P} \in \mathbb{Z}_q^{n\times m}$.
3. Sample $(\mathbf{A}, \mathbf{T_A}) \leftarrow \mathsf{TrapGen}(1^m, 1^n, q)$.
4. Output the public and master secret keys.

$$\mathsf{PH.mpk} = (\{\mathbf{A}_i\}_{i\in[\ell]}, \{\mathbf{B}_j\}_{j\in[t]}, \mathbf{A}, \mathbf{P}), \mathsf{PH.msk} = (\mathbf{T_A}).$$

$\mathsf{PH.KeyGen}(\mathsf{PH.msk}, \widehat{C} \circ \mathsf{IP}_\gamma)$: Given as input a circuit description $\widehat{C} \circ \mathsf{IP}_\gamma$ and the master secret key, the algorithm does the following steps:

1. Let $\mathbf{A}_{\widehat{C}\circ\mathsf{IP}} = \mathsf{Eval}_{\mathsf{pk}}(\{\mathbf{A}_i\}_{i\in[\ell]}, \{\mathbf{B}_j\}_{j\in[t]}, \widehat{C} \circ \mathsf{IP})$.
2. Sample matrix $\mathbf{J} \leftarrow \mathcal{D}_{\mathbb{Z}^{m\times m},\rho}$, and let $\mathbf{U} = \mathbf{P} - \mathbf{AJ}(\text{mod} q)$.
3. Sample $\begin{bmatrix}\mathbf{K}_1\\ \mathbf{K}_2\end{bmatrix} \leftarrow \mathsf{SampleLeft}(\mathbf{A}, \mathbf{A}_{\widehat{C}\circ\mathsf{IP}} + \gamma\mathbf{G}, \mathbf{T_A}, \mathbf{U}, s)$ for parameter s, i.e., the equation holds for $[\mathbf{A}|\mathbf{A}_{\widehat{C}\circ\mathsf{IP}} + \gamma\mathbf{G}] \cdot \begin{bmatrix}\mathbf{K}_1\\ \mathbf{K}_2\end{bmatrix} = \mathbf{U}(\text{mod} q)$.
4. Let $\mathbf{K} = \begin{bmatrix}\mathbf{J} + \mathbf{K}_1\\ \mathbf{K}_2\end{bmatrix}$, and output $\mathsf{sk}_{\widehat{C}\circ\mathsf{IP}_\gamma} = \mathbf{K}$.

$\mathsf{PH.Enc}(\mathsf{PH.mpk}, (\boldsymbol{x}, \boldsymbol{y}), \mu)$: Given as input the master public key, the private attributes $\boldsymbol{x} \in \{0, 1\}^t$, public attributes $\boldsymbol{y} \in \{0, 1\}^\ell$ and message $\mu \in \{0, 1\}$, the algorithm does the following steps:

1. Sample $\boldsymbol{s} \leftarrow D_{\mathbb{Z}^n,s_B}$ and error terms $\boldsymbol{e} \leftarrow \mathcal{D}_{\mathbb{Z}^m,s_B}$ and $\boldsymbol{e}' \leftarrow \mathcal{D}_{\mathbb{Z}^m,s_D}$.
2. Let $\boldsymbol{b} = [0, \cdots, 0, \lceil q/2\rceil\mu]^\top \in \mathbb{Z}_q^m$. Set $\boldsymbol{\beta}_0 = \mathbf{A}^\top \boldsymbol{s} + \boldsymbol{e}$, $\boldsymbol{\beta}_1 = \mathbf{P}^\top \boldsymbol{s} + \boldsymbol{e}' + \boldsymbol{b}$.
3. For $i \in [\ell]$, sample $\mathbf{R}_i \xleftarrow{\$} \{-1, 1\}^{m\times m}$ and set $\boldsymbol{u}_i = (\mathbf{A}_i + y_i \cdot \mathbf{G})^\top \boldsymbol{s} + \mathbf{R}_i^\top \boldsymbol{e}$.
4. For $j \in [t]$, sample $\mathbf{R}'_j \xleftarrow{\$} \{-1, 1\}^{m\times m}$ and set $\boldsymbol{v}_j = (\mathbf{B}_j + x_j \cdot \mathbf{G})^\top \boldsymbol{s} + (\mathbf{R}'_j)^\top \boldsymbol{e}$.
5. Output the ciphertext $\mathsf{ct}_{\boldsymbol{y}} = \big(\boldsymbol{y}, \boldsymbol{\beta}_0, \boldsymbol{\beta}_1, \{\boldsymbol{u}_i\}_{i\in[\ell]}, \{\boldsymbol{v}_j\}_{j\in[t]}\big)$.

$\mathsf{PH.Dec}(\mathsf{sk}_{\widehat{C}\circ\mathsf{IP}_\gamma}, \mathsf{ct}_{\boldsymbol{y}})$: Given as input a secret key and a ciphertext, the algorithm does the following steps:

1. Compute $\boldsymbol{u}_{\widehat{C}\circ\mathsf{IP}} = \mathsf{Eval}_{\mathsf{ct}}\big(\{\mathbf{A}_i, \boldsymbol{u}_i\}_{i\in[\ell]}, \{\mathbf{B}_j, \boldsymbol{v}_j\}_{j\in[t]}, \widehat{C} \circ \mathsf{IP}, \boldsymbol{y}\big)$.
2. Compute $\boldsymbol{\eta} = \boldsymbol{\beta}_1 - \mathbf{K}^\top \begin{pmatrix}\boldsymbol{\beta}_0\\ \boldsymbol{u}_{\widehat{C}\circ\mathsf{IP}}\end{pmatrix}$.
3. Round each coordinate of $\boldsymbol{\eta}$. If $[\mathsf{Round}(\boldsymbol{\eta}[1]), \cdots, \mathsf{Round}(\boldsymbol{\eta}[m-1])] = \mathbf{0}$ then set $\mu = \mathsf{Round}(\boldsymbol{\eta}[m])$ and output μ. Otherwise, output $\perp$.

Theorem 5.3. *Assuming the hardness of* LWE, *then the scheme described in Sect. 5.1 is a* PHPE *for the class $\mathcal{F}$, achieving* $(1, \mathsf{poly})$-Sel-Sim *security that allows at most one 1-key pre-challenge query (and an unbounded polynomial number of 0-keys for both pre and post-challenge queries).*

Due to space limit, we defer the correctness, parameter setting and the detailed proof of Theorem 5.3 to the full version of this paper.

5.2 (Q, poly)-Partially Hiding Predicate Encryption

In this section, we upgrade our basic scheme to handle arbitrary pre- and post-challenge 1-key queries up to Q times (and any unbounded polynomially many 0-keys). Our upgrading technique is similar to that of Agrawal [2] (the Q-bounded PHPE) except that (1) we adopt our new sampling procedure in Sect. 4 for the key generation, (2) we use a simple secret sharing encoding over the message in a novel way, and (3) we take a more efficient way to generate cover-free sets by using a technique of [10]. Our resulting scheme achieves (Q, poly) simulation-based selective security with ciphertext growth additively with $O(Q)$, allowing general 1-key queries up to Q times, whereas the prior scheme of Agrawal [2] requires the adversary to be committed to all the functions of the 1-key queries right after seeing the public parameters, and the ciphertext size grows additively with $O(Q^2)$.

Before presenting the theorem, we first define the following set sampling algorithm.

Lemma 5.4. *Let $N = Qv\kappa^2$ and $v = \Theta(\kappa)$. There exists an efficient sampler $\mathsf{Sampler}_{\mathsf{Set}}(N, Q, v)$ with the following properties: (1) The sampler always outputs a set $\Delta \subset [N]$ with cardinality v; (2) For independent samples $\Delta_1, \ldots, \Delta_Q$ from $\mathsf{Sampler}_{\mathsf{Set}}(N, Q, v)$, the sets are cover-free with probability $(1 - 2^{-\Omega(\kappa)})$, i.e., for all $i \in [Q]$, $\Pr\left[\Delta_i \backslash \left(\bigcup_{j \neq i} \Delta_j\right) \neq \phi\right] \geq 1 - 2Q \cdot 2^{-\Omega(\kappa)}$.*

Proof. We construct $\mathsf{Sampler}_{\mathsf{Set}}(N, Q, v)$ as follows.

- The sampler first defines an (arbitrary) bijection $h : [N] \to [Q] \times [v\kappa^2]$.
- The sampler selects $i \in [Q]$ uniformly random, and a random $\Delta' \subset [v\kappa^2]$ of cardinality v.
- The sampler sets $\Delta = \{h^{-1}(i, j) : j \in \Delta'\}$, and outputs Δ.

The analysis of $\mathsf{Sampler}_{\mathsf{Set}}$ is similar to that in [10], so we just sketch the proof idea. We first observe that the bijection splits $[N]$ into Q buckets, each with $v\kappa^2$ elements. If we randomly throw Q balls to the buckets, then from the Chernoff bound, we have with at least probability $(1 - Q \cdot 2^{-\Omega(\kappa)})$ that all buckets will contain at most κ balls. These buckets correspond to the first index i. Suppose each bucket contains at most κ balls, where each ball corresponds to a random subset in the second index. Then by Lemma 2.2, for certain bucket, the probability that κ random subsets of size v are cover-free is at least $(1 - 2^{-\Omega(\kappa)})$. Furthermore, by union bound, we know that the independent samples $\Delta_1, \ldots, \Delta_Q$ from $\mathsf{Sampler}_{\mathsf{Set}}(N, Q, v)$ are cover free with at least probability $(1 - Q \cdot 2^{-\Omega(\kappa)})$.

The proof of this lemma simply follows from these two facts. □

In general we can choose κ to be $\omega(\log \lambda)$ to achieve $\mathsf{negl}(\lambda)$ security in the asymptotic setting, or say $\lambda^{1/3}$ to achieve $2^{-\Omega(\lambda)}$ security in the concrete setting.

Below we present the construction.

$\mathsf{QPH.Setup}(1^\lambda, 1^t, 1^\ell, 1^d, 1^Q)$: Given as input the security parameter λ, the length of the private and public attributes, t and ℓ respectively, the depth of the circuit family d, and Q as the upper bound of 1-key queries, do the following:

1. Choose public parameters (q, ρ, s, N, v) as described in the following parameter setting paragraph.
2. Choose random matrices $\mathbf{A}_i \in \mathbb{Z}_q^{n\times m}$ for $i \in [\ell], \mathbf{B}_j \in \mathbb{Z}_q^{n\times m}$ for $j \in [t]$, and $\mathbf{P}_k \in \mathbb{Z}_q^{n\times m}$ for $k \in [N]$.
3. Sample $(\mathbf{A}, \mathbf{T_A}) \leftarrow \mathsf{TrapGen}(1^n, q, m)$.
4. Output the public and master secret keys.

$$\mathsf{PH.mpk} = (\{\mathbf{A}_i\}_{i\in[\ell]}, \{\mathbf{B}_j\}_{j\in[t]}, \mathbf{A}, \{\mathbf{P}_k\}_{k\in[N]}), \mathsf{PH.msk} = (\mathbf{T_A})$$

$\mathsf{QPH.KeyGen}(\mathsf{PH.msk}, \widehat{C} \circ \mathsf{IP}_\gamma)$: Given as input a circuit description $\widehat{C} \circ \mathsf{IP}_\gamma$ and the master secret key, do the following:

1. Let $\mathbf{A}_{\widehat{C}\circ\mathsf{IP}} = \mathsf{Eval}_{\mathsf{pk}}(\{\mathbf{A}_i\}_{i\in[\ell]}, \{\mathbf{B}_j\}_{j\in[t]}, \widehat{C} \circ \mathsf{IP})$.
2. Sample a random subset $\Delta \subset [N]$ according sampler $\mathsf{Sampler}_{\mathsf{Set}}(N, Q, v)$ with $|\Delta| = v$, and compute the subset sum $\mathbf{P}_\Delta = \sum_{k\in\Delta} \mathbf{P}_k$.
3. Sample matrix $\mathbf{J} \leftarrow \mathcal{D}_{\mathbb{Z}^{m\times m}, \rho}$, and let $\mathbf{U} = \mathbf{P}_\Delta - \mathbf{A}\mathbf{J}$.
4. Sample $\begin{bmatrix}\mathbf{K}_1\\ \mathbf{K}_2\end{bmatrix} \leftarrow \mathsf{SampleLeft}(\mathbf{A}, \mathbf{A}_{\widehat{C}\circ\mathsf{IP}} + \gamma\mathbf{G}, \mathbf{T_A}, \mathbf{U}, s)$ for Gaussian parameter s, i.e., the equation holds for $[\mathbf{A}|\mathbf{A}_{\widehat{C}\circ\mathsf{IP}} + \gamma\mathbf{G}] \cdot \begin{bmatrix}\mathbf{K}_1\\ \mathbf{K}_2\end{bmatrix} = \mathbf{U}$ mod q.
5. Let $\mathbf{K} = \begin{bmatrix}\mathbf{J} + \mathbf{K}_1\\ \mathbf{K}_2\end{bmatrix}$, and output $\mathsf{sk}_{\widehat{C}\circ\mathsf{IP}_\gamma} = (\Delta, \mathbf{K})$.

$\mathsf{QPH.Enc}(\mathsf{PH.mpk}, (\boldsymbol{x}, \boldsymbol{y}), \mu)$: Given as input the master public key, the private attributes $\boldsymbol{x}$, public attributes $\boldsymbol{y}$ and message μ, do the following:

1. Sample $\boldsymbol{s} \leftarrow \mathcal{D}_{\mathbb{Z}^n, s_B}$ and error terms $\boldsymbol{e} \leftarrow \mathcal{D}_{\mathbb{Z}^m, s_B}$ and $\boldsymbol{e}'_k \leftarrow \mathcal{D}_{\mathbb{Z}^m, s_D}$ for $k \in [N]$.
2. Set $\boldsymbol{\beta}_0 = \mathbf{A}^\top \boldsymbol{s} + \boldsymbol{e}$, $\boldsymbol{b}_k = [0, \cdots, 0, \frac{\lceil q/2 \rceil}{v}\mu] \in \mathbb{Z}_q^m$ for $k \in [N]$, and compute the following vectors as: $\{\boldsymbol{\beta}_{1,k} = \mathbf{P}_k^\top \boldsymbol{s} + \boldsymbol{e}'_k + \boldsymbol{b}_k\}_{k\in[N]}$.
3. For $i \in [\ell]$, sample $\mathbf{R}_i \xleftarrow{\$} \{-1, 1\}^{m\times m}$ and set $\boldsymbol{u}_i = (\mathbf{A}_i + y_i \cdot \mathbf{G})^\top \boldsymbol{s} + \mathbf{R}_i^\top \boldsymbol{e}$.
4. For $j \in [t]$, sample $\mathbf{R}'_j \xleftarrow{\$} \{-1, 1\}^{m\times m}$ and set $\boldsymbol{v}_j = (\mathbf{B}_j + x_j \cdot \mathbf{G})^\top \boldsymbol{s} + (\mathbf{R}'_j)^\top \boldsymbol{e}$.
5. Output the ciphertext $\mathsf{ct}_{\boldsymbol{y}} = \big(\boldsymbol{y}, \boldsymbol{\beta}_0, \{\boldsymbol{\beta}_{1,k}\}_{k\in[N]}, \{\boldsymbol{u}_i\}_{i\in[\ell]}, \{\boldsymbol{v}_j\}_{j\in[t]}\big)$.

$\mathsf{QPH.Dec}(\mathsf{sk}_{\widehat{C}\circ\mathsf{IP}_\gamma}, \mathsf{ct}_{\boldsymbol{y}})$: Given as input a secret key $\mathsf{sk}_{\widehat{C}\circ\mathsf{IP}_\gamma} := (\Delta, \mathbf{K})$ and a ciphertext, do the following:

1. Compute $\boldsymbol{u}_{\widehat{C}\circ\mathsf{IP}} = \mathsf{Eval}_{\mathsf{ct}}\big(\{\mathbf{A}_i, \boldsymbol{u}_i\}_{i\in[\ell]}, \{\mathbf{B}_j, \boldsymbol{v}_j\}_{j\in[t]}, \widehat{C} \circ \mathsf{IP}, \boldsymbol{y}\big)$.
2. Compute $\boldsymbol{\eta} = \sum_{k\in\Delta} \boldsymbol{\beta}_{1,k} - \mathbf{K}^\top \begin{pmatrix}\boldsymbol{\beta}_0\\ \boldsymbol{u}_{\widehat{C}\circ\mathsf{IP}}\end{pmatrix}$.
3. Round each coordinate of $\boldsymbol{\eta}$. If $[\mathsf{Round}(\boldsymbol{\eta}[1]), \cdots, \mathsf{Round}(\boldsymbol{\eta}[m-1])] = \mathbf{0}$ then set $\mu = \mathsf{Round}(\boldsymbol{\eta}[m])$ and output μ. Otherwise, output $\perp$.

Theorem 5.5. *Assuming the hardness of* LWE, *then the* QPHPE *scheme described in Sect. 5.2 is* (Q, poly)-Sel-Sim *secure that allows both pre- and post-challenge* 1*-key queries up to* Q *times and* 0*-key queries for an unbounded polynomial times.*

Due to space limit, we defer the correctness and parameter setting to the full version of this paper. Additionally, we just describe the simulator Sim for Theorem 5.5 here, and defer the detailed proof to the full version.
Simulator$\mathsf{Sim}(1^\lambda, \boldsymbol{y}, 1^{|\boldsymbol{x}|}, b, \mathsf{st})$:

1. $\mathsf{Sim}_1(1^\lambda, \boldsymbol{y}, 1^{|\boldsymbol{x}|})$: It generates all public parameters as in the real PH.Setup, except that it runs $(\mathbf{A}', \mathbf{T}_{\mathbf{A}'}) \leftarrow \mathsf{TrapGen}(1^{n+1}, q, m)$, then parses $\mathbf{A}' = \begin{bmatrix} \mathbf{A} \\ \boldsymbol{z}^\top \end{bmatrix}$, where $\mathbf{A} \in \mathbb{Z}_q^{n \times m}$, and sets $\mathbf{A}$ be the public matrix in PH.mpk.
2. $\mathsf{Sim}_2(1^\lambda, \boldsymbol{y}, 1^{|\boldsymbol{x}|})$: It generates all keys using the real PH.KeyGen.
3. $\mathsf{Sim}_3(1^\lambda, \boldsymbol{y}, 1^{|\boldsymbol{x}|}, b, \mathsf{List})$: It takes as input the public attributes $\boldsymbol{y}$, the size of the private attributes $\boldsymbol{x}$, the message b, and a list List. It constructs the challenge ciphertext as follows.
   - It samples $\boldsymbol{u}_i, \boldsymbol{v}_j$ independently and uniformly from $\mathbb{Z}_q^m$, and sets $\boldsymbol{\beta}_0 = \boldsymbol{z}$, where $\boldsymbol{z}$ is the vector prepared in Sim_1.
   - If $(b, \mathsf{List}) = \perp$, it computes $\{\boldsymbol{\beta}_{1,k}\}_{k \in [N]}$ as follows:
     - Sample random vectors $\widetilde{\boldsymbol{\beta}}_k$ from $\mathbb{Z}_q^m$ for $k \in [N]$.
     - Choose $2Q$ random subsets $\Delta_1, \cdots, \Delta_Q, \Delta'_1, \cdots, \Delta'_Q$ of $[N]$ according sampler $\mathsf{Sampler}_{\mathsf{Set}}(N, Q, v)$, each of which has cardinality v. Note that with an overwhelming probability, the $2Q$ subsets would be cover-free under our parameter selection.
     - Generate random shares $\{b_k\}_{k \in [N]}$ over $\mathbb{Z}_q$ under the following constraints: for $\hat{i} \in [Q]$, (1) $\sum_{k \in \Delta_{\hat{i}}} b_k = 0$, and (2) $\sum_{k \in \Delta'_{\hat{i}}} b_k = \lceil q/2 \rceil$. This can be done efficiently by the cover-freeness of the subsets, using the following standard procedure.
       First, let $\delta_{\hat{i}}$ be a unique index that only appears in $\Delta_{\hat{i}}$ but not the other subsets, and $\delta'_{\hat{i}}$ be a unique index of $\Delta'_{\hat{i}}$. To generate the random shares $\{b_k\}_{k \in [N]}$, we first sample b_k randomly for all $k \in [N] \setminus (\{\delta_{\hat{i}}\}_{\hat{i} \in [Q]} \cup \{\delta'_{\hat{i}}\}_{\hat{i} \in [Q]})$, and then fix $b_{\delta_{\hat{i}}} = -\sum_{k \in \Delta_{\hat{i}} \setminus \{\delta_{\hat{i}}\}} b_k$ for $\hat{i} \in [Q]$, and similarly $b_{\delta'_{\hat{i}}} = \lceil q/2 \rceil - \sum_{k \in \Delta'_{\hat{i}} \setminus \{\delta'_{\hat{i}}\}} b_k$ for $\hat{i} \in [Q]$.
     - Set $\boldsymbol{b}_k = [0, \cdots, 0, b_k] \in \mathbb{Z}_q^m$ for $k \in [N]$, and sample errors $\{\boldsymbol{e}'_k\}_{k \in [N]}$ from the distribution $\mathcal{D}_{\mathbb{Z}_q^m, s_D}$.
     - Set $\boldsymbol{\beta}_{1,k} = \widetilde{\boldsymbol{\beta}}_k + \boldsymbol{b}_k + \boldsymbol{e}'_k$ for $k \in [N]$.
   - If $b = \mu$ and $\mathsf{List} = \{\widehat{C^*_{\hat{i}}} \circ \mathsf{IP}_{\gamma_{\hat{i}}}\}_{\hat{i} \in [Q']}$ for some $Q' \leq Q$, it computes the simulated ciphertext as follows.
     - For $\hat{i} \in [Q']$, compute $\boldsymbol{u}_{\widehat{C^*_{\hat{i}}} \circ \mathsf{IP}} = \mathsf{Eval}_{\mathsf{ct}}(\{\mathbf{A}_i, \boldsymbol{u}_i\}_{i \in [\ell]}, \{\mathbf{B}_j, \boldsymbol{v}_j\}_{j \in [t]}, \widehat{C^*_{\hat{i}}} \circ \mathsf{IP}, \boldsymbol{y})$, and let $\left(\Delta_{\hat{i}}, \mathbf{K}^*_{\hat{i}} = \begin{bmatrix} \mathbf{J}^*_{\hat{i}} + \mathbf{K}^*_{\hat{i},1} \\ \mathbf{K}^*_{\hat{i},2} \end{bmatrix}\right)$ be the keys for $\widehat{C^*_{\hat{i}}} \circ \mathsf{IP}_{\gamma_{\hat{i}}}$, generated by Sim_2 for the pre-challenge 1-key queries.
     - Sample $Q - Q'$ random subsets of cardinality v according sampler $\mathsf{Sampler}_{\mathsf{Set}}(N, Q, v)$, i.e., $\{\Delta_{\hat{i}}\}_{\hat{i} \in [Q'+1, Q]}$, starting with the index $Q'+1$ and ending with Q. We know that by our setting of parameters, the subsets $\{\Delta_{\hat{i}}\}_{\hat{i} \in [Q]}$ are cover-free with an overwhelming probability.

- Compute vectors $\{\boldsymbol{\beta}_{1,k}\}_{k\in[N]}$ as follows:
  * Sample random shares $\{\mu_k\}_{k\in[N]}$ conditioned that $\sum_{k\in\Delta_{\hat{i}}}\mu_k = \lceil q/2\rceil\mu$ for $\hat{i}\in[Q]$. Then set $\boldsymbol{b}_k = [0,\cdots,0,\mu_k]$ for $k\in[N]$.
  * Sample random vectors $\{\widetilde{\boldsymbol{\beta}}_k\}_{k\in[N]}$ condition on the following equations:

$$\sum_{k\in\Delta_{\hat{i}}}\widetilde{\boldsymbol{\beta}}_k = \begin{bmatrix}\mathbf{J}^*_{\hat{i}}+\mathbf{K}^*_{\hat{i},1}\\ \mathbf{K}^*_{\hat{i},2}\end{bmatrix}^\top \cdot \begin{pmatrix}\boldsymbol{\beta}_0\\ \boldsymbol{u}_{\widehat{C^*_{\hat{i}}}\circ\mathsf{IP}}\end{pmatrix} \quad \text{for } \hat{i}\in[Q'].$$

  The above two steps can be done efficiently due to the cover-freeness of the subsets $\{\Delta_{\hat{i}}\}_{\hat{i}\in[Q]}$. The procedure is the same as we have presented in the previous case.
  * Sample errors $\{\boldsymbol{e}_k\}_{k\in[N]}$ according $\mathcal{D}_{\mathbb{Z}_q^m,s_D}$.
  * Set $\boldsymbol{\beta}_{1,k} = \widetilde{\boldsymbol{\beta}}_k + \boldsymbol{b}_k + \boldsymbol{e}'_k$ for $k\in[N]$.
- It outputs the challenge ciphertext

$$\mathsf{ct}^* = \big(\{\boldsymbol{u}_i\}_{i\in[\ell]}, \{\boldsymbol{v}_j\}_{j\in[t]}, \boldsymbol{y}, \boldsymbol{\beta}_0, \{\boldsymbol{\beta}_{1,k}\}_{k\in[N]}\big).$$

4. $\mathsf{Sim}_4(1^\lambda, \boldsymbol{y}, 1^{|\boldsymbol{x}|})$: If the query is a 0-key, then it generates the key using the real QPH.KeyGen. Otherwise, we denote function $\widehat{C^*_{\hat{i}}}\circ\mathsf{IP}_{\gamma_{\hat{i}}}$ be the adversary's 1-key query and $(\mu, \widehat{C^*_{\hat{i}}}\circ\mathsf{IP}_{\gamma_{\hat{i}}})$ be the message received from the oracle $\mathcal{O}$. Here we use index $\hat{i}\in[Q]$ to denote the number of overall 1-key queries up to this point. Then the simulator computes as follows.
   - The simulator first considers the following two cases to determine the parameter Δ:
     - Case 1: $Q' = 0$, i.e., the adversary did not make any 1-key pre-challenge query.
       * If $\mu = 0$, set $\Delta := \Delta_{\hat{i}}$.
       * Else $\Delta := \Delta'_{\hat{i}}$, where $\{\Delta_{\hat{i}}\}_{\hat{i}\in[Q]}$ and $\{\Delta'_{\hat{i}}\}_{\hat{i}\in[Q]}$ are the subsets prepared by Sim_3 in the previous procedure.
     - Case 2: $1 \le Q' < Q$, i.e., the adversary had made Q' 1-key pre-challenge queries.
       * Set $\Delta := \Delta_{\hat{i}}$ where $\Delta_{\hat{i}}$ is the subset prepared by Sim_3 (where μ had been received by Sim_3) in the previous procedure.
   - Compute $\mathbf{P}^*_\Delta = \sum_{k\in\Delta}\mathbf{P}_k$, and compute $\widetilde{\boldsymbol{\beta}}_\Delta = \sum_{k\in\Delta}\widetilde{\boldsymbol{\beta}}_k$, where $\{\widetilde{\boldsymbol{\beta}}_k\}_{k\in[N]}$ are the vectors prepared by Sim_3 in the previous procedure.
   - Compute $\mathbf{A}_{\widehat{C^*_{\hat{i}}}\circ\mathsf{IP}} = \mathsf{Eval}_{\mathsf{pk}}(\{\mathbf{A}_i\}_{i\in[\ell]}, \{\mathbf{B}_j\}_{j\in[t]}, \widehat{C^*_{\hat{i}}}\circ\mathsf{IP})$, and compute $\boldsymbol{u}_{\widehat{C^*_{\hat{i}}}\circ\mathsf{IP}} = \mathsf{Eval}_{\mathsf{ct}}\big(\{\mathbf{A}_i, \boldsymbol{u}_i\}_{i\in[\ell]}, \{\mathbf{B}_j, \boldsymbol{v}_j\}_{j\in[t]}, \widehat{C^*_{\hat{i}}}\circ\mathsf{IP}, \boldsymbol{y}\big)$.
   - Sample $\mathbf{J}^*_{\hat{i}} \leftarrow \mathcal{D}_{\mathbb{Z}^{m\times m},\rho}$, and use $\mathbf{T}_{\mathbf{A}'}$ to sample $\begin{bmatrix}\mathbf{K}^*_{\hat{i},1}\\ \mathbf{K}^*_{\hat{i},2}\end{bmatrix} \leftarrow \mathcal{D}_{\mathbb{Z}^{2m\times m},s}$ such that

$$\begin{bmatrix}\mathbf{A} & \mathbf{A}_{\widehat{C^*_{\hat{i}}}\circ\mathsf{IP}}\\ \boldsymbol{\beta}_0^\top & \boldsymbol{u}^\top_{\widehat{C^*_{\hat{i}}}\circ\mathsf{IP}}\end{bmatrix} \cdot \begin{bmatrix}\mathbf{K}^*_{\hat{i},1}\\ \mathbf{K}^*_{\hat{i},2}\end{bmatrix} = -\begin{bmatrix}\mathbf{A}\\ \boldsymbol{\beta}_0^\top\end{bmatrix}\cdot\mathbf{J}^*_{\hat{i}} + \begin{bmatrix}\mathbf{P}^*_\Delta\\ \widetilde{\boldsymbol{\beta}}^\top_\Delta\end{bmatrix}.$$

– Output $\mathsf{sk}_{\widehat{C_{\hat{i}}^*} \circ \mathsf{IP}_{\gamma_{\hat{i}}}} = \left(\Delta, \begin{bmatrix} \mathbf{J}_{\hat{i}}^* + \mathbf{K}_{\hat{i},1}^* \\ \mathbf{K}_{\hat{i},2}^* \end{bmatrix}\right)$.

5.3 Semi-Adaptively Secure Partially Hiding Predicate Encryption

In this section, we show how to upgrade our PHPE in Sect. 5.2 from (Q, poly)-Sel-SIM security to (Q, poly)-SA-SIM security. Technically, we follow the idea of [17], yet in the case of bounded-length attributes (as used in this work). Below, we present the detailed construction.

Let $\mathsf{PH}_{\mathsf{Sel}} = \{\mathsf{Setup}, \mathsf{KeyGen}, \mathsf{Enc}, \mathsf{Dec}\}$ be a PHPE with private-public attribute space $\{0,1\}^t \times \{0,1\}^\ell$, message space $\mathcal{M}$, and function class $\mathcal{F}$ that is closed under bit-shift on $\{0,1\}^t \times \{0,1\}^\ell$ (i.e., for any $f \in \mathcal{F}$, $(\boldsymbol{r}, \boldsymbol{r}') \in \{0,1\}^t \times \{0,1\}^\ell$, we have $f_{\boldsymbol{r},\boldsymbol{r}'}(\boldsymbol{x}, \boldsymbol{y}) = f(\boldsymbol{x} \oplus \boldsymbol{r}, \boldsymbol{y} \oplus \boldsymbol{r}') \in \mathcal{F}$). Moreover, the encryption algorithm $\mathsf{Enc}((\boldsymbol{x}, \boldsymbol{y}), \mu)$ can be decomposed into three parts: $\mathsf{Enc}_1(\mu; R)$, $\{\mathsf{Enc}_2(x_i; R)\}_{i \in [t]}$, $\{\mathsf{Enc}_3(y_i; R)\}_{i \in [\ell]}$, where R is the common random string among the three algorithms, x_i is the i-th bit of the attribute $\boldsymbol{x}$ whose bit-length is ℓ, and similarly y_i is the i-th bit of $\boldsymbol{y}$. Intuitively, the encryption procedure is done by three different components: with a common random string R, Enc_1 encodes the message, and both Enc_2 and Enc_3 encode the private/public attributes in the bit-by-bit manner.

Additionally, let $\mathsf{PKE} = \{\mathsf{Gen}, \mathsf{Enc}, \mathsf{Dec}\}$ be any semantically secure public-key encryption. Then our transformation is defined as below.

$\mathsf{PH}_{\mathsf{SA}}.\mathsf{Setup}(1^\lambda, 1^t, 1^\ell)$: the algorithm takes the following steps:

- Run the underlying setup $(\mathsf{mpk}_{\mathsf{Sel}}, \mathsf{msk}_{\mathsf{Sel}}) \leftarrow \mathsf{PH}_{\mathsf{Sel}}.\mathsf{Setup}(1^\lambda, 1^\ell)$.
- Generate $\{\mathsf{PKE.pk}_{i,b}, \mathsf{PKE.sk}_{i,b}\}_{i \in [t], b \in \{0,1\}}$, $\{\mathsf{PKE.pk}'_{i,b}, \mathsf{PKE.sk}'_{i,b}\}_{i \in [\ell], b \in \{0,1\}}$ from the scheme PKE.
- Sample a random string $(\boldsymbol{r}, \boldsymbol{r}') \in \{0,1\}^t \times \{0,1\}^\ell$.
- Finally output $\mathsf{mpk}_{\mathsf{SA}} = (\mathsf{mpk}_{\mathsf{Sel}}, \{\mathsf{PKE.pk}_{i,b}\}_{i \in [t], b \in \{0,1\}}, \{\mathsf{PKE.pk}'_{i,b}\}_{i \in [\ell], b \in \{0,1\}})$ as the master public key, and keep private $\mathsf{msk}_{\mathsf{SA}} = (\mathsf{msk}_{\mathsf{Sel}}, \{\mathsf{PKE.sk}_{i,b}\}_{i \in [t], b \in \{0,1\}}, \{\mathsf{PKE.sk}'_{i,b}\}_{i \in [\ell], b \in \{0,1\}}, \boldsymbol{r}, \boldsymbol{r}')$ as the master secret key.

Note: Here Setup might implicitly take input $1^d, 1^Q$ for circuit depth and an upper bound of the 1-key queries. For simplicity, we omit the description.

$\mathsf{PH}_{\mathsf{SA}}.\mathsf{KeyGen}(\mathsf{msk}_{\mathsf{SA}}, f \in \mathcal{F})$: the algorithm defines a related function $f_{\boldsymbol{r},\boldsymbol{r}'}(\boldsymbol{x}, \boldsymbol{y}) := f(\boldsymbol{x} \oplus \boldsymbol{r}, \boldsymbol{y} \oplus \boldsymbol{r}')$, and runs $\mathsf{sk}_{\mathsf{Sel},f} \leftarrow \mathsf{PH}_{\mathsf{Sel}}(\mathsf{msk}_{\mathsf{Sel}}, f_{\boldsymbol{r},\boldsymbol{r}'})$. Then it returns $(\boldsymbol{r}, \boldsymbol{r}', \{\mathsf{PKE.sk}_{i,r_i}\}_{i \in [t]}, \{\mathsf{PKE.sk}'_{i,r'_i}\}_{i \in [\ell]}, \mathsf{sk}_{\mathsf{Sel},f})$ as the secret key.

$\mathsf{PH}_{\mathsf{SA}}.\mathsf{Enc}(\mathsf{mpk}_{\mathsf{SA}}, (\boldsymbol{x}, \boldsymbol{y}), \mu)$: the algorithms runs the following steps:

- Sample a random string R.
- Run $\mathsf{ct}_1 \leftarrow \mathsf{PH}_{\mathsf{Sel}}.\mathsf{Enc}_1(\mu; R)$, $\{L_{i,b} \leftarrow \mathsf{PH}_{\mathsf{Sel}}.\mathsf{Enc}_2(x_i \oplus b; R)\}_{i \in [t], b \in \{0,1\}}$, and $\{L'_{i,b} \leftarrow \mathsf{PH}_{\mathsf{Sel}}.\mathsf{Enc}_3(y_i \oplus b; R)\}_{i \in [\ell], b \in \{0,1\}}$.
- Generate $\{\mathsf{ct}_{i,b} \leftarrow \mathsf{PKE.Enc}(\mathsf{PKE.pk}_{i,b}, L_{i,b})\}_{i \in [t], b \in \{0,1\}}$ and $\{\mathsf{ct}'_{i,b} \leftarrow \mathsf{PKE.Enc}(\mathsf{PKE.pk}'_{i,b}, L'_{i,b})\}_{i \in [\ell], b \in \{0,1\}}$.
- Finally, output the ciphertext as $\mathsf{ct} = (\mathsf{ct}_1, \{\mathsf{ct}_{i,b}\}_{i \in [t], b \in \{0,1\}}, \{\mathsf{ct}'_{i,b}\}_{i \in [\ell], b \in \{0,1\}})$.

$\mathsf{PH}_{\mathsf{SA}}.\mathsf{Dec}(\mathsf{sk}_{\mathsf{SA},f}, \boldsymbol{y}, \mathsf{ct})$:the algorithm runs the following steps:

- Parse $\mathsf{ct} = (\mathsf{ct}_1, \{\mathsf{ct}_{i,b}\}_{i \in [t], b \in \{0,1\}}, \{\mathsf{ct}'_{i,b}\}_{i \in [\ell], b \in \{0,1\}})$.

- Run the PKE decryption on $\{\mathsf{ct}_{i,r_i}\}_{i\in[t]}$ and $\{\mathsf{ct}'_{i,r'_i}\}_{i\in[\ell]}$. Then obtain $\{L_{i,r_i}\}_{i\in[t]}$ and $\{L'_{i,r'_i}\}_{i\in[\ell]}$.
- View $(\mathsf{ct}_1, \{L_{i,r_i}\}_{i\in[t]}, \{L'_{i,r'_i}\}_{i\in[\ell]})$ as the ciphertext of $\mathsf{PH}_{\mathsf{Sel}}$, and decrypt it with $\mathsf{sk}_{\mathsf{Sel},f}$. Output the decrypted outcome.

Theorem 5.6. *Assume that* PKE *is semantically secure, and* $\mathsf{PH}_{\mathsf{Sel}}$ *is* (q_1, q_2)-Sel-SIM *secure for private-public attribute space* $\{0,1\}^t \times \{0,1\}^\ell$, *message space* $\mathcal{M}$, *and function class* $\mathcal{F}$ *that is closed under bit-shift on* $\{0,1\}^t \times \{0,1\}^\ell$. *Then the scheme* $\mathsf{PH}_{\mathsf{SA}}$ *is* (q_1, q_2)-SA-SIM *secure for the same attribute and message spaces and the function class* $\mathcal{F}$.

Due to space limit, we defer the correctness and the proof of Theorem 5.6 to the full version of this paper.

5.4 (Q, poly)-SIM-secure Functional Encryption

In this section, we present the technique from [2], showing that a (Q, poly)-SIM-secure QPHPE with a fully homomorphic encryption scheme implies a (Q, poly)-SIM-secure FE, which is what we desire. Due to space limit, we just describe the theorem from [2], and defer the detailed procedure to the full version.

Theorem 5.7. *Let* $\mathcal{C}$ *be the family of bounded depth circuits,* QPHPE *be a* (Q, poly)-SA-SIM *secure partially-hiding predicate encryption scheme for* $\mathcal{F}$ *as defined in Sect. 5, and* FHE *be a secure fully-homomorphic encryption scheme. Then there exists a functional encryption that is* (Q, poly)-SA-SIM *secure for the class* $\mathcal{C} \times \{I\}$.

We notice that the required QPHPE can be instantiated by Theorems 5.5 and 5.6. Thus, we obtain the following corollary to summarize the final result.

Corollary 5.8. *Assuming the hardness of* LWE *for a sub-exponential modulus-to-noise ratio. Then for any bounded polynomial* $Q = \mathsf{poly}(\lambda)$, *there exists a* (Q, poly)-SA-SIM *secure* FE *for the class* $\mathcal{C} \times \{I\}$.

6 Constructions of FE with Public Index

We notice that our two-stage sampling technique in Sect. 4 can be further used to derived several new feasibilities of FE with public index for the following two function classes.

- The first scheme is IB-FEIP that achieves $(1, \mathsf{poly})$-AD-IND security, i.e., a public-index FE for the class IB $\times$ IP. Detailed definitions are deferred to the full version. This particularly improves the prior analysis of Abdalla et al. [1], who can only achieve the selectively security. As we discussed in full version of this paper, $(1, \mathsf{poly})$-AD-IND is the best we can achieve for the IB predicates as there is only one 1-key corresponding to the challenge index.
  Our construction follows the same design paradigm as [1], except we use the adaptively secure encoding of matrices by [3] and adopt our new sampling algorithm in Sect. 4 for the key generation.

- The second scheme is a generalization of the first scheme that achieves (Q, poly)-SA-IND secure AB-FEIP for any polynomially bounded Q, for general predicate classes (i.e., bounded depth boolean circuits). This new feasibility result is beyond what the prior technique of [1] can achieve.

Due to space limit, we defer the detailed constructions and security proofs of our new IB-FEIP and AB-FEIP to the full version.

Acknowledgements. We would like to thank the anonymous reviewers of Eurocrypt 2021 for their insightful advices. Qiqi Lai is supported by the National Key R&D Program of China (2017YFB0802000), the National Natural Science Foundation of China (61802241, U2001205, 61772326, 61802242), the Natural Science Basic Research Plan in Shaanxi Province of China (2019JQ-360), the National Cryptography Development Foundation during the 13th Five-year Plan Period (MMJJ20180217), and the Fundamental Research Funds for the Central Universities (GK202103093). Feng-Hao Liu and Zhedong Wang are supported by an NSF Award CNS-1657040 and an NSF Career Award CNS-1942400. Any opinions, findings, and conclusions or recommendations expressed in this material are those of the author(s) and do not necessarily reflect the views of the sponsors.

References

1. Abdalla, M., Catalano, D., Gay, R., Ursu, B.: Inner-product functional encryption with fine-grained access control. In: Moriai, S., Wang, H. (eds.) ASIACRYPT 2020, Part III. LNCS, vol. 12493, pp. 467–497. Springer, Cham (2020). https://doi.org/10.1007/978-3-030-64840-4_16
2. Agrawal, S.: Stronger security for reusable garbled circuits, general definitions and attacks. In: Katz, J., Shacham, H. (eds.) CRYPTO 2017, Part I. LNCS, vol. 10401, pp. 3–35. Springer, Cham (2017). https://doi.org/10.1007/978-3-319-63688-7_1
3. Agrawal, S., Boneh, D., Boyen, X.: Efficient lattice (H)IBE in the standard model. In: Gilbert [25], pp. 553–572
4. Agrawal, S., Gorbunov, S., Vaikuntanathan, V., Wee, H.: Functional encryption: new perspectives and lower bounds. In: Canetti, R., Garay, J.A. (eds.) CRYPTO 2013, Part II. LNCS, vol. 8043, pp. 500–518. Springer, Heidelberg (2013). https://doi.org/10.1007/978-3-642-40084-1_28
5. Agrawal, S., Libert, B., Stehlé, D.: Fully secure functional encryption for inner products, from standard assumptions. In: Robshaw, M., Katz, J. (eds.) CRYPTO 2016, Part III. LNCS, vol. 9816, pp. 333–362. Springer, Heidelberg (2016). https://doi.org/10.1007/978-3-662-53015-3_12
6. Agrawal, S., Rosen, A.: Functional encryption for bounded collusions, revisited. In: Kalai, Y., Reyzin, L. (eds.) TCC 2017, Part I. LNCS, vol. 10677, pp. 173–205. Springer, Cham (2017). https://doi.org/10.1007/978-3-319-70500-2_7
7. Albrecht, M.R., Player, R., Scott, S.: On the concrete hardness of learning with errors. J. Math. Cryptology **9**(3), 169–203 (2015)
8. Alwen, J., Peikert, C.: Generating shorter bases for hard random lattices. Theory Comput. Syst. **48**(3), 535–553 (2010)
9. Ananth, P., Brakerski, Z., Segev, G., Vaikuntanathan, V.: From selective to adaptive security in functional encryption. In: Gennaro, R. [23], pp. 657–677

10. Ananth, P., Vaikuntanathan, V.: Optimal bounded-collusion secure functional encryption. In: Hofheinz, D., Rosen, A. (eds.) TCC 2019, Part I. LNCS, vol. 11891, pp. 174–198. Springer, Cham (2019). https://doi.org/10.1007/978-3-030-36030-6_8
11. Boneh, D., et al.: Fully key-homomorphic encryption, arithmetic circuit ABE and compact garbled circuits. In: Nguyen, P.Q., Oswald, E. (eds.) EUROCRYPT 2014. LNCS, vol. 8441, pp. 533–556. Springer, Heidelberg (2014). https://doi.org/10.1007/978-3-642-55220-5_30
12. Boneh, D., Roughgarden, T., Feigenbaum, J. (eds.) 45th ACM STOC. ACM Press, June 2013
13. Boneh, D., Sahai, A., Waters, B.: Functional encryption: definitions and challenges. In: Ishai, Y. (ed.) TCC 2011. LNCS, vol. 6597, pp. 253–273. Springer, Heidelberg (2011). https://doi.org/10.1007/978-3-642-19571-6_16
14. Boyen, X.: Lattice mixing and vanishing trapdoors: a framework for fully secure short signatures and more. In: Nguyen, P.Q., Pointcheval, D. (eds.) PKC 2010. LNCS, vol. 6056, pp. 499–517. Springer, Heidelberg (2010). https://doi.org/10.1007/978-3-642-13013-7_29
15. Brakerski, Z., Döttling, N., Garg, S., Malavolta, G.: Factoring and pairings are not necessary for io: Circular-secure lwe suffices. Cryptology ePrint Archive, Report 2020/1024 (2020). https://eprint.iacr.org/2020/1024
16. Brakerski, Z., Langlois, A., Peikert, C., Regev, O., Stehlé, D.: Classical hardness of learning with errors. In: Boneh et al. [12], pp. 575–584
17. Brakerski, Z., Vaikuntanathan, V.: Circuit-ABE from LWE: unbounded attributes and semi-adaptive security. In: Robshaw, K. [38], pp. 363–384
18. Canetti, R., Chen, Y.: Constraint-hiding constrained PRFs for NC^1 from LWE. In: Coron, J.-S., Nielsen, J.B. (eds.) EUROCRYPT 2017, Part I. LNCS, vol. 10210, pp. 446–476. Springer, Cham (2017). https://doi.org/10.1007/978-3-319-56620-7_16
19. Cash, D. Hofheinz, D., Kiltz, E., Peikert, C.: Bonsai trees, or how to delegate a lattice basis. In: Gilbert [25], pp. 523–552
20. Ducas, L., Micciancio, D.: Improved short lattice signatures in the standard model. In: Garay, J.A., Gennaro, R. (eds.) CRYPTO 2014, Part I. LNCS, vol. 8616, pp. 335–352. Springer, Heidelberg (2014). https://doi.org/10.1007/978-3-662-44371-2_19
21. Garg, S., Gentry, C., Halevi, S., Raykova, M., Sahai, A., Waters, B.: Candidate indistinguishability obfuscation and functional encryption for all circuits. In: 54th FOCS, pp. 40–49. IEEE Computer Society Press, October 2013
22. Gay, R., Pass, R.: Indistinguishability obfuscation from circular security. Cryptology ePrint Archive, Report 2020/1010 (2020). https://eprint.iacr.org/2020/1010
23. Gennaro, R., Robshaw, M.J.B. (eds.): CRYPTO 2015, Part II. LNCS, vol. 9216. Springer, Heidelberg, August 2015
24. Gentry, C., Peikert, C., Vaikuntanathan, V.: Trapdoors for hard lattices and new cryptographic constructions. In: Ladner, R.E., Dwork, C. (eds.), 40th ACM STOC, pp. 197–206. ACM Press, May 2008
25. Gilbert, H., (ed.) EUROCRYPT 2010, volume 6110 of LNCS. Springer, Heidelberg, May/June 2010. https://doi.org/10.1007/978-3-642-13190-5
26. Goldwasser, S., Kalai, Y.T., Popa, R.A., Vaikuntanathan, V., Zeldovich, N.: Reusable garbled circuits and succinct functional encryption. In: Boneh et al. [12], pp. 555–564
27. Gorbunov, S., Vaikuntanathan, V., Wee, H.: Functional encryption with bounded collusions via multi-party computation. In: Safavi-Naini, R., Canetti, R. (eds.) CRYPTO 2012. LNCS, vol. 7417, pp. 162–179. Springer, Heidelberg (2012). https://doi.org/10.1007/978-3-642-32009-5_11

28. Gorbunov, S., Vaikuntanathan, V., Wee, H.: Attribute-based encryption for circuits. In: Boneh et al. [12], pp. 545–554
29. Gorbunov, S., Vaikuntanathan, V., Wee, H.: Predicate encryption for circuits from LWE. In: Gennaro, R. [23], pp. 503–523
30. Goyal, R., Koppula, V., Waters, B.: Semi-adaptive security and bundling functionalities made generic and easy. In: Hirt, M., Smith, A. (eds.) TCC 2016. LNCS, vol. 9986, pp. 361–388. Springer, Heidelberg (2016). https://doi.org/10.1007/978-3-662-53644-5_14
31. Jain, A., Lin, H., Sahai, A.: Indistinguishability obfuscation from well-founded assumptions. Cryptology ePrint Archive, Report 2020/1003 (2020). https://eprint.iacr.org/2020/1003
32. Lai, Q., Liu, F.-H., Wang, Z.: Almost tight security in lattices with polynomial moduli – PRF, IBE, All-but-many LTF, and More. In: Kiayias, A., Kohlweiss, M., Wallden, P., Zikas, V. (eds.) PKC 2020, Part I. LNCS, vol. 12110, pp. 652–681. Springer, Cham (2020). https://doi.org/10.1007/978-3-030-45374-9_22
33. Micciancio, D., Peikert, C.: Trapdoors for lattices: simpler, tighter, faster, smaller. In: Pointcheval, D., Johansson, T. (eds.) EUROCRYPT 2012. LNCS, vol. 7237, pp. 700–718. Springer, Heidelberg (2012). https://doi.org/10.1007/978-3-642-29011-4_41
34. Micciancio, D., Regev, O.: Worst-case to average-case reductions based on Gaussian measures. In: 45th FOCS, pp. 372–381. IEEE Computer Society Press, October 2004
35. O'Neill, A.: Definitional issues in functional encryption. Cryptology ePrint archive, Report 2010/556 (2010). https://eprint.iacr.org/2010/556
36. Peikert, C.: Public-key Cryptosystems from the Worst-case Shortest Vector Problem: Extended Abstract. In: Mitzenmacher, M. (ed.) 41st ACM STOC, pp. 333–342. ACM Press, May/June (2009)
37. Regev, O.: On lattices, learning with errors, random linear codes, and cryptography. In: Gabow, H.N., Fagin, R. (eds.), 37th ACM STOC, pp. 84–93. ACM Press, May 2005
38. Robshaw, M., Katz, J. (eds.): CRYPTO 2016, Part III. LNCS, vol. 9816. Springer, Heidelberg, August 2016

On Bounded Distance Decoding with Predicate: Breaking the "Lattice Barrier" for the Hidden Number Problem

Martin R. Albrecht[1(✉)] and Nadia Heninger[2]

[1] Information Security Group, Royal Holloway, University of London, London, UK
martin.albrecht@royalholloway.ac.uk
[2] University of California, San Diego, USA
nadiah@cs.ucsd.edu

Abstract. Lattice-based algorithms in cryptanalysis often search for a target vector satisfying integer linear constraints as a shortest or closest vector in some lattice. In this work, we observe that these formulations may discard non-linear information from the underlying application that can be used to distinguish the target vector even when it is far from being uniquely close or short.

We formalize lattice problems augmented with a predicate distinguishing a target vector and give algorithms for solving instances of these problems. We apply our techniques to lattice-based approaches for solving the Hidden Number Problem, a popular technique for recovering secret DSA or ECDSA keys in side-channel attacks, and demonstrate that our algorithms succeed in recovering the signing key for instances that were previously believed to be unsolvable using lattice approaches. We carried out extensive experiments using our estimation and solving framework, which we also make available with this work.

1 Introduction

Lattice reduction algorithms [34,51,59,70,71] have found numerous applications in cryptanalysis. These include several general families of cryptanalytic applications including factoring RSA keys with partial information about the secret key via Coppersmith's method [26,62], the (side-channel) analysis of lattice-based schemes [4,8,27,42,55], and breaking (EC)DSA and Diffie-Hellman via side-channel attacks using the Hidden Number Problem.

In the usual statement of the Hidden Number Problem (HNP) [21], the adversary learns some most significant bits of random multiples of a secret integer

N. Heninger—The research of MA was supported by EPSRC grants EP/S020330/1, EP/S02087X/1, by the European Union Horizon 2020 Research and Innovation Program Grant 780701 and Innovate UK grant AQuaSec; NH was supported by the US NSF under grants no. 1513671, 1651344, and 1913210. Part of this work was done while the authors were visiting the Simons Institute for the Theory of Computing. Our experiments were carried out on Cisco UCS equipment donated by Cisco and housed at UCSD. The full version of this work is available at https://ia.cr/2020/1540.

A. Canteaut and F.-X. Standaert (Eds.): EUROCRYPT 2021, LNCS 12696, pp. 528–558, 2021.
https://doi.org/10.1007/978-3-030-77870-5_19

modulo some known integer. This information can be written as integer-linear constraints on the secret. The problem can then be formulated as a variant of the Closest Vector Problem (CVP) known as Bounded Distance Decoding (BDD), which asks one to find a uniquely closest vector in a lattice to some target point $\boldsymbol{t}$. A sufficiently strong lattice reduction will find this uniquely close vector, which can then be used to recover the secret.

The requirement of uniqueness constrains the instances that can be successfully solved with this approach. In short, a fixed instance of the problem is not expected to be solvable when few samples are known, since there are expected to be many spurious lattice points closer to the target than the desired solution. As the number of samples is increased, the expected distance between the target and the lattice shrinks relative to the normalized volume of the lattice, and at some point the problem is expected to become solvable. For some choices of input parameters, however, the problem may be infeasible to solve using these methods if the attacker cannot compute a sufficiently reduced lattice basis to find this solution; if the number of spurious non-solution vectors in the lattice does not decrease fast enough to yield a unique solution; or if simply too few samples can be obtained. In the context of the Hidden Number Problem, the expected infeasibility of lattice-based algorithms for certain parameters has been referred to as the "lattice barrier" in numerous works [12,30,64,73,77].

Nevertheless, the initial cryptanalytic problem may remain well defined even when the gap between the lattice and the target is not small enough to expect a unique closest vector. This is because formulating a problem as a HNP instance omits information: the cryptanalytic applications typically imply non-linear constraints that restrict the solution, often to a unique value. For example, in the most common application of the HNP to side-channel attacks, breaking ECDSA from known nonce bits [18,43], the desired solution corresponds to the discrete logarithm of a public value that the attacker knows. We may consider such additional non-linear constraints as a predicate $h(\cdot)$ that evaluates to true on the unique secret and false elsewhere. Thus, we may reformulate the search problem as a BDD with predicate problem: find a vector $\boldsymbol{v}$ in the lattice within some radius R to the target $\boldsymbol{t}$ such that $f(\boldsymbol{v} - \boldsymbol{t}) := h(g(\boldsymbol{v} - \boldsymbol{t}))$ returns true, where $g(\cdot)$ is a function extracting a candidate secret s from the vector $\boldsymbol{v} - \boldsymbol{t}$.

Contributions

In this work, we define the BDD with predicate problem and give algorithms to solve it. To illustrate the performance of our algorithms, we apply them to the Hidden Number Problem lattices arising from side-channel attacks recovering ECDSA keys from known nonce bits.

In more detail, in Sect. 3, we give a simple refinement of the analysis of the "lattice barrier" and show how this extends the range of parameters that can be solved in practice.

In Sect. 4 we define the Bounded Distance Decoding with predicate ($\mathrm{BDD}_{\alpha,f(\cdot)}$) and the unique Shortest Vector with predicate ($\mathrm{uSVP}_{f(\cdot)}$) problems and mention how Kannan's embedding enables us to solve the former via the latter.

We then give two algorithms for solving the unique Shortest Vector with predicate problem in Sect. 5. One is based on lattice-point enumeration and in principle supports any norm R of the target vector. This algorithm exploits the fact that enumeration *is* exhaustive search inside a given radius. Our other algorithm is based on lattice sieving and is expected to succeed when $R \leq \sqrt{4/3} \cdot \mathrm{gh}(\Lambda)$ where $\mathrm{gh}(\Lambda)$ is the expected norm of a shortest vector in a lattice Λ under the Gaussian heuristic (see below).[1] This algorithm makes use of the fact that a sieve produces a database of short vectors in the lattice, not just a single shortest vector. Thus, the key observation exploited by all our algorithms is that efficient SVP solvers are expected to consider every vector of the lattice within some radius R. Augmenting these algorithms with an additional predicate check then follows naturally. In both algorithms the predicate is checked $(R/\mathrm{gh}(\Lambda))^{d+o(d)}$ times, where d is the dimension of the lattice, which is asymptotically smaller than the cost of the original algorithms.

In Sect. 6, we experimentally demonstrate the performance of our algorithms in the context of ECDSA signatures with partial information about nonce bits. Here, although the lattice-based HNP algorithm has been a well-appreciated tool in the side-channel cryptanalysis community for two decades [17,24,44,53,61,63,68,69,78], we show how our techniques allow us to achieve previous records with fewer samples, bring problem instances previously believed to be intractable into feasible range, maximize the algorithm's success probability when only a fixed number of samples are available, increase the algorithm's success probability in the presence of noisy data, and give new tradeoffs between computation time and sample collection. We also present experimental evidence of our techniques' ability to solve instances given fewer samples than required by the information theoretic limit for lattice approaches. This is enabled by our predicate uniquely determining the secret.

Our experimental results are obtained using a Sage [72]/Python framework for cost-estimating and solving uSVP instances (with predicate). This framework is available at [7] and attached to the electronic version of this work. We expect it to have applications beyond this work.

Related work

There are two main algorithmic approaches to solving the Hidden Number Problem in the cryptanalytic literature. In this work, we focus on lattice-based approaches to solving this problem. An alternative approach, a Fourier analysis-based algorithm due to Bleichenbacher [18], has generally been considered to be more robust to errors, and able to solve HNP instances with fewer bits known, but at the cost of requiring orders of magnitude more samples and a much higher computational cost [12,13,30,73]. Our work can be viewed as extending the applicability of lattice-based HNP algorithms well into parameters believed to be only tractable to Bleichenbacher's algorithm, thus showing how these instances

[1] We note that this technique conflicts with "dimensions for free" [5,32] and thus the expected performance improvement when arbitrarily many samples are available is smaller compared to state-of-the-art sieving (see Sect. 5.3 for details).

can be solved using far fewer samples and less computational time in practice (see Table 3), while gracefully handling input errors (see Fig. 6).

In particular, our work can be considered a systematization, formalization, and generalization of folklore (and often ad hoc) techniques in the literature on lattice-reduction aided side-channel attacks such as examining the entire reduced basis to find the target vector [22,44] or the technique briefly mentioned in [17] of examining candidates after each "tour" of BKZ (BKZ is described below).[2]

More generally, our work can be seen as a continuation of a line of recent works that "open up" SVP oracles, i.e. that forgo treating (approximate) SVP solvers as black boxes inside algorithms. In particular, a series of recent works have taken advantage of the exponentially many vectors produced by a sieve: in [10] the authors use the exponentially many vectors to cost the so-called "dual attack" on LWE [67]; in [5,32,50] the authors exploit the same property to improve sieving algorithms and block-wise lattice reduction; and in [31] the authors use this fact to compute approximate Voronoi cells.

Our work may also be viewed in line with [27], which augments a BDD solver for LWE with "hints" by transforming the input lattice. While these hints must be linear(izable) (with noise), the authors demonstrate the utility of integrating such hints to reduce the cost of finding a solution. On the one hand, our approach allows us to incorporate arbitrary, non-linear hints, as long as these can be expressed as an efficiently computable predicate; this makes our approach more powerful. On the other hand, the scenarios in which our techniques can be applied are much more restricted than [27]. In particular, [27] works for any lattice reduction algorithm and, specifically, for block-wise lattice reduction. Our work, in contrast, does not naturally extend to this setting; this makes our approach less powerful in comparison. We discuss this in Sect. 5.4.

2 Preliminaries

We denote the logarithm with base two by $\log(\cdot)$. We start indexing at zero.

2.1 Lattices

A lattice Λ is a discrete subgroup of $\mathbb{R}^d$. When the rows $\boldsymbol{b}_0, \ldots, \boldsymbol{b}_{d-1}$ of $\boldsymbol{B}$ are linearly independent we refer to it as the basis of the lattice $\Lambda(\boldsymbol{B}) = \{\sum v_i \cdot \boldsymbol{b}_i \mid v_i \in \mathbb{Z}\}$, i.e. we consider row-representations for matrices in this work.

The algorithms considered in this work make use of orthogonal projections $\pi_i : \mathbb{R}^d \mapsto span\left(\boldsymbol{b}_0, \ldots, \boldsymbol{b}_{i-1}\right)^{\perp}$ for $i = 0, \ldots, d-1$. In particular $\pi_0(\cdot)$ is the identity. The *Gram–Schmidt orthogonalization* (GSO) of $\boldsymbol{B}$ is $\boldsymbol{B}^* = (\boldsymbol{b}_0^*, \ldots, \boldsymbol{b}_{d-1}^*)$, where the Gram–Schmidt vector $\boldsymbol{b}_i^*$ is $\pi_i(\boldsymbol{b}_i)$. Then $\boldsymbol{b}_0^* = \boldsymbol{b}_0$ and $\boldsymbol{b}_i^* = \boldsymbol{b}_i - \sum_{j=0}^{i-1} \mu_{i,j} \cdot \boldsymbol{b}_j^*$ for $i = 1, \ldots, d-1$ and $\mu_{i,j} = \frac{\langle \boldsymbol{b}_i, \boldsymbol{b}_j^* \rangle}{\langle \boldsymbol{b}_j^*, \boldsymbol{b}_j^* \rangle}$. Norms in this

[2] For the purposes of this work, the CVP technique used in [17] is not entirely clear from the account given there. We confirmed with the authors that is the analogous strategy to their SVP approach: CVP enumeration interleaved with tours of BKZ.

work are Euclidean and denoted $\|\cdot\|$. We write $\lambda_i(\Lambda)$ for the radius of the smallest ball centred at the origin containing at least i linearly independent lattice vectors, e.g. $\lambda_1(\Lambda)$ is the norm of a shortest vector in Λ.

The Gaussian heuristic predicts that the number $|\Lambda \cap \mathcal{B}|$ of lattice points inside a measurable body $\mathcal{B} \subset \mathbb{R}^n$ is approximately equal to $\mathrm{Vol}(\mathcal{B})/\mathrm{Vol}(\Lambda)$. Applied to Euclidean d-balls, it leads to the following prediction of the length of a shortest non-zero vector in a lattice.

Definition 1 (Gaussian heuristic). *We denote by* $\mathrm{gh}(\Lambda)$ *the expected first minimum of a lattice* Λ *according to the Gaussian heuristic. For a full rank lattice* $\Lambda \subset \mathbb{R}^d$*, it is given by:*

$$\mathrm{gh}(\Lambda) = \left(\frac{\mathrm{Vol}(\Lambda)}{\mathrm{Vol}(\mathfrak{B}_d(1))}\right)^{1/d} = \frac{\Gamma\left(1+\frac{d}{2}\right)^{1/d}}{\sqrt{\pi}} \cdot \mathrm{Vol}(\Lambda)^{1/d} \approx \sqrt{\frac{d}{2\pi e}} \cdot \mathrm{Vol}(\Lambda)^{1/d}$$

where $\mathfrak{B}_d(R)$ *denotes the d-dimensional Euclidean ball with radius R.*

2.2 Hard Problems

A central hard problem on lattices is to find a shortest vector in a lattice.

Definition 2 (Shortest Vector Problem (SVP)). *Given a lattice basis* $\boldsymbol{B}$*, find a shortest non-zero vector in* $\Lambda(\boldsymbol{B})$*.*

In many applications, we are interested in finding closest vectors, and we have the additional guarantee that our target vector is not too far from the lattice. This is known as Bounded Distance Decoding.

Definition 3 (α-Bounded Distance Decoding (BDD_α)). *Given a lattice basis* $\boldsymbol{B}$*, a vector* $\boldsymbol{t}$*, and a parameter* $0 < \alpha$ *such that the Euclidean distance between* $\boldsymbol{t}$ *and the lattice* $\mathrm{dist}(\boldsymbol{t}, \boldsymbol{B}) < \alpha \cdot \lambda_1(\Lambda(\boldsymbol{B}))$*, find the lattice vector* $\boldsymbol{v} \in \Lambda(\boldsymbol{B})$ *which is closest to* $\boldsymbol{t}$*.*

To guarantee a unique solution, it is required that $\alpha < 1/2$. However, the problem can be generalized to $1/2 \leq \alpha < 1$, where we expect a unique solution with high probability. Asymptotically, for any polynomially-bounded $\gamma \geq 1$ there is a reduction from $\mathrm{BDD}_{1/(\sqrt{2}\gamma)}$ to uSVP_γ [14]. The unique shortest vector problem (uSVP) is defined as follows:

Definition 4 (γ-unique Shortest Vector Problem (uSVP_γ)). *Given a lattice* Λ *such that* $\lambda_2(\Lambda) > \gamma \cdot \lambda_1(\Lambda)$ *find a nonzero vector* $\boldsymbol{v} \in \Lambda$ *of length* $\lambda_1(\Lambda)$*.*

The reduction is a variant of the embedding technique, due to Kannan [46], that constructs

$$\boldsymbol{L} = \begin{pmatrix} \boldsymbol{B} & 0 \\ \boldsymbol{t} & \tau \end{pmatrix}$$

where τ is some embedding factor (the reader may think of $\tau = \mathbb{E}\left[\|\boldsymbol{t}-\boldsymbol{v}\|/\sqrt{d}\right]$). If $\boldsymbol{v}$ is the closest vector to $\boldsymbol{t}$ then the lattice $\Lambda(\boldsymbol{L})$ contains $(\boldsymbol{t}-\boldsymbol{v}, \tau)$ which is small.

2.3 Lattice Algorithms

Enumeration

[2,33,45,58,66,71] solves the following problem: Given some matrix $\boldsymbol{B}$ and some bound R, find $\boldsymbol{v} = \sum_{i=0}^{d-1} u_i \cdot \boldsymbol{b}_i$ with $u_i \in \mathbb{Z}$ where at least one $u_i \neq 0$ such that $\|\boldsymbol{v}\|^2 \leq R^2$. By picking the shortest vector encountered, we can use lattice-point enumeration to solve the shortest vector problem. Enumeration algorithms make use of the fact that the vector $\boldsymbol{v}$ can be rewritten with respect to the Gram–Schmidt basis:

$$\boldsymbol{v} = \sum_{i=0}^{d-1} u_i \cdot \boldsymbol{b}_i = \sum_{i=0}^{d-1} u_i \cdot \left(\boldsymbol{b}_i^* + \sum_{j=0}^{i-1} \mu_{i,j} \cdot \boldsymbol{b}_j^* \right) = \sum_{j=0}^{d-1} \left(u_j + \sum_{i=j+1}^{d-1} u_i \cdot \mu_{ij} \right) \cdot \boldsymbol{b}_j^*.$$

Since all the $\boldsymbol{b}_i^*$ are pairwise orthogonal, we can express the norms of projections of $\boldsymbol{v}$ simply as

$$\|\pi_k(\boldsymbol{v})\|^2 = \left\| \sum_{j=k}^{d-1} \left(u_j + \sum_{i=j+1}^{d-1} u_i \, \mu_{i,j} \right) \boldsymbol{b}_j^* \right\|^2 = \sum_{j=k}^{d-1} \left(u_j + \sum_{i=j+1}^{d-1} u_i \, \mu_{i,j} \right)^2 \cdot \|\boldsymbol{b}_j^*\|^2.$$

In particular, vectors do not become longer by projecting. Enumeration algorithms exploit this fact by projecting the problem down to a one dimensional problem of finding candidate $\pi_d(\boldsymbol{v})$ such that $\|\pi_d(\boldsymbol{v})\|^2 \leq R^2$. Each such candidate is then lifted to a candidate $\pi_{d-1}(\boldsymbol{v})$ subject to the constraint $\|\pi_{d-1}(\boldsymbol{v})\|^2 \leq R^2$.

That is, lattice-point enumeration is a depth-first tree search through a tree defined by the u_i. It starts by picking a candidate for u_{d-1} and then explores the subtree "beneath" this choice. Whenever it encounters an empty interval of choices for some u_i it abandons this branch and backtracks. When it reaches the leaves of the tree, i.e. u_0 then it compares the candidate for a full solution to the previously best found and backtracks.

Lattice-point enumeration is expected [40] to consider

$$H_k = \frac{1}{2} \cdot \frac{\mathrm{Vol}(\mathfrak{B}_{d-k}(R))}{\prod_{i=k}^{d-1} \|\boldsymbol{b}_i^*\|}$$

nodes at level k and $\sum_{k=0}^{d-1} H_k$ nodes in total. In particular, enumeration finds the shortest non-zero vector in a lattice in $d^{d/(2e)+o(d)}$ time and polynomial memory [40]. It was recently shown that when enumeration is used as the SVP oracle inside block-wise lattice reduction the time is reduced to $d^{d/8+o(d)}$ [2]. However, the conditions for this improvement are mostly not met in our setting. Significant gains can be made in lower-order terms by considering a different R_i on each level $0 \leq i < d$ instead of a fixed R. Since this prunes branches of the search tree that are unlikely to lead to a solution, this is known as "pruning" in the literature. When the R_i are chosen such that the success probability is exponentially small in d we speak of "extreme pruning" [35].

A state-of-the-art implementation of lattice-point enumeration can be found in FPLLL [74]. This is the implementation we adapt in this work. It visits about $2^{\frac{d \log d}{2e} - 0.995\, d + 16.25}$ nodes to solve SVP in dimension d [2].

Sieving
[1,15,16,41,49,57] takes as input a list of lattice points, $L \subset \Lambda$, and searches for integer combinations of these points that are short. If the initial list is sufficiently large, SVP can be solved by performing this process recursively. Each point in the initial list can be sampled at a cost polynomial in d [48]. Hence the initial list can be sampled at a cost of $|L|^{1+o(1)}$.

Sieves that combine k points at a time are called k-sieves; 2-sieves take integer combinations of the form $\boldsymbol{u} \pm \boldsymbol{v}$ with $\boldsymbol{u}, \boldsymbol{v} \in L$ and $\boldsymbol{u} \neq \pm \boldsymbol{v}$. Heuristic sieving algorithms are analyzed under the heuristic that the points in L are independently and identically distributed uniformly in a thin spherical shell. This heuristic was introduced by Nguyen and Vidick in [65]. As a further simplification, it is assumed that the shell is very thin and normalized such that L is a subset of the unit sphere in $\mathbb{R}^d$. As such, a pair $(\boldsymbol{u}, \boldsymbol{v})$ is reducible if and only if the angle between $\boldsymbol{u}$ and $\boldsymbol{v}$ satisfies $\theta(\boldsymbol{u}, \boldsymbol{v}) < \pi/3$, where $\theta(\boldsymbol{u}, \boldsymbol{v}) = \arccos\left(\langle \boldsymbol{u}, \boldsymbol{v}\rangle / (\|\boldsymbol{u}\| \cdot \|\boldsymbol{v}\|)\right)$, $\arccos(x) \in [0, \pi]$. Under these assumptions, we require $|L| \approx \sqrt{4/3}^{\,d}$ in order to see "collisions", i.e. reductions. Lattice sieves are expected to output a list of $(4/3)^{d/2+o(d)}$ short lattice vectors [5,32]. The asymptotically fastest sieve has a heuristic running time of $2^{0.292\, d+o(d)}$ [15].

We use the performant implementations of lattice sieving that can be found in G6K [5,76] in this work, which includes a variant of [16] ("BGJ1") and [41] (3-Sieve). BGJ1 heuristically runs in time $2^{0.349\, d+o(d)}$ and memory $2^{0.205\, d+o(d)}$. The 3-Sieve heuristically runs in time $2^{0.372\, d+o(d)}$ and memory $2^{0.189\, d+o(d)}$.[3]

BKZ
[70,71] can be used to solve the unique shortest vector problem and thus BDD. BKZ makes use of an oracle that solves the shortest vector problem in dimension β. This oracle can be instantiated using enumeration or sieving. The algorithm then asks the oracle to solve SVP on the first block of dimension β of the input lattice, i.e. of the lattice spanned by $\boldsymbol{b}_0, \ldots, \boldsymbol{b}_{\beta-1}$. This vector is then inserted into the basis and the algorithm asks the SVP oracle to return a shortest vector for the block $\pi_1(\boldsymbol{b}_1), \ldots, \pi_1(\boldsymbol{b}_\beta)$. The algorithm proceeds in this fashion until it reaches $\pi_{d-2}(\boldsymbol{b}_{d-2}), \pi_{d-2}(\boldsymbol{b}_{d-1})$. It then starts again by considering $\boldsymbol{b}_0, \ldots, \boldsymbol{b}_{\beta-1}$. One such loop is called a "tour" and the algorithm will continue with these tours until no more (or only small changes) are made to the basis. For many applications a small, constant number of tours is sufficient for the basis to stabilize.

The key parameter for BKZ is the block size β, i.e. the maximal dimension of the underlying SVP oracle, and we write "BKZ-β". The expected norm of the shortest vector found by BKZ-β and inserted into the basis as $\boldsymbol{b}_0$ for a random

[3] In G6K the 3-Sieve is configured to use a database of size $2^{0.205\, d+o(d)}$ by default, which lowers its time complexity.

lattice is $\|\boldsymbol{b}_0\| \approx \delta_\beta^{d-1} \cdot \mathrm{Vol}(\Lambda)^{1/d}$ for some constant $\delta_\beta \in \mathcal{O}\left(\beta^{1/(2\beta)}\right)$ depending on β.[4]

In [10] the authors formulate a success condition for BKZ-β solving uSVP on a lattice Λ in the language of solving LWE. Let $\boldsymbol{e}$ be the unusually short vector in the lattice and let $\boldsymbol{c}_i^*$ be the Gram–Schmidt vectors of a *typical* BKZ-β reduced basis of a lattice with the same volume and dimension as Λ. Then in [10] it is observed that when BKZ considers the last full block $\pi_{d-\beta}(\boldsymbol{b}_{d-\beta}), \ldots \pi_{d-\beta}(\boldsymbol{b}_{d-1})$ it will insert $\pi_{d-\beta}(\boldsymbol{e})$ at index $d-\beta$ if that projection is the shortest vector in the sublattice spanned by the last block. Thus, when

$$\|\pi_{d-\beta}(\boldsymbol{e})\| < \|\boldsymbol{c}_{d-\beta}^*\| \tag{1}$$

$$\approx$$

$$\sqrt{\beta/d} \cdot \mathbb{E}[\|\boldsymbol{e}\|] < \delta_\beta^{2\beta-d-1} \cdot \mathrm{Vol}(\Lambda)^{1/d} \tag{2}$$

we expect the behavior of BKZ-β on our lattice Λ to deviate from that of a random lattice. This situation is illustrated in Fig. 1. Indeed, in [6] it was shown that once this event happens, the internal LLL calls of BKZ will "lift" and recover $\boldsymbol{e}$. Thus, these works establish a method for estimating the required block size for BKZ to solve uSVP instances. We use this estimate to choose parameters in Sect. 6: given a dimension d, volume $\mathrm{Vol}(\Lambda)$ and $\mathbb{E}[\|\boldsymbol{e}\|]$, we pick the smallest β such that Inequality (2) is satisfied. Note, however, that in small dimensions this reasoning is somewhat complicated by "double intersections" [6] and low "lifting" probability [27]; as a result estimates derived this way are pessimistic for small block sizes. In that case, the model in [27] provides accurate predictions. Instead of only running BKZ, a performance gain can be achieved by following BKZ with one SVP/CVP call in a larger dimension than the BKZ block size [5,53].

2.4 The Hidden Number Problem

In the Hidden Number Problem (HNP) [21], there is a secret integer α and a public modulus n. Information about α is revealed in the form of what we call samples: an oracle chooses a uniformly random integer $0 < t_i < n$, computes $s_i = t_i \cdot \alpha \bmod n$ where the modular reduction is taken as a unary operator so that $0 \leq s_i < n$, and reveals some most significant bits of s_i along with t_i. We will write this as $a_i + k_i = t_i \cdot \alpha \bmod n$, where $k_i < 2^\ell$ for some $\ell \in \mathbb{Z}$ that is a parameter to the problem. For each sample, the adversary learns the pair (t_i, a_i). We may think of the Hidden Number Problem as 1-dimensional LWE [67].

2.5 Breaking ECDSA from Nonce Bits

Many works in the literature have exploited side-channel information about (EC)DSA nonces by solving the Hidden Number Problem (HNP), e.g. [12,19,

[4] The constant is typically defined as $\|\boldsymbol{b}_0\| \approx \delta_\beta^d \cdot \mathrm{Vol}(\Lambda)^{1/d}$ in the literature. From the perspective of the (worst-case) analysis of underlying algorithms, though, normalizing by $d-1$ rather than d is appropriate.

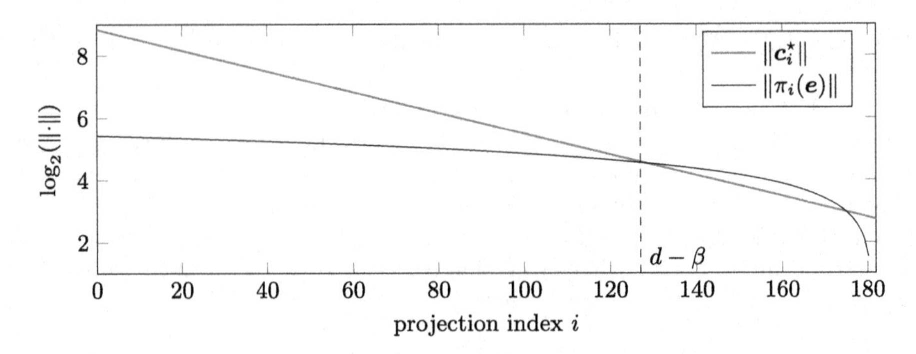

Fig. 1. BKZ$-\beta$ uSVP Success Condition. Expected norms for lattices of dimension $d = 183$ and volume q^{m-n} after BKZ-β reduction for LWE parameters $n = 65, m = 182, q = 521$, standard deviation $\sigma = 8/\sqrt{2\pi}$ and $\beta = 56$. BKZ is expected to succeed in solving a uSVP instance when the two curves intersect at index $d - \beta$ as shown, i.e. when Inequality (1) holds. Reproduced from [6].

44,53,61,63,68,69,73,78], since the seminal works of Bleichenbacher [18] and Howgrave-Graham and Smart [43]. The latter solves HNP using lattice reduction; the former deploys a combinatorial algorithm that can be cast as a variant of the BKW algorithm [3,20,39,47]. The latest in this line of research is [13] which recovers a key from less than one bit of the nonce using Bleichenbacher's algorithm. More recently, in [54] the authors found the first practical attack scenario that was able to make use of Boneh and Venkatesan's [21] original application of the HNP to prime-field Diffie-Hellman key exchange.

Side-channel attacks

Practical side-channel attacks against ECDSA typically run in two stages. First, the attacker collects many signatures while performing side-channel measurements. Next, they run a key recovery algorithm on a suitably chosen subset of the traces. Depending on the robustness of the measurements, the data collection phase can be quite expensive. As examples, in [60] the authors describe having to repeat their attack 10,000 to 20,000 times to obtain one byte of information; in [37] the authors measured 5,000 signing operations, each taking 0.1 s, to obtain 114 usable traces; in [61] the authors describe generating 40,000 signatures in 80 min in order to obtain 35 suitable traces to carry out an attack.

Thus in the side-channel literature, minimizing the amount of data required to mount a successful attack is often an important metric [44,68]. Using our methods as described below will permit more efficient overall attacks.

ECDSA

The global parameters for an ECDSA signature are an elliptic curve $E(\mathbb{F}_p)$ and a generator point G on E of order n. A signing key is an integer $0 \leq d < n$, and the public verifying key is a point dG. To generate an ECDSA signature on a message hash h, the signer generates a random integer nonce $k < n$, and computes the values $r = (kG)_x$ where x subscript is the x coordinate of the point, and $s = k^{-1} \cdot (h + d \cdot r) \bmod n$. The signature is the pair (r, s).

ECDSA as a HNP

In a side-channel attack against ECDSA, the adversary may learn some of the most significant bits of the signature nonce k. Without loss of generality, we will assume that these bits are all 0. Then rearranging the formula for the ECDSA signature s, we have $-s^{-1} \cdot h + k \equiv s^{-1} \cdot r \cdot d \bmod n$, and thus a HNP instance with $a_i = -s^{-1} \cdot h$, $t_i = s^{-1} \cdot r$, and $\alpha = d$.

Solving the HNP with lattices

Boneh and Venkatesan give this lattice for solving the Hidden Number Problem with a BDD oracle:

$$\begin{bmatrix} n & 0 & 0 & \cdots & 0 & 0 \\ 0 & n & 0 & \cdots & 0 & 0 \\ & \vdots & & & \cdots & \\ 0 & 0 & 0 & \cdots & n & 0 \\ t_0 & t_1 & t_2 & \cdots & t_{m-1} & 1/n \end{bmatrix}$$

The target is a vector $(a_0, \dots, a_{m-1}, 0)$ and the lattice vector

$$(t_0 \cdot \alpha \bmod n, \ \dots \ , t_{m-1} \cdot \alpha \bmod n, \ \alpha/n)$$

is within $\sqrt{m+1} \cdot 2^\ell$ of this target when $|k_i| < 2^\ell$.

Most works solve this BDD problem via Kannan's embedding i.e. by constructing the lattice generated by the rows of

$$\begin{bmatrix} n & 0 & 0 & \cdots & 0 & 0 & 0 \\ 0 & n & 0 & \cdots & 0 & 0 & 0 \\ & \vdots & & & \vdots & & \\ 0 & 0 & 0 & \cdots & n & 0 & 0 \\ t_0 & t_1 & t_2 & \cdots & t_{m-1} & 2^\ell/n & 0 \\ a_0 & a_1 & a_2 & \cdots & a_{m-1} & 0 & 2^\ell \end{bmatrix}$$

This lattice contains a vector

$$(k_0, k_1, \dots, k_{m-1}, \ 2^\ell \cdot \alpha/n, \ 2^\ell)$$

that has norm at most $\sqrt{m+2} \cdot 2^\ell$. This lattice also contains $(0, 0, \dots, 0, 2^\ell, 0)$, so the target vector is not generally the shortest vector. There are various improvements we can make to this lattice.

Reducing the size of k by one bit. In an ECDSA input, k is generally positive, so we have $0 \le k_i < 2^\ell$. The lattice works for any sign of k, so we can reduce the bit length of k by one bit by writing $k'_i = k_i - 2^{\ell-1}$. This modification provides a significant improvement in practice and is described in [63], but is not consistently taken advantage of in practical applications.

Eliminating α. Given a set of input equations $a_0 + k_0 \equiv t_0 \cdot \alpha \bmod n, \ldots, a_{m-1} + k_{m-1} = t_{m-1} \cdot \alpha \bmod n$, we can eliminate the variable α and end up with a new set of equations $a'_1 + k_1 \equiv t'_1 \cdot k_0 \bmod n, \ \ldots \ , a'_{m-1} + k_{m-1} \equiv t'_{m-1} \cdot k_0 \bmod n$.

For each relation, $t_i^{-1} \cdot (a_i + k_i) \equiv t_0^{-1} \cdot (a_0 + k_0) \bmod n$; rearranging yields

$$a_i - t_i \cdot t_0^{-1} \cdot a_0 + k_i \equiv t_i \cdot t_0^{-1} \cdot k_0 \bmod n.$$

Thus our new problem instance has $m-1$ relations with $a'_i = a_i - t_i \cdot t_0^{-1} \cdot a_0$ and $t'_i = t_i \cdot t_0^{-1}$.

This has the effect of reducing the dimension of the above lattice by 1, and also making the bounds on all the variables equal-sized, so that normalization is not necessary anymore, and the vector $(0, 0, \ldots, 0, 2^\ell, 0)$ is no longer in the lattice. Thus, the new target $(k_1, k_2, \ldots, k_{m-1}, k_0, 2^\ell)$ is expected to be the unique shortest vector (up to signs) in the lattice for carefully chosen parameters. We note that this transformation is analogous to the normal form transformation for LWE [11]. From a naive examination of the determinant bounds, this transformation would not be expected to make a significant difference in the feasibility of the algorithm, but in the setting of this paper, where we wish to push the boundaries of the unique shortest vector scenario, it is crucial to the success of our techniques.

Let $w = 2^{\ell-1}$. With the above two optimizations, our new lattice Λ is generated by:

$$\begin{bmatrix} n & 0 & 0 & \cdots & 0 & 0 & 0 \\ 0 & n & 0 & \cdots & 0 & 0 & 0 \\ & & \vdots & & \vdots & & \\ 0 & 0 & 0 & \cdots & n & 0 & 0 \\ t'_1 & t'_2 & t'_3 & \cdots & t'_{m-1} & 1 & 0 \\ a'_1 & a'_2 & a'_3 & \cdots & a'_{m-1} & 0 & w \end{bmatrix}$$

and the target vector is $v_t = (k_1 - w, k_2 - w, \ldots, k_{m-1} - w, k_0 - w, w)$.

The expected solution comes from multiplying the second to last basis vector with the secret (in this case, k_0), adding the last vector, and reducing modulo n as necessary. The entries 1 and w are normalization values chosen to ensure that all the coefficients of the short vector will have the same length.

Different-sized k_is. We can adapt the construction to different-sized k_i satisfying $|k_i| < 2^{\ell_i}$ by normalizing each column in the lattice by a factor of $2^{\ell_{max}}/2^{\ell_i}$ [17].

3 The "Lattice Barrier"

It is believed that lattice algorithms for the Hidden Number Problem "become essentially inapplicable when only a very short fraction of the nonce is known for each input sample. In particular, for a single-bit nonce leakage, it is believed that they should fail with high probability, since the lattice vector corresponding to the secret is no longer expected to be significantly shorter than other vectors in the lattice" [13]. Aranha et al. [12] elaborate on this further: "there is a hard limit to what can be achieved using lattice reduction: due to the underlying structure of the HNP lattice, it is impossible to attack (EC)DSA using a single-bit nonce leak with lattice reduction. In that case, the 'hidden lattice point' corresponding to the HNP solution will not be the closest vector even under the Gaussian heuristic (see [64]), so that lattice techniques cannot work." Similar points are made in [30,73,77]; in particular, in [77] it is estimated that a 3-bit bias for a 256-bit curve is not easy and two bits is infeasible, and a 5- or 4-bit bias for a 384-bit curve is not easy and three bits is infeasible.

To see how prior work derived this "lattice barrier", note that the volume of the lattice is

$$\mathrm{Vol}\left(\Lambda\right) = n^{m-1} \cdot w$$

and the dimension is $m+1$. According to the Gaussian heuristic, we expect the shortest vector in the lattice to have norm

$$\begin{aligned} \mathrm{gh}\left(\Lambda\right) &\approx \frac{\Gamma(1+(m+1)/2)^{1/(m+1)}}{\sqrt{\pi}} \cdot \mathrm{Vol}(\Lambda)^{1/(m+1)} \\ &\approx \sqrt{\frac{m+1}{2\,\pi\, e}} \cdot \left(n^{m-1} \cdot w\right)^{1/(m+1)}. \end{aligned}$$

Also, observe that the norm of the target vector $\boldsymbol{v}$ satisfies

$$\|\boldsymbol{v}\| \le \sqrt{m+1} \cdot w. \tag{3}$$

A BDD solver is expected to be successful in recovering $\boldsymbol{v}$ when $\|\boldsymbol{v}\| < \mathrm{gh}(\Lambda)$. We give a representative plot in Fig. 2 comparing the Gaussian heuristic $\mathrm{gh}(\Lambda)$ against the upper bound of the target vectors in Eq. (3) for 1, 2, and 3-bit biases for a 256-bit ECDSA key recovery problem. The resulting lattice dimensions explain the difficulty estimates of [77].

In this work, we make two observations. First, the upper bound for the target vector is a conservative estimate for its length. Since heuristically our problem instances are randomly sampled, we will use the expected norm of a uniformly distributed vector instead. This is only a constant factor different from the upper bound above, but this constant makes a significant difference in the crossover points.

The target vector $\boldsymbol{v}$ we construct after the optimizations above has expected squared norm

$$\mathbb{E}\left[\|\boldsymbol{v}\|^2\right] = \mathbb{E}\left[\left(\sum_{i=1}^{m} (k_i - w)^2\right) + w^2\right] = m \cdot \mathbb{E}\left[(k_i - w)^2\right] + w^2$$

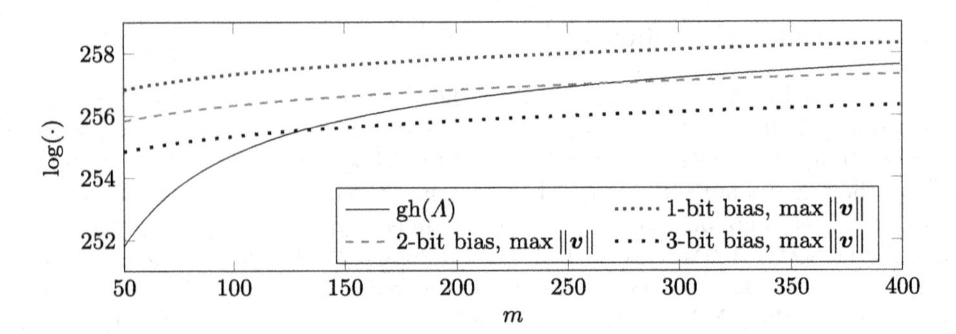

Fig. 2. Illustrating the "lattice barrier". BDD is expected to become feasible when the length of the target vector $\|\boldsymbol{v}\|$ is less than the Gaussian heuristic $\mathrm{gh}(\Lambda)$; we plot the upper bound in Eq. (3) for $\log(n) = 256$ against varying number of samples m.

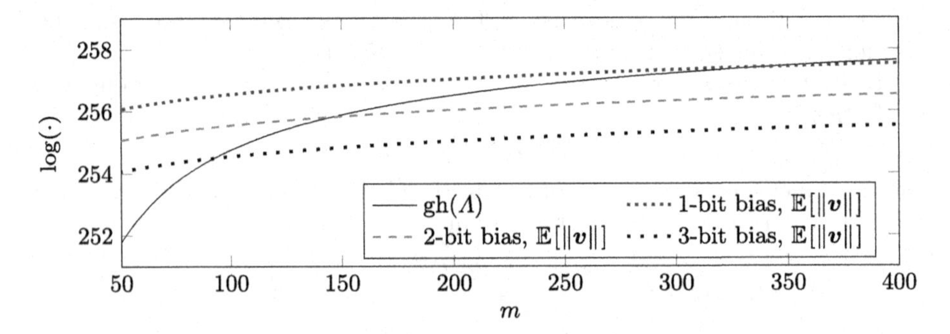

Fig. 3. Updated estimates for feasibility of lattice algorithms. We plot the expected length of the target vector $\|\boldsymbol{v}\|$ against the Gaussian heuristic for varying number of samples m for $\log(n) = 256$. Compared to Fig. 2, the crossover points result in much more tractable instances. We can further decrease the lattice dimension using enumeration and sieving with predicates (see Sect. 4).

with

$$\begin{aligned}
\mathbb{E}\left[(k_i - w)^2\right] &= 1/(2\,w) \cdot \sum_{i=0}^{2\,w-1} (i-w)^2 \\
&= 1/(2\,w) \cdot \sum_{i=0}^{2\,w-1} i^2 - 1/(2\,w) \sum_{i=0}^{2\,w-1} 2\,i \cdot w + 1/(2\,w) \sum_{i=0}^{2\,w-1} w^2 \\
&= w^2/3 + 1/6
\end{aligned}$$

and we arrive at

$$\mathbb{E}\left[\|\boldsymbol{v}\|^2\right] = \mathbb{E}\left[\left(\sum_{i=1}^{m} (k_i - w)^2\right) + w^2\right] = m \cdot w^2/3 + m/6 + w^2. \quad (4)$$

Using this condition, we observe that ECDSA key recovery problems previously believed to be quite difficult to solve with lattices turn out to be within reach, and problems believed to be impossible become merely expensive (see Tables 3 and 4). We illustrate these updated conditions for the example of $\log(n) = 256$ in Fig. 3. The crossover points accurately predict the experimental performance of our algorithms in practice; compare to the experimental results plotted in Fig. 4.

The second observation we make in this work is that we show that lattice algorithms can still be applied when $\|\boldsymbol{v}\| \geq \mathrm{gh}(\Lambda)$, i.e. when the "lattice vector corresponding to the secret is no longer expected to be significantly shorter than other vectors in the lattice" [13]. That is, we observe that the "lattice barrier" is soft, and that violating it simply requires spending more computational time. This allows us to increase the probability of success at the crossover points in Fig. 3 and successfully solve instances with fewer samples than suggested by the crossover points.

An even stronger barrier to the applicability of any algorithm for solving the Hidden Number Problem comes from the amount of information about the secret encoded in the problem itself: each sample reveals $\log(n) - \ell$ bits of information about the secret d. Thus, we expect to require $m \geq \log(n)/(\log(n) - \ell)$ in order to recover d; heuristically, for random instances, below this point we do not expect the solution to be uniquely determined by the lattice, no matter the algorithm used to solve it. We will see below that our techniques allow us to solve instances past both the "lattice barrier" and the information-theoretic limit.

4 Bounded Distance Decoding with Predicate

We now define the key computational problem in this work:

Definition 5 (α-Bounded Distance Decoding with predicate ($\mathrm{BDD}_{\alpha,f(\cdot)}$)). *Given a lattice basis $\boldsymbol{B}$, a vector $\boldsymbol{t}$, a predicate $f(\cdot)$, and a parameter $0 < \alpha$ such that the Euclidean distance $\mathrm{dist}(\boldsymbol{t}, \boldsymbol{B}) < \alpha \cdot \lambda_1(\boldsymbol{B})$, find the lattice vector $\boldsymbol{v} \in \Lambda(\boldsymbol{B})$ satisfying $f(\boldsymbol{v} - \boldsymbol{t}) = 1$ which is closest to $\boldsymbol{t}$.*

We will solve the $\mathrm{BDD}_{\alpha,f(\cdot)}$ using Kannan's embedding technique. However, the lattice we will construct does not necessarily have a unique shortest vector. Rather, uniqueness is expected due to the addition of a predicate $f(\cdot)$.

Definition 6 (unique Shortest Vector Problem with predicate ($\mathrm{uSVP}_{f(\cdot)}$)). *Given a lattice Λ and a predicate $f(\cdot)$ find the shortest nonzero vector $\boldsymbol{v} \in \Lambda$ satisfying $f(\boldsymbol{v}) = 1$.*

Remark 1. Our nomenclature—"BDD" and "uSVP"—might be considered confusing given that the target is neither unusually close nor short. However, the distance to the lattice is still bounded in the first case and the presence of the predicate ensures uniqueness in the second case. Thus, we opted for those names over "CVP" and "SVP".

Explicitly, to solve $\text{BDD}_{\alpha,f(\cdot)}$ using an oracle solving $\text{uSVP}_{f(\cdot)}$, we consider the lattice

$$\boldsymbol{L} = \begin{pmatrix} \boldsymbol{B} & 0 \\ \boldsymbol{t} & \tau \end{pmatrix}$$

where $\tau \approx \mathbb{E}\left[\|\boldsymbol{v} - \boldsymbol{t}\|/\sqrt{d}\right]$ is some embedding factor. If $\boldsymbol{v}$ is the closest vector to $\boldsymbol{t}$ then the lattice $\Lambda(\boldsymbol{L})$ contains $(\boldsymbol{t} - \boldsymbol{v}, \tau)$. Furthermore, we construct the predicate $f'(\cdot)$ given $f(\cdot)$ as in Algorithm 1.

Input: $\boldsymbol{v}$ a vector of dimension d.
Input: $f(\cdot)$ predicate accepting inputs in $\mathbb{R}^{d-1}$.
Output: 0 or 1
1 **if** $|v_{d-1}| \neq \tau$ **then**
2 | **return** *0*;
3 **end**
4 **return** $f((v_0, v_1, \ldots, v_{d-2}))$;

Algorithm 1: uSVP predicate $f'(\cdot)$ from BDD predicate $f()$.

Remark 2. Definitions 5 and 6 are more general than the scenarios used to motivate them in the introduction. That is, both definitions permit the predicate to evaluate to true on more than one vector in the lattice and will return the closest or shortest of those vectors, respectively. In many—but not all—applications, we will additionally have the guarantee that the predicate will only evaluate to true on one vector. Definitions 5 and 6 naturally extend to the case where we ask for a list of all vectors in the lattice up to a given norm satisfying the predicate.

5 Algorithms

We propose two algorithms for solving $\text{uSVP}_{f(\cdot)}$, one based on enumeration—easily parameterized to support arbitrary target norms—and one based on sieving, solving $\text{uSVP}_{f(\cdot)}$ when the norm of the target vector is $\leq \sqrt{4/3} \cdot \text{gh}(\Lambda)$. We will start with recounting the standard uSVP strategy as a baseline to compare against later.

5.1 Baseline

When our target vector $\boldsymbol{v}$ is expected to be shorter than any other vector in the lattice, we may simply use a uSVP solver to recover it. In particular, we may use the BKZ algorithm with a block size β that satisfies the success condition in Eq. (2). Depending on β we may choose enumeration $\beta < 70$ or sieving $\beta \geq 70$ to instantiate the SVP oracle [5]. When $\beta = d$ this computes an HKZ reduced basis and, in particular, a shortest vector in the basis. It is folklore in the literature to search through the reduced basis for the presence of the target vector, that is, to not only consider the shortest non-zero vector in the basis. Thus, when comparing our algorithms against prior work, we will also do this, and consider these algorithms to have succeeded if the target is contained in the reduced basis. We will refer to these algorithms as "BKZ-Enum" and "BKZ-Sieve" depending on the oracle used. We may simply write BKZ-β or BKZ when the SVP oracle or the block size do not need to specified. When $\beta = d$ we will also refer to this approach as the "SVP approach", even though a full HKZ reduced basis is computed and examined. When we need to spell out the SVP oracle used, we will write "Sieve" and "Enum" respectively.

5.2 Enumeration

Our first algorithm is to augment lattice-point enumeration, which is exhaustive search over all points in a ball of a given radius, with a predicate to immediately give an algorithm that exhaustively searches over all points in a ball of a given radius that satisfy a given predicate. In other words, our modification to lattice-point enumeration is simply to add a predicate check whenever the algorithm reaches a leaf node in the tree, i.e. has recovered a candidate solution. If the predicate is satisfied the solution is accepted and the algorithm continues its search trying to improve upon this candidate. If the predicate is not satisfied, the algorithm proceeds as if the search failed. This augmented enumeration algorithm is then used to enumerate all points in a radius R corresponding to the (expected) norm of the target vector. We give pseudocode (adapted from [28]) for this algorithm in in the full version of this work. Our implementation of this algorithm is in the class `USVPPredEnum` in the file `usvp.py` available at [7].

Theorem 1. *Let $\Lambda \subset \mathbb{R}^d$ be a lattice containing vectors $\boldsymbol{v}$ such that $\|\boldsymbol{v}\| \leq R = \xi \cdot \mathrm{gh}(\Lambda)$ and $f(\boldsymbol{v}) = 1$. Assuming the Gaussian heuristic, then enumeration with predicate finds the shortest vector $\boldsymbol{v}$ satisfying $f(\boldsymbol{v}) = 1$ in $\xi^d \cdot d^{d/(2e)+o(d)}$ steps. Enumeration with predicate will make $\xi^{d+o(d)}$ calls to $f(\cdot)$.*

Proof (sketch). Let $R_i = R$. Enumeration runs in

$$\sum_{k=0}^{d-1} \frac{1}{2} \cdot \frac{\mathrm{Vol}(\mathfrak{B}_{d-k}(R))}{\prod_{i=k}^{d-1} \|\boldsymbol{b}_i^*\|}$$

steps [40] which scales by $\xi^{d+o(d)}$ when R scales by ξ. Solving SVP with enumeration takes $d^{d/(2e)+o(d)}$ steps [40]. By the Gaussian heuristic we expect ξ^d points in $\mathfrak{B}_d(R) \cap \Lambda$ on which the algorithm may call the predicate $f(\cdot)$.

Implementation

Modifying FPLLL [74,75] to implement this functionality is relatively straightforward since it already features an `Evaluator` class to validate full solutions—i.e. leaves—with high precision, which we subclassed. We then call this modified enumeration code with a search radius R that corresponds to the expected length of our target. We make use of (extreme) pruned enumeration by computing pruning parameters using FPLLL's `Pruner` module. Here, we make the implicit assumption that rerandomizing the basis means the probability of finding the target satisfying our predicate is independent from previous attempts. We give some example performance figures in Table 1.

Table 1. Enumeration with predicate performance data

		time		#calls to $f(\cdot)$	
ξ	s/r	observed	expected	observed	$(1.01\,\xi)^d$
1.0287	62%	3.1h	2.4 h	1104	30
1.0613	61%	5.1h	5.1 h	2813	483
1.1034	62%	11.8h	15.1 h	15274	15411
1.1384	64%	25.3h	40.1 h	169950	248226

ECDSA instances (see Sect. 6) with $d = 89$ and `USVPPredEnum`. Expected running time is computed using FPLLL's `Pruner` module, assuming 64 CPU cycles are required to visit one enumeration node. Our implementation of enumeration with predicate enumerates a radius of $1.01\cdot\xi\cdot\mathrm{gh}(\Lambda)$. We give the median of 200 experiments. The column "s/r" gives the success rate of recovering the target vector in those experiments.

5.3 Sieving

Our second algorithm is simply a sieving algorithm "as is", followed by a predicate check over the database. That is, taking a page from [5,32], we do not treat a lattice sieve as a black box SVP solver, but exploit that it outputs exponentially many short vectors. In particular, under the heuristic assumptions mentioned in the introduction—all vectors in the database L are on the surface of a d-dimensional ball—a 2-sieve, in its standard configuration, will output all vectors of norm $R \leq \sqrt{4/3}\cdot\mathrm{gh}(\Lambda)$ [32].[5] Explicitly:

Assumption 1. *When a 2-sieve algorithm terminates, it outputs a database L containing all vectors with norm $\leq \sqrt{4/3}\cdot\mathrm{gh}(\Lambda)$.*

Thus, our algorithm simply runs the predicate on each vector of the database. We give pseudocode in the full version of this work. Our implementation of this algorithm is in the class `USVPPredSieve` in the file `usvp.py` available at [7].

Theorem 2. *Let $\Lambda \subset \mathbb{R}^d$ be a lattice containing a vector $\boldsymbol{v}$ such that $\|\boldsymbol{v}\| \leq R = \sqrt{4/3}\cdot\mathrm{gh}(\Lambda)$. Under Assumption 1 sieving with predicate is expected to find the minimal $\boldsymbol{v}$ satisfying $f(\boldsymbol{v}) = 1$ in $2^{0.292\,d+o(d)}$ steps and $(4/3)^{d/2+o(d)}$ calls to $f(\cdot)$.*

[5] The radius $\sqrt{4/3}\cdot\mathrm{gh}(\Lambda)$ can be parameterized in sieving algorithms by adapting the required angle for a reduction and thus increasing the database size. This was used in e.g. [31] to find approximate Voronoi cells.

Table 2. Sieving parameters

Parameter	G6K	This work
BKZ preprocessing	None	$d - 20$
`saturation_ratio`	0.50	0.70
`db_size_factor`	3.20	3.50

Implementation

Implementing this algorithm is trivial using G6K [76]. However, some parameters need to be tuned to make Assumption 1 hold (approximately) in practice. First, since deciding if a vector is a shortest vector is a hard problem, sieve algorithms and implementations cannot use this test to decide when to terminate. As a consequence, implementations of these algorithms such as G6K use a saturation test to decide when to stop: this measures the number of vectors with norm bounded by $C \cdot \mathrm{gh}(\Lambda)$ in the database. In G6K, $C = \sqrt{4/3}$ by default. The required fraction in [76] is controlled by the variable `saturation_ratio`, which defaults to 0.5. Since we are interested in all vectors with norms below this bound, we increase this value. However, increasing this value also requires increasing the variable `db_size_factor`, which controls the size of L. If `db_size_factor` is too small, then the sieve cannot reach the saturation requested by `saturation_ratio`. We compare our final settings with the G6K defaults in Table 2. We justify our choices with the experimental data presented in the full version of this work.

Second, we preprocess our bases with BKZ-$(d-20)$ before sieving. This deviates from the strategy in [5] where such preprocessing is not necessary. Instead, progressive sieving gradually improves the basis there. However, in our experiments we found that this preprocessing step randomized the basis, preventing saturation errors and increasing the success rate. We speculate that this behavior is an artifact of the sampling and replacement strategy used inside G6K.

Conflict with D4F

The performance of sieving in practice benefits greatly from the "dimensions for free" technique introduced in [32]. This technique, which inspired our algorithm, starts from the observation that a sieve will output all vectors of norm $\sqrt{4/3} \cdot \mathrm{gh}(\Lambda)$. This observation is then used to solve SVP in dimension d using a sieve in dimension $d' = d - \Theta(d/\log d)$. In particular, if the projection $\pi_{d-d'}(\boldsymbol{v})$ of the shortest vector $\boldsymbol{v}$ has norm $\|\pi_{d-d'}(\boldsymbol{v})\| \leq \sqrt{4/3} \cdot \mathrm{gh}(\Lambda_{d-d'})$, where $\Lambda_{d-d'}$ is the lattice obtained by projecting Λ orthogonally to the first $d - d'$ vectors of $\boldsymbol{B}$ then it is expected that Babai lifting will find $\boldsymbol{v}$. Clearly, in our setting where the target itself is expected to have norm $> \mathrm{gh}(\Lambda)$ this optimization may not be available. Thus, when there is a choice to construct a uSVP lattice or a $\mathrm{uSVP}_{f(\cdot)}$ lattice in smaller dimension, we should compare the sieving dimension d' of the former against the full dimension of the latter. In [32] an "optimistic" prediction for d' is given as

$$d' = d - \frac{d \log(4/3)}{\log(d/(2\pi e))} \tag{5}$$

which matches the experimental data presented in [32] well. However, we note that G6K achieves a few extra dimensions for free via "on the fly" lifting [5]. We leave investigating an intermediate regime—*fewer* dimensions for free—for future work.

5.4 (No) Blockwise Lattice Reduction with Predicate

Our definitions and algorithms imply two regimes: the traditional BDD/uSVP regime where the target vector is unusually close to/short in the lattice (Sect. 5.1) and our BDD/uSVP with predicate regime where this is not the case and we rely on the predicate to identify it (Sects. 5.2 and 5.3). A natural question then is whether we can use the predicate to improve algorithms in the uSVP regime, that is, when the target vector is unusually short *and* we have a predicate. In other words, can we meaningfully augment the SVP oracle inside block-wise lattice reduction with a predicate?

We first note that the predicate will need to operate on "fully lifted" candidate solutions. That is, when block-wise lattice reduction considers a block $\pi_i(\boldsymbol{b}_i), \ldots, \pi_i(\boldsymbol{b}_{i+\beta-1})$, we must lift any candidate solution to $\pi_0(\cdot)$ to check the predicate. This is because projected sublattices during block-wise lattice reduction are modeled as behaving like random lattices and we have no reason in general to expect our predicate to hold on the projection.

With that in mind, we need to (Babai) lift all candidate solutions before applying the predicate. Now, by assumption, we expect the lifted target to be unusually short with respect to the full lattice. In contrast, we may expect all other candidate solutions to be randomly distributed in the parallelepiped spanned by $\boldsymbol{b}_0^*, \ldots, \boldsymbol{b}_{i-1}^*$ and thus not to be short. In other words, when we lift this way we do not need our predicate to identify the correct candidate. Indeed, the strategy just described is equivalent to picking pruning parameters for enumeration that restrict to the Babai branch on the first i coefficients or to use "dimensions for free" when sieving. Thus, it is not clear that the SVP oracles inside block-wise lattice reduction can be meaningfully be augmented with a predicate.

5.5 Higher-Level Strategies

Our algorithms may fail to find a solution for two distinct reasons. First, our algorithms are randomized: sieving randomly samples vectors and enumeration uses pruning. Second, the gap between the target's norm and the norm of the shortest vector in the lattice might be larger than expected. These two reasons for failure suggest three higher-level strategies:

plain Our "plain" strategy is simply to run enumeration and sieving with predicate as is.

repeat This strategy simply repeats running our algorithms a few times. This addresses failures to solve due to the randomized nature of our algorithms. This strategy is most useful when applied to sieving with predicate as our implementation of enumeration with predicate, which uses extreme pruning [35], already has repeated trials "built-in".

scale This strategy increases the expected radius by some small parameter, say 1.1, and reruns. When the expected target norm $> \sqrt{4/3} \cdot \mathrm{gh}(\Lambda)$ this strategy also switches from sieving with predicate to enumeration with predicate.

6 Application to ECDSA Key Recovery

The source code for the experiments in this section is in the file `ecdsa_hnp.py` available at [7].

Varying the number of samples m

We carried out experiments for common elliptic curve lengths and most significant bits known from the signature nonce to evaluate the success rate of different algorithms as we varied the number of samples, thus varying the expected ratio of the target vector to the shortest vector in the lattice.

As predicted theoretically, the shortest vector technique typically fails when the expected length of the target vector is longer than the Gaussian heuristic, and its success probability rises as the relative length of the target vector decreases. We recall that we considered the shortest vector approach a success if the target vector was contained in the reduced basis. Both the enumeration and sieving algorithms have success rates well above zero when the expected length of the target vector is longer than the expected length of the shortest vector, thus demonstrating the effectiveness of our techniques past the "lattice barrier".

Figure 4 shows the success rate of each algorithm for common parameters of interest as we vary the number of samples. Each data point represents 32 experiments for smaller instances, or 8 experiments for larger instances. The corresponding running times for these algorithms and parameters are plotted in in the full version of this work. We parameterized enumeration with predicate to succeed at a rate of 50%. For some of the larger lattice dimensions, enumeration algorithms were simply infeasible, and we do not report enumeration results for these parameters. These experiments represent more than 60 CPU-years of computation time spread over around two calendar months on a heterogeneous collection of computers with Intel Xeon 2.2 and 2.3GHz E5-2699, 2.4GHz E5-2699A, and 2.5GHz E5-2680 processors.

Table 3 gives representative running times and success rates for sieving with predicate, sieving with predicate, for popular curve sizes and numbers of bits known, and lists similar computations from the literature where we could determine the parameters used. It illustrates how our techniques allow us to solve instances with fewer samples than previous work. We recall that most applications of lattice algorithms for solving ECDSA-HNP instances seem to arbitrarily choose a small block size for BKZ, and experimentally determine the number of

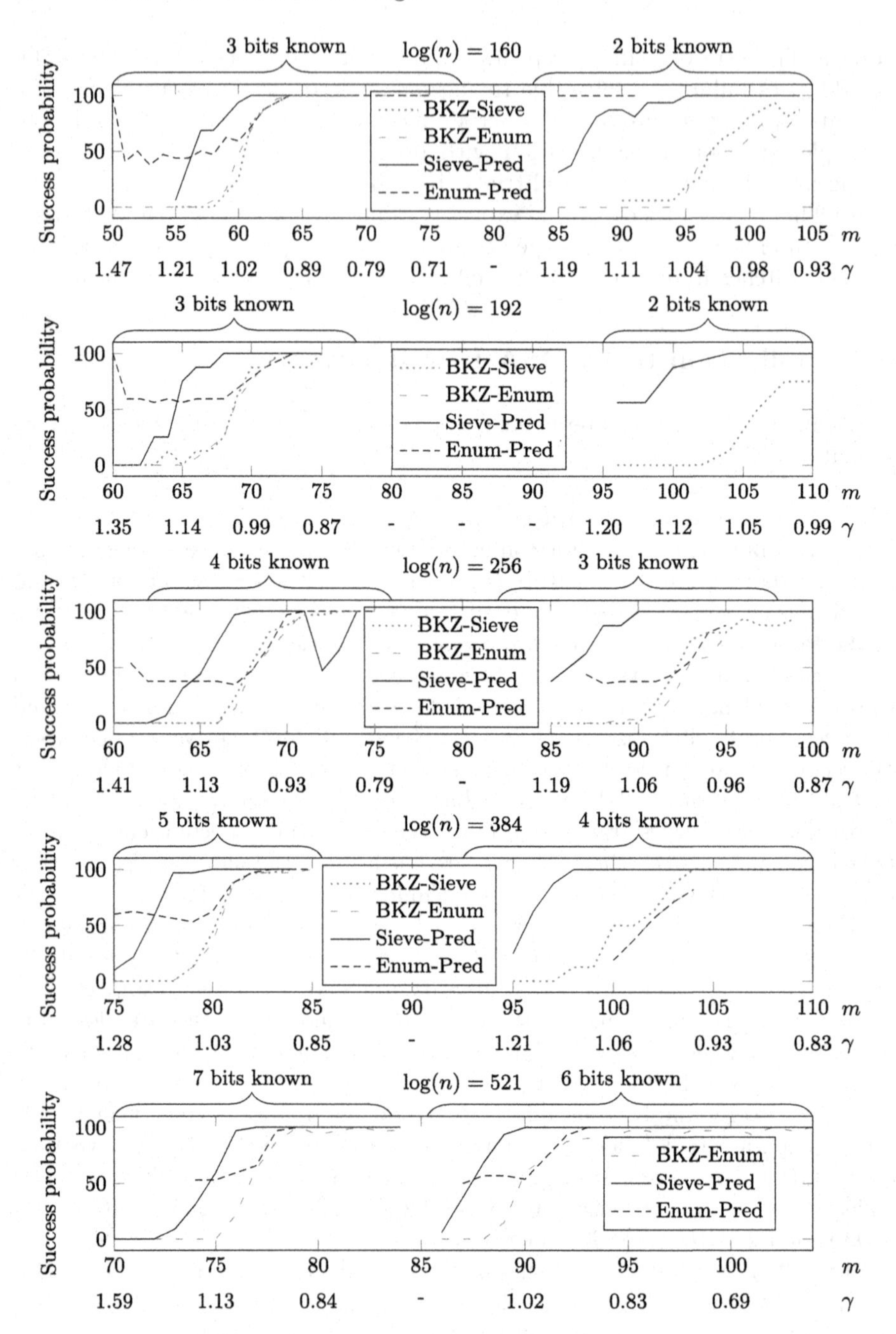

Fig. 4. Comparison of algorithm success rates for ECDSA. We generated HNP instances for common ECDSA parameters and compared the success rates of each algorithm on identical instances. The x-axis labels show the number of samples m and $\gamma = \mathbb{E}[\|\boldsymbol{v}\|] \,/\, \mathbb{E}[\|\boldsymbol{b}_0\|]$, the corresponding ratio between the expected length of the target vector v and the expected length of the shortest vector b_0 in a random lattice.

Table 3. Performance for medium instances

log(n)	bias	m	time	alg.	s/r	previous work
160	3 bits	53	3452 s	E	44%	
160	2 bits	87	4311 s	S	62%	enum, $m \approx 100$, $s/r = 23\%$ in [53]
160	1 bit	–	–	–	–	Bleichenbacher, $m \approx 2^{27}$, in [13]
192	3 bits	63	851s	E	56%	
192	2 bits	98	7500 s	S	56%	
192	1 bit	–	–	–	–	Bleichenbacher, $m \approx 2^{29}$, in [13]
256	4 bits	63	2122 s	E	41%	
256	4 bits	65	76 s	S	66%	BKZ-25, $m \approx 82$, $s/r = 90\%$ in [68]
256	3.6 bits	73	69 s	S	66 %	BKZ-30, $m = 80$, $s/r = 94.5\%$ in [36]
256	3 bits	87	5400 s	S	63%	enum, $m = 100$, $s/r = 21\%$ in [52]
256	2 bits	–	–	–	–	Bleichenbacher, $m \approx 2^{26}$, in [73]
384	5 bits	76	0026 s	E	60%	
384	5 bits	78	412 s	S	91%	BKZ-25, $m \approx 94$, $s/r = 90\%$ in [68]
384	4 bits	97	9200 s	S	88%	BKZ-20, $m = 170$, $s/r = 90\%$ in [9]
521	7 bits	74	6318 s	E	57%	
521	7 bits	75	438s	S	59%	
521	6 bits	88	6643 s	S	77%	

We compare the number of required samples m to previously reported results from the literature, where available. Instances solved using enumeration with predicate are labeled with "E" and are solved using fewer samples than the information-theoretic barrier. Instances solved with using sieving with predicate are labeled "S". Time is in CPU-seconds. The success rate for our experiments is taken over 32 experiments; see Fig. 4 for how the success rate varies with the number of samples.

samples required. For 3 bits known on a 256-bit curve, there are multiple algorithmic results reported in the literature. In [73] the authors report a running time of 238 CPU-hours to run the first phase of Bleichenbacher's algorithm on 2^{23} samples. In [52] the authors report applying BKZ-20 followed by enumeration with linear pruning to achieve a 21% success probability in five hours. Sieving with predicate took 1.5 CPU-hours to solve the same parameters with a 63% success probability using 87 samples.

Table 3 also gives running times and success rates for enumeration with predicate, enumeration with predicate, in solving instances beyond the information-theoretic barrier, that is, when the number of samples available was not large enough to expect the Hidden Number Problem to contain sufficient information to recover the signing key; breaking the "information-theoretic limit". We recall that our techniques can solve these instances because the predicate uniquely determines the target.

We give concrete estimates for the number of required samples and thus the size of the resulting lattice problem in Table 4 for common ECDSA key sizes as

Table 4. Resources required to solve ECDSA with known nonce bits.

$\log(n) = 160$								
Bits known	8	7	6	5	4	3	2	1
Sieve m/d	$21/-2$	25/9	29/15	35/23	45/33	61/49	99/84	258/232
Sieve-Pred m/d	21/22	24/25	28/29	33/34	42/43	57/58	87/88	193/194
Sieve-Pred cost	40.2	38.3	36.4	34.9	33.6	34.4	41.5	80.9
Limit m	20	23	27	32	40	54	80	160
Limit -1 cost	23.5	23.6	24.7	27.9	31.1	36.5	50.6	104.0
$\log(n) = 192$								
Bits known	8	7	6	5	4	3	2	1
Sieve m/d	25/9	29/15	34/21	41/29	51/39	70/57	110/94	255/229
Sieve-Pred m/d	25/26	28/29	33/34	39/40	49/50	65/66	98/99	200/201
Sieve-Pred cost	37.8	36.4	34.9	33.9	33.7	35.7	45.2	83.5
Limit m	24	28	32	39	48	64	96	192
Limit -1 cost	23.7	23.7	26.0	27.2	31.5	38.3	54.2	118.6
$\log(n) = 256$								
Bits known	8	7	6	5	4	3	2	1
Sieve m/d	33/20	38/26	45/33	54/42	69/56	93/79	146/128	341/310
Sieve-Pred m/d	33/34	37/38	43/44	52/53	65/66	87/88	131/132	267/268
Sieve-Pred cost	34.9	34.1	33.6	33.9	35.7	41.5	57.6	108.6
Limit m	32	37	43	52	64	86	128	256
Limit -1 cost	27.2	27.4	29.8	32.3	38.7	48.2	73.7	169.7
$\log(n) = 384$								
Bits known	8	7	6	5	4	3	2	1
Sieve m/d	50/38	57/45	67/54	81/67	103/88	140/122	219/196	512/470
Sieve-Pred m/d	49/50	56/57	65/66	78/79	97/98	130/131	196/197	401/402
Sieve-Pred cost	33.7	34.3	35.7	38.8	44.9	57.2	82.0	158.8
Limit m	48	55	64	77	96	128	192	384
Limit -1 cost	33.7	36.2	39.7	45.2	55.0	74.1	119.0	283.8
$\log(n) = 521$								
Bits known	8	7	6	5	4	3	2	1
Sieve m/d	68/55	78/65	91/77	110/94	139/121	190/169	298/269	696/643
Sieve-Pred m/d	66/67	75/76	88/89	105/106	132/133	176/177	266/267	544/545
Sieve-Pred cost	35.9	38.0	41.8	47.9	58.0	74.5	108.2	212.5
Limit m	66	75	87	105	131	174	261	521
Limit -1 cost	38.0	43.7	50.9	59.4	75.6	105.5	174.1	419.1

Sieve Number of samples m required for solving uSVP as in Sect. 5.1 and sieving dimension according to Eq. (5) (called d' there).
Sieve-Pred Number of samples m required for sieving with predicate and sieving dimension $d = m + 1$.
Sieve-Pred cost Log of expected cost in CPU cycles; cost is estimated as $0.658 \cdot d - 21.11 \log(d) + 119.91$ which does not match the asymptotics but approximates experiments up to dimension 100.
limit Information theoretic limit for m of pure lattice approach: $\lceil \log(n)/\text{bits known} \rceil$.
limit -1 cost Log of expected cost for enumeration with predicate in CPU cycles with $m = \lceil \log(n)/\text{bits known} \rceil - 1$ samples.

the number of known nonce bits varies. These estimates include both instances we are able to solve, as well as problem sizes beyond our current computational ability. When few bits are known, corresponding to large lattices, our approach promises a smaller sieving dimension, but for small (that is, practical) dimensions, "dimensions for free" is more efficient. Thus, when enough samples are available it is still preferable to mount the uSVP attack. We note that Table 4 suggests that there are feasible computations within range for future work with a suitably cluster-parallelized implementation of sieving with predicate, in particular two bits known for a 256-bit modulus, and three bits known for a 384-bit modulus. Furthermore, Table 4 indicates that sieving with predicate allows us to decode at almost the information-theoretic limit for many instances. For comparison, we also give the expected cost of enumeration with predicate when solving with one fewer sample than this limit.

Fixed number of samples m
An implication of Table 4 is that our approach allows us to solve the Hidden Number Problem with fewer samples than the unique SVP bounds would imply. In some attack settings, the attacker may have a hard limit on the number of samples available. Using enumeration and sieving with predicate allows us to increase the probability of a successful attack in this case, and increase the range of parameters for which a feasible attack is possible.

This scenario arose in [22], where the authors searched for flawed ECDSA implementations by applying lattice attacks to ECDSA signatures gathered from public data sources including cryptocurrency blockchains and internet-wide scans of protocols like TLS and SSH. In these cases, the attacker has access to a fixed number of signature samples generated from a given public key, and wishes to maximize the probability of a successful attack against this fixed number of signatures, for as few bits known as possible.

The paper of [22] reported using BKZ in very small dimensions to find 287 distinct keys that used nonce lengths of 160, 128, 110, 64, and less than 32 bits for ECDSA signatures with the 256-bit secp256k1 curve used for Bitcoin. They reported finding two distinct keys using 128-bit nonces in two signatures each.

Experimentally, the BKZ algorithm only has a 70% success rate at recovering the private key for 128-bit nonces with two signature samples, and the success rate drops precipitously as the number of unknown nonce bits increases. In contrast, sieving with predicate has a 100% success rate up to around 132-bit nonces. See Fig. 5 for a comparison of these algorithms as the number of signatures is fixed to two and the number of unknown nonce bits varies.

We hypothesized that this failure rate may have caused the results of [22] to omit some vulnerable keys. Thus, we ran our sieving with predicate approach against the same Bitcoin blockchain snapshot data from September 2018 as used in [22], targeting only 128-bit nonces using pairs of signatures. This snapshot contained 569,396,463 signatures that had been generated by a private key that generated two or more signatures. For the set of m signatures generated by each distinct key, we applied the sieving with predicate algorithm to $2m$ pairs of

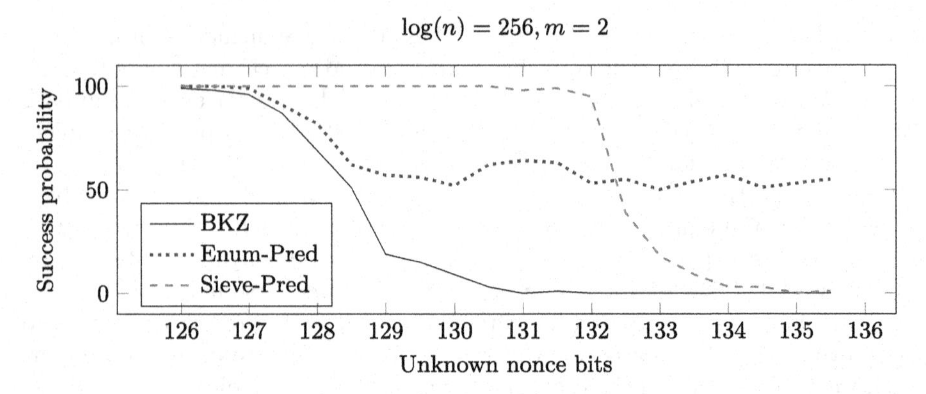

Fig. 5. Algorithm success rates in a small fixed-sample regime. We plot the experimental success rate of each algorithm in recovering a varying number of nonce bits using two samples. Each data point represents the success rate of the algorithm over 100 experiments. Using sieving and enumeration with a predicate allows the attacker to increase the probability of a successful attack even when more samples cannot be collected. We parameterized enumeration with predicate to succeed with probability 1/2.

signatures to check for nonces of length less than 128 bits. Using this approach, we were able to compute the private keys for 9 more distinct secret keys.

Handling errors
In practical side-channel attacks, it is common to have some fraction of measurement errors in the data. In a common setting for ECDSA key recovery from known nonce bits, the side channel leaks the number of leading zeroes of the nonce, but the signal is noisy and thus data may be mislabeled. If the estimate is below the true number, this is not a problem, since the target vector will be even shorter than estimated and thus easier to find. However, if the true number of zero bits is smaller than the estimate, then the desired vector will be larger than estimated which can cause the key recovery algorithm to fail.

It is believed that lattice approaches to the Hidden Number Problem do not deal well with noisy data [68] and "assume that inputs are perfectly correct" [13]. There are a few techniques in the literature to work around these limitations and to deal with noise [44]. The most common approach is simply to repeatedly try running the lattice algorithm on subsamples of the data until one succeeds [23]. Alternatively, one can use more samples in the lattice, in order to increase the expected gap between the target vector and the lattice. For example, it was already observed in [29] that using a lattice construction with more samples increases the success rate in the presence of errors, even using the same block size.

However, the most natural approach does not appear to have been considered in the literature before: Use an estimate of the error rate to compute a new target norm as in Eq. (4) and pick the block size or enumeration radius parameters

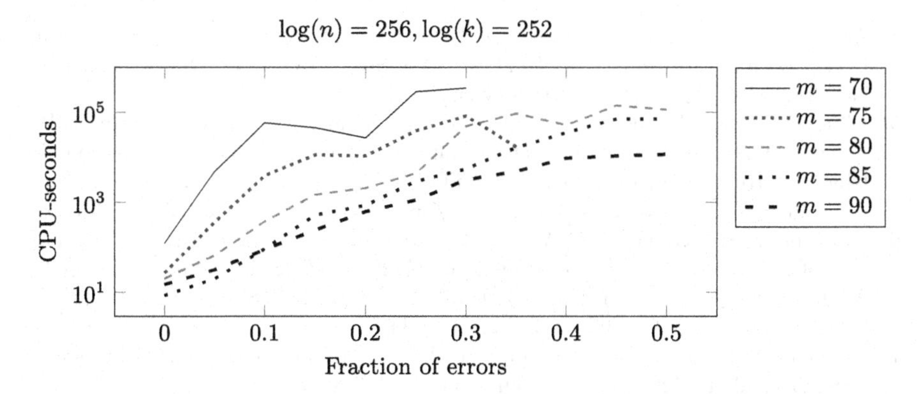

Fig. 6. Search time in the presence of errors. We plot the experimental computation time of the "scale" strategy to find the target vector as we varied the number of errors in the sample. For these experiments, each "error" is a nonce that is one bit longer than the length supplied to the algorithm. Increasing the number of samples decreases the search time.

accordingly. That is, when the error rate can be estimated, this is simply a special case of estimating the norm of the target vector. As before, even if the number m of samples is limited, enumeration with predicate in principle can search out to arbitrarily large target norms.

The most difficult case to handle is when more samples are not available and the error rate is unknown or difficult to estimate properly. In this case, a strategy is to repeatedly increase the expected target norm of the vector, pick an algorithm that solves for this target norm R and attempt to solve the instance: BKZ for $R < \mathrm{gh}(\Lambda)$, sieving with predicate for $R \leq \sqrt{4/3} \cdot \mathrm{gh}(\Lambda)$ and enumeration with predicate for $R > \sqrt{4/3} \cdot \mathrm{gh}(\Lambda)$. We refer to this strategy as "scale" in Sect. 5.5.

Figure 6 illustrates how the running time of the "scale" strategy varies with the fraction of errors and the number of samples used in the lattice.

Acknowledgments. We thank Joe Rowell and Jianwei Li for helpful discussions on an earlier draft of this work, Daniel Genkin for suggesting additional references and a discussion on error resilience, Noam Nissan for feedback on our implementation, and Samuel Neves for helpful suggestions on our characterization of previous work.

References

1. Ajtai, M., Kumar, R., Sivakumar, D.: A sieve algorithm for the shortest lattice vector problem. In: 33rd ACM STOC, pp. 601–610. ACM Press, July 2001
2. Albrecht, M.R., Bai, S., Fouque, P.A., Kirchner, P., Stehlé, D., Wen, W.: Faster enumeration-based lattice reduction: Root hermite factor $k^{1/(2k)}$ time $k^{k/8+o(k)}$. In: Micciancio and Ristenpart [5], pp. 186–212
3. Albrecht, M.R., Cid, C., Faugère, J., Fitzpatrick, R., Perret, L.: On the complexity of the BKW algorithm on LWE. Des. Codes Cryptogr. **74**(2), 325–354 (2015)

4. Albrecht, M.R., Deo, A., Paterson, K.G.: Cold boot attacks on ring and module LWE keys under the NTT. IACR TCHES **2018**(3), 173–213 (2018). https://tches.iacr.org/index.php/TCHES/article/view/7273
5. Albrecht, M.R., Ducas, L., Herold, G., Kirshanova, E., Postlethwaite, E.W., Stevens, M.: The general sieve kernel and new records in lattice reduction. In: Ishai, Y., Rijmen, V. (eds.) EUROCRYPT 2019, Part II. LNCS, vol. 11477, pp. 717–746. Springer, Cham (2019). https://doi.org/10.1007/978-3-030-17656-3_25
6. Albrecht, M.R., Göpfert, F., Virdia, F., Wunderer, T.: Revisiting the expected cost of solving uSVP and applications to LWE. In: Takagi, T., Peyrin, T. (eds.) ASIACRYPT 2017, Part I. LNCS, vol. 10624, pp. 297–322. Springer, Cham (2017). https://doi.org/10.1007/978-3-319-70694-8_11
7. Albrecht, M.R., Heninger, N.: Bounded distance decoding with predicate source code, December 2020. https://github.com/malb/bdd-predicate/
8. Albrecht, M.R., Player, R., Scott, S.: On the concrete hardness of Learning with Errors. J. Math. Cryptol. **9**(3), 169–203 (2015)
9. Aldaya, A.C., Brumley, B.B., ul Hassan, S., García, C.P., Tuveri, N.: Port contention for fun and profit. In: 2019 IEEE Symposium on Security and Privacy, pp. 870–887. IEEE Computer Society Press, May 2019
10. Alkim, E., Ducas, L., Pöppelmann, T., Schwabe, P.: Post-quantum key exchange - a new hope. In: Holz, T., Savage, S. (eds.) USENIX Security 2016, pp. 327–343. USENIX Association, August 2016
11. Applebaum, B., Cash, D., Peikert, C., Sahai, A.: Fast cryptographic primitives and circular-secure encryption based on hard learning problems. In: Halevi, S. (ed.) CRYPTO 2009. LNCS, vol. 5677, pp. 595–618. Springer, Heidelberg (2009). https://doi.org/10.1007/978-3-642-03356-8_35
12. Aranha, D.F., Fouque, P.-A., Gérard, B., Kammerer, J.-G., Tibouchi, M., Zapalowicz, J.-C.: GLV/GLS decomposition, power analysis, and attacks on ECDSA signatures with single-bit nonce bias. In: Sarkar, P., Iwata, T. (eds.) ASIACRYPT 2014, Part I. LNCS, vol. 8873, pp. 262–281. Springer, Heidelberg (2014). https://doi.org/10.1007/978-3-662-45611-8_14
13. Aranha, D.F., Novaes, F.R., Takahashi, A., Tibouchi, M., Yarom, Y.: LadderLeak: reaking ECDSA with less than one bit of nonce leakage. Cryptology ePrint Archive, Report 2020/615 (2020). https://eprint.iacr.org/2020/615
14. Bai, S., Stehlé, D., Wen, W.: Improved reduction from the bounded distance decoding problem to the unique shortest vector problem in lattices. In: Chatzigiannakis, I., Mitzenmacher, M., Rabani, Y., Sangiorgi, D. (eds.) ICALP 2016. LIPIcs, vol. 55, pp. 76:1–76:12. Schloss Dagstuhl, July 2016
15. Becker, A., Ducas, L., Gama, N., Laarhoven, T.: New directions in nearest neighbor searching with applications to lattice sieving. In: Krauthgamer, R. (ed.) 27th SODA, pp. 10–24. ACM-SIAM, January 2016
16. Becker, A., Gama, N., Joux, A.: Speeding-up lattice sieving without increasing the memory, using sub-quadratic nearest neighbor search. Cryptology ePrint Archive, Report 2015/522 (2015). http://eprint.iacr.org/2015/522
17. Benger, N., van de Pol, J., Smart, N.P., Yarom, Y.: "Ooh Aah... just a little bit"?: a small amount of side channel can go a long way. In: Batina, L., Robshaw, M. (eds.) CHES 2014. LNCS, vol. 8731, pp. 75–92. Springer, Heidelberg (2014). https://doi.org/10.1007/978-3-662-44709-3_5
18. Bleichenbacher, D.: On the generation of one-time keys in DL signature schemes. In: Presentation at IEEE P1363 working Group Meeting, p. 81 (2000)
19. Bleichenbacher, D.: Experiments with DSA. CRYPTO 2005-Rump Session (2005)

20. Blum, A., Kalai, A., Wasserman, H.: Noise-tolerant learning, the parity problem, and the statistical query model. In: 32nd ACM STOC, pp. 435–440. ACM Press, May 2000
21. Boneh, D., Venkatesan, R.: Hardness of computing the most significant bits of secret keys in Diffie-Hellman and related schemes. In: Koblitz, N. (ed.) CRYPTO 1996. LNCS, vol. 1109, pp. 129–142. Springer, Heidelberg (1996). https://doi.org/10.1007/3-540-68697-5_11
22. Breitner, J., Heninger, N.: Biased nonce sense: lattice attacks against weak ECDSA signatures in cryptocurrencies. In: Goldberg, I., Moore, T. (eds.) FC 2019. LNCS, vol. 11598, pp. 3–20. Springer, Cham (2019). https://doi.org/10.1007/978-3-030-32101-7_1
23. Brumley, B.B., Tuveri, N.: Remote timing attacks are still practical. In: Atluri, V., Diaz, C. (eds.) ESORICS 2011. LNCS, vol. 6879, pp. 355–371. Springer, Heidelberg (2011). https://doi.org/10.1007/978-3-642-23822-2_20
24. Cabrera Aldaya, A., Pereida García, C., Brumley, B.B.: From A to Z: projective coordinates leakage in the wild. IACR TCHES **2020**(3), 428–453 (2020). https://tches.iacr.org/index.php/TCHES/article/view/8596
25. Capkun, S., Roesner, F. (eds.): USENIX Security 2020. USENIX Association, August 2020
26. Coppersmith, D.: Finding a small root of a bivariate integer equation; factoring with high bits known. In: Maurer, U. (ed.) EUROCRYPT 1996. LNCS, vol. 1070, pp. 178–189. Springer, Heidelberg (1996). https://doi.org/10.1007/3-540-68339-9_16
27. Dachman-Soled, D., Ducas, L., Gong, H., Rossi, M.: LWE with side information: attacks and concrete security estimation. In: Micciancio and Ristenpart [56], pp. 329–358
28. Dagdelen, Ö., Schneider, M.: Parallel enumeration of shortest lattice vectors. In: D'Ambra, P., Guarracino, M., Talia, D. (eds.) Euro-Par 2010. LNCS, vol. 6272, pp. 211–222. Springer, Heidelberg (2010). https://doi.org/10.1007/978-3-642-15291-7_21
29. Dall, F., et al.: CacheQuote: efficiently recovering long-term secrets of SGX EPID via cache attacks. IACR TCHES **2018**(2), 171–191 (2018). https://tches.iacr.org/index.php/TCHES/article/view/879
30. De Mulder, E., Hutter, M., Marson, M.E., Pearson, P.: Using Bleichenbacher's solution to the hidden number problem to attack nonce leaks in 384-Bit ECDSA. In: Bertoni, G., Coron, J.-S. (eds.) CHES 2013. LNCS, vol. 8086, pp. 435–452. Springer, Heidelberg (2013). https://doi.org/10.1007/978-3-642-40349-1_25
31. Doulgerakis, E., Laarhoven, T., de Weger, B.: Finding closest lattice vectors using approximate Voronoi cells. In: Ding, J., Steinwandt, R. (eds.) PQCrypto 2019. LNCS, vol. 11505, pp. 3–22. Springer, Cham (2019). https://doi.org/10.1007/978-3-030-25510-7_1
32. Ducas, L.: Shortest vector from lattice sieving: a few dimensions for free. In: Nielsen, J.B., Rijmen, V. (eds.) EUROCRYPT 2018. LNCS, vol. 10820, pp. 125–145. Springer, Cham (2018). https://doi.org/10.1007/978-3-319-78381-9_5
33. Fincke, U., Pohst, M.: Improved methods for calculating vectors of short length in a lattice, including a complexity analysis. Math. Comput. **44**(170), 463–471 (1985)
34. Gama, N., Nguyen, P.Q.: Finding short lattice vectors within Mordell's inequality. In: Ladner, R.E., Dwork, C. (eds.) 40th ACM STOC, pp. 207–216. ACM Press, May 2008

35. Gama, N., Nguyen, P.Q., Regev, O.: Lattice enumeration using extreme pruning. In: Gilbert, H. (ed.) EUROCRYPT 2010. LNCS, vol. 6110, pp. 257–278. Springer, Heidelberg (2010). https://doi.org/10.1007/978-3-642-13190-5_13
36. García, C.P., Brumley, B.B.: Constant-time callees with variable-time callers. In: Kirda, E., Ristenpart, T. (eds.) USENIX Security 2017, pp. 83–98. USENIX Association, August 2017
37. Genkin, D., Pachmanov, L., Pipman, I., Tromer, E., Yarom, Y.: ECDSA key extraction from mobile devices via nonintrusive physical side channels. In: Weippl, E.R., Katzenbeisser, S., Kruegel, C., Myers, A.C., Halevi, S. (eds.) ACM CCS 2016, pp. 1626–1638. ACM Press (Oct 2016)
38. Gennaro, R., Robshaw, M.J.B. (eds.): CRYPTO 2015, Part I, LNCS, vol. 9215. Springer, Heidelberg (2015). https://doi.org/10.1007/978-3-662-48000-7
39. Guo, Q., Johansson, T., Stankovski, P.: Coded-BKW: Solving LWE using lattice codes. In: Gennaro and Robshaw [38], pp. 23–42
40. Hanrot, G., Stehlé, D.: Improved analysis of Kannan's shortest lattice vector algorithm. In: Menezes, A. (ed.) CRYPTO 2007. LNCS, vol. 4622, pp. 170–186. Springer, Heidelberg (2007). https://doi.org/10.1007/978-3-540-74143-5_10
41. Herold, G., Kirshanova, E.: Improved algorithms for the approximate k-list problem in Euclidean Norm. In: Fehr, S. (ed.) PKC 2017, Part I. LNCS, vol. 10174, pp. 16–40. Springer, Heidelberg (2017). https://doi.org/10.1007/978-3-662-54365-8_2
42. Herold, G., Kirshanova, E., May, A.: On the asymptotic complexity of solving LWE. Des. Codes Cryptogr. **86**(1), 55–83 (2018)
43. Howgrave-Graham, N., Smart, N.P.: Lattice attacks on digital signature schemes. Des. Codes Cryptogr. **23**(3), 283–290 (2001)
44. Jancar, J., Sedlacek, V., Svenda, P., Sys, M.: Minerva: he curse of ECDSA nonces. IACR TCHES **2020**(4), 281–308 (2020). https://tches.iacr.org/index.php/TCHES/article/view/8684
45. Kannan, R.: Improved algorithms for integer programming and related lattice problems. In: 15th ACM STOC, pp. 193–206. ACM Press, April 1983
46. Kannan, R.: Minkowski's convex body theorem and integer programming. Math. Oper. Res. **12**(3), 415–440 (1987)
47. Kirchner, P., Fouque, P.A.: An improved BKW algorithm for LWE with applications to cryptography and lattices. In: Gennaro and Robshaw [38], pp. 43–62
48. Klein, P.N.: Finding the closest lattice vector when it's unusually close. In: Shmoys, D.B. (ed.) 11th SODA. pp. 937–941. ACM-SIAM, January 2000
49. Laarhoven, T.: Search problems in cryptography: from fingerprinting to lattice sieving. Ph.D. thesis, Eindhoven University of Technology (2015)
50. Laarhoven, T., Mariano, A.: Progressive lattice sieving. In: Lange, T., Steinwandt, R. (eds.) PQCrypto 2018. LNCS, vol. 10786, pp. 292–311. Springer, Cham (2018). https://doi.org/10.1007/978-3-319-79063-3_14
51. Lenstra, A.K., Lenstra, H.W., Jr., Lovász, L.: Factoring polynomials with rational coefficients. Math. Ann. **261**, 366–389 (1982)
52. Liu, M., Chen, J., Li, H.: Partially known nonces and fault injection attacks on SM2 signature algorithm. In: Lin, D., Xu, S., Yung, M. (eds.) Inscrypt 2013. LNCS, vol. 8567, pp. 343–358. Springer, Cham (2014). https://doi.org/10.1007/978-3-319-12087-4_22
53. Liu, M., Nguyen, P.Q.: Solving BDD by enumeration: an update. In: Dawson, E. (ed.) CT-RSA 2013. LNCS, vol. 7779, pp. 293–309. Springer, Heidelberg (2013). https://doi.org/10.1007/978-3-642-36095-4_19

54. Merget, R., Brinkmann, M., Aviram, N., Somorovsky, J., Mittmann, J., Schwenk, J.: Raccoon attack: finding and exploiting most-significant-bit-oracles in TLS-DH(E), September 2020. https://raccoon-attack.com/RacoonAttack.pdf. Accessed 11 Sept 2020
55. Micciancio, D., Regev, O.: Lattice-based cryptography. In: Bernstein, D.J., Buchmann, J., Dahmen, E. (eds.) Post-Quantum Cryptography, pp. 147–191. Springer, Heidelberg (2009). https://doi.org/10.1007/978-3-540-88702-7_5
56. Micciancio, D., Ristenpart, T. (eds.): CRYPTO 2020, Part II, LNCS, vol. 12171. Springer, Heidelberg (2020). https://doi.org/10.1007/978-3-030-56880-1
57. Micciancio, D., Voulgaris, P.: Faster exponential time algorithms for the shortest vector problem. In: Charika, M. (ed.) 21st SODA,pp. 1468–1480. ACM-SIAM (2010)
58. Micciancio, D., Walter, M.: Fast lattice point enumeration with minimal overhead. In: Indyk, P. (ed.) 26th SODA, pp. 276–294. ACM-SIAM, January 2015
59. Micciancio, D., Walter, M.: Practical, predictable lattice basis reduction. In: Fischlin, M., Coron, J.-S. (eds.) EUROCRYPT 2016, Part I. LNCS, vol. 9665, pp. 820–849. Springer, Heidelberg (2016). https://doi.org/10.1007/978-3-662-49890-3_31
60. Moghimi, D., Lipp, M., Sunar, B., Schwarz, M.: Medusa: Microarchitectural data leakage via automated attack synthesis. In: Capkun and Roesner [25], pp. 1427–1444
61. Moghimi, D., Sunar, B., Eisenbarth, T., Heninger, N.: TPM-FAIL: TPM meets timing and lattice attacks. In: Capkun and Roesner [25], pp. 2057–2073
62. Nemec, M., Sýs, M., Svenda, P., Klinec, D., Matyas, V.: The return of coppersmith's attack: practical factorization of widely used RSA moduli. In: Thuraisingham, B.M., Evans, D., Malkin, T., Xu, D. (eds.) ACM CCS 2017, pp. 1631–1648. ACM Press (2017)
63. Nguyen, P.Q., Shparlinski, I.: The insecurity of the digital signature algorithm with partially known nonces. J. Cryptol. **15**(3), 151–176 (2002)
64. Nguyen, P.Q., Tibouchi, M.: Lattice-based fault attacks on signatures. In: Joye, M., Tunstall, M. (eds.) Fault Analysis in Cryptography. ISC, pp. 201–220. Springer, Heidelberg (2012). https://doi.org/10.1007/978-3-642-29656-7_12
65. Nguyen, P.Q., Vidick, T.: Sieve algorithms for the shortest vector problem are practical. J. Math. Cryptol. **2**(2), 181–207 (2008)
66. Phost, M.: On the computation of lattice vectors of minimal length, successive minima and reduced bases with applications. SIGSAM Bull. **15**, 37–44 (1981)
67. Regev, O.: On lattices, learning with errors, random linear codes, and cryptography. In: Gabow, H.N., Fagin, R. (eds.) 37th ACM STOC, pp. 84–93. ACM Press, May 2005
68. Ryan, K.: Return of the hidden number problem. IACR TCHES **2019**(1), 146–168 (2018).https://tches.iacr.org/index.php/TCHES/article/view/7337
69. Ryan, K.: Hardware-backed heist: extracting ECDSA keys from qualcomm's TrustZone. In: Cavallaro, L., Kinder, J., Wang, X., Katz, J. (eds.) ACM CCS 2019, pp. 181–194. ACM Press, November 2019
70. Schnorr, C.P.: A hierarchy of polynomial time lattice basis reduction algorithms. Theor. Comput. Sci. **53**, 201–224 (1987)
71. Schnorr, C., Euchner, M.: Lattice basis reduction: improved practical algorithms and solving subset sum problems. Math. Program. **66**, 181–199 (1994)
72. Stein, W., et al.: Sage Mathematics Software Version 9.0. The Sage Development Team (2019). http://www.sagemath.org

73. Takahashi, A., Tibouchi, M., Abe, M.: New Bleichenbacher records: fault attacks on qDSA signatures. IACR TCHES **2018**(3), 331–371 (2018). https://tches.iacr.org/index.php/TCHES/article/view/7278
74. The FPLLL development team: FPLLL, a lattice reduction library (2020). https://github.com/fplll/fplll
75. The FPLLL development team: FPyLLL, a Python interface to FPLLL (2020). https://github.com/fplll/fpylll
76. The G6K development team: G6K (2020). https://github.com/fplll/g6k
77. Tibouchi, M.: Attacks on (ec)dsa with biased nonces (2017). https://ecc2017.cs.ru.nl/slides/ecc2017-tibouchi.pdf, elliptic Curve Cryptography Workshop
78. Weiser, S., Schrammel, D., Bodner, L., Spreitzer, R.: Big numbers - big troubles: Systematically analyzing nonce leakage in (EC)DSA implementations. In: Capkun and Roesner [25], pp. 1767–1784

On the Ideal Shortest Vector Problem over Random Rational Primes

Yanbin Pan[1(✉)], Jun Xu[2], Nick Wadleigh[3], and Qi Cheng[3]

[1] Key Laboratory of Mathematics Mechanization, Academy of Mathematics and Systems Science, Chinese Academy of Sciences, Beijing 100190, China
panyanbin@amss.ac.cn

[2] State Key Laboratory of Information Security, Institute of Information Engineering, Chinese Academy of Sciences, Beijing 100093, China
xujun@iie.ac.cn

[3] School of Computer Science, University of Oklahoma, Norman, OK 73019, USA
{wadleigh,qcheng}@ou.edu

Abstract. Any non-zero ideal in a number field can be factored into a product of prime ideals. In this paper we report a surprising connection between the complexity of the shortest vector problem (SVP) of prime ideals in number fields and their decomposition groups. When applying the result to number fields popular in lattice based cryptosystems, such as power-of-two cyclotomic fields, we show that a majority of rational primes lie under prime ideals admitting a polynomial time algorithm for SVP. Although the shortest vector problem of ideal lattices underpins the security of the Ring-LWE cryptosystem, this work does not break Ring-LWE, since the security reduction is from the worst case ideal SVP to the average case Ring-LWE, and it is one-way.

Keywords: Ring-LWE · Ideal lattice · Average case computational complexity

1 Introduction

Due to their conjectured ability to resist quantum computer attacks, lattice-based cryptosystems have drawn considerable attention. In 1996, Ajtai [1] pioneered the research on worst-case to average-case reduction for the Short Integer Solution problem (SIS). In 2005, Regev [33] presented a worst-case to average-case (quantum) reduction for the Learning With Errors problem (LWE). SIS and LWE became two important cryptographic assumptions, and a large number of cryptographic schemes based on these two problems have been designed. However, the common drawback of such schemes is their limited efficiency.

To improve the efficiency of lattice-based schemes, some special algebraic structures are employed. The first lattice-based scheme with some algebraic structure was the NTRU public key cryptosystem [15], which was introduced by Hoffstein, Pipher and Silverman in 1996. It works in the convolution ring $\mathbb{Z}[x]/(x^p-1)$ where

A. Canteaut and F.-X. Standaert (Eds.): EUROCRYPT 2021, LNCS 12696, pp. 559–583, 2021.
https://doi.org/10.1007/978-3-030-77870-5_20

p is a prime. The cyclic nature of the ring $\mathbb{Z}[x]/(x^p - 1)$ contributes to NTRU's efficiency, and makes NTRU one of the most popular schemes. Later the ring was employed in many other cryptographic primitives, such as [5,21,25,26,31,36].

In 2009, Stehlé *et al.* [37] introduced a structured and more efficient variant of LWE involving the ring $\mathbb{F}_p[x]/(x^N + 1)$ where N is a power of 2 and p is a prime satisfying $p \equiv 3 \pmod 8$. In 2010, Lyubashevsky, Peikert and Regev [22] presented a ring-based variant of LWE, called Ring-LWE. The hardness of problems in [22,37] is based on worst-case assumptions on ideal lattices. Recently, Peikert, Regev and Stephens-Davidowitz [30] presented a polynomial time quantum reduction from (worst-case) ideal lattice problems to Ring-LWE for any modulus and any number field. Lots of schemes employ the ring $\mathbb{Z}[x]/(x^N + 1)$ where N is a power of 2, for example, NewHope [3], Crystals-Kyber [8], and LAC [20] submitted to NIST's post-quantum cryptography standardization. Although solving the ideal SVP does not necessarily break Ring-LWE, understanding the hardness of ideal SVP is no doubt a very important first step to understand the hardness of Ring-LWE.

1.1 Previous Works

Principal ideal lattices are a class of important ideal lattices which can be generated by a single ring element. There is a line of work focusing on the principal ideal SVP. Based on [4,9], solving approx-SVP problems on principal ideal lattices can be divided into the following two steps: Step 1 is finding an ideal generator by using class group computations. In this step, a quantum polynomial time algorithm is presented by Biasse and Song [7], which is based on the work [14]; a classical subexponential time algorithm was given by Biasse, Espitau, Fouque, Gélin and Kirchner [6]. Step 2 is shortening the ideal generator in Step 1 with the log-unit lattice. This step was analyzed by Cramer, Ducas, Peikert and Regev [11]. Then a quantum polynomial time algorithm for approx-SVP, with a $2^{\tilde{O}(\sqrt{N})}$ approximation factor, on principal ideal lattices in cyclotomic number fields was presented in [11].

In 2017, Cramer, Ducas and Wesolowski [12] extended the case of principal ideal lattices in [11] to the case of a general ideal lattice in a cyclotomic ring of prime-power conductor. For approx-SVP on ideal lattices, the result in [12] is better than the BKZ algorithm [34] when the approximation factor is larger than $2^{\tilde{O}(\sqrt{N})}$. Ducas, Plancon and Wesolowski [13] analyzed the approximation factor $2^{\tilde{O}(\sqrt{N})}$ in [11,12] to determine the specific dimension N so that the corresponding algorithms outperform BKZ for an ideal lattice in cyclotomic number fields. Recently, Pellet-Mary, Hanrot and Stehlé [32], inspired by the algorithms in [11,12], proposed an algorithm to solve approx-SVP with the approximation factor $2^{\tilde{O}(\sqrt{N})}$ in ideal lattices for all number fields, aiming to provide trade-offs between the approximation factor and the running time. However, there is an exponential pre-processing phase.

Inspired by Bernstein's logarithm-subfield attack [4], Albrecht, Bai and Ducas [2] and Cheon, Jeong and Lee [10] independently proposed two similar subfield attacks in 2016 against overstretched NTRU that has much larger modulus than in the NTRUEncrypt standard. Later, Kirchner and Fouque [16] proposed a variant of the subfield attacks to improve these two attacks in practice. A typical subfield attack consists of three steps: mapping the lattice to some subfield, solving the lattice problem in the subfield and finally lifting the solution to the full field.

1.2 Our Results

In this paper, we investigate the SVP for lattices corresponding to prime ideals in number fields normal over $\mathbb{Q}$. Every nonzero ideal in a Dedekind Domain can be factored uniquely into a product of prime ideals, so short vectors in prime ideals may help us to find short vectors in general ideals. If, in a general prime ideal $\mathfrak{p}$, we are able to efficiently find a vector with length within the Minkowski bound for $\mathfrak{p}$, then for an ideal $\mathfrak{a}$ with few prime ideal factors, we will be able to approximate the shortest vector in $\mathfrak{a}$ to within a factor much better than what is achieved by the LLL [17] or BKZ [35] algorithms. The most difficult step in factoring an ideal is actually factorization of an integer (the norm of the ideal), which can be done in polynomial time by quantum computers, or in subexponential time by classical computers.

Consider a finite Galois extension $\mathbb{L} \cong \mathbb{Q}[x]/(f(x))$ of $\mathbb{Q}$, and let $\mathfrak{P}$ be a prime ideal in the ring of integers $O_{\mathbb{L}}$ of $\mathbb{L}$. The subgroup of $Gal(\mathbb{L}/\mathbb{Q})$ that stabilizes $\mathfrak{P}$ set-wise is known as the *decomposition group* of $\mathfrak{P}$. Let $\mathbb{K} \subset \mathbb{L}$ be the subfield fixed by the decomposition group of $\mathfrak{P}$. $\mathbb{K}$ is called the *decomposition field* of $\mathfrak{P}$. To find a short vector in $\mathfrak{P}$, we can search for a short vector in the lattice $\mathfrak{P} \cap \mathbb{K}$, which may have smaller rank. More precisely, for a rational prime p, if $pO_{\mathbb{L}}$ is factored into a product of g prime ideals in $O_{\mathbb{L}}$, we can reduce the problem of finding a short vector in any of these prime ideals to a problem of finding a short vector in a rank-g lattice, provided that a basis of $O_{\mathbb{K}}$ can be found efficiently. Equivalently, the fewer the number of irreducible factors of $f(x)$ over $\mathbb{F}_p$, the more efficiently we may solve SVP for prime ideals lying above p. One argues from general facts of algebraic number theory that the determinant of the sublattice is not too large compared to the original lattice in order to relate Minkowski type λ_1 bounds for the two lattices.

We go on to apply the foregoing idea to the rings $\mathbb{Z}[x]/(x^{2^n} + 1)$, which are quite popular in cryptography. We show that there is a hierarchy for the hardness of SVP for these prime ideal lattices. This arises from the observation that the decomposition groups (or their index-two subgroups) form a chain in the subgroup lattice (See Appendices A and B). Roughly speaking, we can classify such prime ideal lattices into n distinct classes, and for a prime ideal lattice in the r-th class, we can find its shortest vector by solving SVP in a dimension-2^r lattice. This suggests that the difficulty of prime ideal SVP can change dramatically from ideal to ideal, an interesting phenomenon that has, to our knowledge, not been pointed out in the literature. By considering some of these classes, we prove that a nontrivial fraction of prime ideals admit an efficient SVP algorithm.

Theorem 1. *Let $N = 2^n$, where n is a positive integer. Let $\mathfrak{p}$ be a prime ideal in the ring $\mathbb{Z}[x]/(x^N + 1)$, and suppose $\mathfrak{p}$ contains a prime number $p \equiv \pm 3 \pmod 8$. Then under the coefficient embedding, the shortest vector in $\mathfrak{p}$ can be found in time poly$(N, \log p)$, and the length of the shortest vector is exactly $\sqrt{p}$.*

Can we conclude from the above result that the *average case* prime ideal SVP is easy? It depends how we define an average prime ideal lattice. As prime ideals are rigid structures, changing distributions gives us totally different complexity results. If prime ideals are selected uniformly at random from the set of those prime ideals whose norms are bounded, then easy cases are rare. Nevertheless our result does show that an average case of the prime ideal SVP in power-of-two cyclotomic fields is not hard, if the prime ideals are selected uniformly at random from the set of all prime ideals whose rational primes are less than some fixed bound. See Subsect. 4.2 for details.

For general (non prime) ideals in $\mathbb{Z}[x]/(x^{2^n} + 1)$, we present an algorithm to confirm that the hierarchy for the hardness of SVP also exists; that is, we can solve SVP for a general ideal lattice by solving SVP in a 2^r-dimensional sublattice, for some positive integer r related to the factorization of the ideal (see Theorem 6). Following Theorem 1, we show how to solve the SVP for ideals all of whose prime factors lie in a certain class. This is a special case of Theorem 6.

Proposition 1. *Let $N = 2^n$, where n is a positive integer. Let $\mathcal{I}$ be an ideal in the ring $\mathbb{Z}[x]/(x^N + 1)$ with prime factorization*

$$\mathcal{I} = \mathfrak{p}_1\mathfrak{p}_2 \cdots \mathfrak{p}_k.$$

If each $\mathfrak{p}_i$ contains a prime integer $\equiv \pm 3 \pmod 8$, the shortest vector in $\mathcal{I}$ can be found in time poly$(N, \log(\mathcal{N}(\mathcal{I})))$.

We would like to stress that the algorithm works by exploiting the multiplicative structure of ideals in the ring of integers of a number field, without factoring the ideal. We regard this as the second contribution of this work, in addition to the algorithm for prime ideals.

Note that a decomposition field is a subfield of the number field. Our algorithm can also be seen as a kind of subfield attack to solve the ideal Hermite-SVP problem. Compared with the previous subfield attacks [2,4,10,16], the main differences are: the previous subfield attacks use relative norm (or trace) to map a lattice into some subfield, while we use the intersection with the decomposition field; The approximation factor in the previous attacks, such as [2], will suffer during the lifting process, while our lifting costs not so much; The previous attacks [2,10,16] work for NTRU with much big modulus, while the instances amenable to our attack must satisfy the condition that the decomposition field is a proper subfield of the number field.

We have to point out that it is still unknown how our result impacts the security of cryptographic schemes. It does not break Ring-LWE, since the security reduction is from the worst case ideal SVP to the average case Ring-LWE, and it is one-way. As pointed out by [4], Smart and Vercauteren [29] proposed an ideal lattice-based fully homomorphic encryption scheme, which generated a prime ideal lattice as the public key. It is enough to break the scheme by finding a short vector in the lattice. To improve the efficiency, they chose ideals of prime determinants, which are not weak instances revealed by our algorithm. Our paper provides a security justification for using such ideal lattices. We should no doubt avoid the weak instances when we construct cryptographic schemes. In addition, our result is a beneficial attempt to solve ideal SVP by exploiting the algebraic structure, and it helps us understand better the hardness of ideal SVP.

1.3 Paper Organization

The remainder of the paper is organized as follows. In Sect. 2, we give some mathematical preliminaries. In Sect. 3 we prove a reduction of approx-SVP in the finite Galois extension of $\mathbb{Q}$. Then in Sect. 4 we present a reduction of SVP for prime ideal lattices and then general ideal lattices in $\mathbb{Z}[\zeta_{2^{n+1}}]$. Finally, a conclusion and some open problems are given in Sect. 5.

2 Mathematical Preliminaries

2.1 Lattices and Some Computational Problems

Lattices are discrete additive subgroups of $\mathbb{R}^N$. Any finite set of linearly independent vectors $b_1, b_2, \cdots, b_m \in \mathbb{R}^N$ generates a lattice:

$$\mathcal{L} = \left\{ \sum_{i=1}^{m} z_i b_i \mid z_i \in \mathbb{Z} \right\}.$$

Denote by B the matrix whose column vectors are the b_i's. We say B is a basis (in matrix form) for $\mathcal{L}$; m and N are the rank and dimension of $\mathcal{L}$, respectively. Denote by $\det(\mathcal{L})$ the determinant of lattice $\mathcal{L}$, which is defined as the (co)volume of $\mathcal{L}$ in the real subspace spanned by $\mathcal{L}$. Note that if $m = N$, the determinant of $\mathcal{L}$ is exactly $|\det(B)|$.

The shortest vector problem (SVP), which refers to the problem of finding a shortest nonzero lattice vector in a given lattice, is one of the most famous hard problems in lattice theory. There are some variants of SVP that are very important for applications.

- Approx-SVP: Given a lattice $\mathcal{L}$ and an approximation factor $\gamma \geq 1$, find a non-zero lattice vector of norm $\leq \gamma \cdot \lambda_1(\mathcal{L})$, where $\lambda_1(\mathcal{L})$ is the length of a shortest non-zero vector in $\mathcal{L}$.
- Hermite-SVP: Given a rank-N lattice $\mathcal{L}$ and an approximation factor $\gamma \geq 1$, find a non-zero lattice vector of norm $\leq \gamma \cdot \det(\mathcal{L})^{\frac{1}{N}}$. Note that Minkowski's theorem [27] tells us that

$$\lambda_1(\mathcal{L}) \leq \frac{2}{\mathcal{V}_N^{1/N}} \cdot \det(\mathcal{L})^{\frac{1}{N}} \leq \sqrt{N} \cdot \det(\mathcal{L})^{\frac{1}{N}},$$

where $\mathcal{V}_N$ is the volume of the N-dimensional ball with radius 1. Thus for any $\gamma \geq \sqrt{N}$, Hermite-SVP is well-defined for all rank-N lattices.
Since $\lambda_1(\mathcal{L})$ is usually hard to determine given a basis of $\mathcal{L}$, it might be very hard to verify a solution returned by an algorithm for Approx-SVP with some approximation factor. However, the solution to Hermite-SVP can be verified efficiently. Hence, many algorithms, such as LLL [17] and BKZ [35], are designed as polynomial-time Hermite-SVP algorithms for some exponential approximation factor.

It is obvious that any algorithm that solves Approx-SVP with factor γ can also solve Hermite-SVP with factor $\gamma\sqrt{N}$ by Minkowski's theorem. Furthermore, based on an idea of Lenstra and Schnorr, Lovász showed that any algorithm solving Hermite-SVP with factor γ can be used to solve Approx-SVP with factor γ^2 in polynomial time [19].

Moreover, a solution to Hermite-SVP with factor $\sqrt{N}$, that is, satisfying the Minkowski bound, is usually taken as a good enough approximation of some shortest vector in a "random" lattice. In addition, when choosing parameters for lattice-based cryptosystems in practice, such as in NewHope[3], Crystals-Kyber [8], and LAC [20], the time complexity of solving Hermite-SVP with some particular factor usually determines the concrete security of these cryptosystems. Therefore, the algorithm for Hermite-SVP is key to both solving Approx-SVP and analyzing the security of lattice-based cryptosystems.

The closest vector problem (CVP) is another famous hard problem in lattice theory. This refers to the problem of finding a lattice vector that is closest to a given vector.

2.2 Some Basic Algebraic Number Theory

We will review some basic algebraic number theory in this section. More details can be found in [23] or [28].

Number fields. An algebraic number $\zeta \in \mathbb{C}$ is any root of a nonzero polynomial $f(x) \in \mathbb{Q}[x]$ and its minimal polynomial is the unique monic irreducible $f(x) \in \mathbb{Q}[x]$ of minimal degree that has ζ as a root. An algebraic number is called an algebraic integer if its minimal polynomial lies in $\mathbb{Z}[x]$.

An algebraic number field is a finite field extension $\mathbb{K}$ of $\mathbb{Q}$. Such a field can be obtained by adjoining a single algebraic integer ζ to $\mathbb{Q}$. That is, $\mathbb{K} = \mathbb{Q}(\zeta)$ for some algebraic integer ζ. The degree N of the minimal polynomial $f(x)$ of ζ is also the degree of $\mathbb{K}$ over $\mathbb{Q}$.

Denote by $O_{\mathbb{K}}$ the ring of algebraic integers in $\mathbb{K}$. It is an integral domain and also a free $\mathbb{Z}$-module with rank N.

For example, let $\zeta_{2^{n+1}}$ be a complex primitive 2^{n+1}-th root of unity, whose minimal polynomial is $f = x^{2^n} + 1$. Then, $\mathbb{K} = \mathbb{Q}(\zeta_{2^{n+1}})$ is the cyclotomic number field of order 2^{n+1} with degree 2^n. Its ring of integers is well known to be $\mathbb{Z}[\zeta_{2^{n+1}}]$.

Embeddings. A number field $\mathbb{K}$ of degree N over $\mathbb{Q}$ has exactly N embeddings into $\mathbb{C}$. Let $\sigma_1, \sigma_2, \cdots, \sigma_{s_1}$ be the real embeddings from $\mathbb{K}$ to $\mathbb{R}$, and let

$$\sigma_{s_1+1}, \sigma_{s_1+2}, \cdots, \sigma_{s_1+s_2},$$

$$\sigma_{s_1+s_2+1} = \overline{\sigma_{s_1+1}}, \quad \sigma_{s_1+s_2+2} = \overline{\sigma_{s_1+2}}, \quad \cdots, \quad \sigma_{s_1+2s_2} = \overline{\sigma_{s_1+s_2}}$$

be the non-real embeddings from $\mathbb{K}$ to $\mathbb{C}$, where $\bar{\cdot}$ denotes complex conjugation.

From these σ_i's we can define the *canonical embedding* $\Sigma_{\mathbb{K}}$ from $\mathbb{K}$ to $\mathbb{C}^N$:

$$\Sigma_{\mathbb{K}} : \mathbb{K} \to \mathbb{C}^N, \quad a \mapsto (\sigma_1(a), \sigma_2(a), \cdots, \sigma_N(a)).$$

It is known that the image of $\Sigma_{\mathbb{K}}$ falls into a subspace in $\mathbb{C}^N$, which is isomorphic to $\mathbb{R}^N$ as an inner product space (see [22]).

Another important embedding from $\mathbb{K}$ to $\mathbb{R}^N$ is the *coefficient embedding,* which is most commonly used in cryptographic constructions. This embedding depends on a choice of generator α for $\mathbb{K}$: write $\mathbb{K} = \mathbb{Q}(\alpha)$ and map $\beta = a_0 + a_1\alpha + ... + a_{N-1}\alpha^{N-1}$ to its coefficient vector, $C(\beta) := (a_0, a_1, ..., a_{N-1})$.

If α may be chosen so that

$$O_{\mathbb{K}} = \mathbb{Z} + \alpha\mathbb{Z} + \alpha^2\mathbb{Z} + ... + \alpha^{N-1}\mathbb{Z}$$

we say $O_{\mathbb{K}}$ is *monogenic.* In this case the coefficient embedding maps $O_{\mathbb{K}}$ to $\mathbb{Z}^N$. Alternatively, via $O_{\mathbb{K}} \cong \mathbb{Z}[x]/(f(x))$, where $f(x)$ is the minimal polynomial of α, we may think of C mapping a polynomial in $\mathbb{Z}[x]/(f(x))$ to its coefficient vector:

$$C(a_0 + a_1x + \cdots + a_{N-1}x^{N-1}) = (a_0, a_1, \cdots, a_{N-1}).$$

Discriminants. If $\mathbb{K} \subset \mathbb{L}$ are number fields, the (relative) discriminant of a $\mathbb{K}$-basis $b_1, b_2, \ldots, b_N$ for $\mathbb{L}$ is defined by

$$d_{\mathbb{L}/\mathbb{K}}(b_1, b_2, \ldots, b_N) = |\det(\sigma_i b_j)|^2,$$

where σ_i varies over the $[\mathbb{L} : \mathbb{K}]$ embeddings $\mathbb{L} \to \mathbb{C}$ which fix all elements of $\mathbb{K}$. The discriminant $\text{disc}(O_{\mathbb{L}}/O_{\mathbb{K}})$, also denoted by $\text{disc}(\mathbb{L}/\mathbb{K})$, is then the ideal of $O_{\mathbb{K}}$ which is generated by the discriminants $d_{\mathbb{L}/\mathbb{K}}(b_1, b_2, \ldots, b_N)$ of all the $\mathbb{K}$-bases $b_1, b_2, \ldots, b_N$ of $\mathbb{L}$ which are contained in $O_{\mathbb{L}}$.

For any number field $\mathbb{K}$, the (absolute) discriminant $\text{disc}(\mathbb{K}/\mathbb{Q})$ becomes the principal ideal generated by $d(b_1, b_2, \ldots, b_N)$ for any basis $b_1, b_2, \ldots, b_N$ of the free $\mathbb{Z}$-module $O_{\mathbb{K}}$. In this case we just write $\text{disc}(\mathbb{K})$ to refer to this ideal or the unique positive integer that generates it. In a sense made precise by the embeddings defined above, the discriminant gives a notion of the co-volume of a ring of integers in its fraction field. Specifically, the discriminant is just the square of this co-volume.

2.3 Ideal Lattices

The ring of integers $O_{\mathbb{K}}$ of $\mathbb{K}$ is a free $\mathbb{Z}$-module, and any ideal $\mathcal{I}$ in $O_{\mathbb{K}}$ is a free $\mathbb{Z}$-submodule since $\mathbb{Z}$ is a principal ideal domain. Under the canonical embedding or the coefficient embedding, any such $\mathcal{I}$ is sent to a lattice in $\mathbb{R}^N$. We call this image the *ideal lattice* associated with $\mathcal{I}$, and we denote it also by $\mathcal{I}$.

Under the canonical embedding $\Sigma_{\mathbb{K}}$ from $\mathbb{K}$ to $\mathbb{C}^N$, the co-volume (i.e. the volume of a fundamental domain) of an ideal lattice $\mathcal{I}$ is given by $N_{\mathbb{K}}(\mathcal{I})\sqrt{|\mathrm{disc}(\mathbb{K})|}$, where $N_{\mathbb{K}}(\mathcal{I})$ is the *norm* of $\mathcal{I}$, defined as the cardinality of $O_{\mathbb{K}}/\mathcal{I}$. Note that when we say the norm of a vector, it refers to the Euclidean norm rather than the algebra norm of an ideal.

Usually it is easier to use the canonical embedding in mathematical analysis, and to use the coefficient embedding in cryptography. For example, under the coefficient embedding of $\mathbb{Z}[\zeta_{2^{n+1}}]$, the lattice associated with the prime ideal $\mathfrak{p}_i = (p, f_i(\zeta_{2^{n+1}}))$ is generated by the coefficient vectors of the following polynomials (modulo $x^N + 1$)

$$f_i, xf_i, \cdots, x^{N-1}f_i \text{ and } p, px, \cdots, px^{N-1},$$

where p is some rational prime, and f_i is some irreducible factor of $x^{2^n} + 1$ modulo p. The minimum generating set should have only N vectors, which can be found by computing the Hermite Normal Form.

Ideals in $\mathbb{Z}[\zeta_{2^n+1}]$. The cyclotomic field of order $2N = 2^{n+1}$ is widely used in cryptography. Its ring of integers is $\mathbb{Z}[\zeta_{2^{n+1}}]$, which is isomorphic to $\mathbb{Z}[x]/(x^N+1)$. Its discriminant is 2^{n2^n}.

Let p be a rational prime, and let

$$x^N + 1 = (f_1 f_2 \cdots f_g)^e$$

be the prime factorization of $x^N + 1$ in the polynomial ring $\mathbb{F}_p[x]$. Then we have

$$(p) = (\mathfrak{p}_1 \mathfrak{p}_2 \cdots \mathfrak{p}_g)^e,$$

where $\mathfrak{p}_i = (p, f_i(\zeta_{2^{n+1}}))$ (here f_i is any integer polynomial which projects to the f_i in the above factorization). We say the prime ideal $\mathfrak{p}_i$ *lies over* the prime p. If e is greater than 1, we say the prime p is *ramified* (in $\mathbb{Z}[\zeta_{2^{n+1}}]$); otherwise we say p is *unramified.* One can verify that 2 is the only ramified rational prime in the cyclotomic field of order $2N$, and that the prime ideal $(2, \zeta_{2^{n+1}} + 1) = (\zeta_{2^{n+1}} + 1)$ lies above the ideal (2).

We are therefore interested in the explicit factorization of the 2^{n+1}-th cyclotomic polynomials, $x^{2^n} + 1$, over $\mathbb{F}_p[x]$. This is computed in [18, Thm. 2.47 and Thm. 3.75] when $p \equiv 1 \pmod 4$ and in [24] when $p \equiv 3 \pmod 4$.

Theorem 2. *Let $p \equiv 1 \pmod 4$, i.e. $p = 2^A \cdot m + 1$, $A \geq 2$, m odd. Denote by U_k the set of all primitive 2^k-th roots of unity modulo p. We have*

- *If $n < A$, then $x^{2^n}+1$ is the product of 2^n irreducible linear factors over $\mathbb{F}_p$:*

$$x^{2^n}+1 = \prod_{u \in U_{n+1}} (x+u).$$

- *If $n \geq A$, then $x^{2^n}+1$ is the product of 2^{A-1} irreducible binomials over $\mathbb{F}_p$ of degree 2^{n-A+1}:*

$$x^{2^n}+1 = \prod_{u \in U_A} (x^{2^{n-A+1}}+u).$$

Theorem 3. *Let $p \equiv 3 \pmod 4$, i.e. $p = 2^A \cdot m - 1$, $A \geq 2$, m odd. Denote by $D_s(x,a)$ the Dickson polynomials*

$$\sum_{i=0}^{\lfloor \frac{s}{2} \rfloor} \frac{s}{s-i}\binom{s-i}{i}(-a)^i x^{s-2i}$$

over $\mathbb{F}_p$. For $n \geq 2$, we have

- *If $n < A$, then $x^{2^n}+1$ is the product of 2^{n-1} irreducible trinomials over $\mathbb{F}_p$:*

$$x^{2^n}+1 = \prod_{\gamma \in \Gamma} (x^2+\gamma x+1),$$

where Γ is the set of all roots of $D_{2^{n-1}}(x,1)$.
- *If $n \geq A$, then $x^{2^n}+1$ is the product of 2^{A-1} irreducible trinomials over $\mathbb{F}_p$ of degree 2^{n-A+1}:*

$$x^{2^n}+1 = \prod_{\delta \in \Delta} (x^{2^{n-A+1}}+\delta x^{2^{n-A}}-1),$$

where Δ is the set of all roots of $D_{2^{A-1}}(x,-1)$.

3 Solving Hermite-SVP for Prime Ideal Lattices in a Galois Extension

In the following, we will consider solving Hermite-SVP for prime ideals of $O_{\mathbb{L}}$ when $\mathbb{L}$ is a finite Galois extension of $\mathbb{Q}$.

A prime ideal $\mathfrak{p}$ in $O_{\mathbb{L}}$ contains a rational prime p, and therefore occurs as one of the prime ideals in the factorization

$$pO_{\mathbb{L}} = (\mathfrak{p}_1\mathfrak{p}_2\cdots\mathfrak{p}_g)^e.$$

Without loss of generality, we assume $\mathfrak{p}_1 = \mathfrak{p}$.

To find a short vector of $\mathfrak{p}_1$, we try to find a short vector in the sublattice given by the intersection of $\mathfrak{p}_1$ with some intermediate field between $\mathbb{Q}$ and $\mathbb{L}$. Since this sublattice has smaller rank, this may lead to a more efficient algorithm than working in $\mathbb{L}$ directly.

More precisely, let G be the Galois group of $\mathbb{L}$ over $\mathbb{Q}$. Recall the decomposition group, D, and decomposition field, $\mathbb{K}$, for $\mathfrak{p}_1$:

$$D := \{\sigma \in G : \sigma(\mathfrak{p}_1) = \mathfrak{p}_1\},$$

$$\mathbb{K} := \{x \in \mathbb{L} : \forall \sigma \in D, \sigma(x) = x\}.$$

Let $O_{\mathbb{K}}$ be the algebraic integer ring of $\mathbb{K}$. It is well known that the degree of $\mathbb{K}$ over $\mathbb{Q}$ is g (see [23, Thm. 28]). This is our desired intermediate field, and we have the following theorem.

Theorem 4. *Suppose $\mathbb{L}/\mathbb{Q}$ is a finite Galois extension with degree N, and suppose $\mathfrak{p}$ is a prime ideal of $O_{\mathbb{L}}$ lying over an unramified rational prime p such that $pO_{\mathbb{L}}$ has g distinct prime ideal factors in $O_{\mathbb{L}}$. If $\mathbb{K}$ is the decomposition field of $\mathfrak{p}$, then a solution to Hermite-SVP with factor γ in the sublattice $\mathfrak{c} = \mathfrak{p} \cap O_{\mathbb{K}}$ under the canonical embedding of $\mathbb{K}$ will also be a solution to Hermite-SVP in $\mathfrak{p}$ with factor $\frac{\sqrt{N/g}}{N_{\mathbb{K}}(disc(\mathbb{L}/\mathbb{K}))^{1/(2N)}} \cdot \gamma$ $(\leq \sqrt{\frac{N}{g}} \cdot \gamma)$ under the canonical embedding of $\mathbb{L}$.*

In particular, when $\gamma = \sqrt{g}$, a vector in the sublattice $\mathfrak{c}$ satisfying the Minkowski bound will produce a vector in the lattice $\mathfrak{p}$ satisfying the Minkowski bound.

Proof. Consider the following diagram

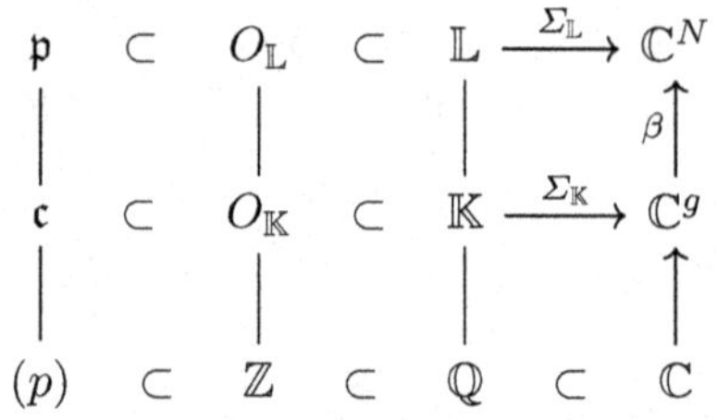

$$\begin{array}{ccccccc}
\mathfrak{p} & \subset & O_{\mathbb{L}} & \subset & \mathbb{L} & \xrightarrow{\Sigma_{\mathbb{L}}} & \mathbb{C}^N \\
| & & | & & | & & \uparrow \beta \\
\mathfrak{c} & \subset & O_{\mathbb{K}} & \subset & \mathbb{K} & \xrightarrow{\Sigma_{\mathbb{K}}} & \mathbb{C}^g \\
| & & | & & | & & \uparrow \\
(p) & \subset & \mathbb{Z} & \subset & \mathbb{Q} & \subset & \mathbb{C}
\end{array}$$

Here β is chosen to be the linear map making the diagram commute.

Note that every embedding of $\mathbb{K}$ in $\mathbb{C}$ can be extended to exactly $\frac{N}{g}$ embeddings of $\mathbb{L}$ in $\mathbb{C}$ [23, Thm. 50]; thus β is (up to permutation) just the linear embedding given by repeating each coordinate N/g times. Thus for any $v \in \mathbb{C}^g$ we have

$$\|\beta(v)\| = \sqrt{\frac{N}{g}} \cdot \|v\|. \quad (1)$$

Note that the norm of $\mathfrak{c}$ is exactly p [23, Thm. 29], so that the determinant of $\mathfrak{c}$ is $p\sqrt{|\mathrm{disc}(\mathbb{K})|}$. Thus, under the canonical embedding of $O_{\mathbb{K}}$ into $\mathbb{C}^g$, any solution $v_0 \in \mathfrak{c}$ to Hermite-SVP with factor γ satisfies

$$\|v_0\| \leq \gamma \cdot p^{\frac{1}{g}} |\mathrm{disc}(\mathbb{K})|^{\frac{1}{2g}}.$$

By Eq. (1) above and the fact that $\text{disc}(\mathbb{L}) = \text{disc}(\mathbb{K})^{N/g} N_{\mathbb{K}}(\text{disc}(\mathbb{L}/\mathbb{K}))$ [28, Corallary (2.10), pp. 202], we therefore have

$$
\begin{aligned}
\|\beta(v_0)\| &\leq \gamma \cdot \sqrt{\frac{N}{g}} p^{\frac{1}{g}} |\text{disc}(\mathbb{K})|^{\frac{1}{2g}} \\
&= \gamma \cdot \frac{\sqrt{N/g}}{N_{\mathbb{K}}(\text{disc}(\mathbb{L}/\mathbb{K}))^{1/(2N)}} p^{\frac{1}{g}} |\text{disc}(\mathbb{L})|^{\frac{1}{2N}} \\
&= \gamma \cdot \frac{\sqrt{N/g}}{N_{\mathbb{K}}(\text{disc}(\mathbb{L}/\mathbb{K}))^{1/(2N)}} (p^{\frac{N}{g}} \sqrt{|\text{disc}(\mathbb{L})|})^{\frac{1}{N}}
\end{aligned}
$$

Note that the norm of $\mathfrak{p}$ is $p^{\frac{N}{g}}$, and thus $p^{\frac{N}{g}} \sqrt{|\text{disc}(\mathbb{L})|}$ is exactly the determinant of the ideal lattice $\mathfrak{p}$ under the canonical embedding of $\mathbb{L}$. Hence v_0 is also a solution to Hermite-SVP with factor $\frac{\sqrt{N/g}}{N_{\mathbb{K}}(\text{disc}(\mathbb{L}/\mathbb{K}))^{1/(2N)}} \cdot \gamma$.

Note that $N_{\mathbb{K}}(\text{disc}(\mathbb{L}/\mathbb{K}))$ is a positive integer. Thus

$$
\frac{\sqrt{N/g}}{N_{\mathbb{K}}(\text{disc}(\mathbb{L}/\mathbb{K}))^{1/(2N)}} \leq \sqrt{\frac{N}{g}}.
$$

In particular, when $\gamma = \sqrt{g}$, $\frac{\sqrt{N/g}}{N_{\mathbb{K}}(\text{disc}(\mathbb{L}/\mathbb{K}))^{1/(2N)}} \cdot \gamma \leq \sqrt{N}$ still holds. The theorem follows. □

Remark 1. To design an algorithm from the theorem, we need to calculate the decomposition field from a prime ideal. In general this is not an easy problem. Fortunately, for power-of-two or prime order cyclotomic fields, the subfield structures have been worked out in the literature. Another technical problem is to compute a basis for $\mathfrak{c} = \mathfrak{p} \cap O_{\mathbb{K}}$. This can be solved if we know a $\mathbb{Q}$-basis of $\mathbb{K}$.

Remark 2. How many prime ideals are vulnerable to this attack? In other words, given an irreducible polynomial over $\mathbb{Z}$, how does its factoring pattern change over $\mathbb{F}_p$ as p varies? This is a central topic of class field theory when the Galois group is solvable. In the general case, it has been studied in the famous Langlands program, where many challenging problems remain. The answer is well known for number fields popular in lattice based cryptography. There exists a set of rational primes, of positive density with non-trivial decomposition group, such that for any p in this set, the decomposition fields of the prime ideals lying above p are never the whole field $\mathbb{L}$. In this case, $\mathfrak{p} \cap O_{\mathbb{K}}$ has rank no more than half that of $\mathfrak{p}$, resulting in a much easier SVP problem.

4 Solving SVP for Ideal Lattices in $\mathbb{Z}[\zeta_{2^{n+1}}]$

In the following, we use the above idea to solve SVP for ideal lattices in $\mathbb{Z}[\zeta_{2^{n+1}}]$, the ring of integers in the cyclotomic field $\mathbb{Q}(\zeta_{2^{n+1}})$, a field which is widely used in lattice-based cryptography. The decomposition field of any prime ideal is either equal to, or a degree-2 subfield of, one of the following

$$\mathbb{Q}[i] \subset \mathbb{Q}[\zeta_8] \subset \cdots \subset \mathbb{Q}[\zeta_{2^n}] \subset \mathbb{Q}[\zeta_{2^{n+1}}].$$

The subfields in this chain are convenient because they are monogenic and their integer rings have $\mathbb{Z}$-bases (powers of $\zeta_{2^{n+1}}$) that are mutually compatible and orthogonal under the canonical embedding. This results in a hierarchy of complexity of prime ideal SVP problems. Furthermore, for a non-prime ideal $\mathcal{I}$, we can approximate the shortest vectors of $\mathcal{I}$ by finding short vectors in $\mathcal{I} \cap O_{\mathbb{K}}$, where $\mathbb{K}$ is the smallest field in the above chain containing all the decomposition fields of the prime factors of $\mathcal{I}$. This allows us to find short vectors for many non-prime ideals. In contrast to the approximation result we achieved in the general setting of Theorem 4, an *exact* SVP solution is possible in power-of-two cyclotomic fields. We will first prove a reduction for SVP for *prime* ideal lattices in $\mathbb{Z}[\zeta_{2^{n+1}}]$, and then we will prove a reduction for general ideals. We would like to point out that in the case of a general ideal lattice $\mathcal{I}$, we do not need to know the prime factorization of $\mathcal{I}$ to run our algorithm.

4.1 Solving SVP for Prime Ideal Lattices in $\mathbb{Z}[\zeta_{2^n+1}]$

For simplicity we let $\zeta = \zeta_{2^{n+1}}$. In the sequel we say goodbye to the canonical embedding and adopt the coefficient embedding C:

$$\mathbb{Q}(\zeta) \to \mathbb{R}^{2^n}, \quad \sum_{i=0}^{2^n-1} a_i \zeta^i \mapsto (a_0,\ a_1, ..., a_{2^n-1}).$$

The coefficient embedding is widely used in cryptographic constructions. For power-of-two cyclotomic fields, the two embeddings are related by scaled-rotations, since for any $v \in \mathbb{Z}[\zeta_{2^{n+1}}]$ it is easy to see that

$$\|\Sigma_{\mathbb{L}}(v)\| = \sqrt{2^n}\|C(v)\|.$$

Hence, the shortest vector under the coefficient embedding of $\mathbb{Q}(\zeta)$ must be the shortest under the canonical embedding.

The prime 2 is the unique ramified prime in $\mathbb{Q}(\zeta)$, and the prime ideal lying over (2) is $(2, \zeta + 1) = (\zeta + 1)$. Hence it is easy to find the shortest vector in the ideal lattice $(\zeta + 1)$, and its length is $\sqrt{2}$.

Below we consider a prime ideal lying over an odd prime and show that there is a hierarchy for the hardness of solving SVP for prime ideal lattices in $\mathbb{Z}[\zeta]$. Roughly speaking, we can classify all the prime ideal lattices into n classes labeled with $1, 2, \cdots, n$, depending on the congruence class of $p \pmod{2^{n+1}}$, and for a prime ideal lattice in the r-th class, we can always find its shortest vector by solving SVP in a 2^r-dimensional lattice. More precisely, we have:

Theorem 5. *For any prime ideal* $\mathfrak{p} = (p, f(\zeta))$ *in* $\mathbb{Z}[\zeta]$*, where* p *is an odd prime and* $f(x)$ *is some irreducible factor of* $x^{2^n} + 1$ *in* $\mathbb{F}_p[x]$*. Write*

$$p = \begin{cases} 2^A \cdot m + 1, & \text{if } p \equiv 1 \pmod 4; \\ 2^A \cdot m - 1, & \text{if } p \equiv 3 \pmod 4, \end{cases}$$

for some odd m *and* $A \geq 2$*, and let*

$$r = \begin{cases} \min\{A-1, n\}, & \text{if } p \equiv 1 \pmod 4; \\ \min\{A, n\}, & \text{if } p \equiv 3 \pmod 4. \end{cases}$$

Then given an oracle that can solve SVP for 2^r*-dimensional lattices, a shortest nonzero vector in* $\mathfrak{p}$ *can be found in poly*$(2^n, \log_2 p)$ *time with the coefficient embedding.*

Proof. It is well known that the Galois group G of $\mathbb{Q}(\zeta)$ over $\mathbb{Q}$ is isomorphic to the multiplicative group $(\mathbb{Z}/2^{n+1}\mathbb{Z})^*$. Let $G = \{\sigma_1, \sigma_3, \cdots, \sigma_{2^{n+1}-1}\}$ where

$$\begin{aligned} \sigma_i : \quad \mathbb{Q}(\zeta) &\to \quad \mathbb{Q}(\zeta); \\ \zeta &\mapsto \quad \zeta^i. \end{aligned}$$

We proceed by considering two separate cases.

Case 1: First we deal with the case when $p \equiv 1 \pmod 4$. The theorem is vacuously true for $n < A$.

If $n \geq A$, we have $r = A - 1$. By Theorem 2, we know that

$$f(x) = x^{2^{n-A+1}} + u = x^{2^{n-r}} + u$$

for some $u \in U_A$. Then the prime ideal lattice $\mathfrak{p}$ can be generated by p and $f(\zeta) = \zeta^{2^{n-r}} + u$. Consider the subgroup $H = \langle \sigma_{2^{r+1}+1} \rangle$ of G generated by $\sigma_{2^{r+1}+1}$. H is a subgroup of the decomposition group of the ideal $\mathfrak{p}$ since

$$\sigma_{2^{r+1}+1}(p) = p, \quad \sigma_{2^{r+1}+1}(f(\zeta)) = f(\zeta).$$

Note that $\mathbb{K} = \mathbb{Q}(\zeta^{2^{n-r}})$ is the fixed field of H and its integer ring $O_{\mathbb{K}}$ has a $\mathbb{Z}$-basis $(1, \zeta^{2^{n-r}}, \zeta^{2 \cdot 2^{n-r}}, \cdots, \zeta^{(2^r-1) \cdot 2^{n-r}})$.

Let $\mathfrak{c} = \mathfrak{p} \bigcap O_{\mathbb{K}}$. We claim that $\mathfrak{p}$ is a direct sum:

$$\mathfrak{p} = \bigoplus_{k=0}^{2^{n-r}-1} \zeta^k \mathfrak{c}. \tag{2}$$

Indeed for any $a \in \mathfrak{p}$, there exist integers z_i's and w_i's such that

$$\begin{aligned} a &= \sum_{i=0}^{2^n-1} z_i \zeta^i f(\zeta) + \sum_{i=0}^{2^n-1} w_i p \zeta^i \\ &= \sum_{k=0}^{2^{n-r}-1} \zeta^k \sum_{j=0}^{2^r-1} (z_{k+j \cdot 2^{n-r}} \zeta^{j \cdot 2^{n-r}} f(\zeta) + w_{k+j \cdot 2^{n-r}} p \zeta^{j \cdot 2^{n-r}}) \\ &= \sum_{k=0}^{2^{n-r}-1} \zeta^k \Big((\sum_{j=0}^{2^r-1} z_{k+j \cdot 2^{n-r}} \zeta^{j \cdot 2^{n-r}}) f(\zeta) + (\sum_{j=0}^{2^r-1} w_{k+j \cdot 2^{n-r}} \zeta^{j \cdot 2^{n-r}}) p \Big). \end{aligned}$$

Let $a^{(k)} = (\sum_{j=0}^{2^r-1} z_{k+j\cdot 2^{n-r}}\zeta^{j\cdot 2^{n-r}})f(\zeta) + (\sum_{j=0}^{2^r-1} w_{k+j\cdot 2^{n-r}}\zeta^{j\cdot 2^{n-r}})p$ for any k. Since $p \in \mathfrak{c}$ and $f(\zeta) \in \mathfrak{c}$, $a^{(k)} \in \mathfrak{c}$. We have established (2).

Since multiplication by ζ is an isometry and for $x \in \mathfrak{c}$, the coefficients of $\zeta^i x$ and $\zeta^j x$ are disjoint for $i \neq j \mod 2^{n-r}$, Eq. (2) implies

$$\lambda_1(\mathfrak{p}) = \lambda_1(\mathfrak{c}),$$

and that to find the shortest vector in the ideal lattice $\mathfrak{p}$, it is enough to find the shortest vector v in the ideal lattice $\mathfrak{c}$, a lattice with dimension 2^r. Indeed $\zeta^k v$ for any $0 \leq k \leq 2^{n-r} - 1$ will be a shortest vector in the ideal lattice $\mathfrak{p}$.

Case 2: For the case when $p \equiv 3 \pmod 4$, everything is similar except that $r = A$.

Algorithm: We can summarize the algorithm to solve SVP in a prime ideal lattice as Algorithm 1.

Algorithm 1. Solve SVP in prime ideal lattice

Input: a prime ideal $\mathfrak{p} = (p, f(\zeta))$ in $\mathbb{Z}[\zeta]$, where p is odd.
Output: a shortest vector in the corresponding prime ideal lattice.

1: Compute the ideal $\mathfrak{c}$ generated by p and $f(\zeta)$ in $O_{\mathbb{K}}$ where $\mathbb{K} = \mathbb{Q}(\zeta^{2^{n-r}})$.
2: Find a shortest vector v in the 2^r-dimensional lattice $\mathfrak{c}$.
3: Output v.

The most time-consuming step in Algorithm 1 is Step 2 and the other steps can be done in poly$(2^n, \log_2 p)$ time. □

Remark 3. By the decomposition (2) above, a similar result will hold for a prime ideal $\mathfrak{p}$ in $O_{\mathbb{L}}$ other than $\mathbb{Q}(\zeta)$, whenever $O_{\mathbb{L}}$ is a free $O_{\mathbb{K}}$-module where $\mathbb{K}$ is the decomposition field of $\mathfrak{p}$, and some $\mathbb{Z}$-basis of $O_{\mathbb{K}}$ can be extended to the $\mathbb{Z}$-basis of $O_{\mathbb{L}}$ that determines the coefficient embedding. If we disregard the last condition that a basis of $O_{\mathbb{K}}$ extends to a basis of $O_{\mathbb{L}}$, there may be a distortion of length, depending on the basis of $O_{\mathbb{K}}$, when we lift the solution from $\mathfrak{c}$ to $\mathfrak{p}$. That is, an approximation factor, which may be much larger than 1, will be involved.

Remark 4. By the remark above, solving the closest vector problem (CVP) for a prime ideal lattice can be also reduced to solving CVP in some 2^r-dimensional sublattice.

SVP of some special prime ideals in $\mathbb{Z}[\zeta_{2^{n+1}}]$. Using Theorem 5, we can prove Theorem 1, which shows that the SVP for prime ideals lying above some special rational primes is very easy.

Proof of Theorem 1

If $p \equiv -3 \pmod 8$, we may write $p = 4m + 1$ with odd m. By Theorem 2, $x^{2^n}+1$ is the product of 2 irreducible binomials over $\mathbb{F}_p$ of degree 2^{n-1}: $x^{2^n}+1 = (x^{2^{n-1}} + u_1) \cdot (x^{2^{n-1}} + u_2)$, where u_i satisfies $u_i^2 \equiv -1 \pmod p$.

For any prime ideal $(p, \zeta^{2^{n-1}} + u_i)$ over (p), by the proof of Theorem 5, the shortest vector can be found by solving the 2-dimensional lattice $\mathcal{L}_i$ generated by $\begin{pmatrix} u_i & 1 \\ -1 & u_i \\ p & 0 \\ 0 & p \end{pmatrix}$. Note that $(-1, u_i) \equiv u_i \cdot (u_i, 1) \pmod{p}$ and $(0, p) = p \cdot (u_i, 1) - u_i \cdot (p, 0)$. The generator matrix can be reduced to the basis of $\mathcal{L}_i$ as $\begin{pmatrix} u_i & 1 \\ p & 0 \end{pmatrix}$, which is exactly the Hermite Normal Form of the lattice basis.

For any vector $v \in \mathcal{L}_i$, there exists an integer vector (z_1, z_2) such that $v = (z_1, z_2)\begin{pmatrix} u_i & 1 \\ p & 0 \end{pmatrix} = (z_1 u_i + z_2 p, z_1)$. Note that

$$\|v\|^2 = (z_1 u_i + z_2 p)^2 + z_1^2 = z_1^2(u_i^2 + 1) + z_2^2 p^2 + 2p z_1 z_2 u_i \equiv 0 \pmod{p}.$$

Then for the nonzero shortest vector v, we have $0 < \|v\|^2 < \frac{4}{\pi} \cdot p < 2p$ (by Minkowski's Theorem [27]) and $\|v\|^2 \equiv 0 \pmod{p}$, which implies that $\|v\|^2 = p$.

In case $p \equiv 3 \pmod{8}$, we may write $p = 4m - 1$ with odd m. By Theorem 3, then $x^{2^n} + 1$ is the product of 2 irreducible binomials over $\mathbb{F}_p$ of degree 2^{n-1}: $x^{2^n} + 1 = (x^{2^{n-1}} + \delta_1 x^{2^{n-2}} - 1) \cdot (x^{2^{n-1}} + \delta_2 x^{2^{n-2}} - 1)$, where δ_i satisfies $\delta_i^2 \equiv -2 \pmod{p}$ since the Dickson polynomial is $D_2(x, -1) = X^2 + 2$.

For any prime ideal $(p, \zeta^{2^{n-1}} + \delta_i \zeta^{2^{n-2}} - 1)$ over (p), we similarly consider the shortest vector in $\mathcal{L}_i$ generated by

$$\begin{pmatrix} -1 & \delta_i & 1 & 0 \\ 0 & -1 & \delta_i & 1 \\ -1 & 0 & -1 & \delta_i \\ -\delta_i & -1 & 0 & -1 \\ p & 0 & 0 & 0 \\ 0 & p & 0 & 0 \\ 0 & 0 & p & 0 \\ 0 & 0 & 0 & p \end{pmatrix}.$$

Similarly, we can easily get the basis for $\mathcal{L}_i$ in the Hermite Normal Form

$$\begin{pmatrix} 0 & -1 & \delta_i & 1 \\ -1 & \delta_i & 1 & 0 \\ 0 & p & 0 & 0 \\ p & 0 & 0 & 0 \end{pmatrix},$$

and prove that for any vector $v \in \mathcal{L}$,

$$\|v\|^2 \equiv 0 \pmod{p}.$$

For the shortest vector v, by Minkowski's Theorem, we know $0 < \|v\|^2 \leq \frac{4\sqrt{2}}{\pi} < 2p$, which implies that $\|v\|^2 = p$. By Theorem 5, the proposition follows. □

4.2 SVP Average-Case Hardness for Prime Ideals in $\mathbb{Z}[\zeta]$

Precisely defining the average-case hardness of SVP for a prime ideal lattice in $\mathbb{Z}[\zeta]$ requires specifying a distribution. We consider the following three distributions.

The first distribution. To select a random prime ideal, one fixes a large M, uniformly randomly selects a prime number in the set

$$\{p \text{ is a prime} : p < M\},$$

and then uniformly randomly selects a prime ideal lying over p. This process provides a reasonable distribution among prime ideals, since every prime ideal in the ring of integers of $\mathbb{Q}[x]/(f(x))$ is of the form $(p, g(x))$, where p is a prime number and $g(x)$ is an irreducible factor of $f(x)$ over $\mathbb{F}_p[x]$. Since roughly half of all primes $p \leq M$ satisfy $p \equiv \pm 3 \pmod 8$, according to Dirichlet's theorem on arithmetic progressions, at least half of all such p have the property that the ideals lying over p admit an efficient algorithm for SVP.

The second distribution. Again fixing a large M, we might alternatively select a prime ideal uniformly at random from the set

$$\{\mathfrak{p} \text{ prime ideal} : p \in \mathfrak{p}, p \text{ is a prime}, p < M\}.$$

In this case, a non-negligible fraction of prime ideals admit efficient SVP algorithm. More precisely, we have

Proposition 2. *Under the distribution above, a random prime ideal of $\mathbb{Z}[\zeta]$ admits an efficient SVP algorithm with probability at least $\frac{1}{1+2^{n-1}}$.*

Proof. For simplicity, we disregard the single prime ideal lying over 2. Note that for $p = 8k \pm 3$, there are exactly two prime ideals over p, and, by Theorem 1, the SVP for the corresponding ideal lattices is easy. For $p = 8k \pm 1$, there are at most 2^n prime ideals lying over p, by Theorems 2 and 3. Then by Dirichlet's prime number theorem, even if we only count the prime ideals lying over $p = 8k \pm 3$, the fraction of easy instances is at least $\frac{1}{1+2^{n-1}}$. □

The third distribution. The third distribution is more common in mathematics. Namely, after fixing a large M, we select uniformly at random a prime ideal from the set

$$\{\mathfrak{p} \text{ prime ideal} : \mathcal{N}(\mathfrak{p}) < M\},$$

where $\mathcal{N}(\mathfrak{p})$ is the norm of the ideal $\mathfrak{p}$.

By Theorem 5, SVP for a prime ideal lattice $\mathfrak{p}$ reduces to SVP for a 2^r-dimensional sub-lattice $\mathfrak{c}$, where r is as defined in the statement of Theorem 5. Note that our algorithm will not improve matters if $r = n$, that is, if p splits completely in $\mathbb{Q}(\zeta)$, or equivalently if $\mathcal{N}(\mathfrak{p}) = p$. By Chebotarev's density theorem [38], there are about $\frac{M}{2^n \log M}$ rational primes which split in $\mathbb{Q}(\zeta)$ and hence $\frac{M}{\log M}$ prime ideals lying above those primes, for which our algorithm cannot provide a reduction for SVP.

If our algorithm is to provide a reduction, the prime ideal under study must lie over a rational prime p with $p \leq \sqrt{M}$, since $\mathcal{N}(\mathfrak{p}) = p^f < M$ where f is some integer greater than 1. Hence there are at most $\sqrt{M}$ such primes and hence at most $2^{n-1}\sqrt{M}$ prime ideals for which our algorithm provides a reduction.

Under such a distribution, therefore, the density of the easy instances for our algorithm is at most $\frac{2^{n-1} \log M}{\sqrt{M}}$, which goes to zero when M tends to infinity.

Remark 5. From a cryptographic perspective, there seems to be no construction relying on the average hardness of ideal SVP in ideals following one of the two first distributions above. However, our algorithm reveals the concrete reason why we should avoid such distributions in the cryptographic constructions although it seems very easy to sample according to the two distributions.

4.3 Solving SVP for a General Ideal Lattice in $\mathbb{Z}[\zeta_{2^{n+1}}]$

For simplicity, we let $\zeta = \zeta_{2^{n+1}}$. We will show that even for a general ideal lattice $\mathcal{I} \subset \mathbb{Z}[\zeta]$, there is a similar hierarchy for the hardness of SVP. We would like to stress that although the following theorem refers to the prime factorization of $\mathcal{I}$, the resulting algorithm does not require it.

Theorem 6. *Let $\mathcal{I}$ be a nonzero ideal of $\mathbb{Z}[\zeta]$ with prime factorization*

$$\mathcal{I} = \mathfrak{p}_1 \cdot \mathfrak{p}_2 \cdots \mathfrak{p}_t,$$

where $\mathfrak{p}_i = (f_i(\zeta), p_i)$ for rational primes p_i, and where the $\mathfrak{p}_i$ are not necessarily distinct. Write $p_i = 2^{A_i} \cdot m_i + 1$ when $p_i \equiv 1 \pmod 4$ and $p_i = 2^{A_i} \cdot m_i - 1$ when $p_i \equiv 3 \pmod 4$ with odd m_i, and let $r = \max\{r_i\}$, where

$$r_i = \begin{cases} \min\{A_i - 1, n\}, & if p_i \equiv 1 \pmod 4; \\ \min\{A_i, n\}, & if p_i \equiv 3 \pmod 4; \\ n, & if p_i = 2. \end{cases}$$

Then the shortest vector in the ideal lattice $\mathcal{L}$ corresponding to $\mathcal{I}$ can be solved via solving SVP in a 2^r-dimensional lattice.

Proof. If $r = n$, then the theorem follows simply.

If $r < n$, W.L.O.G., we assume $r = r_1$. Following the proof of Theorem 5, denote the Galois group $G = \{\sigma_1, \sigma_3, \cdots, \sigma_{2^{n+1}-1}\}$ of $\mathbb{Q}(\zeta)$ over $\mathbb{Q}$, where $\sigma_i(\zeta) = \zeta^i$. Consider the subgroup $H = \langle \sigma_{2^{r+1}+1} \rangle$ of G generated by $\sigma_{2^{r+1}+1}$. For any $\tau \in H$ and every prime ideal $\mathfrak{p}_i = (p_i, f_i(\zeta))$, we have $\tau(\mathfrak{p}_i) = \mathfrak{p}_i$ since $\sigma_{2^{r+1}+1}(p_i) = p_i$, $\sigma_{2^{r+1}+1}(f_i(\zeta)) = f_i(\zeta)$. Note that $\mathbb{K} = \mathbb{Q}(\zeta^{2^{n-r}})$ is the fixed field of H and its integer ring $O_{\mathbb{K}}$ has a $\mathbb{Z}$-basis $(1, \zeta^{2^{n-r}}, \zeta^{2 \cdot 2^{n-r}}, \cdots, \zeta^{(2^r-1) \cdot 2^{n-r}})$.

Let $\mathfrak{c} = \mathcal{I} \bigcap O_{\mathbb{K}}$. We claim that for any $a \in \mathcal{I}$, there exist $a^{(k)} \in \mathfrak{c}$ for $0 \le k < 2^{n-r}$, such that

$$a = \sum_{k=0}^{2^{n-r}-1} \zeta^k a^{(k)}.$$

We proceed by induction. When $t = 1$ the above claim holds by Theorem 5. Suppose the claim holds for $t - 1$. Then setting $\mathcal{I} = \mathfrak{p}_1 \cdot \mathfrak{p}_2 \cdots \mathfrak{p}_t$, and $\overline{\mathcal{I}} = \mathfrak{p}_1 \cdot \mathfrak{p}_2 \cdots \mathfrak{p}_{t-1}$, we have $\mathcal{I} = \overline{\mathcal{I}} \cdot \mathfrak{p}_t$. For any $a \in \mathcal{I}$, we can write $a = \sum x_i y_i$ where $x_i \in \overline{\mathcal{I}}$ and $y_i \in \mathfrak{p}_t$. It suffices to show that for any xy, where $x \in \overline{\mathcal{I}}$ and $y \in \mathfrak{p}_t$, there exist $b^{(k)} \in \mathcal{I} \bigcap O_{\mathbb{K}}$ for $0 \le k < 2^{n-r}$, such that $xy = \sum_{k=0}^{2^{n-r}-1} \zeta^k b^{(k)}$.

By the induction assumption, there exist $x^{(i)} \in \overline{\mathcal{I}} \bigcap O_{\mathbb{K}}$ for $0 \le i < 2^{n-r}$ such that $x = \sum_{i=0}^{2^{n-r}-1} \zeta^i x^{(i)}$, and there exist $y^{(j)} \in \mathfrak{p}_t \bigcap O_{\mathbb{K}}$ for $0 \le j < 2^{n-r}$ such that $y = \sum_{j=0}^{2^{n-r}-1} \zeta^j y^{(j)}$. Hence, we have

$$\begin{aligned}
xy &= \sum_{i=0}^{2^{n-r}-1} \sum_{j=0}^{2^{n-r}-1} \zeta^{i+j} x^{(i)} y^{(j)} \\
&= \sum_{k=0}^{2^{n-r}-1} \zeta^k \sum_{i+j=k} x^{(i)} y^{(j)} + \sum_{k=2^{n-r}}^{2 \cdot 2^{n-r}-2} \zeta^k \sum_{i+j=k} x^{(i)} y^{(j)} \\
&= \sum_{k=0}^{2^{n-r}-1} \zeta^k \sum_{i+j=k} x^{(i)} y^{(j)} + \sum_{k=0}^{2^{n-r}-2} \zeta^k \sum_{i+j=k+2^{n-r}} \zeta^{2^{n-r}} x^{(i)} y^{(j)} \\
&= \sum_{k=0}^{2^{n-r}-2} \zeta^k \Big(\sum_{i+j=k} x^{(i)} y^{(j)} + \sum_{i+j=k+2^{n-r}} \zeta^{2^{n-r}} x^{(i)} y^{(j)} \Big) + \zeta^{2^{n-r}-1} \sum_{i+j=2^{n-r}-1} x^{(i)} y^{(j)}.
\end{aligned}$$

Let $b^{(k)} = \sum_{i+j=k} x^{(i)} y^{(j)} + \sum_{i+j=k+2^{n-r}} \zeta^{2^{n-r}} x^{(i)} y^{(j)}$ for any $0 \le k \le 2^{n-r} - 2$ and $b^{(2^{n-r}-1)} = \sum_{i+j=2^{n-r}-1} x^{(i)} y^{(j)}$. We have that $b^{(k)} \in \mathcal{I} \bigcap O_{\mathbb{K}}$ for $0 \le k < 2^{n-r}$. Hence, for any $a \in \mathcal{I}$, there exist $a^{(k)} \in \mathfrak{c}$ for $0 \le k < 2^{n-r}$, such that $a = \sum_{k=0}^{2^{n-r}-1} \zeta^k a^{(k)}$.

As in the proof of Theorem 5, we can show that $\lambda_1(\mathcal{I}) = \lambda_1(\mathfrak{c})$ and any nonzero shortest vector in $\mathfrak{c}$ will yield 2^{n-r} nonzero shortest vectors in $\mathcal{I}$. □

We would like to point out that in some cases, the r in Theorem 6 can be improved. Consider the case when $n \ge 3$ and $\mathcal{I} = (2, \zeta - 1)^2 = (2, \zeta^2 + 1)$. We need to solve SVP in a 2^n-dimensional lattice by Theorem 6. However, using the intermediate field $\mathbb{Q}(\zeta^2)$ as in the proof of Theorem 6, we can find a shortest vector by solving SVP in a 2^{n-1}-dimensional lattice.

Furthermore, since for any $a \in \mathcal{I}$, there exist $a^{(k)} \in \mathfrak{c}$ for $0 \le k < 2^{n-r}$, such that $a = \sum_{k=0}^{2^{n-r}-1} \zeta^k a^{(k)}$, we conclude that if $(b^{(i)})_{0 \le i < 2^r}$ is a basis of the ideal lattice $\mathfrak{c}$, then $(\zeta^j b^{(i)})_{0 \le i < 2^r, 0 \le j < 2^{n-r}}$ is a basis of the ideal lattice $\mathcal{I}$. Denote by $\mathcal{L}_j$ the lattice generated by $(\zeta^j b^{(i)})_{0 \le i < 2^r}$. Then we have that the ideal lattice $\mathcal{I}$ has an orthogonal decomposition: $\mathcal{L}_0 \oplus \mathcal{L}_1 \oplus \cdots \oplus \mathcal{L}_{2^{n-r}-1}$.

In fact, for any $\bar{r}$, let $\mathfrak{c} = \mathcal{I} \bigcap O_{\mathbb{K}}$ where $\mathbb{K} = \mathbb{Q}(\zeta^{2^{n-\bar{r}}})$. For any basis $(b^{(i)})_{0 \leq i < 2^{\bar{r}}}$ of the ideal lattice $\mathfrak{c}$, if $(\zeta^j b^{(i)})_{0 \leq i < 2^{\bar{r}}, 0 \leq j < 2^{n-\bar{r}}}$ is a basis of the ideal lattice $\mathcal{I}$ (meaning that the ideal lattice $\mathcal{I}$ has an orthogonal decomposition), then the shortest vector in $\mathfrak{c}$ is also a shortest vector in $\mathcal{I}$. Hence we have the following algorithm to solve SVP for a general ideal in $\mathbb{Z}[\zeta]$ without knowing the prime factorization of the ideal.

Algorithm 2. Solve SVP in general ideal lattice

Input: an ideal $\mathcal{I}$;

Output: a shortest vector in the corresponding ideal lattice $\mathcal{L}$.

1: **for** $\bar{r} = 1$ to n **do**

2: Compute a basis $(b^{(i)})_{0 \leq i < 2^{\bar{r}}}$ of the ideal lattice $\mathfrak{c} = \mathcal{I} \bigcap O_{\mathbb{K}}$, where $\mathbb{K} = \mathbb{Q}(\zeta^{2^{n-\bar{r}}})$.

3: **if** $(\zeta^j b^{(i)})_{0 \leq i < 2^{\bar{r}}, 0 \leq j < 2^{n-\bar{r}}}$ is exactly a basis of ideal lattice $\mathcal{I}$ **then**

4: Find a shortest vector v in the $2^{\bar{r}}$-dimensional lattice $\mathfrak{c}$;

5: Output v.

6: **end if**

7: **end for**

Note that Step 2 can be done efficiently by computing the intersection of the lattices $\mathcal{I}$ and $O_{\mathbb{K}}$ under the coefficient embedding.

Remark 6. By the proof of Theorem 6, solving the closest vector problem (CVP) for a general ideal lattice can also be reduced to solving CVP in some $2^{\bar{r}}$-dimensional lattice.

5 Conclusion and Open Problems

We have investigated the SVP of prime ideal lattices in the finite Galois extension of $\mathbb{Q}$, and designed an algorithm exploiting the subfield structure of such fields to solve Hermite-SVP for prime ideal lattices. For the power-of-two cyclotomic fields, we obtained an efficient algorithm for solving SVP in many ideal lattices, either prime or non-prime ideals. We also determined the length of the shortest vector of those prime ideals lying over rational primes congruent to $\pm 3 \pmod 8$. It is an interesting problem to study the length of the shortest vectors in other prime ideals. The worst case hardness of prime ideal lattice SVP for power-of-two cyclotomic fields is also left open.

Acknowledgements. We thank the anonymous referees for their valuable suggestions on how to improve this paper. This work is supported by National Key Research and Development Program of China (No. 2020YFA0712300, 2018YFA0704705), National Natural Science Foundation of China (No. 62032009, 61732021, 61572490) for Y. Pan and J. Xu, and National Science Foundation of USA (CCF-1900820) for N. Wadleigh and Q. Cheng.

A The Subfields of $\mathbb{Q}(\zeta_{2^n})$

Now we sketch the subfield lattice of $\mathbb{Q}(\zeta_{2^{n+1}})$. Consider the three subfields

$$\mathbb{Q}(\zeta_{2^{n+1}} + \zeta_{2^{n+1}}^{-1}),\ \mathbb{Q}(\zeta_{2^n}),\ \mathbb{Q}(\zeta_{2^{n+1}} - \zeta_{2^{n+1}}^{-1}).$$

First we claim $\mathbb{Q}(\zeta_{2^{n+1}})$ is degree two over each. On the one hand, all are proper subfields since $\mathbb{Q}(\zeta_{2^{n+1}}+\zeta_{2^{n+1}}^{-1})$ is contained in the fixed field of the automorphism $\zeta_{2^{n+1}} \mapsto \zeta_{2^{n+1}}^{-1}$, and $\mathbb{Q}(\zeta_{2^{n+1}} - \zeta_{2^{n+1}}^{-1})$ is in the fixed field of the automorphism $\zeta_{2^{n+1}} \mapsto -\zeta_{2^{n+1}}^{-1}$. On the other hand, $\zeta_{2^{n+1}}$ is a root of the quadratic polynomials $x^2 - (\zeta_{2^{n+1}} + \zeta_{2^{n+1}}^{-1})x + 1 \in \mathbb{Q}(\zeta_{2^n} + \zeta_{2^{n+1}}^{-1})[x]$ and $x^2 - (\zeta_{2^{n+1}} - \zeta_{2^{n+1}}^{-1})x - 1 \in \mathbb{Q}(\zeta_{2^{n+1}} - \zeta_{2^{n+1}}^{-1})[x]$.

Moreover, since the involutions

$$\zeta_{2^{n+1}} \mapsto \zeta_{2^{n+1}}^{-1},\ \zeta_{2^{n+1}} \mapsto \zeta_{2^{n+1}}^{2^{n-1}+1},\ \zeta_{2^{n+1}} \mapsto -\zeta_{2^{n+1}}^{-1}$$

are distinct, these three subfields are distinct. Finally it is routine to sketch the subgroup lattice of $\mathbb{Z}_2 \oplus \mathbb{Z}_{2^{n-1}} \cong (\mathbb{Z}/2^{n+1}\mathbb{Z})^* \cong \mathrm{Gal}(\mathbb{Q}(\zeta_{2^{n+1}})/\mathbb{Q})$:

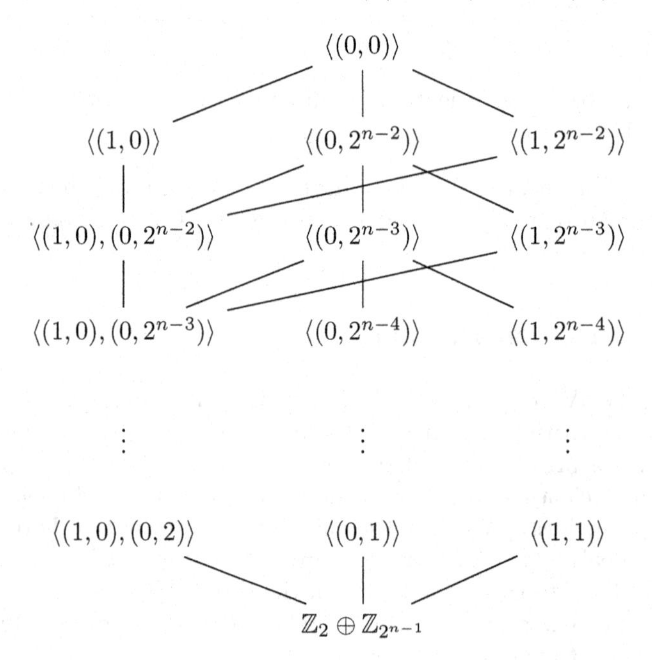

Here all lines indicate extensions of index two. Combining these facts we have the subfield lattice for $\mathbb{Q}(\zeta_{2^n})$:

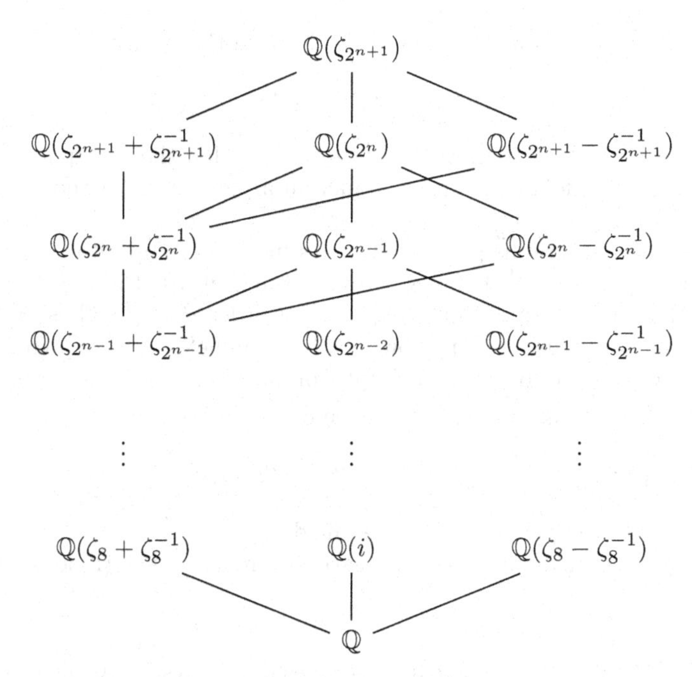

where all lines indicate extensions of order two.

B Decomposition Groups and Fixed Fields

Let $\zeta = \zeta_{2^{n+1}}$, p a rational prime with $p \equiv 3 \pmod 4$, A the natural number with $2^A || p+1$, and let $\mathfrak{p}$ be a prime ideal in $\mathbb{Z}[\zeta]$ containing p. Then

$$\mathfrak{p} = (p,\ \zeta^{2^{n-A+1}} + \delta\zeta^{2^{n-A}} - 1)$$

for some $\delta \in \mathbb{Z}$. Let $\sigma \in Aut(\mathbb{Q}(\zeta)/\mathbb{Q})$ be the automorphism of $\mathbb{Q}(\zeta)$ with $\zeta \mapsto \zeta^{-2^A-1}$. Then we have

$$\begin{aligned}
\sigma\mathfrak{p} &= (p,\ \sigma(\zeta)^{2^{n-A+1}} + \delta\sigma(\zeta)^{2^{n-A}} - 1) \\
&= (p,\ \zeta^{2^{n-A+1}(-2^A-1)} + \delta\zeta^{2^{n-A}(-2^A-1)} - 1) \\
&= (p,\ \zeta^{-2^{n+1}}\zeta^{-2^{n-A+1}} + \delta\zeta^{-2^n}\zeta^{-2^{n-A}} - 1) \\
&= (p,\ \zeta^{-2^{n-A+1}} - \delta\zeta^{-2^{n-A}} - 1) \\
&= (p,\ -\zeta^{-2^{n-A+1}} \cdot (\zeta^{2^{n-A+1}} + \delta\zeta^{2^{n-A}} - 1)) \\
&= \mathfrak{p}.
\end{aligned}$$

We have used the fact that ζ is a unit in $\mathbb{Z}[\zeta]$.

Since $\zeta \mapsto \zeta^{-1}$ is an involution, the order of σ is the order of $\zeta \mapsto \zeta^{2^A+1}$ (denoted by σ') which is the multiplicative order of 2^A+1 in $(\mathbb{Z}/2^{n+1}\mathbb{Z})^*$. We claim that, for $A \geq 2$, this order is 2^{n+1-A}: First note that for $k \equiv 1 \pmod 4$,

$$\mathrm{ord}_{(\mathbb{Z}/2^{n+1}\mathbb{Z})^*}(k) = 2^m$$

if and only if $2^{n+1}||k^{2^m} - 1$. This fact follows easily from the identity

$$k^{2^{g+1}} - 1 = (k^{2^g} - 1)(k^{2^g} + 1)$$

and the fact that for $k = 2^A+1$, we have $2||(k^{2^g}+1)$. Now, that the multiplicative order of $2^A + 1$ is 2^{n+1-A} follows from an induction argument using the above identity.

The preceding two paragraphs prove that σ lies in the decomposition group of $\mathfrak{p}$ and that σ has order 2^{n+1-A}. It follows from a standard result in the theory of number fields that the decomposition group of $\mathfrak{p}$ has order 2^{n+1-A}. Thus $\langle\sigma\rangle$ is precisely the decomposition group of $\mathfrak{p}$. Now recall the subfield/subgroup lattice for $\mathbb{Q}(\zeta)/\mathbb{Q}$ and its Galois group $\mathbb{Z}_{2^{n+1}}^*$. A simple computation shows that σ fixes $\zeta^{2^{n-A}} - \zeta^{-2^{n-A}}$. But from the subfield lattice we can see that

$$[\mathbb{Q}(\zeta) : \mathbb{Q}(\zeta^{2^{n-A}} - \zeta^{-2^{n-A}})] = 2^{n+1-A} = |\langle\sigma\rangle|.$$

Thus $\mathbb{Q}(\zeta^{2^{n-A}} - \zeta^{-2^{n-A}})$ is precisely this fixed field.

A similar, in fact easier, analysis can be carried out for $p \equiv 1 \pmod 4$. In this case

$$\mathfrak{p} = (p,\ \zeta^{2^{n-A+1}} - u)$$

for some $u \in \mathbb{Z}$ and $2^A||p - 1$. Then it is seen that σ' fixes $\mathfrak{p}$. As in the 3 (mod 4) case, we know from a general result of algebraic number theory that the decomposition group of $\mathfrak{p}$ has order 2^{n+1-A}, which matches the order of σ' (computed above). We see that $\mathbb{Q}(\zeta^{2^{n+1-A}})$ is contained in the fixed field of σ', and again, by looking at the subfield lattice to find $[\mathbb{Q}(\zeta) : \mathbb{Q}(\zeta^{2^{n+1-A}})] = 2^{n+1-A}$, we see that $\mathbb{Q}(\zeta^{2^{n+1-A}})$ is precisely the fixed field of the decomposition group of $\mathfrak{p}$.

References

1. Ajtai, M.: Generating hard instances of lattice problems (extended abstract). In: Proceedings of the Twenty-Eighth Annual ACM Symposium on the Theory of Computing - STOC, pp. 99–108 (1996). https://doi.org/10.1145/237814.237838
2. Albrecht, M., Bai, S., Ducas, L.: A subfield lattice attack on overstretched NTRU assumptions. In: Robshaw, M., Katz, J. (eds.) CRYPTO 2016. LNCS, vol. 9814, pp. 153–178. Springer, Heidelberg (2016). https://doi.org/10.1007/978-3-662-53018-4_6
3. Alkim, E., Ducas, L., Pöppelmann, T., Schwabe, P.: Newhope without reconciliation. IACR Cryptology ePrint Archive 2016/1157 (2016). http://eprint.iacr.org/2016/1157
4. Bernstein, D.J.: A subfield-logarithm attack against ideal lattices: computing algebraic number theory tackles lattice-based cryptography. The cr.yp.to blog (2014). https://blog.cr.yp.to/20140213-ideal.html
5. Bernstein, D.J., Chuengsatiansup, C., Lange, T., van Vredendaal, C.: NTRU prime: reducing attack surface at low cost. In: Adams, C., Camenisch, J. (eds.) SAC 2017. LNCS, vol. 10719, pp. 235–260. Springer, Cham (2018). https://doi.org/10.1007/978-3-319-72565-9_12

6. Biasse, J.-F., Espitau, T., Fouque, P.-A., Gélin, A., Kirchner, P.: Computing generator in cyclotomic integer rings. In: Coron, J.-S., Nielsen, J.B. (eds.) EUROCRYPT 2017. LNCS, vol. 10210, pp. 60–88. Springer, Cham (2017). https://doi.org/10.1007/978-3-319-56620-7_3
7. Biasse, J., Song, F.: Efficient quantum algorithms for computing class groups and solving the principal ideal problem in arbitrary degree number fields. In: Proceedings of the Twenty-Seventh Annual ACM-SIAM Symposium on Discrete Algorithms, SODA 2016, pp. 893–902 (2016). https://doi.org/10.1137/1.9781611974331.ch64
8. Bos, J.W., et al.: CRYSTALS - kyber: a CCA-secure module-lattice-based KEM. In: Proceedings of 2018 IEEE European Symposium on Security and Privacy, EuroS&P 2018, pp. 353–367 (2018). https://doi.org/10.1109/EuroSP.2018.00032
9. Campbell, P., Groves, M., Shepherd, D.: Soliloquy: a cautionary tale. In: Proceedings of 2nd ETSI Quantum-Safe Crypto Workshop, vol. 3, no. 9. pp. 1–9 (2014)
10. Cheon, J.H., Jeong, J., Changmin, L.: An algorithm for NTRU problems and cryptanalysis of the GGH multilinear map without a low-level encoding of zero. LMS J. Comput. Math. **19**(A), 255–266 (2016). https://doi.org/10.1112/S1461157016000371
11. Cramer, R., Ducas, L., Peikert, C., Regev, O.: Recovering short generators of principal ideals in cyclotomic rings. In: Fischlin, M., Coron, J.-S. (eds.) EUROCRYPT 2016. LNCS, vol. 9666, pp. 559–585. Springer, Heidelberg (2016). https://doi.org/10.1007/978-3-662-49896-5_20
12. Cramer, R., Ducas, L., Wesolowski, B.: Short stickelberger class relations and application to ideal-SVP. In: Coron, J.-S., Nielsen, J.B. (eds.) EUROCRYPT 2017. LNCS, vol. 10210, pp. 324–348. Springer, Cham (2017). https://doi.org/10.1007/978-3-319-56620-7_12
13. Ducas, L., Plançon, M., Wesolowski, B.: On the shortness of vectors to be found by the ideal-SVP quantum algorithm. In: Boldyreva, A., Micciancio, D. (eds.) CRYPTO 2019. LNCS, vol. 11692, pp. 322–351. Springer, Cham (2019). https://doi.org/10.1007/978-3-030-26948-7_12
14. Eisenträger, K., Hallgren, S., Kitaev, A.Y., Song, F.: A quantum algorithm for computing the unit group of an arbitrary degree number field. In: Proceedings of Symposium on Theory of Computing, STOC 2014, pp. 293–302 (2014). https://doi.org/10.1145/2591796.2591860
15. Hoffstein, J., Pipher, J., Silverman, J.H.: NTRU: a ring-based public key cryptosystem. In: Buhler, J.P. (ed.) ANTS 1998. LNCS, vol. 1423, pp. 267–288. Springer, Heidelberg (1998). https://doi.org/10.1007/BFb0054868
16. Kirchner, P., Fouque, P.-A.: Revisiting lattice attacks on overstretched NTRU parameters. In: Coron, J.-S., Nielsen, J.B. (eds.) EUROCRYPT 2017. LNCS, vol. 10210, pp. 3–26. Springer, Cham (2017). https://doi.org/10.1007/978-3-319-56620-7_1
17. Lenstra, A.K., Lenstra, H.W., Lovász, L.: Factoring polynomials with rational coefficients. Math. Ann. **261**(4), 515–534 (1982). https://doi.org/10.1007/BF01457454
18. Lidl, R., Niederreiter, H.: Finite Fields, Encyclopedia of Mathematics and Its Applications, vol. 20, 2nd edn. Cambridge University Press, Cambridge (1997). https://doi.org/10.1016/s0898-1221(97)84597-x
19. Lovasz, L.: An Algorithmic Theory of Numbers, Graphs, and Convexity. CBMS-NSF Regional Conference Series in Applied Mathematics, vol. 50. Society for Industrial and Applied Mathematics (1986). https://doi.org/10.1137/1.9781611970203

20. Lu, X., et al.: LAC: practical Ring-LWE based public-key encryption with byte-level modulus. IACR Cryptology ePrint Archive 2018/1009 (2018). https://eprint.iacr.org/2018/1009
21. Lyubashevsky, V., Micciancio, D.: Generalized compact knapsacks are collision resistant. In: Bugliesi, M., Preneel, B., Sassone, V., Wegener, I. (eds.) ICALP 2006. LNCS, vol. 4052, pp. 144–155. Springer, Heidelberg (2006). https://doi.org/10.1007/11787006_13
22. Lyubashevsky, V., Peikert, C., Regev, O.: On ideal lattices and learning with errors over rings. In: Gilbert, H. (ed.) EUROCRYPT 2010. LNCS, vol. 6110, pp. 1–23. Springer, Heidelberg (2010). https://doi.org/10.1007/978-3-642-13190-5_1
23. Marcus, D.A.: Number Fields. Universitext, 2nd edn. Springer, New York (2018). https://doi.org/10.1007/978-1-4684-9356-6
24. Meyn, H.: Factorization of the cyclotomic polynomials $x^{2^n}+1$ over finite fields. Finite Fields Appl. **2**, 439–442 (1996). https://doi.org/10.1017/CBO9780511525926
25. Micciancio, D.: Generalized compact knapsacks, cyclic lattices, and efficient one-way functions from worst-case complexity assumptions. In: Proceedings of 43rd Symposium on Foundations of Computer Science (FOCS 2002), pp. 356–365 (2002). https://doi.org/10.1109/SFCS.2002.1181960
26. Micciancio, D.: Generalized compact knapsacks, cyclic lattices, and efficient one-way functions. Comput. Complex. **16**(4), 365–411 (2007). https://doi.org/10.1007/s00037-007-0234-9
27. Micciancio, D., Goldwasser, S.: Complexity of Lattice Problems: A Cryptographic Perspective. The Kluwer International Series in Engineering and Computer Science, vol. 671. Kluwer Academic Publishers (2002). https://doi.org/10.1007/978-1-4615-0897-7
28. Neukirch, J.: Algebraic Number Theory. Grundlehren der mathematischen Wissenschaften, vol. 322, 1st edn. Springer, Heidelberg (1999). https://doi.org/10.1007/978-3-662-03983-0
29. Smart, N.P., Vercauteren, F.: Fully homomorphic encryption with relatively small key and ciphertext sizes. In: Nguyen, P.Q., Pointcheval, D. (eds.) PKC 2010. LNCS, vol. 6056, pp. 420–443. Springer, Heidelberg (2010). https://doi.org/10.1007/978-3-642-13013-7_25
30. Peikert, C., Regev, O., Stephens-Davidowitz, N.: Pseudorandomness of ring-LWE for any ring and modulus. In: Proceedings of the 49th Annual ACM SIGACT Symposium on Theory of Computing, STOC 2017, pp. 461–473 (2017). https://doi.org/10.1145/3055399.3055489
31. Peikert, C., Rosen, A.: Efficient collision-resistant hashing from worst-case assumptions on cyclic lattices. In: Halevi, S., Rabin, T. (eds.) TCC 2006. LNCS, vol. 3876, pp. 145–166. Springer, Heidelberg (2006). https://doi.org/10.1007/11681878_8
32. Pellet-Mary, A., Hanrot, G., Stehlé, D.: Approx-SVP in ideal lattices with pre-processing. In: Ishai, Y., Rijmen, V. (eds.) EUROCRYPT 2019. LNCS, vol. 11477, pp. 685–716. Springer, Cham (2019). https://doi.org/10.1007/978-3-030-17656-3_24
33. Regev, O.: On lattices, learning with errors, random linear codes, and cryptography. J. ACM **56**(6), 1–40 (2009). https://doi.org/10.1145/1568318.1568324. Preliminary version in STOC'05
34. Schnorr, C., Euchner, M.: Lattice basis reduction: improved practical algorithms and solving subset sum problems. Math. Program. **66**, 181–199 (1994). https://doi.org/10.1007/BF01581144

35. Schnorr, C.P., Euchner, M.: Lattice basis reduction: improved practical algorithms and solving subset sum problems. Math. Program. **66**(1–3), 181–199 (1994). https://doi.org/10.1007/BF01581144
36. Stehlé, D., Steinfeld, R.: Making NTRU as secure as worst-case problems over ideal lattices. In: Paterson, K.G. (ed.) EUROCRYPT 2011. LNCS, vol. 6632, pp. 27–47. Springer, Heidelberg (2011). https://doi.org/10.1007/978-3-642-20465-4_4
37. Stehlé, D., Steinfeld, R., Tanaka, K., Xagawa, K.: Efficient public key encryption based on ideal lattices. In: Matsui, M. (ed.) ASIACRYPT 2009. LNCS, vol. 5912, pp. 617–635. Springer, Heidelberg (2009). https://doi.org/10.1007/978-3-642-10366-7_36
38. Tschebotareff, N.: Die bestimmung der dichtigkeit einer menge von primzahlen, welche zu einer gegebenen substitutionsklasse gehören. Math. Ann. **95**(1), 191–228 (1926). https://doi.org/10.1007/BF01206606

Homomorphic Encryption

Efficient Bootstrapping for Approximate Homomorphic Encryption with Non-sparse Keys

Jean-Philippe Bossuat(✉), Christian Mouchet, Juan Troncoso-Pastoriza, and Jean-Pierre Hubaux

École Polytechnique Fédérale de Lausanne, Lausanne, Switzerland
{jean-philippe.bossuat,christian.mouchet,juan.troncoso-pastoriza, jean-pierre.hubaux}@epfl.ch

Abstract. We present a bootstrapping procedure for the full-RNS variant of the approximate homomorphic-encryption scheme of Cheon et al., CKKS (Asiacrypt 17, SAC 18). Compared to the previously proposed procedures (Eurocrypt 18 & 19, CT-RSA 20), our bootstrapping procedure is more precise, more efficient (in terms of CPU cost and number of consumed levels), and is more reliable and 128-bit-secure. Unlike the previous approaches, it does not require the use of sparse secret-keys. Therefore, to the best of our knowledge, this is the first procedure that enables a highly efficient and precise bootstrapping with a low probability of failure for parameters that are 128-bit-secure under the most recent attacks on sparse R-LWE secrets.

We achieve this efficiency and precision by introducing three novel contributions: (i) We propose a generic algorithm for homomorphic polynomial-evaluation that takes into account the approximate rescaling and is optimal in level consumption. (ii) We optimize the key-switch procedure and propose a new technique for linear transformations (*double hoisting*). (iii) We propose a systematic approach to parameterize the bootstrapping, including a precise way to assess its failure probability.

We implemented our improvements and bootstrapping procedure in the open-source Lattigo library. For example, bootstrapping a plaintext in $\mathbb{C}^{32768}$ takes 18 s, has an output coefficient modulus of 505 bits, a mean precision of 19.1 bits, and a failure probability of $2^{-15.58}$. Hence, we achieve 14.1× improvement in bootstrapped throughput (plaintext-bit per second), with respect to the previous best results, and we have a failure probability 468× smaller and ensure 128-bit security.

Keywords: Fully homomorphic encryption · Bootstrapping · Implementation

1 Introduction

Homomorphic encryption (HE) enables computing over encrypted data without decrypting them first; thus, it is becoming increasingly popular as a solution

A. Canteaut and F.-X. Standaert (Eds.): EUROCRYPT 2021, LNCS 12696, pp. 587–617, 2021.
https://doi.org/10.1007/978-3-030-77870-5_21

for processing confidential data in untrustworthy environments. Since Gentry's introduction of the first fully homomorphic-encryption (FHE) scheme over ideal lattices [14], continuous efficiency improvements have brought these techniques closer to practical application domains. As a result, lattice-based FHE schemes are increasingly used in experimental systems [23,26,27], and some of them are now proposed as an industry standard [2].

Cheon et al. [11] introduced a *leveled* encryption scheme for approximate arithmetic (CKKS); the scheme is capable of homomorphically evaluating arbitrary polynomial functions over encrypted complex-number vectors. Although the family of *leveled* cryptosystems enables only a finite multiplicative depth, with each multiplication *consuming* one level, the CKKS scheme enables the homomorphic re-encryption of an exhausted ciphertext into an almost *fresh* one. This capability, commonly called *bootstrapping*, theoretically enables the evaluation of arbitrary-depth circuits. In practice, however, the bootstrapping procedure for CKKS is approximate, and its precision and performance determine the actual maximum depth of a circuit.

Since the initial CKKS bootstrapping procedure by Cheon et al. [10] and until the most recent work by Han and Ki [20] that operates on the full-RNS (residue number systems) version of CKKS, the bootstrapping efficiency has improved by several orders of magnitude. However, this operation remains a bottleneck for its potential applications, and its performance is crucial for the adoption of the scheme. Bootstrapping performance can be improved by following two approaches: (i) adapting the bootstrapping circuit representation by using HE-friendly numerical methods. (ii) optimizing the scheme operations themselves, which also improves the overall scheme performance.

All current CKKS bootstrapping approaches [6,10,20] rely, so far, on sparse secret-keys to reduce the depth of their circuit representation, and none of them has proposed parameters with an equivalent security of at least 128 bits under the recent attacks on sparse R-LWE secrets [9,29]. The lack of stability in the security of sparse R-LWE secrets has lead the standardization initiatives to exclude sparse keys, hence also the bootstrapping operation, from the currently proposed standards [2]. This raises the question about the practicality of a bootstrapping procedure that would not require the use of sparse secret-keys.

1.1 Our Results

We propose an efficient bootstrapping procedure for the full-RNS CKKS scheme; it does not necessarily require the use of sparse secret-keys and provides a greater throughput than the current state of the art (Definition 1 in Sect. 7.1). To achieve this, we make the following contributions:

Homomorphic Polynomial Evaluation (Sect. 3). The full-RNS variant of the CKKS scheme restricts the re-scale operation only to the division by the factors q_i of the ciphertext modulus Q. As the choice of these factors is constrained to those enabling a number theoretic transform (NTT), the rescale cannot be done by a power of two (as in the original CKKS scheme) and it introduces a small scale deviation in the process. For complex circuits, such as polynomial evaluations, additions between ciphertexts of slightly different scales will eventually occur and will introduce errors.

We observe that this problem is trivially solved for linear circuits, by scaling the plaintext constants by the modulus q_i by which the ciphertext will be divided during the next rescale. By doing so, the rescale is exact and the ciphertext scale is unchanged after the operation. As a polynomial can be computed by recursive evaluations of a linear function, the linear case can be generalized. In this work, we propose a generic algorithm that consumes an optimal number of $\lceil \log(d+1) \rceil$ levels to homomorphically evaluate degree-d polynomial functions. Starting from a user-defined output scale, the intermediate scales can be back-propagated in the recursion, thus ensuring that each and every homomorphic addition occurs between ciphertexts of the same scale (hence is errorless). Our algorithm is, to the best of our knowledge, the first general solution for the problem of the approximate rescale arising from the full-RNS variant of the CKKS scheme.

Faster Matrix $\times$ Ciphertext Products (Sect. 4). The most expensive CKKS homomorphic operation is the key-switch. This operation is an integral building block of the homomorphic multiplication, slot rotations, and conjugation. The CKKS bootstrapping requires two linear transformations that involve a large number of rotations (key-switch operations), so minimizing the number of key-switch and/or their complexity has a significant effect on its performance.

Given an $n \times n$ plaintext matrix M and an encrypted vector $\mathbf{v}$, all previous works on the CKKS bootstrapping [6,10,20] use a baby-step giant-step (BSGS) algorithm, proposed by Halevi and Shoup [18], to compute the encrypted product $M\mathbf{v}$ in $\mathcal{O}(\sqrt{n})$ rotations. These works treat the key-switch procedure as a black-box and try to reduce the number of times it is executed. Therefore, they do not exploit the *hoisting* proposed by Halevi and Shoup [19].

We improve this BSGS algorithm by proposing a new format for rotation keys and a modified key-switch procedure that extends the hoisting technique to a second layer. This strategy is generic and it reduces the theoretical minimum complexity (in terms of modular products) of any linear transformation over ciphertexts. In our bootstrapping it reduces the cost of the linear transformations by roughly a factor of *two* compared to the previous hoisting approach.

Improved Bootstrapping Procedure (Sect. 5). We integrate our proposed improvements in the bootstrapping circuit proposed by Cheon et al. [10], Chen et al. [6], Cheon et al. [7] and Han and Ki [20]. We propose a new high-precision and faster bootstrapping circuit with updated parameters that are 128-bit secure, even if considering the most recent attacks on sparse keys [9,29].

Parameterization and Evaluation (Sect. 6). We discuss the parametrization of the CKKS scheme and its bootstrapping circuit, and we propose a procedure to choose and fine-tune the parameters for a given use-case.

We implemented our contributions, as well as our bootstrapping, in the open source library Lattigo: https://github.com/ldsec/lattigo. To the best of our knowledge, this is the first public and open-source implementation of the bootstrapping for the full-RNS variant of the CKKS scheme.

2 Background and Related Work

We now recall the full-RNS variant of the CKKS encryption scheme and review its previously proposed bootstrapping procedures.

2.1 The Full-RNS CKKS Scheme

We consider the CKKS encryption scheme [11] in its full-RNS variant [8]: the polynomial coefficients are always represented in the RNS and NTT domains.

Notation. For a fixed power-of-two N and $L+1$ distinct primes $q_0, \ldots, q_L$, we define $Q_L = \prod_{i=0}^{L} q_i$ and $R_{Q_L} = \mathbb{Z}_{Q_L}[X]/(X^N+1)$, the cyclotomic polynomial ring over the integers modulo Q_L. Unless otherwise stated, we consider elements of R_{Q_L} as their unique representative in the RNS domain: $R_{q_0} \times R_{q_1} \times ... \times R_{q_L} \cong R_{Q_L}$: a polynomial in R_{Q_L} is represented by a $(L+1) \times N$ matrix of coefficients. We denote single elements (polynomials or numbers) in italics, e.g., a, and vectors of such elements in bold, e.g., $\mathbf{a}$, with $\mathbf{a}||\mathbf{b}$ the concatenation of two vectors. We denote $a^{(i)}$ the element at position i of the vector $\mathbf{a}$ or the degree-i coefficient of the polynomial a. We denote $||a||$ the infinity norm of the polynomial (or vector) a in the power basis and $\mathsf{hw}(a)$ the Hamming weight of the polynomial (or vector) a. We denote $\langle \mathbf{a}, \mathbf{b} \rangle$ the inner product between the vectors $\mathbf{a}$ and $\mathbf{b}$. Given two vectors $\mathbf{a}$ and $\mathbf{b}$, each of n values, we denote $\log(\epsilon^{-1})$ the negative log of the L1 norm of their difference: $\epsilon = \frac{1}{n}\sum_{i=0}^{n-1} |a^{(i)} - b^{(i)}|$. $[x]_Q$ denotes reduction of x modulo Q and $\lfloor x \rfloor$, $\lceil x \rceil$, $\lfloor x \rceil$ the rounding of x to the previous, the next, and the closest integer, respectively (if x is a polynomial, the operation is applied coefficient-wise). Unless otherwise stated, logarithms are in base 2.

Plaintext and Ciphertext Space. A plaintext is a polynomial $\mathsf{pt} = m(Y) \in \mathbb{R}[Y]/(Y^{2n}+1)$ with $Y = X^{N/2n}$ and n a power-of-two smaller than N. We define the following plaintext encodings: (i) The *coefficient* encoding for which the message $\mathbf{m} \in \mathbb{R}^{2n}$ is directly encoded as the coefficients of a polynomial in Y. (ii) The *slots* encoding for which the message $\mathbf{m} \in \mathbb{C}^n$ is subjected to the canonical embedding $\mathbb{C}^n \rightarrow Y^{2n}$ for which the negacyclic convolution in $\mathbb{R}[Y]/(Y^{2n}+1)$ results in a Hadamard product in $\mathbb{C}^n$.

We represent plaintexts and ciphertexts, respectively, by the tuples $\{\mathsf{pt}, Q_\ell, \Delta\}$ and $\{\mathsf{ct}, Q_\ell, \Delta\}$, where, for a secret $s \in R_{Q_L}$, pt is a degree-zero polynomial in s, i.e. of R_{Q_ℓ}, and ct is a degree-one polynomial in s, i.e. of $R_{Q_\ell}^2$. We define $Q_\ell = \prod_{i=0}^{\ell} q_i$ as the modulus at level ℓ and Δ as a scaling factor. We denote L as the maximum level and use $0 \leq \ell \leq L$ to represent a level between the smallest level 0 and the highest level L. We refer to the depth of a circuit as the number of levels required for the evaluation of the circuit.

Scheme RNS-CKKS – Basic Operations

- $\mathsf{Setup}(N, h, b, \sigma)$: For a power-of-two ring degree N, a secret-distribution Hamming weight h, a standard deviation σ, and a modulus bit-size b: Select the moduli chains $\{q_0, \ldots, q_L\}$ and $\{p_0, \ldots, p_{\alpha-1}\}$ composed of pairwise different NTT-friendly primes (i.e. $q_i \equiv 1 \mod 2N$) close to powers of two such that $\log(\prod_{i=0}^{L} q_i \times \prod_{j=0}^{\alpha-1} p_j) \leq b$. Set $Q_L = \prod_{i=0}^{L} q_i$, $P = \prod_{j=0}^{\alpha-1} p_j$.
  Define the following distributions over R: χ_{key} with coefficients uniformly distributed over $\{-1, 0, 1\}$ and exactly h non-zero coefficients. χ_{pkenc} with coefficients distributed over $\{-1, 0, 1\}$ with respective probabilities $\{1/4, 1/2, 1/4\}$. χ_{err} with coefficients distributed according to a discrete Gaussian distribution with standard deviation σ and truncated to $[-\lfloor 6\sigma \rfloor, \lfloor 6\sigma \rfloor]$.
- $\mathsf{Encode}\,(\mathbf{m}, \Delta, n, \ell)$ (*coefficients→slots*): For a message $\mathbf{m} \in \mathbb{C}^n$ with $1 \leq n < N$, where n divides N, apply the canonical map $\mathbb{C}^n \to \mathbb{R}[Y]/(Y^{2n}+1) \to R_{Q_\ell}$ with $Y = X^{N/2n}$. Compute $\mathbf{m}' = \mathrm{FFT}_n^{-1}(\mathbf{m})$ and set $\mathbf{m}'_0 || \mathbf{m}'_1 \in \mathbb{R}^{2n}$, with $\mathbf{m}'_0 = \frac{1}{2}(\mathbf{m}' + \overline{\mathbf{m}'})$ and $\mathbf{m}'_1 = \frac{-i}{2}(\mathbf{m}' - \overline{\mathbf{m}'})$, as a polynomial in Y. Finally, scale the coefficients by Δ and round them to the nearest integer, apply the change of variable $Y \to X$ and return $\{\mathsf{pt}, Q_\ell, \Delta\}$.
- $\mathsf{Decode}(\{\mathsf{pt}, Q_\ell, \Delta\}, n)$ (*slots→coefficients*): For $1 \leq n < N$, where n divides N, apply the inverse of the canonical map $R_{Q_\ell} \to \mathbb{R}[Y]/(Y^{2n}+1) \to \mathbb{C}^n$, with $Y = X^{N/2n}$. Map pt to the vector $\mathbf{m}'_0 || \mathbf{m}'_1 \in \mathbb{R}^{2n}$ and return $\mathbf{m} = \mathrm{FFT}_n(\Delta^{-1} \cdot (\mathbf{m}'_0 + i \cdot \mathbf{m}'_1))$.
- $\mathsf{SecKeyGen}(\cdot)$: Sample $s \leftarrow \chi_{\mathsf{key}}$ and return the secret key s.
- $\mathsf{SwitchKeyGen}(s, s', \mathbf{w})$: For $\mathbf{w}$ an integer decomposition basis of β elements, sample $a_i \in R_{PQ_L}$ and $e_i \leftarrow \chi_{\mathsf{err}}$ and return the key-switch key: $\mathsf{swk}_{(s \to s')} = (\mathsf{swk}^{(0)}_{(s \to s')}, \ldots, \mathsf{swk}^{(\beta-1)}_{(s \to s')})$, where $\mathsf{swk}^{(i)}_{(s \to s')} = (-a_i s' + s w^{(i)} P + e_i, a_i)$.
- $\mathsf{PubKeyGen}(s)$: Set the public encryption key $\mathsf{pk} \leftarrow \mathsf{SwitchKeyGen}(0, s, (1))$, the relinearization key $\mathsf{rlk} \leftarrow \mathsf{SwitchKeyGen}(s^2, s, \mathbf{w})$, the rotation keys $\mathsf{rot}_k \leftarrow \mathsf{SwitchKeyGen}(s^{5^k}, s, \mathbf{w})$ (a different key has to be generated for each different k), and the conjugation key $\mathsf{conj} \leftarrow \mathsf{SwitchKeyGen}(s^{-1}, s, \mathbf{w})$ and return: $(\mathsf{pk}, \mathsf{rlk}, \{\mathsf{rot}_k\}_k, \mathsf{conj})$.
- $\mathsf{Enc}(\{\mathsf{pt}, Q_\ell, \Delta\}, s)$: Sample $a \in_u R_{Q_\ell}$ and $e \leftarrow \chi_{\mathsf{err}}$, set $\mathsf{ct} = (-as + e, a) + (\mathsf{pt}, 0)$ and return $\{\mathsf{ct}, Q_\ell, \Delta\}$.
- $\mathsf{PubEnc}(\{\mathsf{pt}, Q_\ell, \Delta\}, \mathsf{pk})$: Sample $u \leftarrow \chi_{\mathsf{pkenc}}$ and $e_0, e_1 \leftarrow \chi_{\mathsf{err}}$, set: $\mathsf{ct} = \mathsf{SwitchKey}(u, \mathsf{pk}) + (\mathsf{pt} + e_0, e_1)$ and return $\{\mathsf{ct}, Q_\ell, \Delta\}$.
- $\mathsf{SwitchKey}(d, \mathsf{swk}_{s \to s'})$: For $d \in R_{Q_\ell}$ a polynomial[1], decompose d base $\mathbf{w}$ such that $d = \langle \mathbf{d}, \mathbf{w} \rangle$ and return $(d_0, d_1) = \lfloor P^{-1} \cdot \langle \mathbf{d}, \mathsf{swk}_{s \to s'} \rangle \rceil \mod Q_\ell$ for $P^{-1} \in \mathbb{R}$.
- $\mathsf{Dec}(\{\mathsf{ct}, Q_\ell, \Delta\}, s)$: For $\mathsf{ct} = (c_0, c_1)$, return $\{\mathsf{pt} = c_0 + c_1 s, Q_\ell, \Delta\}$.

The homomorphic operations of CKKS are detailed in the extended version of the paper [4].

---

[1] SwitchKey does not act directly in a ciphertext; instead, we define it as a generalized intermediate function used as a building block that takes a polynomial as input.

2.2 CKKS Bootstrapping

Let $\mathsf{ct} = (c_0, c_1)$ be a ciphertext at level $\ell = 0$, and s a secret key of Hamming weight h, such that $\mathsf{Decrypt}(\mathsf{ct}, s) = [c_0 + c_1 s]_{Q_0} = \mathsf{pt}$. The goal of the bootstrapping operation is to compute a ciphertext ct' at level $L - k > 0$ (where k is the depth of the bootstrapping circuit) such that $Q_{L-k} \gg Q_0$ and $[c_0' + c_1' s]_{Q_{L-k}} \approx \mathsf{pt}$. Since $[c_0 + c_1 s]_{Q_L} = \mathsf{pt} + Q_0 \cdot I$, where I is an integer polynomial [10], bootstrapping is equivalent to an extension of the CRT basis, followed by a homomorphic reduction modulo Q_0.

Cheon et al. proposed the first procedure [10] to compute this modular reduction, by (i) homomorphically applying the encoding algorithm, to enable the parallel (slot-wise) evaluation, (ii) computing a modular reduction approximated by a scaled sine function on each slot, and (iii) applying the decoding algorithm to retrieve a close approximation of pt without the polynomial I:

$$\underbrace{\mathsf{Encode}(\mathsf{pt} + Q_0 \cdot I) = \mathsf{pt}'}_{\text{(i) }\mathsf{SlotsToCoeffs}(\mathsf{pt}+Q_0 \cdot I)} \Rightarrow \underbrace{\frac{Q_0}{2\pi}\sin\left(\frac{2\pi \mathsf{pt}'}{Q_0}\right) = \mathsf{pt}''}_{\text{(ii) }\mathsf{EvalSine}(\mathsf{pt}')} \Rightarrow \underbrace{\mathsf{Decode}(\mathsf{pt}'') \approx \mathsf{pt}}_{\text{(iii) }\mathsf{CoeffsToSlots}(\mathsf{pt}'')}.$$

The complexity of the resulting bootstrapping circuit is influenced by two parameters: The first one is the secret-key Hamming weight h, which directly impacts the depth of the bootstrapping circuit. Indeed, Cheon et al. show that $||I|| \leq \mathcal{O}(\sqrt{h})$ with very high probability. A denser key will therefore require evaluating a larger-degree polynomial, with a larger depth. The second parameter is the number of plaintext slots n that has a direct impact on the complexity of the circuit (but not on its depth). By scaling down the values to compress them closer to the origin, Cheon et al. are able to evaluate the sine function by using a low-degree Taylor series of the complex exponential and then use repeated squaring (the double angle formula) to obtain the correct result. In their approach, the sine evaluation dominates the circuit's depth, whereas the homomorphic evaluation of the encoding and decoding algorithms, which they express as an $n \times n$ matrix-vector product, dominates its width.

In a subsequent work, Chen et al. [6] propose to compute the encoding by homomorphically evaluating the Cooley-Tukey algorithm. This approach needs $\log(n)$ depth (the number of iterations of the algorithm); to reduce this depth, Chen et al. merge several iterations together, at the cost of an increased complexity. In a concurrent work, Cheon et al. [7] explored techniques to efficiently evaluate DFTs on ciphertexts. They show how to factorize the encoding matrices into a series of $\log_r(n)$ sparse matrices, where r is a power-of-two radix. The contributions in [6,7] enabled the acceleration of the homomorphic evaluation of the encoding functions by two orders of magnitude. Chen et al. [6] also improved the approximation of the scaled sine function by using a Chebyshev interpolant.

More recently, Han and Ki port the bootstrapping procedure to the full-RNS variant of CKKS, with several improvements to the bootstrapping circuit and to the CKKS scheme [20]. They propose a generalization of its key-switch procedure by using an intermediate RNS decomposition that enables a trade-off between the complexity of the key-switch and the homomorphic capacity of a fresh ciphertext. They also give an alternative way to approximate the scaled sine function, which accounts for the magnitude of the underlying plaintext and uses the cosine function and the double angle formula. Combined, these changes yield an acceleration factor of 2.5 to 3, compared to the work of Chen et al. [6].

Both works [6,7] were implemented with HEAAN [21], yet the implementation of only the former was published. The work of [20] was implemented using SEAL [28], but the implementation has still not been published.

2.3 Security of Sparse Keys

One commonality between all the aforementioned works is the use of sparse secret-keys with a Hamming weight $h = 64$. A key with a small Hamming weight enables a low-depth bootstrapping circuit, essential for its practicality. However, recent advances in the cryptanalysis of the R-LWE problem prove that hybrid attacks specifically targeting such sparse keys can severely affect its security [9,29]. In light of the most recent attacks, Curtis and Player [12] estimate that, for a sparse key with $h = 64$ and a ring degree $N = 2^{16}$, the modulus needs to be at most 990 bits to achieve a security of 128 bits. In their initial bootstrapping proposal, Cheon et al. [10] use the parameters $\{N = 2^{16}, \log(Q) = 2480, h = 64, \sigma = 3.2\}$ and estimate the security of these parameters to 80 bits. In their work, Han and Ki [20] propose new parameter sets, one of which they claim has 128-bit of security: $\{N = 2^{16}, \log(Q) = 1450, h = 64, \sigma = 3.2\}$. However, these estimates are based on results obtained using Albrecht's estimator [1] that, at the time, did not take into account the most recent attacks on sparse keys. The security of the parameter set $\{N = 2^{16}, \log(Q) = 1250, h = 64, \sigma = 3.2\}$ is estimated at 113 bits in the more recent work by Son and Cheon [29]. This sets a loose upper bound to security of the parameters (which have a 1450-bit modulus) proposed by Han and Ki [20]. Therefore, the bootstrapping parameters must be updated to comply with the most recent security recommendations, as none of the parameters proposed in the current works achieve a security of 128 bits.

3 Homomorphic Polynomial Evaluation

The main disadvantage of the full-RNS variant of CKKS stems from its rescale operation that does not divide the scale by a power-of-two, as in the original scheme, but by one of the moduli. Those moduli are chosen, for efficiency purposes, as distinct NTT-friendly primes [8]; under this constraint, the power-of-two rescale of the original CKKS scheme can only be approximated. As a result, ciphertexts at the same level can have slightly different scales (depending on the previous homomorphic operations) and additions between such ciphertexts will introduce an error proportional to the difference between their scale. Addressing this issue in a generic and practical way is crucial for the adoption of CKKS.

For a significant step toward this goal, we introduce a homomorphic polynomial-evaluation algorithm that is depth-optimal and ensures that additions are always made between ciphertexts with the exact same scale.

Algorithm 1: BSGS alg. for polynomials in Chebyshev basis

Input: $p(t) = \sum_{i=0}^{d} c_i T_i(t)$.
Output: The evaluation of $p(t)$.
1 $\mathsf{m} \leftarrow \lceil \log(d+1) \rceil$
2 $\mathsf{l} \leftarrow \lfloor \mathsf{m}/2 \rfloor$
3 $T_0(t) = 1$, $T_1(t) = t$
4 Evaluate $T_2(t), T_3(t), \ldots, T_{2^{\mathsf{l}}-1}(t)$ and $T_{2^{\mathsf{l}}}(t), T_{2^{\mathsf{l}+1}}(t), \ldots, T_{2^{\mathsf{m}-1}}(t)$ using $T_{i=a+b}(t) \leftarrow 2T_a(t)T_b(t) - T_{|a-b|}(t)$.
5 Find $q(t)$ and $r(t)$ such that $p(t) = q(t) \cdot T_{2^{\mathsf{m}-1}}(t) + r(t)$.
6 Recurse on step 5 by replacing $p(t)$ by $q(t)$ and $r(t)$ and m by $\mathsf{m}-1$, until the degree of $q(t)$ and $r(t)$ is smaller than 2^{l}.
7 Evaluate $q(t)$ and $r(t)$ using $T_j(t)$ for $0 \leq j \leq 2^{\mathsf{l}} - 1$.
8 Evaluate $p(t)$ using $q(t)$, $r(t)$ and $T_{2^{\mathsf{m}-1}}(t)$.
9 **return** $p(t)$

3.1 The Baby-Step Giant-Step (BSGS) Algorithm

In order to minimize the number of ciphertext-ciphertext multiplications in their bootstrapping circuit, Han and Ki [20] adapt a generic baby-step giant-step (BSGS) polynomial-evaluation algorithm for polynomials expressed in a Chebyshev basis. Algorithm 1 gives a high-level description of the procedure.

For a polynomial $p(t)$ of degree d, with $\mathsf{m} = \lceil \log(d+1) \rceil$ and $\mathsf{l} = \lfloor \mathsf{m}/2 \rfloor$, the algorithm first decomposes $p(t)$ into $\sum_{i=0}^{\lfloor d/\mathsf{l} \rfloor} u_{i,2^{\mathsf{l}}}(t) \cdot T_{2^{i \cdot \mathsf{l}}}(t)$, with $u_{i,2^{\mathsf{l}}}(t) = \sum_{j=0}^{2^{\mathsf{l}}-1} c_{i,j} \cdot T_j(t)$, $c_{i,j} \in \mathbb{C}$ and $T_{0 \leq j < 2^{\mathsf{l}}}$ a pre-computed power basis. We denote $u_{\lfloor d/\mathsf{l} \rfloor, 2^{\mathsf{l}}}(t)$ as u_{max}. The BSGS algorithm then recursively combines the monomials $u_{i,2^{j+1}}(t) = u_{i+1,2^j}(t) \cdot T_{2^j}(t) + u_{i,2^j}(t)$ in a tree-like manner by using a second pre-computed power basis $T_{2^{\mathsf{l}} < i < \mathsf{m}}(t)$ to minimize the number of non-scalar multiplications. The algorithm requires $2^{\mathsf{m}-\mathsf{l}} + 2^{\mathsf{l}} + \mathsf{m} - \mathsf{l} - 3 + \lceil (d+1)/2^{\mathsf{l}} \rceil$ non-scalar products and has, in the best case, depth m.

3.2 Errorless Polynomial Evaluation

We address the errors introduced by the approximate rescale for the evaluation of a polynomial $p(t)$. We scale each of the leaf monomials $u_{i,2^{\mathsf{l}}}(t)$ by some scale Δ such that all evaluations of the subsequent monomials $u_{i,2^{j+1}}(t) = u_{i+1,2^j}(t) \cdot T_{2^j}(t) + u_{i,2^j}(t)$ are done with additions between ciphertexts of the same scale. More formally, let $\Delta_{u_{i,2^{j+1}}(t)}$ be the scale of $u_{i,2^{j+1}}(t)$ (the result of the monomial evaluation), $\Delta_{T_{2^j}(t)}$ the scale of the power-basis element $T_{2^j}(t)$, and $q_{T_{2^j}(t)}$ the modulus by which the product $u_{i+1,2^j}(t) \cdot T_{2^j}(t)$ is rescaled. We set $\Delta_{u_{i+1,2^j}(t)} = \Delta_{u_{i,2^{j+1}}(t)} \cdot q_{T_{2^j}(t)} / \Delta_{T_{2^j}(t)}$ and $\Delta_{u_{i,2^j}(t)} = \Delta_{u_{i,2^{j+1}}(t)}$. Starting from a target scale $\Delta_{p(t)}$ and $p(t) = u_{0,2^{\mathsf{m}}}(t) = u_{1,2^{\mathsf{m}-1}}(t) \cdot T_{2^{\mathsf{m}-1}}(t) + u_{0,2^{\mathsf{m}-1}}(t)$, we recursively compute and propagate down the tree the scale each $u_{i,2^j}(t)$ should have. The recursion ends when reaching $u_{i,2^{\mathsf{l}}}$, knowing the scale that they must have. Since $u_{i,2^{\mathsf{l}}}(t) = \sum_{j=0}^{2^{\mathsf{l}}-1} c_{i,j} T_j(t)$, we can use the same technique to derive by what value

Algorithm 2: EvalRecurse

Input: A target scale Δ, an upper-bound m, a stop factor l, a degree-d polynomial $p(t) = \sum_{i=0}^{d} c_i T_i(t)$, and the power basis $\{T_0, T_1, \ldots, T_{2^l-1}\}$ and $\{T_{2^l}, T_{2^l+1}, \ldots, T_{2^m-1}\}$, pre-computed for a ciphertext ct.

Output: A ciphertext encrypting the evaluation of $p(\mathsf{ct})$.

```
1  if d < 2^l then
2  |  if p(t) = u_max(t) and l > 2^m - 2^(l-1) and l > 1 then
3  |  |  return EvalRecurse(Δ, m = ⌈log(d+1)⌉, l = ⌊⌈log(d+1)⌉/2⌋, p(t), T)
4  |  else
5  |  |  ct ← ⌊c_0 · Δ · q_{T_d}⌉
6  |  |  for i = d; i > 0; i = i − 1 do
7  |  |  |  ct ← Add(ct, MultConst(T_i, ⌊(c_i · Δ · q_{T_d})/Δ_{T_i}⌉))
8  |  |  end
9  |  |  return Rescale(ct)
10 |  end
11 end
12 Express p(t) as q(t) · T_{2^m−1} + r(t)
13 ct_0 ← EvalRecurse((Δ · q_{T_{2^m−2}})/Δ_{T_{2^m−1}}, m − 1, l, q(t), T)
14 ct_1 ← EvalRecurse(Δ, m − 1, l, r(t), T)
15 ct_0 ← Mul(ct_0, T_{2^m−1})
16 if level(ct_0) > level(ct_1) then
17 |  ct_0 ← Add(Rescale(ct_0), ct_1)
18 else
19 |  ct_0 ← Rescale(Add(ct_0, ct_1))
20 end
21 return ct_0
```

each of the coefficients $c_{i,j}$ must be scaled, so that the evaluation of $u_{i,2^l}(t)$ is also done with exact additions and ends up with the desired scale.

Algorithm 2 is our proposed solution: it integrates our scale-propagation technique to the recursive decomposition of $p(t)$ into $q(t)$ and $r(t)$. We compare Algorithms 1 and 2 in Table 1 by evaluating a Chebyshev interpolant of the homomorphic modular reduction done during the bootstrapping circuit. This function plays a central role in the bootstrapping hence is an ideal candidate for

Table 1. Comparison of the homomorphic evaluation of a Chebyshev interpolant of degree d of $\cos(2\pi(x-0.25)/2^r)$ in the interval $(-K/2^r, K/2^r)$ followed by r evaluations of $\cos(2x) = 2\cos^2(x) - 1$. The scheme parameters are $N = 2^{16}$, $n = 2^{15}$, $h = 196$ and $q_i \approx 2^{55}$. $\Delta_\epsilon = |\Delta_{\text{in}} - \Delta_{\text{out}}| \cdot \Delta_{\text{in}}^{-1}$.

		$\log(1/\epsilon)$ for (K, d, r)				
	Δ_ϵ	(12, 34, 2)	(15, 40, 2)	(17, 44, 2)	(21, 52, 2)	(257, 250, 3)
Algorithm 1 ([20])	$2^{-31.44}$	30.36	30.05	29.73	29.19	25.00
Algorithm 2 (ours)	0	37.37	37.16	37.15	37.04	29.46

evaluating the effect of the proposed approaches (see Sect. 5.4). To verify that our algorithm correctly avoids additions between ciphertexts of different scales, we forced both algorithms to always rescale a ciphertext before an addition (in practice, it is better to check the levels of the ciphertexts before an addition, and dynamically assess if a level difference can be used to scale one ciphertext to the scale of the other). We observe that our algorithm yields two advantages: It enables (i) a scale-preserving polynomial evaluation (the output-scale is identical to the input scale), and (ii) a much better precision by successfully avoiding errors due to additions between ciphertexts of different scales.

3.3 Depth-Optimal Polynomial Evaluation

In practice, Algorithm 1 will consume more than the optimal m levels for a specific class of d due to the way the rescale and level management work in the full-RNS variant of the CKKS scheme. This discrepancy arises from the following interactions (recall that Algorithm 1 evaluates each $u_i(t)$ as a linear combination of a pre-computed power-basis $\{T_0(t), T_1(t), \ldots, T_{2^{\mathsf{l}}-1}(t)\}$):

1. If $\mathsf{l} > 1$, then the depth to evaluate $T_{2^{\mathsf{l}}-1}(t)$ is l and evaluating the $u_i(t)$ will necessarily cost $\mathsf{l}+1$ levels due to the constant multiplications.
2. If $\mathsf{l} = 1$, then the depth to evaluate $T_1(t)$ is zero, hence the depth to evaluate the $u_i(t)$ is and remains l.
3. If $d > 8$, then Algorithm 1 sets $\mathsf{l} > 1$.
4. If $2^{\mathsf{m}} - 2^{\mathsf{l}-1} \leq d < 2^{\mathsf{m}}$, then all the elements of the power basis $\{T_{2^{\mathsf{l}}}, T_{2^{\mathsf{l}+1}}, \ldots, T_{2^{\mathsf{m}-1}}\}$ need to be used during the recombination step of Algorithm 1.

Hence, if $\mathsf{l} > 1$ and $d > 2^{\mathsf{m}} - 2^{\mathsf{l}-1}$, the total depth to execute Algorithm 1 is necessarily $\mathsf{m}+1$. This could be avoided by always setting $\mathsf{l} = 1$ regardless of d, but it would lead to a very costly evaluation, as the number of non-scalar multiplications would grow proportionally to d. To mitigate this additional cost, we only enforce $\mathsf{l} = 1$ on the coefficient of $p(t)$ whose degree is $\geq 2^{\mathsf{m}} - 2^{\mathsf{l}-1}$. Hence, Algorithm 2 first splits $p(t)$ into $p(t) = a(t) + b(t) \cdot T_{2^{\mathsf{m}}-2^{\mathsf{l}-1}}(t)$. It then evaluates $a(t)$ with the optimal l and recurses on $b(t)$ until $\mathsf{l} = 1$. The number of additional recursions is bounded by $\log(\mathsf{m})$, because each recursion sets the new degree to half of the square root of the previous one. In practice, these additional recursions add only $\lceil \log(d+1-(2^{\mathsf{m}}-2^{\mathsf{l}-1})) \rceil$ non-scalar multiplications but enable the systematic evaluation of any polynomial by using exactly m levels.

3.4 Conclusions

For an extra cost of $\lceil \log(d+1-(2^{\mathsf{m}}-2^{\mathsf{l}-1})) \rceil$ ciphertext-ciphertext products, our proposed algorithm guarantees an optimal depth hence an optimal-level consumption. This extra cost is negligible, compared to the base cost of Algorithm 1, i.e., $2^{\mathsf{m}-\mathsf{l}}+2^{\mathsf{l}}+\mathsf{m}-\mathsf{l}-3+\lceil (d+1)/2^{\mathsf{l}} \rceil$. It also guarantees exact additions throughout the entire polynomial evaluation, hence preventing the precision loss related to additions between ciphertexts of different scales and making the procedure easier to use. It also enables the possibility to choose the output scale that can be set to the same as the input scale, making the polynomial evaluation scale-preserving. As linear transformations and constant multiplications can already be made to be scale-preserving, our polynomial evaluation is the remaining building block for enabling scale-preserving circuits of arbitrary depth.

4 Key-Switch and Improved Matrix-Vector Product

The key-switch procedure is the generic public-key operation of the CKKS scheme. By generating specific public *key-switch keys* derived from secret keys s' and s, it is possible to enable the public re-encryption of ciphertexts from key s' to s. Beyond the public encryption procedure (switching from $s' = 0$ to s), a key-switch is required by most homomorphic operations to cancel the effect of encrypted arithmetic on the decryption circuit, thus ensuring the compactness of the scheme. In particular, homomorphic multiplications require the re-encryption from key s^2 back to s, whereas slot-rotations require the re-encryption from the equivalent rotation of s back to s. The cost of the key-switch dominates the cost of these operations by one to two orders of magnitude because it requires many NTTs and CRT reconstructions. Hence, optimizations of the key-switch algorithm have a strong effect on the overall efficiency of the scheme.

We propose an optimized key-switch key format and key-switch algorithm (Sect. 4.1). We then apply them to rotation-keys and further improve the hoisted-rotation technique (Sect. 4.2) introduced by Halevi and Shoup [19]. Finally, we propose a modified procedure for matrix-vector multiplications over packed ciphertexts (Sect. 4.3) which features a novel *double-hoisting* optimization.

4.1 Improved Key-Switch Keys

Given a ciphertext modulus $Q_L = \prod_{j=0}^{L} q_j$, we use a basis $\mathbf{w}$ composed of products among the q_j, as described by Han and Ki [20]. We also include the entire basis $\mathbf{w}$ in the keys, as done by Bajard et al. and Halevi et al. [3,16]; this saves one constant multiplication during the key-switch and enables a simpler key-switch keys generation. A more detailed overview of these works can be found in the extended version of the paper [4].

We propose a simpler and more efficient hybrid approach. Specifically, we use the basis $w^{(i)} = \frac{Q_L}{q_{\alpha_i}}[(\frac{Q_L}{q_{\alpha_i}})^{-1}]_{q_{\alpha_i}}$ with $q_{\alpha_i} = \prod_{j=\alpha_i}^{\min(\alpha(\beta+1)-1,L)} q_j$ for $0 \leq i < \beta$, $\beta = \lceil (L+1)/\alpha \rceil$ and α a positive integer. In other words, Q is factorized into β equally-sized composite-numbers q_{α_i}, each composed of up to α different primes. Thus, our key-switch keys have the following format:

$$\left(\mathsf{swk}^0_{q_{\alpha_i}}, \mathsf{swk}^1_{q_{\alpha_i}}\right) = \left([-a_i s + s' \cdot P \cdot \tfrac{Q_L}{q_{\alpha_i}} \cdot [(\tfrac{Q_L}{q_{\alpha_i}})^{-1}]_{q_{\alpha_i}} + e_i]_{PQ_L}, [a_i]_{PQ_L}\right).$$

We set $P = \prod_{j=0}^{\alpha-1} p_j$, and the bit-size of P such that $q_{\alpha_i} \leq P, \forall \alpha_i$. As shown by Gentry et al. [15], this leads to a negligible error introduced by the key-switch operation. Algorithm 3 describes the associated key-switch procedure that corresponds to the standard one adapted to our keys.

Algorithm 3: Key-switch

Input: $c \in R_{Q_\ell}$, the key-switch key $\mathsf{swk}_{s \to s'}$.
Output: $(a, b) \in R^2_{Q_\ell}$.
1 $\mathbf{d} \leftarrow \left[[c]_{q_{\alpha_{0 \leq i < \beta}}}\right]_{PQ_\ell}$
2 $(a, b) \leftarrow (\langle \mathbf{d}, \mathsf{swk}^0 \rangle, \langle \mathbf{d}, \mathsf{swk}^1 \rangle)$
3 $(a, b) \leftarrow (\lfloor P^{-1} \cdot a \rceil, \lfloor P^{-1} \cdot b \rceil)$
4 **return** (a, b)

4.2 Improved Hoisted-Rotations

The slot-rotation operation in CKKS is defined by the automorphism $\phi_k : X \rightarrow X^{5^k} \pmod{X^N + 1}$. It rotates the message slots by k positions to the left. After a rotation, the secret under which the ciphertext is encrypted is changed from s to $\phi_k(s)$, and a key-switch $\phi_k(s) \rightarrow s$ is applied to return to the original key.

Halevi and Shoup [19] show that as ϕ_k is an automorphism, it distributes over addition and multiplication, and commutes with the power-of-two base decomposition. As ϕ_k acts individually on the coefficients by permuting them without changing their norm (the modular reduction by $X^N + 1$ at most induces a sign change), it also commutes with the special RNS decomposition (see Supplementary material in the extended version of the paper [4]): $[\phi_k(a)]_{q_{\alpha_i}} = \phi_k([a]_{q_{\alpha i}})$.

Hence, when several rotations have to be applied on the same ciphertext, $[a]_{q_{\alpha_i}}$ can be pre-computed and re-used for each subsequent rotation: $\sum \phi_k([a]_{q_{\alpha_i}}) \cdot \mathsf{rot}_{k,q_{\alpha_i}}$. This technique proposed by Halevi et al., called *hoisting*, significantly reduces the number of NTTs and CRT reconstructions, at the negligible cost of having to compute the automorphism for each of the $[a]_{q_{\alpha_i}}$.

We further exploit the properties of the automorphism to reduce its execution cost, by observing that ϕ_k^{-1} can be directly pre-applied on the rotation keys:

$$\left(\widetilde{\mathsf{rot}}^0_{k,q_{\alpha_i}}, \widetilde{\mathsf{rot}}^1_{k,q_{\alpha_i}}\right) = \left([-a_i\phi_k^{-1}(s) + s \cdot P \cdot \tfrac{Q_L}{q_{\alpha_i}} \cdot [(\tfrac{Q_L}{q_{\alpha_i}})^{-1}]_{q_{\alpha_i}} + e_i]_{PQ_L}, [a_i]_{PQ_L}\right)$$

Compared to a $\mathsf{rot}_{k,q_{\alpha_i}}$, a traditional rotation-key as defined in Sect. 2.1, this reduces the number of automorphisms per-rotation to only one:

$$\langle \phi_k(\mathbf{a}), \mathsf{rot}_k \rangle = \phi_k\left(\langle \mathbf{a}, \widetilde{\mathsf{rot}}_k \rangle\right).$$

Our improved algorithm for hoisted rotations is detailed in Algorithm 4.

4.3 Faster Matrix-Vector Operations

We now discuss the application of homomorphic slot-rotations to the computation of matrix-vector products on packed ciphertexts. The ability to efficiently apply generic linear transformations to encrypted vectors is pivotal for a wide variety of applications of homomorphic encryption. In particular, the homomorphic evaluation of the CKKS encoding and decoding procedures, which are linear transformations, dominates the cost in the original bootstrapping procedure.

Algorithm 4: Optimized Hoisting-Rotations

Input: $\mathsf{ct} = (c_0, c_1) \in R^2_{Q_\ell}$ and a set of r rotation keys $\widetilde{\mathsf{rot}}_{r_k}$.
Output: $\mathbf{v}$ a list containing each r_k rotation of ct.

```
1 d ← [[c_1]_{q_{α_{0≤i<β}}}]_{PQ_ℓ} // (Decompose)
2 foreach r_k do
3     (a, b) ← (⟨d, rot~^0_{r_k}⟩, ⟨d, rot~^1_{r_k}⟩) // (MultSum)
4     (a, b) ← (⌊P^{-1} · a⌉, ⌊P^{-1} · b⌉) // (ModDown)
5     v_{r_k} ← (φ_{r_k}(c_0 + a), φ_{r_k}(b)) // (Permute)
6 end
7 return v
```

Halevi and Shoup propose to express an $n \times n$ matrix M in diagonal form and to use a baby-step giant-step (BSGS) algorithm (Algorithm 5) to evaluate the matrix product in $\mathcal{O}(\sqrt{n})$ rotations [17,18]. At the time of this writing, all the existing bootstrapping procedures for the CKKS scheme are based on this approach and are not reported to use *hoisting*, unlike done for BGV [18,19]. We now break down the cost of this BSGS algorithm, analyze its components and, using our observations, we present our improvements to this approach.

Algorithm 5: BSGS Algorithm of [19] For Matrix × Vector Multiplication

Input: ct a ciphertext encrypting $\mathbf{m} \in \mathbb{C}^n$, $\mathbf{M}_{diag}$ the diagonal rows of $\mathbf{M}$ a $n \times n$ matrix with $n = n_1 n_2$.
Output: The evaluation $\mathsf{ct}' = \mathbf{M} \times \mathsf{ct}$.

```
1  for i = 0; i < n_1; i = i + 1 do
2      ct_i ← Rotate_i(ct)
3  end
4  ct' ← (0, 0)
5  for j = 0; j < n_2; j = j + 1 do
6      r ← (0, 0)
7      for i = 0; i < n_1; i = i + 1 do
8          r ← Add(r, Mul(ct_i, Rotate_{-n_1·j}(M_diag^{(n_1·j+i)})))
9      end
10     ct' ← Add(ct', Rotate_{n_1·j}(r))
11 end
12 ct' ← Rescale(ct')
13 return ct'
```

Dominant Complexity of Rotations. The dominant cost factor of Algorithm 5 is the number of rotations, as each rotation requires key-switch operations. These rotations comprise four steps (see Algorithm 4):

1. *Decompose*: Decompose a polynomial of R_{Q_ℓ} in base $\mathbf{w}$ and return the result in R_{PQ_ℓ}. This operation requires NTTs and CRT basis extensions.
2. *MultSum*: Compute a sum of products of polynomials in R_{PQ_ℓ}. This operation only requires coefficient-wise additions and multiplications.

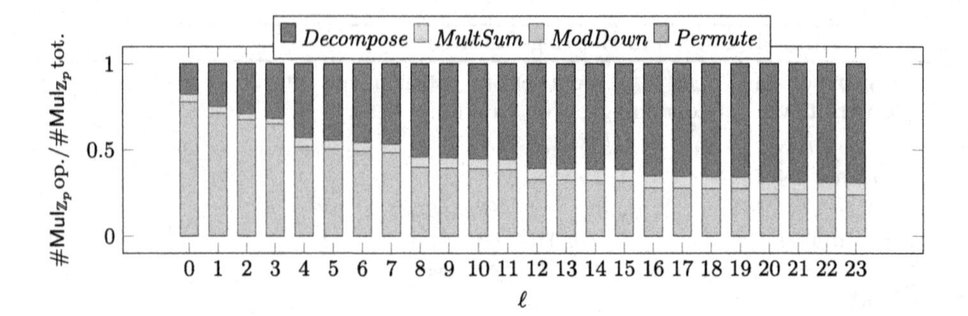

Fig. 1. Normalized complexity of each step (op.) of a rotation. The complexity for each operation was computed with $N = 2^{16}$, $0 \leq \ell \leq 23$ and $\alpha = 4$. The complexity derivation can be found in the extended version of the paper [4].

3. *ModDown*: Divide a polynomial of R_{PQ_ℓ} by P and return the result in R_{Q_ℓ}. This operation requires NTTs and CRT basis extensions.
4. *Permute*: Apply the automorphism ϕ_k on a polynomial of R_{Q_ℓ}. It represents a permutation of the coefficients and has in theory no impact on complexity.

Let n be the number of non-zero diagonals of M, and n_1, n_2 be two integers such that $n = n_1 n_2$; the complexity of the original BSGS algorithm (Algorithm 5) is $n_1 + n_2$ rotations and it is minimized when $n_1 \approx n_2$:

$$(n_2 + n_1) \cdot (\textit{Decompose} + \textit{MultSum} + \textit{ModDown} + \textit{Permute}),$$

to which $2n_2n_1$ multiplications in R_{Q_ℓ} should also be added (line 8 of Algorithm 5). We denote *inner-loop* and *outer-loop* the lines that depend, respectively, on n_1 and n_2. Figure 1 shows the weight of each of the four steps in the total complexity. The complexity of the steps *MultSum* and *Permute* is negligible compared to the complexity of *Decompose* and *ModDown*, as products and additions are very inexpensive compared to NTTs and CRT basis extensions. We base our optimization on this observation.

Improved BSGS Algorithm. We propose a new optimization that we refer to as *double-hoisting*. This optimization improves the *hoisting* technique proposed by Halevi and Shoup [19] and further reduces the complexity related to the *inner-loop* rotations by adding a second layer of *hoisting*.

The *first level*, proposed by Halevi and Shoup [19], applies to the *inner-loop* rotations (line 8 of Algorithm 5). This renders the computation devoted to *Decompose* independent of the value n_1, so the complexity is reduced to

$$\begin{aligned} &n_2 \cdot (\textit{Decompose} + \textit{MultSum} + \textit{ModDown} + \textit{Permute}) \\ &+ n_1 \cdot (\textit{MultSum} + \textit{ModDown} + \textit{Permute}) + \textit{Decompose}. \end{aligned}$$

The *second level*, which we propose, introduces an additional hoisting for the *inner-loop* rotations, as the *ModDown* step is a coefficient-wise operation.

Algorithm 6: Double-hoisting BSGS matrix×vector algorithm

Input: ct $= (c_0, c_1) \in R^2_{Q_\ell}$, $\mathbf{M}_{diag} \in R_{PQ_\ell}$ the pre-rotated diagonals of $\mathbf{M}_{n\times n}$, $n_1 n_2 = n$, $\text{rot}_i \in R^2_{PQ_\ell}$ the set of necessary rotations keys.

Output: The evaluation of $\mathbf{M} \times$ ct.

1 $\mathbf{d} \leftarrow \left[[c_1]_{q_{\alpha_{0\le i<\beta}}}\right]_{PQ_\ell}$ // $Q_\ell \rightarrow PQ_\ell$
2 **for** $i = 0; i < n_1; i = i + 1$ **do**
3 $\quad (a_i, b_i) \leftarrow (P \cdot c_0 + \langle \mathbf{d}, \widetilde{\text{rot}}_i^0\rangle, \langle \mathbf{d}, \widetilde{\text{rot}}_i^1\rangle)$ // $\in PQ_\ell$
4 **end**
5 $(r_0, r_1), r_2 \leftarrow (0, 0), (0)$
6 **for** $j = 0; j < n_2; j = j + 1$ **do**
7 $\quad (u_0, u_1) \leftarrow (0, 0)$
8 $\quad$**for** $i = 0; i < n_1; i = i + 1$ **do**
9 $\quad\quad (u_0, u_1) \leftarrow (u_0, u_1) + (a_i, b_i) \cdot \mathbf{M}_{diag}^{(n_1 \cdot j + i)}$ // $\in PQ_\ell$
10 $\quad$**end**
11 $\quad (u_0, u_1) \leftarrow (\lfloor P^{-1} \cdot u_0\rceil, \lfloor P^{-1} \cdot u_1)\rceil$ // $(PQ_\ell \rightarrow Q_\ell)$
12 $\quad \mathbf{d} \leftarrow \left[[u_1]_{q_{\alpha_{0\le i<\beta}}}\right]_{PQ_\ell}$ // $Q_\ell \rightarrow PQ_\ell$
13 $\quad (r_0, r_1) \leftarrow (r_0, r_1) + \left(\phi_{n_1 \cdot j}\left(\langle \mathbf{d}, \widetilde{\text{rot}}^0_{n_1\cdot j}\rangle\right), \phi_{n_1 \cdot j}\left(\langle \mathbf{d}, \widetilde{\text{rot}}^1_{n_1\cdot j}\rangle\right)\right)$ // $\in PQ_\ell$
14 $\quad r_2 \leftarrow r_2 + \phi_{n_1 \cdot j}(u_0)$ // $\in Q_\ell$
15 **end**
16 $(r_0, r_1) \leftarrow (\lfloor P^{-1} \cdot r_0\rceil, \lfloor P^{-1} \cdot r_1\rceil)$ // $(PQ_\ell \rightarrow Q_\ell)$
17 **return** $(r_0 + r_2, r_1)$

Similarly to the *Decompose* step, this operation commutes with the *Permute* step and the ciphertext-plaintext multiplications (line 8 of Algorithm 5). Therefore, we need to apply it only once after the entire *inner-loop* of n_1 rotations. Applying the same reasoning for the *ModDown* step of the *outer-loop* rotations we can reduce the number of key-switch operations from $n_1 + n_2$ to $n_2 + 1$:

$$n_2 \cdot (\textit{Decompose} + \textit{MultSum} + \textit{ModDown} + \textit{Permute})$$
$$+\, n_1 \cdot (\textit{MultSum} + \textit{Permute}) + \textit{Decompose} + \textit{ModDown}.$$

Algorithm 6 describes our double-hoisting BSGS for matrix-vector products.

Discussion. In addition to benefiting from our improved key-switch (Sect. 4.1) and rotation (Sect. 4.2) procedures, Algorithm 6 introduces a trade-off: The *ModDown* step in the *inner-loop* now depends on the value n_2, and the *ModDown* step of the *outer-loop* is performed only once. However, the $2n_1n_2$ multiplications and additions are performed in R_{PQ_ℓ} instead of R_{Q_ℓ}. Hence, the complexity dependency on n_1 is significantly reduced at the cost of slightly increasing the dependency on n_1n_2. Applying the *ModDown* step at the end of each loop has the additional benefit of introducing the rounding error only once.

Table 2 compares the complexity of a non-hoisted, single-hoisted (Algorithm 5) and double-hoisted (Algorithm 6) BSGS, each with its optimal ratio n_1/n_2. Our approach minimizes the complexity when $2^3 \le n_1/n_2 \le 2^4$.

Table 2. Complexity of Algorithm 5 [17], *1-hoisted* Algorithm 5 [19] and our *2-hoisted* Algorithm 6. M is a $2^{15} \times 2^{15}$ matrix with $n = n_1 n_2$ non zero diagonals. The used parameters are $N = 2^{16}$, $n = 2^{15}$, $\ell = 18$, $\alpha = 4$. The speed-up factor is the ratio between the $\#\mathsf{Mul}_{\mathbb{Z}_p}$, taking as baseline the 1-hoisted approach.

	No hoisting [17]			1-hoisted [19]		2-hoisted (proposed)		
n	n_1/n_2	$\log(\#\mathsf{Mul}_{\mathbb{Z}_p})$	Speed-up	n_1/n_2	$\log(\#\mathsf{Mul}_{\mathbb{Z}_p})$	n_1/n_2	$\log(\#\mathsf{Mul}_{\mathbb{Z}_p})$	Speed-up
32768	2	37.276	0.777×	2	36.913	8	36.813	1.071×
16384	1	36.500	0.765×	4	36.114	16	35.903	1.157×
8192	2	35.865	0.706×	2	35.364	8	35.055	1.238×
4096	1	35.152	0.705×	4	34.648	16	34.205	1.359×
2048	2	34.597	0.652×	2	33.981	8	33.446	1.448×
1024	1	33.927	0.664×	4	33.337	16	32.672	1.585×
512	2	33.422	0.619×	2	32.732	8	32.014	1.644×
256	1	32.769	0.645×	4	32.137	16	31.318	1.764×
128	2	32.282	0.609×	2	31.568	8	30.753	1.759×
64	1	31.614	0.649×	4	30.992	16	30.127	1.821×
32	2	31.112	0.623×	2	30.430	8	29.637	1.732×
16	1	30.375	0.682×	4	29.842	16	29.311	1.445×
8	2	29.792	0.685×	2	29.248	2	29.116	1.094×

This shows that the strategy of the previously proposed bootstrapping procedures [6,10,20], which minimize the number of rotations by setting $n_1 \approx n_2$, is not optimal anymore. The maximum gain occurs when n (the number of non zero diagonals) is around 128. This can be exploited by factorizing the linear transforms, used during the bootstrapping, into several sparse matrices (see Sect. 5.3).

Increasing the ratio from $n_2/n_1 \approx 1$ to $n_2/n_1 \approx 16$ in our bootstrapping parameters (Sect. 6) increases the number of keys by a factor around 1.6 and reduces the computation time by 20%. Hence, Algorithm 6 reduces the overall complexity of matrix-vector products, by introducing a time-memory trade-off.

We also observe that these improvements are not restricted to plaintext matrices or to the CKKS scheme and can be applied to other R-LWE scheme, such as BGV [5] or BFV [13], as long as the scheme (or its implementation) allows for the factorization of an expensive operation. For example, in the BFV scheme, the quantization (division by Q/t) (as well as the re-linearization if the matrix is in ciphertext) can be delayed to the *outer-loop*.

5 Bootstrapping for the Full-RNS CKKS Scheme

We present our improved bootstrapping procedure for the full-RNS variant of the CKKS scheme. We follow the high-level procedure of Cheon et al. [10] and adapt each step by relying on the techniques proposed in Sects. 3 and 4.

The purpose of the CKKS bootstrapping [10] is, in contrast with BFV's [13], not to reduce the error. Instead, and similarly to BGV [5] bootstrapping, it is meant to reset the ciphertext modulus to a higher level in order to enable further homomorphic multiplications. The approximate nature of CKKS, due to the plaintext and ciphertext error being mixed together, implies that each homomorphic operation decreases the output precision. As a result, all the currently proposed bootstrapping circuits only approximate the ideal bootstrapping operation, and their output precision also determines their practical utility.

5.1 Circuit Overview

Let $\{\mathsf{ct} = (c_0, c_1), Q_0, \Delta\}$ be a ciphertext that encrypts an n-slot message under a secret-key s with Hamming weight h, such that $\mathsf{Decrypt}(\mathsf{ct}, s) = c_0 + sc_1 = \lfloor \Delta \cdot m(Y) \rceil + e \in \mathbb{Z}[Y]/(Y^{2n}+1)$, where $Y = X^{N/2n}$. The bootstrapping operation outputs a ciphertext $\{\mathsf{ct}' = (c_0', c_1'), Q_{L-k}, \Delta\}$ such that $c_0' + sc_1' = \lfloor \Delta \cdot m(Y) \rceil + e' \in \mathbb{Z}[Y]/(Y^{2n} + 1)$, where $k < L$ is the number of levels consumed by the bootstrapping and $||e'|| \geq ||e||$ is the error that results from the combination of the initial error e and the error induced by the bootstrapping circuit.

The bootstrapping circuit is divided into the five steps detailed below. For the sake of conciseness, we describe the plaintext circuit and omit the error terms.

1. $\mathsf{ModRaise}$: ct is raised to the modulus Q_L by applying the CRT map $R_{q_0} \rightarrow R_{q_0} \times R_{q_1} \times \cdots \times R_{q_L}$. This yields a ciphertext $\{\mathsf{ct}, Q_L, \Delta\}$ for which
$$[c_0 + sc_1]_{Q_L} = Q_0 \cdot I(X) + \lfloor \Delta \cdot m(Y) \rceil = m',$$
where $Q_0 \cdot I(X) = \left[- [sc_1]_{Q_0} + sc_1 \right]_{Q_L}$ is an integer polynomial for which $||I(X)||$ is $\mathcal{O}(\sqrt{h})$ [10]. The next four steps remove this unwanted $Q_0 \cdot I(X)$ polynomial by homomorphically evaluating an approximate modular reduction by Q_0.
2. SubSum: If $2n \neq N$, then $Y \neq X$ and $I(X)$ is not a polynomial in Y. SubSum maps $Q_0 \cdot I(X) + \lfloor \Delta \cdot m(Y) \rceil$ to $(N/2n) \cdot (Q_0 \cdot \tilde{I}(Y) + \lfloor \Delta \cdot m(Y) \rceil)$, a polynomial in Y [10].
3. $\mathsf{CoeffsToSlots}$: The message $m' = Q_0 \cdot \tilde{I}(Y) + \lfloor \Delta \cdot m(Y) \rceil$ is in the *coefficient* domain, which prevents slot-wise evaluation of the modular reduction. This step homomorphically evaluates the inverse discrete-Fourier-transform (DFT) and produces a ciphertext encrypting $\mathsf{Encode}(m')$ that enables the slot-wise evaluation of the approximated modular reduction.
Remark: This step returns two ciphertexts, each encrypting $2n$ real values. If $4n \leq N$, these ciphertexts can be repacked into one. Otherwise, the next step is applied separately on both ciphertexts.
4. $\mathsf{EvalSine}$: The modular reduction $f(x) = x \bmod 1$ is homomorphically evaluated on the ciphertext(s) encrypting $\mathsf{Encode}(m')$. This function is approximated by $\dfrac{Q_0}{2\pi\Delta} \cdot \sin\left(\dfrac{2\pi\Delta x}{Q_0}\right)$, which is tight when $Q_0/\Delta \gg ||m(Y)||$. As the range of x is determined by $||\tilde{I}(Y)||$, the approximation needs to account for the secret-key density.
5. $\mathsf{SlotsToCoeffs}$: This step homomorphically evaluates the DFT on the ciphertext encrypting $f(\mathsf{Encode}(m'))$. It returns a ciphertext at level $L - k$ that

encrypts $\mathsf{Decode}(f(\mathsf{Encode}(m'))) \approx f(m') \approx \lfloor \Delta \cdot m(Y) \rceil$, which is a close approximation of the original message.

We now detail our approach for each step.

5.2 ModRaise and SubSum

We base the ModRaise and SubSum operations directly on the initial bootstrapping of Cheon et al. [10]. The SubSum step multiplies the encrypted message by a factor $N/2n$ that needs to be subsequently cancelled. We take advantage of the following CoeffsToSlot step, a linear transformation, to scale the corresponding matrices by $2n/N$. As we also use this trick for grouping other constants, we elaborate more on the matrices scaling in Sect. 5.5.

5.3 CoeffsToSlots and SlotsToCoeffs

Let n be a power-of-two integer such that $1 \leq n < N$; the following holds for any two vectors $\mathbf{m}, \mathbf{m}' \in \mathbb{C}^n$ due to the convolution property of the complex DFT

$$\mathsf{Decode}_n(\mathsf{Encode}_n(\mathbf{m}) \otimes \mathsf{Encode}_n(\mathbf{m}')) \approx \mathbf{m} \odot \mathbf{m}',$$

where $\otimes$ and $\odot$ respectively denote the nega-cyclic convolution and Hadamard multiplication. I.e., the encoding and decoding algorithms define an isomorphism between $\mathbb{R}[Y]/(Y^{2n}+1)$ and $\mathbb{C}^n$ [11]. The goal of the CoeffsToSlots and SlotsToCoeffs steps is to homomorphically evaluate this isomorphism on a ciphertext.

Let $\psi = e^{i\pi/n}$ be a $2n$-th primitive root of unity. As 5 and $-1 \bmod 2n$ span $\mathbb{Z}_{2n}$, $\{\psi^{5^k}, \overline{\psi^{5^k}}, 0 \leq k < n\}$ is the set of all $2n$-th primitive roots of unity. Given a polynomial $m(Y) \in \mathbb{R}[Y]/(Y^{2n}+1)$ with $Y = X^{N/2n}$, the decoding algorithm is defined as the evaluation of this polynomial at each root of unity $\mathsf{Decode}_n(m(Y)) = (m(\psi), m(\psi^5), \ldots, m(\psi^{5^{2n-1}}))$. The decoding isomorphism is fully defined by the $n \times n$ special Fourier transform matrix $\mathrm{SF}_{n,(j,k)} = \psi^{j5^k}$, with inverse (the encoding matrix) $\mathrm{SF}_n^{-1} = \frac{1}{n}\overline{\mathrm{SF}}_n^T$ [7]. Its homomorphic evaluation can be expressed in terms of plaintext matrix-vector products:

1. $\mathsf{CoeffsToSlots}(\mathbf{m}) : t_0 = \frac{1}{2}(\mathrm{SF}_n^{-1} \times \mathbf{m} + \overline{\mathrm{SF}_n^{-1} \times \mathbf{m}}), t_1 = -\frac{1}{2}i(\mathrm{SF}_n^{-1} \times \mathbf{m} - \overline{\mathrm{SF}_n^{-1} \times \mathbf{m}})$
2. $\mathsf{SlotsToCoeffs}(t_0, t_1) : \mathbf{m} = \mathrm{SF}_n \times (t_0 + i \cdot t_1)$.

DFT Evaluation. In their initial bootstrapping proposal, Cheon et al. [10] homomorphically compute the DFT as a single matrix-vector product in $\mathcal{O}(\sqrt{n})$ rotations and depth 1, by using the baby-step giant-step (BSGS) approach of Halevi and Shoup [18] (Algorithm 5 in Sect. 4.3). To further reduce the complexity, two recent works from Cheon et al. [7] and Chen et al. [6] exploit the structure of the equivalent FFT algorithm by recursively merging its iterations,

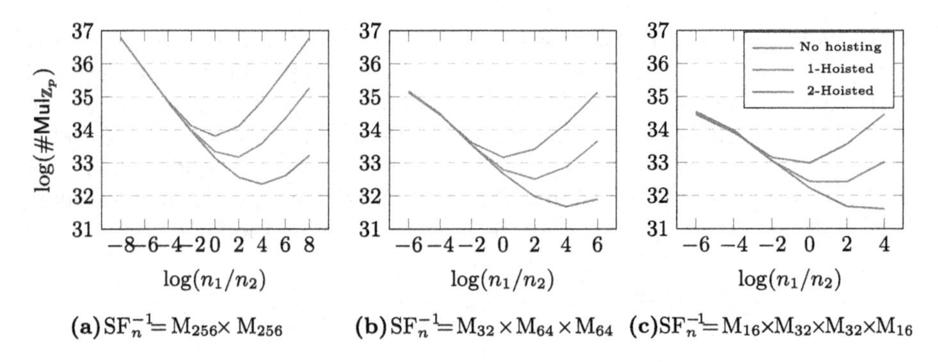

Fig. 2. Theoretical complexity of CoeffToSlots for different $\rho_{\mathrm{SF}_n^{-1}}$ using Algorithm 5 with no hoisting, single hoisting, and double hoisting (Algorithm 6).

reducing the complexity to $\mathcal{O}(\sqrt{r}\log_r(n))$ rotations at the cost of increasing the depth to $\mathcal{O}(\log_r(n))$, for r a power-of-two radix between 2 and n.

We base our approach on the work of [7] and [6], and we use our *double hoisting* BSGS to evaluate the matrix-vector products (see Sect. 4.3 and Algorithm 6). This step is parameterized by $\rho = \lceil \log_r(n) \rceil$, the depth of the linear transformation (i.e., the number of matrices that we need to evaluate).

Figure 2 shows the effect of our algorithm on the CoeffsToSlots step, compared with the original BSGS algorithm for $\rho_{\mathrm{SF}_n^{-1}} = \{2, 3, 4\}$. The complexity is computed as the number of products in $\mathbb{Z}_p$, with parameters $N = 2^{16}$, a target $\ell = 17$ (the level after CoeffsToSlots) and $n = 2^{15}$ slots.

Each level of hoisting reduces the total complexity by a noticeable amount. Regular *hoisting*, as proposed by Halevi and Shoup [19], achieves its minimum complexity when $n_1 \approx 2^2 n_2$ instead of $n_1 \approx n_2$. Using our *double hoisting*, the minimum complexity is further shifted to $n_1 \approx 2^4 n_2$. On average, our method reduces the complexity of the linear transformations in the bootstrapping by a factor of $2\times$ compared to the single *hoisting* technique of Halevi and Shoup.

Efficient Repacking of Sparse Plaintexts. The first part of CoeffsToSlots is a DFT that outputs a vector of $\mathbb{C}^n$ values; the second part of CoeffsToSlots applies the map $\mathbb{C}^n \rightarrow \mathbb{R}^{2n}$ to this vector. During the decoding, the inverse mapping $\mathbb{R}^{2n} \rightarrow \mathbb{C}^n$ is used. This map can be computed with simple operations, e.g., conjugation, multiplication by $-i$, and additions. If the original ciphertext is not fully packed ($0 < n < N/2$ slots), the two resulting ciphertexts can be merged into one, requiring one evaluation of EvalSine instead of two.

We observe that decoding a plaintext $\mathbf{m} \in \mathbb{C}^n$ by using the decoding algorithm for a plaintext of $\mathbb{C}^{2^k n}$ slots (assuming that $2^k n < N$) outputs a vector comprising 2^k concatenated replicas of $\mathbf{m}$. Therefore, a ciphertext that encrypts $\mathbf{m} \in \mathbb{C}^n$ can also be seen as a ciphertext encrypting $\mathbf{m}' \in \mathbb{C}^{2n}$ for $\mathbf{m}' = \mathbf{m}||\mathbf{m}$. This property can be used to save two levels when repacking and unpacking ciphertexts before and after the EvalSine:

- **Repacking before the EvalSine** ($\mathbb{C}^n \rightarrow \mathbb{R}^{2n}$): Repacking into one ciphertext is done by extending the domain of the plaintext vectors of the last matrix of the CoeffsToSlots step from $\mathbb{C}^n$ to $\mathbb{C}^n||0^n$. Thus, the last n slots are set to zero and can be used to store the imaginary part of the first n slots. This repacking involves one additional rotation and it does not consume any additional levels.
- **Unpacking after the EvalSine** ($\mathbb{R}^{2n} \rightarrow \mathbb{C}^n$): For this operation, we evaluate the following $2n \times 2n$ matrix on the ciphertext before the DFT

$$\begin{bmatrix} I_n & i \cdot I_n \\ I_n & i \cdot I_n \end{bmatrix},$$

where I_n is the $n \times n$ identity matrix. Its effect is to homomorphically apply the map $\mathbb{R}^{2n} \rightarrow \mathbb{C}^n||\mathbb{C}^n$, which is a valid encoding of $\mathbb{C}^n$, due to the properties of the encoding algorithm. This additional matrix (transformation) is combined with the first group of the SlotsToCoeffs matrices, thus slightly increasing its density.

5.4 EvalSine

EvalSine implements the homomorphic modular reduction of the message $m' = Q_0 \cdot \tilde{I}(Y) + \Delta \cdot m(Y)$ modulo Q_0. The modular reduction is approximated by

$$f(x) = \frac{Q_0}{\Delta}\frac{1}{2\pi}\sin\left(2\pi x\frac{\Delta}{Q_0}\right) \approx \frac{Q_0}{\Delta} \cdot \left(\frac{\Delta}{Q_0}x \bmod 1\right),$$

which scales the message m' down to $\tilde{I}(Y) + (\Delta/Q_0) \cdot m(Y)$, removes the $\tilde{I}(Y)$ polynomial by reducing the message modulo 1, and scales the message back to $\Delta \cdot m(Y)$. As $\tilde{I}(Y)$ determines the range and degree of the approximation, the EvalSine step has to account for the secret-key density h. In particular, the range of the approximation $(-K, K)$ is chosen such that $\Pr[||\tilde{I}(Y)|| > K] \leq \kappa$ for a user-defined κ. We elaborate more on how we parameterize K, in Sect. 6.2.

Previous Work. Chen et al. [6] directly approximate the function $\frac{1}{2\pi} \cdot \sin(2\pi x)$ by using a standard Chebyshev interpolant of degree $d = 119$ in an interval of $(-K, K)$ for $K = 12$ (using a sparse key with $h = 64$). Han and Ki [20] approximate $\cos(2\pi\frac{1}{2^r}(x - 0.25))$ followed by r iterations of the double angle formula $\cos(2x) = 2\cos(x)^2 - 1$ to obtain $\sin(2\pi x)$. The factor $1/2^r$ reduces the range of the approximation to $(-K/2^r, K/2^r)$, enabling the use of a smaller-degree interpolant. They combine it with a specialized Chebyshev interpolation that places the node around the expected intervals of the input. This reduces the degree of the approximation and the cost of its evaluation. In their work, they use an interpolant of degree 30 with a scaling factor $r = 2$ (they also use a sparse key with $h = 64$).

In a recent work, Lee et al. [25] propose to compose the sine/cosine function with a low degree arcsine. This additional step corrects the error introduced by the sine, especially if Q_0/Δ is small (when the values are not close to the origin). This improves the overall precision of the bootstrapping and enables bootstrapping messages with larger values. However, this comes at the cost of increasing the depth of the EvalSine step, as a second polynomial must be evaluated.

Our Work. Both the methods of Chen et al. and Han and Ki have $d = \mathcal{O}(K)$, therefore doubling K requires at most doubling d, and the evaluation will require at most one additional level, as the Chebyshev interpolant can be evaluated in $\mathcal{O}(\log(K))$ levels. Hence, precision put aside, the level consumption should not be a fundamental problem when evaluating the large degree interpolant (as required by dense keys). However, the effects of the approximate rescale procedure, if not properly managed, can significantly reduce the output precision. Our EvalSine makes use of our novel polynomial evaluation technique (Sect. 3).

We propose a more compact expression of the modular reduction function $f(x) = \frac{1}{2\pi}\sin(2\pi x)$, which is approximated by $g_r(x)$, a modified scaled cosine followed by r iterations of the double-angle formula:

$$g_0(x) = \frac{1}{\sqrt[2^r]{2\pi}} \cos\left(2\pi \tfrac{1}{2^r}(x - 0.25)\right) \text{ and } g_{i+1} = 2g_i^2 - \left(\frac{1}{\sqrt[2^r]{2\pi}}\right)^{2^i}.$$

We include the $1/2\pi$ factor directly in the function we approximate, even when using the double angle formula, without consuming an additional level, impacting the precision, or fundamentally changing its evaluation. We observed that even though the approximation technique of Han and Ki is well suited for small K, the standard Chebyshev interpolation technique, as used by Chen et al., remains more efficient when K is large. The reason is that Han and Ki's interpolant has a minimum degree of $2K - 1$, so it grows in degree with respect to K much faster than the standard Chebyshev interpolation. Hence, we use the approximation method of Han and Ki when K is small (for sparse keys) and the standard Chebyshev approximation, as done by Chen et al., for dense keys.

As suggested by Lee et al. [25], we can further improve this step by composing it with $\arcsin(x)$, i.e., $\frac{1}{2\pi}\arcsin(\sin(2\pi x))$, which corrects the error $e_{g_r(x)} = |g_r(x) - x \bmod 1|$. Unlike Lee et al., we do not interpolate the arcsine, rather we choose to use a low degree Taylor polynomial and show in our results (see Sect. 7.2) that it is sufficient to achieve similar results.

Algorithm 7 details our implementation of the EvalSine procedure. The ciphertext must be multiplied by several constants, before and after the polynomial evaluation. For efficiency, we merge these constants with the linear transformations. See Sect. 5.5 for further details.

Algorithm 7: EvalSine

```
   Input: {ct, Q_ℓ, Δ} a ciphertext, p(t) a Chebyshev interpolant of degree d of
          f(x) = x mod 1, K the range of interpolation, r a scaling factor.
   Output: The evaluation ct' = ⌊Q_0/Δ⌉ · p(⌊Q_0/Δ⌉^{-1} · ct).
 1 Δ ← Δ · ⌊Q_0/Δ⌉ // Division by ⌊Q_0/Δ⌉
 2 T_0 ← 1
 3 T_1 ← AddConst(ct, −0.5/(2^{r+1}K))
 4 m ← ⌈log(d + 1)⌉
 5 l ← ⌊m/2⌋
 6 T ← {T_0, T_1, ..., T_{2^l}; T_{2^l+1}, ..., T_{2^m−1}} // Compute the power basis
 7 for i = 0; i < r; i = i + 1 do
 8 |  Δ ← sqrt(Δ · q_{L−CtS depth−EvalSine depth−r+i}) // Pre-compute target Δ
 9 end
10 ct' ← EvalRecurse(Δ, m, l, p(t), T) (Algorithm 2) // Outputs ct' with target
       Δ scale
11 for i = 0; i < r; i = i + 1 do
12 |  ct' ← AddConst(2 · Mul(ct', ct'), −(1/2π)^{1/2^{r−i}})
13 |  ct' ← Rescale(ct') // Δ ← Δ^2/q_{L−CtS depth−EvalSine depth−i}
14 end
15 Δ ← Δ · ⌊Q_0/Δ⌉^{-1} // Multiplication by ⌊Q_0/Δ⌉
16 return ct'
```

5.5 Matrix Scaling

Several steps of the bootstrapping circuit require the ciphertexts to be multiplied by constant plaintext values. This is most efficiently done by merging them and pre-multiplying the resulting constants to the SF_n^{-1} and SF_n matrices.

Before EvalSine, the ciphertext has to be multiplied (i) by $1/N$ to cancel the $N/2n$ and $2n$ factors introduced by the SubSum and CoeffsToSlots steps, (ii) by $1/(2^r K)$ for the scaling by $1/2^r$ and change of variable for the polynomial evaluation in Chebyshev basis, and (iii) by $Q_0/2^{\lceil \log(Q_0) \rceil}$ to compensate for the error introduced by the approximate division by $\lfloor Q_0/\Delta \rceil$. Therefore, the matrices resulting from the factorization of SF_n^{-1} are scaled by

$$\mu_{\mathsf{CtS}} = \left(\frac{1}{2^r KN} \cdot \frac{Q_0}{2^{\lfloor \log(Q_0) \rceil}} \right)^{\frac{1}{\rho_{\mathrm{SF}_n^{-1}}}},$$

where $\rho_{\mathrm{SF}_n^{-1}}$ is the degree of factorization of SF_n^{-1}. Evenly spreading the scaling factors across all matrices ensures that they are scaled by a value as close as possible to 1.

After EvalSine, the ciphertext has to be multiplied (i) by $2^{\lceil \log(q_0) \rceil}/Q_0$ to compensate for the error introduced by the approximate multiplication by $\lfloor Q_0/\Delta \rceil$, and (ii) by Δ/δ, where Δ is the scale of the ciphertext after the EvalSine step and δ is the desired ciphertext output scale. Therefore, the matrices resulting

Table 3. Modulus size $\log(QP)$ for different secret-key densities h ($\lambda \geq 128$).

h	$\log(QP)$		
	$\log(QP, N), \lambda \geq 128$	$N = 2^{15}$	$N = 2^{16}$
64	$0.015121N - 8.248756$	496	982
96	$0.018896N - 3.671642$	619	1234
128	$0.021370N - 3.601990$	699	1396
192	$0.023448N - 3.611940$	767	1533
$N/2$	[12]	881	1782

from the factorization of SF_n are scaled by

$$\mu_{\mathsf{StC}} = \left(\frac{\Delta}{\delta} \cdot \frac{2^{\lfloor \log(Q_0) \rceil}}{Q_0}\right)^{\frac{1}{\rho_{\mathrm{SF}_n}}},$$

where ρ_{SF_n} is the degree of factorization of SF_n.

6 Parameter Selection

A proper parameterization is paramount to the security and correctness of the bootstrapping procedure. Whereas security is based on traditional hardness assumptions, setting the correctness-related parameters is accomplished mostly through experimental processes for finding appropriate trade-offs between the performance and the probability of decryption errors. In this Section, we discuss various constraints and inter-dependencies in the parameter selection. Then, we propose a generic procedure for finding appropriate parameter sets.

6.1 Security

For each parameter set, we select a modulus size with an estimated security of 128 bits. These values are shown in Table 3 for several choices of the secret-key Hamming weight h, and are based on the work of Curtis and Player [12]. According to the authors, these parameters result from conservative estimations, and account for hypothetical future improvements to the most recent attacks of Cheon et al. [9] and Son et al. [29]. Therefore, their actual security is underestimated.

6.2 Choosing K for EvalSine

Each coefficient of the polynomial $\tilde{I}(Y) \in \mathbb{R}[Y]/(Y^{2n}+1)$ is the result of the sum of $h+1$ uniformly distributed variables in $\mathbb{Z}_{Q_0}$ [10], hence it follows an Irwin–Hall distribution [25]. By centering and normalizing the coefficients of $\tilde{I}(Y)$, we get instead the sum of $h+1$ uniformly distributed variables in $(-0.5, 0.5)$. The probability $\Pr[||\tilde{I}(Y)|| > K]$ can be computed by adapting the cumulative probability function of the Irwin-Hall distribution:

Table 4. $\Pr[||\tilde{I}(Y)|| > K] \approx 2^{-16}$ for $n = 2^{15}$ and variable h.

$\log_2(h)$	6	7	8	9	10	11	12	13	14	15
K	14	20	29	41	58	82	116	163	232	328
$\log_2(\Pr[\lVert\tilde{I}(Y)\rVert > K])$	-14.6	-14.6	-15.7	-15.6	-15.5	-15.4	-15.4	-15.4	-15.4	-15.4
$K/\sqrt{h}$	1.75	1.76	1.81	1.81	1.81	1.81	1.81	1.81	1.81	1.81

$$1 - \left(\left(\frac{2}{(h+1)!}\sum_{i=0}^{\lfloor K+0.5(h+1)\rfloor}(-1)^i\binom{h+1}{i}(K+0.5(h+1)-i)^{h+1}\right)-1\right)^{2n}. \tag{1}$$

The previous works [6,7,10,20] use a sparse key with $h = 64$ and $K = 12$, which regardless of the security, gives a failure probability of $2^{-14.7}$ and $2^{-6.7}$ for $n = 2^7$ and $n = 2^{15}$ respectively, according to Eq. (1). Clearly, these parameters were not chosen for large n and are most likely an artifact of the first proposal for a bootstrapping for CKKS [10], for which only a small number of slots was practical. In our work, we increase h to ensure an appropriate security and use a much larger number of slots (e.g., $n = 2^{15}$), hence we need to adapt K. Table 4 shows that if we target a failure probability $\leq 2^{-15.0}$ for $n = 2^{15}$ slots and take h as a parameter, then $K \approx 1.81\sqrt{h}$.

6.3 Finding Parameters

We describe a general heuristic procedure for selecting and fine-tuning bootstrapping parameters. Each operation of the bootstrapping requires a different scaling and a different precision, therefore different moduli. Choosing each modulus optimally for each operation not only leads to a better performance and a better final precision but also optimizes the bit consumption of each operation and increases the remaining homomorphic capacity after the bootstrapping.

We describe our procedure to find suitable parameters for the bootstrapping in Algorithm 8 and propose five reference parameter sets that result from this algorithm. The parameter sets were selected for their performance and similarity with those in previous works, thus enabling a comparison. For each set, Table 5 shows the parameters related to CKKS and to the bootstrapping circuit.

7 Evaluation

We implemented the improved algorithm of Sects. 3 and 4, along with the bootstrapping procedure of Sect. 5 in the Lattigo library [24]. We evaluated it by using the parameters of Sect. 6.3. Lattigo is an open-source library that implements the RNS variants of the BFV [3,13,16] and CKKS [8] schemes in Golang [30]. All experiments were conducted single-threaded on an i5-6600k at 3.5 GHz with 32 GB of RAM running Windows 10 (Go version 1.15.6, GOARCH = amd64, GOOS = windows).

Table 5. The sets of parameters of the full-RNS CKKS used to evaluate the performance of our bootstrapping code. + means concatenation in the chain and $a \cdot b$ denotes the consecutive concatenation of a different moduli of size b. Moduli with fractional a are only partially used by the step they are allocated to.

Parameters										
Set	h	N	Δ	$\log(QP)$	L	$\log(q_i)$				$\log(p_j)$
						$q_{0\leq i\leq(L-k)}$	StC	Sine	CtS	
I	192	2^{16}	2^{40}	1546	25	$60+9\cdot 40$	$3\cdot 39$	$8\cdot 60$	$4\cdot 56$	$5\cdot 61$
II	192		2^{45}	1547	24	$60+5\cdot 45$	$3\cdot 42$	$11\cdot 60$	$4\cdot 58$	$4\cdot 61$
III	192		2^{30}	1553	21	$55+7.5\cdot 60$	$1.5\cdot 60$	$8\cdot 55$	$4\cdot 53$	$5\cdot 61$
IV	32768		2^{45}	1792	28	$50+9\cdot 40$	$56+28$	$12\cdot 60$	$4\cdot 53$	$6\cdot 61$
V	192	2^{15}	2^{25}	768	14	$33+50+25$	60	$8\cdot 50$	$2\cdot 49$	$2\cdot 50$

7.1 The Bootstrapping Metrics

Although CPU costs are an important aspect when evaluating a bootstrapping procedure, these factors have to be considered together with other performance-related metrics such as the size of the output plaintext space, the failure probability, the precision, and the remaining multiplicative depth. To compare our bootstrapping procedure with the existing ones, we use the same concept of a *bootstrapping utility metric*, as introduced by Han and Ki [6].

Definition 1 (Bootstrapping Throughput). *For n a number of plaintext slots, $\log(\epsilon^{-1})$ the output precision, $\log(Q_{L-k})$ the output coefficient-modulus size after the bootstrapping (remaining homomorphic capacity) and complexity a measure of the computational cost (in CPU time), the bootstrapping throughput is defined as:*

$$\textit{throughput} = \frac{n \times \log(\epsilon^{-1}) \times \log(Q_{L-k})}{\textit{complexity}}.$$

Note that we express the remaining homomorphic capacity in terms of the modulus size, instead of the number of levels, because Q_{L-k} can be re-allocated differently at each bootstrapping call, e.g., a small number of moduli with a large plaintext scale or a large number of moduli with a small plaintext scale.

As κ, the bootstrapping failure probability, is a probability and not a metric, we chose to not include it directly in Definition 1. However, we still believe it should be taken into account as an opportunity-cost variable. Indeed, the event of a bootstrapping failure will likely result in the need to re-run the entire circuit. Hence, the probability of failure should be weighed vs. the cost of having to re-run a circuit to determine if κ is in an acceptable range.

7.2 Results

We run our benchmarks and report the bootstrapping performance for each parameter set of Table 5, and we compare them with the previous works of Chen et al. [6], Han and Ki [20], and the recent and concurrent work of Lee et al.

Algorithm 8: Heuristic Parameter Selection

Input: λ a security parameter.

Output: The parameters $(N, n, h, Q_L, P, \kappa, \alpha, d_{\text{sin}}, r, d_{\text{arcsin}}, \rho_{\text{SF}_n^{-1}}, \rho_{\text{SF}_n})$.

1 Select n, N and h and derive $\log(PQ_L)$ according to λ.

2 Given Δ (the scale of the message), compute the ratio Q_0/Δ and select the bootstrapping output precision δ.

3 Given a target failure probability κ, estimate K using Equation (1).

4 Given the bootstrapping output precision δ, find d_{sin} (the degree of the sine polynomial), r (the number of double angle) and d_{arcsin} (the degree of the arcsine polynomial) such that the polynomial approximation of x mod 1 of the EvalSine step in the interval $(-K/2^r, K/2^r)$ gives a precision greater than $\log(Q_0/\Delta) + \delta$ bits.

5 Select $\rho_{\text{SF}_n^{-1}}$ and ρ_{SF_n} (the depth of the CoeffsToSlots and SlotsToCoeffs steps).

6 Allocate the q_j of the CoeffsToSlots, EvalSine and SlotsToCoeffs steps, with the maximum possible bit-size for all q_j.

7 Select α and allocate $P = \prod_{j=0}^{\alpha-1} p_j$, ensuring that $P \approx \beta ||q_{\alpha_i}||$.

8 Run the bootstrapping and find the minimum values for d_{sin}, r and d_{arcsin} such that the output has δ bits of precision.

9 Run the bootstrapping and find the minimum bit-size for the q_j of the EvalSine such that the output reaches the desired precision or until it plateaus.

10 Run the bootstrapping and find the minimum bit-size for the q_j of the CoeffsToSlots such that the output precision is not affected.

11 Run the bootstrapping and find the minimum bit-size for the q_j of the SlotsToCoeffs such that the output precision is not affected.

12 Allocate the rest of the moduli of Q_L such that $\log(PQ_L)$ ensures a security of at least λ and check again step 7.

13 If additional residual homomorphic capacity is needed or the security λ cannot be achieved

1. Reduce α, $\rho_{\text{SF}_n^{-1}}$ and/or ρ_{SF_n} and check again line 6.
2. Increase h to increase $\log(PQ_L)$ and restart at line 1.
3. Increase N to increase $\log(PQ_L)$ and restart at line 1.

return $(N, n, h, Q_L, P, \kappa, \alpha, d_{\text{sin}}, r, d_{\text{arcsin}}, \rho_{\text{SF}_n^{-1}}, \rho_{\text{SF}_n})$

[25]. Unfortunately, the implementations of these works have not been publicly released and we were not able to reproduce their results on our own hardware for a fair comparison. The parameters and results are summarized in Table 6 and 7, respectively. Reports on experiments that demonstrate the numerical stability of our bootstrapping can be found in the extended version of the paper [4].

Focusing only on the overall performance, our most performing set (Set III) achieves throughput 14.1× and 28.4× larger than the best result reported by Han and Ki [20] and Lee et al. [25] respectively. Our Set IV uses dense keys and achieves a throughput 4.6× and 9× larger than the work of Han and Ki and Lee et al. respectively. Both these works use SEAL [28] and are evaluated on similar hardware. Our sets III and IV achieve a throughput 54.2× and 17.4× larger than the best result reported by Chen et al. [6], implemented using the

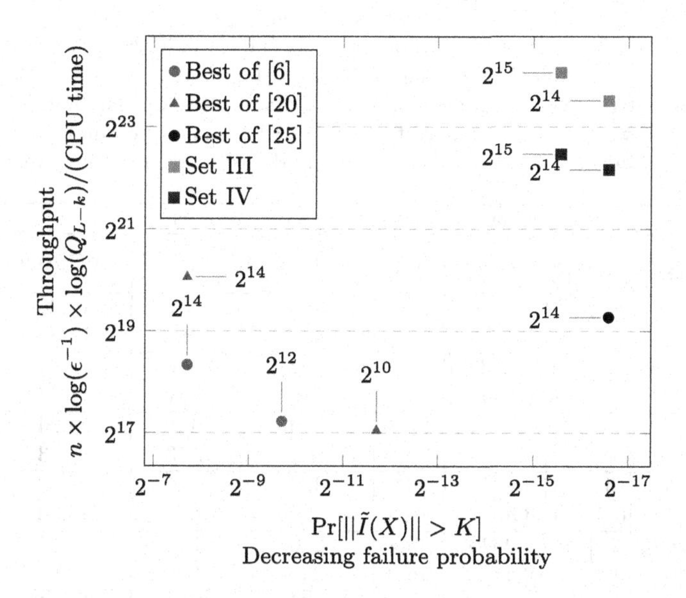

Fig. 3. Bootstrapping throughput comparison. We plot the results for our best performing parameter set against the state of the art. Nodes are labeled with n, the number of plaintext slots.

Table 6. Parameter comparison of [6,20,25] and our work. "–" means that value was not reported. Lee et al.'s [25] parameters are based on our Set III.

Bootstrapping Parameters											
Set	N	$\log(QP)$	h	λ	$\rho_{\text{SF}_n^{-1}}$	ρ_{SF_n}	Q_0/Δ	K	$d_{\sin(x)}$	r	$d_{\arcsin(x)}$
[25]	2^{16}	1553	192	≈ 128	2	2	256	25	66	2	0
							8		68		5
[6]		2480	64	< 80	4	4	1024	12	119	0	0
[20]		1452	64	< 100	-	-	1024	12	31	2	0
I		1546	192	≈ 128	4	3	256	25	63	2	0
II		1547	192	≈ 128	4	3	256	25	63	2	0
III		1553	192	≈ 128	4	3	4	25	63	2	7
IV		1792	32768	≈ 128	4	3	256	325	255	4	0
[6]	2^{15}	1240	64	< 80	2	2	1024	12	119	0	0
[20]		910	64	< 90	-	-	1024	12	31	2	0
V		768	192	≈ 128	2	2	256	25	63	2	0

HEAAN library [21]. HEAAN does not implement the full-RNS variant of CKKS, hence the latter comparison shows the significant performance gains that can be achieved by combining optimized algorithms with a full-RNS implementation.

The implementation of Lee et al. makes use of the recent work of Kim et al. [22] which proposes new techniques to minimize the error during computation, notably a *delayed rescaling* that consists in rescaling the ciphertext before a multiplication and not after, so that the error is as small as possible when doing

Table 7. Comparison of the bootstrapping performances of [6,20,25] and our proposed bootstrapping for the full-RNS variant of CKKS with parameter sets I, II, III, IV and V. MU, SS, CtS, StC designate ModUp, SubSum, SlotstoCoeffs, CoeffstoSlots. "–" indicates that the prior work did not report the value. All timings are single threaded. The plaintext real and imaginary part are uniformly distributed in the interval −1 and 1.

Bootstrapping Performances												
Set	n	Timing(s)						$\log(Q_{L-k})$	$\log(\epsilon^{-1})$	$\log(\text{bits}/s)$	$\log(\kappa)$
		MU	SS	CtS	StC	Sine	**Total**				
[25]	2^{14}	-	-	-	-	-	461.5	**653**	27.2	19.26	-16.58
[25]	2^{14}	-	-	-	-	-	451.5	533	**32.6**	19.26	-16.58
[6]	2^{14}	119.8				38.5	158.3	172	18.6	18.33	-7.70
[6]	2^{12}	127.5				40.4	167.9	301	20.9	17.22	-9.70
[20]	2^{14}	-	-	-	-	-	52.8	370	10.8	20.24	-7.70
[20]	2^{10}	-	-	-	-	-	37.6	370	15.3	17.23	-11.70
I	2^{15}	0.06	0	6.5	3.7	12.8	23.0	420	25.7	23.87	-15.58
I	2^{14}	0.06	0.3	6.3	3.8	6.3	16.9	420	26.0	23.33	-16.58
II	2^{15}	0.06	0	6.8	2.2	14.2	23.4	240	31.5	23.33	-15.58
II	2^{14}	0.06	0.3	6.0	2.4	7.1	16.0	240	31.6	22.88	-16.58
III	2^{15}	0.06	0	5.4	2.4	10.1	**18.1**	505	19.1	**24.06**	-15.58
III	2^{14}	0.06	0.3	5.0	2.6	5.0	13.1	505	18.9	23.50	-16.58
IV	2^{15}	0.07	0	7.9	28.2	3.0	39.2	410	16.8	22.45	-14.90
IV	2^{14}	0.07	0.4	7.1	14.1	3.2	24.9	410	17.3	22.15	-15.90
[6]	2^{10}	28.8				9.5	38.3	150	6.9	14.75	-11.70
[6]	2^{8}	16.9				9.2	26.0	75	10.03	12.85	-13.70
[20]	2^{2}	-	-	-	-	-	7.5	185	15.0	10.53	-19.70
[20]	2^{1}	-	-	-	-	-	7.0	185	16.8	9.79	-20.70
V	2^{14}	0.02	0	3.7	0.7	2.9	7.5	110	15.5	21.82	-16.58
V	2^{13}	0.02	0.4	1.6	0.4	1.5	3.9	110	15.4	21.76	-17.58

the multiplication. This enables Lee et al. to achieve a slightly higher precision than ours (our implementation does not use the work of Kim et al.). Lee et al. results are also the ones with the most residual homomorphic capacity. The primary reason is the implementation of the CKKS scheme in SEAL, which can only use one special prime ($\alpha = 1$, see Sect. 4) during the key-switching. This increases the ciphertext homomorphic capacity, but at the cost of an increased key-switch complexity. The second reason is that they allocate less levels to the linear transformations (in total, three less than our parameters). This enables them to reduce the depth of the bootstrapping, at the cost of increasing its complexity, which shows in their timings.

We observe that there is a correlation between the value Q_0/Δ and the precision. A better precision is achieved when using a smaller ratio, even when the arcsin is not composed with the scaled sine. Previous works usually assume that $||m|| \approx ||\text{FFT}^{-1}(m)||$ to set Q_0/Δ and derive the expected precision of the scaled sine. In practice, since each coefficient of $\text{FFT}^{-1}(m)$ is a dot product between the vector m and a complex vector of roots of unity (zero-mean and small variance), if the mean of m is close to zero, then $||\text{FFT}^{-1}(m)|| \ll ||m||$ with overwhelming probability. For example, given m uniform in $(-1, 1)$ and $n = 2^{15}$ slots, then $||m||/||\text{FFT}^{-1}(m)|| \approx 100$. Hence the message is much closer to the origin than expected, which reduces the inherent error of the scaled sine and amplifies the effectiveness of the arcsine. We note that even if the distribution of m is not known, it is possible to enforce this behavior with a single plaintext multiplication by homomorphically negating half of its coefficients before the bootstrapping. One could even homomorphically split m in half and create two symmetric vectors to enforce a zero mean. A more detailed analysis of this behavior and how to efficiently exploit it or integrate it into the linear transforms of the bootstrapping could be a interesting future research line.

All our sets have a failure probability that is two to three orders magnitude smaller than previous works, except for the results of Lee et al. which use our suggested parameters. For example, following Eq. (1), if successive bootstrappings are carried out with $n = 2^{15}$ slots, then [6] and [20] would reach a 1/2 failure probability after 52 bootstrappings, whereas ours would reach the same probability after 24,656 bootstrappings.

Figure 3 plots the best performing instances of Table 7.

8 Conclusion

In this work, we have introduced a secure, reliable, precise and efficient bootstrapping procedure for the full-RNS CKKS scheme that does not require the use of sparse secret-keys. To the best of our knowledge, this is the first reported instance of a practical bootstrapping parameterized for at least 128-bit security.

To achieve this, we have proposed a generic algorithm for the homomorphic evaluation of polynomials with reduced error and optimal in level consumption. In addition to the increase in precision and efficiency, our algorithm also improves the usability of the full-RNS variant of CKKS (for which managing a changing scale in large circuits is known to be a difficult task).

We have also proposed an improved key-switch format that we apply to the homomorphic matrix-vector multiplication. Our novel *double hoisting* algorithm reduces the complexity of the CoeffsToSlots and SlotsToCoeffs by roughly a factor of 2 compared to previous works. The performance gain for these procedures enables their use outside of the bootstrapping, for applications where the conversion between coefficient- and slot-domains would enable much more efficient homomorphic circuits (e.g., in the training of convolutional neural networks or R-LWE to LWE ciphertext conversion).

We have also proposed a systematic approach to parameterize the bootstrapping, including a way to precisely assess its failure probability. We have evaluated our bootstrapping procedure and have shown that its throughput with "dense" secret-keys ($h = N/2$) is up to 4.6× larger than the best state-of-the-art results with sparse keys ($h = 64$). When the sparse-keys-adjusted parameters of Curtis and Player [12] for $h = 192$ and 128-bits of security are considered, our procedure has a 14.1× larger throughput than the previous work that uses a sparse key with $h = 64$ with insecure parameters. Additionally, all our parameters lead to a more reliable instance than the previous works, with a failure probability orders of magnitude lower.

We have implemented our contributions in the Lattigo library [24]. This is, to the best of our knowledge, the first open-source implementation of a bootstrapping procedure for the full-RNS variant of the CKKS scheme.

Acknowledgments. We would like to thank Anamaria Costache, Mariya Georgieva and the anonymous reviewers for their valuable feedback. We also thank Lee et al. (authors of [25]) for the insightful discussions. This work was supported in part by the grant #2017-201 of the ETH Domain PHRT Strategic Focal Area.

References

1. Albrecht, M.R., Player, R., Scott, S.: On the concrete hardness of learning with errors. J. Math. Cryptol. **9**(3), 169–203 (2015)
2. Albrecht, M., et al.: Homomorphic encryption security standard. Technical report, HomomorphicEncryption.org, Toronto, Canada, November 2018
3. Bajard, J.-C., Eynard, J., Hasan, M.A., Zucca, V.: A full RNS variant of FV like somewhat homomorphic encryption schemes. In: Avanzi, R., Heys, H. (eds.) SAC 2016. LNCS, vol. 10532, pp. 423–442. Springer, Cham (2017). https://doi.org/10.1007/978-3-319-69453-5_23
4. Bossuat, J.-P., et al.: Efficient Bootstrapping for Approximate Homomorphic Encryption with Non-Sparse Keys. Cryptology ePrint Archive, Report 2020/1203 (2020). https://eprint.iacr.org/2020/1203
5. Brakerski, Z., Gentry, C., Vaikuntanathan, V.: (Leveled) fully homomorphic encryption without bootstrapping. ACM Trans. Comput. Theory (TOCT) **6**(3), 1–36 (2014)
6. Chen, H., Chillotti, I., Song, Y.: Improved bootstrapping for approximate homomorphic encryption. In: Ishai, Y., Rijmen, V. (eds.) EUROCRYPT 2019. LNCS, vol. 11477, pp. 34–54. Springer, Cham (2019). https://doi.org/10.1007/978-3-030-17656-3_2
7. Cheon, J.H., Han, K., Hhan, M.: Faster Homomorphic Discrete Fourier Transforms and Improved FHE Bootstrapping. IACR Cryptology ePrint Archive 2018/1073 (2018)
8. Cheon, J.H., Han, K., Kim, A., Kim, M., Song, Y.: A full RNS variant of approximate homomorphic encryption. In: Cid, C., Jacobson, M. (eds.) SAC 2018. LNCS, vol. 11349, pp. 347–368. Springer, Cham (2018). https://doi.org/10.1007/978-3-030-10970-7_16
9. Cheon, J.H., et al.: A hybrid of dual and meet-in-the-middle attack on sparse and ternary secret LWE. IEEE Access **7**, 89497–89506 (2019)
10. Cheon, J.H., Han, K., Kim, A., Kim, M., Song, Y.: Bootstrapping for approximate homomorphic encryption. In: Nielsen, J.B., Rijmen, V. (eds.) EUROCRYPT 2018. LNCS, vol. 10820, pp. 360–384. Springer, Cham (2018). https://doi.org/10.1007/978-3-319-78381-9_14
11. Cheon, J.H., Kim, A., Kim, M., Song, Y.: Homomorphic encryption for arithmetic of approximate numbers. In: Takagi, T., Peyrin, T. (eds.) ASIACRYPT 2017. LNCS, vol. 10624, pp. 409–437. Springer, Cham (2017). https://doi.org/10.1007/978-3-319-70694-8_15

12. Curtis, B.R., Player, R.: On the feasibility and impact of standardising sparse-secret LWE parameter sets for homomorphic encryption. In: Proceedings of the 7th Workshop on Encrypted Computing and Applied Homomorphic Cryptography (2019)
13. Fan, J., Vercauteren, F.: Somewhat practical fully homomorphic encryption. IACR Cryptology ePrint Archive 2012/144 (2012)
14. Gentry, C.: Fully homomorphic encryption using ideal lattices. In: Proceedings of the Forty-First Annual ACM Symposium on Theory of Computing, pp. 169–178 (2009)
15. Gentry, C., Halevi, S., Smart, N.P.: Homomorphic evaluation of the AES circuit. In: Safavi-Naini, R., Canetti, R. (eds.) CRYPTO 2012. LNCS, vol. 7417, pp. 850–867. Springer, Heidelberg (2012). https://doi.org/10.1007/978-3-642-32009-5_49
16. Halevi, S., Polyakov, Y., Shoup, V.: An improved RNS variant of the BFV homomorphic encryption scheme. In: Matsui, M. (ed.) CT-RSA 2019. LNCS, vol. 11405, pp. 83–105. Springer, Cham (2019). https://doi.org/10.1007/978-3-030-12612-4_5
17. Halevi, S., Shoup, V.: Algorithms in HElib. In: Garay, J.A., Gennaro, R. (eds.) CRYPTO 2014. LNCS, vol. 8616, pp. 554–571. Springer, Heidelberg (2014). https://doi.org/10.1007/978-3-662-44371-2_31
18. Halevi, S., Shoup, V.: Bootstrapping for HElib. In: Oswald, E., Fischlin, M. (eds.) EUROCRYPT 2015. LNCS, vol. 9056, pp. 641–670. Springer, Heidelberg (2015). https://doi.org/10.1007/978-3-662-46800-5_25
19. Halevi, S., Shoup, V.: Faster homomorphic linear transformations in HElib. In: Shacham, H., Boldyreva, A. (eds.) CRYPTO 2018. LNCS, vol. 10991, pp. 93–120. Springer, Cham (2018). https://doi.org/10.1007/978-3-319-96884-1_4
20. Han, K., Ki, D.: Better bootstrapping for approximate homomorphic encryption. In: Jarecki, S. (ed.) CT-RSA 2020. LNCS, vol. 12006, pp. 364–390. Springer, Cham (2020). https://doi.org/10.1007/978-3-030-40186-3_16
21. HEAAN. https://github.com/snucrypto/HEAAN
22. Kim, A., Papadimitriou, A., Polyakov, Y.: Approximate Homomorphic Encryption with Reduced Approximation Error. Cryptology ePrint Archive, Report 2020/1118 (2020). https://eprint.iacr.org/2020/1118
23. Kim, M., et al.: Ultra-fast homomorphic encryption models enable secure outsourcing of genotype imputation. bioRxiv (2020). https://doi.org/10.1101/2020.07.02.183459
24. Lattigo 2.0.0. EPFL-LDS, September 2020. https://github.com/ldsec/lattigo
25. Lee, J.-W., et al.: High-Precision Bootstrapping of RNS-CKKS Homomorphic Encryption Using Optimal Minimax Polynomial Approximation and Inverse Sine Function. Cryptology ePrint Archive, Report 2020/552 (2020). https://eprint.iacr.org/2020/552. Accepted to Eurocrypt 2021
26. Masters, O., et al.: Towards a Homomorphic Machine Learning Big Data Pipeline for the Financial Services Sector. IACR Cryptology ePrint Archive 2019/1113 (2019)
27. Sav, S., et al.: POSEIDON: Privacy-Preserving Federated Neural Network Learning. arXiv preprint (2020). arXiv:2009.00349
28. Microsoft SEAL (release 3.6). Microsoft Research, Redmond, WA, November 2020. https://github.com/Microsoft/SEAL
29. Son, Y., Cheon, J.H.: Revisiting the Hybrid attack on sparse and ternary secret LWE. In: IACR Cryptology ePrint Archive 2019/1019 (2019)
30. The Go Programming Language, September 2020. https://golang.org/

High-Precision Bootstrapping of RNS-CKKS Homomorphic Encryption Using Optimal Minimax Polynomial Approximation and Inverse Sine Function

Joon-Woo Lee[1(✉)], Eunsang Lee[1], Yongwoo Lee[1], Young-Sik Kim[2], and Jong-Seon No[1]

[1] Department of Electrical and Computer Engineering, INMC, Seoul National University, Seoul, Republic of Korea
{joonwoo42,shaeunsang,jsno}@snu.ac.kr, yongwool@ccl.snu.ac.kr

[2] Department of Information and Communication Engineering, Chosun University, Gwangju, Republic of Korea
iamyskim@chosun.ac.kr

Abstract. Approximate homomorphic encryption with the residue number system (RNS), called RNS-variant Cheon-Kim-Kim-Song (RNS-CKKS) scheme [12,13], is a fully homomorphic encryption scheme that supports arithmetic operations for real or complex number data encrypted. Although the RNS-CKKS scheme is a fully homomorphic encryption scheme, most of the applications with the RNS-CKKS scheme use it as the only leveled homomorphic encryption scheme because of the lack of the practicality of the bootstrapping operation of the RNS-CKKS scheme. One of the crucial problems of the bootstrapping operation is its poor precision. While other basic homomorphic operations ensure sufficiently high precision for practical use, the bootstrapping operation only supports about 20-bit fixed-point precision at best, which is not high precision enough to be used for the reliable large-depth homomorphic computations until now.

In this paper, we improve the message precision in the bootstrapping operation of the RNS-CKKS scheme. Since the homomorphic modular reduction process is one of the most important steps in determining the precision of the bootstrapping, we focus on the homomorphic modular reduction process. Firstly, we propose a fast algorithm of obtaining the optimal minimax approximate polynomial of modular reduction function and the scaled sine/cosine function over the union of the approximation regions, called an improved multi-interval Remez algorithm. In fact, this algorithm derives the optimal minimax approximate polynomial of any continuous functions over any union of the finite number of intervals. Next, we propose the composite function method using the inverse sine function to reduce the difference between the scaling factor used in the bootstrapping and the default scaling factor. With these methods, we reduce the approximation error in the bootstrapping of the RNS-CKKS

This work is supported by Samsung Advanced Institute of Technology.

A. Canteaut and F.-X. Standaert (Eds.): EUROCRYPT 2021, LNCS 12696, pp. 618–647, 2021.
https://doi.org/10.1007/978-3-030-77870-5_22

scheme by 1/1176–1/42 (5.4–10.2-bit precision improvement) for each parameter setting. While the bootstrapping without the composite function method has 27.2–30.3-bit precision at maximum, the bootstrapping with the composite function method has 32.6–40.5-bit precision.

Keywords: Approximate homomorphic encryption · Bootstrapping · Composite function approximation · Fully homomorphic encryption (FHE) · Improved multi-interval Remez algorithm · Inverse sine function · Minimax approximate polynomial · RNS-variant Cheon-Kim-Kim-Song (RNS-CKKS) scheme

1 Introduction

Fully homomorphic encryption (FHE) is the encryption scheme enabling any logical operations [6,14,16,19,30] or arithmetic operations [12,13] with encrypted data. The FHE scheme makes it possible to preserve security in data processing. However, in the traditional encryption schemes, they are not encrypted to enable the processing of encrypted data, which causes clients to be dissuaded from receiving services and prevents companies from developing various related systems because of the lack of clients' privacy. FHE solves this problem clearly so that clients can receive many services by ensuring their privacy.

First, Gentry constructed the FHE scheme by coming up with the idea of bootstrapping [18]. After this idea was introduced, cryptographers constructed many FHE schemes using bootstrapping. Approximate homomorphic encryption, which is also called a Cheon-Kim-Kim-Song (CKKS) scheme [13], is one of the promising FHE schemes, which deals with any real and complex numbers. The CKKS scheme is particularly in the spotlight for much potential power in many applications such as machine learning [2,3,5,7,15,23], in that data is usually represented by real numbers. Lots of research for the optimization of the CKKS scheme have been done actively for practical use. Cheon et al. proposed the residue number system (RNS) variant CKKS scheme (RNS-CKKS) [12] so that the necessity of arbitrary precision library can be removed and only use the word-size operations. The running time of the homomorphic operations in the RNS-CKKS scheme is 10 times faster than that of the original CKKS scheme with the single thread, and further, the RNS-CKKS scheme has an advantage in parallel computation, which leads to much better running time performance with the multi-core environment. Because of the fast homomorphic operations, most homomorphic encryption libraries, including `SEAL` [29] and `PALISADE` [1], are implemented using the RNS-CKKS scheme. Thus, we focus on the RNS-CKKS scheme in this paper.

Since the CKKS scheme includes noises used to ensure security as the approximate error in the message, the use of the RNS-CKKS scheme requires more sensitivity to the precision of the message than other homomorphic encryption schemes that support accurate decryption and homomorphic evaluation. This can be more sensitive for large-depth homomorphic operations because errors are likely to be amplified by the operations and distort the data significantly.

Fortunately, the basic homomorphic operations in the RNS-CKKS scheme can ensure sufficiently high precision for practical use, but this is not the case for the bootstrapping operation. Ironically, while the bootstrapping operation in other homomorphic encryption schemes reduces the effect of the errors on messages so that they do not distort messages, the bootstrapping operation in the CKKS scheme amplifies the errors, which makes it the most major cause of data distortion among any other homomorphic operations in the RNS-CKKS scheme. Since advanced operations with large depth may require bootstrapping operation many times, the message precision problem in the bootstrapping operation is a crucial obstacle to applying the RNS-CKKS scheme to advanced applications.

Although the RNS-CKKS scheme is currently one of the most potential solutions to implement privacy-preserving machine learning (PPML) system [2,3,15], the methods for the PPML studied so far have mainly been applied to simple models such as MNIST, which has such a low depth that bootstrapping is not required. Thus, the message precision problem in the bootstrapping operation in the RNS-CKKS scheme did not need to be considered in the PPML model until now. However, the advanced machine learning model currently presented requires a large depth, and thus we should introduce the bootstrapping operation and cannot avoid the message precision problem in the bootstrapping operation. Of course, the fact that bootstrapping requires longer running time and larger depth than other homomorphic operations is also pointed out as a major limitation of bootstrapping. While these points may be improved by simple parameter adjustments and using hardware optimization, the message precision problem in bootstrapping is difficult to solve with these simple methods.

Most of the works about PPML with FHE focused on the inference process rather than the training process because of the large running time. However, training neural networks with encrypted data is actually more important from a long-term perspective for solving the real security problem in machine learning, in that the companies cannot gather sufficiently many important but sensitive data, such as genetic or financial information so that they cannot construct the deep learning model for them because of the privacy of the data owners. While the inference process does not need a high precision number system, the training process is affected sensitively by the precision of the number system. Chen et al. [9] showed that convolutional neural networks (CNN) learning MNIST could not converge when the model is trained using a 16-bit fixed-point number system. When the 32-bit fixed-point number system is used to train the CNN with MNIST, the training performance was slightly lower than the case of using the single-precision floating-point number system, although all bits except one bit representing the sign are used to represent the data in 32-bit fixed-point number system, which is much better precision than the single-precision floating-point number system, which is 23-bit precision. Although many works proposed to use low-precision fixed-point numbers in the training procedure, they used additional special techniques, such as stochastic rounding [20] or the dynamic fixed-point number system [21], which cannot be supported by the RNS-CKKS scheme until now.

While most of the deep learning systems use single-precision floating-point numbers, the maximum precision achieved with the bootstrapping of the CKKS scheme in the previous papers was about only 20 bits. Considering that the CKKS scheme only supports fixed-point arithmetic, the 20-bit precision is not large enough to be applied wholly to the deep learning system. Thus, to apply the RNS-CKKS scheme to deep learning systems, it is necessary to achieve a precision sufficiently better than the 32-bit fixed-point precision, which requires a breakthrough for the bootstrapping in the RNS-CKKS scheme concerning its precision.

1.1 Our Contribution

In this paper, we propose two methods to improve the bootstrapping operation of the RNS-CKKS scheme. Firstly, we devise a fast algorithm, called an improved multi-interval Remez algorithm, obtaining the optimal minimax approximate polynomial of any continuous functions over any union of the finite number of intervals, which include the modular reduction function and the scaled sine/cosine function over the union of the approximation regions. Although the previous works have suggested methods to obtain polynomials that approximate the scaled sine/cosine function well from the minimax perspective, which are used to approximate the modular reduction function, these methods cannot obtain the optimal minimax approximate polynomial.

The original multi-interval Remez algorithm is not theoretically proven to obtain the minimax approximate polynomial, and it is only practically used for two or three approximation regions in the finite impulse response filter design, while we need to approximate functions over the union of tens of intervals. Furthermore, it takes impractically much time if this algorithm is used without further improvement to obtain a polynomial that can be used for the bootstrapping. To make the multi-interval Remez algorithm practical, we modify the multi-interval Remez algorithm as the improved multi-interval Remez algorithm. Then we prove the correctness of the improved multi-interval Remez algorithm, including the original multi-interval Remez algorithm, for the union of any finite number of intervals. Since it can obtain the optimal minimax approximate polynomial in seconds, we can even adaptively obtain the polynomial when we abruptly change some parameters on processing the ciphertexts so that we have to update the approximate polynomial. All polynomial approximation methods proposed in previous works for bootstrapping in the CKKS scheme can be replaced with the improved multi-interval Remez algorithm, which ensures the best quality of the approximation. It ensures to use the least degree of the approximate polynomial for a given amount of error.

Next, we propose the composite function method to enlarge the approximation region in the homomorphic modular reduction process using the inverse sine function. The crucial point in the bootstrapping precision is that the difference between the modular reduction function and the sine/cosine function gives a significant precision loss. All previous works have used methods that approximate the modular reduction function as a part of the sine/cosine functions. This approximation has an inherent approximation error so that the limitation of the precision occurs. Besides, to ensure that these two functions are significantly close to each other, the approximation region has to be reduced significantly. They set the half-width of one interval in the approximation region as 2^{-10}, which is equal to the ratio of default scaling factor to the scaling factor used in the bootstrapping. The message has to be scaled by multiplying 2^{-10} to make the message into the approximation region, and it is scaled by multiplying 2^{10} at the end of the bootstrapping. Thus, the precision error in the computation for bootstrapping is amplified by 2^{10}, and the 10-bit precision loss occurs. If we try to reduce this precision loss by enlarging the approximation region, the approximation error by the sine/cosine function becomes large, and thus the overall precision becomes lower than before.

Therefore, we propose to compose the optimal approximate polynomial of the inverse sine function to the sine/cosine function, since composing the inverse sine function to the sine/cosine function extends the approximation region of the modular reduction function, which makes it possible to improve the precision of the bootstrapping. Note that the inverse sine function we use has only one interval in the approximation region, and thus we can reach the small approximate error with relatively low degree polynomials. We obtain the minimax approximate polynomials for the scaled cosine function and the inverse sine function with sufficiently small minimax error by the improved multi-interval Remez algorithm. We apply these polynomials in the homomorphic modular reduction process by homomorphically evaluating the approximate polynomial for the scaled cosine function, several double-angle formulas, and the approximate polynomial for the inverse sine function. This enables us to minimize the inevitable precision loss by approximating the modular reduction function to the sine/cosine function.

Since the previous works do not focus on the maximum precision of the bootstrapping of the RNS-CKKS scheme, we check the maximum precision of the bootstrapping with the previous techniques. The detailed relation with the precision of the bootstrapping and various parameters is analyzed with `SEAL` library. With the proposed methods, we reduce the approximation error in the bootstrapping of the RNS-CKKS scheme by 1/1176–1/42 (5.4–10.2-bit precision improvement) for each parameter setting. While the bootstrapping without the composite function method has 27.2–30.3-bit precision at maximum, the bootstrapping with the proposed composite function method has 32.6–40.5-bit precision, which are better precision than 32-bit fixed-point precision.

1.2 Related Works

The CKKS scheme [13] was firstly proposed without bootstrapping as a somewhat homomorphic encryption scheme supporting only the finite number of multiplications. Cheon et al. [11] firstly suggested bootstrapping operation with the homomorphic linear transformation enabling transformation between slots and coefficients, and approximation of homomorphic modulus reduction function as the sine function with Taylor approximation and the double-angle formula. Chen et al.

[8] applied a modified fast Fourier transform (FFT) algorithm to evaluate homomorphic linear transformation and used Chebyshev interpolation and Paterson-Stockmeyer algorithm to approximate the sine function efficiently in terms of the running time and the depth consumption. Han et al. [22] improved the homomorphic modular reduction in the bootstrapping operation. While Chen et al. approximated the sine function in one interval, Han et al. approximated the cosine function only in the separated approximation regions, reducing the degree of polynomials and using simpler double-angle formula than that of the sine function. Still, their approximate polynomial is also not optimal in the minimax aspect.

On the other hand, the RNS-CKKS scheme was proposed. Since big integers used to represent the ciphertexts in the CKKS scheme cannot be stored with the basic data type, the original CKKS scheme had to resort to the arbitrary precision data type libraries, such as the number theory library (NTL). To remove the reliance on the external libraries for performance improvement, Cheon et al. applied the RNS system in the CKKS scheme. Most practical homomorphic encryption libraries, such as `SEAL` and `PALISADE`, implement the RNS-CKKS scheme. The approximate rescaling procedure, which enables using the RNS system in the RNS-CKKS scheme, causes more approximation error in the homomorphic multiplication of the RNS-CKKS scheme than in that of the original CKKS scheme. Kim et al. [24] recently suggested the management method for the scaling factor in the RNS-CKKS scheme. Thus the approximation error in the homomorphic multiplication of the RNS-CKKS scheme was made the same as that of the original CKKS scheme.

Bossuat et al. [4] optimized various performances of the bootstrapping of the RNS-CKKS scheme. Their two main techniques are the scale-invariant polynomial evaluation and the double hoisting. In the scale-invariant polynomial evaluation, the coefficients of an approximate polynomial are slightly adjusted by multiplication with some adjustment factor so that the messages in the output ciphertext are not affected by the approximate rescaling. Also, it always ensures optimal depth consumption by introducing additional recursive loops. The double hoisting technique optimized the homomorphic evaluation of a linear combination of several rotated ciphertexts from the same ciphertext with different rotation steps. Bossuat et al.'s techniques are compatible with our techniques; that is, their techniques and our techniques can be applied simultaneously in the RNS-CKKS scheme.

1.3 Outline

The outline of the paper is given as follows. Section 2 deals with some preliminaries for the RNS-CKKS scheme, approximation theory, and the Remez algorithm. In Sect. 3, we propose an improved multi-interval Remez algorithm for obtaining the optimal minimax approximate polynomial. The numerical relation between the message precision and several parameters in the RNS-CKKS scheme is dealt with in Sect. 4, and the upper bound of the message precision in the bootstrapping of the RNS-CKKS scheme is also included. In Sect. 5, we propose the composite function method, which makes it possible to reduce the difference of the two scaling factors in default operations and in bootstrapping operations, and numerically shows the improvement of the message precision in the proposed bootstrapping operation in the RNS-CKKS scheme. Section 6 concludes the paper.

2 Preliminary

2.1 Notation

Let $\mathsf{round}(x)$ be the function that outputs the integer nearest to x, and we do not have to consider the case of tie in this paper. The Chebyshev polynomials $T_n(x)$ are defined by $\cos n\theta = T_n(\cos\theta)$. The remainder of a divided by q is denoted as $[a]_q$. If $\mathcal{C} = \{q_0, q_1, \cdots, q_{\ell-1}\}$ is the set of positive integers coprime each other and $a \in \mathbb{Z}_Q$ where $Q = \prod_{i=0}^{\ell-1} q_i$, the RNS representation of a with regard to $\mathcal{C}$ is denoted by $[a]_{\mathcal{C}} = ([a]_{q_0}, [a]_{q_1}, \cdots, [a]_{q_{\ell-1}}) \in \mathbb{Z}_{q_0} \times \cdots \times \mathbb{Z}_{q_{\ell-1}}$. The base of logarithm in this paper is two.

2.2 CKKS Scheme and RNS-CKKS Scheme

It is known that the CKKS scheme supports several operations for encrypted data of real numbers or complex numbers. Since it usually deals with real numbers, the noise that ensures the security of the CKKS scheme can be embraced outside of the significant figures of the data, which is the crucial concept of the CKKS scheme.

The RNS-CKKS scheme [12] uses the RNS form to represent the ciphertexts and to perform the homomorphic operations efficiently. While the power-of-two modulus is used in the CKKS scheme, the product of large primes is used for ciphertext modulus in the RNS-CKKS scheme so that the RNS system can be applied. These large primes are chosen to be similar to the scaling factor, which is some power-of-two integer. There is a crucial difference in the rescaling operation between the CKKS scheme and the RNS-CKKS scheme. While the CKKS scheme can rescale the ciphertext by the exact scaling factor, the RNS-CKKS scheme has to rescale the ciphertext by one of the RNS moduli, which is not equal to the scaling factor. Thus, the RNS-CKKS scheme allows approximation in the rescaling procedure. The specific procedure is not needed in this paper, and thus we omit the detailed procedures. Detailed procedures in the CKKS scheme and the RNS-CKKS scheme are found in [13] and [12], respectively.

2.3 Kim-Papadimitriou-Polyakov (KPP) Scaling Factor Management

Kim et al. [24] suggested a method of eliminating the large rescaling error in the RNS-CKKS scheme. Instead of using the same power-of-two scaling factor for each level, they used different scaling factors in different levels. If the maximum level is L, and the ciphertext modulus for level i is denoted as q_i, the scaling factor for each level is given as follows: $\Delta_L = q_L$ and $\Delta_i = \Delta_{i+1}^2/q_{i+1}$ for $i = 0, \cdots, L-1$.

If the two ciphertexts are at the same level, it does not introduce the approximate rescaling error when they are multiplied homomorphically. If the two ciphertexts are in the different level, that is, in the levels i and j such that $i > j$, the moduli $q_i, \cdots, q_{j+1}$ in the first ciphertext are dropped, the first ciphertext is multiplied by a constant $\lceil \frac{\Delta_j q_{j+1}}{\Delta_i} \rfloor$, and it is rescaled by q_{j+1}. Then we perform the conventional homomorphic multiplication with the two ciphertexts, which are now at the same level, together with rescaling in the RNS-CKKS scheme. The approximate rescaling error is also not introduced in this case.

2.4 Bootstrapping for CKKS Scheme

The framework of the bootstrapping of the CKKS scheme was introduced in [13], which is the same as the case of the RNS-CKKS scheme. The purpose of bootstrapping is to refresh the ciphertext of level 0, whose multiplication cannot be performed anymore, to the fresh ciphertext of level L having the same messages. Bootstrapping is composed of the following four steps:

i) Modulus raising
ii) Homomorphic linear transformation; CoeffToSlot
iii) Homomorphic modular reduction
iv) Homomorphic linear transformation; SlotToCoeff

Modulus Raising: The starting point of bootstrapping is modulus raising, where we simply consider the ciphertext of level 0 as an element of $\mathcal{R}_Q^2$, instead of $\mathcal{R}_{q_0}^2$. Since the ciphertext of level 0 is supposed to be $\langle \mathsf{ct}, \mathsf{sk} \rangle \approx m \mod q_0$, we have $\langle \mathsf{ct}, \mathsf{sk} \rangle \approx m + q_0 I \mod Q$ for some $I \in \mathcal{R}$ when we try to decrypt it. We are assured that the absolute values of coefficients of I are rather small, for example, usually smaller than 12, because coefficients of sk consist of small numbers [11]. The crucial part of the bootstrapping of the CKKS scheme is to make ct' such that $\langle \mathsf{ct}', \mathsf{sk} \rangle \approx m \mod q_L$. This is divided into two parts: homomorphic linear transform and homomorphic evaluation of modular reduction function.

Homomorphic Linear Transformation: The ciphertext ct after modulus raising can be considered as the ciphertext encrypting $m + q_0 I$, and thus we now have to perform modular reduction to coefficients of message polynomial homomorphically. However, the operations we have are all for slots, not coefficients of the message polynomial. Thus, to perform some meaningful operations on coefficients, we have to convert ct into a ciphertext that encrypts coefficients of $m + q_0 I$ as its slots. After evaluation of homomorphic modular reduction function, we have to reversely convert this ciphertext into the other ciphertext ct' that encrypts the slots of the previous ciphertext as the coefficients of its message. These two operations are called CoeffToSlot and SlotToCoeff operations. These operations are regarded as homomorphic evaluation of encoding and decoding of messages, which are a linear transformation by some variants of Vandermonde matrix for roots of $\Phi_M(x)$. This can be performed by general homomorphic matrix multiplication [11], or FFT-like operation [8].

Homomorphic Modular Reduction Function: After CoeffToSlot is performed, we now have to perform modular reduction homomorphically on each slot in modulus q_0. This procedure is called EvalMod. This modular reduction function is not an arithmetic function and even not a continuous function. Fortunately, by restricting the range of the messages such that m/q_0 is small enough, the approximation region can be given only near multiples of q_0. This allows us to approximate the modular reduction function more effectively. Since the operations that the CKKS supports are arithmetic operations, most of the works [8,11,22] dealing with CKKS bootstrapping approximate the modular reduction function with some polynomials, which are sub-optimal approximate polynomials.

The scaling factor is increased when the bootstrapping is performed because m/q_0 needs to be very small in the homomorphic modular reduction function. In this paper, the default scaling factor means the scaling factor used in the intended applications, and the bootstrapping scaling factor means the scaling factor used in the bootstrapping. The bit-length difference between these two scaling factors is usually 10.

2.5 Approximation Theory

There are many theorems for the minimax approximate polynomials of a function defined on a compact set in approximation theory. Before introducing these theorems, we refer to a definition of the Haar condition of a set of functions that deals with the generalized version of power bases used in polynomial approximation and its equivalent statement. It is a well-known fact that the power basis $\{1, x, x^2, \cdots, x^d\}$ satisfies the Haar condition. Thus, if an argument deals with the polynomials concerning a set of basis functions satisfying the Haar condition, it naturally includes the case of polynomials.

Definition 2.1 ([10] **Haar's Condition).** *A set of functions* $\{g_1, g_2, \cdots, g_n\}$ *satisfies the Haar condition if each* g_i *is continuous and if each determinant*

$$D[x_1, \cdots, x_n] = \begin{vmatrix} g_1(x_1) & \cdots & g_n(x_1) \\ \vdots & \ddots & \vdots \\ g_1(x_n) & \cdots & g_n(x_n) \end{vmatrix}$$

for any n *distinct points* $x_1, \cdots, x_n$ *is not zero.*

Lemma 2.2 ([10]). *A set of functions* $\{g_1, \cdots, g_n\}$ *satisfies the Haar condition if and only if the zero function is the only function of the form* $\sum_i c_i g_i$ *that has more than* $n-1$ *roots.*

We now introduce the core property of the minimax approximate polynomial for a function on D.

Theorem 2.3 ([10] **Chebyshev Alternation Theorem).** *Let $\{g_1, \cdots, g_n\}$ be a set of continuous functions defined on $[a, b]$ satisfying the Haar condition, and let D be a closed subset of $[a, b]$. A polynomial $p = \sum_i c_i g_i$ is the minimax approximate polynomial on D to any given continuous function f defined on D if and only if there are $n+1$ distinct elements $x_0 < \cdots < x_n$ in D such that for the error function $r = f - p$ restricted on D,*

$$r(x_i) = -r(x_{i-1}) = \pm \sup_{x \in D} |r(x)|.$$

This condition is also called the equioscillation condition. This means that if we find a polynomial satisfying the equioscillation condition, this is the unique minimax approximate polynomial. It is needless to compare with the maximum approximation error of any polynomials.

2.6 Algorithms for Minimax Approximation

Remez Algorithm. Remez algorithm [10,27,28] is an iterative algorithm that always returns the minimax approximate polynomial for any continuous function on an interval of $[a, b]$. This algorithm strongly uses the Chebyshev alternation theorem [10] in that its purpose is finding the polynomial satisfying equioscillation condition. In fact, the Remez algorithm can be applied to obtain the minimax approximate polynomial, whose basis function $\{g_1, \cdots, g_n\}$ satisfies the Haar condition. The specific algorithm is shown in Algorithm 1.

Multi-interval Remez Algorithm. Since the Remez algorithm works only when the approximation region is one interval, we need another multi-interval Remez algorithm that works when the approximation region is the union of several intervals. The above Remez algorithm can be extended to the multiple sub-intervals of an interval [17,26,28]. The multi-interval Remez algorithm is the same as Algorithm 1, except Steps 3 and 4. For each iteration, firstly, we find all of the local extreme points of the error function $p - f$ whose absolute error values are larger than the absolute error values at the current reference points. Then, we choose $n+1$ new extreme points among these points satisfying the following two criteria:

i) The error values alternate in sign.
ii) A new set of extreme points includes the global extreme point.

These two criteria are known to ensure the convergence to the minimax polynomial, even though there is no exact proof of its convergence to the best of our knowledge. However, it is noted that there are many choices of sets of extreme points satisfying these criteria. In the next section, we modify the multi-interval Remez algorithm, where one of the two criteria is changed.

Algorithm 1: Remez Algorithm [10,27,28]

Input : An input domain $[a, b]$, a continuous function f on $[a, b]$, an approximation parameter δ, and a basis $\{g_1, \cdots, g_n\}$.
Output: The minimax approximate polynomial p for f

1 Select $x_1, x_2, \cdots, x_{d+2} \in [a, b]$ in strictly increasing order.
2 Find the polynomial $p(x) = \sum_{i=1}^{n} c_i g_i(x)$ with $p(x_i) - f(x_i) = (-1)^i E$ for $i = 1, \cdots, d+2$ and some E by solving the system of linear equations with variables c_i's and E.
3 Divide the interval into $n+1$ sections $[z_{i-1}, z_i]$, $i = 1, \cdots, n+1$, from zeros $z_1, \cdots, z_n$ of $p(x) - f(x)$, where $x_i < z_i < x_{i+1}$, and boundary points $z_0 = a$, $z_{n+1} = b$.
4 Find the maximum (resp. minimum) points for each section when $p(x_i) - f(x_i)$ has positive (resp. negative) value. Denote these extreme points $y_1, \cdots, y_{n+1}$.
5 $\epsilon_{\mathsf{max}} \leftarrow \max_i |p(y_i) - f(y_i)|$
6 $\epsilon_{\mathsf{min}} \leftarrow \min_i |p(y_i) - f(y_i)|$
7 **if** $(\epsilon_{\mathsf{max}} - \epsilon_{\mathsf{min}})/\epsilon_{\mathsf{min}} < \delta$ **then**
8 | **return** $p(x)$
9 **else**
10 | Replace x_i's with y_i's and go to line 2.
11 **end**

3 Efficient Algorithm for Optimal Minimax Approximate Polynomial

In this section, we propose an improved multi-interval Remez algorithm for obtaining the optimal minimax approximate polynomial. With this proposed algorithm, we can obtain the optimal minimax approximate polynomial for continuous function on the union of finitely many closed intervals to apply the Remez algorithm to the bootstrapping of the CKKS scheme. The function we are going to approximate is the normalized modular reduction function defined in only near finitely many integers given as

$$\mathsf{normod}(x) = x - \mathsf{round}(x), \quad x \in \bigcup_{i=-(K-1)}^{K-1} [i - \epsilon, i + \epsilon],$$

where K determines the number of intervals in the domain. normod function corresponds to the modular reduction function scaled for both its domain and range.

In addition, Han et al. [22] uses the cosine function to approximate $\mathsf{normod}(x)$ to use double-angle formula for efficient homomorphic evaluation. If we use double-angle formula ℓ times, we have to approximate the following cosine function

$$\cos\left(\frac{2\pi}{2^\ell}\left(x - \frac{1}{4}\right)\right), \quad x \in \bigcup_{i=-(K-1)}^{K-1} [i - \epsilon, i + \epsilon].$$

To design an approximation algorithm that deals with the above two functions, we assume the general continuous function defined on an union of finitely many closed intervals, which is given as

$$D = \bigcup_{i=1}^{t} [a_i, b_i] \subset [a, b] \subset \mathbb{R},$$

where $a_i < b_i < a_{i+1} < b_{i+1}$ for all $i = 1, \cdots, t-1$.

When we propose the improved multi-interval Remez algorithm to approximate a given continuous function on D with a polynomial having a degree less than or equal to d, we have to consider two crucial points. One is to establish an efficient criterion for choosing new $d+2$ reference points among several extreme points. The other is to make efficient some steps in the improved multi-interval Remez algorithm. We deal with these two issues for the improved multi-interval Remez algorithm in Sects. 3.1 and 3.3, respectively.

3.1 Improved Multi-interval Remez Algorithm with Criteria for Choosing Extreme Points

Assume that we apply the multi-interval Remez algorithm on D and use $\{g_1, \cdots, g_n\}$ satisfying Haar condition on $[a, b]$ as the basis of polynomials. After obtaining the minimax approximate polynomial regarding the set of reference points for each iteration, we have to choose a new set of reference points for the next iteration. However, there are many boundary points in D, and all these boundary points have to be considered as extreme points of the error function. For this reason, there are many cases of selecting $n+1$ points among these extreme points. For bootstrapping in the CKKS scheme, there are many intervals to be considered, and thus there are lots of candidate extreme points. Since the criterion of the original multi-interval Remez algorithm cannot determine the unique new set of reference points for each iteration, it is necessary to make how to choose $n+1$ points for each iteration to reduce the number of iterations as small as possible. Otherwise, it requires a large number of iteration for convergence to the minimax approximate polynomial. On the other hand, if the criterion is not designed properly, the improved multi-interval Remez algorithm may not converge into a single polynomial in some cases.

In order to set the criterion for selecting $n+1$ reference points, we need to define a simple function for extreme points, $\mu_{p,f} : D \to \{-1, 0, 1\}$ as follows,

$$\mu_{p,f}(z) = \begin{cases} 1 & p(x) - f(x) \text{ is concave at z on D} \\ -1 & p(x) - f(x) \text{ is convex at z on D} \\ 0 & \text{otherwise,} \end{cases}$$

where $p(x)$ is a polynomial obtained in that iteration and $f(x)$ is a continuous function on D to be approximated. We abuse the notation $\mu_{p,f}$ as μ.

Assume that the number of extreme points of $p(x) - f(x)$ on D is finite, and the set of extreme points is denoted by $B = \{w_1, w_2, \cdots, w_m\}$. Assume that B

is ordered in increasing order, $w_1 < w_2 < \cdots < w_m$, and then the values of μ at these points are always 1 or -1. Let $\mathcal{S}$ be a set of functions defined as

$$\mathcal{S} = \{\sigma : [n+1] \to [m] \mid \sigma(i) < \sigma(i+1) \text{ for all } i = 1, \cdots, n\},$$

which means all the ways of choosing $n+1$ points of the m points. Clearly, $\mathcal{S}$ has only the identity function if $n+1 = m$.

Then, we set three criteria for selecting $n+1$ extreme points as follows:

i) *Local extreme value condition.* If E is the absolute value of error at points in the set of reference points, then we have

$$\min_i \mu(x_{\sigma(i)})(p(x_{\sigma(i)}) - f(x_{\sigma(i)})) \geq E.$$

ii) *Alternating condition.* $\mu(x_{\sigma(i)}) \cdot \mu(x_{\sigma(i+1)}) = -1$ for $i = 1, \cdots, n$.

iii) *Maximum absolute sum condition.* Among σ's satisfying the above two conditions, choose σ maximizing the following value

$$\sum_{i=1}^{n+1} |p(x_{\sigma(i)}) - f(x_{\sigma(i)})|.$$

It is noted that the local extreme value condition in i) means in particular that the extreme points are discarded if the local maximum value of $p(x) - f(x)$ is negative or the local minimum of $p(x) - f(x)$ is positive.

Note that the first two conditions are also included in the original multi-interval Remez algorithm. The third condition, the maximum absolute sum condition, is the replacement of the condition that the new set of reference points includes the global extreme point. The numerical analysis will show that the third condition makes the proposed improved multi-interval Remez algorithm converge to the optimal minimax approximate polynomial fast. Although there are some cases in which the global maximum point is not included in the new set of reference points chosen by the maximum absolute sum condition, we prove that the maximum absolute sum condition is enough for the improved multi-interval Remez algorithm to converge to the minimax approximate polynomial in the next subsection.

We propose the improved multi-interval Remez algorithm for the continuous function on the union of finitely many closed intervals as in Algorithm 2. The local extreme value condition is reflected in Step 3, and the alternating condition and the maximum absolute sum condition are reflected in Step 4.

3.2 Correctness of Improved Multi-interval Remez Algorithm

We now have to prove that the improved multi-interval Remez algorithm always converges to the minimax approximate polynomial for a given continuous function on the union of finite intervals D. This proof is similar to the convergence proof of the original Remez algorithm on one closed interval [10,27], but there

Algorithm 2: Improved Multi-interval Remez Algorithm

Input : An input domain $D = \bigcup_{i=1}^{t} [a_i, b_i] \subset \mathbb{R}$, a continuous function f on D, an approximation parameter δ, and a basis $\{g_1, \cdots, g_n\}$
Output: The minimax approximate polynomial p for f

1 Select $x_1, x_2, \cdots, x_{n+1} \in D$ in strictly increasing order.
2 Find the polynomial $p(x)$ with $p(x_i) - f(x_i) = (-1)^i E$ for some E.
3 Gather all extreme and boundary points such that $\mu_{p,f}(x)(p(x) - f(x)) \geq |E|$ into a set B.
4 Find $n+1$ extreme points $y_1 < y_2 < \cdots < y_{n+1}$ with alternating condition and maximum absolute sum condition in B.
5 $\epsilon_{\mathsf{max}} \leftarrow \max_i |p(y_i) - f(y_i)|$
6 $\epsilon_{\mathsf{min}} \leftarrow \min_i |p(y_i) - f(y_i)|$
7 **if** $(\epsilon_{\mathsf{max}} - \epsilon_{\mathsf{min}})/\epsilon_{\mathsf{min}} < \delta$ **then**
8 | **return** $p(x)$
9 **else**
10 | Replace x_i's with y_i's and go to line 2.
11 **end**

are a few more general arguments than the original proof. This convergence proof includes the proof for both the variant of the Remez algorithm and the modified Remez algorithm. Theorem 3.1 is the exact statement of the correctness of the improved multi-interval Remez algorithm. We include the sketch of the proof of Theorem 3.1. The full proof is shown in the full version of the paper [25].

Theorem 3.1. *Let $\{g_1, \cdots, g_n\}$ be a set of functions satisfying the Haar condition on $[a, b]$, D be the multiple sub-intervals of $[a, b]$, and f be a continuous function on D. Let p_k be an approximate polynomial generated in the k-th iteration of the modified Remez algorithm, and p^* be the optimal minimax approximate polynomial of f. Then, as k increases, p_k converges uniformly to p^* as in the following inequality*

$$\|p_k - p^*\|_\infty \leq A\theta^k,$$

where A is a non-negative constant and $0 < \theta < 1$.

Proof. (Sketch) Let $\{x_1^{(0)}, \cdots, x_{n+1}^{(0)}\}$ be the initial set of reference points and $\{x_1^{(k)}, \cdots, x_{n+1}^{(k)}\}$ be the new set of reference points chosen at the end of iteration k. Let $r_k = p_k - f$ be the error function of p_k and $r^* = p^* - f$ be the error function of p^*. Since p_k is generated such that the absolute values of the error function r_k at the reference points $x_i^{(k-1)}$, $i = 1, 2, \cdots, n+1$ are the same. For $k \geq 1$, we define

$$\alpha_k = \min_i |r_k(x_i^{(k-1)})| = \max_i |r_k(x_i^{(k-1)})|,$$

$$\beta_k = \|r_k\|_\infty,$$

$$\gamma_k = \min_i |r_k(x_i^{(k)})|.$$

Let $\beta^* = \|r^*\|_\infty$. Then, we can prove the following facts, which are proven in the full version of the paper.

i) $\alpha_k \leq \gamma_k \leq \alpha_{k+1} \leq \beta^* \leq \beta_k$ for $k \geq 1$.
ii) α_{k+1} is a weighted average of $|r_k(x_i^{(k)})|$ for $i = 1, \cdots, n+1$. In other words, for all $k \geq 1$ there are weights $\theta_i^{(k)} \geq 0$ such that $\alpha_{k+1} = \sum_{i=1}^{n+1} \theta_i^{(k)} |r_k(x_i^{(k)})|$ where $\sum_{i=1}^{n+1} \theta_i^{(k)} = 1$.
iii) All weights for the weighted average is larger than some positive constant throughout all iterations. In other words, there is a global constant $\theta' > 0$ such that $\theta_i^{(k)} \geq \theta'$.
iv) $\sum_{i=1}^{n+1} |r_k(x_i^{(k)})| \geq \beta_k$ for $k \geq 1$.

For convenience, we set $\theta = 1 - \theta'$. It is enough to show that $\beta_k - \beta^* \leq C\theta^k$ for some positive constant C to prove the theorem, and it is also proven in the full paper that this is a sufficient condition for the theorem.

From the facts i)–iv), we have

$$\begin{aligned}
\gamma_{k+1} - \gamma_k &\geq \alpha_{k+1} - \gamma_k \\
&= \sum_{i=1}^{n+1} \theta_i^{(k)} (|r_k(x_i^{(k)})| - \gamma_k) \\
&\geq (1-\theta)(\beta_k - \gamma_k) \qquad (1) \\
&\geq (1-\theta)(\beta^* - \gamma_k). \qquad (2)
\end{aligned}$$

From (2), we have

$$\begin{aligned}
\beta^* - \gamma_{k+1} &= (\beta^* - \gamma_k) - (\gamma_{k+1} - \gamma_k) \\
&\leq (\beta^* - \gamma_k) - (1-\theta)(\beta^* - \gamma_k) \\
&= \theta(\beta^* - \gamma_k).
\end{aligned}$$

Then, we obtain the following inequality for some nonnegative B as

$$\beta^* - \gamma_k \leq B\theta^k. \qquad (3)$$

From (1) and (3), we have

$$\begin{aligned}
\beta_k - \beta^* &\leq \beta_k - \gamma_k \\
&\leq \frac{1}{1-\theta}(\gamma_{k+1} - \gamma_k) \\
&\leq \frac{1}{1-\theta}(\beta^* - \gamma_k) \\
&\leq \frac{1}{1-\theta} B\theta^k \\
&\leq C\theta^k.
\end{aligned}$$

□

Remark. Note that i)–iv) in the above proof can be satisfied if we include the global extreme point to the new set of reference points as in the original multi-interval Remez algorithm, instead of the maximum absolute sum condition in the improved multi-interval Remez algorithm. Thus, this proof naturally includes the convergence proof of the original variant of the Remez algorithm.

From the sketch of the proof, we know that the convergence rate of α_k determines the convergence rate of the algorithm. Since α_k is always lower than β^* and non-decreasing sequence, it is desirable to obtain α_k as large as possible for each iteration. The maximum sum condition is more effective than the global extreme point inclusion condition; The global extreme point inclusion condition cannot care about the reference points other than the global extreme point, but the maximum sum condition cares for all the reference points to be large. This can give some intuition for the effectiveness of the maximum sum condition.

3.3 Efficient Implementation of Improved Multi-interval Remez Algorithm

In this section, we have to consider the issues in each step of Algorithm 2 and suggest how to implement Steps 1, 2, 3, and 4 of Algorithm 2 as follows.

Initialization: Depending on the initialization method, there can be a large difference in the number of iterations required. Therefore, the closer the polynomial produced by initializing the initial reference points to the optimal minimax approximation polynomial, the fewer iterations are required. We use the node setting method of Han et al. [22] to effectively set the initial reference points in the improved multi-interval Remez algorithm. Since Han et al.'s node setting method was for polynomial interpolation, it chooses the $d+1$ number of nodes when we need the approximate polynomial of degree d. Instead, if we need to obtain the optimal minimax approximate polynomial of degree d, we choose the $d+2$ number of nodes with Han et al.'s method as if we need the approximate polynomial of degree $d+1$, and uses them for the initial reference points.

Finding Approximate Polynomial: A naive approach is finding coefficients of the approximate polynomial with power basis at the current reference points for the continuous function $f(x)$, i.e., we can obtain c_j's in the following equation

$$\sum_{j=0}^{d} c_j x_i^j - f(x_i) = (-1)^i E,$$

where E is also an unknown variable in this system of linear equations. However, this method suffers from the precision problem for the coefficients. It is known that as the degree of the basis of approximate polynomial increases, the coefficients usually decrease, and we have to set higher precision for the coefficients of the higher degree basis. Han et al. [22] use the Chebyshev basis for this coefficient precision problem since the coefficients of a polynomial with the Chebyshev basis usually have the almost same order. Thus, we also use the Chebyshev basis instead of the power basis.

Obtaining Extreme Points: Since we are dealing with a tiny minimax approximation error, we have to obtain the extreme points as precisely as possible. Otherwise, we cannot reach the extreme point for the minimax approximate polynomial precisely, and then the minimax approximation error obtained with this algorithm becomes large. Basically, to obtain the extreme points, we can scan $p(x) - f(x)$ with a small scan step and obtain the extreme points where the increase and decrease are exchanged. A small scan step increases the accuracy of the extreme point but causes a long scan time accordingly. To be more specific, it takes approximately 2^{ℓ} proportional time to find the extreme points with the accuracy of ℓ-bit. Therefore, it is necessary to devise a method to obtain high accuracy extreme points more quickly.

In order to obtain the exact point of the extreme value, we use a method of finding the points where the increase and decrease are exchanged and then finding the exact extreme point using a kind of binary search. Let $r(x) = p(x) - f(x)$ and sc be the scan step. If we can find $x_{i,0}$ where $\mu(x_{i,0})r(x_{i,0}) \geq |E|$, and $(r(x_{i,0}) - r(x_{i,0} - \mathsf{sc}))(r(x_{i,0} + \mathsf{sc}) - r(x_{i,0})) \leq 0$, we obtain the i-th extreme points using the following process successively ℓ times,

$$x_{i,k} = \underset{x \in \{x_{i,k-1} - \mathsf{sc}/2^k, x_{i,k-1}, x_{i,k-1} + \mathsf{sc}/2^k\}}{\arg\max} |r(x)|, k = 1, 2, \cdots, \ell,$$

where the i-th extreme point x_i is set to be $x_{i,\ell}$. Then, we obtain the extreme point with $O(\log(\mathsf{sc}) + \ell)$-bit precision. Since sc needs not to be a too small value, we can find the extreme point with arbitrary precision with linear time to precision ℓ. In summary, we propose that the ℓ-bit precision of the extreme points can be obtained by the linear time of ℓ instead of 2^{ℓ}.

This procedure for each interval in the approximation region can be performed independently with each other, and thus it can be performed effectively with several threads. Since this step is the slowest step among any other steps in the improved multi-interval Remez algorithm, the parallel processing for this procedure is desirable to make the whole algorithm much fast.

One can say that the Newton method is more efficient than the binary search method in finding the extreme points because we may just find the roots of the derivative of $p(x) - f(x)$. However, the extreme points are very densely distributed in our situation, and thus the Newton method may not be stably performed. Even if we miss only one extreme point, the algorithm can act in an undefined manner. The binary search method is fast enough and finds all of the extreme points very robustly, and thus we use the binary search instead of the Newton method.

Obtaining New Reference Points: When we find the new reference points satisfying the local extreme value condition, the alternating condition, and maximum absolute sum condition, there is a naive approach: among local extreme points which satisfy the local extreme value condition, find all $d+2$ points satisfying the alternating condition and choose the $n+1$ points which have the maximum absolute sum value. If we have m local extreme points, we have to investigate $\binom{m}{d+2}$ points, and this value is too large, making this algorithm impractical. Thus, we have to find a more efficient method than this naive approach.

We propose a very efficient and provable algorithm for finding the new reference points. The proposed algorithm always gives the $d+2$ points satisfying the three criteria. It can be considered as an elimination method in that we eliminate some elements for each iteration in the proposed algorithm until we obtain $n+1$ points. It is clear that as long as $m > d+2$, we can find at least one element which may not be included in the new reference points. This proposed algorithm is given in Algorithm 3. Algorithm 3 takes $O(m \log m)$ running time, which is a quasi-linear time algorithm.

We note that there are always some points in all situations such that we can ensure that if we choose a set of $d+2$ points including these points satisfying the alternating condition, there exists the other set of $d+2$ points without these points which satisfies the alternating condition and whose absolute sum is larger. Algorithm 3 finds these points until the number of the remaining points is $d+2$. The correctness proof follows the above basic principle, and the full proof can be found in the full paper [25].

To understand the last part of Algorithm 3, the example can be given that if the extreme point x_2 is removed, $T = \{|r(x_1)|+|r(x_2)|, |r(x_2)|+|r(x_3)|, |r(x_3)|+|r(x_4)|, \cdots\}$ is changed to $T = \{|r(x_1)| + |r(x_3)|, |r(x_3)| + |r(x_4)|, \cdots\}$. It is assumed that whenever we remove an element in the ordered set B in Algorithm 3, the remaining points remain sorted and indices are relabeled in increasing order. When we compare the values to remove some extreme points, there are the cases that the compared values are equal or the smallest element is more than one. In such cases, we randomly remove one of these elements.

3.4 Numerical Analysis with Improved Multi-interval Remez Algorithm

This subsection shows the numerical analysis of the improved multi-interval Remez algorithm for its efficiency and the optimal minimax approximation error.

Maximum Sum Condition: Table 1 shows the number of iterations required to converge to the optimal minimax approximate polynomial in the multi-interval Remez algorithm and the improved multi-interval Remez algorithm. The initial set of reference points is selected uniformly in each interval since we want to observe their performances in the worst case. While selecting new reference points is not unique for each iteration in the multi-interval Remez algorithm, the improved multi-interval Remez algorithm selects the new reference points uniquely for each iteration. Thus, when we analyze the multi-interval Remez algorithm, we randomly select the new reference points for each iteration among the possible sets of reference points that satisfy the local extreme value condition and the alternating condition and have the global extreme point. We set the

Algorithm 3: New Reference

Input : An increasing ordered set of extreme points $B = \{t_1, t_2, \cdots, t_m\}$ with $m \geq d+2$, and the degree of the approximate polynomial d.

Output: $d+2$ points in B satisfying alternating condition and maximum absolute sum condition.

```
1  i ← 1
2  while t_i is not the last element of B do
3      if μ(t_i)μ(t_{i+1}) = 1 then
4          Remove from B one of two points t_i, t_{i+1} having the smaller value
           among {|r(t_i)|, |r(t_{i+1})|}.
5      else
6          i ← i + 1
7      end
8  end
9  if |B| > d + 3 then
10     Calculate all |r(t_i)| + |r(t_{i+1})| for i = 1, ··· , |B| − 1 and sort and store these
       values into the array T.
11 while |B| > d + 2 do
12     if |B| = d + 3 then
13         Remove from B one of two points t_1, t_{|B|} having less value among
           {|r(t_1)|, |r(t_{|B|})|}.
14     else if |B| = d + 4 then
15         Insert |r(t_1)| + |r(t_{|B|})| into T and sort T. Remove from B the two
           elements having the smallest value in T.
16     else
17         if t_1 or t_{|B|} is included in the smallest element in T then
18             Remove from B only t_1 or t_{|B|}.
19         else
20             Remove from B the two elements having the smallest element in T.
21         end
22         Remove from T all elements related to the removed extreme points, and
           insert into T the sum of absolute error values of the two newly adjacent
           extreme points.
23     end
24 end
```

approximation parameter δ in Algorithm 2 as 2^{-40} and repeat this simulation 100 times. It shows that the improved multi-interval Remez algorithm is much better to reduce the iteration number of the Remez algorithm.

Note that the number of iterations depends on the initial set of reference points. In fact, the uniformly distributed reference points are not desirable as an initial set of reference points because these reference points are far from the converged reference points. In fact, the improved multi-interval Remez algorithm with the initialization method explained in the previous subsection only needs

4–14 iterations. The overall running time of the improved multi-interval Remez algorithm with the method in the previous subsection is 1–3 s by PC with AMD Ryzen Threadripper 1950X 16-core CPU @ 3.40 GHz.

Table 1. Comparison of iteration numbers between the improved multi-interval Remez algorithm and the multi-interval Remez algorithm for $\delta = 2^{-40}$

Degree of approx. poly.	Modified Remez algorithm	Multi-interval Remez algorithm			
		Average	Standard deviation	Max	Min
79	**28**	60.0	9.38	82	41
99	**8**	17.1	3.34	28	11
119	**26**	53.4	8.10	79	37
139	**39**	60.3	4.71	79	48
159	**39**	72.1	9.71	98	42
179	**48**	72.3	9.72	105	53
199	**56**	80.4	7.28	94	60

Minimax Error: We obtain the optimal minimax approximate polynomials for the modular reduction function and the scaled cosine function with the scaling number two. Figure 1(a) shows the minimax approximation error of the approximate polynomial of the modular reduction function derived by the improved multi-interval Remez algorithm and the minimax approximation error of the previous homomorphic modular reduction method with scaling number zero in [22], compared to the modular reduction function. That is, let $p_1(x)$ be the optimal minimax approximate polynomial of the normod function and let $q_1(x)$ be the approximate polynomial obtained by Han et al.'s method with scaling number zero when the half-width of approximation region is 2^{-10}. Then, $\max_{x\in D}|p_1(x) - \mathsf{normod}(x)|$ and $\max_{x\in D}|q_1(x) - \mathsf{normod}(x)|$ are compared in Fig. 1(a). Note that while the minimax approximation error of the approximate polynomial of the modular reduction function decreases steadily as the degree of the approximate polynomial increases, the minimax approximation error of the previous method does not decrease when the degree is larger than 76 because of the approximation error between the modular reduction function and the sine/cosine function.

Figure 1(b) shows the minimax approximation error of the composition of the approximate polynomial of the scaled cosine function with scaling number two derived by the improved multi-interval Remez algorithm and two double angle formulas and the minimax approximation error of method in [22], compared to the cosine function. That is, let $p_2(x)$ be the optimal minimax approximate polynomial of $\cos\left(\frac{\pi}{2}(x - 1/4)\right)$ and let $q_2(x)$ be the approximate polynomial obtained by Han et al.'s method with scaling number two when the half-width of approximation region is 2^{-3}. If $r(x) = 2x^2 - 1$, then $\max_{x\in D}|r \circ r \circ p_2(x) - \sin(2\pi x)|$ and $\max_{x\in D}|r \circ r \circ q_2(x) - \sin(2\pi x)|$ are compared in Fig. 1(b). The

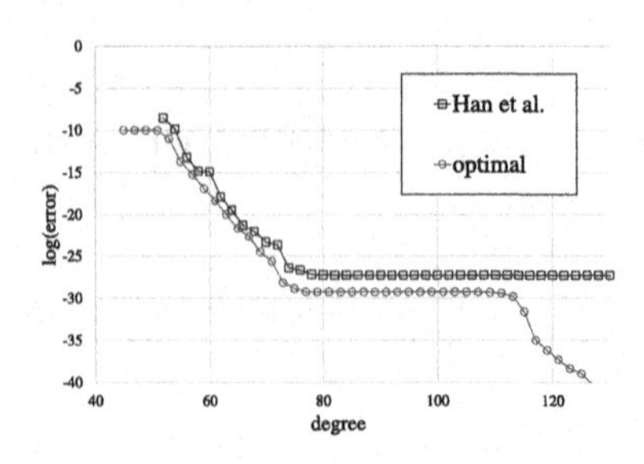

(a) Modular reduction function

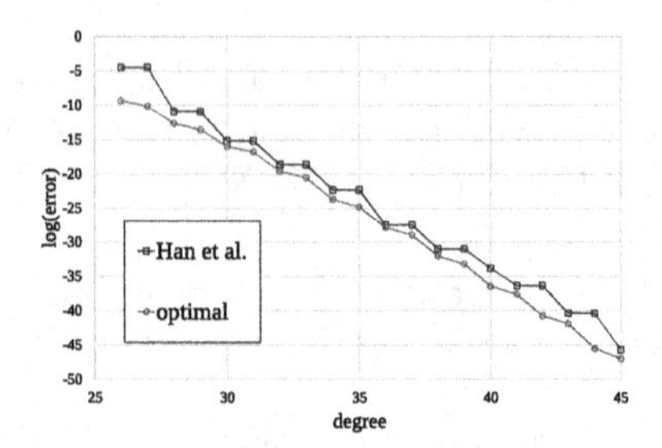

(b) Cosine function with scaling number two

Fig. 1. Comparison of minimax approximatio error between the previous approximation method and the improved multi-interval Remez algorithm.

proposed method improves the minimax approximation error by 2.3 bits on average, and by 5 bits at maximum for the same degree of the approximate polynomial. This improvement leads to a reduction of 1–2 degrees for the given minimax approximation error.

In fact, the approximate polynomial for the modular reduction function cannot yet be used in the bootstrapping of the RNS-CKKS scheme because of the huge coefficients. It is a very unstable polynomial to evaluate in the RNS-CKKS scheme in that these large coefficients amplify the approximation error in the message. It is an interesting open problem to stably use the minimax approximate polynomial of the modular reduction function in the RNS-CKKS scheme. Instead of using the unstable minimax approximate polynomial of the modular reduction function, we approximate the modular reduction function with a composition of several stable polynomials in Sect. 5.

4 Numerical Analysis of Message Precision in Bootstrapping with Improved Multi-Interval Remez Algorithm in SEAL Library

Since the previous researches for the bootstrapping of the RNS-CKKS scheme did not deal deeply with its message precision, we numerically analyze the message precision for the bootstrapping with improved multi-interval Remez algorithm of the RNS-CKKS scheme by changing several parameters: the degree d of the approximate polynomial of the scaled cosine function, the bit-length difference $\delta_{\mathsf{diff}} = \log \Delta_{\mathsf{boot}} - \log \Delta$ between the default scaling factor and the bootstrapping scaling factor, the bootstrapping scaling factor Δ_{boot}, and the number of the slots. We assume that the range of the real part and the imaginary part of the messages to be bootstrapped is $[-1, 1]$. The bootstrapping precision is measured as $-\log_2(e_r + e_i)/2$, where e_r and e_i are the average error of the real part and the imaginary part of all slots, respectively.

Numerical analysis in this section is conducted in PC with Intel(R) Xeon(R) Silver 4210 CPU @ 2.20 GHz single-threaded, and the SEAL library version 3.5.9 [29] is used. The double angle formula for the cosine function is assumed to be used twice. The improved multi-interval Remez algorithm is used to obtain the optimal minimax approximate polynomial in all simulations, rather than polynomial approximation methods in the previous papers, [8,11,22]. The polynomial modulus degree is set to be 2^{16}, the secret key Hamming weight is set to be 192, the value of K is set to be 25, and the maximum modulus for the ciphertext is set to be 2^{1553}, which satisfies the 128-bit security as in [4]. The CoeffToSlot and SlotToCoeff procedures in [8] with two level consumption are used in all simulations. The scaling factor management method and the delayed rescaling method in [24] is applied, and the depth consumption of the polynomial evaluation is optimized by Bossuat et al.'s evaluation method [4]. The input messages are sampled by the uniform distribution over the bootstrapping range.

Degree of Approximate Polynomial: Table 2 shows the message precision of the bootstrapping with the improved multi-interval Remez algorithm when the degree of the approximate polynomial for the scaled cosine function is changed. The value of $\log \delta_{\mathsf{diff}}$ is 12, $\log \Delta_{\mathsf{boot}}$ is 60, and the number of the slots is 2^8 in this simulation. The approximation error means the minimax error of the approximate polynomial for the scaled cosine function, and the bootstrapping error means the average error for each slot when the bootstrapping is performed with the library. When the scaling factor is changed from the bootstrapping scaling factor to the default scaling factor, the message and its error are multiplied by δ_{diff}. We show both the bootstrapping error before changing the scaling factor and that after changing the scaling factor. Although the approximation error continues to decrease as the degree of the approximate polynomial increases, the bootstrapping error does not decrease below a certain value. This bound is caused by either the difference between the modular reduction function and the cosine function or the rescaling error, depending on the situation. Thus, we cannot raise the message precision infinitely by using a high degree approximate polynomial. The actual lower bound of the bootstrapping error before changing the scaling factor is denoted by e_{min} in this section, and then the lower bound of the bootstrapping error after changing the scaling factor is $e_{\mathsf{min}}\delta_{\mathsf{diff}}$.

Table 2. Message precision of the bootstrapping with the improved multi-interval Remez algorithm for various degrees of the approximate polynomials

Degree of approx. poly.	Approximation error by the optimal minimax polynomial	Bootstrapping error		Message precision (bits)
		Before changing Scaling factor e_{min}	After changing Scaling factor $e_{\mathsf{min}}\delta_{\mathsf{diff}}$	
60	1.77×10^{-11}	2.83×10^{-10}	1.16×10^{-6}	19.7
62	5.26×10^{-13}	8.50×10^{-12}	3.48×10^{-8}	24.8
64	3.07×10^{-14}	6.17×10^{-13}	2.53×10^{-9}	28.6
66	1.56×10^{-15}	3.76×10^{-13}	1.54×10^{-9}	29.2
68	6.59×10^{-17}	3.76×10^{-13}	1.54×10^{-9}	29.2

Value of δ_{diff}: The bit length difference between the default scaling factor Δ and the bootstrapping scaling factor Δ_{boot}, which will be denoted as $\delta_{\mathsf{diff}} = \log \Delta_{\mathsf{boot}} - \log \Delta$, is closely related to the message precision. The value of δ_{diff} is usually chosen as 10 bits to lower the difference between the modular reduction function and the sine/cosine function since the half-width of each interval in the approximation region is $2^{-\delta_{\mathsf{diff}}}$.

It causes loss to the message precision of the bootstrapping. In the bootstrapping procedure, we need to divide the message by $2^{\delta_{\mathsf{diff}}}$ so that it can be included in the approximation region and multiply $2^{\delta_{\mathsf{diff}}}$ at the end of the bootstrapping. If the precision error until multiplying $2^{\delta_{\mathsf{diff}}}$ is e, the final error becomes $2^{\delta_{\mathsf{diff}}}e$. e cannot be reduced below a certain error value because of the rescaling error dealt with in the previous subsection. If we denote this lower bound as $e_b = 2^{-\delta_b}$, the message precision will be $\delta_b - \delta_{\mathsf{diff}}$.

Because δ_{diff} has a significant effect on both the message precision of the bootstrapping and the message precision of the intended operation in the application, it is desirable to reduce the δ_{diff} to prevent this precision loss. However, the difference between the sine/cosine function and the modular reduction function is somewhat dominant, and this difference becomes more dominant as δ_{diff} increases.

Table 3 shows the maximum message precision of the bootstrapping with improved multi-interval Remez algorithm for various δ_{diff}. The degree of the approximate polynomials for each case is set to be large enough to reach the minimum approximate error e_{min}, and the scaling factor and the number of slots are fixed to be 60 and 2^8, respectively.

The bootstrapping error after changing the scaling factor is $e_{\mathsf{min}}\delta_{\mathsf{diff}}$. As δ_{diff} decreases, e_{min} increases rapidly so that the $e_{\mathsf{min}}\delta_{\mathsf{diff}}$ grows. This is because the difference between the modular reduction function and the cosine function becomes larger when the approximation region is enlarged. We can naively expect that the bootstrapping error can be decreased infinitely when $\log \delta_{\mathsf{diff}}$ is increased, because $|\epsilon - \sin \epsilon| = O(\epsilon^3)$. However, if the δ_{diff} is larger than 16, the value of e_{min} does not decrease, and thus $e_{\mathsf{min}}\delta_{\mathsf{diff}}$ increases. This lower bound of e_{min} is caused by the rescaling error and the homomorphic linear transform in the bootstrapping. This bound of e_{min} is related to the bootstrapping scaling factor Δ_{boot} and the number of slots, which will be dealt with in the following paragraphs.

Note that we do not need to use the scaling factor $\Delta_{\mathsf{boot}}/\delta_{\mathsf{diff}}$ after bootstrapping. If we use a scaling factor $2^{-\ell}\Delta_{\mathsf{boot}}/\delta_{\mathsf{diff}}$, the bootstrapping error is amplified by ℓ-bit and the range of the message becomes $[-2^{\ell}, 2^{\ell}]$. Indeed, there are many cases that large range of the message is more important than low bootstrapping error, and thus users can control the scaling factor after bootstrapping concerning the range and the bootstrapping error they want to use.

Bootstrapping Scaling Factor: Table 4 shows the maximum message precision for various bootstrapping scaling factors when the number of slots is 2^8. The degree of the approximate polynomial and the value of δ_{diff} are set to reach the lower bound of e_{min} for each bootstrapping scaling factor and to minimize the value of $e_{\mathsf{min}}\delta_{\mathsf{diff}}$, which determines the actual message precision of the boot-

Table 3. Message precision of the bootstrapping with the improved multi-interval Remez algorithm for various values of $\log \delta_{\text{diff}}$

$\log \delta_{\text{diff}}$	Bootstrapping error		Message precision (bits)
	Before changing scaling factor e_{min}	After changing scaling factor $e_{\text{min}}\delta_{\text{diff}}$	
3	2.55×10^{-5}	2.04×10^{-4}	12.3
7	7.45×10^{-9}	9.53×10^{-7}	20.0
10	1.32×10^{-11}	1.35×10^{-8}	26.1
11	1.64×10^{-12}	3.36×10^{-9}	28.1
12	3.76×10^{-13}	1.54×10^{-9}	29.2
13	2.88×10^{-13}	2.36×10^{-9}	28.7
14	2.77×10^{-13}	4.54×10^{-9}	27.7

strapping. The second column in Table 4 shows the lower bound of e_{min}, the third column shows the value of δ_{diff} which minimizes $e_{\text{min}}\delta_{\text{diff}}$, and the last column shows the maximum message precision with the corresponding bootstrapping scaling factor.

The maximum message precision of the bootstrapping in the RNS-CKKS scheme decreases as the bootstrapping scaling factor decreases. This means that we have to use as large a bootstrapping scaling factor as possible when we need precise bootstrapping. Since the bootstrapping scaling factor is related to the multiplicative depth, this gives the trade-off between the depth and the precision.

Note that the bit-length of scaling factors can be different for each level, and thus we do not need to use the same scaling factor throughout the bootstrapping. This fact is used in Bossuat et al.'s work [4].

Table 4. Maximum message precision of the bootstrapping with the improved multi-interval Remez algorithm for various bootstrapping scaling factors

$\log \Delta_{\text{boot}}$	$\log \delta_{\text{diff}}$	Bootstrapping error		Message precision (bits)
		Before changing scaling factor e_{min}	After changing scaling factor $e_{\text{min}}\delta_{\text{diff}}$	
50	9	3.30×10^{-10}	1.69×10^{-7}	22.5
54	10	2.21×10^{-11}	2.27×10^{-8}	25.4
57	11	2.88×10^{-12}	5.90×10^{-9}	27.3
60	12	3.71×10^{-13}	1.52×10^{-9}	29.3

Number of Slots: Table 5 shows the maximum message precision for various numbers of slots when the bootstrapping scaling factor is 60, the maximum scaling factor. The degree of the approximate polynomials and δ_{diff} are set to the same as Table 4. The error analysis in [11] shows that the approximation error in SlotToCoeff step is amplified more as we use more slots. The result of Table 5 corresponds to this error analysis. This gives the trade-off between the number of slots and the message precision. Note that all precision is less than 32-bit precision. We will improve these precision results in the next section by using the composite function method with the inverse sine function.

Note that $\log \delta_{\mathsf{diff}}$ is not high when the number of slots is large, although this $\log \delta_{\mathsf{diff}}$ value does not ensure the high precision as the difference between the modular reduction function and the sine/cosine function is rather high. This phenomenon is because the coefficients of the message polynomial is very small when the number of slots is large as discussed in [4]. Thus, the approximation region for the cosine function can be generally much less than $1/\delta_{\mathsf{diff}}$.

Table 5. Maximum message precision of the bootstrapping with improved multi-interval Remez algorithm for various numbers of slots

$\log n$	Degree of approx. poly.	$\log \delta_{\mathsf{diff}}$	Bootstrapping error		Message precision (bits)	Remaining modulus	Running time (s)
			Before changing scaling factor e_{min}	After changing scaling factor $e_{\mathsf{min}}\delta_{\mathsf{diff}}$			
5	67	14	4.52×10^{-14}	7.42×10^{-10}	30.3	653	91.9
8	66	12	3.71×10^{-13}	1.52×10^{-9}	29.3	653	133.6
10	66	11	1.34×10^{-12}	2.75×10^{-9}	28.4	653	189.3
12	66	9	8.46×10^{-12}	4.33×10^{-9}	27.8	653	287.0
14	66	8	2.46×10^{-11}	6.31×10^{-9}	27.2	653	461.0

5 Improvement of Message Precision by Composite Function Approximation of Modular Reduction Function

At first glance, it seems to be the best method to use the optimal minimax approximate polynomials for the modular reduction function. However, we can see that some of the coefficients of the optimal minimax approximate polynomials with regard to the Chebyshev basis are so large that the amplified approximate errors by these coefficients totally distort the messages in the ciphertext. On the other hand, the optimal minimax approximate polynomial coefficients of the scaled sine/cosine functions with more than one scale number are small enough not to distort the messages. Thus, the approximation of the modular reduction function by the sine/cosine function is essential for the correctness of the RNS-CKKS scheme.

When we adhere to the approximation by the scaled sine/cosine function, the difference of the modular reduction function and the sine/cosine function is a crucial obstacle, which is mentioned as an important open problem in Han et al.'s paper

[22]. This difference is sharply increased as the approximation region of the modular reduction function becomes longer, and this prevents us from reducing δ_{diff}.

5.1 Composite Function Approximation of Modular Reduction Function by Inverse Sine Function

We propose a simple and novel method for solving this problem, which is called the composite function approximation method. In short, we compose the optimal minimax approximate polynomial of the sine/cosine function and the approximate polynomial of the inverse sine function. It is easy to check that if we have two functions f and g for $0 < \epsilon < \frac{1}{4}$ as

$$f : \bigcup_{k=-\infty}^{\infty} [2\pi(k-\epsilon), 2\pi(k+\epsilon)] \to [-\sin 2\pi\epsilon, \sin 2\pi\epsilon], \quad f(x) = \sin x$$

$$g : [-\sin 2\pi\epsilon, \sin 2\pi\epsilon] \to [-2\pi\epsilon, 2\pi\epsilon], \quad g(x) = \arcsin x,$$

then the following equation holds as

$$x - 2\pi \cdot \mathsf{round}\left(\frac{x}{2\pi}\right) = (g \circ f)(x), \quad x \in \bigcup_{k=-\infty}^{\infty} [2\pi(k-\epsilon), 2\pi(k+\epsilon)].$$

If we substitute $t = \frac{x}{2\pi}$, then we have

$$\mathsf{normod}(t) = \frac{1}{2\pi}(g \circ f)(2\pi t), \quad t \in \bigcup_{k=-\infty}^{\infty} [k-\epsilon, k+\epsilon]. \tag{4}$$

If we approximate both f and g with the optimal minimax approximate polynomials derived by the improved multi-interval Remez algorithm, we can approximate the modular reduction function with any small approximate error by the composition of f and g. Note that $g(x)$ can be approximated very well with some approximate polynomials of a small degree since the domain of $g(x)$ is only one interval. Indeed, the cosine approximation with double-angle formula in [22] can be regarded as the special case of the proposed composite function approximation, in that they approximate $g(x)$ with x, that is, the identity function. Note that the cosine function in [22] is merely a parallel shift of the sine function. Thus, it is said that they approximate the sine function instead of the cosine function.

The sine function f was evaluated by composing the scaled cosine function and several double-angle formulas in [22]. If the number of the used double-angle formula is ℓ, then the functions h_1, h_2, and h_3 are defined as

$$h_1(x) = \cos\left(\frac{2\pi}{2^\ell}\left(x - \frac{1}{4}\right)\right), \; h_2(x) = 2x^2 - 1, \; h_3(x) = \frac{1}{2\pi}\arcsin(x).$$

Then, the normod function, which is equivalent to the modular reduction function, can be represented as

$$\mathsf{normod}(x) = h_3 \circ h_2^\ell \circ h_1(x).$$

Thus, if $\tilde{h}_1$ is the optimal minimax approximate polynomial of h_1 and $\tilde{h}_3$ is that of h_3, we can approximate normod function by the composition of several polynoimals as

$$\mathsf{normod}(x) \approx \tilde{h}_3 \circ h_2^{\ell} \circ \tilde{h}_1(x).$$

With this method, we can approximate the modular reduction function by the composition of several polynomials at arbitrary precision. This enables us to reduce δ_{diff} to 3 and reach the message precision of $\delta_b - \delta_{\mathsf{diff}}$, which is the best precision mentioned in the previous section. The next section shows that we can indeed reach this high precision in the SEAL library.

5.2 Simulation Result with SEAL Library

This subsection demonstrates that the composite function method can improve the message precision in the RNS-CKKS scheme. The simulation environment is the same as the simulation in Sect. 4.

Table 6 shows that the value of e_{min} with the composite function method does not change. The degrees of approximate polynomials of the scaled cosine function and inverse sine function are set to minimize e_{min}, and these degrees are shown in Table 6. In contrast to the result in Table 3, all of the values of e_{min} in Table 6 are almost the same as the minimum value of e_{min} in Table 3 regardless of δ_{diff}. Since e_{min} is fixed with the minimum value, the bootstrapping precision, which is determined by $e_{\mathsf{min}}\delta_{\mathsf{diff}}$, is increased as δ_{diff} decreases.

Table 6. Maximum message precision of the bootstrapping with improved multi-interval Remez algorithm and composite function method for various δ_{diff}

$\log \delta_{\mathsf{diff}}$	Bootstrapping error		Message precision (bits)
	Before changing scaling factor e_{min}	After changing scaling factor $e_{\mathsf{min}}\delta_{\mathsf{diff}}$	
3	2.93×10^{-13}	2.34×10^{-12}	38.6
7	2.90×10^{-13}	3.71×10^{-11}	34.6
10	2.85×10^{-13}	2.92×10^{-10}	31.7
11	3.21×10^{-13}	6.58×10^{-10}	30.5
12	2.88×10^{-13}	1.18×10^{-9}	29.7

Table 7 shows the maximum precision of the bootstrapping with the improved multi-interval Remez algorithm and composite function method for various slots. The δ_{diff} value and $\log \Delta_{\mathsf{diff}}$ value are set to be 3 and 60, respectively. The maximum message precision is increased by 5.4–10.2 bits and becomes 32.6–40.5 bit precision. All of the message precision is larger than the 32-bit precision. Thus, we make the bootstrapping of the RNS-CKKS scheme more reliable enough to be used in practical applications.

The half-width of the approximation region has to be $2^{-\delta_{\mathsf{diff}}}$ when the range of real and imaginary part of the messages is assumed to be the same as that of the coefficients of the message polynomial. However, when we sample the messages from the uniform distribution over the range, the coefficients are significantly reduced so that we may reduce the approximation region as discussed in [4]. ϵ denotes the half-width of each approximation region we set for each number of slots, and this value is numerically set to have no effect on the bootstrapping. If one wants to be conservative on the distribution of the message and the range of the coefficients, they may set ϵ to be $2^{-\delta_{\mathsf{diff}}}$.

Table 7. Maximum message precision of the bootstrapping with composite function method for various number of slots

$\log n$	$\log 1/\epsilon$	Deg. of app. poly. of cos.	Deg. of app. poly. of inv. sine	Bootstrapping error		Message precision (bits)	Remaining modulus	Running time (s)
				Before changing scaling factor e_{min}	After changing scaling factor $e_{\mathsf{min}}\delta_{\mathsf{diff}}$			
5	4	71	15	7.93×10^{-14}	6.34×10^{-13}	40.5	473	94.7
8	6	70	9	2.93×10^{-13}	2.34×10^{-12}	38.6	473	133.2
10	9	69	7	1.14×10^{-12}	9.13×10^{-12}	36.7	533	188.9
12	10	69	5	4.84×10^{-12}	3.87×10^{-11}	34.5	533	273.9
14	10	68	5	1.97×10^{-11}	1.53×10^{-10}	32.6	533	451.5

Although we add the inverse sine approximation procedure, the overall running time of the bootstrapping is similar or reduced. Note that the more depth level left in a ciphertext, the more time homomorphic evaluation takes. Since we consume more depth level in the inverse sine approximation procedure, the ciphertexts in the SlotToCoeff procedure have less remaining depth level. Thus, the running time of the SlotToCoeff procedure in the new bootstrapping is less than that in the original one. The remaining modulus bit length is reduced because of the additional depth consumption of the inverse sine approximation procedure. This additional depth consumption can be seen as a trade-off for the high precision.

6 Conclusion

We proposed the algorithm for obtaining the optimal minimax approximate polynomial for any continuous function on the union of the finite set, including the scaled cosine function on separate approximation regions. Then we analyzed the message precision of the bootstrapping with the improved multi-interval Remez algorithm in RNS-CKKS, and its maximum message precision is measured in the SEAL library. We proposed the composite function method with inverse sine function to improve the message precision of the bootstrapping significantly, and thus the improved message precision bootstrapping has the precision higher than the precision of the 32-bit fixed-point number system, even when lots of slots are used. Thus, the large-depth operations in advanced applications, such as training a convolutional neural network for encrypted data, are needed to be implemented by the RNS-CKKS scheme with the improved message precision bootstrapping.

Acknowledgement. We thank Jean-Philippe Bossuat for his help with optimizing the approximation for the inverse sine function by observing the distribution of the message polynomial coefficients.

References

1. PALISADE Lattice Cryptography Library (release 1.10.4), September 2020. https://palisade-crypto.org/
2. Boemer, F., Costache, A., Cammarota, R., Wierzynski, C.: nGraph-HE2: a high-throughput framework for neural network inference on encrypted data. In: Proceedings of the 7th ACM Workshop on Encrypted Computing & Applied Homomorphic Cryptography, pp. 45–56 (2019)
3. Boemer, F., Lao, Y., Cammarota, R., Wierzynski, C.: nGraph-HE: a graph compiler for deep learning on homomorphically encrypted data. In: Proceedings of the 16th ACM International Conference on Computing Frontiers, pp. 3–13 (2019)
4. Bossuat, J.P., Mouchet, C., Troncoso-Pastoriza, J., Hubaux, J.P.: Efficient bootstrapping for approximate homomorphic encryption with non-sparse keys. Cryptology ePrint Archive, Report 2020/1203 (2020). https://eprint.iacr.org/2020/1203. Accepted to Eurocrypt 2021
5. Bourse, F., Minelli, M., Minihold, M., Paillier, P.: Fast homomorphic evaluation of deep discretized neural networks. In: Shacham, H., Boldyreva, A. (eds.) CRYPTO 2018. LNCS, vol. 10993, pp. 483–512. Springer, Cham (2018). https://doi.org/10.1007/978-3-319-96878-0_17
6. Brakerski, Z., Gentry, C., Vaikuntanathan, V.: (Leveled) fully homomorphic encryption without bootstrapping. ACM Trans. Comput. Theory **6**(3), 13 (2014)
7. Brutzkus, A., Gilad-Bachrach, R., Elisha, O.: Low latency privacy preserving inference. In: International Conference on Machine Learning, pp. 812–821. PMLR (2019)
8. Chen, H., Chillotti, I., Song, Y.: Improved bootstrapping for approximate homomorphic encryption. In: Ishai, Y., Rijmen, V. (eds.) EUROCRYPT 2019. LNCS, vol. 11477, pp. 34–54. Springer, Cham (2019). https://doi.org/10.1007/978-3-030-17656-3_2
9. Chen, Y., et al.: Dadiannao: a machine-learning supercomputer. In: 2014 47th Annual IEEE/ACM International Symposium on Microarchitecture, pp. 609–622. IEEE (2014)
10. Cheney, E.: Introduction to Approximation Theory. McGraw-Hill, New York (1966)
11. Cheon, J.H., Han, K., Kim, A., Kim, M., Song, Y.: Bootstrapping for approximate homomorphic encryption. In: Nielsen, J.B., Rijmen, V. (eds.) EUROCRYPT 2018. LNCS, vol. 10820, pp. 360–384. Springer, Cham (2018). https://doi.org/10.1007/978-3-319-78381-9_14
12. Cheon, J., Han, K., Kim, A., Kim, M., Song, Y.: A full RNS variant of approximate homomorphic encryption. In: Cid, C., Jacobson, M. (eds.) SAC 2018. LNCS, vol. 11349, pp. 347–368. Springer, Cham (2018). https://doi.org/10.1007/978-3-030-10970-7_16

13. Cheon, J.H., Kim, A., Kim, M., Song, Y.: Homomorphic encryption for arithmetic of approximate numbers. In: Takagi, T., Peyrin, T. (eds.) ASIACRYPT 2017. LNCS, vol. 10624, pp. 409–437. Springer, Cham (2017). https://doi.org/10.1007/978-3-319-70694-8_15
14. Chillotti, I., Gama, N., Georgieva, M., Izabachène, M.: TFHE: fast fully homomorphic encryption over the torus. J. Cryptol. **33**(1), 34–91 (2020)
15. Dathathri, R., et al.: Chet: an optimizing compiler for fully-homomorphic neural-network inferencing. In: Proceedings of the 40th ACM SIGPLAN Conference on Programming Language Design and Implementation, pp. 142–156 (2019)
16. Fan, J., Vercauteren, F.: Somewhat practical fully homomorphic encryption. Cryptology ePrint Archive, Report 2012/144 (2012). https://eprint.iacr.org/2012/144
17. Filip, S.: A robust and scalable implementation of the Parks-McClellan algorithm for designing FIR filters. ACM Trans. Math. Softw. **43**(1), 1–24 (2016)
18. Gentry, C.: Fully homomorphic encryption using ideal lattices. In: Proceedings of the Forty-First Annual ACM Symposium on Theory of Computing, pp. 169–178 (2009)
19. Gentry, C., Sahai, A., Waters, B.: Homomorphic encryption from learning with errors: conceptually-simpler, asymptotically-faster, attribute-based. In: Canetti, R., Garay, J.A. (eds.) CRYPTO 2013. LNCS, vol. 8042, pp. 75–92. Springer, Heidelberg (2013). https://doi.org/10.1007/978-3-642-40041-4_5
20. Gupta, S., Agrawal, A., Gopalakrishnan, K., Narayanan, P.: Deep learning with limited numerical precision. In: International Conference on Machine Learning, pp. 1737–1746 (2015)
21. Gysel, P., Motamedi, M., Ghiasi, S.: Hardware-oriented approximation of convolutional neural networks. In: International Conference on Learning Representation (2016)
22. Han, K., Ki, D.: Better bootstrapping for approximate homomorphic encryption. In: Jarecki, S. (ed.) CT-RSA 2020. LNCS, vol. 12006, pp. 364–390. Springer, Cham (2020). https://doi.org/10.1007/978-3-030-40186-3_16
23. Jiang, X., Kim, M., Lauter, K., Song, Y.: Secure outsourced matrix computation and application to neural networks. In: Proceedings of the 2018 ACM SIGSAC Conference on Computer and Communications Security, pp. 1209–1222 (2018)
24. Kim, A., Papadimitriou, A., Polyakov, Y.: Approximate homomorphic encryption with reduced approximation error. Cryptology ePrint Archive, Report 2020/1118 (2020). https://eprint.iacr.org/2020/1118
25. Lee, J.W., Lee, E., Lee, Y., Kim, Y.S., No, J.S.: High-precision bootstrapping of RNS-CKKS homomorphic encryption using optimal minimax polynomial approximation and inverse sine function. Cryptology ePrint Archive, Report 2020/552 (2020). https://eprint.iacr.org/2020/552
26. McClellan, J., Parks, T.: A personal history of the Parks-McClellan algorithm. IEEE Signal Process. Mag. **22**(2), 82–86 (2005)
27. Powell, M.: Approximation Theory and Methods. Cambridge University Press, Cambridge (1981)
28. Remez, E.: Sur la détermination des polynômes d'approximation de degré donnée. Commun. Kharkov Math. Soc. **10**(196), 41–63 (1934)
29. Microsoft SEAL (release 3.5), Microsoft Research, Redmond, April 2020. https://github.com/Microsoft/SEAL
30. van Dijk, M., Gentry, C., Halevi, S., Vaikuntanathan, V.: Fully homomorphic encryption over the integers. In: Gilbert, H. (ed.) EUROCRYPT 2010. LNCS, vol. 6110, pp. 24–43. Springer, Heidelberg (2010). https://doi.org/10.1007/978-3-642-13190-5_2

On the Security of Homomorphic Encryption on Approximate Numbers

Baiyu Li(✉) and Daniele Micciancio

University of California, San Diego, USA
{baiyu,daniele}@cs.ucsd.edu

Abstract. We present passive attacks against CKKS, the homomorphic encryption scheme for arithmetic on approximate numbers presented at Asiacrypt 2017. The attack is both theoretically efficient (running in expected polynomial time) and very practical, leading to complete key recovery with high probability and very modest running times. We implemented and tested the attack against major open source homomorphic encryption libraries, including HEAAN, SEAL, HElib and PALISADE, and when computing several functions that often arise in applications of the CKKS scheme to machine learning on encrypted data, like mean and variance computations, and approximation of logistic and exponential functions using their Maclaurin series.

The attack shows that the traditional formulation of IND-CPA security (or indistinguishability against chosen plaintext attacks) achieved by CKKS does not adequately capture security against passive adversaries when applied to approximate encryption schemes, and that a different, stronger definition is required to evaluate the security of such schemes.

We provide a solid theoretical basis for the security evaluation of homomorphic encryption on approximate numbers (against passive attacks) by proposing new definitions, that naturally extend the traditional notion of IND-CPA security to the approximate computation setting. We propose both indistinguishability-based and simulation-based variants, as well as restricted versions of the definitions that limit the order and number of adversarial queries (as may be enforced by some applications). We prove implications and separations among different definitional variants, and discuss possible modifications to CKKS that may serve as a countermeasure to our attacks.

1 Introduction

Fully homomorphic encryption (FHE) schemes allow to perform arbitrary computations on encrypted data (without knowing the decryption key), and, at least in theory, can be a very powerful tool to address a wide range of security problems, especially in the area of distributed or outsourced computation. Since the discovery of Gentry's bootstrapping technique [23] and the construction of the

Research supported by Global Research Cluster program of Samsung Advanced Institute of Technology and NSF Award 1936703.

A. Canteaut and F.-X. Standaert (Eds.): EUROCRYPT 2021, LNCS 12696, pp. 648–677, 2021.
https://doi.org/10.1007/978-3-030-77870-5_23

first FHE schemes based on standard lattice assumptions [9–12], improving the efficiency of these constructions has been one of the main challenges in the area, both in theory and in practice.

The main source of inefficiency in FHE constructions is the fact that these cryptosystems (or, more generally, encryption schemes based on lattice problems [38,45]) are inherently noisy: encrypting (say) an integer message m, and then applying the *raw* decryption function produces a perturbed message $m+e$, where e is a small error term added for security purposes during the encryption process. This is not much of a problem when using only encryption and decryption operations: the error can be easily removed by scaling the message m by an appropriate factor $B > 2|e|$ (e.g., as already done in [45]), or applying some other form of error correction to m before encryption. Then, if the raw decryption function outputs a perturbed value $v = m \cdot B + e$, the original message m can be easily recovered by rounding v to the closest multiple of B. However, when computing on encrypted messages using a homomorphic encryption scheme, the errors can grow very quickly, making the resulting ciphertext undecryptable, or requiring such a large value of B (typically exponential or worse in the depth of the computation) that the cost of encryption becomes prohibitive. The size of the encryption noise e can be reduced using the bootstrapping technique introduced by Gentry in [23], thereby allowing to perform arbitrary computations with a fixed value of B. However, all known bootstrapping methods are very costly, making them the main efficiency bottleneck for general purpose computation on encrypted data. So, reducing the growth rate of the noise e during encrypted computations is of primary importance to either use bootstrapping less often, or avoid the use of bootstrapping altogether by employing a sufficiently large (but not too big) scaling factor B. In fact, controlling the error growth during homomorphic computations has been the main objective of much research work, starting with [9–12].

Homomorphic encryption for arithmetic on approximate numbers. One of the most recent and interesting contributions along these lines is the approach suggested in [14,16–18,33] based on the idea that in many practical scenarios, computations are performed on real-world data which is already approximate, and the result of the computation inherently contains small errors even when carried out in the clear (without any encryption), due to statistical noise or measurement errors. If the goal of encryption is to secure these *approximate* real-world computations, requiring the decryption function to produce exact results may seem an overkill, and rightly so: if the decryption algorithm simply outputs $m+e$, the application can treat e just like the noise already present in the input and output of the (unencrypted) computation. Interestingly, [18] shows that the resulting "approximate encryption" scheme produces results that are almost as accurate as floating point computations on plaintext data. But the practical impact on the concrete efficiency of the scheme is substantial: by avoiding the large scaling factor B, the scheme achieves much slower error growth than "exact" homomorphic encryption schemes. This allows to perform much deeper computations before the need to invoke a costly bootstrapping procedure, and, in many settings, completely avoid the use of bootstrapping while still delivering results that are sufficiently accurate for the application.

Not surprisingly, the scheme of [18] and its improved variants [14,16,17,33] (generically called CKKS after the authors of [18]) have attracted much attention as a potentially more practical method to apply homomorphic computation on the encryption of real data. The CKKS paper [18] already provided an open source implementation in the "Homomorphic Encryption for Arithmetic on Approximate Numbers" (HEAAN) library [30]. Subsequently, other implementations of the scheme have been included in pretty much all mainstream libraries for secure computation on encrypted data, like Microsoft's "Simple Encrypted Arithmetic Library" SEAL [15], IBM's "Homomorphic Encryption" library HElib [26–28], and NJIT's lattice cryptography library PALISADE [41]. Some of these libraries are used as a backend for other tools, like Intel's nGraph-HE compiler [6,7] for secure machine learning applications, and a wide range of other applications, including the encrypted computation of logistic regression [29], security-preserving support vector machines [42], homomorphic training of logistic regression models [5], homomorphic evaluation of neural networks and tensor programs [20,21], compiling ngraph programs for deep learning [7], private text classification [2], and clustering over encrypted data [19] just to name a few.

Our contribution. While, as argued in much previous work, approximate computations have little impact on the correctness of many applications, we bring into question their impact on security. In particular, we show that the traditional formulation of indistinguishability under chosen plaintext attack (IND-CPA, [4,25], see Definition 1) is inadequate to capture security against passive adversaries when applied to approximate encryption schemes. In fact, as our work shows, an approximate homomorphic encryption scheme can satisfy IND-CPA security and still be completely insecure from both a theoretical and practical standpoint. In order to put the study of approximate homomorphic encryption schemes on a sound theoretical basis, we propose a new, more refined formulation of passive security which properly captures the capabilities of a passive adversary when applied to approximate (homomorphic) encryption schemes. We call this notion IND-CPA^{D} security, or "indistinguishability under chosen plaintext attacks with decryption oracles", for reasons that will soon be clear. Our new IND-CPA^{D} security definition is a conservative extension of IND-CPA, in the sense that (1) it implies IND-CPA security, and (2) when applied to standard (exact, possibly homomorphic) encryption schemes, it is perfectly equivalent to IND-CPA. However, when applied to approximate encryption, it is strictly stronger: there are approximate encryption schemes that are IND-CPA secure, but not IND-CPA^{D}.

This is not just a theoretical problem: we show (both by means of theoretical analysis and practical experimentation) that the definitional shortcomings highlighted by our investigation directly affect concrete homomorphic encryption schemes proposed and implemented in the literature. In particular, we show that the CKKS FHE scheme for arithmetics on approximate numbers (both as described in the original paper [18], and as implemented in all major FHE software libraries [30,31,41,47]) is subject to a devastating *key recovery attack* that can be carried out by a passive adversary, accessing the encryption function only through the public interfaces provided by the libraries. We remark that there is no contradiction between our results and the formal security claims made in [18]: the CKKS scheme satisfies IND-CPA security under standard assumptions on the hardness of the (Ring) LWE problem. The problem is with the technical definition of IND-CPA used in [18], which does not offer any reasonable level of security against passive adversaries when applied to approximate schemes.

The ideas behind the new $\mathsf{IND\text{-}CPA}^{\mathsf{D}}$ definition and the attacks to CKKS are easily explained. The traditional formulation of IND-CPA security lets the adversary choose the messages being encrypted, in order to model a-priori knowledge about the message distribution, or even the possibility of the adversary influencing the choice of the messages encrypted by honest parties. This is good, but not enough. When using a homomorphic encryption scheme, a passive adversary may also choose/know the homomorphic computation being performed[1]. Finally, a passive adversary may observe the decrypted result of some homomorphic computations. (See Fig. 1 for an illustration.) So, our $\mathsf{IND\text{-}CPA}^{\mathsf{D}}$ definition provides the adversary with encryption, evaluation, and a severely restricted decryption oracle[2] that model the input/output interfaces of encryption and evaluation algorithms and the output interface of the decryption algorithm. We chose the name $\mathsf{IND\text{-}CPA}^{\mathsf{D}}$ to indicate its close relationship to IND-CPA, but with some emphasis on the adversarial ability to observe the decryption results[3]. It is easy to check (see Lemma 1) that as long as the definition is applied to a standard (exact) encryption scheme, observing the decryption of the final result of the homomorphic computation provides no additional power to the adversary: since the adversary already knows the initial message m and the function f, it can also compute the final result $f(m)$ on its own. So, there is no need to explicitly give to the adversary access to a decryption (or homomorphic evaluation) oracle.

However, for *approximate* encryption schemes, seeing the result of decryption may provide additional information, which the adversary cannot easily compute (or simulate) on its own. In particular, this additional data may provide useful information about other ciphertexts, or even the secret key material. This possibility is quite real, as we demonstrate it can be used to attack all the major libraries implementing the CKKS scheme. The attack is very simple. It involves encrypting a collection of messages, optionally performing some homomorphic computations on them, and finally observing the decryption of the result. Then, using only the information available to a passive adversary (i.e., the input values,

[1] This computation may or may not be secret, depending on whether the scheme is "circuit-hiding".

[2] We remark that this use of decryption oracle is only a technical detail of our formulation, and it is quite different from the decryption oracle used for defining active (chosen ciphertext) attacks: Our decryption oracle only provides access to the plaintext output interface of a decryption algorithm, and does not allow to apply the decryption algorithm on adversarially chosen ciphertexts.

[3] The name $\mathsf{IND\text{-}CPA}^{+}$ was used in earlier versions of this paper. An alternative notation could be IND-CPA-D.

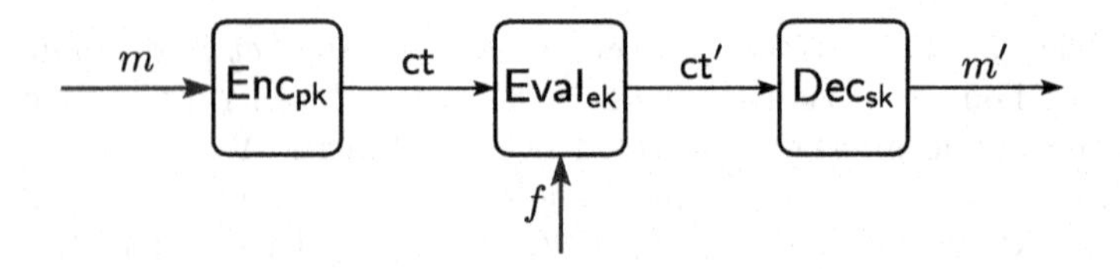

Fig. 1. A passive attacker against a homomorphic encryption scheme may choose/know the plaintext m and the homomorphic computation f (thick blue interfaces), and it can read from black interfaces to learn the ciphertexts $\mathsf{ct}, \mathsf{ct}'$ and the decryption results m'. The adversary has only passive access to the communication and final output channels, i.e., it can eavesdrop, but is not allowed to tamper with (or inject) ciphertexts or alter the final result of the computation. (Color figure online)

encrypted ciphertexts, and final decrypted result of the computation), the attack attempts to recover the secret key using standard linear algebra or lattice reduction techniques. We demonstrate the attack on a number of simple, but representative computations: the computation of the mean or variance of a large data set, and the approximate computation of the logistic and exponential functions using their Maclaurin series. These are all common computations that arise in the application of CKKS to secure machine learning, the primary target area for approximate homomorphic encryption. We implemented and tested the attack against all main open source libraries implementing approximate homomorphic encryption (HEAAN, RNS-HEAAN, PALISADE, SEAL and HElib), showing that they are all vulnerable. We stress that this is due not to an implementation bug in the libraries (which faithfully implement CKKS encryption), but to the shortcomings of the theoretical security definition originally used to evaluate the CKKS scheme. Still, our key recovery attack works very well both in theory and in practice, provably running in expected polynomial time and with success probability 1, and recovering the key in practice, even for large values of the security parameter, in just a few seconds. So, the attack may pose a real threat to applications using the libraries. It immediately follows from the attack that the CKKS scheme is not $\mathsf{IND\text{-}CPA}^{\mathsf{D}}$ secure. In practice, such an attack can be carried out in systems where the decryption results are made publicly available, or, more generally, they may be disclosed to selected parties. As an example, consider privacy-preserving data sharing and aggregation services for medical data [44]. In this setting, individual hospitals encrypt their own sensitive medical records using a public key approximate homomorphic encryption scheme and upload the ciphertexts to a cloud computing service; the cloud service accepts queries from an investigator, perhaps from one of the hospitals, and homomorphically computes the requested statistics. Finally, it decrypts or re-encrypts the final computation result (possibly with the help of a third party that holds the secret decryption key) and sends it to the investigator. We may assume that the service checks that the query issued by the investigator is legitimate, and does not reveal sensitive information about individual patient records. Still, our attack shows that the result of the query may be enough to recover in full the secret decryption key, exposing the entire medical record database of all participating hospitals. Similar attacks are also feasible in homomorphic encryption based vehicular ad-hoc networks [48] where homomorphically evaluated data analytics (both ciphertexts and decrypted results) can be accessed by a passive attacker.

On the theoretical side, we consider several restricted versions of $\mathsf{IND\text{-}CPA}^{\mathsf{D}}$, showing implications and separations among them. For example, one may consider adversaries that perform only a bounded number k of decryption queries, as may be enforced by an application that chooses a new key every k homomorphic computations. ($\mathsf{IND\text{-}CPA}$ may be considered a special case where $k = 0$.) Interestingly, we show that for every k there are approximate encryption schemes that are secure up to k decryption queries, but completely insecure for $k + 1$.

Relations to other attacks to homomorphic encryption schemes. It is well known that homomorphic encryption schemes cannot be secure under adaptive chosen ciphertext attacks (CCA2). In [35], Li, Galbraith, and Ma presented adaptive key recovery attacks against the GSW homomorphic encryption scheme as well as modifications to GSW to prevent such attacks. We remark that both attacks considered in [35] are active attacks that require calling a decryption oracle on ciphertexts formed by the adversary. So these attacks are outside of the $\mathsf{IND\text{-}CPA}^{\mathsf{D}}$ security model that we consider in this paper.

Organization. The rest of the paper is organized as follows. In Sect. 2 we provide some mathematical background about the LWE problem and lattice-based (homomorphic) encryption. In Sect. 3 we present our $\mathsf{IND\text{-}CPA}^{\mathsf{D}}$ security definition, and initiate its theoretical study, proving implication and separation results between different variants of the definition. In Sect. 4 we give a detailed description and rigorous analysis of our attack. Practical experiments using our implementation of the attack are described in Sect. 5. Section 6 concludes with some general remarks and a discussion of possible countermeasures to our attack.

2 Preliminaries

Notation. We use the notation $\mathbf{a} = (a_0, \ldots, a_{n-1})$ for column vectors, and $\mathbf{a}^t = [a_0, \ldots, a_{n-1}]$ for rows. Vector entries are indexed starting from 0, and denoted by a_i or $\mathbf{a}[i]$. The dot product between two vectors (with entries in a ring) is written $\langle \mathbf{a}, \mathbf{b} \rangle$ or $\mathbf{a}^t \cdot \mathbf{b}$. Scalar functions $f(\mathbf{a}) = (f(a_0), \ldots, f(a_{n-1}))$ are applied to vectors componentwise.

For any finite set A, we write $x \leftarrow A$ for the operation of selecting x uniformly at random from A. More generally, if χ is a probability distribution, $x \leftarrow \chi$ selects x according to χ.

Standard cryptographic definitions. In all our definitions, we denote the security parameter by κ. A function f in κ is *negligible* if $f(\kappa) = \kappa^{-\omega(1)}$. We use $\mathsf{negl}(\kappa)$ to denote an arbitrary negligible function in κ.

We recall the standard notions of public-key encryption scheme and homomorphic encryption scheme. A *public-key encryption scheme* with a message space $\mathcal{M}$ is a tuple (KeyGen, Enc, Dec) consisting of three algorithms:

- a randomized key generation algorithm KeyGen that takes the security parameter 1^κ and outputs a secret key sk and a public key pk,
- a randomized encryption algorithm Enc that takes pk and a message $m \in \mathcal{M}$ and outputs a ciphertext ct, and
- a deterministic decryption algorithm Dec that takes sk and a ciphertext ct and outputs a message m' or a special symbol $\perp$ indicating decryption failure.

We usually parameterize Enc with pk and write $\mathsf{Enc}_{\mathsf{pk}}(\cdot)$ to denote the function $\mathsf{Enc}(\mathsf{pk}, \cdot)$, and similarly we write $\mathsf{Dec}_{\mathsf{sk}}(\cdot)$ for the function $\mathsf{Dec}(\mathsf{sk}, \cdot)$. A public-key encryption scheme is *correct* if for all $m \in \mathcal{M}$ and keys (sk, pk) in the support of $\mathsf{KeyGen}(1^\kappa)$, $\Pr\{\mathsf{Dec}_{\mathsf{sk}}(\mathsf{Enc}_{\mathsf{pk}}(m)) = m\} = 1 - \mathsf{negl}(\kappa)$, where the probability is over the randomness of Enc.

A public-key *homomorphic* encryption scheme is a public-key encryption scheme with an additional, possibly randomized, *(homomorphic) evaluation* algorithm Eval, and such that KeyGen outputs an additional *evaluation key* ek besides sk and pk. The algorithm Eval takes ek, a circuit $g : \mathcal{M}^l \to \mathcal{M}$ for some $l \geq 1$, and a sequence of l ciphertexts ct_i, and it outputs a ciphertext ct'. The *correctness* of a homomorphic encryption scheme requires that, for all keys (sk, pk, ek) in the support of $\mathsf{KeyGen}(1^\kappa)$, for all circuits $g : \mathcal{M}^l \to \mathcal{M}$ and for all $m_i \in \mathcal{M}$, $1 \leq i \leq l$, it holds that

$$\Pr\left\{\begin{array}{l}\mathsf{ct}_i \leftarrow \mathsf{Enc}_{\mathsf{pk}}(m_i) \text{ for } 1 \leq i \leq l, \\ \mathsf{Dec}_{\mathsf{sk}}(\mathsf{Eval}_{\mathsf{ek}}(g, (\mathsf{ct}_i)_{i=1}^l)) = g((m_i)_{i=1}^l)\end{array}\right\} = 1 - \mathsf{negl}(\kappa),$$

where the probability is over the randomness of Enc and Eval. We also require that the complexity of Dec is independent (or a slow growing function) of the size of the circuit g.

In terms of security, we recall the standard security notion of *indistinguishability under chosen plaintext attack*, or IND-CPA, for public-key (homomorphic) encryption schemes.

Definition 1 (IND-CPA Security). *Let* (KeyGen, Enc, Dec, Eval) *be a homomorphic encryption scheme. We define an experiment* $\mathsf{Expr}_b^{\mathrm{cpa}}[\mathcal{A}]$ *parameterized by a bit* $b \in \{0, 1\}$ *and an efficient adversary* $\mathcal{A}$*:*

$$\begin{aligned}\mathsf{Expr}_b^{\mathrm{cpa}}[\mathcal{A}](1^\kappa): \quad & (\mathsf{sk}, \mathsf{pk}, \mathsf{ek}) \leftarrow \mathsf{KeyGen}(1^\kappa) \\ & (x_0, x_1) \leftarrow \mathcal{A}(1^\kappa, \mathsf{pk}, \mathsf{ek}) \\ & \mathsf{ct} \leftarrow \mathsf{Enc}_{\mathsf{pk}}(x_b) \\ & b' \leftarrow \mathcal{A}(\mathsf{ct}) \\ & \mathsf{return}(b')\end{aligned}$$

We say that the scheme is IND-CPA secure *if for any efficient adversary* $\mathcal{A}$*, it holds that*

$$\mathsf{Adv}_{\mathsf{cpa}}[\mathcal{A}](\kappa) = |\Pr\{\mathsf{Expr}_0^{\mathrm{cpa}}[\mathcal{A}](1^\kappa) = 1\} - \Pr\{\mathsf{Expr}_1^{\mathrm{cpa}}[\mathcal{A}](1^\kappa) = 1\}| = \mathsf{negl}(\kappa).$$

Lattices and rings. A lattice is a (typically full rank) discrete subgroup of $\mathbb{R}^n$. Lattices $L \subset \mathbb{R}^n$ can be represented by a basis, i.e., a matrix $\mathbf{B} \in \mathbb{R}^{n\times k}$ with linearly independent columns such that $L = \mathbf{B}\mathbb{Z}^k$. The length of the shortest nonzero vector in a lattice L is denoted by $\lambda(L)$. The Shortest Vector Problem, given a lattice L, asks to find a lattice vector of length $\lambda(L)$. The Approximate SVP relaxes this condition to finding a nonzero lattice vector of length at most $\gamma\cdot\lambda(L)$, where the approximation factor $\gamma \geq 1$ may be a function of the dimension n or other lattice parameters.

We write $\mathbb{Z}, \mathbb{Q}, \mathbb{R}, \mathbb{C}$ for the sets of integer, rational, real and complex numbers. For any positive $q > 0$, we write $\mathbb{R}_q = \mathbb{R}/(q\mathbb{Z})$ for the set of reals modulo q (as a quotient of additive groups), uniquely represented as values in the centered interval $[-q/2, q/2)$. Similarly, for any positive integer $q > 0$, we write $\mathbb{Z}_q = \mathbb{Z}/(q\mathbb{Z})$ for the ring of integers modulo q, uniquely represented as values in $[-q/2, q/2) \cap \mathbb{Z} = \{-\lceil\frac{q-1}{2}\rceil, \ldots, \lfloor\frac{q-1}{2}\rfloor\}$.

Let $N = 2^k$ be a power of 2, $\zeta_{2N} = e^{\pi\imath/N}$ the principal $(2N)$th complex root of unity. We write $\mathcal{K}^{(2N)} = \mathbb{Q}[X]/(X^N+1)$ for the cyclotomic field of order $2N$, and $\mathcal{O}^{(2N)} = \mathbb{Z}[X]/(X^N+1)$ for its ring of integers. The primitive roots of unity ζ_{2N}^{2j+1}, for $j = 0, \ldots, N-1$, are precisely the roots of the cyclotomic polynomial $X^N + 1$. We omit the index $2N$ and simply write $\mathcal{K}, \mathcal{O}$ and ζ when the value of N is clear from the context. Elements of $\mathcal{K}$ (and $\mathcal{O}$) are uniquely represented as polynomials $a(X) = a_0 + a_1 \cdot X + \ldots + a_{N-1} \cdot X^{N-1}$ of degree less than N, and identified with their vectors of coefficients $\mathbf{a} = (a_0, \ldots, a_{N-1}) \in \mathbb{Q}^N$ (and $\mathbb{Z}^N$). For any positive integer $q > 0$, we write $\mathcal{K}_q = \mathcal{K}/(q\mathcal{K}) \equiv \mathbb{Q}_q^N$ for the set of vectors/polynomials with entries/coefficients reduced modulo q. Similarly for $\mathcal{O} \equiv \mathbb{Z}^N$ and $\mathcal{O}_q \equiv \mathbb{Z}_q^N$.

LWE and homomorphic encryption. The *(Ring) Learning With Errors (LWE)* distribution $\mathsf{RLWE}_s(N, q, \chi)$ with *secret* $s \in \mathcal{O}^{(2N)}$ and *error distribution* χ (over $\mathcal{O}^{(2N)}$), produces pairs $(a, b) \in \mathcal{O}_q^{(2N)}$ where $a \leftarrow \mathcal{O}_q^{(2N)}$ is chosen uniformly at random, and $b = s\cdot a+e$ for $e \leftarrow \chi$. The *(decisional) Ring LWE* assumption over $\mathcal{O}^{(2N)}$ with error distribution χ and secret distribution χ' and m samples, states that when $s \leftarrow \chi'$, the product distribution $\mathsf{RLWE}_s(N, q, \chi)^m$ is pseudorandom, i.e., it is computationally indistinguishable from the uniform distribution over $(\mathcal{O}_q \times \mathcal{O}_q)^m$.

For appropriate choices of χ, χ' and q, the Ring LWE problem is known to be computationally hard, based on (by now) standard assumptions on the worst-case complexity of computing approximately shortest vectors in ideal lattices. Theoretical work supports setting the error distribution χ to a discrete Gaussian of standard deviation $O(\sqrt{N})$, and setting the secret distribution χ' to either the uniform distribution over $\mathcal{O}_q$, or the same distribution as the errors χ. For the sake of efficiency, the Ring LWE problem is often employed by homomorphic encryption schemes also for narrower secret and error distributions, that lack the same theoretical justifications, but for which no efficient attack is known, e.g., distributions over vectors with binary $\{0, 1\}$ or ternary $\{-1, 0, 1\}$ coefficients.

The *raw* (Ring) LWE encryption scheme works as follows:

- The key generation algorithm picks $s \leftarrow \chi'$, $e \leftarrow \chi$, $a \leftarrow \mathcal{O}_q$, and outputs secret key $\mathsf{sk} = (-s, 1) \in \mathcal{O}_q^2$ and public key $\mathsf{pk} = (a, b) \in \mathcal{O}_q^2$ where $b = s \cdot a + e$ follows the LWE distribution.
- The encryption algorithm, $\mathsf{Enc}_{\mathsf{pk}}(m)$ picks random $u \leftarrow \{0,1\}^N$ and $\mathbf{e} = (e_0, e_1) \leftarrow \chi^2$, and outputs $\mathsf{ct} = u \cdot \mathsf{pk} + \mathbf{e} + (0, m) \in \mathcal{O}_q^2$
- The *raw* decryption algorithm $\mathsf{Dec}_{\mathsf{sk}}(\mathsf{ct})$ outputs $\langle \mathsf{sk}, \mathsf{ct} \rangle \bmod q$.

The secret and public keys satisfy the property that $\langle \mathsf{sk}, \mathsf{pk} \rangle = e$ equals the short error vector chosen during key generation. We qualified this scheme and the decryption algorithm as "raw" because applying the encryption algorithm, and subsequently decrypting the result (with a matching pair of public and secret keys) does not recover the original message, but only a value close to it. In fact, for any $(\mathsf{sk}, \mathsf{pk})$ produced by the key generation algorithm, we have

$$\mathsf{Dec}_{\mathsf{sk}}(\mathsf{Enc}_{\mathsf{pk}}(m)) = u \cdot \langle \mathsf{sk}, \mathsf{pk} \rangle + \langle \mathsf{sk}, \mathbf{e} \rangle + m = m + (ue - se_0 + e_1) \pmod q$$

where the perturbation $\tilde{e} = (ue - se_0 + e_1)$ is small because it is a combination of short vectors u, e, s, e_0, e_1. (The size of these vectors is best quantified with respect to the message encoding used by the application, and it is discussed below.) In order to obtain a proper encryption scheme that meets the correctness requirement, the message m must be preprocessed, by encoding it with an appropriate error correcting code, which allows to recover from the error $\tilde{e}$. For example, if m has binary entries, one can multiply m by a scaling factor $\lfloor q/2 \rceil$, and then round (each coefficient of) the output of the raw decryption algorithm to the closest multiple of $\lfloor q/2 \rceil$. For the sake of improving the efficiency of homomorphic computations, the CKKS encryption scheme [18] gets away without applying error correction, and directly using the raw decryption algorithm to produce "approximate" decryptions of the ciphertexts. So, in the following we focus on the "raw" LWE scheme, and postpone the discussion of error correction to later.

By linearity of Enc, LWE encryption directly supports (bounded) addition of ciphertexts: if $\mathsf{ct}_0 = (a_0, b_0)$ and $\mathsf{ct}_1 = (a_1, b_1)$ are encryptions of m_0 and m_1 with noise e_0 and e_1 respectively, then the vector sum

$$\mathsf{ct}_0 + \mathsf{ct}_1 = (a_0 + a_1, b_0 + b_1) \bmod q$$

is an encryption of $m_0 + m_1$ with noise $e_0 + e_1$.

There are several ways to perform homomorphic multiplication on LWE ciphertexts. As in [18], here we focus on the "tensoring" technique of [10] implemented using the "raising the modulus" multiplication method of [24]. This multiplication method uses an appropriate multiple pq of the ciphertext modulus q, and requires an "evaluation key", produced during key generation, which is computed and used as follows:

- $\mathsf{ek} = (a, b) \in \mathcal{O}_{pq}^2$ where $a \leftarrow \mathcal{O}_{pq}$, $e \leftarrow \chi_e$ and $b = as + e + ps^2 \pmod{pq}$.
- Using ek, the product of two ciphertexts $\mathsf{ct}_0 = (a_0, b_0), \mathsf{ct}_1 = (a_1, b_1)$ is computed as

$$\mathsf{ct}_0 \times \mathsf{ct}_1 = (a_0 b_1 + a_1 b_0, b_0 b_1) + \lfloor (a_0 a_1 \bmod q) \cdot \mathsf{ek}/p \rceil .$$

In order to approximately evaluate deep arithmetic circuits, the CKKS scheme combines these addition and multiplication procedures with a *rescaling operation* RS, implemented using the *key switching* technique of [10]. Rescaling requires the use of a sequence of moduli q_l, which for simplicity we assume to be of the form $q_l = q_0 \cdot p^l$ for some base p, e.g., $p = 2$. Ciphertexts may live at different levels, with level l ciphertexts encrypted using modulus q_l. The key generation algorithm takes as auxiliary input the highest number of desired levels L, and produces public and evaluation keys with respect to the largest modulus q_L. CKKS directly supports addition and multiplication only between ciphertexts at the same level. Rescaling is used to map ciphertexts $\mathsf{ct} \in \mathcal{O}^2_{q_{l+l'}}$ to a lower level l with the operation

$$\mathsf{RS}_{l'}(\mathsf{ct}) = \left\lfloor \mathsf{ct}/p^{l'} \right\rceil \in \mathcal{O}^2_{q_l}$$

where the division and rounding are performed componentwise.

The CKKS message encoding. The CKKS scheme considers a vector of complex numbers (or Gaussian integers) $\tilde{\mathbf{a}}$ as the set of evaluation points $\tilde{a}_j = a(x_j)$ of a real (in fact, integer) polynomial $a(X) \in \mathbb{Z}[X]$. This allows to perform pointwise addition and multiplication of vectors (SIMD style) by means of addition and multiplication of polynomials as $(a(X) \circ b(X))(x_j) = a(x_j) \circ b(x_j)$ for any x_j, where $\circ \in \{+, \times\}$. The evaluation points are chosen among the primitive $(2N)$th roots of unity ζ^{2j+1}, so that the cyclotomic polynomial $X^N + 1$ evaluates to zero at all those points, and reduction modulo $X^N + 1$ does not affect the value of $a(x_j)$. This allows to operate on the polynomials modulo $X^N + 1$, i.e., as elements of the cyclotomic ring $\mathcal{O}$. Since $a(X)$ has real coefficients and primitive roots come in complex conjugate pairs $\zeta^{2j+1}, \zeta^{2(N-j)-1}$, the value of $a(X)$ can be freely chosen only for half of the roots, with the value of $a(\zeta^{2(N-j)-1})$ uniquely determined as the complex conjugate of $a(\zeta^{2j+1})$. So, $a(X)$ is used to represent a vector $\tilde{\mathbf{a}}$ of $N/2$ complex values. Setting the evaluation points to $x_j = \zeta^{4j+1}$ (for $j = 0, \ldots, N/2 - 1$), and using the fact that these points are primitive roots of unity, interpolation and evaluation can be efficiently computed (in $O(N \log N)$ time) using the Fast Fourier Transform.

Let $\varphi \colon \mathcal{O} \to \mathbb{C}^{N/2}$ be the transformation mapping $a(X) \in \mathcal{O} \equiv \mathbb{Z}^N$ to $\varphi(a) = \tilde{\mathbf{a}} = (a(\zeta^{4j+1}))_{j=0}^{N/2-1} \in \mathbb{C}^{N/2}$, and its extension $\varphi \colon \mathcal{S} \to \mathbb{C}^{N/2}$ to arbitrary real polynomials, where $\mathcal{S} = \mathbb{R}[X]/(X^N + 1) \equiv \mathbb{R}^N$. We can identify any polynomial $a \in \mathcal{S}$ by its coefficient vector $(a_0, a_1, \ldots, a_{N-1})$, and we set $\|a\|_2 = \|(a_0, a_1, \ldots, a_{N-1})\|_2$. Similarly we can define $\|a\|_1$ and $\|a\|_\infty$ as the corresponding norms on the coefficient vector. So the transformation $\varphi \colon \mathcal{S} \to \mathbb{C}^{N/2}$ is a scaled isometry, satisfying $\|\varphi(a)\|_2 = \sqrt{N}\|a\|_2$ and $\|\varphi(a)\|_\infty \leq \|a\|_1$. In what follows, we assume, as a message space, the set of complex vectors $\tilde{\mathbf{a}} \in \varphi(\mathcal{O}) \subset \mathbb{C}^{N/2}$ which are the evaluation of polynomials $a(X) \in \mathcal{O}$ with integer coefficients much smaller than the ciphertext modulus q. Arbitrary vectors $\mathbf{z} \in \mathbb{C}^{N/2}$ can be encrypted (approximately) by taking the inverse transform φ^{-1} on a scaled vector $\Delta \cdot \mathbf{z}$, for some scaling factor $\Delta \in \mathbb{R}$, such that $\|\varphi^{-1}(\Delta \cdot \mathbf{z})\| \ll q$ and rounding $\varphi^{-1}(\Delta \cdot \mathbf{z})$ to a nearby point of the form $\varphi(a)$ for some $a(X) \in \mathcal{O}$.

The complete message encoding and decoding functions in CKKS are defined as

- $\mathsf{Encode}(\mathbf{z} \in \mathbb{C}^{N/2}; \Delta) = \lfloor \Delta \cdot \varphi^{-1}(\mathbf{z}) \rceil \in \mathcal{O}$.
- $\mathsf{Decode}(a \in \mathcal{O}; \Delta) = \varphi(\Delta^{-1} \cdot a) \in \mathbb{C}^{N/2}$.

Once encoded, the scaling factor Δ is usually implicitly tied to a plaintext polynomial, so we sometimes omit it when its value is clear from the context.

Since these encoding and decoding operations can be performed without any knowledge of the secret or public keys, sometimes we assume they are performed at the outset, at the application level, before invoking the encryption or decryption algorithms. More specifically, we may assume messages $\varphi(\Delta^{-1} \cdot m) \in \mathbb{C}^{N/2}$ are provided to the encryption algorithm by specifying the integer polynomial $m \in \mathcal{O}$, and the decryption algorithm returns a message $\tilde{\mathbf{m}}' = \mathsf{Decode}(m'; \Delta)$ represented as the underlying polynomial $m' \in \mathcal{O}$ that is an approximation of m. All this is only for the sake of theoretical analysis, and all concrete implementations (of the scheme and our attacks to it) include encoding and decoding procedures as part of the encryption and decryption algorithms. Message encoding can be quite relevant to quantify the amount of noise in a ciphertext. We say that a ciphertext ct *approximately encrypts* message $\tilde{\mathbf{m}}$ with scaling factor Δ and noise $\tilde{\mathbf{e}}$ if $\mathsf{Decode}(\mathsf{Dec}_{\mathsf{sk}}(\mathsf{ct}); \Delta) = \tilde{\mathbf{m}} + \tilde{\mathbf{e}}$.

3 Security Notions for Approximate Encryption

In this section we present general definitions in the public-key setting that accurately capture passive attacks against a (possibly approximate, homomorphic) encryption scheme. We recall that in a passive attack the adversary may control which messages get encrypted, what homomorphic computations are performed on them, and may observe all ciphertexts produced in the process, as well as the decrypted result of the computations (as illustrated in Fig. 1).

We present an indistinguishability-based definition (similar in spirit to the standard IND-CPA notion described in Definition 1). A simulation-based notion is presented in the full version of this paper. Then, we explore restricted and extended variants of these basic definitions.

3.1 Indistinguishability-Based Definition

Our first definition is indistinguishability-based: the adversary chooses a number of pairs of plaintext messages, and its goal is to determine whether the ciphertexts it receives are encryptions of the first or the second plaintext in the pairs. In contrast to Definition 1, our new definition allows an adversary to make *multiple* challenge queries (m_0, m_1), rather than a single one. Our adversary can also issue homomorphic evaluation and decryption queries. We now give the formal definition. For simplicity, and as common in homomorphic encryption schemes, we assume all messages belong to a fixed message space $\mathcal{M}$. In particular, all messages have (or can be padded to) the same length. We refer to our definition as $\mathsf{IND\text{-}CPA^D}$, as it includes IND-CPA (see Definition 1) as a special case, where the adversary makes only one encryption query, and no homomorphic evaluation or decryption queries, whereas our definition explicitly provides the adversary with a restricted decryption oracle which allows to observe decryption results of honestly generated ciphertexts.

Definition 2 ($\mathsf{IND\text{-}CPA^D}$ Security). *Let $\mathcal{E} = (\mathsf{KeyGen}, \mathsf{Enc}, \mathsf{Dec}, \mathsf{Eval})$ be a public-key homomorphic (possibly approximate) encryption scheme with plaintext space $\mathcal{M}$ and ciphertext space $\mathcal{C}$. We define an experiment $\mathsf{Expr}_b^{\mathrm{indcpa^D}}[\mathcal{A}]$, parameterized by a bit $b \in \{0,1\}$ and involving an efficient adversary $\mathcal{A}$ that is given access to the following oracles, sharing a common state $S \in (\mathcal{M} \times \mathcal{M} \times \mathcal{C})^*$ consisting of a sequence of message-message-ciphertext triplets:*

- *An encryption oracle $\mathsf{E}_{\mathsf{pk}}(m_0, m_1)$ that, given a pair of plaintext messages m_0, m_1, computes $c \leftarrow \mathsf{Enc}_{\mathsf{pk}}(m_b)$, extends the state*
$$S := [S; (m_0, m_1, c)]$$
with one more triplet, and returns the ciphertext c to the adversary.
- *An evaluation oracle $\mathsf{H}_{\mathsf{ek}}(g, J)$ that, given a function $g : \mathcal{M}^k \to \mathcal{M}$ and a sequence of indices $J = (j_1, \ldots, j_k) \in \{1, \ldots, |S|\}^k$, computes the ciphertext $c \leftarrow \mathsf{Eval}_{\mathsf{pk}}(g, S[j_1].c, \ldots, S[j_k].c)$, extends the state*
$$S := [S; (g(S[j_1].m_0, \ldots, S[j_k].m_0),\ g(S[j_1].m_1, \ldots, S[j_k].m_1),\ c)]$$
with one more triplet, and returns the ciphertext c to the adversary. Here and below $|S|$ denotes the number of triplets in the sequence S, and $S[j].m_0$, $S[j].m_1$ and $S[j].c$ denote the three components of the jth element of S.
- *A decryption oracle $\mathsf{D}_{\mathsf{sk}}(j)$ that, given an index $j \le |S|$, checks whether $S[j].m_0 = S[j].m_1$, and, if so, returns $\mathsf{Dec}_{\mathsf{sk}}(S[j].c)$ to the adversary. (If the check fails, a special error symbol $\perp$ is returned.)*

The experiment is defined as

$$\begin{aligned}\mathsf{Expr}_b^{\mathrm{indcpa^D}}[\mathcal{A}](1^\kappa): \quad & (\mathsf{sk}, \mathsf{pk}, \mathsf{ek}) \leftarrow \mathsf{KeyGen}(1^\kappa)\\ & S := [\,]\\ & b' \leftarrow \mathcal{A}^{\mathsf{E}_{\mathsf{pk}}, \mathsf{H}_{\mathsf{ek}}, \mathsf{D}_{\mathsf{sk}}}(1^\kappa, \mathsf{pk}, \mathsf{ek})\\ & \mathsf{return}(b')\end{aligned}$$

The advantage of adversary $\mathcal{A}$ against the $\mathsf{IND\text{-}CPA^D}$ security of the scheme is

$$\mathsf{Adv}_{\mathrm{indcpa^D}}[\mathcal{A}](\kappa) = |\Pr\{\mathsf{Expr}_0^{\mathrm{indcpa^D}}[\mathcal{A}](1^\kappa) = 1\} - \Pr\{\mathsf{Expr}_1^{\mathrm{indcpa^D}}[\mathcal{A}](1^\kappa) = 1\}|,$$

where the probability is over the randomness of $\mathcal{A}$ and the experiment. The scheme $\mathcal{E}$ is $\mathsf{IND\text{-}CPA^D}$-secure if for any efficient (probabilistic polynomial time) $\mathcal{A}$, the advantage $\mathsf{Adv}_{\mathrm{indcpa^D}}[\mathcal{A}]$ is negligible in κ.

As a standard convention, if at any point in an experiment the adversary makes an invalid query (e.g., a circuit g not supported by the scheme, or indices out of range), the oracle simply returns an error symbol $\perp$.

We remark that, while the adversary in Definition 2 is given access to a decryption oracle, this should not be confused with indistinguishability under a *chosen ciphertext attack* (IND-CCA), which models active adversaries with the capability of tampering with (or injecting) arbitrary ciphertexts. Definition 2 only allows for decryption queries on valid ciphertexts that have been honestly computed using the correct encryption and homomorphic evaluation algorithms (modeled by the oracles E and H). Furthermore, the requirement that $S[j].m_0 = S[j].m_1$ is to eliminate trivial attacks where the adversary can distinguish between two computations that lead to different results when computed on exact values.

Exact encryption schemes can be seen as a special case of approximate encryption, with the added correctness requirement. So, Definition 2 can be applied to exact as well as approximate encryption schemes. As a sanity check, we compare our new definition with the traditional formulation of IND-CPA security (Definition 1) modeling passive attacks against exact encryption schemes. Perhaps not surprisingly, for the case of exact encryption schemes, our new security definition coincides with the standard notion of IND-CPA security.

Lemma 1. *Any exact homomorphic encryption scheme $\mathcal{E}$ is IND-CPA secure if and only if it is IND-CPA^{D} secure.*

Proof. It is easy to see that IND-CPA^{D} security implies IND-CPA security, as an adversary making only one E query but no other queries in the IND-CPA^{D} experiment is also an IND-CPA adversary. So we consider the reverse direction.

Assume $\mathcal{E}$ is IND-CPA secure. Let $\mathcal{A}$ be any adversary breaking the IND-CPA^{D} security of $\mathcal{E}$, and assume $\mathcal{A}$ makes at most l queries in total to E and H. We build adversaries $\mathcal{B}^{(i)}$, for $0 \le i < l$, to break the IND-CPA security of $\mathcal{E}$.

$\mathcal{B}^{(i)}$ takes input 1^κ, pk, ek, and it then runs $\mathcal{A}(1^\kappa, \mathsf{pk}, \mathsf{ek})$. It maintains a state $S \in (\mathcal{M} \times \mathcal{M} \times \mathcal{C})^*$ just like $\mathsf{Expr}^{\mathrm{indcpa}^{\mathrm{D}}}$, and it answers oracle queries made by $\mathcal{A}$ as follows:

- For each query (m_0, m_1) to E, if $|S| < i$, then let $c \leftarrow \mathsf{Enc}_{\mathsf{pk}}(m_1)$; if $|S| > i$, then let $c \leftarrow \mathsf{Enc}_{\mathsf{pk}}(m_0)$; and if $|S| = i$, $\mathcal{B}^{(i)}$ sends (m_0, m_1) to $\mathsf{Expr}_b^{\mathrm{cpa}}$ and receives c. The state S is extended by one more triplet (m_0, m_1, c), and c is returned to $\mathcal{A}$.
- For each query (g, J) to H, where $g : \mathcal{M}^k \to \mathcal{M}$ and $J = (j_1, \ldots, j_k)$, let $c \leftarrow \mathsf{Eval}_{\mathsf{ek}}(g, S[j_1].c, \ldots, S[j_k].c)$, extend S by one more triplet

$$(g(S[j_1].m_0, \ldots, S[j_k].m_0),\ g(S[j_1].m_1, \ldots, S[j_k].m_1), c),$$

  and return c to $\mathcal{A}$.
- For each query j to D, if $j \le |S|$ and $S[j].m_0 = S[j].m_1$, then return $S[j].m_0$ to $\mathcal{A}$; otherwise return an error symbol $\perp$.

Finally, when $\mathcal{A}$ halts with a bit b', $\mathcal{B}^{(i)}$ outputs this bit.

Since $\mathcal{B}^{(i)}$ does not depend on the secret key sk to answer the D queries, it is a valid adversary in the $\mathsf{IND\text{-}CPA}$ experiment. Now, let $\mathcal{H}^{(i)} = \mathsf{Expr}_0^{\mathrm{cpa}}[\mathcal{B}^{(i)}]$ for $0 \leq i < l$, and let $\mathcal{H}^{(l)} = \mathsf{Expr}_1^{\mathrm{cpa}}[\mathcal{B}^{(l-1)}]$. For $1 \leq i < l$, note that $\mathcal{H}^{(i)}$ is exactly the same distribution as $\mathsf{Expr}_1^{\mathrm{cpa}}[\mathcal{B}^{(i-1)}]$. Furthermore, by the correctness of exact homomorphic encryption schemes, the D responses from $\mathcal{B}^{(i)}$ to $\mathcal{A}$ are indistinguishable from those in the $\mathsf{IND\text{-}CPA^D}$ experiment; so $\mathcal{H}^{(0)}$ and $\mathsf{Expr}_0^{\mathrm{indcpa^D}}[\mathcal{A}]$ are indistinguishable, and the same holds true for $\mathcal{H}^{(l)}$ and $\mathsf{Expr}_1^{\mathrm{indcpa^D}}[\mathcal{A}]$. So $\mathsf{Adv}^{\mathrm{indcpa^D}}[\mathcal{A}] \leq \sum_{0 \leq i < l} \mathsf{Adv}^{\mathrm{cpa}}[\mathcal{B}^{(i)}] + \mathsf{negl}(\kappa)$, which is negligible since $\mathcal{E}$ is $\mathsf{IND\text{-}CPA}$ secure. □

Notice that the above lemma makes essential use of the correctness of exact encryption schemes, and the proof does not extend to approximate encryption schemes. In fact, for approximate encryption schemes, the result of decryption is not a simple function of the encrypted messages (and the computations performed on them), and may potentially depend (in an indirect, unspecified way) on the scheme's secret key and encryption randomness. So the information provided by decryption queries is not easily computed by the adversary on its own, and, at least in principle, $\mathsf{IND\text{-}CPA^D}$ may be a stronger security notion than $\mathsf{IND\text{-}CPA}$ when applied to approximate encryption schemes. We will make this intuition clear in the following sections, proving formal separation results, and providing concrete attacks to actual approximate encryption schemes.

Also note that the above definition does not guarantee circuit privacy in the homomorphically evaluated ciphertexts, as the circuit to be evaluated in a query to oracle H does not depend on the bit b of the $\mathsf{IND\text{-}CPA^D}$ experiment. In the full version we extend our definitions with circuit privacy. Here we focus on the basic definition (without circuit privacy) which is the most common in cryptography.

3.2 Restricted Security Notions and Separations Between Them

We have observed that, for exact encryption schemes, $\mathsf{IND\text{-}CPA^D}$ security is equivalent to the traditional $\mathsf{IND\text{-}CPA}$ security. (See Lemma 1.) We now show that $\mathsf{IND\text{-}CPA^D}$ is strictly stronger than $\mathsf{IND\text{-}CPA}$, i.e., there are approximate encryption schemes that are provably $\mathsf{IND\text{-}CPA}$ secure (under standard complexity assumptions) but are not $\mathsf{IND\text{-}CPA^D}$ secure. In order to get a more refined understanding of the gap between these notions, we introduce a natural parameterization of $\mathsf{IND\text{-}CPA^D}$ security, that smoothly interpolates between $\mathsf{IND\text{-}CPA}$ and $\mathsf{IND\text{-}CPA^D}$. Then, we define a number of restricted notions of security, and show separations between them, showing that there is an infinite chain of (strictly) increasingly stronger definitions, ranging from $\mathsf{IND\text{-}CPA}$ all the way to $\mathsf{IND\text{-}CPA^D}$.

Restricting the numbers of queries. We parameterize the definition by imposing a bound on the number of queries that may be asked by the adversary.

Definition 3 (**((q, ℓ)-IND-CPA^{D} Security).** *For any two functions $q(\kappa)$ and $\ell(\kappa)$ of the security parameter κ, we say that a homomorphic encryption scheme is (q, ℓ)-IND-CPA^{D} secure if it satisfies Definition 2 for all adversaries $\mathcal{A}$ that make at most $\ell(\kappa)$ queries to oracles* E, H, *and at most $q(\kappa)$ queries to oracle* D.

We combined the encryption (E) and evaluation (H) queries into a single bound $\ell(\kappa)$ for simplicity, and because both types of queries produce ciphertexts. The bound ℓ could be significant for approximate encryption schemes as security with respect ℓ queries to E and H does not appear to imply security with respect to $\ell+1$ such queries. This is in contrast to proper (exact) encryption schemes in the public-key setting where one-message security implies multi-message security. It remains an interesting open question to find out the relationship between (q, ℓ)-IND-CPA^{D} and $(q, \ell+1)$-IND-CPA^{D} securities.

The definition is easily extended to more general formulations, but we will be primarily interested in the bound q on the number of decryption queries, which are the distinguishing feature of approximate encryption schemes. When ℓ is an arbitrary polynomial, and only the number of decryption queries $q(\kappa)$ is restricted, we say that a scheme is q-IND-CPA^{D} secure.

Now, we can think of IND-CPA security as a special case of (q, ℓ)-IND-CPA^{D}, for $q = 0$ and $\ell = 1$, as the only query to E/H must be an encryption query. (Oracle E must be called at least once before one can use H to homomorphically evaluate a function on a ciphertext.) So, bounding the number of queries allows to smoothly transition from the traditional IND-CPA definition (i.e., $(0, 1)$-IND-CPA^{D} security), to our IND-CPA^{D} (i.e., $(\mathsf{poly}, \mathsf{poly})$-IND-CPA^{D} security).

Naturally, for proper (exact) encryption schemes, all these definitions are equivalent, and it is only in the approximate encryption setting that the definitions can be separated.

In the following proposition we show that there exists some scheme that is secure for up to some fixed number q of decryption queries but insecure for just $q+1$ decryption queries. We remark that the encryption scheme described in the proof is presented for the sole purpose of separating the two definitions. More natural examples that separate IND-CPA and IND-CPA^{D} will be described in Sect. 4, where we present attacks to approximate encryption schemes from the literature.

Proposition 1. *Assume there exist a pseudorandom function and an IND-CPA-secure exact homomorphic encryption scheme. Then, for any fixed $q \geq 2$, there exists a homomorphic approximate encryption scheme that is (q, ℓ)-IND-CPA^{D}-secure but not $(q+1, \ell)$-IND-CPA^{D}-secure.*

Proof (sketch). Let $\mathcal{E} = (\mathsf{KeyGen}, \mathsf{Enc}, \mathsf{Dec}, \mathsf{Eval})$ be an exact, IND-CPA-secure, HE scheme. The main idea is to construct a new scheme $\mathcal{E}'$ that consists of the same encryption and evaluation algorithms, but with new key generation KeyGen' and decryption algorithms Dec'. $\mathsf{KeyGen}'(1^\kappa)$ samples keys $(\mathsf{sk}, \mathsf{pk}, \mathsf{ek}) \leftarrow \mathsf{KeyGen}(1^\kappa)$ as in $\mathcal{E}$, and then it samples a PRF key K to form the new secret key $\mathsf{sk}' = (\mathsf{sk}, K)$, while keeping the public key pk and the evaluation key ek the same. Given a ciphertext c, $\mathsf{Dec}'_{\mathsf{sk}'}$ first runs $\mathsf{Dec}_{\mathsf{sk}}(c)$ to obtain the exact plaintext m, and then it adds to m a secret share (encoded as a small number) of sk produced by the PRF on m. Specifically, if $m \pmod{q+1} \equiv 0$, then the share is $\mathsf{sk} \oplus r$ for $r = \oplus_{i=1}^{q} \mathsf{PRF}_K(i)$; otherwise, the share is $\mathsf{PRF}_K(m \bmod (q+1))$. Here a PRF is used to keep Dec' deterministic. Since $\mathcal{E}$ is IND-CPA secure, and since any q or less shares of sk are pseudorandom, our new approximate encryption scheme is (q, ℓ)-IND-CPA^{D}-secure. However, an adversary can fully recover sk using $q+1$ decryption queries, breaking $(q+1, \ell)$-IND-CPA^{D} security of the new scheme. □

Restricting the query ordering. In the definition of IND-CPA^{D} security, we did not state any restriction on the relative order of queries made by the adversary. In particular, queries can be made in many rounds, and a later query can depend on the responses from earlier queries. Such notion is called *security with adaptively chosen queries*, or simply *adaptive* security.

There are several other natural query orderings that can be imposed on the adversary, and enforced by an application. For example, it is often the case that inputs are encrypted and collected in advance, before any homomorphic evaluation or decryption operation takes place. As an extreme situation, one can consider a fully non-adaptive setting, where the adversary specifies all its queries in advance after seeing the public/evaluation key. We call this the *(fully) non-adaptive* model. Non-adaptive security is much easier to formulate, and we fully spell out its definition now.

Definition 4 (Non-Adaptive (q,ℓ)-IND-CPA^{D} Security). *Let $\mathcal{E}$ be a homomorphic (possibly approximate) encryption scheme $\mathcal{E} = (\mathsf{KeyGen}, \mathsf{Enc}, \mathsf{Dec}, \mathsf{Eval})$. Let q and ℓ be two polynomial bounds in κ. We say that $\mathcal{E}$ is* non-adaptively (q,ℓ)-IND-CPA^{D}-secure *if for all efficient adversary $\mathcal{A} = (\mathcal{A}_0, \mathcal{A}_1)$ consisting of two steps such that*

$$(\{m_0^{(i)}\}_{i=1}^{k}, \{m_1^{(i)}\}_{i=1}^{k}, \{(g_i, J_i)\}_{i=k+1}^{\ell}, \{j_i\}_{i=1}^{q}, \mathsf{st}) \leftarrow \mathcal{A}_0(1^\kappa, \mathsf{pk}, \mathsf{ek}),$$

where $(\mathsf{sk}, \mathsf{pk}, \mathsf{ek}) \leftarrow \mathsf{KeyGen}(1^\kappa)$, $m_0^{(i)} = g_i(m_0^{(J_i)})$, $m_1^{(i)} = g_i(m_1^{(J_i)})$ *for* $i = k+1, \ldots, \ell$, *and all g_i are valid circuits with indices $J_i \in \{1, \ldots, \ell\}^*$, the following two distributions are indistinguishable to $\mathcal{A}_1(1^\kappa, \mathsf{st})$:*

$$\{\ \{c_i \leftarrow \mathsf{Enc}_{\mathsf{pk}}(m_0^{(i)})\}_{i=1}^{k},\ \{c_i \leftarrow \mathsf{Eval}_{\mathsf{ek}}(g_i, c_{(J_i)})\}_{i=k+1}^{\ell},\ \{\mathsf{Dec}_{\mathsf{sk}}(c_i) \mid m_0^{j_i} = m_1^{j_i}\}_{i=1}^{q}\ \},$$

and

$$\{\ \{c_i \leftarrow \mathsf{Enc}_{\mathsf{pk}}(m_1^{(i)})\}_{i=1}^{k},\ \{c_i \leftarrow \mathsf{Eval}_{\mathsf{ek}}(g_i, c_{(J_i)})\}_{i=k+1}^{\ell},\ \{\mathsf{Dec}_{\mathsf{sk}}(c_i) \mid m_0^{j_i} = m_1^{j_i}\}_{i=1}^{q}\ \},$$

where the probability is over the randomness of $\mathcal{A}$ and in Enc *and* Eval.

Typically the same security notion is weaker in the non-adaptive model than in the adaptive model, as some attacks are only feasible in the latter model. We show that this is also the case for homomorphic approximate encryption schemes. As before, the encryption scheme described in the following proof is not intended to be used. It is just a theoretical construction, provided simply for the purpose of showing that a scheme may satisfy one definition but not the other.

Proposition 2. *Assume there exist an IND-CPA-secure exact homomorphic encryption scheme and a secure pseudorandom permutation. Then there exists a homomorphic approximate encryption scheme that is non-adaptively $\mathsf{IND\text{-}CPA}^{\mathsf{D}}$-secure, but it is not adaptively $(2,2)$-$\mathsf{IND\text{-}CPA}^{\mathsf{D}}$-secure.*

Proof (sketch). Let $\mathcal{E}$ be an IND-CPA-secure exact HE scheme, and let $H : \{0,1\}^\kappa \times \{0,1\}^\kappa \to \{0,1\}^\kappa$ be a pseudorandom permutation. We can define another pseudorandom permutation F:

$$\forall x \in \{0,1\}^\kappa.\ F_K(x) = H_K^{-1}(H_K(x) \oplus 1).$$

Notice that $F_K(F_K(x)) = x$ for all $x \in \{0,1\}^\kappa$.

We now modify $\mathcal{E}$ to obtain a homomorphic approximate encryption scheme $\mathcal{E}'$ with the same encryption and evaluation algorithms but modified key generation and decryption algorithms:

- $\mathsf{KeyGen}'(1^\kappa)$: Sample $(\mathsf{sk}, \mathsf{pk}, \mathsf{ek}) \leftarrow \mathcal{E}.\mathsf{KeyGen}(1^\kappa)$ and $K \leftarrow \{0,1\}^\kappa$ (for the pseudorandom permutation F). Return $(\mathsf{sk}', \mathsf{pk}, \mathsf{ek})$, where $\mathsf{sk}' = (\mathsf{sk}, K)$.
- $\mathsf{Dec}'_{\mathsf{sk}'}(c) = m + \pi(r)$, where $m = \mathcal{E}.\mathsf{Dec}_{\mathsf{sk}}(c)$, $r = F_K(\mathsf{sk})$ if $m = 0$, and $r = F_K(m)$ otherwise, and (π, π^{-1}) is an encoding scheme from $\{0,1\}^\kappa$ to small numbers.

One can check that F_K is a pseudorandom permutation in the non-adaptive model. So $\mathsf{Dec}'_{\mathsf{sk}'}$ can be simulated without knowing the secret key (sk, K) in the non-adaptive model, and hence our new scheme is non-adaptively $\mathsf{IND\text{-}CPA}^{\mathsf{D}}$-secure. However, an adaptive adversary $\mathcal{A}$ can first ask to encrypt 0 and then ask to decrypt the corresponding ciphertext to get $e = \pi(F_K(\mathsf{sk}))$. Next, $\mathcal{A}$ asks to encrypt $\pi^{-1}(e) = F_K(\mathsf{sk})$ and then asks to decrypt its ciphertext. At this point $\mathcal{A}$ can fully recover sk using the decryption result $\pi^{-1}(e) + \pi(\mathsf{sk})$. So $\mathcal{E}'$ is not adaptively $(2,2)$-$\mathsf{IND\text{-}CPA}^{\mathsf{D}}$-secure. □

4 Attacks to Homomorphic Encryption for Arithmetics on Approximate Numbers

In this section we describe a key recovery attack against the CKKS scheme, including both theoretical and practical analysis. Based on such attack, we can conclude that the CKKS scheme is not $\mathsf{IND\text{-}CPA}^{\mathsf{D}}$ secure. Note that our attack is **much stronger** than a simple indistinguishability attack: we show how to efficiently recover the secret (decryption) key of the scheme! Clearly, once the secret key has been recovered, it is easy to break the formal $\mathsf{IND\text{-}CPA}^{\mathsf{D}}$ security definition. While recovering the secret key makes our attacks stronger, any security analysis of improved variants of CKKS or other approximate encryption schemes should still target $\mathsf{IND\text{-}CPA}^{\mathsf{D}}$ as a security goal, and not simply protect the scheme against full key recovery.

4.1 Theoretical Outline

The technical idea behind the attack is easily explained by exemplifying it on a symmetric key version of LWE encryption. (Breaking the CKKS scheme involves additional complications due to the details of the encoding/decoding functions discussed below.) We recall that in a passive attack (against a symmetric key encryption scheme $E_{\mathbf{s}}(m)$), the adversary can observe the encryption $E_{\mathbf{s}}(m)$ of any message m of its choice. In LWE encryption, the key is a random vector $\mathbf{s} \in \mathbb{Z}_q^n$, and a (possibly encoded) message $m \in \mathbb{Z}_q$ is encrypted as $E_{\mathbf{s}}(m) = (\mathbf{a}, b)$ where $\mathbf{a} \in \mathbb{Z}_q^n$ is chosen at random, and $b = \langle \mathbf{s}, \mathbf{a} \rangle + m + e \pmod q$ for a small random integer perturbation $e \in \mathbb{Z}$. If the encryption scheme works on "approximate numbers", $(m + e)$ is treated as an approximation of m, and the decryption algorithm outputs $D_{\mathbf{s}}(\mathbf{a}, b) = b - \langle \mathbf{s}, \mathbf{a} \rangle = m + e$.

Our most basic attack involves an adversary that asks for an encryption of $m = 0$, so to obtain a ciphertext $\mathsf{ct} = (\mathbf{a}, b)$ where $b = \langle \mathbf{s}, \mathbf{a} \rangle + e \pmod q$. The adversary then asks to compute the identity function $id(x) = x$ on it. (This is the same as performing no computation at all.) Finally, it asks for an approximate decryption of the result, and computes

$$c = b - \mathsf{Dec}_{\mathbf{s}}(\mathsf{ct}) = (\langle \mathbf{s}, \mathbf{a} \rangle + e) - (m + e) = \langle \mathbf{s}, \mathbf{a} \rangle \pmod q. \tag{1}$$

This provides a linear equation $\langle \mathbf{s}, \mathbf{a} \rangle = c \pmod q$ in the secret key. Collecting n such linear equations and solving the resulting system (e.g., by Gaussian elimination) recovers the secret key $\mathbf{s}$ with high probability.

It is easy to see that there is nothing special about the message 0, or the fact that no computation is performed: as long as the adversary knows the ciphertext ct (possibly the result of a homomorphic computation) and gets to see the approximate decryption of ct, the same attack goes through. However, the actual scheme described in [18] and subsequent papers, and their open source implementations include several modifications of the above scheme, introduced to make the scheme more useful in practice, but which also make the attack less straightforward. We briefly describe each of these modifications, and how the attack is adapted. In the most general case, our attack requires not just the solution of a linear system of equations, but the use of lattice reduction for the (polynomial time) solution of a lattice approximation problem.

Public key. First, CKKS is a public key encryption scheme, where, as standard in lattice based encryption, the public key can be seen as a collection of encryptions of 0 values. This makes no difference in the attack, as the ciphertexts still have the same structure with respect to the secret key, and the (approximate) decryption algorithm is unmodified. Switching to a public key system has the only effect of producing larger noise vectors e in ciphertexts.

Ring lattices. In order to achieve practical performance, all instantiations of the CKKS scheme make use of cyclic/ideal lattices [40] and the Ring LWE problem [38,39]. Specifically, the vectors $\mathbf{a}, \mathbf{s}$ are interpreted as (coefficients of) polynomials a, s in the power-of-two cyclotomic rings $\mathcal{O}^{(2N)}$ popularized by the

SWIFFT hash function [36,37,43] and widely used in the implementation of lattice cryptography since then. In a sense, switching to ideal lattices makes the attack only more efficient: the linear equation $\langle \mathbf{s}, \mathbf{a} \rangle = \mathbf{c} \pmod q$ becomes an equation $a \cdot s = c \in \mathcal{O}_q^{(2N)}$ in the cyclotomic ring modulo q, which can be solved (even using a single ciphertext) by computing the (ring) inverse of a, and recovering s as

$$s' = a^{-1} \cdot c \in \mathcal{O}_q. \tag{2}$$

A little difficulty arises due to the choice of q. The first implementation of CKKS, the HEAAN library [30] sets q to a power of 2 to simplify the treatment of floating point numbers. Subsequent instantiations of CKKS use a prime (or square-free) q of the form $h \cdot 2^n + 1$ together with the Number Theoretic Transform for very fast ring operations[37]. For a (sufficiently large) prime q, the probability of a random element a being invertible is very close to 1, but this is not the case when q is a power of two. If a is not invertible, we can still recover partial information about the secret key s, and completely recover s by using multiple ciphertexts.

Euclidean embedding. In order to conveniently apply the CKKS scheme on practical problems, the input message space is set to $\mathbb{C}^{N/2}$ for some N that is a power of 2, the set of vectors with complex entries, or, more precisely, their floating point approximations. A message $\mathbf{z} \in \mathbb{C}^k$, for some integer $1 \leq k \leq N/2$, can be considered as a vector in $\mathbb{C}^{N/2}$ (by padding it with 0 entries), and it is then encoded to

$$m = \mathsf{Encode}(\mathbf{z}; \Delta) = \lfloor \Delta \cdot \varphi^{-1}(\mathbf{z}) \rceil \in \mathbb{Z}^N \equiv \mathcal{O},$$

where Δ is some precision factor. The "decode" operation $\mathsf{Decode} : \mathcal{O} \to \mathbb{C}^k$ sends an integer polynomial m to

$$\mathsf{Decode}(m; \Delta) = \varphi(\Delta^{-1} \cdot m) \in \mathbb{C}^k,$$

where the entries corresponding to the 0-paddings are dropped. Decode is an approximate inverse of Encode as $\mathbf{z}' = \mathsf{Decode}(\mathsf{Encode}(\mathbf{z}; \Delta); \Delta)$ is close (but not exactly equal) to $\mathbf{z}$.

This is slightly more problematic for our attack, because a passive adversary only gets to see the result of final decryption $\mathbf{z}' \in \mathbb{C}^k$, rather than the ring element $m' = a \cdot s + b \in \mathcal{O}$ that is required by our attack, in addition to the ciphertext $\mathsf{ct} = (a, b)$. Moreover, given the approximate nature of the encoding/decoding process, $\mathsf{Decode}(m')$ is not even the exact (mathematical) transformation $\varphi(\Delta^{-1} \cdot m')$, but only the result of an approximate floating point computation. We address this by setting $k = N/2$ (so, at least the vector $\mathsf{Decode}(m')$ has the right dimension over $\mathbb{C}$), and re-encoding the message output by the decryption algorithm to obtain $\mathsf{Encode}(\mathsf{Decode}(m'))$.

At this point, depending on the concrete choice of parameters of the scheme, we may have $\mathsf{Encode}(\mathsf{Decode}(m')) = m'$, in which case we can carry out the above attack by setting up a system of linear equations or computing inverses in the cyclotomic ring. We summarize this case in the following theorem.

Theorem 1 (Linear Key-Recovery Attack against CKKS). *Fix a particular instantiation of the CKKS scheme under the Ring-LWE assumption of dimension N and modulus q, and fix a key tuple* $(\mathsf{sk}, \mathsf{pk}, \mathsf{ek}) \leftarrow \mathsf{KeyGen}(1^\kappa)$. *Given $k = O(N)$ ciphertext ct_i for $1 \leq i \leq k$, that are either encryptions under* pk *or homomorphic evaluations under* ek*, and given their approximate decryption results $\mathbf{z}'_i = \mathsf{Decode}(\mathsf{Dec}_{\mathsf{sk}}(\mathsf{ct}_i); \Delta)$ with a scaling factor Δ, if $\mathsf{Encode}(\mathbf{z}'_i; \Delta) = \mathsf{Dec}_{\mathsf{sk}}(\mathsf{ct}_i)$ for all $1 \leq i \leq k$, then we can efficiently recover the secret key* sk *with high probability.*

Moreover, if the ciphertext modulus q is a prime or a product of distinct primes, then the above holds for all $k \geq 1$.

4.2 Analysis of Encoding/Decoding Errors

To see for what concrete parameters the linear attack can be applied, we take a closer look at the error introduced by the encoding and decoding computation. In practice, since N is a power of 2, the classical Cooley-Tukey FFT algorithm is used to implement the transformation φ and its inverse φ^{-1}, and the computation is done using floating-point arithmetic that could cause round-off errors.

Fix a ciphertext ct, and let $m' = \mathsf{Dec}_{\mathsf{sk}}(\mathsf{ct}) \in \mathcal{O}$ be its approximate decryption (before decoding) with a scaling factor Δ. Let $\hat{\mathbf{z}}' = \mathsf{Decode}(m'; \Delta)$ be the computed value of $\mathbf{z}' = \varphi(\Delta^{-1} \cdot m')$. To carry out the attack, we compute the encoding of $\hat{\mathbf{z}}'$ with the scaling factor Δ: first we apply inverse FFT to compute $\mathbf{u} = \Delta \cdot \varphi^{-1}(\hat{\mathbf{z}}')$, and then we round its computed value $\hat{\mathbf{u}}$ to $m'' = \lfloor \hat{\mathbf{u}} \rceil \in \mathcal{O}$. Let $\varepsilon = \hat{\mathbf{u}} - \mathbf{m}'$ be the *encoding error*, where $\mathbf{m}'$ is the coefficient vector of m'. We see that $\mathsf{Encode}(\mathsf{Decode}(m'; \Delta); \Delta) = m'$ if and only if $\|\varepsilon\|_\infty = \|\hat{\mathbf{u}} - \mathbf{m}'\|_\infty < \frac{1}{2}$.

Assume the relative error in computing the Cooley-Tukey FFT in dimension N is at most μ in l_2 norm. Then $\|\hat{\mathbf{z}}' - \mathbf{z}'\|_2 \leq \mu \cdot \frac{\sqrt{N}}{\Delta}\|\mathbf{m}'\|_2$, $\|\hat{\mathbf{u}} - \mathbf{u}\|_2 \leq \mu(1 + \mu) \cdot \|\mathbf{m}'\|_2$, and $\|\mathbf{u} - \mathbf{m}'\|_2 \leq \mu \cdot \|\mathbf{m}'\|_2$. It follows that

$$\|\varepsilon\|_\infty = \|\hat{\mathbf{u}} - \mathbf{m}'\|_\infty \leq \|\hat{\mathbf{u}} - \mathbf{m}'\|_2 \leq (2\mu + \mu^2)\|\mathbf{m}'\|_2.$$

In [13], Brisebarre et al. presented tight bounds on the relative error μ in applying the Cooley-Tukey FFT algorithm on IEEE-754 floating-point numbers. According to their estimate, $\mu \approx 53 \cdot 2^{-53}$ for $N = 2^{16}$ and double-precision floating-point numbers. So, we expect to see $\mathsf{Encode}(\mathsf{Decode}(m'; \Delta); \Delta) \neq m'$ in such setting, i.e., $\|\varepsilon\|_\infty > \frac{1}{2}$, when $\|m'\|_2 > 2^{45}$. (As we will see in the next section, our experimental results using existing CKKS implementations suggest this is a very conservative estimation.) The rescaling operation can be used to reduce the size of the approximate plaintext m', which is already used to maximize the capacity of homomorphic computation in CKKS.

Lattice attack. In case $\mathsf{Encode}(\mathsf{Decode}(m')) \approx m'$ is only an approximation of what we want for the linear key recovery attack, it is still possible to recover sk by solving a (polynomial time) lattice approximation problem.

Theorem 2 (Lattice Attack against CKKS). *Fix a particular instantiation of the CKKS scheme under the Ring-LWE assumption of dimension N and* *modulus* q, *and fix a key tuple* $(\mathsf{sk}, \mathsf{pk}, \mathsf{ek}) \leftarrow \mathsf{KeyGen}(1^\kappa)$. *Given a ciphertext* $\mathsf{ct} \in \mathcal{O}_q^2$ *with a scaling factor* Δ, *and given an approximate decryption* $\mathbf{z}' = \mathsf{Decode}(\mathsf{Dec}_{\mathsf{sk}}(\mathsf{ct}); \Delta)$ *of* ct, *if the encoding error* $\varepsilon = \Delta \cdot \varphi^{-1}(\mathbf{z}') - \mathsf{Dec}_{\mathsf{sk}}(\mathsf{ct})$ *satisfies* $\|\varepsilon\|_2 \leq 2^{-\frac{N}{2}} \cdot (q\sqrt{N} - h)$, *where* $h = \mathsf{HW}(s) \leq N$ *is the Hamming weight of* s, *then the secret key* sk *can be efficiently recovered.*

Proof (sketch). Let $\mathsf{ct} = (a, b)$ for some $a, b \in \mathcal{O}_q$. We consider the following approximate CVP instance. Let $A = \phi(a) \in \mathbb{Z}^{N \times N}$ be the negacyclic matrix representation of a. Consider the following matrix

$$B = \begin{pmatrix} A & qI_N \\ \mathbf{1}^t & \mathbf{0}^t \end{pmatrix} \in \mathbb{Z}^{(N+1)\times(2N)},$$

where $\mathbf{1}^t = [1, \ldots, 1]$ is a N-dimensional row vector of all 1 entries. Let $\mathcal{L} = \mathcal{L}(B)$ be the integer lattice generated by B, let $\mathbf{u} = \Delta \cdot \varphi^{-1}(\mathbf{z}') \in \mathbb{R}^N$, and let $\mathbf{t} = (\mathbf{u} - \mathbf{b}, 0)^t \in \mathbb{R}^{N+1}$, where $\mathbf{b}$ is the coefficient vector of b. Our CVP instance asks to find $\mathbf{v} \in \mathcal{L}$ such that $\|\mathbf{v} - \mathbf{t}\|_2 \leq \delta$ for some $\delta > 0$.

To set the parameter δ, notice that $\mathbf{v}_0 = (m' - b, \langle \mathbf{1}, \mathbf{s} \rangle)$ is a lattice point, and $\|\mathbf{v}_0 - \mathbf{t}\|_2^2 = \|\varepsilon\|_2^2 + \langle \mathbf{1}, \mathbf{s} \rangle^2$. On the other hand, if $m'' - b = A\mathbf{r} + q\mathbf{w}$ for some $\mathbf{r}, \mathbf{w} \in \mathbb{Z}^N$, then $\mathbf{v}_1 = (m'' - b, \langle \mathbf{1}, \mathbf{r} \rangle) \in \mathcal{L}$ is also a lattice point. We have $\|\mathbf{v}_1 - \mathbf{t}\|_2^2 = \|\varepsilon - \lfloor \varepsilon \rceil\|_2^2 + \langle \mathbf{1}, \mathbf{r} \rangle^2$. Note that $\mathbf{r} = A^{-1}(m' - b) + A^{-1} \lfloor \varepsilon \rceil \pmod q = \mathbf{s} + A^{-1} \lfloor \varepsilon \rceil \pmod q$. In CKKS, $\mathbf{s}$ is chosen from a uniform distribution on ternary coefficients $\{\pm 1, 0\}$ with Hamming weight $h \leq N$, so $|\langle \mathbf{1}, \mathbf{r} \rangle| \geq |\langle \mathbf{1}, A^{-1}\varepsilon \rangle - h|$. We can assume that $\lfloor \varepsilon \rceil$ is independent of $m' - b$, so $A^{-1} \lfloor \varepsilon \rceil \pmod q$ is close to uniform, and so it holds with high probability that $|\langle \mathbf{1}, A^{-1}\varepsilon \rangle| \leq 2\sqrt{3} \cdot q\sqrt{N}$. When $\|\varepsilon\|_2 \leq 2^{-\frac{N}{2}} \cdot (q\sqrt{N} - h)$, we can set $\delta = 2\sqrt{3} \cdot q\sqrt{N}$ and obtain $m' - b$ with high probability by solving such CVP instance in polynomial time. Then, we can mount the linear attack as in Theorem 1. □

5 Experiments

The basic idea of our linear attack is so simple that it requires no validation. However, as described in the previous section, a concrete instantiation of the CKKS scheme may include a number of details that make the attack more difficult in practice. Given the simplicity of our attack, we also considered the possibility that the implementations of CKKS may not correspond too closely to the theoretical scheme described in the papers, and included some additional countermeasures to defend against the attack.

To put our linear attack to a definitive test, we implemented it against publicly available libraries HEAAN [30], PALISADE [41], SEAL [47], and HElib [31] that implement the CKKS scheme, and we ran our attack over some homomorphic computations that are commonly used in real world privacy-preserving machine-learning applications. Our experimental results against the libraries are summarized in Tables 1 and 2. For most of the parameter settings, our attack can successfully and quite efficiently recover the secret key, showing it is widely

Algorithm 1: The pseudocode outlining our key recovery attack experiments.

Input: Lattice parameters $(N, \log q)$, initial scaling factor Δ_0, plaintext bound B, and circuit g.

1 Sample $(\mathsf{sk}, \mathsf{pk}, \mathsf{ek}) \leftarrow \mathsf{KeyGen}(N, \log q, \Delta_0)$, where $(1, s) = \mathsf{sk}$
2 Sample $\mathbf{z} \leftarrow \mathbb{C}^{N/2}$ such that $|z_i| \leq B$ for all $1 \leq i \leq N/2$
3 Encrypt $\mathsf{ct}_{\mathsf{in}} \leftarrow \mathsf{Enc}_{\mathsf{pk}}(\mathsf{Encode}(\mathbf{z}; \Delta_0))$
4 Evaluate $\mathsf{ct}_{\mathsf{out}} \leftarrow \mathsf{Eval}_{\mathsf{ek}}(g, \mathsf{ct}_{\mathsf{in}})$
5 Decrypt $\mathbf{z}' \leftarrow \mathsf{Decode}(\mathsf{Dec}_{\mathsf{sk}}(\mathsf{ct}_{\mathsf{out}}); \Delta)$, where Δ is the scaling factor in $\mathsf{ct}_{\mathsf{out}}$
6 Encode $m'' \leftarrow \mathsf{Encode}(\mathbf{z}'; \Delta)$
7 Compute $s' \leftarrow a^{-1} \cdot (m'' - b) \in \mathcal{O}_q$, where $(b, a) = \mathsf{ct}_{\mathsf{out}}$
8 **return** $s' = s$

applicable to these CKKS implementations. In the following, we discuss our experiment and the relevant implementation details of these libraries, and we briefly analyze the results. We also consider RNS-HEAAN [46], an alternative implementation similar to HEAAN that includes RNS (residue number system) optimizations, obtaining similar results.

We did not implement the lattice based attack. The main difficulty in running the lattice attack in our experiment is that it requires lattice reduction in very large dimension, beyond what is currently supported by state of the art lattice reduction libraries. However, the theoretical running time of the attack is polynomial, and the corresponding parameter settings should still be considered insecure. In the following, we refer to our linear attack as *the* attack.

5.1 Implementation of Our Attack and Experiments

A pseudocode outline of our experiment programs is presented in Algorithm 1. Such programs model the situations where an attacker can influence an honest user to perform certain homomorphic computations and can obtain both the final ciphertexts and the decrypted approximate numbers. A successful run indicates that the target CKKS implementation is not $\mathsf{IND\text{-}CPA^D}$-secure.

For concrete homomorphic computations, we choose to compute the variance of a wide range of input data to exemplify how our attack may be affected by large underlying plaintexts in extreme cases. Specifically, our program encrypts the input data to a single ciphertext $\mathsf{ct}_{\mathsf{in}}$ in the full packing mode, and then it performs one homomorphic squaring, followed by several homomorphic rotations and summations to homomorphically compute the sum of squares, and finally it does a homomorphic multiplication by a constant $2/N$ to obtain $\mathsf{ct}_{\mathsf{out}}$ that encrypts the variance. We also compute the logistic function $(1 + e^{-x})^{-1}$ and exponential functions e^x using their Maclaurin series up to a degree d, to check whether our attack may be affected by the bigger noises and the possibly adjusted scaling factors due to multi-level homomorphic computations. Once the homomorphic computation is done, our program decrypts $\mathsf{ct}_{\mathsf{out}}$ to approximate numbers $\mathbf{z}'$, and mounts our linear attack as in Steps 6 and 7 of Algorithm 1.

Table 1. The results of applying our attack on homomorphically computed variance of $N/2 = 2^{15}$ random complex numbers of magnitude $1 \leq B \leq 2^9$. We carried out the attack against all main open source implementations of CKKS, obtaining similar results. Numbers are packed into all slots, and are encoded using various initial scaling factors Δ_0. For each parameter combination (Δ_0, B), we ran our programs 100 times against each library. A "✓" indicates that, for *all* these libraries with the particular parameters, the attack *always* succeeded to recover sk. A few cells where a number is shown, correspond to extreme parameters where some runs failed to recover sk, and the number is the maximum (over all libraries) of the average l_∞ norms of the encoding error ε. These settings are still subject to attacks based on lattice reduction, see Sects. 4.2 and 5.3 for details.

	Attack applied to HEAAN, PALISADE, SEAL, HElib										
	B	1	2	2^2	2^3	2^4	2^5	2^6	2^7	2^8	2^9
Variance	$\log \Delta_0 = 30$	✓	✓	✓	✓	✓	✓	✓	✓	✓	✓
	$\log \Delta_0 = 40$	✓	✓	✓	✓	✓	✓	✓	✓	✓	✓
	$\log \Delta_0 = 50$	✓	✓	✓	✓	✓	✓	1.21	5.41	20.65	80.19

We remark that all these homomorphic computations are very common in applications of the CKKS scheme.

In our programs, we use the data structures and public APIs provided by each library to carry out the key recovery computation[4]. Note that an attacker is free to use any method, not necessarily these public interfaces, to carry out the attack.

5.2 Details on Different Implementations of CKKS

We considered the latest versions of all these libraries: HEAAN version 2.1 [30], PALISADE version 1.10.4 [41], SEAL version 3.5 [47], and HElib version 1.1.0 [31] and RNS-HEAAN [46]. All these libraries implement the transformation φ and its inverse using the classical Cooley-Tukey FFT algorithm on double-precision floating-point numbers. Still, they contain several distinct implementation details relevant to our attack.

Multi-precision integers vs. double-CRT representation. All versions of HEAAN (version 1.0 as in [18], version 1.1 as in [16], and the most recent version 2.1) use multi-precision integers to represent key materials and ciphertexts. Consequently, HEAAN achieves very good accuracy in approximate decryption, but at the same time it rarely introduces any encoding error, resulting in a great success rate in our key recovery experiment.

To improve efficiency, the residual number system, as known as double-CRT representation, is adopted to the CKKS scheme in [17], and it is implemented

[4] The source code of our attack implementations are available at https://github.com/ucsd-crypto/CKKSKeyRecovery.

Table 2. The results of applying our attack to homomorphically computed logistic and exponential functions on random real numbers of magnitude $B \in \{1, 2, 8\}$ packed into full $N/2 = 2^{15}$ slots, evaluated using their Maclaurin series of degree $d \in \{5, 10\}$. For each parameter setting, we ran our experimental program 100 times for each library, and here "✓" indicates sk was recovered in all these runs against a particular library. A few cells where a number is shown, correspond to extreme parameters when some runs failed to recover sk, and the number is the average l_∞ norm of the encoding error ε in these runs. For HElib, "n/a" indicates the parameters are not supported by the library.

			Attack applied to HEAAN, PALISADE, SEAL, HElib							
			HEAAN		PALISADE		SEAL		HElib	
	Δ_0	B	$d = 5$	$d = 10$	$d = 5$	$d = 10$	$d = 5$	$d = 10$	$d = 5$	$d = 10$
Logistic	2^{30}	1	✓	✓	✓	✓	✓	✓	✓	✓
	2^{40}	1	✓	✓	✓	✓	✓	✓	3.1	6.7
	2^{50}	1	✓	✓	✓	✓	✓	✓	8.2	8.2
Exponential	2^{30}	1	✓	✓	✓	✓	✓	✓	✓	✓
		2	✓	✓	✓	✓	✓	✓	n/a	n/a
		8	✓	✓	✓	✓	✓	✓	n/a	n/a
	2^{40}	1	✓	✓	✓	✓	✓	✓	1.9	8.2
		2	✓	✓	✓	✓	✓	✓	n/a	n/a
		8	✓	✓	✓	✓	✓	✓	n/a	n/a
	2^{50}	1	✓	✓	✓	✓	✓	✓	8.1	8.2
		2	✓	✓	✓	✓	✓	✓	n/a	n/a
		8	7.6	15.2	8.1	18.2	2.2	4.3	n/a	n/a

in RNS-HEAAN. Other libraries also implement the RNS variant of CKKS, with some different details:

- During decryption, RNS-HEAAN uses only the first RNS tower of ciphertexts; so it expects the scaled plaintext to be much smaller than the 60-bit prime modulus in the first tower. Other libraries convert the double-CRT format to multi-precision integers before applying the canonical embedding; so they support a larger plaintext space and are more accurate.
- During rescaling, RNS-HEAAN uses a power-of-2 rescaling factor, while the other libraries' rescaling factors are the primes or close to primes in the moduli chain. In particular, PALISADE optimizes the rescaling factors to reduce the errors and precision loss in many homomorphic operations [32].

As observed in our experiment, among the RNS implementations of CKKS, our attack was more successful against the libraries using more accurate element representations and scaling factors.

PALISADE. In addition, PALISADE uses extended precision floating-point arithmetic in Decode, which has 64-bit precision on X86 CPUs. This further improves the accuracy of approximate decryption, but perhaps unintentionally making our attack more successful by a tiny margin (comparing to other libraries).

HElib. Unlike other libraries, HElib adjusts the scaling factor used in Encode and many homomorphic operations according to the estimated noise size and the magnitude of the plaintext. It expects the input numbers to have magnitude at most 1 for optimal precisions. So our experiment with HElib chooses random input only within the unit circle.

RNS-HEAAN. Looking back to RNS-HEAAN, its implementation of Decode introduces a small round-off error in a conversion from `uint64_t` to `double`. As a result, such (seemingly unexpected) implementation choice may lead to reduced precision (by only a few bits), but it also results in more failed runs in our experiment. Still, when our attack fails, the encoding errors are quite small, and so RNS-HEAAN is still subject to the lattice reduction attack. We tried to "fix" this by more carefully converting between number systems, and we immediately see a much better success rate for our attack.

5.3 Experiment Results

We set up all libraries with the highest supported lattice dimension $N = 2^{16}$, which also corresponds to the highest security level. By the analysis in Sect. 4.2 (and also observed in our experiment), the larger the dimension is, the higher the chance an encoding error may show up (leading to failed attack runs). On the other hand, since the claimed security decreases with larger values of the modulus q, we set it to around 350 bit, which is a secure, yet realistic value for FHE schemes. According to common evaluation methodologies [1], the associated LWE problem provides a level of security well above 256 bits. (Specifically, in dimension $N = 2^{16}$, it is estimated that 256-bits of security are achieved even for moduli q with over 700 bits.)

In all our experiments, we use the full packing mode with $N/2$ slots. For the variance computation, we generate random input numbers with magnitude $B \leq 2^9$. For the experiment on the logistic and the exponential functions, we set the maximal degree of their Maclaurin series to $d \leq 10$, which provides good approximation for inputs smaller than 1.

Our experiments are executed in a 64-bit Linux environment running on an Intel i7-4790 CPU. The attack is very efficient, especially for the RNS-CKKS implementations, as the key recovery computation can benefit from using NTT and parallelization. Each individual run in our experiment finishes within several seconds to just one minute, with most of the running time taken by the key generation and encryption/homomorphic evaluation operations, rather than the attack itself. For each homomorphic computation task, for each parameter setting, and for each library, we run our attack 100 times to record the success rate and the encoding error ε. The results of the experiments with HEAAN, PALISADE, SEAL, and HElib are presented in Tables 1 and 2. As shown in these tables, our attack *always* succeeded to recover the secret key in *most* parameter settings against *all* the libraries, especially for typical input sizes and scaling factors. The failed cases in both tables correspond to the extreme parameters where the l_2 norm of the underlying plaintext exceeds 2^{52}, showing better practical performance than the worst case analysis in Sect. 4.2. (There are more failed cases with HElib because its adjusted scaling factors are typically larger and so are the plaintexts.) Comparing the results on the logistic and the exponential functions, we conclude that a deeper level of homomorphic computation has no significant effect on our attack, and the runs in the last row of Table 2 failed due to larger plaintext sizes. In particular, the encoding error ε with SEAL is smaller than other libraries because its implementation of Decode incurs less round-off errors in scaling by Δ^{-1}.

We did a limited number of experiments with RNS-HEAAN because it has a small plaintext space. Nonetheless, we see a consistent but small encoding error of size $\|\varepsilon\|_\infty \leq 2^7$ in our RNS-HEAAN experiments when $B^2\Delta_0 \approx 2^{50}$.

6 Conclusion

We proposed new security definitions, extending the traditional IND-CPA security notion, that properly capture the passive security requirement for (homomorphic) *approximate* encryption schemes. The necessity of adequate security notions for approximate encryption reminds us that correctness and security are two essential issues for cryptographic systems that must be considered at the same time. From a theoretical perspective, we initiated the study of IND-CPA^D security for approximate computation, by presenting implication and separation results between variants of the definition. There are still many very interesting research directions and open questions regarding IND-CPA^D security as well as simulation-based security. We leave further study of these new security notions to future work.

For our attack against the CKKS scheme, as it essentially recovers the encryption noise $\tilde{e}$ of the ciphertext, a natural countermeasure to harden the CKKS scheme is to modify the decryption algorithm, so that it does not output $m + \tilde{e}$, but only an approximation that does not depend on the secret key and encryption randomness. Below we discuss some specific ways to do that. We remark that our suggestions are just simple countermeasures to mitigate the effect of the attacks described in our work. Finding a more solid solution, provably achieving the notion of IND-CPA^D security proposed in this paper, is left as an open problem.

Gaussian noise. Perhaps the most natural way to do that (from an LWE perspective) is to add Gaussian noise to the result of the decryption function, similar to the noise introduced by the encryption algorithm. While this makes the schemes perhaps more robust, it does not seem an adequate countermeasure. The reason is that an attacker may repeatedly request decryptions of the same ciphertext. If the noise is unbiased, it can be easily reduced by taking several decryptions, and computing their average. The result will not be exact, but the noise can be made arbitrarily small by using a sufficiently large (still polynomial) number of calls, so that it can be eliminated either using rounding and Gaussian elimination, or applying the theoretical lattice-based attacks described in Sect. 4.2.

Deterministic noise or rounding. To avoid the above weakness, one can effectively limit the number of decryption calls to 1 per ciphertext by adding a deterministic noise in the decryption algorithm, e.g., as a pseudorandom function applied to the ciphertext (and key derived from the decryption key.) This is similar to the noise flooding techniques used in many lattice cryptographic contexts such as bootstrapping and circuit privacy of homomorphic encryption. It is perhaps not practical to apply noise flooding generically to achieve unbounded $\mathsf{IND\text{-}CPA^D}$ security, as it requires a superpolynomial modulus, but it might be feasible to achieve a bounded q-$\mathsf{IND\text{-}CPA^D}$ security for any a-priori fixed q, using techniques similar to [3,8,22]. A rigorous analysis is required not only for security but also for practical efficiency and tradeoffs in the accuracy of the approximate computation, and we leave this to future work.

Exact decryption. One can set up parameters in such a way that $\lfloor (m + \tilde{e})/\Delta \rceil = \lfloor m/\Delta \rceil$, at least with high probability, where the rounding operation is taken to certain precision. This effectively replaces the idea of approximate decryption with an exact decryption algorithm, but for a modified message. This could be a more promising direction to enhance the CKKS scheme, and it requires a careful analysis of encryption noises together with rounding errors. Intuitively, instead of interpreting a ciphertext as encoding an approximate number $m + \tilde{e}$, we regard them as encryptions of an approximate value $\lfloor (m + \tilde{e})/\Delta \rceil$. We can then define the operations supported by the homomorphic encryption scheme in such a way to ensure exact, deterministic behavior, both for homomorphic computations and final decryption.

Since the resulting scheme satisfies the standard notion of correctness for encryption (even if, perhaps, for a less standard set of operations than simple addition and multiplication), it can be easily analyzed using the traditional definition of security, and it is immediate to show that the scheme is secure under passive attacks based on a standard (Ring) LWE assumption.

Responsible Disclosure

We disclosed details of our attack to the developers of HEAAN, SEAL, HElib, and PALISADE at the beginning of October 2020 (and also to Lattigo [34] at a later time, after porting our attack to the GO programming language used by that library), before making our paper public. All teams were very responsive, and they quickly acknowledged that our attack works and it represents a serious threat that needs to be addressed. They have taken various actions, addressing the vulnerability to different degrees, ranging from warning the users that any use of the decryption function (except in very controlled environments where the result of decryption is kept private,) to implementing some mitigation strategy along the lines discussed in the previous section. In particular, we have tested the latest development versions of HElib and PALISADE, and we can confirm our attack is no longer effective against them. Developing and implementing a variant of CKKS which provably achieves $\mathsf{IND\text{-}CPA}^{\mathsf{D}}$ with only a modest decrease in performance is left to future work.

Acknowledgment. We would like to thank Mark Schultz, Jessica Sorrell, and the SAIT research team for useful discussions. We would like to thank Victor Shoup and Jingwei Chen for pointing out errors in an earlier version of this paper.

References

1. Albrecht, M., et al.: Homomorphic encryption security standard. Technical report, HomomorphicEncryption.org, Toronto, Canada, November 2018. https://homomorphicencryption.org/standard/
2. Badawi, A.A., Hoang, L., Mun, C.F., Laine, K., Aung, K.M.M.: Privft: private and fast text classification with homomorphic encryption. CoRR, abs/1908.06972 (2019)
3. Bai, S., Lepoint, T., Roux-Langlois, A., Sakzad, A., Stehlé, D., Steinfeld, R.: Improved security proofs in lattice-based cryptography: using the Rényi divergence rather than the statistical distance. J. Cryptol. **31**(2), 610–640 (2018)
4. Bellare, M., Desai, A., Jokipii, E., Rogaway, P.: A concrete security treatment of symmetric encryption. In: 38th Annual Symposium on Foundations of Computer Science, FOCS 1997, pp. 394–403. IEEE Computer Society (1997)
5. Bergamaschi, F., Halevi, S., Halevi, T.T., Hunt, H.: Homomorphic training of 30,000 logistic regression models. In: Deng, R.H., Gauthier-Umaña, V., Ochoa, M., Yung, M. (eds.) ACNS 2019. LNCS, vol. 11464, pp. 592–611. Springer, Cham (2019). https://doi.org/10.1007/978-3-030-21568-2_29
6. Boemer, F., Costache, A., Cammarota, R., Wierzynski, C.: nGraph-HE2: a high-throughput framework for neural network inference on encrypted data. CoRR, abs/1908.04172 (2019)
7. Boemer, F., Lao, Y., Cammarota, R., Wierzynski, C.: nGraph-HE: a graph compiler for deep learning on homomorphically encrypted data. In: CF 2019, pp. 3–13. ACM (2019)
8. Bogdanov, A., Guo, S., Masny, D., Richelson, S., Rosen, A.: On the hardness of learning with rounding over small modulus. In: Kushilevitz, E., Malkin, T. (eds.) TCC 2016, Part I. LNCS, vol. 9562, pp. 209–224. Springer, Heidelberg (2016). https://doi.org/10.1007/978-3-662-49096-9_9
9. Brakerski, Z.: Fully homomorphic encryption without modulus switching from classical GapSVP. In: Safavi-Naini, R., Canetti, R. (eds.) CRYPTO 2012. LNCS, vol. 7417, pp. 868–886. Springer, Heidelberg (2012). https://doi.org/10.1007/978-3-642-32009-5_50
10. Brakerski, Z., Gentry, C., Vaikuntanathan, V.: (Leveled) fully homomorphic encryption without bootstrapping. ACM Trans. Comput. Theory **6**(3), 13:1–13:36 (2014)
11. Brakerski, Z., Vaikuntanathan, V.: Fully homomorphic encryption from ring-LWE and security for key dependent messages. In: Rogaway, P. (ed.) CRYPTO 2011. LNCS, vol. 6841, pp. 505–524. Springer, Heidelberg (2011). https://doi.org/10.1007/978-3-642-22792-9_29

12. Brakerski, Z., Vaikuntanathan, V.: Efficient fully homomorphic encryption from (standard) LWE. SIAM J. Comput. **43**(2), 831–871 (2014)
13. Brisebarre, N., Joldes, M., Muller, J., Nanes, A., Picot, J.: Error analysis of some operations involved in the Cooley-Tukey fast fourier transform. ACM Trans. Math. Softw. **46**(2), 11:1–11:27 (2020)
14. Chen, H., Chillotti, I., Song, Y.: Improved bootstrapping for approximate homomorphic encryption. In: Ishai, Y., Rijmen, V. (eds.) EUROCRYPT 2019, Part II. LNCS, vol. 11477, pp. 34–54. Springer, Cham (2019). https://doi.org/10.1007/978-3-030-17656-3_2
15. Chen, H., Laine, K., Player, R.: Simple encrypted arithmetic library - SEAL v2.1. In: Brenner, M., et al. (eds.) FC 2017. LNCS, vol. 10323, pp. 3–18. Springer, Cham (2017). https://doi.org/10.1007/978-3-319-70278-0_1
16. Cheon, J.H., Han, K., Kim, A., Kim, M., Song, Y.: Bootstrapping for approximate homomorphic encryption. In: Nielsen, J.B., Rijmen, V. (eds.) EUROCRYPT 2018, Part I. LNCS, vol. 10820, pp. 360–384. Springer, Cham (2018). https://doi.org/10.1007/978-3-319-78381-9_14
17. Cheon, J.H., Han, K., Kim, A., Kim, M., Song, Y.: A full RNS variant of approximate homomorphic encryption. In: Cid, C., Jacobson, M. (eds.) SAC 2018. LNCS, vol. 11349, pp. 347–368. Springer, Cham (2018). https://doi.org/10.1007/978-3-030-10970-7_16
18. Cheon, J.H., Kim, A., Kim, M., Song, Y.: Homomorphic encryption for arithmetic of approximate numbers. In: Takagi, T., Peyrin, T. (eds.) ASIACRYPT 2017, Part I. LNCS, vol. 10624, pp. 409–437. Springer, Cham (2017). https://doi.org/10.1007/978-3-319-70694-8_15
19. Cheon, J.H., Kim, D., Park, J.H.: Towards a practical clustering analysis over encrypted data. IACR Cryptology ePrint Archive 2019/465 (2019)
20. Dathathri, R., et al.: CHET: compiler and runtime for homomorphic evaluation of tensor programs. CoRR, abs/1810.00845 (2018)
21. Dathathri, R., et al.: CHET: an optimizing compiler for fully-homomorphic neural-network inferencing. In: PLDI 2019, pp. 142–156. ACM (2019)
22. Ducas, L., Stehlé, D.: Sanitization of FHE ciphertexts. In: Fischlin, M., Coron, J.-S. (eds.) EUROCRYPT 2016, Part I. LNCS, vol. 9665, pp. 294–310. Springer, Heidelberg (2016). https://doi.org/10.1007/978-3-662-49890-3_12
23. Gentry, C.: Fully homomorphic encryption using ideal lattices. In: STOC 2009, pp. 169–178. ACM (2009)
24. Gentry, C., Halevi, S., Smart, N.P.: Homomorphic evaluation of the AES circuit. In: Safavi-Naini, R., Canetti, R. (eds.) CRYPTO 2012. LNCS, vol. 7417, pp. 850–867. Springer, Heidelberg (2012). https://doi.org/10.1007/978-3-642-32009-5_49
25. Goldwasser, S., Micali, S.: Probabilistic encryption. J. Comput. Syst. Sci. **28**(2), 270–299 (1984)
26. Halevi, S., Shoup, V.: Algorithms in HElib. In: Garay, J.A., Gennaro, R. (eds.) CRYPTO 2014, Part I. LNCS, vol. 8616, pp. 554–571. Springer, Heidelberg (2014). https://doi.org/10.1007/978-3-662-44371-2_31
27. Halevi, S., Shoup, V.: Bootstrapping for `HElib`. In: Oswald, E., Fischlin, M. (eds.) EUROCRYPT 2015, Part I. LNCS, vol. 9056, pp. 641–670. Springer, Heidelberg (2015). https://doi.org/10.1007/978-3-662-46800-5_25
28. Halevi, S., Shoup, V.: Faster homomorphic linear transformations in HElib. In: Shacham, H., Boldyreva, A. (eds.) CRYPTO 2018, Part I. LNCS, vol. 10991, pp. 93–120. Springer, Cham (2018). https://doi.org/10.1007/978-3-319-96884-1_4
29. Han, K., Hong, S., Cheon, J.H., Park, D.: Logistic regression on homomorphic encrypted data at scale. In: AAAI 2019, pp. 9466–9471. AAAI Press (2019)

30. HEAAN (release 2.1). SNUCRYPTO (2018). https://github.com/snucrypto/HEAAN
31. HElib (release 1.1.0). IBM (2020). https://github.com/homenc/HElib
32. Kim, A., Papadimitriou, A., Polyakov, Y.: Approximate homomorphic encryption with reduced approximation error. Cryptology ePrint Archive, Report 2020/1118 (2020). https://eprint.iacr.org/2020/1118
33. Kim, D., Song, Y.: Approximate homomorphic encryption over the conjugate-invariant ring. In: Lee, K. (ed.) ICISC 2018. LNCS, vol. 11396, pp. 85–102. Springer, Cham (2019). https://doi.org/10.1007/978-3-030-12146-4_6
34. Lattigo 2.0.0. EPFL-LDS, October 2020. http://github.com/ldsec/lattigo
35. Li, Z., Galbraith, S.D., Ma, C.: Preventing adaptive key recovery attacks on the Gentry-Sahai-Waters leveled homomorphic encryption scheme. IACR Cryptology ePrint Archive 2016/1146 (2016)
36. Lyubashevsky, V., Micciancio, D.: Generalized compact knapsacks are collision resistant. In: Bugliesi, M., Preneel, B., Sassone, V., Wegener, I. (eds.) ICALP 2006, Part II. LNCS, vol. 4052, pp. 144–155. Springer, Heidelberg (2006). https://doi.org/10.1007/11787006_13
37. Lyubashevsky, V., Micciancio, D., Peikert, C., Rosen, A.: SWIFFT: a modest proposal for FFT hashing. In: Nyberg, K. (ed.) FSE 2008. LNCS, vol. 5086, pp. 54–72. Springer, Heidelberg (2008). https://doi.org/10.1007/978-3-540-71039-4_4
38. Lyubashevsky, V., Peikert, C., Regev, O.: On ideal lattices and learning with errors over rings. J. ACM **60**(6), 43:1–43:35 (2013)
39. Lyubashevsky, V., Peikert, C., Regev, O.: A toolkit for ring-LWE cryptography. In: Johansson, T., Nguyen, P.Q. (eds.) EUROCRYPT 2013. LNCS, vol. 7881, pp. 35–54. Springer, Heidelberg (2013). https://doi.org/10.1007/978-3-642-38348-9_3
40. Micciancio, D.: Generalized compact knapsacks, cyclic lattices, and efficient one-way functions. Comput. Complex. **16**(4), 365–411 (2007)
41. PALISADE lattice cryptography library (release 1.10.4). PALISADE Project (2020). https://gitlab.com/palisade/
42. Park, S., Lee, J., Cheon, J.H., Lee, J., Kim, J., Byun, J.: Security-preserving support vector machine with fully homomorphic encryption. In: SafeAI@AAAI 2019. CEUR Workshop Proceedings, vol. 2301. CEUR-WS.org (2019)
43. Peikert, C., Rosen, A.: Efficient collision-resistant hashing from worst-case assumptions on cyclic lattices. In: Halevi, S., Rabin, T. (eds.) TCC 2006. LNCS, vol. 3876, pp. 145–166. Springer, Heidelberg (2006). https://doi.org/10.1007/11681878_8
44. Raisaro, J.L., Klann, J.G., Wagholikar, K.B., Estiri, H., Hubaux, J.-P., Murphy, S.N.: Feasibility of homomorphic encryption for sharing I2B2 aggregate-level data in the cloud. AMIA Summits Transl. Sci. Proc. **2017**, 176–185 (2018)
45. Regev, O.: On lattices, learning with errors, random linear codes, and cryptography. J. ACM **56**(6), 34:1–34:40 (2009)
46. RNS-HEAAN. SNUCRYPTO (2018). https://github.com/KyoohyungHan/FullRNS-HEAAN
47. Microsoft SEAL (release 3.5). Microsoft Research, Redmond, April 2020. https://github.com/Microsoft/SEAL
48. Sun, X., Yu, F.R., Zhang, P., Xie, W., Peng, X.: A survey on secure computation based on homomorphic encryption in vehicular ad hoc networks. Sensors **20**(15), 4253 (2020)

The Rise of Paillier: Homomorphic Secret Sharing and Public-Key Silent OT

Claudio Orlandi(✉), Peter Scholl, and Sophia Yakoubov

Aarhus University, Aarhus, Denmark
{orlandi,peter.scholl,sophia.yakoubov}@cs.au.dk

Abstract. We describe a simple method for solving the distributed discrete logarithm problem in Paillier groups, allowing two parties to locally convert multiplicative shares of a secret (in the exponent) into additive shares. Our algorithm is perfectly correct, unlike previous methods with an inverse polynomial error probability. We obtain the following applications and further results.

- **Homomorphic secret sharing.** We construct homomorphic secret sharing for branching programs with *negligible* correctness error and supporting *exponentially large* plaintexts, with security based on the decisional composite residuosity (DCR) assumption.
- **Correlated pseudorandomness.** Pseudorandom correlation functions (PCFs), recently introduced by Boyle et al. (FOCS 2020), allow two parties to obtain a practically unbounded quantity of correlated randomness, given a pair of short, correlated keys. We construct PCFs for the oblivious transfer (OT) and vector oblivious linear evaluation (VOLE) correlations, based on the quadratic residuosity (QR) or DCR assumptions, respectively. We also construct a pseudorandom correlation generator (for producing a bounded number of samples, all at once) for general degree-2 correlations including OLE, based on a combination of (DCR or QR) and the learning parity with noise assumptions.
- **Public-key silent OT/VOLE.** We upgrade our PCF constructions to have a *public-key setup*, where after independently posting a public key, each party can locally derive its PCF key. This allows completely *silent generation* of an arbitrary amount of OTs or VOLEs, without any interaction beyond a PKI, based on QR, DCR, a CRS and a random oracle. The public-key setup is based on a novel non-interactive vector OLE protocol, which can be seen as a variant of the Bellare-Micali oblivious transfer protocol.

1 Introduction

Homomorphic secret sharing, or HSS, allows two parties to non-interactively perform computations on secret-shared private inputs. In contrast to homomorphic encryption, where a single party carries out the computation on encrypted data, HSS can be viewed as a distributed variant where several servers are each given a share of the inputs, and then (without further interaction) can

A. Canteaut and F.-X. Standaert (Eds.): EUROCRYPT 2021, LNCS 12696, pp. 678–708, 2021.
https://doi.org/10.1007/978-3-030-77870-5_24

homomorphically evaluate these to obtain a share of the desired output. Useful applications of HSS include succinct forms of secure multi-party computation [BGI16a, BGI17, BGMM20], private querying to public databases [GI14, BGI15, WYG+17] and generating correlated randomness in secure computation protocols [BCGI18, BCG+19]. In this work, we will focus on a strong flavour of HSS with *additive reconstruction*, meaning that the server's shares of the output can be simply added together (in an abelian group) to give the result of the computation.[1]

At Crypto 2016, Boyle, Gilboa and Ishai [BGI16a] constructed two-server HSS for the class of polynomial-size branching programs based on the decisional Diffie-Hellman (DDH) assumption. Branching programs are a class of computations that cover restricted classes of circuits such as NC^1 and logspace computations. One application of their construction is succinct secure computation protocols for these types of computation, where the communication complexity is proportional only to the input and output lengths [BGI17]. However, Boyle et al. also managed to achieve secure computation for general, leveled circuits with a communication cost that is *sublinear in the circuit size* by a logarithmic factor. Previously, breaking this circuit-size barrier was only known to be possible using fully homomorphic encryption, so this result positioned HSS as an alternative path towards secure computation with low communication.

At the heart of the DDH-based construction [BGI16a] is a *distributed discrete log* procedure, where two parties are given multiplicative shares of a secret g^x (for some fixed base g), and wish to locally convert these into *additive shares* of x. Their method of solving this unfortunately has an inverse polynomial probability ε of correctness error, which is expensive to keep small, since the workload in homomorphic evaluation scales with $O(1/\varepsilon)$.

Their original HSS construction has been extended in several works, including a simpler "public-key" style sharing phase [BGI17], a variant based on Paillier encryption [FGJS17], improved efficiency of the distributed discrete log step [BCG+17, DKK18] and techniques for mitigating leakage that can arise from the non-negligible correctness error [BCG+17].

Despite this progress, all these constructions still have the limitation of a non-negligible chance of an incorrect computation, which requires a large amount of extra work to keep small. In fact, Dinur et al. [DKK18] showed a conditional lower bound that solving the distributed discrete log protocol with correctness probability ε *requires* $\Omega(1/\sqrt{\varepsilon})$ computation, unless the discrete logarithm problem in an interval can be solved more efficiently. They also gave a matching upper bound.

On the other hand, if we rely on the learning with errors (LWE) assumption, instead of discrete log- or factoring-based assumptions, it is possible to obtain HSS for arbitrary circuits [DHRW16, BGI+18], and with a *negligible* probability of correctness error, when using LWE with a superpolynomial modulus. This construction builds on fully homomorphic encryption [Gen09, BV11], and despite much recent progress, this still involves a significant computational overhead.

[1] This leads to a form of optimally succinct reconstruction that even fully homomorphic cannot achieve on its own.

When restricting computations to branching programs instead of circuits, and limiting the number of servers to two, there is a much specialized construction that reduces these costs [BKS19].

Pseudorandom Correlation Generators. A recent, promising application of techniques based on HSS is to build *pseudorandom correlation generators* (PCGs) [BCGI18,BCG+19], which are a way of expanding short, correlated seeds into a large amount of correlated randomness. This correlated randomness might be, for instance, a batch of oblivious transfers (OTs) on random inputs, which can be used to obtain cheap, information-theoretic protocols for secure computation of Boolean circuits. Other correlations can be used to securely compute arithmetic circuits over a ring R, for example, in oblivious linear evaluation (OLE), each sample has the form $(a, b), (x, ax + b)$ for random $a, b, x \in R$. Another example is vector oblivious linear evaluation (VOLE), which has the restriction that x is fixed for each sample of the correlation.

More concretely, a PCG is a pair of algorithms $(\mathsf{Gen}, \mathsf{Expand})$, where Gen outputs a pair of short, correlated seeds, while Expand takes one of these seeds and expands it into a longer output string. The security requirements are that the joint distribution of both outputs is indistinguishable from the desired correlation, and also that each seed preserves privacy of the other party's output.

While PCGs can be constructed from suitably expressive HSS [BCG+19], this requires homomorphic evaluation of a pseudorandom generator inside HSS and typically leads to poor concrete efficiency. Instead, the most promising constructions so far are based on variants of the learning parity with noise (LPN) assumption, and build upon practical constructions of HSS for point functions (or, function secret sharing) [GI14,BGI15,BGI16b]. Using LPN, we can obtain PCGs for the VOLE [BCGI18], OT [BCG+19] and OLE [BCG+19,BCG+20b] correlations, and these can even be concretely efficient when relying on structured variants of LPN such as ring-LPN, or using quasi-cyclic codes.

Pseudorandom Correlation Functions. Very recently, Boyle et al. [BCG+20a] extended the notion of PCG to a *pseudorandom correlation function* (PCF), which allows generating an unbounded number of correlated outputs in an on-the-fly manner, given a pair of correlated keys. This is similar to how a pseudorandom function extends the concept of a pseudorandom generator. While PCFs can be constructed in a generic but inefficient manner based on LWE, Boyle et al. also gave constructions based on new flavours of *variable-density* LPN assumptions. The practical security and efficiency of these constructions has yet to be determined, although their initial results suggest that the PCFs for the OT and VOLE correlations could be concretely efficient.

1.1 Our Contributions

In this work, we present new constructions of homomorphic secret sharing and pseudorandom correlation functions based on standard, number-theoretic assumptions related to factoring. At the heart of most of our constructions is a single, key technique, namely, an efficient algorithm for solving the distributed discrete logarithm problem when using the Paillier cryptosystem over $\mathbb{Z}_{N^2}^*$ (where N is an RSA modulus). Unlike previous algorithms [BGI16a, FGJS17, DKK18], which always incurred an inverse polynomial probability of error, our method is very simple and has *perfect correctness.*

Building on this technique, we obtain the following results.

Homomorphic Secret Sharing. We construct homomorphic secret sharing for branching programs with *negligible correctness error* and supporting computations on an *exponentially large* plaintext space. We present several variants. The first two are based on circular security assumptions (of Paillier encryption and of a Paillier-ElGamal hybrid, respectively); however, the second has the advantage of allowing for a public-key style setup. The third variant also allows for a public-key setup, and additionally relies solely on the decisional composite residuosity (DCR) assumption. However, it is less efficient.

This gives the first construction of negligible-error HSS for branching programs without relying on LWE with a superpolynomial modulus. Compared with previous constructions based on discrete log-type assumptions [BGI16a, BGI17], as well as the Paillier-based construction of Fazio et al. [FGJS17], we avoid their limitations of a $1/\mathsf{poly}$ correctness error and polynomial-sized plaintext space. We also obtain smaller share sizes and much better computational efficiency.

Pseudorandom Correlation Functions and Pseudorandom Correlation Generators. We construct PCFs for producing an arbitrary number of random instances of vector oblivious linear evaluation, based on the DCR assumption, and oblivious transfer, based on the quadratic residuosity (QR) assumption. These constructions are very simple and have relatively small key sizes, consisting of $O(1)$ and $O(\lambda)$ group elements, respectively (for security parameter λ). We also construct a weaker pseudorandom correlation generator (for producing a bounded number of samples, or, all at once) for general degree-2 correlations, when assuming $\{\text{DCR} \vee \text{QR}\} \wedge \text{LPN}$. Compared with a previous construction based on only LPN [BCG+19], we reduce the key size by a factor $O(\lambda)$ and reduce the computational cost from quadratic to linear in the output length. This can also be upgraded to a PCF, when assuming a recent, variable-density version of LPN [BCG+20a].

Public-Key Silent OT and VOLE. We show how to upgrade our PCF constructions to have a *public-key setup.* After independently posting a public key, which uses a CRS, each party can use the other party's key, together with the private randomness for its own key, to derive a key for the PCF. Using their PCF keys, the parties can then silently compute an arbitrary quantity of OT or VOLE correlations, all *without any interaction* beyond the PKI. To our knowledge, these are the first such constructions that allow producing non-trivial correlations from a reusable public-key setup, without relying on lattice-based assumptions and homomorphic evaluation of PRGs inside multi-key FHE [DHRW16,BCG+19]. Note that we assume both a CRS and the random oracle model.

The public-key PCF for OT can be plugged into an existing construction of two-round multi-party computation with an OT correlations setup [GIS18], and reduces the complexity of its setup phase. This leads to a passively secure, two-round MPC protocol based on the DCR and QR assumptions, which makes a black-box use of a PRG, and has a PKI setup that scales independently of the circuit size. This is in contrast to the *strong PKI* setup from [GIS18], where the size of each public key scales linearly with the circuit size.

1.2 Comparison with Previous Results

We now give a more detailed comparison of our results with those from previous work, and discuss some efficiency metrics.

Homomorphic Secret Sharing. As already mentioned, we avoid the $1/\mathsf{poly}$ correctness error and small message space associated with previous constructions based on DDH [BGI16a,BCG+17,BGI17] or Paillier [FGJS17]. This brings us the additional benefit that the share size of our HSS is smaller, since in all these constructions, each share of an input x contains encryptions of $x \cdot d_i$, where d_i are the bits of the secret key. Since we support a large message space, we can instead choose d_i to be a much larger chunk of the secret key, so that each share only contains a *constant* number of group elements, instead of $O(\lambda)$. The smaller share size and improved share conversion step in our construction also give us much lower computational costs, since previous works require a workload that scales with $\Omega(1/\sqrt{\varepsilon})$, where ε is the correctness error probability [DKK18].

We can also compare our HSS with constructions based on LWE. Using LWE with a superpolynomial modulus, these can also support negligible error and a superpolynomial plaintext space [BKS19], and can even go beyond branching programs to evaluate general circuits [DHRW16,BGI+18]. When restricted to branching programs or low-depth circuits, and using ring-LWE, homomorphic evaluation would likely be more efficient than our HSS, due to fast algorithms for polynomial arithmetic in ring-LWE, compared with exponentiations in Paillier. On the other hand, our scheme has much smaller shares, since ring-LWE ciphertexts with a superpolynomial modulus are orders of magnitude larger than Paillier ciphertexts (ranging from 100 kB–3 MB in [BKS19] vs under 1 kB for Paillier).

Finally, we remark that LWE is a very different assumption to DCR. On the one hand, it can plausibly resist attacks by quantum computers; however, in a purely classical setting, factoring-based assumptions are arguably better studied than ring-LWE with a superpolynomial modulus.

PCFs and PCGs. Compared with PCGs for OT and VOLE based on LPN [BCGI18,BCG+19], our PCFs have the advantages of a public-key setup and the ability to incrementally produce an unbounded number of outputs (which comes with being a PCF and not just a PCG). In VOLE, our PCF has the additional benefit of much smaller keys, since each party's key only contains two Paillier group elements, compared with $\tilde{O}(\lambda^2)$ bits using LPN. The key size in our PCF for OT is around λ elements of $\mathbb{Z}_N$, which is comparable to the LPN-based PCG keys at typical security levels. The main drawback of our constructions is their computational efficiency, since our PCF for VOLE requires one exponentiation in $\mathbb{Z}_{N^2}^*$ to produce a single output in $\mathbb{Z}_N$, while the PCF for OT requires ≈ 128 exponentiations in $\mathbb{Z}_N^*$ to obtain one string-OT (at the 128-bit security level). Similarly, our PCG for OLE (based on both LPN and DCR) reduces the key size of previous LPN-based PCGs for OLE by a factor of $O(\lambda)$, at the cost of requiring exponentiations and limited to OLEs over $\mathbb{Z}_N$ or $\mathbb{Z}_2$, rather than more general rings or fields.

The recent PCFs for OT and VOLE from variable-density LPN [BCG+20a] overcome the PCG limitation of standard LPN-based constructions, although still do not have a public-key setup. They also come with much larger keys than their PCG counterparts, as well as our VOLE and OT constructions. Their computational efficiency has not yet been demonstrated, although they may be faster than our number-theoretic constructions due to being based on lightweight primitives like distributed point functions.

1.3 Overview of Techniques

We start by recalling the share conversion procedure used by Boyle et al. [BGI16a]. The basic idea of their scheme allows two parties to locally multiply secrets $x, y \in \mathbb{Z}$, where x is encrypted and y is secret shared, obtaining a secret sharing of the result $z = xy$. However, z is now *multiplicatively* (or, rather, *divisively*) shared; that is, the parties have group elements $g_0, g_1 \in \mathbb{G}$, such that $g_1 = g_0 \cdot g^z$. To continue evaluating a program, they would like to convert these into *linear* (*subtractive*) shares, so they can be used in another multiplication (with a ciphertext).

Boyle et al. described an ingenious protocol for converting such divisive shares to subtractive shares. To obtain subtractive shares of z, it is enough that the parties *agree upon* some distinguished element h that is not too far away from g_0, g_1 in terms of multiplications by g. If they find such an h, then party σ can compute the distance of g_σ from h by brute force: by multiplying h by g repeatedly, and seeing how many such multiplications it takes to get to g_σ. If we're guaranteed that h isn't too far away, this should not be too inefficient. Let d_σ be the distance of g_σ from h—that is, $hg^{d_\sigma} = g_\sigma$. Then,

$$\begin{aligned} g_1 = g_0 \cdot g^z &\Leftrightarrow hg^{d_1} = hg^{d_0} g^z \\ &\Leftrightarrow g^{d_1} = g^{d_0} g^z \end{aligned}$$

We can conclude that $d_1 \equiv d_0 + z$ modulo the order of the subgroup generated by g, and if d_0, d_1 are small then these shares can be recovered efficiently.

The major challenge is agreeing upon a common point h. Boyle et al. did so by having the parties first fix a set of random, distinguished points in the group;

party σ then finds the closest point in this set to g_σ. As long as both parties find the same point, this will lead to a correct share conversion. To make this process efficient, the distance t between successive points can't be too large, since the running time will be $O(t)$. However, there is then an inherent $\approx 1/t$ probability of failure, in case a point lies between the original two shares and they fail to agree.

This leads to a tradeoff between running time and correctness of the share conversion procedure. Dinur, Keller and Klein [DKK18] described an improved conversion algorithm, which achieves $1/t$ error probability in only $O(\sqrt{t})$ steps. On the negative side, they showed that any algorithm which beats this could be used to improve the cost of finding discrete logarithms in an interval, a well-studied problem that is believed to be hard.

Despite the correctness difficulty, this weaker form of HSS still suffices for many applications including sublinear secure two-party computation, with some additional work to correct for errors [BGI16a,BGI17].

Share Conversion in Paillier. By moving to a Paillier group ($\mathbb{Z}_{N^2}^*$), we can overcome the challenges of (a) agreeing on a distinguished point and (b) efficiently finding the distance of a multiplicative share from that point, *without requiring a correctness/efficiency tradeoff.*

The parties' multiplicative shares will still have the form g_0, g_1 such that $g_1 = g_0 \cdot g^z$; however, now we take $g = (1+N)$, which has order N in $\mathbb{Z}_{N^2}^*$. To find the distinguished point h, the parties simply *reduce their shares mod* N. Remarkably, this leads to the parties always agreeing upon the same value h, which is also guaranteed to be the *smallest* value in the coset $X = (g_0, g_0 \cdot g, \ldots, g_0 \cdot g^{N-1})$. To see that parties agree on h, notice that since $(1+N)^x \equiv 1 \pmod N$, we have $g_0 \equiv g_1 \pmod N$. To see that h lies in X, write $g_0 = (h + h'N)$ and suppose that $h = g_0(1+N)^s$ for some s. Then, since $(1+N)^s \equiv (1+sN) \pmod{N^2}$, we have $h \equiv (h+h'N)(1+sN) \pmod{N^2}$, which we can easily solve to get $s \equiv -h'h^{-1} \pmod N$.

Given this, party σ can compute the distance from their multiplicative shares to h *without the use of brute force.* They simply take $h/g_\sigma = (1+N)^{z_\sigma}$, and then take the discrete logarithm of this (exploiting the fact that discrete logs are easy with base $1+N$) to find their additive share z_σ satisfying $z_0 - z_1 = z \bmod N$.

As well as removing the correctness error, moving to Paillier groups has removed the limitation that messages must be small, since we can efficiently apply share conversion to shares of any message in $\mathbb{Z}_N$. We are still missing one step, however; to be able to continue the HSS computation, we need shares of z *over the integers*, rather than modulo N. Using a trick previously used in an LWE-based scheme [BKS19], if z is sufficiently smaller than N, with high probability subtractive shares of z modulo N are *already* valid shares over the integers. Assuming z to be much smaller than N is not very limiting, since we can still have, say, z around $\sqrt{N}$ and achieve both negligible failure probability and exponentially large plaintexts.

HSS Variants. We use the trick described above to get homomorphic secret sharing for branching programs. We subtractively share (digits of) the Paillier decryption key d (where $d \equiv 1 \pmod{N}$ and $d \equiv 0 \pmod{\phi(N)}$) between the two parties. Similarly to [FGJS17], we can use such a sharing of the secret key to go from a Paillier ciphertext to a divisive sharing of the underlying message x of the form $g_1 = g_0 \cdot (1+N)^x$. Once we have that, we can obtain a subtractive sharing of x, as described above. Given subtractive shares of d times some $y \in [N]$, we can similarly get a divisive sharing—and then a subtractive sharing—of xy. Using encryptions of digits of the key d, we can maintain the invariant that we always have subtractive shares of d times our intermediate values available, so we can continue the computation and multiply more encrypted values by our intermediate values.

There are two downsides to our Paillier-based HSS scheme: (1) it requires trusted setup (to distribute shares of the key d), and (2) since we use encryptions of digits of d, we need to assume the *circular security* of Paillier. We can avoid trusted setup by instead using Paillier-ElGamal encryption [CS02, DGS03, BCP03]. When using Paillier-ElGamal, once a modulus N is generated and published, the parties need only do a public-key style setup, where each party independently generates a secret/public key pair, and publishes its public key, following a previous ElGamal-based method [BGI17]. We can additionally avoid assuming circular security by using the Brakerski-Goldwasser scheme [BG10], which is *provably* circular-secure. The downside of using this scheme is much larger ciphertexts.

Pseudorandom Correlation Functions. Our pseudorandom correlation functions use techniques similar to our Paillier-based homomorphic secret sharing construction, with the difference that the Paillier decryption key d is known to one of the parties (whereas before, it was secret shared). Our PCF constructions also crucially rely on the fact that Paillier ciphertexts can be obliviously sampled; any element in $\mathbb{Z}_{N^2}^*$ is in fact a valid ciphertext! In our PCF for the VOLE correlation, one party knows d, the other party knows a value x, and dx is subtractively secret shared between the two. Given a random ciphertext, the party who knows d can decrypt it to learn a, and both parties can recover shares of xa using the trick from our HSS construction.

We get a PCF for OT from similar techniques, but using the Goldwasser-Micali bit-encryption scheme [GM82], which admits a simple distributed discrete log procedure (as also observed in [DGI+19]). We also construct the weaker notion of a PCG for OLE, by essentially generating many instances of the VOLE PCF setup, but compressing them in clever ways using the LPN assumption together with function secret sharing, inspired by previous PCG constructions [BCGI18]. This construction also generalizes in several ways, to give secret-shared degree-2 correlations over $\mathbb{Z}_N$ or $\mathbb{Z}_2$, and to give PCFs when relying on the variable-density LPN assumption [BCG+20a] instead of standard LPN.

Public-Key Setup for PCFs. Our PCF for VOLE requires a setup where one party knows d, the other knows x, and both hold subtractive shares of dx. (Our PCF for OT uses a similar setup.) We show that such setup can be instantiated *non-interactively*; each party locally generates a secret/public key pair, and extracts the setup information it needs from its own key pair and the other party's public key.

The PCF setup itself can be seen as an OLE instance for values x and d. So, our public-key setup is based on a novel non-interactive vector-OLE protocol that can be seen as a variant of Bellare-Micali oblivious transfer [BM90] with a CRS. Their original non-interactive oblivious transfer protocol allows two parties to use a (non-reusable) PKI setup to non-interactively obtain an OT. In our Paillier-based variant, instead of only producing OT – and crucially thanks to our distributed discrete log procedure – we show how the parties can obtain a *vector OLE*, where the sender's input is x, the receiver's input are some values $a_1, \ldots, a_n$, and the parties end up with additive sharings of the product $x \cdot a_i$ for all i's. This suffices to generate the keys for our PCF constructions non-interactively.

2 Preliminaries

We work with Blum integers of the form $N = pq$, where p and q are safe primes.[2] We let $(N, p, q) \leftarrow \mathsf{GenPQ}(1^\lambda)$ be a randomized algorithm which, on input the security parameter λ, samples two such random primes p, q of length $\ell = \ell(\lambda)$, and outputs (N, p, q). In some of our constructions, $N = pq$ will be a public modulus generated by a trusted setup algorithm (such that no-one knows the factorization p and q), while in other cases the factorization will be known to one party.

2.1 Assumptions

Assumption 1 (Decisional Composite Residuosity (DCR) Assumption). *For $(N, p, q) \leftarrow \mathsf{GenPQ}(1^\lambda)$, let $g_0 \leftarrow \mathbb{Z}^*_{N^2}$, and $g_1 = g_0^N \bmod N^2$. For $b \leftarrow \{0, 1\}$, for all PPT algorithms $\mathcal{A}$,*

$$\Pr[\mathcal{A}(N, g_b) = b] \leq \frac{1}{2} + \mathsf{negl}(\lambda).$$

Assumption 2 (Quadratic Residuosity (QR) Assumption). *For $(N, p, q) \leftarrow \mathsf{GenPQ}(1^\lambda)$, let $g_0 \leftarrow \mathbb{Z}^*_N$, and $g_1 = g_0^2 \bmod N$. For $b \leftarrow \{0, 1\}$, for all PPT algorithms $\mathcal{A}$,*

$$\Pr[\mathcal{A}(N, g_b) = b] \leq \frac{1}{2} + \mathsf{negl}(\lambda).$$

[2] A safe prime p is equal to $2p' + 1$ where p' is also prime. This is not actually required by all our constructions, but for simplicity we use the same group generation algorithm through the paper.

We also leverage a lemma from Brakerski and Goldwasser [BG10] that refers to the *interactive vector game.* We rephrase the lemma here in terms of Paillier groups only. Consider the decomposition $\mathbb{Z}_{N^2}^* = \mathbb{G}_R \times \mathbb{G}_M$, where $\mathbb{G}_R$ is the group of Nth residues modulo N^2 and $\mathbb{G}_M$ is the group of elements of orders that divide N. The DCR assumption (Assumption 1) states that a random element from $\mathbb{G}_R$ is indistinguishable from a random element in $\mathbb{Z}_{N^2}^*$. In the interactive vector game, the challenger samples a bit $b \leftarrow \{0,1\}$, and $(g_1, \ldots, g_l) \leftarrow \mathbb{G}_R^l$ (for a parameter l). It sends $(g_1, \ldots, g_l)$ to the adversary. The adversary then makes adaptive queries of the form $(a_1, \ldots, a_l) \in \mathbb{G}_M^l$, to which the challenger responds by sampling r from $[N^2]$ and returns $(a_1^b g_1^r, \ldots, a_l^b g_l^r)$. Finally, the adversary returns a guess b' at the value of b.

Lemma 2.1 (Rephrased Lemma B.1 From [BG10]**).** *Assuming the DCR assumption, for all efficient adversaries $\mathcal{A}$, the probability that $\mathcal{A}$ guesses b correctly in the interactive vector game is at most negligibly greater than half.*

2.2 Encryption

KDM Security. In some of our constructions, we assume that (variants of) the Paillier encryption scheme are *key-dependent message* (KDM) secure. The definition we use is similar to the one given by Brakerski and Goldwasser [BG10], with the differences that we only consider one key pair, and do not consider adaptive adversary queries. (This makes for a weaker definition, and thus a milder assumption.)

Definition 2.2 (KDM Security). *An encryption scheme* ($\mathsf{ES.Gen}, \mathsf{ES.Enc}, \mathsf{ES.Dec}$) *is* KDM secure *over the set of programs F if for all security parameters $\lambda \in \mathbb{N}$, for all polynomial sets of fixed output length programs $f_1, \ldots, f_\rho \in F$ and for all PPT adversaries $\mathcal{A}$,*

$$\left| \Pr\left[\mathcal{A}(\mathsf{pk}, \mathsf{ct}_{\beta,1}, \ldots, \mathsf{ct}_{\beta,\rho}) = \beta \;\middle|\; \begin{array}{l} (\mathsf{pk}, \mathsf{sk}) \leftarrow \mathsf{ES.Gen}(1^\lambda), \\ x_{0,i} \leftarrow f_i(\mathsf{sk}) \text{ for } i \in [\rho], \\ x_{1,i} \leftarrow 0^{|x_{0,i}|} \text{ for } i \in [\rho], \\ \mathsf{ct}_{b,i} \leftarrow \mathsf{ES.Enc}(\mathsf{pk}, x_b) \text{ for } b \in \{0,1\}, i \in [\rho], \\ \beta \leftarrow \{0,1\} \end{array} \right] - \frac{1}{2} \right| \leq \mathsf{negl}(\lambda).$$

Paillier Encryption. While there are many known variants of the Paillier cryptosystem [Pai99], we use the variant where the decryption key is an integer d such that raising any ciphertext to the power d gives $(1+N)^m \pmod{N^2}$, where m is the message and N the public modulus. Since it is easy to compute discrete logarithms with base $1+N$ in $\mathbb{Z}_{N^2}^*$, this gives an efficient decryption procedure. We describe the Paillier cryptosystem below; its security is based on the DCR assumption (Assumption 1).

$\mathsf{Paillier.Gen}(1^\lambda)$:

1. Sample $(N, p, q) \leftarrow \mathsf{GenPQ}(1^\lambda)$.

2. Compute $d \in \mathbb{Z}$ such that $d \equiv 0 \pmod{\phi(N)}$ and $d \equiv 1 \pmod{N}$.
3. Output $\mathsf{pk} = N$, $\mathsf{sk} = d$.

$\mathsf{Paillier.Enc}(\mathsf{pk}, x)$:
1. Sample a random $r \leftarrow [N^2]$.
2. Output $\mathsf{ct} = r^N(1+N)^x \bmod N^2$.

Observe that, since $(1+N)^x \equiv 1 + xN \pmod{N^2}$, $(1+N)$ has order N in $\mathbb{Z}^*_{N^2}$. Additionally, observe that since the order of r in $\mathbb{Z}^*_{N^2}$ must divide $N\phi(N)$,

$$\begin{aligned}
\mathsf{ct}^d \pmod{N^2} &\equiv r^{Nd}(1+N)^{dx} && \pmod{N^2} \\
&\equiv r^{Nd \pmod{N\phi(N)}}(1+N)^{dx \pmod{N}} && \pmod{N^2} \\
&\equiv (1+N)^x && \pmod{N^2} \\
&\equiv 1 + xN.
\end{aligned}$$

$\mathsf{Paillier.Dec}(\mathsf{sk}, \mathsf{ct})$:
1. Output $x = \frac{(\mathsf{ct}^d \bmod N^2) - 1}{N}$.

We will also use the following fact, namely, that the encryption function is a bijection. In particular, this implies that a randomly chosen element of $\mathbb{Z}_{N^2}$ defines a valid ciphertext with overwhelming probability.

Proposition 2.3 ([Pai99]). *The following map is a bijection:*

$$\begin{aligned}
\mathbb{Z}_N \times \mathbb{Z}^*_N &\to \mathbb{Z}^*_{N^2} \\
(x, r) &\mapsto r^N(1+N)^x
\end{aligned}$$

In our homomorphic secret sharing constructions (Sect. 4), we use two other flavors of Paillier encryption: a Paillier-ElGamal hybrid, and the KDM-secure scheme due to Brakerski and Goldwasser [BG10]. In Sect. 5, we also use the Goldwasser–Micali cryptosystem [GM82].

2.3 Secret Sharing

We work with subtractive secret sharing. We let $\langle x\rangle_0^{(m)}, \langle x\rangle_1^{(m)}$ denote a subtractive sharing of x modulo m, such that $\langle x\rangle_1^{(m)} - \langle x\rangle_0^{(m)} \equiv x \pmod{m}$. If one share is chosen uniformly at random from $[m]$ (and the other is chosen to satisfy the equation above), each share alone perfectly hides x, while the two together allow the reconstruction of x.

Similarly, we let $\langle x\rangle_0^{(\mathbb{Z})}, \langle x\rangle_1^{(\mathbb{Z})}$ denote a subtractive sharing of x over the integers, such that $\langle x\rangle_1^{(\mathbb{Z})} - \langle x\rangle_0^{(\mathbb{Z})} = x$. For $x \in \{0, \dots, m-1\}$, in order for each share alone to statistically hide x, $\langle x\rangle_1^{(\mathbb{Z})}$ can be chosen uniformly at random from the range $\{0, \dots, m2^\kappa - 1\}$, where κ is the statistical security parameter. If $\langle x\rangle_0^{(\mathbb{Z})}$ is then defined as $\langle x\rangle_1^{(\mathbb{Z})} - x$, then it is within statistical distance $2^{-\kappa}$ of the uniform distribution.

3 Share Conversion for Paillier Encryption

Suppose two parties hold respective values g_0 and g_1 in $\mathbb{Z}_{N^2}^*$, such that $g_1 \equiv g_0(1+N)^x \pmod{N^2}$ for some $x \in \mathbb{Z}_N$. The parties wish to locally convert these *multiplicative* (or, rather, *divisive*) shares into *subtractive* shares of x.

We can view g_0 and g_1 as elements of the coset

$$X_{g_0} := \{g_0, g_0(1+N), g_0(1+N)^2, \cdots, g_0(1+N)^{N-1}\}.$$

If both parties can *agree upon* a distinct element of this set, say h, without communicating, then they can each calculate the distance (in terms of powers of $1+N$) between g_i and h to obtain a subtractive share of x. In particular, if they obtain $h = g_0(1+N)^z$ for some z, then P_1 can compute the discrete logarithm of $g_1/h = (1+N)^{x-z}$ and output $z_1 := x - z$, while P_0 uses $g_0/h = (1+N)^{-z}$ to get $z_0 := -z$, giving $z_1 - z_0 \equiv x \pmod N$.

To agree upon such a representative h, we have the parties compute the smallest element from X_{g_0}, defined by viewing elements of $\mathbb{Z}_{N^2}^*$ as integers in $\{0, \ldots, N^2-1\}$. Surprisingly, this can be done by simply computing $h = g_i \bmod N$. Since $(1+N)^x \equiv 1 \pmod N$, it is clear that this gives the same h for both g_0 and g_1. It remains to show that h lies in the same coset.

Proposition 3.1. *Let $g \in \mathbb{Z}_{N^2}^*$, $h = g \bmod N$ and $h' = \lfloor g/N \rfloor$. Then h can be written as $g(1+N)^{-z}$, where $z = h'h^{-1} \bmod N$.*

Proof. Suppose that we can write $h = g(1+N)^{-z} \bmod N^2$, for some $z \in \mathbb{Z}$. Since $g = h + h'N$, this is equivalent to

$$\begin{aligned} h &= (h + h'N) \cdot (1 - zN) \mod N^2 \\ &= h + (h' - zh)N \mod N^2 \end{aligned}$$

The above is satisfied if and only if $h'N \equiv zhN \pmod{N^2}$, or equivalently, $z \equiv h'h^{-1} \pmod N$. □

This gives us a direct way to solve the distributed discrete log problem, which we present in Algorithm 3.2. Instead of computing g_i/h and then finding the discrete logarithm with respect to $1+N$, we can simply compute z as in Proposition 3.1.

Algorithm 3.2: $\mathsf{DDLog}_N(g)$

1. Write $g = h + h'N$, where $h, h' < N$, using the division algorithm.
2. Output $z = h'h^{-1} \mod N$.

Lemma 3.3. *Let $g_0, g_1 \in \mathbb{Z}_{N^2}^*$ such that $g_1 = g_0(1+N)^x \bmod N^2$. If $z_b = \mathsf{DDLog}_N(g_b)$, then $z_1 - z_0 \equiv x \pmod N$.*

Proof. First, observe that since each g_i is in $\mathbb{Z}_{N^2}^*$, it must have an inverse modulo N, so DDLog will not fail.

From Proposition 3.1, each z_i satisfies $h \equiv g_i(1+N)^{-z_i} \pmod{N^2}$, where $h = g_0 \bmod N = g_1 \bmod N$. This gives

$$\begin{aligned} g_1(1+N)^{-z_1} &\equiv g_0(1+N)^{-z_0} &\pmod{N^2} \\ \Leftrightarrow (1+N)^{x-z_1} &\equiv (1+N)^{-z_0} &\pmod{N^2} \\ \Leftrightarrow x &\equiv z_1 - z_0 &\pmod{N}. \end{aligned}$$

□

Remark 3.4. We can alternatively interpret the share conversion procedure by viewing each input g_i as a Paillier ciphertext $g_i = (1+N)^{x_i} r^N$ for some (unknown) message x_i, and the *same randomness* r. Under this condition, share conversion allows each party to locally obtain a subtractive share of $x = x_1 - x_0$. Note that this does not violate the security of Paillier, because we were given two ciphertexts with the same randomness.

3.1 Using a Secret Shared Decryption Key

Getting g_0, g_1 such that $g_1 = g_0(1+N)^x$ given a Paillier encryption $g = (1+N)^x r^N$ of x can be done using subtractive shares (over the integers) $\langle d \rangle_0^{(\mathbb{Z})}, \langle d \rangle_1^{(\mathbb{Z})}$ of the Paillier decryption key d (where $\langle d \rangle_1^{(\mathbb{Z})} - \langle d \rangle_0^{(\mathbb{Z})} = d$, $d \equiv 1 \pmod{N}$, and $d \equiv 0 \pmod{\phi(N)}$).

Using our ciphertext $g = (1+N)^x r^N$, we compute

$$g_0 = g^{\langle d \rangle_0^{(\mathbb{Z})}} = (1+N)^{x \langle d \rangle_0^{(\mathbb{Z})}} (r^{\langle d \rangle_0^{(\mathbb{Z})}})^N \bmod N^2,$$

and

$$g_1 = g^{\langle d \rangle_1^{(\mathbb{Z})}} = (1+N)^{x \langle d \rangle_1^{(\mathbb{Z})}} (r^{\langle d \rangle_1^{(\mathbb{Z})}})^N \bmod N^2.$$

Since $d \equiv 0 \pmod{\phi(N)}$, it follows that $d = \langle d \rangle_1^{(\mathbb{Z})} - \langle d \rangle_0^{(\mathbb{Z})} \Rightarrow \langle d \rangle_0^{(\mathbb{Z})} \equiv \langle d \rangle_1^{(\mathbb{Z})} \pmod{\phi(N)}$ and therefore

$$r^{\langle d \rangle_1^{(\mathbb{Z})} N} \equiv r^{\langle d \rangle_0^{(\mathbb{Z})} N} \pmod{N^2}.$$

Then, as desired,

$$\begin{aligned} \frac{g_1}{g_0} &\equiv (1+N)^{x(\langle d \rangle_1^{(\mathbb{Z})} - \langle d \rangle_0^{(\mathbb{Z})})} \pmod{N^2} \\ &\equiv (1+N)^x \pmod{N^2}. \end{aligned}$$

Remark 3.5. If, instead of subtractive shares of d, we have shares of yd (that is, $\langle yd \rangle_0^{(\mathbb{Z})}, \langle yd \rangle_1^{(\mathbb{Z})}$ such that $\langle yd \rangle_1^{(\mathbb{Z})} - \langle yd \rangle_0^{(\mathbb{Z})} = yd$), we can use these as described above to get g_0, g_1 such that $g_1 \equiv g_0(1+N)^{xy} \pmod{N^2}$.

$\mathsf{Exp}^{\mathsf{HSS,sec}}_{\mathcal{A},\sigma,b}(\lambda)$:
$(x_0, x_1, \mathsf{state}) \leftarrow \mathcal{A}(1^\lambda)$
$(\mathsf{pk}, (\mathsf{ek}_0, \mathsf{ek}_1)) \leftarrow \mathsf{HSS.Setup}(1^\lambda)$
$(\mathsf{I}_0, \mathsf{I}_1) \leftarrow \mathsf{HSS.Input}(\mathsf{pk}, x_b)$
$b' \leftarrow \mathcal{A}(\mathsf{state}, \mathsf{pk}, \mathsf{ek}_\sigma, \mathsf{I}_\sigma)$
return b'

Fig. 1. Security of HSS.

3.2 Getting Shares over Integers

The previous sections describe how to use $g_1 = g_0(1+N)^x \mod N^2$ to get subtractive shares of x over $\mathbb{Z}_N$. However, we often want subtractive shares of x over the integers. This can be done as long as x is sufficiently smaller than N.

Observe that, if $z_1 - z_0 \equiv x \pmod{N}$, then $z_1 - z_0 = x$ as long as $z_1 - x \geq 0$. There are only x values of z_1 such that this isn't true; so, for $x < N/2^\kappa$ and uniform choice of $z_1 \in \mathbb{Z}_N$, $z_1 - z_0 = x$ over $\mathbb{Z}$, except with probability $\leq 2^{-\kappa}$.

4 Homomorphic Secret Sharing

4.1 Definitions

We base our definitions of homomorphic secret sharing (HSS) on those given by Boyle *et al.* [BKS19].

Definition 4.1 (Homomorphic Secret Sharing). *A (2-party, public-key)* Homomorphic Secret Sharing (HSS) *scheme for a class of programs* $\mathcal{P}$ *over a ring* R *with input space* $\mathcal{I} \subseteq R$ *consists of PPT algorithms* $(\mathsf{HSS.Setup}, \mathsf{HSS.Input}, \mathsf{HSS.Eval})$ *with the following syntax:*

- $\mathsf{HSS.Setup}(1^\lambda) \rightarrow (\mathsf{pk}, (\mathsf{ek}_0, \mathsf{ek}_1))$: *Given a security parameter* 1^λ, *the setup algorithm outputs a public key* pk *and a pair of evaluation keys* $(\mathsf{ek}_0, \mathsf{ek}_1)$.
- $\mathsf{HSS.Input}(\mathsf{pk}, x) \rightarrow (\mathsf{I}_0, \mathsf{I}_1)$: *Given public key* pk *and private input value* $x \in \mathcal{I}$, *the input algorithm outputs input information* $(\mathsf{I}_0, \mathsf{I}_1)$.
- $\mathsf{HSS.Eval}(\sigma, \mathsf{ek}_\sigma, (\mathsf{I}^{(1)}_\sigma, \ldots, \mathsf{I}^{(\rho)}_\sigma), P) \rightarrow y_\sigma$: *On input a party index* $\sigma \in \{0,1\}$, *evaluation key* ek_σ, *vector of* ρ *input values and a program* $P \in \mathcal{P}$ *with* ρ *input values, the homomorphic evaluation algorithm outputs* $y_\sigma \in R$, *which is party* σ*'s share of an output* $y \in R$.

Note that, in the constructions we consider, we have $\mathsf{I}_0 = \mathsf{I}_1$. *We say that* $(\mathsf{HSS.Setup}, \mathsf{HSS.Input}, \mathsf{HSS.Eval})$ *is a* homomorphic secret sharing scheme *for the class of programs* $\mathcal{P}$ *if the following conditions hold:*

- ***Correctness.*** *For all security parameters* $\lambda \in \mathbb{N}$, *for all programs* $P \in \mathcal{P}$, *for all* $x^{(1)}, \ldots, x^{(\rho)} \in \mathcal{I}$ *(where* $\mathcal{I}$ *is the input space of* P*), for* $(\mathsf{pk}, \mathsf{ek}_0, \mathsf{ek}_1) \leftarrow \mathsf{HSS.Setup}(1^\lambda)$ *and for* $(\mathsf{I}^{(i)}_0, \mathsf{I}^{(i)}_1) \leftarrow \mathsf{HSS.Input}(\mathsf{pk}, x^{(i)})$, *we have*

$$\Pr\left[y_0 + y_1 = P(x^{(1)}, \ldots, x^{(\rho)})\right] \geq 1 - \mathsf{negl}(\lambda),$$

where

$$y_\sigma \leftarrow \mathsf{HSS.Eval}(\sigma, \mathsf{ek}_\sigma, (\mathsf{I}_\sigma^{(1)}, \ldots, \mathsf{I}_\sigma^{(\rho)}), P)$$

for $\sigma \in \{0,1\}$ *where the probability is taken over the random coins of* HSS.Setup, HSS.Input *and* HSS.Eval.

- ***Security.*** *For each* $\sigma \in \{0,1\}$ *and non-uniform adversary* $\mathcal{A}$ *(of size polynomial in the security parameter* λ*), it holds that*

$$\left|\Pr[\mathsf{Exp}_{\mathcal{A},\sigma,0}^{\mathsf{HSS,sec}}(\lambda) = 1] - \Pr[\mathsf{Exp}_{\mathcal{A},\sigma,1}^{\mathsf{HSS,sec}}(\lambda) = 1]\right| \leq \varepsilon(\lambda)$$

for all sufficiently large λ*, where* $\mathsf{Exp}_{\mathcal{A},\sigma,b}^{\mathsf{HSS,sec}}(\lambda)$ *for* $b \in \{0,1\}$ *is as defined in Fig. 1.*

Restricted Multiplication Straight-Line Programs. Our HSS schemes support homomorphic evaluation for a class of programs called *Restricted Multiplication Straight-line (RMS)* programs [Cle91,BGI16a]. An RMS program is an arithmetic circuit, with the restriction that every multiplication must be between an input value and an intermediate value of the computation, called a *memory value.* This class of programs captures polynomial-size branching programs, which includes arbitrary logspace computations and NC1 circuits.

Definition 4.2 (RMS programs). *An RMS program consists of a magnitude bound* B_{msg} *and a sequence of instructions of the four types described below. The inputs to the program are initially provided as a set of* input values I_x*, for each input* $x \in \mathbb{Z}$*. We consider the class of programs where the absolute value of all* memory values *during the computation is bounded above by* B_{msg}*.*

- $\mathsf{ConvertInput}(\mathsf{I}_x) \rightarrow \mathsf{M}_x$*: Load an input* x *into memory.*
- $\mathsf{Add}(\mathsf{M}_x, \mathsf{M}_y) \rightarrow \mathsf{M}_z$*: Add two memory values, obtaining* $z = x + y$*.*
- $\mathsf{Add}(\mathsf{I}_x, \mathsf{I}_y) \rightarrow \mathsf{I}_z$*: Add two input values, obtaining* $z = x + y$*.*
- $\mathsf{Mul}(\mathsf{I}_x, \mathsf{M}_y) \rightarrow \mathsf{M}_z$*: Multiply a memory value by an input, obtaining* $z = x \cdot y$*.*
- $\mathsf{Output}(\mathsf{M}_x, n_{\mathsf{out}}) \rightarrow x \bmod n_{\mathsf{out}}$*: Output a memory value, reduced modulo* n_{out} *(for some* $n_{\mathsf{out}} \leq B_{\mathsf{msg}}$*).*

We additionally assume that each instruction is implicitly assigned a unique identifier $\mathsf{id} \in \{0,1\}^*$*.*

If at any step of execution the size of a memory value exceeds the bound B_{msg} *or becomes negative (i.e.* $z > B_{\mathsf{msg}}$ *or* $z < 0$*), the output of the program on the corresponding input is undefined. Otherwise, the output is the sequence of* Output *values. Note that we consider addition of input values merely for the purpose of efficiency.*

4.2 HSS from Paillier

We follow the blueprint from Fazio *et al.* [FGJS17], based on Boyle *et al.* [BGI16a] to build an HSS scheme for RMS programs. Our scheme requires that an encryption of the secret decryption key be available. However, for correctness, our scheme also requires that all ciphertexts encrypt values much smaller than N; so, we are forced to decompose our secret key into digits before encrypting it. We use B_{sk} to refer to the base used for this decomposition, or, in other words, as an upper bound on the size of each digit. B_{sk} affects the efficiency of our scheme, since it will take $\ell = \log_{B_{\mathsf{sk}}}(N^2)$ ciphertexts to contain our secret key. B_{sk} is also related to the bound B_{msg} on our message space, since we require that all our input and memory values *times a digit of the secret key* be at least 2^κ times smaller than N: we get $B_{\mathsf{msg}} = \frac{N}{B_{\mathsf{sk}} 2^\kappa}$.

If we want $B_{\mathsf{msg}} = 2^\kappa$, then we get $B_{\mathsf{sk}} = \frac{N}{2^{2\kappa}}$. As long as N is at least 3κ bits long, B_{sk} will be at least κ bits long; so, we will need around 6 ciphertexts to contain our secret key.

As in RMS programs, we consider *input values* and *memory values*. Input values, denoted I, are the inputs to the computation, consisting of Paillier encryptions. Memory values, denoted M, are subtractively secret-shared intermediate values. More concretely, let $d^{(0)}, \ldots, d^{(\ell-1)}$ denote the digits of d (modulo some base B_{sk}), where d is the Paillier decryption key.

- An *input value* I_x consists of X, which is a Paillier encryption of x, and $X^{(0)}, \ldots, X^{(\ell-1)}$, which are Paillier encryptions of $d^{(0)}x, \ldots, d^{(\ell-1)}x$.
- A *memory value* $\mathsf{M}_x = (\mathsf{M}_{x,0}, \mathsf{M}_{x,1})$ consists of subtractive sharings of x and $d^{(0)}x, \ldots, d^{(\ell-1)}x$ over the integers. That is, party σ's memory value for x is $\mathsf{M}_{x,\sigma} = (\langle x \rangle_\sigma^{(\mathbb{Z})}, \langle xd^{(0)} \rangle_\sigma^{(\mathbb{Z})}, \ldots, \langle xd^{(\ell-1)} \rangle_\sigma^{(\mathbb{Z})})$.

We describe our HSS scheme for RMS programs in Construction 4.4. We defer the proof of Theorem 4.3 to the full version of this paper.

Theorem 4.3. *Construction 4.4 is a secure HSS scheme assuming the KDM security of the Paillier encryption scheme, and assuming that $F^{(N)}$ is a secure PRF.*

Construction 4.4: Construction $\mathsf{HSS}_{\mathsf{Paillier}}$

$\mathsf{Setup}(1^\lambda)$: Set up the scheme.

1. Sample $(\mathsf{pk}_{\mathsf{Paillier}} = N, \mathsf{sk} = d) \leftarrow \mathsf{Paillier.Gen}(1^\lambda)$. Let $d^{(0)}, \ldots, d^{(\ell-1)}$ denote the digits of d base B_{sk}.
2. Subtractively secret share the digits of d as $\langle d^{(i)} \rangle_0^{(\mathbb{Z})}, \langle d^{(i)} \rangle_1^{(\mathbb{Z})}$ such that $\langle d^{(i)} \rangle_1^{(\mathbb{Z})} - \langle d^{(i)} \rangle_0^{(\mathbb{Z})} = d^{(i)}$. Each $\langle \cdot \rangle_1^{(\mathbb{Z})}$ is drawn uniformly at random from $[2^\kappa B_{\mathsf{sk}}]$; $\langle \cdot \rangle_0^{(\mathbb{Z})}$ is selected to complete the subtractive sharing.
3. Sample a key $\mathsf{k}_{\mathsf{prf}}$ for the prf $F^{(2^\kappa)}$ which outputs values in $[2^\kappa]$.
4. For $\sigma \in \{0, 1\}$, let $\mathsf{ek}_\sigma = (\mathsf{k}_{\mathsf{prf}}, \langle d^{(0)} \rangle_\sigma^{(\mathbb{Z})}, \ldots, \langle d^{(\ell-1)} \rangle_\sigma^{(\mathbb{Z})})$.

5. Encrypt the digits of d as $D^{(0)} \leftarrow \mathsf{Paillier.Enc}(\mathsf{pk}_{\mathsf{Paillier}}, d^{(0)}), \ldots, D^{(\ell-1)} \leftarrow \mathsf{Paillier.Enc}(\mathsf{pk}_{\mathsf{Paillier}}, d^{(\ell-1)})$.
6. Let $\mathsf{pk} = (\mathsf{pk}_{\mathsf{Paillier}}, D^{(0)}, \ldots, D^{(\ell-1)})$.
7. Output $(\mathsf{pk}, (\mathsf{ek}_0, \mathsf{ek}_1))$.

$\mathsf{Input}(\mathsf{pk}, x)$: Generate an input value for x.

1. Generate a Paillier ciphertext $X \leftarrow \mathsf{Paillier.Enc}(\mathsf{pk}_{\mathsf{Paillier}}, x)$.
2. For $i \in [0, \ldots, \ell-1]$, generate an encryption $X^{(i)}$ of $d^{(i)}x$ by homomorphically multiplying $D^{(i)}$ by x, and then re-randomizing. Concretely using Paillier, $X^{(i)} = r_i^N (D^{(i)})^x$ for a randomly sampled $r_i \leftarrow \mathbb{Z}^*_{N^2}$.
3. Let $\mathsf{I} = (X, X^{(0)}, \ldots, X^{(\ell-1)})$.
4. Output $(\mathsf{I}_0 = \mathsf{I}, \mathsf{I}_1 = \mathsf{I})$.

$\mathsf{ConvertInput}(\sigma, \mathsf{ek}_\sigma, \mathsf{I} = (X, X^{(0)}, \ldots, X^{(\ell-1)}))$: Convert an input to a memory value. First, we take a canonical secret sharing of 1 as $\langle 1\rangle_1^{(\mathbb{Z})} = F^{(2^\kappa)}_{\mathsf{k}_{\mathsf{prf}}}(1) + 1 \bmod N$, and $\langle 1\rangle_0^{(\mathbb{Z})} = F^{(2^\kappa)}_{\mathsf{k}_{\mathsf{prf}}}(1)$. Then we create a memory value for 1 as $\mathsf{M}_{1,\sigma} = (\langle 1\rangle_\sigma^{(\mathbb{Z})}, \langle d^{(0)}\rangle_\sigma^{(\mathbb{Z})}, \ldots, \langle d^{(\ell-1)}\rangle_\sigma^{(\mathbb{Z})})$ for $\sigma \in \{0,1\}$, and evaluate $\mathsf{Mul}(\sigma, \mathsf{ek}_\sigma, \mathsf{I}_x, \mathsf{M}_{1,\sigma})$.[a]

$\mathsf{Add}(\sigma, \mathsf{ek}_\sigma, \mathsf{M}_{x,\sigma}, \mathsf{M}_{y,\sigma})$: Add two memory values.

1. Parse $\mathsf{M}_{x,\sigma} = (\langle x\rangle_\sigma^{(\mathbb{Z})}, \langle xd^{(0)}\rangle_\sigma^{(\mathbb{Z})}, \ldots, \langle xd^{(\ell-1)}\rangle_\sigma^{(\mathbb{Z})})$, and $\mathsf{M}_{y,\sigma} = (\langle y\rangle_\sigma^{(\mathbb{Z})}, \langle yd^{(0)}\rangle_\sigma^{(\mathbb{Z})}, \ldots, \langle yd^{(\ell-1)}\rangle_\sigma^{(\mathbb{Z})})$.
2. Let $\langle z\rangle_\sigma^{(\mathbb{Z})} = \langle x\rangle_\sigma^{(\mathbb{Z})} + \langle y\rangle_\sigma^{(\mathbb{Z})}$, and $\langle zd^{(i)}\rangle_\sigma^{(\mathbb{Z})} = \langle xd^{(i)}\rangle_\sigma^{(\mathbb{Z})} + \langle yd^{(i)}\rangle_\sigma^{(\mathbb{Z})}$ for $i \in [0, \ldots, \ell-1]$.
3. Output $\mathsf{M}_{z,\sigma} = (\langle z\rangle_\sigma^{(\mathbb{Z})}, \langle zd^{(0)}\rangle_\sigma^{(\mathbb{Z})}, \ldots, \langle zd^{(\ell-1)}\rangle_\sigma^{(\mathbb{Z})})$.

$\mathsf{Add}(\mathsf{pk}, \mathsf{I}_x = (X, X^{(0)}, \ldots, X^{(\ell-1)}), \mathsf{I}_y = (Y, Y^{(0)}, \ldots, Y^{(\ell-1)}))$: Add two input values by homomorphically evaluating addition on the ciphertexts to get $\mathsf{I}_z = (Z, Z^{(0)}, \ldots, Z^{(\ell-1)})$. Concretely using Paillier, $Z = XY \bmod N^2$, and $Z^{(i)} = X^{(i)}Y^{(i)} \bmod N^2$. Output I_z.

$\mathsf{Mul}(\sigma, \mathsf{ek}_\sigma, \mathsf{I}_x, \mathsf{M}_{y,\sigma}))$: Multiply an input value and a memory value. We let id be the index of this multiplication; all such indices are assumed to be unique.

1. Parse $\mathsf{I}_x = (X, X^{(0)}, \ldots, X^{(\ell-1)})$.
2. Parse $\mathsf{M}_{y,\sigma} = (\langle y\rangle_\sigma^{(\mathbb{Z})}, \langle yd^{(0)}\rangle_\sigma^{(\mathbb{Z})}, \ldots, \langle yd^{(\ell-1)}\rangle_\sigma^{(\mathbb{Z})})$.
3. Let $\langle yd\rangle_\sigma^{(\mathbb{Z})} = \sum_{i=0}^{\ell-1} B_{\mathsf{sk}}{}^i \langle yd^{(i)}\rangle_\sigma^{(\mathbb{Z})}$ be a subtractive share of yd over the integers.
4. Let
$$\langle z\rangle_\sigma^{(N)} = \mathsf{DDLog}_N((X)^{\langle yd\rangle_\sigma^{(\mathbb{Z})}}) + F^{(N)}_{\mathsf{k}_{\mathsf{prf}}}(\mathsf{id}) \pmod N.$$
This yields a subtractive sharing of $z = xy \pmod N$. Since $z \ll N$, we can take this to be a share of z over the integers; that is,
$$\langle z\rangle_\sigma^{(\mathbb{Z})} = \langle z\rangle_\sigma^{(N)}.$$
5. Similarly, let
$$\langle zd^{(i)}\rangle_\sigma^{(N)} = \mathsf{DDLog}_N((X^{(i)})^{\langle yd\rangle_\sigma^{(\mathbb{Z})}}) + F^{(N)}_{\mathsf{k}_{\mathsf{prf}}}(\mathsf{id}, i),$$
and
$$\langle zd^{(i)}\rangle_\sigma^{(\mathbb{Z})} = \langle zd^{(i)}\rangle_\sigma^{(N)}.$$
6. Output $\mathsf{M}_{z,\sigma} = (\langle z\rangle_\sigma^{(\mathbb{Z})}, \langle zd^{(0)}\rangle_\sigma^{(\mathbb{Z})}, \ldots, \langle zd^{(\ell-1)}\rangle_\sigma^{(\mathbb{Z})})$.

$\mathsf{Output}(\sigma, \mathsf{ek}_\sigma, \mathsf{M}_{x,\sigma} = (\langle x\rangle_\sigma^{(\mathbb{Z})}, \langle xd^{(0)}\rangle_\sigma^{(\mathbb{Z})}, \dots, \langle xd^{(\ell-1)}\rangle_\sigma^{(\mathbb{Z})}), n_{\mathsf{out}})$: Output $\langle x\rangle_\sigma^{(\mathbb{Z})} \bmod n_{\mathsf{out}}$.

[a] Note that in our HSS construction based on Paillier, we do not actually use $\langle 1\rangle_\sigma^{(\mathbb{Z})}$ in the multiplication; it is only necessary for Output. However, in HSS constructions based on PaillierEG and BG described in the full version of this paper, this will be needed.)

4.3 HSS Variants

The HSS construction in the previous section has two drawbacks: (1) it requires a local trusted setup for each pair of parties, and (2) its security relies on the assumption that Paillier is KDM secure. We address both these issues by giving two alternative HSS constructions. In the first one we replace Paillier encryption with the Paillier-ElGamal encryption scheme [CS02, DGS03, BCP03], which is essentially ElGamal over the group $\mathbb{Z}_{N^2}^*$. In this variant multiple users can share the same modulus N, and the decryption key is a random exponent d (as in ElGamal). This has the advantage of only requiring a public-key style setup, where each party publishes a public key, and each can then non-interactively derive their shared public key and their own evaluation key. Note that the trusted setup now only contains the modulus N, and can be used by any number of parties. In the last construction we replace Paillier encryption with the provably KDM secure encryption scheme of Brakerski and Goldwasser [BG10]. This has the unexpected advantage that generating encryptions of the digits of the secret key can trivially be done having access to the public key only. While both alternative constructions follow the same blueprint as the one from "regular" Paillier, several details need to be taken care of. The details of the constructions are deferred to the full version of this paper.

5 Pseudorandom Correlation Functions

In this section, we present our constructions of pseudorandom correlation functions (PCFs). We first recap the syntax and definitions of a PCF in Sect. 5.1. Then, in Sect. 5.2, we give our PCF for the vector oblivious linear evaluation (VOLE) correlation, based on the DCR assumption, and in Sect. 5.3, our PCF for the oblivious transfer (OT) correlation based on quadratic residuosity. Our public-key variants of these PCFs are deferred until Sect. 6. Finally, in Sect. 5.4, we also construct the weaker notion of a pseudorandom correlation generator (PCG) for the oblivious linear evaluation (OLE) correlation, based on a combination of the DCR and learning parity with noise assumptions.

5.1 Definitions

To formalize our constructions for VOLE and OT, we use the concept of a *pseudorandom correlation function* (PCF) by Boyle *et al.* [BCG+20a]. Informally, a pseudorandom correlation function enables two parties to sample an arbitrary amount of correlated randomness, given a one-time setup that outputs a pair of short, correlated keys. This extends the previous notion of a pseudorandom correlation generator [BCG+19], analogously to how a PRF extends a PRG, where in the latter, the outputs are typically of bounded length and/or must be computed all at once.

One example of desirable correlated randomness is an instance of random oblivious transfer (OT), where one party obtains (s_0, s_1) uniform over $\{0,1\}^2$, and the other obtains (b, s_b) for b uniform over $\{0,1\}$. Another example is vector oblivious linear evaluation (VOLE) over a ring R, where the parties obtain respective outputs $(u, v) \in R^2$ and $(x, w) \in R^2$, where u, v are random, $w = ux + v$, and x is sampled at random, but fixed for all samples from the correlation.

We model a target correlation as a probabilistic algorithm $\mathcal{Y}$, which produces a pair of outputs (y_0, y_1) for the two parties. To define security, we additionally require the correlation to be *reverse-sampleable*, meaning that given an output y_σ, there is an efficient algorithm which produces a $y_{1-\sigma}$ from the distribution of $\mathcal{Y}$ conditioned on y_σ. Note that in the case of VOLE, due to the fixed x, we also use a master secret key msk which parametrizes the algorithm $\mathcal{Y}$. Such a correlation with a master secret key is called a *correlation with setup*, which we focus on below.

Definition 5.1 (Reverse-sampleable correlation with setup). *Let* $1 \leq \ell_0(\lambda), \ell_1(\lambda) \leq \mathsf{poly}(\lambda)$ *be output-length functions. Let* $(\mathsf{Setup}, \mathcal{Y})$ *be a tuple of probabilistic algorithms, such that*

- Setup, *on input* 1^λ, *returns a master key* msk, *and*
- $\mathcal{Y}$, *on input* 1^λ *and* msk, *returns a pair of outputs* $(y_0, y_1) \in \{0,1\}^{\ell_0(\lambda)} \times \{0,1\}^{\ell_1(\lambda)}$.

We say that the tuple $(\mathsf{Setup}, \mathcal{Y})$ *defines a* reverse sampleable correlation with setup *if there exists a probabilistic polynomial time algorithm* $\mathsf{RSample}$ *such that*

- $\mathsf{RSample}$, *on input* 1^λ, msk, $\sigma \in \{0,1\}$ *and* $y_\sigma \in \{0,1\}^{\ell_\sigma(\lambda)}$, *returns* $y_{1-\sigma} \in \{0,1\}^{\ell_{1-\sigma}(\lambda)}$ *such that for all* $\mathsf{msk}, \mathsf{msk}'$ *in the image of* Setup *and all* $\sigma \in \{0,1\}$, *the following distributions are statistically close:*

$$\{(y_0, y_1) \mid (y_0, y_1) \leftarrow \mathcal{Y}(1^\lambda, \mathsf{msk})\}$$
$$\{(y_0, y_1) \mid (y_0', y_1') \leftarrow \mathcal{Y}(1^\lambda, \mathsf{msk}'), y_\sigma \leftarrow y_\sigma', y_{1-\sigma} \leftarrow \mathsf{RSample}(1^\lambda, \mathsf{msk}, \sigma, y_\sigma)\}$$

A PCF for a correlation $\mathcal{Y}$ consists of a key generation algorithm, Gen, which outputs a pair of correlated keys, together with an evaluation algorithm, Eval, which is given one of the keys and a public input, and produces a correlated output. In a *weak* PCF, we only consider running Eval with randomly chosen

$\mathsf{Exp}^{\mathsf{pr}}_{\mathcal{A},Q,0}(\lambda)$:	$\mathsf{Exp}^{\mathsf{pr}}_{\mathcal{A},Q,1}(\lambda)$:
$\mathsf{msk} \leftarrow \mathsf{Setup}(1^\lambda)$	$(\mathsf{k}_0, \mathsf{k}_1) \leftarrow \mathsf{PCF.Gen}(1^\lambda)$
for $i = 1$ **to** $Q(\lambda)$:	**for** $i = 1$ **to** $Q(\lambda)$:
$\quad x^{(i)} \leftarrow \{0,1\}^{n(\lambda)}$	$\quad x^{(i)} \leftarrow \{0,1\}^{n(\lambda)}$
$\quad (y_0^{(i)}, y_1^{(i)}) \leftarrow \mathcal{Y}(1^\lambda, \mathsf{msk})$	$\quad$ **for** $\sigma \in \{0,1\}$: $y_\sigma^{(i)} \leftarrow \mathsf{PCF.Eval}(\sigma, \mathsf{k}_\sigma, x^{(i)})$
$b \leftarrow \mathcal{A}(1^\lambda, (x^{(i)}, y_0^{(i)}, y_1^{(i)})_{i \in [Q(\lambda)]})$	$b \leftarrow \mathcal{A}(1^\lambda, (x^{(i)}, y_0^{(i)}, y_1^{(i)})_{i \in [Q(\lambda)]})$
return b	**return** b

Fig. 2. Pseudorandom $\mathcal{Y}$-correlated outputs of a PCF.

$\mathsf{Exp}^{\mathsf{sec}}_{\mathcal{A},Q,\sigma,0}(\lambda)$:	$\mathsf{Exp}^{\mathsf{sec}}_{\mathcal{A},Q,\sigma,1}(\lambda)$:
$(\mathsf{k}_0, \mathsf{k}_1) \leftarrow \mathsf{PCF.Gen}(1^\lambda)$	$(\mathsf{k}_0, \mathsf{k}_1) \leftarrow \mathsf{PCF.Gen}(1^\lambda)$
for $i = 1$ **to** $Q(\lambda)$:	$\mathsf{msk} \leftarrow \mathsf{Setup}(1^\lambda)$
$\quad x^{(i)} \leftarrow \{0,1\}^{n(\lambda)}$	**for** $i = 1$ **to** $Q(\lambda)$:
$\quad y_{1-\sigma}^{(i)} \leftarrow \mathsf{PCF.Eval}(1-\sigma, \mathsf{k}_{1-\sigma}, x^{(i)})$	$\quad x^{(i)} \leftarrow \{0,1\}^{n(\lambda)}$
$b \leftarrow \mathcal{A}(1^\lambda, \sigma, k_\sigma, (x^{(i)}, y_{1-\sigma}^{(i)})_{i \in [Q(\lambda)]})$	$\quad y_\sigma^{(i)} \leftarrow \mathsf{PCF.Eval}(\sigma, \mathsf{k}_\sigma, x^{(i)})$
return b	$\quad y_{1-\sigma}^{(i)} \leftarrow \mathsf{RSample}(1^\lambda, \mathsf{msk}, \sigma, y_\sigma^{(i)})$
	$b \leftarrow \mathcal{A}(1^\lambda, \sigma, k_\sigma, (x^{(i)}, y_{1-\sigma}^{(i)})_{i \in [Q(\lambda)]})$
	return b

Fig. 3. Security of a PCF. Here, RSample is the algorithm for reverse sampling $\mathcal{Y}$ as in Definition 5.1.

inputs, whereas in a *strong* PCF, the inputs can be chosen arbitrarily. Boyle *et al.* [BCG+20a] show that any weak PCF can be used together with a programmable random oracle to obtain a strong PCF, so from here on, our default notion of PCF will be a weak PCF.

There are two security requirements for a PCF: firstly, a pseudorandomness requirement, meaning that the joint distribution of both parties' outputs of Eval are indistinguishable from outputs of $\mathcal{Y}$. Secondly, there is a security property, which intuitively requires that pseudorandomness still holds even when given one of the parties' keys.

Definition 5.2 (Pseudorandom correlation function (PCF)). *Let* $(\mathsf{Setup}, \mathcal{Y})$ *fix a reverse-sampleable correlation with setup which has output length functions* $\ell_0(\lambda), \ell_1(\lambda)$*, and let* $\lambda \leq n(\lambda) \leq \mathsf{poly}(\lambda)$ *be an input length function. Let* $(\mathsf{PCF.Gen}, \mathsf{PCF.Eval})$ *be a pair of algorithms with the following syntax:*

- $\mathsf{PCF.Gen}(1^\lambda)$ *is a probabilistic polynomial time algorithm that on input* 1^λ*, outputs a pair of keys* $(\mathsf{k}_0, \mathsf{k}_1)$*;*
- $\mathsf{PCF.Eval}(\sigma, \mathsf{k}_\sigma, x)$ *is a deterministic polynomial-time algorithm that on input* $\sigma \in \{0,1\}$*, key* k_σ *and input value* $x \in \{0,1\}^{n(\lambda)}$*, outputs a value* $y_\sigma \in \{0,1\}^{\ell_\sigma(\lambda)}$*.*

We say $(\mathsf{PCF.Gen}, \mathsf{PCF.Eval})$ *is a (weak) pseudorandom correlation function (PCF) for* $\mathcal{Y}$*, if the following conditions hold:*

- ***Pseudorandom $\mathcal{Y}$-correlated outputs.*** *For every $\sigma \in \{0,1\}$ and non-uniform adversary $\mathcal{A}$ of size* $\mathsf{poly}(\lambda)$, *and every $Q = \mathsf{poly}(\lambda)$, it holds that*

$$\left|\Pr[\mathsf{Exp}^{\mathsf{pr}}_{\mathcal{A},Q,0}(\lambda) = 1] - \Pr[\mathsf{Exp}^{\mathsf{pr}}_{\mathcal{A},Q,1}(\lambda) = 1]\right| \leq \mathsf{negl}(\lambda)$$

for all sufficiently large λ, where $\mathsf{Exp}^{\mathsf{pr}}_{\mathcal{A},Q,b}(\lambda)$ for $b \in \{0,1\}$ is as defined in Fig. 2. (In particular, where the adversary is given access to $Q(\lambda)$ samples.)
- ***Security.*** *For each $\sigma \in \{0,1\}$ and non-uniform adversary $\mathcal{A}$ of size $B(\lambda)$, and every $Q = \mathsf{poly}(\lambda)$, it holds that*

$$\left|\Pr[\mathsf{Exp}^{\mathsf{sec}}_{\mathcal{A},Q,\sigma,0}(\lambda) = 1] - \Pr[\mathsf{Exp}^{\mathsf{sec}}_{\mathcal{A},Q,\sigma,1}(\lambda) = 1]\right| \leq \mathsf{negl}(\lambda)$$

for all sufficiently large λ, where $\mathsf{Exp}^{\mathsf{sec}}_{\mathcal{A},Q,\sigma,b}(\lambda)$ for $b \in \{0,1\}$ is as defined in Fig. 3 (again, with $Q(\lambda)$ samples).

5.2 PCF for Vector-OLE from Paillier

Vector oblivious linear evaluation, or VOLE, over a ring $R = R(\lambda)$, is a correlation defined by an algorithm Setup, which outputs $\mathsf{msk} = x$ for a random $x \in R$, and an algorithm $\mathcal{Y}_{\mathsf{VOLE}}$, which on input msk, samples random elements $u, v \in R$, computes $w = ux + v$ and outputs the pair $((u,v),(w,x))$. Note that w, v can be viewed as a subtractive secret sharing of the product ux. Since x is fixed, this means that a batch of VOLE samples can be used to perform scalar-vector multiplications on secret-shared inputs, as part of, for instance, a secure two-party computation protocol.

The main idea behind our PCF for VOLE is the following. In the standard Paillier cryptosystem, every element of $\mathbb{Z}^*_{N^2}$ defines a valid ciphertext, which makes it possible to obliviously sample an encryption of a random message, without knowing the underlying message. We exploit this by having both parties locally generate the same random ciphertexts[3], which are then viewed as encryptions of random inputs a in the HSS construction. Then, given a subtractive secret sharing of xd, where d is the secret key and $x \in \mathbb{Z}_N$ is some fixed value, the parties can use the distributed multiplication protocol from the HSS scheme to obtain shares z_0, z_1 such that $z_1 = z_0 + ax$. If one party is additionally given the secret key d (and hence learns the a's) and the other party is given x, then this process can be repeated to produce an arbitrarily long VOLE correlation.

In the PCF construction, shown in Construction 5.4, the values x, d and shares of xd are distributed by the PCF Gen algorithm, while the random ciphertexts are given as public inputs to the Eval algorithm, since we are only building a weak PCF and not a strong one. Additionally, the parties use a PRF to randomize their output shares and ensure that these are uniformly distributed. We defer the proof of Theorem 5.3 to the full version of this paper.

[3] E.g. with a random oracle, or some other public source of randomness.

Theorem 5.3. *Suppose the DCR assumption holds, and that F is a secure PRF. Then Construction 5.4 is a secure PCF for the VOLE correlation, $\mathcal{Y}_{\mathsf{VOLE}}$, over the ring $\mathbb{Z}_N$.*

Construction 5.4: PCF for Vector Oblivious Linear Evaluation

Let $F : \{0,1\}^\lambda \times \{0,1\}^\lambda \to \mathbb{Z}_N$ be a pseudorandom function.

Gen: On input 1^λ:

1. Sample $(N, p, q) \leftarrow \mathsf{GenPQ}(1^\lambda)$.
2. Compute $d \in \mathbb{Z}$ such that $d \equiv 0 \pmod{\varphi(N)}$ and $d \equiv 1 \pmod{N}$.
3. Sample $x \leftarrow [N], y_0 \leftarrow [N^3 2^\kappa]$, and let $y_1 = y_0 + x \cdot d$ over the integers.
4. Sample $\mathsf{k}_{\mathsf{prf}} \leftarrow \{0,1\}^\lambda$.
5. Output the keys $\mathsf{k}_0 = (N, \mathsf{k}_{\mathsf{prf}}, y_0, d)$ and $\mathsf{k}_1 = (N, \mathsf{k}_{\mathsf{prf}}, y_1, x)$.

Eval: On input $(\sigma, \mathsf{k}_\sigma, c)$, for a random input $c \in \mathbb{Z}^*_{N^2}$:

- If $\sigma = 0$, parse $\mathsf{k}_0 = (N, \mathsf{k}_{\mathsf{prf}}, y_0, d)$:
  1. Compute $a = \mathsf{Paillier.Dec}(d, c)$.
  2. Compute $z_0 = \mathsf{DDLog}_N(c^{y_0}) + F_{\mathsf{k}_{\mathsf{prf}}}(c) \bmod N$.
  3. Output (z_0, a)
- If $\sigma = 1$, parse $\mathsf{k}_1 = (N, \mathsf{k}_{\mathsf{prf}}, y_1, x)$:
  1. Compute $z_1 = \mathsf{DDLog}_N(c^{y_1}) + F_{\mathsf{k}_{\mathsf{prf}}}(c) \bmod N$.
  2. Output (z_1, x)

5.3 PCF for Oblivious Transfer from Quadratic Residuosity

To build a PCF for OT, we will first build a PCF for *XOR-correlated OT*, where the sender's messages are all of the form $z_1, z_1 \oplus s$ for some fixed string s. This is formally defined by a correlation $\mathcal{Y}_{\oplus\text{-}\mathsf{OT}}$, where the setup algorithm Setup picks a random $\mathsf{msk} = s \leftarrow \{0,1\}^\lambda$, and then each call to $\mathcal{Y}_{\oplus\text{-}\mathsf{OT}}(\mathsf{msk})$ first samples $b \leftarrow \{0,1\}, z_0 \leftarrow \{0,1\}^\lambda$, lets $z_1 = z_0 \oplus b \cdot s$, and outputs the pair $(z_0, b), (z_1, s)$.

Our PCF construction proceeds analogously to the VOLE case, except we rely on the Goldwasser-Micali cryptosystem instead of Paillier.

GM Encryption. We use the Goldwasser–Micali (GM) cryptosystem [GM82], with the simplified decryption procedure by Katz and Yung [KY02], which allows threshold decryption when p, q are both $3 \pmod{4}$.[4]

$\mathsf{GM.Gen}(1^\lambda)$:

1. Sample $(N, p, q) \leftarrow \mathsf{GenPQ}(1^\lambda)$.

[4] One can also obtain a similar threshold-compatible decryption under more general requirements for the modulus; see Desmedt and Kurosawa [DK07].

2. Let $d = \phi(N)/4 = (N - p - q + 1)/4$.
3. Output $\mathsf{pk} = N$, $\mathsf{sk} = d$.

$\mathsf{GM.Enc}(\mathsf{pk}, x \in \{0,1\})$:
1. Sample a random $r \leftarrow \mathbb{Z}_N$.
2. Output $\mathsf{ct} = r^2(-1)^x \bmod N$.

Observe that, if $x = 0$, ct will be a random quadratic residue modulo N; if $x = 1$, ct will be a random non-residue.

$\mathsf{GM.Dec}(\mathsf{sk}, \mathsf{ct})$:
1. Compute $y = \mathsf{ct}^d \bmod N$, which is in $\{1, -1\}$, and output $x = 0$ if $y = 1$, or $x = 1$ if $y = -1$.

Notice that in the GM cryptosystem, $\mathbb{J}_N$ (the elements of $\mathbb{Z}_N$ with Jacobi symbol 1) defines the set of valid ciphertexts. This allows us to sample a random ciphertext without knowing the corresponding message, by generating a random element of $\mathbb{Z}_N$ and testing that it has Jacobi symbol 1, which can be done efficiently.

We also use the distributed discrete log procedure $\mathsf{DDLog}^{\mathsf{GM}}$, shown in Algorithm 5.5. By inspection, it can be seen that for any two inputs $a_0, a_1 \in \mathbb{Z}_N^*$ satisfying $a_1/a_0 = (-1)^b$ for a bit b, we have $\mathsf{DDLog}^{\mathsf{GM}}(a_0) \oplus \mathsf{DDLog}^{\mathsf{GM}}(a_1) = b$. Note that this procedure was previously used to construct trapdoor hash functions [DGI+19].

Algorithm 5.5: $\mathsf{DDLog}^{\mathsf{GM}}(a \in \mathbb{Z}_N)$

1. Map a to an integer in $\{0, \ldots, N-1\}$.
2. If $a < N/2$ then output $z = 1$, otherwise, output $z = 0$.

PCF for Oblivious Transfer. The construction proceeds similarly to the VOLE case, except instead of one sharing, the Gen algorithm samples λ subtractive sharings of $s_j \cdot d$, where d is the GM secret key and s_j is one bit of the sender's fixed correlated OT offset. Then, given a random encryption of a bit b in Eval, the parties run $\mathsf{DDLog}^{\mathsf{GM}}$ λ times to obtain XOR shares of the string $b \cdot s \in \{0,1\}^\lambda$, giving a correlated OT as required. We defer the proof of Theorem 5.7 to the full version of this paper.

Construction 5.6: PCF for Oblivious Transfer

Let $F : \{0,1\}^\lambda \times \mathbb{Z}_N \to \{0,1\}^\lambda$ be a pseudorandom function.

Gen: On input 1^λ:

1. Sample $(N, p, q) \leftarrow \mathsf{GenPQ}(1^\lambda)$, and let $d = \varphi(N)/4$.
2. Sample $\mathsf{k}_{\mathsf{prf}} \leftarrow \{0,1\}^\lambda$.

3. For $j = 1, \ldots, \lambda$, sample $s_j \leftarrow \{0,1\}$, $y_{0,j} \leftarrow [N2^\kappa]$, and let $y_{1,j} = y_{0,j} + s_j \cdot d$ over the integers. Write $s = (s_1, \ldots, s_\lambda)$.
4. Output the keys $\mathsf{k}_0 = (N, \mathsf{k}_{\mathsf{prf}}, \{y_{0,j}\}_{j\in[\lambda]}, d)$ and $\mathsf{k}_1 = (N, \mathsf{k}_{\mathsf{prf}}, \{y_{1,j}\}_{j\in[\lambda]}, s)$.

Eval: On input $(\sigma, \mathsf{k}_\sigma, c)$, for a random input $c \in \mathbb{J}_N$:

- If $\sigma = 0$, parse $\mathsf{k}_0 = (N, \mathsf{k}_{\mathsf{prf}}, \{y_{0,j}\}_{j\in[\lambda]}, d)$:
  1. Compute $b = \mathsf{GM.Dec}(d, c)$ in $\{0,1\}$.
  2. For $j = 1, \ldots, \lambda$, compute $z_{0,j} = \mathsf{DDLog}^{\mathsf{GM}}(c^{y_{0,j}})$.
  3. Let $z_0 = (z_{0,0}, \ldots, z_{0,\lambda}) \oplus F_{\mathsf{k}_{\mathsf{prf}}}(c)$.
  4. Output (z_0, b).
- If $\sigma = 1$, parse $\mathsf{k}_1 = (N, \mathsf{k}_{\mathsf{prf}}, \{y_{1,j}\}_{j\in[\lambda]}, s)$:
  1. For $j = 1, \ldots, \lambda$, compute $z_{1,j} = \mathsf{DDLog}^{\mathsf{GM}}(c^{y_{1,j}})$.
  2. Let $z_1 = (z_{1,1}, \ldots, z_{1,\lambda}) \oplus F_{\mathsf{k}_{\mathsf{prf}}}(c)$.
  3. Output (z_1, s).

Theorem 5.7. *Suppose the QR assumption holds, and that F is a secure PRF. Then Construction 5.6 is a secure PCF for the correlated OT correlation, $\mathcal{Y}_{\oplus\text{-}\mathsf{OT}}$.*

Extension to Random Oblivious Transfer. A correlated OT can be locally coverted into a *random OT*, where both of the sender's messages are independently random, using a hash function and the technique of Ishai *et al.* [IKNP03]. The sender simply applies the hash function to compute its outputs $\mathsf{H}(z_1), \mathsf{H}(z_1 \oplus s)$, while the receiver outputs $\mathsf{H}(z_0) = \mathsf{H}(z_1 \oplus b \cdot s)$. Assuming a suitable correlation robustness property of H, the resulting OT messages are pseudorandom. It was shown by Boyle et al. [BCG+19,BCG+20a] that this transformation can be used to convert any PCF or PCG for correlated OT into one for the random OT correlation. Hence, we obtain the following.

Corollary 5.8. *Suppose the QR assumption holds, and there is a secure correlation-robust hash function. Then, there exists a secure PCF for the random oblivious transfer correlation.*

5.4 PCG for OLE and Degree-2 Correlations from LPN and Paillier

In Sect. 5.2, we showed how to build a PCF for VOLE, where the parties obtain $(u, v) \in R^2$ and $(x, w) \in R^2$, respectively, such that u, v are random, $w = ux + v$, and x is fixed for all samples from the correlation. In this section we show how to upgrade this to more general degree-2 correlations, including OLE, where x is sampled freshly at random for each instance. Of course, if we could get many VOLE PCF setups, each one of those could yield one OLE instance (if we only use it once!). Here, we show how to get m setups for the VOLE construction *all at once* from a smaller amount of correlated randomness, in what amounts to a PCG for OLE. (We emphasize that this is a pseudorandom correlation *generator*, not *function*, since it produces a fixed number of correlations.)

In the main construction we fix N, as well as the associated Paillier decryption key d, which we give to party 0, across all m instances. Our goal is run m copies of Construction 5.4 so we would *like* to give party 1 m random values $x_1, \ldots, x_m$, and secret share each dx_i over the integers between the two parties. However, this doesn't give us a PCG, because the size of our setup would be the same as the number of correlations we are able to produce. In order to keep our setup size much smaller than m, we instead produce the *setup* with a variant of a PCG based on the LPN assumption [BCGI18]. We give party 1 a *sparse* n-element vector $\boldsymbol{e}$ of elements in $[N]$, for $m < n$, which only contains $t = \mathsf{poly}(\lambda)$ non-zero elements. (Since it is sparse, it can actually be represented in $t \log(n) \log(N) \ll n$ bits.) By the dual form of the LPN assumption, $H \cdot \boldsymbol{e}$ for such a sparse $\boldsymbol{e}$ and some public $H \in \mathbb{Z}^{m \times n}$ looks pseudorandom (if $\boldsymbol{e}$ is unknown), so we can expand $\boldsymbol{e}$ to give m psuedorandom elements. In order to similarly compress a sharing of $d \cdot \boldsymbol{e}$, we use a *function secret sharing of the multi-point function* defined by $d \cdot \boldsymbol{e}$. (Note that we need to use a large enough modulus in the function secret sharing so that the output shares the parties obtain are shares over the integers with overwhelming probability.) This allows both parties to obtain shares of $d \cdot \boldsymbol{e}$, and then to compute shares of $H \cdot (d \cdot \boldsymbol{e})$. This completes the setup of m instances of our VOLE PCF; the parties then use each of those instances once, to get m instances of the OLE correlation.

We present the complete PCG in Construction 5.10. Preliminaries on FSS and the proof of Theorem 5.9 are deferred to the full version of this paper.

Theorem 5.9. *Let* H *be a random oracle,* F *a secure PRF, and suppose that both the LPN and DCR assumptions hold. Then Construction 5.10 is a secure PCG for the OLE correlation.*

Construction 5.10: $\mathsf{PCG_{OLE}}$

Let $\mathsf{H} : \{0,1\}^* \to \mathbb{Z}_{N^2}$ be a hash function, modelled as a random oracle, and $F : \{0,1\}^\lambda \times \{0,1\}^\lambda \to \mathbb{Z}_N$ be a PRF.

Let m, n be length parameters with $m < n$, and $H \in \mathbb{Z}^{m \times n}$ be a matrix for which the dual-LPN problem is hard over $\mathbb{Z}_N$.

Gen: On input 1^λ:

1. Sample $(N, p, q) \leftarrow \mathsf{GenPQ}(1^\lambda)$.
2. Compute $d \in \mathbb{Z}$ such that $d = 0 \bmod \varphi(N)$ and $d = 1 \bmod N$.
3. Sample $\mathsf{k_{prf}} \leftarrow \{0,1\}^\lambda$.
4. Sample a vector $\boldsymbol{e} \in \mathbb{Z}_N^n$ with t random, non-zero entries, and zero elsewhere.
5. Generate FSS keys $\mathsf{k}_0^{\mathsf{fss}}, \mathsf{k}_1^{\mathsf{fss}}$ for the multi-point function defined by $d \cdot \boldsymbol{e}$, with domain size n and range $\mathbb{Z}_{N'}$, for $N' = N^3 2^\kappa$.[a]
6. Output the seeds $\mathsf{k}_0 = (N, \mathsf{k_{prf}}, \mathsf{k}_0^{\mathsf{fss}}, d)$ and $\mathsf{k}_1 = (N, \mathsf{k_{prf}}, \mathsf{k}_1^{\mathsf{fss}}, \boldsymbol{e})$.

Expand: On input $(\sigma, \mathsf{k}_\sigma)$:

- If $\sigma = 0$, parse $\mathsf{k}_0 = (N, \mathsf{k_{prf}}, \mathsf{k}_0^{\mathsf{fss}}, d)$:

1. Let $\boldsymbol{y}_0' = \mathsf{FSS.FullEval}(0, \mathsf{k}_0^{\mathsf{fss}})$ in $\mathbb{Z}^n$.
2. Compute $\boldsymbol{y}_0 = H \cdot \boldsymbol{y}_0'$ in $\mathbb{Z}^m$.
3. For $j = 1, \ldots, m$:
   (a) Let $c_j = \mathsf{H}(sid, j)$ in $\mathbb{Z}_{N^2}$
   (b) Compute $a_j = (c^d - 1)/N$ in $\mathbb{Z}_N$.
   (c) Compute $z_{0,j} = \mathsf{DDLog}_N(c_j^{y_{0,j}}) + F_{\mathsf{k}_{\mathsf{prf}}}(j)$.
4. Output $\boldsymbol{a} = (a_1, \ldots, a_m)$ and $\boldsymbol{z}_0 = (z_{0,1}, \ldots, z_{0,m})$.

– If $\sigma = 1$, parse $\mathsf{k}_1 = (N, \mathsf{k}_{\mathsf{prf}}, \mathsf{k}_1^{\mathsf{fss}}, \boldsymbol{e})$:
1. Let $\boldsymbol{y}_1' = \mathsf{FSS.FullEval}(1, \mathsf{k}_1^{\mathsf{fss}})$ in $\mathbb{Z}^n$.
2. Compute $\boldsymbol{y}_1 = H \cdot \boldsymbol{y}_1'$ in $\mathbb{Z}^m$ and $\boldsymbol{b} = H \cdot \boldsymbol{e}$ in $\mathbb{Z}_N^m$.
3. For $j = 1, \ldots, m$:
   (a) Let $c_j = \mathsf{H}(sid, j)$ in $\mathbb{Z}_{N^2}$
   (b) Compute $z_{1,j} = \mathsf{DDLog}_N(c_j^{y_{1,j}}) + F_{\mathsf{k}_{\mathsf{prf}}}(j)$.
4. Output $\boldsymbol{b} = H \cdot \boldsymbol{e}$ and $\boldsymbol{z}_1 = (z_{1,1}, \ldots, z_{1,m})$, both in $\mathbb{Z}_N^m$.

---

[a] This ensures that $d \cdot \boldsymbol{e}$ is always much less than N', so we get shares over the integers.

6 Public-Key Setup for PCFs

6.1 Non-interactive VOLE

In this section, we present our protocol for non-interactive VOLE with semi-honest security, based on Paillier. Party P_A has input values $a_1, \ldots, a_n$, while party P_B has a single input value x, and the goal is to obtain additive shares of $a_i \cdot x$ modulo N, by exchanging just one simultaneous message. We assume that the modulus N has been generated as a trusted setup, and no-one knows its factorization.

Our protocol starts off in the spirit of Bellare-Micali OT [BM90], where P_B sends g^s for a random s, and P_A sends $g^{r_i} \cdot C^{a_i}$, for random r_i, where g and C are some fixed random group elements. Note that in Bellare-Micali, a_i is a bit, whereas here it is in $\mathbb{Z}_N$. At this point (where we depart slightly from [BM90]), P_A can compute the keys $g^{r_i s}$, while P_B can compute, $g^{r_i s} \cdot C^{a_i s}$ (without knowing a_i). We then have that the *ratio* of each of the two parties' keys is $C^{a_i s}$. Next, we additionally have P_B send a correction value $D = C^s \cdot (1+N)^x$, which allows P_A to adjust its key so that the ratios become $(1+N)^{a_i x}$. Finally, each party locally applies the distributed discrete log procedure to convert each key into an additive share of $a_i \cdot x$ modulo N. To allow for simulation, both parties randomize their output shares. Since we are dealing with passive security, it is enough for one of the parties (P_A) to sample those values. The full protocol is specified in Construction 6.3. We defer the proof of Theorem 6.1 to the full version of this paper.

Theorem 6.1. *The protocol in Construction 6.3 securely implements Functionality 6.2 in the presence of passive, static corruptions under the DCR and QR assumptions.*

Functionality 6.2: $\mathcal{F}_{\mathbb{Z}_N}$-VOLE

The functionality interacts with parties $\mathsf{P_B}$, $\mathsf{P_A}$ and an adversary $\mathcal{A}$.

On input $a_1, \ldots, a_n \in \mathbb{Z}_N$ from $\mathsf{P_A}$ and $x \in \mathbb{Z}_N$ from $\mathsf{P_B}$, the functionality does the following:

- Sample $y_{0,i} \leftarrow \mathbb{Z}_N$, for $i = 1, \ldots, n$, and set $y_{1,i} = y_{0,i} + a_i \cdot x$.
- Output $(y_{0,1}, \ldots, y_{0,n})$ to $\mathsf{P_B}$ and $(y_{1,1}, \ldots, y_{1,n})$ to $\mathsf{P_A}$.

Construction 6.3: Non-interactive VOLE protocol

CRS: The algorithms below implicitly have access to $\mathsf{crs} = (N, g, C)$, where $(N, p, q) \leftarrow \mathsf{GenPQ}(1^\lambda)$, and $g, C \leftarrow \mathbb{Z}^*_{N^2}$.

Message from $\mathsf{P_A}$: On input $(a_1, \ldots, a_n) \in \mathbb{Z}^n_N$, sample $r_i \leftarrow [N^2], t_i \leftarrow \mathbb{Z}_N$, compute $A_i = g^{r_i} \cdot C^{a_i}$ and send (A_i, t_i) for $i = 1, \ldots, n$.

Message from $\mathsf{P_B}$: On input $x \in \mathbb{Z}_N$, sample $s \leftarrow [N^2]$ and send (B, D) where $B = g^s$, $D = C^s \cdot (1 + N)^x$.

Output of $\mathsf{P_A}$: On receiving (B, D), compute $K'_i = B^{r_i} \cdot D^{a_i}$ and output $(y_{1,1}, \ldots, y_{1,n})$, where $y_{1,i} = \mathsf{DDLog}_N(K'_i) + t_i$.

Output of $\mathsf{P_B}$: On receiving $(A_1, \ldots, A_n, t_1, \ldots, t_n)$, compute $K_i = A_i^s$ and output $(y_{0,1}, \ldots, y_{0,n})$, where $y_{0,i} = \mathsf{DDLog}_N(K_i) + t_i$.

6.2 Public-Key Silent PCFs

We can plug our non-interactive VOLE protocol into the PCFs of Sect. 5 to obtain a public-key variant of those protocols where, after independently posting a public key, each party can locally derive its PCF key. Using their PCF keys, together with a random oracle to generate the public random inputs, the parties can then silently compute an arbitrary quantity of OT or VOLE correlations, without any interaction beyond the PKI.

Formally speaking, we can model this by defining a *public-key PCF* the same way as a standard PCF, except we replace the Gen algorithm with two separate algorithms GenA and GenB, which output key pairs $(\mathsf{sk_A}, \mathsf{pk_A})$ and $(\mathsf{sk_B}, \mathsf{pk_B})$. After running these algorithms, we define the two parties' PCF keys to be $(\mathsf{sk_A}, \mathsf{pk_A}, \mathsf{pk_B})$ and $(\mathsf{sk_B}, \mathsf{pk_A}, \mathsf{pk_B})$, respectively, and the rest of the definition follows the same way as before.

We sketch the constructions below. To distinguish between the different moduli involved, we refer to the modulus in the CRS (needed for the NIVOLE protocol) as $\tilde{N}$, and we refer to the modulus in the PCF as N.

Public-Key Silent VOLE. We replace the Gen algorithm of Construction 5.4 with the following: Both parties generate the first message of a non-interactive key exchange (NIKE) protocol (this will be used to derive the PRF key $\mathsf{k_{prf}}$). Party 0 runs $(N, p, q) \leftarrow \mathsf{GenPQ}(1^\lambda)$ and computes d as in Gen, while party 1 picks a random x, and they run the protocol in Construction 6.3 with $n = 1$ (that is, they implement a single OLE). They both include the message from the NIOLE and NIKE protocol in their public key, while party 0 includes the modulus N too. Upon receiving the public key of the other party, they can compute their PCF key k_σ by completing the NIKE and NIOLE protocols. Note that we require y_0, y_1 to be a share of $x \cdot d$ over the integers, so this requires the modulus $\tilde{N}$ in the CRS for the NIOLE to be sufficiently large i.e., $\tilde{N} > N^3 2^\kappa$.

Public-Key Silent OT. We replace the Gen algorithm of Construction 5.6 with the following: As above both parties generates the first message of a NIKE. Party 0 runs $(N, p, q) \leftarrow \mathsf{GenPQ}(1^\lambda)$ and computes d as in Gen, while party 1 picks random bits $s_j \in \{0, 1\}$ for $j \in [\lambda]$, and they run the protocol in Construction 6.3 with $n = \lambda$ (that is, this is a "proper" instance of VOLE). They both include the message from the NIVOLE and NIKE protocol in their public key, while party 0 includes the modulus N too. Upon receiving the public key of the other party, they can compute their PCF key k_σ by completing the NIKE and NIOLE protocol. Note that we require y_0, y_1 to be a share of $s_j \cdot d$ over the integers, so this requires the modulus $\tilde{N}$ in the CRS for the NIVOLE to be sufficiently large i.e., $\tilde{N} > N 2^\kappa$.

Acknowledgements. We would like to thank Damiano Abram for pointing out a bug in Theorem 6.1, and the anonymous reviewers for their time and feedback.

This work was supported by: the Concordium Blockhain Research Center, Aarhus University, Denmark; the Carlsberg Foundation under the Semper Ardens Research Project CF18-112 (BCM); the European Research Council (ERC) under the European Unions's Horizon 2020 research and innovation programme under grant agreement No 669255 (MPCPRO) and grant agreement No 803096 (SPEC); a starting grant from Aarhus University Research Foundation.

This material is based upon work supported by the Defense Advanced Research Projects Agency (DARPA) under Contract No. HR001120C0085. Any opinions, findings and conclusions or recommendations expressed in this material are those of the author(s) and do not necessarily reflect the views of the Defense Advanced Research Projects Agency (DARPA). Distribution Statement "A" (Approved for Public Release, Distribution Unlimited).

References

[BCG+17] Boyle, E., Couteau, G., Gilboa, N., Ishai, Y., Orrù, M.: Homomorphic secret sharing: optimizations and applications. In: ACM CCS 2017. ACM Press, October/November 2017

[BCG+19] Boyle, E., Couteau, G., Gilboa, N., Ishai, Y., Kohl, L., Scholl, P.: Efficient pseudorandom correlation generators: silent OT extension and more. In: Boldyreva, A., Micciancio, D. (eds.) CRYPTO 2019, Part III. LNCS, vol. 11694, pp. 489–518. Springer, Cham (2019). https://doi.org/10.1007/978-3-030-26954-8_16

[BCG+20a] Boyle, E., Couteau, G., Gilboa, N., Ishai, Y., Kohl, L., Scholl, P.: Correlated pseudorandom functions from variable-density LPN. In: FOCS (2020)

[BCG+20b] Boyle, E., Couteau, G., Gilboa, N., Ishai, Y., Kohl, L., Scholl, P.: Efficient pseudorandom correlation generators from ring-LPN. In: Micciancio, D., Ristenpart, T. (eds.) CRYPTO 2020, Part II. LNCS, vol. 12171, pp. 387–416. Springer, Cham (2020). https://doi.org/10.1007/978-3-030-56880-1_14

[BCGI18] Boyle, E., Couteau, G., Gilboa, N., Ishai, Y.: Compressing vector OLE. In: ACM CCS 2018. ACM Press, October 2018

[BCP03] Bresson, E., Catalano, D., Pointcheval, D.: A simple public-key cryptosystem with a double trapdoor decryption mechanism and its applications. In: Laih, C.-S. (ed.) ASIACRYPT 2003. LNCS, vol. 2894, pp. 37–54. Springer, Heidelberg (2003). https://doi.org/10.1007/978-3-540-40061-5_3

[BG10] Brakerski, Z., Goldwasser, S.: Circular and leakage resilient public-key encryption under subgroup indistinguishability. In: Rabin, T. (ed.) CRYPTO 2010. LNCS, vol. 6223, pp. 1–20. Springer, Heidelberg (2010). https://doi.org/10.1007/978-3-642-14623-7_1

[BGI15] Boyle, E., Gilboa, N., Ishai, Y.: Function secret sharing. In: Oswald, E., Fischlin, M. (eds.) EUROCRYPT 2015, Part II. LNCS, vol. 9057, pp. 337–367. Springer, Heidelberg (2015). https://doi.org/10.1007/978-3-662-46803-6_12

[BGI16a] Boyle, E., Gilboa, N., Ishai, Y.: Breaking the circuit size barrier for secure computation under DDH. In: Robshaw, M., Katz, J. (eds.) CRYPTO 2016, Part I. LNCS, vol. 9814, pp. 509–539. Springer, Heidelberg (2016). https://doi.org/10.1007/978-3-662-53018-4_19

[BGI16b] Boyle, E., Gilboa, N., Ishai, Y.: Function secret sharing: improvements and extensions. In: ACM CCS 2016. ACM Press, October 2016

[BGI17] Boyle, E., Gilboa, N., Ishai, Y.: Group-based secure computation: optimizing rounds, communication, and computation. In: Coron, J.-S., Nielsen, J.B. (eds.) EUROCRYPT 2017, Part II. LNCS, vol. 10211, pp. 163–193. Springer, Cham (2017). https://doi.org/10.1007/978-3-319-56614-6_6

[BGI+18] Boyle, E., Gilboa, N., Ishai, Y., Lin, H., Tessaro, S.: Foundations of homomorphic secret sharing. In: ITCS 2018. LIPIcs, January 2018

[BGMM20] Bartusek, J., Garg, S., Masny, D., Mukherjee, P.: Reusable two-round MPC from DDH. Cryptology ePrint Archive, Report 2020/170 (2020). https://eprint.iacr.org/2020/170

[BKS19] Boyle, E., Kohl, L., Scholl, P.: Homomorphic secret sharing from lattices without FHE. In: Ishai, Y., Rijmen, V. (eds.) EUROCRYPT 2019, Part II. LNCS, vol. 11477, pp. 3–33. Springer, Cham (2019). https://doi.org/10.1007/978-3-030-17656-3_1

[BM90] Bellare, M., Micali, S.: Non-interactive oblivious transfer and applications. In: Brassard, G. (ed.) CRYPTO 1989. LNCS, vol. 435, pp. 547–557. Springer, New York (1990). https://doi.org/10.1007/0-387-34805-0_48

[BV11] Brakerski, Z., Vaikuntanathan, V.: Efficient fully homomorphic encryption from (standard) LWE. In: 52nd FOCS. IEEE Computer Society Press, October 2011

[Cle91] Cleve, R.: Towards optimal simulations of formulas by bounded-width programs. Comput. Complex. **1**, 91–105 (1991)

[CS02] Cramer, R., Shoup, V.: Universal hash proofs and a paradigm for adaptive chosen ciphertext secure public-key encryption. In: Knudsen, L.R. (ed.) EUROCRYPT 2002. LNCS, vol. 2332, pp. 45–64. Springer, Heidelberg (2002). https://doi.org/10.1007/3-540-46035-7_4

[DGI+19] Döttling, N., Garg, S., Ishai, Y., Malavolta, G., Mour, T., Ostrovsky, R.: Trapdoor hash functions and their applications. In: Boldyreva, A., Micciancio, D. (eds.) CRYPTO 2019, Part III. LNCS, vol. 11694, pp. 3–32. Springer, Cham (2019). https://doi.org/10.1007/978-3-030-26954-8_1

[DGS03] Damgård, I., Groth, J., Salomonsen, G.: The theory and implementation of an electronic voting system. In: Gritzalis, D.A. (ed.) Secure Electronic Voting, pp. 77–98. Springer, Boston (2003). https://doi.org/10.1007/978-1-4615-0239-5_6

[DHRW16] Dodis, Y., Halevi, S., Rothblum, R.D., Wichs, D.: Spooky encryption and its applications. In: Robshaw, M., Katz, J. (eds.) CRYPTO 2016, Part III. LNCS, vol. 9816, pp. 93–122. Springer, Heidelberg (2016). https://doi.org/10.1007/978-3-662-53015-3_4

[DK07] Desmedt, Y., Kurosawa, K.: A generalization and a variant of two threshold cryptosystems based on factoring. In: Garay, J.A., Lenstra, A.K., Mambo, M., Peralta, R. (eds.) ISC 2007. LNCS, vol. 4779, pp. 351–361. Springer, Heidelberg (2007). https://doi.org/10.1007/978-3-540-75496-1_23

[DKK18] Dinur, I., Keller, N., Klein, O.: An optimal distributed discrete log protocol with applications to homomorphic secret sharing. In: Shacham, H., Boldyreva, A. (eds.) CRYPTO 2018, Part III. LNCS, vol. 10993, pp. 213–242. Springer, Cham (2018). https://doi.org/10.1007/978-3-319-96878-0_8

[FGJS17] Fazio, N., Gennaro, R., Jafarikhah, T., Skeith, W.E.: Homomorphic secret sharing from paillier encryption. In: Okamoto, T., Yu, Y., Au, M.H., Li, Y. (eds.) ProvSec 2017. LNCS, vol. 10592, pp. 381–399. Springer, Cham (2017). https://doi.org/10.1007/978-3-319-68637-0_23

[Gen09] Gentry, C.: Fully homomorphic encryption using ideal lattices. In: 41st ACM STOC. ACM Press, May/June 2009

[GI14] Gilboa, N., Ishai, Y.: Distributed point functions and their applications. In: Nguyen, P.Q., Oswald, E. (eds.) EUROCRYPT 2014. LNCS, vol. 8441, pp. 640–658. Springer, Heidelberg (2014). https://doi.org/10.1007/978-3-642-55220-5_35

[GIS18] Garg, S., Ishai, Y., Srinivasan, A.: Two-round MPC: information-theoretic and black-box. In: Beimel, A., Dziembowski, S. (eds.) TCC 2018, Part I. LNCS, vol. 11239, pp. 123–151. Springer, Cham (2018). https://doi.org/10.1007/978-3-030-03807-6_5

[GM82] Goldwasser, S., Micali, S.: Probabilistic encryption and how to play mental poker keeping secret all partial information. In: 14th ACM STOC. ACM Press, May 1982

[IKNP03] Ishai, Y., Kilian, J., Nissim, K., Petrank, E.: Extending oblivious transfers efficiently. In: Boneh, D. (ed.) CRYPTO 2003. LNCS, vol. 2729, pp. 145–161. Springer, Heidelberg (2003). https://doi.org/10.1007/978-3-540-45146-4_9

[KY02] Katz, J., Yung, M.: Threshold cryptosystems based on factoring. In: Zheng, Y. (ed.) ASIACRYPT 2002. LNCS, vol. 2501, pp. 192–205. Springer, Heidelberg (2002). https://doi.org/10.1007/3-540-36178-2_12

[Pai99] Paillier, P.: Public-key cryptosystems based on composite degree residuosity classes. In: Stern, J. (ed.) EUROCRYPT 1999. LNCS, vol. 1592, pp. 223–238. Springer, Heidelberg (1999). https://doi.org/10.1007/3-540-48910-X_16

[WYG+17] Wang, F., Yun, C., Goldwasser, S., Vaikuntanathan, V., Zaharia, M.: Splinter: practical private queries on public data. In: 14th USENIX Symposium on Networked Systems Design and Implementation, NSDI 2017, pp. 299–313 (2017)

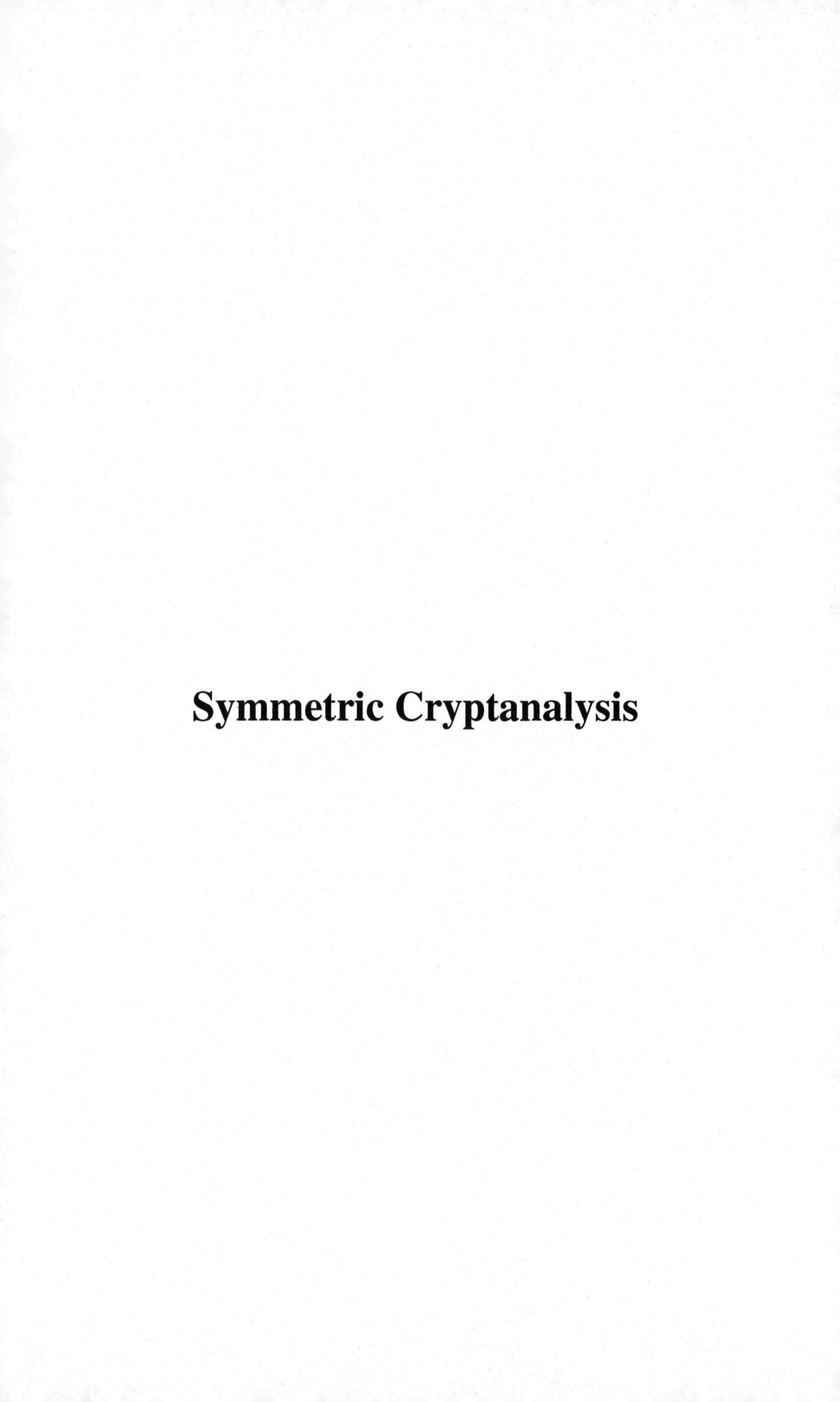

Symmetric Cryptanalysis

Improved Linear Approximations to ARX Ciphers and Attacks Against ChaCha

Murilo Coutinho(✉) and Tertuliano C. Souza Neto

Research and Development Center for Communications Security (CEPESC), Rio de Janeiro, Brazil
murilo.coutinho@redes.unb.br

Abstract. In this paper, we present a new technique which can be used to find better linear approximations in ARX ciphers. Using this technique, we present the first explicitly derived linear approximations for 3 and 4 rounds of ChaCha and, as a consequence, it enables us to improve the recent attacks against ChaCha. Additionally, we present new differentials for 3 and 3.5 rounds of ChaCha that, when combined with the proposed technique, lead to further improvement in the complexity of the Differential-Linear attacks against ChaCha.

Keywords: Differential-linear cryptanalysis · ARX-Ciphers · ChaCha

1 Introduction

Symmetric cryptographic primitives are heavily used in a variety of contexts. In particular, ARX-based design is a major building block of modern ciphers due to its efficiency in software. ARX stands for addition, word-wise rotation and XOR. Indeed, ciphers following this framework are composed of those operations and avoid the computation of smaller S-boxes through look-up tables. The ARX-based design approach is used to design stream ciphers (e.g., Salsa20 [7] and ChaCha [6]), efficient block ciphers (e.g., Sparx [15]), cryptographic permutations (e.g., Sparkle [3]) and hash functions (e.g., Blake [2]).

ARX-based designs are not only efficient but provide good security properties. The algebraic degree of ARX ciphers is usually high after only a very few rounds as the carry bit within one modular addition already reaches almost maximal degree. For differential and linear attacks, ARX-based designs show weaknesses for a small number of rounds. However, after some rounds the differential and linear probabilities decrease rapidly. Thus, the probabilities of differentials and the absolute correlations of linear approximations decrease very quickly as we increase the number of rounds. In fact, this property led to the long-trail strategy for designing ARX-based ciphers [15].

Ciphers and primitives based on Salsa20 and ChaCha families are heavily used in practice. In 2005, Bernstein proposed the stream cipher Salsa20 [7] as a contender to the eSTREAM [26], the ECRYPT Stream Cipher Project. As

A. Canteaut and F.-X. Standaert (Eds.): EUROCRYPT 2021, LNCS 12696, pp. 711–740, 2021.
https://doi.org/10.1007/978-3-030-77870-5_25

outlined by the author, Salsa20 is an ARX type family of algorithms which can be ran with several number of rounds, including the well known Salsa20/12 and Salsa20/8 versions. Latter, in 2008, Bernstein proposed some modifications to Salsa20 in order to provide better diffusion per round and higher resistance to cryptanalysis. These changes originated a new stream cipher, a variant which he called ChaCha [6]. Although Salsa20 was one of the winners of the eSTREAM competition, ChaCha has received much more attention through the years. Nowadays, we see the usage of this cipher in several projects and applications.

ChaCha, along with Poly1305 [5], is in one of the cipher suits of the new TLS 1.3 [21], which has been used by Google on both Chrome and Android. Not only has ChaCha been used in TLS but also in many other protocols such as SSH, Noise and S/MIME 4.0. In addition, the RFC 7634 proposes the use of ChaCha in IKE and IPsec. ChaCha has been used not only for encryption, but also as a pseudo-random number generator in any operating system running Linux kernel 4.8 or newer [25,28]. Additionally, ChaCha has been used in several applications such as WireGuard (VPN) (see [18] for a huge list of applications, protocols and libraries using ChaCha).

Related Work. Since ChaCha is so heavily used, it is very important to understand its security. Indeed, the cryptanalysis of ChaCha is well understood and several authors studied its security [1,9,11–14,16,17,19,22–24,27,29] which show weaknesses in the reduced round versions of the cipher.

The cryptanalysis of Salsa20 was introduced by Crowley [11] in 2005. Crowley developed a differential attack against Salsa20/5, namely the 5-round version of Salsa20, and received the $1000 prize offered by Bernstein for the most interesting Salsa20 cryptanalysis in that year. In 2006, Fischer et al. [16] improved the attack against Salsa20/5 and presented their attack against Salsa20/6.

Probably the most important cryptanalysis in this regard was proposed by Aumasson et al. at FSE 2008 [1] with the introduction of Probabilistic Neutral Bits (PNBs), showing attacks against Salsa20/7, Salsa20/8, ChaCha20/6 and ChaCha20/7. After that, several authors proposed small enhancements on the attack of Aumasson et al. The work by Shi et al. [27] introduced the concept of Column Chaining Distinguisher (CCD) to achieve some incremental advancements over [1] for both Salsa and ChaCha.

Maitra, Paul and Meier [22] studied an interesting observation regarding round reversal of Salsa, but no significant cryptanalytic improvement could be obtained using this method. Maitra [23] used a technique of Chosen IVs to obtain certain improvements over existing results. Dey and Sarkar [13] showed how to choose values for the PNB to further improve the attack.

In a paper presented in FSE 2017, Choudhuri and Maitra [9] significantly improved the attacks by considering the mathematical structure of both Salsa and ChaCha in order to find differential characteristics with much higher correlations. Recently, Coutinho and Souza [10] proposed new multi-bit differentials using the mathematical framework of Choudhuri and Maitra. In Crypto 2020, Beierle et al. [4] proposed improvements to the framework of differential-linear cryptanalysis against ARX-based designs and further improved the attacks against ChaCha.

Our Contribution. In this work, we provide a new framework to find linear approximations for ARX ciphers. Using this framework we provide the first explicitly derived linear approximations for 3 and 4 rounds of ChaCha. Exploring these linear approximations, we can improve the attacks for 6 and 7 rounds of ChaCha. Additionally, we present new differentials for 3 and 3.5 rounds of ChaCha which also improve the attacks. We summarize our findings along with other significant attacks for comparison in Table 1. Also, we verified all theoretical results with random experiments. We provide the source code to reproduce this paper in Github https://github.com/MurCoutinho/cryptanalysisChaCha.git, which is, for the best of our knowledge, the first implementation of cryptanalysis against ChaCha available to the public. We should note that it is possible to find attacks with less complexity for related key attacks, but we do not consider them in this work.

Table 1. The best attacks against ChaCha with 256-bit key.

Rounds	Time Complexity	Data Complexity	Reference
4	2^{6}	2^{6}	[9]
4.5	2^{12}	2^{12}	[9]
5	2^{16}	2^{16}	[9]
6	2^{139}	2^{30}	[1]
	2^{136}	2^{28}	[27]
	2^{130}	2^{35}	[9]
	$2^{127.5}$	$2^{37.5}$	[9]
	2^{116}	2^{116}	[9]
	$2^{102.2}$	2^{56}	[10]
	$2^{77.4}$	2^{58}	[4]
	2^{75}	2^{75}	[10]
	2^{51}	2^{51}	This work
7	2^{248}	2^{27}	[1]
	$2^{246.5}$	2^{27}	[27]
	$2^{238.9}$	2^{96}	[23]
	$2^{237.7}$	2^{96}	[9]
	$2^{231.9}$	2^{50}	[10]
	$2^{230.86}$	$2^{48.8}$	[4]
	$2^{228.51}$	$2^{80.51}$	This work
	2^{224}	2^{224}	This work
	2^{218}	2^{218}	This work

Organization of the Paper. In Sect. 2, we provide an overview of previous results, including a description of ChaCha, a summary of differential-linear cryptanalysis and a review of the techniques developed by Choudhuri and Maitra in

[9]. In Sect. 3, we present a new technique which can be used to find better linear approximations in ARX ciphers and theoretically develop new linear relations between bits of different rounds for ChaCha. Then, in Sect. 4, we show that these new linear approximations lead to a better distinguisher and key recovery attacks of ChaCha reduced to 6 and 7 rounds. Finally, Sect. 5 presents the conclusion and future work.

2 Specifications and Preliminaries

The main notation we will use throughout the paper is defined in Table 2. Next we define the algorithm ChaCha.

Table 2. Notation

Notation	Description
X	a 4×4 state matrix of ChaCha
$X^{(0)}$	initial state matrix of ChaCha
$X^{(R)}$	state matrix after application of R round functions
Z	output of ChaCha, $Z = X^{(0)} + X^{(R)}$
$x_i^{(R)}$	i^{th} word of the state matrix $X^{(R)}$ (words arranged in row major)
$x_{i,j}^{(R)}$	j^{th} bit of ith word of the state matrix $X^{(R)}$
$x_i^{(R)}[j_0, j_1, ..., j_t]$	the sum $x_{i,j_0}^{(R)} \oplus x_{i,j_1}^{(R)} \oplus \cdots \oplus x_{i,j_t}^{(R)}$
$x + y$	addition of x and y modulo 2^{32}
$x - y$	subtraction of x and y modulo 2^{32}
$x \oplus y$	bitwise XOR of x and y
$x \lll n$	rotation of x by n bits to the left
$x \ggg n$	rotation of x by n bits to the right
Δx	XOR difference of x and x'. $\Delta x = x \oplus x'$
$\Delta X^{(R)}$	XOR difference of $X^{(R)}$ and $X'^{(R)}$. $\Delta X^{(R)} = X^{(R)} \oplus X'^{(R)}$
$\Delta x_i^{(R)}$	differential $\Delta x_i^{(R)} = x_i^{(R)} \oplus x'^{(R)}_i$
$\Delta x_{i,j}^{(R)}$	differential $\Delta x_{i,j}^{(R)} = x_{i,j}^{(R)} \oplus x'^{(R)}_{i,j}$
$\Pr(E)$	probability of occurrence of an event E
$\mathcal{ID}$	input difference
$\mathcal{OD}$	output difference

2.1 ChaCha

The stream cipher Salsa20 was proposed by Bernstein [7] to the *eSTREAM* competition and later Bernstein proposed ChaCha [6] as an improvement of Salsa20. ChaCha consists of a series of ARX (addition, rotation, and XOR) operations on 32-bit words, being highly efficient in software and hardware.

Each round of ChaCha has a total of 16 bitwise XOR, 16 addition modulo 2^{32} and 16 constant-distance rotations.

ChaCha operates on a state of 64 bytes, organized as a 4×4 matrix with 32-bit integers, initialized with a 256-bit key $k_0, k_1, ..., k_7$, a 64-bit nonce v_0, v_1 and a 64-bit counter t_0, t_1 (we may also refer to the nonce and counter words as IV words), and 4 constants $c_0 = $ 0x61707865, $c_1 = $ 0x3320646e, $c_2 = $ 0x79622d32 and $c_3 = $ 0x6b206574. For ChaCha, we have the following initial state matrix:

$$X^{(0)} = \begin{pmatrix} x_0^{(0)} & x_1^{(0)} & x_2^{(0)} & x_3^{(0)} \\ x_4^{(0)} & x_5^{(0)} & x_6^{(0)} & x_7^{(0)} \\ x_8^{(0)} & x_9^{(0)} & x_{10}^{(0)} & x_{11}^{(0)} \\ x_{12}^{(0)} & x_{13}^{(0)} & x_{14}^{(0)} & x_{15}^{(0)} \end{pmatrix} = \begin{pmatrix} c_0 & c_1 & c_2 & c_3 \\ k_0 & k_1 & k_2 & k_3 \\ k_4 & k_5 & k_6 & k_7 \\ t_0 & t_1 & v_0 & v_1 \end{pmatrix}. \tag{1}$$

The state matrix is modified in each round by a *Quarter Round Function* (QRF), denoted by $QR\left(x_a^{(r-1)}, x_b^{(r-1)}, x_c^{(r-1)}, x_d^{(r-1)}\right)$, which receives and updates 4 integers in the following way:

$$\begin{array}{llll} x_{a'}^{(r-1)} = x_a^{(r-1)} + x_b^{(r-1)}; & x_{d'}^{(r-1)} = (x_d^{(r-1)} \oplus x_{a'}^{(r-1)}) \lll 16; \\ x_{c'}^{(r-1)} = x_c^{(r-1)} + x_{d'}^{(r-1)}; & x_{b'}^{(r-1)} = (x_b^{(r-1)} \oplus x_{c'}^{(r-1)}) \lll 12; \\ x_a^{(r)} = x_{a'}^{(r-1)} + x_{b'}^{(r-1)}; & x_d^{(r)} = (x_{d'}^{(r-1)} \oplus x_a^{(r)}) \lll 8; \\ x_c^{(r)} = x_{c'}^{(r-1)} + x_d^{(r)}; & x_b^{(r)} = (x_{b'}^{(r-1)} \oplus x_c^{(r)}) \lll 7; \end{array} \tag{2}$$

One round of ChaCha is defined as 4 applications of the QRF. There is, however, a difference between odd and even rounds. For odd rounds, i.e. $r \in \{1, 3, 5, 7, ...\}$, $X^{(r)}$ is obtained from $X^{(r-1)}$ by applying

$$\begin{array}{l} \left(x_0^{(r)}, x_4^{(r)}, x_8^{(r)}, x_{12}^{(r)}\right) = QR\left(x_0^{(r-1)}, x_4^{(r-1)}, x_8^{(r-1)}, x_{12}^{(r-1)}\right) \\ \left(x_1^{(r)}, x_5^{(r)}, x_9^{(r)}, x_{13}^{(r)}\right) = QR\left(x_1^{(r-1)}, x_5^{(r-1)}, x_9^{(r-1)}, x_{13}^{(r-1)}\right) \\ \left(x_2^{(r)}, x_6^{(r)}, x_{10}^{(r)}, x_{14}^{(r)}\right) = QR\left(x_2^{(r-1)}, x_6^{(r-1)}, x_{10}^{(r-1)}, x_{14}^{(r-1)}\right) \\ \left(x_3^{(r)}, x_7^{(r)}, x_{11}^{(r)}, x_{15}^{(r)}\right) = QR\left(x_3^{(r-1)}, x_7^{(r-1)}, x_{11}^{(r-1)}, x_{15}^{(r-1)}\right) \end{array}.$$

On the other hand, for even rounds, i.e. $r \in \{2, 4, 6, 8, , ...\}$, $X^{(r)}$ is calculated from $X^{(r-1)}$ by applying

$$\begin{array}{l} \left(x_0^{(r)}, x_5^{(r)}, x_{10}^{(r)}, x_{15}^{(r)}\right) = QR\left(x_0^{(r-1)}, x_5^{(r-1)}, x_{10}^{(r-1)}, x_{15}^{(r-1)}\right) \\ \left(x_1^{(r)}, x_6^{(r)}, x_{11}^{(r)}, x_{12}^{(r)}\right) = QR\left(x_1^{(r-1)}, x_6^{(r-1)}, x_{11}^{(r-1)}, x_{12}^{(r-1)}\right) \\ \left(x_2^{(r)}, x_7^{(r)}, x_8^{(r)}, x_{13}^{(r)}\right) = QR\left(x_2^{(r-1)}, x_7^{(r-1)}, x_8^{(r-1)}, x_{13}^{(r-1)}\right) \\ \left(x_3^{(r)}, x_4^{(r)}, x_9^{(r)}, x_{14}^{(r)}\right) = QR\left(x_3^{(r-1)}, x_4^{(r-1)}, x_9^{(r-1)}, x_{14}^{(r-1)}\right) \end{array}.$$

The output of ChaCha20/R is then defined as the sum of the initial state with the state after R rounds $Z = X^{(0)} + X^{(R)}$. One should note that it is possible to parallelize each application of the QRF on each round and also that each round is reversible. Hence, we can compute $X^{(r-1)}$ from $X^{(r)}$. For more information on ChaCha, we refer to [6].

2.2 Differential-Linear Cryptanalysis

In this section, we describe the technique of Differential-Linear cryptanalysis as used to attack ChaCha. Let E be a cipher and suppose we can write $E = E_2 \circ E_1$, where E_1 and E_2 are sub ciphers, covering m and l rounds of the main cipher, respectively. We can apply an input difference $\mathcal{ID}$ $\Delta X^{(0)}$ in the sub cipher E_1 obtaining an output difference $\mathcal{OD}$ $\Delta X^{(m)}$ (see the left side of Fig. 1). The next step is to apply Linear Cryptanalysis to the second sub cipher E_2. Using masks Γ_m and Γ_{out}, we attempt to find good linear approximations covering the remaining l rounds of the cipher E. Applying this technique we can construct a differential-linear distinguisher covering all $m + l$ rounds of the cipher E. This is the main idea in Langford and Hellman's classical approach [20].

Sometimes, however, it can be useful to divide the cipher E into three other ciphers, i.e. $E = E_3 \circ E_2 \circ E_1$. In this scenario, we can explore properties of the cipher in the first part E_1, and then apply a differential linear attack where we divide the differential part of the attack in two (see the right side of Fig. 1). Here, the $\mathcal{OD}$ from the sub cipher E_1 after r rounds, namely $\Delta X^{(r)}$, is the $\mathcal{ID}$ for the sub cipher E_2 which produces an output difference $\Delta X^{(m)}$. For more information in this regard, see [4].

It is important to understand how to compute the complexity of a differential-linear attack. We denote the differential of the state matrix as $\Delta X^{(r)} = X^{(r)} \oplus X'^{(r)}$ and the differential of individual words as $\Delta x_i^{(r)} = x_i^{(r)} \oplus x_i'^{(r)}$. Let $x_{i,j}^{(r)}$ denote the j-th bit of the i-th word of the state matrix after r rounds and let $\mathcal{J}$ be a set of bits. Also, let σ and σ' be linear combinations of bits in the set $\mathcal{J}$

$$\sigma = \left(\bigoplus_{(i,j) \in \mathcal{J}} x_{i,j}^{(r)} \right), \quad \sigma' = \left(\bigoplus_{(i,j) \in \mathcal{J}} x_{i,j}'^{(r)} \right).$$

Then

$$\Delta\sigma = \left(\bigoplus_{(i,j) \in \mathcal{J}} \Delta x_{i,j}^{(r)} \right)$$

is the linear combination of the differentials. We can write

$$\Pr\left[\Delta\sigma = 0 | \Delta X^{(0)}\right] = \frac{1}{2}(1 + \varepsilon_d), \tag{3}$$

where ε_d is the differential correlation.

Using linear cryptanalysis, it is possible to go further and find new relations between the initial state matrix and the state matrix after $R > r$ rounds. To do so, let $\mathcal{L}$ denote another set of bits and define

$$\rho = \left(\bigoplus_{(i,j) \in \mathcal{L}} x_{i,j}^{(R)} \right), \quad \rho' = \left(\bigoplus_{(i,j) \in \mathcal{L}} x_{i,j}'^{(R)} \right).$$

Then, as before,

$$\Delta\rho = \left(\bigoplus_{(i,j)\in\mathcal{L}} \Delta x_{i,j}^{(R)} \right).$$

We can define $\Pr[\sigma = \rho] = \frac{1}{2}(1+\varepsilon_L)$, where ε_L is the linear correlation. We want to find γ such that $\Pr\left[\Delta\rho = 0 | \Delta X^{(0)}\right] = \frac{1}{2}(1+\gamma)$.

To compute γ, we write (to simplify the notation we make the conditional to $\Delta X^{(0)}$ implicit):

$$\begin{aligned}\Pr[\Delta\sigma = \Delta\rho] &= \Pr[\sigma = \rho] \cdot \Pr\left[\sigma' = \rho'\right] + \Pr[\sigma = \bar{\rho}] \cdot \Pr\left[\sigma' = \overline{\rho'}\right] \\ &= \frac{1}{2}\left(1 + \varepsilon_L^2\right).\end{aligned}$$

Then,

$$\begin{aligned}\Pr[\Delta\rho = 0] &= \Pr[\Delta\sigma = 0] \cdot \Pr[\Delta\sigma = \Delta\rho] + \Pr[\Delta\sigma = 1] \cdot \Pr[\Delta\sigma = \overline{\Delta\rho}] \\ &= \frac{1}{2}\left(1 + \varepsilon_d \cdot \varepsilon_L^2\right).\end{aligned}$$

Therefore, the differential-linear correlation is given by $\gamma = \varepsilon_d \cdot \varepsilon_L^2$, which defines a distinguisher with complexity $\mathcal{O}\left(\frac{1}{\varepsilon_d^2 \varepsilon_L^4}\right)$. For further information on differential-linear cryptanalysis we refer to [8].

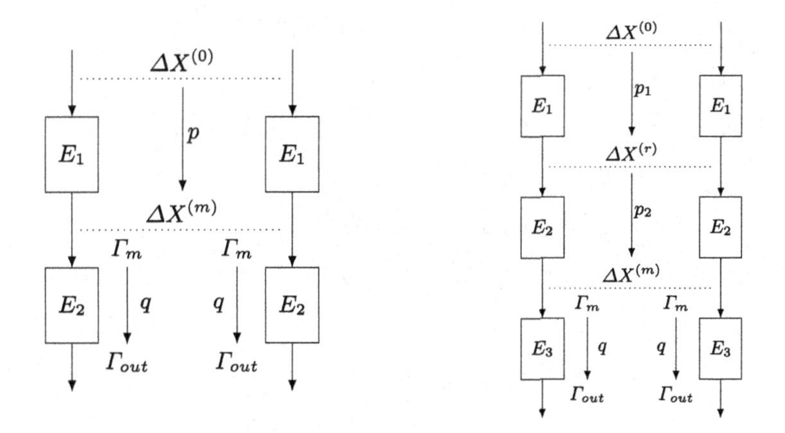

Fig. 1. A classical differential-linear distinguisher (on the left) and a differential-linear distinguisher with experimental evaluation of the correlation p_2 (on the right).

2.3 Multi-bit Differential for Reduced Round ChaCha

In this section, we review the work presented in [9] and in [10]. In these works, the authors developed the theory for selecting specific combination of bits to give high correlations for Chacha. To do that, in both papers the authors analyzed the QRF directly, representing each equation in its bit level. In the following, we change the original notation of the referred papers in order to create a notation that will be better for the purposes of this work.

Thus, let $\Theta(x, y) = x \oplus y \oplus (x + y)$ be the carry function of the sum $x + y$. Define $\Theta_i(x, y)$ as the i-th bit of $\Theta(x, y)$. By definition, we have $\Theta_0(x, y) = 0$. We can write the QRF equations of ChaCha (Eq. 2) as

$$\begin{aligned}
x'^{(m-1)}_{a,i} &= x^{(m-1)}_{a,i} \oplus x^{(m-1)}_{b,i} \oplus \Theta_i(x^{(m-1)}_a, x^{(m-1)}_b) \\
x'^{(m-1)}_{d,i+16} &= x^{(m-1)}_{d,i} \oplus x'^{(m-1)}_{a,i} \\
x'^{(m-1)}_{c,i} &= x^{(m-1)}_{c,i} \oplus x'^{(m-1)}_{d,i} \oplus \Theta_i(x^{(m-1)}_c, x'^{(m-1)}_d) \\
x'^{(m-1)}_{b,i+12} &= x^{(m-1)}_{b,i} \oplus x'^{(m-1)}_{c,i} \\
x^{(m)}_{a,i} &= x'^{(m-1)}_{a,i} \oplus x'^{(m-1)}_{b,i} \oplus \Theta_i(x'^{(m-1)}_a, x'^{(m-1)}_b) \\
x^{(m)}_{d,i+8} &= x'^{(m-1)}_{d,i} \oplus x^{(m)}_{a,i} \\
x^{(m)}_{c,i} &= x'^{(m-1)}_{c,i} \oplus x^{(m)}_{d,i} \oplus \Theta_i(x'^{(m-1)}_c, x^{(m)}_d) \\
x^{(m)}_{b,i+7} &= x'^{(m-1)}_{b,i} \oplus x^{(m)}_{c,i}
\end{aligned} \tag{4}$$

Inverting these equations, we get:

$$x'^{(m-1)}_{b,i} = x^{(m)}_{b,i+7} \oplus x^{(m)}_{c,i} \tag{5}$$

$$x'^{(m-1)}_{c,i} = x^{(m)}_{c,i} \oplus x^{(m)}_{d,i} \oplus \Theta_i(x'^{(m-1)}_c, x^{(m)}_d) \tag{6}$$

$$x'^{(m-1)}_{d,i} = x^{(m)}_{a,i} \oplus x^{(m)}_{d,i+8} \tag{7}$$

$$x'^{(m-1)}_{a,i} = x^{(m)}_{a,i} \oplus x^{(m)}_{b,i+7} \oplus x^{(m)}_{c,i} \oplus \Theta_i(x'^{(m-1)}_a, x'^{(m-1)}_b) \tag{8}$$

$$x^{(m-1)}_{b,i} = \mathcal{L}^{(m)}_{b,i} \oplus \Theta_i(x'^{(m-1)}_c, x^{(m)}_d) \tag{9}$$

$$x^{(m-1)}_{c,i} = \mathcal{L}^{(m)}_{c,i} \oplus \Theta_i(x'^{(m-1)}_c, x^{(m)}_d) \oplus \Theta_i(x^{(m-1)}_c, x'^{(m-1)}_d) \tag{10}$$

$$x^{(m-1)}_{d,i} = \mathcal{L}^{(m)}_{d,i} \oplus \Theta_i(x'^{(m-1)}_a, x'^{(m-1)}_b) \tag{11}$$

$$\begin{aligned}
x^{(m-1)}_{a,i} = \mathcal{L}^{(m)}_{a,i} &\oplus \Theta_i(x'^{(m-1)}_a, x'^{(m-1)}_b) \oplus \\
&\Theta_i(x'^{(m-1)}_c, x^{(m)}_d) \oplus \Theta_i(x^{(m-1)}_a, x^{(m-1)}_b)
\end{aligned} \tag{12}$$

where

$$\mathcal{L}^{(m)}_{a,i} = x^{(m)}_{a,i} \oplus x^{(m)}_{b,i+7} \oplus x^{(m)}_{b,i+19} \oplus x^{(m)}_{c,i+12} \oplus x^{(m)}_{d,i} \tag{13}$$

$$\mathcal{L}^{(m)}_{b,i} = x^{(m)}_{b,i+19} \oplus x^{(m)}_{c,i} \oplus x^{(m)}_{c,i+12} \oplus x^{(m)}_{d,i} \tag{14}$$

$$\mathcal{L}^{(m)}_{c,i} = x^{(m)}_{a,i} \oplus x^{(m)}_{c,i} \oplus x^{(m)}_{d,i} \oplus x^{(m)}_{d,i+8} \tag{15}$$

$$\mathcal{L}^{(m)}_{d,i} = x^{(m)}_{a,i} \oplus x^{(m)}_{a,i+16} \oplus x^{(m)}_{b,i+7} \oplus x^{(m)}_{c,i} \oplus x^{(m)}_{d,i+24} \tag{16}$$

Lemma 1. *It holds that* $x^{(m-1)}_{l,0} = \mathcal{L}^{(m)}_{l,0}$, *for* $l \in \{a, b, c, d\}$.

Proof. This result follows directly from Eqs. (9)–(12) by using the fact that $\Theta_0(x, y) = 0$. □

From this equations, we can derive the following lemma:

Lemma 2. *(Lemma 3 of [9]) Let*

$$\begin{aligned}
\Delta A^{(m)} &= \Delta x_{\alpha,0}^{(m)} \oplus \Delta x_{\beta,7}^{(m)} \oplus \Delta x_{\beta,19}^{(m)} \oplus \Delta x_{\gamma,12}^{(m)} \oplus \Delta x_{\delta,0}^{(m)}\\
\Delta B^{(m)} &= \Delta x_{\beta,19}^{(m)} \oplus \Delta x_{\gamma,0}^{(m)} \oplus \Delta x_{\gamma,12}^{(m)} \oplus \Delta x_{\delta,0}^{(m)}\\
\Delta C^{(m)} &= \Delta x_{\delta,0}^{(m)} \oplus \Delta x_{\gamma,0}^{(m)} \oplus \Delta x_{\delta,8}^{(m)} \oplus \Delta x_{\alpha,0}^{(m)}\\
\Delta D^{(m)} &= \Delta x_{\delta,24}^{(m)} \oplus \Delta x_{\alpha,16}^{(m)} \oplus \Delta x_{\alpha,0}^{(m)} \oplus \Delta x_{\gamma,0}^{(m)} \oplus \Delta x_{\beta,7}^{(m)}
\end{aligned}$$

After m rounds of ChaCha, the following holds:

$$\left|\varepsilon_{(A(m))}\right| = \left|\varepsilon_{\left(x_{\alpha,0}^{(m-1)}\right)}\right|, \left|\varepsilon_{\left(B^{(m)}\right)}\right| = \left|\varepsilon_{\left(x_{\beta,0}^{(m-1)}\right)}\right|$$

$$\left|\varepsilon_{\left(C^{(m)}\right)}\right| = \left|\varepsilon_{\left(x_{\gamma,0}^{(m-1)}\right)}\right|, \left|\varepsilon_{\left(D^{(m)}\right)}\right| = \left|\varepsilon_{\left(x_{\delta,0}^{(m-1)}\right)}\right|$$

The tuples $(\alpha, \beta, \gamma, \delta)$ vary depending on whether m is odd or even.

- *Case I. m is odd:*

$$(\alpha, \beta, \gamma, \delta) \in \{(0,4,8,12), (1,5,9,13), (2,6,10,14), (3,7,11,15)\}.$$

- *Case II. m is even:*

$$(\alpha, \beta, \gamma, \delta) \in \{(0,5,10,15), (1,6,11,12), (2,7,8,13), (3,4,9,14)\}.$$

Proof. See [9]. □

Lemma 3. *(Lemma 9 of [9]) For one active input bit in round $m-1$ and multiple active output bits in round m, the following holds for $i > 0$.*

$$x_{b,i}^{(m-1)} = \mathcal{L}_{b,i}^{(m)} \oplus x_{d,i-1}^{(m)}, \quad w.p.\ \tfrac{1}{2}\left(1+\tfrac{1}{2}\right)$$

$$x_{a,i}^{(m-1)} = \mathcal{L}_{a,i}^{(m)} \oplus x_{b,i+18}^{(m)} \oplus x_{c,i+11}^{(m)} \oplus x_{d,i-2}^{(m)} \oplus x_{d,i+6}^{(m)}, \quad w.p.\ \tfrac{1}{2}\left(1+\tfrac{1}{2^4}\right)$$

$$x_{c,i}^{(m-1)} = \mathcal{L}_{c,i}^{(m)} \oplus x_{a,i-1}^{(m)} \oplus x_{d,i+7}^{(m)} \oplus x_{d,i-1}^{(m)}, \quad w.p.\ \tfrac{1}{2}\left(1+\tfrac{1}{2^2}\right)$$

$$x_{d,i}^{(m-1)} = \mathcal{L}_{d,i}^{(m)} \oplus x_{c,i-1}^{(m)} \oplus x_{b,i+6}^{(m)}, \quad w.p.\ \tfrac{1}{2}\left(1+\tfrac{1}{2}\right)$$

Proof. See [9]. □

Finally, using Lemma 2 and Lemma 3, it is possible to find linear approximations for two rounds of ChaCha.

Lemma 4. *(Lemma 10 of [9]) The following holds with probability $\frac{1}{2}\left(1+\frac{1}{2}\right)$*

$$\begin{aligned}
x_{11,0}^{(3)} = {}& x_0^{(5)}[0,8,16,24] \oplus x_{1,0}^{(5)} \oplus x_{3,0}^{(5)} \oplus x_{4,7}^{(5)} \oplus x_4^{(5)}[14,15] \oplus x_5^{(5)}[7,19] \oplus\\
& x_8^{(5)}[0,7,8] \oplus x_{9,12}^{(5)} \oplus x_{11,0}^{(5)} \oplus x_{12}^{(5)}[0,24] \oplus x_{13,0}^{(5)} \oplus x_{15}^{(5)}[0,8].
\end{aligned}$$

Proof. See [9]. □

Recently, Coutinho and Souza [10] found linear approximations with fewer terms using the same techniques.

Lemma 5. *(Lemma 5 of [10]) When m is odd, each of the following also holds with probability* $\frac{1}{2}(1+\frac{1}{2})$

$$x_{0,0}^{(m-2)} \oplus x_{5,0}^{(m-2)} = x_{0,0}^{(m)} \oplus x_{2,0}^{(m)} \oplus x_{4,7}^{(m)} \oplus x_{4,19}^{(m)} \oplus x_{5,26}^{(m)} \oplus x_{8,12}^{(m)} \oplus x_{9,7}^{(m)} \oplus x_{9,19}^{(m)} \oplus x_{10,0}^{(m)} \oplus x_{12,0}^{(m)} \oplus x_{13,6}^{(m)} \oplus x_{13,7}^{(m)} \oplus x_{14,0}^{(m)} \oplus x_{14,8}^{(m)}$$

$$x_{1,0}^{(m-2)} \oplus x_{6,0}^{(m-2)} = x_{1,0}^{(m)} \oplus x_{3,0}^{(m)} \oplus x_{5,7}^{(m)} \oplus x_{5,19}^{(m)} \oplus x_{6,26}^{(m)} \oplus x_{9,12}^{(m)} \oplus x_{10,7}^{(m)} \oplus x_{10,19}^{(m)} \oplus x_{11,0}^{(m)} \oplus x_{13,0}^{(m)} \oplus x_{14,6}^{(m)} \oplus x_{14,7}^{(m)} \oplus x_{15,0}^{(m)} \oplus x_{15,8}^{(m)}$$

$$x_{2,0}^{(m-2)} \oplus x_{7,0}^{(m-2)} = x_{0,0}^{(m)} \oplus x_{2,0}^{(m)} \oplus x_{6,7}^{(m)} \oplus x_{6,19}^{(m)} \oplus x_{7,26}^{(m)} \oplus x_{8,0}^{(m)} \oplus x_{10,12}^{(m)} \oplus x_{11,7}^{(m)} \oplus x_{11,19}^{(m)} \oplus x_{12,0}^{(m)} \oplus x_{12,8}^{(m)} \oplus x_{14,0}^{(m)} \oplus x_{15,6}^{(m)} \oplus x_{15,7}^{(m)}$$

$$x_{3,0}^{(m-2)} \oplus x_{4,0}^{(m-2)} = x_{1,0}^{(m)} \oplus x_{3,0}^{(m)} \oplus x_{4,26}^{(m)} \oplus x_{7,7}^{(m)} \oplus x_{7,19}^{(m)} \oplus x_{8,7}^{(m)} \oplus x_{8,19}^{(m)} \oplus x_{9,0}^{(m)} \oplus x_{11,12}^{(m)} \oplus x_{12,6}^{(m)} \oplus x_{12,7}^{(m)} \oplus x_{13,0}^{(m)} \oplus x_{13,8}^{(m)} \oplus x_{15,0}^{(m)}$$

Proof. See [10]. □

In [9], the authors showed that using as $\mathcal{ID}$ a single bit at $x_{13,13}^{(0)}$ and $\mathcal{OD}$ at $x_{11,0}^{(3)}$, it is possible to obtain $\varepsilon_d = -0.0272 \approx -\frac{1}{2^{5.2}}$, experimentally. And from Lemma 2 it is possible to extend to a 4-round differential-linear correlation with $\varepsilon_L = 1$ when the $\mathcal{OD}$ is $x_{1,0}^{(4)} \oplus x_{11,0}^{(4)} \oplus x_{12,8}^{(4)} \oplus x_{12,0}^{(4)}$. Further, it is possible to extend to a 5-round differential-linear correlation using the last equation from Lemma 4 with probability $\frac{1}{2}\left(1+\frac{1}{2}\right)$. This gives a total differential-linear 5th round correlation of $\varepsilon_d \cdot \varepsilon_L^2 \approx -0.0068 = -\frac{1}{2^{7.2}}$. This leads to a 5 round distinguisher with complexity approximately 2^{16}.

Extending the linear approximation for 3 rounds comes at a cost. As discussed prior to the above lemma, for ChaCha, setting $i=0$ in Lemma 2 allows linear approximation of probability 1 for LSB variables. The cost is thus determined by the non LSB variables. A simple count of the non LSB variables in the form (Variable Type, # non LSB occurrence) gives $(x_a, 3), (x_b, 5), (x_c, 3)$, and $(x_d, 2)$. Now, using the probabilities of Lemma 3 and Lemma 4, the linear correlation is $\varepsilon_L = 1/2^{1+3\cdot4+5\cdot1+3\cdot2+2\cdot1} = 2^{-26}$. This leads to a 6 round correlation of $\varepsilon_L^2 \varepsilon_d \approx \frac{1}{2^{57.2}}$. The distinguisher for this correlation has a complexity of 2^{116}.

In [10], the authors used Lemma 5 to derive a distinguisher for 6 rounds. To do that, they found a differential with correlation $\varepsilon_d = 0.00048$ for $(a,b) = (3,4)$ when the input difference is given by $\Delta x_{14,6}^{(0)} = 1$, and 0 for all remaining bits. Therefore, expanding for 6 rounds from Lemma 5 with weights 4, 1, 2, 1 for x_a, x_b, x_c and x_d, respectively, they got $\varepsilon_L = 1/2^{1+0\cdot4+3\cdot1+3\cdot2+3\cdot1} = 2^{-13}$. Then we have $\varepsilon_d \varepsilon_L^2 \approx 2^{-37.02}$, which leads to an attack against 6 rounds of ChaCha with complexity 2^{75}. This is the currently best known 6 round attack on ChaCha.

3 Improved Linear Approximations for ARX-Based Ciphers

The challenge of finding good linear approximations in ARX-based designs comes from the addition operation which is responsible for the non-linearity of the design. In 2003, Wallén [30] published a very important paper where a mathematical framework for finding linear approximations of addition modulo 2^n was developed. Since then, several authors used these technique to find linear approximations in ARX-based designs [9].

Therefore, as before, let $\Theta(x, y) = x \oplus y \oplus (x + y)$ be the carry function of the sum $x + y$. Define $\Theta_i(x, y)$ as the i-th bit of $\Theta(x, y)$. By definition, we have $\Theta_0(x, y) = 0$. Using Theorem 3 of [30], we can generate all possible linear approximations with a given correlation. In particular, we will use the following linear approximations:

$$\Pr(\Theta_i(x, y) = y_{i-1}) = \frac{1}{2}\left(1 + \frac{1}{2}\right), i > 0. \tag{17}$$

$$\Pr(\Theta_i(x, y) \oplus \Theta_{i-1}(x, y) = 0) = \frac{1}{2}\left(1 + \frac{1}{2}\right), i > 0. \tag{18}$$

In previous works of cryptanalysis of ARX ciphers, authors concentrated in finding approximations for particular bits in one round and then repeating the same equations to expand the linear approximation to further rounds (see [9] and [10] for some examples). However, by combining Eqs. 17 and 18 when attacking ARX ciphers we can create a strategy to improve linear approximations when considering more rounds. The main idea is that when using Eq. 17 in one round we will create consecutive terms that can be expanded together using Eq. 18.

For example, consider the sum $z = x + y$. If we want a linear approximation for the bit z_7, we can use Eq. 17 to obtain $z_7 = x_7 \oplus y_7 \oplus \Theta_7(x, y) = x_7 \oplus y_7 \oplus y_6$ with probability 0.75. Since the XOR operation will not change the indexes and the rotation will probably keep y_6 and y_7 adjacent, we can use Eq. 18 in the subsequent round to cancel out the non-linear terms rather than expanding them, leading to a linear equation with higher correlation and fewer terms to be expanded further. Next, we will use this technique to find new linear approximations for ChaCha.

3.1 Linear Approximations for the Quarter Round Function

The first step is to find linear approximations for the QRF of ChaCha. Of course, we already know some of them from previous works (Sect. 2.3). However, here we will consider adjacent bits and several other combinations that cancel out non-linear terms or use Eq. (18). At first glance, these results may seem innocuous, but latter they will prove themselves useful when deriving linear approximations for multiple rounds of ChaCha.

We start with a better linear approximation for $x_{a,i}^{(m-1)}$.

Lemma 6. *The following holds for $i > 0$*

$$x_{a,i}^{(m-1)} = \mathcal{L}_{a,i}^{(m)} \oplus x_{b,i+6}^{(m)} \oplus x_{b,i+18}^{(m)} \oplus x_{c,i+11}^{(m)} \oplus x_{d,i-1}^{(m)}, \ w.p.\ \tfrac{1}{2}\left(1 + \tfrac{1}{2^3}\right).$$

Proof. From Eq. (12) we have

$$x_{a,i}^{(m-1)} = \mathcal{L}_{a,i}^{(m)} \oplus \Theta_i(x_a'^{(m-1)}, x_b'^{(m-1)}) \oplus \Theta_i(x_c'^{(m-1)}, x_d^{(m)}) \oplus \Theta_i(x_a^{(m-1)}, x_b^{(m-1)}).$$

Using Eq. (17) and the Piling-up Lemma we can write

$$x_{a,i}^{(m-1)} = \mathcal{L}_{a,i}^{(m)} \oplus x_{b,i-1}'^{(m-1)} \oplus \Theta_i(x_c'^{(m-1)}, x_d^{(m)}) \oplus x_{b,i-1}^{(m-1)},$$

with probability $\frac{1}{2}\left(1 + \frac{1}{2^2}\right)$. Using Eq. (9) we get

$$x_{a,i}^{(m-1)} = \mathcal{L}_{a,i}^{(m)} \oplus x_{b,i-1}'^{(m-1)} \oplus \Theta_i(x_c'^{(m-1)}, x_d^{(m)}) \oplus \mathcal{L}_{b,i-1}^{(m)} \oplus \Theta_{i-1}(x_c'^{(m-1)}, x_d^{(m)}).$$

Using the approximation of Eq. (18) and the Piling-up Lemma we can write

$$x_{a,i}^{(m-1)} = \mathcal{L}_{a,i}^{(m)} \oplus x_{b,i-1}'^{(m-1)} \oplus \mathcal{L}_{b,i-1}^{(m)},$$

with probability $\frac{1}{2}\left(1 + \frac{1}{2^3}\right)$. Finally, using Eqs. (5) and (14) we get

$$x_{a,i}^{(m-1)} = \mathcal{L}_{a,i}^{(m)} \oplus x_{b,i+6}^{(m)} \oplus x_{b,i+18}^{(m)} \oplus x_{c,i+11}^{(m)} \oplus x_{d,i-1}^{(m)},$$

which completes the proof. □

Lemma 7. *For two active input bits in round $m-1$ and multiple active output bits in round m, the following holds for $i > 0$*

$$x_{\lambda,i}^{(m-1)} \oplus x_{\lambda,i-1}^{(m-1)} = \mathcal{L}_{\lambda,i}^{(m)} \oplus \mathcal{L}_{\lambda,i-1}^{(m)}, \ w.p.\ \frac{1}{2}\left(1 + \frac{1}{2^\sigma}\right),$$

where $(\lambda, \sigma) \in \{(a, 3), (b, 1), (c, 2), (d, 1)\}$.

Proof. This proof follows directly from Eqs. (9)–(12) using the approximation of Eq. (18) and the Piling-up Lemma. □

Lemma 8. *Suppose that $(\lambda, \sigma) \in \{(i, i-2), (i-1, i-1)\}$, $i > 1$. Then for three active input bits in round $m-1$ and multiple active output bits in round m, the following holds*

$$x_{b,\lambda}^{(m-1)} \oplus x_{c,i}^{(m-1)} \oplus x_{c,i-1}^{(m-1)} = \mathcal{L}_{b,i-1}^{(m)} \oplus \mathcal{L}_{c,i}^{(m)} \oplus \mathcal{L}_{c,i-1}^{(m)} \oplus x_{d,\sigma}^{(m)}, \ w.p.\ \frac{1}{2}\left(1 + \frac{1}{2^2}\right).$$

Proof. Using Eq. (9) and Eq. (10) we get

$$\begin{aligned} x_{b,\lambda}^{(m-1)} \oplus x_{c,i}^{(m-1)} \oplus x_{c,i-1}^{(m-1)} = {} & \mathcal{L}_{b,\lambda}^{(m)} \oplus \mathcal{L}_{c,i}^{(m)} \oplus \mathcal{L}_{c,i-1}^{(m)} \oplus \Theta_\lambda(x_c'^{(m-1)}, x_d^{(m)}) \oplus \\ & \Theta_i(x_c'^{(m-1)}, x_d^{(m)}) \oplus \Theta_i(x_c^{(m-1)}, x_d'^{(m-1)}) \oplus \\ & \Theta_{i-1}(x_c'^{(m-1)}, x_d^{(m)}) \oplus \Theta_{i-1}(x_c^{(m-1)}, x_d'^{(m-1)}). \end{aligned}$$

Canceling out common factors and using the approximation of Eq. (18) we can write

$$x_{b,\lambda}^{(m-1)} \oplus x_{c,i}^{(m-1)} \oplus x_{c,i-1}^{(m-1)} = \mathcal{L}_{b,i}^{(m)} \oplus \mathcal{L}_{c,i}^{(m)} \oplus \mathcal{L}_{c,i-1}^{(m)} \oplus \Theta_{\sigma+1}(x_c'^{(m-1)}, x_d^{(m)}).$$

with probability $\frac{1}{2}\left(1+\frac{1}{2}\right)$. Using Eq. (17) we get

$$x_{b,\lambda}^{(m-1)} \oplus x_{c,i}^{(m-1)} \oplus x_{c,i-1}^{(m-1)} = \mathcal{L}_{b,i}^{(m)} \oplus \mathcal{L}_{c,i}^{(m)} \oplus \mathcal{L}_{c,i-1}^{(m)} \oplus x_{d,\sigma}^{(m)},$$

with probability $\frac{1}{2}\left(1+\frac{1}{2^2}\right)$. □

Lemma 9. *For multiple active input bits in round $m-1$ and multiple active output bits in round m, the following linear approximations hold for ChaCha with probability $\frac{1}{2}\left(1+\frac{1}{2^k}\right)$:*

$$x_{b,i}^{(m-1)} \oplus x_{c,i}^{(m-1)} = \mathcal{L}_{b,i}^{(m)} \oplus \mathcal{L}_{c,i}^{(m)} \oplus x_{a,i-1}^{(m)} \oplus x_{d,i+7}^{(m)} \qquad k=1, i>0 \qquad (19)$$

$$x_{a,i}^{(m-1)} \oplus x_{b,i}^{(m-1)} = \mathcal{L}_{a,i}^{(m)} \oplus \mathcal{L}_{b,i-1}^{(m)} \oplus \mathcal{L}_{b,i}^{(m)} \oplus x_{b,i+6}^{(m)} \oplus x_{c,i-1}^{(m)} \oplus x_{d,i-2}^{(m)} \qquad k=3, i>1 \qquad (20)$$

$$x_{a,1}^{(m-1)} \oplus x_{b,1}^{(m-1)} = \mathcal{L}_{a,1}^{(m)} \oplus \mathcal{L}_{b,0}^{(m)} \oplus \mathcal{L}_{b,1}^{(m)} \oplus x_{b,7}^{(m)} \oplus x_{c,0}^{(m)} \qquad k=2 \qquad (21)$$

$$x_{a,i}^{(m-1)} \oplus x_{c,i}^{(m-1)} = \mathcal{L}_{a,i}^{(m)} \oplus \mathcal{L}_{b,i-1}^{(m)} \oplus \mathcal{L}_{c,i}^{(m)} \oplus x_{a,i-1}^{(m)} \oplus x_{b,i+6}^{(m)} \oplus x_{c,i-1}^{(m)} \oplus x_{d,i-2}^{(m)} \oplus x_{d,i+7}^{(m)} \qquad k=4, i>1 \qquad (22)$$

$$x_{a,1}^{(m-1)} \oplus x_{c,1}^{(m-1)} = \mathcal{L}_{a,1}^{(m)} \oplus \mathcal{L}_{b,0}^{(m)} \oplus \mathcal{L}_{c,1}^{(m)} \oplus x_{a,0}^{(m)} \oplus x_{b,7}^{(m)} \oplus x_{c,0}^{(m)} \oplus x_{d,8}^{(m)} \qquad k=3 \qquad (23)$$

$$x_{a,i}^{(m-1)} \oplus x_{d,i}^{(m-1)} = \mathcal{L}_{a,i}^{(m)} \oplus \mathcal{L}_{d,i}^{(m)} \oplus \mathcal{L}_{b,i-1}^{(m)} \qquad k=2, i>1 \qquad (24)$$

$$x_{a,i-1}^{(m-1)} \oplus x_{a,i}^{(m-1)} \oplus x_{c,i}^{(m-1)} = \mathcal{L}_{a,i-1}^{(m)} \oplus \mathcal{L}_{a,i}^{(m)} \oplus \mathcal{L}_{c,i}^{(m)} \oplus x_{d,i-2}^{(m)} \oplus x_{a,i-1}^{(m)} \oplus x_{d,i+7}^{(m)} \qquad k=4, i>1 \qquad (25)$$

$$x_{a,i}^{(m-1)} \oplus x_{a,i-1}^{(m-1)} \oplus x_{b,i}^{(m-1)} = \mathcal{L}_{a,i}^{(m)} \oplus \mathcal{L}_{a,i-1}^{(m)} \oplus \mathcal{L}_{b,i}^{(m)} \oplus x_{d,i-2}^{(m)}, \qquad k=3, i>1 \qquad (26)$$

$$x_{b,i-1}^{(m-1)} \oplus x_{a,i}^{(m-1)} \oplus x_{d,i}^{(m-1)} = \mathcal{L}_{a,i}^{(m)} \oplus \mathcal{L}_{d,i}^{(m)} \oplus x_{d,i-1}^{(m)}, \qquad k=2, i>1 \qquad (27)$$

$$x_{b,i-1}^{(m-1)} \oplus x_{b,i}^{(m-1)} \oplus x_{c,i-1}^{(m-1)} \oplus x_{c,i}^{(m-1)} = \mathcal{L}_{b,i-1}^{(m)} \oplus \mathcal{L}_{b,i}^{(m)} \oplus \mathcal{L}_{c,i-1}^{(m)} \oplus \mathcal{L}_{c,i}^{(m)}, \qquad k=1, i>1 \qquad (28)$$

$$x_{a,i}^{(m-1)} \oplus x_{a,i-1}^{(m-1)} \oplus x_{b,i}^{(m-1)} \oplus x_{c,i-1}^{(m-1)} = \mathcal{L}_{a,i}^{(m)} \oplus \mathcal{L}_{a,i-1}^{(m)} \oplus \mathcal{L}_{b,i}^{(m)} \oplus \mathcal{L}_{c,i-1}^{(m)} \oplus x_{a,i-2}^{(m)} \oplus x_{d,i+6}^{(m)}, \qquad k=3, i>1 \qquad (29)$$

$$x_{a,i}^{(m-1)} \oplus x_{a,i-1}^{(m-1)} \oplus x_{c,i-1}^{(m-1)} \oplus x_{d,i}^{(m-1)} \oplus x_{d,i-1}^{(m-1)} = \mathcal{L}_{a,i-1}^{(m)} \oplus \mathcal{L}_{a,i}^{(m)} \oplus \mathcal{L}_{c,i-1}^{(m)} \oplus \mathcal{L}_{d,i-1}^{(m)} \oplus \mathcal{L}_{d,i}^{(m)} \oplus x_{d,i-1}^{(m)} \oplus x_{a,i-2}^{(m)} \oplus x_{d,i+6}^{(m)}, \qquad k=3, i>2 \qquad (30)$$

Proof. The proof for each equation follows the same basic steps: (1) cancel common factors; (2) cancel adjacent non-linear terms using Eq. (18), updating the probability using the Piling-Up Lemma; (3) substitute the remaining non-linear terms using Eq. (17), updating the probability using the Piling-Up Lemma. For completeness, we list all proofs in Appendix A. □

3.2 Linear Approximations for Multiple Rounds of ChaCha

In this section, we use the proposed technique to construct several new linear approximations for the stream cipher ChaCha which will prove useful to construct better distinguishers. We developed a program (available in https://github.com/MurCoutinho/cryptanalysisChaCha.git) that makes the process of finding linear approximations partly automatic. Our program is capable of expanding the equations and, after statistically verifying the correlation, it outputs the resulting linear approximation in LATEXcode.

We start using the result of Coutinho and Souza [10]. We will only consider the equation for $x_{3,0}^{(3)} \oplus x_{4,0}^{(3)}$ of Lemma 5 but the same reasoning could be applied to any other equation in that lemma. Then, we have

$$\begin{aligned} x_{3,0}^{(3)} \oplus x_{4,0}^{(3)} = x_{1,0}^{(5)} \oplus x_{3,0}^{(5)} \oplus x_{4,26}^{(5)} \oplus x_{7,7}^{(5)} \oplus x_{7,19}^{(5)} \oplus x_{8,7}^{(5)} \oplus x_{8,19}^{(5)} \oplus \\ x_{9,0}^{(5)} \oplus x_{11,12}^{(5)} \oplus x_{12,6}^{(5)} \oplus x_{12,7}^{(5)} \oplus x_{13,0}^{(5)} \oplus x_{13,8}^{(5)} \oplus x_{15,0}^{(5)} \end{aligned} \tag{31}$$

with probability $\frac{1}{2}\left(1+\frac{1}{2}\right)$.

As presented in Sect. 2.3, to expand the equation to the 6-th round, we could use only Lemma 3 as proposed in [9]. In this case, we have weights 4, 1, 2, 1 for x_a, x_b, x_c and x_d, respectively, and a count of $(x_a, 0)$, $(x_b, 3)$, $(x_c, 3)$ e $(x_d, 3)$. Thus, the linear correlation is $\varepsilon_L = 1/2^{1+0\cdot4+3\cdot1+3\cdot2+3\cdot1} = 2^{-13}$. However, we can do better with the new technique proposed in Sect. 3. This will lead us to the following lemma

Lemma 10. *The following linear approximation holds with probability* $\frac{1}{2}\left(1+\frac{1}{2^8}\right)$

$$\begin{aligned} x_{3,0}^{(3)} \oplus x_{4,0}^{(3)} = {} & x_0^{(6)}[0,16] \oplus x_1^{(6)}[0,6,7,11,12,22,23] \oplus x_2^{(6)}[0,6,7,8,16,18, \\ & 19,24] \oplus x_4^{(6)}[7,13,19] \oplus x_5^{(6)}[7] \oplus x_6^{(6)}[7,13,14,19] \oplus \\ & x_7^{(6)}[6,7,14,15,26] \oplus x_8^{(6)}[0,7,8,19,31] \oplus x_9^{(6)}[0,6,12,26] \oplus \\ & x_{10}^{(6)}[0] \oplus x_{11}^{(6)}[6,7] \oplus x_{12}^{(6)}[0,11,12,19,20,30,31] \oplus \\ & x_{13}^{(6)}[0,14,15,24,26,27] \oplus x_{14}^{(6)}[8,25,26] \oplus x_{15}^{(6)}[24]. \end{aligned}$$

Proof. First, from Eq. (31) we can use Lemma 1 to replace $x_{1,0}^{(5)}, x_{3,0}^{(5)}, x_{9,0}^{(5)}, x_{13,0}^{(5)}, x_{15,0}^{(5)}$ by $\mathcal{L}_{1,0}^{(6)}, \mathcal{L}_{3,0}^{(6)}, \mathcal{L}_{9,0}^{(6)}, \mathcal{L}_{13,0}^{(6)}, \mathcal{L}_{15,0}^{(6)}$ with probability 1. Next, note that, since we are transitioning from round 5 to 6, we have

$$(a,b,c,d) \in \{(0,5,10,15),(1,6,11,12),(2,7,8,13),(3,4,9,14)\}.$$

We have already considered the case $(a,b,c,d) = (0,5,10,15)$. Then we still have 3 cases left to consider.

– **Case 1:** When $(a, b, c, d) = (1, 6, 11, 12)$, we have the factors $x^{(5)}_{11,12}$, $x^{(5)}_{12,6}$, $x^{(5)}_{12,7}$. Then we can use Lemma 3 and Lemma 7 in order to get

$$\Pr\left(x^{(5)}_{11,12} = \mathcal{L}^{(6)}_{11,12} \oplus x^{(6)}_{1,11} \oplus x^{(6)}_{12,19} \oplus x^{(6)}_{12,11}\right) = \frac{1}{2}\left(1 + \frac{1}{2^2}\right)$$

and

$$\Pr\left(x^{(5)}_{12,7} \oplus x^{(5)}_{12,6} = \mathcal{L}^{(6)}_{12,7} \oplus \mathcal{L}^{(6)}_{12,6}\right) = \frac{1}{2}\left(1 + \frac{1}{2}\right).$$

– **Case 2:** If $(a, b, c, d) = (2, 7, 8, 13)$, we have the factors $x^{(5)}_{7,7}$, $x^{(5)}_{7,19}$, $x^{(5)}_{8,7}$, $x^{(5)}_{8,19}$, $x^{(5)}_{13,8}$ and we can use Lemma 3 and Eq. (19) of Lemma 9 to get

$$\Pr\left(x^{(5)}_{13,8} = \mathcal{L}^{(6)}_{13,8} \oplus x^{(6)}_{8,7} \oplus x^{(6)}_{7,i+6}\right) = \frac{1}{2}\left(1 + \frac{1}{2}\right),$$

$$\Pr\left(x^{(5)}_{7,7} \oplus x^{(5)}_{8,7} = \mathcal{L}^{(6)}_{7,7} \oplus \mathcal{L}^{(6)}_{8,7} \oplus x^{(6)}_{2,6} \oplus x^{(6)}_{13,14}\right) = \frac{1}{2}\left(1 + \frac{1}{2}\right),$$

$$\Pr\left(x^{(5)}_{7,19} \oplus x^{(5)}_{8,19} = \mathcal{L}^{(6)}_{7,19} \oplus \mathcal{L}^{(6)}_{8,19} \oplus x^{(6)}_{2,18} \oplus x^{(6)}_{13,26}\right) = \frac{1}{2}\left(1 + \frac{1}{2}\right).$$

– **Case 3:** When considering $(a, b, c, d) = (3, 4, 9, 14)$, we have $x^{(5)}_{4,26}$ and we can use Lemma 3 to obtain

$$\Pr\left(x^{(5)}_{4,26} = \mathcal{L}^{(6)}_{4,26} \oplus x^{(6)}_{14,26}\right) = \frac{1}{2}\left(1 + \frac{1}{2}\right).$$

By the Piling-up Lemma, we have that all these changes result in a probability of $\frac{1}{2}\left(1 + \frac{1}{2^8}\right)$. Expanding the linear terms using Eqs. (13)–(16) and canceling out common factors completes the proof. □

Computational Result 1 *The linear approximation of Lemma 10 holds computationally with* $\varepsilon_{L_0} = 0.006942 \approx 2^{-7.17}$. *This correlation was verified using* 2^{38} *random samples.*

In [9], the authors remarked that an expansion of this method to 7 rounds would be unlikely to be useful. Indeed, if we only apply Lemma 3 (which are the linear approximations proposed by Choudhuri and Maitra) we would have $(x_a, 14)$, $(x_b, 13)$, $(x_c, 9)$, $(x_d, 15)$. Therefore, the aggregated correlation would be $\varepsilon_L = 1/2^{7+14\cdot4+13\cdot1+9\cdot2+15\cdot1} = 2^{-109}$. Thus, using this linear expansion in a differential-linear attack would lead to a distinguisher with complexity no less then 2^{436}. However, using our new linear approximations we can get a much better result.

Lemma 11. *The following linear approximation holds with probability* $\frac{1}{2}\left(1+\frac{1}{2^{55}}\right)$

$$x_{3,0}^{(3)} \oplus x_{4,0}^{(3)} = x_0^{(7)}[0,3,4,7,8,11,12,14,15,18,20,27,28] \oplus x_1^{(7)}[0,5,7,8,10, 11,14,15,16,22,23,24,25,27,30,31] \oplus x_2^{(7)}[6,7,9,10,16,18,19, 25,26] \oplus x_3^{(7)}[6,7,8,24] \oplus x_4^{(7)}[0,2,3,5,18,22,23,27] \oplus x_5^{(7)}[1,2, 9,10,13,14,18,21,22,25,29,30] \oplus x_6^{(7)}[2,3,5,7,10,11,13,14,19, 22,23,27,30,31] \oplus x_7^{(7)}[1,2,13,25,26,30,31] \oplus x_8^{(7)}[8,11,13,20, 25,27,28,30,31] \oplus x_9^{(7)}[2,3,6,7,14,15,18,27] \oplus x_{10}^{(7)}[0,3,4,8,12, 13,14,18,20,27,28,30] \oplus x_{11}^{(7)}[6,14,15,18,19,23,24,27] \oplus x_{12}^{(7)}[3,4,6,11,13,22,23,24,26,27,30,31] \oplus x_{13}^{(7)}[1,2,6,7,8,10, 11,13,14,16,18,20,22,23,24,25,26] \oplus x_{14}^{(7)}[0,6,13,14,15,16, 23,24] \oplus x_{15}^{(7)}[16,25,26].$$

Proof. If we start from Lemma 10 then we want to expand the equation one more round. To do so, first note that since we are transitioning from round 6 to 7, we have $(a,b,c,d) \in \{(0,4,8,12),(1,5,9,13),(2,6,10,14),(3,7,11,15)\}$. Therefore, we can divide the factors of the equation in 4 distinct groups:

- Group I - $x_0^{(6)}[0,16], x_4^{(6)}[7,13,19], x_8^{(6)}[0,7,8,19,31]$, $x_{12}^{(6)}[0,11,12,19,20, 30,31]$.
- Group II - $x_1^{(6)}[0,6,7,11,12,22,23]$, $x_5^{(6)}[7]$, $x_9^{(6)}[0,6,12,26]$, $x_{13}^{(6)}[0,14,15, 24,26,27]$.
- Group III - $x_2^{(6)}[0,6,7,8,16,18,19,24], x_6^{(6)}[7,13,14,19], x_{10}^{(6)}[0], x_{14}^{(6)}[8,25,26]$.
- Group IV - $x_7^{(6)}[6,7,14,15,26]$, $x_{11}^{(6)}[6,7], x_{15}^{(6)}[24]$.

The procedure to expand and compute the correlation is similar to that in the proof of Lemma 10. To simplify the notation we will compute the probability given by the Piling-up Lemma by summing values k where the probability of a particular linear equation will be given by $\frac{1}{2}\left(1+\frac{1}{2^k}\right)$.

In Group I, the factors $x_{0,0}^{(6)}, x_{8,0}^{(6)}, x_{12,0}^{(6)}$ can be expanded using Lemma 1 with probability 1. Next, we can combine the following factors: $x_{4,7}^{(6)}, x_{8,7}^{(6)}, x_{8,8}^{(6)}$ using Lemma 8 ($k=2$); $x_{4,19}^{(6)}, x_{8,19}^{(6)}$ using Eq. (19) of Lemma 9 ($k=1$); $x_{12,11}^{(6)}, x_{12,12}^{(6)}$ using Lemma 7 with ($k=1$); $x_{12,19}^{(6)}, x_{12,20}^{(6)}$ using Lemma 7 with ($k=1$); $x_{12,30}^{(6)}, x_{12,31}^{(6)}$ using Lemma 7 with ($k=1$). Finally, it remains some single terms to be expanded: $x_{0,16}^{(6)}$ using Lemma 6 ($k=3$); $x_{4,13}^{(6)}$ using Lemma 3 ($k=1$); $x_{8,31}^{(6)}$ using Lemma 3 ($k=2$). By the Piling-up Lemma, we can combine these linear relations to obtain

$$\begin{aligned} &x_0^{(6)}[0,16] \oplus x_4^{(6)}[7,13,19] \oplus x_8^{(6)}[0,7,8,19,31] \oplus x_{12}^{(6)}[0,11,12,19,20,30,31] = \\ &x_0^{(7)}[0,3,4,7,8,11,12,14,15,18,20,27,28] \oplus x_4^{(7)}[0,2,3,5,18,22,23,27] \oplus \\ &x_8^{(7)}[8,11,13,20,25,27,28,30,31] \oplus x_{12}^{(7)}[3,4,6,11,13,22,23,24,26,27,30,31] \end{aligned} \tag{32}$$

with probability $\frac{1}{2}\left(1+\frac{1}{2^{12}}\right)$.

In Group II, the factors $x^{(6)}_{1,0}, x^{(6)}_{9,0}, x^{(6)}_{13,0}$ can be expanded using Lemma 1 with probability 1. Next, we can combine the following factors: $x^{(6)}_{1,6}, x^{(6)}_{1,7}, x^{(6)}_{5,7}, x^{(6)}_{9,6}$ using Eq. (29) of Lemma 9 ($k = 3$); $x^{(6)}_{1,11}, x^{(6)}_{1,12}, x^{(6)}_{9,12}$ using Eq. (25) of Lemma 9 ($k = 4$); $x^{(6)}_{1,22}, x^{(6)}_{1,23}$ using Lemma 7 ($k = 3$); $x^{(6)}_{13,14}, x^{(6)}_{13,15}$ using Lemma 7 ($k = 1$); $x^{(6)}_{13,26}, x^{(6)}_{13,27}$ using Lemma 7 ($k = 1$). Finally, it remains some single terms to be expanded: $x^{(6)}_{9,26}$ using Lemma 3 ($k = 2$); $x^{(6)}_{13,24}$ using Lemma 3 ($k = 1$). By the Piling-up Lemma, we can combine these linear relations to obtain

$$\begin{aligned}&x_1^{(6)}[0,6,7,11,12,22,23] \oplus x_5^{(6)}[7] \oplus x_9^{(6)}[0,6,12,26] \oplus x_{13}^{(6)}[0,14,15,24,26,\\&27] = x_1^{(7)}[0,5,7,8,10,11,14,15,16,22,23,24,25,27,30,31] \oplus x_5^{(7)}[1,2,9,10,\\&13,14,18,21,22,25,29,30] \oplus x_9^{(7)}[2,3,6,7,14,15,18,27] \oplus x_{13}^{(7)}[1,2,6,7,8,10,\\&11,13,14,16,18,20,22,23,24,25,26]\end{aligned} \tag{33}$$

with probability $\frac{1}{2}\left(1 + \frac{1}{2^{15}}\right)$.

In Group III, the factors $x^{(6)}_{2,0}$ and $x^{(6)}_{10,0}$ can be expanded using Lemma 1 with probability 1. Next, we can combine the following factors: $x^{(6)}_{2,6}, x^{(6)}_{2,7}$ using Lemma 7 ($k = 3$); $x^{(6)}_{6,13}, x^{(6)}_{6,14}$ using Lemma 7 ($k = 1$); $x^{(6)}_{14,25}, x^{(6)}_{14,26}$ using Lemma 7 ($k = 1$); $x^{(6)}_{2,18}, x^{(6)}_{2,19}, x^{(6)}_{6,19}$ using Eq. (26) of Lemma 9 ($k = 3$); $x^{(6)}_{2,8}, x^{(6)}_{6,7}, x^{(6)}_{14,8}$ using Eq. (27) of Lemma 9 ($k = 2$). Finally, it remains some single terms to be expanded: $x^{(6)}_{2,16}$ using Lemma 6 ($k = 3$); $x^{(6)}_{2,24}$ using Lemma 6 ($k = 3$). By the Piling-up Lemma, we can combine these linear relations to obtain

$$\begin{aligned}&x_2^{(6)}[0,6,7,8,16,18,19,24] \oplus x_6^{(6)}[7,13,14,19] \oplus x_{10}^{(6)}[0] \oplus x_{14}^{(6)}[8,25,26] =\\&x_2^{(7)}[6,7,9,10,16,18,19,25,26] \oplus x_6^{(7)}[2,3,5,7,10,11,13,14,19,22,23,27,30,\\&31] \oplus x_{10}^{(7)}[0,3,4,8,12,13,14,18,20,27,28,30] \oplus x_{14}^{(7)}[0,6,13,14,15,16,23,24]\end{aligned} \tag{34}$$

with probability $\frac{1}{2}\left(1 + \frac{1}{2^{16}}\right)$.

In Group IV, we can combine the following factors: $x^{(6)}_{7,14}, x^{(6)}_{7,15}$ using Lemma 7 ($k = 1$); $x^{(6)}_{7,6}, x^{(6)}_{7,7}, x^{(6)}_{11,6}, x^{(6)}_{11,7}$ using Eq. (28) of Lemma 9 ($k = 1$). It remains some single terms to be expanded: $x^{(6)}_{7,26}$ using Lemma 3 ($k = 1$); $x^{(6)}_{15,24}$ using Lemma 3 ($k = 1$). By the Piling-up Lemma, we can combine these linear relations to obtain

$$\begin{aligned}&x_7^{(6)}[6,7,14,15,26] \oplus x_{11}^{(6)}[6,7] \oplus x_{15}^{(6)}[24] = x_3^{(7)}[6,7,8,24] \oplus x_7^{(7)}[1,2,\\&13,25,26,30,31] \oplus x_{11}^{(7)}[6,14,15,18,19,23,24,27] \oplus x_{15}^{(7)}[16,25,26]\end{aligned} \tag{35}$$

with probability $\frac{1}{2}\left(1 + \frac{1}{2^4}\right)$.

Finally, using the Piling-up Lemma we can combine the results from Lemma 10 and Eqs. (32)–(35), which leads to a correlation of $\varepsilon_L = 1/2^{8+12+15+16+4} = 2^{-55}$. □

Computational Result 2 *The linear approximation of Eq.* (32) *holds computationally with* $\varepsilon_{L_1} = 0.000301 \approx 2^{-11.70}$. *This correlation was verified using* 2^{42} *random samples.*

Computational Result 3 *The linear approximation of Eq.* (33) *holds computationally with* $\varepsilon_{L_2} = 0.000100 \approx 2^{-13.29}$. *This correlation was verified using* 2^{42} *random samples.*

Computational Result 4 *The linear approximation of Eq.* (34) *holds computationally with* $\varepsilon_{L_3} = 0.000051 \approx 2^{-14.26}$. *This correlation was verified using* 2^{42} *random samples.*

Computational Result 5 *The linear approximation of Eq.* (35) *holds computationally with* $\varepsilon_{L_4} = 0.0625 \approx 2^{-4}$. *This correlation was verified using* 2^{38} *random samples.*

Next, we will only work with a linear approximation for the bit $x_{5,0}^{(3.5)}$. As we will see in the next session, we are able to find a differential correlation to this bit (as introduced in [9], half a round of ChaCha consists in applying half the operations of the QRF. Thus, from Eq. (2) we can write $x_a^{(r-1/2)} = x_{a'}^{(r-1)}$, $\dots$ $x_d^{(r-1/2)} = x_{d'}^{(r-1)}$). Using Eq. (5) it is easy to see that we have $x_{5,0}^{(3.5)} = x_{5,7}^{(4)} \oplus x_{10,0}^{(4)}$. Additionally, using Lemma 3 we can expand one more round and we get

$$x_{5,0}^{(3.5)} = x_{2,0}^{(5)} \oplus x_{5,26}^{(5)} \oplus x_{9,7}^{(5)} \oplus x_{9,19}^{(5)} \oplus x_{10,0}^{(5)} \oplus x_{13,6}^{(5)} \oplus x_{13,7}^{(5)} \oplus x_{14,0}^{(5)} \oplus x_{14,8}^{(5)}, \quad (36)$$

with probability $\frac{1}{2}\left(1+\frac{1}{2}\right)$.

Lemma 12. *The following linear approximation holds with probability* $\frac{1}{2}\left(1+\frac{1}{2^8}\right)$

$$\begin{aligned} x_{5,0}^{(3.5)} = {} & x_0^{(6)}[0] \oplus x_2^{(6)}[0,6,7,22,23] \oplus x_3^{(6)}[0,6,7,8,16,18,19,24] \oplus \\ & x_4^{(6)}[7,14,15] \oplus x_5^{(6)}[13] \oplus x_7^{(6)}[7,13,14,19] \oplus x_8^{(6)}[6,7,12] \oplus \\ & x_9^{(6)}[0,8,19] \oplus x_{10}^{(6)}[0,6,26] \oplus x_{13}^{(6)}[0,30,31] \oplus \\ & x_{14}^{(6)}[0,6,7,14,15,18,19,24,26,27] \oplus x_{15}^{(6)}[0,8,25,26]. \end{aligned}$$

Proof. We start from Eq. (36) and we can use Lemma 1 $x_{2,0}^{(5)}, x_{10,0}^{(5)}, x_{14,0}^{(5)}$ by $\mathcal{L}_{2,0}^{(6)}, \mathcal{L}_{10,0}^{(6)}, \mathcal{L}_{14,0}^{(6)}$ with probability 1. Next, note that, since we are transitioning from round 5 to 6, we have $(a,b,c,d) \in \{(0,5,10,15), (1,6,11,12), (2,7,8,13), (3,4,9,14)\}$. Considering $(a,b,c,d) = (0,5,10,15)$, we have the factor $x_{5,26}^{(5)}$ and we can apply Lemma 3 to get

$$\Pr\left(x_{5,26}^{(5)} = \mathcal{L}_{5,26}^{(6)} \oplus x_{15,25}^{(6)}\right) = \frac{1}{2}\left(1+\frac{1}{2}\right).$$

Considering $(a,b,c,d) = (2,7,8,13)$, we have the factors $x_{13,6}^{(5)}$ and $x_{13,7}^{(5)}$. Then we can use Lemma 7 to get

$$\Pr\left(x_{13,6}^{(5)} \oplus x_{13,7}^{(5)} = \mathcal{L}_{13,6}^{(6)} \oplus \mathcal{L}_{13,7}^{(6)}\right) = \frac{1}{2}\left(1+\frac{1}{2}\right).$$

Considering $(a,b,c,d) = (3,4,9,14)$ we have $x_{9,7}^{(5)}, x_{9,19}^{(5)}$ and $x_{14,8}^{(5)}$, and then we can apply Lemma 3 to obtain

$$\Pr\left(x_{9,7}^{(5)} = \mathcal{L}_{9,7}^{(6)} \oplus x_{3,6}^{(6)} \oplus x_{14,14}^{(6)} \oplus x_{14,6}^{(6)}\right) = \frac{1}{2}\left(1 + \frac{1}{2^2}\right),$$

$$\Pr\left(x_{9,19}^{(5)} = \mathcal{L}_{9,19}^{(6)} \oplus x_{3,18}^{(6)} \oplus x_{14,26}^{(6)} \oplus x_{14,18}^{(6)}\right) = \frac{1}{2}\left(1 + \frac{1}{2^2}\right),$$

$$\Pr\left(x_{14,8}^{(5)} = \mathcal{L}_{14,8}^{(6)} \oplus x_{9,7}^{(6)} \oplus x_{4,14}^{(6)}\right) = \frac{1}{2}\left(1 + \frac{1}{2}\right).$$

By the Piling-up Lemma, we have that all these changes result in a probability of $\frac{1}{2}\left(1 + \frac{1}{2^8}\right)$. Expanding the linear terms using Eqs. (13)–(16) and canceling out common factors completes the proof. □

Computational Result 6 *The linear approximation of Lemma 12 holds computationally with* $\varepsilon_{L_0} = 0.00867 \approx 2^{-6.85}$.

Lemma 13. *The following linear approximation holds with probability* $\frac{1}{2}\left(1 + \frac{1}{2^{47}}\right)$

$$\begin{aligned} x_{5,0}^{(3.5)} = {} & x_0^{(7)}[0,6,7,11,12] \oplus x_1^{(7)}[7,8,14,15,16,18,19,30,31] \oplus \\ & x_2^{(7)}[0,2,3,5,6,8,10,11,14,15,16,18,19,24,25,27,30,31] \oplus \\ & x_3^{(7)}[6,7,9,10,18,19,25,26] \oplus x_4^{(7)}[1,2,7,19,26] \oplus x_5^{(7)}[0,5,6,7] \oplus \\ & x_6^{(7)}[1,2,9,10,19,21,22,29,31] \oplus x_7^{(7)}[2,3,5,10,11,13,14,19,22,23, \\ & 27,30,31] \oplus x_8^{(7)}[6,14,15,19,26,27] \oplus x_9^{(7)}[8,13,19,25,30,31] \oplus \\ & x_{10}^{(7)}[2,3,7,12,14,15,23,24,27] \oplus x_{11}^{(7)}[0,3,4,8,12,13,14,18,20,27, \\ & 28,30] \oplus x_{12}^{(7)}[0,5,6,11,12,19,20] \oplus x_{13}^{(7)}[0,7,12,13,15,16,18,19,22, \\ & 23,24,26,27] \oplus x_{14}^{(7)}[1,2,8,10,11,13,14,16,18,19,22,23,24,25,26, \\ & 30,31] \oplus x_{15}^{(7)}[5,6,7,8,13,14,15,16,23]. \end{aligned}$$

Proof. If we start from Lemma 12 we want to expand the equation one more round. To do so, first note that since we are transitioning from round 6 to 7, we have $(a,b,c,d) \in \{(0,4,8,12),(1,5,9,13),(2,6,10,14),(3,7,11,15)\}$. Therefore, we can divide the factors of the equation in 4 distinct groups:

- Group I - $x_0^{(6)}[0], x_4^{(6)}[7,14,15], x_8^{(6)}[6,7,12]$.
- Group II - $x_5^{(6)}[13], x_9^{(6)}[0,8,19], x_{13}^{(6)}[0,30,31]$.
- Group III - $x_2^{(6)}[0,6,7,22,23]$, $x_{10}^{(6)}[0,6,26]$, $x_{14}^{(6)}[0,6,7,14,15,18,19,24,26,27]$.
- Group IV - $x_3^{(6)}[0,6,7,8,16,18,19,24], x_7^{(6)}[7,13,14,19], x_{15}^{(6)}[0,8,25,26]$.

Here, we follow the same strategy as in the proof of Lemma 11. In Group I, the factor $x_{0,0}^{(6)}$ can be expanded using Lemma 1 with probability 1. Next, we can combine the following factors: $x_{4,7}^{(6)}, x_{8,6}^{(6)}, x_{8,7}^{(6)}$ using Lemma 8 ($k=2$); $x_{4,14}^{(6)}, x_{4,15}^{(6)}$

using Lemma 7 with ($k = 1$). Finally, we expand $x^{(6)}_{8,12}$ using Lemma 3 ($k = 2$). By the Piling-up Lemma, we can combine these linear relations to obtain

$$\begin{aligned} x^{(6)}_0[0] \oplus x^{(6)}_4[7,14,15] \oplus x^{(6)}_8[6,7,12] = x^{(7)}_0[0,6,7,11,12] \oplus \\ x^{(7)}_4[1,2,7,19,26] \oplus x^{(7)}_8[6,14,15,19,26,27] \oplus x^{(7)}_{12}[0,5,6,11,12,19,20] \end{aligned} \tag{37}$$

with probability $\frac{1}{2}\left(1 + \frac{1}{2^5}\right)$.

In Group II, the factors $x^{(6)}_{9,0}, x^{(6)}_{13,0}$ can be expanded using Lemma 1 with probability 1. Next, we can combine $x^{(6)}_{13,30}, x^{(6)}_{13,31}$ using Lemma 7 ($k = 1$). The remaining terms can be expanded with Lemma 3: $x^{(6)}_{9,8}$ ($k = 2$); $x^{(6)}_{9,19}$ ($k = 2$); $x^{(6)}_{5,13}$ ($k = 1$). By the Piling-up Lemma, we can combine these linear relations to obtain

$$\begin{aligned} x^{(6)}_5[13] \oplus x^{(6)}_9[0,8,19] \oplus x^{(6)}_{13}[0,30,31] = x^{(7)}_1[7,8,14,15,16,18,19,30,31] \oplus \\ x^{(7)}_5[0,5,6,7] \oplus x^{(7)}_9[8,13,19,25,30,31] \oplus x^{(7)}_{13}[0,7,12,13,15,16,18,19,22, \\ 23,24,26,27] \end{aligned} \tag{38}$$

with probability $\frac{1}{2}\left(1 + \frac{1}{2^6}\right)$.

In Group III, the factors $x^{(6)}_{2,0}, x^{(6)}_{10,0}$ and $x^{(6)}_{14,0}$ can be expanded using Lemma 1 with probability 1. Next, we can combine the following factors: $x^{(6)}_{2,6}, x^{(6)}_{2,7}, x^{(6)}_{10,6}, x^{(6)}_{14,6}, x^{(6)}_{14,7}$ using Eq. (30) of Lemma 9 ($k = 3$); $x^{(6)}_{2,22}, x^{(6)}_{2,23}$ using Lemma 7 ($k = 3$); $x^{(6)}_{14,14}, x^{(6)}_{14,15}$ using Lemma 7 ($k = 1$); $x^{(6)}_{14,18}, x^{(6)}_{14,19}$ using Lemma 7 ($k = 1$); $x^{(6)}_{14,26}, x^{(6)}_{14,27}$ using Lemma 7 ($k = 1$). Finally, it remains some single terms to be expanded: $x^{(6)}_{10,26}$ using Lemma 3 ($k = 2$); $x^{(6)}_{14,24}$ using Lemma 6 ($k = 1$). By the Piling-up Lemma, we can combine these linear relations to obtain

$$\begin{aligned} x^{(6)}_2[0,6,7,22,23], x^{(6)}_{10}[0,6,26], x^{(6)}_{14}[0,6,7,14,15,18,19,24,26,27] = \\ x^{(7)}_2[0,2,3,5,6,8,10,11,14,15,16,18,19,24,25,27,30,31] \oplus \\ x^{(7)}_6[1,2,9,10,19,21,22,29,31] \oplus x^{(7)}_{10}[2,3,7,12,14,15,23,24,27] \oplus \\ x^{(7)}_{14}[1,2,8,10,11,13,14,16,18,19,22,23,24,25,26,30,31] \end{aligned} \tag{39}$$

with probability $\frac{1}{2}\left(1 + \frac{1}{2^{12}}\right)$.

In Group IV, the factors $x^{(6)}_{3,0}$ and $x^{(6)}_{15,0}$ can be expanded using Lemma 1 with probability 1. Then we can combine the following factors: $x^{(6)}_{3,6}, x^{(6)}_{3,7}, x^{(6)}_{7,7}$ using Eq. (26) of Lemma 9 ($k = 3$); $x^{(6)}_{3,18}, x^{(6)}_{3,19}, x^{(6)}_{7,19}$ using Eq. (26) of Lemma 9 ($k = 3$); $x^{(6)}_{3,8}, x^{(6)}_{15,8}$ using Eq. (24) of Lemma 9 ($k = 2$); $x^{(6)}_{15,25}, x^{(6)}_{15,26}$ using Lemma 7 ($k = 1$); $x^{(6)}_{7,13}, x^{(6)}_{7,14}$ using Lemma 7 ($k = 1$). It remains some single terms to be expanded: $x^{(6)}_{3,16}$ using Lemma 6 ($k = 3$); $x^{(6)}_{3,24}$ using Lemma 6 ($k = 3$). By the Piling-up Lemma, we can combine these linear relations to obtain

$$x_3^{(6)}[0,6,7,8,16,18,19,24], x_7^{(6)}[7,13,14,19], x_{15}^{(6)}[0,8,25,26] = \\ x_3^{(7)}[6,7,9,10,18,19,25,26] \oplus x_7^{(7)}[2,3,5,10,11,13,14,19,22,23,27,30,31] \oplus \\ x_{11}^{(7)}[0,3,4,8,12,13,14,18,20,27,28,30] \oplus x_{15}^{(7)}[5,6,7,8,13,14,15,16,23] \tag{40}$$

with probability $\frac{1}{2}\left(1+\frac{1}{2^{16}}\right)$.

Finally, using the Piling-up Lemma we can combine the results from Lemma 12 and Eqs. (37)-(40), which leads to a correlation of $\varepsilon_L = 1/2^{8+5+6+12+16} = 2^{-47}$. □

Computational Result 7 *The linear approximation of Eq.* (37) *holds computationally with* $\varepsilon_{L_1} = 0.0416 \approx 2^{-4.59}$. *This correlation was verified using* 2^{38} *random samples.*

Computational Result 8 *The linear approximation of Eq.* (38) *holds computationally with* $\varepsilon_{L_2} = 0.0278 \approx 2^{-5.19}$. *This correlation was verified using* 2^{38} *random samples.*

Computational Result 9 *The linear approximation of Eq.* (39) *holds computationally with* $\varepsilon_{L_3} = 0.000398 \approx 2^{-11.29}$. *This correlation was verified using* 2^{42} *random samples.*

Computational Result 10 *The linear approximation of Eq.* (40) *holds computationally with* $\varepsilon_{L_4} = 0.000047 \approx 2^{-14.38}$. *This correlation was verified using* 2^{42} *random samples.*

It is interesting to note that the experimental correlation is higher than expected in several cases. Of course, since the hypothesis of independence for the Piling-Up Lemma does not hold, it is expected to see deviations between what is predicted theoretically and what we see in practice. The fact that the correlation is usually higher indicates a positive correlation between some equations. In future works, it may be interesting to try to understand why ChaCha has this behavior.

4 Improved Differential-Linear Attacks Against ChaCha

4.1 New Differentials

In this section, we present new differentials for 3.5 rounds of ChaCha. As in previous works, these differential correlations were found experimentally. To find these correlations we used the technique proposed by Beierle et al. at Crypto 2020 [4], and described in Sect. 2.2. Here, the cipher is divided into the sub ciphers E_1 covering 1 round and E_2 covering 2.5 rounds to find a differential path for 3.5 rounds. Thus we want a particular differential characteristic of the form

$$\Delta X^{(0)} \xrightarrow{\text{1 round}} \Delta X^{(1)} \xrightarrow{\text{2.5 rounds}} \Delta X^{(3.5)}.$$

The idea is to generate consistent $\Delta X^{(1)}$ whose Hamming weight is minimized. In [4], the authors showed that the following differential characteristic occurs with probability 2^{-5} on average for the QRF of ChaCha

$$\Delta X^{(0)} = (([]),([]),([]),([i])) \rightarrow \Delta X^{(1)} = (([i+28]),([i+31,i+23,i+11,\\ i+3]),([i+24,i+16,i+4]),\\ ([i+24,i+4])). \tag{41}$$

From there we computed $\Delta X^{(3.5)}$ by generating random states $X^{(1)}$ and $X'^{(1)}$ and statistically testing for correlations in particular bits of $\Delta X^{(3.5)}$. We note that this procedure is computationally intensive as some of the correlations are very small. For some bits, we executed this procedure up to 2^{50} pairs of random states in the first round. To achieve this amount of computation we used 8 NVIDIA GPUs (RTX 2080ti). As in the referred paper, we used $i = 6$. Also, we fixed the differential of Eq. (41) in the third column of the state matrix. Table 3 shows the results.

Table 3. New differentials after 3.5 rounds, starting from $\Delta X^{(1)}$ in the third column of the state matrix with $i = 6$ in Eq. (41).

$\mathcal{OD}$	$\lvert\varepsilon_d\rvert$
$\Delta x_{0,0}^{(3.5)}$	0.000307
$\Delta x_{1,0}^{(3.5)}$	0.000124
$\Delta x_{12,0}^{(3.5)}$	0.000017
$\Delta x_{13,0}^{(3.5)}$	0.000016
$\Delta x_{5,0}^{(3.5)}$	0.0000002489

4.2 Distinguishers

Using the linear approximations of Lemma 10 and Lemma 11, the differential correlation $\varepsilon_d = 0.00048$ for $(a, b) = (3, 4)$ described in [10], and the estimated correlations from the Computational Results 1–5, we get $\varepsilon_d \varepsilon_{L_0}^2 \approx 2^{-25.37}$ which gives us a distinguisher for 6 rounds of ChaCha with complexity less than 2^{51}. Also, we get $\varepsilon_d(\varepsilon_{L_0}\varepsilon_{L_1}\varepsilon_{L_2}\varepsilon_{L_3}\varepsilon_{L_4})^2 \approx 2^{-111.86}$ which gives us a distinguisher for 7 rounds of ChaCha with complexity less than 2^{224}.

Using the linear approximations of Lemma 12 and Lemma 13, the differential correlation for $\Delta_{5,0}^{(3.5)}$ presented in Table 3, and the estimated correlations from the Computational Results 6–10, we get $\varepsilon_d \varepsilon_{L_0}^2 \approx 2^{-35.64}$ which gives us a distinguisher for 6 rounds of ChaCha with complexity less than $2^{72+5} = 2^{77}$ (here we have to repeat the procedure 2^5 times on average as in [4]). Also, we get $\varepsilon_d(\varepsilon_{L_0}\varepsilon_{L_1}\varepsilon_{L_2}\varepsilon_{L_3}\varepsilon_{L_4})^2 \approx 2^{-106.5}$ which gives us a distinguisher for 7 rounds of ChaCha with complexity less than $2^{213+5} = 2^{218}$.

4.3 New Attack Using Probabilistic Neutral Bits (PNBs)

One of the most important attacks against ChaCha is the proposal of Aumasson [1]. The attack first identifies good choices of truncated differentials, then it uses probabilistic backwards computation with the notion of PNBs, estimating the complexity of the attack. This attack is described in several previous works [1,22,23], thus, in our description, we skip several details.

The PNB-based key recovery is a fully experimental approach. We summarize the technique as follows:

- Let the correlation in the forward direction (a.k.a, differential-linear distinguisher) after r rounds be ε_d.
- Let n be the number of PNBs given by a correlation γ. Namely, even if we flip one bit in PNBs, we still observe correlation γ.
- Let the correlation in the backward direction, where all PNB bits are fixed to 0 and non-PNB bits are fixed to the correct ones, is ε_a

Then, the time complexity of the attack is estimated as $2^{256-n}N + 2^{256-\alpha}$, where the data complexity N is given as

$$N = \left(\frac{\sqrt{\alpha \log(4)} + 3\sqrt{1 - \varepsilon_a^2 \varepsilon_d^2}}{\varepsilon_a \varepsilon_d} \right)^2,$$

where α is a parameter that the attacker can choose.

We can improve previous attacks for 7 rounds of ChaCha using this technique by considering the new differential correlation for $\Delta_{5,0}^{(3.5)}$ presented in Table 3. Using Eq. (5) it is easy to see that we have $x_{5,0}^{(3.5)} = x_{5,7}^{(4)} \oplus x_{10,0}^{(4)}$. Therefore, we consider $\mathcal{ID}$ given by Eq. (41) with $i = 6$ and $\mathcal{OD}$ $x_{5,7}^{(4)} \oplus x_{10,0}^{(4)}$. Using $\gamma = 0.35$ we found 108 PNBs, and we obtained $\varepsilon_a = 0.000169$. From that, we get an attack with data complexity of $2^{75.51}$ and time complexity $2^{223.51}$. As in [4], we have to repeat this attack 2^5 times on average. Thus, the final attack has data complexity of $2^{80.51}$ and time complexity $2^{228.51}$. Bellow we list all PNBs:

PNB = (2, 3, 4, 7, 8, 9, 10, 11, 12, 13, 14, 15, 16, 19, 20, 21, 22, 23, 26, 27, 28, 29, 30, 31, 32, 33, 39, 40, 41, 42, 43, 44, 45, 46, 47, 48, 49, 50, 51, 52, 53, 54, 55, 56, 57, 58, 59, 60, 61, 62, 63, 66, 67, 68, 69, 78, 79, 80, 102, 103, 111, 112, 115, 128, 129, 130, 135, 136, 143, 144, 147, 148, 149, 150, 151, 155, 156, 157, 158, 159, 160, 161, 162, 163, 168, 169, 170, 173, 174, 175, 176, 179, 180, 181, 182, 185, 186, 191, 199, 200, 201, 219, 220, 221, 222, 223, 232, 255).

5 Conclusion

In this paper, we presented a new technique to find linear approximations for ARX ciphers. Applying this technique we presented new linear approximations to the stream cipher ChaCha which gave us new and improved distinguishers. In addition, we presented new differential characteristics for 3.5 rounds of ChaCha and use them to improve the attacks based on Probabilistic Neutral Bits. For future works, we expect that the proposed technique can be used to improve attacks against similar ARX-based designs, as the stream cipher Salsa and the hash function Blake.

Acknowledgements. The authors would like to thank the anonymous reviewers for their valuable comments and suggestions which helped us to improve our work.

A Proofs

In this appendix, we expand the proof of Lemma 9 for each individual linear approximation.

A.1 Eq. (19)

Proof. Using Eqs. (9) and (10) we can write

$$x_{b,i}^{(m-1)} \oplus x_{c,i}^{(m-1)} = \mathcal{L}_{b,i}^{(m)} \oplus \Theta_i(x_c'^{(m-1)}, x_d^{(m)}) \oplus \\ \mathcal{L}_{c,i}^{(m)} \oplus \Theta_i(x_c'^{(m-1)}, x_d^{(m)}) \oplus \Theta_i(x_c^{(m-1)}, x_d'^{(m-1)}).$$

Using the approximation of Eq. (17) we can write $\Theta_i(x_c^{(m-1)}, x_d'^{(m-1)}) = x_{d,i-1}'^{(m-1)}$ with probability $\frac{1}{2}\left(1+\frac{1}{2}\right)$. Thus, using Eq. (7) and canceling out common factors we get

$$x_{b,i}^{(m-1)} \oplus x_{c,i}^{(m-1)} = \mathcal{L}_{b,i}^{(m)} \oplus \mathcal{L}_{c,i}^{(m)} \oplus x_{a,i-1}^{(m)} \oplus x_{d,i+7}^{(m)},$$

with probability $\frac{1}{2}\left(1+\frac{1}{2}\right)$, which concludes the proof. □

A.2 Eqs. (20) and (21)

Proof. Using Eqs. (9) and (12) we can write

$$x_{a,i}^{(m-1)} \oplus x_{b,i}^{(m-1)} = \mathcal{L}_{a,i}^{(m)} \oplus \mathcal{L}_{b,i}^{(m)} \oplus \Theta_i(x_c'^{(m-1)}, x_d^{(m)}) \oplus \Theta_i(x_c'^{(m-1)}, x_d^{(m)}) \oplus \\ \Theta_i(x_a'^{(m-1)}, x_b'^{(m-1)}) \oplus \Theta_i(x_a^{(m-1)}, x_b^{(m-1)}).$$

Cancelling out common factors, using the approximation of Eq. (17) and the Piling-up Lemma we can write

$$x_{a,i}^{(m-1)} \oplus x_{b,i}^{(m-1)} = \mathcal{L}_{a,i}^{(m)} \oplus \mathcal{L}_{b,i}^{(m)} \oplus x_{b,i-1}'^{(m-1)} \oplus x_{b,i-1}^{(m-1)}$$

with probability $\frac{1}{2}\left(1+\frac{1}{2^2}\right)$. Now we can replace $x_{b,i-1}'^{(m-1)}$ using Eq. (5) and $x_{b,i-1}^{(m-1)}$ using Lemma 3, which leads to

$$x_{a,i}^{(m-1)} \oplus x_{b,i}^{(m-1)} = \mathcal{L}_{a,i}^{(m)} \oplus \mathcal{L}_{b,i}^{(m)} \oplus x_{b,i+6}^{(m)} \oplus x_{c,i-1}^{(m)} \oplus \mathcal{L}_{b,i-1}^{(m)} \oplus x_{d,i-2}^{(m)},$$

with probability $\frac{1}{2}\left(1+\frac{1}{2^3}\right)$ by the Piling-up Lemma. We can also use Lemma 1 in order to obtain

$$x_{a,1}^{(m-1)} \oplus x_{b,1}^{(m-1)} = \mathcal{L}_{a,1}^{(m)} \oplus \mathcal{L}_{b,1}^{(m)} \oplus x_{b,7}^{(m)} \oplus x_{c,0}^{(m)} \oplus \mathcal{L}_{b,0}^{(m)},$$

with probability $\frac{1}{2}\left(1+\frac{1}{2^2}\right)$. □

A.3 Eqs. (22) and (23)

Proof. Combining Eq. (10) and Eq. (12) we have

$$x_{a,i}^{(m-1)} \oplus x_{c,i}^{(m-1)} = \mathcal{L}_{a,i}^{(m)} \oplus \mathcal{L}_{c,i}^{(m)} \oplus \Theta_i(x_c^{(m-1)}, x_d'^{(m-1)}) \oplus \Theta_i(x_a'^{(m-1)}, x_b'^{(m-1)}) \oplus \Theta_i(x_a^{(m-1)}, x_b^{(m-1)}).$$

Using the approximation of Eq. (17) and the Piling-up Lemma we can write

$$x_{a,i}^{(m-1)} \oplus x_{c,i}^{(m-1)} = \mathcal{L}_{a,i}^{(m)} \oplus \mathcal{L}_{c,i}^{(m)} \oplus x_{d,i-1}'^{(m-1)} \oplus x_{b,i-1}'^{(m-1)} \oplus x_{b,i-1}^{(m-1)}$$

with probability $\frac{1}{2}\left(1+\frac{1}{2^3}\right)$. Now we can replace $x_{d,i-1}'^{(m-1)}$ using Eq. (7), $x_{b,i-1}'^{(m-1)}$ using Eq. (5) and $x_{b,i-1}^{(m-1)}$ using Lemma 3 if $i > 1$ or 1 if $i = 1$, which leads to

$$x_{a,i}^{(m-1)} \oplus x_{c,i}^{(m-1)} = \mathcal{L}_{a,i}^{(m)} \oplus \mathcal{L}_{c,i}^{(m)} \oplus x_{a,i-1}^{(m)} \oplus x_{d,i+7}^{(m)} \oplus x_{b,i+6}^{(m)} \oplus x_{c,i-1}^{(m)} \oplus \mathcal{L}_{b,i-1}^{(m)} \oplus x_{d,i-2}^{(m)},$$

with probability $\frac{1}{2}\left(1+\frac{1}{2^4}\right)$ by the Piling-up Lemma or

$$x_{a,1}^{(m-1)} \oplus x_{c,1}^{(m-1)} = \mathcal{L}_{a,1}^{(m)} \oplus \mathcal{L}_{c,1}^{(m)} \oplus x_{a,0}^{(m)} \oplus x_{d,8}^{(m)} \oplus x_{b,7}^{(m)} \oplus x_{c,0}^{(m)} \oplus \mathcal{L}_{b,0}^{(m)},$$

with probability $\frac{1}{2}\left(1+\frac{1}{2^3}\right)$. □

A.4 Eq. (24)

Proof. Using Eq. (11) and Eq. (12) we can write

$$x_{a,i}^{(m-1)} \oplus x_{d,i}^{(m-1)} = \mathcal{L}_{a,i}^{(m)} \oplus \mathcal{L}_{d,i}^{(m)} \oplus \Theta_i(x_c'^{(m-1)}, x_d^{(m)}) \oplus \Theta_i(x_a^{(m-1)}, x_b^{(m-1)}).$$

Using Eq. (17) we get

$$x_{a,i}^{(m-1)} \oplus x_{d,i}^{(m-1)} = \mathcal{L}_{a,i}^{(m)} \oplus \mathcal{L}_{d,i}^{(m)} \oplus \Theta_i(x_c'^{(m-1)}, x_d^{(m)}) \oplus x_{b,i-1}^{(m-1)},$$

and from Eq. (9)

$$x_{a,i}^{(m-1)} \oplus x_{d,i}^{(m-1)} = \mathcal{L}_{a,i}^{(m)} \oplus \mathcal{L}_{d,i}^{(m)} \oplus \Theta_i(x_c'^{(m-1)}, x_d^{(m)}) \oplus \mathcal{L}_{b,i-1}^{(m)} \oplus \Theta_{i-1}(x_c'^{(m-1)}, x_d^{(m)}),$$

with probability $\frac{1}{2}\left(1+\frac{1}{2}\right)$. Thus, using the approximation of Eq. (18) and the Piling-up Lemma we can write

$$x_{a,i}^{(m-1)} \oplus x_{d,i}^{(m-1)} = \mathcal{L}_{a,i}^{(m)} \oplus \mathcal{L}_{d,i}^{(m)} \oplus \mathcal{L}_{b,i-1}^{(m)},$$

with probability $\frac{1}{2}\left(1+\frac{1}{2^2}\right)$. □

A.5 Eq. (25)

Proof. Using Eq. (12) and Eq. (10) and canceling out common factors we get

$$\begin{aligned} x_{a,i-1}^{(m-1)} \oplus x_{a,i}^{(m-1)} \oplus x_{c,i}^{(m-1)} = & \mathcal{L}_{a,i-1}^{(m)} \oplus \mathcal{L}_{a,i}^{(m)} \oplus \mathcal{L}_{c,i}^{(m)} \oplus \\ & \Theta_{i-1}(x_a^{\prime(m-1)}, x_b^{\prime(m-1)}) \oplus \Theta_{i-1}(x_c^{\prime(m-1)}, x_d^{(m)}) \oplus \\ & \Theta_{i-1}(x_a^{(m-1)}, x_b^{(m-1)}) \oplus \Theta_i(x_a^{\prime(m-1)}, x_b^{\prime(m-1)}) \oplus \\ & \Theta_i(x_a^{(m-1)}, x_b^{(m-1)}) \oplus \Theta_i(x_c^{(m-1)}, x_d^{\prime(m-1)}) \end{aligned}$$

Using the approximation of Eq. (18) and the Piling-up Lemma we obtain

$$\begin{aligned} x_{a,i-1}^{(m-1)} \oplus x_{a,i}^{(m-1)} \oplus x_{c,i}^{(m-1)} = & \mathcal{L}_{a,i-1}^{(m)} \oplus \mathcal{L}_{a,i}^{(m)} \oplus \mathcal{L}_{c,i}^{(m)} \oplus \\ & \Theta_{i-1}(x_c^{\prime(m-1)}, x_d^{(m)}) \oplus \Theta_i(x_c^{(m-1)}, x_d^{\prime(m-1)}) \end{aligned}$$

with probability $\frac{1}{2}\left(1+\frac{1}{2^2}\right)$. Using Eq. (17) and Eq. (7) we get

$$\begin{aligned} x_{a,i-1}^{(m-1)} \oplus x_{a,i}^{(m-1)} \oplus x_{c,i}^{(m-1)} = & \mathcal{L}_{a,i-1}^{(m)} \oplus \mathcal{L}_{a,i}^{(m)} \oplus \mathcal{L}_{c,i}^{(m)} \oplus \\ & x_{d,i-2}^{(m)} \oplus x_{a,i-1}^{(m)} \oplus x_{d,i+7}^{(m)} \end{aligned}$$

with probability $\frac{1}{2}\left(1+\frac{1}{2^4}\right)$. □

A.6 Eq. (26)

Proof. Using Eq. (9) and Eq. (12) and canceling out common factors we can write

$$\begin{aligned} & x_{a,i}^{(m-1)} \oplus x_{a,i-1}^{(m-1)} \oplus x_{b,i}^{(m-1)} = \mathcal{L}_{a,i}^{(m)} \oplus \mathcal{L}_{a,i-1}^{(m)} \oplus \mathcal{L}_{b,i}^{(m)} \oplus \\ & \Theta_{i-1}(x_a^{\prime(m-1)}, x_b^{\prime(m-1)}) \oplus \Theta_{i-1}(x_c^{\prime(m-1)}, x_d^{(m)}) \oplus \Theta_{i-1}(x_a^{(m-1)}, x_b^{(m-1)}) \oplus \\ & \Theta_i(x_a^{\prime(m-1)}, x_b^{\prime(m-1)}) \oplus \Theta_i(x_a^{(m-1)}, x_b^{(m-1)}). \end{aligned}$$

Using the approximation of Eq. (18) and the Piling-up Lemma we can write

$$\begin{aligned} x_{a,i}^{(m-1)} \oplus x_{a,i-1}^{(m-1)} \oplus x_{b,i}^{(m-1)} = & \mathcal{L}_{a,i}^{(m)} \oplus \mathcal{L}_{a,i-1}^{(m)} \\ & \oplus \mathcal{L}_{b,i}^{(m)} \oplus \Theta_{i-1}(x_c^{\prime(m-1)}, x_d^{(m)}). \end{aligned}$$

with probability $\frac{1}{2}\left(1+\frac{1}{2^2}\right)$. Using the approximation of Eq. (17) we get

$$x_{a,i}^{(m-1)} \oplus x_{a,i-1}^{(m-1)} \oplus x_{b,i}^{(m-1)} = \mathcal{L}_{a,i}^{(m)} \oplus \mathcal{L}_{a,i-1}^{(m)} \oplus \mathcal{L}_{b,i}^{(m)} \oplus x_{d,i-2}^{(m)}.$$

with probability $\frac{1}{2}\left(1+\frac{1}{2^3}\right)$. □

A.7 Eq. (27)

Proof. Using Eq. (11) and Eq. (12), and canceling out common factors we have

$$x_{b,i-1}^{(m-1)} \oplus x_{a,i}^{(m-1)} \oplus x_{d,i}^{(m-1)} = x_{b,i-1}^{(m-1)} \oplus \mathcal{L}_{a,i}^{(m)} \oplus \Theta_i(x_c'^{(m-1)}, x_d^{(m)}) \oplus \Theta_i(x_a^{(m-1)}, x_b^{(m-1)}) \oplus \mathcal{L}_{d,i}^{(m)}.$$

Using the approximation of Eq. (17) we have $\Theta_i(x_a^{(m-1)}, x_b^{(m-1)}) = x_{b,i-1}^{(m-1)}$ occurring with probability $\frac{1}{2}\left(1 + \frac{1}{2^2}\right)$. Then

$$x_{b,i-1}^{(m-1)} \oplus x_{a,i}^{(m-1)} \oplus x_{d,i}^{(m-1)} = \mathcal{L}_{a,i}^{(m)} \oplus \mathcal{L}_{d,i}^{(m)} \oplus \Theta_i(x_c'^{(m-1)}, x_d^{(m)}).$$

with probability $\frac{1}{2}\left(1 + \frac{1}{2}\right)$. Finally, using the approximation of Eq. (17) and the Piling-up Lemma we get

$$x_{b,i-1}^{(m-1)} \oplus x_{a,i}^{(m-1)} \oplus x_{d,i}^{(m-1)} = \mathcal{L}_{a,i}^{(m)} \oplus \mathcal{L}_{d,i}^{(m)} \oplus x_{d,i-1}^{(m)}.$$

with probability $\frac{1}{2}\left(1 + \frac{1}{2^2}\right)$. □

A.8 Eq. (28)

Proof. Using Eq. (9) and Eq. (10), we can write

$$x_{b,i-1}^{(m-1)} \oplus x_{b,i}^{(m-1)} \oplus x_{c,i-1}^{(m-1)} \oplus x_{c,i}^{(m-1)} = \mathcal{L}_{b,i-1}^{(m)} \oplus \Theta_{i-1}(x_c'^{(m-1)}, x_d^{(m)}) \oplus \mathcal{L}_{b,i}^{(m)} \oplus \Theta_i(x_c'^{(m-1)}, x_d^{(m)}) \oplus \mathcal{L}_{c,i-1}^{(m)} \oplus \Theta_{i-1}(x_c'^{(m-1)}, x_d^{(m)}) \oplus \Theta_{i-1}(x_c^{(m-1)}, x_d'^{(m-1)}) \oplus \mathcal{L}_{c,i}^{(m)} \oplus \Theta_i(x_c'^{(m-1)}, x_d^{(m)}) \oplus \Theta_i(x_c^{(m-1)}, x_d'^{(m-1)}).$$

Canceling out common factors we get

$$x_{b,i-1}^{(m-1)} \oplus x_{b,i}^{(m-1)} \oplus x_{c,i-1}^{(m-1)} \oplus x_{c,i}^{(m-1)} = \mathcal{L}_{b,i-1}^{(m)} \oplus \mathcal{L}_{b,i}^{(m)} \oplus \mathcal{L}_{c,i-1}^{(m)} \oplus \mathcal{L}_{c,i}^{(m)} \oplus \Theta_{i-1}(x_c^{(m-1)}, x_d'^{(m-1)}) \oplus \Theta_i(x_c^{(m-1)}, x_d'^{(m-1)}).$$

Thus, using the approximation of Eq. (18) we get

$$x_{b,i-1}^{(m-1)} \oplus x_{b,i}^{(m-1)} \oplus x_{c,i-1}^{(m-1)} \oplus x_{c,i}^{(m-1)} = \mathcal{L}_{b,i-1}^{(m)} \oplus \mathcal{L}_{b,i}^{(m)} \oplus \mathcal{L}_{c,i-1}^{(m)} \oplus \mathcal{L}_{c,i}^{(m)}.$$

with probability $\frac{1}{2}\left(1 + \frac{1}{2}\right)$. □

A.9 Eq. (29)

Proof. Using Eqs. (9), (10) and (12)

$$x_{a,i}^{(m-1)} \oplus x_{a,i-1}^{(m-1)} \oplus x_{b,i}^{(m-1)} \oplus x_{c,i-1}^{(m-1)} = \mathcal{L}_{a,i}^{(m)} \oplus \mathcal{L}_{a,i-1}^{(m)} \oplus \mathcal{L}_{b,i}^{(m)} \oplus \mathcal{L}_{c,i-1}^{(m)} \oplus \Theta_i(x_a'^{(m-1)}, x_b'^{(m-1)}) \oplus \Theta_i(x_a^{(m-1)}, x_b^{(m-1)}) \oplus \Theta_{i-1}(x_a'^{(m-1)}, x_b'^{(m-1)}) \oplus \Theta_{i-1}(x_a^{(m-1)}, x_b^{(m-1)}) \oplus \Theta_{i-1}(x_c^{(m-1)}, x_d'^{(m-1)}).$$

Using the approximation of Eq. (18) and the Piling-up Lemma we can write

$$x_{a,i}^{(m-1)} \oplus x_{a,i-1}^{(m-1)} \oplus x_{b,i}^{(m-1)} \oplus x_{c,i-1}^{(m-1)} = \mathcal{L}_{a,i}^{(m)} \oplus \mathcal{L}_{a,i-1}^{(m)} \oplus \mathcal{L}_{b,i}^{(m)} \oplus \mathcal{L}_{c,i-1}^{(m)} \oplus \Theta_{i-1}(x_c^{(m-1)}, x_d^{\prime(m-1)}).$$

with probability $\frac{1}{2}\left(1+\frac{1}{2^2}\right)$. Therefore, Eqs. (17) and (7) give us

$$x_{a,i}^{(m-1)} \oplus x_{a,i-1}^{(m-1)} \oplus x_{b,i}^{(m-1)} \oplus x_{c,i-1}^{(m-1)} = \mathcal{L}_{a,i}^{(m)} \oplus \mathcal{L}_{a,i-1}^{(m)} \oplus \mathcal{L}_{b,i}^{(m)} \oplus \mathcal{L}_{c,i-1}^{(m)} \oplus x_{a,i-2}^{(m)} \oplus x_{d,i+6}^{(m)}.$$

with probability $\frac{1}{2}\left(1+\frac{1}{2^3}\right)$. □

A.10 Eq. (30)

Proof. Using Eqs. (10), (11) and (12), we can write

$$x_{a,i}^{(m-1)} \oplus x_{a,i-1}^{(m-1)} \oplus x_{c,i-1}^{(m-1)} \oplus x_{d,i}^{(m-1)} \oplus x_{d,i-1}^{(m-1)} = \mathcal{L}_{a,i-1}^{(m)} \oplus \mathcal{L}_{a,i}^{(m)} \oplus \mathcal{L}_{c,i-1}^{(m)} \oplus \mathcal{L}_{d,i-1}^{(m)} \oplus \mathcal{L}_{d,i}^{(m)} \oplus \Theta_{i-1}(x_a^{(m-1)}, x_b^{(m-1)}) \oplus \Theta_i(x_c^{\prime(m-1)}, x_d^{(m)}) \oplus \Theta_i(x_a^{(m-1)}, x_b^{(m-1)}) \oplus \Theta_{i-1}(x_c^{(m-1)}, x_d^{\prime(m-1)}).$$

Using the approximation of Eq. (18) we have

$$x_{a,i}^{(m-1)} \oplus x_{a,i-1}^{(m-1)} \oplus x_{c,i-1}^{(m-1)} \oplus x_{d,i}^{(m-1)} \oplus x_{d,i-1}^{(m-1)} = \mathcal{L}_{a,i-1}^{(m)} \oplus \mathcal{L}_{a,i}^{(m)} \oplus \mathcal{L}_{c,i-1}^{(m)} \oplus \mathcal{L}_{d,i-1}^{(m)} \oplus \mathcal{L}_{d,i}^{(m)} \oplus \Theta_i(x_c^{\prime(m-1)}, x_d^{(m)}) \oplus \Theta_{i-1}(x_c^{(m-1)}, x_d^{\prime(m-1)})$$

with probability $\frac{1}{2}\left(1+\frac{1}{2}\right)$. Finally, by the Piling-up Lemma and using the approximation of Eq. (17) and Eq. (7), we get

$$x_{a,i}^{(m-1)} \oplus x_{a,i-1}^{(m-1)} \oplus x_{c,i-1}^{(m-1)} \oplus x_{d,i}^{(m-1)} \oplus x_{d,i-1}^{(m-1)} = \mathcal{L}_{a,i-1}^{(m)} \oplus \mathcal{L}_{a,i}^{(m)} \oplus \mathcal{L}_{c,i-1}^{(m)} \oplus \mathcal{L}_{d,i-1}^{(m)} \oplus \mathcal{L}_{d,i}^{(m)} \oplus x_{d,i-1}^{(m)} \oplus x_{a,i-2}^{(m)} \oplus x_{d,i+6}^{(m)}$$

with probability $\frac{1}{2}\left(1+\frac{1}{2^3}\right)$. □

References

1. Aumasson, J.-P., Fischer, S., Khazaei, S., Meier, W., Rechberger, C.: New features of Latin dances: analysis of Salsa, ChaCha, and Rumba. In: Nyberg, K. (ed.) FSE 2008. LNCS, vol. 5086, pp. 470–488. Springer, Heidelberg (2008). https://doi.org/10.1007/978-3-540-71039-4_30
2. Aumasson, J.P., Henzen, L., Meier, W., Phan, R.C.W.: SHA-3 proposal blake. Submission to NIST 92 (2008)
3. Beierle, C., et al.: Schwaemm and Esch: lightweight authenticated encryption and hashing using the Sparkle permutation family (2019)
4. Beierle, C., Leander, G., Todo, Y.: Improved differential-linear attacks with applications to ARX ciphers. In: Micciancio, D., Ristenpart, T. (eds.) CRYPTO 2020. LNCS, vol. 12172, pp. 329–358. Springer, Cham (2020). https://doi.org/10.1007/978-3-030-56877-1_12

5. Bernstein, D.J.: The poly1305-AES message-authentication code. In: Gilbert, H., Handschuh, H. (eds.) FSE 2005. LNCS, vol. 3557, pp. 32–49. Springer, Heidelberg (2005). https://doi.org/10.1007/11502760_3
6. Bernstein, D.J.: ChaCha, a variant of Salsa20. In: Workshop Record of SASC, vol. 8, 3–5 (2008)
7. Bernstein, D.J.: The Salsa20 family of stream ciphers. In: Robshaw, M., Billet, O. (eds.) New Stream Cipher Designs. LNCS, vol. 4986, pp. 84–97. Springer, Heidelberg (2008). https://doi.org/10.1007/978-3-540-68351-3_8
8. Blondeau, C., Leander, G., Nyberg, K.: Differential-linear cryptanalysis revisited. J. Cryptol. **30**(3), 859–888 (2016). https://doi.org/10.1007/s00145-016-9237-5
9. Choudhuri, A.R., Maitra, S.: Significantly improved multi-bit differentials for reduced round Salsa and Chacha. IACR Transa. Symmetric Cryptol. 261–287 (2016)
10. Coutinho, M., Neto, T.S.: New multi-bit differentials to improve attacks against ChaCha. IACR Cryptology ePrint Archive 2020, 350 (2020)
11. Crowley, P.: Truncated differential cryptanalysis of five rounds of Salsa20. In: The State of the Art of Stream Ciphers SASC 2006, pp. 198–202 (2006)
12. Dey, S., Roy, T., Sarkar, S.: Revisiting design principles of Salsa and ChaCha. Adv. Math. Commun. **13**(4), 689 (2019)
13. Dey, S., Sarkar, S.: Improved analysis for reduced round Salsa and Chacha. Discrete Appl. Math. **227**, 58–69 (2017)
14. Ding, L.: Improved related-cipher attack on Salsa20 stream cipher. IEEE Access **7**, 30197–30202 (2019)
15. Dinu, D., Perrin, L., Udovenko, A., Velichkov, V., Großschädl, J., Biryukov, A.: Design strategies for ARX with provable bounds: SPARX and LAX. In: Cheon, J.H., Takagi, T. (eds.) ASIACRYPT 2016. LNCS, vol. 10031, pp. 484–513. Springer, Heidelberg (2016). https://doi.org/10.1007/978-3-662-53887-6_18
16. Fischer, S., Meier, W., Berbain, C., Biasse, J.-F., Robshaw, M.J.B.: Non-randomness in eSTREAM Candidates Salsa20 and TSC-4. In: Barua, R., Lange, T. (eds.) INDOCRYPT 2006. LNCS, vol. 4329, pp. 2–16. Springer, Heidelberg (2006). https://doi.org/10.1007/11941378_2
17. Hernandez-Castro, J.C., Tapiador, J.M.E., Quisquater, J.-J.: On the Salsa20 core function. In: Nyberg, K. (ed.) FSE 2008. LNCS, vol. 5086, pp. 462–469. Springer, Heidelberg (2008). https://doi.org/10.1007/978-3-540-71039-4_29
18. IANIX: ChaCha usage & deployment (2020). https://ianix.com/pub/chacha-deployment.html. Accessed 13 Jan 2020
19. Ishiguro, T., Kiyomoto, S., Miyake, Y.: Latin dances revisited: new analytic results of Salsa20 and ChaCha. In: Qing, S., Susilo, W., Wang, G., Liu, D. (eds.) ICICS 2011. LNCS, vol. 7043, pp. 255–266. Springer, Heidelberg (2011). https://doi.org/10.1007/978-3-642-25243-3_21
20. Langford, S.K., Hellman, M.E.: Differential-linear cryptanalysis. In: Desmedt, Y.G. (ed.) CRYPTO 1994. LNCS, vol. 839, pp. 17–25. Springer, Heidelberg (1994). https://doi.org/10.1007/3-540-48658-5_3
21. Langley, A., Chang, W., Mavrogiannopoulos, N., Strombergson, J., Josefsson, S.: ChaCha20-Poly1305 cipher suites for transport layer security (TLS). RFC 7905 (10) (2016)
22. Maitra, S., Paul, G., Meier, W.: Salsa20 cryptanalysis: new moves and revisiting old styles. In: The Ninth International Workshop on Coding and Cryptography (2015)
23. Maitra, S.: Chosen IV cryptanalysis on reduced round ChaCha and Salsa. Discrete Appl. Math. **208**, 88–97 (2016)

24. Mouha, N., Preneel, B.: A proof that the ARX cipher Salsa20 is secure against differential cryptanalysis. IACR Cryptology ePrint Archive 2013, 328 (2013)
25. Muller, S.: Documentation and analysis of the Linux random number generator - federal office for information security (Germany's) (2019). https://www.bsi.bund.de/SharedDocs/Downloads/EN/BSI/Publications/Studies/LinuxRNG/LinuxRNG_EN.pdf;jsessionid=6B0F8D7795B80F5EADA3DB3DB3E4043B.1_cid360?__blob=publicationFile&v=19
26. Robshaw, M., Billet, O. (eds.): New Stream Cipher Designs. LNCS, vol. 4986. Springer, Heidelberg (2008). https://doi.org/10.1007/978-3-540-68351-3
27. Shi, Z., Zhang, B., Feng, D., Wu, W.: Improved key recovery attacks on reduced-round Salsa20 and ChaCha. In: Kwon, T., Lee, M.-K., Kwon, D. (eds.) ICISC 2012. LNCS, vol. 7839, pp. 337–351. Springer, Heidelberg (2013). https://doi.org/10.1007/978-3-642-37682-5_24
28. Torvalds, L.: Linux kernel source tree (2016). https://git.kernel.org/pub/scm/linux/kernel/git/torvalds/linux.git/commit/?id=818e607b57c94ade9824dad63a96c2ea6b21baf3
29. Tsunoo, Y., Saito, T., Kubo, H., Suzaki, T., Nakashima, H.: Differential cryptanalysis of Salsa20/8. In: Workshop Record of SASC, vol. 28 (2007)
30. Wallén, J.: Linear approximations of addition modulo 2^n. In: Johansson, T. (ed.) FSE 2003. LNCS, vol. 2887, pp. 261–273. Springer, Heidelberg (2003). https://doi.org/10.1007/978-3-540-39887-5_20

Rotational Cryptanalysis from a Differential-Linear Perspective

Practical Distinguishers for Round-Reduced `FRIET`, `Xoodoo`, and `Alzette`

Yunwen Liu[1,2,3], Siwei Sun[2,3](✉), and Chao Li[1]

[1] College of Liberal Arts and Science, National University of Defense Technology, Changsha, China
[2] State Key Laboratory of Information Security, Institute of Information Engineering, Chinese Academy of Sciences, Beijing, China
[3] University of Chinese Academy of Sciences, Beijing, China

Abstract. The differential-linear attack, combining the power of the two most effective techniques for symmetric-key cryptanalysis, was proposed by Langford and Hellman at CRYPTO 1994. From the exact formula for evaluating the bias of a differential-linear distinguisher (JoC 2017), to the differential-linear connectivity table (DLCT) technique for dealing with the dependencies in the switch between the differential and linear parts (EUROCRYPT 2019), and to the improvements in the context of cryptanalysis of ARX primitives (CRYPTO 2020), we have seen significant development of the differential-linear attack during the last four years. In this work, we further extend this framework by replacing the differential part of the attack by rotational-xor differentials. Along the way, we establish the theoretical link between the rotational-xor differential and linear approximations, revealing that it is nontrivial to directly apply the closed formula for the bias of ordinary differential-linear attack to rotational differential-linear cryptanalysis. We then revisit the rotational cryptanalysis from the perspective of differential-linear cryptanalysis and generalize Morawiecki et al.'s technique for analyzing `Keccak`, which leads to a practical method for estimating the bias of a (rotational) differential-linear distinguisher in the special case where the output linear mask is a unit vector. Finally, we apply the rotational differential-linear technique to the permutations involved in `FRIET`, `Xoodoo`, `Alzette`, and `SipHash`. This gives significant improvements over existing cryptanalytic results, or offers explanations for previous experimental distinguishers without a theoretical foundation. To confirm the validity of our analysis, all distinguishers with practical complexities are verified experimentally.

Keywords: Differential-linear cryptanalysis · Rotational cryptanalysis · ARX · `FRIET` · `Xoodoo` · `Alzette` · `SipHash`

A. Canteaut and F.-X. Standaert (Eds.): EUROCRYPT 2021, LNCS 12696, pp. 741–770, 2021.
https://doi.org/10.1007/978-3-030-77870-5_26

1 Introduction

The practical security of a symmetric-key primitive is determined by evaluating its resistance against an almost exhaustive list of known cryptanalytic techniques. Therefore, it is of essential importance to generalize existing cryptanalytic methods or develop new techniques. Sometimes the boundary between the two can be quite blurred. For example, the development of the invariant attacks [23,24,35], ploytopic cryptanalysis [33], division properties [34,36], rotational cryptanalysis [1,17], etc. in recent years belongs to these two approaches.

Another approach is to employ known techniques in combination to enhance the effectiveness of the individual attacks. The boomerang [37] and differential-linear cryptanalysis are the best examples. In particular, during the past four years, we have seen significant advancements in the development of the differential-linear cryptanalysis introduced by Langford and Hellman at CRYPTO 1994 [22], which combines the power of the two most important techniques (differential and linear attacks) for symmetric-key cryptanalysis. Our work starts with an attempt to further extend the differential-linear framework by replacing the differential part of this cryptanalytic technique with rotational-xor differentials.

Rotational and Rotational-xor Cryptanalysis. Rotational cryptanalysis was first formally introduced in [17] by Khovratovich and Nikolic, where the evolution of the so-called rotational pair $(x, x \lll t)$ through a target cipher was analyzed. The rotational properties of the building blocks of ARX primitives were then applied to the rotational rebound attack on the hash function Skein [19], and later were refined to consider a chain of modular additions [18]. Recently, cryptanalytic results of ARX-based permutations Chaskey and Chacha with respect to rotational cryptanalysis were reported [5,21]. Apart from the ARX constructions, permutations built with logical operations without modular additions, also known as AND-RX or LRX [3] primitives, are particularly interesting with respect to rotational attacks. In 2010, Morawiecki *et al.* applied this technique to distinguish the round-reduced `Keccak`-$f[1600]$ permutation by feeding in rotational pairs and observing the bias of the XOR of the $(i+t)$-th and i-th bits of the corresponding outputs, where t is the rotation offset and the addition should be taken modulo the size of the rotated word [31]. We will come back to Morawiecki *et al.*'s technique and show that it has an intimate relationship with the so-called rotational differential-linear cryptanalysis we proposed in Sect. 3. To thwart rotational attacks, constants which are not rotation-invariant can be injected into the data path. Still, in certain cases, it is possible to overcome this countermeasure with some ad-hoc techniques.

Later, Ashur and Liu [1] generalized the concept of rotational pair by considering the propagation of a data pair (x, x') that is related by the so-called rotational-xor (RX) difference $(x \lll t) \oplus x' = \delta$. The cryptanalytic technique based on RX-difference was named as rotational-xor cryptanalysis. Note that when the RX-difference of the pair (x, x') is zero, it degenerates to a rotational pair. RX cryptanalysis integrates the effect of constants into the analysis and it has been successfully applied to many ARX or AND-RX designs [26,28]. Hereafter, we refer both rotational and rotational-xor cryptanalysis as rotational cryptanalysis, or in a general sense, rotational cryptanalysis contains all the statistical attacks requiring chosen data (e.g., plaintexts) with certain rotational relationships.

Differential-linear Cryptanalysis. Given an encryption function E, we divide it into two consecutive subparts E_0 and E_1. Let $\delta \to \Delta$ be a differential for E_0 with probability p, and $\Gamma \to \gamma$ be a linear approximation for E_1 with bias $\epsilon_{\Gamma,\gamma} = \Pr[\Gamma \cdot y \oplus \gamma \cdot E_1(y) = 0] - \frac{1}{2}$. Then, the overall bias $\mathcal{E}_{\delta,\gamma}$ of the differential-linear distinguisher can be estimated with the piling-up lemma [30] as

$$\mathcal{E}_{\delta,\gamma} = \Pr[\gamma \cdot (E(x) \oplus E(x \oplus \delta)) = 0] - \frac{1}{2} = (-1)^{\Gamma \cdot \Delta} \cdot 2p\epsilon_{\Gamma,\gamma}^2, \tag{1}$$

since $\gamma \cdot (E(x) \oplus E(x \oplus \delta))$ can be decomposed into the XOR sum of the following three terms

$$\begin{cases} \Gamma \cdot (E_0(x) \oplus E_0(x \oplus \delta)), \\ \Gamma \cdot E_0(x \oplus \delta) \oplus \gamma \cdot E(x \oplus \delta), \\ \Gamma \cdot E_0(x) \oplus \gamma \cdot E(x). \end{cases}$$

The derivation of Eq. (1) not only relies on the independence of E_0 and E_1, but also the assumption

$$\Pr[\Gamma \cdot (E_0(x) \oplus E_0(x \oplus \delta)) = 0 \mid E_0(x) \oplus E_0(x \oplus \delta) \neq \Delta] = \frac{1}{2}, \tag{2}$$

under which we have $\Pr[\Gamma \cdot (E_0(x) \oplus E_0(x \oplus \delta)) = 0] = \frac{1}{2} + \frac{(-1)^{\Gamma \cdot \Delta}}{2}p$.

However, it has long been observed that Eq. (2) may fail in many cases and multiple linear approximations have to be taken into account to make the estimates more accurate [22,27,29]. In [9], Blondeau, Leander, and Nyberg presented a closed formula for the overall bias $\mathcal{E}_{\delta,\gamma}$ based on the link between differential and linear attacks [12] under the sole assumption that E_0 and E_1 are independent. However, this closed formula is generally not applicable in practice even if E_0 and E_1 are independent, since it requires the computation of the exact bias $\epsilon_{\delta,v} = \Pr[v \cdot (E_0(x) \oplus E_0(x \oplus \delta)) = 0] - \frac{1}{2}$ for all v. [1] Moreover, in some cases the dependency between E_0 and E_1 can be significant. Inspired by the boomerang-connectivity table (BCT) and its successful applications in the context of boomerang attacks [13], Bar-On, Dunkelman, Keller, and Weizman introduced the differential-linear connectivity table (DLCT) [4], where the target cipher is decomposed as $E = E_1 \circ E_m \circ E_0$ and the actual differential-linear probability of the middle part E_m is determined by experiments, fully addressing the issue of dependency in the switch between E_0 and E_1 (The effect of multiple

[1] Unlike the estimation of the probability of a differential with a large number of characteristics, a partial evaluation of the differential-linear distinguisher without the full enumeration of intermediate masks can be inaccurate, since both positive and negative biases occur.

characteristics and approximations still has to be handled by the framework of Blondeau *et al.* [9]). Most recently, Beierle, Leander, and Todo presented several improvements to the framework of differential-linear attacks with a special focus on ARX ciphers at CRYPTO 2020 [7].

Our Contribution. We start from the natural idea to extend the framework of differential-linear attacks by replacing the differential part with rotational-xor differentials. Specifically, given a pair of data with RX-difference $\delta = (x \lll t) \oplus x'$ and a linear mask γ, a *rotational differential-linear* distinguisher of a cipher E exploits the bias of $\gamma \cdot (\texttt{rot}(E(x)) \oplus E(\texttt{rot}(x) \oplus \delta))$, where $\texttt{rot}(\cdot)$ is some rotation-like operation.

We then present an informal formula similar to Eq. (1) to estimate the bias of a rotational differential-linear distinguisher by the probability of the rotational-xor differential covering E_0 and the biases of the linear approximation and its rotated version covering E_1, where $E = E_1 \circ E_0$. This formula, as in the case of ordinary differential-linear cryptanalysis, requires certain assumptions that may not hold in practice.

Consequently, we try to derive a closed formula for computing the bias of a rotational differential-linear distinguisher, which we expect to be analogous to Blondeau *et al.*'s result [9]. Although we failed to achieve this goal, we manage to establish a general link between the rotational-xor cryptanalysis and linear cryptanalysis as a by-product of this failed endeavour. From a practical point of view, we do not lose much due to the absence of a closed formula, since this kind of formula will inevitably involve the correlations of exponentially many trails which are hard to evaluate in most situations.

Then, we focus our attention on the special case of rotational differential-linear cryptanalysis where the output linear mask γ is a unit vector. In this case, the bias $\Pr[e_i \cdot (\texttt{rot}(f(x)) \oplus f(\texttt{rot}(x) \oplus \delta)) = 0] - \frac{1}{2}$ is

$$\Pr[(E(x))_j \oplus (E(x'))_i = 0] - \frac{1}{2} = \frac{1}{2} - \Pr[(E(x))_j \neq (E(x'))_i], \tag{3}$$

for some i and j, where $x' = \texttt{rot}(x) \oplus \delta$. With this formulation, we immediately realize that Morawiecki *et al.*'s approach [31] gives rise to an efficient method for evaluating the biases of rotational differential-linear distinguishers, as well as ordinary differential-linear distinguishers whose output linear masks are unit vectors. We generalize some results from Morawiecki *et al.*'s work and arrive at formulas which are able to predict $\Pr[(f(x))_j \neq f(x')_i]$ based on the information $\Pr[x_j \neq x_i]$ for many common operations f appearing in ARX designs. In particular, we give the explicit formula for computing the differential-linear and rotational differential-linear probability for an n-bit modular addition with $O(n)$ operations, while a direct application of Bar-On *et al.*'s approach [4] based on the Fast Fourier Transformation (FFT) by treating the modular addition as an $2n \times n$ S-box would require a complexity of $\mathcal{O}(2^{2n})$. The probability evaluation can be iteratively applied for an ARX or AND-RX construction. Nevertheless, we note that the accuracy of the probability evaluation is affected by the dependency among the neighbour bits.

Finally, we apply the technique of rotational differential-linear cryptanalysis to the cryptographic permutations involved in FRIET, Xoodoo and Alzette. For FRIET, we find a 6-round rotational differential-linear distinguisher with a correlation $2^{-5.81}$, and it can be extended to a practical 8-round rotational differential-linear distinguisher with a correlation of $2^{-17.81}$. As a comparison, the correlation of the best known 8-round linear trail of FRIET is 2^{-40}. Moreover, our 6-round distinguisher for FRIET can be further extended to a 13-round one. For Xoodoo, we identify a 4-round rotational differential-linear distinguisher with a correlation 1, while previous best result for Xoodoo is a 3-round differential with a probability 2^{-36}. For Alzette, the 64-bit ARX-box, we find a 4-round differential-linear distinguisher with a correlation $2^{-0.27}$ and a 4-round rotational differential-linear distinguisher with a correlation $2^{-11.37}$. A summary of the results is shown in Table 1, where all distinguishers with practical complexities are experimentally verified.

Table 1. A summary of the results. R-DL = rotational differential-linear, DL = differential-linear, LC = linear characteristic, DC = differential characteristic. We show differentials with probabilities and LC/DL/R-DL with correlations.

Permutation	Type	# Round	Probability/Correlation		Ref.
			Theoretical	Experimental	
FRIET	R-DL	6	$2^{-5.81}$	$2^{-5.12}$	Sect. 5
	R-DL	7	$2^{-9.81}$	$2^{-9.12}$	Sect. 5
	LC	7	2^{-29}	–	[32]
	R-DL	8	$2^{-17.81}$	$2^{-17.2}$	Sect. 5
	LC	8	2^{-40}	–	[32]
	R-DL	13	$2^{-117.81}$	–	Sect. 5
Xoodoo	DC	3	2^{-36}	–	[14]
	R-DL	4	1	1	Sect. 5
Alzette	DC	4	2^{-6}	–	[6]
	R-DL	4	$2^{-11.37}$	$2^{-7.35}$	Sect. 6
	DL	4	$2^{-0.27}$	$2^{-0.1}$	Sect. 6

Outline. Section 2 introduces the notations and preliminaries for rotational-xor and linear cryptanalysis. We propose the rotational differential-linear cryptanalysis and establish the theoretical link between the rotational-xor cryptanalysis and linear cryptanalysis in Sect. 3. This is followed by Sect. 4 where we explore the methods for evaluating the biases of rotational differential-linear distinguishers. In Sect. 5 and Sect. 6, we apply the techniques developed in previous sections to AND-RX and ARX primitives. Section 7 concludes the paper with some open problems.

2 Notations and Preliminaries

Let $\mathbb{F}_2 = \{0, 1\}$ be the field with two elements. We denote by x_i the i-th bit of a bit string $x \in \mathbb{F}_2^n$. For a vectorial Boolean function $F : \mathbb{F}_2^n \to \mathbb{F}_2^m$ with $y = F(x) \in \mathbb{F}_2^m$, its i-th output bit y_i is denoted by $(F(x))_i$. For an n-bit string x, we use the indexing scheme $x = (x_{n-1}, \cdots, x_1, x_0)$. In addition, concrete values in $\mathbb{F}_2^n$ are specified in hexadecimal notations. For example, we use `1111` to denote the binary string $(\texttt{0001 0001 0001 0001})_2$.

The XOR-difference and rotational-xor difference with offset t of two bit strings x and x' in $\mathbb{F}_2^n$ are defined as $x \oplus x'$ and $(x \lll t) \oplus x'$, respectively. For the rotational-xor difference $\delta = (x \lll t) \oplus x'$, we may omit the rotation offset and write $\delta = \overleftarrow{x} \oplus x'$ or $\delta = \texttt{rot}(x) \oplus x'$ to make the notation more compact when it is clear from the context. Moreover, by abusing the notation, $\overleftarrow{x}$ and $\texttt{rot}(x)$ may rotate the entire string x or rotate the substrings of x to the left separately with a common offset, depending on the context. For instance, in the analysis of Keccak-f, we rotate each lane of the state by certain amount [31]. Correspondingly, $\overrightarrow{x}$ and $\texttt{rot}^{-1}(x)$ rotate x or its substrings to the right. Similar to differential cryptanalysis with XOR-difference, we can define the probability of an RX-differential as follows.

Definition 1 (RX-differential probability). *Let $f : \mathbb{F}_2^n \to \mathbb{F}_2^n$ be a vectorial boolean function. Let α and β be n-bit words. Then, the RX-differential probability of the RX-differential $\alpha \to \beta$ for f is defined as*

$$\Pr[\alpha \to \beta] = 2^{-n}\#\{x \in \mathbb{F}_2^n : \texttt{rot}(f(x)) \oplus f(\texttt{rot}(x) \oplus \alpha) = \beta\}$$

Finally, the definitions of correlation, bias, and some lemmas concerning Boolean functions together with the piling-up lemma are needed.

Definition 2 ([10,11]). *The correlation of a Boolean function $f : \mathbb{F}_2^n \to \mathbb{F}_2$ is defined as* $\mathrm{cor}(f) = 2^{-n}(\#\{x \in \mathbb{F}_2^n : f(x) = 0\} - \#\{x \in \mathbb{F}_2^n : f(x) = 1\})$.

Definition 3 ([10,11]).] *The bias $\epsilon(f)$ of a Boolean function $f : \mathbb{F}_2^n \to \mathbb{F}_2$ is defined as $2^{-n}\#\{x \in \mathbb{F}_2^n : f(x) = 0\} - \frac{1}{2}$.*

From Definition 2 and Definition 3 we can see that $\mathrm{cor}(f) = 2\epsilon(f)$.

Definition 4. *Let $f : \mathbb{F}_2^n \to \mathbb{F}_2$ be a Boolean function. The Walsh-Hadamard transformation takes in f and produces a real-valued function $\hat{f} : \mathbb{F}_2^n \to \mathbb{R}$ such that*

$$\forall w \in \mathbb{F}_2^n, \quad \hat{f}(w) = \sum_{x \in \mathbb{F}_2^n} f(x)(-1)^{x \cdot w}.$$

Definition 5. *Let $f : \mathbb{F}_2^n \to \mathbb{F}_2$ and $g : \mathbb{F}_2^n \to \mathbb{F}_2$ be two Boolean functions. The convolutional product of f and g is a Boolean function defined as*

$$\forall y \in \mathbb{F}_2^n, \quad (f \star g)(y) = \sum_{x \in \mathbb{F}_2^n} g(x)f(x \oplus y).$$

Lemma 1 (**[11], Corollary 2**). *Let $\hat{f}$ be the Walsh-Hadamard transformation of f. Then the Walsh-Hadamard transformation of $\hat{f}$ is $2^n f$.*

Lemma 2 [11], **Proposition 6**). $\widehat{(f \star g)}(z) = \hat{f}(z)\hat{g}(z)$ *and thus* $\widehat{(f \star f)} = (\hat{f})^2$.

Lemma 3 (Piling-up Lemma [30]**).** *Let $Z_0, \cdots, Z_{m-1}$ be m independent binary random variables with $\Pr[Z_i = 0] = p_i$. Then we have that*

$$\Pr[Z_0 \oplus \cdots \oplus Z_{m-1} = 0] = \frac{1}{2} + 2^{m-1} \prod_{i=0}^{m-1} (p_i - \frac{1}{2}),$$

or alternatively, $2\Pr[Z_0 \oplus \cdots \oplus Z_{m-1} = 0] - 1 = \prod_{i=0}^{m-1}(2p_i - 1)$.

3 Rotational Differential-Linear Cryptanalysis

A natural extension of the differential-linear cryptanalysis is to replace the differential part of the attack by rotational-xor (RX) differentials. Let $E = E_1 \circ E_0$ be an encryption function. Assume that we have an RX-differential $\delta \to \Delta$ covering E_0 with $\Pr[\mathtt{rot}(E_0(x)) \oplus E_0(\mathtt{rot}(x) \oplus \delta) = \Delta] = p$ and a linear approximation $\Gamma \to \gamma$ of E_1 such that

$$\begin{cases} \epsilon_{\Gamma,\gamma} = \Pr[\Gamma \cdot y \oplus \gamma \cdot E_1(y) = 0] - \frac{1}{2}, \\ \epsilon_{\mathtt{rot}^{-1}(\Gamma),\mathtt{rot}^{-1}(\gamma)} = \Pr[\mathtt{rot}^{-1}(\Gamma) \cdot y \oplus \mathtt{rot}^{-1}(\gamma) \cdot E_1(y) = 0] - \frac{1}{2}. \end{cases}$$

Let $x' = \mathtt{rot}(x) \oplus \delta$. If the assumption

$$\Pr[\Gamma \cdot (\mathtt{rot}(E_0(x)) \oplus E_0(x')) = 0 \mid \mathtt{rot}(E_0(x)) \oplus E_0(x') \neq \Delta] = \frac{1}{2} \tag{4}$$

holds. We have

$$\Pr[\Gamma \cdot (\mathtt{rot}(E_0(x)) \oplus E_0(x')) = 0] = \frac{1}{2} + \frac{(-1)^{\Gamma \cdot \Delta}}{2} p.$$

Since

$$\begin{aligned} \gamma \cdot (\mathtt{rot}(E(x)) \oplus E(x')) &= \gamma \cdot \mathtt{rot}(E(x)) \oplus \Gamma \cdot \mathtt{rot}(E_0(x)) \\ &\oplus \Gamma \cdot (\mathtt{rot}(E_0(x)) \oplus E_0(x')) \\ &\oplus \Gamma \cdot E_0(x') \oplus \gamma \cdot E(x') \\ &= \mathtt{rot}(\mathtt{rot}^{-1}(\gamma) \cdot E(x) \oplus \mathtt{rot}^{-1}(\Gamma) \cdot E_0(x)) \\ &\oplus \Gamma \cdot (\mathtt{rot}(E_0(x)) \oplus E_0(x')) \\ &\oplus \Gamma \cdot E_0(x') \oplus \gamma \cdot E(x'), \end{aligned}$$

the bias of the rotational differential-linear distinguisher can be estimated by piling-up lemma as

$$\mathcal{E}^{\text{R-DL}}_{\delta,\gamma} = \Pr[\gamma \cdot (\overleftarrow{E}(x) \oplus E(x')) = 0] - \frac{1}{2} = (-1)^{\Gamma \cdot \Delta} \cdot 2p\epsilon_{\Gamma,\gamma}\epsilon_{\mathtt{rot}^{-1}(\Gamma),\mathtt{rot}^{-1}(\gamma)}, \tag{5}$$

and the corresponding correlation of the distinguisher is

$$\mathcal{C}_{\delta,\gamma}^{\text{R-DL}} = 2\mathcal{E}_{\delta,\gamma}^{\text{R-DL}} = (-1)^{\Gamma\cdot\Delta} \cdot 4p\epsilon_{\Gamma,\gamma}\epsilon_{\texttt{rot}^{-1}(\Gamma),\texttt{rot}^{-1}(\gamma)}. \tag{6}$$

We can distinguish E from random permutations if the absolute value of $\mathcal{E}_{\delta,\gamma}^{\text{R-DL}}$ or $\mathcal{C}_{\delta,\gamma}^{\text{R-DL}}$ is sufficiently high. Note that if we set the rotation offset to zero, the rotational differential-linear attack is exactly the ordinary differential-linear cryptanalysis. Therefore, the rotational differential-linear attack is a strict generalization of the ordinary differential-linear cryptanalysis.

A rotational differential-linear distinguisher can be extended by appending linear approximations at the end. Given a rotational differential-linear distinguisher of a function f with a bias

$$\epsilon_{\delta,\gamma} = \Pr[\gamma \cdot (\texttt{rot}(f(x)) \oplus f(\texttt{rot}(x) \oplus \delta)) = 0] - \frac{1}{2},$$

and a linear approximation (γ, μ) over a function g with

$$\begin{cases} \epsilon_{\gamma,\mu} = \Pr[\gamma \cdot x \oplus \mu \cdot g(x) = 0] - \frac{1}{2}, \\ \epsilon_{\texttt{rot}^{-1}(\gamma),\texttt{rot}^{-1}(\mu)} = \Pr[\texttt{rot}^{-1}(\gamma) \cdot x \oplus \texttt{rot}^{-1}(\mu) \cdot g(x) = 0] - \frac{1}{2}, \end{cases}$$

we can compute the bias of the rotational differential-linear distinguisher of $h = g \circ f$ with input RX-difference δ and output linear mask μ by the piling-up lemma. Since

$$\begin{aligned} \mu \cdot (\texttt{rot}(h(x)) \oplus h(\texttt{rot}(x) \oplus \delta)) = {} & \gamma \cdot (\texttt{rot}(f(x)) \oplus f(\texttt{rot}(x) \oplus \delta)) \\ & \oplus \gamma \cdot \texttt{rot}(f(x)) \oplus \mu \cdot \texttt{rot}(h(x)) \\ & \oplus \gamma \cdot f(\texttt{rot}(x) \oplus \delta) \oplus \mu \cdot h(\texttt{rot}(x) \oplus \delta) \end{aligned},$$

the bias of the rotational differential-linear distinguisher can be estimated as

$$\Pr[\mu \cdot (\texttt{rot}(h(x)) \oplus h(\texttt{rot}(x) \oplus \delta)) = 0] - \frac{1}{2} = 4\epsilon_{\delta,\gamma}\epsilon_{\gamma,\mu}\epsilon_{\texttt{rot}^{-1}(\gamma),\texttt{rot}^{-1}(\mu)}. \tag{7}$$

However, as in ordinary differential-linear attacks, the assumption described by Eq. (4) may not hold in practice, and we prefer a closed formula for the bias $\mathcal{E}_{\delta,\gamma}^{\text{R-DL}}$ without this assumption for much the same reasons leading to Blondeau *et al.*'s work [9]. Also, we would like to emphasize that if Eq. (5) and (7) are used to estimate the bias, we should verify the results experimentally whenever possible.

3.1 Towards a Closed Formula for the Bias of the Rotational Differential-Linear Distinguisher

In [9], Blondeau *et al.* proved the following theorem based on the general link between differential and linear cryptanalysis [12].

Theorem 1 ([9]). *If E_0 and E_1 are independent, the bias of a differential-linear distinguisher with input difference δ and output linear mask γ can be computed as*

$$\mathcal{E}_{\delta,\gamma} = \sum_{v \in \mathbb{F}_2^n} \epsilon_{\delta,v} c_{v,\gamma}^2, \tag{8}$$

for all $\delta \neq 0$ and $\gamma \neq 0$, where

$$\begin{cases} \epsilon_{\delta,v} = \Pr[v \cdot (E_0(x) \oplus E_0(x \oplus \delta)) = 0] - \frac{1}{2} \\ c_{v,\gamma} = \mathrm{cor}(v \cdot y \oplus \gamma \cdot E_1(y)) \end{cases}.$$

To replay Blondeau *et al.*'s technique in an attempt to derive the rotational differential-linear counterpart of Eq. (8), we have to first establish the relationship between rotational differential-linear cryptanalysis and linear cryptanalysis.

Link Between RX-cryptanalysis and Linear Cryptanalysis. Let $F : \mathbb{F}_2^n \to \mathbb{F}_2^n$ be a vectorial Boolean function. The cardinality of the set

$$\{x \in \mathbb{F}_2^n : \overleftarrow{F}(x) \oplus F(\overleftarrow{x} \oplus a) = b\}$$

is denoted by $\xi_F(a, b)$, and the correlation of $u \cdot x \oplus v \cdot F(x)$ is $\mathrm{cor}(u \cdot x \oplus v \cdot F(x))$. Let $\overleftarrow{\underrightarrow{F}} : \mathbb{F}_2^n \to \mathbb{F}_2^n$ be the vectorial Boolean function mapping x to $\overleftarrow{F}(\overrightarrow{x})$. It is easy to show that

$$\mathrm{cor}(u \cdot x \oplus v \cdot \overleftarrow{\underrightarrow{F}}(x)) = \mathrm{cor}(\overrightarrow{u} \cdot x \oplus \overrightarrow{v} \cdot F(x)).$$

In what follows, we are going to establish the relationship between

$$\xi_F(a, b), \quad \mathrm{cor}(u \cdot x \oplus v \cdot F(x)), \quad \text{and} \quad \mathrm{cor}(\overrightarrow{u} \cdot x \oplus \overrightarrow{v} \cdot F(x)).$$

Definition 6. *Given a vectorial Boolean function $F : \mathbb{F}_2^n \to \mathbb{F}_2^n$, the Boolean function $\theta_F : \mathbb{F}_2^{2n} \to \mathbb{F}_2$ is defined as*

$$\theta_F(x, y) = \begin{cases} 1 & \text{if} \quad y = F(x), \\ 0 & \text{otherwise.} \end{cases} \tag{9}$$

Lemma 4. *Let $F : \mathbb{F}_2^n \to \mathbb{F}_2^n$ be a vectorial Boolean function. Then for any $(a, b) \in \mathbb{F}_2^{2n}$, we have $\xi_F(a, b) = (\theta_{\overleftarrow{\underrightarrow{F}}} \star \theta_F)(a, b)$.*

Proof. According to Definition 5, we have

$$\begin{aligned}
(\theta_{\overleftarrow{\underrightarrow{F}}} \star \theta_F)(a,b) &= \sum_{x||y \in \mathbb{F}_2^{2n}} \theta_{\overleftarrow{\underrightarrow{F}}}(x,y)\theta_F(a \oplus x, b \oplus y) \\
&= \sum_{x \in \mathbb{F}_2^n} \sum_{y \in \mathbb{F}_2^n} \theta_{\overleftarrow{\underrightarrow{F}}}(x,y)\theta_F(a \oplus x, b \oplus y) \\
&= \sum_{x \in \mathbb{F}_2^n} \theta_{\overleftarrow{\underrightarrow{F}}}(x, \overleftarrow{\underrightarrow{F}}(x))\theta_F(a \oplus x, b \oplus \overleftarrow{\underrightarrow{F}}(x)) \\
&= \sum_{x \in \mathbb{F}_2^n} \theta_F(a \oplus x, b \oplus \overleftarrow{\underrightarrow{F}}(x)) \\
&= \#\{x \in \mathbb{F}_2^n : b \oplus \overleftarrow{\underrightarrow{F}}(x) = F(a \oplus x)\} \\
&= \xi_F(a,b).
\end{aligned}$$

□

Lemma 5. *Let $F : \mathbb{F}_2^n \to \mathbb{F}_2^n$ be a vectorial Boolean function. Then for any $(a,b) \in \mathbb{F}_2^{2n}$, we have $\mathrm{cor}(a \cdot x \oplus b \cdot F(x)) = 2^{-n}\hat{\theta}_F(a,b)$.*

Proof. According to Definition 4, we have

$$\begin{aligned}
\hat{\theta}_F(a,b) &= \sum_{x||y \in \mathbb{F}_2^{2n}} \theta_F(x,y)(-1)^{(x||y)\cdot(a||b)} \\
&= \sum_{x \in \mathbb{F}_2^n} \sum_{y \in \mathbb{F}_2^n} \theta_F(x,y)(-1)^{a\cdot x \oplus b \cdot y} \\
&= \sum_{x \in \mathbb{F}_2^n} (-1)^{a \cdot x \oplus b \cdot F(x)} \\
&= 2^n \mathrm{cor}(a \cdot x \oplus b \cdot F(x)).
\end{aligned}$$

□

In addition, applying Lemma 5 to $\overleftarrow{\underrightarrow{F}}$ gives $\mathrm{cor}(a \cdot x \oplus b \cdot \overleftarrow{\underrightarrow{F}}(x)) = \frac{1}{2^n}\hat{\theta}_{\overleftarrow{\underrightarrow{F}}}(a,b)$.

Theorem 2. *The link between RX-differentials and linear approximations can be summarized as*

$$\xi_F(a,b) = \sum_{u \in \mathbb{F}_2^n} \sum_{v \in \mathbb{F}_2^n} (-1)^{u \cdot a \oplus v \cdot b} \mathrm{cor}(\overrightarrow{u} \cdot x \oplus \overrightarrow{v} \cdot F(x)) \mathrm{cor}(u \cdot x \oplus v \cdot F(x)). \quad (10)$$

Proof. According to Lemma 4 and Lemma 2, we have

$$2^{2n}\xi_F(a,b) = (\widehat{\theta_{\overleftarrow{\underrightarrow{F}}} \star \theta_F})(a,b) = \hat{\theta}_{\overleftarrow{\underrightarrow{F}}}\hat{\theta}_F(a,b).$$

Since $\hat{\theta}_{\overleftarrow{\underrightarrow{F}}}\hat{\theta}_F = 2^{2n}\mathrm{cor}(u \cdot x \oplus v \cdot \overleftarrow{\underrightarrow{F}}(x))\mathrm{cor}(u \cdot x \oplus v \cdot F(x))$ due to Lemma 5,

$$
\begin{aligned}
\widehat{\hat{\theta}_{\overleftarrow{\underrightarrow{F}}}\hat{\theta}_F}(a,b) &= 2^{2n} \sum_{u||v \in \mathbb{F}_2^{2n}} (-1)^{(u||v)\cdot(a||b)} \mathrm{cor}(u \cdot x \oplus v \cdot \overleftarrow{\underrightarrow{F}}(x))\mathrm{cor}(u \cdot x \oplus v \cdot F(x)) \\
&= 2^{2n} \sum_{u,v \in \mathbb{F}_2^{n}} (-1)^{u\cdot a \oplus v \cdot b} \mathrm{cor}(u \cdot x \oplus v \cdot \overleftarrow{\underrightarrow{F}}(x))\mathrm{cor}(u \cdot x \oplus v \cdot F(x)) \\
&= 2^{2n} \sum_{u,v \in \mathbb{F}_2^{n}} (-1)^{u\cdot a \oplus v \cdot b} \mathrm{cor}(\overrightarrow{u} \cdot x \oplus \overrightarrow{v} \cdot F(x))\mathrm{cor}(u \cdot x \oplus v \cdot F(x))
\end{aligned}
$$

□

If the function F is rotation invariant, i.e., $\overleftarrow{F(x)} = F(\overleftarrow{x})$, then we have $\mathrm{cor}(\overrightarrow{u} \cdot x \oplus \overrightarrow{v} \cdot F(x)) = \mathrm{cor}(u \cdot x \oplus v \cdot F(x))$. As a result, the theoretical link between rotational-xor and linear cryptanalysis degenerates to the link between ordinary differential cryptanalysis and linear cryptanalysis. Moreover, based on the link between differential and linear cryptanalysis, Blondeau *et al.* derive a closed formula for the bias of an ordinary differential-linear distinguisher as shown in Eq. (8). We try to mimic Blondeau *et al.*'s approach to obtain a closed formula for the biases of rotational differential-linear distinguishers. However, we failed in this attempt due to a fundamental difference between rotational-xor differentials and ordinary differentials: the output RX-difference is not necessarily zero when the input RX-difference $\mathtt{rot}(x) \oplus x'$ is zero. We leave it as an open problem to derive a closed formula for the bias of a rotational differential-linear distinguisher. From a practical point of view, we do not lose much due to the absence of a closed formula since this kind of formula will inevitably involve the correlations of exponentially many trails which are hard to evaluate in most situations.

3.2 Morawiecki *et al.*'s Technique Revisited

In [31], Morawiecki *et al.* performed a rotational cryptanalysis on the Keccak-f permutation E. In this attack, the probability of

$$\Pr[(E(x))_{i-t} \neq (E(x \lll t))_i]$$

was exploited to distinguish the target. In what follows, we show that Morawiecki *et al.*'s technique can be regarded as a special case of the rotational differential-linear framework.

Eventually, what we exploit in a rotational differential-linear attack associated with an input RX-difference $\delta \in \mathbb{F}_2^n$ and an output linear mask $\gamma \in \mathbb{F}_2^n$ is the abnormally high absolute bias or correlation of the Boolean function

$$\gamma \cdot (\mathtt{rot}(E(x)) \oplus E(\mathtt{rot}(x) \oplus \delta)).$$

Following the notation of [9], let $\mathrm{sp}(\gamma) \subseteq \mathbb{F}_2^n$ be the linear space spanned by γ, and $\mathrm{sp}(\gamma)^{\perp} = \{u \in \mathbb{F}_2^n : \forall v \in \mathrm{sp}(\gamma), u \cdot v = 0\}$ be the orthogonal space of $\mathrm{sp}(\gamma)$.

We then define two sets $\mathbb{D}_0$ and $\mathbb{D}_1$ which form a partition of $\mathbb{F}_2^n$:

$$\begin{cases} \mathbb{D}_0 = \{x \in \mathbb{F}_2^n : \texttt{rot}(E(x)) \oplus E(\texttt{rot}(x) \oplus \delta) \in \mathrm{sp}(\gamma)^\perp\} \\ \mathbb{D}_1 = \{x \in \mathbb{F}_2^n : \texttt{rot}(E(x)) \oplus E(\texttt{rot}(x) \oplus \delta) \in \mathbb{F}_2^n - \mathrm{sp}(\gamma)^\perp\} \end{cases}.$$

Under the above notations, for any $x \in \mathbb{D}_0$, $\gamma \cdot (\texttt{rot}(E(x)) \oplus E(\texttt{rot}(x) \oplus \delta)) = 0$ and for any $x \in \mathbb{D}_1$, $\gamma \cdot (\texttt{rot}(E(x)) \oplus E(\texttt{rot}(x) \oplus \delta)) = 1$.

Thus, the higher the absolute value of

$$\#\mathbb{D}_0 - \#\mathbb{D}_1 = 2^n \mathrm{cor}(\gamma \cdot (\texttt{rot}(E(x)) \oplus E(\texttt{rot}(x) \oplus \delta))),$$

the more effective the attack is.

If $\gamma = e_i$ is the i-th unit vector, we have $\mathrm{sp}(\gamma) = \{0, e_i\}$ and $\mathrm{sp}(\gamma)^\perp$ contains all vectors whose i-th bit is 0. In this case,

$$\begin{aligned} \#\mathbb{D}_0 - \#\mathbb{D}_1 &= 2^n - 2\#\mathbb{D}_1 \\ &= 2^n - 2^{n+1} (\Pr[e_i \cdot (\texttt{rot}(E(x)) \oplus E(\texttt{rot}(x) \oplus \delta)) = 1]) \\ &= 2^n - 2^{n+1} (\Pr[(E(x))_j \neq (E(\texttt{rot}(x) \oplus \delta))_i]) \\ &= 2^n - 2^{n+1} (\Pr[(E(x))_j \neq (E(x')_i])\,. \end{aligned}$$

Therefore, the effectiveness of the rotational differential-linear attack can be completely characterized by $\Pr[(E(x))_{i-t} \neq (E(x'))_i]$. In the next section, we show how to compute this type of probabilities for the target cipher.

4 Evaluate the Bias of Rotational Differential-Linear Distinguishers

According to the previous section, for a rotational differential-linear distinguisher with an input RX-difference δ and output linear mask e_i, the bias of the distinguisher can be completely determined by

$$\Pr[(E(x))_{i-t} \neq (E(x'))_i], \text{ where } x' = x \lll t \oplus \delta,$$

and we call it the rotational differential-linear probability or R-DL probability. Note that for a random pair $(x, x' = x \lll t \oplus \delta)$ with rotational-xor difference $\delta \in \mathbb{F}_2^n$, we have

$$\Pr[x_{i-t} \neq x'_i] = \frac{1 + (-1)^{1-\delta_i}}{2},$$

for $0 \leq i < n$. Therefore, what we need is a method to evaluate the probability

$$\Pr[(F(x))_{i-t} \neq (F(x'))_i]$$

for $0 \leq i < m - 1$, where $F : \mathbb{F}_2^n \to \mathbb{F}_2^m$ is a vectorial Boolean function that represents a component of E. Then, with certain independence assumptions, we can iteratively determine the probability $\Pr[(E(x))_{i-t} \neq (E(x'))_i]$.

Observation 1 *Let $F : \mathbb{F}_2^n \to \mathbb{F}_2^m$ be a vectorial Boolean function. Assume that the input pair (x, x') satisfies $\Pr[x_{i-t} \neq x'_i] = p_i$ for $0 \leq i < n$, where $x, x' \in \mathbb{F}_2^n$. For $u \in \mathbb{F}_2^n$, we define the set $\mathcal{S}_u = \{(x, x') \in \mathbb{F}_2^n \times \mathbb{F}_2^n : (x \lll t) \oplus x' = u\}$ with $\#\mathcal{S}_u = 2^n$. Let y_i and y'_i be the i-th bit of $F(x)$ and $F(x')$ respectively for $0 \leq i < m$. Then we have*

$$\begin{aligned}\Pr[y_{i-t} \neq y_i] &= \sum_{u \in \mathbb{F}_2^n} \Pr[y_{i-t} \neq y_i | (x, x') \in \mathcal{S}_u] \Pr[(x, x') \in \mathcal{S}_u] \\ &= \sum_{u \in \mathbb{F}_2^n} \Pr[y_{i-t} \neq y_i | (x, x') \in \mathcal{S}_u] \prod_{i=0}^{n-1} ((1 - u_i) - (-1)^{u_i} p_i) \\ &= \frac{1}{2^n} \sum_{u \in \mathbb{F}_2^n} \#\{(x, x') \in \mathcal{S}_u : y_{i-t} \neq y_i\} \prod_{i=0}^{n-1} ((1 - u_i) - (-1)^{u_i} p_i).\end{aligned}$$

The observation is inspired by Morawiecki *et al.*'s work on rotational cryptanalysis [31] where, given a rotational pair, the bias of the output pair being unequal at certain bit is calculated for one-bit AND, NOT and XOR. In the following, we reformulate and generalize their propagation rules in terms of rotational differential-linear probability. Note that all these rules can be derived from Observation 1.

Proposition 1 (AND-rule). *Let a, b, a', and b' be n-bit strings with $\Pr[a_{i-t} \neq a'_i] = p_i$ and $\Pr[b_{i-t} \neq b'_i] = q_i$. Then*

$$\Pr[(a \wedge b)_{i-t} \neq (a' \wedge b')_i] = \frac{1}{2}(p_i + q_i - p_i q_i).$$

Proposition 2 (XOR-rule). *Let a, b, a', and b' be n-bit strings with $\Pr[a_{i-t} \neq a'_i] = p_i$ and $\Pr[b_{i-t} \neq b'_i] = q_i$. Then*

$$\Pr[(a \oplus b)_{i-t} \neq (a' \oplus b')_i] = p_i + q_i - 2p_i q_i.$$

Proposition 3 (NOT-rule). *Let a and b be n-bit strings with $\Pr[a_{i-t} \neq b_i] = p_i$. Then $\Pr[\bar{a}_{i-t} \neq \bar{b}_i] = p_i$.*

Next, we consider constant additions. Let $(x, x') \in \mathbb{F}_2^{2n}$ be a data pair with $\Pr[x_{i-t} \neq x'_i] = p_i$ for some integer t and $c \in \mathbb{F}_2^n$ be a constant. Then $\Pr[(x \oplus c)_{i-t} \neq (x' \oplus c)_i] = \Pr[x_{i-t} \oplus x'_i \neq c_{i-t} \oplus c_i]$. In [31], only the cases where $c_{i-t} \oplus c_i = 1$ or $c_{i-t} = c_i = 0$ are considered. We generalize the rule for constant addition from [31] to the following proposition with all possibilities taken into account.

Proposition 4 (Adjusted C-rule). *Let a and a' be n-bit strings with $\Pr[a_{i-t} \neq a'_i] = p_i$ and $c \in \mathbb{F}_2^n$ be a constant. Then we have*

$$\Pr[(a \oplus c)_{i-t} \neq (a' \oplus c)_i] = \begin{cases} 1 - p_i, & c_{i-t} \oplus c_i = 1 \\ p_i, & c_{i-t} \oplus c_i = 0 \end{cases}$$

4.1 Propagation of R-DL Probabilities in Arithmetic Operations

For functions with AND-RX or LRX construction, such as the permutation Keccak-f, the propagation of the R-DL probability can be evaluated by the propositions previously shown, under the independency assumptions on the neighbouring bits. However, when dependency takes over, even if a function can be expressed as a boolean circuit, a direct applications of the AND, XOR, NOT and adjusted C-rule may lead to errors that accumulated during the iterated evaluation. One such example is the modular addition. In the following, we will derive the propagation rules of the differential-linear (DL) probability and R-DL probability for an n-bit modular addition.

Lemma 6 (carry-rule). *Let $\varsigma : \mathbb{F}_2^3 \to \mathbb{F}_2$ be the carry function*

$$\varsigma(x_0, x_1, x_2) = x_0x_1 \oplus x_1x_2 \oplus x_0x_2.$$

Let a, b, c, a', b', and c' be binary random variables with

$$p_0 = \Pr[a \neq a'], p_1 = \Pr[b \neq b'], p_2 = \Pr[c \neq c'].$$

Then, we have that

$$\Pr[\varsigma(a,b,c) \neq \varsigma(a',b',c')] = p_0p_1p_2 - \frac{p_0p_1 + p_0p_2 + p_1p_2}{2} + \frac{p_0 + p_1 + p_2}{2}.$$

Proof. We prove the carry-rule with Observation 1 by enumerating $u \in \mathbb{F}_2^3$. For $u = (0,0,0)$, $\Pr[\varsigma(a,b,c) \neq \varsigma(a',b',c')|a = a', b = b', c = c'] = 0$. For $u = (0,0,1)$, $\Pr[\varsigma(a,b,c) \neq \varsigma(a',b',c')|a = a', b = b', c \neq c'] = \Pr[a \oplus b = 1] = 1/2$ and $\prod_{i=0}^{2}((1-u_i) + (-1)^{1-u_i}p_i) = (1-p_a)(1-p_b)p_c$.

Similarly, one can derive the expression for all $u \in \mathbb{F}_{2^3}$, and we omit the details.The overall probability of the event $ab \oplus ac \oplus bc \neq a'b' \oplus a'c' \oplus b'c'$ is $p_ap_bp_c - (p_ap_b + p_ap_c + p_bp_c)/2 + (p_a + p_b + p_c)/2$. □

Based on the carry-rule, we can immediately prove the following two theorems on the DL and R-DL probabilities for n-bit modulo additions.

Theorem 3 ($\boxplus$-rule for DL). *Let x, y and x', y' be n-bit string, such that $\Pr[x_i \neq x_i'] = p_i$ and $\Pr[y_i \neq y_i'] = q_i$. Then, the differential-linear probability for modular addition can be computed as*

$$\Pr[(x \boxplus y)_i \neq (x' \boxplus y')_i] = p_i + q_i - 2p_iq_i - 2p_is_i - 2q_is_i + 4p_iq_is_i$$

where $s_0 = 0$ and

$$s_{i+1} = p_iq_is_i - \frac{p_iq_i + p_is_i + q_is_i}{2} + \frac{p_i + q_i + s_i}{2}, i \leq n-1$$

Proof. For inputs x and y, denote the carry by

$$c = (x \boxplus y) \oplus x \oplus y = (c_{n-1}, \cdots, c_1, c_0),$$

where $c_0 = 0, c_{i+1} = x_i y_i \oplus x_i c_i \oplus y_i c_i$. Similarly, for x' and y', denote the carry by $c' = (c'_{n-1}, \cdots, c'_1, c'_0)$. Let s_i denote the probability $\Pr[c_i \neq c'_i]$. Then, $s_0 = 0$ and for $i \geq 1$, the event $c_i \neq c'_i$ is equivalent to

$$x_{i-1}y_{i-1} \oplus x_{i-1}c_{i-1} \oplus y_{i-1}c_{i-1} \neq x'_{i-1}y'_{i-1} \oplus x'_{i-1}c'_{i-1} \oplus y'_{i-1}c'_{i-1}.$$

Therefore, s_i can be computed as

$$p_{i-1}q_{i-1}s_{i-1} - (p_{i-1}q_{i-1} + p_{i-1}q_{i-1} + q_{i-1}s_{i-1})/2 + (p_{i-1} + q_{i-1} + s_{i-1})/2$$

according to Lemma 6. Since $x \boxplus y = x \oplus y \oplus c$, and $x' \boxplus y' = x' \oplus y' \oplus c'$, with the XOR-rule, we have

$$\Pr[(x \boxplus y)_i \neq (x' \boxplus y')_i] = p_i + q_i - 2p_iq_i - 2p_is_i - 2q_is_i + 4p_iq_is_i.$$

□

Example 1. Consider an 8-bit modular addition with input difference being $a = 7$ and $b = 7$. Then, we have for $0 \leq i \leq 7$,

$$p_i = \frac{1 + (-1)^{1-a_i}}{2}, q_i = \frac{1 + (-1)^{1-b_i}}{2},$$

so

$$p_0 = p_1 = p_2 = 1, p_3 = p_4 = p_5 = p_6 = p_7 = 0,$$
$$q_0 = q_1 = q_2 = 1, q_3 = q_4 = q_5 = q_6 = q_7 = 0.$$

The $\boxplus$-rule gives the output DL-probabilities in Table 2. The probabilities predicted in the table are verified by running through the 16-bit input space. In addition, we verified the $\boxplus$-rule in DL with all input differences on an 8-bit modular addition. Under the precision level given in Table 2, the experiments match the theoretical prediction perfectly.

Table 2. The DL-probabilities of an 8-bit modular addition with input differences $a = b = 7$ by theoretical evaluation, which are confirmed by experiments.

i	0	1	2	3	4	5	6	7
p_i	0	2^{-1}	$2^{-0.415037}$	$2^{-0.192645}$	$2^{-1.19265}$	$2^{-2.19265}$	$2^{-3.19265}$	$2^{-4.19265}$

As for the rotational differential-linear cryptanalysis of an n-bit modular addition, a left rotation by t bits is applied to the operands. Firstly, we present the $\boxplus$-rule for RX-difference with a rotation offset $t = 1$.

Theorem 4 ($\boxplus$-rule for RL, $t = 1$). *Given random n-bit strings x, y and x', y' such that $x' = (x \lll 1) \oplus a, y' = (y \lll 1) \oplus b$, where $\Pr[x_{i-1} \neq x'_i] = p_i, \Pr[y_{i-1} \neq y'_i] = q_i$. Then, the rotational differential-linear probability of the modular addition can be computed as*

$$\Pr[(x \boxplus y)_{i-1} \neq (x' \boxplus y')_i] = p_i + q_i - 2p_iq_i - 2p_is_i - 2q_is_i + 4p_iq_is_i,$$

where $s_0 \approx 1/2, s_1 = 1/4$,

$$s_{i+1} = p_iq_is_i - \frac{p_iq_i + p_is_i + q_is_i}{2} + \frac{p_i + q_i + s_i}{2}, 2 \leq i \leq n-1.$$

Proof. Denote $x = (x_{n-1}, \cdots, x_1, x_0)$, $y = (y_{n-1}, \cdots, y_1, y_0)$. Then

$$x' = ((x'_{n-1}, \cdots, x'_1, x'_0) = (x_{n-2} \oplus a_{n-1}, \cdots, x_0 \oplus a_1, x_{n-1} \oplus a_0)$$

$$y' = ((y'_{n-1}, \cdots, y'_1, y'_0) = (y_{n-2} \oplus b_{n-1}, \cdots, y_0 \oplus b_1, y_{n-1} \oplus b_0)$$

Let $c = (c_{n-1}, \cdots, c_0) = (x \boxplus y) \oplus x \oplus y$ and $c' = (c'_{n-1}, \cdots, c'_0) = (x' \boxplus y') \oplus x' \oplus y'$ be the two carries.

Let s_i denote the probability $\Pr[c_{i-1} \neq c'_i]$. When $i = 0$, $s_0 = \Pr[c_{n-1} \neq c'_0] = \Pr[x_{n-2}y_{n-2} \oplus x_{n-2}c_{n-2} \oplus y_{n-2}c_{n-2} = 0] \approx 1/2$, because the LHS term is balanced for independent random variables x and y. For $i = 1$, $s_1 = \Pr[c_0 \neq c'_1] = \Pr[x'_0y'_0 \neq 0] = 1/4$. For $i > 1$, s_i is equal to

$$\begin{aligned}\Pr[c_{i-1} \neq c'_i] &= \Pr[x_{i-2}y_{i-2} \oplus x_{i-2}c_{i-2} \oplus y_{i-2}c_{i-2} \neq x'_{i-1}y'_{i-1} \oplus x'_{i-1}c'_{i-1} \oplus y'_{i-1}c'_{i-1}] \\ &= p_{i-1}q_{i-1}s_{i-1} - \frac{p_{i-1}q_{i-1} + p_{i-1}s_{i-1} + q_{i-1}s_{i-1}}{2} + \frac{p_{i-1} + q_{i-1} + s_{i-1}}{2}\end{aligned}$$

For $x \boxplus y$ and $x' \boxplus y'$, applying the XOR-rule on the inputs and the carry vector gives

$$\Pr[(x \boxplus y)_{i-1} \neq (x' \boxplus y')_i] = p_i + q_i - 2p_iq_i - 2p_is_i - 2q_is_i + 4p_iq_is_i$$

□

Example 2. Consider an 8-bit modular addition with input RX-difference (left rotate by 1-bit) being $a = 7$ and $b = 7$, which implies that

$$p_0 = p_1 = p_2 = 1, p_3 = p_4 = p_5 = p_6 = p_7 = 0,$$
$$q_0 = q_1 = q_2 = 1, q_3 = q_4 = q_5 = q_6 = q_7 = 0.$$

The R-DL probability of the i-th output bit, $0 \leq i < 8$ is given in Table 3. The probabilities predicted for $i \geq 2$ are verified by running through the 16-bit input space, and the probability for $i = 0$ is $2^{-1.01132}$ by experiment.

The experiments on an 8-bit modular addition show that the theoretical estimation of the DL and R-DL probabilities match the experiments well, except that the approximation in R-DL probability for the least significant bit has a marginal error in precision.

With a similar deduction, we give the following theorem for computing the R-DL probability through a modular addition under the condition that $\mathtt{rot}(x) = x \lll t$, for an integer $2 \leq t \leq n-1$.

Table 3. The RL-probabilities of an 8-bit modular addition with input differences $a, b = 7$. $\texttt{rot}(x) = x \lll 1$. The index i represents the position of the output bit.

i	0	1	2	3	4	5	6	7
p	2^{-1}	2^{-2}	$2^{-0.678072}$	$2^{-0.29956}$	$2^{-1.29956}$	$2^{-2.29956}$	$2^{-3.29956}$	$2^{-4.29956}$

Theorem 5 ($\boxplus$-rule for RL for arbitrary $t > 1$). *Given random n-bit strings x, y and x', y' such that $x' = x \lll t \oplus a, y' = y \lll t \oplus b$, where $\Pr[x_{i-1} \neq x'_i] = p_i, \Pr[y_{i-1} \neq y'_i] = q_i$. Then, the rotational differential-linear probability of the modular addition for $i \geq 0$ can be computed as*

$$\Pr[(x \boxplus y)_{i-1} \neq (x' \boxplus y')_i] = p_i + q_i - 2p_iq_i - 2p_is_i - 2q_is_i + 4p_iq_is_i,$$

where $s_0 \approx 1/2, s_t = 1/2$,

$$s_{i+1} = p_iq_is_i - \frac{p_iq_i + p_is_i + q_is_i}{2} + \frac{p_i + q_i + s_i}{2}, 1 \leq i \leq n-1, i \neq t$$

Proof. Denote $x = (x_{n-1}, \cdots, x_1, x_0)$, $y = (y_{n-1}, \cdots, y_1, y_0)$, then

$$x' = ((x'_{n-1}, \cdots, x'_1, x'_0) = (x_{n-1-t} \oplus a_{n-1}, \cdots, x_{n-t+1} \oplus a_1, x_{n-t} \oplus a_0)$$

$$y' = ((y'_{n-1}, \cdots, y'_1, y'_0) = (y_{n-1-t} \oplus b_{n-1}, \cdots, y_0 \oplus b_1, y_{n-1} \oplus b_0).$$

Let $c = (c_{n-1}, \cdots, c_1, c_0)$ and $c' = (c'_{n-1}, \cdots, c'_1, c'_0)$ be the carries. Let s_i denote the probability $\Pr[c_{i-t} \neq c'_i]$. When $i = 0$,

$$s_0 = \Pr[c_{n-t} \neq c'_0] = \Pr[x_{n-t-1}y_{n-t-1} \oplus x_{n-t-1}c_{n-t-1} \oplus y_{n-t-1}c_{n-t-1} \neq 0] \approx 1/2$$

When $i = t$, $s_t = \Pr[c_0 \neq c'_t] = \Pr[x'_{t-1}y'_{t-1} \oplus x'_{t-1}c'_{t-1} \oplus y'_{t-1}c'_{t-1} \neq 0] \approx 1/2$
For all i, $i \neq 0, t$,

$$\begin{aligned} s_i &= \Pr[c_{i-t} \neq c'_i] \\ &= \Pr[x'_{i-1}y'_{i-1} \oplus x'_{i-1}c'_{i-1} \oplus y'_{i-1}c'_{i-1} \\ &\qquad \neq x_{n-t+i-1}y_{n-t+i-1} \oplus x_{n-t+i-1}c_{n-t+i-1} \oplus c_{n-t+i-1}y_{n-t+i-1}] \\ &= p_{i-1}q_{i-1}s_{i-1} - \frac{p_{i-1}q_{i-1} + p_{i-1}s_{i-1} + q_{i-1}s_{i-1}}{2} + \frac{p_{i-1} + q_{i-1} + s_{i-1}}{2}. \end{aligned}$$

Then, we have

$$\Pr[(x \boxplus y)_{i-t} \neq (x' \boxplus y')_i] = p_i + q_i - 2p_iq_i - 2p_is_i - 2q_is_i + 4p_iq_is_i.$$

□

The $\boxplus$-rules for DL and R-DL allows us to compute the partial DLCT of an n-bit modular addition accurately and efficiently. A naive application of Bar-On *et al.*'s approach [4] based on the Fast Fourier Transformation (FFT) by treating the modular addition as an $2n \times n$ S-box would require a complexity of

$\mathcal{O}(2^{2n})$, where it requires a complexity of $O(n2^{2n})$ to obtain the n rows of the DLCT whose output masks are the unit vectors. In contrast, with the $\boxplus$-rule for DL, given the input difference, the DL-probability for all output masks that are unit vectors can be evaluated in $\mathcal{O}(n)$ operations, which achieves an exponential speed-up.

4.2 Finding Input Differences for Local Optimization

According to Proposition 1 and Proposition 2, for x and y in $\mathbb{F}_2$, if $\Pr[x \neq x'] = p_1, \Pr[y \neq y'] = p_2$, we have

$$\Pr[xy \neq x'y'] = \frac{1}{2}(p_1 + p_2 - p_1 p_2), \quad \Pr[x \oplus y \neq x' \oplus y'] = p_1 + p_2 - 2p_1 p_2.$$

Obviously, $\Pr[xy \neq x'y']$ is in the interval $[0, 0.5]$ and $\Pr[x \oplus y \neq x' \oplus y']$ is in the interval $[0, 1]$. Moreover, a behaviour of $\Pr[x \oplus y \neq x' \oplus y']$ is that it collapses to $\frac{1}{2}$ (e.g., correlation zero) whenever one of p_1 and p_2 is $\frac{1}{2}$. This observation suggests that the input probabilities should be biased from $\frac{1}{2}$ as much as possible. Otherwise, the probabilities will rapidly collapse to $\frac{1}{2}$ for all one-bit output masks after a few iterative evaluations of the round function.

In order to find distinguishers that cover as many rounds of a function F as possible, our strategy is to look for an input RX-difference δ, such that the DL or R-DL probability after one or a few propagations still has a relatively large imbalance for all the output masks whose Hamming weights are one. Therefore, we can define the objective function to maximize the summation of the absolute biases:

$$\sum_i (|\Pr[e_i \cdot (\texttt{rot}(f(x)) \oplus f(\texttt{rot}(x) \oplus \delta)) = 0] - 1/2|). \tag{11}$$

For 8-bit modular additions, we observed that the absolute DL and R-DL bias are relatively large when the input RX-differences are either with a large Hamming weight or a small weight. For instance, with RX-difference $(x \lll 1) \oplus x'$, when the input differences are $a = \texttt{0}$ and $b = \texttt{1}$, the RL-probabilities are given as follows for $e_i, i = 0, 1, \ldots, 7$.

$$2^{-1}, 2^{-2}, 2^{-3}, 2^{-4}, 2^{-5}, 2^{-6}, 2^{-7}, 2^{-8}.$$

Whereas for $a = \texttt{ff}$ and $b = \texttt{ff}$, the RL-probabilities are given as follows for $e_i, i = 0, 1, \ldots, 7$.

$$2^{-1}, 2^{-2}, 2^{-0.678072}, 2^{-0.29956}, 2^{-0.142019}, 2^{-0.0692627}, 2^{-0.0342157}, 2^{-0.0170064}.$$

When the size of the operands are large (e.g., $n = 32$), it is difficult to find the optimal input difference manually. Next, we show the optimal input RX-difference with respect to the objective function given by Eq. (11) in a 32-bit modular addition. See the full version of this paper [25] for the search of such differences.

Example 3. Consider the R-DL probability for a 32-bit modular addition with $\mathtt{rot}(x) = x \lll 1$. With input RX-differences

$$a = \mathtt{7ffffffc}, b = \mathtt{7ffffffe},$$

the objective function in Eq. 11 is maximized, and the R-DL probabilities $\Pr[e_i \cdot (\mathtt{rot}(x \boxplus y) \oplus ((\mathtt{rot}(x) \oplus a) \boxplus (\mathtt{rot}(y) \oplus b))) = 1]$ for $0 \leq i \leq 31$ are shown as follows.

i	0	1	2	3	4	5	6	7
p_i	0.5	0.75	0.5	0.75	0.875	0.9375	0.96875	0.984375
i	8	9	10	11	12	13	14	15
p_i	0.992188	0.996094	0.998047	0.999023	0.999512	0.999756	0.999878	0.999939
i	16	17	18	19	20 – 31			
p_i	0.999969	0.999985	0.999992	0.999996	1			

5 Applications to AND-RX Primitives

In this section, we apply the rotational differential-linear technique to the AND-RX permutations involved in FRIET and Xoodoo, and significant improvements are obtained. To confirm the validity of the results, all distinguishers with practical complexities are experimentally verified, and the source code is available[1].

5.1 Distinguishers for Round-Reduced FRIET

FRIET is an authenticated encryption scheme with built-in fault detection mechanisms proposed by Simon et al. at EUROCRYPT 2020 [32]. Its fault detection ability comes from its underlying permutation, which is designed based on the so-called *code embedding* approach.

The core permutation FRIET-P employed in FRIET operates on a $4 \times 128 = 512$-bit state arranged into a rectangular with 4 rows (called limbs) and 128 columns (called slices) as shown in Fig. 1. The permutation FRIET-P is an iterative design with its round function g_{rc_i} visualized in Fig. 2, where a, b, and $c \in \mathbb{F}_2^{128}$ are the four limbs (see Fig. 1) of the input state and rc_i is the round constant for the i-th round.

By design, the round function g_{rc_i} is slice-wise *code-abiding* for the parity code $[4, 3, 2]_{\mathbb{F}_2}$, meaning that every slice of the output state is a code word if every slice of the input state is a code word. Mathematically, it means that $a + b + c = d$ implies $a' + b' + c' = d'$. This slice-wise *code-abiding* property is

[1] https://github.com/YunwenL/Rotational-cryptanalysis-from-a-differential-linear-perspective.

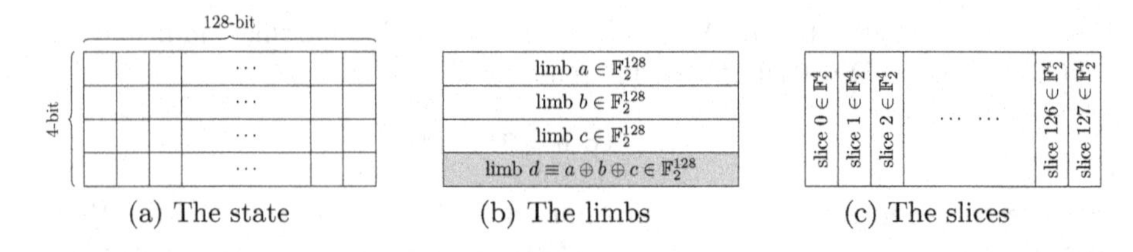

(a) The state (b) The limbs (c) The slices

Fig. 1. The view of the state

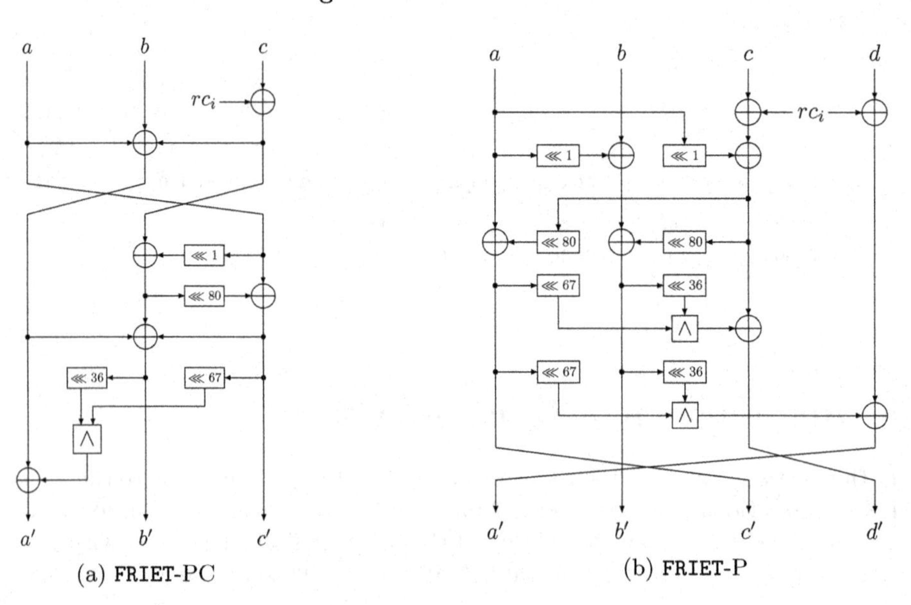

(a) FRIET-PC (b) FRIET-P

Fig. 2. The round functions of Friet-PC and Friet-P

inherited by the permutation FRIET-P $= g_{rc_{t-1}} \circ \cdots \circ g_{rc_1} \circ g_{rc_0}$. Consequently, faults will be detected if some output slice is not a code word when all of the slices of the input state are code words. Note that the behavior of the permutation FRIET-PC is identical to FRIET-P by design if we ignore the limb d.

Practical Distinguishers for FRIET-PC. Since a distinguisher for the permutation FRIET-PC directly translates to a distinguisher for FRIET-P, we focus on the permutation FRIET-PC. Let (a, b, c) and (a', b', c') in $\mathbb{F}_2^{128\times 3}$ be the input pair of the permutation with RX-differences

$$\Delta_a = (a \lll t) \oplus a', \quad \Delta_b = (b \lll t) \oplus b', \quad \Delta_c = (c \lll t) \oplus c'.$$

In our analysis, we only consider input RX-differences such that $wt(\Delta_a) + wt(\Delta_b) + wt(\Delta_c) \leq 1$.

According to the adjusted C-rule (see Proposition 4), the constant addition injects an RX-difference $c \oplus (c \lll t)$ to the state, and alters the R-DL-probabilities when the corresponding bits in $c \oplus (c \lll t)$ is nonzero. A rule-of-thumb for choosing the rotational amount is to minimize the weight of the

RX-difference introduced by the round constants, so that the effect of the constants on destroying the rotational propagation is presumably decreased. The first 6 round constants of FRIET-PC are (in Hexadecimal)

$$\texttt{1111}, \texttt{11100000}, \texttt{1101}, \texttt{10100000}, \texttt{101}, \texttt{10110000}.$$

To minimize the Hamming weight of the RX-differences from the round constants, one of the best rotational operations is to left rotate by 4 bits, such that the consecutive nonzero nibbles cancel themselves as many as possible. Then, the injected RX-differences due to the round constants are

$$\texttt{10001}, \texttt{100100000}, \texttt{10111}, \texttt{111100000}, \texttt{1111}, \texttt{111010000}.$$

With the AND-rule, XOR-rule and adjusted C-rule, the R-DL probability can be evaluated given the input RX-differences with $w_h(\Delta_a)+w_h(\Delta_b)+w_h(\Delta_c) \leq 1$ and the output linear mask e_i. Table 4 shows the rotational differential-linear distinguishers with the largest absolute correlation we found in reduced-round FRIET-PC, where $\Delta_a, \Delta_b, \Delta_c$ are the input RX-differences, and $\gamma_a, \gamma_b, \gamma_c$ are the output masks for the limbs a, b, c, respectively.

Table 4. Distinguishers for reduced-round FRIET-PC with rotation offset $t = 4$.

Round	Δ_a	Δ_b	Δ_c	γ_a	γ_b	γ_c	Correlation	
							Theoretical	Experimental
1	0	0	0	1	0	0	1	1
2	0	0	0	1	0	0	1	1
3	0	0	0	1	0	0	1	1
4	0	0	0	0	1	0	1	1
5	0	0	1	0	0	400000000000000000000	$2^{-0.96}$	$2^{-0.83}$
6	0	0	10000	0	0	40000	$2^{-5.81}$	$2^{-5.12}$

For FRIET-PC reduced to 4-round, an R-DL distinguisher with correlation 1 is detected, with input RX-differences $(0, 0, 0)$ and output masks $(0, 1, 0)$. For 5, 6-round FRIET-PC, we found practical rotational differential-linear distinguishers with correlation $2^{-0.96}$ and $2^{-5.81}$, respectively. All the distinguishers shown in Table 4 are verified experimentally with 2^{24} random plaintexts.

Extending the Practical Distinguishers. According to the discussion of Sect. 3, we can extend a rotational differential-linear distinguisher by appending a linear approximation $\gamma \rightarrow \mu$, and the bias of the extended distinguisher can be computed with Eq. (7). Consequently, this extension is optimal when $\epsilon_{\gamma,\mu}$ and $\epsilon_{\texttt{rot}^{-1}(\gamma),\texttt{rot}^{-1}(\mu)}$ reach their largest possible absolute values simultaneously. For FRIET-PC, we always have $\epsilon_{\gamma,\mu} = \epsilon_{\texttt{rot}^{-1}(\gamma),\texttt{rot}^{-1}(\mu)}$, and thus we can focus on finding an optimal linear approximation $\gamma \rightarrow \mu$.

Here we take the 6-round R-DL distinguisher presented in Table 4 and append optimal linear approximations to extend it. The output linear mask of the 6-round distinguisher is $(\texttt{0}, \texttt{0}, \texttt{40000})$. In Table 5, we list the correlations of the optimal linear approximations for round-reduced FRIET-PC whose input masks are $(\texttt{0}, \texttt{0}, \texttt{40000})$, which are found with the SMT-based approach [20].

Table 5. The correlation of optimal linear trails found in round-reduced FRIET-PC with the input masks $(\texttt{0}, \texttt{0}, \texttt{40000})$

# Round	1	2	3	4	5	6	7
Correlation	2^{-2}	2^{-6}	2^{-12}	2^{-20}	2^{-30}	2^{-42}	2^{-56}

The optimal 1-round linear trail we found has output masks

$$\mu_a = \texttt{00000000000000020000000000040000}$$
$$\mu_b = \texttt{00004000000000020000000000040000}$$
$$\mu_c = \texttt{00000000000080020000000000060000}.$$

Thus a 7-round distinguisher can be built by concatenating the 6-round distinguisher with a 1-round linear approximation, and the estimated correlation is $2^{-5.81} \times 2^{-2\times 2} = 2^{-9.81}$. With 2^{24} pairs of inputs satisfying the input RX-difference, the output difference under the specified mask are biased with a correlation approximately $2^{-9.12}$. Similarly, by appending a 2-round linear trail with output masks

$$\mu_a = \texttt{00000000000000030000000000060000}$$
$$\mu_b = \texttt{00006000000000010000000030020000}$$
$$\mu_c = \texttt{600000000000c0010000000000030000}.$$

at the end of the 6-round rotational differential-linear distinguisher, we get a 8-round RL-distinguisher with a correlation $2^{-17.81}$. And with 2^{40} pairs of inputs satisfying the input RX-difference, we find the experimental correlation of the 8-round distinguisher is $2^{-17.2}$. As a comparison, the 7-,8-round linear trails presented in the specification of FRIET-PC have correlation 2^{-29} and 2^{-40}, respectively. With the linear trails shown in Table 5, the concatenated distinguisher can reach up to 13 rounds, with an estimated correlation $2^{-117.81}$.

5.2 Distinguishers for Round-Reduced Xoodoo

Xoodoo [14] is a 384-bit lightweight cryptographic permutation whose primary target application is in the Farfalle construction [8]. The state of Xoodoo is arranged into a $4 \times 3 \times 32$ cuboid and the bit at a specific position is accessed as $a[x][y][z]$. One round of Xoodoo consists of the following operations.

$$a[x][y][z] = a[x][y][z] \oplus \sum_y a[x-1][y][z-5] \oplus \sum_y a[x-1][y][z-14]$$
$$a[x][1][z] = a[x-1][1][z], a[x][2][z] = a[x][2][z-11]$$
$$a[0][0] = a[0][0] \oplus RC_i$$
$$a[x][y][z] = a[x][y][z] \oplus ((a[x][y+1][z]+1) * (a[x][y+2][z]))$$
$$a[x][1][z] = a[x][1][z-1], a[x][2][z] = a[x-1][2][z-8]$$

The total number of rounds in Xoodoo is 12, and in some modes (Farfalle [8] for instance), the core permutation calls a 6-round Xoodoo permutation. The round constants of Xoodoo are shown in the following, and for Xoodoo reduced to r rounds, the round constants are $c_{-(r-1)}, \cdots, c_0$.

$$\begin{array}{llll} c_{-11} = \texttt{00000058}, & c_{-8} = \texttt{000000D0}, & c_{-5} = \texttt{00000060}, & c_{-2} = \texttt{000000F0} \\ c_{-10} = \texttt{00000038}, & c_{-7} = \texttt{00000120}, & c_{-4} = \texttt{0000002C}, & c_{-1} = \texttt{000001A0} \\ c_{-9} = \texttt{000003C0}, & c_{-6} = \texttt{00000014}, & c_{-3} = \texttt{00000380}, & c_0 = \texttt{00000012} \end{array}$$

Given input difference being all-zero, *i.e.*, the input pair is exactly a rotational pair, let the rotation amount be left-rotate by 1-bit. We find that after 3 rounds of Xoodoo, there are still many output bits that are highly biased, with the largest correlation being 1 and the one-bit mask at position $(1, 0, 16)$. This suggests a nonzero mask 10000 at the lane $(1, 0)$. However, extending one extra round, we no longer see any significant correlation.

Noticing that the round constant is XORed into the state right after the first two linear operations, one can control the input RX-difference such that the difference is cancelled by the injection of the first-round constant. As a result, it gains one round free at the beginning, and we are able to construct a 4-round distinguishers for Xoodoo. When the left-rotational amount is set to 1-bit, the RX-difference of the first constant c_{-3} is 00000480. This suggests that if we take input RX-differences

$$\begin{array}{l} a[0][0] = \texttt{484ccc80}; \texttt{a}[0][1] = \texttt{484cc800}; \texttt{a}[0][2] = \texttt{484cc800}; \\ a[1][0] = \texttt{3ab9821a}; \texttt{a}[1][1] = \texttt{3ab9821a}; \texttt{a}[1][2] = \texttt{3ab9821a}; \\ a[2][0] = \texttt{37b6cde9}; \texttt{a}[2][1] = \texttt{37b6cde9}; \texttt{a}[2][2] = \texttt{37b6cde9}; \\ a[3][0] = \texttt{45a3f0cb}; \texttt{a}[3][1] = \texttt{45a3f0cb}; \texttt{a}[3][2] = \texttt{45a3f0cb}. \end{array}$$

The RX-difference after the first round of Xoodoo will be all zero. Hence, we are able to find a 4-round distinguishers with significant correlations. We find a rotational differential-linear distinguishers with correlation 1 with the output mask being 10000 at lane $(1, 0)$ and zero for the rest lanes. Another two distinguishers with the same correlation are found with output mask 20000 at lane $(1, 1)$ and 1000000 at lane $(3, 2)$.

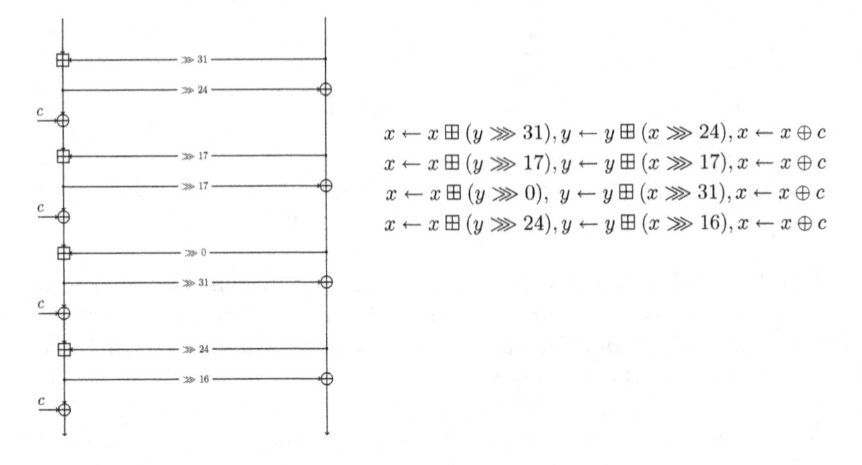

Fig. 3. The `Alzette` instance.

6 Applications to ARX Primitives

In this section, we apply the rotational differential-linear technique to the ARX permutations involved in `Alzette` and `SipHash`, and the source code for experimental verifications is available[2].

6.1 Application in the 64-Bit ARX-box `Alzette`

At CRYPTO 2020, Beierle et al. presented a 64-bit ARX-box `Alzette` [6] that is efficient for software implementation. The design is along the same research line with a previous design called SPARX [15] with a 32-bit ARX-box where a long trail argument was proposed for deriving a security bound in ARX ciphers. Figure 3 shows an instance of `Alzette` with an input $(x, y) \in \mathbb{F}_2^{32} \times \mathbb{F}_2^{32}$.

The differential and linear properties of `Alzette` is comparable to the 8-bit S-box of AES. The optimal differential characteristic in `Alzette` has a probability of 2^{-6}. In addition, because of the modular additions in `Alzette` and the diffusion, the designers showed by division property that the `Alzette` may have full degree in all its coordinates.

In the following, we present the rotational differential-linear and differential-linear distinguishers of `Alzette` found with the techniques in Sect. 4. The constant $c =$ `B7E15162` (the first constant in SPARX-based design Sparkle-128) is considered for illustration.

Rotational Differential-Linear Distinguisher. In Sect. 4.2, (`7ffffffc`, `7ffffffe`) is found to be optimal in 32-bit modular addition under the objective function

[2] https://github.com/YunwenL/Rotational-cryptanalysis-from-a-differential-linear-perspective.

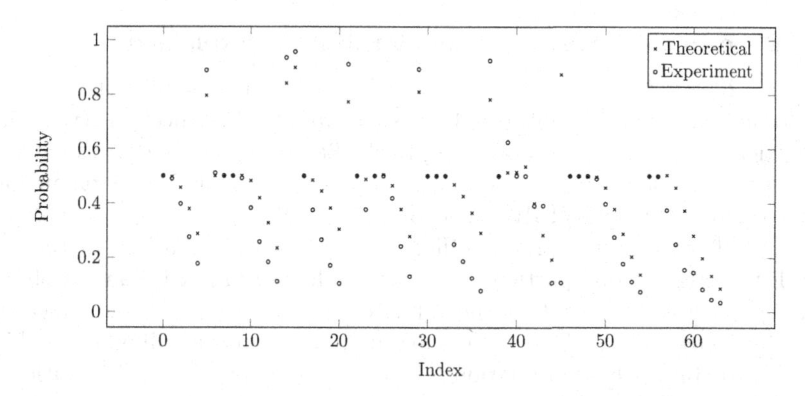

Fig. 4. A comparison between the differential-linear probability in `Alzette` by theoretical computation and by experiment. The index shows the index of the nonzero bit in the unit-vector output mask. For instance, when the index is 0, the output mask is (0,1), and when the index is 63, it is (80000000,0).

considered in Example 3. Here, the difference can be used as the input difference of the first modular addition in `Alzette`. Because of the right rotation by 31 bits before the modular addition, the input RX-difference to `Alzette` is (`7ffffffc`, `3fffffff`). With an iterative evaluation on the steps in `Alzette`, we found that the second least significant bit is biased. Specifically, with an output mask $(2, 0)$, the RL-probability is 0.500189, that is a correlation $2^{-11.37}$. By taking 2^{28} pairs of random plaintexts, the experimental correlation of the distinguisher is $2^{-7.35}$. In addition, we checked all input RX-differences (a, b) with Hamming weight $wt(a) + wt(b) = 1$, but no rotational differential-linear distinguisher is found.

Differential-linear distinguisher. For all input differences with Hamming weight 1, we compute the differential-linear probability of `Alzette` with the technique in Sect. 4. The best found distinguisher has an input difference $(80000000, 0)$ and output mask $(80000000, 0)$, with a probability of 0.086, equivalently, a correlation of $2^{-0.27}$. By experiment verification with 2^{28} pairs of random plaintexts, the correlation is $2^{-0.1}$.

The following Fig. 4 shows a comparison of the probability for an input difference $(80000000, 0)$ and output masks $(1 \lll t, 0)$ (for all integer $t \in [0, 31]$), by our evaluation technique and the experiment with 2^{24} pairs of random plaintexts. The theoretical evaluation matches the experiment within a tolerable fluctuation.

Comparing with RL-distinguishers and DL-distinguisher found in `Alzette`, the latter is significantly stronger. Also, it is interesting to notice that input differences with low Hamming weight often lead to good differential-linear distinguishers in `Alzette`, whereas we didn't find any rotational differential-linear distinguisher with low-weight RX-differences when the rotational offset is greater than zero. The influence of the constants in RL-distinguishers may be the main cause.

6.2 Experimental Distinguishers for SipHash Explained

SipHash [2], designed by Aumasson and Bernstein, is a family of ARX-based pseudorandom functions optimized for short inputs. Instances of SipHash are widely deployed in practice. For example, SipHash-2-4 is used in the dnscache instances of all OpenDNS resolvers and employed as hash() in Python for all major platforms (https://131002.net/siphash/#us).

In [16], from a perspective of differential cryptanalysis, a bias of the difference distribution of one particular output bit for 3-round SipHash is observed when the Hamming weight of the input difference is one. For instance, with input difference $a = 1$, He and Yu showed that the output difference is biased at the 27-th bit with a correlation 2^{-6} by experiments. This observation was obtained through extensive experiments and the theoretical reason behind these distinguishers is unclear as stated by He and Yu:

> "*... we are not concerned about why it shows a rotation property or why it reaches such a bias level. However, a great number of experiments can support those observations.* (see [16, Section 4.2, Page 11])"

According to the discussion of Sect. 3.2, the bias of $E(x) \oplus E(x \oplus \delta)$ observed in [16] is equivalent to the bias of

$$e_i \cdot (E(x) \oplus E(x \oplus \delta)).$$

It can be interpreted in the differential-linear framework and analyzed with the theoretical approach presented in Sect. 4. Here, we apply the rules for modular addition and XOR, and compute the DL-probability of the 3-round distinguisher found in SipHash. With our technique, we confirm that the 3-round differential-linear distinguisher with the aforementioned difference and mask, the predicted correlation is $2^{-6.6}$ which is close to He and Yu's experiments.

In addition, we can explain the observation on the rotation property with the $\boxplus$-rule in differential-linear. We will adopt the notations that are used in Theorem 3.

Because the input difference in their experiment has only one nonzero bit, we consider the DL-probability of an n-bit modular addition where the input difference is $(e_k, 0)$, for an integer k.

Then, for a pair of inputs (x, y) and (x', y'), the probability $p_k = \Pr[x_k \neq x'_k] = 1$. And for the remaining bits, $p_i = \Pr[x_i \neq x'_i], i \neq k$ and $q_i = \Pr[y_i, y'_i]$ are equal to zero.

Let $s_i = \Pr[\varsigma(x, y)_i \neq \varsigma(x', y')]$. We have $s_0, \cdots, s_k = 0, s_{k+t} = 2^{-t}, 1 \leq t \leq n - 1 - k$. As a result, the DL-probabilities through the modular addition at the i-th bit is given by $P_i = \Pr[(x \boxplus y)_i \neq (x' \boxplus y')_i], 0 \leq i \leq n - 1$, where

$$\Pr[(x \boxplus y)_i \neq (x' \boxplus y')_i] = \begin{cases} 0, & i \leq k \\ 2^{-i+k}, & \text{otherwise} \end{cases} \tag{12}$$

By rotating the input difference $(\mathtt{1} \lll k, 0)$ to the left by one bit, the differential-linear probability for the i-th bit of the output $\overleftarrow{P_i}$ is equal to 2^{-i+k+1} for $k + 1 < i \leq n - 1$, and to zero for $i \leq k + 1$.

It is obvious that the by rotating the differential-linear probability in Eq. (12), we obtain the probabilities $\overleftarrow{P_i}$ for all but the least significant bit, where $\overleftarrow{P_0} = 0$ and $P_{n-1} = 2^{-n-1+k}$. Nevertheless, the error is negligible if $n - k$ is large, and it holds for large modular additions such as the 64-bit one adopted in SipHash.

For input differences with Hamming weight more than 1, a similar rotational property can be observed for the ⊞-rule in differential-linear. And it gives a straightforward intuition on the rotational property observed in the differential-linear distinguishers of SipHash.

7 Conclusion and Open Problems

We extend the differential-linear framework by using rotational-xor differentials in the differential part of the framework and we name the resulting cryptanalytic technique as rotational differential-linear cryptanalysis. We give an informal formula to estimate the bias of rotational differential-linear distinguisher under certain assumptions. In particular, we show Morawiecki et al.'s technique can be generalized to estimate the bias of a rotational differential-linear distinguisher whose output linear mask is a unit vector. We apply our method to the permutations involved in `FRIET`, `Xoodoo`, `Alzette`, and `SipHash`, which leads to significant improvements over existing cryptanalytic results or explanations for previous experimental distinguishers without a theoretical foundation. Finally, we would like to mention that we failed to derive a closed formula for the bias of a rotational differential-linear distinguisher under the sole assumption of the independence between the rotational-xor differential part and linear part. This is left open and the link between rotational-xor differential and linear cryptanalysis we presented in this work can be seen as a first step towards solving this problem.

A natural extension of rotational differential-linear cryptanalysis is to the SPN-type primitives, where one aims at finding a rotational relation that is preserved with a significant probability through the nonlinear Sbox layer. Especially, it is feasible to check all the rotational differences for their transition probabilities in a small-scale Sbox. Comparing to binary and arithmetic operations, our observation is that rotational relations are less likely to preserve in Sboxes, so it is challenging to find good distinguishers in Sbox-based designs. We leave it as an interesting future work.

Acknowledgement. We would like to thank the reviewers of Eurocrypt 2021 for their comments and suggestions to improve this paper. This work is supported by National Key R&D Program of China (2017YFB0802000, 2018YFA0704704), Natural Science Foundation of China (NSFC) under Grants 61902414, 61722213, 62032014, 61772519, and 61772545 and the Chinese Major Program of National Cryptography Development Foundation (MMJJ20180102).

References

1. Ashur, T., Liu, Y.: Rotational cryptanalysis in the presence of constants. IACR Trans. Symmetric Cryptol. **2016**(1), 57–70 (2016)
2. Aumasson, J.-P., Bernstein, D.J.: SipHash: a fast short-input PRF. In: Galbraith, S., Nandi, M. (eds.) INDOCRYPT 2012. LNCS, vol. 7668, pp. 489–508. Springer, Heidelberg (2012). https://doi.org/10.1007/978-3-642-34931-7_28
3. Aumasson, J.-P., Jovanovic, P., Neves, S.: Analysis of NORX: investigating differential and rotational properties. In: Aranha, D.F., Menezes, A. (eds.) LATINCRYPT 2014. LNCS, vol. 8895, pp. 306–324. Springer, Cham (2015). https://doi.org/10.1007/978-3-319-16295-9_17
4. Bar-On, A., Dunkelman, O., Keller, N., Weizman, A.: DLCT: a new tool for differential-linear cryptanalysis. In: Ishai, Y., Rijmen, V. (eds.) EUROCRYPT 2019. Part I. LNCS, vol. 11476, pp. 313–342. Springer, Cham (2019). https://doi.org/10.1007/978-3-030-17653-2_11
5. Barbero, S., Bellini, E., Makarim, R.H.: Rotational analysis of ChaCha permutation. CoRR abs/2008.13406 (2020). https://arxiv.org/abs/2008.13406
6. Beierle, C., et al.: Alzette: a 64-Bit ARX-box. In: Micciancio, D., Ristenpart, T. (eds.) CRYPTO 2020. Part III. LNCS, vol. 12172, pp. 419–448. Springer, Cham (2020). https://doi.org/10.1007/978-3-030-56877-1_15
7. Beierle, C., Leander, G., Todo, Y.: Improved differential-linear attacks with applications to ARX ciphers. In: Micciancio, D., Ristenpart, T. (eds.) CRYPTO 2020. Part III. LNCS, vol. 12172, pp. 329–358. Springer, Cham (2020). https://doi.org/10.1007/978-3-030-56877-1_12
8. Bertoni, G., Daemen, J., Hoffert, S., Peeters, M., Assche, G.V., Keer, R.V.: Farfalle: parallel permutation-based cryptography. IACR Trans. Symmetric Cryptol. **2017**(4), 1–38 (2017)
9. Blondeau, C., Leander, G., Nyberg, K.: Differential-linear cryptanalysis revisited. J. Cryptol. **30**(3), 859–888 (2017). https://doi.org/10.1007/s00145-016-9237-5
10. Canteaut, A.: Lecture notes on cryptographic Boolean functions (2016). https://www.rocq.inria.fr/secret/Anne.Canteaut/
11. Carlet, C.: Boolean functions for cryptography and error correcting codes (2006). https://www.rocq.inria.fr/secret/Anne.Canteaut/
12. Chabaud, F., Vaudenay, S.: Links between differential and linear cryptanalysis. In: De Santis, A. (ed.) EUROCRYPT 1994. LNCS, vol. 950, pp. 356–365. Springer, Heidelberg (1995). https://doi.org/10.1007/BFb0053450
13. Cid, C., Huang, T., Peyrin, T., Sasaki, Y., Song, L.: Boomerang connectivity table: a new cryptanalysis tool. In: Nielsen, J.B., Rijmen, V. (eds.) EUROCRYPT 2018. Part II. LNCS, vol. 10821, pp. 683–714. Springer, Cham (2018). https://doi.org/10.1007/978-3-319-78375-8_22
14. Daemen, J., Hoffert, S., Assche, G.V., Keer, R.V.: The design of Xoodoo and Xoofff. IACR Trans. Symmetric Cryptol. **2018**(4), 1–38 (2018)
15. Dinu, D., Perrin, L., Udovenko, A., Velichkov, V., Großschädl, J., Biryukov, A.: Design strategies for ARX with provable bounds: SPARX and LAX. In: Cheon, J.H., Takagi, T. (eds.) ASIACRYPT 2016. Part I. LNCS, vol. 10031, pp. 484–513. Springer, Heidelberg (2016). https://doi.org/10.1007/978-3-662-53887-6_18
16. He, L., Yu, H.: Cryptanalysis of reduced-round SipHash. IACR Cryptology ePrint Archive 2019/865 (2019)
17. Khovratovich, D., Nikolić, I.: Rotational cryptanalysis of ARX. In: Hong, S., Iwata, T. (eds.) FSE 2010. LNCS, vol. 6147, pp. 333–346. Springer, Heidelberg (2010). https://doi.org/10.1007/978-3-642-13858-4_19

18. Khovratovich, D., Nikolic, I., Pieprzyk, J., Sokolowski, P., Steinfeld, R.: Rotational cryptanalysis of ARX revisited. In: Fast Software Encryption - 22nd International Workshop, FSE 2015, Istanbul, Turkey, 8–11 March 2015, Revised Selected Papers, pp. 519–536 (2015)
19. Khovratovich, D., Nikolić, I., Rechberger, C.: Rotational rebound attacks on reduced skein. In: Abe, M. (ed.) ASIACRYPT 2010. LNCS, vol. 6477, pp. 1–19. Springer, Heidelberg (2010). https://doi.org/10.1007/978-3-642-17373-8_1
20. Kölbl, S., Leander, G., Tiessen, T.: Observations on the SIMON block cipher family. In: Gennaro, R., Robshaw, M. (eds.) CRYPTO 2015. Part I. LNCS, vol. 9215, pp. 161–185. Springer, Heidelberg (2015). https://doi.org/10.1007/978-3-662-47989-6_8
21. Kraleva, L., Ashur, T., Rijmen, V.: Rotational cryptanalysis on MAC algorithm Chaskey. In: Conti, M., Zhou, J., Casalicchio, E., Spognardi, A. (eds.) ACNS 2020. Part I. LNCS, vol. 12146, pp. 153–168. Springer, Cham (2020). https://doi.org/10.1007/978-3-030-57808-4_8
22. Langford, S.K., Hellman, M.E.: Differential-linear cryptanalysis. In: Desmedt, Y.G. (ed.) CRYPTO 1994. LNCS, vol. 839, pp. 17–25. Springer, Heidelberg (1994). https://doi.org/10.1007/3-540-48658-5_3
23. Leander, G., Abdelraheem, M.A., AlKhzaimi, H., Zenner, E.: A cryptanalysis of PRINTcipher: the invariant subspace attack. In: Rogaway, P. (ed.) CRYPTO 2011. LNCS, vol. 6841, pp. 206–221. Springer, Heidelberg (2011). https://doi.org/10.1007/978-3-642-22792-9_12
24. Leander, G., Minaud, B., Rønjom, S.: A generic approach to invariant subspace attacks: cryptanalysis of Robin, iSCREAM and Zorro. In: Oswald, E., Fischlin, M. (eds.) EUROCRYPT 2015. Part I. LNCS, vol. 9056, pp. 254–283. Springer, Heidelberg (2015). https://doi.org/10.1007/978-3-662-46800-5_11
25. Liu, Y., Sun, S., Li, C.: Rotational cryptanalysis from a differential-linear perspective, practical distinguishers for round-reduced Friet, Xoodoo, and Alzette. IACR Cryptology ePrint Archive 2021/189 (2021)
26. Liu, Y., Witte, G.D., Ranea, A., Ashur, T.: Rotational-XOR cryptanalysis of reduced-round SPECK. IACR Trans. Symmetric Cryptol. **2017**(3), 24–36 (2017)
27. Liu, Z., Gu, D., Zhang, J., Li, W.: Differential-multiple linear cryptanalysis. In: Bao, F., Yung, M., Lin, D., Jing, J. (eds.) Inscrypt 2009. LNCS, vol. 6151, pp. 35–49. Springer, Heidelberg (2010). https://doi.org/10.1007/978-3-642-16342-5_3
28. Lu, J., Liu, Y., Ashur, T., Sun, B., Li, C.: Rotational-XOR cryptanalysis of Simon-like block ciphers. In: Liu, J.K., Cui, H. (eds.) ACISP 2020. LNCS, vol. 12248, pp. 105–124. Springer, Cham (2020). https://doi.org/10.1007/978-3-030-55304-3_6
29. Lu, J.: A methodology for differential-linear cryptanalysis and its applications. Des. Codes Cryptogr. **77**(1), 11–48 (2014). https://doi.org/10.1007/s10623-014-9985-x
30. Matsui, M.: Linear cryptanalysis method for DES cipher. In: Helleseth, T. (ed.) EUROCRYPT 1993. LNCS, vol. 765, pp. 386–397. Springer, Heidelberg (1994). https://doi.org/10.1007/3-540-48285-7_33
31. Morawiecki, P., Pieprzyk, J., Srebrny, M.: Rotational cryptanalysis of round-reduced Keccak. In: Moriai, S. (ed.) FSE 2013. LNCS, vol. 8424, pp. 241–262. Springer, Heidelberg (2014). https://doi.org/10.1007/978-3-662-43933-3_13
32. Simon, T., et al.: Friet: an authenticated encryption scheme with built-in fault detection. In: Canteaut, A., Ishai, Y. (eds.) EUROCRYPT 2020. Part I. LNCS, vol. 12105, pp. 581–611. Springer, Cham (2020). https://doi.org/10.1007/978-3-030-45721-1_21

33. Tiessen, T.: Polytopic cryptanalysis. In: Fischlin, M., Coron, J.-S. (eds.) EUROCRYPT 2016. Part I. LNCS, vol. 9665, pp. 214–239. Springer, Heidelberg (2016). https://doi.org/10.1007/978-3-662-49890-3_9
34. Todo, Y.: Structural evaluation by generalized integral property. In: Oswald, E., Fischlin, M. (eds.) EUROCRYPT 2015. Part I. LNCS, vol. 9056, pp. 287–314. Springer, Heidelberg (2015). https://doi.org/10.1007/978-3-662-46800-5_12
35. Todo, Y., Leander, G., Sasaki, Y.: Nonlinear invariant attack: practical attack on full SCREAM, iSCREAM, and Midori64. J. Cryptol. **32**(4), 1383–1422 (2019). https://doi.org/10.1007/s00145-018-9285-0
36. Todo, Y., Morii, M.: Bit-based division property and application to SIMON family. In: Peyrin, T. (ed.) FSE 2016. LNCS, vol. 9783, pp. 357–377. Springer, Heidelberg (2016). https://doi.org/10.1007/978-3-662-52993-5_18
37. Wagner, D.: The boomerang attack. In: Knudsen, L. (ed.) FSE 1999. LNCS, vol. 1636, pp. 156–170. Springer, Heidelberg (1999). https://doi.org/10.1007/3-540-48519-8_12

Automatic Search of Meet-in-the-Middle Preimage Attacks on AES-like Hashing

Zhenzhen Bao[2(✉)], Xiaoyang Dong[3(✉)], Jian Guo[2(✉)], Zheng Li[4,5(✉)], Danping Shi[1,6(✉)], Siwei Sun[1,6(✉)], and Xiaoyun Wang[3,7(✉)]

[1] State Key Laboratory of Information Security, Institute of Information Engineering, Chinese Academy of Sciences, Beijing, China
{shidanping,sunsiwei}@iie.ac.cn
[2] Division of Mathematical Sciences, School of Physical and Mathematical Sciences, Nanyang Technological University, Singapore, Singapore
{zzbao,guojian}@ntu.edu.sg
[3] Institute for Advanced Study, BNRist, Tsinghua University, Beijing, China
{xiaoyangdong,xiaoyunwang}@tsinghua.edu.cn
[4] Faculty of Information Technology, Beijing University of Technology, Beijing, China
[5] Beijing Key Laboratory of Trusted Computing, Beijing University of Technology, Beijing, China
lizhengcn@bjut.edu.cn
[6] University of Chinese Academy of Sciences, Beijing, China
[7] Key Laboratory of Cryptologic Technology and Information Security, Ministry of Education, Shandong University, Shandong, China

Abstract. The Meet-in-the-Middle (MITM) preimage attack is highly effective in breaking the preimage resistance of many hash functions, including but not limited to the full MD5, HAVAL, and Tiger, and reduced SHA-0/1/2. It was also shown to be a threat to hash functions built on block ciphers like AES by Sasaki in 2011. Recently, such attacks on AES hashing modes evolved from merely using the freedom of choosing the internal state to also exploiting the freedom of choosing the message state. However, detecting such attacks especially those evolved variants is difficult. In previous works, the search space of the configurations of such attacks is limited, such that manual analysis is practical, which results in sub-optimal solutions. In this paper, we remove artificial limitations in previous works, formulate the essential ideas of the construction of the attack in well-defined ways, and translate the problem of searching for the best attacks into optimization problems under constraints in Mixed-Integer-Linear-Programming (MILP) models. The MILP models capture a large solution space of valid attacks; and the objectives of the MILP models are attack configurations with the minimized computational complexity. With such MILP models and using the off-the-shelf solver, it is efficient to search for the best attacks exhaustively. As a result, we obtain the first attacks against the full (5-round) and an extended (5.5-round) version of Haraka-512 v2, and 8-round AES-128 hashing modes, as well as improved attacks covering more rounds of Haraka-256 v2 and other members of AES and Rijndael hashing modes.

A. Canteaut and F.-X. Standaert (Eds.): EUROCRYPT 2021, LNCS 12696, pp. 771–804, 2021.
https://doi.org/10.1007/978-3-030-77870-5_27

Keywords: AES · Haraka v2 · MITM · Preimage · Automatic search · MILP

1 Introduction

Hash function is one of the most important cryptographic primitives, due to its wide and crucial applications such as digital signatures, verification of message integrity and passwords etc. To support these applications, collision resistance, preimage resistance, and second-preimage resistance form the three basic security requirements for cryptographic hash functions. Unlike many public-key cryptographic systems, whose security can be usually reduced to some hard mathematical problems, most of the hash function standards in use could not enjoy such a security reduction. The confidence of the security strength of many symmetric-key primitives mainly relies on intensive and persistent cryptanalysis from the research community. Hence, such effort is of utmost importance, especially against the basic security properties of the standards and the ones used in practice. In this paper, we mainly focus on preimage resistance of hash functions built on the block cipher Advanced Encryption Standard (AES) [16] and the like (we call them AES-like hashing for short). Typical examples are the three PGV-modes [48] – Davies-Meyer (DM), Matyas-Meyer-Oseas (MMO), and Miyaguchi-Preneel (MP), instantiated with AES. Both PGV-modes and AES have long-standing security supported by rigorous and massive cryptanalysis, including the recent quantum collision attacks [18,30]. The MMO-mode instantiated with AES is standardized by Zigbee [1] and also suggested by ISO [32] as a standard way of building hash function based on block ciphers. Furthermore, many feature-rich cryptographic protocols, *e.g.*, multi-party computation protocols [22,34], use hash functions as building blocks and their instances adopt AES-MMO due to its high efficiency when implemented with AES-NI. Besides, since the standardization of AES, many new ciphers follow a similar design strategy or using AES round function directly as building blocks to share the security proof and implementation benefits, *e.g.*, hash functions Grindahl [38], ECHO [13], Grøstl [21], and Haraka v2 [39], and authenticated encryption [15].

THE MITM PREIMAGE ATTACKS. Informally, preimage resistance refers to the property that, for a hash function H and a target T given at random, it is *computationally* difficult to find a message x, such that $H(x) = T$. Theoretically, for a secure hash function with a digest of n bits in size, the expected number of H evaluations required to find such an x is 2^n. Any such algorithm with a time complexity lower than 2^n is considered as a *preimage attack.*

In [7], Aumasson *et al.* devised preimage attacks on step-reduced MD5 and full 3-pass HAVAL [64], in which the key technique can be viewed as the application of local-collision combined with the Meet-in-the-Middle (MITM) approach. Sasaki and Aoki in [51] formally proposed to combine the MITM and local-collision approaches and successfully devised preimage attacks on full versions of 3, 4, and 5-pass HAVAL. Further, they in [6] proposed the *splice-and-cut* technique and in [53] invented the concept of the *initial structure*, which add more strength to the MITM attack, and successfully broke the preimage resistance of the full MD5. These techniques were then formalized as *bicliques* [14,35,36], and further evolved to differential views [19,37]. Since these pioneering works, the MITM preimage attack turned out to be very powerful and found many applications in the last decade. It broke the theoretical preimage security claims of MD4 [26], MD5 [53], Tiger [26,59], HAVAL [27,51] and round-reduced variants of many other hash functions such as SHA-0 and SHA-1 [5,19,37], SHA-2 [4], BLAKE [19], HAS-160 [29], RIPEMD and RIPEMD-160 [60], Stribog [2], Whilwind [3], and AES hashing modes [10,49,62]. Interestingly, the idea of MITM preimage attack also leads to the progress of collision attacks against reduced SHA-2 [42] and key-recovery attack against full KTANTAN [61], which are technically different from the DS-MITM attacks [23–25].

The core of a MITM preimage attack on the hash function is generally a MITM pseudo-preimage attack on its compression function (denoted by CF). The basic idea of the attack on the CF is as follows (take the DM-mode as an example). First, the iterative round-based computation of the CF is divided at an intermediate round (starting point) into two chunks. One chunk is computed forward (named as *forward chunk*), the other is computed backward (named as *backward chunk*), and one of them is computed across the first and last rounds via the feed-forward mechanism of the hashing mode, and they end at a common intermediate round (matching point). In each of the chunks, the computation involves at least one distinct message word (or a few bits of it), such that they can be computed over all possible values of the involved message word(s) independently from the message word(s) involved in the other chunk (the distinct words are called *neutral words*). When an initial structure is used, it covers few consecutive rounds at the starting point, within which the two chunks overlapped and the neutral words for both chunks appear simultaneously, but still, the computations of the two chunks on the neutral words are independent.

In [49], Sasaki applied such MITM preimage attack to AES-hashing modes. Together with the partial matching technique, the attack successfully penetrated 7 out of the 10, 12, 14 rounds respectively for AES-128, AES-192, and AES-256. Later, Wu *et al.* in [62] improved the complexities in multi-target setting. Different from early MITM attacks on the MD-SHA family, their attacks select the neutral bytes from the internal state and fix the material fed into the key/message-schedule to an arbitrary constant. Recently, such attacks on AES hashing modes evolved to not only using the freedom of selecting the internal state but also exploiting the freedom of selecting the message state (key materials of the block ciphers), and improved results are achieved in [10]. Due to the fact that there are too many possible configurations (selection of neutral words, position of initial structure and matching rounds, extra conditions imposed to limit propagation of neutral words, etc.) to test out by bruteforce, all existing attacks cover only a small portion of configurations, which were believed to potentially give better cryptanalysis results according to the attackers' intuition and experiences.

AUTOMATIC TOOLS. In the last decade, cryptanalysis has also made significant progress from manual methods to those aided by dedicated computer programs searching for best differential/linear paths etc. [46] and best attacks [17], then to automatic tools for solving Constraint Satisfaction Problems (CSP), such as Mixed Integer Linear Programming (MILP), Constrained Programming (CP), Satisfiability Solvers (SAT), and Satisfiability Modulo Theories (SMT). These automatic tools convert the problem of finding better cryptanalytic attacks to optimization problems solvable by the tools, under certain constraints, which ensure the validity of the attacks. They not only enlarge the possible solution space covered by previous manual methods and dedicated search programs, but also helped generalize and even re-define the attack models which in turn further enlarge solution spaces. As a result, these tools have made significant advances in cryptanalysis, such as differential/linear path search [20,40,45,47,57,58], cube (-like) attacks [28,43,44,55,56], integral attacks based on division properties [63], three-subset and Demirci-Selçuk meet-in-the-middle attacks [50,54]. These usually lead to attacks for more rounds and/or lower time/memory complexities. With these available capacities, a more accurate security assessment is possible, and many recent primitive designs [9,12,39] benefited from these tools in determining the round number and the security margin with better confidence.

It is important to note that, literally every problem in cryptanalysis, complex or simple, can be converted into one under automatic tools. However, when the problem is complex, tools may not be able to output solutions in real time. Hence, different from the traditional manual cryptanalysis, the difficulty of tool-aided cryptanalysis is to find a proper model, which balances the problem solving time and size of solution space the model covers (number of attack configurations in case of AES-like hashing). Obviously, a model covering larger solution space comes with lesser constraints, which is harder to solve by the tools, but has bigger chances to offer better cryptanalysis results. All our effort in this paper is to convert the preimage finding problem into one under the MILP language, by a model covering largest possible solution space, while keeping the model solvable in practical time within our computation capacity in hands.

Our Contributions. In this paper, we manage to automatize the search for the best MITM preimage attacks with MILP models. We focus ourselves on hash functions built on AES and AES-like ciphers.

We extend the construction of attacks by removing the limitations taken by previous works [10,49,62]. That includes releasing the boundaries of the initial structure by applying the essential idea to every possible round; considering the possibility of imposing degree of freedom both from the internal state and from the message, which is done by allowing selecting neutral bytes from both of the encryption state and key state, and for both directions of computation; considering a desynchronized selection of neutral bytes in the encryption computation flow and the key-schedule flow (meaning that we allow the key state, from which the neutral bytes be selected, be at any possible round, instead of adhering to the round at where neutral bytes are selected in the encryption state) as appeared already *e.g.*, in [10,26].

We formalize the essential idea behind the advanced techniques used in the MITM preimage attack, including the above mentioned extended form of initial

Table 1. Results of applications of our tool compared with previous best results

Target	#Round	Time-1	Time-2	(DoF^+, DoF^-, DoM) in bits	Ref.
AES-128	7/10	2^{120}	2^{125}	(8, 8, 32)	[49]
	7/10	$2^{120-\min(t,24)}$	2^{123}	(8, 32, 32)	[62]
	7/10	2^{104}	2^{117}	(24, 32, 24)	[10]
	8/10	2^{120}	2^{125}	(16, 8, 8)	Fig. 7
AES-192	7/12	2^{120}	2^{125}	(8, 8, 32)	[49]
	7/12	2^{96}	2^{113}	(32, 32, 32)	[10]
	8/12	$2^{112-\min(t,16)}$	2^{116}	(16, 32, 32)	[10]
	9/12	2^{120}	2^{125}	(8, 8, 8)	*Fig. 9
AES-256	7/14	2^{120}	2^{125}	(8, 8, 32)	[49]
	8/14	2^{96}	2^{113}	(32, 32, 32)	[10]
	9/14	$2^{120-\min(t,24)}$	2^{123}	(8, 32, 32)	*Fig. 10
Rijndael-256	9/14	2^{248}	2^{253}	(16, 16, 8)	*Fig. 12
Haraka-256 v2	7/10	2^{248}	2^{248}	(8, 8, 96)	[39]
	9/10	2^{224}	2^{224}	(32, 32, 64)	*Fig. 13
Haraka-512 v2	8/10	2^{248}	2^{248}	(8, 8, 64)	[39]
	10/10	$2^{224-\min(t,32)}$	2^{224}	(128, 32, 64)	*Fig. 14
	11/10	2^{240}	2^{240}	(128, 128, 16)	Fig. 8

–* Please refer to the full version of this paper [11].
– Following [10], we use Time-1 to represent the time complexity of pseudo-preimage. Here, 2^t is the number of available targets for multi-target pseudo-preimage attacks; use Time-2 to represent the complexity of using the (multi-target) pseudo-preimage attacks to do (second-)preimage attacks when requiring an upper layer of meet-in-the-middle procedure of conversion for some PGV-modes, and here a single target is given. For Haraka-512 v2, the conversion is not needed and Time-2 should be the same with Time-1.
– The unit of complexity is one computation of the compression function.
– #Round is the number of AES-like round (one Haraka v2 round consists of two AES-like rounds).
– (DoF^+, DoF^-, DoM) is (the degree of freedom for forward computation, the degree of freedom for backward computation, the degree of matching), please refer to Sect. 3.

structure and the partial matching, using explicit-defined rules. In our formulation of the MITM preimage attacks, we do not pre-set any hard boundaries for the initial structures (*i.e.*, the number of rounds and which rounds are covered), but allow it to evolve automatically according to certain rules from well-defined and potentially desynchronized starting states towards a clear objective. Thanks to this formulation, the MITM preimage attack is ready to be transformed into MILP models covering a larger solution space than previous works.

We refine the MILP model for the operations involved in AES-like round functions to accurately capture all possible effects of them on the forward and backward computation paths. For example, instead of separately treating the `AddRoundKey` and `MixColumns`, we treat them as a whole (a composition transformation) and formalize constraints that can result in all possible impacts from the input states to the output state. In doing that, the models can capture the solutions where the difference in the active cells in the key state and that in encryption state be mutually (partially) canceled, which is impossible when treat the two operations separately. Such treatment further enlarges the search space to capture more potentially better attacks.

With such MILP models and using off-the-shelf solver, we apply the automatic search to AES-like hashing. Improved attacks than the previous ones were obtained. That includes the first preimage attacks on 8-round AES-128, 9-round AES-192, 9-round AES-256, 9-round Rijndael hashing modes, 4.5-round (9 AES-rounds) Haraka-256 v2 and the full 5-round (10 AES-rounds) version and extended 5.5-round (11 AES-rounds) version of Haraka-512 v2. The detailed results, together with a comparison to the previous related works, are summarized in Table 1.

Note that the modified versions of Haraka v2 are used in instantiations of SPHINCS$^+$ (which replaces the DM-mode with Sponge-based construction) [31] and Gravity-SPHINCS (which extends one round on top of the 5-round version) [8]. Gravity-SPHINCS is one of the first round, and SPHINCS$^+$ is one of the third round alternate candidates of digital signatures in the NIST Post-Quantum Cryptography Standardization Process. Our attacks on Haraka v2 do not directly break the security of SPHINCS$^+$-Haraka and Gravity-SPHINCS-Haraka. For SPHINCS$^+$-Haraka, the security relies on a preimage resistance of 128-bit rather than 256-bit. For Gravity-SPHINCS-Haraka, the security relies on a collision resistance of 128-bit rather than preimage resistance, besides, the underlying Haraka v2 variants have increased the AES-like rounds from 10 to 12, while our attacks cover at most 11 rounds.

2 AES-like Hashing and MITM Preimage Attacks

Most current hash functions are based on compression functions (CF) with fixed length input and output; and the support for variable-length messages can be achieved through domain extenders. Here, we focus on the challenge of inverting the CF, i.e., given one or multiple targets T, find input chaining value h and message block M, such that $\mathsf{CF}(h, M) = T$. Such attacks are called pseudo-preimage attacks, in which the chaining value is free of choice. Pseudo-preimages can be converted to (second-)preimages of hash functions using generic methods (details can be found in Appendix C of the full version [11]).

2.1 AES-like Hashing

Typically, the compression function of hash functions can be constructed from block ciphers applying the secure PGV-modes [48]. When the underlying block ciphers are AES-like, we call the hash functions as AES-like hashing. Concretely, in AES-like hashing, the underlying compression function is based on AES-like round functions as depicted in Fig. 2, where the state being manipulated is organized into an $\mathrm{N}_{\mathtt{row}} \times \mathrm{N}_{\mathtt{col}}$ two-dimensional array of c-bit cells. One AES-like round function typically consists of the following operations:

- **SubBytes.** Substitute each cell according to an S-boxes $S : \mathbb{F}_{2^c} \rightarrow \mathbb{F}_{2^c}$.

- **ShiftRows$_{\pi_t}$.** Permute the cell positions according to the permutation π_t.
- **MixColumns.** Update each column by left-multiplying an $\mathtt{N_{row}} \times \mathtt{N_{row}}$ MDS matrix (maximal distance separable matrix, with branch number $\mathtt{B_n} = \mathtt{N_{row}} + 1$, *i.e.*, as long as the input/output of the MDS matrix is non-zero, the sum of non-zero elements in the input and output is at least $\mathtt{N_{row}} + 1$).
- **AddRoundKey.** XOR a round key or a round-dependent constant into the state depending on whether the intended construction is keyed or not.

2.2 Advanced Techniques in Meet-in-the-Middle Preimage Attacks

Since the pioneering works on preimage attacks on MD4, MD5, and HAVAL [7,41, 51,52], the MITM approach has been applied and further developed for preimage attacks on many other hash functions. This method develops into *splice-and-cut* [6] MITM preimage attacks with support from *initial structure* [53] and (indirect) *partial matching* techniques.

Initial Structures [53]. From the idea of local-collision, Sasaki and Aoki proposed a novel concept – *initial structure*. The purpose of the initial structure is to skip several steps/rounds at the beginning of chunks in a MITM attack so that the attack covers more steps/rounds. It is a few consecutive starting steps, where the two chunks overlapped. Although the two sets of neutral words, denoted by N^+ and N^-, appear simultaneously at these steps, they are only involved in the computation of one chunk each. Besides, one can add constraints to the values of neutral words of one chunk, such that different values lead to constant impact on the computation of the opposite chunk. Thus, a proper initial structure should satisfy that, steps after the initial structure (forward chunk) can be computed independently of N^- and steps before the initial structure (backward chunk) can be computed independently of N^+.

Remark 1 (Related work – the formalism of Biclique). Notably, the initial structure was viewed as the most promising and underutilized technique for MITM preimage attack in the subsequent years since its invention. In [36], authors replaced the idea of initial structure with a more formal and general concept, which is named biclique. With this formalism, one can view the structure in a differential view, and built it by applying various tools available for collision search and differential attacks. This concept of biclique has been applied to both preimage attacks on hash functions (*e.g.*, SHA-1, SHA-2 and Skein-512 [36,37]) and key-recovery attacks on block ciphers (*e.g.*, AES and IDEA [14,35]).

In this paper, independent of the formalism using concept of biclique, and instead of adhering to a formal definition, we apply the essential idea behind the original concept of initial structure. We formalize the basic idea using explicit rules and extend the initial structure to be less structured.

(Indirect-) Partial matching [6,53]. In the two ending states for matching, as long as there remain one common word of which the value can be computed independently between the forward and backward chunks, the matching can be performed. Further, apart from directly matching values of common words, any

determined relations between words in the states at the matching point can be exploited to filter out miss-matched computations. For example, Sasaki in [49] exploited the following property of the AES MixColumns to do indirect matching: knowing any b bytes ($b > 4$) among the input and output of MixColumns on one column, one can built a filter of $b - 4$ bytes. For example, in Fig. 1d, it is possible to do partial matching between states $\#\mathrm{MC}^1$ and $\#\mathrm{AK}^1$, and each column provides $2 + 3 - 4 = 1$ byte filter, as exemplified in Fig. 1b.

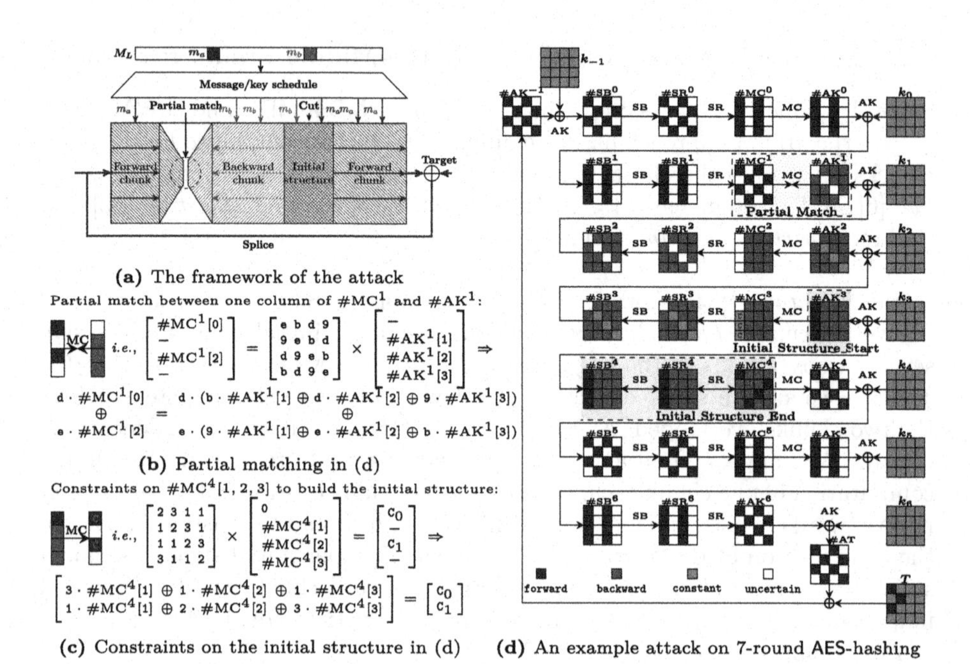

Fig. 1. The MITM pseudo-preimage attack [49,62] (Color figure online)

Multi-targets [26,62]. When multiple targets are available, it adds the degree of freedom to the chunk where the targets are added to.

The Attack Framework. The procedure (Fig. 1) and complexities of the MITM pseudo-preimage attack depend on the following configurations:

1. Chunk separation – the position of initial structure and matching points.
2. The neutral bytes – the selection and the constraints on the neutral bytes, which determine the degrees of freedom for each chunk.
3. The bytes for matching – the deterministic relation used for matching, which determines the filtering ability (degree of matching).

After setting up the configuration, the basic attack procedure goes as follows. Denote the neutral bytes for the forward and backward chunk by N^+ and N^-:

1. Assign arbitrary compatible values to all bytes except those that depend on the neutral bytes (*e.g.*, the Gray cells in Fig. 1d).

2. Obtain possible values of neutral bytes N^+ and N^- under the constraints on them (*e.g.*, in Fig. 1c). Suppose there are 2^{d_1} values for N^+, and 2^{d_2} for N^-.
3. For all 2^{d_1} values of N^+, compute forward from the initial structure to the matching point to get a table L^+, whose indices are the values for matching, and the elements are the values of N^+.
4. For all 2^{d_2} values of N^-, compute backward from the initial structure to the matching point to get a table L^-, whose indices are the values for matching, and the elements are the values of N^-.
5. Check whether there is a match on indices between L^+ and L^-.
6. In case of partial-matching exist in the above step, for the surviving pairs, check for a full-state match. In case none of them are fully matched, repeat the procedure by changing values of fixed bytes till find a full match.

The Attack Complexity. Denote the size of the internal state by n, the degree of freedom in the forward and backward chunks by d_1 and d_2, and the number of bits for the match by m, the time complexity of the attack is [10]:

$$2^{n-(d_1+d_2)} \cdot (2^{\max(d_1,d_2)} + 2^{d_1+d_2-m}) \simeq 2^{n-\min(d_1,d_2,m)}. \quad (1)$$

2.3 Basic Rules Applied to MITM Attacks on AES-like Hashing

Sources of Degrees of Freedom. Shown by the complexity analysis, the MITM attack benefits from larger degrees of freedom in both chunks and matching. In early MITM preimage attacks on the MD-SHA family, the degree of freedom comes from the message words. Whereas, in early MITM preimage attacks on AES-like hashing [49,62], the degree of freedom comes from the bytes in encryption states[1], and the attacks set the material fed into the key-schedule as arbitrary constant. In [10], the authors proposed to introduce neutral bytes not only from the encryption state but also from the key state. The principle is that, for one chunk, one adds as much degree of freedom as possible to improve the computational complexity, and at the same time, keeps their impacts on the opposite chunk as little as possible to cover as many rounds as possible. To keep the analysis manually doable, the authors in [10] proposed that the neutral bytes in key states are all introduced for merely one chunk.

Ways to Control Impacts on the Opposite Chunk. For the ways to cancel impacts from neutral words for one chunk on the opposite chunk, recall that early preimage attacks on MD-SHA used the (cross) absorption properties of Boolean functions by setting an input variable to a special value to absorb the difference in another input variable. In the attack on AES-like hashing, the ways to control the impacts of the neutral bytes is to add constraints on those neutral bytes

[1] In a hash function, there is no encryption and key-schedule. Here, focusing on hash functions built on block ciphers, we use them to represent the two algorithms updating the chaining values and updating the message words. For different mode-of-operations, the correspondence might be different.

when they are inputs to the following operations. Note that adding constraints means consuming the degree of freedom.

- `AddRoundKey` and `XOR`: one can restrict that the `XOR` of two neutral bytes be constant. The rationale is to use the difference in one neutral byte (*e.g.*, in the key state) to absorb the difference in another neutral byte (*e.g.*, in the encryption state). That will consume one-byte degree of freedom.
- `MixColumns` (MC): Even if the input contains neutral bytes (active) for one chunk, one can add restriction on their values, such that their impacts on some output bytes of the `MC` be constant. Therefore, the opposite chunk can be computed independently as long as the constant impacts are known. Take the attack in Fig. 1d for example. In the computation from $\#\text{MC}^4$ to $\#\text{AK}^4$, the values of `Red` cells in state $\#\text{MC}^4$ are restricted such that changing them does not change impact on the `Blue` cells marked by `C` in $\#\text{AK}^4$ (exemplified in Fig. 1c). This restriction consumes the degree of freedom that lies in neutral bytes for backward chunk, but enables the independent forward computation. Explicitly, if there are i neutral bytes for one chunk involved in the input of `MC` , then we can control their impacts on j bytes of the output be constant by consuming j bytes degree of freedom. For AES-like hashing, because the matrix `MC` in `MixColumns` is MDS, there is a limitation for applying this control, that is $i + \text{N}_{\texttt{row}} - j \geq \text{N}_{\texttt{row}} + 1$, *i.e.*, $i \geq j + 1$.
- `MixColumns` $\circ$ `AddRoundKey` (XOR-MC): in backward chunk, when there are forward neutral bytes in both the key and the encryption state, to control their impacts, one may first apply the above-mentioned way of restriction on `AddRoundKey` and then on `MixColumns`. Besides that, we apply restriction on the composition transformation of `AddRoundKey` and `MixColumns`. The rationale is that, the `XOR` operation in `AddRoundKey` is byte-wise. Only when two bytes being *at the same position* in two states, the difference in one byte can absorb the difference in the other byte. As for `MixColumns`, only when two bytes being *in the same state*, the difference in one byte can absorb the difference in the other byte. However, when considering the composition `MixColumns`$\circ$`AddRoundKey`, even when the neutral bytes for the forward chunk lie in different states (some in the key state and some in the encryption state) and in different byte positions, we can still use the difference of some neutral bytes to absorb the difference of others. Section 4.1 and the listed attacks will provide formal descriptions and concrete examples.
  Explicitly, suppose that there are i forward neutral bytes in the key state, and j forward neutral bytes in the encryption state, and they lie in columns with a common index. Let k be the number of different byte positions considering these neutral bytes together (*i.e.*, k equals the Hamming weight of the 'OR' between the indicator vector of whether a position has a neutral byte in the key state and that in the encryption state). Then, considering the MDS property of `MC` in `MixColumns`, we can control the impacts of neutral bytes on t bytes of the output by consuming t bytes degree of freedom as long as $k + \text{N}_{\texttt{row}} - t \geq \text{N}_{\texttt{row}} + 1$, *i.e.*, $k \geq t + 1$.

Remark 2. (Relation with previous MITM attacks on AES hashing modes). Note that the ways to control the impacts have already been used in previous MITM preimage attacks on AES-like hashing [10,49,62], which is an essential element for constructing the initial structure. In this paper, we consider the possibility to impose such constraints to any round, and in this sense, the boundaries of the initial structure disappear. Besides, as has been mentioned above, the ways to select the neutral bytes were limited in previous works to make the analysis doable manually. In this paper, we remove these restrictions by allowing the selection of neutral bytes in both encryption state and key state, and for both forward and backward chunks.

In the subsequent sections, we will use these ideas to get explicit rules for selecting neutral bytes, consuming degree of freedom on neutral bytes to control their impacts. Incorporating with other optimization techniques (*e.g.*, partial matching and multi-targets), we convert the problem of searching for the best configurations into optimization problems under constraints in MILP-models. With the obtained MILP-models and the off-the-shelf solver, we can search for the best MITM attacks on AES-like hashing exhaustively.

Remark 3 (Relation with another work on using MILP to searching MITM attack). In [50], Sasaki already applied the MILP formalization to search the three-subset MITM attack on GIFT-64. In the tool, which rounds covered by an initial structure are predefined. Neutral bits are all from the key state because the goal is a key-recovery attack. Besides, because it is dedicated to GIFT-64 (with a bit-permutation linear layer), the previously mentioned rules for optimizing MITM attacks on AES-like hashing are not included, which is essentially the most challenging parts in our formalization.

3 Formulate the MITM Attack on AES-like Hashing

To search for MITM attacks on AES-like hashing, we now formulate the attack with the general construction shown in Fig. 2.

Denote the *starting states* in the encryption data path and key-schedule data path by $S^{\mathtt{ENC}}$ and $S^{\mathtt{KSA}}$, respectively (corresponding to the location of an initial structure previously); and denote the *ending states* for the forward computation and backward computation by E^+ and E^-, respectively (corresponding to the previous matching). In the formalized attack, partial knowledge of E^+ and E^- that is used for matching is supposed to be obtained by computing from $S^{\mathtt{ENC}}$ and $S^{\mathtt{KSA}}$ forward and backward, respectively[2].

Without loss of generality, we assume that the states in the encryption data paths and the key-schedule both have n c-bit cells (with $n = \mathrm{N}_{\mathtt{row}} \cdot \mathrm{N}_{\mathtt{col}}$). To reference the cells of certain n-cell states, denote by $\mathcal{B}^{\mathtt{ENC}}$, $\mathcal{B}^{\mathtt{KSA}}$, $\mathcal{R}^{\mathtt{ENC}}$, $\mathcal{R}^{\mathtt{KSA}}$, $\mathcal{C}$, and $\mathcal{D}$ the ordered subsets of $\mathcal{N} = \{0, 1, \cdots, n-1\}$ whose elements are increasingly

[2] Note that after finding out a formalized attack, adaptation will be made manually to launch a concrete attack; the forward and backward computations may start from the most decisive states instead of $S^{\mathtt{ENC}}$ and $S^{\mathtt{KSA}}$ while keeping the complexity.

ordered. Here, the $\mathcal{B}^{\text{ENC}}$ and $\mathcal{B}^{\text{KSA}}$ refer to the neutral cells from the internal state and message (or key state of the underlying block cipher) for the forward chunk, and $\mathcal{R}^{\text{ENC}}$ and $\mathcal{R}^{\text{KSA}}$ for the backward chunk. The $\mathcal{C}$ and $\mathcal{D}$ refer to the known and active cells in the ending states E^+ and E^- of the forward and backward chunks, respectively. For example, we may have $\mathcal{C} = \{0, 2, 7\}$, and for a 16-cell state S, $S[\mathcal{C}]$ is defined to be $(S[0], S[2], S[7])$ or $S[0, 2, 7]$.

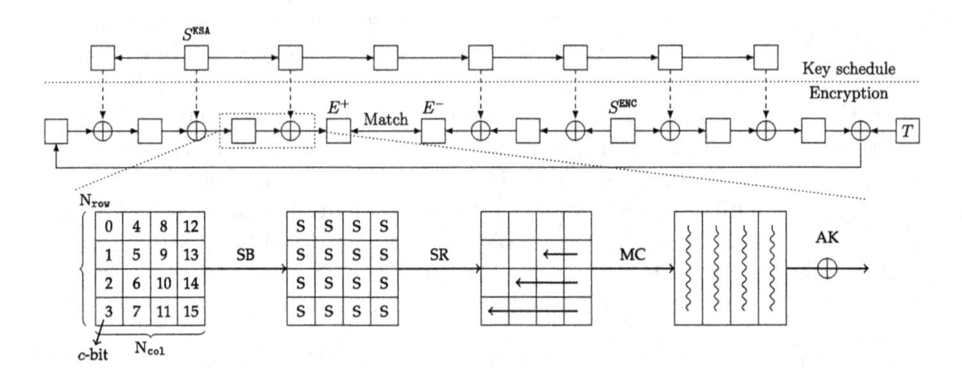

Fig. 2. A high-level overview of the MITM preimage attack

Before one can mount a MITM preimage attack, these four states: S^{ENC}, S^{KSA}, E^+, E^-, and six subsets $\mathcal{B}^{\text{ENC}}$, $\mathcal{B}^{\text{KSA}}$, $\mathcal{R}^{\text{ENC}}$, $\mathcal{R}^{\text{KSA}}$, $\mathcal{C}$, $\mathcal{D}$ ($\mathcal{B}^{\text{ENC}} \cap \mathcal{R}^{\text{ENC}} = \emptyset$ and $\mathcal{B}^{\text{KSA}} \cap \mathcal{R}^{\text{KSA}} = \emptyset$ for independence between chunks) must be specified.

Note that to visualize these subsets and the attack, we will introduce a coloring system in Sect. 4, where cells referenced by $\mathcal{B}^{\text{ENC}}$ and $\mathcal{B}^{\text{KSA}}$ are `Blue`, and cells referenced by $\mathcal{R}^{\text{ENC}}$ and $\mathcal{R}^{\text{KSA}}$ are `Red`. The remaining cells in the starting states referenced by $\mathcal{G}^{\text{ENC}}$ and $\mathcal{G}^{\text{KSA}}$ are `Gray`, where $\mathcal{G}^{\text{ENC}} = \mathcal{N} - \mathcal{B}^{\text{ENC}} \cup \mathcal{R}^{\text{ENC}}$ and $\mathcal{G}^{\text{KSA}} = \mathcal{N} - \mathcal{B}^{\text{KSA}} \cup \mathcal{R}^{\text{KSA}}$. Moreover, $\mathcal{C}$ references the `Blue` cells in the ending state E^+, and $\mathcal{D}$ the `Red` cells in E^-.

In what follows, the degree of freedom (DoF) refers to number of cells, rather than bits. We call $\lambda^+ = |\mathcal{B}^{\text{ENC}}| + |\mathcal{B}^{\text{KSA}}|$ the initial DoF for the forward chunk, and $\lambda^- = |\mathcal{R}^{\text{ENC}}| + |\mathcal{R}^{\text{KSA}}|$ the initial DoF for the backward chunk. For forward and backward chunks being computed independently, these initial DoFs might be consumed by adding constraints on neutral cells in S^{ENC} and S^{KSA}. Thus, neutral cells in the starting states may not take all $2^{c \cdot \lambda^+}$ and $2^{c \cdot \lambda^-}$ values.

If the forward neutral cells $(S^{\text{ENC}}[\mathcal{B}^{\text{ENC}}], S^{\text{KSA}}[\mathcal{B}^{\text{KSA}}])$ (in `Blue`) in the starting states can only take values in $\mathbb{X} \subseteq \mathbb{F}_{2^c}^{|\mathcal{B}^{\text{ENC}}|+|\mathcal{B}^{\text{KSA}}|}$ with $|\mathbb{X}| = (2^c)^{d_1} \leq (2^c)^{|\mathcal{B}^{\text{ENC}}|+|\mathcal{B}^{\text{KSA}}|}$, and the backward neutral cells $(S^{\text{ENC}}[\mathcal{R}^{\text{ENC}}], S^{\text{KSA}}[\mathcal{R}^{\text{KSA}}])$ (in `Red`) in the starting states can only take values in $\mathbb{Y} \subseteq \mathbb{F}_{2^c}^{|\mathcal{R}^{\text{ENC}}|+|\mathcal{R}^{\text{KSA}}|}$ with $|\mathbb{Y}| = (2^c)^{d_2} \leq (2^c)^{|\mathcal{R}^{\text{ENC}}|+|\mathcal{R}^{\text{KSA}}|}$, then after fixing the `Gray` cells $(S^{\text{ENC}}[\mathcal{G}^{\text{ENC}}], S^{\text{KSA}}[\mathcal{G}^{\text{KSA}}])$ in the starting states to some constant in $\mathbb{F}_{2^c}^{(n-|\mathcal{B}^{\text{ENC}}|-|\mathcal{R}^{\text{ENC}}|)+(n-|\mathcal{B}^{\text{KSA}}|-|\mathcal{R}^{\text{KSA}}|)}$, the attacker can compute $(2^c)^{d_1}$ different values of $E^+[\mathcal{C}]$ in the forward direction which only depend on $(S^{\text{ENC}}[\mathcal{B}^{\text{ENC}}], S^{\text{KSA}}[\mathcal{B}^{\text{KSA}}])$. The attacker stores these $(2^c)^{d_1}$

values in a list L^+. Similarly, the attacker can compute $(2^c)^{d_2}$ different values of $E^-[\mathcal{D}]$ in the backward direction which only depend on $(S^{\texttt{ENC}}[\mathcal{R}^{\texttt{ENC}}], S^{\texttt{KSA}}[\mathcal{R}^{\texttt{KSA}}])$. The attacker stores these $(2^c)^{d_2}$ values in a list L^-. For the two lists L^+ and L^-, the attacker can perform an m-cell matching. Then, $|L^+ \times L^-|/(2^c)^m$ pairs from $L^+ \times L^-$ are expected to pass the test.

We call m the degrees of matching (denoted by DoM). Note that $\mathcal{B}^{\texttt{ENC}}$ and $\mathcal{B}^{\texttt{KSA}}$ indicate the sources of the degrees of freedom for the forward computation, and $\mathcal{R}^{\texttt{ENC}}$ and $\mathcal{R}^{\texttt{KSA}}$ indicate the sources of the degrees of freedom for the backward computation. Since in the forward computation and backward computation, $(S^{\texttt{ENC}}[\mathcal{B}^{\texttt{ENC}}], S^{\texttt{KSA}}[\mathcal{B}^{\texttt{KSA}}])$ and $(S^{\texttt{ENC}}[\mathcal{R}^{\texttt{ENC}}], S^{\texttt{KSA}}[\mathcal{R}^{\texttt{KSA}}])$ are restricted to $\mathbb{X}$ and $\mathbb{Y}$ respectively, with $|\mathbb{X}| = (2^c)^{d_1}$ and $|\mathbb{Y}| = (2^c)^{d_2}$, we call d_1 the degrees of freedom for the forward computation (denoted by DoF^+) and d_2 the degrees of freedom for the backward computation (denoted by DoF^-).

With this configuration, it is shown that the time complexity to find a full n-cell match between the two ending states is $(2^c)^{n-\min\{d_1,d_2,m\}}$. Therefore, for a valid MITM preimage attack, we must have $\text{DoF}^+ \geq 1$, $\text{DoF}^- \geq 1$, and $\text{DoM} \geq 1$. In the following section, we will show how to automatically determine $\mathcal{B}^{\texttt{ENC}}$, $\mathcal{B}^{\texttt{KSA}}$, $\mathcal{R}^{\texttt{ENC}}$, $\mathcal{R}^{\texttt{KSA}}$, $\mathcal{C}$, and $\mathcal{D}$ with MILP such that the complexity $(2^c)^{n-\min\{\text{DoF}^+,\text{DoF}^-,\text{DoM}\}}$ of the corresponding attack is minimized *when the starting states and ending states are given.* Note that the choices of the starting states and ending states are quite limited and thus can be enumerated automatically.

Remark 4. Our program enumerates all combinations of the locations of starting and ending points in encryption, and all combinations of the locations of starting points in the encryption and key-schedule algorithm. That is, for an N-round targeted cipher, our program generates MILP-models for each of the possible combinations $\{(\texttt{init}_r^{\text{E}}, \texttt{init}_r^{\text{K}}, \texttt{match}_r) \mid 0 \leq \texttt{init}_r^{\text{E}} < N,\ -1 \leq \texttt{init}_r^{\text{K}} < N,\ 0 \leq \texttt{init}_r^{\text{E}} < N,\ \texttt{init}_r^{\text{E}} \neq \texttt{match}_r\}$, where $\texttt{init}_r^{\text{E}}$ is the location of starting point in encryption, $\texttt{init}_r^{\text{K}}$ is that in key-schedule, and $\texttt{match}_r$ is the location of the matching point. To find the optimal attacks, the MILP solver solves them all. Note that for each individual model, the locations of the matching and the initial states are set, but the states are not set.

Note 1 (Tricks for matching the ending states as indirect matching and matching through MixColumns used in [4,10,26]*).* Note that in the MITM preimage attack on AES-like hash functions, the last sub-key addition leading to E^- is close to the boundary of the forward and backward computation as illustrated in Fig. 15a in the full version [11].

Therefore, to perform matching, one can decompose state as $K = K^+ + K^-$, and translate the computation in Fig. 15a in the full version [11] into its equivalent form shown in Fig. 15b in the full version [11], since $\text{MC}(E^+) \oplus K = \text{MC}(E^+ \oplus \text{MC}^{-1}(K^+)) \oplus K^-$. Full explanation can be found in Appendix C of the full version [11].

In the following description of our modeling method, for simplicity, we let the number of rows of the state $\text{N}_{\texttt{row}}$ be 4, and thus, the branch number of the

MixColumns $B_n = N_{row} + 1$ be 5. However, the modeling method can be directly applied to other AES-like hashing that formalized in Sect. 2.1.

4 Programming the MITM Preimage Attacks with MILP

To facilitate the visualization of our analysis, each cell can take one of the four colors (Gray, Red, Blue, and White) according to certain rules, and a valid coloring scheme in our model corresponds to a MITM pseudo-preimage attack. The semantics of the colors of cells are listed as follows.

- Gray (G): known constant in both forward and backward chunk.
- Red (R): known and active in the backward chunk but unknown in the forward.
- Blue (B): known and active in the forward chunk but unknown in the backward.
- White (W): unknown in both the forward and backward chunk.

For the ith cell of a state S, we introduce two 0–1 variables x_i^S and y_i^S to encode its color, where $(x_i^S, y_i^S) = (0, 0)$ represents W, $(x_i^S, y_i^S) = (0, 1)$ represents R, $(x_i^S, y_i^S) = (1, 0)$ represents B, and $(x^S, y^S) = (1, 1)$ represents G. The encoding scheme is chosen such that $x_i^S = 1$ if and only if $S[i]$ is a known cell for the forward computation, and $y_i^S = 1$ if and only if $S[i]$ is a known cell for the backward computation. Under this encoding scheme, the number of Blue cells and Gray cells (known cells for the forward computation) in S can be computed as $\sum_i x_i^S$. Similarly, the number of Red cells and Gray cells (known cells in the backward computation) in S can be computed as $\sum_i y_i^S$. We also introduce an indicator 0-1 variable β_i^S for each cell such that $\beta_i^S = 1$ if and only if the cell $S[i]$ is Gray, which can be described by the constraints in Eq. (2).

Under these constraints, the number of Blue cells in S can be computed as $\sum_i x_i^S - \sum_i \beta_i^S$, and the number of Red cells in S can be computed as $\sum_i y_i^S - \sum_i \beta_i^S$. Moreover, the Blue cells in the starting states are used to capture $(S^{\text{ENC}}[\mathcal{B}^{\text{ENC}}], S^{\text{KSA}}[\mathcal{B}^{\text{KSA}}])$, and the Red cells in the starting states are used to capture $(S^{\text{ENC}}[\mathcal{R}^{\text{ENC}}], S^{\text{KSA}}[\mathcal{R}^{\text{KSA}}])$.

Constraints for the Starting States. For the starting states, we introduce two additional variables λ^+ and λ^- that compute the so-called *initial degrees of freedom*, where λ^+ (the initial DoF for the forward computation) is defined as the number of Blue cells in S^{ENC} and S^{KSA}, and λ^- (the initial DoF for the backward computation) is defined as the number of Red cells in S^{ENC} and S^{KSA}. Putting the definitions into equations, we have Eq. (3).

$$\begin{cases} x_i^S - \beta_i^S \geq 0; \\ y_i^S - \beta_i^S \geq 0; \\ x_i^S + y_i^S - 2\beta_i^S \leq 1. \end{cases} \quad (2)$$

$$\begin{cases} \lambda^+ = \sum_i x_i^{S^{\text{ENC}}} - \sum_i \beta_i^{S^{\text{ENC}}} + \sum_i x_i^{S^{\text{KSA}}} - \sum_i \beta_i^{S^{\text{KSA}}}; \\ \lambda^- = \sum_i y_i^{S^{\text{ENC}}} - \sum_i \beta_i^{S^{\text{ENC}}} + \sum_i y_i^{S^{\text{KSA}}} - \sum_i \beta_i^{S^{\text{KSA}}}. \end{cases} \quad (3)$$

Constraints for the Ending States. To be concrete, we describe the constraints for matching through the `MixColumns` operation of AES.

Property 1. Let $(E^-[4j], E^-[4j+1], E^-[4j+2], E^-[4j+3])^T$ and $(E^+[4j], E^+[4j+1], E^+[4j+2], E^+[4j+3])^T$ be the jth columns of the ending states E^- and E^+ that are linked by the `MixColumns` operation. When t ($t \geq 5$) out of the 8 bytes of the two columns are known, there is a filter of $t-4$ bytes.

Since the time complexity of the attack is $(2^c)^{n-\min\{\text{DoF}^+, \text{DoF}^-, \text{DoM}\}}$, we must impose the constraint DoM ≥ 1 to ensure a valid attack. The known bytes of the jth column of the ending state E^+ for the forward computation path from the starting states to E^+ can be computed in our model as $\sum_{i=0}^{3}(x_{4j+i}^{E^+} + y_{4j+i}^{E^+} - \beta_{4j+i}^{E^+})$. Similarly, the known bytes of the jth column of the ending state E^- for the backward computation path from the starting states to E^- can be computed as $\sum_{i=0}^{3}(x_{4j+i}^{E^-} + y_{4j+i}^{E^-} - \beta_{4j+i}^{E^-})$. Therefore, according to Property 1, we have the constraints for DoM in Eq. (4) (suppose each state has four columns).

$$\begin{cases} \text{DoM} = \sum_{j=0}^{3} \max\{0, \left(\sum_{i=0}^{3}(x_{4j+i}^{E^+} + y_{4j+i}^{E^+} - \beta_{4j+i}^{E^+}) + \sum_{i=0}^{3}(x_{4j+i}^{E^-} + y_{4j+i}^{E^-} - \beta_{4j+i}^{E^-}) - 4\right)\}; \\ \text{DoM} \geq 1. \end{cases} \tag{4}$$

Constraints for the States in the Computation Paths. This is an essential part of this work. In this part, we extend the construction of attacks on the basis of previous works. We refine and apply the critical idea behind the initial structure to a greater extent, and explicitly describe more possible ways to propagate the attributes (expressed in the four colors) of the cells that are involved in computation paths in both the encryption and the key-schedule. Therefore, we would like to devote one separate whole section (Sect. 4.1) for the details of this part. Here we only give some high-level descriptions.

Let f be an operation that transforms a state $S_{\texttt{IN}}$ into a state $S_{\texttt{OUT}}$. Then the coloring scheme of $(S_{\texttt{IN}}, S_{\texttt{OUT}})$ must obey certain rules associated with f and the direction of the computation in which f is involved, such that the semantics of the colors are respected.

If we restrict the `Red` cells $(S^{\texttt{ENC}}[\mathcal{R}^{\texttt{ENC}}], S^{\texttt{KSA}}[\mathcal{R}^{\texttt{KSA}}])$ in the starting states to some carefully constructed set $\mathbb{Y}$ defined in Sect. 3, it may be valid to transform certain `Red` cells in $S_{\texttt{IN}}$ to `Gray` cells (or even `Blue` cells) in $S_{\texttt{OUT}}$ by some operations along the forward computation path (starting from the starting states to the ending state E^+). By doing so, impacts from the `Red` cells on the forward computation are limited, meanwhile, the degrees of freedom of the `Red` cells in the starting states should be reduced from λ^-; similar situations happen along the backward computation path (starting from the starting states to the ending state E^-). In our MILP model, we must keep track of how much degrees of freedom are consumed to ensure the remaining degrees of freedom for the forward computation (DoF$^+$) and for the backward computation (DoF$^-$) always greater or equal to one. The variables and constraints introduced for the above purpose are detailed in Sect. 4.1.

The Objective Function. To minimize the time complexity of the attack, $\min\{\mathrm{DoF}^+, \mathrm{DoF}^-, \mathrm{DoM}\}$ should be maximized. To this end, we can introduce an auxiliary variable $v_{\mathtt{Obj}}$, impose the constraints in Eq. (5) and set the objective function to maximize $v_{\mathtt{Obj}}$.

In the multi-target setting, we suppose that the degree of freedom for the chunk to which the targets are added can be directly increased. Thus, for models where the starting point (resp. matching point) is at the upper round than the matching point (resp. starting point), DoF^- (resp. DoF^+) can be directly increased, the objective is to maximize $\min\{\mathrm{DoF}^+, \mathrm{DoM}\}$ (resp. $\min\{\mathrm{DoF}^-, \mathrm{DoM}\}$).

$$\begin{cases} v_{\mathtt{Obj}} \leq \mathrm{DoF}^+; \\ v_{\mathtt{Obj}} \leq \mathrm{DoF}^-; \\ v_{\mathtt{Obj}} \leq \mathrm{DoM}. \end{cases} \quad (5) \qquad \begin{cases} \mathrm{DoF}^+ = \lambda^+ - \sigma^+; \\ \mathrm{DoF}^- = \lambda^- - \sigma^-. \end{cases} \quad (6)$$

4.1 MILP Constraints for the States in the Computation Paths and the Consumption of Degrees of Freedom

Recalling the formalized framework of MITM attack in Sect. 3, before we perform the attack on a given target with predefined positions of starting states and ending states, we have to determine $\mathcal{B}^{\mathtt{ENC}}$, $\mathcal{B}^{\mathtt{KSA}}$, $\mathcal{R}^{\mathtt{ENC}}$, and $\mathcal{R}^{\mathtt{KSA}}$ for the starting states $\mathcal{S}^{\mathtt{ENC}}$ and $\mathcal{S}^{\mathtt{KSA}}$. In our visualizations of the attacks, the `Blue` cells in the starting states $\mathcal{S}^{\mathtt{ENC}}$ and $\mathcal{S}^{\mathtt{KSA}}$ are meant to capture $\mathcal{B}^{\mathtt{ENC}}$ and $\mathcal{B}^{\mathtt{KSA}}$ respectively. Similarly, the `Red` cells in the starting states are used to capture $\mathcal{R}^{\mathtt{ENC}}$ and $\mathcal{R}^{\mathtt{KSA}}$, and the `Gray` cells in the starting states are used to capture $\mathcal{G}^{\mathtt{ENC}}$, and $\mathcal{G}^{\mathtt{KSA}}$.

Therefore, according to Eq. (3), the number of `Blue` cells and the number of `Red` cells in the starting states correspond to the initial degrees of freedom λ^+ and λ^-, respectively. To control the impacts from neutral cells in one direction on the opposite direction, along the computation paths leading to the ending states, the initial degrees of freedom are consumed according to the coloring schemes.

Basically, forward computation consumes λ^-, and backward computation consumes λ^+. The consumption of degrees of freedom is counted in cells. Let σ^+ and σ^- be the accumulated degrees of freedom that have been consumed in the backward and forward computation paths, respectively. We have Eq. (6) for calculating the remaining degrees of freedom. That is, the remaining DoF for the forward computation is computed as the initial DoF of the forward computation minus the DoF consumed by the backward computation (from the starting state to the ending state E^-), and the remaining DoF of the backward computation is computed as the initial DoF of the backward computation minus the

DoF consumed by the forward computation (from the starting state to the ending state E^+). Since the complexity of the attack is $(2^c)^{n-\min\{\text{DoF}^+,\text{DoF}^-,\text{DoM}\}}$, we always require $\text{DoF}^+ \geq 1$ and $\text{DoF}^- \geq 1$. Moreover, σ^+ is computed as $\sum \sigma^+(S_{\texttt{IN}} \rightarrow S_{\text{out}})$ along the computation path that consumes DoF for the forward computation, where $\sigma^+(S_{\texttt{IN}} \rightarrow S_{\texttt{OUT}})$ is the DoF for the forward computation consumed by the transition from state $S_{\texttt{IN}}$ to $S_{\texttt{OUT}}$, and σ^- is computed as $\sum \sigma^-(S_{\texttt{IN}} \rightarrow S_{\texttt{OUT}})$ along the computation path that consumes the DoF for the backward computation. To show how to compute σ^+ in our model, we will take the most complicated XOR-MC operation as an example. For other operations, one can obtain the constraints similarly.

According to the semantics of the colors, the rules for coloring the input and output states of an operation, and how they consume the degree of freedom to limit the impacts should be different for the forward and the backward computation paths. Therefore, for each type of operations, we will give two sets of rules for different directions of the computation.

First of all, an invertible S-box preserves the color of the input cell, and the ShiftRows permutes the coloring scheme of the input state according to the permutations associated with the ShiftRows in both forward and backward computations. Both S-box and ShiftRows operations can not be used to reduce the impacts via consuming the degree of freedom. In the sequel, we will focus on more nontrivial operations.

XOR. The XOR operations exist in the `AddRoundKey` and the key/message-schedule (if any). Here we need to distinguish two different directions. If the XOR to be modeled is involved in the forward computation path from the starting states to the ending state E^+, the coloring scheme of the input and output cells of the XOR operation obeys the set of rules (denoted by $\texttt{XOR}^+\texttt{-RULE}$, where a "+" sign signifies the forward computation) shown in Fig. 3a. Similarly, if the XOR to be modeled is involved in the backward computation path from the starting states to the ending state E^-, the coloring scheme of the input and output cells of the XOR operation obeys the set of rules named as $\texttt{XOR}^-\texttt{-RULE}$, which is visualized in Fig. 3b. Note that $\texttt{XOR}^-\texttt{-RULE}$ (Fig. 3b) can be obtained from $\texttt{XOR}^+\texttt{-RULE}$ (Fig. 3a) by exchanging the `Red` cells and `Blue` cells, since the meanings of `Red` and `Blue` are dual for the forward and backward computations.

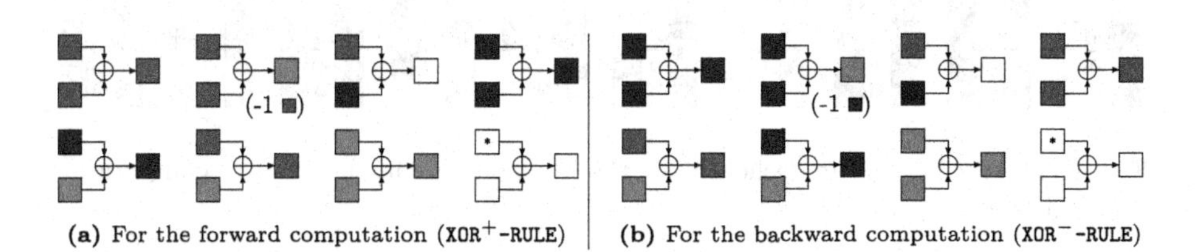

(a) For the forward computation ($\texttt{XOR}^+\texttt{-RULE}$) **(b)** For the backward computation ($\texttt{XOR}^-\texttt{-RULE}$)

Fig. 3. Rules for XOR operations, where a "*" means that the cell can be any color (Color figure online)

Let $A[0]$, $B[0]$ be the input cells and $C[0]$ be the output cell. The set of rules XOR$^+$-RULE restricts $(x_0^A, y_0^A, x_0^B, y_0^B, x_0^C, y_0^C)$ to a subset of $\mathbb{F}_2^6$, which can be described by a system of linear inequalities by using the convex hull computation method [58], and the set of rules XOR$^-$-RULE can be described similarly.

Within each of the two sets of rules for XOR operations, only one coloring scheme consumes the degree of freedom, *e.g.*, the ■ ⊕ ■ → ■ in Fig. 3a, which describes the possibility that the difference in one cell cancels that in another.

MixColumns. For the MixColumns operation in the forward computation, we have the following set of rules (denoted by MC$^+$-RULE) for the coloring schemes of the input and output columns. Examples of valid coloring schemes are shown in Fig. 4.

- ▶ MC$^+$-RULE-1. If there is at least one White cell in the input column, all the output cells are White (one unknown cell in the input causes all cells in the output be unknown);
- ▶ MC$^+$-RULE-2. If there are Blue cells but no White cells and no Red cell in the input column, then all the output cells are Blue (can perform full forward computations);
- ▶ MC$^+$-RULE-3. If all the input cells are Gray, then all the output cells are Gray (can perform bi-direction computations on fixed constants);
- ▶ MC$^+$-RULE-4. If there are Red and Blue cells but no White cells in the input column, each output cell must be Blue or White. Moreover, a condition should be fulfilled, that is, the sum of the numbers of Blue and Gray cells in the input and output columns must be no more than 3 (*i.e.*, $8-5$) (can partially cancel the impacts from ■ on ■ within an input column by consuming λ^-, and perform partial forward computations. Because of the MDS property of MixColumns, this is possible only when the condition is fulfilled);
- ▶ MC$^+$-RULE-5. If there are Red cells but no White cells and no Blue cells in the input column, then each output cell must be Red or Gray. Moreover, a condition should be fulfilled, that is, the number of Gray cells in the input and output columns must be no more than 3 (*i.e.*, $8-5$) (can partially cancel the difference within an input column by consuming λ^-. Because of the MDS property of MixColumns, this is possible only when the condition is fulfilled).

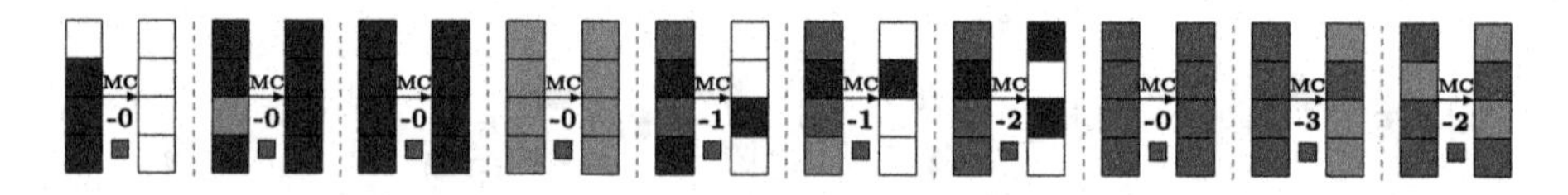

Fig. 4. Some valid coloring schemes for the MixColumns in the forward computation (Color figure online)

All the above rules can be described by linear inequalities.

First, we introduce three 0-1 indicator variables μ, υ, ω for the input column and necessary constraints into the model to satisfy the following cases.

- $\mu = 1, \upsilon = 0, \omega = 0$ if and only if MC$^+$-RULE-1 is fulfilled;
- $\mu = 0, \upsilon = 1, \omega = 0$ if and only if MC$^+$-RULE-2 is fulfilled;
- $\mu = 0, \upsilon = 1, \omega = 1$ if and only if MC$^+$-RULE-3 is fulfilled;
- $\mu = 0, \upsilon = 0, \omega = 0$ if and only if MC$^+$-RULE-4 is fulfilled;
- $\mu = 0, \upsilon = 0, \omega = 1$ if and only if MC$^+$-RULE-5 is fulfilled.

This can be done as follows.

Let $(A[0], A[1], A[2], A[3])^T$ and $(B[0], B[1], B[2], B[3])^T$ be the input and output columns. Without any restriction, there are 2^8 possible coloring schemes for the input column since $(x_0^A, y_0^A, \cdots, x_3^A, y_3^A) \in \mathbb{F}_2^8$. We define the set of vectors

$$\{(x_0^A, y_0^A, \cdots, x_3^A, y_3^A, \mu) : (x_0^A, y_0^A, \cdots, x_3^A, y_3^A) \in \mathbb{F}_2^8\}, \tag{7}$$

where $\mu = 1$ if and only if there exists $i \in \{0, 1, 2, 3\}$ such that $(x_i^A, y_i^A) = (0, 0)$. This subset can be described by linear inequalities with the convex hull computation method [58].

The indicator variable $\upsilon = 1$ if and only if $x_i^A = 1$ for each $i \in \{0, 1, 2, 3\}$. This can be done by linear inequalities listed in Eq. (8). The indicator variable $\omega = 1$ if and only if $y_i^A = 1$ for each $i \in \{0, 1, 2, 3\}$. This can be done by similar inequalities as Eq. (8).

Now, with the help of these variables μ, υ, ω, we can convert MC$^+$-RULE into a system of inequalities shown in Eq. (9).

$$\begin{cases} \sum_{i=0}^{3} x_i^A - 4\upsilon \geq 0; \\ \sum_{i=0}^{3} x_i^A - \upsilon \leq 3. \end{cases} \tag{8}$$

$$\begin{cases} \sum_{i=0}^{3} x_i^B + 4\mu \leq 4; \\ \sum_{i=0}^{3} y_i^B + 4\mu \leq 4; \\ \sum_{i=0}^{3} y_i^B - 4\omega = 0; \end{cases} \qquad \begin{cases} \sum_{i=0}^{3} (x_i^A + x_i^B) - 5\upsilon \leq 3; \\ \sum_{i=0}^{3} (x_i^A + x_i^B) - 8\upsilon \geq 0. \end{cases} \tag{9}$$

Since the semantics of the Red cells and Blue cells are dual in the forward and backward computation, the set of rules for backward computation (denoted by MC$^-$-RULE) can be obtained from MC$^+$-RULE by exchanging the words Blue and Red. We omit the details to save spaces.

XOR Then MixColumns (XOR-MC). For the operation which maps the two input columns $(A[0], A[1], A[2], A[3])^T$ and $(B[0], B[1], B[2], B[3])^T$ to $C[0, 1, 2, 3] = \text{MC}^{-1}(A[0, 1, 2, 3] + B[0, 1, 2, 3])$, we have the following rules for the coloring schemes of the input and output columns. Note that this operation only appears in the backward computation for all the targets in this paper. Therefore, we only specify the set of rules for XOR-MC for the backward computation.

- XOR-MC-RULE-1. If there is at least one White cell in the input columns, all the output cells are White (one unknown cell in the input causes all cells in the output be unknown);

- ▶ XOR-MC-RULE-2. If there are Red cells but no White cells and no Blue cells in the input columns, all output cells are Red (can perform full backward computations);
- ▶ XOR-MC-RULE-3. If all input cells are Gray, then all output cells are Gray (can perform bi-direction computations on fixed constants);
- ▶ XOR-MC-RULE-4. If there are Blue cells and Red cells but no White cells in the input columns, each output cell must be Red or White. Moreover, when combining the two input columns as a 4×2 matrix, the number of rows with one or two Blue cells plus the number of White cells in the output column must be greater or equal to 5 (can partially cancel the impacts from ■ on ■ within two input columns by consuming λ^+, and perform partial backward computations. Because of the MDS property of inverse MixColumns, this is possible only when the condition is fulfilled);
- ▶ XOR-MC-RULE-5. If there are Blue cells but no Red cells and no White cells in the input columns, each output cell must be Blue or Gray. Moreover, when combining the two input columns as a 4×2 matrix, the number of rows with one or two Blue cells plus the number of Blue cells in the output column must be greater or equal to 5 (can partially cancel the difference within two input columns by consuming λ^+. Because of the MDS property of MixColumns, this is possible only when the condition is fulfilled).

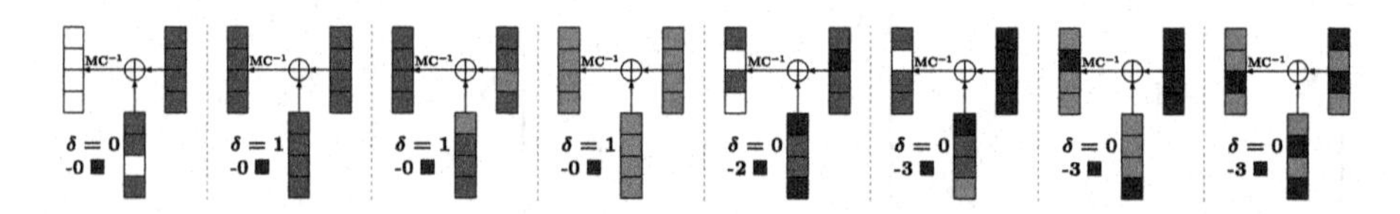

Fig. 5. Some valid coloring schemes for the XOR-MC in the backward computation (Color figure online)

All the above rules can be described by similar linear inequalities for MC$^-$-RULE. Three 0-1 indicator variables μ, υ, ω also be introduced for the input columns. $\mu = 1$ if and only if there exists $i \in \{0,1,2,3\}$ such that $(x_i^A, y_i^A) = (0,0)$ or $(x_i^B, y_i^B) = (0,0)$. $\upsilon = 1$ if and only if $x_i^A = 1$ and $x_i^B = 1$ for each $i \in \{0,1,2,3\}$. $\omega = 1$ if and only if $y_i^A = 1$ and $y_i^B = 1$ for each $i \in \{0,1,2,3\}$. These constraints can be generated from that of MC-RULE. For example, introduce μ^A (resp μ^B) for input column $(A[0], A[1], A[2], A[3])^T$ (resp $(B[0], B[1], B[2], B[3])^T$) and necessary constraints as Eq. (7). Then $\mu = 1$ if and only if $\mu^A = 1$ or $\mu^B = 1$. Then

- ▶ $\mu = 1, \upsilon = 0, \omega = 0$ if and only if XOR-MC-RULE-1 is fulfilled;
- ▶ $\mu = 0, \upsilon = 0, \omega = 1$ if and only if XOR-MC-RULE-2 is fulfilled;
- ▶ $\mu = 0, \upsilon = 1, \omega = 1$ if and only if XOR-MC-RULE-3 is fulfilled;
- ▶ $\mu = 0, \upsilon = 0, \omega = 0$ if and only if XOR-MC-RULE-4 is fulfilled;
- ▶ $\mu = 0, \upsilon = 1, \omega = 0$ if and only if XOR-MC-RULE-5 is fulfilled.

Another four 0-1 variables $\tau_0, \tau_1, \tau_2, \tau_3$ are introduced for each row, $\tau_i = 1$ if and only if $A[i]$ or $B[i]$ is `Blue` cell.

Now, with the help of these variables $\mu, \epsilon, \omega, \tau_i$ for $i \in \{0, 1, 2, 3\}$, we can convert `XOR-MC-RULE` into a system of inequalities as listed in Eq. (10).

$$\begin{cases} \sum_{i=0}^{3} x_i^C + 4\mu \leq 4; \\ \sum_{i=0}^{3} y_i^C + 4\mu \leq 4; \\ \sum_{i=0}^{3} x_i^C - 4v = 0; \end{cases} \quad \begin{cases} \sum_{i=0}^{3} (y_i^C - \tau_i) - 5\omega - \mu \leq -1; \\ \sum_{i=0}^{3} (y_i^C - \tau_i) - 8\omega \geq -4. \end{cases} \tag{10}$$

Remark 5. One may attempt to model the XOR-MC operation by applying `XOR⁻-RULE` and `MC⁻-RULE` separately. This approach is valid but misses important coloring schemes that may lead to better attacks. For example, considering the input columns shown in Fig. 6, applying `XOR⁻-RULE` results in `White` cells after the XOR operation. Subsequently, applying `MC⁻-RULE`, we will end up with a full column of `White` cells. However, if we model the XOR-MC operation as a whole, we can still preserve some `Red` cells from impact according to the sixth sub-figure in Fig. 5. This coloring scheme can be explained by the equation shown in Fig. 6, where the second term of the right-hand side of the equation is known for the backward computation. Therefore, we can restrict the values of $(B[0], A[0], A[1], A[2], A[3])$ such that

$$\begin{aligned} \mathtt{e} \cdot (A[0] \oplus B[0]) \oplus \mathtt{b} \cdot A[1] \oplus \mathtt{d} \cdot A[2] \oplus 9 \cdot A[3] = \mathtt{C}_0 \\ \mathtt{d} \cdot (A[0] \oplus B[0]) \oplus 9 \cdot A[1] \oplus \mathtt{e} \cdot A[2] \oplus \mathtt{b} \cdot A[3] = \mathtt{C}_2 \\ \mathtt{b} \cdot (A[0] \oplus B[0]) \oplus \mathtt{d} \cdot A[1] \oplus 9 \cdot A[2] \oplus \mathtt{e} \cdot A[3] = \mathtt{C}_3 \end{aligned} \tag{11}$$

where $\mathtt{C}_0$, $\mathtt{C}_2$, and $\mathtt{C}_3$ are constants, which implies that only $C[1]$ is unknown for the backward computation (see the sixth sub-figure in Fig. 5). The principle is to let the differences of multiple cells in two input columns mutually canceled at particular output cells.

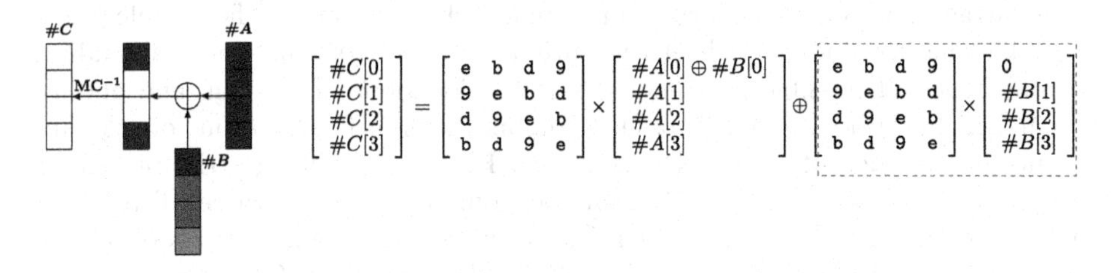

Fig. 6. The inaccuracy of modeling XOR-MC in the backward computation by applying `XOR⁻-RULE` and `MC⁻-RULE` separately. (Color figure online)

Compute consumed DoF in `XOR-MC-RULE`*s.* In all of our applications, the XOR-MC operation only appears in the backward computation and thus only consumes the DoF for the forward computation. Let $(A[0], \cdots, A[3])$ and $(B[0], \cdots, B[3])$ be the two input columns and $(C[0], \cdots, C[3])$ be the output column. Given a valid coloring scheme of A, B, and C, the consumed DoF (measured in cells) $\sigma^+((A[0,\cdots,3], B[0,\cdots,3]) \rightarrow C[0,\cdots,3])$ equals the number of `Red` and `Gray` cells (known cells of the output column in the backward computation) when there is at least one `Blue` cell in the input columns. Otherwise, the consumed DoF is zero.

Let δ be a 0-1 indicator variable such that $\delta = 1$ if and only if there are no `Blue` cells and no `White` cells in the input columns, which can be achieved by imposing the following constraints on δ:

$$\begin{cases} -\delta + \sum_{i=0}^{3} y_i^A + \sum_{i=0}^{3} y_i^B \leq 7; \\ y_i^A \geq \delta, \quad y_i^B \geq \delta, \quad \text{for } i \in \{0,1,2,3\}. \end{cases} \tag{12}$$

Then we have $\sigma^+((A[0,\cdots,3], B[0,\cdots,3]) \rightarrow C[0,\cdots,3]) = -4\delta + \sum_{i=0}^{3} y_i^C$. In Fig. 5 we give some example coloring schemes of the XOR-MC operation together with their consumed DoF. Similarly, the constraints describing how the XOR and MC operations consume DoF can be deduced.

5 Applications

Equipped with the presented tool, we evaluated the security of hash functions built on AES and AES-like ciphers, including all members of AES and the members of Rijndael with 256-bit block-size [16] in PGV-modes (note the equivalence among PGV-modes for the attacks as shown in [10]) and Haraka v2 [39].

For all targets, improved attacks are identified. In particular, our tool found the first preimage attacks on 8-round AES-128 hashing modes, and on the full 5-round and the extended 5.5-round (10 and 11 AES-rounds) Haraka-512 v2. Due to the page limit, we only describe two attacks in detail. The list of optimal attacks we found is presented in Table 1. With the help of the visualizations of these attacks, one can reconstruct concrete attacks and confirm the complexities.

The time for finding each of the optimal attacks is within hours, including enumerating all possible combinations of the locations of starting and ending points in encryption, and all possible combinations of the locations of starting points in the encryption and key-schedule. For example, to get the presented attack on 8-round AES-128 hashing modes, our program generated all possible MILP-models and the MILP solver Gurobi solved them all, which took about two hours on a PC with an Intel Core i7-7500U CPU and 8 GB memory.

5.1 Improved Attacks on AES and Rijndael Hashing Modes

Searching the attacks. We apply our method to AES hashing modes. With our tool, many new attacks are found automatically. We list some examples for each member of AES and also the members of Rijndael with 256-bit block-size [16] (denoted by Rijndael-256) in Fig. 7, 9, 10, 11, and 12 in the full version [11].

Notably, apart from new attacks with better complexities, an 8-round attack on AES-128 and 9-round attacks on AES-192 and AES-256 hashing mode were found, which extend one more round compared with previous attacks [10,49,62].

To be clear, in the figures, some information are presented, such as which states are the starting states (in the searching for the attacks, not necessarily in the concrete attacks), how independent computation flows propagated in the states, and where the two chunks meet. Besides, which rules are applied to the states and how the degrees of freedom are consumed by the specific coloring scheme in our MILP models are also exhibited. Furthermore, the initial degrees of freedom (λ^+, λ^-), and the final configuration $(\mathrm{DoF}^+, \mathrm{DoF}^-, \mathrm{DoM})$ which determines the attack complexity are summarized at the bottom.

For example, from Fig. 7, it can be seen that, in the searching of our model, the starting states are $\#\mathrm{SB}^4$ and k_4, and the ending states are $\#\mathrm{MC}^1$ and $\#\mathrm{SB}^2$. Also, we have $\mathcal{B}^{\mathtt{ENC}} = [0, 5, 10, 15]$, $\mathcal{B}^{\mathtt{KSA}} = [0, 1, 2, 3, 4, 6, 7, 8, 9, 11, 12, 13, 14]$, $\mathcal{R}^{\mathtt{ENC}} = [1, 2, 3, 4, 6, 7, 8, 9, 11, 12, 13, 14]$, $\mathcal{R}^{\mathtt{KSA}} = \emptyset$, $\mathcal{C} = [0, 2, 5, 7, 8, 10, 13, 15]$, and $\mathcal{D} = [1, 2, 3]$. Accordingly, the initial degrees of freedom for the forward computation and backward computation are 17 and 12 respectively, and the degree of matching is $2 + 3 - 4 = 1$. The states $\#\mathrm{SB}^4$, k_3, and $\#\mathrm{MC}^3$ are enclosed by a dashed light-green frame, which means that `XOR-MC-RULE` is applied to them, and the specific coloring scheme consumes 12 cells of degrees of freedom for the forward computation. Similarly, the `XOR-MC-RULE` is applied to states $\#\mathrm{SB}^3$, k_2, and $\#\mathrm{MC}^2$, and that consumes 3 cells of degrees of freedom for the forward computation. The states $\#\mathrm{MC}^4$ and $\#\mathrm{AK}^4$ are enclosed by a dashed light-purple frame, which means $\mathtt{MC}^+$`-RULE` is applied to them, and that consumes 9 cells of degrees of freedom for the backward computation. Similarly, the $\mathtt{MC}^+$`-RULE` is applied to states $\#\mathrm{MC}^5$ and $\#\mathrm{AK}^5$, and that consumes 2 cells of degrees of freedom for the backward computation. Accordingly, in the solution of our model, $\mathrm{DoF}^+ = 17 - 12 - 3 = 2$ and $\mathrm{DoF}^- = 12 - 9 - 2 = 1$, which indicates that the values of $(\#\mathrm{SB}^4[\mathcal{B}^{\mathtt{ENC}}], k_4[\mathcal{B}^{\mathtt{KSA}}])$ are restricted to a subset $\mathbb{X}$ of $\mathbb{F}_{2^8}^{17}$ with $(2^8)^2$ elements, and the values of $(\#\mathrm{SB}^4[\mathcal{R}^{\mathtt{ENC}}], k_4[\mathcal{R}^{\mathtt{KSA}}])$ are restricted to a subset $\mathbb{Y}$ of $\mathbb{F}_{2^8}^{12}$ with 2^8 elements. To be more concrete, $\mathbb{X}$ and $\mathbb{Y}$ should be chosen such that the forward computation is irrelevant of $(\#\mathrm{SB}^4[\mathcal{R}^{\mathtt{ENC}}], k_4[\mathcal{R}^{\mathtt{KSA}}])$, and the backward computation is irrelevant of $(\#\mathrm{SB}^4[\mathcal{B}^{\mathtt{ENC}}], k_4[\mathcal{B}^{\mathtt{KSA}}])$. Since the degrees of freedom for the forward and backward computations (DoF^+ and DoF^-) are derived rather formally without giving the actual contents of $\mathbb{X}$ and $\mathbb{Y}$, some readers may doubt whether such $\mathbb{X}$ and $\mathbb{Y}$ really exist.

In the following (in the precomputation phase and more details in Appendix B.1 in the full version [11]), we explicitly show in this example, how to obtain $\mathbb{X}$ and $\mathbb{Y}$ such that the required properties are fulfilled, and under the configuration obtained by the MILP model, how to launch the concrete attack.

The attack on 8-round AES -128 hashing (refer to Fig. 7)

The Precomputation Phase (precompute possible initial values of neutral bytes)

1. To be able to compute backward chunk independently of forward neutral bytes, the forward neutral bytes should have constant impacts on the 12 C-marked Red bytes in $\#\mathrm{MC}^3$ and on the 3 C-marked Red bytes in $\#\mathrm{MC}^2$. Therefore, denote the 12 constant impacts on 12 bytes in $\#\mathrm{MC}^3$ by $\mathtt{C}_{1,0}$, $\mathtt{C}_{1,1}$, $\cdots$, $\mathtt{C}_{1,11}$, we derive constraints on forward neutral bytes, which is a system linear equation Eq. (13) in [11]. Similarly, denote the 3 constant impacts on 3 bytes in $\#\mathrm{MC}^2$ by $\mathtt{C}_{2,0}, \mathtt{C}_{2,1}, \mathtt{C}_{2,2}$, we derive constraints on forward neutral bytes, which is a system of linear equation Eq. (14) in [11]. In total, requiring impacts to be constant will impose 15 bytes constraints on forward neutral bytes (20 bytes) as shown in the system of linear equation Eq. (17) in [11]. Solving Eq. (17) in [11], one gets 2^{40} solutions.
   In the following main procedure, the values of $\mathtt{C}_{1,0}$, $\mathtt{C}_{1,1}$,..., $\mathtt{C}_{1,11}$, and $\mathtt{C}_{2,0}$, $\mathtt{C}_{2,1}$, $\mathtt{C}_{2,2}$ are fixed such that we only need to solve Eq. (17) in [11] once. However, the main procedure will need to trail on many values of Gray bytes in k_4 (*i.e.*, $k_4[5, 10, 15]$) to find full match. So here, we precompute values of forward neutral bytes that correspond to each value of $k_4[5, 10, 15]$. That can be done as follows. For each of the 2^{40} solution, k_3 and $\#\mathrm{SB}^4[0, 5, 10, 15]$ are determined. Compute k_4 using k_3, and store k_4 and the values of $\#\mathrm{SB}^4[0, 5, 10, 15]$ in table T_1 indexed by the values of 3 Gray bytes $k_4[5, 10, 15]$.
   • Note that there are 2^{24} entries in T_1, and the total size of T_1 is about 2^{40}. Under each index, there are about 2^{16} elements. We can either use 2^{16} or 2^8 of them. The total complexity of the full attack will be the same (because DoF^+ and DoM are all one byte). Thus, we use 2^8. Therefore, the complexity of this procedure is 2^{32}, and the memory requirement is 2^{32}.
2. To be able to compute forward chunk independently of backward neutral bytes, the backward neutral bytes should have constant impacts on the 2 C-marked Blue bytes in $\#\mathrm{AK}^5$. Therefore, denote the 2 constant impacts on 2 bytes in $\#\mathrm{AK}^5$ by $\mathtt{C}_{4,0}$ and $\mathtt{C}_{4,1}$, we derive constraints on backward neutral bytes, which is a linear equation system Eq. (18) in [11]. For each possible $\mathtt{C}_{4,0}$ and $\mathtt{C}_{4,1}$, when solve Eq. (18) in [11], one gets 2^8 solutions.
   In the following main procedure, we need to trail on many values of ($\mathtt{C}_{4,0}$, $\mathtt{C}_{4,1}$) to find a full match. So here, we precompute values of backward neutral bytes that correspond to each value of ($\mathtt{C}_{4,0}$, $\mathtt{C}_{4,1}$), store values of $\#\mathrm{MC}^5[1, 2, 3]$ fulfilling Eq. (18) in [11] in table T_2 indexed by the values of ($\mathtt{C}_{4,0}$, $\mathtt{C}_{4,1}$).
   • There are 2^{16} entries in T_2, and the total size of T_2 is 2^{24}. Under each index, there are 2^8 elements.

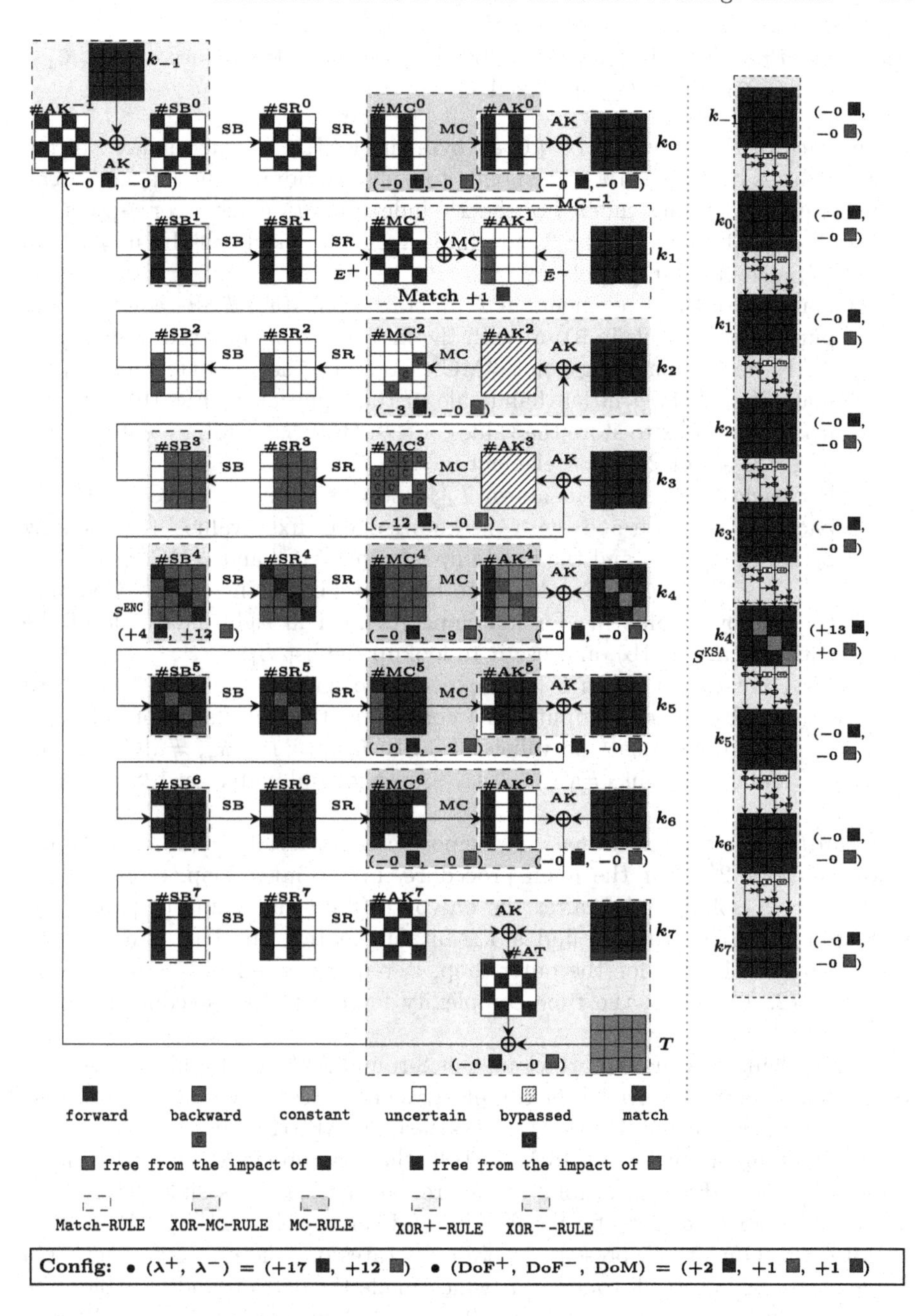

Fig. 7. An MITM pseudo-preimage attack on 8-round AES-128 hashing. Note that, because the use of XOR-MC-RULE, we do not introduce any variable in our MILP model for states #AK^2 and #AK^3, and thus we bypass them. (Color figure online)

The Main Procedure. During the following procedure, the values of $\mathtt{C}_{1,0}$, $\mathtt{C}_{1,1}$, ..., $\mathtt{C}_{1,11}$, and $\mathtt{C}_{2,0}$, $\mathtt{C}_{2,1}$, $\mathtt{C}_{2,2}$ are fixed.

1. For each of the 2^x values of 9 `Gray` bytes in #AK^4, for each index i of the 2^{24} indexes of T_1 (each i corresponds to each candidate value of the 3 `Gray` bytes in k_4), for each index j of the 2^{16} indexes of T_2 (each j corresponds to each candidate value of the 2-byte impact on `C`-marked cells by in #AK^5), do: Initialize an empty table L_1.
   (a) For each of the 2^8 elements in $T_1[i]$, start from state #SB^4 and k_4, compute forward (cells in `Blue`) with the knowledge of the fixed impact j on #AK^5 to the matching point #MC^1. Compute the one-byte value m_1 for matching (defined in left-hand side of the equation in Fig. 1b), and use m_1 as the index to store the values of (#SB$^4[0, 5, 10, 15]$, k_4) into $L_1[m_1]$ (there is about $2^{8-8} = 1$ element in each $L_1[m_1]$).
   (b) For each of the 2^8 elements in $T_2[j]$, start from state #MC^5, compute backward (cells in `Red`) with the knowledge of fixed value i (*i.e.*, 3 `Gray` bytes) in state k_4 and the fixed impacts on #MC^3 and #MC^2 (*i.e.*, $\mathtt{C}_{1,0}$, $\mathtt{C}_{1,1}$, $\cdots$, $\mathtt{C}_{1,11}$, $\mathtt{C}_{2,0}$, $\mathtt{C}_{2,1}$, $\mathtt{C}_{2,2}$) to the matching point #AK^1. Compute the one-byte value m_2 for matching (defined in right-hand side of the equation in Fig. 1b), and use it to lookup the list L_1:
      i. For each element in $L_1[m_2]$ (expected to exist one): restart the forward and backward computations combining the knowledge of values in both directions (the values of #SB$^4[0, 5, 10, 15]$, k_4, #MC^5) to the matching point (#MC^1, #AK^1), test for full match on 128-bit state.

Complexity. The computational and memory complexity of the precomputation phase is about 2^{32}. For the main procedure, in the inner loop, there will be $2^{(8+8-8)} = 2^8$ solutions left after the one-byte (8-bit) matching (m_1 and m_2) in Step 1 (b) i. In order to find a 128-bit full match, one has to match the other 120 bits. Hence, for the outer loop, it requires $x + 24 + 16 = 120 - 8$, *i.e.*, $x = 72$. Therefore, the time complexity for the main procedure is about $2^{(x+24+16)+8} = 2^{x+40} = 2^{120}$.

We implemented the full attack on this 8-round AES-128-hashing (with partial matchings), which verified the complexity. The codes and results are available via https://github.com/MITM-AES-like-Hashing/AES128_8R.

Apart from the biclique attacks in [14], the best previous pseudo-preimage attacks against AES-128 hashing modes remain as 7 rounds since 2011, with a time complexity of 2^{120} by Sasaki [49] and improved to 2^{112} by Bao *et al.* in 2019 [10]. Our attack presented here penetrates one more round. There is a unique features observed from Fig. 7, which made the extra round possible. The backward chunk covers one more round compared with that in [10,49]. This is only possible after the consumption of 12 and 3 `Blue` bytes of freedom degrees (forward neutral bytes) in consecutive two rounds. Without the introduction of DoF from key bytes in [10], this would not be possible. Note that the backward chunk only outputs 3 bytes, which are just sufficient to form a filter of one byte together with the 2 `Blue` bytes before the `MixColumns` at the matching point.

As depicted in Fig. 9, 10, 11, 12 in the full version of this paper [11] and summarized in Table 1, when our search models are applied to hashing modes based on other AES variants, they are also able to improve by one round against AES-192 and AES-256 hashing as in [10]. Some configurations (*e.g.*, Fig. 9, 11 in [11]) are more involved, in which the key states have neutral bytes for both forward and backward chunks. That might be hard to be found by manual.

5.2 Improved Attacks on Haraka V2

Haraka v2 [39] is a family of hash functions designed to be efficient for short-input and for post-quantum applications. It includes two versions, denoted by Haraka-256 v2 and Haraka-512 v2, both output 256-bit hash digests and claim 256-bit security against (second)-preimage attacks. They only process short-input (s-bit string, denoted by x) and thus employ s-bit permutation (denoted by π_s) in the DM-mode as follows:

$$\text{Haraka-256 v2}(x) = \pi_s(x) \oplus x \text{ and } \text{Haraka-512 v2}(x) = \text{trunc}(\pi_s(x) \oplus x)$$

where trunc truncates 512-bit state to 256-bit output. To achieve high performance on platforms supporting AES-NI and share security analysis of AES, the round function of the permutation π_s first applies two layers of b AES-round-functions in parallel on a state that can be evenly divided into b sub-states (each of which is identical to the state of AES), then it applies a shuffle (denoted by $\mathbf{mix}_s$) among the columns of the state. For Haraka-256 v2, $s = 256, b = 2$, and for Haraka-512 v2, $s = 512, b = 4$. For both of them, the number of rounds is 5 that involves 10 AES-rounds in sequential.

The former version of Haraka (named as Haraka v1) was broken by Jean [33] due to its weak round constants. Then an updated version Haraka v2 [39] was published. The designers provide MITM preimage attacks on 3.5-round Haraka-256 v2 and on 4-round Haraka-512 v2.

Searching the Attacks. For both versions of Haraka v2, our tool produced improved MITM preimage attacks. In particular, for Haraka-256 v2, our tool found attacks that cover up to 4.5-round (9 AES-rounds). An example that has the optimal complexity is visualized in Fig. 13 in the full version [11], of which the complexity is $2^{256-8\times\min\{\text{DoF}^+,\ \text{DoF}^-,\ \text{DoM}\}} = 2^{256-8\times\min\{4,\ 4,\ 8\}} = 2^{224}$. Note that this attack directly implies an attack covering 4-round (8 AES-rounds) with the same complexity. For Haraka-512 v2, our tool finds attacks that penetrate the full 5-round (10 AES-rounds) and the extended 5.5-round (11 AES-rounds) version. The detailed configuration of one of the attacks on the full 5-round (10 AES-rounds) is visualized in Fig. 14 in [11]. In the following, we present one of the searching results on the extended 5.5-round (11 AES-rounds) and the concrete attack corresponding to the configuration visualized in Fig. 8.

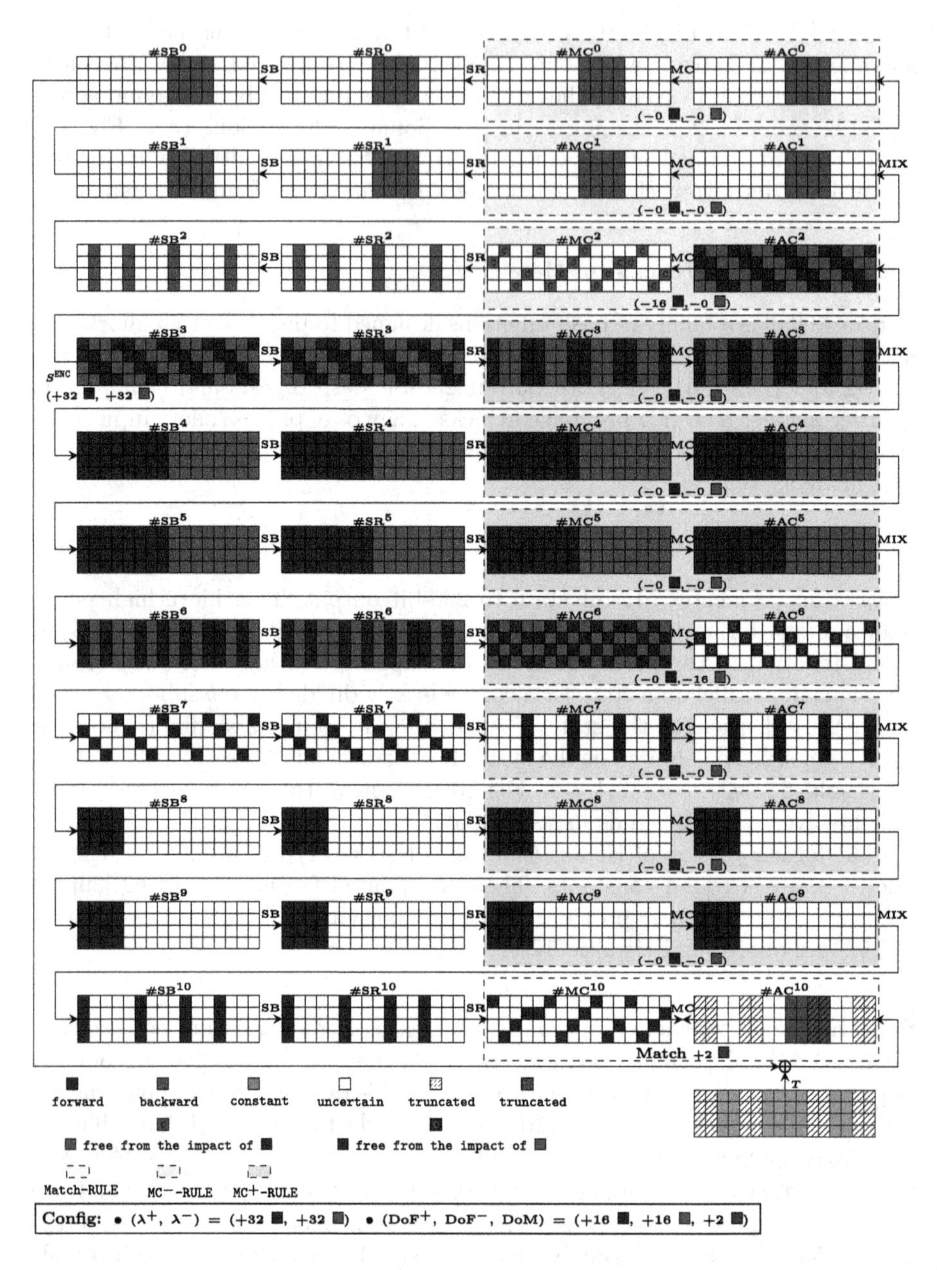

Fig. 8. An MITM preimage attack on the extended 5.5-round (11 AES-rounds) Haraka-512 v2. Note that in our MILP-models, the position of the used hash bits are treated and used as constant in gray cell of the target T, and the bits discarded are treated as 'uncertain' although we distinct them using hatched pattern. However, in the attack procedure, the discarded bits are free of choice such that the state cells in hatched pattern are free of matching. (Color figure online)

From Fig. 8, it can be seen that in the searching of our model, the starting state is $\#\mathrm{SB}^3$, and the ending states are $\#\mathrm{MC}^{10}$ and $\#\mathrm{AC}^{10}$. Also, we have $\mathcal{B}^{\mathrm{ENC}} = [16 \cdot i + j \mid i \in \{0,1,2,3\},\ j \in \{0,1,5,6,10,11,12,15\}]$, $\mathcal{R}^{\mathrm{ENC}} = [16 \cdot i + j \mid i \in \{0,1,2,3\},\ j \in \{2,3,4,7,8,9,13,14\}]$, $\mathcal{C} = [16 \cdot i + j \mid i \in \{0,3\},\ j \in \{0,7,10,13\}] \cup [16 \cdot i + j \mid i \in \{1,2\},\ j \in \{1,4,11,14\}]$, and $\mathcal{D} = [16 \cdot i + j \mid i \in \{2\},\ j \in \{0,1,\ldots,7\}]$. Therefore, both of the initial degrees of freedom for the forward computation and backward computation are 32, *i.e.*, $\lambda^+ = \lambda^- = 32$, and the degree of matching is $\mathrm{DoM} = (1+4-4) \times 2 = 2$. The $\texttt{MC}^-\texttt{-RULE}$ applied to states $\#\mathrm{MC}^2$ and $\#\mathrm{AC}^2$ consumes 16 cells of degrees of freedom for the forward computation. And the $\texttt{MC}^+\texttt{-RULE}$ applied to states $\#\mathrm{MC}^6$ and $\#\mathrm{AC}^6$ consumes 16 cells of degrees of freedom for the backward computation. Accordingly, in the solution of our model, $\mathrm{DoF}^+ = 32 - 16 = 16$ and $\mathrm{DoF}^- = 32 - 16 = 16$. This indicates that the values of $\#\mathrm{SB}^3[\mathcal{B}^{\mathrm{ENC}}]$ are restricted to a subset $\mathbb{X}$ of $\mathbb{F}_{2^8}^{32}$ with $2^{8\times16}$ elements, and the values of $\#\mathrm{SB}^3[\mathcal{R}^{\mathrm{ENC}}]$ are restricted to a subset $\mathbb{Y}$ of $\mathbb{F}_{2^8}^{32}$ with $2^{8\times16}$ elements. To be more concrete, $\mathbb{X}$ and $\mathbb{Y}$ should be chosen such that the forward computation is irrelevant of $\#\mathrm{SB}^3[\mathcal{R}^{\mathrm{ENC}}]$ and the backward computation is irrelevant of $\#\mathrm{SB}^3[\mathcal{B}^{\mathrm{ENC}}]$. In summary, the decisive parameters for the obtained attack is $(\mathrm{DoF}^+, \mathrm{DoF}^-, \mathrm{DoM}) = (16, 16, 2)$. From these parameters, one can directly obtain that the time complexity of the corresponding pseudo-preimage attack is $(2^8)^{32-\min\{\mathrm{DoF}^+,\mathrm{DoF}^-,\mathrm{DoM}\}} = 2^{240}$.

The concrete attack on 11-AES -round Haraka-512 v2 (refer to Fig. 8)

1. For each of the 2^x values of impacts (16-byte impacts on the C-marked `Red` cells in $\#\mathrm{MC}^2$ and 16-byte impacts on the C-marked `Blue` cells in $\#\mathrm{AC}^6$), do: Initialize two empty tables L_1 and L_2
   (a) With the knowledge of the value of 16-byte impacts on the C-marked `Red` cells in $\#\mathrm{MC}^2$, we can collect $2^{16\times8} = 2^{128}$ possible values of `Blue` bytes (neutral bytes for the forward) in $\#\mathrm{AC}^2$ by solving sets of linear equations column-by-column. For example, in the first column of $\#\mathrm{MC}^2$ and $\#\mathrm{AC}^2$, the two `Blue` bytes and 1-byte impact (denoted by $\mathtt{C}_0$) on the C-marked cell have to meet: $\mathtt{9} \cdot \#\mathrm{AC}^2[0] \oplus \mathtt{e} \cdot \#\mathrm{AC}^2[1] = \mathtt{C}_0$. There are 16 sets of such linear equations, one set per column. For each column, we obtain 2^8 solutions. Hence, it is expected to get 2^{128} solutions by solving 16 sets of linear equations with 32 variables in total. The number 2^{128} is also the degrees of freedom for forward chunk.
   (b) For each of the 2^{128} solutions for `Blue` bytes in $\#\mathrm{AC}^2$, compute forward with the knowledge of the 16-byte impacts on the C-marked cells in $\#\mathrm{AC}^6$ to the matching point $\#\mathrm{MC}^{10}$, extract the two-byte value m_1 for matching, store the values of `Blue` bytes in $\#\mathrm{AC}^2$ in $L_1[m_1]$.
   (c) Similarly, collect 2^{128} possible values for `Red` bytes in state $\#\mathrm{MC}^6$ and compute backward to the matching point $\#\mathrm{AC}^{10}$, extract the two-byte value m_2 for matching, store the values of `Red` bytes in $\#\mathrm{MC}^6$ in $L_2[m_2]$.
   (d) For entries with common index between L_1 and L_2, form pairs of values of `Blue` bytes in $\#\mathrm{AC}^2$ and `Red` bytes in $\#\mathrm{MC}^6$; for each pair, restart the forward and backward computations combining the knowledge of values in both direction, test for full match on 256 bits.

Complexity. In Step 1 (d), it is expected to find $2^{128+128-16} = 2^{240}$ matches on 16 bits. Among them, it is expected to left 1 solution that also match on the other 240 bits, that implies a full match on 256 bits. Hence, to find a full match, it is expected to need 2^x outer loops where $x = 0$. The memory requirement is $2 \cdot 2^{128}$ to store L_1 and L_2. The time complexity of Step 1 (a) is no more than 2^{128}. The same complexity also applies to Step 1 (b) and Step 1 (c). The time complexity of Step 1 (d) is approximately $2^{16} \times 2^{2\times 112} = 2^{240}$ (L_1 and L_2 contains 2^{16} entries each; each entry is expected to contain 2^{112} values. Under a common 16-bit index, there are $2^{2\times 112}$ pairs to check for full match.) Therefore, the total time complexity is 2^{240}.

6 Conclusions

We modeled the MITM preimage attack into the language of MILP, generalized the attack model, and obtained better results in terms of number of attacked rounds against AES-like hashing including the 8-round AES-128, 9-round AES-192, 9-round AES-256, and 9-round Rijndael-256 hashing modes, 4.5-round Haraka-256 v2, the full version (5-round) and extended version (5.5-round) of Haraka-512 v2.

Acknowledgements. We thank the anonymous reviewers for the helpful comments. This research is partially supported by the National Natural Science Foundation of China (Grant No. 61802400, 62032014, 61772519, 61961146004), the National Key Research and Development Program of China (Grant No. 2018YFA0704701, 2018YFA0704704), the Chinese Major Program of National Cryptography Development Foundation (No. MMJJ20180101, MMJJ20180102), the Major Program of Guangdong Basic and Applied Research (Grant No. 2019B030302008); Nanyang Technological University in Singapore under Grant 04INS000397C230, Singapore's Ministry of Education under Grants RG18/19, RG91/20, and MOE2019-T2-1-060; the Gopalakrishnan – NTU Presidential Postdoctoral Fellowship 2020.

References

1. Alliance, Z.: ZigBee 2007 specification (2007). http://www.zigbee.org/
2. AlTawy, R., Youssef, A.M.: Preimage attacks on reduced-round stribog. In: Pointcheval, D., Vergnaud, D. (eds.) AFRICACRYPT 2014. LNCS, vol. 8469, pp. 109–125. Springer, Cham (2014). https://doi.org/10.1007/978-3-319-06734-6_7
3. AlTawy, R., Youssef, A.M.: Second preimage analysis of whirlwind. In: Lin, D., Yung, M., Zhou, J. (eds.) Inscrypt 2014. LNCS, vol. 8957, pp. 311–328. Springer, Cham (2015). https://doi.org/10.1007/978-3-319-16745-9_17
4. Aoki, K., Guo, J., Matusiewicz, K., Sasaki, Yu., Wang, L.: Preimages for step-reduced SHA-2. In: Matsui, M. (ed.) ASIACRYPT 2009. LNCS, vol. 5912, pp. 578–597. Springer, Heidelberg (2009). https://doi.org/10.1007/978-3-642-10366-7_34
5. Aoki, K., Sasaki, Yu.: Meet-in-the-middle preimage attacks against reduced SHA-0 and SHA-1. In: Halevi, S. (ed.) CRYPTO 2009. LNCS, vol. 5677, pp. 70–89. Springer, Heidelberg (2009). https://doi.org/10.1007/978-3-642-03356-8_5

6. Aoki, K., Sasaki, Yu.: Preimage attacks on one-block MD4, 63-Step MD5 and more. In: Avanzi, R.M., Keliher, L., Sica, F. (eds.) SAC 2008. LNCS, vol. 5381, pp. 103–119. Springer, Heidelberg (2009). https://doi.org/10.1007/978-3-642-04159-4_7
7. Aumasson, J.-P., Meier, W., Mendel, F.: Preimage attacks on 3-pass HAVAL and step-reduced MD5. In: Avanzi, R.M., Keliher, L., Sica, F. (eds.) SAC 2008. LNCS, vol. 5381, pp. 120–135. Springer, Heidelberg (2009). https://doi.org/10.1007/978-3-642-04159-4_8
8. Aumasson, J.P., Endignoux, G.: Gravity-SPHINCS. Technical report, National Institute of Standards and Technology (2017). https://csrc.nist.gov/projects/post-quantum-cryptography/round-1-submissions
9. Banik, S., Pandey, S.K., Peyrin, T., Sasaki, Y., Sim, S.M., Todo, Y.: GIFT: a small present - towards reaching the limit of lightweight encryption. In: Fischer, W., Homma, N. (eds.) CHES 2017. LNCS, vol. 10529, pp. 321–345. Springer, Heidelberg (2017)
10. Bao, Z., Ding, L., Guo, J., Wang, H., Zhang, W.: Improved meet-in-the-middle preimage attacks against AES hashing modes. IACR Trans. Symm. Cryptol. **2019**(4), 318–347 (2019)
11. Bao, Z., et al.: Automatic search of meet-in-the-middle preimage attacks on AES-like hashing. Cryptology ePrint Archive, Report 2020/467 (2020). https://eprint.iacr.org/2020/467
12. Beierle, C., et al.: The SKINNY family of block ciphers and its low-latency variant MANTIS. In: Robshaw, M., Katz, J. (eds.) CRYPTO 2016, Part II. LNCS, vol. 9815, pp. 123–153. Springer, Heidelberg (2016). https://doi.org/10.1007/978-3-662-53008-5_5
13. Benadjila, R., et al.: SHA-3 proposal: ECHO. Submission to NIST (updated), p. 113 (2009)
14. Bogdanov, A., Khovratovich, D., Rechberger, C.: Biclique cryptanalysis of the full AES. In: Lee, D.H., Wang, X. (eds.) ASIACRYPT 2011. LNCS, vol. 7073, pp. 344–371. Springer, Heidelberg (2011). https://doi.org/10.1007/978-3-642-25385-0_19
15. Canteaut, A., et al.: Saturnin: a suite of lightweight symmetric algorithms for post-quantum security. Submission to NIST (2019)
16. Daemen, J., Rijmen, V.: The Design of Rijndael: AES - The Advanced Encryption Standard. Information Security and Cryptography. Springer, Heidelberg (2002). https://doi.org/10.1007/978-3-662-04722-4
17. Derbez, P., Fouque, P.-A.: Automatic search of meet-in-the-middle and impossible differential attacks. In: Robshaw, M., Katz, J. (eds.) CRYPTO 2016, Part II. LNCS, vol. 9815, pp. 157–184. Springer, Heidelberg (2016). https://doi.org/10.1007/978-3-662-53008-5_6
18. Dong, X., Sun, S., Shi, D., Gao, F., Wang, X., Hu, L.: Quantum Collision Attacks on AES-Like Hashing with Low Quantum Random Access Memories. In: Moriai, S., Wang, H. (eds.) ASIACRYPT 2020, Part II. LNCS, vol. 12492, pp. 727–757. Springer, Cham (2020). https://doi.org/10.1007/978-3-030-64834-3_25
19. Espitau, T., Fouque, P.-A., Karpman, P.: Higher-order differential meet-in-the-middle preimage attacks on SHA-1 and BLAKE. In: Gennaro, R., Robshaw, M. (eds.) CRYPTO 2015, Part I. LNCS, vol. 9215, pp. 683–701. Springer, Heidelberg (2015). https://doi.org/10.1007/978-3-662-47989-6_33
20. Fu, K., Wang, M., Guo, Y., Sun, S., Hu, L.: MILP-based automatic search algorithms for differential and linear trails for speck. In: Peyrin, T. (ed.) FSE 2016. LNCS, vol. 9783, pp. 268–288. Springer, Heidelberg (2016). https://doi.org/10.1007/978-3-662-52993-5_14

21. Gauravaram, P., et al.: Grøstl – a SHA-3 candidate, March 2011. http://www.groestl.info/Groestl.pdf
22. Guo, C., Katz, J., Wang, X., Yu, Y.: Efficient and secure multiparty computation from fixed-key block ciphers. In: 2020 IEEE Symposium on Security and Privacy, San Francisco, CA, USA, 18–21 May 2020, pp. 825–841 (2020)
23. Guo, J., Jean, J., Nikolić, I., Sasaki, Yu.: Meet-in-the-middle attacks on generic feistel constructions. In: Sarkar, P., Iwata, T. (eds.) ASIACRYPT 2014, Part I. LNCS, vol. 8873, pp. 458–477. Springer, Heidelberg (2014). https://doi.org/10.1007/978-3-662-45611-8_24
24. Guo, J., Jean, J., Nikolic, I., Sasaki, Y.: Extended meet-in-the-middle attacks on some Feistel constructions. Des. Codes Crypt. **80**(3), 587–618 (2016)
25. Guo, J., Jean, J., Nikolic, I., Sasaki, Y.: Meet-in-the-middle attacks on classes of contracting and expanding Feistel constructions. IACR Trans. Symm. Cryptol. **2016**(2), 307–337 (2016). http://tosc.iacr.org/index.php/ToSC/article/view/576
26. Guo, J., Ling, S., Rechberger, C., Wang, H.: Advanced meet-in-the-middle preimage attacks: first results on full tiger, and improved results on MD4 and SHA-2. In: Abe, M. (ed.) ASIACRYPT 2010. LNCS, vol. 6477, pp. 56–75. Springer, Heidelberg (2010). https://doi.org/10.1007/978-3-642-17373-8_4
27. Guo, J., Su, C., Yap, W.: An improved preimage attack against HAVAL-3. Inf. Process. Lett. **115**(2), 386–393 (2015)
28. Hao, Y., Leander, G., Meier, W., Todo, Y., Wang, Q.: Modeling for three-subset division property without unknown subset. In: Canteaut, A., Ishai, Y. (eds.) EUROCRYPT 2020, Part I. LNCS, vol. 12105, pp. 466–495. Springer, Cham (2020). https://doi.org/10.1007/978-3-030-45721-1_17
29. Hong, D., Koo, B., Sasaki, Yu.: Improved preimage attack for 68-step HAS-160. In: Lee, D., Hong, S. (eds.) ICISC 2009. LNCS, vol. 5984, pp. 332–348. Springer, Heidelberg (2010). https://doi.org/10.1007/978-3-642-14423-3_22
30. Hosoyamada, A., Sasaki, Yu.: Finding hash collisions with quantum computers by using differential trails with smaller probability than birthday bound. In: Canteaut, A., Ishai, Y. (eds.) EUROCRYPT 2020, Part II. LNCS, vol. 12106, pp. 249–279. Springer, Cham (2020). https://doi.org/10.1007/978-3-030-45724-2_9
31. Hulsing, A., et al.: SPHINCS+. Technical report, National Institute of Standards and Technology (2019). https://csrc.nist.gov/projects/post-quantum-cryptography/round-2-submissions
32. ISO/IEC: 10118-2:2010 Information technology — Security techniques - Hash-functions - Part 2: Hash-functions using an n-bit block cipher, 3rd edn. International Organization for Standardization, Geneve, Switzerland, October 2010
33. Jean, J.: Cryptanalysis of Haraka. IACR Trans. Symmetric Cryptol. **2016**(1), 1–12 (2016)
34. Keller, M., Orsini, E., Scholl, P.: MASCOT: faster malicious arithmetic secure computation with oblivious transfer. In: ACM SIGSAC 2016, Vienna, Austria, 24–28 October 2016, pp. 830–842 (2016)
35. Khovratovich, D., Leurent, G., Rechberger, C.: Narrow-bicliques: cryptanalysis of full IDEA. In: Pointcheval, D., Johansson, T. (eds.) EUROCRYPT 2012. LNCS, vol. 7237, pp. 392–410. Springer, Heidelberg (2012). https://doi.org/10.1007/978-3-642-29011-4_24
36. Khovratovich, D., Rechberger, C., Savelieva, A.: Bicliques for preimages: attacks on skein-512 and the SHA-2 family. In: Canteaut, A. (ed.) FSE 2012. LNCS, vol. 7549, pp. 244–263. Springer, Heidelberg (2012). https://doi.org/10.1007/978-3-642-34047-5_15

37. Knellwolf, S., Khovratovich, D.: New preimage attacks against reduced SHA-1. In: Safavi-Naini, R., Canetti, R. (eds.) CRYPTO 2012. LNCS, vol. 7417, pp. 367–383. Springer, Heidelberg (2012). https://doi.org/10.1007/978-3-642-32009-5_22
38. Knudsen, L.R., Rechberger, C., Thomsen, S.S.: The Grindahl hash functions. In: Biryukov, A. (ed.) FSE 2007. LNCS, vol. 4593, pp. 39–57. Springer, Heidelberg (2007). https://doi.org/10.1007/978-3-540-74619-5_3
39. Kölbl, S., Lauridsen, M.M., Mendel, F., Rechberger, C.: Haraka v2 - efficient short-input hashing for post-quantum applications. IACR Trans. Symm. Cryptol. **2016**(2), 1–29 (2016). http://tosc.iacr.org/index.php/ToSC/article/view/563
40. Kölbl, S., Leander, G., Tiessen, T.: Observations on the SIMON block cipher family. In: Gennaro, R., Robshaw, M. (eds.) CRYPTO 2015, Part I. LNCS, vol. 9215, pp. 161–185. Springer, Heidelberg (2015). https://doi.org/10.1007/978-3-662-47989-6_8
41. Leurent, G.: MD4 is not one-way. In: Nyberg, K. (ed.) FSE 2008. LNCS, vol. 5086, pp. 412–428. Springer, Heidelberg (2008). https://doi.org/10.1007/978-3-540-71039-4_26
42. Li, J., Isobe, T., Shibutani, K.: Converting meet-in-the-middle preimage attack into pseudo collision attack: application to SHA-2. In: Canteaut, A. (ed.) FSE 2012. LNCS, vol. 7549, pp. 264–286. Springer, Heidelberg (2012). https://doi.org/10.1007/978-3-642-34047-5_16
43. Li, Z., Bi, W., Dong, X., Wang, X.: Improved conditional cube attacks on Keccak keyed modes with MILP method. In: Takagi, T., Peyrin, T. (eds.) ASIACRYPT 2017, Part I. LNCS, vol. 10624, pp. 99–127. Springer, Cham (2017). https://doi.org/10.1007/978-3-319-70694-8_4
44. Li, Z., Dong, X., Bi, W., Jia, K., Wang, X., Meier, W.: New conditional cube attack on Keccak keyed modes. IACR Trans. Symm. Cryptol. **2019**(2), 94–124 (2019)
45. Liu, G., Ghosh, M., Song, L.: Security analysis of SKINNY under related-tweakey settings (long paper). IACR Trans. Symm. Cryptol. **2017**(3), 37–72 (2017)
46. Matsui, M.: On correlation between the order of S-boxes and the strength of DES. In: De Santis, A. (ed.) EUROCRYPT 1994. LNCS, vol. 950, pp. 366–375. Springer, Heidelberg (1995). https://doi.org/10.1007/BFb0053451
47. Mouha, N., Wang, Q., Gu, D., Preneel, B.: Differential and linear cryptanalysis using mixed-integer linear programming. In: Wu, C.-K., Yung, M., Lin, D. (eds.) Inscrypt 2011. LNCS, vol. 7537, pp. 57–76. Springer, Heidelberg (2012). https://doi.org/10.1007/978-3-642-34704-7_5
48. Preneel, B., Govaerts, R., Vandewalle, J.: Hash functions based on block ciphers: a synthetic approach. In: Stinson, D.R. (ed.) CRYPTO 1993. LNCS, vol. 773, pp. 368–378. Springer, Heidelberg (1994). https://doi.org/10.1007/3-540-48329-2_31
49. Sasaki, Yu.: Meet-in-the-middle preimage attacks on AES hashing modes and an application to whirlpool. In: Joux, A. (ed.) FSE 2011. LNCS, vol. 6733, pp. 378–396. Springer, Heidelberg (2011). https://doi.org/10.1007/978-3-642-21702-9_22
50. Sasaki, Yu.: Integer linear programming for three-subset meet-in-the-middle attacks: application to GIFT. In: Inomata, A., Yasuda, K. (eds.) IWSEC 2018. LNCS, vol. 11049, pp. 227–243. Springer, Cham (2018). https://doi.org/10.1007/978-3-319-97916-8_15
51. Sasaki, Yu., Aoki, K.: Preimage attacks on 3, 4, and 5-Pass HAVAL. In: Pieprzyk, J. (ed.) ASIACRYPT 2008. LNCS, vol. 5350, pp. 253–271. Springer, Heidelberg (2008). https://doi.org/10.1007/978-3-540-89255-7_16
52. Sasaki, Yu., Aoki, K.: Preimage attacks on step-reduced MD5. In: Mu, Y., Susilo, W., Seberry, J. (eds.) ACISP 2008. LNCS, vol. 5107, pp. 282–296. Springer, Heidelberg (2008). https://doi.org/10.1007/978-3-540-70500-0_21

53. Sasaki, Yu., Aoki, K.: Finding preimages in full MD5 faster than exhaustive search. In: Joux, A. (ed.) EUROCRYPT 2009. LNCS, vol. 5479, pp. 134–152. Springer, Heidelberg (2009). https://doi.org/10.1007/978-3-642-01001-9_8
54. Shi, D., Sun, S., Derbez, P., Todo, Y., Sun, B., Hu, L.: Programming the Demirci-Selçuk meet-in-the-middle attack with constraints. In: Peyrin, T., Galbraith, S. (eds.) ASIACRYPT 2018, Part II. LNCS, vol. 11273, pp. 3–34. Springer, Cham (2018). https://doi.org/10.1007/978-3-030-03329-3_1
55. Song, L., Guo, J.: Cube-attack-like cryptanalysis of round-reduced KECCAK using MILP. IACR Trans. Symm. Cryptol. **2018**(3), 182–214 (2018)
56. Song, L., Guo, J., Shi, D., Ling, S.: New MILP modeling: improved conditional cube attacks on Keccak-based constructions. In: Peyrin, T., Galbraith, S. (eds.) ASIACRYPT 2018, Part II. LNCS, vol. 11273, pp. 65–95. Springer, Cham (2018). https://doi.org/10.1007/978-3-030-03329-3_3
57. Sun, S., Gerault, D., Lafourcade, P., Yang, Q., Todo, Y., Qiao, K., Hu, L.: Analysis of AES, SKINNY, and others with constraint programming. IACR Trans. Symm. Cryptol. **2017**(1), 281–306 (2017)
58. Sun, S., Hu, L., Wang, P., Qiao, K., Ma, X., Song, L.: Automatic security evaluation and (related-key) differential characteristic search: application to SIMON, PRESENT, LBlock, DES(L) and other bit-oriented block ciphers. In: Sarkar, P., Iwata, T. (eds.) ASIACRYPT 2014, Part I. LNCS, vol. 8873, pp. 158–178. Springer, Heidelberg (2014). https://doi.org/10.1007/978-3-662-45611-8_9
59. Wang, L., Sasaki, Yu.: Finding preimages of tiger up to 23 steps. In: Hong, S., Iwata, T. (eds.) FSE 2010. LNCS, vol. 6147, pp. 116–133. Springer, Heidelberg (2010). https://doi.org/10.1007/978-3-642-13858-4_7
60. Wang, L., Sasaki, Yu., Komatsubara, W., Ohta, K., Sakiyama, K.: (Second) preimage attacks on step-reduced RIPEMD/RIPEMD-128 with a new local-collision approach. In: Kiayias, A. (ed.) CT-RSA 2011. LNCS, vol. 6558, pp. 197–212. Springer, Heidelberg (2011). https://doi.org/10.1007/978-3-642-19074-2_14
61. Wei, L., Rechberger, C., Guo, J., Wu, H., Wang, H., Ling, S.: Improved meet-in-the-middle cryptanalysis of KTANTAN (poster). In: Parampalli, U., Hawkes, P. (eds.) ACISP 2011. LNCS, vol. 6812, pp. 433–438. Springer, Heidelberg (2011). https://doi.org/10.1007/978-3-642-22497-3_31
62. Wu, S., Feng, D., Wu, W., Guo, J., Dong, L., Zou, J.: (Pseudo) preimage attack on round-reduced `Grøstl` hash function and others. In: Canteaut, A. (ed.) FSE 2012. LNCS, vol. 7549, pp. 127–145. Springer, Heidelberg (2012). https://doi.org/10.1007/978-3-642-34047-5_8
63. Xiang, Z., Zhang, W., Bao, Z., Lin, D.: Applying MILP method to searching integral distinguishers based on division property for 6 lightweight block ciphers. In: Cheon, J.H., Takagi, T. (eds.) ASIACRYPT 2016, Part I. LNCS, vol. 10031, pp. 648–678. Springer, Heidelberg (2016). https://doi.org/10.1007/978-3-662-53887-6_24
64. Zheng, Y., Pieprzyk, J., Seberry, J.: HAVAL — a one-way hashing algorithm with variable length of output (extended abstract). In: Seberry, J., Zheng, Y. (eds.) AUSCRYPT 1992. LNCS, vol. 718, pp. 81–104. Springer, Heidelberg (1993). https://doi.org/10.1007/3-540-57220-1_54

A Deeper Look at Machine Learning-Based Cryptanalysis

Adrien Benamira[1(✉)], David Gerault[1,2], Thomas Peyrin[1], and Quan Quan Tan[1]

[1] Nanyang Technological University, Singapore, Singapore
{adrien002,quanquan001}@e.ntu.edu.sg, thomas.peyrin@ntu.edu.sg
[2] University of Surrey, Guildford, UK

Abstract. At CRYPTO'19, Gohr proposed a new cryptanalysis strategy based on the utilisation of machine learning algorithms. Using deep neural networks, he managed to build a neural based distinguisher that surprisingly surpassed state-of-the-art cryptanalysis efforts on one of the versions of the well studied NSA block cipher SPECK (this distinguisher could in turn be placed in a larger key recovery attack). While this work opens new possibilities for machine learning-aided cryptanalysis, it remains unclear how this distinguisher actually works and what information is the machine learning algorithm deducing. The attacker is left with a black-box that does not tell much about the nature of the possible weaknesses of the algorithm tested, while hope is thin as interpretability of deep neural networks is a well-known difficult task.

In this article, we propose a detailed analysis and thorough explanations of the inherent workings of this new neural distinguisher. First, we studied the classified sets and tried to find some patterns that could guide us to better understand Gohr's results. We show with experiments that the neural distinguisher generally relies on the differential distribution on the ciphertext pairs, but also on the differential distribution in penultimate and antepenultimate rounds. In order to validate our findings, we construct a distinguisher for SPECK cipher based on pure cryptanalysis, without using any neural network, that achieves basically the same accuracy as Gohr's neural distinguisher and with the same efficiency (therefore improving over previous non-neural based distinguishers).

Moreover, as another approach, we provide a machine learning-based distinguisher that strips down Gohr's deep neural network to a bare minimum. We are able to remain very close to Gohr's distinguishers' accuracy using simple standard machine learning tools. In particular, we show that Gohr's neural distinguisher is in fact inherently building a very good approximation of the Differential Distribution Table (DDT) of the cipher during the learning phase, and using that information to directly classify ciphertext pairs. This result allows a full interpretability of the distinguisher and represents on its own an interesting contribution towards interpretability of deep neural networks.

Finally, we propose some method to improve over Gohr's work and possible new neural distinguishers settings. All our results are confirmed with experiments we have been conducted on SPECK block cipher (source code available online).

A. Canteaut and F.-X. Standaert (Eds.): EUROCRYPT 2021, LNCS 12696, pp. 805–835, 2021.
https://doi.org/10.1007/978-3-030-77870-5_28

Keywords: Differential cryptanalysis · SPECK · Machine learning · Deep neural networks · Interpretability

1 Introduction

While modern symmetric-key cryptography designs are heavily relying on security by construction with strong security arguments (resistance against simple differential/linear attacks, study of algebraic properties, etc.), cryptanalysis remains a crucial part of a cipher's validation process. Only a primitive that went through active and thorough scrutiny of third-party cryptanalysts should gain enough trust by the community to be considered as secure. However, there has been more and more cipher proposals in the past decade (especially with the recent rise of lightweight cryptography) and cryptanalysis effort could not really keep up the pace: conducting cryptanalysis remains a very tough and low-rewarding task.

In order to partially overcome this shortage in cryptanalysts manpower, a recent trend arose of automating as much as possible the various tasks of an attacker. Typically, searching for differential and linear characteristics can now be modeled as Satisfiability/Satisfiability Modulo Theories [17] (SAT/SMT), Mixed Linear Integer Programming [18] (MILP) or Constraint Programming [25] (CP) problems, which can in turn simply be handled by an appropriate solver. The task of the cryptanalyst is therefore reduced to only providing an efficient modeling of the problem to be studied. Due to the impressive results considering the simplicity of the process, a lot of advances have been made in the past decade in this very active research field and this even improved the ciphers designs themselves (how to choose better cryptographic bricks and how to assemble them has been made much easier thanks to these new automated tools). One is then naturally tempted to push this idea further by even getting rid of the modeling part. More generally, can a tool recognize possible weaknesses/patterns in a cipher by just interacting with it, with as little input as possible from the cryptanalysts? One does not expect such a tool to replace a cryptanalyst's job, but it might come in handy for easily pre-checking a cipher (or reduced versions of it) for possible weaknesses.

Machine learning and particularly deep learning have recently attracted a lot of attention, due to impressive advances in important research areas such as computer vision, speech recognition, etc. Some possible connections between cryptography and machine learning were already identified in [21] and we have seen many applications of machine learning for side-channels analysis [16]. However, machine learning for black-box cryptanalysis remained mostly unexplored until Gohr's article presented at CRYPTO'19 [11].

In his work, Gohr trained a deep neural network on labeled data composed of ciphertext pairs: half the data coming from ciphering plaintexts pairs with a fixed input difference with the cipher studied, half from random values. He then checks if the trained neural network is able to classify accurately random from real ciphertext pairs. Quite surprisingly, when applying his framework to the block cipher SPECK-32/64 (the 32-bit block 64-bit key version of SPECK [2]), he managed to obtain a good accuracy for a non-negligible number of rounds. He even managed to mount a key recovery process on top of his neural distinguisher, eventually leading to the current best known key recovery attack for this number of rounds (improving over works on SPECK-32/64 such as [6,24]). Even if his distinguisher/key recovery attack had not been improving over the state-of-the-art, the prospect of a generic tool that could pre-scan for vulnerabilities in a cryptographic primitive (while reaching an accuracy close to exiting cryptanalysis) would have been very attractive anyway.

Yet, Gohr's paper actually opened many questions. The most important, listed by the author as an open problem, is the interpretability of the distinguisher. An obvious issue with a neural distinguisher is that its black-box nature is not really telling us much about the actual weakness of the cipher analyzed. More generally, interpretability for deep neural networks has been known to be a very complex problem and represents a key challenge for the machine learning community. At first sight, it seems therefore very difficult to make any advances in this direction.

Another interesting aspect to explore is to try to match Gohr's neural distinguisher/key recovery attack with classical cryptanalysis tools. It remains very surprising that a trained deep neural network can perform better than the scrutiny of experienced cryptanalysts. As remarked by Gohr, his neural distinguisher is mostly differential in nature (on the ciphertext pairs), but some unknown extra property is exploited. Indeed, as demonstrated by one of his experiments, the neural distinguisher can still distinguish between a real and a random set that have the exact same differential distribution on the ciphertext pairs. Since we know there is some property that researchers have not seen or exploited, what is it?

Finally, a last natural question is: can we do better? Are there some better settings that could improve the accuracy of Gohr's distinguishers?

Our Contributions. In this article, we analyze the behavior of Gohr's neural distinguishers when working on SPECK-32/64 cipher. We first study in detail the classified sets of real/random ciphertext pairs in order to get some hints on what criterion the neural network is actually basing its decisions on. Looking for patterns, we observe that the neural distinguisher is very probably deducing some differential conditions not on the ciphertext pairs directly, but on the penultimate or antepenultimate rounds. We then conduct some experiments to validate our hypothesis.

In order to further confirm our findings, we construct for 5, 6 and 7-round reduced SPECK-32/64 a new distinguisher purely based on cryptanalysis, without any neural network or machine learning algorithm, that matches Gohr's neural distinguisher's accuracy while actually being faster and using the same amount of precomputation/training data. In short, our distinguisher relies on selective partial decryption: in order to attack nr rounds, some hypothesis is made on some bits of the last round subkey and partial decryption is performed, eventually filtered by a precomputed approximated DDT on $nr - 1$ rounds.

We then take a different approach by tackling the problem not from the cryptanalysis side, but the machine learning side. More precisely, as a deep learning model learns high-level features by itself, in order to reach full interpretability we need to discover what these features are. By analyzing the components of Gohr's neural network, we managed to identify a procedure to model these features, while retaining almost the same accuracy as Gohr's neural distinguishers. Moreover, we also show that our method performs similarly on other primitives by applying it on the SIMON block cipher. This result is interesting from a cryptography perspective, but also from a machine learning perspective, showing an example of interpretability by transformation of a deep neural network.

Finally, we explore possible improvements over Gohr's neural distinguishers. By using batches of ciphertexts instead of pairs, we are able to significantly improve the accuracy of the distinguisher, while maintaining identical experimental conditions.

Outline. In Sect. 2, we introduce notations as well as basic cryptanalysis and machine learning concepts that will be used in the rest of the paper. In Sect. 3, we describe in more detail the various experiments conducted by Gohr and the corresponding results. We provide in Sect. 4 an explanation of his neural distinguishers as well as the description of an actual cryptanalysis-only distinguisher that matches Gohr's accuracy. We propose in Sect. 5 a machine learning approach to enable interpretability of the neural distinguishers. Finally, we studied possible improvements in Sect. 6.

2 Preliminaries

Basic notations. In the rest of this article, $\oplus$, $\wedge$ and $\boxplus$ will denote the eXclusive-OR operation, the bitwise AND operation and the modular addition[1] respectively. A right/left bit rotation will be denoted as $\ggg$ and $\lll$ respectively, while $a||b$ will represent the concatenation of two bit strings a and b.

2.1 A Brief Description of SPECK

The lightweight family of ARX block ciphers SPECK was proposed by the US National Security Agency (NSA) [2] in 2013, targeting mainly good performances on micro-controllers. Several versions of the cipher have been proposed within its family, but in this article (and in Gohr's work [11]) we will focus mainly on SPECK-32/64, the 32-bit block 64-bit key version of SPECK, which is composed of 22 rounds (for simplicity, SPECK-32/64 will be referred to as SPECK in the rest of the article).

The 32-bit internal state is divided into a 16-bit left and a 16-bit right part, that we will generally denote l_i and r_i at round i respectively, and is initialised with the plaintext $(l_0||r_0) \leftarrow P$. The round function of the cipher is then a very simple Feistel structure combining bitwise XOR operation and 16-bit modular

[1] The modulo will be stated explicitly if it is not clear from the context.

addition. See Fig. 1 where k_i represents the 16-bit subkey at round i and where $\alpha = 7$, $\beta = 2$. The final ciphertext C is then obtained as $C \leftarrow (l_{22}||r_{22})$. The subkeys are generated with a key schedule that is very similar to the round function (we refer to [2] for a complete description, as we do not make use of the details of the key schedule in this article).

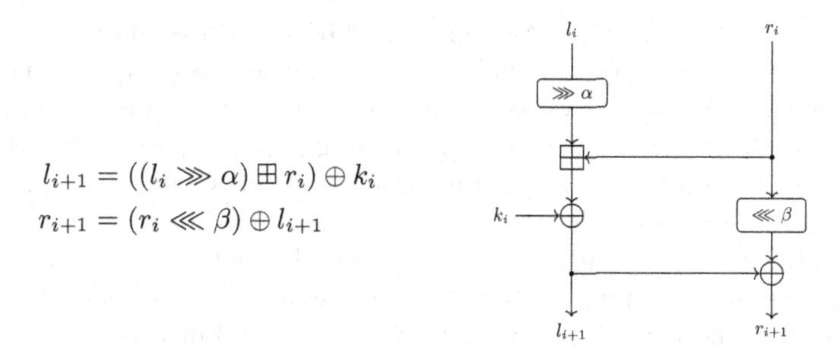

Fig. 1. The SPECK-32/64 round function.

2.2 Differential Cryptanalysis

Differential cryptanalysis studies the propagation of a difference through a cipher. Let a function $f : \mathbb{F}_2^b \rightarrow \mathbb{F}_2^b$ and x, x' be two different inputs for f with a difference $\Delta x = x \oplus x'$. Let $y = f(x)$ and $y' = f(x')$ and a difference $\Delta y = y \oplus y'$. Then, we are interested in the transition probability from Δx to Δy ($\Delta x \xrightarrow{f} \Delta y$):

$$\mathbb{P}(\Delta x \xrightarrow{f} \Delta y) := \frac{\#\{x | f(x) \oplus f(x \oplus \Delta x) = \Delta y\}}{2^b}$$

One classical tool for differential cryptanalysis is the Difference Distribution Table (DDT), which simply lists the differential transition probabilities for each possible input/output difference pairs $(\Delta x, \Delta y)$. The studied function f is usually some Sbox, or some small cipher sub-component, as the DDT of an entire 64-bit or 128-bit cipher would obviously be too large to store.

Since SPECK is internally composed of a left and right part, for a ciphertext C we will denote by C_l and C_r its 16-bit left and right parts respectively. Then, for two ciphertexts C and C', we will denote ΔL the XOR difference $C_l \oplus C'_l$ between the left parts of the two ciphertexts (respectively $\Delta R = C_r \oplus C'_r$ for the right parts). Moreover, for a round i, we will denote by V_i the difference between the two parts of the internal state $V_i = l_i \oplus r_i$.

2.3 Deep Neural Networks

Deep Neural Networks (DNN) are a family of non-linear machine learning classifiers that have gained popularity since their success in addressing a variety of data-driven tasks, such as computer vision, speech recognition, etc.

The main problem tackled by DNN is, given a dataset $D = \{(x_0, y_0)... (x_n, y_n)\}$, with $x_i \in X$ being samples and $y_i \in [0, \ldots, l]$ being labels, to find the optimal parameters θ^* for the DNN_θ model, with the parameters θ such that:

$$\theta^* = \underset{\theta}{\operatorname{argmin}} \sum_{i=0}^{n} L(y_i, DNN_\theta(x_i)) \quad (1)$$

with L being the loss function. As there is no literal expression of θ^*, the approximate solution will depend on the chosen optimization algorithm such as the stochastic gradient descent. Moreover, hyper-parameters of the problem (parameters whose value is used to control the learning process) need to be adjusted as they play an important role in the final quality of the solution.

DNN are powerful enough to derive accurate non-linear features from the training data, but these features are not robust. Indeed, adding a small amount of noise at the input can cause these features to deviate and confuse the model. In other words, the DNN is a very unbiased classifier, but has a high variance.

Different blocks can be used to implement these complex models. However, in this paper, we will be using four types of blocks: the linear neural network, the one-dimensional convolutional neural network, the activation functions (ReLU and sigmoid) and the batch normalization.

Linear neural network. Linear neural networks applies a linear transformation to the incoming data: $out = in.A^T + b$. Here we have $\theta = (A, b)$. The linear neural network is also commonly named perceptron layer or dense layer.

One-dimensional convolutional neural network. The 1D-CNN applies a convolution over a fixed (multi-)temporal input signal. The 1D-CNN operation can be seen as multiple linear neural networks (one per filter) where each one is applied to a sub-part of the input. This sub-part is sliding, its size is kernel size, its pitch is the stride and its start and end points depend on the padding.

Activation functions. The three activation functions that we discuss here are the Rectified Linear Unit (ReLU), defined as $\text{ReLU}(x) = \max(0, x)$, the sigmoid, defined as $\text{Sigmoid}(x) = \sigma(x) = \frac{1}{1+\exp(-x)}$ and the Heaviside step function, defined as $H(x) = \frac{1}{2} + \frac{sgn(x)}{2}$. This block, added between each layer of the DNN, introduces the non-linear part of the model.

Batch normalization. Training samples are typically randomly collected in batches to speed up the training process. It is thus usual to normalize the overall tensor according to the batch dimension.

3 A First Look at Gohr's CRYPTO 2019 Results

Since its release, the lightweight block cipher SPECK attracted a lot of external cryptanalysis, together with its sibling SIMON (this was amplified by the fact that no cryptanalysis was reported in the original specifications document [2]). Many different aspects of SPECK have been covered by these efforts, but the works from Dinur [6] and Song *et al.* [24] are the most successful advances on its differential cryptanalysis aspect so far. Dinur [6] studied all versions of SPECK, improving the best known differential characteristics (from [1,3]) as well as describing a new key recovery strategy for this cipher. In particular, he devised a 4-round attack for 11 rounds of SPECK-32/64 using a 7 round differential characteristic, that has a time complexity of 2^{46} and data complexity of 2^{22} (chosen plaintexts).

Later, at CRYPTO'19, Gohr published a cryptanalysis work on SPECK-32/64 that is based on deep learning [11]. Gohr proposed a key-recovery attack on 11-round SPECK-32/64 with estimated time complexity 2^{38}, improving the previous best attack [6] in 2^{46}, albeit with a slightly higher data complexity: $2^{14.5}$ ciphertext pairs required. In this section, we will briefly review Gohr's results [11].

Overview. In his article, Gohr proposes multiple differential cryptanalysis of SPECK, focusing on the input difference Δ_{in} = 0x0040/0000. In this setting, the aim is to distinguish *real pairs*, *i.e.*, encryptions of plaintext pairs P, P' such that $P \oplus P' = \Delta_{in}$, from *random pairs*, which are the encryptions of random pairs of plaintext with no fixed input difference. Gohr compares a traditional (pure) differential distinguisher with a distinguisher based on a DNN for 5 to 8 rounds of SPECK-32/64 and showed that the DNN performs better.

Pure differential distinguishers. Gohr computed the full DDT for the input difference Δ_{in}, using the Markov assumption. Then, to classify a ciphertext pair (C, C'), the probability p of the output difference $C \oplus C'$ is read from the DDT and compared to the uniform probability. Let $\Delta_{out} = C \oplus C'$, then

$$\text{Classification} = \begin{cases} Real & \text{if } DDT(\Delta_{in} \to \Delta_{out}) > \frac{1}{2^{32}-1} \\ Random & \text{otherwise} \end{cases}$$

These distinguishers for reduced-round SPECK-32/64 are denoted D_{nr}, where $nr \in \{5, 6, 7, 8\}$ represents the number of rounds. The neural distinguishers are denoted as N_{nr}.

Gohr's neural distinguisher. We provide in Fig. 2 a representation of Gohr's neural distinguisher. It is a deep neural network, whose main components are:

1. Block 1: a 1D-CNN with kernel size of 1, a batch normalization and a ReLU activation function
2. Blocks 2-i: one to ten layers with each layer consisting of two 1D-CNN with kernel size of 3, each followed by batch normalization and a ReLU activation function.
3. Block 3: a non-linear final classification block, composed of three perceptron layers separated by two batch normalization and ReLU functions, and finished with a sigmoid function.

The input to the initial convolution block (Block 1) is a 4×16 matrix, where each row corresponds to each 16-bit value in this order: C_l, C_r, C'_l, C'_r,

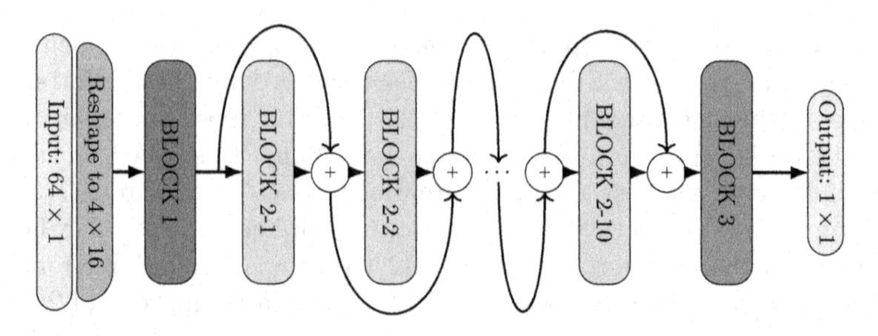

Fig. 2. The whole pipeline of Gohr's deep neural network. Block 1 refers to the initial convolution block, Block 2-1 to 2-10 refer to the residual block and Block 3 refers to the classification block.

a convolution layer with 32 filters is then applied. The kernel size of this 1D-CNN is 1, thus, it maps (C_l, C_r, C'_l, C'_r) to $(filter_1, filter_2, ..., filter_{32})$. Each *filter* is a non-linear combination of the features (C_l, C_r, C'_l, C'_r) after the ReLU activation function depending on the value of the inputs and weights learned by the 1D-CNN. The output of the first block is connected to the input and added to the output of the subsequent layer in the residual block (see Fig. 3).

In the residual blocks (Blocks 2-i), both 1D-CNNs have a kernel of size 3, a padding of size 1 and a stride of size 1 which make the temporal dimension invariant across layers. At the end of each layer, the output is connected to the input and added to the output of the subsequent layer to prevent the relevant input signal from being wiped out across layers. The output of a residual block is a (32×16) feature tensor (see Fig. 4).

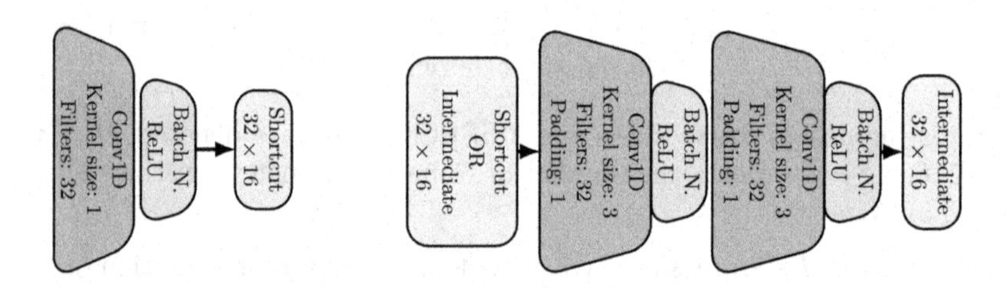

Fig. 3. Initial convolution block (Block 1).

Fig. 4. The residual block (Blocks 2-i).

The final classification block takes as input the flattened output tensor of the residual block. This 512×1 vector is passed into three perceptron layers (Multi-Layer Perceptron or MLP) with batch normalization and ReLU activation functions for the first two layers and a final sigmoid activation function performs the binary classification (see Fig. 5).

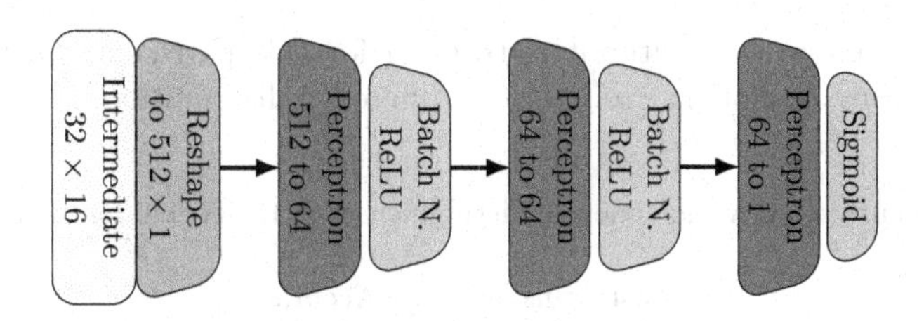

Fig. 5. The classification block (Block 3).

Accuracy and efficiency of the neural distinguishers. For each pair, the neural distinguishers outputs a real-valued score between 0 and 1. If this score is greater than or equal to 0.5, the sample is classified as a real pair, and as a random pair otherwise. The results given by Gohr are presented in Table 1. Note that N_7 and N_8 are trained using some sophisticated methods (we refer to [11] for more details on the training). We remark that Gohr's neural distinguisher has about 100,000 floating parameters, which is size efficient considering the accuracies obtained.

Table 1. Accuracies of neural distinguishers for 5, 6, 7 and 8 rounds (taken from Table 2 of [11]). TPR and TNR denote true positive and true negative rates respectively.

Rds	Distinguisher	Accuracy	TPR	TNR
5	D_5	0.911	0.877	0.947
	N_5	$0.929 \pm 5.13 \times 10^{-4}$	$0.904 \pm 8.33 \times 10^{-4}$	$0.954 \pm 5.91 \times 10^{-4}$
6	D_6	0.758	0.680	0.837
	N_6	$0.788 \pm 8.17 \times 10^{-4}$	$0.724 \pm 1.26 \times 10^{-3}$	$0.853 \pm 1.00 \times 10^{-3}$
7	D_7	0.591	0.543	0.640
	N_7	$0.616 \pm 9.7 \times 10^{-4}$	$0.533 \pm 1.41 \times 10^{-3}$	$0.699 \pm 1.30 \times 10^{-3}$
8	D_8	0.512	0.496	0.527
	N_8	$0.514 \pm 1.00 \times 10^{-3}$	$0.519 \pm 1.41 \times 10^{-3}$	$0.508 \pm 1.42 \times 10^{-3}$

Real differences experiment. The neural distinguishers performed better than the distinguishers using the full DDT, indicating that the neural distinguishers may learn something more than pure differential cryptanalysis. Gohr explores this effect with the *real differences experiment.* In this experiment, instead of distinguishing a real pair from a random pair, the challenge is to distinguish real pairs from masked real pairs, computed as $(C \oplus M, C' \oplus M)$, where M is a random 32-bit value. These experiments use the N_{nr} distinguishers directly, without retraining them for this new task. Table 2 shows the accuracies of these distinguishers. Notice that this operation does not affect $\Delta_{out} = C \oplus C' = (C \oplus M) \oplus (C' \oplus M)$ and thus the output difference distribution. However, the neural distinguishers are still able to distinguish real pairs from masked pairs even without re-training for this particular purpose, which shows that they do not just rely on the difference distribution.

Table 2. Accuracies of various neural distinguishers in the real differences experiment.

Rds	Distinguisher	Accuracy
5	N_5	$0.707 \pm 9.10 \times 10^{-4}$
6	N_6	$0.606 \pm 9.77 \times 10^{-4}$
7	N_7	$0.551 \pm 9.95 \times 10^{-4}$
8	N_8	$0.507 \pm 1.00 \times 10^{-3}$

4 Interpretation of Gohr's Neural Network: A Cryptanalysis Perspective

Interpretability of neural networks remains a highly researched area in machine learning, but the focus has always been on improving the model and computational efficiency. We will discuss more about the interpretability in a machine learning sense in Sect. 5. In this section, we want to find out why and how the neural distinguishers work in a cryptanalysis sense. In essence, we want to answer the following question:

What type of cryptanalysis is Gohr's neural distinguisher learning?

If the neural distinguisher is learning some currently-unknown form of cryptanalysis, then we would like to extrapolate the additional statistics that it exploits. If not, then we want to find out what causes Gohr's neural distinguishers to perform better than pure differential attacks, and even improve state-of-the-art attacks. With this question in mind, we perform a series of experiments and analyses in order to come up with a reasonable guess, later validated by the creation of a pure cryptanalysis-based distinguisher that matches the accuracy of Gohr's one.

Gohr's neural distinguishers are able to correctly identify approximately 90.4%, 68.0% and 54.3% of the real ciphertext pairs (given by the true positive rates) for 5, 6 and 7 rounds of SPECK-32/64 respectively (see Table 1). We will try to find out what these ciphertext pairs are if there are any common patterns and see whether we are able to identify and isolate them.

Choice of input difference. As a start, we looked into Gohr's choice of input difference: 0x0040/0000. This difference is part of a 9-round differential characteristics from Table 7 of [1]. The reason given by Gohr is that this difference deterministically transits to a difference with low Hamming weight after one round. Using constraint programming and techniques similar to [10], we found that the best differential characteristics with a **fixed** input difference of 0x0040/0000 for 5 rounds is 0x0040/0000 → 0x802a/d4a8, with probability of 2^{-13}. In contrast, when we do not restrict the input difference, the best differential characteristics for 5 rounds is 0x2800/0010 → 0x850a/9520, with probability of 2^{-9}. However, when we trained the neural distinguishers to recognize ciphertext pairs with the input difference of 0x2800/0010, the neural distinguishers performed worse (an accuracy of 75.85% for 5 rounds). This is surprising as it is generally natural for a cryptanalyst to maximize the differential probability when choosing a differential characteristic. We believe this is explained by the fact that 0x0040/0000 is the input difference maximizing the differential probability for 3 or 4 rounds of SPECK-32/64 (verified with constraint programming), which has the most chances to provide a biased distribution one or two rounds later. Generally, we believe that when using such neural distinguisher, a good method to choose an input difference is to simply use the input difference leading to the highest differential probability for $nr-1$ or $nr-2$ rounds.

Changing the inputs to the neural network. Gohr's neural distinguishers are trained using the actual ciphertext pairs (C, C') whereas the pure differential distinguishers are only using the difference between the two ciphertexts $C \oplus C'$. Thus, it is unfair to compare both as they are not exploiting the same amount of information. To have a fair comparison of the capability of neural distinguishers and pure differential distinguishers, we trained new neural distinguishers using $C \oplus C'$, instead of (C, C'). The results are an accuracy of 90.6% for 5 rounds, 75.4% for 6 rounds and 58.3% for 7 rounds. This shows us that when the neural distinguishers are restricted to only have access to the difference distribution, they do not perform as well as their respective N_{nr}, and similarly to D_{nr}[2] as can be seen in Table 1. Therefore, this is another confirmation (on top of the real differences experiment conducted in [11]), that Gohr's neural distinguishers are learning more than just the distribution of the differences on the ciphertext. With that information, we therefore naturally looked beyond just the difference distribution at round nr.

4.1 Analyzing Ciphertext Pairs

In this section, we limit and focus the discussions and results mostly to 5 rounds of SPECK-32/64. We recall that the last layer of the neural distinguisher is a sigmoid activation function. Thus, its output is a value between 0 and 1. When the score is 0.5 or more, the neural distinguisher predicts it as a real pair or otherwise, random pair.

The closer a score is to 0.5, the least certain the neural distinguisher is on the classification. In order to know what are the traits that the neural distinguisher is looking for, we segregate the ciphertext pairs that yield extreme scores, *i.e.* scores that are either less than 0.1 (bad score) or more than 0.9 (good score). For the rest of this section, we label the ciphertext pairs as "bad" and "good" ciphertext pairs and refer to the sets as B and G respectively. As we were experimenting with them, we kept the keys (unique to each pair) that are used to generate the

[2] Note that the new neural distinguishers are trained with 10^7 pairs, the same number as in [11].

ciphertext pairs. The goal now is to find similarities and differences in these two groups separately.

As we believe that most of the features the neural distinguishers learned is differential in nature, we focus on the differentials of these ciphertext pairs. To start, we did the following experiment (Experiment A):

1. Using 10^5 real 5-round SPECK-32/64 ciphertext pairs, extract the set G.
2. Obtain the differences of the ciphertext pairs and sort them by frequency
3. For each of the differences δ:
   (a) Generate 10^4 random 32-bit numbers and apply the difference, δ to get 10^4 different ciphertext pairs.
   (b) Feed the pairs to the neural distinguisher N_5 to obtain the scores.
   (c) Note down the number of pairs that yield a score ≥ 0.5

In Table 3, we show the top 25 differences for 5 rounds of SPECK-32/64 with their respective score from the above experiment. Out of the first 1000 differences, each records about 75% of the pairs scoring more than 0.5. Also, there exist multiple pairs of differences such that one is more probable than the other, and yet, it has a lower number of pairs classifying as real (*e.g.* No. 21 in Table 3). Thus, there is little evidence showing that if a difference is more probable, then the neural distinguisher is necessarily more likely to recognize it.

Table 3. The top 25 differences (5 rounds of SPECK-32/64) in G with their respective results for Experiment A as a percentage of how many pairs having a score of ≥ 0.5 out of 10^4 pairs. Cnt refers to the number of differences obtained in G.

No.	Difference	Cnt	Percent.	No.	Difference	Cnt	Percent.
1	0x802a/d4a8	116	75	14	0x883a/dcb8	45	75
2	0x802e/d4ac	81	76	15	0x801e/d49c	45	75
3	0x803a/d4b8	73	74	16	0xa026/f4a4	42	75
4	0x8e2a/daa8	73	75	17	0xbe1a/ea98	41	75
5	0x822a/d6a8	72	75	18	0x821a/d698	41	76
6	0xb82a/eca8	67	75	19	0xbe26/eaa4	41	75
7	0x882a/dca8	65	75	20	0x83ea/d768	40	75
8	0x801a/d498	62	75	21	0x8626/caa4	40	38
9	0xa02a/f4a8	62	75	22	0x886a/dce8	40	75
10	0xbe2a/eaa8	62	75	23	0xa06a/f4e8	40	75
11	0x806a/d4e8	59	74	24	0x8e1a/da98	39	75
12	0x8e26/daa4	47	75	25	0x8226/cea4	38	37
13	0x8026/d4a4	46	74				

Since the neural distinguishers outperform the ones with just the XOR input, we started to look beyond just the differences at 5 rounds. We decided to partially decrypt the ciphertext pairs from G for a few rounds and re-run Experiment A on these partially decrypted pairs: for each pair, we compute the difference and for each difference, we created 10^4 random plaintext pairs with these differences and encrypted them to round nr using random keys. The results are very intriguing, as compared to that of Table 3: almost all of the (top 1000) unique differences obtained in this experiment achieved 99% or 100% of ciphertext pairs having a score of $\geq$0.5.

We can see that the differences at rounds 3 and 4 (after decrypting 2 and 1 round respectively) start to show some strong biases. In fact, for all of the top 1000 differences at rounds 3 and 4, all 10^4 pairs $\times$ 1000 differences returned a score of $\geq$0.5[3]. With that, we conduct yet another experiment (Experiment B):

1. For all the ciphertext pairs in G, decrypt i rounds with their respective keys and compute the corresponding difference. Denote the set of differences as $\texttt{Diff}_{\texttt{5-i}}$.
2. Generate 10^5 plaintext pairs with a difference of `0x0040/0000` with random keys, encrypt to 4 rounds
3. If the pair's difference is in $\texttt{Diff}_{\texttt{5-i}}$, keep the pair. Otherwise, discard.
4. Encrypt the remaining pairs to 5 rounds and evaluate them using N_5.

When $i = 2$, we obtain 1669 unique differences with a dataset size of 89,969. 97.86% of these ciphertext pairs yielded a score $\geq$0.5 (*i.e.* by this method, we can isolate 88.04% of the true positive ciphertexts pair). Using $i = 1$, we have 128,039 unique differences and the size of the dataset is 74,077. While we could get a cleaner set with 99.98% of these ciphertext pairs obtaining a score of $\geq$0.5, we only managed to isolate 74.06% of the true positive pairs. Comparing with the true positive rate of N_5 from Table 1, which is $0.904 \pm 8.33 \times 10^{-4}$, the case when $i = 2$ seems to be closer.

We also looked into the bias of the difference bits (the jth difference bit refers to the jth bit index of $C_{5-2} \oplus C'_{5-2}$ where C_{nr-i} refers to the nr round ciphertext decrypted by i rounds. Table 4 shows the difference bit biases of the first 1000 (most common) unique differences of ciphertext pairs in G and B after decrypting two rounds. We assume that the neural distinguisher is able to identify some bits at these rounds because they are significantly more biased, though both the set B and G are from the real distribution.

Now, we state the assumption required for our conjecture, which we will verify experimentally in Sect. 4.3.

Assumption 1 *Given a 5-round* SPECK-*32/64 ciphertext pair,* N_5 *is able to determine the difference of certain bits at rounds 3 and 4 with high accuracy.*

Conjecture 1. Given a 5-round SPECK-32/64 ciphertext pair, N_5 finds the difference of certain bits at round 3 and decides if the ciphertext pair is real or random.

---

[3] The differences were obtained experimentally.

Table 4. Difference bit bias of ciphertext pairs in G and B after decrypting 2 rounds. A negative (resp. positive) value indicates a bias towards '0' (resp. '1').

bit position	31	30	29	28	27	26	25	24	23	22	21	20	19	18	17	16
G	$\underline{0.476}$	$\underline{-0.454}$	−0.355	−0.135	0.045	0.084	−0.009	$\underline{0.487}$	$\underline{-0.473}$	−0.426	−0.300	−0.050	0.006	0.019	$\underline{0.500}$	$\underline{-0.500}$
B	−0.002	0.018	0.008	−0.011	0.044	0.002	0.023	−0.022	0.010	−0.002	0.013	−0.004	0.006	−0.005	0.103	0.072
bit position	**15**	**14**	**13**	**12**	**11**	**10**	**9**	**8**	**7**	**6**	**5**	**4**	**3**	**2**	**1**	**0**
G	$\underline{0.476}$	$\underline{-0.454}$	−0.142	−0.006	0.025	0.084	−0.009	$\underline{0.487}$	$\underline{-0.473}$	−0.426	0.165	0.094	−0.006	0.019	$\underline{-0.500}$	$\underline{-0.500}$
B	0.031	−0.009	−0.015	−0.007	−0.014	−0.024	0.025	0.026	0.034	−0.005	−0.018	−0.021	0.006	0.009	0.079	−0.065

Interestingly, the difference bit biases after decrypting 1 and 2 rounds are very similar (in their positions). We will provide an explanation in Sect. 4.2. The exact truncated differentials are ($*$ denotes no specific constraint, while 0 or 1 denotes the expected bit difference):

$$\text{3 rounds: } 10*****00*****00\ 10*****00*****10$$
$$\text{4 rounds: } 10*****10*****10\ 10*****10*****00$$

We refer to these particular truncated differential masks as TD_3 and TD_4 for the following discussion. Using constraint programming, we evaluate that the probabilities for these truncated differentials are 87.86% and 49.87% respectively. In order to verify how much the neural distinguisher is relying on these bits, we perform the following experiment (Experiment C):

1. Generate 10^6 plaintext pairs with initial difference `0x0040/0000` and 10^6 random keys.
2. Encrypt all 10^6 plaintext pairs to $5-i$ rounds. If a plaintext pair satisfies the TD_{5-i}, then we keep it. Otherwise, it will be discarded.
3. Encrypt the remaining pairs to 5 rounds and evaluate them using N_5.

Table 5. Results of Experiment C with TD_3 and TD_4. Proport. refers to the number of true positive ciphertext pairs captured by the experiment.

5-i	Trunc. Diff.	Dataset size	Acc.	Proport.
3	TD3	87741	99.277%	87.11%
4	TD4	50063	99.996%	50.06%

Table 5 shows the statistics of the above experiment with 5 rounds of SPECK-32/64. The true positive rates for ciphertext pairs that follow these are closer to that of Gohr's neural distinguisher. Now, there remains about 3% of the ciphertext pairs yet to be explained (comparing the results of TD_{5-2} with N_5). The important point to note here is that the pairs we have identified are exactly the ones verified by the neural distinguisher as well, by the nature of these experiments. In other words, we managed to find what the neural distinguisher is looking for and not just another distinguisher that would achieve a good accuracy by identifying a different set of ciphertext pairs.

4.2 Deriving TD_3 and TD_4

With an input difference of 0x0040/0000, which has a deterministic transition to 0x8000/8000 in round 1, the difference will only start to spread after round 1 due to the modular addition in the SPECK-32/64 round function. The inputs to the modular addition at round 2 are 0x0100 and 0x8000 (cf. Fig. 6). While there are two active bits, only one of them will propagate the carry (as the other is the MSB), resulting in multiple differences. Assuming a uniform distribution, the carry has a probability $\frac{1}{2}$ of propagating to the left. This causes the probability of the various differentials to reduce by $\frac{1}{2}$ as the carry bit propagates until b_{31} (bit position 31) is reached and any further carry will be removed by the modular addition.

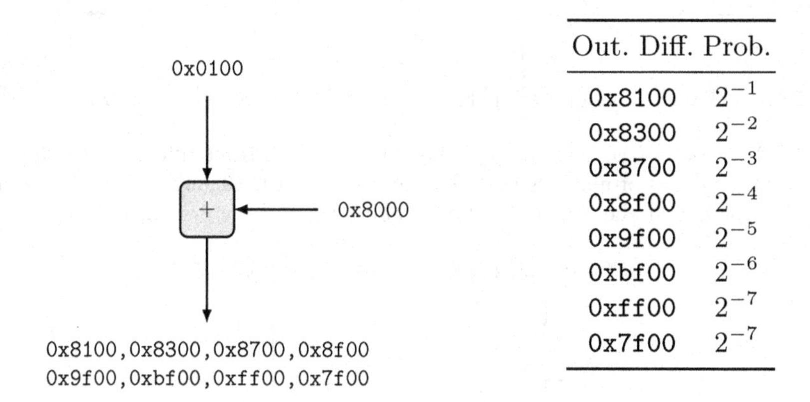

Out. Diff.	Prob.
0x8100	2^{-1}
0x8300	2^{-2}
0x8700	2^{-3}
0x8f00	2^{-4}
0x9f00	2^{-5}
0xbf00	2^{-6}
0xff00	2^{-7}
0x7f00	2^{-7}

Fig. 6. The distribution of the possible output differences after passing through the modular addition operation.

In Fig. 7 and Fig. 8, we show how the bits evolve along the most probable differential path from round 1 (0x8000/8000) to round 4 (0x850a/9520). As it passes through the modular addition operation, we highlight the bits that have a relatively higher probability of being different from the most probable differential. The darker the color, the higher the probability of the difference being toggled.

Figure 7 and Fig. 8 show us why TD_3 is important at round 3, and how the active bits shift in SPECK-32/64 when we start with the input difference of 0x0040/0000. In every round, b_{31}, (the leftmost bit) has a high probability of staying active. This bit is then rotated to b_{24} before it goes into the modular addition operation. In each round, b_{26} has a $\frac{1}{2}$ chance of switching from $1 \rightarrow 0$ or the other way round. b_{27} and b_{28} have a $\frac{1}{4}$ and $\frac{1}{8}$ chance respectively of switching. This makes them highly volatile and therefore, unreliable. On the other hand, the right part of SPECK-32/64 rotates by 2 to the left at the end of each round. Because of the high rotation value in the left part of SPECK-32/64, low rotation

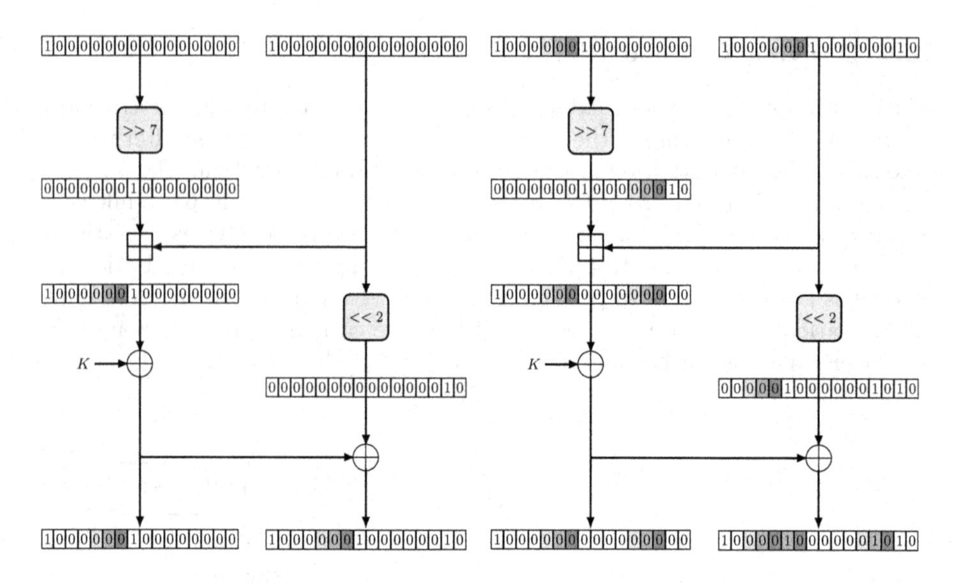

Fig. 7. The left (resp. right) part shows how the active bit from difference `0x8000/8000` (resp. `0x8100/8102`) propagates to difference `0x8100/8102` (`0x8000/820a`). The darker the color, the higher the probability ($\geq \frac{1}{4}$) that it has a carry propagated to.

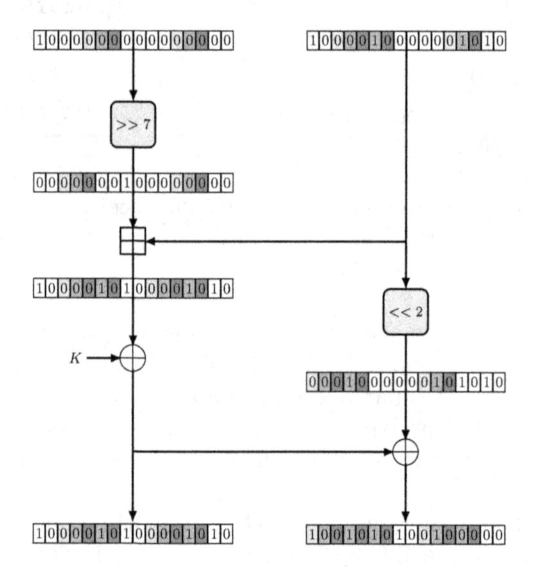

Fig. 8. Showing how the active bit from difference `0x8000/820a` propagates to difference `0x850a/9520`. The darker the color, the higher the probability ($\geq \frac{1}{4}$) that it has a carry propagated to.

value of the right part of SPECK-32/64, and the fact that the left part is added into the right part after the rotation, it takes about 3 to 4 rounds for the volatile and unreliable bits to spread.

4.3 Verifying Assumption 1

To verify if Gohr's neural distinguisher is able to recognize the truncated differential, we retrain the neural distinguisher with a slight difference (Experiment D):

1. Generate 10^7 plaintext pairs such that about $\frac{1}{2}$ of the pairs satisfy TD_3 (these are the positive pairs)
2. Encrypt the plaintext pairs for two rounds
3. Train the neural network to distinguish the two distributions, and validate with the same hyper-parameters as in [11], with a depth of 1 in the residual block.

After retraining, the neural distinguisher has an accuracy of 96.57% (TPR: 99.95%, TNR: 93.19%) This shows that the neural distinguisher has the capabilities to actually recognize the truncated differential with an outstanding accuracy.

4.4 SPECK-32/64 Reduced to 6 Rounds

We perform Experiments C for 6 rounds of SPECK-32/64 as well. Table 6 shows the comparison of the true positive results of rounds 5 and 6. While the results are not as obvious as for the case of 5 rounds, we can still observe a similar trend for 6 rounds.

Table 6. Results of SPECK-32/64 reduced to six rounds for Experiment C. Proportion refers to the number of true positive ciphertext pairs captured by the experiment.

6-i rds	Truncated Differential	Size of dataset	Accuracy	Proport.
4	10*****10*****10 10*****10*****00	49902	99.41%	49.6%
5	10*****00*****10 11*****01*****00	6884	99.927%	6.88%

4.5 Average Key Rank Differential Distinguisher

Taking into consideration the observations we presented in this section, we introduce a new average key rank distinguisher that is not based on machine learning and almost matches the accuracy as Gohr's neural network for 5, 6 and 7 rounds of SPECK-32/64. Here are the key considerations used in our distinguisher:

- The training set of Gohr's neural network consists of 10^7 ciphertext pairs. Thus, we restrict our distinguisher to only use 10^7 ciphertext pairs as well.
- If we do an exhaustive key search for two rounds, the time complexity will be extremely high. Instead, we may need to limit ourselves to only one round to match the complexity of the neural distinguishers.

– If we know the difference at round i, the $i-1$ round difference for the right part is known as well, since $r_{i-1} = (l_i \oplus r_i) \ggg 2$

With those pointers in mind, we created a distinguisher that uses an approximated DDT (aDDT); that is, a truncated DDT that is experimentally constructed based on n ciphertext pairs. In this distinguisher, we use $n = 10^7$ to ensure that both our distinguisher and the neural distinguishers have the same amount of information. The idea of the distinguisher is to decrypt the last round, nr, using all possible subkey bits that are relevant to the bits we are interested in. Then, we compute the average of the probabilities of all partial decryptions for a given pair, read from $aDDT(nr-1)$, to get a score. If the score is greater than that of the random distribution, the distinguisher will return 1 (Real) and 0 (Random) otherwise. The bits we are interested in can be represented as an AND mask, that is, a mask that has '1' in the bit positions that we want to consider the bit and '0' for those we want to ignore. The mask value we have chosen is `0xff8f/ff8f` rather than the expected `0xc183/c183` as we believe the truncated differential they are detecting is at $nr-2$ rounds. Thus, other than the bits that are identified earlier in this section, we decided to include more bits to improve the accuracy. With the look-up table to the aDDT, we do not just only match the data complexity (of the offline training) of the Gohr's neural distinguishers, but at the same time, include the correlations between bits as well.

The pseudocodes for creating the aDDT and the average key rank distinguisher can be found in the long version of the paper that can be found on eprint. We applied the distinguisher for 5, 6 and 7 rounds of SPECK-32/64 and the results are given in Table 7. It shows that our distinguisher closely matches the accuracies of Gohr's neural distinguishers.

Degree of closeness. We now study the similarity between our distinguishers and Gohr's neural distinguishers. In particular, we are interested in whether the classifications of the ciphertext pairs are the same for both distinguishers. To verify this, we gave a set of 10^5 5-round ciphertext pairs (approx. 50,000 from real and random distribution each) to both our average key rank distinguisher and N_5, and measured how many times did they have the same output. The results for $nr = 5$ are shown in Table 8. We can see that about 97.6% of the ciphertext pairs tested have the same classification in both distinguishers. For $nr = 6$, we achieved 94.98% of the pairs with the same classification.

Complexity comparison. In our average key rank distinguisher, for each pair, we perform the partial decryption of two ciphertexts, and a table lookup in aDDT. In the partial decryption, we enumerate the 2^{12} keys affecting the rightmost 13 bits of δl_{nr-1} covered by our mask. Therefore, the complexity of our distinguisher is 2^{13} one-round SPECK-32/64 decryptions, and 2^{13} table lookups. Comparing its complexity with Gohr's distinguishers is not trivial, as the operations involved are different. Gohr evaluates the complexity of his neural key recovery by their runtime and an estimation of the number of speck encryptions that could be performed at the same time on a GPU implementation. We propose to use the number of floating point multiplications performed by the neural network instead. Let I and O respectively denote the number of inputs and outputs to one layer. The computational cost of going through a dense layer is $I \cdot O$ multiplications. For 1D-CNN with kernel size $ks = 1$, a null padding, a stride equal to 1 and F filters, with input size (I, T) the cost is computed as $I \cdot F \cdot T$ multiplications. With the same input but with kernel size $ks = 3$, a padding equal to 1, the cost is $I \cdot ks \cdot F \cdot T$ Applying these formulas to Gohr's neural network, we obtain a total of $137280 \approx 2^{17.07}$ multiplications. Note that we do not account for batch normalizations and additions, which are dominated by the cost of the multiplications. Using this estimation, it seems that our distinguisher is slightly better in terms of complexity.

Table 7. Accuracy of the average key rank distinguisher with a mask value of `0xff8f/ff8f`.

N_5	Accuracy	TPR	TNR
5	92.98%	90.76%	95.22%
6	78.79%	72.53%	85.07%
7	60.28%	55.31%	65.24%

Table 8. Closeness of the outputs of N_5 and average key rank distinguisher.

		N_r output	
		≥0.5	<0.5
AKR Dist	1	46.6%	1.48%
	0	0.953%	51.0%

4.6 Discussion

Even though Gohr trained a neural distinguisher with a fixed input difference, it is unfair to compare the accuracy of neural distinguisher to that of a pure differential cryptanalysis (with the use of DDT), since there are alternative cryptanalysis methods that the neural distinguisher may have learned. The experiments performed indicate that while Gohr's neural distinguishers did not rely much on difference at the nr round, they rely strongly on the differences at round $nr - 1$ and even more strongly at round $nr - 2$. These results support the hypothesis that the neural distinguisher may learn differential-linear cryptanalysis [13] in the case of SPECK. While we did not present any attacks here, using the MILP model shown in [9], we verified that there are indeed many linear relations with large biases for 2 to 3 rounds.

Unlike traditional linear cryptanalysis, which usually use independent characteristics or linear hull involving the same plaintext and ciphertext bits, a well-trained neural network is able to learn and exploit several linear characteristics while taking into account their dependencies and correlations.

We believe that neural networks find the easiest way to achieve the best accuracy. In the case of SPECK, it seems that differential-linear cryptanalysis would be a good fit since it requires less data and the truncated differential has a very high probability. Thus, we think that neural networks have the ability to efficiently learn short but strong differential, linear or differential-linear characteristics for small block ciphers for a small number of rounds.

4.7 Application to AES-2-2-4 [7]

We are also interested in the capabilities of the neural distinguishers on a Substitution-Permutation Network (SPN) cipher. We chose a small scale variant of AES from [7] with the parameters: $r = 2, c = 2, e = 4$. We chose this cipher as it has a small state size, which could be exhaustively searched through. AES-2-2-8 would be a good choice as it also has a state size of 32-bit, however, our distinguishers are not able to learn anything significant. We trained AES-2-2-4 with 2^{15} pairs, starting with an input difference of $(1, 0, 0, 1)$. This input difference was chosen such that only after two rounds, all S-boxes will be active. We trained them for 3 rounds and obtained an accuracy of 61.0%. In contrast, we use the same number of pairs, we trained an aDDT distinguisher and we obtained an accuracy of 62.3%.

To show the possibilities of relying purely on differences, we perform an experiment similar to Experiment A. With the trained neural distinguisher, we exhaust all possible 16-bit differences and we generate 100 random pairs for each difference. Next, we feed the pairs to the neural distinguisher and count the number of pairs in each basket of score: $[0.0 - 0.1), [0.1 - 0.2), ..., [0.9 - 1.0]$. Our result shows that for each differential, the 100 random pairs form a cluster about a center similar to a Gaussian distribution. These results seem to suggest the nature of the neural distinguisher for AES-2-2-4 is one that relies fully on differential: giving a confidence interval based on just the difference.

5 Interpretation of Gohr's Neural Network: A Machine Learning Perspective

In this section, we are exploring the following practical question:

Can Gohr's neural network be replaced by a strategy inspired by both differential cryptanalysis and machine learning?

We will demonstrate here that this is possible. First of all, it should be emphasized that DNNs often outperform mathematical modeling or standard machine learning approaches in supervised data-driven settings, especially on high-dimensional data. It seems to be the case because correlations found between input and output pairs during DNN training lead to more relevant characteristics than those found by experts. In other words, Gohr's neural distinguisher seems to be capable of finding a property $\mathcal{P}$ currently unknown by cryptanalysts. One may ask if we could experimentally approach this unknown property $\mathcal{P}$ that encodes the neural distinguisher behavior, using both machine learning and cryptanalysis expertise. With this question in mind, we propose our best estimate with a focus on 5 and 6 SPECK-32/64 rounds where the DNN achieves accuracies of 92.9% and 78.8% in a real/random distinction setting and where the full DDT approach can achieve accuracies of 91.1% and 75.8%. In our best setting, we reach accuracy values of 92.3% and 77.9%.

Section 3 discusses in detail how Gohr's neural distinguisher is modeled in three blocks. Our objective here is to replace each of these individual blocks by a more interpretable one, coming either from machine learning or from the cryptanalysts' point of view. This work is thus the result of the collaboration between two worlds addressing the open question of deep learning interpretability. In the course of the study, we set forth and challenged four conjectures to estimate the property $\mathcal{P}$ learned by the DNN as detailed below.

5.1 Four Conjectures

Conjectures 2 & 3 aim to uncover Block 3 behavior. Conjecture 4 tackles Block 1 while Conjecture 5 concerns Block 2-i.

The DNN can not be entirely replaced by another machine learning model. Ensemble-based machine learning models such as random forests [4] and gradient boosting decision trees [8] are accurate and easier to interpret than DNNs [14]. Nevertheless, DNNs outperform ensemble-based machine learning models for most tasks on high-dimensional data such as images. However, with only 64 bits of input, we could legitimately wonder whether the DNN could be replaced by another ensemble-based machine learning model. Despite our small size problem, our experiments reveal that other models significantly decrease the accuracy.

Conjecture 2. Gohr's neural network outperforms other non-neuronal network machine learning models.

Experiment. To challenge this conjecture, we tested multiple machine learning models, such as Random Forest (RF), Light Gradient Boosting Machine (LGBM), Multi-Layer Perceptron (MLP), Support Vector Machine (SVM) and Linear Regression (LR). They all performed equally. For the rest of this paper, we will only consider LGBM [12] as an alternative ensemble classifier for DNN and MLP. LGBM is an extension of Gradient Boosting Decision Tree (GBDT) [8] and we fixed our choice on it because it is accurate, interpretable and faster to train than RF or GBDT. In support of our conjecture, we established that the accuracy for the LGBM model is significantly lower than the one of the DNN when the inputs are (C_l, C_r, C'_l, C'_r), see third column of Table 9.

Table 9. A comparison of the neural distinguisher and LGBM model for 5 round, for 10^6 samples generated of type (C_l, C_r, C'_l, C'_r).

N_5	D_5	LGBM as classifier for theoriginal input	LGBM as classifier for the 512-feature	LGBM as classifier for the 64-feature
92.9%	91.1%	76.34% $\pm$ 2.62	91.49% $\pm$ 0.09	92.36% $\pm$ 0.07

The final MLP block is not essential. As described above, we can not replace the entire DNN with another non-neuronal machine learning model that is easier to interpret. However, we may be able to replace the last block (Block 3) of the neural distinguisher performing the final classification, by an ensemble model.

Conjecture 3. The MLP block of Gohr's neural network can be replaced by another ensemble classifier.

Experiment. We successfully exchanged the final MLP block for a LGBM model. The reasons for choosing LGBM as a non-linear classifier were detailed in the previous experiment paragraph. The first attempt is a complete substitution of Block 3, taking the 512-dimension output of Block 2–10 as input. In the fourth column of Table 9, we observe that this experiment leads to much better results than the one from Conjecture 2, and even better results than the classical DDT method D_5 (+0.39%). To further improve the accuracy, we implemented a partial substitution, taking only the 64-dimension output of the first layer of the MLP as input. As can be seen in the fifth column from Table 9, the accuracy with those inputs is now much closer to the DNN accuracy. In both cases, the accuracy is close to the neural distinguisher, supporting our conjecture. At this point, in order to grasp the unknown property $\mathcal{P}$, one needs to understand the feature vector at the residuals' output.

The linear transformation on the inputs. We saw in Sect. 3 that Block 1 performs a linear transformation on the input. By looking at the weights of the DNN first convolution, we observe that it contains many opposite values. This indicates that the DNN is looking for differences between the input features. Consequently, we propose the following conjecture.

Conjecture 4. The first convolution layer of Gohr's neural network transforms the input (C_l, C_r, C'_l, C'_r) into $(\Delta L, \Delta V, V_0, V_1)$ and a linear combination of those terms.

Experiments. As the inputs of the first convolution are binary, we could formally verify our conjecture. By forcing to one all non-zero values of the output of this layer, we calculated the truth-table of the first convolution. We thus obtained the boolean expression of the first layer for the 32 filters. We observed that eight filters were empty and the remaining twenty-four filters were simple. The filter expressions are provided in the long version of the paper that can be found on eprint.

However, one may argue that setting all non-zero values to one is an over-simplified approach. Therefore, we replaced the first ReLU activation function by the Heaviside activation function, and then we retrained the DNN. Since the Heaviside function binarizes the intermediate value (as in [28]), we can establish the formal expression of the first layer of the retrained DNN. This second

DNN had the same accuracy as the first one and almost the same filter boolean expression.

Finally, we trained the same DNN with the following entries $(\Delta L, \Delta V, V_0, V_1)$. Using the same method as before, we established the filters' boolean expressions. This time, we obtained twenty five null filters and seven non-null filters, with the following expressions: ΔL, $\overline{V_0} \wedge V_1$, $\overline{\Delta L}$, ΔL, $\overline{V_0} \wedge \overline{V_1}$, $\Delta L \wedge \Delta V$, $\overline{\Delta L} \wedge \overline{\Delta V}$. These observations support conjecture 4. Therefore, we kept only $(\Delta L, \Delta V, V_0, V_1)$ as inputs for our pipeline.

The masked output distribution table. With regards to the remaining residual block replacement, our first assumption is that the DNN calculates a shape close to the DDT in that residual block. However, two major properties of the neural distinguisher prevent us from assuming that it is a DDT in the classical sense of the term. The first property, as explained in Sect. 3, is that the neural distinguisher does not only rely on the difference distribution to distinguish real pairs as presented in Table 2. The second specificity is that the DNN has only approximately 100,000 floating parameters to perform classification, which can be considered as size efficient. Our second assumption is therefore that the DNN is able to compress the distribution table. We introduce the following definitions.

Output Distribution Table (ODT). We propose to compute a distribution table on the values $(\Delta L, \Delta V, V_0, V_1)$ directly, instead of doing so on the difference of the ciphertext pair $(C_l \oplus C'_l, C_r \oplus C'_r)$. We call this new table an Output Distribution Table (ODT) and it can be seen as a generalization of the DDT. The entries of the ODT are 64 bits, which is not tractable for 10^7 samples. Also, the DNN has only 100,000 parameters. The DNN is therefore able to compress the ODT.

Masked Output Distribution Table (M-ODT). A compressed ODT means that the input is not 64 bits, but instead hw bits, where hw represents the Hamming weight of the mask. Let us consider a mask $M \in \mathcal{M}_{hw}$ with $\mathcal{M}_{hw}$ the ensemble of 64-bits masks with Hamming weight hw and $M = (M_1, M_2, M_3, M_4)$, with M_i a 16-bit mask. Compressing the ODT therefore means applying the M mask to all inputs. In our case, with $I = (\Delta L, \Delta V, V_0, V_1)$, we get $I_M = (\Delta L \wedge M_1, \Delta V \wedge M_2, V_0 \wedge M_3, V_1 \wedge M_4) = I \wedge M$, before computing the ODT. By calculating that way, the number of ODT entries per mask decreases. It becomes a function that depends only on hw and on the bit positions in the masks. It is therefore a more compact representation of the complete ODT. However, it turns out that if we consider only one mask, we get only one value per sample to perform the classification: $P(\texttt{Real}|I_M)$, while the DNN has a final vector size of 512. We considered several masks. Thus, by defining the ensemble $R_M \in \mathcal{M}_{hw}$, the set of relevant masks of $\mathcal{M}_{hw}$, we can calculate for a specific input $I = (\Delta L, \Delta V, V_0, V_1)$ the probability $P(\texttt{Real}|I_M), \forall M \in R_M$. Then, we concatenate all the probabilities into a feature vector of size $m = |R_M|$. We get the feature F for the input I: $F = \left(P(\texttt{Real}|I_{M1}) \; P(\texttt{Real}|I_{M2}) \cdots P(\texttt{Real}|I_{Mm}) \right)^T$. We are now able to propose the final conjecture.

Conjecture 5. The neural distinguisher internal data processing of Block 2-i can be approached by:

1. Computing a distribution table for input $(\Delta L, \Delta V, V_0, V_1)$.
2. Finding several relevant masks and applying them to the input in order to compress the output distribution table.

We abbreviate M-ODT this Masked-Output Distribution Table. Thus, the feature vector of the DNN can be replaced by a vector where each value represents the probability stored in the M-ODT for each mask.

This approach enables us to replace Block 2-i of the DNN. Though, we still need to clarify how to get the R_M ensemble.

Extracting masks. Based on local interpretation methods, we can extract these masks from the DNN. Indeed, these methods consist of highlighting the most important bits of the entries for classification. Thus, by sorting the entries according to their score and by applying these local interpretation methods, we can obtain the relevant masks.

5.2 Approximating the Expression of the Property $\mathcal{P}$

From our conjectures, we hypothesized that we can approximate the unknown property $\mathcal{P}$ that encodes the neural distinguisher behavior by the following:

- Changing (C, C') into $I = (\Delta L, \Delta V, V_0, V_1)$.
- Changing the 512-feature vector of the DNN by the feature vector of probabilities $F = \left(P(\texttt{Real}|I_{M1}) \; P(\texttt{Real}|I_{M2}) \cdots P(\texttt{Real}|I_{Mm}) \right)^T$.
- Changing the final MLP block by the ensemble machine learning model LGBM.

These points stand respectively for Block 1, Block 2-i and Block 3.

5.3 Implementation

In this section and based on the verified conjectures, we are describing the stepwise implementation of our method. We consider that we have a DNN formed with 10^7 data of type $(\Delta L, \Delta V, V_0, V_1)$ for 5 and 6 rounds of SPECK-32/64. We developed a three-step approach:

1. Extraction of the masks from the DNN with a first dataset.
2. Construction of the M-ODT with a second dataset.
3. Training of the final classifier from the probabilities stored in the M-ODT with a third dataset.

Mask extraction from the DNN. We first ranked 10^4 real samples according to DNN score, as described in Sect. 4.1, in order to estimate the masks from these entries. We used multiple local interpretation methods: Integrated Gradients [26], DeepLift [22], Gradient Shap [15], Saliency maps [23], Shapley Value [5], and Occlusion [27]. These methods score each bit according to their importance for the classification. Following averaging by batch and by method, there were two possible ways to move forward. We could either assign a Hamming weight or else set a threshold above which all bits would be set to one. After a wide range of experiments, we chose the first option and set the Hamming weight to sixteen and eighteen (which turned out to be the best values in our testing). This approach allowed us to build the ensemble R_M of the relevant masks.

Implementation details. We used the captum library[4] which brings together multiple methods on local interpretation. The dataset is divided into batches of size about 2,500 and grouped by scores. The categories we used were: scores from 1 to 0.9 (about 2,000 samples), scores from 0.9 to 0.5 (about 500 samples), scores from 1 to 0.8 (about 2,100 samples) and scores from 1 to 0.5 (about 2,500 samples). This way, one score per method could be derived for each bit of each sample. We then proposed several methods to average these importance scores by bit of category: the sum of absolute values, the median of absolute values and the average of absolute values. Then, we took the sixteen and eighteen best values and we obtained a mask. There is one mask per score, one per local interpretation method and one per averaging method. On average, for 5,000 samples we generate about 100 relevant masks. Finally, with the methods available in scikit-learn [20], we ranked the features and so the masks according to their performance. After multiple repetitions of mask generation and selection at every time, we obtained 50 masks that are effective: they are provided in the long version of the paper. The final ensemble of masks is the addition of those 50 effective masks and the generated relevant masks.

Constructing the M-ODT. Once the ensemble R_M of relevant masks is determined, we compute the M-ODT. (Algorithm can be found in long version of the paper) describes our construction method which is similar to that of the DDT. The inputs of the algorithm include a second dataset composed of $n = 10^7$ real samples of type $I = (\Delta L, \Delta V, V_0, V_1)$, and the set of relevant masks R_M. The output is the M-ODT dictionary with the mask as first key, the masked input as second key, and $P(\texttt{Real}|I \wedge M) = P(\texttt{Real}|I_M)$ as value.

The M-ODT dictionary is constructed as follow: first, for each mask M in R_M, we compute the corresponding masked-dataset $\mathcal{D}_M$ which is simply the operation $I_M = I \wedge M$ for all I in $\mathcal{D}$. Secondly we compute a dictionary U with key the element of D_M and with value the occurrences number of that element in D_M. Then, we compute for all element I_M in $\mathcal{D}_M$ the probability:

[4] https://github.com/pytorch/captum.

$$P(\texttt{Real}|I_M) = \frac{P(I_M|\texttt{Real})P(\texttt{Real})}{P(I_M|\texttt{Real})P(\texttt{Real}) + P(I_M|\texttt{Random})P(\texttt{Random})}$$

with $P(\texttt{Real}) = P(\texttt{Random}) = 0.5$, $P(I_M|\texttt{Random}) = 2^{-HW(M)}$, $HW(M)$ being the Hamming weight of M and $P(I_M|\texttt{Real}) = \frac{1}{n} \times U[I_M]$. Finally we update M-ODT as follow: M-ODT$[M][I_M] = P(\texttt{Real}|I_M)$.

Training the classifier on probabilities. Upon building the M-ODT, we can start training the classifier. Given a third dataset $\mathcal{D} = \{(input_0, y_0)...(input_n, y_n)\}$, with $input_j$ a sample of type (C, C'), transformed into $(\Delta L, \Delta V, V_0, V_1)$ and the label $y_j \in [0, 1]$, with $n = 10^6$, we first compute the feature vector $F_j = \left(P(\texttt{Real}|I_j \wedge M1) \; P(\texttt{Real}|I_j \wedge M2) \cdots P(\texttt{Real}|I_j \wedge Mm) \right)^T$ for all inputs and for $m = |R_M|$. Next, we determined the optimal θ parameters for the g_θ model according to Eq. 1, with L being the square loss. Here, the g_θ classifier is Light Gradient Boosting Machine (LGBM) [12].

Implementation details. Feature vectors are standardized. Model hyper-parameters fine-tuning has been achieved by grid search. Results were obtained by cross-validation on 20% of the train set and the test set had 10^5 samples. Finally, results are obtained on the complete pipeline for three different seeds, five times for every seed.

5.4 Results

The M-ODT pipeline was implemented with numpy, scikit-learn [20] and pytorch [19]. The project code can be found at this URL address[5]. Our work station is constituted of a GPU Nvidia GeForce GTX 970 with 4043 MiB memory and four Intel core i5-4460 processors clocked at 3.20 GHz.

General results. Table 10 shows accuracies of the DDT, the DNN and our M-ODT pipeline on 5 and 6-round reduced SPECK-32/64 for 1.1×10^7 generated samples. When compared to DNN and DDT, our M-ODT pipeline reached an intermediate performance right below DNN. The main difference is the true positive rate which is higher in our pipeline (this can be explained by the fact that our M-ODT preprocessing only considers real samples). All in all, our M-ODT pipeline successfully models the property $\mathcal{P}$.

Matching. Table 11 summarizes the results of the quantitative correspondence studies for the prediction between the two models. We compared the DNN trained on samples type $(\Delta L, \Delta V, V_0, V_1)$ to our M-ODT pipeline. On 5 rounds, we obtained a rate of 97.5% identical predictions. In addition, 91.3% were both

[5] https://github.com/AnonymousSubmissionEuroCrypt2021/A-Deeper-Look-at-Machine-Learning-Based-Cryptanalysis.

Table 10. A comparison of Gohr's neural network, the DDT and our M-ODT pipeline accuracies for around 150 masks generated each time, with input $(\Delta L, \Delta V, V_0, V_1)$, LGBM as classifier and 1.1×10^7 samples generated in total. TPR and TNR refers to true positive and true negative rate respectively.

Rd	Distinguisher	Accuracy	TPR	TNR
5	D_5	91.1%	87.7%	94.7%
	N_5	92.9% ± 0.05	90.4% ± 0.08	95.4% ± 0.06
	M-ODT (Ours)	92.3% ± 0.08	95.5% ± 0.09	89.1% ± 0.2
6	D_6	75.8%	68.0%	83.7%
	N_6	78.8% ± 0.08	72.4% ± 0.01	85.3% ± 0.1
	M-ODT (Ours)	77.9% ± 0.1	85.2% ± 0.1	70.6% ± 0.2

identical and equal to the label. On 6 rounds, matching prediction reduces down to 93.1%.

We thus demonstrated that our method advantageously approximates the performance of the neural distinguisher. With an initial linear transformation on the inputs, computing a M-ODT for a set of masks extracted from the DNN and then classifying the resulting feature vector with LGBM, we achieved an efficient yet more easily interpretable approach than Gohr distinguishers. Indeed, DNN obscure features are simply approached in our pipeline by $F = \left(P(\texttt{Real}|I_{M1})\ P(\texttt{Real}|I_{M2}) \cdots P(\texttt{Real}|I_{Mm}) \right)^T$. Finally, we interpret the performance of the classifier globally (i.e. retrieving the decision tree) and locally (i.e. deducing which feature played the greatest role in the classification for each sample) as in [14]. Those results are not displayed as they are beyond the scope of the present work, but they can be found in the project code.

Table 11. A comparison of Gohr's neural network predictions and our M-ODT pipeline predictions for around 150 masks generated each time, with input $(\Delta L, \Delta V, V_0, V_1)$, LGBM as classifier and 1.1×10^7 samples generated in total.

Nr	Model	Accuracy	Matching	Matching & equal to label
5	N_5	92.9%	97.5% ± 0.06	91.3% ± 0.08
	M-ODT (Ours)	92.3%		
6	N_6	78.8%	93.1% ± 0.07	75.3% ± 0.11
	M-ODT (Ours)	77.9%		

5.5 Application to SIMON Cipher

In order to check whether our approach could be generalized to other cryptographic primitives, we evaluated our M-ODT method on 8 rounds of SIMON-

32/64 block cipher. Implementing the same pipeline, we enjoyed a 82.2% accuracy for the classification, whereas the neural distinguisher achieves 83.4% accuracy. In addition, the matching rate between the two models was up to 92.4%. The slight deterioration in the results of our pipeline for SIMON can be explained by the lack of efficient masks as introduced in Sect. 5.3 for SPECK.

5.6 Discussions

From the cryptanalysts' standpoint, one important aspect of using the neural distinguisher is to uncover the property $\mathcal{P}$ learned by the DNN. Unfortunately, while being powerful and easy to use, Gohr's neural network remains opaque.

Our main conjecture is that the 10-layer residual blocks, considered as the core of the model, are acting as a compressed DDT applied on the whole input space. We model our idea with a Masked Output Distribution Table (M-ODT). The M-ODT can be seen as a distribution table applied on masked outputs, in our case $(\Delta L, \Delta V, V_0, V_1)$, instead of only the difference $(C_l \oplus C'_l, C_r \oplus C'_r)$. By doing so, features are no longer abstract as in the neural distinguisher. In our pipeline, each one of the features is a probability for the sample to be real knowing the mask and the input. In the end, with our M-ODT pipeline, we successfully obtained a model which has only -0.6% difference accuracy with the DNN and a matching of 97.3% on 5 rounds of SPECK-32/64. Additional analysis of our pipeline (e.g. masks independence, inputs influence, classifiers influence) are available into the project code. To the best of our knowledge, this work is the first successful attempt to exhibit the underlying mechanism of the neural distinguisher. However, we note that a minor limitation of our method is that it still requires the DNN to extract the relevant masks during the preparation of the distinguisher. Since it is only during preparation, this does not remove anything with regards to the interpretability of the distinguisher. Future work will aim at computing these masks without DNN. All in all, our findings represent an opportunity to guide the development of a novel, easy-to-use and interpretable cryptanalysis method.

6 Improved Training Models

While in the two previous sections we focused on understanding how the neural distinguisher works, here we will explain how one can outperform Gohr's results. The main idea is to create batches of ciphertext inputs instead of pairs.

We refer to batch input of size B, a group of B ciphertexts that are constructed from the same key. Here, we can distinguish two ways to train and evaluate the neural distinguisher pipeline with batch input. The straightforward one is to evaluate the neural distinguisher score for each element of the batch and then to take the median of the results. The second is to consider the whole batch as a single input for a neural distinguisher. In order to do so, we used 2-dimensional CNN (2D-CNN) where the channel dimension is the features $(\Delta L, \Delta V, V_0, V_1)$. We should point out that, for sake of comparability with

Gohr's work, we maintained the product of the training set size by the batch size to be equal to 10^7. Both batch size-based challenging methods yielded similar accuracy values (see Table 12). Notably, in both cases, we enjoyed 100% accuracy on 5 and 6 rounds with batch sizes 10 and 50 respectively.

Table 12. Study of the batch size methods on the accuracies with (ΔL, ΔV, V_0, V_1) as input for 5 and 6 rounds.

Rounds	5			6			
Batch input size	1	5	10	1	5	10	50
Averaging Method	92.9%	99.8%	100%	78.6%	95.41%	99.0%	100%
2D-CNN Method	–	99.4%	100%	–	93.27%	97.7%	100%

Considering these encouraging outcomes, we extended the method to 7 rounds. As the 7-round training is more sophisticated and the two previous methods are equivalent, we decided to only apply the first method (the averaging one), because it requires to train only one neural distinguisher. Results given in Table 13 confirm our previous findings: with a batch size of 100, we obtain 99.7% accuracy on 7 rounds. This remarkable outcome demonstrates the major improvement of our batch strategy over those from earlier Gohr's work.

Table 13. Study of the averaging batch size method on the 7-round accuracies with (ΔL, ΔV, V_0, V_1) as input.

Batch input size	1	5	10	50	100
Averaging Method	61.2%	73.5%	80.8%	96.7%	99.7%

Conclusion

In this article, we proposed a thorough analysis of Gohr's deep neural network distinguishers of SPECK-32/64 from CRYPTO'19. By carefully studying the classified sets, we managed to uncover that these distinguishers are not only basing their decisions on the ciphertext pair difference, but also the internal state difference in penultimate and antepenultimate rounds. We confirmed our findings by proposing pure cryptanalysis-based distinguishers on SPECK-32/64 that match Gohr's accuracy. Moreover, we also proposed a new simplified pipeline for Gohr's distinguishers, that could reach the same accuracy while allowing a complete interpretability of the decision process. We finally gave possible directions to even improve over Gohr's accuracy.

Our results indicate that Gohr's neural distinguishers are not really producing novel cryptanalysis attacks, but more like optimizing the information extraction with the low-data constraints. Many more distinguisher settings, machine learning pipelines, types of ciphers should be studied to have a better understanding of what machine learning-based cryptanalysis might be capable of. Yet, we foresee that such tools could become of interest for cryptanalysts and designers to easily and generically pre-test a primitive for simple weaknesses.

Our work also opens interesting directions with regards to interpretability of deep neural networks and we believe our simplified pipeline might lead to better interpretability in other areas than cryptography.

Acknowledgements. The authors are grateful to the anonymous reviewers for their insightful comments that improved the quality of the paper. The authors are supported by the Temasek Laboratories NTU grant DSOCL17101. We would like to thank Aron Gohr for pointing out that the differential characteristics mentioned in the attacks of Dinur's[6] have been extended by one free round, thus, our previous suggestion of extending Dinur's attack by one round is invalid.

References

1. Abed, F., List, E., Lucks, S., Wenzel, J.: Differential cryptanalysis of round-reduced SIMON and SPECK. In: Cid, C., Rechberger, C. (eds.) FSE 2014. LNCS, vol. 8540, pp. 525–545. Springer, Heidelberg (2015). https://doi.org/10.1007/978-3-662-46706-0_27
2. Beaulieu, R., Shors, D., Smith, J., Treatman-Clark, S., Weeks, B., Wingers, L.: The SIMON and SPECK families of lightweight block ciphers. IACR Cryptol. ePrint Arch. 2013, 404 (2013). http://eprint.iacr.org/2013/404
3. Biryukov, A., Roy, A., Velichkov, V.: Differential analysis of block ciphers SIMON and SPECK. In: Cid, C., Rechberger, C. (eds.) FSE 2014. LNCS, vol. 8540, pp. 546–570. Springer, Heidelberg (2015). https://doi.org/10.1007/978-3-662-46706-0_28
4. Breiman, L.: Random forests. Mach. Learn. **45**(1), 5–32 (2001)
5. Castro, J., Gómez, D., Tejada, J.: Polynomial calculation of the shapley value based on sampling. Comput. Oper. Res. **36**(5), 1726–1730 (2009)
6. Dinur, I.: Improved differential cryptanalysis of round-reduced speck. In: Joux, A., Youssef, A. (eds.) SAC 2014. LNCS, vol. 8781, pp. 147–164. Springer, Cham (2014). https://doi.org/10.1007/978-3-319-13051-4_9
7. Duan, X., Yue, C., Liu, H., Guo, H., Zhang, F.: Attitude tracking control of small-scale unmanned helicopters using quaternion-based adaptive dynamic surface control. IEEE Access **9**, 10153–10165 (2021). https://doi.org/10.1109/ACCESS.2020.3043363
8. Friedman, J.H.: Greedy function approximation: a gradient boosting machine. Ann. Stat. 1189–1232 (2001)
9. Fu, K.,Wang, M., Guo, Y., Sun, S., Hu, L.: MILP-based automatic search algorithms for differential and linear trails for speck. IACR Cryptol. ePrint Arch. 407 (2016)
10. Gerault, D., Minier, M., Solnon, C.: Constraint programming models for chosen key differential cryptanalysis. In: Rueher, M. (ed.) CP 2016. LNCS, vol. 9892, pp. 584–601. Springer, Cham (2016). https://doi.org/10.1007/978-3-319-44953-1_37

11. Gohr, A.: Improving attacks on round-reduced speck32/64 using deep learning. In: Boldyreva, A., Micciancio, D. (eds.) CRYPTO 2019, Part II. LNCS, vol. 11693, pp. 150–179. Springer, Cham (2019). https://doi.org/10.1007/978-3-030-26951-7_6
12. Ke, G., et al.: Lightgbm: A highly efficient gradient boosting decision tree. Adv. Neural Inf. Process. Syst. 3146–3154 (2017)
13. Langford, S.K., Hellman, M.E.: Differential-linear cryptanalysis. In: Desmedt, Y.G. (ed.) CRYPTO 1994. LNCS, vol. 839, pp. 17–25. Springer, Heidelberg (1994). https://doi.org/10.1007/3-540-48658-5_3
14. Lundberg, S.M., et al.: Explainable AI for trees: From local explanations to global understanding. arXiv preprint arXiv:1905.04610 (2019)
15. Lundberg, S.M., Lee, S.I.: A unified approach to interpreting model predictions. In: Advances in Neural Information Processing systems, pp. 4765–4774 (2017)
16. Maghrebi, H., Portigliatti, T., Prouff, E.: Breaking cryptographic implementations using deep learning techniques. In: Carlet, C., Hasan, M.A., Saraswat, V. (eds.) SPACE 2016. LNCS, vol. 10076, pp. 3–26. Springer, Cham (2016). https://doi.org/10.1007/978-3-319-49445-6_1
17. Mouha, N., Preneel, B.: A proof that the ARX cipher salsa20 is secure against differential cryptanalysis. IACR Cryptol. ePrint Arch. 328 (2013). http://eprint.iacr.org/2013/328
18. Mouha, N., Wang, Q., Gu, D., Preneel, B.: Differential and linear cryptanalysis using mixed-integer linear programming. Inf. Secur. Cryptology - Inscrypt **2011**, 57–76 (2011)
19. Paszke, A., et al.: Pytorch: An imperative style, high-performance deep learning library. In: Advances in Neural Information Processing Systems, pp. 8026–8037 (2019)
20. Pedregosa, F., et al.: Scikit-learn: machine learning in python. J. Mach. Learn. Res. **12**, 2825–2830 (2011)
21. Rivest, R.L.: Cryptography and machine learning. In: Imai, H., Rivest, R.L., Matsumoto, T. (eds.) ASIACRYPT 1991. LNCS, vol. 739, pp. 427–439. Springer, Heidelberg (1993). https://doi.org/10.1007/3-540-57332-1_36
22. Shrikumar, A., Greenside, P., Kundaje, A.: Learning important features through propagating activation differences. arXiv preprint arXiv:1704.02685 (2017)
23. Simonyan, K., Vedaldi, A., Zisserman, A.: Deep inside convolutional networks: Visualising image classification models and saliency maps. arXiv preprint arXiv:1312.6034 (2013)
24. Song, L., Huang, Z., Yang, Q.: Automatic differential analysis of ARX block ciphers with application to SPECK and LEA. In: Liu, J.K., Steinfeld, R. (eds.) ACISP 2016. LNCS, vol. 9723, pp. 379–394. Springer, Cham (2016). https://doi.org/10.1007/978-3-319-40367-0_24
25. Sun, S., Gérault, D., Lafourcade, P., Yang, Q., Todo, Y., Qiao, K., Hu, L.: Analysis of AES, skinny, and others with constraint programming. IACR Trans. Symmetric Cryptol. **2017**(1), 281–306 (2017)
26. Sundararajan, M., Taly, A., Yan, Q.: Axiomatic attribution for deep networks. arXiv preprint arXiv:1703.01365 (2017)
27. Zeiler, M.D., Fergus, R.: Visualizing and understanding convolutional networks. In: Fleet, D., Pajdla, T., Schiele, B., Tuytelaars, T. (eds.) ECCV 2014, Part I. LNCS, vol. 8689, pp. 818–833. Springer, Cham (2014). https://doi.org/10.1007/978-3-319-10590-1_53
28. Zhou, S., Wu, Y., Ni, Z., Zhou, X., Wen, H., Zou, Y.: Dorefa-net: Training low bitwidth convolutional neural networks with low bitwidth gradients. CoRR abs/1606.06160 (2016). http://arxiv.org/abs/1606.06160

Author Index

Printed in the United States
by Baker & Taylor Publisher Services